D0212494

A HISTORY OF
MATHEMATICAL
NOTATIONS

FLORIAN CAJORI

Two Volumes Bound As One

Volume I:
Notations in Elementary Mathematics

Volume II:
Notations Mainly in Higher Mathematics

DOVER PUBLICATIONS, INC.
New York

Bibliographical Note

This Dover edition, first published in 1993, is an unabridged
and unaltered republication in one volume of the work first
published in two volumes by The Open Court Publishing Company, La Salle, Illinois, in 1928 and 1929.

Library of Congress Cataloging-in-Publication Data

Cajori, Florian, 1859–1930.
    A history of mathematical notations / by Florian Cajori.
        p.   cm.
    Originally published: Chicago : Open Court Pub. Co., 1928–
1929.
    "Two volumes bound as one."
    Includes indexes.
    Contents: v. 1. Notations in elementary mathematics — v.
2. Notations mainly in higher mathematics.
    ISBN 0-486-67766-4 (pbk.)
    1. Mathematical notation—History. 2. Mathematics—His-
tory. 3. Numerals—History. I. Title.
QA41.C32 1993
510'.148—dc20   93-29211
                                                                         CIP

Manufactured in the United States of America
Dover Publications, Inc., 31 East 2nd Street, Mineola, N.Y. 11501

PREFACE

The study of the history of mathematical notations was suggested to me by Professor E. H. Moore, of the University of Chicago. To him and to Professor M. W. Haskell, of the University of California, I am indebted for encouragement in the pursuit of this research. As completed in August, 1925, the present history was intended to be brought out in one volume. To Professor H. E. Slaught, of the University of Chicago, I owe the suggestion that the work be divided into two volumes, of which the first should limit itself to the history of symbols in elementary mathematics, since such a volume would appeal to a wider constituency of readers than would be the case with the part on symbols in higher mathematics. To Professor Slaught I also owe generous and vital assistance in many other ways. He examined the entire manuscript of this work in detail, and brought it to the sympathetic attention of the Open Court Publishing Company. I desire to record my gratitude to Mrs. Mary Hegeler Carus, president of the Open Court Publishing Company, for undertaking this expensive publication from which no financial profits can be expected to accrue.

I gratefully acknowledge the assistance in the reading of the proofs of part of this history rendered by Professor Haskell, of the University of California; Professor R. C. Archibald, of Brown University; and Professor L. C. Karpinski, of the University of Michigan.

FLORIAN CAJORI

UNIVERSITY OF CALIFORNIA

TABLE OF CONTENTS

I. INTRODUCTION

PARAGRAPHS

II. NUMERAL SYMBOLS AND COMBINATIONS OF SYMBOLS . . . 1–99

    Babylonians 1–15
    Egyptians 16–26
    Phoenicians and Syrians 27–28
    Hebrews 29–31
    Greeks 32–44
    Early Arabs 45
    Romans 46–61
    Peruvian and North American Knot Records 62–65
    Aztecs 66–67
    Maya 68
    Chinese and Japanese 69–73
    Hindu-Arabic Numerals 74–99
        Introduction 74–77
        Principle of Local Value 78–80
        Forms of Numerals 81–88
        Freak Forms 89
        Negative Numerals 90
        Grouping of Digits in Numeration 91
        The Spanish Calderón 92–93
        The Portuguese Cifrão 94
        Relative Size of Numerals in Tables 95
        Fanciful Hypotheses on the Origin of Numeral Forms . 96
        A Sporadic Artificial System 97
        General Remarks 98
        Opinion of Laplace 99

III. SYMBOLS IN ARITHMETIC AND ALGEBRA (ELEMENTARY PART) 100

    A. Groups of Symbols Used by Individual Writers . . . 101
        Greeks—Diophantus, Third Century A.D. 101–5
        Hindu—Brahmagupta, Seventh Century 106–8
        Hindu—The Bakhshālī Manuscript 109
        Hindu—Bhāskara, Twelfth Century 110–14
        Arabic—al-Khowârizmî, Ninth Century 115
        Arabic—al-Karkhî, Eleventh Century 116
        Byzantine—Michael Psellus, Eleventh Century . . 117
        Arabic—Ibn Albanna, Thirteenth Century . . . 118
        Chinese—Chu Shih-Chieh, Fourteenth Century . . 119, 120

PARAGRAPHS

Byzantine—Maximus Planudes, Fourteenth Century 121
Italian—Leonardo of Pisa, Thirteenth Century . . 122
French—Nicole Oresme, Fourteenth Century . . . 123
Arabic—al-Qalasâdî, Fifteenth Century 124
German—Regiomontanus, Fifteenth Century . . . 125–27
Italian—Earliest Printed Arithmetic, 1478 128
French—Nicolas Chuquet, 1484 129–31
French—Estienne de la Roche, 1520 132
Italian—Pietro Borgi, 1484, 1488 133
Italian—Luca Pacioli, 1494, 1523 134–38
Italian—F. Ghaligai, 1521, 1548, 1552 139
Italian—H. Cardan, 1532, 1545, 1570 . . . . . 140, 141
Italian—Nicolo Tartaglia, 1506–60 . . . . . . 142, 143
Italian—Rafaele Bombelli, 1572 . . . . 144, 145
German—Johann Widman, 1489, 1526 146
Austrian—Grammateus, 1518, 1535 147
German—Christoff Rudolff, 1525 . . . . . . 148, 149
Dutch—Gielis van der Hoecke, 1537 150
German—Michael Stifel, 1544, 1545, 1553 151–56
German—Nicolaus Copernicus, 1566 157
German—Johann Scheubel, 1545, 1551 . . . . . 158, 159
Maltese—Wil. Klebitius, 1565 160
German—Christophorus Clavius, 1608 161
Belgium—Simon Stevin, 1585 . . . . 162, 163
Lorraine—Albert Girard, 1629 164
German-Spanish—Marco Aurel, 1552 165
Portuguese-Spanish—Pedro Nuñez, 1567 166
English—Robert Recorde, 1543(?), 1557 167–68
English—John Dee, 1570 169
English—Leonard and Thomas Digges, 1579 . . . 170
English—Thomas Masterson, 1592 171
French—Jacques Peletier, 1554 172
French—Jean Buteon, 1559 173
French—Guillaume Gosselin, 1577 174
French—Francis Vieta, 1591 176–78
Italian—Bonaventura Cavalieri, 1647 179
English—William Oughtred, 1631, 1632, 1657 . . . 180–87
English—Thomas Harriot, 1631 188
French—Pierre Hérigone, 1634, 1644 189
Scot-French—James Hume, 1635, 1636 190
French—René Descartes 191
English—Isaac Barrow 192
English—Richard Rawlinson, 1655–68 193
Swiss—Johann Heinrich Rahn 194

PARAGRAPHS

English—John Wallis, 1655, 1657, 1685 . . . . 195, 196
Extract from *Acta eruditorum*, Leipzig, 1708 . . . 197
Extract from *Miscellanea Berolinensia*, 1710 (Due to
   G. W. Leibniz) 198
Conclusions 199

B. Topical Survey of the Use of Notations 200–356
  Signs of Addition and Subtraction 200–216
    Early Symbols 200
    Origin and Meaning of the Signs 201–3
    Spread of the + and − Symbols 204
    Shapes of the + Sign 205–7
    Varieties of − Signs 208, 209
    Symbols for "Plus or Minus" 210, 211
    Certain Other Specialized Uses of + and − . . 212–14
    Four Unusual Signs 215
    Composition of Ratios 216
  Signs of Multiplication 217–34
    Early Symbols 217
    Early Uses of the St. Andrew's Cross, but Not as the
      Symbol of Multiplication of Two Numbers . . 218–30
      The Process of Two False Positions 219
      Compound Proportions with Integers 220
      Proportions Involving Fractions 221
      Addition and Subtraction of Fractions . . . 222
      Division of Fractions 223
      Casting Out the 9's, 7's, or 11's 225
      Multiplication of Integers 226
      Reducing Radicals to Radicals of the Same Order 227
      Marking the Place for "Thousands" 228
      Place of Multiplication Table above 5×5 . . 229
    The St. Andrew's Cross Used as a Symbol of Multi-
      plication 231
    Unsuccessful Symbols for Multiplication 232
    The Dot for Multiplication 233
    The St. Andrew's Cross in Notation for Transfinite
      Ordinal Numbers 234
  Signs of Division and Ratio 235–47
    Early Symbols 235, 236
    Rahn's Notation 237
    Leibniz's Notations 238
    Relative Position of Divisor and Dividend . . . 241
    Order of Operations in Terms Containing Both ÷
      and × 242
    A Critical Estimate of : and ÷ as Symbols . . 243

PARAGRAPHS

Notations for Geometric Ratio   244
Division in the Algebra of Complex Numbers . . 247
Signs of Proportion 248–59
  Arithmetical and Geometrical Progression . . . 248
  Arithmetical Proportion 249
  Geometrical Proportion 250
  Oughtred's Notation 251
  Struggle in England between Oughtred's and Wing's
    Notations before 1700 252
  Struggle in England between Oughtred's and Wing's
    Notations during 1700–1750 253
  Sporadic Notations 254
  Oughtred's Notation on the European Continent . 255
  Slight Modifications of Oughtred's Notation . . 257
  The Notation : :: : in Europe and America . . 258
  The Notation of Leibniz 259
Signs of Equality 260–70
  Early Symbols 260
  Recorde's Sign of Equality 261
  Different Meanings of = ... 262
  Competing Symbols 263
  Descartes' Sign of Equality 264
  Variations in the Form of Descartes' Symbol . . 265
  Struggle for Supremacy 266
  Variation in the Form of Recorde's Symbol . . . 268
  Variation in the Manner of Using It 269
  Nearly Equal 270
Signs of Common Fractions 271–75
  Early Forms 271
  The Fractional Line 272
  Special Symbols for Simple Fractions 274
  The Solidus 275
Signs of Decimal Fractions 276–89
  Stevin's Notation 276
  Other Notations Used before 1617 278
  Did Pitiscus Use the Decimal Point? 279
  Decimal Comma and Point of Napier 282
  Seventeenth-Century Notations Used after 1617 . 283
  Eighteenth-Century Discard of Clumsy Notations . 285
  Nineteenth Century : Different Positions for Point
    and for Comma 286
  Signs for Repeating Decimals 289
Signs of Powers 290–315
  General Remarks 290

PARAGRAPHS

Double Significance of R and l 291
Facsimiles of Symbols in Manuscripts 293
Two General Plans for Marking Powers 294
Early Symbolisms: Abbreviative Plan, Index Plan 295
Notations Applied Only to an Unknown Quantity,
    the Base Being Omitted 296
Notations Applied to Any Quantity, the Base Being
    Designated 297
Descartes' Notation of 1637 298
Did Stampioen Arrive at Descartes' Notation Inde-
    pendently? 299
Notations Used by Descartes before 1637 . . . 300
Use of Hérigone's Notation after 1637 301
Later Use of Hume's Notation of 1636 302
Other Exponential Notations Suggested after 1637 . 303
Spread of Descartes' Notation 307
Negative, Fractional, and Literal Exponents . . 308
Imaginary Exponents 309
Notation for Principal Values 312
Complicated Exponents 313
D. F. Gregory's $(+)^r$ 314
Conclusions 315
Signs for Roots 316–38
Early Forms, General Statement 316, 317
The Sign R, First Appearance 318
Sixteenth-Century Use of R 319
Seventeenth-Century Use of R 321
The Sign l 322
Napier's Line Symbolism 323
The Sign $\sqrt{}$ 324–38
    Origin of $\sqrt{}$ 324
    Spread of the $\sqrt{}$ 327
    Rudolff's Signs outside of Germany 328
    Stevin's Numeral Root-Indices 329
    Rudolff and Stifel's Aggregation Signs . . . 332
    Descartes' Union of Radical Sign and Vinculum . 333
    Other Signs of Aggregation of Terms 334
    Redundancy in the Use of Aggregation Signs . 335
    Peculiar Dutch Symbolism 336
    Principal Root-Values 337
    Recommendation of the U.S. National Committee 338
Signs for Unknown Numbers 339–41
Early Forms 339

                                                    PARAGRAPHS
Crossed Numerals Representing Powers of Un-
    knowns   340
Descartes' z, y, x   340
Spread of Descartes' Signs   341
Signs of Aggregation   342–56
    Introduction   342
    Aggregation Expressed by Letters   343
    Aggregation Expressed by Horizontal Bars or Vincu-
        lums   344
    Aggregation Expressed by Dots   348
    Aggregation Expressed by Commas   349
    Aggregation Expressed by Parentheses   350
    Early Occurrence of Parentheses   351
    Terms in an Aggregate Placed in a Vertical Column   353
    Marking Binomial Coefficients   354
    Special Uses of Parentheses   355
    A Star to Mark the Absence of Terms   356

IV. SYMBOLS IN GEOMETRY (ELEMENTARY PART)   357–85

A. Ordinary Elementary Geometry   357
    Early Use of Pictographs   357
    Signs for Angles   360
    Signs for "Perpendicular"   364
    Signs for Triangle, Square, Rectangle, Parallelogram .   365
    The Square as an Operator   366
    Sign for Circle   367
    Signs for Parallel Lines   368
    Signs for Equal and Parallel   369
    Signs for Arcs of Circles   370
    Other Pictographs   371
    Signs for Similarity and Congruence 372
    The Sign ⇌ for Equivalence   375
    Lettering of Geometric Figures   376
    Sign for Spherical Excess   380
    Symbols in the Statement of Theorems 381
    Signs for Incommensurables   382
    Unusual Ideographs in Elementary Geometry . . . 383
    Algebraic Symbols in Elementary Geometry   . . . 384

B. Past Struggles between Symbolists and Rhetoricians in
    Elementary Geometry   385

INDEX

ILLUSTRATIONS

FIGURE PARAGRAPHS

1. Babylonian Tablets of Nippur 4

2. Principle of Subtraction in Babylonian Numerals . . . 9

3. Babylonian Lunar Tables 11

4. Mathematical Cuneiform Tablet CBS 8536 in the Museum of the University of Pennsylvania 11

5. Egyptian Numerals 17

6. Egyptian Symbolism for Simple Fractions 18

7. Algebraic Equation in Ahmes . . . . 2?

8. Hieroglyphic, Hieratic, and Coptic Numerals 24

9. Palmyra (Syria) Numerals 27

10. Syrian Numerals 28

11. Hebrew Numerals 30

12. Computing Table of Salamis 36

13. Account of Disbursements of the Athenian State, 418–415 b.c. 36

14. Arabic Alphabetic Numerals 45

15. Degenerate Forms of Roman Numerals 56

16. Quipu from Ancient Chancay in Peru 65

17. Diagram of the Two Right-Hand Groups 65

18. Aztec Numerals 66

19. Dresden Codex of Maya 67

20. Early Chinese Knots in Strings, Representing Numerals . 70

21. Chinese and Japanese Numerals 74

22. Hill's Table of Boethian Apices 80

23. Table of Important Numeral Forms 80

24. Old Arabic and Hindu-Arabic Numerals 83

25. Numerals of the Monk Neophytos 88

26. Chr. Rudolff's Numerals and Fractions 89

27. A Contract, Mexico City, 1649 93

FIGURE	PARAGRAPHS
28. REAL ESTATE SALE, MEXICO CITY, 1718	94
29. FANCIFUL HYPOTHESES	96
30. NUMERALS DESCRIBED BY NOVIOMAGUS	98
31. SANSKRIT SYMBOLS FOR THE UNKNOWN	108
32. BAKHSHĀLĪ ARITHMETIC	109
33. ŚRĪDHARA'S *Trisátikā*	112
34. ORESME'S *Algorismus Proportionum*	123
35. al-QALASÂDÎ'S ALGEBRAIC SYMBOLS	125
36. COMPUTATIONS OF REGIOMONTANUS	127
37. CALENDAR OF REGIOMONTANUS	128
38. FROM EARLIEST PRINTED ARITHMETIC	128
39. MULTIPLICATIONS IN THE "TREVISO" ARITHMETIC	128
40. DE LA ROCHE'S *Larismethique*, FOLIO 60*B*	132
41. DE LA ROCHE'S *Larismethique*, FOLIO 66*A*	132
42. PART OF PAGE IN PACIOLI'S *Summa*, 1523	138
43. MARGIN OF FOLIO 123*B* IN PACIOLI'S *Summa*	139
44. PART OF FOLIO 72 OF GHALIGAI'S *Practica d'arithmetica*, 1552	139
45. GHALIGAI'S *Practica d'arithmetica*, FOLIO 198	139
46. CARDAN, *Ars magna*, ED. 1663, PAGE 255	141
47. CARDAN, *Ars magna*, ED. 1663, PAGE 297	141
48. FROM TARTAGLIA'S *General Trattato*, 1560	143
49. FROM TARTAGLIA'S *General Trattato*, FOLIO 4	144
50. FROM BOMBELLI'S *Algebra*, 1572	144
51. BOMBELLI'S *Algebra* (1579 IMPRESSION), PAGE 161	145
52. FROM THE MS OF BOMBELLI'S *Algebra* IN THE LIBRARY OF BOLOGNA	145
53. FROM PAMPHLET NO. 595*N* IN THE LIBRARY OF THE UNIVERSITY OF BOLOGNA	146
54. WIDMAN'S *Rechnung*, 1526	146
55. FROM THE ARITHMETIC OF GRAMMATEUS	146
56. FROM THE ARITHMETIC OF GRAMMATEUS, 1535	147
57. FROM THE ARITHMETIC OF GRAMMATEUS, 1518(?)	147
58. FROM CHR. RUDOLFF'S *Coss*, 1525	148

FIGURE   PARAGRAPHS

59. From Chr. Rudolff's *Coss*, Ev. 148

60. From Van der Hoecke' *In arithmetica* 150

61. Part of Page from Stifel's *Arithmetica integra*, 1544 . . . 150

62. From Stifel's *Arithmetica integra*, Folio 31*B* 152

63. From Stifel's Edition of Rudolff's *Coss*, 1553 156

64. Scheubel, Introduction to Euclid, Page 28 159

65. W. Klebitius, Booklet, 1565 161

66. From Clavius' *Algebra*, 1608 161

67. From S. Stevin's *Le Thiende*, 1585 162

68. From S. Stevin's *Arithmetiqve* 162

69. From S. Stevin's *Arithmetiqve* 164

70. From Aurel's *Arithmetica* 165

71. R. Recorde, *Whetstone of Witte*, 1557 168

72. Fractions in Recorde 168

73. Radicals in Recorde 168

74. Radicals in Dee's Preface 169

75. Proportion in Dee's Preface 169

76. From Digges's *Stratioticos* 170

77. Equations in Digges 172

78. Equality in Digges 172

79. From Thomas Masterson's *Arithmeticke*, 1592 172

80. J. Peletier's *Algebra*, 1554 172

81. Algebraic Operations in Peletier's *Algebra* 172

82. From J. Buteon, *Arithmetica*, 1559 173

83. Gosselin's *De arte magna*, 1577 174

84. Vieta, *In artem analyticam*, 1591 176

85. Vieta, *De emendatione aeqvationvm* 178

86. B. Cavalieri, *Exercitationes*, 1647 179

87. From Thomas Harriot, 1631, Page 101 180

88. From Thomas Harriot, 1631, Page 65 180

89. From Hérigone, *Cursus mathematicus*, 1644 189

90. Roman Numerals for *x* in J. Hume, 1635 191

FIGURE
PARAGRAPHS

91. RADICALS IN J. HUME, 1635 191

92. R. DESCARTES, *Géométrie* 191

93. I. BARROW'S *Euclid*, LATIN EDITION. NOTES BY ISAAC NEWTON . 193

94. I. BARROW'S *Euclid*, ENGLISH EDITION 193

95. RICH. RAWLINSON'S SYMBOLS 194

96. RAHN'S *Teutsche Algebra*, 1659 195

97. BRANCKER'S TRANSLATION OF RAHN, 1668 195

98. J. WALLIS, 1657 195

99. FROM THE HIEROGLYPHIC TRANSLATION OF THE AHMES PAPYRUS 200

100. MINUS SIGN IN THE GERMAN MS C. 80, DRESDEN LIBRARY . . 201

101. PLUS AND MINUS SIGNS IN THE LATIN MS C. 80, DRESDEN LIBRARY 201

102. WIDMANS' MARGINAL NOTE TO MS C. 80, DRESDEN LIBRARY . 201

103. FROM THE ARITHMETIC OF BOETHIUS, 1488 250

104. SIGNS IN GERMAN MSS AND EARLY GERMAN BOOKS 294

105. WRITTEN ALGEBRAIC SYMBOLS FOR POWERS FROM PEREZ DE MOYA'S *Arithmetica* 294

106. E. WARING'S REPEATED EXPONENTS 313

Volume I

NOTATIONS IN ELEMENTARY MATHEMATICS

I

INTRODUCTION

In this history it has been an aim to give not only the first appearance of a symbol and its origin (whenever possible), but also to indicate the competition encountered and the spread of the symbol among writers in different countries. It is the latter part of our program which has given bulk to this history.

The rise of certain symbols, their day of popularity, and their eventual decline constitute in many cases an interesting story. Our endeavor has been to do justice to obsolete and obsolescent notations, as well as to those which have survived and enjoy the favor of mathematicians of the present moment.

If the object of this history of notations were simply to present an array of facts, more or less interesting to some students of mathematics—if, in other words, this undertaking had no ulterior motive—then indeed the wisdom of preparing and publishing so large a book might be questioned. But the author believes that this history constitutes a mirror of past and present conditions in mathematics which can be made to bear on the notational problems now confronting mathematics. The successes and failures of the past will contribute to a more speedy solution of the notational problems of the present time.

NUMERAL SYMBOLS AND COMBINATIONS OF SYMBOLS

BABYLONIANS

1. In the Babylonian notation of numbers a vertical wedge Υ stood for 1, while the characters $\langle$ and $\Upsilon\!\!\succ$ signified 10 and 100, respectively. Grotefend[1] believes the character for 10 originally to have been the picture of two hands, as held in prayer, the palms being pressed together, the fingers close to each other, but the thumbs thrust out. Ordinarily, two principles were employed in the Babylonial notation—the additive and multiplicative. We shall see that limited use was made of a third principle, that of subtraction.

2. Numbers below 200 were expressed ordinarily by symbols whose respective values were to be *added*. Thus, $\Upsilon\!\!\succ\!\!\langle\langle\Upsilon\Upsilon\Upsilon$ stands for 123. The principle of multiplication reveals itself in $\langle\,\Upsilon\!\!\succ$ where the smaller symbol 10, placed before the 100, is to be multiplied by 100, so that this symbolism designates 1,000.

3. These cuneiform symbols were probably invented by the early Sumerians. Their inscriptions disclose the use of a decimal scale of numbers and also of a *sexagesimal* scale.[2]

Early Sumerian clay tablets contain also numerals expressed by circles and curved signs, made with the blunt circular end of a stylus, the ordinary wedge-shaped characters being made with the pointed end. A circle ● stood for 10, a semicircular or lunar sign stood for 1. Thus, a "round-up" of cattle shows ●●DDD, or 36, cows.[3]

4. The sexagesimal scale was first discovered on a tablet by E. Hincks[4] in 1854. It records the magnitude of the illuminated portion

---

[1] His first papers appeared in *Göttingische Gelehrte Anzeigen* (1802), Stück 149 und 178; *ibid.* (1803), Stück 60 und 117.

[2] In the division of the year and of the day, the Babylonians used also the duodecimal plan.

[3] G. A. Barton, *Haverford Library Collection of Tablets*, Part I (Philadelphia, 1905), Plate 3, HCL 17, obverse; see also Plates 20, 26, 34, 35. Allotte de la Fuye, "En-e-tar-zi patési de Lagaš," *H. V. Hilprecht Anniversary Volume* (Chicago, 1909), p. 128, 133.

[4] "On the Assyrian Mythology," *Transactions of the Royal Irish Academy.* "Polite Literature," Vol. XXII, Part 6 (Dublin, 1855), p. 406, 407.

of the moon's disk for every day from new to full moon, the whole disk being assumed to consist of 240 parts. The illuminated parts during the first five days are the series 5, 10, 20, 40, 1.20, which is a geometrical progression, on the assumption that the last number is 80. From here on the series becomes arithmetical, 1.20, 1.36, 1.52, 2.8, 2.24, 2.40, 2.56, 3.12, 3.28, 3.44, 4, the common difference being 16. The last number is written in the tablet ⌄⌄⌄, and, according to Hincks's interpretation, stood for $4 \times 60 = 240$.

Obverse.   *Reverse.*

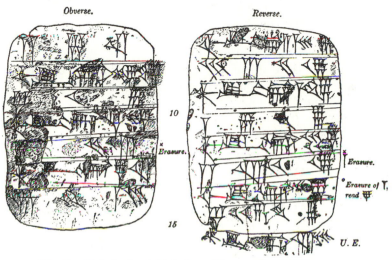

Fig. 1.—Babylonian tablets from Nippur, about 2400 B.C.

5. Hincks's explanation was confirmed by the decipherment of tablets found at Senkereh, near Babylon, in 1854, and called the *Tablets of Senkereh*. One tablet was found to contain a table of square numbers, from 1^2 to 60^2, a second one a table of cube numbers from 1^3 to 32^3. The tablets were probably written between 2300 and 1600 B.C. Various scholars contributed toward their interpretation. Among them were George Smith (1872), J. Oppert, Sir H. Rawlinson, Fr. Lenormant, and finally R. Lepsius.[1] The numbers 1, 4, 9, 16, 25, 36,

[1] George Smith, *North British Review* (July, 1870), p. 332 n.; J. Oppert, *Journal asiatique* (August–September, 1872; October–November, 1874); J. Oppert, *Étalon des mesures assyr. fixé par les textes cunéiformes* (Paris, 1874); Sir H. Rawlinson and G. Smith, "The Cuneiform Inscriptions of Western Asia," Vol. IV: *A Selection from the Miscellaneous Inscriptions of Assyria* (London, 1875), Plate 40; R. Lepsius, "Die Babylonisch-Assyrischen Längenmaasse nach der Tafel von Senkereh," *Abhandlungen der Königlichen Akademie der Wissenschaften zu Berlin* (aus dem Jahre 1877 [Berlin, 1878], Philosophisch-historische Klasse), p. 105–44.

and 49 are given as the squares of the first seven integers, respectively.
We have next $1.4 = 8^2$, $1.21 = 9^2$, $1.40 = 10^2$, etc. This clearly indicates
the use of the sexagesimal scale which makes $1.4 = 60+4$, $1.21 = 60+$
21, $1.40 = 60+40$, etc. This sexagesimal system marks the earliest
appearance of the all-important "principle of position" in writing
numbers. In its general and systematic application, this principle requires a symbol for zero. But no such symbol has been found on early
Babylonian tablets; records of about 200 B.C. give a symbol for zero,
as we shall see later, but it was not used in calculation. The earliest
thorough and systematic application of a symbol for zero and the
principle of position was made by the Maya of Central America, about
the beginning of the Christian Era.

6. An extension of our knowledge of Babylonian mathematics
was made by H. V. Hilprecht who made excavations at Nuffar (the
ancient Nippur). We reproduce one of his tablets[1] in Figure 1.

Hilprecht's transliteration, as given on page 28 of his text is
as follows:

Line 1.	125	720		Line 9.	2,000	18
Line 2.	IGI-GAL-BI	103,680		Line 10.	IGI-GAL-BI	6,480
Line 3.	250	360		Line 11.	4,000	9
Line 4.	IGI-GAL-BI	51,840		Line 12.	IGI-GAL-BI	3,240
Line 5.	500	180		Line 13.	8,000	18
Line 6.	IGI-GAL-BI	25,920		Line 14.	IGI-GAL-BI	1,620
Line 7.	1,000	90		Line 15.	16,000	9
Line 8.	IGI-GAL-BI	12,960		Line 16.	IGI-GAL-BI	810

7. In further explanation, observe that in

Line 1.	$125 = 2 \times 60 + 5$,	$720 = 12 \times 60 + 0$
Line 2.	Its denominator,	$103,680 = [28 \times 60 + 48(?)] \times 60 + 0$
Line 3.	$250 = 4 \times 60 + 10$,	$360 = 6 \times 60 + 0$
Line 4.	Its denominator,	$51,840 = [14 \times 60 + 24] \times 60 + 0$
Line 5.	$500 = 8 \times 60 + 20$,	$180 = 3 \times 60 + 0$
Line 6.	Its denominator,	$25,920 = [7 \times 60 + 12] \times 60 + 0$
Line 7.	$1,000 = 16 \times 60 + 40$,	$90 = 1 \times 60 + 30$
Line 8.	Its denominator,	$12,960 = [3 \times 60 + 36] \times 60 + 0$

[1] *The Babylonian Expedition of the University of Pennsylvania.* Series A:
"Cuneiform Texts," Vol. XX, Part 1. *Mathematical, Metrological and Chrono-
logical Tablets from the Temple Library of Nippur* (Philadelphia, 1906), Plate 15,
No. 25.

Line 9. $2,000 = 33 \times 60 + 20$, $18 = 10 + 8$
Line 10. Its denominator, $6,480 = [1 \times 60 + 48] \times 60 + 0$
Line 11. $4,000 = [1 \times 60 + 6] \times 60 + 40$,   9
Line 12. Its denominator, $3,240 = 54 \times 60 + 0$
Line 13. $8,000 = [2 \times 60 + 13] \times 60 + 20$,   18
Line 14. Its denominator, $1,620 = 27 \times 60 + 0$
Line 15. $16,000 = [4 \times 60 + 26] \times 60 + 40$,   9
Line 16. Its denominator, $810 = 13 \times 60 + 30$
$IGI\text{-}GAL =$ Denominator, $BI =$ Its, i.e., the number 12,960,000 or 60^4.

We quote from Hilprecht (*op. cit.*, pp. 28–30):

"We observe (*a*) that the first numbers of all the odd lines (1, 3, 5, 7, 9, 11, 13, 15) form an increasing, and all the numbers of the even lines (preceded by *IGI-GAL-BI* = 'its denominator') a descending geometrical progression; (*b*) that the first number of every odd line can be expressed by a fraction which has 12,960,000 as its numerator and the closing number of the corresponding even line as its denominator, in other words,

$$125 = \frac{12,960,000}{103,680} \; ; \qquad 250 = \frac{12,960,000}{51,840} \; ; \qquad 500 = \frac{12,960,000}{25,920} \; ;$$

$$1,000 = \frac{12,960,000}{12,960} \; ; \qquad 2,000 = \frac{12,960,000}{6,480} \; ; \qquad 4,000 = \frac{12,960,000}{3,240} \; ;$$

$$8,000 = \frac{12,960,000}{1,620} \; ; \qquad 16,000 = \frac{12,960,000}{810} \; .$$

But the closing numbers of all the odd lines (720, 360, 180, 90, 18, 9, 18, 9) are still obscure to me. . . . . .

"The question arises, what is the meaning of all this? What in particular is the meaning of the number 12,960,000 ($= 60^4$ or $3,600^2$) which underlies all the mathematical texts here treated . . . . ? . . . . This 'geometrical number' (12,960,000), which he [Plato in his *Republic* viii. 546*B–D*] calls 'the lord of better and worse births,' is the arithmetical expression of a great law controlling the Universe. According to Adam this law is 'the Law of Change, that law of inevitable degeneration to which the Universe and all its parts are subject'—an interpretation from which I am obliged to differ. On the contrary, it is the Law of Uniformity or Harmony, i.e. that fundamental law which governs the Universe and all its parts, and which cannot be ignored and violated without causing an anomaly, i.e. without resulting in a degeneration of the race." The nature of the "Platonic number" is still a debated question.

8. In the reading of numbers expressed in the Babylonian sexagesimal system, uncertainty arises from the fact that the early Babylonians had no symbol for zero. In the foregoing tablets, how do we know, for example, that the last number in the first line is 720 and not 12? Nothing in the symbolism indicates that the 12 is in the place where the local value is "sixties" and not "units." Only from the study of the entire tablet has it been inferred that the number intended is 12×60 rather than 12 itself. Sometimes a horizontal line was drawn following a number, apparently to indicate the absence of units of lower denomination. But this procedure was not regular, nor carried on in a manner that indicates the number of vacant places.

9. To avoid confusion some Babylonian documents even in early times contained symbols for 1, 60, 3,600, 216,000, also for 10, 600, 36,000.[1] Thus · was 10, ● was 3,600, ⊙ was 36,000.

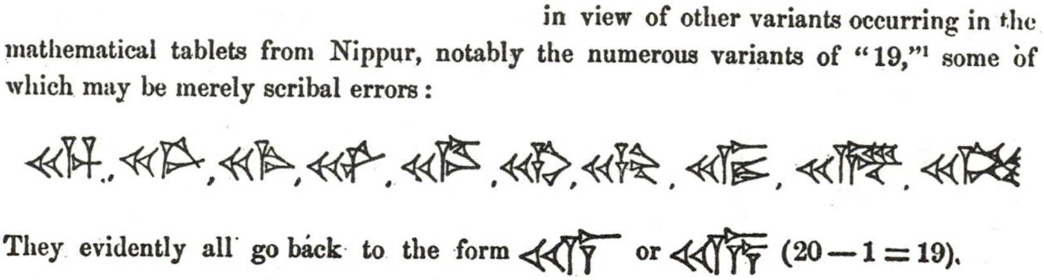

in view of other variants occurring in the mathematical tablets from Nippur, notably the numerous variants of "19,"[1] some of which may be merely scribal errors:

They evidently all go back to the form ⟪⟪𝌆 or ⟪⟪𝌆 (20−1=19).

Fig. 2.—Showing application of the principle of subtraction

10. Besides the principles of addition and multiplication, Babylonian tablets reveal also the use of the principle of subtraction, which is familiar to us in the Roman notation XIX (20−1) for the number 19. Hilprecht has collected ideograms from the Babylonian tablets which he has studied, which represent the number 19. We reproduce his symbols in Figure 2. In each of these twelve ideograms (Fig. 2), the two symbols to the left signify together 20. Of the symbols immediately to the right of the 20, one vertical wedge stands for "one" and the remaining symbols, for instance ⟨, for *LAL* or "minus"; the entire ideogram represents in each of the twelve cases the number 20−1 or 19.

One finds the principle of subtraction used also with curved signs;[2] D ●● ⟨ D meant 60+20−1, or 79.

---

[1] See François Thureau-Dangin, *Recherches sur l'origine de l'écriture cunéiforme* (Paris, 1898), Nos. 485–91, 509–13. See also G. A. Barton, *Haverford College Library Collection of Cuneiform Tablets*, Part I (Philadelphia, 1905), where the forms are somewhat different; also the *Hilprecht Anniversary Volume* (Chicago, 1909), p. 128 ff.

[2] G. A. Barton, *op. cit.*, Plate 3, obverse.

11. The symbol used about the second century B.C. to designate the *absence* of a number, or a blank space, is shown in Figure 3, containing numerical data relating to the moon.[1] As previously stated, this symbol, $\leqq$, was not used in computation and therefore performed

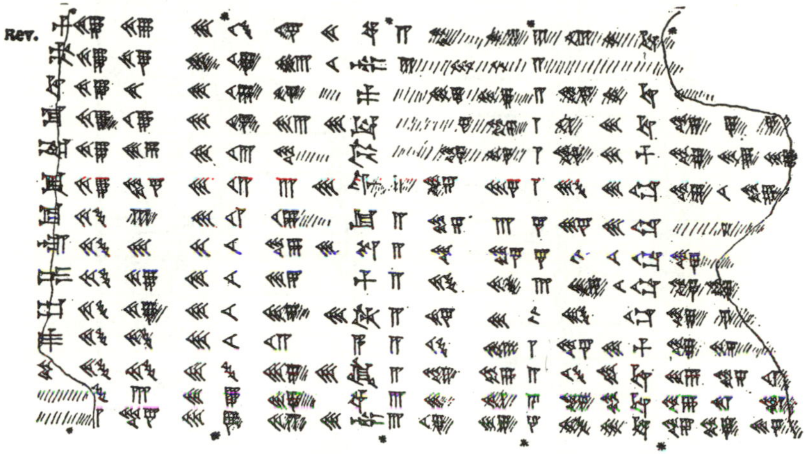

Fig. 3.—Babylonian lunar tables, reverse; full moon for one year, about the end of the second century B.C.

only a small part of the functions of our modern zero. The symbol is seen in the tablet in row 10, column 12; also in row 8, column 13. Kugler's translation of the tablet, given in his book, page 42, is shown below. Of the last column only an indistinct fragment is preserved; the rest is broken off.

REVERSE

1...	*Nisannu*	28°56′30″	19°16′ ″	Librae	3ᶻ 6°45′	4ᴵ74ᴵᴵ10ᴵᴵᴵ	sik
2...	Airu	28 38 30	17 54 30	Scorpii	3 21 28	6 20 30	sik
3...	*Simannu*	28 20 30	16 15	Arcitenentis	3 31 39	3 45 30	sik
4...	*Dûzu*	28 18 30	14 33 30	Capri	3 34 41	1 10 30	sik
5...	*Âbu*	28 36 30	13 9	Aquarii	3 27 56	1 24 30	bar
6...	Ulûlu	29 54 30	13 3 30	Piscium	3 15 34	1 59 30	num
7...	Tišrîtu	29 12 30	11 16	Arietis	2 58 3	4 34 30	num
8...	Araḫ-s.	29 30 30	10 46 30	Tauri	2 40 54	6 0 10	num
9...	Kislimu	29 48 30	10 35	Geminorum	2 29 29	3 25 10	num
10...	Tebitu	29 57 30	10 32 30	Cancri	2 24 30	0 57 10	num
11...	Šabâtu	29 39 30	10 12	Leonis	2 30 53	1 44 50	bar
12...	*Adâru* I	29 21 30	9 33 30	Virginis	2 42 56	2 19 50	sik
13...	*Adâru* II	29 3 30	8 36	Librae	3 0 21	4 54 50	sik
14...	*Nisannu*	28 45 30	7 21 30	Scorpii	3 17 36	5 39 50	sik

[1] Franz Xaver Kugler, S.J., *Die babylonische Mondrechnung* (Freiburg im Breisgau, 1900), Plate IV, No. 99 (81–7–6), lower part.

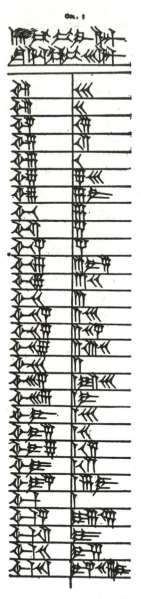

Fig. 4.—Mathematical cuneiform tablet, CBS 8536, in the Museum of the University of Pennsylvania.

12. J. Oppert pointed out the Babylonian use of a designation for the sixths, viz., $\frac{1}{6}$, $\frac{1}{3}$, $\frac{1}{2}$, $\frac{2}{3}$, $\frac{5}{6}$. These are unit fractions or fractions whose numerators are one less than the denominators.[1] He also advanced evidence pointing to the Babylonian use of sexagesimal *fractions* and the use of the sexagesimal system in weights and measures. The occurrence of sexagesimal fractions is shown in tablets recently examined. We reproduce in Figure 4 two out of twelve columns found on a tablet described by H. F. Lutz.[2] According to Lutz, the tablet "cannot be placed later than the Cassite period, but it seems more probable that it goes back even to the First Dynasty period, *ca.* 2000 B.C."

13. To mathematicians the tablet is of interest because it reveals operations with sexagesimal fractions resembling modern operations with decimal fractions. For example, 60 is divided by 81 and the quotient expressed sexagesimally. Again, a sexagesimal number with two fractional places, 44(26)(40), is multiplied by itself, yielding a product in four fractional places, namely, [32]55(18)(31)(6)(40). In this notation the [32] stands for 32×60 units, and to the (18), (31), (6), (40) must be assigned, respectively, the denominators 60, 60^2, 60^3, 60^4.

The tablet contains twelve columns of figures. The first column (Fig. 4) gives the results of dividing 60 in succession by twenty-nine different divisors from 2 to 81. The eleven other columns contain tables of multiplication; each of the numbers 50, 48, 45, 44(26)(40), 40, 36, 30, 25, 24, 22(30), 20 is multiplied by integers up to 20, then by the numbers 30, 40, 50, and finally by itself. Using our modern numerals, we interpret on page 10 the first and the fifth columns. They exhibit a larger number of fractions than do the other columns. The Babylonians had no mark separating the fractional from the integral parts of a number. Hence a number like 44(26)(40) might be interpreted in different ways; among the possible meanings are $44 \times 60^2 + 26 \times 60 + 40$, $44 \times 60 + 26 + 40 \times 60^{-1}$, and $44 + 26 \times 60^{-1} + 40 \times 60^{-2}$. Which interpretation is the correct one can be judged only by the context, if at all.

The exact meaning of the first two lines in the first column is uncertain. In this column 60 is divided by each of the integers written on the left. The respective quotients are placed on the right.

[1] Symbols for such fractions are reproduced also by Thureau-Dangin, *op. cit.*, Nos. 481–84, 492–508, and by G. A. Barton, *Haverford College Library Collection of Cuneiform Tablets*, Part I (Philadelphia, 1905).

[2] "A Mathematical Cuneiform Tablet," *American Journal of Semitic Languages and Literatures*, Vol. XXXVI (1920), p. 249–57.

In the fifth column the multiplicand is 44(26)(40) or $44\frac{4}{9}$. The last two lines seem to mean "$60^2 \div 44(26)(40) = 81$, $60^2 \div 81 = 44(26)(40)$."

First Column
. . . . gal (?) -bi 40 -ám
šu *a- na* gal-bi 30 -ám

igi 2	30
igi 3	20
igi 4	15
igi 5	12
igi 6	10
igi 8	7(30)
igi 9	6(40)
igi 10	6
igi 12	5
igi 15	4
igi 16	3(45)
igi 18	3(20)
igi 20	3
igi 24	2(30)
igi 25	2(24)
igi 28*	2(13)(20)
igi 30	2
igi 35*	1(52)(30)
igi 36	1(40)
igi 40	1(30)
igi 45	1(20)
igi 48	1(15)
igi 50	1(12)
igi 54	1(6)(40)
igi 60	1
igi 64	(56)(15)
igi 72	(50)
igi 80	(45)
igi 81	(44)(26)(40)

Fifth Column

	44(26)(40)
1	44(26)(40)
2	[1]28(53)(20)
3	[2]13(20)
4	[2]48(56)(40)*
5	[3]42(13)(20)
6	[4]26(40)
7	[5]11(6)(40)
9	[6]40
10	[7]24(26)(40)
11	[8]8(53)(20)
12	[8]53(20)
13	[9]27(46)(40)*
14	[10]22(13)(20)
15	[11]6(40)
16	[11]51(6)(40)
17	[12]35(33)(20)
18	[13]20
19	[14]4(26)(40)
20	[14]48(53)(20)
30	[22]13(20)
40	[29]37(46)(40)
50	[38]2(13)(20)*
44(26)(40)a-na 44(26)(40)	
[32]55(18)(31)(6)(40)	
44(26)(40) square	
igi 44(26)(40)	81
igi 81	44(26)(40)

Numbers that are incorrect are marked by an asterisk (*).

14. The Babylonian use of sexagesimal fractions is shown also in a clay tablet described by A. Ungnad.[1] In it the diagonal of a rectangle whose sides are 40 and 10 is computed by the approximation

[1] *Orientalische Literaturzeitung* (ed. Peise, 1916), Vol. XIX, p. 363–68. See also Bruno Meissner, *Babylonien und Assyrien* (Heidelberg, 1925), Vol. II, p. 393.

$40+2\times40\times10^2\div60^2$, yielding $42(13)(20)$, and also by the approximation $40+10^2\div\{2\times40\}$, yielding $41(15)$. Translated into the decimal scale, the first answer is $42.22+$, the second is 41.25, the true value being $41.23+$. These computations are difficult to explain, except on the assumption that they involve sexagesimal *fractions*.

15. From what has been said it appears that the Babylonians had ideograms which, transliterated, are *Igi-Gal* for "denominator" or "division," and *Lal* for "minus." They had also ideograms which, transliterated, are *Igi-Dua* for "division," and *A-Du* and *Ara* for "times," as in $Ara-1$ 18, for "$1\times18=18$," $Ara-2$ 36 for "$2\times18=36$"; the *Ara* was used also in "squaring," as in 3 Ara 3   9 for "$3\times3=9$." They had the ideogram $Ba-Di-E$ for "cubing," as in 27-E 3 $Ba-Di-E$ for "$3^3=27$"; also $Ib-Di$ for "square," as in 9-E 3 $Ib-Di$ for "$3^2=9$." The sign $A-An$ rendered numbers "distributive."[1]

EGYPTIANS

16. The Egyptian number system is based on the scale of 10, although traces of other systems, based on the scales of 5, 12, 20, and 60, are believed to have been discovered.[2] There are three forms of Egyptian numerals: the hieroglyphic, hieratic, and demotic. Of these the hieroglyphic has been traced back to about 3300 B.C.;[3] it is found mainly on monuments of stone, wood, or metal. Out of the hieroglyphic sprang a more cursive writing known to us as hieratic. In the beginning the hieratic was simply the hieroglyphic in the rounded forms resulting from the rapid manipulation of a reed-pen as contrasted with the angular and precise shapes arising from the use of the chisel. About the eighth century B.C. the demotic evolved as a more abbreviated form of cursive writing. It was used since that time down to the beginning of the Christian Era. The important mathematical documents of ancient Egypt were written on papyrus and made use of the hieratic numerals.[4]

[1] Hilprecht, *op. cit.*, p. 23; Arno Poebel, *Grundzüge der sumerischen Grammatik* (Rostock, 1923), p. 115; B. Meissner, *op. cit.*, p. 387–89.

[2] Kurt Sethe, *Von Zahlen und Zahlworten bei den alten Ägyptern* (Strassburg, 1916), p. 24–29.

[3] J. E. Quibell and F. W. Green, *Hierakonopolis* (London, 1900–1902), Part I, Plate 26B, who describe the victory monument of King *Ncr-mr;* the number of prisoners taken is given as 120,000, while 400,000 head of cattle and 1,422,000 goats were captured.

[4] The evolution of the hieratic writing from the hieroglyphic is explained in G. Möller, *Hieratische Paläographie*, Vol. I, Nos. 614 ff. The demotic writing

17. The hieroglyphic symbols were ‖ for 1, ∩ for 10, ℭ for 100, ⸂ for 1,000, ⎮ for 10,000, 🐟 for 100,000, 𓁨 for 1,000,000, ◯ for 10,000,000. The symbol for 1 represents a vertical staff; that for 1,000 a lotus plant; that for 10,000 a pointing finger; that for 100,000 a burbot; that for 1,000,000 a man in astonishment, or, as more recent

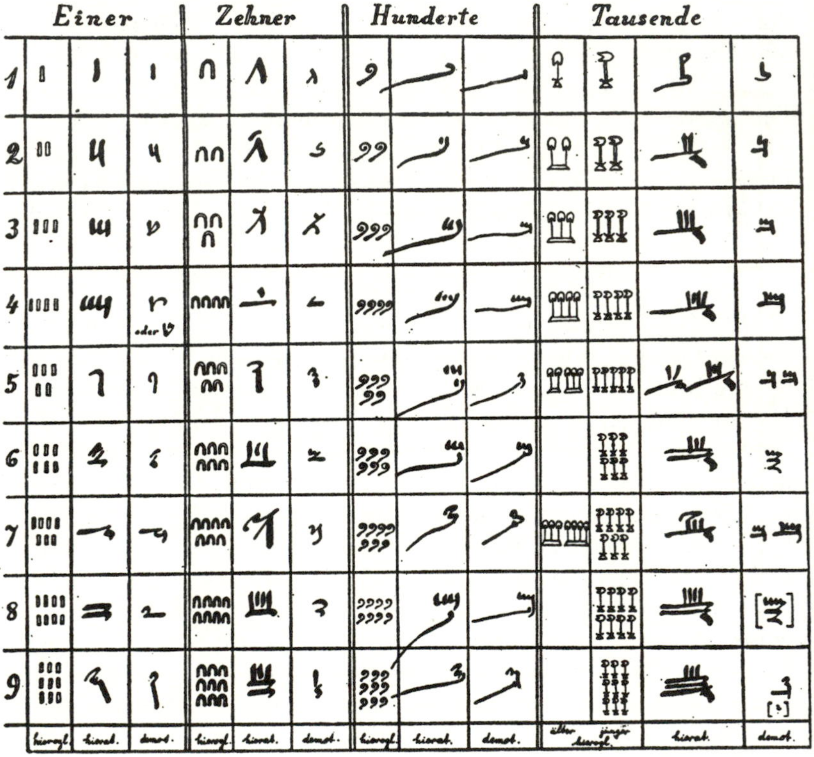

Fig. 5.—Egyptian numerals. Hieroglyphic, hieratic, and demotic numeral symbols. (This table was compiled by Kurt Sethe.)

Egyptologists claim, the picture of the cosmic deity Hh.[1] The symbols for 1 and 10 are sometimes found in a horizontal position.

18. We reproduce in Figures 5 and 6 two tables prepared by Kurt

is explained by F. L. Griffith, *Catalogue of the Demotic Papyri in the John Rylands Library* (Manchester, 1909), Vol. III, p. 415 ff., and by H. Brugsch, *Grammaire démotique*, §§ 131 ff.

[1] Sethe, *op. cit.*, p. 11, 12.

Sethe. They show the most common of the great variety of forms which are found in the expositions given by Möller, Griffith, and Brugsch.

Observe that the old hieratic symbol for $\frac{1}{4}$ was the cross $\times$, signifying perhaps a part obtainable from two sections of a body through the center.

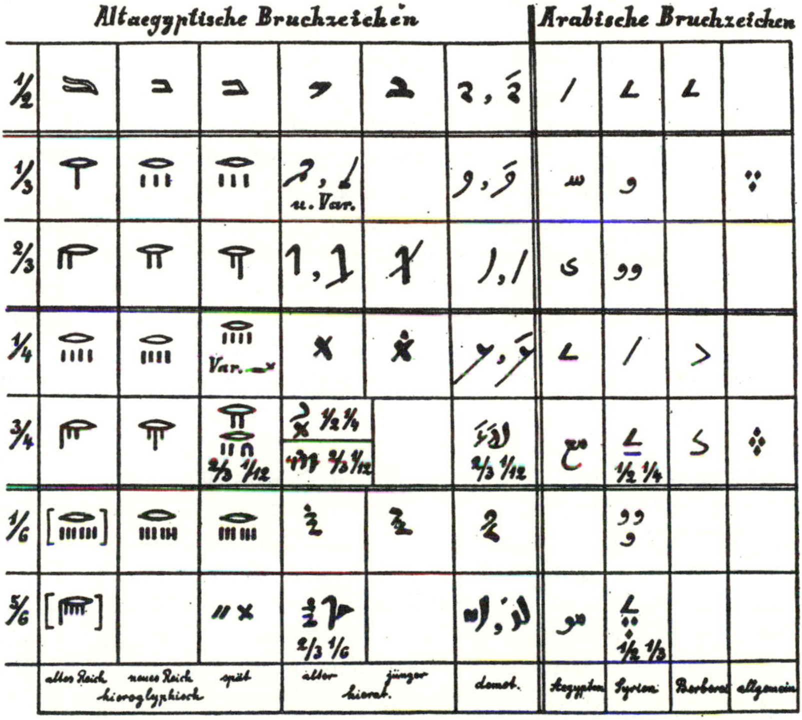

FIG. 6.—Egyptian symbolism for simple fractions. (Compiled by Kurt Sethe)

19. In writing numbers, the Egyptians used the principles of addition and multiplication. In applying the additive principle, not more than four symbols of the same kind were placed in any one group. Thus, 4 was written in hieroglyphs ||||; 5 was not written |||||, but either ||| || or $\begin{smallmatrix}|||\\||\end{smallmatrix}$. There is here recognized the same need which caused the Romans to write V after IIII, L=50 after XXXX=40, D=500 after CCCC=400. In case of two unequal groups, the Egyptians always wrote the larger group before, or above the smaller group; thus, seven was written $\begin{smallmatrix}||||\\|||\end{smallmatrix}$.

20. In the *older* hieroglyphs 2,000 or 3,000 was represented by two or three lotus plants grown *in one bush*. For example, 2,000 was 𝖄; correspondingly, 7,000 was designated by 𝖘𝖑𝖞 𝖘𝖑𝖞. The later hieroglyphs simply place two lotus plants together, to represent 2,000, without the appearance of springing from one and the same bush.

21. The multiplicative principle is not so old as the additive; it came into use about 1600–2000 B.C. In the oldest example hitherto known,[1] the symbols for 120, placed before a lotus plant, signify 120,000. A smaller number written before or below or above a symbol representing a larger unit designated multiplication of the larger by the smaller. Möller cites a case where 2,800,000 is represented by one burbot, with characters placed beneath it which stand for 28.

22. In hieroglyphic writing, unit fractions were indicated by placing the symbol ⌒ over the number representing the denominator. Exceptions to this are the modes of writing the fractions ½ and ⅔; the old hieroglyph for ⅓ was ⌒, the later was ⌒⌒; of the slightly varying hieroglyphic forms for ⅔, ⧎ was quite common.[2]

23. We reproduce an algebraic example in hieratic symbols, as it occurs in the most important mathematical document of antiquity known at the present time—the *Rhind papyrus*. The scribe, Ahmes, who copied this papyrus from an older document, used black and red ink, the red in the titles of the individual problems and in writing auxiliary numbers appearing in the computations. The example which, in the Eisenlohr edition of this papyrus, is numbered 34, is hereby shown.[3] Hieratic writing was from right to left. To facilitate the study of the problem, we write our translation from right to left and in the same relative positions of its parts as in the papyrus, except that numbers are written in the order familiar to us; i.e., 37 is written in our translation 37, and not 73 as in the papyrus. Ahmes writes unit fractions by placing a dot over the denominator, except in case of

[1] *Ibid.*, p. 8.

[2] *Ibid.*, p. 92–97, gives detailed information on the forms representing ⅔. The Egyptian procedure for decomposing a quotient into unit fractions is explained by V. V. Bobynin in *Abh. Gesch. Math.*, Vol. IX (1899), p. 3.

[3] *Ein mathematisches Handbuch der alten Ägypter* (*Papyrus Rhind des British Museum*) *übersetzt und erklärt* (Leipzig, 1877; 2d ed., 1891). The explanation of Problem 34 is given on p. 55, the translation on p. 213, the facsimile reproduction on Plate XIII of the first edition. The second edition was brought out without the plates. A more recent edition of the Ahmes papyrus is due to T. Eric Peet and appears under the title *The Rhind Mathematical Papyrus*, British Museum, Nos. 10057 and 10058, Introduction, Transcription, and Commentary (London, 1923).

$\frac{1}{2}$, $\frac{1}{3}$, $\frac{2}{3}$, $\frac{1}{4}$, each of which had its own symbol. Some of the numeral symbols in Ahmes deviate somewhat from the forms given in the two preceding tables; other symbols are not given in those tables. For the reading of the example in question we give here the following symbols:

Four	—	One-fourth	×	
Five	"|	Heap	 ᗱ ᚎ†	See Fig. 7
Seven	ᒾ	The whole	13	See Fig. 7
One-half	⊅	It gives	3⚡	See Fig. 7

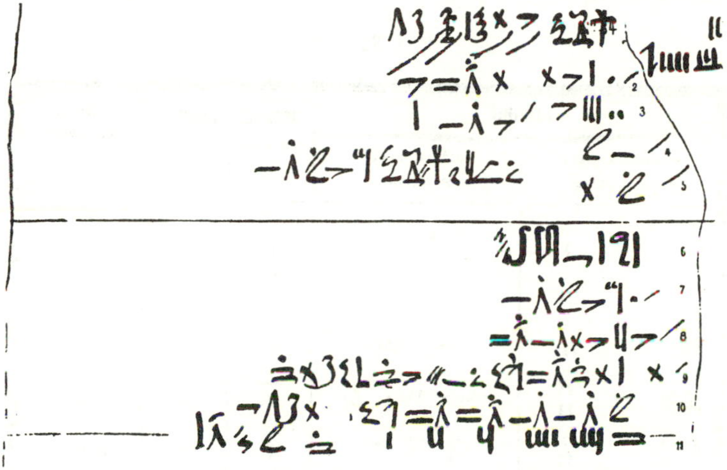

FIG. 7.—An algebraic equation and its solution in the Ahmes papyrus, 1700 B.C., or, according to recent authorities, 1550 B.C. (Problem 34, Plate XIII in Eisenlohr; p. 70 in Peet; in chancellor Chace's forthcoming edition, p. 76, as R. C. Archibald informs the writer.)

Translation (reading from right to left):

"10 gives it, whole its, $\frac{1}{4}$ its, $\frac{1}{2}$ its, Heap   No. 34

| $\frac{1}{2}$ | $\frac{1}{28}$ $\frac{1}{4}$ | $\frac{1}{4}$ $\frac{1}{2}$ 1 |
| 1 | $\frac{1}{14}$ $\frac{1}{2}$ | $\frac{1}{2}$ 3 .. |

$\frac{1}{14}$ $\frac{1}{7}$ $\frac{1}{2}$ 5 is heap the together   7   4

$\frac{1}{4}$ $\frac{1}{7}$

Proof the of Beginning

$\frac{1}{14}$ $\frac{1}{7}$ $\frac{1}{2}$ 5

$\frac{1}{28}$ $\frac{1}{14}$ $\frac{1}{4}$ $\frac{1}{2}$ 2 $\frac{1}{2}$

$\frac{1}{8}$ $\frac{1}{4}$ Remainder $\frac{1}{8}$ $\frac{1}{2}$ 9 together $\frac{1}{56}$ $\frac{1}{28}$ $\frac{1}{8}$ $\frac{1}{4}$ 1 $\frac{1}{4}$

14 gives $\frac{1}{4}$ $\frac{1}{6}$ $\frac{1}{28}$ $\frac{1}{28}$ $\frac{1}{14}$ $\frac{1}{14}$ $\frac{1}{7}$

21 Together .7 gives $\frac{1}{8}$ 1   2   2   4   4   8"

24. Explanation:

$$\text{The algebraic equation is } \frac{x}{2}+\frac{x}{4}+x=10$$

$$\text{i.e.,} \qquad (1+\tfrac{1}{2}+\tfrac{1}{4})x=10$$

The solution answers the question, By what must $(1\ \tfrac{1}{2}\ \tfrac{1}{4})$ be multiplied to yield the product 10? The four lines 2–5 contain on the right the following computation:

Twice $(1\ \tfrac{1}{2}\ \tfrac{1}{4})$ yields $3\tfrac{1}{2}$.

Four times $(1\ \tfrac{1}{2}\ \tfrac{1}{4})$ yields 7.

One-seventh of $(1\ \tfrac{1}{2}\ \tfrac{1}{4})$ is $\tfrac{1}{4}$.

↑° UNITES.

- SIGNES		LETTRES NUMÉRALES coptes.	VALEUR des SIGNES.	NOMS DE NOMBRE en dialecte thébain.
HIÉROGLYPHIQUES, creux et pleins.	HIÉRATIQUES, avec variantes.			
𝕀 ı	) ι ? ?	Ⲁ̄	1	oua.
𝕀𝕀 ıı	५ ५	Ⲃ̄	2	snau.
𝕀𝕀𝕀 ııı	ⱳ ⱳ	Ⲅ̄	3	choment.
𝕀𝕀 𝕀𝕀 ıı/ıı	ⱳⱳ ⱳⱳ Ⳑ	Ⲇ̄	4	ftoou.
𝕀𝕀𝕀 𝕀𝕀 ııı/ıı	⁊ ⅔ ⁊	Ⲉ̄	5	tiou.
𝕀𝕀𝕀 𝕀𝕀𝕀 ııı/ııı	⁄⁄ ⁄⁄	Ⲋ̄	6	soou.
𝕀𝕀𝕀𝕀 𝕀𝕀𝕀 ıııı/ııı	⌐ ⁊⁄ ⁊⁄	Ⲍ̄	7	sachf.
𝕀𝕀𝕀𝕀 𝕀𝕀𝕀𝕀 ıııı/ıııı	⇒ ⇒	Ⲏ̄	8	chmoun.
𝕀𝕀𝕀 𝕀𝕀𝕀 𝕀𝕀𝕀 ııı/ııı/ııı	⁊ ⁊	Ⲑ̄	9	psis.

[Continued on facing page]

[i.e., taking $(1\ \tfrac{1}{2}\ \tfrac{1}{4})$ once, then four times, together with $\tfrac{1}{7}$ of it, yields only 9; there is lacking 1. The remaining computation is on the four lines 2–5, on the left. Since $\tfrac{1}{7}$ of $(1\ \tfrac{1}{2}\ \tfrac{1}{4})$ yields $(\tfrac{1}{4}\ \tfrac{1}{14}\ \tfrac{1}{28})$ or $\tfrac{1}{4}$, $\tfrac{2}{7}$ or]

$(\tfrac{1}{4}\ \tfrac{1}{28})$ of $(1\ \tfrac{1}{2}\ \tfrac{1}{4})$, yields $\tfrac{1}{2}$.

And the double of this, namely, $(\tfrac{1}{2}\ \tfrac{1}{14})$ of $(1\ \tfrac{1}{2}\ \tfrac{1}{4})$ yields 1.

Adding together 1, 4, $\tfrac{1}{7}$ and $(\tfrac{1}{2}\ \tfrac{1}{14})$, we obtain Heap=$5\tfrac{1}{2}$

$\tfrac{1}{7}\ \tfrac{1}{14}$ or $5\tfrac{5}{7}$, the answer.

Proof.—5 $\frac{1}{2}$ $\frac{1}{7}$ $\frac{1}{14}$ is multiplied by (1 $\frac{1}{2}$ $\frac{1}{4}$) and the partial products are added. In the first line of the proof we have 5 $\frac{1}{2}$ $\frac{1}{7}$ $\frac{1}{14}$, in the second line half of it, in the third line one-fourth of it. Adding at first only the integers of the three partial products and the simpler fractions $\frac{1}{2}$, $\frac{1}{2}$, $\frac{1}{4}$, $\frac{1}{4}$, $\frac{1}{8}$, the partial sum is 9 $\frac{1}{2}$ $\frac{1}{8}$. This is $\frac{1}{4}$ $\frac{1}{8}$ short of 10. In the fourth line of the proof (l. 9) the scribe writes the remaining fractions and, reducing them to the common denominator 56, he writes (in

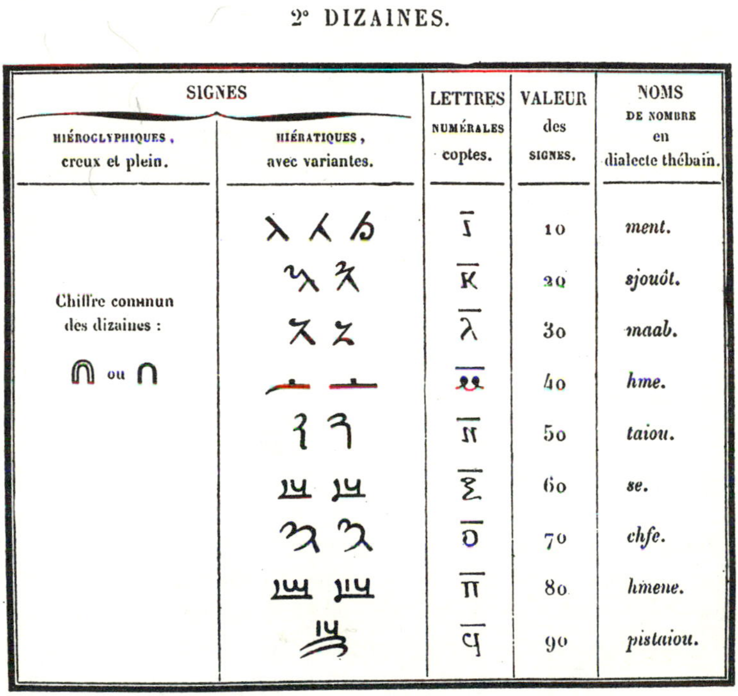

2° DIZAINES.

SIGNES		LETTRES NUMÉRALES coptes.	VALEUR des SIGNES.	NOMS DE NOMBRE en dialecte thébain.
HIÉROGLYPHIQUES, creux et plein.	HIÉRATIQUES, avec variantes.			
Chiffre commun des dizaines : ⋒ ou ⋒	⋋ ⋏ ♭	$\overline{\text{Ⲓ}}$	10	*ment.*
		$\overline{\text{Ⲕ}}$	20	*sjouôt.*
		$\overline{\lambda}$	30	*maab.*
		$\overline{\text{ⲙ}}$	40	*hme.*
		$\overline{\text{Ⲛ}}$	50	*taiou.*
		$\overline{\text{Ⲝ}}$	60	*se.*
		$\overline{\text{Ⲟ}}$	70	*chfe.*
		$\overline{\text{Ⲡ}}$	80	*hmene.*
		$\overline{\text{Ⳁ}}$	90	*pistaiou.*

FIG. 8.—Hieroglyphic, hieratic, and Coptic numerals. (Taken from A. P. Pihan, *Exposé des signes de numération* [Paris, 1860], p. 26, 27.)

red color) in the last line the numerators 8, 4, 4, 2, 2, 1 of the reduced fractions. Their sum is 21. But $\frac{21}{56} = \frac{14+7}{56} = \frac{1}{4} \frac{1}{8}$, which is the exact amount needed to make the total product 10.

A pair of legs symbolizing addition and subtraction, as found in impaired form in the Ahmes papyrus, are explained in § 200.

25. The Egyptian Coptic numerals are shown in Figure 8. They are of comparatively recent date. The hieroglyphic and hieratic **are**

the oldest Egyptian writing; the demotic appeared later. The Coptic writing is derived from the Greek and demotic writing, and was used by Christians in Egypt after the third century. The Coptic numeral symbols were adopted by the Mohammedans in Egypt after their conquest of that country.

26. At the present time two examples of the old Egyptian solution of problems involving what we now term "quadratic equations"[1] are known. For square root the symbol ⌐ has been used in the modern hieroglyphic transcription, as the interpretation of writing in the two papyri; for quotient was used the symbol ⅜ .

PHOENICIANS AND SYRIANS

27. The Phoenicians[2] represented the numbers 1–9 by the respective number of vertical strokes. Ten was usually designated by a horizontal bar. The numbers 11–19 were expressed by the juxtaposition of a horizontal stroke and the required number of vertical ones.

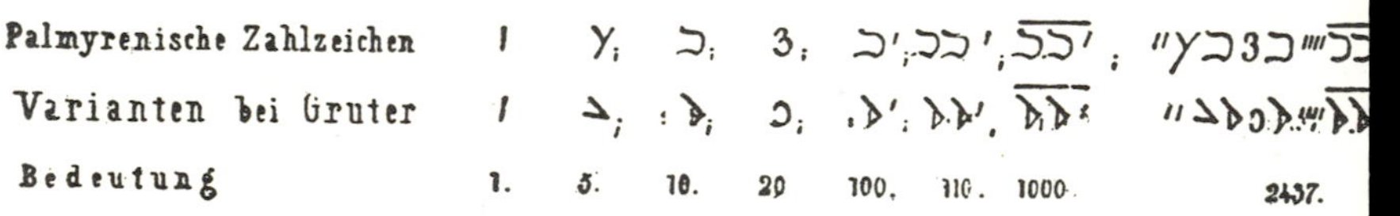

Palmyrenische Zahlzeichen	I	Y;	כ;	3;	כ',;ככ',;ככ'	, ΣΣ' ;	"Yככ3ככ ""ΣΣ
Varianten bei Gruter	I	�;	: ʖ;	כ;	,ʖ'; ʖʖ',	ʖʖ:	" ᐸʖכʖ.""ʖʖ
Bedeutung	1.	5.	10.	20	100.	110.	1000								2437.

Fig. 9.—Palmyra (Syria) numerals. (From M. Cantor, *Kulturleben, etc.*, Fig. 48)

As Phoenician writing proceeded from right to left, the horizontal stroke signifying 10 was placed farthest to the right. Twenty was represented by two parallel strokes, either horizontal or inclined and sometimes connected by a cross-line as in Η, or sometimes by two strokes, thus Λ. One hundred was written thus |<| or thus |ᴐ| . Phoenician inscriptions from which these symbols are taken reach back several centuries before Christ. Symbols found in Palmyra (modern Tadmor in Syria) in the first 250 years of our era resemble somewhat the numerals below 100 just described. New in the Palmyra numer-

[1] See H. Schack-Schackenburg, "Der Berliner Papyrus 6619," *Zeitschrift für ägyptische Sprache und Altertumskunde*, Vol. XXXVIII (1900), p. 136, 138, and Vol. XL (1902), p. 65–66.

[2] Our account is taken from Moritz Cantor, *Vorlesungen über Geschichte der Mathematik*, Vol. I (3d ed.; Leipzig, 1907), p. 123, 124; *Mathematische Beiträge zum Kulturleben der Völker* (Halle, 1863), p. 255, 256, and Figs. 48 and 49.

als is γ for 5. Beginning with 100 the Palmyra numerals contain new forms. Placing a | to the right of the sign for 10 (see Fig. 9) signifies multiplication of 10 by 10, giving 100. Two vertical strokes || mean 10×20, or 200; three of them, 10×30, or 300.

28. Related to the Phoenician are numerals of Syria, found in manuscripts of the sixth and seventh centuries A.D. Their shapes and their mode of combination are shown in Figure 10. The Syrians employed also the twenty-two letters of their alphabet to represent the numbers 1–9, the tens 10–90, the hundreds 100–400. The following hundreds were indicated by juxtaposition: $500 = 400 + 100$, $600 = 400 + 200, \ldots, 900 = 400 + 400 + 100$, or else by writing respectively 50–90 and placing a dot over the letter to express that its value is to be taken tenfold. Thousands were indicated by the letters for 1–9, with a stroke annexed as a subscript. Ten thousands were expressed

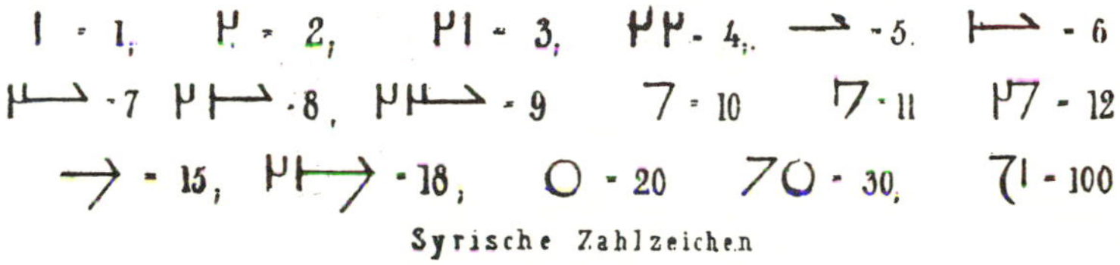

Syrische Zahlzeichen

FIG. 10.—Syrian numerals. (From M. Cantor, *Kulturleben*, etc., Fig. 49)

by drawing a small dash below the letters for one's and ten's. Millions were marked by the letters 1–9 with two strokes annexed as subscripts (i.e., $1,000 \times 1,000 = 1,000,000$).

❧ HEBREWS

29. The Hebrews used their alphabet of twenty-two letters for the designation of numbers, on the decimal plan, up to 400. Figure 11 shows three forms of characters: the Samaritan, Hebrew, and Rabbinic or cursive. The Rabbinic was used by commentators of the Sacred Writings. In the Hebrew forms, at first, the hundreds from 500 to 800 were represented by juxtaposition of the sign for 400 and a second number sign. Thus, תק stood for 500, תר for 600, תש for 700, תת for 800.

30. Later the end forms of five letters of the Hebrew alphabet came to be used to represent the hundreds 500–900. The five letters representing 20, 40, 50, 80, 90, respectively, had two forms; one of

LETTRES			NOMS ET TRANSCRIPTION DES LETTRES		VALEURS.	NOMS DE NOMBRE.
SAMARITAINES.	HÉBRAÏQUES.	RABBINIQUES.				
ᴀ	א	פ	aleph,	a	1	ekhâd.
ᴅ	ב	ב	bet,	b	2	chenaïm.
ᴅ	ג	ב	ghimel,	gh	3	chelochâh.
ᴄ	ד	ל	dalet,	d	4	arbâ´âh.
ᴇ	ה	ק	hé,	h	5	khamichâh.
ᴢ	ו	ז	waw,	w	6	chichâh.
ᴘ	ז	ז	zaïn,	z	7	chib´âh.
ᴨ	ח	ק	khet,	kh	8	chemonâh.
ᴧ	ט	ʋ	t´et´,	t´	9	tich´âh.
ᴔ	י	כ	iod,	i	10	´asârâh.
ᴣ	כ	כ	kaph,	k	20	´esrîm.
ᴣ	ל	ל	lamed,	l	30	chelochîm.
ᴤ	מ	ק	mem,	m	40	arbâ´îm.
ᴤ	נ	ɔ	noun,	n	50	khamichîm.
ᴪ	ס	ꭉ	s´amek	ś	60	chichîm.
ᴠ	ע	ʋ	´aïn,	´a	70	chib´îm.
ᴐ	פ	פ	phé,	ph	80	chemonîm.
ᴙ	צ	ק	tsadé,	ts	90	tich´îm.
ᴘ	ק	ק	qoph,	q	100	méâh.
ᴐ	ר	ꭓ	rech,	r	200	mâtaïm.
ᴟ	ש	ʃ	chin,	ch	300	châlôch méôt.
ᴧ	ת	ק	tau,	t	400	arba´ méôt.

Fig. 11.—Hebrew numerals. (Taken from A. P. Pihan, *Exposé des signes de numération* [Paris, 1860], p. 172, 173.)

the forms occurred when the letter was a terminal letter of a word.
These end forms were used as follows:

ץ	ף	ז	ם	ך
900	800	700	600	500 .

To represent thousands the Hebrews went back to the beginning of
their alphabet and placed two dots over each letter. Thereby its
value was magnified a thousand fold. Accordingly, א represented
1,000. Thus any number less than a million could be represented by
their system.

31. As indicated above, the Hebrews wrote from right to left.
Hence, in writing numbers, the numeral of highest value appeared on
the right; הא meant 5,001, אה meant 1,005. But 1,005 could be
written also הא, where the two dots were omitted, for when א meant
unity, it was always placed to the left of another numeral. Hence
when appearing on the right it was interpreted as meaning 1,000.
With a similar understanding for other signs, one observes here the
beginning of an imperfect application in Hebrew notation of the
principle of local value. By about the eighth century A.D., one finds
that the signs התמה signify 5,845, the number of verses in the laws
as given in the Masora. Here the sign on the extreme right means
5,000; the next to the left is an 8 and must stand for a value less than
5,000, yet greater than the third sign representing 40. Hence the
sign for 8 is taken here as 800.[1]

GREEKS

32. On the island of Crete, near Greece, there developed, under
Egyptian influence, a remarkable civilization. Hieroglyphic writing
on clay, of perhaps about 1500 B.C., discloses number symbols as
follows:) or | for 1,))))) or ||| || or $\begin{smallmatrix} ||| \\ || \end{smallmatrix}$ for 5, · for 10, \ or / for
100, ◇ for 1,000, V for $\frac{1}{4}$ (probably), \\\\::::))) for 483.[2] In this
combination of symbols only the additive principle is employed.
Somewhat later,[3] 10 is represented also by a horizontal dash; the

[1] G. H. F. Nesselmann, *Die Algebra der Griechen* (Berlin, 1842), p. 72, 494;
M. Cantor, *Vorlesungen über Geschichte der Mathematik*, Vol. I (3d ed.), p. 126, 127.

[2] Arthur J. Evans, *Scripta Minoa*, Vol. I (1909), p. 258, 256.

[3] Arthur J. Evans, *The Palace of Minos* (London, 1921). Vol. I, p 646; see
also p. 279.

sloping line indicative of 100 and the lozenge-shaped figure used for 1,000 were replaced by the forms ○ for 100, and ◇ for 1,000.

◇◇○° ° ═ ═ ═ |‖ stood for 2,496 .

33. The oldest strictly Greek numeral symbols were the so-called *Herodianic signs,* named after Herodianus, a Byzantine grammarian of about 200 A.D., who describes them. These signs occur frequently in Athenian inscriptions and are, on that account, now generally called Attic. They were the initial letters of numeral adjectives.[1] They were used as early as the time of Solon, about 600 B.C., and continued in use for several centuries, traces of them being found as late as the time of Cicero. From about 470 to 350 B.C. this system existed in competition with a newer one to be described presently. The Herodianic signs were

Ⱶ Iota for 1	H Eta for 100
Π or ⲅ or Ⲅ Pi for 5	X Chi for 1,000
Δ Delta for 10	M My for 10,000

34. Combinations of the symbols for 5 with the symbols for 10,100, 1,000 yielded symbols for 50, 500, 5,000. These signs appear on an abacus found in 1847, represented upon a Greek marble monument on the island of Salamis.[2] This computing table is represented in Figure 12.

The four right-hand signs I C T X, appearing on the horizontal line below, stand for the fractions $\frac{1}{6}$, $\frac{1}{12}$, $\frac{1}{24}$, $\frac{1}{48}$, respectively. Proceeding next from right to left, we have the symbols for 1, 5, 10, 50, 100, 500, 1,000, 5,000, and finally the sign Ⱶ for 6,000. The group of symbols drawn on the left margin, and that drawn above, do not contain the two symbols for 5,000 and 6,000. The pebbles in the columns represent the number 9,823. The four columns represented by the five vertical lines on the right were used for the representation of the fractional values $\frac{1}{6}$, $\frac{1}{12}$, $\frac{1}{24}$, $\frac{1}{48}$, respectively.

35. Figure 13 shows the old Herodianic numerals in an Athenian state record of the fifth century B.C. The last two lines are: Κεφάλαιον

---

[1] See, for instance, G. Friedlein, *Die Zahlzeichen und das elementare Rechnen der Griechen und Römer* (Erlangen, 1869), p. 8; M. Cantor, *Vorlesungen über Geschichte der Mathematik,* Vol. I (3d ed.), p. 120; H. Hankel, *Zur Geschichte der Mathematik im Alterthum und Mittelalter* (Leipzig, 1874), p. 37.

[2] Kubitschek, "Die Salaminische Rechentafel," *Numismatische Zeitschrift* (Vienna, 1900), Vol. XXXI, p. 393–98; A. Nagl, *ibid.,* Vol. XXXV (1903), p. 131–43; M. Cantor, *Kulturleben der Völker* (Halle, 1863), p. 132, 136; M. Cantor, *Vorlesungen über Geschichte der Mathematik,* Vol. I (3d ed.), p. 133.

ἀνα[λώατοςτ] οὗ ἐπὶ τ[ης] ἀρχῆς ΗΗΗΓ ΤΤΤ . . . . ; i.e., "Total of expenditures during our office three hundred and fifty-three talents. . . . ."

36. The exact reason for the displacement of the Herodianic symbols by others is not known. It has been suggested that the commercial intercourse of Greeks with the Phoenicians, Syrians, and Hebrews brought about the change. The Phoenicians made one important contribution to civilization by their invention of the alphabet. The Babylonians and Egyptians had used their symbols to represent whole syllables or words. The Phoenicians borrowed hieratic

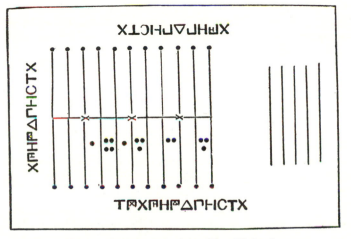

Fig. 12.—The computing table of Salamis

signs from Egypt and assigned them a more primitive function as letters. But the Phoenicians did not use their alphabet for numerical purposes. As previously seen, they represented numbers by vertical and horizontal bars. The earliest use of an entire alphabet for designating numbers has been attributed to the Hebrews. As previously noted, the Syrians had an alphabet representing numbers. The Greeks are supposed by some to have copied the idea from the Hebrews. But Moritz Cantor[1] argues that the Greek use is the older and that the invention of alphabetic numerals must be ascribed to the Greeks. They used the twenty-four letters of their alphabet, together with three strange and antique letters, ϛ (old *van*), ϙ (*koppa*), ϡ (*sampi*), and the symbol M. This change was decidedly for the worse, for the old Attic numerals were less burdensome on the memory inas-

[1] *Vorlesungen über Geschichte der Mathematik*, Vol. I (3d ed., 1907), p. 25.

Fig. 13.—Account of disbursements of the Athenian state, 418–415 B.C., British Museum, Greek Inscription No. 23. (Taken from R. Brown, *A History of Accounting and Accountants* [Edinburgh, 1905], p. 26.)

much as they contained fewer symbols. The following are the Greek alphabetic numerals and their respective values:

α	β	γ	δ	ϵ	ς	ζ	η	θ	ι	κ	λ	μ	ν	ξ	o	π	φ
1	2	3	4	5	6	7	8	9	10	20	30	40	50	60	70	80	90

ρ	σ	τ	υ	ϕ	χ	ψ	ω	$\backepsilon$	$,\alpha$	$,\beta$	$,\gamma,$
100	200	300	400	500	600	700	800	900	1,000	2,000	3,000

etc.

M	$\overset{\beta}{M}$	$\overset{\gamma}{M,}$	etc.
10,000	20,000	30,000	

37. A horizontal line drawn over a number served to distinguish it more readily from words. The coefficient for M was sometimes placed before or behind instead of over the M. Thus 43,678 was written $\overline{\delta M,\gamma\chi o\eta}$. The horizontal line over the Greek numerals can hardly be considered an essential part of the notation; it does not seem to have been used except in manuscripts of the Byzantine period.[1] For 10,000 or myriad one finds frequently the symbol M or Mν, sometimes simply the dot · , as in $\beta\cdot o\delta$ for 20,074. Often[2] the coefficient of the myriad is found written above the symbol μ^ν.

38. The paradox recurs, Why did the Greeks change from the Herodianic to the alphabet number system? Such a change would not be made if the new did not seem to offer some advantages over the old. And, indeed, in the new system numbers could be written in a more compact form. The Herodianic representation of 1,739 was χ ⊓HHΔΔΔΠ||||; the alphabetic was $,\alpha\psi\lambda\theta$. A scribe might consider the latter a great innovation. The computer derived little aid from either. Some advantage lay, however, on the side of the Herodianic, as Cantor pointed out. Consider HHHH + HH = ⊓ H, ΔΔΔΔ + ΔΔ = ⊿ Δ; there is an analogy here in the addition of hundred's and of ten's. But no such analogy presents itself in the alphabetic numerals, where the corresponding steps are $\upsilon+\sigma=\chi$ and $\mu+\kappa=\xi$; adding the hundred's expressed in the newer notation affords no clew as to the sum of the corresponding ten's. But there was another still more important consideration which placed the Herodianic far above the alphabetical numerals. The former had only six symbols, yet they afforded an easy representation of numbers below 100,000; the latter demanded twenty-seven symbols for numbers below 1,000! The mental effort

[1] *Encyc. des scien. math.*, Tome I, Vol. I (1904), p. 12.　　[2] *Ibid.*

of remembering such an array of signs was comparatively great. We are reminded of the centipede having so many legs that it could hardly advance.

39. We have here an instructive illustration of the fact that a mathematical topic may have an amount of symbolism that is a hindrance rather than a help, that becomes burdensome, that obstructs progress. We have here an early exhibition of the truth that the movements of science are not always in a forward direction. Had the Greeks not possessed an abacus and a finger symbolism, by the aid of which computations could be carried out independently of the numeral notation in vogue, their accomplishment in arithmetic and algebra might have been less than it actually was.

40. Notwithstanding the defects of the Greek system of numeral notation, its use is occasionally encountered long after far better systems were generally known. A Calabrian monk by the name of Barlaam,[1] of the early part of the fourteenth century, wrote several mathematical books in Greek, including arithmetical proofs of the second book of Euclid's *Elements*, and six books of *Logistic*, printed in 1564 at Strassburg and in several later editions. In the *Logistic* he develops the computation with integers, ordinary fractions, and sexagesimal fractions; numbers are expressed by Greek letters. The appearance of an arithmetical book using the Greek numerals at as late a period as the close of the sixteenth century in the cities of Strassburg and Paris is indeed surprising.

41. Greek writers often express fractional values in words. Thus Archimedes says that the length of a circle amounts to three diameters and a part of one, the size of which lies between one-seventh and ten-seventy-firsts.[2] Eratosthenes expresses $\frac{11}{83}$ of a unit arc of the earth's meridian by stating that the distance in question "amounts to eleven parts of which the meridian has eighty-three."[3] When expressed in symbols, fractions were often denoted by first writing the numerator marked with an accent, then the denominator marked with two accents and written twice. Thus,[4] $ιζ'\ κα''\ κα'' = \frac{17}{21}$. Archimedes, Eutocius, and Diophantus place the denominator in the position of the

[1] All our information on Barlaam is drawn from M. Cantor, *Vorlesungen über Geschichte der Mathematik*, Vol. I (3d ed.), p. 509, 510; A. G. Kästner, *Geschichte der Mathematik* (Göttingen, 1796), Vol. I, p. 45; J. C. Heilbronner, *Historia matheseos universae* (Lipsiae, 1742), p. 488, 489.

[2] *Archimedis opera omnia* (ed. Heiberg; Leipzig, 1880), Vol. I, p. 262.

[3] Ptolemäus, Μεγάλη σύνταξις (ed. Heiberg), Pars I, Lib. 1, Cap. 12, p. 68.

[4] Heron, *Stereometrica* (ed. Hultsch; Berlin, 1864), Pars I, Par. 8, p. 155.

modern exponent; thus[1] Archimedes and Eutocius use the notation $\overset{\kappa\alpha'}{\iota\zeta}$ or $\iota\zeta$ for $\frac{17}{21}$, and Diophantus (§§ 101–6), in expressing large numbers, writes (*Arithmetica*, Vol. IV, p. 17), $\dfrac{\beta_1\psi\delta\,\sim}{\gamma\cdot,\,\zeta\,\chi\kappa\alpha}$ for $\dfrac{36,621}{2,704}$.

Here the sign $\sim$ takes the place of the accent. Greek writers, even as late as the Middle Ages, display a preference for unit fractions, which played a dominating rôle in old Egyptian arithmetic.[2] In expressing such fractions, the Greeks omitted the α' for the numerator and wrote the denominator only once. Thus $\mu\delta'' = \frac{1}{44}$. Unit fractions in juxtaposition were added,[3] as in $\zeta'' \kappa\eta'' \rho\iota\beta'' \sigma\kappa\delta'' = \frac{1}{7} + \frac{1}{28} + \frac{1}{112} + \frac{1}{224}$. One finds also a single accent,[4] as in $\delta' = \frac{1}{4}$. Frequent use of unit fractions is found in Geminus (first century B.C.), Diophantus (third century A.D.), Eutocius and Proclus (fifth century A.D.). The fraction $\frac{1}{2}$ had a mark of its own,[5] namely, $\angle$ or $\angle'$, but this designation was no more adopted generally among the Greeks than were the other notations of fractions. Ptolemy[6] wrote 38°50′ (i.e., $38°\frac{1}{2}\,\frac{1}{3}$) thus, $\lambda\eta'$ $\angle'\gamma'''$. Hultsch has found in manuscripts other symbols for $\frac{1}{2}$, namely, the semicircles C^{VI}, C, and the sign $\mathcal{S}$; the origin of the latter is uncertain. He found also a symbol for $\frac{2}{3}$, resembling somewhat the small omega (ω).[7] Whether these symbols represent late practice, but not early usage, it is difficult to determine with certainty.

42. A table for reducing certain ordinary fractions to the sum of unit fractions is found in a Greek papyrus from Egypt, described by

[1] G. H. F. Nesselmann, *Algebra der Griechen* (Berlin, 1842), p. 114.

[2] J. Baillet describes a papyrus, "Le papyrus mathématique d'Akhmîm," in *Mémoires publiés par les membres de la Mission archéologique française au Caire* (Paris, 1892), Vol. IX, p. 1–89 (8 plates). This papyrus, found at Akhmîm, in Egypt, is written in Greek, and is supposed to belong to the period between 500 and 800 A.D. It contains a table for the conversion of ordinary fractions into unit fractions.

[3] Fr. Hultsch, *Metrologicorum scriptorum reliquiae* (1864–66), p. 173–75; M. Cantor, *Vorlesungen über Geschichte der Mathematik*, Vol. I (3d ed.), p. 129.

[4] Nesselmann, *op. cit.*, p. 112.

[5] *Ibid.*; James Gow, *Short History of Greek Mathematics* (Cambridge, 1884), p. 48, 50.

[6] *Geographia* (ed. Carolus Müllerus; Paris, 1883), Vol. I, Part I, p. 151.

[7] *Metrologicorum scriptorum reliquiae* (Leipzig, 1864), Vol. I, p. 173, 174. On p. 175 and 176 Hultsch collects the numeral symbols found in three Parisian manuscripts, written in Greek, which exhibit minute variations in the symbolism. For instance, 700 is found to be ψ^π, ψ, ψ'.

L. C. Karpinski,[1] and supposed to be intermediate between the Ahmes papyrus and the Akhmim papyrus. Karpinski (p. 22) says: "In the table no distinction is made between integers and the corresponding unit fractions; thus γ' may represent either 3 or $\frac{1}{3}$, and actually $\gamma'\gamma'$ in the table represents $3\frac{1}{3}$. Commonly the letters used as numerals were distinguished in early Greek manuscripts by a bar placed above the letters but not in this manuscript nor in the Akhmim papyrus." In a third document dealing with unit fractions, a Byzantine table of fractions, described by Herbert Thompson,[2] $\frac{2}{3}$ is written $Ϟ$; $\frac{1}{2}$, ϵ; $\frac{1}{3}$, $ᚌ$ (from $Ϟ$ '); $\frac{1}{4}$, $\triangle$ (from $\triangle'$); $\frac{1}{6}$, $ℰ$ (from ϵ'); $\frac{1}{8}$, $ᚹ$ (from H'). As late as the fourteenth century, Nicolas Rhabdas of Smyrna wrote two letters in the Greek language, on arithmetic, containing tables for unit fractions.[3] Here letters of the Greek alphabet used as integral numbers have bars placed above them.

43. About the second century before Christ the Babylonian sexagesimal numbers were in use in Greek astronomy; the letter omicron, which closely resembles in form our modern zero, was used to designate a vacant space in the writing of numbers. The Byzantines wrote it usually $\bar{o}$, the bar indicating a numeral significance as it has when placed over the ordinary Greek letters used as numerals.[4]

44. The division of the circle into 360 equal parts is found in Hypsicles.[5] Hipparchus employed sexagesimal fractions regularly, as did also C. Ptolemy[6] who, in his *Almagest*, took the approximate value of π to be $3+\dfrac{8}{60}+\dfrac{30}{60\times60}$. In the Heiberg edition this value is written $\bar{\gamma}\ \bar{\eta}\ \bar{\lambda}$, purely a *notation of position*. In the tables, as printed by Heiberg, the dash over the letters expressing numbers is omitted. In the edition of N. Halma[7] is given the notation $\bar{\gamma}\ \eta'\ \lambda''$, which is

---

[1] "The Michigan Mathematical Papyrus No. 621," *Isis*, Vol. V (1922), p. 20–25.

[2] "A Byzantine Table of Fractions," *Ancient Egypt*, Vol. I (1914), p. 52–54.

[3] The letters were edited by Paul Tannery in *Notices et extraits des manuscrits de la Bibliothèque Nationale*, Vol. XXXII, Part 1 (1886), p. 121–252.

[4] C. Ptolemy, *Almagest* (ed. N. Halma; Paris, 1813), Book I, chap. ix, p. 38 and later; J. L. Heiberg, in his edition of the *Almagest* (*Syntaxis mathematica*) (Leipzig, 1898; 2d ed., Leipzig, 1903), Book I, does not write the bar over the o but places it over all the significant Greek numerals. This procedure has the advantage of distinguishing between the o which stands for 70 and the o which stands for zero. See *Encyc. des scien. math.*, Tome I, Vol. I (1904), p. 17, n. 89.

[5] Ἀναφορικός (ed. K. Manitius), p. xxvi.

[6] *Syntaxis mathematica* (ed. Heiberg), Vol. I, Part 1, p. 513.

[7] *Composition math. de Ptolémée* (Paris, 1813), Vol. I, p. 421; see also *Encyc. des scien. math.*, Tome I, Vol. I (1904), p. 53, n. 181.

probably the older form. Sexagesimal fractions were used during the whole of the Middle Ages in India, and in Arabic and Christian countries. One encounters them again in the sixteenth and seventeenth centuries. Not only sexagesimal fractions, but also the sexagesimal notation of integers, are explained by John Wallis in his *Mathesis universalis* (Oxford, 1657), page 68, and by V. Wing in his *Astronomia Britannica* (London, 1652, 1669), Book I.

EARLY ARABS

45. At the time of Mohammed the Arabs had a script which did not differ materially from that of later centuries. The letters of the early Arabic alphabet came to be used as numerals among the Arabs

1	ا	10	ى	100	ق	1000	غ	10 000	يغ	100 000	قغ
2	ب	20	ك	200	ر	2000	بغ	20 000	كغ	200 000	رغ
3	ج	30	ل	300	ش	3000	جغ	30 000	لغ	300 000	شغ
4	د	40	م	400	ت	4000	دغ	40 000	مغ	400 000	تغ
5	ه	50	ن	500	ث	5000	هغ	50 000	نغ	500 000	ثغ
6	و	60	س	600	خ	6000	وغ	60 000	سغ	600 000	خغ
7	ز	70	ع	700	ذ	7000	زغ	70 000	عغ	700 000	ذغ
8	ح	80	ف	800	ض	8000	حغ	80 000	فغ	800 000	ضغ
9	ط	90	ص	900	ظ	9000	طغ	90 000	صغ	900 000	ظغ

FIG. 14.—Arabic alphabetic numerals used before the introduction of the Hindu-Arabic numerals.

as early as the sixth century of our era.[1] After the time of Mohammed, the conquering Moslem armies coming in contact with Greek culture acquired the Greek numerals. Administrators and military leaders used them. A tax record of the eighth century contains numbers expressed by Arabic letters and also by Greek letters.[2] Figure 14 is a table given by Ruska, exhibiting the Arabic letters and the numerical values which they represent. Taking the symbol for 1,000 twice, on the multiplicative principle, yielded 1,000,000. The Hindu-Arabic

[1] Julius Ruska, "Zur ältesten arabischen Algebra und Rechenkunst," *Sitzungsberichte d. Heidelberger Akademie der Wissensch.* (Philos.-histor. Klasse, 1917; 2. Abhandlung), p. 37.

[2] *Ibid.*, p. 40.

numerals, with the zero, began to spread among the Arabs in the ninth and tenth centuries, and they slowly displaced the Arabic and Greek numerals.[1]

ROMANS

46. We possess little definite information on the origin of the Roman notation of numbers. The Romans never used the successive letters of their alphabet for numeral purposes in the manner practiced by the Syrians, Hebrews, and Greeks, although (as we shall see) an alphabet system was at one time proposed by a late Roman writer. Before the ascendancy of Rome the Etruscans, who inhabited the country nearly corresponding to modern Tuscany and who ruled in Rome until about 500 B.C., used numeral signs which resembled letters of their alphabet and also resembled the numeral signs used by the Romans. Moritz Cantor[2] gives the Etrurian and the old Roman signs, as follows: For 5, the Etrurian $\wedge$ or V, the old Roman V; for 10 the Etrurian X or $+$, the old Roman X; for 50 the Etrurian $\uparrow$ or $\downarrow$, the old Roman $\curlyvee$ or $\downarrow$ or $\perp$ or $\lfloor$ or L; for 100 the Etrurian $\oplus$, the old Roman $\ominus$; for 1,000 the Etrurian 8, the old Roman $\oplus$. The resemblance of the Etrurian numerals to Etrurian letters of the alphabet is seen from the following letters: V, $+$, $\downarrow$, $\subset$, 8. These resemblances cannot be pronounced accidental. "Accidental, on the other hand," says Cantor, "appears the relationship with the later Roman signs, I V, X, L, C, M, which from their resemblance to letters transformed themselves by popular etymology into these very letters." The origins of the Roman symbols for 100 and 1,000 are uncertain; those for 50 and 500 are generally admitted to be the result of a bisection of the two former. "There was close at hand," says G. Friedlein,[3] "the abbreviation of the word *centum* and *mille* which at an early age brought about for 100 the sign C, and for 1,000 the sign $\Lambda\Lambda$ and after Augustus[4] M." A view held by some Latinists[5] is that "the signs for 50, 100, 1,000 were originally the three Greek aspirate letters which the Romans did not require, viz., Ψ, $\odot$, $\oplus$, i.e., χ, θ, Φ. The Ψ was written $\perp$ and abbreviated into L; $\odot$ from a false notion of its origin made like

[1] *Ibid.*, p. 47.

[2] *Vorlesungen über Geschichte der Mathematik*, Vol. I (3d ed.), p. 523, and the table at the end of the volume.

[3] *Die Zahlzeichen und das elementare Rechnen der Griechen und Römer* (Erlangen, 1869), p. 28.

[4] Theodor Mommsen, *Die unteritalischen Dialekte* (Leipzig, 1840), p. 30.

[5] Ritschl, *Rhein. Mus.*, Vol. XXIV (1869), p. 12.

the initial of centum; and ① assimilated to ordinary letters CI⊃. The half of ①, viz., D, was taken to be ½ 1,000, i.e., 500; X probably from the ancient form of ⊖, viz., ⊗, being adopted for 10, the half of it V was taken for 5."[1]

47. Our lack of positive information on the origin and early history of the Roman numerals is not due to a failure to advance working hypotheses. In fact, the imagination of historians has been unusually active in this field.[2] The dominating feature in the Roman notation is the principle of addition, as seen in II, XII, CC, MDC, etc.

48. Conspicuous also is the frequent use of the principle of subtraction. If a letter is placed before another of greater value, its value is to be subtracted from that of the greater. One sees this in IV, IX, XL. Occasionally one encounters this principle in the Babylonian notations. Remarks on the use of it are made by Adriano Cappelli in the following passage:

"The well-known rule that a smaller number, placed to the left of a larger, shall be subtracted from the latter, as ①|⊃⊃=4,000, etc., was seldom applied by the old Romans and during the entire Middle Ages one finds only a few instances of it. The cases that I have found belong to the middle of the fifteenth century and are all cases of IX, never of IV, and occurring more especially in French and Piedmontese documents. Walther, in his *Lexicon diplomaticum*, Göttingen, 1745–47, finds the notation LXL=90 in use in the eighth century. On the other hand one finds, conversely, the numbers IIIX, VIX with the meaning of 13 and 16, in order to conserve, as Lupi remarks, the Latin terms *tertio decimo* and *sexto decimo*."[3] L. C. Karpinski points out that the subtractive principle is found on some early tombstones and on a signboard of 130 B.C., where at the crowded end of a line 83 is written XXCIII, instead of LXXXIII.

---

[1] H. J. Roby, *A Grammar of the Latin Language from Plautus to Suetonius* (4th ed.; London, 1881), Vol. I, p. 441.

[2] Consult, for example, Friedlein, *op. cit.*, p. 26–31; Nesselmann, *op. cit.*, p. 86–92; Cantor, *Mathematische Beiträge zum Kulturleben der Völker*, p. 155–67; J. C. Heilbronner, *Historia Matheseos universae* (Lipsiae, 1742), p. 732–35; Grotefend, *Lateinische Grammatik* (3d ed.; Frankfurt, 1820), Vol. II, p. 163, is quoted in the article "Zahlzeichen" in G. S. Klügel's *Mathematisches Wörterbuch*, continued by C. B. Mollweide and J. A. Grunert (Leipzig, 1831); Mommsen, *Hermes*, Vol. XXII (1887), p. 596; Vol. XXIII (1888), p. 152. A recent discussion of the history of the Roman numerals is found in an article by Ettore Bortolotti in *Bolletino della Mathesis* (Pavia, 1918), p. 60–66, which is rich in bibliographical references, as is also an article by David Eugene Smith in *Scientia* (July–August, 1926).

[3] *Lexicon Abbreviaturarum* (Leipzig, 1901), p. xlix.

49. Alexander von Humboldt[1] makes the following observations: "Summations by juxtaposition one finds everywhere among the Etruscans, Romans, Mexicans and Egyptians; subtraction or lessening forms of speech in Sanskrit among the Indians: in 19 or *unavinsati;* 99 *unusata;* among the Romans in *undeviginti* for 19 (*unus de viginti*), *undeoctoginta* for 79; *duo de quadraginta* for 38; among the Greeks *eikosi deonta henos* 19, and *pentekonta düoin deontoin* 48, i.e., 2 missing in 50. This lessening form of speech has passed over in the graphics of numbers when the group signs for 5, 10 and even their multiples, for example, 50 or 100, are placed to the left of the characters they modify (IV and IΛ, XL and XT for 4 and 40) among the Romans and Etruscans (Otfried Müller, *Etrusker*, II, 317-20), although among the latter, according to Otfried Müller's new researches, the numerals descended probably entirely from the alphabet. In rare Roman inscriptions which Marini has collected (*Iscrizioni della Villa di Albano*, p. 193; Hervas, *Aritmetica delle nazioni* [1786], p. 11, 16), one finds even 4 units placed before 10, for example, IIIIX for 6."

50. There are also sporadic occurrences in the Roman notations of the principle of multiplication, according to which VM does not stand for 1,000−5, but for 5,000. Thus, in Pliny's *Historia naturalis* (about 77 A.D.), VII, 26; XXXIII, 3; IV praef., one finds[2] LXXXIII.M, XCII.M, CX.M for 83,000, 92,000, 110,000, respectively.

51. The thousand-fold value of a number was indicated in some instances by a horizontal line placed above it. Thus, Aelius Lampridius (fourth century A.D.) says in one place, "$\overline{CXX}$, equitum Persarum fudimus: et mox $\overline{X}$ in bello interemimus," where the numbers designate 120,000 and 10,000. Strokes placed on top and also on the sides indicated hundred thousands; e.g., $|\overline{X}|\overline{CLXXXDC}$ stood for 1,180,600. In more recent practice the strokes sometimes occur only on the sides, as in $|X| \cdot DC.XC.$, the date on the title-page of Sigüenza's *Libra astronomica*, published in the city of Mexico in 1690. In antiquity, to prevent fraudulent alterations, XXXM was written for 30,000, and later still CIϽ took the place of M.[3] According to

---

[1] "Über die bei verschiedenen Völkern üblichen Systeme von Zahlzeichen, etc.," *Crelle's Journal für die reine und angewandte Mathematik* (Berlin, 1829), Vol. IV, p. 210, 211.

[2] Nesselmann, *op. cit.*, p. 90.

[3] Confer, on this point, Theodor Mommsen and J. Marquardt, *Manuel des antiquités romaines* (trans. G. Humbert), Vol. X by J. Marquardt (trans. A. Vigié; Paris, 1888), p. 47, 49.

Cappelli[1] "one finds, often in French documents of the Middle Ages, the multiplication of 20 expressed by two small x's which are placed as exponents to the numerals III, VI, VIII, etc., as in IIII^{xx}=80, VI^{xx}XI = 131."

52. A Spanish writer[2] quotes from a manuscript for the year 1392 the following:

M   C
"IIII, IIII, LXXIII florins" for 4,473 florins.

M   XX
"III C IIII III florins" for 3,183 (?) florins.

In a Dutch arithmetic, printed in 1771, one finds[3]

c   c m c
i ꝛꝛiiȷ for 123, i ꝛꝛiiȷ iiiȷ lbȷ for 123,456.

53. For 1,000 the Romans had not only the symbol M, but also I, ∞ and CIↃ. According to Priscian, the celebrated Latin grammarian of about 500 A.D., the ∞ was the ancient Greek sign X for 1,000, but modified by connecting the sides by curved lines so as to distinguish it from the Roman X for 10. As late as 1593 the ∞ is used by C. Dasypodius[4] the designer of the famous clock in the cathedral at Strasbourg. The CIↃ was a | inclosed in parentheses (or *apostrophos*). When only the right-hand parenthesis is written, IↃ, the value represented is only half, i.e., 500. According to Priscian,[5] "quinque milia per I et duas in dextera parte apostrophos, IↃↃ. decem milia per supra dictam formam additis in sinistra parte contrariis duabus notis quam sunt apostrophi, CCIↃↃ." Accordingly, IↃↃ stood for 5,000, CCIↃↃ for 10,000; also IↃↃↃ represented 50,000; and CCCIↃↃↃ, 100,000; (∞), 1,000,000. If we may trust Priscian, the symbols that look like the letters C, or those letters facing in the opposite direction, were not really letters C, but were apostrophes or what we have called

---

[1] *Op. cit.*, p. xlix.

[2] Liciniano Saez, *Demostración Histórica del verdadero valor de Todas Las Monedas que corrían en Castilla durante el reynado del Señor Don Enrique III* (Madrid, 1796).

[3] *De Vernieuwde Cyfferinge* van M.ʳ Willem Bartjens. Herstelt, ... door M.ʳ Jan van Dam, ... en van alle voorgaande Fauten gezuyvert door . . . . Klaas Bosch (Amsterdam, 1771), p. 8.

[4] *Cunradi Dasypodii Institutionum Mathematicarum voluminis primi Erotemata* (1593), p. 23.

[5] "De figuris numerorum," *Henrici Keilii Grammatici Latini* (Lipsiae, 1859), Vol. III, 2, p. 407.

parentheses. Through Priscian it is established that this notation is at least as old as 500 A.D.; probably it was much older, but it was not widely used before the Middle Ages.

54. While the Hindu-Arabic numerals became generally known in Europe about 1275, the Roman numerals continued to hold a commanding place. For example, the fourteenth-century banking-house of Peruzzi in Florence—Compagnia Peruzzi—did not use Arabic numerals in their account-books. Roman numerals were used, but the larger amounts, the thousands of *lira*, were written out in words; one finds, for instance, "lb. quindicimilia CXV / V ⅃ VI in fiorini" for 15,115 *lira* 5 *soldi* 6 *denari;* the specification being made that the *lira* are *lira a fiorino d'oro* at 20 *soldi* and 12 *denari*. There appears also a symbol much like ⟩, for thousand.[1]

Nagl states also: "Specially characteristic is . . . . during all the Middle Ages, the regular prolongation of the last | in the units, as VI|=VII, which had no other purpose than to prevent the subsequent addition of a further unit."

55. In a book by H. Giraua Tarragones[2] at Milan the Roman numerals appear in the running text and are usually underlined; in the title-page, the date has the horizontal line *above* the numerals. The Roman four is IIII. In the tables, columns of degrees and minutes are headed "G.M."; of hour and minutes, "H.M." In the tables, the Hindu-Arabic numerals appear; the five is printed ⅚, without the usual upper stroke. The vitality of the Roman notation is illustrated further by a German writer, Sebastian Frank, of the sixteenth century, who uses Roman numerals in numbering the folios of his book and in his statistics: "Zimmet kumpt von Zailon .CC.VÑ LX. teütscher meil von Calicut weyter gelegen. . . . . Die Nägelin kummen von Meluza / für Calicut hinaussgelegen vij·c. vnd XL. deutscher meyl."[3] The two numbers given are 260 and 740 German miles. Peculiar is the insertion of *vnd* ("and"). Observe also the use of the principle of multiplication in vij·c. (=700). In Jakob Köbel's *Rechenbiechlin* (Augsburg, 1514), fractions appear in Roman numerals;

thus, $\dfrac{\text{II}^{\text{C}}}{\text{IIII}^{\text{C}}.\text{LX}}$ stands for $\frac{200}{460}$.

[1] Alfred Nagl, *Zeitschrift für Mathematik und Physik*, Vol. XXXIV (1889), Historisch-literarische Abtheilung, p. 164.

[2] *Dos Libros de Cosmographie*, compuestos nueuamente por Hieronymo Giraua Tarragones (Milan, M.D.LVI).

[3] *Weltbuch / spiegel vnd bildtnis des gantzen Erdtbodens . . . . von Sebastiano Franco Wördensi . . . . (M.D. XXXIIII), fol. ccxx.

56. In certain sixteenth-century Portuguese manuscripts on navigation one finds the small letter *b* used for 5, and the capital letter *R* for 40. Thus, *xb*iij stands for 18, *R*iij for 43.[1]

Fig. 15.—Degenerate forms of Roman numerals in English archives (Common Pleas, Plea Rolls, 637, 701, and 817; also Recovery Roll 1). (Reduced.)

A curious development found in the archives of one or two English courts of the fifteenth and sixteenth centuries[2] was a special Roman

[1] J. I. de Brito Rebello, *Livro de Marinharia* (Lisboa, 1903), p. 37, 85–91, 193, 194.

[2] *Antiquaries Journal* (London, 1926), Vol. VI, p. 273, 274.

numeration for the membranes of their Rolls, the numerals assuming a degraded form which in its later stages is practically unreadable. In Figure 15 the first three forms show the number 147 as it was written in the years 1421, 1436, and 1466; the fourth form shows the number 47 as it was written in 1583.

57. At the present time the Roman notation is still widely used in marking the faces of watches and clocks, in marking the dates of books on title-pages, in numbering chapters of books, and on other occasions calling for a double numeration in which confusion might arise from the use of the same set of numerals for both. Often the Roman numerals are employed for aesthetic reasons.

58. A striking feature in Roman arithmetic is the partiality for duodecimal fractions. Why duodecimals and not decimals? We can only guess at the answer. In everyday affairs the division of units into two, three, four, and six equal parts is the commonest, and duodecimal fractions give easier expressions for these parts. Nothing definite is known regarding the time and place or the manner of the origin of these fractions. Unlike the Greeks, the Romans dealt with concrete fractions. The Roman *as*, originally a copper coin weighing one pound, was divided into 12 *unciae*. The abstract fraction $\frac{11}{12}$ was called *deuna* ($= de\ uncia$, i.e., *as* [1] less *uncia* [$\frac{1}{12}$]). Each duodecimal subdivision had its own name and symbol. This is shown in the following table, taken from Friedlein,[1] in which S stands for *semis* or "half" of an *as*.

TABLE

as.	1	. . . .	. . . .
deunx	$\frac{11}{12}$	$S = = -$ or $S :: \cdot$	(de uncia $1 - \frac{1}{12}$)
dextans ⎱ (decunx) ⎰	$\frac{5}{6}$	$S = =$ or $S ::$	$\left\{\begin{array}{l}\text{(de sextans } 1-\frac{1}{6})\\ \text{(decem unciae)}\end{array}\right.$
dodrans	$\frac{3}{4}$	$S = -$ or $S = 1$ or $S : \cdot$	(de quadrans $1 - \frac{1}{4}$)
bes	$\frac{2}{3}$	$S =$ or $- S -$ or $S :$	(duae assis *sc.* partes)
septunx	$\frac{7}{12}$	$S -$ or $S \cdot$	(septem unciae)
semis	$\frac{1}{2}$	S	. . . .
quincunx	$\frac{5}{12}$	$= = -$ or $= - =$ or $:: \cdot$	(quinque unciae)
triens	$\frac{1}{3}$	$= =$ or $\sum \sum$ or $::$	. . . .
quadrans	$\frac{1}{4}$	$= -$ or $= 1$ or $: \cdot$	. . . .
sextans	$\frac{1}{6}$	$=$ or Z or $:$	. . . .
sescuncia $1\frac{1}{2}$	$\frac{1}{12} = \frac{1}{8}$	$- \ L \ \vdash \ \mathcal{L} \ \Sigma \ \mathcal{L}$	. . . .
uncia	$\frac{1}{12}$	$-$ or $\cdot$ or on bronze abacus $\ominus$	

In place of straight lines $-$ occur also curved ones $\sim$.

---

[1] *Op. cit.*, Plate 2, No. 13; see also p. 35.

59. Not all of these names and signs were used to the same extent. Since $\frac{1}{2}+\frac{1}{3}=\frac{5}{6}$, there was used in ordinary life $\frac{1}{2}$ and $\frac{1}{3}$ (*semis et triens*) in place of $\frac{5}{6}$ or $\frac{10}{12}$ (*decunx*). Nor did the Romans confine themselves to the duodecimal fractions or their simplified equivalents $\frac{1}{2}$, $\frac{1}{3}$, $\frac{1}{4}$, $\frac{1}{6}$, etc., but used, for instance, $\frac{1}{10}$ in measuring silver, a *libella* being $\frac{1}{10}$ *denarius*. The *uncia* was divided in 4 *sicilici*, and in 24 *scripuli* etc.[1] In the *Geometry* of Boethius the Roman symbols are omitted and letters of the alphabet are used to represent fractions. Very probably this part of the book is not due to Boethius, but is an interpolation by a writer of later date.

60. There are indeed indications that the Romans on rare occasions used letters for the expression of integral numbers.[2] Theodor Mommsen and others discovered in manuscripts found in Bern, Einsiedeln, and Vienna instances of numbers denoted by letters. Tartaglia gives in his *General trattato di nvmeri*, Part I (1556), folios 4, 5, the following:

A	500	I	1	R	80
B	300	K	51	S	70
C	100	L	50	T	160
D	500	M	1,000	V	5
E	250	N	90	X	10
F	40	O	11	Y	150
G	400	P	400	Z	2,000
H	200	Q	500		

61. Gerbert (Pope Sylvestre II) and his pupils explained the Roman fractions. As reproduced by Olleris,[3] Gerbert's symbol for $\frac{1}{2}$ does not resemble the capital letter S, but rather the small letter ς.

---

[1] For additional details and some other symbols used by the Romans, consult Friedlein, p. 33–46 and Plate 3; also H. Hankel, *op. cit.*, p. 57–61, where computations with fractions are explained. Consult also Fr. Hultsch, *Metrologic. scriptores Romani* (Leipzig, 1866).

[2] Friedlein, *op. cit.*, p. 20, 21, who gives references. In the *Standard Dictionary of the English Language* (New York, 1896), under *S*, it is stated that $\bar{S}$ stood for 7 or 70.

[3] *Œuvres de Gerbert* (Paris, 1867), p. 343–48, 393–96, 583, 584.

PERUVIAN AND NORTH AMERICAN KNOT RECORDS[1]
ANCIENT *QUIPU*

62. "The use of knots in cords for the purpose of reckoning, and recording numbers" was practiced by the Chinese and some other ancient people; it had a most remarkable development among the Inca of Peru, in South America, who inhabited a territory as large as the United States east of the Rocky Mountains, and were a people of superior mentality. The period of Inca supremacy extended from about the eleventh century A.D. to the time of the Spanish conquest in the sixteenth century. The *quipu* was a twisted woolen cord, upon which other smaller cords of different colors were tied. The color, length, and number of knots on them and the distance of one from another all had their significance. Specimens of these ancient *quipu* have been dug from graves.

63. We reproduce from a work by L. Leland Locke a photograph of one of the most highly developed *quipu*, along with a line diagram of the two right-hand groups of strands. In each group the top strand usually gives the sum of the numbers on the four pendent strands. Thus in the last group, the four hanging strands indicate the numbers 89, 258, 273, 38, respectively. Their sum is 658; it is recorded by the top string. The repetition of units is usually expressed by a long knot formed by tying the overhand knot and passing the cord through the loop of the knot as many times as there are units to be denoted. The numbers were expressed on the decimal plan, but the *quipu* were not adopted for calculation; pebbles and grains of maize were used in computing.

64. Nordenskiöld shows that, in Peru, 7 was a magic number; for in some *quipu*, the sums of numbers on cords of the same color, or the numbers emerging from certain other combinations, are multiples of 7 or yield groups of figures, such as 2777, 777, etc. The *quipu* disclose also astronomical knowledge of the Peruvian Indians.[2]

65. Dr. Leslie Spier, of the University of Washington, sends me the following facts relating to Indians in North America: "The data that I have on the *quipu*-like string records of North-American Indians indicate that there are two types. One is a long cord with knots and

[1] The data on Peru knot records given here are drawn from a most interesting work, *The Ancient Quipu or Peruvian Knot Record*, by L. Leland Locke (American Museum of Natural History, 1923). Our photographs are from the frontispiece and from the diagram facing p. 16. See Figs. 16 and 17.

[2] Erland Nordenskiöld, *Comparative Ethnographical Studies*, No. 6, Part 1 (1925), p. 36.

bearing beads, etc., to indicate the days. It is simply a string record. This is known from the Yakima of eastern Washington and some Interior Salish group of Nicola Valley,[1] B.C.

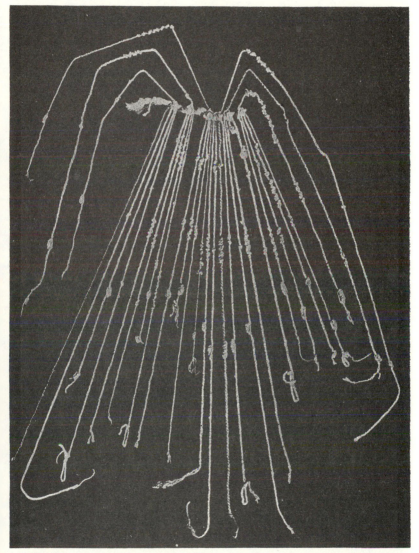

Fig. 16.—A *quipu*, from ancient Chancay in Peru, now kept in the American Museum of Natural History (Museum No. B8713) in New York City.

[1] J. D. Leechman and M. R. Harrington, *String Records of the Northwest, Indian Notes and Monographs* (1921).

"The other type I have seen in use among the Havasupai and Walapai of Arizona. This is a cord bearing a number of knots to indicate the days until a ceremony, etc. This is sent with the messenger who carries the invitation. A knot is cut off or untied for each day that elapses; the last one indicating the night of the dance. This is also used by the Northern and Southern Maidu and the Miwok of California.[1] There is a mythical reference to these among the Zuñi of New Mexico.[2] There is a note on its appearance in San Juan Pueblo in the same state in the seventeenth century, which would indicate that its use was widely known among the Pueblo Indians. 'They directed him (the leader of the Pueblo rebellion of 1680) to make a rope of the palm leaf and tie in it a number of knots to represent the number of days before the rebellion was to take place; that he must send the rope to all the Pueblos in the Kingdom, when each should signify its approval of, and union with, the conspiracy by untying one of the knots.'[3] The Huichol of Central Mexico also have knotted strings to keep count of days, untieing them as the days elapse. They also keep records of their lovers in the same way.[4] The Zuñi also keep records of days worked in this fashion.[5]

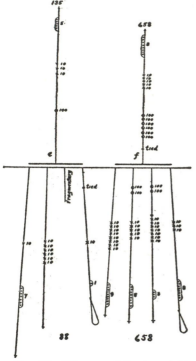

Fig. 17.—Diagram of the two right-hand groups of strands in Fig. 16.

[1] R. B. Dixon, "The Northern Maidu," *Bulletin of the American Museum of Natural History*, Vol. XVII (1905), p. 228, 271; P.-L. Faye, "Notes on the Southern Maidu," *University of California Publications of American Archaeology and Ethnology*, Vol. XX (1923), p. 44; Stephen Powers, "Tribes of California," *Contributions to North American Ethnology*, Vol. III (1877), p. 352.

[2] F. H. Cushing, "Zuñi Breadstuff," *Indian Notes and Monographs*, Vol. VIII (1920), p. 77.

[3] Quoted in J. G. Bourke, "Medicine-Men of the Apache," *Ninth Annual Report, Bureau of American Ethnology* (1892), p. 555.

[4] K. Lumholtz, *Unknown Mexico*, Vol. II, p. 218–30.

[5] Leechman and Harrington, *op. cit.*

"Bourke[1] refers to medicine cords with olivella shells attached among the Tonto and Chiricahua Apache of Arizona and the Zuñi. This may be a related form.

"I think that there can be no question the instances of the second type are historically related. Whether the Yakima and Nicola Valley usage is connected with these is not established."

AZTECS

66. "For figures, one of the numerical signs was the dot ($\cdot$), which marked the units, and which was repeated either up to 20 or up to the figure 10, represented by a lozenge. The number 20 was represented by a flag, which, repeated five times, gave the number 100, which was

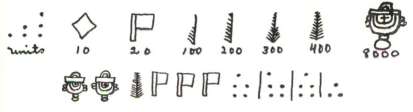

Fig. 18.—Aztec numerals

marked by drawing quarter of the barbs of a feather. Half the barbs was equivalent to 200, three-fourths to 300, the entire feather to 400. Four hundred multiplied by the figure 20 gave 8,000, which had a purse for its symbol."[2] The symbols were as shown in the first line of Figure 18.

The symbols for 20, 400, and 8,000 disclose the number 20 as the base of Aztec numeration; in the juxtaposition of symbols the additive principle is employed. This is seen in the second line[3] of Figure 18, which represents

$$2\times8,000+400+3\times20+3\times5+3=16,478 .$$

67. The number systems of the Indian tribes of North America, while disclosing no use of a symbol for zero nor of the principle of

---

[1] *Op. cit.*, p. 550 ff.

[2] Lucien Biart, *The Aztecs* (trans. J. L. Garner; Chicago, 1905), p. 319.

[3] Consult A. F. Pott, *Die quinäre und vigesimale Zählmethode bei Völkern aller Welttheile* (Halle, 1847).

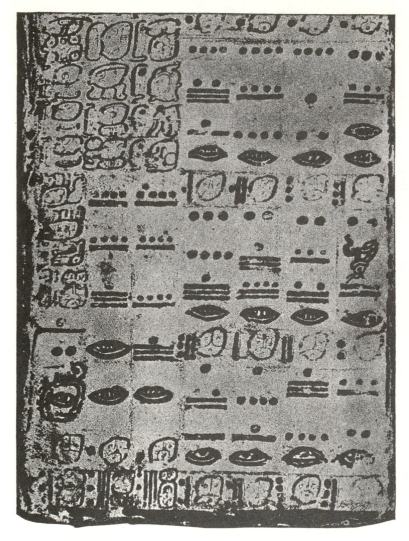

Fig. 19.—From the Dresden Codex, of the Maya, displaying numbers. The second column on the left, from above down, displays the numbers 9, 9, 16, 0, 0, which stand for $9 \times 144{,}000 + 9 \times 7{,}200 + 16 \times 360 + 0 + 0 = 1{,}366{,}560$. In the third column are the numerals 9, 9, 9, 16, 0, representing 1,364,360. The original appears in black and red colors. (Taken from Morley, *An Introduction to the Study of the Maya Hieroglyphs*, p. 266.)

local value, are of interest as exhibiting not only quinary, decimal, and vigesimal systems, but also ternary, quaternary, and octonary systems.[1]

MAYA

68. The Maya of Central America and Southern Mexico developed hieroglyphic writing, as found on inscriptions and codices, dating apparently from about the beginning of the Christian Era, which discloses the use of a remarkable number system and chronology.[2] The number system discloses the application of the principle of local value, and the use of a symbol for zero centuries before the Hindus began to use their symbol for zero. The Maya system was vigesimal, except in one step. That is, 20 units (*kins*, or "days") make 1 unit of the next higher order (*uinals*, or 20 days), 18 *uinals* make 1 unit of the third order (*tun*, or 360 days), 20 *tuns* make 1 unit of the fourth order (*Katun*, or 7,200 days), 20 *Katuns* make 1 unit of the fifth order (*cycle*, or 144,000 days), and finally 20 *cycles* make 1 *great cycle* of 2,880,000 days. In the Maya codices we find symbols for 1–19, expressed by bars and dots. Each bar stands for 5 units, each dot for 1 unit. For instance,

·	··	··	——	··	·	····
1	2	4	5	7	11	19

The zero is represented by a symbol that looks roughly like a half-closed eye. In writing 20 the principle of local value enters. It is expressed by a dot placed over the symbol for zero. The numbers are written vertically, the lowest order being assigned the lowest position (see Fig. 19). The largest number found in the codices is 12,489,781.

CHINA AND JAPAN

69. According to tradition, the oldest Chinese representation of number was by the aid of *knots in strings*, such as are found later among the early inhabitants of Peru. There are extant two Chinese tablets[3] exhibiting knots representing numbers, odd numbers being designated by white knots (standing for the complete, as day, warmth,

[1] W. C. Eells, "Number-Systems of North-American Indians," *American Mathematical Monthly*, Vol. XX (1913), p. 263–72, 293–99; also *Bibliotheca mathematica* (3d series, 1913), Vol. XIII, p. 218–22.

[2] Our information is drawn from S. G. Morley, *An Introduction to the Study of the Maya Hieroglyphs* (Washington, 1915).

[3] Paul Perny, *Grammaire de la langue chinoise orale et écrite* (Paris, 1876), Vol. II, p. 5–7; Cantor, *Vorlesungen über Geschichte der Mathematik*, Vol. 1 (3d ed.), p. 674.

the sun) while even numbers are designated by black knots (standing for the incomplete, as night, cold, water, earth). The left-hand tablet shown in Figure 20 represents the numbers 1–10. The right-hand tablet pictures the magic square of nine cells in which the sum of each row, column, and diagonal is 15.

70. The Chinese are known to have used three other systems of writing numbers, the Old Chinese numerals, the mercantile numerals, and what have been designated as scientific numerals. The time of the introduction of each of these systems is uncertain.

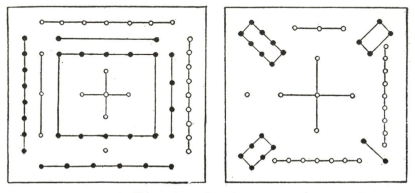

Fɪɢ. 20.—Early Chinese knots in strings, representing numerals

71. The *Old Chinese numerals* were written vertically, from above down. Figure 21 shows the Old Chinese numerals and mercantile numerals, also the Japanese cursive numerals.[1]

72. The *Chinese scientific numerals* are made up of vertical and horizontal rods according to the following plan: The numbers 1–9 are represented by the rods |, ||, |||, ||||, |||||, T, TT, TTT, TTTT; the numbers 10–90 are written thus — = ≡ ≣ ≣ ⊥ ⊥ ⊥ ⊥. According to the Chinese author Sun-Tsu, units are represented, as just shown, by vertical rods, ten's by horizontal rods, hundred's again by vertical rods, and so on. For example, the number 6,728 was designated by ⊥ TT = TTT.

73. The Japanese make use of the Old Chinese numerals, but have two series of names for the numeral symbols, one indigenous, the other derived from the Chinese language, as seen in Figure 21.

[1] See also Ed. Biot, *Journal asiatique* (December, 1839), p. 497–502; Cantor, *Vorlesungen über Geschichte der Mathematik*, Vol. I, p. 673; Biernatzki, *Crelle's Journal*, Vol. LII (1856), p. 59–94.

HINDU-ARABIC NUMERALS

74. *Introduction.*—It is impossible to reproduce here all the forms of our numerals which have been collected from sources antedating 1500 or 1510 A.D. G. F. Hill, of the British Museum, has devoted a

CHIFFRES			VALEURS.	NOMS DE NOMBRE	
CHINOIS *kidï-chôu.*	JAPONAIS cursifs.	DU COMMERCE.		EN JAPONAIS PUR.	EN SINICO-JAPONAIS.
一	一	۱	1	*fitots.*	*itsi.*
二	二	۱۱	2	*foutats.*	*ni.*
三	三	۱۱۱	3	*mits.*	*san.*
四	四	メ	4	*yots.*	*si.*
五	五	४	5	*itsouts.*	*go.*
六	六	上	6	*mouts.*	*rok.*
七	七	士	7	*nanats.*	*sitsi.*
八	八	丰	8	*yats.*	*fats.*
九	九	夂	9	*kokonots.*	*kou.*
十	十	十	10	*towo.*	*zyou.*
百	百	θ	100	*momo.*	*fak ou fyak.*
千	千	千	1,000	*tsidzi.*	*sen.*
萬	萬	万	10,000	*yorodz.*	*man.*

FIG. 21.—Chinese and Japanese numerals. (Taken from A. P. Pihan, *Exposé les signes de numération* [Paris, 1860], p. 15.)

whole book[1] of 125 pages to the early numerals in Europe alone. Yet even Hill feels constrained to remark: "What is now offered, in the shape of just 1,000 classified examples, is nothing more than a *vinde-*

[1] *The Development of Arabic Numerals in Europe* (exhibited in 64 tables; Oxford, 1915).

miatio prima." Add to the Hill collection the numeral forms, or supposedly numeral forms, gathered from other than European sources, and the material would fill a volume very much larger than that of Hill. We are compelled, therefore, to confine ourselves to a few of the more important and interesting forms of our numerals.[1]

75. One feels the more inclined to insert here only a few tables of numeral forms because the detailed and minute study of these forms has thus far been somewhat barren of positive results. With all the painstaking study which has been given to the history of our numerals we are at the present time obliged to admit that we have not even settled the time and place of their origin. At the beginning of the present century the Hindu origin of our numerals was supposed to have been established beyond reasonable doubt. But at the present time several earnest students of this perplexing question have expressed grave doubts on this point. Three investigators—G. R. Kaye in India, Carra de Vaux in France, and Nicol. Bubnov in Russia—working independently of one another, have denied the Hindu origin.[2] However, their arguments are far from conclusive, and the hypothesis of the Hindu origin of our numerals seems to the present writer to explain the known facts more satisfactorily than any of the substitute hypotheses thus far advanced.[3]

[1] The reader who desires fuller information will consult Hill's book which is very rich in bibliographical references, or David Eugene Smith and Louis Charles Karpinski's *The Hindu-Arabic Numerals* (Boston and London, 1911). See also an article on numerals in English archives by H. Jenkinson in *Antiquaries Journal*, Vol. VI (1926), p. 263–75. The valuable original researches due to F. Woepcke should be consulted, particularly his great "Mémoire sur la propagation des chiffres indiens" published in the *Journal asiatique* (6th series; Paris, 1863), p. 27–79, 234–90, 442–529. Reference should be made also to a few other publications of older date, such as G. Friedlein's *Zahlzeichen und das elementare Rechnen der Griechen und Römer* (Erlangen, 1869), which touches questions relating to our numerals. The reader will consult with profit the well-known histories of mathematics by H. Hankel and by Moritz Cantor.

[2] G. R. Kaye, "Notes on Indian Mathematics," *Journal and Proceedings of the Asiatic Society of Bengal* (N.S., 1907), Vol. III, p. 475–508; "The Use of the Abacus in Ancient India," *ibid.*, Vol. IV (1908), p. 293–97; "References to Indian Mathematics in Certain Mediaeval Works," *ibid.*, Vol. VII (1911), p. 801–13; "A Brief Bibliography of Hindu Mathematics," *ibid.*, p. 679–86; *Scientia*, Vol. XXIV (1918), p. 54; "Influence grecque dans le développement des mathématiques hindoues," *ibid.*, Vol. XXV (1919), p. 1–14; Carra de Vaux, "Sur l'origine des chiffres," *ibid.*, Vol. XXI (1917), p. 273–82; Nicol. Bubnov, *Arithmetische Selbstständigkeit der europäischen Kultur* (Berlin, 1914) (trans. from Russian ed.; Kiev, 1908).

[3] F. Cajori, "The Controversy on the Origin of Our Numerals," *Scientific Monthly*, Vol. IX (1919), p. 458–64. See also B. Datta in *Amer. Math. Monthly*, Vol. XXXIII, p. 449; *Proceed. Benares Math. Soc.*, Vol. VII.

76. Early Hindu mathematicians, Aryabhata (b. 476 A.D.) and Brahmagupta (b. 598 A.D.), do not give the expected information about the Hindu-Arabic numerals.

Āryabhaṭa's work, called *Aryabhatiya*, is composed of three parts, in only the first of which use is made of a special notation of numbers. It is an alphabetical system[1] in which the twenty-five consonants represent 1–25, respectively; other letters stand for 30, 40, ... , 100, etc.[2] The other mathematical parts of Āryabhata consists of rules without examples. Another alphabetic system prevailed in Southern India, the numbers 1–19 being designated by consonants, etc.[3]

In Brahmagupta's *Pulverizer*, as translated into English by H. T. Colebrooke,[4] numbers are written in our notation with a zero and the principle of local value. But the manuscript of Brahmagupta used by Colebrooke belongs to a late century. The earliest commentary on Brahmagupta belongs to the tenth century; Colebrooke's text is later.[5] Hence this manuscript cannot be accepted as evidence that Brahmagupta himself used the zero and the principle of local value.

77. Nor do inscriptions, coins, and other manuscripts throw light on the origin of our numerals. Of the old notations the most important is the Brahmi notation which did not observe place value and in which 1, 2, and 3 are represented by $—$, $=$, $\equiv$. The forms of the Brahmi numbers do not resemble the forms in early place-value notations[6] of the Hindu-Arabic numerals.

Still earlier is the Kharoshthi script,[7] used about the beginning of the Christian Era in Northwest India and Central Asia. In it the first three numbers are | || |||, then X = 4, |X = 5, ||X = 6, XX = 8, ꓶ = 10, 3 = 20, 33 = 40, ꓶ33 = 50, ⅄| = 100. The writing proceeds from right to left.

78. *Principle of local value.*—Until recently the preponderance of authority favored the hypothesis that our numeral system, with its concept of local value and our symbol for zero, was wholly of Hindu origin. But it is now conclusively established that the principle of

[1] M. Cantor, *Vorlesungen über Geschichte der Mathematik*, Vol. I (3d ed.), p. 606.

[2] G. R. Kaye, *Indian Mathematics* (Calcutta and Simla, 1915), p. 30, gives full explanation of Aryabhata's notation.

[3] M. Cantor, *Math. Beiträge z. Kulturleben der Völker* (1863), p. 68, 69.

[4] *Algebra with Arithmetic and Mensuration from the Sanscrit* (London, 1817), p. 326 ff.

[5] *Ibid.*, p. v, xxxii.

[6] See forms given by G. R. Kaye, *op. cit.*, p. 29.   [7] *Ibid.*

local value was used by the Babylonians much earlier than by the Hindus, and that the Maya of Central America used this principle and symbols for zero in a well-developed numeral system of their own and at a period antedating the Hindu use of the zero (§ 68).

79. The earliest-known reference to Hindu numerals outside of India is the one due to Bishop Severus Sebokht of Nisibis, who, living in the convent of Kenneshre on the Euphrates, refers to them in a fragment of a manuscript (MS Syriac [Paris], No. 346) of the year 662 A.D. Whether the numerals referred to are the ancestors of the modern numerals, and whether his Hindu numerals embodied the principle of local value, cannot at present be determined. Apparently hurt by the arrogance of certain Greek scholars who disparaged the Syrians, Sebokht, in the course of his remarks on astronomy and mathematics, refers to the Hindus, "their valuable methods of calculation; and their computing that surpasses description. I wish only to say that this computation is done by means of nine signs."[1]

80. Some interest attaches to the earliest dates indicating the use of the perfected Hindu numerals. That some kind of numerals with a zero was used in India earlier than the ninth century is indicated by Brahmagupta (b. 598 A.D.), who gives rules for computing with a zero.[2] G. Bühler[3] believes he has found definite mention of the decimal system and zero in the year 620 A.D. These statements do not necessarily imply the use of a decimal system based on the principle of local value. G. R. Kaye[4] points out that the task of the antiquarian is complicated by the existence of forgeries. In the eleventh century in India "there occurred a specially great opportunity to regain confiscated endowments and to acquire fresh ones." Of seventeen citations of inscriptions before the tenth century displaying the use of place value in writing numbers, all but two are eliminated as forgeries; these two are for the years 813 and 867 A.D.; Kaye is not sure of the reliability even of these. According to D. E. Smith and L. C. Karpinski,[5] the earliest authentic document unmistakably containing the numerals with the zero in India belongs to the year 876 A.D. The earli-

[1] See M. F. Nau, *Journal asiatique* (10th ser., 1910), Vol. XVI, p. 255; L. C. Karpinski, *Science*, Vol. XXXV (1912), p. 969-70; J. Ginsburg, *Bulletin of the American Mathematical Society*, Vol. XXIII (1917), p. 368.

[2] Colebrooke, *op. cit.*, p. 339, 340.

[3] "Indische Paläographie," *Grundriss d. indogerman. Philologie u. Altertumskunde*, Band I, Heft 11 (Strassburg, 1896), p. 78.

[4] *Journal of the Asiatic Society of Bengal* (N.S., 1907), Vol. III, p. 482-87.

[5] *The Hindu-Arabic Numerals* (New York, 1911), p. 52.

est Arabic manuscripts containing the numerals are of 874[1] and 888 A.D. They appear again in a work written at Shiraz in Persia[2] in 970 A.D. A church pillar[3] not far from the Jeremias Monastery in Egypt has

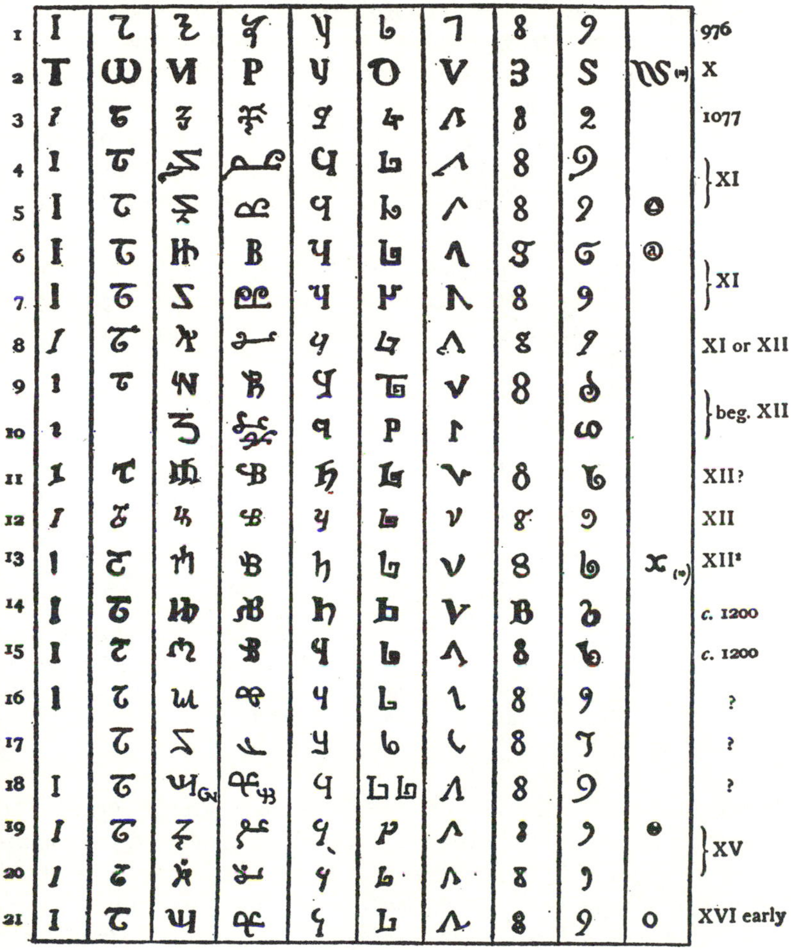

Fig. 22.—G. F. Hill's table of early European forms and Boethian apices. (From G. F. Hill, *The Development of Arabic Numerals in Europe* [Oxford, 1915], p. 28. Mr. Hill gives the MSS from which the various sets of numerals in this table are derived: [1] Codex Vigilanus; [2] St. Gall MS now in Zürich; [3] Vatican MS 3101, etc. The Roman figures in the last column indicate centuries.)

[1] Karabacek, *Wiener Zeitschrift für die Kunde des Morgenlandes*, Vol. II (1897), p. 56.

[2] L. C. Karpinski, *Bibliotheca mathematica* (3d ser., 1910–11), p. 122.

[3] Smith and Karpinski, *op. cit.*, p. 138–43.

the date 349 A.H. (= 961 A.D.). The oldest definitely dated European manuscript known to contain the Hindu-Arabic numerals is the Codex Vigilanus (see Fig. 22, No. 1), written in the Albelda Cloister in Spain in 976 A.D. The nine characters without the zero are given, as an addition, in a Spanish copy of the *Origines* by Isidorus of Seville, 992 A.D. A tenth-century manuscript with forms differing materially from those in the Codex Vigilanus was found in the St. Gall manuscript (see Fig. 22, No. 2), now in the University Library at Zürich. The numerals are contained in a Vatican manuscript of 1077 (see Fig. 22, No. 3), on a Sicilian coin of 1138, in a Regensburg (Bavaria)

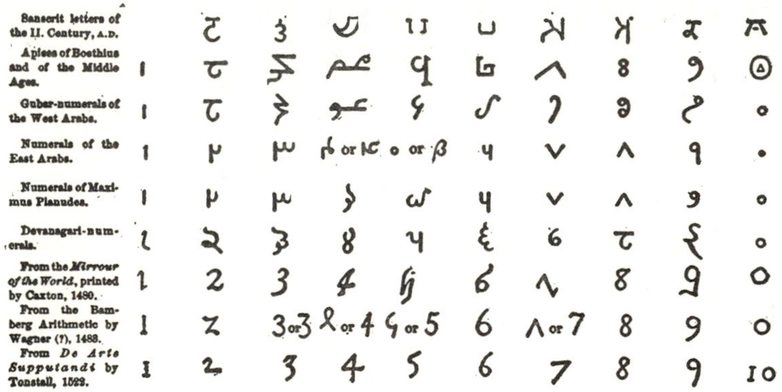

FIG. 23.—Table of important numeral forms. (The first six lines in this table are copied from a table at the end of Cantor's *Vorlesungen über Geschichte der Mathematik*, Vol. I. The numerals in the Bamberg arithmetic are taken from Friedrich Unger, *Die Methodik der praktischen Arithmetik in historischer Entwickelung* [Leipzig, 1888], p. 39.)

chronicle of 1197. The earliest manuscript in French giving the numerals dates about 1275. In the British Museum one English manuscript is of about 1230–50; another is of 1246. The earliest undoubted Hindu-Arabic numerals on a gravestone are at Pforzheim in Baden of 1371 and one at Ulm of 1388. The earliest coins outside of Italy that are dated in the Arabic numerals are as follows: Swiss 1424, Austrian 1484, French 1485, German 1489, Scotch 1539, English 1551.

81. *Forms of numerals.*—The Sanskrit letters of the second century A.D. head the list of symbols in the table shown in Figure 23. The implication is that the numerals have evolved from these letters. If such a connection could be really established, the Hindu origin of our numeral forms would be proved. However, a comparison of the forms appearing in that table will convince most observers that an origin

from Sanskrit letters cannot be successfully demonstrated in that way; the resemblance is no closer than it is to many other alphabets.

The forms of the numerals varied considerably. The 5 was the most freakish. An upright 7 was rare in the earlier centuries. The symbol for zero first used by the Hindus was a dot.[1] The symbol for zero (0) of the twelfth and thirteenth centuries is sometimes crossed by a horizontal line, or a line slanting upward.[2] The Boethian apices, as found in some manuscripts, contain a triangle inscribed in the circular zero. In Athelard of Bath's translation of Al-Madjrītī's revision of Al-Khowarizmi's astronomical tables there are in different manuscripts three signs for zero,[3] namely, the $\ominus$ (= theta?) referred to above, then $\check{\tau}$ (= *teca*),[4] and $\bar{0}$. In one of the manuscripts 38 is written several times XXXO, and 28 is written XXO, the O being intended most likely as the abbreviation for *octo* ("eight").

82. The symbol τ for zero is found also in a twelfth-century manuscript[5] of N. Ocreatus, addressed to his master Athelard. In that century it appears especially in astronomical tables as an abbreviation for *teca*, which, as already noted, was one of several names for zero;[6] it is found in those tables by itself, without connection with other numerals. The symbol occurs in the *Algorismus vulgaris* ascribed to Sacrobosco.[7] C. A. Nallino found $\bar{o}$ for zero in a manuscript of Escurial, used in the preparation of an edition of Al-Battani. The symbol $\ominus$ for zero occurs also in printed mathematical books.

The one author who in numerous writings habitually used θ for zero was the French mathematician Michael Rolle (1652–1719). One finds it in his *Traité d'algèbre* (1690) and in numerous articles in the publications of the French Academy and in the *Journal des sçavans*.

[1] Smith and Karpinski, *op. cit.*, p. 52, 53.

[2] Hill, *op. cit.*, p. 30–60.

[3] H. Suter, *Die astronomischen Tafeln des Muḥammed ibn Mūsā Al-Khwārizmī in der Bearbeitung des Maslama ibn Aḥmed Al-Madjrītī und der lateinischen Uebersetzung des Athelhard von Bath* (København, 1914), p. xxiii.

[4] See also M. Curtze, *Petri Philomeni de Dacia in Algorismum vulgarem Johannis de Sacrobosco Commentarius* (Hauniae, 1897), p. 2, 26.

[5] "Prologus N. Ocreati in Helceph ad Adelardum Batensem Magistrum suum. Fragment sur la multiplication et la division publié pour la première fois par Charles Henry," *Abhandlungen zur Geschichte der Mathematik*, Vol. III (1880), p. 135–38.

[6] M. Curtze, *Urkunden zur Geschichte der Mathematik im Mittelalter und der Renaissance* (Leipzig, 1902), p. 182.

[7] M. Curtze, *Abhandlungen zur Geschichte der Mathematik*, Vol. VIII (Leipzig, 1898), p. 3–27.

Manuscripts of the fifteenth century, on arithmetic, kept in the Ashmolean Museum[1] at Oxford, represent the zero by a circle, crossed by a vertical stroke and resembling the Greek letter ϕ. Such forms for zero are reproduced by G. F. Hill[2] in many of his tables of numerals.

83. In the fifty-six philosophical treatises of the brothers Iḫwān aṣ-ṣafā (about 1000 A.D.) are shown Hindu-Arabic numerals and the corresponding Old Arabic numerals.

The forms of the Hindu-Arabic numerals, as given in Figure 24, have maintained themselves in Syria to the present time. They appear with almost identical form in an Arabic school primer, printed

Fig. 24.—In the first line are the Old Arabic numerals for 10, 9, 8, 7, 6, 5, 4, 3, 2, 1. In the second line are the Arabic names of the numerals. In the third line are the Hindu-Arabic numerals as given by the brothers Iḫwān aṣ-ṣafā. (Reproduced from J. Ruska, *op. cit.*, p. 87.)

at Beirut (Syria) in 1920. The only variation is in the 4, which in 1920 assumes more the form of a small Greek epsilon. Observe that 0 is represented by a dot, and 5 by a small circle. The forms used in modern Arabic schoolbooks cannot be recognized by one familiar only with the forms used in Europe.

84. In fifteenth-century Byzantine manuscripts, now kept in the Vienna Library,[3] the numerals used are the Greek letters, but the principle of local value is adopted. Zero is γ or in some places $\cdot$; aa means 11, $\beta\gamma$ means 20, $a\gamma\gamma\gamma$ means 1,000. "This symbol γ for zero means elsewhere 5," says Heiberg, "conversely, o stands for 5 (as now among the Turks) in Byzantine scholia to Euclid. . . . . In Constantinople the new method was for a time practiced with the retention of

[1] Robert Steele, *The Earliest Arithmetics in English* (Oxford, 1922), p. 5.

[2] *Op. cit.*, Tables III, IV, V, VI, VIII, IX, XI, XV, XVII, XX, XXI, XXII. See also E. Wappler, *Zur Geschichte der deutschen Algebra im XV. Jahrhundert* (Zwickauer Gymnasialprogramm von 1887), p. 11–30.

[3] J. L. Heiberg, "Byzantinische Analekten," *Abhandlungen zur Geschichte der Mathematik*, Vol. IX (Leipzig, 1899), p. 163, 166, 172. This manuscript in the Vienna Library is marked "Codex Phil. Gr. 65."

the old letter-numerals, mainly, no doubt, in daily intercourse." At
the close of one of the Byzantine manuscripts there is a table of
numerals containing an imitation of the Old Attic numerals. The table
gives also the Hindu-Arabic numerals, but apparently without recognition of the principle of local value; in writing 80, the 0 is placed over
the 8. This procedure is probably due to the ignorance of the scribe.

85. A manuscript[1] of the twelfth century, in Latin, contains the
symbol Ⱶ for 3 which Curtze and Nagl[2] declare to have been found
only in the twelfth century. According to Curtze, the foregoing
strange symbol for 3 is simply the symbol for *tertia* used in the notation for sexagesimal fractions which receive much attention in this
manuscript.

86. Recently the variations in form of our numerals have been sum-
marized as follows: "The form[3] of the numerals 1, 6, 8 and 9 has not
varied *much* among the [medieval] Arabs nor among the Christians
of the Occident; the numerals of the Arabs of the Occident for 2, 3 and
5 have forms offering some analogy to ours (the 3 and 5 are originally
reversed, as well among the Christians as among the Arabs of the
Occident); but the form of 4 and that of 7 have greatly modified
themselves. The numerals 5, 6, 7, 8 of the Arabs of the Orient differ
distinctly from those of the Arabs of the Occident (Gobar numerals).
For *five* one still writes 5 and ﺱ." The use of *i* for 1 occurs in the first
printed arithmetic (Treviso, 1478), presumably because in this early
stage of printing there was no type for 1. Thus, 9,341 was printed
934*i*.

87. Many points of historical interest are contained in the following quotations from the writings of Alexander von Humboldt.
Although over a century old, they still are valuable.

"In the Gobar[4] the group signs are dots, that is zeroes, for in
India, Tibet and Persia the zeroes and dots are identical. The Gobar
symbols, which since the year 1818 have commanded my whole attention, were discovered by my friend and teacher, Mr. Silvestre de
Sacy, in a manuscript from the Library of the old Abbey St. Germain
du Près. This great orientalist says: 'Le Gobar a un grand rapport

---

[1] Algorithmus-MSS Clm 13021, fols. 27–29, of the Munich Staatsbibliothek.
Printed and explained by Maximilian Curtze, *Abhandlungen zur Geschichte der
Mathematik*, Vol. VIII (Leipzig, 1898), p. 3–27.

[2] *Zeitschrift für Mathematik und Physik* (Hist. Litt. Abth.), Vol. XXXIV
(Leipzig, 1889), p. 134.

[3] *Encyc. des Scien. math.*, Tome I, Vol. I (1904), p. 20, n. 105, 106.

[4] Alexander von Humboldt, *Crelle's Journal*, Vol. IV (1829), p. 223, 224.

avec le chiffre indien, mais il n'a pas de zéro (S. *Gramm. arabe*, p. 76, and the note added to Pl. 8).' I am of the opinion that the zero-symbol is present, but, as in the Scholia of Neophytos on the units, it stands over the units, not by their side. Indeed it is these very zero-symbols or dots, which give these characters the singular name *Gobar* or *dust-writing*. At first sight one is uncertain whether one should recognize therein a transition between numerals and letters of the alphabet. One distinguishes with difficulty the Indian 3, 4, 5 and 9. *Dal* and *ha* are perhaps ill-formed Indian numerals 6 and 2. The notation by dots is as follows:

$$3 \overset{\cdot}{} \quad \text{for } 30 \ ,$$
$$4 \overset{\cdot\cdot}{} \quad \text{for } 400 \ ,$$
$$6 \overset{\cdot\cdot}{\cdot} \quad \text{for } 6{,}000 \ .$$

These dots remind one of an old-Greek but rare notation (Ducange, *Palaeogr.*, p. xii), which begins with the myriad: $a \overset{\cdot\cdot}{}$ for 10,000, $\beta \overset{\cdot\cdot}{\cdot}$ for 200 millions. In this system of geometric progressions a single dot, which however is not written down, stands for 100. In Diophantus and Pappus a dot is placed between letter-numerals, instead of the initial $M\upsilon$ (myriad). A dot multiplies what lies to its left by 10,000. . . . . A real zero symbol, standing for the absence of some unit, is applied by Ptolemy in the descending sexagesimal scale for missing degrees, minutes or seconds. Delambre claims to have found our symbol for zero also in manuscripts of Theon, in the Commentary to the Syntaxis of Ptolemy.[1] It is therefore much older in the Occident than the invasion of the Arabs and the work of Planudes on *arithmoi indikoi.*" L. C. Karpinski[2] has called attention to a passage in the Arabic biographical work, the *Fihrist* (987 A.D.), which describes a Hindu notation using dots placed *below* the numerals; one dot indicates tens, two dots hundreds, and three dots thousands.

88. There are indications that the magic power of the principle of local value was not recognized in India from the beginning, and that our perfected Hindu-Arabic notation resulted from gradual evolution. Says Humboldt: "In favor of the successive perfecting of the designation of numbers in India testify the Tamul numerals which, by means

---

[1] J. B. J. Delambre, *Histoire de l'astron. ancienne*, Vol. I, p. 547; Vol. II, p. 10. The alleged passage in the manuscripts of Theon is not found in his printed works. Delambre is inclined to ascribe the Greek sign for zero either as an abbreviation of *ouden* or as due to the special relation of the numeral omicron to the sexagesimal fractions (*op. cit.*, Vol. II, p. 14, and *Journal des sçavans* [1817], p. 539).

[2] *Bibliotheca mathematica*, Vol. XI (1910–11), p. 121–24.

of the nine signs for the units and by signs of the groups 10, 100, or 1,000, express all values through the aid of multipliers placed on the

FIG. 25.—The numerals of the monk Neophytos

left. This view is supported also by the singular *arithmoi indikoi* in the scholium of the monk *Neophytos*, which is found in the Parisian

A HISTORY OF MATHEMATICAL NOTATIONS

Wie die visier ziffer er= kendt werden.

Die Visier ziffer werden gewonlich mit jren Chaͤ
ractern wie hernach volgt also geschriben/habe gleich
wol nit vil sondere verwandlung gegen den gemei
nen ziffern/außgenoͤmen das fünfft vnd sibend.

Auch solt du sonderlich mercken/ wenn bey einer
ziffer drey punct stehn/ so helt dasselbig daß gerad so
vil Eymer/vnd kein viertheil minder noch mehr.

Die halben Eymer werden allein mit einer lini oder
strichlin vnterscheiden.Deñ als offt ein strichlin durch
ein ziffer geht/benimpt es ein halben Eymer/vnd das
geschicht allein bey den Eymern vnnd nicht bein vier
tein.

1 Eins　　　·|, Ein halber Eymer.
2 Zwey　　　·|· Ein gantzer Eymer.
3 Drey　　　·⚹ Anderthalber Eymer
4 Vier　　　·⚹· Zwen Eymer.
5 Fünff　　　·⚹· Dritthalber Eymer.
6 Sechs　　　⚹· Drey Eymer.
7 Siben

8 Acht　　　⚹ Vierdthalber Eymer.
9 Neune　　·⚹· Vier Eymer.
10 Zehen　　·⚹· Fünffthalb Eymer.
　　　　　　·⚹· Fünff Eymer.
　　　　　　·⚹· Sechshalb Eymer
　　　　　　·⚹· Sechs Eymer.
So aber ein vaß et＝　⚹· Sibenthalber Eymer
lich vierteil mehr: oder　⚹· Siben Eymer.
minder vber die ge＝　·⚹· Achthalb Eymer
funden Eymer helt /　
das wirde durch die　·8· Acht Eymer.
zwey volgenden zey＝　⚹· Neundhalb Eymer.
chen geschriben / vnd　
die nachfolgende zif＝　⚹· Neun Eymer.
fer bedeut die vier＝　·10· Zehenthalber Eymer.
theil.　　　　　·10· Zehen Eymer.
　　　　　　·11· Eilffthalber Eymer.

Bedent der vier＝
tel minder.
Bedeut der vier＝
tel mehr.
Exempel auff
minder.
Vier aymer min＝
der vier viertel.
Zwen aymer min
der zwey vierteil.
Drey aymer min＝
der sechs vierteil.
Siben aymer
minder fünff vier
theil.
Exempel auff
mehr.
Siben aymer vnd
vier viertheil.
Sechs aymer
vnd drey viertel.
Fünff aymer vnd
zwey viertheyl.
Sechzehen ty＝
mer vnd ein viers
theil.
Achthalber ay＝
mer vñ drey vier＝
theil.

Fig. 26.—From Christoff Rudolff's *Künstliche Rechnung mit der Ziffer* (Augsburg, 1574[?]).

Library (Cod. Reg., fol. 15), for an account of which I am indebted to Prof. Brandis. The nine digits of Neophytos wholly resemble the Persian, except the 4. The digits 1, 2, 3 and 9 are found even in Egyptian number inscriptions (Kosegarten, *de Hierogl. Aegypt.*, p. 54). The nine units are enhanced tenfold, 100 fold, 1,000 fold by writing above them one, two or three zeros, as in: $\overset{\circ}{2}=20$, $\overset{\circ}{24}=24$, $\overset{\circ\circ}{5}=500$, $\overset{\circ\circ\circ}{6}=6,000$. If we imagine dots in place of the zero symbols, then we have the arabic Gobar numerals."[1] Humboldt copies the scholium of Neophytos. J. L. Heiberg also has called attention to the scholium of Neophytos and to the numbering of scholia to Euclid in a Greek manuscript of the twelfth century (Codex Vindobonensis, Gr. 103), in which numerals resembling the Gobar numerals occur.[2] The numerals of the monk Neophytos (Fig. 25), of which Humboldt speaks, have received the special attention of P. Tannery.[3]

89. *Freak forms.*—We reproduce herewith from the Augsburg edition of Christoff Rudolff's *Künstliche Rechnung* a set of our numerals, and of symbols to represent such fractions

[1] *Op. cit.*, p. 227.

[2] See J. L. Heiberg's edition of *Euclid* (Leipzig, 1888), Vol. V; P. Tannery, *Revue archéol.* (3d ser., 1885), Vol. V, p. 99, also (3d ser., 1886), Vol. VII, p. 355; *Encycl des scien. math.*, Tome I, Vol. I (1904), p. 20, n. 102.

[3] *Mémoires scientifiques*, Vol. IV (Toulouse and Paris, 1920), p. 22.

and mixed numbers as were used in Vienna in the measurement of wine. We have not seen the first edition (1526) of Rudolff's book, but Alfred Nagl[1] reproduces part of these numerals from the first edition. "In the Viennese wine-cellars," says Hill, "the casks were marked according to their contents with figures of the forms given."[2] The symbols for fractions are very curious.

90. *Negative numerals.*—J. Colson[3] in 1726 claimed that, by the use of negative numerals, operations may be performed with "more ease and expedition." If 8605729398715 is to be multiplied by 389175836438, reduce these to small numbers $1\overline{4}1\overline{4}33\overline{1}40\overline{1}315$ and $4\overline{1}\overline{1}2\overline{2}4\overline{2}4444\overline{2}$. Then write the multiplier on a slip of paper and place it in an inverted position, so that its first figure is just over the left-hand figure of the multiplicand. Multiply $4 \times 1 = 4$ and write down 4. Move the multiplier a place to the right and collect the two products, $4 \times \overline{1} + \overline{1} \times 1 = \overline{5}$; write down $\overline{5}$. Move the multiplier another place to the right, then $4 \times \overline{4} + \overline{1} \times \overline{1} + \overline{1} \times 1 = \overline{1}6$; write the $\overline{1}$ in the second line. Similarly, the next product is 11, and so on. Similar processes and notations were proposed by A. Cauchy,[4] E. Selling,[5] and W. B. Ford,[6] while J. P. Ballantine[7] suggests 1 inverted, thus ┰, as a sign for negative 1, so that ┰ $\times 7 = $ ┰3 and the logarithm $9.69897 - 10$ may be written ┰9.69897 or ┰.69897. Negative logarithmic characteristics are often marked with a negative sign placed over the numeral (Vol. II, § 476).

91. *Grouping digits in numeration.*—In the writing of numbers containing many digits it is desirable to have some symbol separating the numbers into groups of, say, three digits. Dots, vertical bars, commas, arcs, and colons occur most frequently as signs of separation.

In a manuscript, Liber algorizmi,[8] of about 1200 A.D., there appear

---

[1] *Monatsblatt der numismatischen Gesellschaft in Wien*, Vol. VII (December, 1906), p. 132.

[2] G. F. Hill, *op. cit.*, p. 53.

[3] *Philosophical Transactions*, Vol. XXXIV (1726), p. 161–74; *Abridged Transactions*, Vol. VI (1734), p. 2–4. See also G. Peano, *Formulaire mathématique*, Vol. IV (1903), p. 49.

[4] *Comptes rendus*, Vol. XI (1840), p. 796; *Œuvres* (1st ser.), Vol. V, p. 434–55.

[5] *Eine neue Rechenmaschine* (Berlin, 1887), p. 16; see also *Encyklopädie d. Math. Wiss.*, Vol. I, Part 1 (Leipzig, 1898–1904), p. 944.

[6] *American Mathematical Monthly*, Vol. XXXII (1925), p. 302.

[7] *Op. cit.*, p. 302.

[8] M. Cantor, *Zeitschrift für Mathematik*, Vol. X (1865), p. 3; G. Eneström, *Bibliotheca mathematica* (3d ser., 1912–13), Vol. XIII, p. 265.

dots to mark periods of three. Leonardo of Pisa, in his *Liber Abbaci* (1202), directs that the hundreds, hundred thousands, hundred millions, etc., be marked with an accent above; that the thousands, millions, thousands of millions, etc., be marked with an accent below.

In the 1228 edition,[1] Leonardo writes 678 935 784 105 296. Johannes de Sacrobosco (d. 1256), in his *Tractatus de arte numerandi*, suggests that every third digit be marked with a dot.[2] His commentator, Petrus de Dacia, in the first half of the fourteenth century, does the same.[3] Directions of the same sort are given by Paolo Dagomari[4] of Florence, in his *Regoluzze di Maestro Paolo dall Abbaco* and Paolo of Pisa,[5] both writers of the fourteenth century. Luca Pacioli, in his *Summa* (1494), folio 19b, writes 8 659 421 635 894 676; Georg Peurbach (1505),[6] 3790528614. Adam Riese[7] writes 86789325178. M. Stifel (1544)[8] writes 2329089562800. Gemma Frisius[9] in 1540 wrote 24 456 345 678. Adam Riese (1535)[10] writes 86·7·89·3·25·178. The Dutch writer, Martinus Carolus Creszfeldt,[11] in 1557 gives in his *Arithmetica* the following marking of a number:

"Exempel. || 5 8 7 4 9 3 6 2 5 3 4 || ."

---

[1] *El liber abbaci di Leonardo Pisano* .... da B. Boncompagni (Roma, 1857), p. 4.

[2] J. O. Halliwell, *Rara mathematica* (London, 1839), p. 5; M. Cantor, *Vorlesungen*, Vol. II (2d ed., 1913), p. 89.

[3] *Petri Philomeni de Dacia in Algorismum vulgarem Iohannis de Sacrobosco commentarius* (ed. M. Curtze; Kopenhagen, 1897), p. 3, 29; J. Tropfke, *Geschichte der Elementarmathematik* (2d ed., 1921), Vol. I, p. 8.

[4] Libri, *Histoire des sciences mathématiques en Italie*, Vol. III, p. 296–301 (Rule 1).

[5] *Ibid.*, Vol. II, p. 206, n. 5, and p. 526; Vol. III, p. 295; see also Cantor, *op. cit.*, Vol. II (2d ed., 1913), p. 164.

[6] *Opus algorithmi* (Herbipoli, 1505). See Wildermuth, "Rechnen," *Encyklopaedie des gesammten Erziehungs- und Unterrichtswesens* (Dr. K. A. Schmid, 1885).

[7] *Rechnung auff der Linien vnnd Federn* (1544); Wildermuth, "Rechnen," *Encyklopaedie* (Schmid, 1885), p. 739.

[8] Wildermuth, *op. cit.*, p. 739.

[9] *Arithmeticae practicae methodus facilis* (1540); F. Unger, *Die Methodik der praktischen Arithmetik in historischer Entwickelung* (Leipzig, 1888), p. 25, 71.

[10] *Rechnung auff d. Linien u. Federn* (1535). Taken from H. Hankel, *op. cit.* (Leipzig, 1874), p. 15.

[11] *Arithmetica* (1557). Taken from Bierens de Haan, *Bouwstoffen voor de Geschiedenis der Wis-en Natuurkundige Wetenschappen*, Vol. II (1887), p. 3.

Thomas Blundeville (1636)[1] writes 5|936|649. Tonstall[2] writes

. . . .   4 3 2 1 0

3210987654321. Clavius[3] writes 42329089562800. Chr. Rudolff[4] writes

$2\overset{\cdot}{3}40\overset{\cdot}{5}6\overset{\cdot}{3}9567$. Johann Caramuel[5] separates the digits, as in "34:252,-

Integri.   Partes.

341;154,329"; W. Oughtred,[6] 9|876|543|$210\overline{12}$|$\overline{345}$|$\overline{678}$|9; K. Schott[7],

$\overset{|||}{7}69\overset{||}{74}3\overset{|}{2}329089562436$; N. Barreme,[8] 254.567.804.652; W. J. G.

     III   II   I   0

Karsten,[9] 872 094,826 152,870 364,008; I. A. de Segner,[10] 5|329″|870|

325′|743|297°, 174; Thomas Dilworth,[11] 789 789 789; Nicolas Pike,[12]

    3   2   1

356;809,379;120,406;129,763; Charles Hutton,[13] 281,427,307; E.

Bézout,[14] 23, 456, 789, 234, 565, 456.

In M. Lemos' *Portuguese encyclopedia*[15] the population of New

---

[1] *Mr. Blundevil, His Exercises contayning eight Treatises* (7th ed., Ro. Hartwell; London, 1636), p. 106.

[2] *De Arte Svppvtandi, libri qvatvor Cvtheberti Tonstalli* (Argentorati), Colophon 1544, p. 5.

[3] *Christophori Clavii epitome arithmeticae practicae* (Romae, 1583), p. 7.

[4] *Künstliche Rechnung mit der Ziffer* (Augsburg, 1574[?]), Aiij B.

[5] *Joannis Caramvelis mathesis biceps, vetus et nova* (Companiae [southeast of Naples], 1670), p. 7. The passage is as follows: "Punctum finale (.) est, quod ponitur post unitatem: ut cùm scribimus 23. viginti tria. Comma (,) post millenarium scribitur . . . . ut cùm scribimus, 23,424. . . . . Millenarium à centenario distinguere alios populos docent Hispani, qui utuntur hoc charactere $\backslash f$, ... Hypocolon (;) millionem à millenario separat, ut cùm scribimus 2;041,311. Duo puncta ponuntur post billionem, seu millionem millionum, videlicet 34:252,341;154,329." Caramuel was born in Madrid. For biographical sketch see *Revista matemática Hispano-American*, Vol. I (1919), p. 121, 178, 203.

[6] *Clavis mathematicae* (London, 1652), p. 1 (1st ed., 1631).

[7] *Cursus mathematicus* (Herbipoli, 1661), p. 23.

[8] *Arithmétique* (new ed.; Paris, 1732), p. 6.

[9] *Mathesis theoretica elementaris atqve svblimior* (Rostochii, 1760), p. 195.

[10] *Elementa arithmeticae geometriae et calcvli geometrici* (2d ed.; Halle, 1767), p. 13.

[11] *Schoolmaster's Assistant* (22d ed.; London, 1784), p. 3.

[12] *New and Complete System of Arithmetic* (Newburyport, 1788), p. 18.

[13] "Numeration," *Mathematical and Philosophical Dictionary* (London, 1795).

[14] *Cours de mathématiques* (Paris, 1797), Vol. I, p. 6.

[15] "Portugal," *Encyclopedia Portugueza Illustrada* . . . . de Maximiano Lemos (Porto).

York City is given as "3.437:202"; in a recent Spanish encyclopedia,[1] the population of America is put down as "150·979,995."

In the process of extracting square root, two early commentators[2] on Bhāskara's *Lilavati*, namely Rama-Crishna Deva and Gangad'hara (*ca.* 1420 A.D.), divide numbers into periods of two digits in this manner, $\overset{\shortmid}{8}\ \overset{-}{8}\ \overset{\shortmid}{2}\ \overset{-}{0}\ \overset{\shortmid}{9}$. In finding cube roots Rama-Crishna Deva writes $\overset{\shortmid}{1}\ \overset{-}{9}\ \overset{\shortmid}{5}\ \overset{\shortmid}{3}\ \overset{-}{1}\ \overset{-}{2}\ \overset{\shortmid}{5}$.

92. *The Spanish "calderón."*—In Old Spanish and Portuguese numeral notations there are some strange and curious symbols. In a contract written in Mexico City in 1649 the symbols "7U291e" and "VIIUCCXCIps" each represent 7,291 *pesos*. The U, which here resembles an O that is open at the top, stands for "thousands."[3] I. B. Richman has seen Spanish manuscripts ranging from 1587 to about 1700, and Mexican manuscripts from 1768 to 1855, all containing symbols for "thousands" resembling U or D, often crossed by one or two horizontal or vertical bars. The writer has observed that after 1600 this U is used freely both with Hindu-Arabic and with Roman numerals; *before 1600* the U occurs more commonly with Roman numerals. Karpinski has pointed out that it is used with the Hindu-Arabic numerals as early as 1519, in the accounts of the Magellan voyages. As the Roman notation does not involve the principle of local value, U played in it a somewhat larger rôle than merely to afford greater facility in the reading of numbers. Thus VIUCXV equals $6 \times 1,000 + 115$. This use is shown in manuscripts from Peru of 1549 and 1543,[4] in manuscripts from Spain of 1480[5] and 1429.[6]

We have seen the corresponding *type symbol* for 1,000 in Juan Perez de Moya,[7] in accounts of the coining in the Real Casa de Moneda de

[1] "América," *Enciclopedia illustrada segui Diccionario universal* (Barcelona).

[2] Colebrooke, *op. cit.*, p. 9, 12, xxv, xxvii.

[3] F. Cajori, "On the Spanish Symbol U for 'thousands,' " *Bibliotheca mathematica*, Vol. XII (1912), p. 133.

[4] *Cartas de Indias publicalas por primera vez el Ministerio de Fomento* (Madrid, 1877), p. 502, 543, *facsimiles X* and *Y*.

[5] José Gonzalo de las Casas, *Anales de la Paleografia Española* (Madrid, 1857), Plates 87, 92, 109, 110, 113, 137.

[6] Liciniano Saez, *Demostración Histórica del verdadero valor de todas las monedas que corrían en Castilla durante el Reynado del Señor Don Enrique III* (Madrid, 1796), p. 447. See also Colomera y Rodríguez, Venancio, *Paleografia castellana* (1862).

[7] *Aritmetica practica* (14th ed.; Madrid, 1784), p. 13 (1st ed., 1562).

Mexico (1787), in eighteenth-century books printed in Madrid,[1] in the *Gazetas de Mexico* of 1784 (p. 1), and in modern reprints of seventeenth-century documents.[2] In these publications the printed symbol resembles the Greek sampi ꝯ for 900, but it has no known connection with it. In books printed in Madrid[3] in 1760, 1655, and 1646, the symbol is a closer imitation of the written U, and is curiously made up of the two small printed letters, *l*, *f*, each turned halfway around. The two inverted letters touch each other below, thus ʋſ. Printed symbols representing a distorted U have been found also in some Spanish arithmetics of the sixteenth century, particularly in that of Gaspard de Texeda[4] who writes the number 103,075,102,300 in the Castellanean form c.iijU.75q̊s c.ijU300 and also in the algoristic form 103U075q̊s 102U300. The Spaniards call this symbol and also the sampi-like symbol a *calderón*.[5] A non-Spanish author who explains the *calderón* is Johann Caramuel,[6] in 1670.

93. The present writer has been able to follow the trail of this curious symbol U from Spain to Northwestern Italy. In Adriano Cappelli's *Lexicon* is found the following: "In the liguric documents of the second half of the fifteenth century we found in frequent use, to indicate the multiplication by 1,000, in place of M, an O crossed by a horizontal line."[7] This closely resembles some forms of our Spanish symbol U. Cappelli gives two facsimile reproductions[8] in

---

[1] Liciniano Saez, *op. cit.*

[2] Manuel Danvila, *Boletin de la Real Academia de la Historia* (Madrid, 1888), Vol. XII, p. 53.

[3] *Cuentas para todas, compendio arithmético, e Histórico* .... su autor D. Manuel Recio, Oficial de la contaduría general de postos del Reyno (Madrid, 1760); *Teatro Eclesiástico de la primitiva Iglesia de las Indias Occidentales* .... el M. Gil Gonzalez Davila, su Coronista Mayor de las Indias, y de los Reynos de las dos Castillas (Madrid, 1655), Vol. II; *Memorial, y Noticias Sacras, y reales del Imperio de las Indias Occidentales* .... Escriuiale por el año de 1646, Juan Diez de la Calle, Oficial Segundo de la Misma Secretaria.

[4] *Suma de Arithmetica pratica* (Valladolid, 1546), fol. iiijr.; taken from D. E. Smith, *History of Mathematics*, Vol. II (1925), p. 88. The q̊s means *quentos* (*cuentos*, "millions").

[5] In Joseph Aladern, *Diccionari popular de la Llengua Catalana* (Barcelona, 1905), we read under "Caldero": "Among ancient copyists a sign (ʋſ) denoted a thousand."

[6] *Joannis Caramvelis Mathesis biceps vetus et nova* (Companiae, 1670), p. 7.

[7] *Lexicon Abbreviaturarum* (Leipzig, 1901), p. l.

[8] *Ibid.*, p. 436, col. 1, Nos. 5 and 6.

which the sign in question is small and is placed in the position of an exponent to the letters XL, to represent the number 40,000. This corresponds to the use of a small *c* which has been found written to the right of and above the letters XI, to signify 1,100. It follows, therefore, that the modified U was in use during the fifteenth century in Italy, as well as in Spain, though it is not known which country had the priority.

What is the origin of this *calderón?* Our studies along this line make it almost certain that it is a modification of one of the Roman

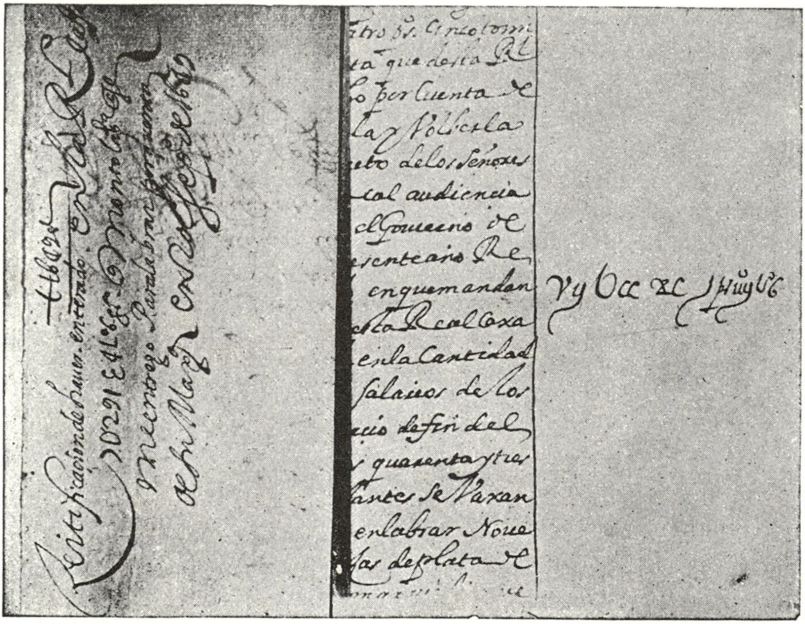

Fig. 27.—From a contract (Mexico City, 1649). The right part shows the sum of 7,291 *pesos,* 4 *tomines,* 6 *granos,* expressed in Roman numerals and the *calderón.* The left part, from the same contract, shows the same sum in Hindu-Arabic numerals and the *calderón.*

symbols for 1,000. Besides M, the Romans used for 1,000 the symbols CIꓛ, T, ∞, and ↑. These symbols are found also in Spanish manuscripts. It is easy to see how in the hands of successive generations of amanuenses, some of these might assume the forms of the *calderón.* If the lower parts of the parentheses in the forms CIꓛ or CIIꓛ are united, we have a close imitation of the U, crossed by one or by two bars.

94. *The Portuguese "cifrão."*—Allied to the distorted Spanish U is the Portuguese symbol for 1,000, called the *cifrão*.[1] It looks somewhat like our modern dollar mark, $. But its function in writing numbers was identical with that of the *calderón*. Moreover, we have seen forms of this Spanish "thousand" which need only to be turned through a right angle to appear like the Portuguese symbol for 1,000. Changes of that sort are not unknown. For instance, the Arabic numeral 5 appears upside down in some Spanish books and manuscripts as late as the eighteenth and nineteenth centuries.

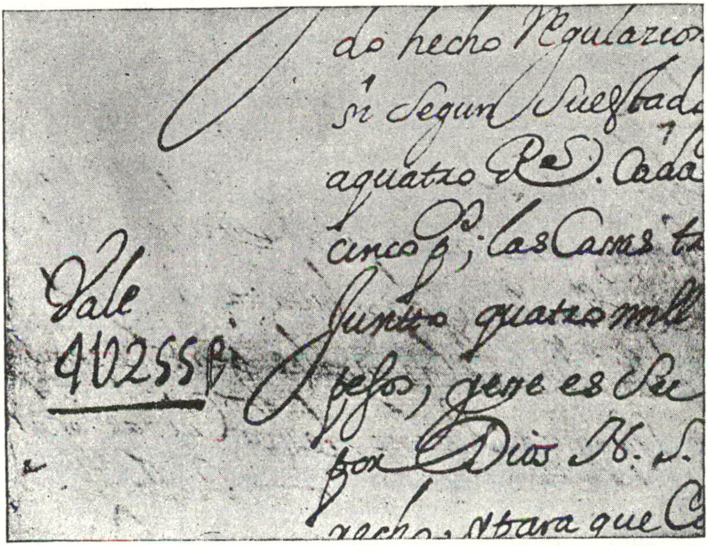

Fig. 28.—Real estate sale in Mexico City, 1718. The sum written here is 4,255 *pesos*.

95. *Relative size of numerals in tables.*—André says on this point: "In certain numerical tables, as those of Schrön, all numerals are of the same height. In certain other tables, as those of Lalande, of Callet, of Houël, of Dupuis, they have unequal heights: the 7 and 9 are prolonged downward; 3, 4, 5, 6 and 8 extend upward; while 1 and 2 do not reach above nor below the central body of the writing. . . . . The unequal numerals, by their very inequality, render the long train of numerals easier to read; numerals of uniform height are less legible."[2]

[1] See the word *cifrão* in Antonio de Moraes Silva, *Dicc. de Lingua Portuguesa* (1877); in Vieira, *Grande Dicc. Portuguez* (1873); in *Dicc. Comtemp. da Lingua Portuguesa* (1881).

[2] D. André, *Des notations mathématiques* (Paris, 1909), p. 9.

96. Fanciful hypotheses on the origin of the numeral forms.—A problem as fascinating as the puzzle of the origin of language relates to the evolution of the forms of our numerals. Proceeding on the tacit assumption that each of our numerals contains within itself, as a skeleton so to speak, as many dots, strokes, or angles as it represents units, imaginative writers of different countries and ages have advanced hypotheses as to their origin. Nor did these writers feel that they were indulging simply in pleasing pastime or merely contributing to mathematical recreations. With perhaps only one exception, they were as convinced of the correctness of their explanations as are circle-squarers of the soundness of their quadratures.

The oldest theory relating to the forms of the numerals is due to the Arabic astrologer Aben Ragel[1] of the tenth or eleventh century. He held that a circle and two of its diameters contained the required forms as it were in a nutshell. A diameter represents 1; a diameter and the two terminal arcs on opposite sides furnished the 2. A glance at Part I of Figure 29 reveals how each of the ten forms may be evolved from the fundamental figure.

On the European Continent, a hypothesis of the origin from dots is the earliest. In the seventeenth century an Italian Jesuit writer, Mario Bettini,[2] advanced such an explanation which was eagerly accepted in 1651 by Georg Philipp Harsdörffer[3] in Germany, who said: "Some believe that the numerals arose from points or dots," as in Part II. The same idea was advanced much later by Georges Dumesnil[4] in the manner shown in the first line of Part III. In cursive writing the points supposedly came to be written as dashes, yielding forms resembling those of the second line of Part III. The two horizontal dashes for 2 became connected by a slanting line yielding the modern form. In the same way the three horizontal dashes for 3 were joined by two slanting lines. The 4, as first drawn, resembled the 0; but confusion was avoided by moving the upper horizontal stroke into a

[1] J. F. Weidler, *De characteribus numerorum vulgaribus dissertatio mathematica-critica* (Wittembergae, 1737), p. 13; quoted from M. Cantor, *Kulturleben der Völker* (Halle, 1863), p. 60, 373.

[2] *Apiaria universae philosophiae, mathematicae*, Vol. II (1642), Apiarium XI, p. 5. See Smith and Karpinski, *op. cit.*, p. 36.

[3] *Delitae mathematicae et physicae* (Nürnberg, 1651). Reference from M. Sterner, *Geschichte der Rechenkunst* (München and Leipzig [1891]), p. 138, 524.

[4] "Note sur la forme des chiffres usuels," *Revue archéologique* (3d ser.; Paris, 1890), Vol. XVI, p. 342–48. See also a critical article, "Prétendues notations Pythagoriennes sur l'origine de nos chiffres," by Paul Tannery, in his *Mémoires scientifiques*, Vol. V (1922), p. 8.

vertical position and placing it below on the right. To avoid con-
founding the 5 and 6, the lower left-hand stroke of the first 5 was

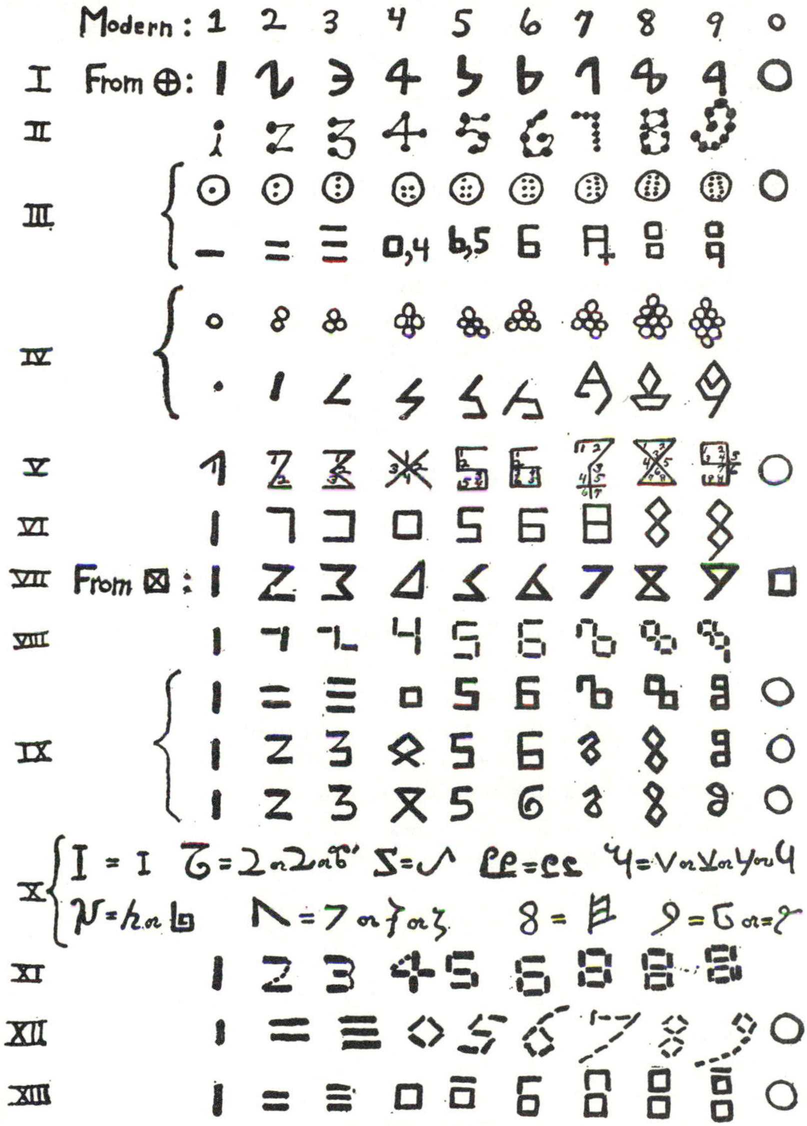

FIG. 29.—Fanciful hypotheses

changed from a vertical to a horizontal position and placed at the
top of the numeral. That all these changes were accepted as historical,

without an atom of manuscript evidence to support the different steps in the supposed evolution, is an indication that Baconian inductive methods of research had not gripped the mind of Dumesnil. The origin from dots appealed to him the more strongly because points played a rôle in Pythagorean philosophy and he assumed that our numeral system originated with the Pythagoreans.

Carlos le-Maur,[1] of Madrid, in 1778 suggested that lines joining the centers of circles (or pebbles), placed as shown in the first line of Part IV, constituted the fundamental numeral forms. The explanation is especially weak in accounting for the forms of the first three numerals.

A French writer, P. Voizot,[2] entertained the theory that originally a numeral contained as many angles as it represents units, as seen in Part V. He did not claim credit for this explanation, but ascribed it to a writer in the Genova Catholico Militarite. But Voizot did originate a theory of his own, based on the number of strokes, as shown in Part VI.

Édouard Lucas[3] entertains readers with a legend that Solomon's ring contained a square and its diagonals, as shown in Part VII, from which the numeral figures were obtained. Lucas may have taken this explanation from Jacob Leupold[4] who in 1727 gave it as widely current in his day.

The historian Moritz Cantor[5] tells of an attempt by Anton Müller[6] to explain the shapes of the digits by the number of strokes necessary to construct the forms as seen in Part VIII. An eighteenth-century writer, Georg Wachter,[7] placed the strokes differently, somewhat as in Part IX. Cantor tells also of another writer, Piccard,[8] who at one time had entertained the idea that the shapes were originally deter-

[1] *Elementos de Matématica pura* (Madrid, 1778), Vol. I, chap. i.

[2] "Les chiffres arabes et leur origine," *La nature* (2d semestre, 1899), Vol. XXVII, p. 222.

[3] *L'Arithmétique amusante* (Paris, 1895), p. 4. Also M. Cantor, *Kulturleben der Völker* (Halle, 1863), p. 60, 374, n. 116; P. Treutlein, *Geschichte unserer Zahlzeichen* (Karlsruhe, 1875), p. 16.

[4] *Theatrvm Arithmetico-Geometricvm* (Leipzig, 1727), p. 2 and Table III.

[5] *Kulturleben der Völker*, p. 59, 60.

[6] *Arithmetik und Algebra* (Heidelberg, 1833). See also a reference to this in P. Treutlein, *op. cit.* (1875), p. 15.

[7] *Naturae et Scripturae Concordia* (Lipsiae et Hafniae, 1752), chap. iv.

[8] Mémoire sur la forme et de la provenance des chiffres, *Société Vaudoise des sciences naturelles* (séances du 20 Avril et du 4 Mai, 1859), p. 176, 184. M. Cantor reproduces the forms due to Piccard; see Cantor, *Kulturleben, etc.*, Fig. 44.

mined by the number of strokes, straight or curved, necessary to express the units to be denoted. The detailed execution of this idea, as shown in Part IX, is somewhat different from that of Müller and some others. But after critical examination of his hypothesis, Piccard candidly arrives at the conclusion that the resemblances he pointed out are only accidental, especially in the case of 5, 7, and 9, and that his hypothesis is not valid.

This same Piccard offered a special explanation of the forms of the numerals as found in the geometry of Boethius and known as the "Apices of Boethius." He tried to connect these forms with letters in the Phoenician and Greek alphabets (see Part X). Another writer whose explanation is not known to us was J. B. Reveillaud.[1]

The historian W. W. R. Ball[2] in 1888 repeated with apparent approval the suggestion that the nine numerals were originally formed by drawing as many strokes as there are units represented by the respective numerals, with dotted lines added to indicate how the writing became cursive, as in Part XI. Later Ball abandoned this explanation. A slightly different attempt to build up numerals on the consideration of the number of strokes is cited by W. Lietzmann.[3] A still different combination of dashes, as seen in Part XII, was made by the German, David Arnold Crusius, in 1746.[4] Finally, C. P. Sherman[5] explains the origin by numbers of short straight lines, as shown in Part XIII. "As time went on," he says, "writers tended more and more to substitute the easy curve for the difficult straight line and not to lift the pen from the paper between detached lines, but to join the two—which we will call cursive writing."

These hypotheses of the origin of the forms of our numerals have been barren of results. The value of any scientific hypothesis lies in co-ordinating known facts and in suggesting new inquiries likely to advance our knowledge of the subject under investigation. The hypotheses here described have done neither. They do not explain the very great variety of forms which our numerals took at different times

[1] *Essai sur les chiffres arabes* (Paris, 1883). Reference from Smith and Karpinski, *op. cit.*, p. 36.

[2] *A Short Account of the History of Mathematics* (London, 1888), p. 147.

[3] *Lustiges und Merkwürdiges von Zahlen und Formen* (Breslau, 1922), p. 73, 74. He found the derivation in Raether, *Theorie und Praxis des Rechenunterrichts* (1. Teil, 6. Aufl.; Breslau, 1920), p. 1, who refers to H. von Jacobs, *Das Volk der Siebener-Zähler* (Berlin, 1896).

[4] *Anweisung zur Rechen-Kunst* (Halle, 1746), p. 3.

[5] *Mathematics Teacher*, Vol. XVI (1923), p. 398–401.

and in different countries. They simply endeavor to explain the numerals as they are printed in our modern European books. Nor have they suggested any fruitful new inquiry. They serve merely as entertaining illustrations of the operation of a pseudo-scientific imagination, uncontrolled by all the known facts.

97. *A sporadic artificial system.*—A most singular system of numeral symbols was described by Agrippa von Nettesheim in his *De occulta philosophia* (1531) and more fully by Jan Bronkhorst of Nimwegen in Holland who is named after his birthplace Noviomagus.[1] In 1539 he published at Cologne a tract, *De numeris*, in which he describes numerals composed of straight lines or strokes which, he claims, were used by *Chaldaei et Astrologi*. Who these Chaldeans are whom he mentions it is difficult to ascertain; Cantor conjectures that they were late Roman or medieval astrologers. The symbols are given again in a document published by M. Hostus in 1582 at Antwerp. An examination of the symbols indicates that they enable one to write numbers up into the millions in a very concise form. But this conciseness is attained at a great sacrifice of simplicity; the burden on the memory is great. It does not appear as if these numerals grew by successive steps of time; it is more likely that they are the product of some inventor who hoped, perhaps, to see his symbols supersede the older (to him) crude and clumsy contrivances.

An examination, in Figure 30, of the symbols for 1, 10, 100, and 1,000 indicates how the numerals are made up of straight lines. The same is seen in 4, 40, 400, and 4,000 or in 5, 50, 500, and 5,000.

98. *General remarks.*—Evidently one of the earliest ways of recording the small numbers, from 1 to 5, was by writing the corresponding number of strokes or bars. To shorten the record in expressing larger numbers new devices were employed, such as placing the bars representing higher values in a different position from the others, or the introduction of an altogether new symbol, to be associated with the primitive strokes on the additive, or multiplicative principle, or in some cases also on the subtractive principle.

After the introduction of alphabets, and the observing of a fixed sequence in listing the letters of the alphabets, the use of these letters

[1] See M. Cantor, *Vorlesungen über Geschichte der Mathematik*, Vol. II (2d ed.; Leipzig, 1913), p. 410; M. Cantor, *Mathemat. Beiträge zum Kulturleben der Völker* (Halle, 1863), p. 166, 167; G. Friedlein, *Die Zahlzeichen und das elementare Rechnen der Griechen und Römer* (Erlangen, 1869), p. 12; T. H. Martin, *Annali di matematica* (B. Tortolini; Rome, 1863), Vol. V, p. 298; J. C. Heilbronner, *Historia Matheseos universae* (Lipsiae, 1742), p. 735–37; J. Ruska, *Archiv für die Geschichte der Naturwissenschaften und Technik*, Vol. IX (1922), p. 112–26.

for the designation of numbers was introduced among the Syrians, Greeks, Hebrews, and the early Arabs. The alphabetic numeral systems called for only very primitive powers of invention; they made

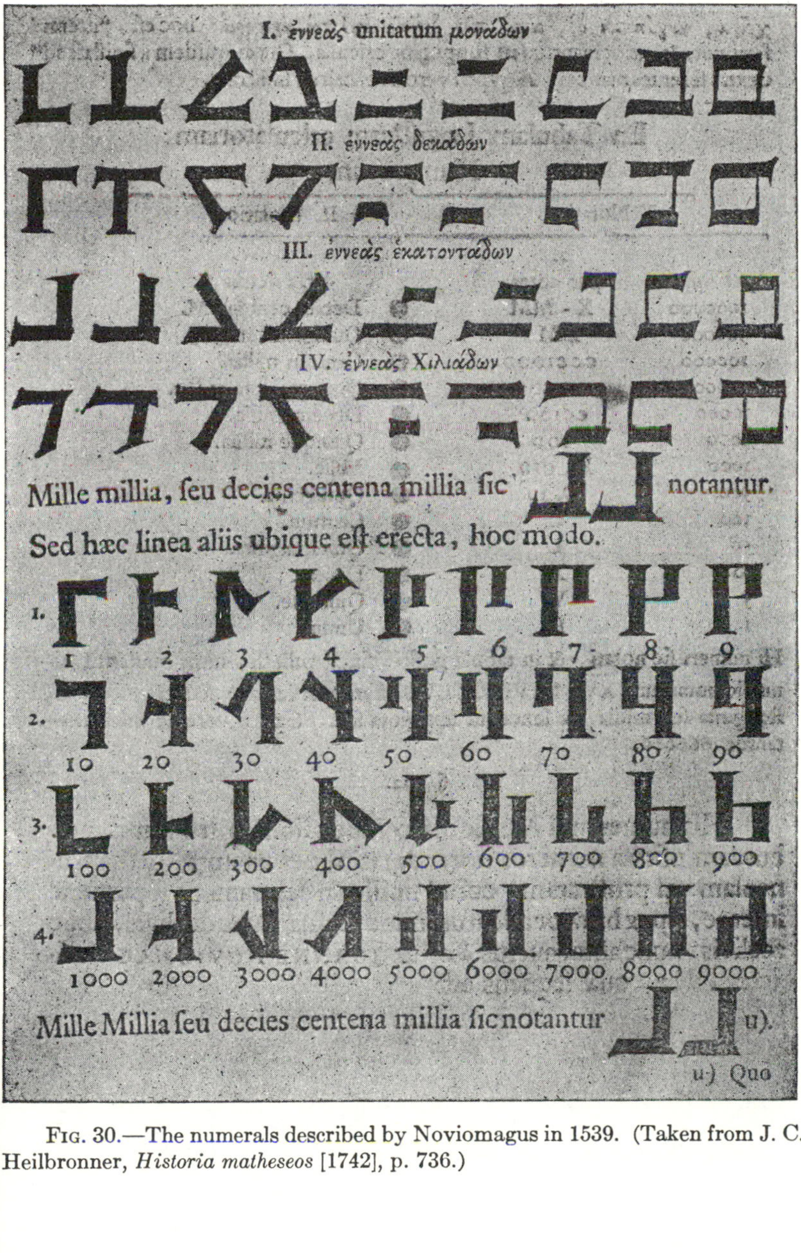

FIG. 30.—The numerals described by Noviomagus in 1539. (Taken from J. C. Heilbronner, *Historia matheseos* [1742], p. 736.)

unnecessarily heavy demands on the memory and embodied no attempt to aid in the processes of computation.

The highest powers of invention were displayed in the systems employing the principle of local value. Instead of introducing new symbols for units of higher order, this principle cleverly utilized the position of one symbol relative to others, as the means of designating different orders. Three important systems utilized this principle: the Babylonian, the Maya, and the Hindu-Arabic systems. These three were based upon different scales, namely, 60, 20 (except in one step), and 10, respectively. The principle of local value applied to a scale with a small base affords magnificent adaptation to processes of computation. Comparing the processes of multiplication and division which we carry out in the Hindu-Arabic scale with what the alphabetical systems or the Roman system afforded places the superiority of the Hindu-Arabic scale in full view. The Greeks resorted to abacal computation, which is simply a primitive way of observing local value in computation. In what way the Maya or the Babylonians used their notations in computation is not evident from records that have come down to us. The scales of 20 or 60 would call for large multiplication tables.

The origin and development of the Hindu-Arabic notation has received intensive study. Nevertheless, little is known. An outstanding fact is that during the past one thousand years no uniformity in the shapes of the numerals has been reached. An American is sometimes puzzled by the shape of the number 5 written in France. A European traveler in Turkey would find that what in Europe is a 0 is in Turkey a 5.

99. *Opinion of Laplace.*—Laplace[1] expresses his admiration for the invention of the Hindu-Arabic numerals and notation in this wise: "It is from the Indians that there has come to us the ingenious method of expressing all numbers, in ten characters, by giving them, at the same time, an absolute and a place value; an idea fine and important, which appears indeed so simple, that for this very reason we do not sufficiently recognize its merit. But this very simplicity, and the extreme facility which this method imparts to all calculation, place our system of arithmetic in the first rank of the useful inventions. How difficult it was to invent such a method one can infer from the fact that it escaped the genius of Archimedes and of Apollonius of Perga, two of the greatest men of antiquity."

[1] *Exposition du système du monde* (6th ed.; Paris, 1835), p. 376.

III

SYMBOLS IN ARITHMETIC AND ALGEBRA
(ELEMENTARY PART)

100. In ancient Babylonian and Egyptian documents occur certain ideograms and symbols which are not attributable to particular individuals and are omitted here for that reason. Among these signs is ⌐ for square root, occurring in a papyrus found at Kahun and now at University College, London,[1] and a pair of walking legs for squaring in the Moscow papyrus.[2] These symbols and ideograms will be referred to in our "Topical Survey" of notations.

A. GROUPS OF SYMBOLS USED BY INDIVIDUAL WRITERS

GREEK: DIOPHANTUS, THIRD CENTURY A.D.

101. The unknown number in algebra, defined by Diophantus as containing an undefined number of units, is represented by the Greek letter s with an accent, thus s', or in the form $s^{o'}$. In plural cases the symbol was doubled by the Byzantines and later writers, with the addition of case endings. Paul Tannery holds that the evidence is against supposing that Diophantus himself duplicated the sign.[3] G. H. F. Nesselmann[4] takes this symbol to be final sigma and remarks that probably its selection was prompted by the fact that it was the only letter in the Greek alphabet which was not used in writing numbers. Heath favors "the assumption that the sign was a mere tachygraphic abbreviation and not an algebraical symbol like our x, though discharging much the same function."[5] Tannery suggests that the sign is the ancient letter koppa, perhaps slightly modified. Other views on this topic are recorded by Heath.

[1] Moritz Cantor, *Vorlesungen über Geschichte der Mathematik*, Vol. I, 3d ed., Leipzig, p. 94.

[2] B. Touraeff, *Ancient Egypt* (1917), p. 102.

[3] *Diophanti Alexandrini opera omnia cum Graecis commentariis* (Lipsiae, 1895), Vol. II, p. xxxiv–xlii; Sir Thomas L. Heath, *Diophantus of Alexandria* (2d ed.; Cambridge, 1910), p. 32, 33.

[4] *Die Algebra der Griechen* (Berlin, 1842), p. 290, 291.

[5] *Op. cit.*, p. 34–36.

A square, x^2, is in Diophantus' *Arithmetica* Δ^Y

A cube, x^3, is in Diophantus' *Arithmetica* K^Y

A square-square, x^4, is in Diophantus' *Arithmetica* $\Delta^Y\Delta$

A square-cube, x^5, is in Diophantus' *Arithmetica* ΔK^Y

A cube-cube, x^6, is in Diophantus' *Arithmetica* $K^Y K$

In place of the capital letters kappa and delta, small letters are some-times used.[1] Heath[2] comments on these symbols as follows: "There is no obvious connection between the symbol Δ^Y and the symbol s of which it is the square, as there is between x^2 and x, and in this lies the great inconvenience of the notation. But upon this notation no advance was made even by late editors, such as Xylander, or by Bachet and Fermat. They wrote N (which was short for *Numerus*) for the s of Diophantus, Q (*Quadratus*) for Δ^Y, C (*Cubus*) for K^Y, so that we find, for example, $1Q+5N=24$, corresponding to $x^2+5x=24$.[3] Other symbols were however used even before the publication of Xylander's *Diophantus*, e.g., in Bombelli's *Algebra*."

102. Diophantus has no symbol for multiplication; he writes down the numerical results of multiplication without any preliminary step which would necessitate the use of a symbol. Addition is expressed

---

[1] From Fermat's edition of Bachet's *Diophantus* (Toulouse, 1670), p. 2, Definition II, we quote: "Appellatvr igitur Quadratus, Dynamis, & est illius nota δ' superscriptum habens $\bar{v}$ sic $\delta^{\bar{v}}$. Qui autem sit ex quadrato in suum latus cubus est, cuius nota est $\overset{\prime}{\kappa}$, superscriptum habens $\bar{v}$ hoc pacto $\kappa^{\bar{v}}$. Qui autem sit ex quad-rato in seipsum multiplicato, quadrato-quadratus est, cuius nota est geminum δ' habens superscriptum $\bar{v}$, hac ratione $\delta\delta^{\bar{v}}$. Qui sit quadrato in cubum qui ab eodem latere profectus est, ducto, quadrato-cubus nominatur, nota eius $\delta\bar{\kappa}$ superscriptum habens $\bar{v}$ sic $\delta\kappa^{\bar{v}}$. Qui ex cubo in se ducto nascitur, cubocubus vocatur, & est eius nota geminum $\bar{\kappa}$ superscriptum habens $\bar{v}$, hoc pacto $\kappa\kappa^{\bar{v}}$. Cui vero nulla harum proprietatum obtigit, sed constat multitudine vnitatem rationis experte, numerus vocatur, nota eius $\overset{\prime}{s}$. Est et aliud signum immutabile definitorum, vnitas, cuius nota $\bar{\mu}$ superscriptum habens $\bar{o}$ sic $\mu^{\bar{o}}$." The passage in Bachet's edition of 1621 is the same as this.

[2] *Op. cit.*, p. 38.

[3] In Fermat's edition of Bachet's *Diophantus* (Toulouse, 1670), p. 3, Definition II, we read: "Haec ad verbum exprimenda esse arbitratus sum potiùs quàm cum Xilandro nescio quid aliud comminisci. Quamuis enim in reliqua versione nostra notis ab eodem Xilandro excogitatis libenter vsus sim, quas tradam infrà. Hîc tamen ab ipso Diophanto longiùs recedere nolui, quòd hac definitione notas ex-plicet quibus passim libris istis vtitur ad species omnes compendio designandas, & qui has ignoret ne quidem Graeca Diophanti legere possit. Porrò quadratum Dy-namin vocat, quae vox potestatem sonat, quia videlicet quadratus est veluti potestas cuius libet lineae, & passim ab Euclide, per id quod potest linea, quadratus illius designatur. Itali, Hispanique eadem ferè de causa Censum vocant, quasi

by mere juxtaposition. Thus the polynomial $X^3 + 13x^2 + 5x + 2$ would be in Diophantine symbols $K^Y\bar{a}\Delta^Y\overline{\iota\gamma}s\bar{\epsilon}\overset{o}{M}\bar{\beta}$, where $\overset{o}{M}$ is used to represent units and shows that $\bar{\beta}$ or 2 is the absolute term and not a part of the coefficient of s or x. It is to be noted that in Diophantus' "square-cube" symbol for x^5, and "cube-cube" symbol for x^6, the *additive* principle for exponents is employed, rather than the *multiplicative* principle (found later widely prevalent among the Arabs and Italians), according to which the "square-cube" power would mean x^6 and the "cube-cube" would mean x^9.

103. Diophantus' symbol for subtraction is "an inverted Ψ with the top shortened, $\wedge$." Heath pertinently remarks: "As Diophantus used no distinct sign for $+$, it is clearly necessary, in order to avoid confusion, that all the negative terms in an expression, should be placed together after all the positive terms. And so in fact he does place them."[1] As regards the origin of this sign $\pitchfork$, Heath believes that the explanation which is quoted above from the Diophantine text as we have it is not due to Diophantus himself, but is "an explanation made by a scribe of a symbol which he did not understand." Heath[2] advances the hypothesis that the symbol originated by placing a | within the uncial form $\wedge$, thus yielding $\wedge$. Paul Tannery,[3] on the other hand, in 1895 thought that the sign in question was adapted from the old letter sampi $?$, but in 1904 he[4] concluded that it was rather a conventional abbreviation associated with the root of a certain Greek verb. His considerations involve questions of Greek grammar and were prompted by the appearance of the Diophantine sign

---

dicas redditum, prouentúmque, quòd à latere seu radice, tanquam à feraci solo quadratus oriatur. Inde factum vt Gallorum nonnulli & Germanorum corrupto vocabulo zenzum appellarint. Numerum autem indeterminatum & ignotum, qui & aliarum omnium potestatum latus esse intelligitur, Numerum simpliciter Diophantus appellat. Alij passim Radicem, vel latus, vel rem dixerunt, Itali patrio vocabulo Cosam. Caeterùm nos in versione nostra his notis *N. Q. C. QQ. QC. CC.* designabimus Numerum, Quadratum, Cubum, Quadratoquadratum, Quadratocubum, Cubocubum. Nam quod ad vnitates certas & determinatas spectat, eis notam aliquam adscribere superuacaneum duxi, quòd hae seipsis absque vlla ambiguitate sese satis indicent. Ecquis enim cùm audit numerum 6. non statim cogitat sex vnitates? Quid ergo necesse est sex vnitates dicere, cùm sufficiat dicere, sex? . . . . " This passage is the same as in Bachet's edition of 1621.

[1] Heath, *op. cit.*, p. 42.

[2] *Ibid.*, p. 42, 43.

[3] Tannery, *op. cit.*, Vol. II, p. xli.

[4] *Bibliotheca mathematica* (3d ser.), Vol. V, p. 5–8.

of subtraction in the critical notes to Schöne's edition[1] of the *Metrica* of Heron.

For equality the sign in the archetypal manuscripts seems to have been $ι^σ$; "but copyists introduced a sign which was sometimes confused with the sign ᛃ" (Heath).

104. The notation for division comes under the same head as the notation for fractions (see § 41). In the case of unit fractions, a double accent is used with the denominator: thus $γ'' = \frac{1}{3}$. Sometimes a simple accent is used; sometimes it appears in a somewhat modified form as $\wedge$, or (as Tannery interprets it) as $\curlyvee$: thus $y^{\curlyvee} = \frac{1}{10}$. For $\frac{1}{2}$ appear the symbols $\angle'$ and $\smile$, the latter sometimes without the dot. Of fractions that are not unit fractions, $\frac{2}{3}$ has a peculiar sign ω of its own, as was the case in Egyptian notations. "Curiously enough," says Heath, "it occurs only four times in Diophantus." In some old manuscripts the denominator is written above the numerator, in some rare cases. Once we find $ιε^δ = \frac{15}{4}$, the denominator taking the position where we place exponents. Another alternative is to write the numerator first and the denominator after it in the same line, marking the denominator with a submultiple sign in some form: thus, $\overline{γ}δ' = \frac{3}{4}$.[2] The following are examples of fractions from Diophantus:

From v. 10: $\dfrac{ιβ}{ιζ} = \dfrac{17}{12}$ From v. 8, Lemma: $\barβ\angle's' = 2\frac{1}{2}\frac{1}{6}$

From iv. 3: $ς\curlyvee\barη = \dfrac{8}{x}$ From iv. 15: $Δ^Y\curlyvee\overline{σν} = \dfrac{250}{x^2}$

From vi. 12: $Δ^Y\barξ\overset{\circ}{M},\overline{βφκ}$ ἐν γορίῳ $Δ^YΔ\bar{α}\overset{\circ}{M} ⌐ \wedge Δ^Y\barξ$

$$= (60x^2 + 2,520)/(x^4 + 900 - 60x^2) \ .$$

105. The fact that Diophantus had only one symbol for unknown quantity affected considerably his mode of exposition. Says Heath: "This limitation has made his procedure often very different from our modern work." As we have seen, Diophantus used but few symbols. Sometimes he ignored even these by describing an operation in words, when the symbol would have answered as well or better. Considering the amount of symbolism used, Diophantus' algebra may be designated as "syncopated."

[1] *Heronis Alexandrini opera*, Vol. III (Leipzig, 1903), p. 156, l. 8, 10. The manuscript reading is μονάδων οδτιδ', the meaning of which is $74 - \frac{1}{14}$.

[2] Heath, *op. cit.*, p. 45, 47.

106. We begin with a quotation from H. T. Colebrooke on Hindu algebraic notation:[1] "The Hindu algebraists use abbreviations and initials for symbols: they distinguish negative quantities by a dot, but have not any mark, besides the absence of the negative sign, to discriminate a positive quantity. No marks or symbols (other than abbreviations of words) indicating operations of addition or multiplication, etc., are employed by them: nor any announcing equality[2] or relative magnitude (greater or less). . . . . A fraction is indicated by placing the divisor under the dividend, but without a line of separation. The two sides of an equation are ordered in the same manner, one under the other. . . . . The symbols of unknown quantity are not confined to a single one: but extend to ever so great a variety of denominations: and the characters used are the initial syllables of the names of colours, excepting the first, which is the initial of *yávat-távat*, as much as."

107. In Brahmagupta,[3] and later Hindu writers, abbreviations occur which, when transliterated into our alphabet, are as follows:

> *ru* for *rupa*, the absolute number
> *ya* for *yávat-távat*, the (first) unknown
> *ca* for *calaca* (black), a second unknown
> *ní* for *nílaca* (blue), a third unknown
> *pí* for *pítaca* (yellow), a fourth unknown
> *pa* for *pandu* (white), a fifth unknown
> *lo* for *lohita* (red), a sixth unknown
> *c* for *caraní*, surd, or square root
> *ya v* for x^2, the *v* being the contraction for
>   *varga*, square number

108. In Brahmagupta,[4] the division of *ru* 3 *c* 450 *c* 75 *c* 54 by *c* 18 *c* 3 (i.e., $3+\sqrt{450}+\sqrt{75}+\sqrt{54}$ by $\sqrt{18}+\sqrt{3}$) is carried out as follows: "Put *c* 18 *c* 3. The dividend and divisor, multiplied by this, make *ru* 75 *c* 625. The dividend being then divided by the single surd *ru* 15 constituting the divisor, the quotient is *ru* 5 *c* 3."

[1] H. T. Colebrooke, *Algebra, with Arithmetic and Mensuration from the Sanscrit of Bramegupta and Bháscara* (London, 1817), p. x, xi.

[2] The Bakhshālī MS (§ 109) was found after the time of Colebrooke and has an equality sign.

[3] *Ibid.*, p. 339 ff.

[4] *Brahme-sphuta-sidd'hánta*, chap. xii. Translated by H. T. Colebrooke in *op. cit.* (1817), p. 277–378; we quote from p. 342.

In modern symbols, the statement is, substantially: Multiply dividend and divisor by $\sqrt{18}-\sqrt{3}$; the products are $75+\sqrt{675}$ and 15; divide the former by the latter, $5+\sqrt{3}$.

"Question 16.[1] When does the residue of revolutions of the sun, less one, fall, on a Wednesday, equal to the square root of two less than the residue of revolutions, less one, multiplied by ten and augmented by two?

"The value of residue of revolutions is to be here put square of *yávat-távat* with two added: *ya v* 1 *ru* 2 is the residue of revolutions.

Sanskrit characters or letters, by which the Hindus denote the unknown quantities in their notation, are the following : पा, का, नी, पी, लो.

Fig. 31.—Sanskrit symbols for unknowns. (From Charles Hutton, *Mathematical Tracts*, II, 167.) The first symbol, *pa*, is the contraction for "white"; the second, *ca*, the initial for "black"; the third, *ni*, the initial for "blue"; the fourth, *pi*, the initial for "yellow"; the fifth, *lo*, for "red."

This less two is *ya v* 1; the square root of which is *ya* 1. Less one, it is *ya* 1 *ru* $\dot{1}$; which multiplied by ten is *ya* 10 *ru* $\dot{10}$; and augmented by two, *ya* 10 *ru* $\dot{8}$. It is equal to the residue of revolutions *ya v* 1 *ru* 2 less one; viz. *ya v* 1 *ru* 1. Statement of both sides $\dfrac{ya\ v\ 0\ ya\ 10\ ru\ \dot{8}}{ya\ v\ 1\ ya\ \ 0\ ru\ 1}$. Equal subtraction being made conformably to rule 1 there arises *ya v* 1 $\dfrac{ru\ \dot{9}}{ya\ \dot{10}}$.

Now, from the absolute number ($\dot{9}$), multiplied by four times the [coefficient of the] square ($3\dot{6}$), and added to (100) the square of the [coefficient of the] middle term (making consequently 64), the square root being extracted (8), and lessened by the [coefficient of the] middle term ($1\dot{0}$), the remainder is 18 divided by twice the [coefficient of the] square (2), yields the value of the middle term 9. Substituting with this in the expression put for the residue of revolutions, the answer comes out, residue of revolutions of the sun 83. Elapsed period of days deduced from this, 393, must have the denominator in least terms added so often until it fall on Wednesday."

[1] Colebrooke, *op. cit.*, p. 346. The abbreviations *ru, c, ya, ya v, ca, ni*, etc., are transliterations of the corresponding letters in the Sanskrit alphabet.

Notice that $\begin{smallmatrix} ya & v & 0 & ya & 10 & ru & \overset{\bullet}{8} \\ ya & v & 1 & ya & 0 & ru & 1 \end{smallmatrix}$ signifies $0x^2+10x-8=x^2+0x+1.$

Brahmagupta gives[1] the following equation in three unknown quantities and the expression of one unknown in terms of the other two:

"ya 197 ca 1$\overset{\bullet}{6}$44 ni $\overset{\bullet}{1}$ ru 0

ya 0 ca 0 ni 0 ru 6302.

Equal subtraction being made, the value of $y\acute{a}vat$-$t\acute{a}vat$ is

ca 1644 ni 1 ru 6302 ."

(ya) 197

In modern notation:

$$197x-1644y-z+0=0x+0y+0z+6302 ,$$

whence,

$$x=\frac{1644y+z+6302}{197} .$$

HINDU: THE BAKHSHĀLĪ MS

109. The so-called Bakhshālī MS, found in 1881 buried in the earth near the village of Bakhshālī in the northwestern frontier of India, is an arithmetic written on leaves of birch-bark, but has come down in mutilated condition. It is an incomplete copy of an older manuscript, the copy having been prepared, probably about the eighth, ninth, or tenth century. "The system of notation," says A. F. Rudolph Hoernle,[2] "is much the same as that employed in the arithmetical works of Brahmagupta and Bhāskara. There is, however, a very important exception. The *sign for the negative* quantity is a cross (+). It looks exactly like our modern sign for the positive quantity, but it is placed after the number which it qualifies. Thus $\begin{smallmatrix} 12 & 7+ \\ 1 & 1 \end{smallmatrix}$ means $12-7$ (i.e. 5). This is a sign which I have not met with in any other Indian arithmetic. . . . . The sign now used is a dot placed over the number to which it refers. Here, therefore, there appears to be a mark of great antiquity. As to its origin I am not able to suggest any satisfactory explanation. . . . . A *whole* number, when it occurs in an arithmetical operation, as may be seen from the above given example, is indicated by placing the number 1 under it. This, however, is

[1] Colebrooke, *op. cit.*, p. 352.

[2] "The Bakhshālī Manuscript," *Indian Antiquary*, Vol. XVII (Bombay, 1888), p. 33–48, 275–79; see p. 34.

a practice which is still occasionally observed in India. . . . . The following statement from the first example of the twenty-fifth *sutra* affords a good example of the system of notation employed in the Bakhshālī arithmetic:

$$\left| \begin{array}{c} \cdot \\ 1 \end{array} \begin{array}{ccc} 1 & 1 & 1 \quad bh\hat{a} \ 32 \\ 1 & 1 & 1 \\ 3+ & 3+ & 3+ \end{array} \right| \quad phala\dot{m} \ 108$$

Here the initial dot is used much in the same way as we use the letter x to denote the unknown quantity, the value of which is sought. The number 1 under the dot is the sign of the whole (in this case, unknown) number. A fraction is denoted by placing one number under the other without any line of separation; thus $\begin{smallmatrix}1\\3\end{smallmatrix}$ is $\frac{1}{3}$, i.e. one-third. A mixed number is shown by placing the three numbers under one another; thus $\begin{smallmatrix}1\\1\\3\end{smallmatrix}$ is $1+\frac{1}{3}$ or $1\frac{1}{3}$, i.e. one and one-third. Hence $\begin{smallmatrix}1\\1\\3+\end{smallmatrix}$ means $1-\frac{1}{3}$ $\left(\text{i.e. } \frac{2}{3}\right)$. Multiplication is usually indicated by placing the numbers side by side; thus

$$\left| \begin{array}{cc} 5 & 32 \\ 8 & 1 \end{array} \right| \quad phala\dot{m} \ 20$$

means $\frac{5}{8}\times 32=20$. Similarly $\begin{smallmatrix}1 & 1 & 1\\1 & 1 & 1\\3+ & 3+ & 3+\end{smallmatrix}$ means $\frac{2}{3}\times\frac{2}{3}\times\frac{2}{3}$ or $\left(\frac{2}{3}\right)^3$, i.e. $\frac{8}{27}$. *Bhâ* is an abbreviation of *bhâga*, 'part,' and means that the number preceding it is to be treated as a denominator. Hence $\begin{smallmatrix}1 & 1 & 1\\1 & 1 & 1\\3+ & 3+ & 3+\end{smallmatrix}$ *bhâ* means $1\div\frac{8}{27}$ or $\frac{27}{8}$. The whole statement, therefore,

$$\left| \begin{array}{c} \cdot \\ 1 \end{array} \begin{array}{ccc} 1 & 1 & 1 \\ 1 & 1 & 1 \quad bh\hat{a} \ 32 \\ 3+ & 3+ & 3+ \end{array} \right| \quad phala\dot{m} \ 108 \ ,$$

means $\frac{27}{8}\times 32=108$, and may be thus explained,—'a certain number is found by dividing with $\frac{8}{27}$ and multiplying with 32; that number is 108.' The dot is also used for another purpose, namely as one of the ten fundamental figures of the decimal system of notation, or the zero (0123456789). It is still so used in India for both purposes, to indicate the unknown quantity as well as the naught. . . . . The Indian dot, unlike our modern zero, is not properly a numerical figure at all. It is simply a sign to indicate an empty place or a hiatus. This is clearly shown by its name *sûnya*, 'empty.' . . . . Thus the two figures 3 and 7, placed in juxtaposition (37), mean 'thirty-seven,' but with an 'empty space' interposed between them (3 7), they mean 'three hundred and seven.' To prevent misunderstanding the presence of the 'empty space' was indicated by a dot (3.7); or by what is now the zero (307). On the other hand, occurring in the statement of a problem, the 'empty place' could be filled up, and here the dot which marked its presence signified a 'something' which was to be discovered and to be put in the empty place. . . . . In its double signification, which still survives in India, we can still discern an indication of that country as its birthplace. . . . . The operation of multiplication alone is not indicated by any special sign. Addition is indicated by *yu* (for *yuta*), subtraction by $+$ (*ka* for *kanita?*) and division by *bhâ* (for *bhâga*). The whole operation is commonly enclosed between lines (or sometimes double lines), and the result is set down outside, introduced by *pha* (for *phala*)." Thus, *pha* served as a sign of equality.

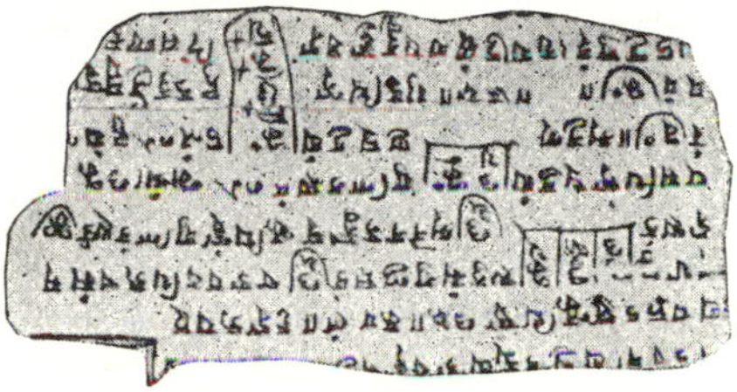

Fig. 32.—From Bakhshālī arithmetic (G. R. Kaye, *Indian Mathematics* [1915], p. 26; R. Hoernle; *op. cit.*, p. 277).

The problem solved in Figure 32 appears from the extant parts to have been: Of a certain quantity of goods, a merchant has to pay, as duty, $\frac{1}{3}$, $\frac{1}{4}$, and $\frac{1}{5}$ on three successive occasions. The total duty is 24. What was the original quantity of his goods? The solution appears in the manuscript as follows: "Having subtracted the series from one," we get $\frac{2}{3}$, $\frac{3}{4}$, $\frac{4}{5}$; these multiplied together give $\frac{2}{5}$; that again, subtracted from 1 gives $\frac{3}{5}$; with this, after having divided (i.e., inverted, $\frac{5}{3}$), the total duty (24) is multiplied, giving 40; that is the original amount. Proof: $\frac{2}{5}$ multiplied by 40 gives 16 as the remainder. Hence the original amount is 40. Another proof: 40 multiplied by $1-\frac{1}{3}$ and $1-\frac{1}{4}$ and $1-\frac{1}{5}$ gives the result 16; the deduction is 24; hence the total is 40.

110. Bhāskara speaks in his *Lilavati*[1] of squares and cubes of numbers and makes an allusion to the raising of numbers to higher powers than the cube. Ganesa, a sixteenth-century Indian commentator of Bhāskara, specifies some of them. Taking the words *varga* for square of a number, and *g'hana* for cube of a number (found in Bhāskara and earlier writers), Ganesa explains[2] that the product of four like numbers is the square of a square, *varga-varga;* the product of six like numbers is the cube of a square, or square of a cube, *varga-g'hana* or *g'hana-varga;* the product of eight numbers gives *varga-varga-varga;* of nine, gives the cube of a cube, *g'hana-g'hana.* The fifth power was called *varga-g'hana-gháta;* the seventh, *varga-varga-g'hana-gháta.*

111. It is of importance to note that the higher powers of the unknown number are built up on the principle of involution, except the powers whose index is a prime number. According to this principle, indices are multiplied. Thus *g'hana-varga* does not mean $n^3 \cdot n^2 = n^5$, but $(n^3)^2 = n^6$. Similarly, *g'hana-g'hana* does not mean $n^3 \cdot n^3 = n^6$, but $(n^3)^3 = n^9$. In the case of indices that are prime, as in the fifth and seventh powers, the multiplicative principle became inoperative and the additive principle was resorted to. This is indicated by the word *gháta* ("product"). Thus, *varga-g'hana-gháta* means $n^2 \cdot n^3 = n^5$.

In the application, whenever possible, of the multiplicative principle in building up a symbolism for the higher powers of a number, we see a departure from Diophantus. With Diophantus the symbol for x^2, followed by the symbol for x^3, meant $x^5;$ with the Hindus it meant x^6. We shall see that among the Arabs and the Europeans of the thirteenth to the seventeenth centuries, the practice was divided, some following the Hindu plan, others the plan of Diophantus.

112. In Bhāskara, when unlike colors (dissimilar unknown quantities, like x and y) are multiplied together, the result is called *bhavita* ("product"), and is abbreviated *bha.* Says Colebrooke: "The product of two unknown quantities is denoted by three letters or syllables, as *ya.ca bha, ca.ni bha*, etc. Or, if one of the quantities be a higher power, more syllables or letters are requisite; for the square, cube, etc., are likewise denoted by the initial syllables, *va, gha, va-va, va-gha, gha-gha,*[3] etc. Thus *ya va · ca gha bha* will signify the square of the

---

[1] Colebrooke, *op. cit.*, p. 9, 10.

[2] *Ibid.*, p. 10, n. 3; p. 11.

[3] *Gha-gha* for the sixth, instead of the ninth, power, indicates the use here of the additive principle.

first unknown quantity multiplied by the cube of the second. A dot is, in some copies of the text and its commentaries, interposed between the factors, without any special direction, however, for this notation."[1] Instead of *ya va* one finds in Brahmagupta and Bhāskara also the severer contraction *ya v;* similarly, one finds *cav* for the square of the second unknown.[2]

It should be noted also that "equations are not ordered so as to put all the quantities positive; nor to give precedence to a positive term in a compound quantity: for the negative terms are retained, and even preferably put in the first place."[3]

According to N. Ramanujacharia and G. R. Kaye,[4] the content of the part of the manuscript shown in Figure 33 is as follows: The

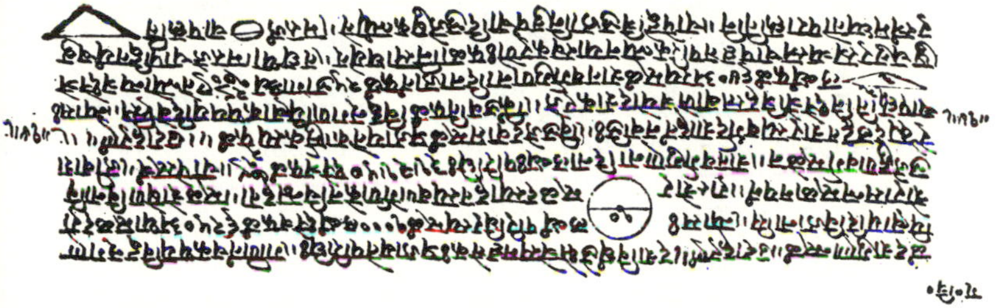

Fig. 33.—Śridhara's *Triśátiká*. Śridhara was born 991 A.D. He is cited by Bhāskara; he explains the "Hindu method of completing the square" in solving quadratic equations.

circumference of a circle is equal to the square root of ten times the square of its diameter. The area is the square root of the product of ten with the square of half the diameter. Multiply the quantity whose square root cannot be found by any large number, take the square root of the product, leaving out of account the remainder. Divide it by the square root of the factor. To find the segment of a circle, take the sum of the chord and arrow, multiply it by the arrow, and square the product. Again multiply it by ten-ninths and extract its square root. Plane figures other than these areas should be calculated by considering them to be composed of quadrilaterals, segments of circles, etc.

[1] *Op. cit.*, p. 140, n. 2; p. 141. In this quotation we omitted, for simplicity, some of the accents found in Colebrooke's transliteration from the Sanskrit.

[2] *Ibid.*, p. 63, 140, 346.

[3] *Ibid.*, p. xii.

[4] *Bibliotheca mathematica* (3d ser.), Vol. XIII (1912–13), p. 206, 213, 214.

113. *Bhāskara Āchábrya, "Lilavati,"*[1] 1150 *A.D.*—"Example: Tell me the fractions reduced to a common denominator which answer to three and a fifth, and one-third, proposed for addition; and those which correspond to a sixty-third and a fourteenth offered for subtraction. Statement:

$$\begin{array}{ccc} 3 & 1 & 1 \\ 1 & 5 & 3 \end{array}$$

Answer: Reduced to a common denominator

$$\begin{array}{ccc} 45 & 3 & 5 \\ 15 & 15 & 15 \end{array} \cdot \quad \text{Sum } \dfrac{53}{15} \cdot$$

Statement of the second example:

$$\begin{array}{cc} 1 & 1 \\ 63 & 14 \end{array} \cdot$$

Answer: The denominator being abridged, or reduced to least terms, by the common measure seven, the fractions become

$$\begin{array}{cc} 1 & 1 \\ 9 & 2 \end{array} \cdot$$

Numerator and denominator, multiplied by the abridged denominators, give respectively $\dfrac{2}{126}$ and $\dfrac{9}{126}$. Subtraction being made, the difference is $\dfrac{7}{126}$."

114. *Bhāskara Āchábrya, "Vija-Ganita."*[2]—"Example: Tell quickly the result of the numbers three and four, negative or affirmative, taken together: . . . . The characters, denoting the quantities known and unknown, should be first written to indicate them generally; and those, which become negative, should be then marked with a dot over them. Statement:[3] 3·4. Adding them, the sum is found 7. Statement: 3̇·4̇. Adding them, the sum is 7̇. Statement: 3·4̇. Taking the difference, the result of addition comes out 1̇.

" 'So much as' and the colours 'black, blue, yellow and red,'[4] and others besides these, have been selected by venerable teachers for names of values of unknown quantities, for the purpose of reckoning therewith.

[1] Colebrooke, *op. cit.*, p. 13, 14.   [2] *Ibid.*, p. 131.

[3] In modern notation, $3+4=7$, $(-3)+(-4)=-7$, $3+(-4)=-1$.

[4] Colebrooke, *op. cit.*, p. 139.

"Example:[1] Say quickly, friend, what will affirmative one unknown with one absolute, and affirmative pair unknown less eight absolute, make, if addition of the two sets take place? .... Statement:[2]

$$ya\ 1 \quad ru\ 1$$
$$ya\ 2 \quad ru\ \overset{\centerdot}{8}$$

Answer: The sum is $ya\ 3 \quad ru\ \overset{\centerdot}{7}$.

"When absolute number and colour (or letter) are multiplied one by the other, the product will be colour (or letter). When two, three or more homogeneous quantities are multiplied together, the product will be the square, cube or other [power] of the quantity. But, if unlike quantities be multiplied, the result is their (*bhávita*) 'to be' product or factum.

"23. Example:[3] Tell directly, learned sir, the product of the multiplication of the unknown (*yávat-távat*) five, less the absolute number one, by the unknown (*yávat-távat*) three joined with the absolute two: .... Statement:[4]

$$ya\ 5 \quad ru\ \overset{\centerdot}{1}$$
$$ya\ 3 \quad ru\ 2$$

Product: $ya\ v\ 15 \quad ya\ 7 \quad ru\ \overset{\centerdot}{2}$.

"Example:[5] 'So much as' three, 'black' five, 'blue' seven, all affirmative: how many do they make with negative two, three, and one of the same respectively, added to or subtracted from them? Statement:[6]

$$ya\ 3 \quad ca\ 5 \quad ni\ 7$$
$$ya\ \overset{\centerdot}{2} \quad ca\ \overset{\centerdot}{3} \quad ni\ \overset{\centerdot}{1}$$

Answer: Sum $ya\ 1 \quad ca\ 2 \quad ni\ 6$.

Difference $ya\ 5 \quad ca\ 8 \quad ni\ 8$.

"Example:[7] Say, friend, [find] the sum and difference of two irrational numbers eight and two: .... after full consideration, if thou be acquainted with the sixfold rule of surds. Statement:[8] $c\ 2 \quad c\ 8$.

---

[1] *Ibid.* [2] In modern notation, $x+1$ and $2x-8$ have the sum $3x-7$

[3] Colebrooke, *op. cit.*, p. 141, 142.

[4] In modern notation $(5x-1)(3x+2)=15x^2+7x-2$.

[5] Colebrooke, *op. cit.*, p. 144.

[6] In modern symbols, $3x+5y+7z$ and $-2x-3y-z$ have the sum $x+2y+6z$, and the difference $5x+8y+8z$.

[7] Colebrooke, *op. cit.*, p. 146.

[8] In modern symbols, the example is $\sqrt{8}+\sqrt{2}=\sqrt{18}$, $\sqrt{8}-\sqrt{2}=\sqrt{2}$. The same example is given earlier by Brahmagupta in his *Brahme-spuṭa-sidd'hánta*, chap. xviii, in Colebrooke, *op. cit.*, p. 341.

Answer: Addition being made, the sum is c 18. Subtraction taking place, the difference is c 2."

ARABIC: aL-KHOWÂRIZMÎ, NINTH CENTURY A.D.

115. In 772 Indian astronomy became known to Arabic scholars. As regards algebra, the early Arabs failed to adopt either the Diophantine or the Hindu notations. The famous *Algebra* of al-Khowârizmî of Bagdad was published in the original Arabic, together with an English translation, by Frederic Rosen,[1] in 1831. He used a manuscript preserved in the Bodleian Collection at Oxford. An examination of this text shows that the exposition was altogether rhetorical, i.e., devoid of all symbolism. "Numerals are in the text of the work always expressed by words: [Hindu-Arabic] figures are only used in some of the diagrams, and in a few marginal notes."[2] As a specimen of al-Khowârizmî's exposition we quote the following from his *Algebra*, as translated by Rosen:

"What must be the amount of a square, which, when twenty-one dirhems are added to it, becomes equal to the equivalent of ten roots of that square? Solution: Halve the number of the roots; the moiety is five. Multiply this by itself; the product is twenty-five. Subtract from this the twenty-one which are connected with the square; the remainder is four. Extract its root; it is two. Subtract this from the moiety of the roots, which is five; the remainder is three. This is the root of the square which you required, and the square is nine. Or you may add the root to the moiety of the roots; the sum is seven; this is the root of the square which you sought for, and the square itself is forty-nine."[3]

By way of explanation, Rosen indicates the steps in this solution, expressed in modern symbols, as follows: Example:

$$x^2 + 21 = 10x \; ; \quad x = \tfrac{10}{2} \pm \sqrt{[(\tfrac{10}{2})^2 - 21]} = 5 \pm \sqrt{25 - 21} = 5 \pm \sqrt{4} = 5 \pm 2 \,.$$

ARABIC: aL-KARKHÎ, EARLY ELEVENTH CENTURY A.D.

116. It is worthy of note that while Arabic algebraists usually build up the higher powers of the unknown quantity on the multiplicative principle of the Hindus, there is at least one Arabic writer, al-Karkhî of Bagdad, who followed the Diophantine additive principle.[4]

[1] *The Algebra of Mohammed Ben Musa* (ed. and trans. Frederic Rosen; London, 1831). See also L. C. Karpinski, *Robert of Chester's Latin Translation of the Algebra of Al-Khowarizmi* (1915).

[2] Rosen, *op. cit.*, p. xv.   [3] *Ibid.*, p. 11.

[4] See Cantor, *op. cit.*, Vol. I (3d ed.), p. 767, 768; Heath, *op. cit.*, p. 41.

In al-Kharkī's work, the *Fakhrī*, the word *mal* means x^2, *ka͏ᶜb* means x^3; the higher powers are *māl māl* for x^4, *māl ka͏ᶜb* for x^5 (not for x^6), *ka͏ᶜb ka͏ᶜb* for x^6 (not for x^9), *māl māl ka͏ᶜb* for x^7 (not for x^{12}), and so on.

Cantor[1] points out that there are cases among Arabic writers where *māl* is made to stand for x, instead of x^2, and that this ambiguity is reflected in the early Latin translations from the Arabic, where the word *census* sometimes means x, and not x^2.[2]

117. Michael Psellus, a Byzantine writer of the eleventh century who among his contemporaries enjoyed the reputation of being the first of philosophers, wrote a letter[3] about Diophantus, in which he gives the names of the successive powers of the unknown, used in Egypt, which are of historical interest in connection with the names used some centuries later by Nicolas Chuquet and Luca Pacioli. In Psellus the successive powers are designated as the first number, the second number (square), etc. This nomenclature appears to have been borrowed, through the medium of the commentary by Hypatia, from Anatolius, a contemporary of Diophantus.[4] The association of the successive powers of the unknown with the series of natural numbers is perhaps a partial recognition of exponential values, for which there existed then, and for several centuries that followed Psellus, no adequate notation. The next power after the fourth, namely, x^5, the Egyptians called "the first undescribed," because it is neither a square nor a cube; the sixth power they called the "cube-cube"; but the seventh was "the second undescribed," as being the product of the square and the "first undescribed." These expressions for x^5 and x^7 are closely related to Luca Pacioli's *primo relato* and *secondo relato*, found in his *Summa* of 1494.[5] Was Pacioli directly or indirectly influenced by Michael Psellus?

118. While the early Arabic algebras of the Orient are characterized by almost complete absence of signs, certain later Arabic works on

---

[1] *Op. cit.*, p. 768. See also Karpinski, *op. cit.*, p. 107, n. 1.

[2] Such translations are printed by G. Libri, in his *Histoire des sciences mathématiques*, Vol. I (Paris, 1838), p. 276, 277, 305.

[3] Reproduced by Paul Tannery, *op. cit.*, Vol. II (1895), p. 37–42.

[4] See Heath, *op. cit.*, p. 2, 18.

[5] See *ibid.*, p. 41; Cantor, *op. cit.*, Vol. II (2d ed.), p. 317.

algebra, produced in the Occident, particularly that of al-Qalasâdî of Granada, exhibit considerable symbolism. In fact, as early as the thirteenth century symbolism began to appear; for example, a notation for continued fractions in al-Ḥaṣṣâr (§ 391). Ibn Khaldûn[1] states that Ibn Albanna at the close of the thirteenth century wrote a book when under the influence of the works of two predecessors, Ibn Almunᶜim and Alaḥdab. "He [Ibn Albanna] gave a summary of the demonstrations of these two works and of other things as well, concerning the technical employment of symbols[2] in the proofs, which serve at the same time in the abstract reasoning and the representation to the eye, wherein lies the secret and essence of the explication of theorems of calculation with the aid of signs." This statement of Ibn Khaldûn, from which it would seem that symbols were used by Arabic mathematicians before the thirteenth century, finds apparent confirmation in the translation of an Arabic text into Latin, effected by Gerard of Cremona (1114–87). This translation contains symbols for x and x^2 which we shall notice more fully later. It is, of course, quite possible that these notations were introduced into the text by the translator and did not occur in the original Arabic. As regards Ibn Albanna, many of his writings have been lost and none of his extant works contain algebraic symbolism.

CHINESE: CHU SHIH-CHIEH

(1303 A.D.)

119. Chu Shih-Chieh bears the distinction of having been "instrumental in the advancement of the Chinese abacus algebra to the highest mark it has ever attained."[3] The Chinese notation is interesting as being decidedly unique. Chu Shih-Chieh published in 1303 a treatise, entitled *Szu-yuen Yü-chien*, or "The Precious Mirror of the Four Elements," from which our examples are taken. An expression like $a+b+c+d$, and its square, $a^2+b^2+c^2+d^2+2ab+2ac+2ad+$

---

[1] Consult F. Woepcke, "Recherches sur l'histoire des sciences mathématiques chez les orientaux," *Journal asiatique* (5th ser.), Vol. IV (Paris, 1854), p. 369–72; Woepcke quotes the original Arabic and gives a translation in French. See also Cantor, *op. cit.*, Vol. I (3d ed.), p. 805.

[2] Or, perhaps, letters of the alphabet.

[3] Yoshio Mikami, *The Development of Mathematics in China and Japan* (Leipzig, 1912), p. 89. All our information relating to Chinese algebra is drawn from this book, p. 89–98.

$2bc+2bd+2cd$, were represented as shown in the following two illustrations:

```
                                        1
                    1            2   0   2
                                           2
            1   *   1        1 0   *   0 1
                                   2
                    1            2   0   2
                                        1
```

Where we have used the asterisk in the middle, the original has the character *t'ai* ("great extreme"). We may interpret this symbolism by considering *a* located one space to the right of the asterisk (✳), *b* above, *c* to the left, and *d* below. In the symbolism for the square of $a+b+c+d$, the 0's indicate that the terms *a*, *b*, *c*, *d* do not occur in the expression. The squares of these letters are designated by the 1's two spaces from ✳. The four 2's farthest from ✳ stand for $2ab$, $2ac$, $2bc$, $2bd$, respectively, while the two 2's nearest to ✳ stand for $2ac$ and $2bd$. One is impressed both by the beautiful symmetry and by the extreme limitations of this notation.

120. Previous to Chu Shih-Chieh's time algebraic equations of only one unknown number were considered; Chu extended the process to as many as four unknowns. These unknowns or elements were called the "elements of heaven, earth, man, and thing." Mikami states that, of these, the heaven element was arranged below the known quantity (which was called "the great extreme"), the earth element to the left, the man element to the right, and the thing element above. Letting ✳ stand for the great extreme, and *x*, *y*, *z*, *u*, for heaven, earth, man, thing, respectively, the idea is made plain by the following representations:

$$\boxed{\begin{array}{c}✳\\\hline 1\end{array}}=x\,,\qquad \boxed{\begin{array}{c|c}1&✳\end{array}}=y\,,\qquad \boxed{\begin{array}{c|c}✳&1\end{array}}=z\,,\qquad \boxed{\begin{array}{c}1\\\hline ✳\end{array}}=u\,.$$

Mikami gives additional illustrations:

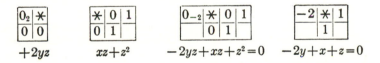

$+2yz$ $\qquad$ $xz+z^2$ $\qquad$ $-2yz+xz+z^2=0$ $\qquad$ $-2y+x+z=0$

Using the Hindu-Arabic numerals in place of the Chinese calculating pieces or rods, Mikami represents three equations, used by Chu, in the following manner:

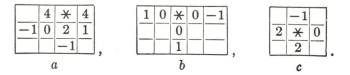

In our notation, the four equations are, respectively,

$$a)\ 2x - x^2 + 4y - xy^2 + 4z + xy = 0 \,,$$
$$b)\ x^2 + y^2 - z^2 = 0 \,,$$
$$c)\ 2x + 2y - u = 0 \,.$$

No sign of equality is used here. All terms appear on one side of the equation. Notwithstanding the two-dimensional character of the notation, which permits symbols to be placed above and below the starting-point, as well as to left and right, it made insufficient provision for the representation of complicated expressions and for easy methods of computation. The scheme does not lend itself easily to varying algebraic forms. It is difficult to see how, in such a system, the science of algebra could experience a rapid and extended growth. The fact that Chinese algebra reached a standstill after the thirteenth century may be largely due to its inelastic and faulty notation.

BYZANTINE: MAXIMUS PLANUDES, FOURTEENTH CENTURY A.D.

121. Maximus Planudes, a monk of the first half of the fourteenth century residing in Constantinople, brought out among his various compilations in Greek an arithmetic,[1] and also scholia to the first two books of Diophantus' *Arithmetica*.[2] These scholia are of interest to us, for, while Diophantus evidently wrote his equations in the running text and did not assign each equation a separate line, we find in Planudes the algebraic work broken up so that each step or each equation is assigned a separate line, in a manner closely resembling modern practice. To illustrate this, take the problem in Diophantus (i. 29),

[1] *Das Rechenbuch des Maximus Planudes* (Halle: herausgegeben von C. I. Gerhardt, 1865).

[2] First printed in Xylander's Latin translation of Diophantus' *Arithmetica* (Basel, 1575). These scholia in Diophantus are again reprinted in P. Tannery, *Diophanti Alexandrini opera omnia* (Lipsiae, 1895), Vol. II, p. 123–255; the example which we quote is from p. 201.

"to find two numbers such that their sum and the difference of their squares are given numbers." We give the exposition of Planudes and its translation.

	Planudes				Translation
	$\bar{\kappa}$		$\bar{\pi}$	[Given the numbers],	20, 80
ἔκϑ ·	$s\bar{a}\mu^{\circ}\bar{\iota}$		$\mu^{\circ}\bar{\iota} \wedge s\bar{a}$.......	Putting for the numbers,	$x+10,$
					$10-x$
τετρ ·	$\Delta^{Y}\bar{a}ss\bar{\kappa}\mu^{\circ}\overline{\rho}$		$\Delta^{Y}\bar{a}\mu^{\circ}\rho \wedge ss\bar{\kappa}$...	Squaring,	$x^2+20x+100,$
					$x^2+100-20x$
ὑπεροχ·	$ss\bar{\mu}$	ι^{σ}·	$\mu^{\circ}\overline{\pi}$	Taking the difference,	$40x=80$
μερ ·	$s\bar{a}$	ι^{σ}·	$\mu^{\circ}\overline{\beta}$	Dividing,	$x=2$
ὑπ ·	$\mu^{\circ}\overline{\iota\beta}$		$\mu^{\circ}\overline{\eta}$·	Result,	12, 8

ITALIAN: LEONARDO OF PISA
(1202 A.D.)

122. Leonardo of Pisa's mathematical writings are almost wholly rhetorical in mode of exposition. In his *Liber abbaci* (1202) he used the Hindu-Arabic numerals. To a modern reader it looks odd to see expressions like $\frac{1}{13}\ \frac{2}{11}\ \frac{3}{5}$ 42, the fractions written before the integer in the case of a mixed number. Yet that mode of writing is his invariable practice. Similarly, the coefficient of x is written after the name for x, as, for example,[1] —"radices $\frac{1}{2}$12" for $12\frac{1}{2}x$. A computation is indicated, or partly carried out, on the margin of the page, and is inclosed in a rectangle, or some irregular polygon whose angles are right angles. The reason for the inverted order of writing coefficients or of mixed numbers is due, doubtless, to the habit formed from the study of Arabic works; the Arabic script proceeds from right to left. Influenced again by Arabic authors, Leonardo uses frequent geometric figures, consisting of lines, triangles, and rectangles to illustrate his arithmetic or algebraic operations. He showed a partiality for unit fractions; he separated the numerator of a fraction from its denominator by a fractional line, but was probably not the first to do this (§ 235). The product of a and b is indicated by *factus ex.a.b.* It has been stated that he denoted multiplication by juxtaposition,[2] but G. Eneström shows by numerous quotations from the *Liber abbaci* that such is not the case.[3] Cantor's quotation from the *Liber abbaci*,

[1] *Il liber abbaci di Leonardo Pisano* (ed. B. Boncompagni), Vol. I (Rome, 1867), p. 407.

[2] Cantor, *op. cit.*, Vol. II (2d ed.), p. 62.

[3] *Bibliotheca mathematica* (3d ser.), Vol. XII (1910–11), p. 335, 336.

"sit numerus .a.e.c. quaedam coniunctio quae uocetur prima, numeri vero .d.b.f. sit coniunctio secunda,"[1] is interpreted by him as a product, the word *coniunctio* being taken to mean "product." On the other hand, Eneström conjectures that *numerus* should be *numeri*, and translates the passage as meaning, "Let the numbers a, e, c be the first, the numbers d, b, f the second combination." If Eneström's interpretation is correct, then a.e.c and d.b.f are not products. Leonardo used in his *Liber abbaci* the word *res* for x, as well as the word *radix*. Thus, he speaks, "et intellige pro re summam aliquam ignotam, quam inuenire uis."[2] The following passage from the *Liber abbaci* contains the words *numerus* (for a given number), *radix* for x, and *census* for x^2: "Primus enim modus est, quando census et radices equantur numero. . . . . Verbi gratia: duo census, et decem radices equantur denariis 30,"[3] i.e., $2x^2+10x=30$. The use of *res* for x is found also in a Latin translation of al-Khowârizmî's algebra,[4] due perhaps to Gerard of Cremona, where we find, "res in rem fit census," i.e., $x.x=x^2$. The word *radix* for x as well as *res*, and *substantia* for x^2, are found in Robert of Chester's Latin translation of al-Khowârizmî's algebra.[5] Leonardo of Pisa calls x^3 *cubus*, x^4 *census census*, x^6 *cubus cubus*, or else *census census census;* he says, " . . . . est multiplicare per cubum cubi, sicut multiplicare per censum census census."[6] He goes even farther and lets x^8 be *census census census census*. Observe that this phraseology is based on the additive principle $x^2 \cdot x^2 \cdot x^2 \cdot x^2 = x^8$. Leonardo speaks also of *radix census census*.[7]

The first appearance of the abbreviation R or $\not{R}$ for *radix* is in his *Practica geometriae* (1220),[8] where one finds the $\not{R}$ meaning "square root" in an expression "et minus $\not{R}$. 78125 dragme, et diminuta radice 28125 dragme." A few years later, in Leonardo's *Flos*,[9] one finds marginal notes which are abbreviations of passages in the text relating to square root, as follows:

[1] *Op. cit.*, Vol. I (3d ed.), p. 132.

[2] *Ibid.*, Vol. I, p. 191.

[3] *Ibid.*, Vol. I, p. 407.

[4] Libri, *Histoire des sciences mathématiques en Italie*, Vol. I (Paris, 1838), p. 268.

[5] L. C. Karpinski, *op. cit.*, p. 68, 82.

[6] *Il liber abbaci*, Vol. I, p. 447.

[7] *Ibid.*, Vol. I, p. 448.

[8] *Scritti di Leonardo Pisano* (ed. B. Boncompagni), Vol. II (Rome, 1862), p. 209.

[9] *Op. cit.*, Vol. II, p. 231. For further particulars of the notations of Leonardo of Pisa, see our §§ 219, 220, 235, 271–73, 290, 292, 318, Vol. II, § 389.

$.R.^x p^i. Bino.^{ij}$ for *primi [quidem] binomij radix*
$2.^i B. R.^x$ for *radix [quippe] secundi binomij*
$.Bi. 3^i. R.^x$ for *Tertij [autem] binomij radix*
$.Bi. 4^i. R.^x$ for *Quarti [quoque] binomij radix*

FRENCH: NICOLE ORESME, FOURTEENTH CENTURY A.D.

123. Nicole Oresme (*ca.* 1323–82), a bishop in Normandy, prepared a manuscript entitled *Algorismus proportionum*, of which several
copies are extant.[1] He was the first to conceive the notion of fractional
powers which was afterward rediscovered by Stevin. More than this,
he suggested a notation for fractional powers. He considers powers of
ratios (called by him *proportiones*). Representing, as does Oresme
himself, the ratio 2:1 by 2, Oresme expresses $2^{\frac{1}{2}}$ by the symbolism

$$\boxed{\begin{array}{c} 1.p \\ \hline 2.2 \end{array}}$$ and reads this *medietas [proportionis] duplae;* he expresses

$(2\frac{1}{2})^{\frac{1}{4}}$ by the symbolism $\boxed{\begin{array}{c} 1.p.1 \\ \hline 4.2.2 \end{array}}$ and reads it *quarta pars [proportionis]*

duplae sesquialterae. The fractional exponents $\frac{1}{2}$ and $\frac{1}{4}$ are placed to the
left of the ratios affected.

H. Wieleitner adds that Oresme did not use these symbols in computation. Thus, Oresme expresses in words, ". . . . proponatur proportio, que sit due tertie quadruple; et quia duo est numerator, ipsa
erit vna tertia quadruple duplicate, sev sedecuple,"[2] i.e., $4^{\frac{2}{3}}=(4^2)^{\frac{1}{3}}=16^{\frac{1}{3}}$.
Oresme writes[3] also: "Sequitur quod *.a.* moueatur velocius *.b.* in pro-
portione, que est medietas proportionis .50. ad .49.," which means,
"the velocity of a:velocity of $b = \sqrt{50}:\sqrt{49}$," the word *medietas* meaning "square root."[4]

The transcription of the passage shown in Figure 34 is as follows:

"Una media debet sic scribi $\boxed{\begin{array}{c} 1 \\ \hline 2 \end{array}}$, una tertia sic $\boxed{\begin{array}{c} 1 \\ \hline 3. \end{array}}$ et due tertie

sic $\boxed{\begin{array}{c} 2 \\ \hline 3. \end{array}}$; et sic de alijs. et numerus, qui supra uirgulam, dicitur

[1] Maximilian Curtze brought out an edition after the MS R. 4° 2 of the Gym-
nasiat-Bibliothek at Thorn, under the title *Der Algorithmus Proportionum des
Nicolaus Oresme* (Berlin, 1868). Our photographic illustration is taken from that
publication.

[2] Curtze, *op. cit.*, p. 15.   [3] *Ibid.*, p. 24.

[4] See Eneström, *op. cit.*, Vol. XII (1911–12), p. 181. For further details see
also Curtze, *Zeitschrift für Mathematik und Physik*, Vol. XIII (Suppl. 1868),
p. 65 ff.

numerator, iste uero, qui est sub uirgula, dicitur denominator. 2. Proportio dupla scribitur isto modo 2.^{la}, et tripla isto modo 3.^{la}; et sic
de alijs. Proportio sesquialtera sic scribitur $\boxed{\frac{p\ 1.}{1\ 2.}}$, et sesquitert^{e}

$\boxed{\frac{p\ 1}{1\ 3}}$. Proportio superpartiens duas tertias scribitur sic $\boxed{\frac{p\ 2}{1\ 3.}}$.

Proportio dupla superpartiens duas quartas scribitur sic $\boxed{\frac{p\ 2}{2\ 4}}$; et

sic de alijs. 3. Medietas duple scribitur sic $\boxed{\frac{1\ p}{2\cdot 2}}$, quarta pars

duple sesquialtere scribitur sic $\boxed{\frac{1\cdot p\ 1}{4\cdot 2\cdot 2}}$; et sic de alijs."

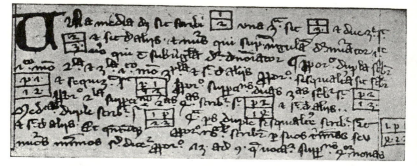

Fig. 34.—From the first page of Oresme's *Algorismus proportionum* (four-teenth century).

A free translation is as follows:

"Let a half be written $\boxed{\frac{1}{2}}$, a third $\boxed{\frac{1}{3}}$, and two-thirds $\boxed{\frac{2}{3}}$,
and so on. And the number above the line is called the 'numerator,'
the one below the line is called the 'denominator.' 2. A double ratio
is written in this manner 2.^{la}, a triple in this manner 3.^{la}, and thus in

other cases. The ratio one and one-half is written $\boxed{\frac{p\ 1.}{1\ 2.}}$, and one and

one-third is written $\boxed{\frac{p\ 1}{1\ 3}}$. The ratio one and two-thirds is written

$\boxed{\frac{p\ 2}{1\ 3.}}$. A double ratio and two-fourths are written $\boxed{\frac{p\ 2}{2\ 4}}$, and thus in other cases. 3. The square root of two is written thus $\boxed{\dfrac{1.\ p}{2\ \ 2}}$, the

fourth root of two and one-half is written thus $\boxed{\dfrac{1.\ p\ 1}{4.\ 2.\ 2}}$, and thus in other cases."

ARABIC: AL-QALASÂDÎ, FIFTEENTH CENTURY A.D.

124. Al-Qalasâdî's *Raising of the Veil of the Science of Gubar* appeared too late to influence the progress of mathematics on the European Continent. Al-Qalasâdî used ⌐, the initial letter in the Arabic word *jidr*, "square root"; the symbol was written above the number whose square root was required and was usually separated from it by a horizontal line. The same symbol, probably considered this time as the first letter in *jahala* ("unknown"), was used to represent the unknown term in a proportion, the terms being separated by the sign ∴ . But in the part of al-Qalasâdî's book dealing more particularly with algebra, the unknown quantity x is represented by the letter ش, x^2 by the letter ص, x^3 by the letter ک ; these are written above their respective coefficients. Addition is indicated by juxtaposition. Subtraction is ⅄ʃ ; the equality sign, ⅃ , is seen to resemble the Diophantine ι, if we bear in mind that the Arabs wrote from right to left, so that the curved stroke faced in both cases the second member of the equation. We reproduce from Woepcke's article a few samples of al-Qalasâdî's notation. Observe the peculiar shapes of the Hindu-Arabic numerals (Fig. 35).

Woepcke[1] reproduces also symbols from an anonymous Arabic manuscript of unknown date which uses symbols for the powers of x and for the powers of the reciprocal of x, built up on the additive principle of Diophantus. The total absence of data relating to this manuscript diminishes its historic value.

GERMAN: REGIOMONTANUS

(ca. 1473)

125. Regiomontanus died, in the prime of life, in 1476. After having studied in Rome, he prepared an edition of Ptolemy[2] which was issued in 1543 as a posthumous publication. It is almost purely rhetorical, as appears from the following quotation on pages 21 and 22.

[1] *Op. cit.*, p. 375–80.

[2] *Ioannis de Monte Regio et Georgii Pvrbachii epitome, in Cl. Ptolemaei magnam compositionem* (Basel, 1543). The copy examined belongs to Mr. F. E. Brasch.

By the aid of a quadrant is determined the angular elevation ACE, "que erit altitudo tropici hiemalis," and the angular elevation ACF, "que erit altitudo tropici aestivalis," it being required to find the arc EF between the two. "Arcus itaque EF, fiet distantia duorum tropi-

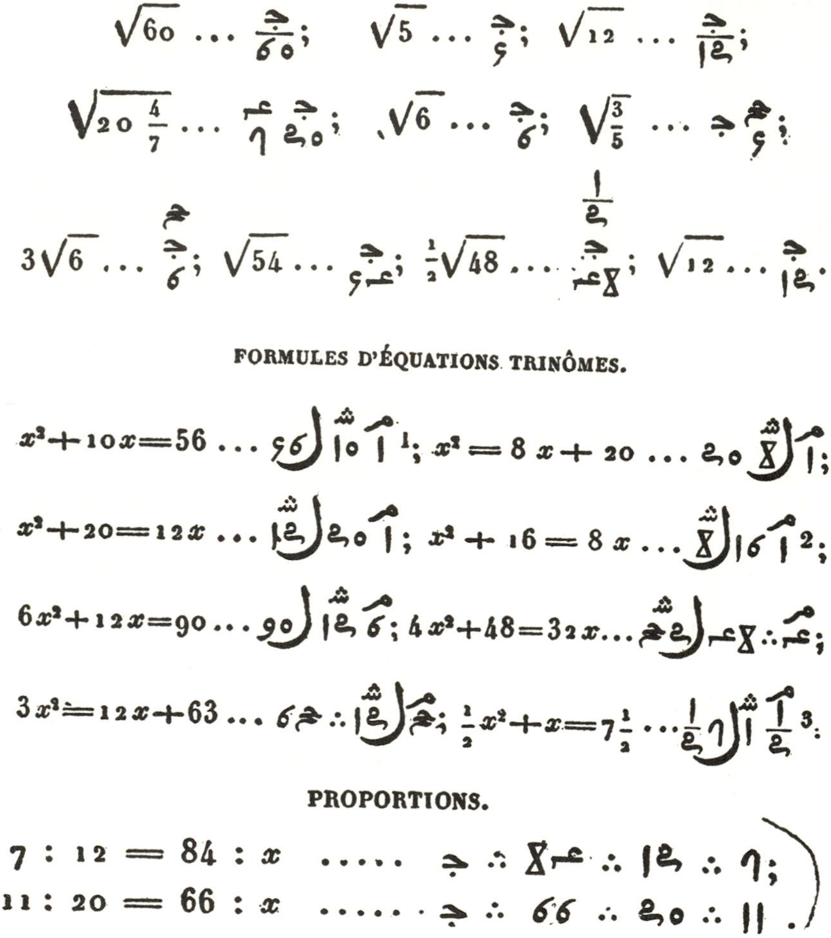

$$\sqrt{60} \dots \frac{\Rightarrow}{60}; \quad \sqrt{5} \dots \frac{\Rightarrow}{\varsigma}; \quad \sqrt{12} \dots \frac{\Rightarrow}{|\varsigma|};$$

$$\sqrt{20\tfrac{4}{7}} \dots \overset{c}{\eta}\overset{\Rightarrow}{\varsigma_0}; \quad \sqrt{6} \dots \frac{\Rightarrow}{6}; \quad \sqrt{\tfrac{3}{5}} \dots \Rightarrow\overset{\curvearrowright}{\varsigma};$$

$$3\sqrt{6} \dots \overset{\Rightarrow}{\varsigma}; \quad \sqrt{54} \dots \overset{\Rightarrow}{\varsigma\equiv}; \quad \tfrac{1}{2}\sqrt{48} \dots \overset{\tfrac{1}{2}}{\overset{\Rightarrow}{\equiv X}}; \quad \sqrt{12} \dots \overset{\Rightarrow}{|\varsigma|}.$$

FORMULES D'ÉQUATIONS TRINÔMES.

$$x^2 + 10x = 56 \dots \varsigma 6 \big) |0 \overset{\curvearrowleft}{1}{}^1; \quad x^2 = 8x + 20 \dots 20 \underset{X}{\overset{\curvearrowleft}{\big)} \overset{\frown}{1}};$$

$$x^2 + 20 = 12x \dots |\overset{\curvearrowleft}{\equiv} \big) 20 \overset{\frown}{1}; \quad x^2 + 16 = 8x \dots \overset{\curvearrowleft}{X} \big) 16 \overset{\frown}{1}{}^2;$$

$$6x^2 + 12x = 90 \dots 90 \big) |\overset{\curvearrowleft}{\equiv} \overset{\frown}{6}; \quad 4x^2 + 48 = 32x \dots \overset{\curvearrowleft}{\equiv 2} \big) \equiv X \therefore \overset{\frown}{\equiv};$$

$$3x^2 = 12x + 63 \dots 6 \Rightarrow \therefore |\overset{\curvearrowleft}{\equiv} \big) \overset{\frown}{\equiv}; \quad \tfrac{1}{2}x^2 + x = 7\tfrac{1}{2} \dots |\overset{\curvearrowleft}{\equiv} \eta | \overset{\curvearrowleft}{1} \overset{\frown}{\tfrac{1}{2}}{}^3.$$

PROPORTIONS.

$$7 : 12 = 84 : x \quad \dots \dots \quad \Rightarrow \therefore X \overset{c}{\equiv} \therefore |\equiv \therefore \eta;$$

$$11 : 20 = 66 : x \quad \dots \dots \quad \Rightarrow \therefore 66 \therefore 20 \therefore ||.$$

Fig. 35.—Al-Qalasâdî's algebraic symbols. (Compiled by F. Woepcke, *Journal asiatique* [Oct. and Nov., 1854], p. 363, 364, 366.)

corum quęsita. Hâc Ptolemaeus reperit 47. graduum 42. minutorum 40. secundorum. Inuenit enim proportionem eius ad totum circulū sicut 11. ad 83, postea uerò minorem inuenerunt. Nos autem inuenimus arcum AF 65. graduum 6. minutorum, & arcum AE 18. graduum 10.

minutorum. Ideoq. nunc distantia tropicorum est 46. graduum 56. minutorum, ergo declinatio solis maxima nostro tempore est 23. graduum 28. minutorum."

126. We know, however, that in some of his letters and manuscripts symbols appear. They are found in letters and sheets containing computations, written by Regiomontanus to Giovanni Bianchini, Jacob von Speier, and Christian Roder, in the period 1463–71. These documents are kept in the Stadtbibliothek of the city of Nürnberg.[1] Regiomontanus and Bianchini designate angles thus: $\overset{u}{gr}$ 35 $\overset{u}{m}$ 17; Regiomontanus writes also: $\overline{44}$. 42′. 4″ (see also § 127).

In one place[2] Regiomontanus solves the problem: Divide 100 by a certain number, then divide 100 by that number increased by 8; the sum of the quotients is 40. Find the first divisor. Regiomontanus writes the solution thus:

In Modern Symbols

"$\dfrac{100}{1\,\gamma}$ $\qquad$ $\dfrac{100}{1\,\gamma\ \text{et}\ 8}$ $\qquad\qquad$ $\dfrac{100}{x}$ $\qquad$ $\dfrac{100}{x+8}$

$100\,\gamma$ et 800 $\qquad\qquad\qquad\qquad$ $100x + 800$
$100\,\gamma$ $\qquad\qquad\qquad\qquad\qquad\ $ $100x$

$\dfrac{200\,\gamma\ \text{et}\ 800}{1\,c\!\!\!/\ \text{et}\ 8\,\gamma} - 40$ $\qquad\qquad$ $\dfrac{200x + 800}{x^2 + 8x} = 40$

$40\,c\!\!\!/$ et $320\,\gamma$ — $200\,\gamma$ et 800 $\qquad$ $40x^2 + 320x = 200x + 800.$
$40\,c\!\!\!/$ et $120\,\gamma$ — 800 $\qquad\qquad\ $ $40x^2 + 120x = 800$
$\ \ 1\,c\!\!\!/$ et $\ \ \ 3\,\gamma$ — $\ \ 20$ $\qquad\qquad$ $x^2 + \ \ \ 3x = \ \ 20$

$\tfrac{3}{2} \cdot \tfrac{9}{4}$ addo numerum $20\tfrac{9}{4}$—$\tfrac{8.9}{4}$ $\qquad$ $\tfrac{3}{2} \mid \tfrac{9}{4}$ add the no. $20\tfrac{9}{4} = \tfrac{8.9}{4}$
Radix-quadrata de $\tfrac{8.9}{4}$ minus $\tfrac{3}{2}$—$1\,\gamma$ $\qquad$ $\sqrt{\tfrac{8.9}{4}} - \tfrac{3}{2} = x$
Primus ergo divisor fuit $R\!\!\!/$ de $22\tfrac{1}{4}$ $\qquad$ Hence the first divisor was
$\quad\overline{i9}\ 1\tfrac{1}{2}$." $\qquad\qquad\qquad\qquad\qquad$ $\sqrt{22\tfrac{1}{4}} - 1\tfrac{1}{2}$.

Note that "plus" is indicated here by *et;* "minus" by $\overline{i9}$, which is probably a ligature or abbreviation of "minus." The unknown quantity is represented by γ and its square by $c\!\!\!/$. Besides, he had a sign for equality, namely, a horizontal dash, such as was used later in Italy by Luca Pacioli, Ghaligai, and others. See also Fig. 36.

[1] Curtze, *Urkunden zur Geschichte der Mathematik im Mittelalter und der Renaissance* (Leipzig, 1902), p. 185–336 = *Abhandlungen zur Geschichte der Mathematik*, Vol. XII. See also L. C. Karpinski, *Robert of Chester's Translation of the Algebra of Al-Khowarizmi* (1915), p. 36, 37.

[2] Curtze, *op. cit.*, p. 278.

127. Figure 37[1] illustrates part of the first page of a calendar issued by Regiomontanus. It has the heading *Janer* ("January"). Farther to the right are the words *Sunne—Monde—Stainpock* ("Sun—Moon—Capricorn"). The first line is *1 A. Kl. New Jar* (i.e., "first day, A. calendar, New Year"). The second line is *2. b. 4. nō. der achtet S. Stephans.* The seven letters *A,b,c,d, e,F, g,* in the second column on the left, are the dominical letters of the calendars. Then come the days of the Roman calendar. After the column of saints' days comes a double column for the place of the sun. Then follow two double columns for the moon's longitude; one for the mean, the other for the

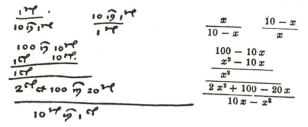

$$\frac{x}{10-x} \qquad \frac{10-x}{x}$$

$$\begin{array}{c} 100-10\,x \\ x^2-10\,x \\ \hline x^2 \\ \hline 2\,x^2+100-20\,x \\ \hline 10\,x-x^2 \end{array}$$

Fig. 36.—Computations of Regiomontanus, in letters of about 1460. (From manuscript, Nürnberg, fol. 23. (Taken from J. Tropfke, *Geschichte der Elementar-Mathematik* (2d ed.), Vol. II [1921], p. 14.)

true. The *S* signifies *signum* (i.e., 30°); the *G* signifies *gradus*, or "degree." The numerals, says De Morgan, are those facsimiles of the numerals used in manuscripts which are totally abandoned before the end of the fifteenth century, except perhaps in reprints. Note the shapes of the 5 and 7. This almanac of Regiomontanus and the *Compotus* of Anianus are the earliest almanacs that appeared in print.

ITALIAN: THE EARLIEST PRINTED ARITHMETIC

(1478)

128. The earliest aritnmetic was printed anonymously at Treviso, a town in Northeastern Italy. Figure 38 displays the method of solving proportions. The problem solved is as follows: A courier travels from Rome to Venice in 7 days; another courier starts at the same time and travels from Venice to Rome in 9 days. The distance between Rome and Venice is 250 miles. In how many days will the

[1] Reproduced from Karl Falkenstein, *Geschichte der Buchdruckerkunst* (Leipzig, 1840), Plate XXIV, between p. 54 and 55. A description of the almanac of Regiomontanus is given by A. de Morgan in the *Companion to the British Almanac*, for 1846, in the article, "On the Earliest Printed Almanacs," p. 18–25.

couriers meet, and how many miles will each travel before meeting? Near the top of Figure 38 is given the addition of 7 and 9, and the

FIG. 37.—"Calendar des Magister Johann von Kunsperk (Johannes Regiomontanus) Nürnberg um 1473."

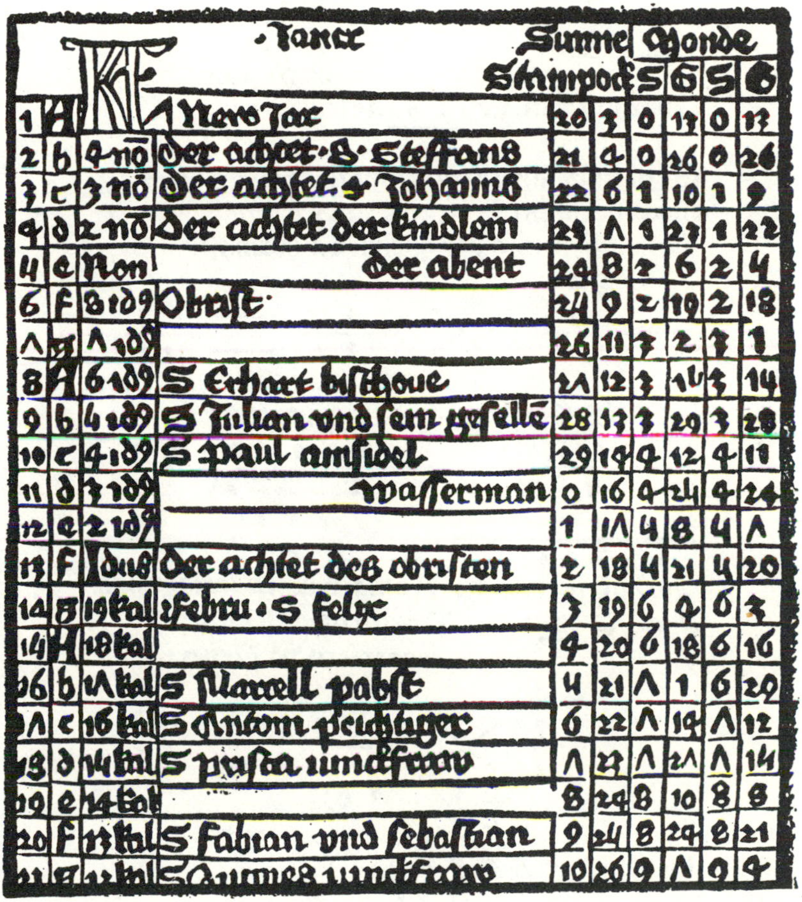

division of 63 by 16, by the scratch method.[1] The number of days is $3\frac{15}{16}$. The distance traveled by the first courier is found by the pro-

[1] Our photograph is taken from the *Atti dell'Accademia Pontificia de' nuovi Lincei*, Vol. XVI (Roma, 1863), p. 570.

portion $7:250 = \frac{68}{18}:x$. The mode of solution is interesting. The 7 and 250 are written in the form of fractions. The two lines which cross

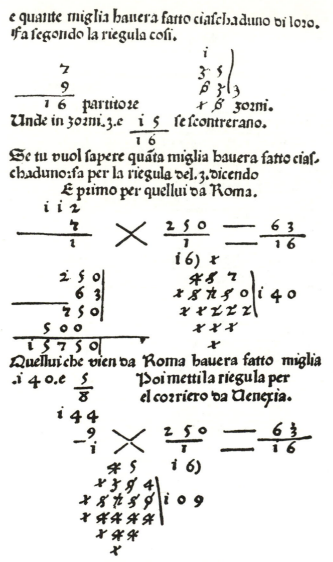

FIG. 38.—From the earliest printed arithmetic, 1478

and the two horizontal lines on the right, connecting the two numerators and the two denominators, respectively, indicate what numbers are to be multiplied together: $7 \times 1 \times 16 = 112$; $1 \times 250 \times 63 = 15,750$. The multiplication of 250 and 63 is given; also the division of 15,750

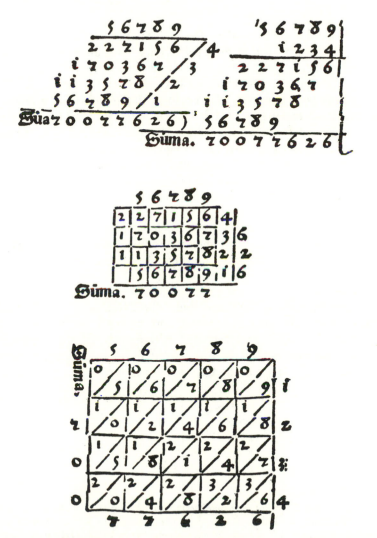

Fig. 39.—Multiplications in the Treviso arithmetic; four multiplications of 56,789 by 1,234 as given on one page of the arithmetic.

by 112, according to the scratch method. Similarly is solved the proportion $9:250 = \frac{6}{1}\frac{3}{8}:x$. Notice that the figure 1 is dotted in the same way as the Roman I is frequently dotted. Figure 39 represents other examples of multiplication.[1]

FRENCH: NICOLAS CHUQUET

(1484)

129. Over a century after Oresme, another manuscript of even greater originality in matters of algebraic notation was prepared in France, namely, *Le triparty en la science des nombres* (1484), by Nicolas Chuquet, a physician in Lyons.[2] There are no indications that he had seen Oresme's manuscripts. Unlike Oresme, he does not use fractional exponents, but he has a notation involving integral, zero, and negative exponents. The only possible suggestion for such exponential notation known to us might have come to Chuquet from the Gobar numerals, the Fihrist, and from the scholia of Neophytos (§§ 87, 88) which are preserved in manuscript in the National Library at Paris. Whether such connection actually existed we are not able to state. In any case, Chuquet elaborates the exponential notation to a completeness apparently never before dreamed of. On this subject Chuquet was about one hundred and fifty years ahead of his time; had his work been printed at the time when it was written, it would, no doubt, have greatly accelerated the progress of algebra. As it was, his name was known to few mathematicians of his time.

Under the head of "Numeration," the *Triparty* gives the Hindu-Arabic numerals in the inverted order usual with the Arabs: ".0.9.8.7.6.5.4.3.2.1." and included within dots, as was customary in late manuscripts and in early printed books. Chuquet proves addition by "casting out the 9's," arranging the figures as follows:

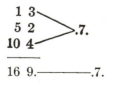

[1] *Ibid.*, p. 550.

[2] *Op. cit.* (publié d'après le manuscrit fonds Français N. 1346 de la Bibliothèque nationale de Paris et précédé d'une notice, par M. Aristide Marre), *Bullettino di Bibliog. e di Storia delle scienze mat. et fisiche*, Vol. XIII (1880), p. 555–659, 693–814; Vol. XIV, p. 413–60.

The addition of $\frac{2}{3}$ and $\frac{3}{4}$ is explained in the text, and the following arrangement of the work is set down by itself:[1]

130. In treating of roots he introduces the symbol $\mathbb{R}$, the first letter in the French word *racine* and in the Latin *radix*. A number, say 12, he calls *racine premiere*, because 12, taken as a factor once, gives 12; 4 is a *racine seconde* of 16, because 4, taken twice as a factor, gives 16. He uses the notations $\mathbb{R}^1.12.$ equal .12., $\mathbb{R}^2.16.$ equal .4., $\mathbb{R}^4.16.$ equal .2., $\mathbb{R}^5.32.$ equal .2. To quote: "Il conuiendroit dire que racine p̄miere est entendue pour tous nombres simples Cōme qui diroit la racine premiere de .12. que lon peult ainsi noter en mettant .1. dessus $\mathbb{R}$. en ceste maniere $\mathbb{R}^1.12.$ cest .12. Et $\mathbb{R}^1.9.$ est .9. et ainsi de tous aultres nobres. Racine seconde est celle qui posee en deux places lune soubz laultre et puys multipliee lune par laultre pduyt le nombre duquel elle est racine seconde Comme 4. et .4. qui multipliez lung par laultre sont .16. ainsi la racine seconde de .16. si est .4. ... on le peult ainsi mettre $\mathbb{R}^2 16.$ ... Et $\mathbb{R}^5.32.$ si est .2. Racine sixe se doit ainsi mettre $\mathbb{R}^6.$ et racine septiesme ainsi $\mathbb{R}^7.$ ... Aultres manieres de racines sont que les simples devant dictes que lon peult appeller racines composees Cōme de 14. plus $\mathbb{R}^2 180.$ dont sa racine seconde si est .3. $\overline{p}$. $\mathbb{R}^2 5.$ [i.e., $\sqrt{14+\sqrt{180}} = 3+\sqrt{5}$] ... cōe la racine seconde de .14. $\overline{p}$ $\mathbb{R}^2.180.$ se peult ainsi mettre $\mathbb{R}^2.\underline{14.\overline{p}.\mathbb{R}^2.180.}$"[2]

Not only have we here a well-developed notation for roots of integers, but we have also the horizontal line, drawn underneath the binomial $14+\sqrt{180}$, to indicate aggregation, i.e., to show that the square root of the entire binomial is intended.

Chuquet took a position in advance of his time when he computed with zero as if it were an actual quantity. He obtains,[3] according to his rule, $x = 2 \pm \sqrt{4-4}$ as the roots of $3x^2+12=12x$. He adds: "... reste .0. Donc $\mathbb{R}^2.0.$ adioustee ou soustraicte avec .2. ou de .2. monte .2. qui est le nōbe que lon demande."

131. Chuquet uses $\overline{p}$ and $\overline{m}$ to designate the words *plus* and *moins*. These abbreviations we shall encounter among Italian writers. Proceeding to the development of his exponential theory and notation,

[1] Boncompagni, *Bullettino*, Vol. XIII, p. 636.

[2] *Ibid.*, p. 655.

[3] *Ibid.*, p. 805; Eneström, *Bibliotheca mathematica*, Vol. VIII (1907-8), p. 203.

he states first that a number may be considered from different points of view.[1] One is to take it without any denomination (*sans aulcune denomīacion*), or as having the denomination 0, and mark it, say, .12? and .13? Next a number may be considered the primary number of a continuous quantity, called "linear number" (*nombre linear*), designated .12¹ .13¹ .20¹, etc. Third, it may be a secondary or superficial number, such as 12². 13². 19²., etc. Fourth, it may be a cubical number, such as .12³. 15³. 1³., etc. "On les peult aussi entendre estre nombres quartz ou quarrez de quarrez qui seront ainsi signez . 12⁴. 18⁴. 30⁴., etc." This nomenclature resembles that of the Byzantine monk Psellus of the eleventh century (§ 117).

Chuquet states that the ancients called his primary numbers "things" (*choses*) and marked them .ρ.; the secondary numbers they called "hundreds" and marked them .ɕ.; the cubical numbers they indicated by □; the fourth they called "hundreds of hundreds" (*champs de champ*), for which the character was t ɕ. This ancient nomenclature and notation he finds insufficient. He introduces a symbolism "que lon peult noter en ceste maniere $R^2.12^1$. $R^2.12^2$. $R^2.12^3$. $R^2.12^4$. etc. $R^3.12^1$. $R^3.12^2$. $R^3.12^3$. $R^3.12^4$. etc. $R^4.13^5$. $R^6.12^6$. etc." Here "$R^4.13^5$." means $\sqrt[4]{13x^5}$. He proceeds further and points out "que lon peult ainsi noter .12¹ ⁻ᵐ ou moins 12.," thereby introducing the notion of an exponent "minus one." As an alternative notation for this last he gives ".$\tilde{m}.12^1$," which, however, is not used again in this sense, but is given another interpretation in what follows.

From what has been given thus far, the modern reader will probably be in doubt as to what the symbolism given above really means. Chuquet's reference to the ancient names for the unknown and the square of the unknown may have suggested the significance that he gave to his symbols. His 12² does not mean 12×12, but our $12x^2$; *the exponent is written without* its base. Accordingly, his ".12.¹ ⁻ᵐ" means $12x^{-1}$. This appears the more clearly when he comes to "adiouster 8¹ avec $\tilde{m}.5^1$ monte tout .3.¹ Ou .10.¹ avec .$\tilde{m}.16.^1$ mõte tout $\tilde{m}.6.^1$," i.e., $8x - 5x = 3x$, $10x - 16x = -6x$. Again, ".8.² avec .12.² montent .20.²" means $8x^2 + 12x^2 = 20x^2$; subtracting ".$\tilde{m}.16^0$." from ".12.²" leaves "12.² $\tilde{m}$. $\tilde{m}$. 16? qui valent autant cõme .12.² $\bar{p}$. 16?"[2] The meaning of Chuquet's ".12?" appears from his "Example. qui multiplie .12⁰ par .12? montent .144. puis qui adiouste .0. avec .0. monte 0. ainsi monte ceste multiplicacion .144?,"[3] i.e., $12x^0 \times 12x^0 =$

[1] Boncompagni, *op. cit.*, Vol. XIII, p. 737.

[2] *Ibid.*, p. 739.   [3] *Ibid.*, p. 740.

$144x^0$. Evidently, $x^0 = 1$; he has the correct interpretation of the exponent zero. He multiplies $.12^?$ by $.10.^2$ and obtains $120.^2$; also $.5.^1$ times $.8.^1$ yields $.40.^2$; $.12.^3$ times $.10.^5$ gives $.120.^8$; $.8.^1$ times $.7^{1.\tilde{m}.}$ gives $.56^?$ or $.56.$; $.8^3$ times $.7^{1.\tilde{m}.}$ gives $.56.^2$ Evidently algebraic multiplication, involving the product of the coefficients and the sum of the exponents, is a familiar process with Chuquet. Nevertheless, he does not, in his notation, apply exponents to given numbers, i.e., with him "3^2" never means 9, it always means $3x^2$. He indicates (p. 745) the division of $30 - x$ by $x^2 + x$ in the following manner:

$$\frac{30.\ \tilde{m}.\ 1^1}{1^2\ \bar{p}.\ 1^1}\ .$$

As a further illustration, we give $R^2 1.\frac{1}{2}.\bar{p} R^2 24.\bar{p}.R^2 1.\frac{1}{2}.$ multiplied by $R^2 1.\frac{1}{2}\ \bar{p}\ R^2 24.\ \ \tilde{m} R^2 1.\frac{1}{2}.$ gives $R^2 24.$ This is really more compact and easier to print than our $\sqrt{1\frac{1}{2} + \sqrt{24}} + \sqrt{1\frac{1}{2}}$ *times* $\sqrt{1\frac{1}{2} + \sqrt{24}} - \sqrt{1\frac{1}{2}}$ *equals* $\sqrt{24}$.

FRENCH: ÉSTIENNE DE LA ROCHE
(1520)

132. Éstienne de la Roche, Villefranche, published *Larismethique,* at Lyon in 1520, which appeared again in a second edition at Lyon in 1538, under the revision of Gilles Huguetan. De la Roche mentions Chuquet in two passages, but really appropriates a great deal from his distinguished predecessor, without, however, fully entering into his spirit and adequately comprehending the work. It is to be regretted that Chuquet did not have in De la Roche an interpreter acting with sympathy and full understanding. De la Roche mentions the Italian Luca Pacioli.

De la Roche attracted little attention from writers antedating the nineteenth century; he is mentioned by the sixteenth-century French writers Buteo and Gosselin, and through Buteo by John Wallis. He employs the notation of Chuquet, intermixed in some cases, by other notations. He uses Chuquet's $\bar{p}$ and $\tilde{m}$ for *plus* and *moins*, also Chuquet's radical notation R^2, R^3, R^4, ... , but gives an alternative notation: $R\ \square$ for R^3, $R R$ for R^4, $R R\ \square$ for R^6. His strange uses of the geometric square are shown further by his writing $\square$ to indicate the cube of the unknown, an old procedure mentioned by Chuquet.

The following quotation is from the 1538 edition of De la Roche, where, as does Chuquet, he calls the unknown and its successive powers by the names of primary numbers, secondary numbers, etc.:

"... vng chascun nombre est considere comme quantité continue que aultrement on dit nombre linear qui peult etre appelle chose ou premier: et telz nombres seront notez apposition de une unite au dessus deulx en ceste maniere 12^1 ou 13^1, etc., ou telz nombres seront signes dung tel characte apres eux comme 12.ρ. ou 13.ρ. ... cubes que lon peut ainsi marquer $12.^3$ ou $13.^3$ et ainsi 12 □ ou 13 □."[1]

The translation is as follows:

"And a number may be considered as a continuous quantity, in other words, a linear number, which may be designated a thing or as primary, and such numbers are marked by the apposition of unity above them in this manner 12^1 or 13^1, etc., or such numbers are indicated also by a character after them, like 12.ρ. or 13.ρ. ... Cubes one

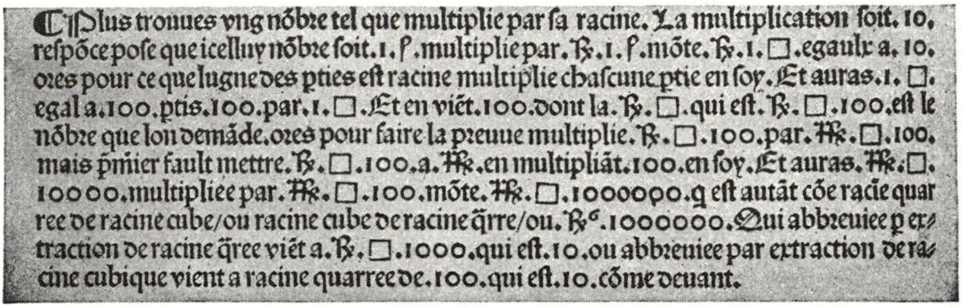
Plus trouues vng nõbre tel que multiplie par sa racine. La multiplication soit. 10. respõce pose que icelluy nõbre soit. 1. ρ. multiplie par. ℞. 1. ρ. mõte. ℞. 1. □. egaulx a. 10. ores pour ce que lugne des ptíes est racine multiplie chascune ptie en soy. Et auras. 1. □. egal a. 100. ptis. 100. par. 1. □. Et en viẽt. 100. dont la. ℞. □. qui est. ℞. □. 100. est le nõbre que lon demãde. ores pour faire la preuue multiplie. ℞. □. 100. par. ℞℞. □. 100. mais pmier fault mettre. ℞. □. 100. a. ℞℞. en multipliãt. 100. en soy. Et auras. ℞℞. □. 10000. multipliee par. ℞℞. □. 100. mõte. ℞℞. □. 1000000. q est autãt cõe racie quarree de racine cube/ou racine cube de racine q̃rre/ou. ℞⁶. 1000000. Qui abbreuiee p extraction de racine q̃ree viẽt a. ℞. □. 1000. qui est. 10. ou abbreuiee par extraction de racine cubique vient a racine quarree de. 100. qui est. 10. cõme deuant.

FIG. 40.—Part of fol. 60*B* of De la Roche's *Larismethique* of 1520

may mark $12.^3$ or $13.^3$ and also 12 □. or 13□." (We have here $12^1 = 12x$, $12.^3 = 12x^3$, etc.)

A free translation of the text shown in Figure 40 is as follows:

"Next find a number such that, multiplied by its root, the product is 10. Solution: Let the number be x. This multiplied by $\sqrt{x}$ gives $\sqrt{x^3} = 10$. Now, as one of the sides is a radical, multiply each side by itself. You obtain $x^3 = 100$. Solve. There results the cube root of 100, i.e., $\sqrt[3]{100}$ is the required number. Now, to prove this, multiply $\sqrt[3]{100}$ by $\sqrt[6]{100}$. But first express $\sqrt[3]{100}$ as $\sqrt[6]{}$, by multiplying 100 by itself, and you have $\sqrt[6]{10{,}000}$. This multiplied by $\sqrt[6]{100}$ gives $\sqrt[6]{1{,}000{,}000}$, which is the square root of the cube root, or the cube root of the square root, or $\sqrt[6]{1{,}000{,}000}$. Extracting the square root gives $\sqrt[3]{1{,}000}$ which is 10, or reducing by the extraction of the cube root gives the square root of 100, which is 10, as before."

[1] See an article by Terquem in the *Nouvelles annales de mathématiques* (Terquem et Gerono), Vol. VI (1847), p. 41, from which this quotation is taken. For extracts from the 1520 edition, see Boncompagni, *op. cit.*, Vol. XIV (1881), p. 423.

The end of the solution of the problem shown in Figure 41 is in modern symbols as follows:

$$\begin{array}{lll}
 & & \text{first } \sqrt{34}+7\,. \\
x \text{———} x+1 & & x+4 \\
\quad 2\tfrac{1}{2}x & & 1\tfrac{1}{2}x+4\tfrac{1}{2} \\
\hline
2\tfrac{1}{2}x^2+2\tfrac{1}{2}x \text{———} & & 1\tfrac{1}{2}x^2+10\tfrac{1}{2}x+18 \\
\quad x^2 \text{——————————} & & 8x+18 \\
 & & 4 \quad 16 \\
\hline
 & & 16 \quad \sqrt{34}+4 \text{ second.}
\end{array}$$

$$[\text{i.e., } x=\sqrt{34}+4]$$

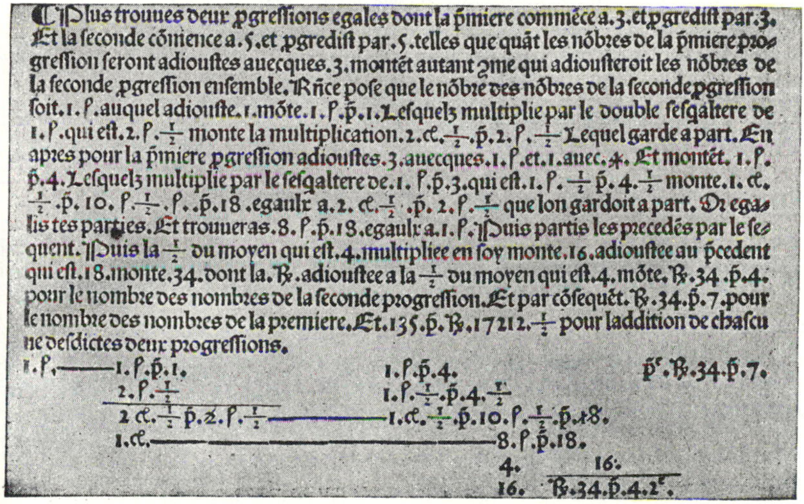
¶ Plus trouues deux pgressions egales dont la pmiere commẽce a.3.et pgredist par.3. Et la seconde cõmence a.5.et pgredist par.5.telles que quãt les nõbres de la pmiere progression seront adioustes auecques.3.montẽt autant cõme qui adiousteroit les nõbres de la seconde pgression ensemble. Rñce pose que le nõbre des nõbres de la seconde pgression soit.1.ρ.auquel adiouste.1.mõte.1.ρ.p̄.1.Lesquelz multiplie par le double sesqaltere de 1.ρ.qui est.2.ρ.½ monte la multiplication.2.ce.½.p̄.2.ρ.½ Lequel garde a part. En apres pour la pmiere pgression adioustes.3.auecques.1.ρ.et.1.auec.4. Et montẽt.1.ρ.p̄.4.Lesquelz multiplie par le sesqaltere de.1.ρ.p̄.3.qui est.1.ρ.½ p̄.4.½ monte.1.ce.½.p̄.10.ρ.½.ρ..p̄.18.egaulx a.2.ce.½.p̄.2.ρ.½ que lon gardoit a part. Or egalis tes parties. Et trouueras.8.ρ.p̄.18.egaulx a.1.ρ. Puis partis les precedẽs par le sequent. Puis la ½ du moyen qui est.4.multipliee en soy monte.16.adioustee au pcedent qui est.18.monte.34.dont la.℞.adioustee a la ½ du moyen qui est.4.mõte.℞.34.p̄.4. pour le nombre des nombres de la seconde progression. Et par cõsequẽt.℞.34.p̄.7.pour le nombre des nombres de la premiere. Et.135.p̄.℞.17212.½ pour laddition de chascune desdictes deux progressions.

1.ρ.——1.ρ.p̄.1. 1.ρ.p̄.4. p^e.℞.34.p̄.7.
2.ρ.½ 1.ρ.½.p̄.4.½
2 ce.½ p̄.2.ρ.½ ————1.ce.½.p̄.10.ρ.½.p̄.18.
1.ce.——————————8.ρ.p̄.18.
4. 16.
16. ℞.34.p̄.4.2^e.

FIG. 41.—Part of fol. 66 of De la Roche's *Larismethique* of 1520

ITALIAN: PIETRO BORGI (OR BORGHI)
(1484, 1488)

133. Pietro Borgi's *Arithmetica* was first printed in Venice in 1484; we use the edition of 1488. The book contains no algebra. It displays the scratch method of division and the use of dashes in operating with fractions (§§ 223, 278). We find in this early printed *Arithmetica* the use of curved lines in the solution of problems in alligation. Such graphic aids became frequent in the solution of the indeterminate problems of alligation, as presented in arithmetics. Pietro Borgi, on the unnumbered folio 79*B*, solves the following problem: Five sorts of spirits, worth per *ster*, respectively, 44, 48, 52, 60, 66 *soldi*,

are to be mixed so as to obtain 50 *ster*, worth each 56 *soldi*. He solves this by taking the qualities of wine in pairs, always one quality dearer and the other cheaper than the mixture, as indicated by the curves in the example.

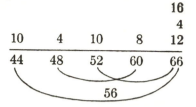

Then $56-44=12$; $66-56=10$; write 12 above 66 and 10 above 44. Proceed similarly with the pairs 48 and 60, 52 and 66. This done, add 10, 4, 10, 8, 16. Their sum is 48, but should be 50. Hence multiply each by $\frac{50}{48}$ and you obtain $10\frac{5}{12}$ as the number of *ster* of wine worth 44 *soldi* to be put into the mixture, etc.

ITALIAN: LUCA PACIOLI

(1494, 1523)

134. *Introduction.*—Luca Pacioli's *Summa de arithmetica geometria proportioni et proportionalita* (Venice, 1494)[1] is historically important because in the first half of the sixteenth century it served in Italy as the common introduction to mathematics and its influence extended to other European countries as well. The second edition (1523) is a posthumous publication and differs from the first edition

[1] Cosmo Gordon ("Books on Accountancy, 1494–1600," *Transactions of the Bibliographical Society* [London], Vol. XIII, p. 148) makes the following remarks on the edition of 1494: "The *Summa de arithmetica* occurs in two states. In the first the body of the text is printed in Proctor's type 8, a medium-sized gothic. On sig. *a* 1, on which the text begins, there is the broad wood-cut border and portrait-initial *L* already described. In the second state of the *Summa*, of which the copy in the British Museum is an example, not only do the wood-cut border and initial disappear from *a* 1, but sigs. *a–c* with the two outside leaves of sigs. *d* and *e*, and the outside leaf of sig. *a*, are printed in Proctor's type 10**, a type not observed by him in any other book from Paganino's press. There are no changes in the text of the reprinted pages, but that they are reprinted is clear from the fact that incorrect head-lines are usually corrected, and that the type of the remaining pages in copies which contain the reprints shows signs of longer use than in copies where the text type does not vary. It may be supposed that a certain number of the sheets of the signatures in question were accidentally destroyed, and that type 8 was already in use. The sheets had, therefore, to be supplied in the nearest available type." The copy of the 1494 edition in the Library of the University of California exhibits the type 10.

only in the spelling of some of the words. References to the number of the folio apply to both editions.

In the *Summa* the words "plus" and "minus," in Italian *più* and *meno*, are indicated by $\bar{p}$ and $\tilde{m}$. The unknown quantity was called "thing," in the Italian *cosa*, and from this word were derived in Germany and England the words *Coss* and "cossic art," which in the sixteenth and seventeenth centuries were synonymous with "algebra." As pointed out more fully later, *co.* (*cosa*) meant our *x; ce.* (*censo*) meant our x^2; *cu.* (*cubo*) meant our x^3. Pacioli used the letter $\mathcal{R}$ for *radix. Censo* is from the Latin *census* used by Leonardo of Pisa and Regiomontanus. Leonardo of Pisa used also the word *res* ("thing").

135. *Different uses of the symbol $\mathcal{R}$.*—The most common use of $\mathcal{R}$, the abbreviation for the word *radix* or *radici*, was to indicate roots. Pacioli employs for the same purpose the small letter $\curlywedge\gamma$, sometimes in the running text,[1] but more frequently when he is pressed for space in exhibiting algebraic processes on the margin.[2] He writes in Part I of his *Summa:*

(Fol. 70*B*)	$\mathcal{R}$.200. for $\sqrt{200}$
(Fol. 119*B*)	$\mathcal{R}$.*cuba. de* .64. for $\sqrt[3]{64}$
(Fol. 182*A*)[3]	$\mathcal{R}$.*relato.* for fifth root
(Fol. 182*A*)	$\mathcal{R}$ $\mathcal{R}$ $\mathcal{R}$. *cuba.* for seventh root
(Fol. 86*A*)	$\mathcal{R}$ .6.$\tilde{m}$.$\mathcal{R}$.2. for $\sqrt{6}-\sqrt{2}$
(Fol. 131*A*)	$\mathcal{R}$ $\mathcal{R}$.120. for $\sqrt[4]{120}$
(Fol. 182*A*)	$\mathcal{R}$. *cuba. de $\mathcal{R}$. cuba.* for sixth root
(Fol. 182*A*)	$\mathcal{R}$ $\mathcal{R}$. *cuba. de $\mathcal{R}$. cuba.* for eighth root.

The use of the $\mathcal{R}v$. for the designation of the roots of expressions containing two or more terms is shown in the following example:

(Fol. 149*A*) $\mathcal{R}v$. $\mathcal{R}$.20$\frac{1}{4}$.$\tilde{m}$.$\frac{1}{2}$. for $\sqrt{\sqrt{20\frac{1}{4}}-\frac{1}{2}}$.

The following are probably errors in the use of $\mathcal{R}v.$:

(Fol. 93*A*) $\mathcal{R}v$. 50000.$\tilde{m}$.200. for $\sqrt{50,000-200}$,

(Fol. 93*A*) R Rv. 50000.$\tilde{m}$.200. for $\sqrt{\sqrt{50,000-200}}$.

In combining symbols to express the higher roots, Pacioli uses the additive principle of Diophantus, while in expressing the higher powers

[1] Part I (1523), fol. 86 *A*.

[2] *Ibid.*, fol. 124 *A*.

[3] On the early uses of *radix relata* and *primo relato* see Eneström, *Bibliotheca mathematica*, Vol. XI (1910–11), p. 353.

he uses the multiplication principle of the Hindus. Thus Pacioli indicates the seventh root by $R̶$ $R̶$ $R̶$. *cuba*. (2+2+3), but the eighth power by *ce.ce.ce*. (2×2×2). For the fifth, seventh, and eleventh powers, which are indicated by prime numbers, the multiplication principle became inapplicable. In that case he followed the notation of wide prevalence at that time and later: $p°r°$ (*primo relato*) for the fifth power, $2°r°$ (*secundo relato*) for the seventh power, $3°r°$ (*terzo relato*) for the eleventh power.[1] Whenever the additive principle was used in marking powers or roots, these special symbols became superfluous. Curiously, Pacioli applies the additive principle in his notation for roots, yet does not write $R.R̶$ *cuba* (2+3) for the fifth root, but $R̶$. *relata*. However, the seventh root he writes R R $R̶$. *cuba* (2+2+3) and not $R2°r°.$[2]

136. In other parts of Pacioli's *Summa* the sign $R̶$ is assigned altogether different meanings. Apparently, his aim was to describe the various notations of his day, in order that readers might select the symbols which they happened to prefer. Referring to the prevailing diversity, he says, "tante terre: tante vsanze."[3] Some historians have noted only part of Pacioli's uses of $R̶$, while others have given a fuller account but have fallen into the fatal error of interpreting certain powers as being roots. Thus far no one has explained all the uses of the sign $R̶$ in Pacioli's *Summa*. It was Julius Rey Pastor and Gustav Eneström who briefly pointed out an inaccuracy in Moritz Cantor, when he states that Pacioli indicated by $R̶$ 30 the thirtieth *root*, when Pacioli really designated by $R̶$.$30°$ the twenty-ninth *power*. This point is correctly explained by J. Tropfke.[4]

We premise that Pacioli describes two notations for representing powers of an unknown, x^2, x^3, ... , and three notations for x. The one most commonly used by him and by several later Italian writers of the sixteenth century employs for x, x^2, x^3, x^4, x^5, x^6, x^7, ... , the abbreviations *co.* (*cosa*), *ce.* (*censo*), *cu.* (*cubo*), *ce.ce.* (*censo de censo*), $p°r°$ (*primo relato*), *ce.cu.* (*censo de cubo*), $2°r°$ (*secundo relato*), ...[5]

Pacioli's second notation for powers involves the use of $R̶$, as already indicated. He gives: $R̶.p^a$ (*radix prima*) for x^0, $R̶.2^a$ (*radix secunda*) for x, $R̶.3^a$ (*radix terza*) for x^2, ... , $R̶.30^a$ (*nono relato*) for x^{29}.[6] When Eneström asserts that folio 67B deals, not with roots,

[1] Part I, fol. 67*B*.

[2] *Ibid.*, fol. 182*A*.

[3] *Ibid.*, fol. 67*B*.   [4] *Op. cit.* (2d ed.), **Vol. II (1921), p. 109.**

[5] *Op. cit.*, Part I, fol. 67*B*.   [6] *Ibid.*

but exclusively with the powers x^0, x, x^2, ... , x^{29}, he is not quite accurate, for besides the foregoing symbols placed on the margin of the page, he gives on the margin also the following: "Rx. Radici; R R. Radici de Radici; Rv. Radici vniuersale. Ouer radici legata. O voi dire radici vnita; R. cu. Radici cuba; $\tilde{q}\beta^a$ quantita." These expressions are used by Pacioli in dealing with roots as well as with powers, except that Rv. is employed with roots only; as we have seen, it signifies the root of a binomial or polynomial. In the foregoing two uses of R, how did Pacioli distinguish between roots and powers? The ordinal number, *prima*, *secunda*, *terza*, etc., placed after the R, always signifies a "power," or a *dignita*. If a root was intended, the number affected was written after the R; for example, R.200. for $\sqrt{200}$. In folio 143AB Pacioli dwells more fully on the use of R in the designation of powers and explains the multiplication of such expressions as R. 5^a *via*. R. 11^a *fa* R. 15^a, i.e., $x^4 \times x^{10} = x^{14}$. In this notation one looks in vain for indications of the exponential concepts and recognition of the simple formula $a^m \cdot a^n = a^{m+n}$. Pacioli's results are in accordance with the formula $a^m \cdot a^n = a^{m+n-1}$. The ordinal numbers in R 11^a, etc., exceed by unity the power they represent. This clumsy designation made it seem necessary to Pacioli to prepare a table of products, occupying one and one-half pages, and containing over two hundred and sixty entries; the tables give the various combinations of factors whose products do not exceed x^{29}. While Eneström and Rey Pastor have pointed out that expressions like R.28^a mark powers and not roots, they have failed to observe that Pacioli makes no use whatever of this curious notation in the working of problems. Apparently his aim in inserting it was encyclopedial.

137. In working examples in the second part of the *Summa*, Pacioli exhibits a third use of the sign R not previously noted by historians. There R is used to indicate powers of numbers, but in a manner different from the notation just explained. We quote from the *Summa* a passage[1] in which R refers to powers as well as to roots. Which is meant appears from the mode of phrasing: "... R.108. e questo mca con laxis ch' R.16. fa. R.1728 piglià el .$\frac{1}{3}$. cioe recca .3. a. R. fa .9. parti .1728 in. 9. neuien. 192. e. R.192. . . . ." ($\therefore$ $\sqrt{108}$ and multiplying this with the axis which is $\sqrt{16}$ gives $\sqrt{1{,}728}$. Take $\frac{1}{3}$, i.e., raising 3 to the second power gives 9; dividing 1,728 by 9 gives 192, and the $\sqrt{192}$. . . . .) Here "recca. 3. a. R. fa. 9." identifies R with a power. In Part I, folio 186A, one reads, "quando fia recata prima. 1.

[1] *Ibid.*, Part II, fol. 72 *B*.

co. a. R. fa. 1. ce" ("raising the x to the second power gives x^2"). Such phrases are frequent as, Part II, folio 72B, "reca. 2. a. R. cu. fa. 8" ("raise 2 to the third power; it gives 8"). Observe that $R.$ $cu.$ means the "third" power, while[1] R. 3^a and $R.$ $terza.$ refer to the "second" power. The expression of powers by the Diophantine additive plan $(2+3)$ is exhibited in "reca. 3. a. R R. cuba fa. 729" ("raise 3 to the fifth power; it gives 729").[2]

A fourth use of R is to mark the unknown x. We have previously noted Pacioli's designation of x by co. ($cosa$) and by R. 2^a. In Part II, folio 15B, he gives another way: "la mita dun censo e .12. dramme: sonno equali a .5.R. E questo cõme a dire .10. radici sonno equali a vn censo e. 24. dramme" ("Half of x^2 and the number 12 are equal to $5x$; and this amounts to saying $10x$ are equal to x^2 and the number 24").

In Part I, folio 60B, the sign R appears on the margin twice in a fifth rôle, namely, as the abbreviation for $rotto$ ("fraction"), but this use is isolated. From what we have stated it is evident that Pacioli employed R in five different ways; the reader was obliged to watch his step, not to get into entanglements.

138. *Sign of equality.*—Another point not previously noted by historians is that Pacioli used the dash (—) as a symbol for equality. In Part I, folio 91A, he gives on the margin algebraic expressions relating to a problem that is fully explained in the body of the page. We copy the marginal notes and give the modern equivalents:

	Summa (Part I, fol. 91A)		Modern Equivalents	
$\bar{p}^a$	1. *co.* $\tilde{m}$. 1. $\cancel{\beta}^a$	1st	$x - y$
3^a	1. *co.* $\bar{p}$. 1. $\bar{\cancel{\beta}}^a$	3d	$x + y$

1. *co.* $\tilde{m}$. 1. *ce. de.* $\bar{\cancel{\beta}}^a$	36		$x^2 - y^2 = 36$

$Rv.$ 1. *ce.* $\tilde{m}$ 36 ___ 1. *ce.* $\bar{de}$ $\cancel{\beta}^a$ $\sqrt{x^2 - 36} = y$,

Valor quantitatis.   the value of y .

p^a	1. *co.* $\tilde{m}$ $Rv.$ 1. *ce.* $\tilde{m}$ 36	1st	$x - \sqrt{x^2 - 36}$
2^a	6	2d	6
3^a	1. *co.* $\bar{p}$ $Rv.$ 1. *ce.* $\tilde{m}$ 36 .	3d	$x + \sqrt{x^2 - 36}$

2. *co.* $\bar{p}$. 6.——216		$2x + 6$	$= 216$
2. *co.* ___ 210		$2x$	$= 210$
Valor rei. 105		Value of x	105

[1] Part I, fol. 67B.   [2] Part II, fol. 72B.

Notice that the *co.* in the third expression should be *ce.*, and that the .1. *ce. de* $\bar{\bar{\gamma}}\beta^a$ in the fourth expression should be .1. *co. de* $\bar{\bar{\gamma}}\beta^a$. Here, the short lines or dashes express equality. Against the validity of this interpretation it may be argued that Pacioli uses the dash for several different purposes. The long lines above are drawn to separate the sum or product from the parts which are added or multiplied. The short line or dash occurs merely as a separator in expressions like

in Part I, folio 39A. The dash is used in Part I, folio 54 B, to indicate multiplication, as in

where the dash between 5 and 7 expresses 5×7, one slanting line means 2×7, the other slanting line 5×3. In Part II, folio 37*A*, the dash represents some line in a geometrical figure; thus $d\underline{\quad 3 \quad}k$ means that the line *dk* in a complicated figure is 3 units long. The fact that Pacioli uses the dash for several distinct purposes does not invalidate the statement that one of those purposes was to express equality. This interpretation establishes continuity of notation between writers preceding and following Pacioli. Regiomontanus,[1] in his correspondence with Giovanni Bianchini and others, sometimes used a dash for equality. After Pacioli, Francesco Ghaligai, in his *Pratica d'arithmetica*, used the dash for the same purpose. Professor E. Bortolotti informs me that a manuscript in the Library of the University of Bologna, probaby written between 1550 and 1568, contains two parallel dashes (=) as a symbol of equality. The use of two dashes was prompted, no doubt, by the desire to remove ambiguity arising from the different interpretations of the single dash.

Notice in Figure 42 the word *cosa* for the unknown number, and its abbreviation, *co.; censo* for the square of the unknown, and its contraction, *ce.; cubo* for the cube of the unknown; also *.$\tilde{p}$.* for "plus" and *.$\tilde{m}$.* for "minus." The explanation given here of the use of *cosa, censo, cubo*, is not without interest.

[1] See Maximilian Curtze, *Urkunden zur Geschichte der Mathematik im Mittelalter und der Renaissance* (Leipzig, 1902), p. 278.

The first part of the extract shown in Figure 43 gives $\sqrt{\sqrt{40}+6}+\sqrt{\sqrt{40}-6}$ and the squaring of it. The second part gives $\sqrt{\sqrt{20}+2}+\sqrt{\sqrt{20}-2}$ and the squaring of it; the simplified result is given as $\sqrt{80}+4$, but it should be $\sqrt{80}+8$. Remarkable in this second example is the omission of the *v* to express *vniversale.* From the computation as well as from the explanation of the text it appears that the first ℞ was intended to express universal root, i.e., $\sqrt{\sqrt{20}+2}$ and not $\sqrt[4]{20}+2$.

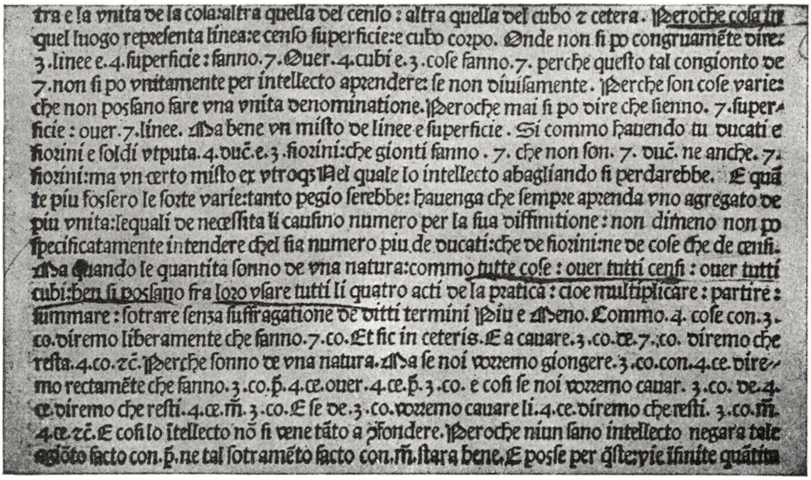
tra e la vnita de la cosa: altra quella del censo: altra quella del cubo z cetera. Peroche cosa in
quel luogo representa linea: e censo superficie: e cubo corpo. Onde non si po congruamēte dire:
3. linee e. 4. superficie: fanno. 7. Ouer. 4. cubi e. 3. cose fanno. 7. perche questo tal congionto de
7. non si po vnitamente per intellecto aprendere: se non diuisamente. Perche son cose varie:
che non posano fare vna vnita denominatione. Peroche mai si po dire che sienno. 7. superficie: ouer. 7. linee. Ma bene vn misto de linee e superficie. Si commo hauendo tu ducati e
fiorini e soldi vtputa. 4. duc̄. e. 3. fiorini: che gionti fanno. 7. che non son. 7. duc̄. ne anche. 7.
fiorini: ma vn certo misto ex vtroq3 Nel quale lo intellecto abagliando si perdarebbe. E quā
te piu fosero le sorte varie: tanto pegio serebbe: hauenga che sempre aprenda vno agregato de
piu vnita: lequali de necessita li causino numero per la sua diffinitione: non dimeno non po
specificatamente intendere chel sia numero piu de ducati: che de fiorini: ne de cose che de censi.
Ma quando le quantita sonno de vna natura: commo tutte cose: ouer tutti censi: ouer tutti
cubi: ben si posano fra loro vsare tutti li quatro acti de la pratica: cioe multiplicare: partire:
summare: sotrare senza suffragatione de ditti termini Piu e Meno. Commo. 4. cose con. 3.
co. diremo liberamente che fanno. 7. co. Et sic in ceteris. E a cauare. 3. co. de. 7. co. diremo che
resta. 4. co. zc̄. Perche sonno de vna natura. Ma se noi voremo giongere. 3. co. con. 4. ce. diremo rectamēte che fanno. 3. co. p̄. 4. ce. ouer. 4. ce. p̄. 3. co. e cosi se noi voremo cauar. 3. co. de. 4.
ce. diremo che resti. 4. ce. m̄. 3. co. E se de. 3. co. voremo cauare li. 4. ce. diremo che resti. 3. co. m̄.
4. ce. zc̄. E cosi lo itellecto nō si vene tāto a confondere. Peroche niun sano intellecto negara tale
agiōto facto con. p̄. ne tal sotramēto facto con. m̄. stara bene. E posse per queste vie īfinite quātita

FIG. 42.—Part of a page in Luca Pacioli's *Summa*, Part I (1523), fol. 112*A*

ITALIAN: F. GHALIGAI

(1521, 1548, 1552)

139. Ghaligai's *Pratica d'arithmetica*[1] appeared in earlier editions, which we have not seen, in 1521 and 1548. The three editions do not differ from one another according to Riccardi's *Biblioteca matematica italiana* (I, 500–502). Ghaligai writes (fol. 71*B*): $x = cosa = c^{\circ}$, $x^2 = censo =$, $x^3 = cubo =$, $x^5 = relato =$, $x^7 = pronico =$ 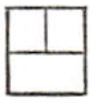 , $x^{11} = tronico =$ 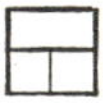, $x^{13} = dromico =$. He uses the m° for "minus" and the $\tilde{p}$ and *e* for "plus," but frequently writes in full *piu* and *meno.*

[1] *Pratica d'arithmetica di Francesco Ghaligai Fiorentino* (Nuouamente Riuista, & con somma Diligenza Ristampata. In Firenze. M.D.LII).

Equality is expressed by dashes (— — —); a single dash (—) is used also to separate factors. The repetition of a symbol, simply to fill up an interval, is found much later also in connection with the sign of equality (=). Thus, John Wallis, in his *Mathesis universalis* ([Oxford, 1657], p. 104) writes: $1+2-3= = =0$.

Ghaligai does not claim these symbols as his invention, but ascribes them to his teacher, Giovanni del Sodo, in the statement (folio 71*B*): "Dimostratione di 8 figure, le quale Giovanni del Sodo pratica la sua Arciba & perche in parte terro 'el suo stile le dimostreto.' "[1] The page shown (Fig. 45) contains the closing part of the

[1] *Op. cit.* (1552), fols. 2*B*, 65; Eneström, *Bibliotheca mathematica*, 3. S., Vol. VIII, 1907–8, p. 96.

solution of the problem to find three numbers, P, S, T, in continued proportion, such that $S^2 = P + T$, and, each number being multiplied

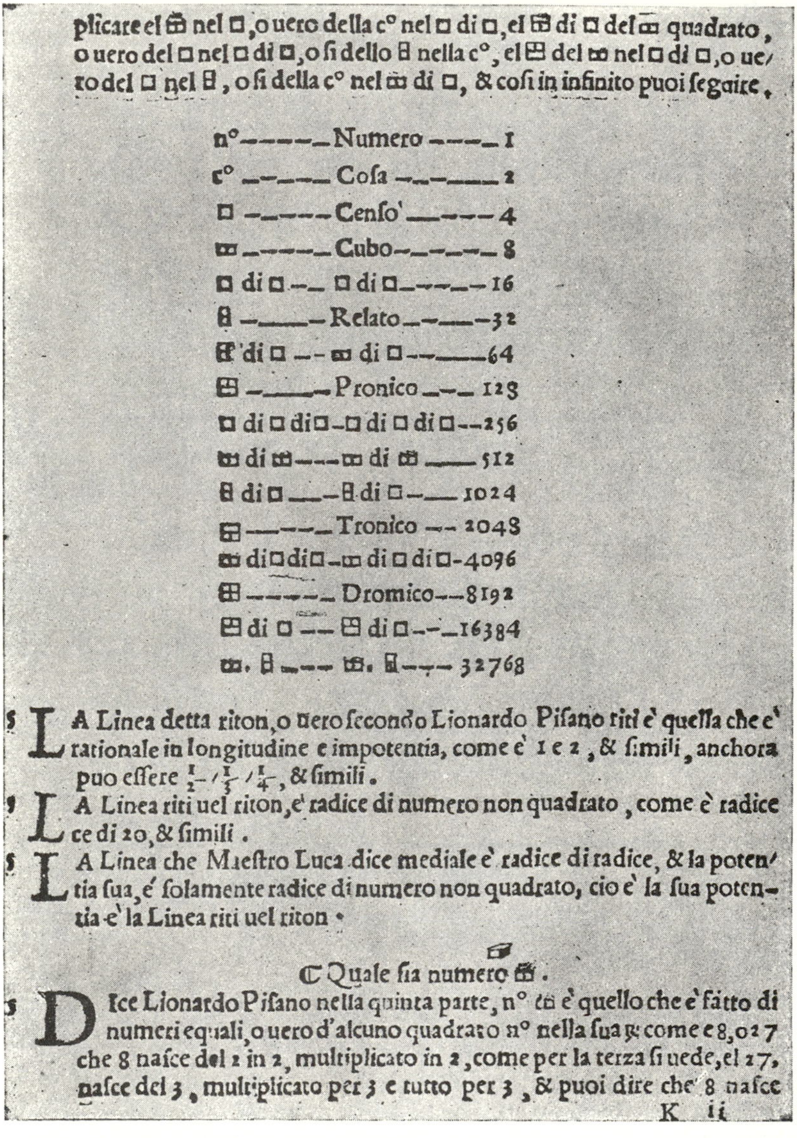

plicare el ⊞ nel □, o uero della c° nel □ di □, el ⊞ di □ del ⊠ quadrato, o uero del □ nel □ di □, o fi dello ⊟ nella c°, el ⊞ del ⊠ nel □ di □, o ue/ ro del □ nel ⊟, o fi della c° nel ⊠ di □, & cofi in infinito puoi feguire.

n°	Numero	1
c°	Cofa	2
□	Cenfo'	4
⊠	Cubo	8
□ di □	□ di □	16
⊟	Relato	32
⊞ di □	⊠ di □	64
⊞	Pronico	128
□ di □ di □	□ di □ di □	256
⊠ di ⊠	⊠ di ⊠	512
⊟ di □	⊟ di □	1024
⊞	Tronico	2048
⊠ di □ di □	⊠ di □ di □	4096
⊞	Dromico	8192
⊞ di □	⊞ di □	16384
⊠. ⊟	⊠. ⊟	32768

§ **L**A Linea detta riton, o uero fecondo Lionardo Pifano riti è quella che è rationale in longitudine e impotentia, come è 1 e 2, & fimili, anchora puo effere $\frac{1}{2}$, $\frac{1}{3}$, $\frac{1}{4}$, & fimili.

¶ **L**A Linea riti uel riton, è radice di numero non quadrato, come è radice ce di 20, & fimili.

§ **L**A Linea che Maeftro Luca dice mediale è radice di radice, & la potentia fua, è folamente radice di numero non quadrato, cio è la fua potentia è la Linea riti uel riton.

☞ ¶ Quale fia numero ⊞.

§ **D**Ice Lionardo Pifano nella quinta parte, n° ⊠ è quello che è fatto di numeri equali, o uero d'alcuno quadrato n° nella fua ℞ come e 8,027 che 8 nafce del 2 in 2, multiplicato in 2, come per la terza fi uede, el 27, nafce del 3, multiplicato per 3 e tutto per 3, & puoi dire che 8 nafce

K ii

FIG. 44.—Part of fol. 72 of Ghaligai's *Pratica d'arithmetica* (1552). This exhibits more fully his designation of powers.

by the sum of the other two, the sum of these products is equal to twice the second number multiplied by the sum of the other two, plus 72. Ghaligai lets $S = 3co$ or $3x$. He has found $x = 2$, and the root of x^2 equal to $\sqrt{4}$.

The translation of the text in Figure 45 is as follows: "equal to $\sqrt{4}$, and the $\sqrt{x^4}$ is equal to $\sqrt{16}$, hence the first quantity was $18 - \sqrt{288}$, and the second was 6, and the third $18 + \sqrt{288}$.

$S. \quad 3x \quad P. \text{ and } T. \quad 9x^2$ $\qquad\qquad 18x^2 + 6x, \times 3x$.

$P. \quad 4\frac{1}{2}x^2 - \sqrt{20\frac{1}{4}x^4 - 9x^2}$

$T. \quad 4\frac{1}{2}x^2 + \sqrt{20\frac{1}{4}x^4 - 9x^2}$, $\qquad\qquad = 54x^3 + 18x^2 = 54x^3 + 72$

$S. \quad 3x$ $\qquad\qquad\qquad\qquad\qquad\qquad 18x^2 = 72$

$\qquad P. \quad 4\frac{1}{2}x^2 - \sqrt{20\frac{1}{4}x^4 - 9x^2}$ $\qquad\qquad \sqrt{4}$

$= 9x^2 + 3x, \times 2$

$\qquad\qquad \backslash - 4 - - - /$ $\qquad\qquad\qquad$ Value of x which is 2

$\qquad\qquad 18 \quad \sqrt{324}$ $\qquad\qquad P. \text{ was } 18 - \sqrt{288}$

$\qquad\qquad\qquad 36$ $\qquad\qquad\qquad S. \text{ was } 6$

$\qquad\qquad\qquad \overline{\sqrt{288}}$ $\qquad\qquad T. \text{ was } 18 + \sqrt{288}$

Proof

$24 + \sqrt{288}$ $\qquad\qquad\qquad 24 - \sqrt{288}$

$18 - \sqrt{288}$ $\qquad\qquad\qquad 18 + \sqrt{288}$

$432 + \sqrt{93,312} - 288$ $\qquad 432 + \sqrt{165,888} - 288$

$288 - \sqrt{165,888}$ $\qquad\qquad 288 - \sqrt{93,312}$

144 $\qquad\qquad\qquad\qquad\quad 144$

$\qquad\qquad 144 + \sqrt{93,312}$ $\qquad\qquad\qquad 18 - \sqrt{288}$

$\qquad\qquad\quad - \sqrt{165,888}$ $\qquad\qquad\qquad 18 + \sqrt{288}$

$\qquad\qquad\quad + \sqrt{165,888}$

$\qquad\qquad 144 - \sqrt{93,312}$ $\qquad\qquad\qquad = 36, \times 6$

$\qquad\qquad \text{Gives } 288$ $\qquad\qquad\qquad\qquad 216, \times 2$

$\qquad\qquad\qquad 216$

$\qquad\qquad\qquad\overline{\quad}$ $\qquad\qquad\qquad\qquad\qquad 432$

$\qquad\qquad \text{Gives } 504$ $\qquad\qquad\qquad\qquad\quad 72$

As it should 504."

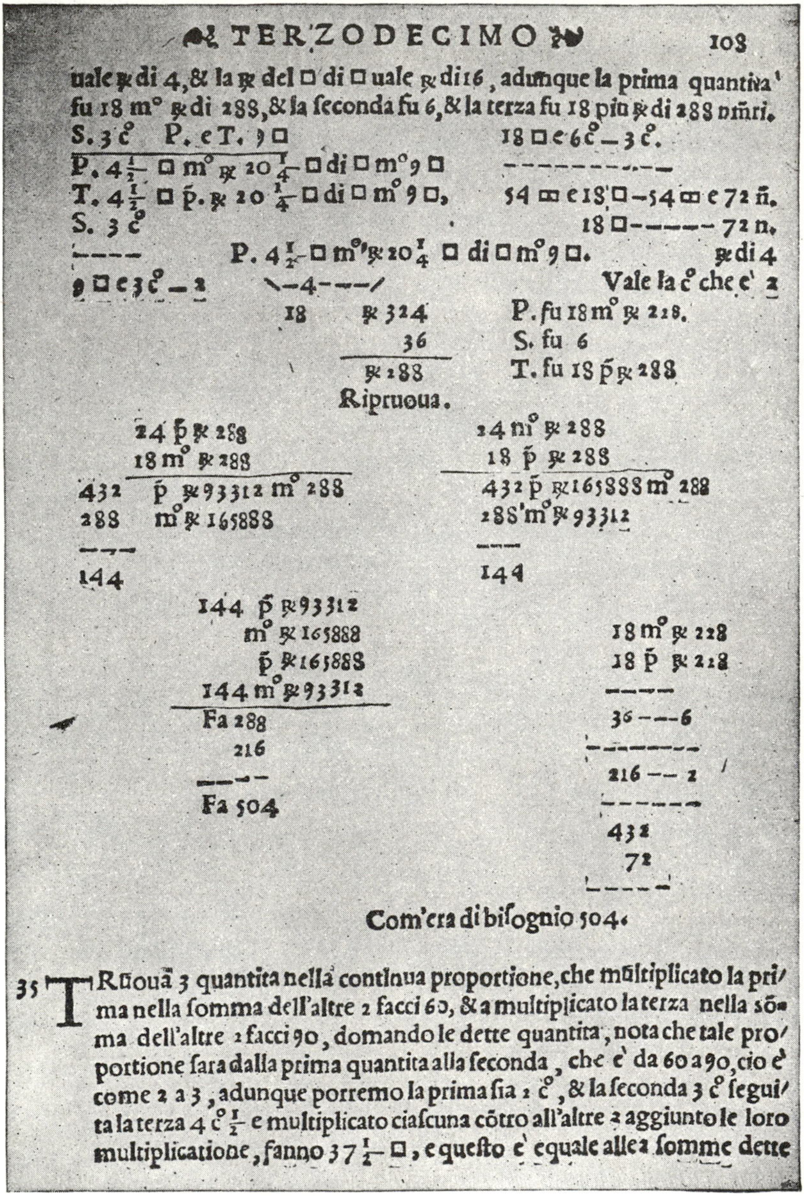

ɑ̨ TERZODECIMO ⅔̨   103

uale ℞ di 4,& la ℞ del □ di □ uale ℞ di 16 , adunque la prima quantita'
fu 18 m° ℞ di 288,& la seconda fu 6,& la terza fu 18 piu ℞ di 288 nmri.

S. 3 c̃   P. e T. 9 □   18 □ e 6 c̃ — 3 c̃.

P. 4½ □ m° ℞ 20¼ □ di □ m° 9 □   — — — — — — —

T. 4½ □ p̃. ℞ 20¼ □ di □ m° 9 □,   54 ⫿ e 18 □ — 54 ⫿ e 72 ñ.

S. 3 c̃   18 □ — — — — 72 n,

— — — —   P. 4½ □ m° ℞ 20¼ □ di □ m° 9 □.   ℞ di 4

9 □ e 3 c̃ — 2   \— 4 — —/   Vale la c̃ che e' 2

                    18   ℞ 324   P. fu 18 m° ℞ 218.

                            36   S. fu 6

                          ℞ 288   T. fu 18 p̃ ℞ 288

                        Ripruoua.

24 p̃ ℞ 288   24 m̃ ℞ 288
18 m° ℞ 288   18 p̃ ℞ 288

432   p̃ ℞ 93312 m° 288   432 p̃ ℞ 165888 m° 288
288   m° ℞ 165888   288 m° ℞ 93312

— — —   — — —

144   144

        144 p̃ ℞ 93312
            m° ℞ 165888
            p̃ ℞ 165888   18 m° ℞ 228
        144 m° ℞ 93312   18 p̃ ℞ 228

        Fa 288   — — — —
        216   36 — — 6

        — — —   — — — —
        Fa 504   216 — — 2

                                                 — — — —
                                                 432
                                                 72
                                                 — — — —

                Com'era di bisognio 504.

35 T Ruouã 3 quantita nella continua proportione, che multiplicato la pri-
      ma nella somma dell'altre 2 facci 60, & a multiplicato la terza nella sõ-
      ma dell'altre 2 facci 90, domando le dette quantita, nota che tale pro-
      portione sara dalla prima quantita alla seconda , che e' da 60 a 90, cio e'
      come 2 a 3 , adunque porremo la prima sia 2 c̃ , & la seconda 3 c̃ segui-
      ta la terza 4 c̃ ½ e multiplicato ciascuna cõtro all'altre 2 aggiunto le loro
      multiplicatione , fanno 37 ½ □ , e questo e' equale alle 2 somme dette

FIG. 45.—Ghaligai's *Pratica d'arithmetica* (1552), fol. 108

The following equations are taken from the same edition of 1552:

Translation

(Folio 110) $\frac{1}{4}$ □ di □ $\tilde{m}$ $\frac{1}{4}$ di □ — 1 □ $\frac{1}{4}x^4 - \frac{1}{4}x^2 = x^2$

$\frac{1}{4}$ □ di □ — $1\frac{1}{4}$ di □ $\frac{1}{4}x^4 = 1\frac{1}{4}x^2$

$\frac{1}{4}$ □ — — — — — $1\frac{1}{4}$ $\tilde{n}$ $\frac{1}{4}x^2 = 1\frac{1}{4}$

(Folio 113) $\frac{1}{4}$ □ □ $\mathring{m}$ 4 □ — — — 4 □ $\frac{1}{4}x^4 - 4x^2 = 4x^2$

$\frac{1}{4}$ □ □ — — 8 □ $\frac{1}{4}x^4 = 8x^2$

Ghaligai uses his combinations of little squares to mark the orders of roots. Thus, folio 84B, R □ di 3600 — che $è$ 60, i.e., $\sqrt{3,600}=60$; folio 72B, la R ⬜⬜ di 8 $diciamo$ 2, i.e., $\sqrt[3]{8}=2$; folio 73B, R ⊟ di 7776 for $\sqrt[5]{7,776}$; folio 73B, R ⬜⬜ di □ di 262144 for $\sqrt[6]{262,144}$.

ITALIAN: HIERONYMO CARDAN
(1539, 1545, 1570)

140. Cardan uses $\tilde{p}$ and $\tilde{m}$ for "plus" and "minus" and R for "root." In his *Practica arithmeticae generalis* (Milano, 1539) he uses Pacioli's symbols *nu.*, *co.*, *ce.*, *cu.*, and denotes the successive higher powers, *ce.ce.*, *Rel. p.*, *cu.ce.*, *Rel. 2.*, *ce.ce.ce.*, *cu.cu.*, *ce. Rel.*[1] However, in his *Ars magna* (1545) Cardan does not use *co.* for x, *ce.* for x^2, etc., but speaks of "rem ignotam, quam vocamus positionem,"[2] and writes $60+20x=100$ thus: "60. $\tilde{p}$. 20. positionibus aequalia 100." Farther on[3] he writes $x^2+2x=48$ in the form "1. quad. $\tilde{p}$. 2. pos. aeq. 48.," x^4 in the form[4] "1. quadr. quad.," $x^5+6x^3=80$ in the form[5] "r. p^m $\tilde{p}$. 6. cub.   80," $x^5=7x^2+4$ in the form[6] "r^mp^m 7. quad. $\tilde{p}$. 4." Observe that in the last two equations there is a blank space where we write the sign of equality (=). These equations appear in the text in separate lines; in the explanatory text is given *aequale* or *aequatur*. For the representation of a second unknown he follows Pacioli in using the word *quantitas*, which he abbreviates to *quan.* or *qua.* Thus[7] he writes $7x+3y=122$ in the form "7. pos. $\tilde{p}$. 3. qua. aequal. 122."

Attention should be called to the fact that in place of the $\tilde{p}$ and $\tilde{m}$, given in Cardan's *Opera*, Volume IV (printed in 1663), one finds in Cardan's original publication of the *Ars magna* (1545) the signs p:

---

[1] *Hieronymi Cardani operum tomvs quartvs* (Lvgdvni, 1663), p. 14.

[2] *Ibid.*, p. 227.

[3] *Ibid.*, p. 231.   [5] *Ibid.*

[4] *Ibid.*, p. 237.   [6] *Ibid.*, p. 239.

[7] *Ars magna* in *Operum tomvs quartvs*, p. 241, 242.

and $m:$. For example, in 1545 one finds $(5+\sqrt{-15})\ (5-\sqrt{-15})=25-(-15)=40$ printed in this form:

$$\text{`` }5p:\ \cancel{R}\ m:\ 15$$
$$5m:\ \cancel{R}\ m:\ 15$$
$$\overline{25m:m:\ 15\ \tilde{q}d\ est\ 40\ ,\text{''}}$$

while in 1663 the same passage appears in the form:

$$\text{`` }5.\ \tilde{p}.\ \cancel{R}.\ \tilde{m}.\ 15.$$
$$5.\ \tilde{m}.\ \cancel{R}.\ m.\ 15.$$
$$\overline{25.\ m.\ m.\ 15.\ quad.\ est\ 40.\ \text{''}}{}^{1}$$

141. Cardan uses $\cancel{R}$ to mark square root. He employs[2] Pacioli's *radix vniversalis* to binomials and polynomials, thus "$R.V.7.\ \tilde{p}\ \cancel{R}.\ 4.$ vel sic $(\cancel{R})\ 13.\tilde{p}\ \cancel{R}.\ 9.$" for $\sqrt{7+\sqrt{4}}$ or $\sqrt{13+\sqrt{9}}$; "$\cancel{R}.V.10.p.\cancel{R}.16.p.3.p\ \cancel{R}.64.$" for $\sqrt{10+\sqrt{16}+3+\sqrt{64}}$. Cardan proceeds to new notations. He introduces the *radix ligata* to express the roots of each of the terms of a binomial; he writes: "$L\cancel{R}.\ 7.\ \tilde{p}R.\ 10.$"[3] for $\sqrt{7}+\sqrt{10}$. This L would seem superfluous, but was introduced to distinguish between the foregoing form and the *radix distincta*, as in "$\cancel{R}.D.\ 9\ p.\ \cancel{R}.\ 4.$," which signified 3 and 2 taken separately. Accordingly, "$\cancel{R}.D.\ 4.\ p.\ \cancel{R}.\ 9.$," multiplied into itself, gives $4+9$ or 13, while the "$R.L.\ 4.\ p.\ R.\ 9.$," multiplied into itself, gives $13+\sqrt{144}=25$. In later passages Cardan seldom uses the *radix ligata* and *radix distincta*.

In squaring binomials involving radicals, like "$\cancel{R}.V.L.\ \cancel{R}.\ 5.\ \tilde{p}.\ \cancel{R}.\ 1.\ \tilde{m}\ \cancel{R}.V.L.\ \cancel{R}.\ 5.\ \tilde{m}\ \cancel{R}.\ 1.$," he sometimes writes the binomial a second time, beneath the first, with the capital letter X between the two binomials, to indicate cross-multiplication.[4] Of interest is the following passage in the *Regula aliza* which Cardan brought out in 1570: "$\cancel{R}p:$ est $p:\ \cancel{R}\ m:$ quadrata nulla est iuxta usum communem" ("The square root of a positive number is positive; the square root of a negative number is not proper, according to the common acceptation").[5]

[1] See Tropfke, *op. cit.*, Vol. III (1922), p. 134, 135.

[2] Cardan, *op. cit.*, p. 14, 16, of the *Practica arithmeticae* of 1539.

[3] *Ibid.*, p. 16.

[4] *Ibid.*, p. 194.

[5] *Op. cit.* (Basel, 1570), p. 15. Reference taken from Eneström, *Bibliotheca mathematica*, Vol. XIII (1912–13), p. 163.

However, in the *Ars magna*[1] Cardan solves the problem, to divide 10 into two parts, whose product is 40, and writes (as shown above):

" 5. $\bar{p}$ ℞. $\tilde{m}$. 15.
5. $\tilde{m}$ ℞. m.. 15.

25 *m.m.*. 15. *quad. est* 40 . "

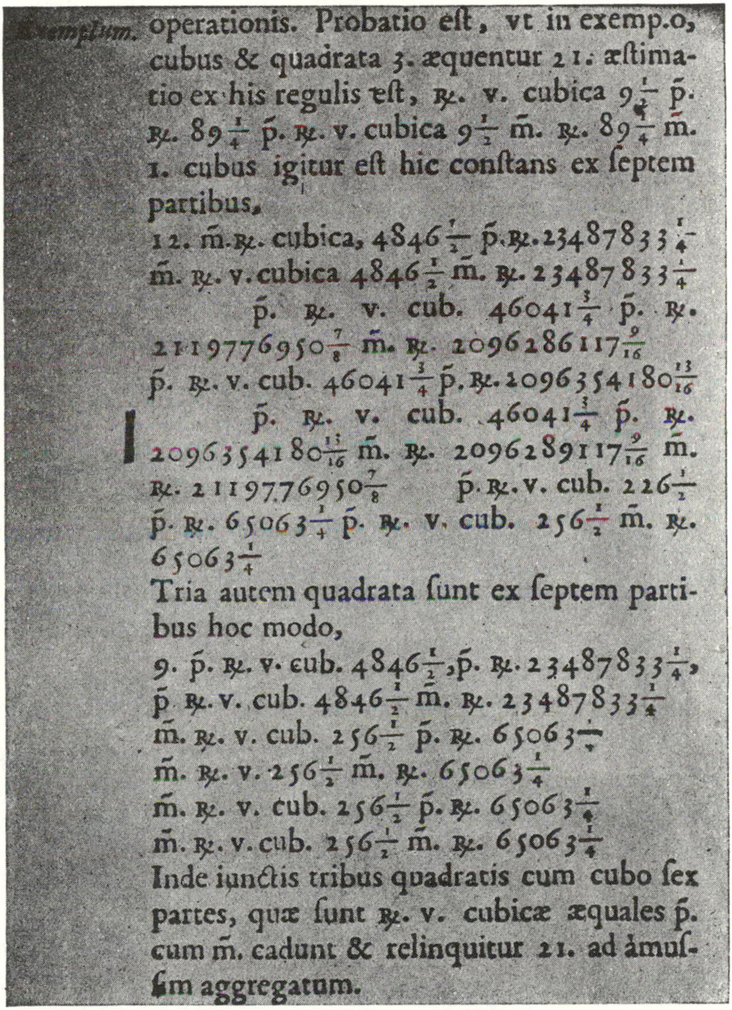

exemplum. operationis. Probatio eſt, vt in exemp.o, cubus & quadrata 3. æquentur 21. æſtima-tio ex his regulis eſt, ℞. v. cubica $9\frac{1}{4}$ $\bar{p}$. ℞. $89\frac{1}{4}$ $\bar{p}$. ℞. v. cubica $9\frac{1}{2}$ $\tilde{m}$. ℞. $89\frac{1}{4}$ $\tilde{m}$. 1. cubus igitur eſt hic conſtans ex ſeptem partibus.

12. $\tilde{m}$. ℞. cubica, $4846\frac{1}{2}$ $\bar{p}$. ℞. $23487833\frac{1}{4}$. $\tilde{m}$. ℞. v. cubica $4846\frac{1}{2}$ $\tilde{m}$. ℞. $23487833\frac{1}{4}$

$\bar{p}$. ℞. v. cub. $46041\frac{3}{4}$ $\bar{p}$. ℞. $2119776950\frac{7}{8}$ $\tilde{m}$. ℞. $2096286117\frac{9}{16}$
$\bar{p}$. ℞. v. cub. $46041\frac{3}{4}$ $\bar{p}$. ℞. $2096354180\frac{13}{16}$

$\bar{p}$. ℞. v. cub. $46041\frac{3}{4}$ $\bar{p}$. ℞. $2096354180\frac{13}{16}$ $\tilde{m}$. ℞. $2096289117\frac{9}{16}$ $\tilde{m}$. ℞. $2119776950\frac{7}{8}$ $\bar{p}$. ℞. v. cub. $226\frac{1}{4}$ $\bar{p}$. ℞. $65063\frac{1}{4}$ $\bar{p}$. ℞. v. cub. $256\frac{1}{2}$ $\tilde{m}$. ℞. $65063\frac{1}{4}$

Tria autem quadrata ſunt ex ſeptem parti-bus hoc modo,

9. $\bar{p}$. ℞. v. cub. $4846\frac{1}{2}$, $\bar{p}$. ℞. $23487833\frac{1}{4}$,
$\bar{p}$. ℞. v. cub. $4846\frac{1}{2}$ $\tilde{m}$. ℞. $23487833\frac{1}{4}$
$\tilde{m}$. ℞. v. cub. $256\frac{1}{2}$ $\bar{p}$. ℞. $65063\frac{1}{4}$
$\tilde{m}$. ℞. v. $256\frac{1}{2}$ $\tilde{m}$. ℞. $65063\frac{1}{4}$
$\tilde{m}$. ℞. v. cub. $256\frac{1}{2}$ $\bar{p}$. ℞. $65063\frac{1}{4}$
$\tilde{m}$. ℞. v. cub. $256\frac{1}{2}$ $\tilde{m}$. ℞. $65063\frac{1}{4}$

Inde iunctis tribus quadratis cum cubo ſex partes, quæ ſunt ℞. v. cubicæ æquales $\bar{p}$. cum $\tilde{m}$. cadunt & relinquitur 21. ad àmuſ-ſem aggregatum.

FIG. 46.—Part of a page (255) from the *Ars magna* as reprinted in H. Cardan's *Operum tomvs quartvs* (Lvgdvni, 1663). The *Ars magna* was first published in 1545.

[1] *Operum tomvs quartvs*, p. 287.

In one place Cardan not only designates known numbers by letters, but actually operates with them. He lets a and b stand for any given numbers and then remarks that $R \dfrac{a}{b}$ is the same as $\dfrac{R a}{R b}$, i.e., $\sqrt{\dfrac{a}{b}}$ is the same as $\dfrac{\sqrt{a}}{\sqrt{b}}$.[1]

Figure 46 deals with the cubic $x^3+3x^2=21$. As a check, the value of x, expressed in radicals, is substituted in the given equation. There are two misprints. The $226\frac{1}{4}$ should be $256\frac{1}{4}$. Second, the two lines which we have marked with a stroke on the left should be omitted, except the $\tilde{m}$ at the end. The process of substitution is unnecessarily complicated. For compactness of notation, Cardan's symbols rather surpass the modern symbols, as will be seen by comparing his passage with the following translation:

"The proof is as in the example $x^3+3x^2=21$. According to these rules, the result is $\sqrt[3]{9\frac{1}{2}+\sqrt{89\frac{1}{4}}}+\sqrt[3]{9\frac{1}{2}-\sqrt{89\frac{1}{4}}}.-1$. The cube [i.e., x^3] is made up of seven parts:

$$12-\sqrt[3]{4,846\frac{1}{2}+\sqrt{23,487,833\frac{1}{4}}}-\sqrt[3]{4,846\frac{1}{2}-\sqrt{23,487,833\frac{1}{4}}}$$
$$+\sqrt[3]{46,041\frac{3}{4}+\sqrt{2,119,776,950\frac{7}{8}}}-\sqrt{2,096,286,117\frac{9}{16}}$$
$$+\sqrt[3]{46,041\frac{3}{4}+\sqrt{2,096,354,180\frac{1}{8}}}-\sqrt{2,119,776,950\frac{7}{8}}$$
$$+\sqrt[3]{256\frac{1}{2}+\sqrt{65,063\frac{1}{4}}}+\sqrt[3]{256\frac{1}{2}-\sqrt{65,063\frac{1}{4}}}\ .$$

"The three squares [i.e., $3x^2$] are composed of seven parts in this manner:

$$9+\sqrt[3]{4,846\frac{1}{2}+\sqrt{23,487,833\frac{1}{4}}}$$
$$+\sqrt[3]{4,846\frac{1}{2}-\sqrt{23,487,833\frac{1}{4}}}$$
$$-\sqrt[3]{256\frac{1}{2}+\sqrt{65,063\frac{1}{4}}}$$
$$-\sqrt[3]{256\frac{1}{2}-\sqrt{65,063\frac{1}{4}}}$$
$$-\sqrt[3]{256\frac{1}{2}+\sqrt{65,063\frac{1}{4}}}$$
$$-\sqrt[3]{256\frac{1}{2}-\sqrt{65,063\frac{1}{4}}}\ .$$

Now, adding the three squares with the six parts in the cube, which are equal to the general cube root, there results 21, for the required aggregate."

[1] *De regula aliza* (1570), p. 111. Quoted by Eneström, *op. cit.*, Vol. VII (1906–7), p. 387.

In translation, Figure 47 is as follows:

"The *Quaestio VIII.*

"Divide 6 into three parts, in continued proportion, of which the sum of the squares of the first and second is 4. We let the first be the

QVÆSTIO VIII.

Fac ex 6. tres partes, in continua proportione, quarum quadrata primæ & ſecundæ iuncta ſimul faciant 4. ponemus primam 1. poſitionem, quadratum eius eſt 1. quadratum, reſiduum igitur ad 4. eſt quadratum ſecundæ quantitatis, id eſt 4. m̃. 1. quadrato, huius radicem, & 1 poſitionem detrahe ex 6. habebis tertiam quantitatem, vt vides, quare ducta prima in teriam, ha-

1. poſ. | v. ℞. 4. m̃. 1. quad. | 6. m̃. 1. poſ. m̃. ℞. v. 4. m̃. 1. quad.
6. poſ. m̃. 1. quad. m̃. ℞. v. 4. quad. m̃. 1. quad. quad.

4. | 6. poſ. m̃. ℞. v. 4. quad. m̃. 1. quad. quad.
6. poſ. m̃. 4. æqual. ℞. v. 4. quad. m̃. 1. quad. quad.
36. quad. p̃. 16. m̃. 48. poſ. æquantur 4. quad. m̃. 1. quad. quad.

32. quad. p̃. 16. p̃. 1. quad. quad. æqualia 48. poſ.

1. quad. quad. p̃. 32. quad. p̃. 256. æqualia 48. poſ. p̃. 240.

FIG. 47.—Part of p. 297, from the *Ars magna*, as reprinted in H. Cardan's *Operum tomvs quartvs* (Lvgdvni, 1663).

1. position [i.e., x]; its square is 1. square [i.e., x^2]. Hence 4 minus this is the square of the second quantity, i.e., $4-1$. square [i.e., $4-x^2$]. Subtract from 6 the square root of this and also 1. position, and you will have the third quantity [i.e., $6-x-\sqrt{4-x^2}$], as you see, because the first multiplied by the third . . . . :

$$x \mid \sqrt{4-x^2} \mid 6-x-\sqrt{4-x^2} \mid$$
$$6x-x^2-\sqrt{4x^2-x^4}$$

$$4. \mid 6x-\sqrt{4x^2-x^4}$$
$$6x-4=\sqrt{4x^2-x^4}$$
$$36x^2+16-48x=4x^2-x^4$$

$$32x^2+16+x^4=48x$$

$$1x^4+32x^2+256=48x+240$$.,,

ITALIAN: NICOLO TARTAGLIA
(1537, 1543, 1546, 1556–60)

142. Nicolo Tartaglia's first publication, of 1537, contains little algebraic symbolism. He writes: "Radice .200. censi piu .10. cose" for $\sqrt{200x^2+10x}$, and "trouamo la cosa ualer Radice .200. men. 10." for "We find $x=\sqrt{200}-10$."[1] In his edition of Euclid's *Elements*[2] he writes "ℛ ℛ ℛ ℛ" for the sixteenth root. In his *Qvesiti*[3] of 1546 one reads, "Sia .1. cubo de censo piu .48. equal à 14. cubi" for "Let $x^6+48=14x^3$," and "la ℛ. cuba de .8. ualera la cosa, cioe. 2." for "The $\sqrt[3]{8}$ equals x, which is 2."

More symbolism appeared ten years later. Then he used the $\bar{p}$ and $\widetilde{m}$ of Pacioli to express "plus" and "minus," also the *co., ce., cu.,* etc., for the powers of numbers. Sometimes his abbreviations are less intense than those of Pacioli, as when he writes[4] *men* instead of $\widetilde{m}$, or[5] *cen* instead of *ce*. Tartaglia uses ℛ for *radix* or "root." Thus "la ℛ ℛ di $\frac{1}{16}$ è $\frac{1}{2}$,"[6] "la ℛ cu. di $\frac{1}{8}$ è $\frac{1}{2}$,"[7] "la ℛ rel. di $\frac{1}{32}$ è $\frac{1}{2}$,"[8] "la ℛ cen. cu. di $\frac{1}{64}$ è $\frac{1}{2}$,"[9] "la ℛ cu. cu. di $\frac{1}{512}$ è $\frac{1}{2}$,"[10] "la ℛ terza rel. di $\frac{1}{2048}$ è $\frac{1}{2}$."[11]

143. Tartaglia writes proportion by separating the three terms which he writes down by two slanting lines. Thus,[12] he writes "9// 5//100," which means in modern notation $9:5=100:x$. For his occasional use of parentheses, see § 351.

---

[1] *Nova scientia* (Venice, 1537), last two pages of "Libro secondo."

[2] *Evclide Megarense* (Venice, 1569), fol. 229 (1st ed., 1543).

[3] *Qvesiti, et inventioni* (Venice, 1546), fol. 132.

[4] *Seconda parte del general trattato di nvmeri, et misvri de Nicolo Tartaglia* (Venice, 1556), fol. 88B.

[5] *Ibid.*, fol. 73.

[6] *Ibid.*, fol. 38.

[7] *Ibid.*, fol. 34.

[8] *Ibid.*, fol. 43.

[9] *Ibid.*, fol. 47B.

[10] *Ibid.*, fol. 60.

[11] *Ibid.*, fol. 68.

[12] *Ibid.*, fol. 162.

On the margin of the page shown in Figure 48 are given the symbols of powers of the unknown number, viz., *co.*, *ce.*, etc., up to the twenty-ninth power. In the illustrations of multiplication, the absolute number 5 is marked "5 $\mathring{n}/0$"; the 0 after the solidus indicates the *dignità* or power 0, as shown in the marginal table. His

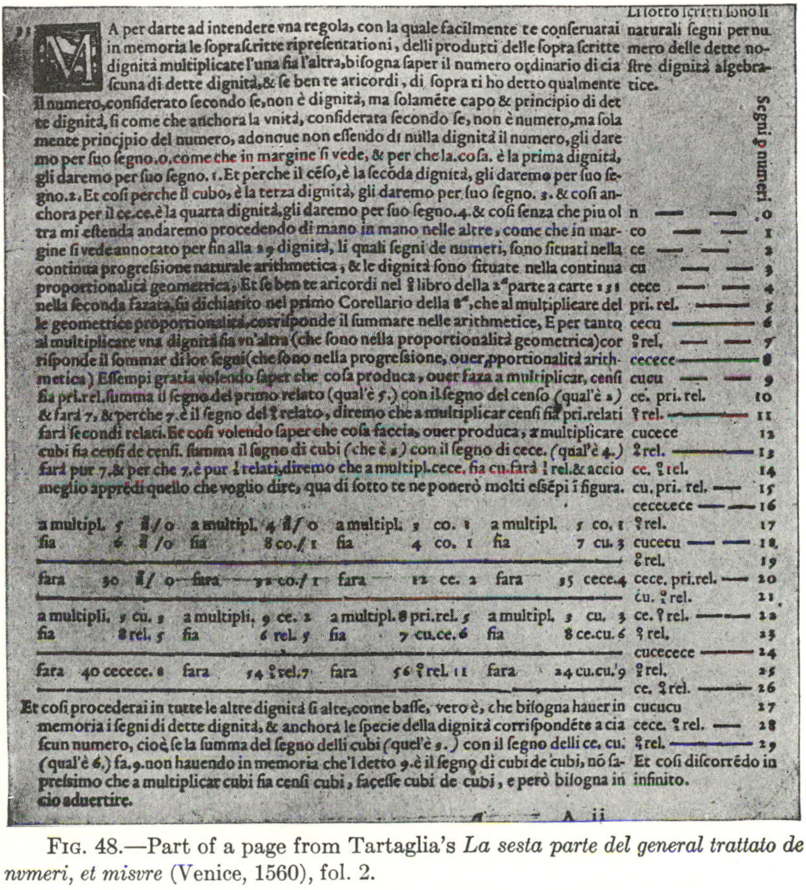

MA per darte ad intendere vna regola, con la quale facilmente te conſeruarai in memoria le ſopraſcritte ripreſentationi, delli produtti delle ſopra ſcritte dignità multiplicate l'una fia l'altra, biſogna ſaper il numero ordinario di ciaſcuna di dette dignità, & ſe ben te aricordi, di ſopra ti ho detto qualmente il numero, conſiderato ſecondo ſe, non è dignità, ma ſolamẽte capo & principio di dette dignità, ſi come che anchora la vnità, conſiderata ſecondo ſe, non è numero, ma ſolamente principio del numero, adonque non eſſendo di nulla dignità il numero, gli daremo per ſuo ſegno .o. come che in margine ſi vede, & per che la .coſa. è la prima dignità, gli daremo per ſuo ſegno. 1. Et perche il cẽſo, è la ſecõda dignità, gli daremo per ſuo ſegno. 2. Et coſi perche il cubo, è la terza dignità, gli daremo per ſuo ſegno. 3. & coſi anchora per il ce.ce. è la quarta dignità, gli daremo per ſuo ſegno. 4. & coſi ſenza che piu oltra mi eſtenda andaremo procedendo di mano in mano nelle altre, come che in margine ſi vede annotato per fin alla 29 dignità, li quali ſegni de numeri, ſono ſituati nella continua progreſsione naturale arithmetica, & le dignità ſono ſituate nella continua proportionalità geometrica, Et ſe ben te aricordi nel ° libro della 2ª parte a carte 131 nella ſeconda fazata, fu dichiarito nel primo Corellario della 8ª, che al multiplicare delle geometrice proportionalità, corriſponde il ſummare nelle arithmetice, E per tanto al multiplicare vna dignità fia vn'altra (che ſono nella proportionalità geometrica) corriſponde il ſommar di lor ſegni (che ſono nella progreſsione, ouer proportionalità arithmetica) Eſſempi gratia volendo ſaper che coſa produca, ouer faza a multiplicar, cenſi fia pri.rel. ſumma il ſegno del primo relato (qual'è 5.) con il ſegno del cenſo (qual'è 2.) & farà 7. & perche 7. è il ſegno del 2° relato, diremo che a multiplicar cenſi fia pri.relati farà ſecondi relati. Et coſi volendo ſaper che coſa faccia, ouer produca, a multiplicare cubi fia cenſi de cenſi. ſumma il ſegno di cubi (che è 3) con il ſegno di cece. (qual'è 4.) farà pur 7. & per che 7. è pur 2 relati, diremo che a multipl. cece. fia cu. farà 2 rel. & accio meglio apprẽdi quello che voglio dire, qua di ſotto te ne ponerò molti eſſẽpi ĩ figura.

a multipl.	5	n/o	a multipl.	4	n/o	a multipl.	3	co. 1	a multipl.	5	co. 1
fia	6	n/o	fia	8	co./1	fia	4	co. 1	fia	7	cu. 3
fara	30	n/o	fara	32	co./1	fara	12	ce. 2	fara	35	cece. 4
a multipli.	5	cu. 3	a multipli.	9	ce. 2	a multipl.	8	pri.rel. 5	a multipl.	3	cu. 3
fia	8	rel. 5	fia	6	rel. 5	fia	7	cu.ce. 6	fia	8	ce.cu. 6
fara	40	cecece. 8	fara	54	2° rel. 7	fara	56	3° rel. 11	fara	24	cu.cu. 9

Et coſi procederai in tutte le altre dignità ſi alte, come baſſe, vero è, che biſogna hauer in memoria i ſegni di dette dignità, & anchora le ſpecie della dignità corriſpondẽte a ciaſcun numero, cioè ſe la ſumma del ſegno delli cubi (quel'è 3.) con il ſegno delli ce. cu. (qual'è 6.) fa. 9. non hauendo in memoria che'l detto 9. è il ſegno di cubi de cubi, nõ ſapreſsimo che a multiplicar cubi fia cenſi cubi, faceſſe cubi de cubi, e però biſogna in cio aduertire.

A ii

Li ſotto ſcritti ſono li naturali ſegni per numero delle dette noſtre dignità algebratice.

Segni p numeri

n	0
co	1
ce	2
cu	3
cece	4
pri. rel.	5
cecu	6
2° rel.	7
cecece	8
cucu	9
ce. pri. rel.	10
3° rel.	11
cucece	12
4° rel.	13
ce. 2° rel.	14
cu. pri. rel.	15
cecececе	16
5° rel.	17
cucecu	18
6° rel.	19
cece. pri. rel.	20
cu. 2° rel.	21
ce. 3° rel.	22
7° rel.	23
cucecece	24
° rel.	25
ce. 4° rel.	26
cucucu	27
cece. 2° rel.	28
8° rel.	29

Et coſi diſcorrẽdo in infinito.

FIG. 48.—Part of a page from Tartaglia's *La sesta parte del general trattato de nvmeri, et misvre* (Venice, 1560), fol. 2.

illustrations stress the rule that in multiplication of one *dignità* by another, the numbers expressing the *dignità* of the factors must be added.

ITALIAN: RAFAELE BOMBELLI

(1572, 1579)

144. Bombelli's *L'algebra* appeared at Venice in 1572 and again at Bologna in 1579. He used *p.* and *m.* for "plus" and "minus."

Following Cardan, Bombelli used almost always *radix legata* for a root affecting only one term. To write two or more terms into one, Bombelli wrote an L right after the ℞ and an inverted ⅃ at the end of the expression to be radicated. Thus he wrote: ℞ L 7 $p.$℞ 14 ⅃ for our modern $\sqrt{7+\sqrt{14}}$, also[1] Rq L Rc L Rq 68 $p.2$ ⅃ m Rc L R q 68 m 2 ⅃⅃ for the modern $\sqrt{\left\{ \sqrt[3]{(\sqrt{68}+2)} - \sqrt[3]{(\sqrt{68}-2)} \right\}}$.

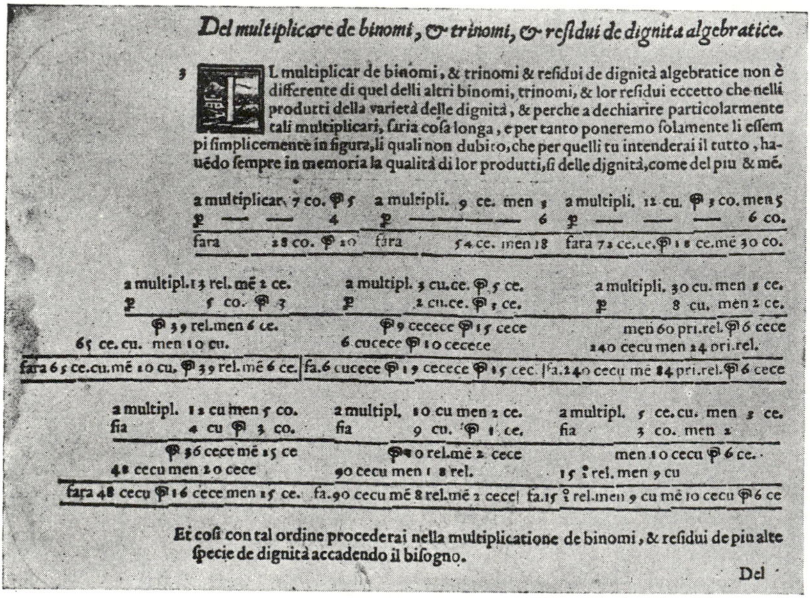

Fig. 49.—Part of a page from Tartaglia's *La sesta parte del general trattato de nvmeri, et misvre* (Venice, 1560), fol. 4. Shows multiplication of binomials. Observe the fancy .$\bar{p}$. for "plus." For "minus" he writes here $m\bar{e}$ or *men*.

An important change in notation was made for the expression of powers which was new in Italian algebras. The change is along the line of what is found in Chuquet's manuscript of 1484. It is nothing less than the introduction of positive integral exponents, but without writing the base to which they belonged. As long as the exponents were applied only to the unknown x, there seemed no need of writing the x. The notation is shown in Figure 50.

[1] Copied by Cantor, *op. cit.*, Vol. II (2d ed., 1913), p. 624, from Bombelli's *L'algebra*, p. 99.

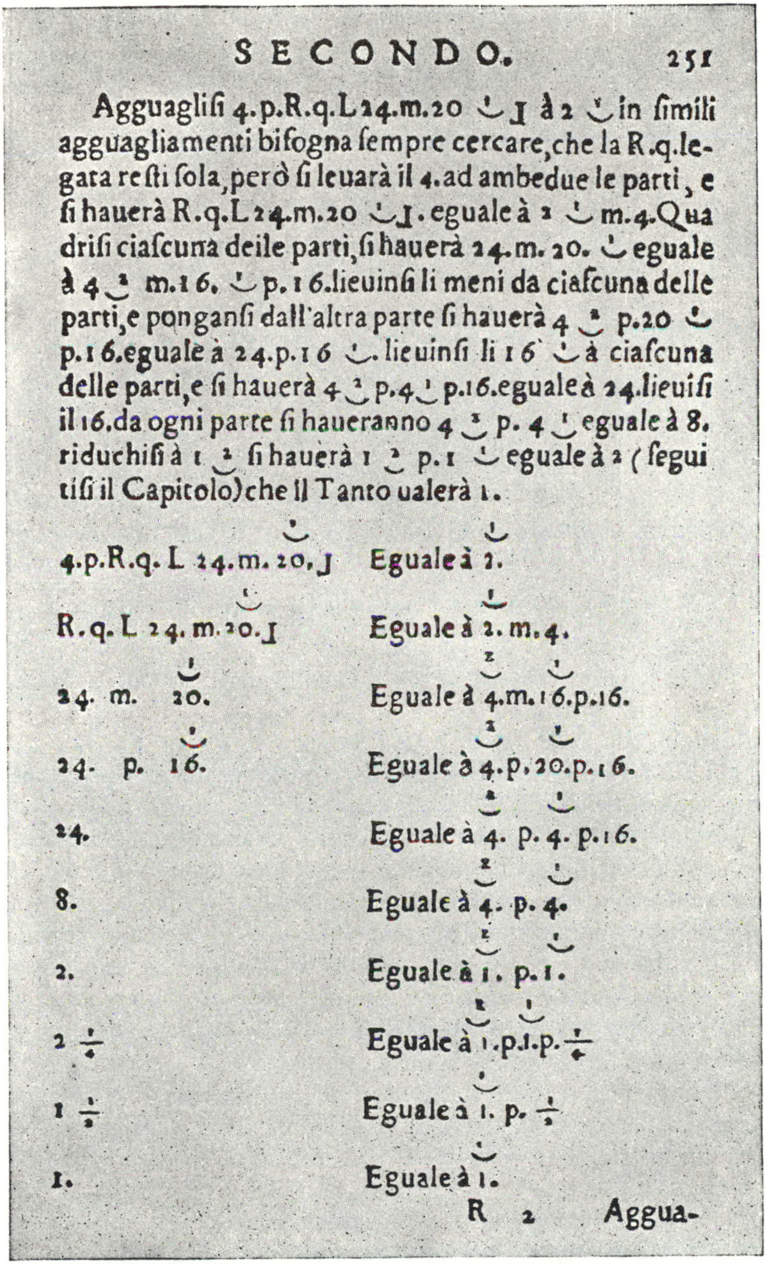

SECONDO.   251

Agguaglifi 4.p.R.q.L24.m.20 ‿ 1̑ à 2 ‿ in fimili agguagliamenti bifogna fempre cercare, che la R.q. legata refti fola, però fi leuarà il 4. ad ambedue le parti, e fi hauerà R.q.L24.m.20 ‿ 1̑. eguale à 2 ‿ m.4. Quadrifi ciafcuna deile parti, fi hauerà 24. m. 20. ‿ eguale à 4 ‿ m.16. ‿ p. 16. lieuinfi li meni da ciafcuna delle parti, e ponganfi dall'altra parte fi hauerà 4 ‿ p.20 ‿ p.16. eguale à 24. p.16 ‿. lieuinfi li 16 ‿ à ciafcuna delle parti, e fi hauerà 4 ‿ p.4 ‿ p.16. eguale à 24. lieuifi il 16. da ogni parte fi haueranno 4 ‿ p. 4 ‿ eguale à 8. riduchifi à 1 ‿ fi hauerà 1 ‿ p.1 ‿ eguale à 2 (feguitifi il Capitolo) che Il Tanto ualerà 1.

4.p.R.q.L 24.m. 20. 1̑	Eguale à 2̑.
R.q.L 24. m. 20. 1̑	Eguale à 2̑. m. 4.
24. m. 20.	Eguale à 4.m.16.p.16.
24. p. 16.	Eguale à 4.p.20.p.16.
24.	Eguale à 4. p. 4. p.16.
8.	Eguale à 4. p. 4.
2.	Eguale à 1. p. 1.
2 ÷	Eguale à 1. p.1.p. ÷
1 ÷	Eguale à 1. p. ÷
1.	Eguale à 1.

R 2   Aggua-

Fig. 50.—From Bombelli's *L'algebra* (1572)

In Figure 50 the equations are:

$$4+\sqrt{24-20x}=2x ,$$
$$\sqrt{24-20x}=2x-4 ,$$
$$24-20x=4x^2-16x+16 ,$$
$$24+16x=4x^2+20x+16 ,$$
$$24=4x^2+4x+16 ,$$
$$8=4x^2+4x ,$$
$$2=x^2+x ,$$
$$2\tfrac{1}{4}=x^2+x+\tfrac{1}{4} ,$$
$$1\tfrac{1}{2}=x+\tfrac{1}{2} ,$$
$$1=x .$$

Bombelli expressed square root by *R. q.*, cube root by *R. c.*, fourth root by *R R. q.*, fifth root (*Radice prima incomposta, ouer relata*) by *R. p. r.*, sixth root by *R. q. c.*, seventh root by *R. s. r.*, the square root of a polynomial (*Radice quadrata legata*) by *R. q. L ⌐*; the cube root of a polynomial (*Radice cubica legata*) by *R. c. L ⌐*. Some of these symbols are shown in Figure 51. He finds the sum of $\sqrt[3]{72-\sqrt{1,088}}$ and $\sqrt[3]{\sqrt{4,352}+16}$ to be $\sqrt[3]{232+\sqrt{53,312}}$.

The first part of the sentence preceding page 161 of Bombelli's *Algebra*, as shown in Figure 51, is "Sommisi *R. c. L R. q.* 4352 *.p.* 16.⌐ con *R. c. L* 72. *m. R. q.* 1088.⌐."

145. Bombelli's *Algebra* existed in manuscript about twenty years before it was published. The part of a page reproduced in Figure 52 is of interest as showing that the mode of expressing aggregation of terms is different from the mode in the printed texts. We have here the expression of the radicals representing x for the cubic $x^3=32x+24$. Note the use of horizontal lines with cross-bars at the ends; the lines are placed below the terms to be united, as was the case in Chuquet. Observe also that here a negative number is not allowed to stand alone: -1069 is written $0-1069$. The cube root is designated by R^3, as in Chuquet.

A manuscript, kept in the Library of the University of Bologna, contains data regarding the sign of equality ($=$). These data have been communicated to me by Professor E. Bortolotti and tend to show that ($=$) as a sign of equality was developed at Bologna independently of Robert Recorde and perhaps earlier.

The problem treated in Figure 53 is to divide 900 into two parts, one of which is the cube root of the other. The smaller part is desig-

R.q. 1088. ɪ Queſte due R.ſi poſſono ſommare, perche il lato di R. c. L 72. m. R. q. 1088. ch'è R. c. L
68. m. 2. ɪ è in proportione dupla à R. c. L R. q. 43 -
52. m. 16. ɪ reſiduo di R. c. L R. q. 4352. p. 16. ɪ però ſi poſſono ſommare (com'è detto) partendo la mag
giore per la minore, cioè per R. c. L 72. m. R. q. 1088, ɪ
che moltiplicata uia il ſuo Binomio (come ſi uede nel-

R.c.L 72.m.R.q.1088.   R.c.L R.q.4352.p.16.ɪ
R.c.L 72.p.R.q.1088.   R.c.L 72.p.R.q.1088.ɪ
────────────────   ────────────────
R.c.4096.   R.c.L.R.q.18415616.p.3328.ɪ
Lato   16.partitore.   Lato R.q.272.p.4.

Auenimento R.q.1 $\frac{1}{16}$ p. $\frac{1}{4}$, giontoli 1,
fa R.q. 1 $\frac{1}{16}$ p. 1 $\frac{1}{4}$ Via R.q. 1 $\frac{1}{16}$ p. 1 $\frac{1}{4}$

Fà R.q.4. $\frac{1}{6}$ $\frac{1}{4}$ p.2. $\frac{1}{8}$
R.q.1 $\frac{1}{16}$ p. 1 $\frac{1}{4}$
────────────────

R.c.L 5 $\frac{1}{1}$ $\frac{5}{6}$ p.R.q.35. $\frac{2}{2}$ $\frac{1}{6}$ ɪ·
R.c.L 72.m.R.q.1088.ɪ
────────────────

Somma.   R.c.L 232.p R.q.53312.ɪ

la figura.) fà 16, e queſto è il partitore , e moltiplicato
R. c. L R. q. 4352. p. 16. ɪ uia Rad. c. L 72. p. Rad.
1088. ɪ Binomio del partitore fà R.c. L R.q.18415616.
p. 3328. ɪ , che il ſuo lato è R. q. 272. p. 4, che parti-
to per 16. ne uiene R. q. 1 $\frac{1}{16}$ p. $\frac{1}{4}$, che aggiontoli 1
per regola fà 1 $\frac{1}{4}$ p. Rad. q. 1 $\frac{1}{6}$, e queſto ſi hà da
moltiplicare uia R.c. L 72. m. R. q. 1088. ɪ però riducaſi à
L   R.c.

Fɪɢ. 51.—Bombelli's *Algebra*, p. 161 of the 1579 impression, exhibiting the
calculus of radicals. In the third line of the computation, instead of 18,415,616
there should be 27,852,800. Notice the broken fractional lines, indicating difficulty
in printing fractions with large numerators and denominators.

nated by a symbol consisting of *c* and a flourish (probably intended for *co*). Then follows the equation 900 $\tilde{m}$ 1*co*① = 1*cu*③. (our $900 - x = x^3$). One sees here a mixture of two notations for x and x^3: the notation *co* and *cu* made familiar by Luca Pacioli, and Bombelli's exponential notation, with the 1 and 3, placed above the line, each exponent resting in a cup. It is possible that the part of the algebra here photographed may go back as far as about 1550. The cross-writing in the photograph begins: "in libro vecchio a carte 82: quella di far di 10 due parti: dice messer Nicolo che l'ona e ℞ 43 *p* 5 $\tilde{m}$ *R*18: et l'altra il resto sino a 10, cioe 5 $\tilde{m}$ ℞ 43 $\tilde{p}$. ℞ 18." This Nicolo is supposed to be Nicolo Tartaglia who died in 1557. The phrasing "Messer Nicolo" implies, so Bortolotti argues, that Nicolo was a living contemporary. If these contentions are valid, then the manuscript in question was written in 1557 or earlier.[1]

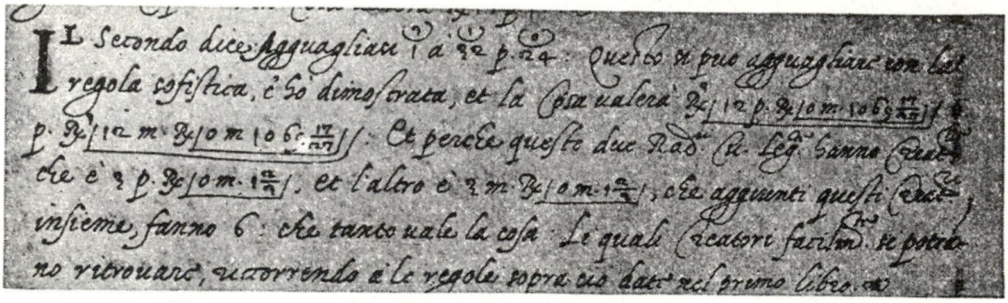
IL Secondo dice: Agguagliasi 1 a 32 p. 24: Questo si puo agguagliare con la
regola sofistica, c'ho dimostrata, et la Cosa valera ...
... Et perche queste due Rad. ... hanno ...
che è 3 p. ... et l'altro è 3 m. ..., che aggiunti questi ...
insieme, fanno 6: che tanto vale la cosa. Le quali ... si potra-
no ritrovare, ricorrendo a le regole sopra cio dati nel primo libro.

FIG. 52.—From the manuscript of the *Algebra* of Bombelli in the Comunale Library of Bologna. (Courtesy of Professor E. Bortolotti, of Bologna.)

The novel notations of Bombelli and of Ghaligai before him did not find imitators in Italy. Thus, in 1581 there appeared at Brescia the arithmetic and mensuration of Antonio Maria Visconti,[2] which follows the common notation of Pacioli, Cardan, and Tartaglia in designating powers of the unknown.

GERMAN: IOHANN WIDMAN
(1489, 1526)

146. Widman's *Behennde vnnd hübsche Rechnūg auff allen Kauffmanschafften* is the earliest printed arithmetic which contains the signs plus (+) and minus (−) (see §§ 201, 202).

[1] Since the foregoing was written, E. Bortolotti has published an article, on mathematics at Bologna in the sixteenth century, in the *Periodico di Matematiche* (4th ser., Vol. V, 1925), p. 147–84, which contains much detailed information, and fifteen facsimile reproductions of manuscripts exhibiting the notations then in use at Bologna, particularly the use of a dash (—) and the sign (=) to express equality.

[2] *Antonii Mariae Vicecomitis Civis Placentini practica numerorum & mensurarum* (Brixiae, 1581).

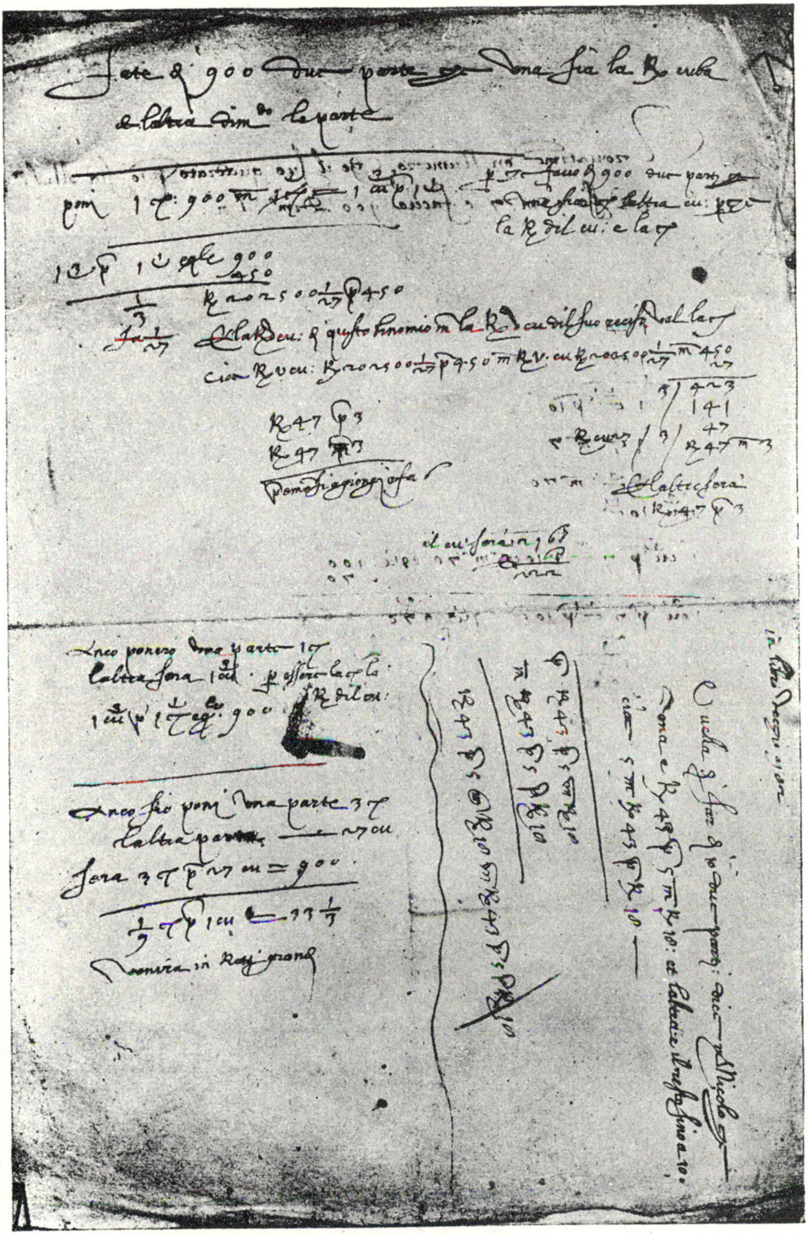

Fig. 53.—From a pamphlet (marked No. 595N, in the Library of the University of Bologna) containing studies and notes which Professor Bortolotti considers taken from the lessons of Pompeo Eolognetti ([Bologna?]–1568).

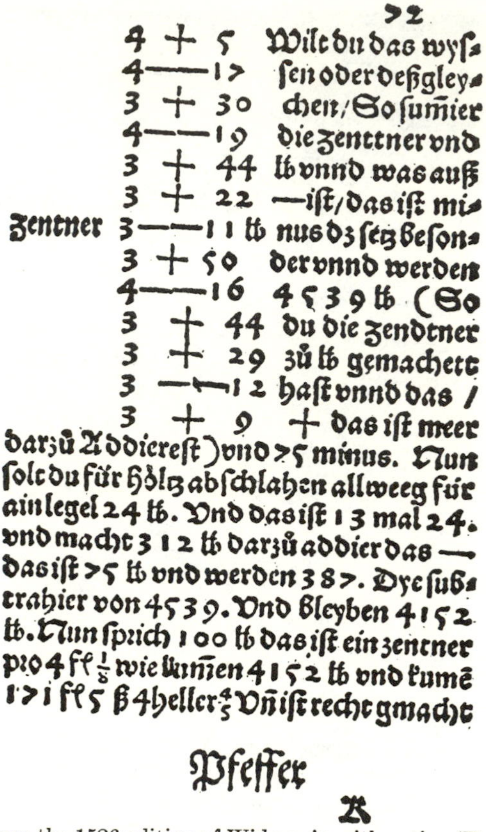

72

4	+	5	Wilt du das wyſ-
4	—	17	ſen oder deßgley-
3	+	30	chen/ So ſumier
4	—	19	die zenttner vnd
3	+	44	lb vnnd was auß
3	+	22	—iſt/das iſt mi-
3	—	11 lb	nus dz ſetz beſon-
3	+	50	der vnnd werden
4	—	16	4539 lb (So
3	✝	44	du die zendtner
3	+	29	zů lb gemachett
3	—	12	haſt vnnd das /
3	+	9	+ das iſt meer

Zentner

darzů Addiereſt) vnd 75 minus. Nun
ſolt du für Holtz abſchlahen allweeg für
ain legel 24 lb. Vnd das iſt 13 mal 24.
vnd macht 312 lb darzů addier das —
das iſt 75 lb vnd werden 387. Dye ſub-
trahier von 4539. Vnd bleyben 4152.
lb. Nun ſprich 100 lb das iſt ein zentner
pro 4 fl ½ wie kũmen 4152 lb vnd kumẽ
171 fl 5 ß 4 heller ⅓ Vnd iſt recht gmacht

Pfeffer

₰

Fig. 54.—From the 1526 edition of Widman's arithmetic. (Taken from D. E. Smith, *Rara arithmetica*, p. 40.)

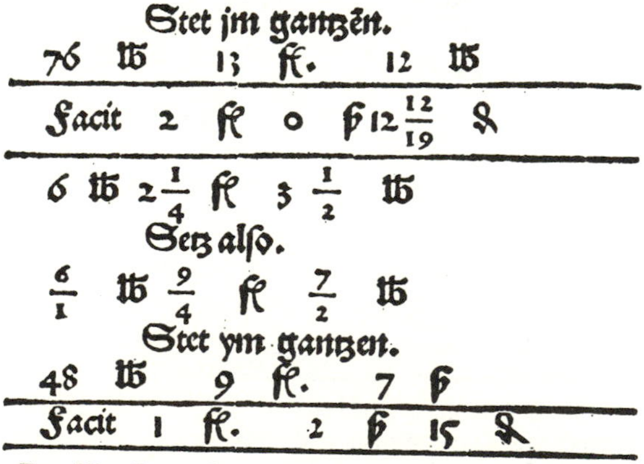

Stet im gantzen.

76	lb	13	fl.		12	lb
Facit	2	fl	0	ß	12 $\frac{12}{19}$	₰

6	lb	2 $\frac{1}{4}$	fl	3	$\frac{1}{2}$	lb

Setz also.

$\frac{6}{1}$	lb	$\frac{9}{4}$	fl		$\frac{7}{2}$	lb

Stet ym gantzen.

48	lb	9	fl.		7	ß
Facit	1	fl.		2	ß	15	₰

Fig. 55.—From the arithmetic of Grammateus (1518)

AUSTRIAN: HEINRICH SCHREIBER (GRAMMATEUS)
(1518, 1535)

147. Grammateus published an arithmetic and algebra, entitled *Ayn new Kunstlich Buech* (Vienna), printed at Nürnberg (1518), of which the second edition appeared in 1535. Grammateus used the

¶Additio.

Albie fein zu addiren die quantitet eines na=
mens/als N. mit N: prima mit prima/ secunda
mit secūda/tertia mit tertia zc. Vnd man brau=
chet solche zeichen als ─┼─ ist mehr/vnd ─┬─/min
der/in welcher sein zu mercken drei Regel.

¶Die Erst Regel.

Wann ein quantitet hat an beyden orten ─┼─
oder ─ so sol mann solche quantitet addirn hin
zū gesatzt das zeychen ─┼─ oder ─
als   9 pri. ─┼─ 7 N.   6 pri. ─ 4 N.
      6 pri. ─┼─ 5 N.   8 pri. ─ 10 N.
Facit 15 pri. ─┼─ 12 N.   14 pri. ─ 14 N.

¶Die ander Regel.

Ist in der obern quantitet ─┼─ vnd in der vn
dern ─/vnd ─┼─ übertrifft ─/ so sol die vnder
quantitet von der obern subtrahirt werden/vñ
zu dem übrigen setz ─┼─ So aber die vnder quā
titet ist grösser/so subtrahir die kleinern vō der
grössern/vñ zu dem das do bleibend ist/setze ─
als   6 pri. ─┼─ 6 N: 4 pri. ─┼─ 2 N.
     12 pri. ─ 4 N.   6 pri: ─ 6 N.
     18 pri. ─┼─ 2 N.   10 pri. ─ 4 N.

¶Die dritt Regel.

So in der obgesatzten quantitet würt fundē
─ vnd in der vndern ─┼─/vnd ─ übertrifft ─┼─/
so subtrahir eins von dem andern/vnd zum ü=
brigen schreib ─ Ist es aber/das die vnder quā
titet übertrifft die obern/so ziehe eins von dem
andern/vnd zu dem ersten setze ─┼─ als

FIG. 56.—From the arithmetic of Grammateus (1535). (Taken from D. E. Smith, *Rara arithmetica*, p. 125.)

plus and minus signs in a technical sense for addition and subtraction. Figure 55 shows his mode of writing proportion: $76lb. : 13fl. = 12lb. : x$. He finds $x = 2fl.$ 0 s. $12\frac{1}{13}\vartheta$. $[1fl. = 8s., 1s. = 30\vartheta]$.

The unknown quantity x and its powers x^2, x^3, ... , were called, respectively, *pri* (*prima*), 2a. or *se.* (*seconda*), 3a. or *ter.* (*terza*), 4a.

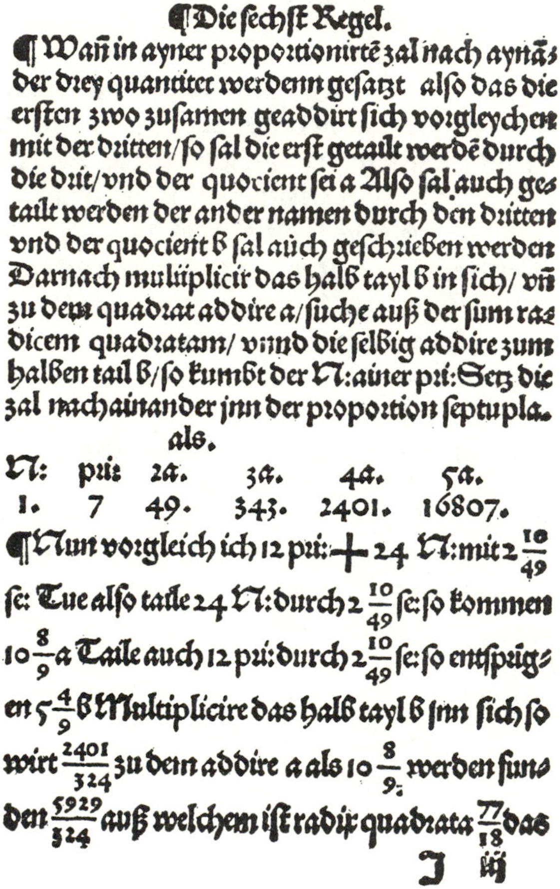

FIG. 57.—From the arithmetic of Grammateus (1518)

or *quart.* (*quarta*), 5a. or *quit.* (*quinta*), 6a. or *sex.* (*sexta*); N. stands for absolute number.

Fig. 56 shows addition of binomials. Figure 57 amounts to the solution of a quadratic equation. In translation: "The sixth rule: When in a proportioned number [i.e., in 1, x, x^2] three quantities are taken so that the first two added together are equal to the third [i.e.,

$d+ex=fx^2$], then the first shall be divided by [the coefficient of] the third and the quotient designated a. In the same way, divide the [coefficient of] the second by the [coefficient of] the third and the quotient designated b. Then multiply the half of b into itself and to the square add a; find the square root of the sum and add that to half of b. Thus is found the N. of 1 *pri.* [i.e., the value of x]. Place the number successively in the seven-fold proportion

N:	x	x^2	x^3	x^4	x^5
1.	7	49	343.	2,401.	16,807.

Now I equate $12x+24$ with $2\frac{10}{49}x^2$. Proceed thus: Divide 24 by $2\frac{10}{49}x^2$; there is obtained $10\frac{8}{9}a$. Divide also $12x$ by $2\frac{10}{49}x^2$; thus arises $5\frac{4}{9}b$. Multiplying the half of b by itself gives $\frac{2401}{324}$, to which adding a, i.e., $10\frac{8}{9}$, will yield $\frac{5929}{324}$, the square root of which is $\frac{77}{18}$; add this to half of the part b or $\frac{49}{18}$, and there results the number 7 as the number 1 *pri.* [i.e., x]."

The following example is quoted from Grammateus by Treutlein:[1]

" 6 *pri.*$+8$ *N.*	Modern Symbols
Durch	$6x+8$
5 *pri.*-7 *N.*	$5x-7$
30 *se.*$+40$ *pri.*	$30x^2+40x$
$\quad-42$ *pri.*-56 *N.*	$\quad-42x-56$
30 *se.*$-\ 2$ *pri.*-56 *N.* "	$30x^2-2x-56$.

In the notation of Grammateus, 9 *ter.*$+30$ *se.*-6 *pri.* $+48N$. stands for $9x^3+30x^2-6x+48$.[2]

We see in Grammateus an attempt to discard the old cossic symbols for the powers of the unknown quantity and to substitute in their place a more suitable symbolism. The words *prima, seconda*, etc., remind one of the nomenclature in Chuquet. His notation was adopted by Gielis van der Hoecke.

GERMAN: CHRISTOFF RUDOLFF
(1525)

148. Rudolff's *Behend vnnd Hubsch Rechnung durch die kunstreichen regeln Algebre so gemeincklich die Coss genent werden* (Strass-

[1] P. Treutlein, *Abhandlungen zur Geschichte der Mathematik*, Vol. II (Leipzig, 1879), p. 39.

[2] For further information on Grammateus, see C. I. Gerhardt, "Zur Geschichte der Algebra in Deutschland," *Monatsbericht d. k. Akademie der Wissenschaften zu Berlin* (1867), p. 51.

burg, 1525) is based on algebras that existed in manuscript (§ 203). Figure 58 exhibits the symbols for indicating powers up to the ninth. The symbol for *cubus* is simply the letter *c* with a final loop resembling the letter *e*, but is not intended as such. What appears below the symbols reads in translation: "*Dragma* or *numerus* is taken here as 1. It is no number, but assigns other numbers their kind. *Radix* is the

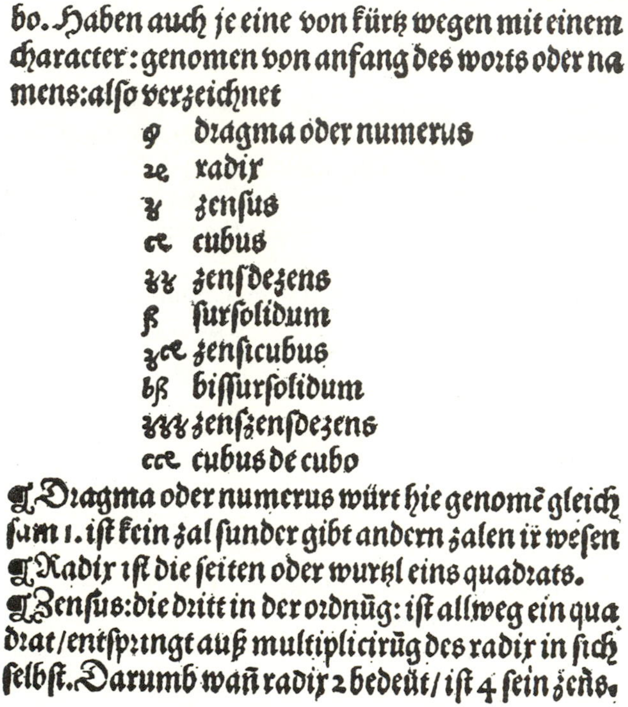
bo. Haben auch je eine von kürtz wegen mit einem character: genomen von anfang des worts oder na mens: also verzeichnet
dragma oder numerus
radix
zensus
cubus
zensdezens
sursolidum
zensicubus
bissursolidum
zenszensdezens
cubus de cubo
Dragma oder numerus würt hie genomē gleich sam 1. ist kein zal sunder gibt andern zalen ir wesen
Radix ist die seiten oder wurtzl eins quadrats.
Zensus: die dritt in der ordnūg: ist allweg ein quadrat / entspringt auß multiplicirūg des radix in sich selbst. Darumb wañ radix 2 bedeüt / ist 4 sein zēns.

FIG. 58.—From Rudolff's *Coss* (1525)

side or root of a square. *Zensus*, the third in order, is always a square; it arises from the multiplication of the *radix* into itself. Thus, when *radix* means 2, then 4 is the *zensus*." Adam Riese assures us that these symbols were in general use ("zeichen ader benennung Di in gemeinen brauch teglich gehandelt werdenn").[1] They were adopted by Adam

[1] Riese's *Coss* was found, in manuscript, in the year 1855, in the Kirchen- und Schulbibliothek of Marienberg, Saxony; it was printed in 1892 in the following publication: *Adam Riese, sein Leben, seine Rechenbücher und seine Art zu rechnen. Die Coss von Adam Riese*, by Realgymnasialrektor Bruno Berlet, in Annaberg i. E., 1892.

Riese, Apian, Menher, and others. The addition of radicals is shown in Figure 59. Cube root is introduced in Rudolff's *Coss* of 1525 as follows: "Würt radix cubica in diesem algorithmo bedeut durch solchen character $\mathsf{w}\!\surd$, als $\mathsf{w}\!\surd$ 8 is zu versteen radix cubica aufs 8." ("In this algorithm the cubic root is expressed by this character $\mathsf{w}\!\surd$, as $\mathsf{w}\!\surd 8$ is to be understood to mean the cubic root of 8.") The fourth root Rudolff indicated by $\mathsf{v}\!\surd$; the reader naturally wonders why two strokes should signify fourth root when three strokes indicate cube root. It is not at once evident that the sign for the fourth

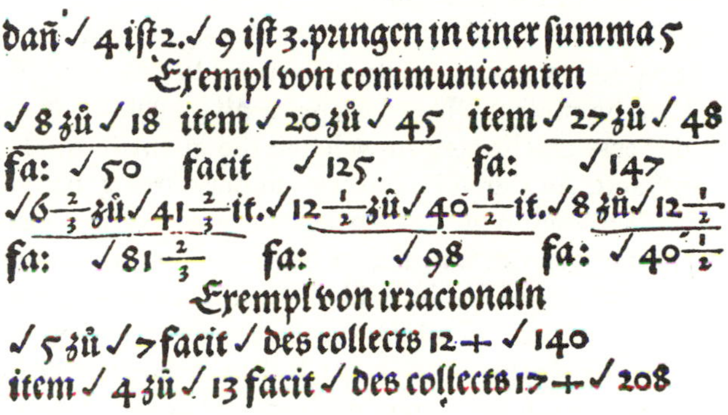
dañ √ 4 ist 2. √ 9 ist 3. bringen in einer summa 5
Exempl von communicanten
√8 zů √18 item √20 zů √45 item √27 zů √48
fa: √50 facit √125. fa: √147
√6 2/3 zů √41 2/3 it. √12 1/2 zů √40 1/2 it. √8 zů √12 1/2
fa: √81 2/3 fa: √98 fa: √40 1/2
Exempl von irracionaln
√5 zů √7 facit √ des collects 12 + √140
item √4 zů √13 facit √ des collects 17 + √208

FIG. 59.—From Rudolff's *Coss* (1525)

root represented two successive square-root signs, thus, $\surd\surd$. This crudeness in notation was removed by Michael Stifel, as we shall see later.

The following example illustrates Rudolff's subtraction of fractions:[1]

$$\text{“}\frac{1\,\mathcal{X}-2}{12}\text{ von }\frac{12}{1\,\mathcal{X}+2}\text{ Rest }\frac{148-1\mathfrak{z}}{12\,\mathcal{X}+24}\,.\text{”}$$

On page 141 of his *Coss*, Rudolff indicates aggregation by a dot;[2] i.e., the dot in "$\surd.12+\surd140$" indicates that the expression is $\sqrt{12+\sqrt{140}}$, and not $\sqrt{12}+\sqrt{140}$. In Stifel sometimes a second dot appears at the end of the expression (§ 348). Similar use of the dot we shall find in Ludolph van Ceulen, P. A. Cataldi, and, in form of the colon (:), in William Oughtred.

When dealing with two unknown quantities, Rudolff represented

[1] Treutlein, "Die deutsche Coss," *op. cit.*, Vol. II, p. 40.

[2] G. Wertheim, *Abhandlungen zur Geschichte der Mathematik*, Vol. VIII (Leipzig, 1898), p. 153.

the second one by the small letter q, an abbreviation for *quantita*, which Pacioli had used for the second unknown.[1]

Interesting at this early period is the following use of the letters a, c, and d to represent ordinary numbers (folio Giij[a]): "Nim ½ solchs collects | setz es auff ein ort | dz werd von lere wegen c genennt. Darnach subtrahier das c vom a | das übrig werd gesprochen d. Nun sag ich dz $\sqrt{c}+\sqrt{d}$ ist quadrata radix des ersten binomij." ("Take ½ this sum, assume for it a position, which, being empty, is called c. Then subtract c from a, what remains call d. Now I say that $\sqrt{c}+\sqrt{d}$ is the square root of the first binomial.")[2]

149. Rudolff was convinced that development of a science is dependent upon its symbols. In the Preface to the second part of Rudolff's *Coss* he states: "Das bezeugen alte bücher nit vor wenig jaren von der coss geschriben, in welchen die quantitetn, als dragma, res, substantia etc. nit durch character, sunder durch gantz geschribne wort dargegeben sein, vnd sunderlich in practicirung eines' yeden exempels die frag gesetzt, ein ding, mit solchen worten, ponatur vna res." In translation: "This is evident from old books on algebra, written many years ago, in which quantities are represented, not by characters, but by words written out in full, 'drachm,' 'thing,' 'substance,' etc., and in the solution of each special example the statement was put, 'one thing,' in such words as *ponatur, una res*, etc."[3]

In another place Rudolff says: "Lernt die zalen der coss aussprechen vnnd durch ire charakter erkennen vnd schreiben."[4] ("Learn to pronounce the numbers of algebra and to recognize and write them by their characters.")

DUTCH: GIELIS VAN DER HOECKE
(1537)

150. An early Dutch algebra was published by Gielis van der Hoecke which appeared under the title, *In arithmetica een sonderlinge excellēt boeck* (Antwerp [1537]).[5] We see in this book the early appear-

---

[1] Chr. Rudolff, *Behend vnnd Hubsch Rechnung* (Strassburg, 1525), fol. R1[a]. Quoted by Eneström, *Bibliotheca mathematica*, Vol. XI (1910–11), p. 357.

[2] Quoted from Rudolff by Eneström, *ibid.*, Vol. X (1909–10), p. 61.

[3] Quoted by Gerhardt, *op. cit.* (1870), p. 153. This quotation is taken from the second part of Gerhardt's article; the first part appeared in the same publication, for the year 1867, p. 38–54.

[4] *Op. cit.*, Buch I, Kap. 5, Bl. Dijr°; quoted by Tropfke, *op. cit.* (2. ed.), Vol. II, p. 7.

[5] On the date of publication, see Eneström, *op. cit.*, Vol. VII (1906–7), p. 211; Vol. X (1909–10), p. 87.

ance of the plus and minus signs in Holland. As the symbols for powers one finds here the notation of Grammateus, *N.*, *pri.*, *se.*, 3^a, 4^a, 5^a, etc., though occasionally, to fill out a space on a line, one en-

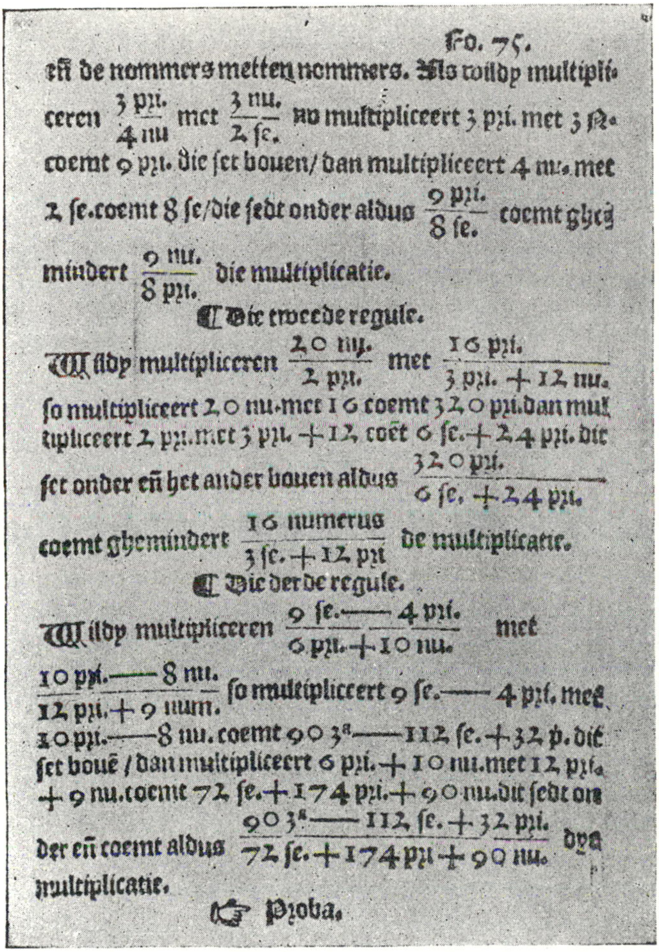
Fo. 75.

ēn de nommers metten nommers. Als wildy multipliceren 3 pri./4 nu met 3 nu./2 se. so multipliceert 3 pri. met 3 N. coemt 9 pri. die set bouen/ dan multipliceert 4 nu. met 2 se. coemt 8 se/die sedt onder aldus 9 pri./8 se. coemt [illegible] minderd 9 nu./8 pri. die multiplicatie.

Die tweede regule.

Wildy multipliceren 20 nu./2 pri. met 16 pri./3 pri. + 12 nu. so multipliceert 20 nu. met 16 coemt 320 pri. dan multipliceert 2 pri. met 3 pri. + 12 coēt 6 se. + 24 pri. die set onder ēn het ander bouen aldus 320 pri./6 se. + 24 pri. coemt ghemindert 16 numerus/3 se. + 12 pri de multiplicatie.

Die derde regule.

Wildy multipliceren 9 se. — 4 pri./6 pri. + 10 nu. met 10 pri. — 8 nu./12 pri. + 9 num. so multipliceert 9 se. — 4 pri. met 10 pri. — 8 nu. coemt 90 3ª — 112 se. + 32 p. die set bouē / dan multipliceert 6 pri. + 10 nu. met 12 pri. + 9 nu. coemt 72 se. + 174 pri. + 90 nu. die sedt on der ēn coemt aldus 90 3ª — 112 se. + 32 pri./72 se. + 174 pri + 90 nu. dye multiplicatie.

Proba.

FIG. 60.—From Gielis van der Hoecke's *In arithmetica* (1537). Multiplication of fractions by *regule cos.*

counters *numerus*, *num.*, or *nu.* in place of *N.;* also *secu.* in place of *se.* For *pri.* he uses a few times $\dot{p}$.

The translation of matter shown in Figure 60 is as follows: "[In order to multiply fractions simply multiply numerators by numerators] and denominators. Thus, if you wish to multiply $\frac{3x}{4}$ by $\frac{3}{2x^2}$, you multiply $3x$ by $\dot{3}$, this gives $9x$, which you write down. Then multiply 4 by $2x^2$, this gives $8x^2$, which you write under the other $\frac{9x}{8x^2}$. Simplified this becomes $\frac{9}{8x}$, the product. *Second rule:* If you wish to multiply $\frac{20}{2x}$ by $\frac{16x}{3x+12}$, multiply 20 by 16 [*sic*] which

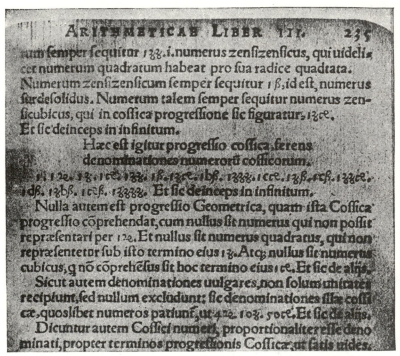

Fig. 61.—Part of a page from M. Stifel's *Arithmetica integra* (1544), fol. 235

gives $320x$, then multiply $2x$ by $3x+12$, which gives $6x^2+24x$. Place this under the other obtained above $\frac{320x}{6x^2+24x}$, this simplified gives:[1] $\frac{16}{3x^2+12x}$, the product. . . . ."

As radical sign Gielis van der Hoecke does not use the German symbols of Rudolff, but the capital R of the Italians. Thus he writes (fol. 90B) "$6+R8$" for $6+\sqrt{8}$, "$-R\ 32\ pri.$" for $-\sqrt{32x}$.

[1] The numerator should be 160, the denominator $3x+12$.

GERMAN: MICHAEL STIFEL
(1544, 1545, 1553)

151. Figure 61 is part of a page from Michael Stifel's important work on algebra, the *Arithmetica integra* (Nürnberg, 1544). From the ninth and the tenth lines of the text it will be seen that he uses the same symbols as Rudolff had used to designate powers, up to and including x^9. But Stifel carries here the notation as high as x^{16}. As Tropfke remarks,[1] the b in the symbol $b\beta$ of the seventh power leads Stifel to the happy thought of continuing the series as far as one may choose. Following the alphabet, his *Arithmetica integra* (1544) gives $c\beta = x^{11}$, $d\beta = x^{13}$, $e\beta = x^{17}$, etc.; in the revised *Coss* of Rudolff (1553), Stifel writes $\mathfrak{B}\beta$, $\mathfrak{C}\beta$, $\mathfrak{D}\beta$, $\mathfrak{G}\beta$. He was the first[2] who in print discarded the symbol for *dragma* and wrote a given number by itself. Where Rudolff, in his *Coss* of 1525 wrote 4ϕ, Stifel, in his 1553 edition of that book, wrote simply 4.

A multiplication from Stifel (*Arithmetica integra*, fol. 236*v*)[3] follows:

[Concluding part "$6\mathfrak{z} + 8\mathcal{R} - 6$
of a problem:] $2\mathfrak{z} - 4$

$$12\mathfrak{zz} + 16c\!\!c - 12\mathfrak{z}$$
$$- 24\mathfrak{z} - 32\mathcal{R} + 24$$

$$12\mathfrak{zz} + 16c\!\!c - 36\mathfrak{z} - 32\mathcal{R} + 24 \text{ "}$$

In Modern Symbols

$$6x^2 + 8x - 6$$
$$2x^2 - 4$$

$$12x^4 + 16x^3 - 12x^2$$
$$- 24x^2 - 32x + 24$$

$$12x^4 + 16x^3 - 36x^2 - 32x + 24$$

We give Stifel's treatment of the quartic equation, $1\mathfrak{zz} + 2c\!\!c + 6\mathfrak{z} + 5\mathcal{R} + 6$ *aequ.* 5550: "Quaeritur numerus ad quem additum suum quadratum faciat 5550. Pone igitur quod quadratum illud faciat $1A\mathfrak{z}$. tunc radix eius quadrata faciet $1A$. Et sic $1A\mathfrak{z} + 1A$. aequabitur

---

[1] *Op. cit.*, Vol. II, p. 120.

[2] Rudolff, *Coss* (1525), Signatur Hiiij (Stifel ed. [1553], p. 149); see Tropfke, *op. cit.*, Vol. II, p. 119, n. 651.

[3] Treutlein, *op. cit.*, p. 39.

5550. Itacq 1$A_{\mathfrak{z}}$. aequabit 5550−1A. Facit 1A. 74. Ergo cum. 1$_{\mathfrak{zz}}$+ 2α+6$_{\mathfrak{z}}$+5$\mathcal{R}$+6, aequetur. 5550. Sequitur quod. 74. aequetur 1$_{\mathfrak{z}}$+1$\mathcal{R}$+2. . . . . Facit itacq. 1$\mathcal{R}$.8."[1]

Translation:

$$\text{``}x^4+2x^3+6x^2+5x+6=5{,}550\ .$$

Required the number which, when its square is added to it, gives 5,550. Accordingly, take the square, which it makes, to be A^2. Then the square root of that square is A. Then $A^2+A=5{,}550$ and $A^2=5{,}550-A$. A becomes 74. Hence, since $x^4+2x^3+6x^2+5x+6=5{,}550$, it follows that $74=1x^2+x+2$. . . . . Therefore x becomes 8."

152. When Stifel uses more than the one unknown quantity $\mathcal{R}$, he at first follows Cardan in using the symbol q (abbreviation for *quantita*),[2] but later he represents the other unknown quantities by A, B, C. . . . . In the last example in the book he employs five unknowns, $\mathcal{R}$, A, B, C, D. In the example solved in Figure 62 he represents the unknowns by $\mathcal{R}$, A, B. The translation is as follows:

"Required three numbers in continued proportion such that the multiplication of the [sum of] the two extremes and the difference by which the extremes exceed the middle number gives 4,335. And the multiplications of that same difference and the sum of all three gives 6,069.

$A+x$ is the sum of the extremes,

$A-x$ the middle number,

$2A$ the sum of all three,

$2x$ the difference by which the extremes exceed the

middle. Then $2x$ multiplied into the sum of the extremes, i.e., in $A+x$, yields $2xA+2x^2=4{,}335$. Then $2x$ multiplied into $2A$ or the sum of all make $4xA=6{,}069$.

"Take these two equations together. From the first it follows that $xA=\dfrac{4{,}335-2x^2}{2}$. But from the second it follows that $1xA=\dfrac{6{,}069}{4}$. Hence $\dfrac{4{,}335-2x^2}{2}=\dfrac{6{,}069}{4}$, for, since they are equal to one and the same, they are equal to each other. Therefore [by reduction] $17{,}340-8x^2=12{,}138$, which gives $x^2=650\frac{1}{4}$ and $x=25\frac{1}{2}$.

[1] *Arithmetica integra*, fol. 307 B.

[2] *Ibid.*, III, vi, 252A. This reference is taken from H. Bosmans, *Bibliotheca mathematica* (3d ser., 1906–7), Vol. VII, p. 66.

DE ARITHMET. CARDANI 313

Quærantur tres numeri continue proportionales, ita ut
multiplicatio duorum extremorum, per differentiam, quam
habent extremi simul, ultra numerum medium, faciant 4335.
Et multiplicatio eiusdem differentiæ, in summam omnium
trium faciat 6069.

1 A ╾ 1 ℞. Est summa extremorum.
1 A ╾ 1 ℞. Est summa medij.
2 A. Est summa omnium trium.
2 ℞ Est differentia quam habent extremi ultra mediū.
Itaqȝ 2 ℞ multiplicatæ in summam extremorum, id est, in 1 A
╾ 1 ℞. faciunt 2 ℞ A ╾ 2 ℥. æquata. 4335. Deinde 2 ℞ multi
plicatæ in 2 A seu in summam omnium, faciunt 4 ℞ A æquata 6069.

Confer iam duas æquationes illas. Nam ex priore sequitur
quòd 1 ℞ A faciat $\frac{4335}{2}$ — 1 ℥ Ex posteriore autē sequitur quod
1 ℞ A. faciat $\frac{6069}{4}$. Sequitur ergo quod $\frac{4335}{2}$ — 1 ℥ & $\frac{6069}{4}$ inter se æquentur. Quia quæ uni & eidem sunt æqualia, etiam si-
bi inuicem sunt æqualia. Ergo (per reductionem) 17340 — 8 ℥
æquantur. 12138. facit 1 ℥. 650 $\frac{1}{4}$ Et 1 ℞. facit 25 $\frac{1}{2}$.

Restat iam ut 1 A. etiam resoluatur, facit autem (ut paulo su
perius uidimus) 1 ℞ A. $\frac{6069}{4}$. Cum igitur duo illa inter se sint æ-
qualia, Diuide utrunqȝ per 1 ℞. tunc inuenies 1 A. æquari, seu
facere. $\frac{6069}{4℞}$. Cum autem 1 ℞ faciat 25 $\frac{1}{2}$. facient 4 ℞. 102. Itaqȝ
6069. diuisa per 102. faciunt 59 $\frac{1}{2}$. Et tantum facit 1 A. Quare
1 A — 1 ℞. id est, medius numerus facit 34. Et 1 A ╾ 1 ℞. id est,
summa duorum extremorum facit 85. Iam igitur oritur noua
quæstio hæc.

Diuidantur 85 in duas partes, ita ut 34 mediet inter eas
proportionaliter.

Sic stant numeri.
1 B. 34. 85 — 1 B.
Vnde 85 B — 1 B ℥ æquatur 1156. facit 1 B 17. Et sic stant
numeri exempli. 17. 34. 68.
KK Exem

FIG. 62.—From Stifel's *Arithmetica integra* (1544), fol. 313

"It remains to find also $1A$. One has [as we saw just above] $1xA = \dfrac{6,069}{4}$. Since these two are equal to each other, divide each by x, and there follows $A = \dfrac{6,069}{4x}$. But as $x = 25\frac{1}{2}$, one has $4x = 102$, and 6,069 divided by 102 gives $59\frac{1}{2}$. And that is what A amounts to. Since $A - x$, i.e., the middle number equals 34, and $A + x$, i.e., the sum of the two extremes is 85, there arises this new problem:

"Divide 85 into two parts so that 34 is a mean proportional between them. These are the numbers:

$$B \, , \qquad 34 \, , \qquad 85 - B \, .$$

Since $85B - B^2 = 1,156$, there follows $B = 17$. And the numbers of the example are 17, 34, 68."

Observe the absence of a sign of equality in Stifel, equality being expressed in words or by juxtaposition of the expressions that are equal; observe also the designation of the square of the unknown B by the sign B_3. Notice that the fractional line is very short in the case of fractions with binomial (or polynomial) numerators—a singularity found in other parts of the *Arithmetica integra*. Another oddity is Stifel's designation of the multiplication of fractions.[1] They are written as we write ascending continued fractions. Thus

$$\frac{1}{7}\frac{2}{3}\frac{3}{4}$$

means "Tres quartae, duarum tertiarum, unius septimae," i.e., $\frac{3}{4}$ of $\frac{2}{3}$ of $\frac{1}{7}$.

The example in Fig. 62 is taken from the closing part of the *Arithmetica integra* where Cardan's *Ars magna*, particularly the solutions of cubic and quartic equations, receive attention. Of interest is Stifel's suggestion to his readers that, in studying Cardan's *Ars magna*, they should translate Cardan's algebraic statements into the German symbolic language: "Get accustomed to transform the signs used by him into our own. Although his signs are the older, ours are the more commodious, at least according to my judgment."[2]

---

[1] *Arithmetica integra* (1548), p. 7; quoted by S. Günther, *Vermischte Untersuchungen* (Leipzig, 1876), p. 131.

[2] *Arithmetica integra* (Nürnberg, 1544), Appendix, p. 306. The passage, as quoted by Tropfke, *op. cit.*, Vol. II (2. ed.), p. 7, is as follows: "Assuescas, signa eius, quibus ipse utitur, transfigurare ad signa nostra. Quamvis enim signa quibus ipse utitur, uetustiora sint nostris, tamen nostra signa (meo quidē iudicio) illis sunt commodiora."

153. Stifel rejected Rudolff's symbols for radicals of higher order and wrote $\sqrt{3}$ for $\sqrt{}$, $\sqrt{cc}$ for $\sqrt[\beta]{}$, etc., as will be seen more fully later.

But he adopts Rudolff's dot notation for indicating the root of a binomial:[1]

"$\sqrt{3}\cdot 12+\sqrt{3}6\cdot -\cdot\sqrt{3}\cdot 12-\sqrt{3}6$ has for its square $12+\sqrt{3}6+12-\sqrt{3}6-\sqrt{3}138-\sqrt{3}138$"; i.e., "$\sqrt{\sqrt{12+\sqrt{6}}}-\sqrt{\sqrt{12-\sqrt{6}}}$ has for its square $12+\sqrt{6}+12-\sqrt{6}-\sqrt{138}-\sqrt{138}$." Again:[2] "Tertio vide, utrũ $\sqrt{3}\cdot\sqrt{3}\,12500-50$ addita ad $\sqrt{3}\cdot\sqrt{3}\,12500+50$. faciat $\sqrt{3}\cdot\sqrt{3}\,50000+200$" ("Third, see whether $\sqrt{\sqrt{12,500-50}}$ added to $\sqrt{\sqrt{12,500+50}}$ makes $\sqrt{\sqrt{50,000+200}}$"). The dot is employed to indicate that the root of all the terms following is required.

154. Apparently with the aim of popularizing algebra in Germany by giving an exposition of it in the German language, Stifel wrote in 1545 his *Deutsche arithmetica*[3] in which the unknown x is expressed by *sum*, x^2 by *"sum: sum,"* etc. The nature of the book is indicated by the following equation:

"Der Algorithmus meiner deutschen Coss braucht zum ersten schlecht vnd ledige zalē | wie der gemein Algorithmus | als da sind 1 2 3 4 5 etc. Zum audern braucht er die selbigen zalen vnder diesem namen | Sum̄a. Vnd wirt dieser nam Suma | also verzeichnet | Sum̄: Als hie | 1 sum̄: 2 sum̄: 3 sum̄ etc. . . . . . So ich aber 2 sum: Multiplicir mit 3 sum: so kom̄en mir 6 sum: sum: Das mag ich also lesen | 6 summē summarum | wie man deñ im Deutschē offt findet | sum̄a sum̄arum. . . . . . Soll ich multipliciren 6 sum: sum: sum: mit 12 sum: sum: sum: So sprich ich | 12 mal 6. macht 72 sum: sum: sum: sum sum . . . ."[4] Translation: "The algorithm of my Deutsche Coss uses, to start with, simply the pure numbers of the ordinary algorithm, namely, 1, 2, 3, 4, 5, etc. Besides this it uses these same numbers under the name of *summa*. And this name *summa* is marked *sum̄:*, as in 1 *sum̄:* 2 *sum̄:* 3 *sum̄*, etc. . . . . . But when I multiply 2 *sum:* by 3 *sum:* I obtain 6 *sum: sum:*. This I may read | 6 *summē summarum* | for in German one encounters often *sum̄a sum̄arum*. . . . . . When I am to multiply 6 *sum: sum: sum:* by 12 *sum: sum: sum:*, I say | 12 times 6 makes 72 *sum: sum: sum: sum: sum: sum:* . . . ."

---

[1] *Op. cit.*, fol. 138a.   [2] *Ibid.*, fol. 315a.

[3] *Op. cit. Inhaltend. Die Hauszrechnung. Deutsche Coss. Rechnung* (1545).

[4] Treutlein, *op. cit.*, Vol. II, p. 34. For a facsimile reproduction of a page of Stifel's *Deutsche arithmetica*, see D. E. Smith, *Rara arithmetica* (1898), p. 234.

The inelegance of this notation results from an effort to render the subject easy; Stifel abandoned the notation in his later publications, except that the repetition of factors to denote powers reappears in 1553 in his "Cossische Progresʒ" (§ 156).

In this work of 1545 Stifel does not use the radical signs found in his *Arithmetica integra;* now he uses $\surd$, $\overline{\surd}$, $\overline{\overline{\surd}}$, for square, cube, and fourth root, respectively. He gives (fol. 74) the German capital letter 𝔐 as the sign of multiplication, and the capital letter 𝔇 as the sign of division, but does not use either in the entire book.[1]

155. In 1553 Stifel brought out a revised edition of Rudolff's *Coss*. Interesting is Stifel's comparison of Rudolff's notation of radicals with his own, as given at the end of page 134 (see Fig. 63a), and his declaration of superiority of his own symbols. On page 135 we read: "How much more convenient my own signs are than those of Rudolff, no doubt everyone who deals with these algorithms will notice for himself. But I too shall often use the sign $\surd$ in place of the $\surd_{\mathfrak{z}}$, for brevity.

"But if one places this sign before a simple number which has not the root which the sign indicates, then from that simple number arises a surd number.

"Now my signs are much more convenient and clearer than those of Christoff. They are also more complete for they embrace all sorts of numbers in the arithmetic of surds. They are [here he gives the symbols in the middle of p. 135, shown in Fig. 63b]. Such a list of surd numbers Christoff's symbols do not supply, yet they belong to this topic.

"Thus my signs are adapted to advance the subject by putting in place of so many algorithms a single and correct algorithm, as we shall see.

"In the first place, the signs (as listed) themselves indicate to you how you are to name or pronounce the surds. Thus, $\surd\beta 6$ means the sursolid root of 6, etc. Moreover, they show you how they are to be reduced, by which reduction the declared unification of many (indeed all such) algorithms arises and is established."

156. Stifel suggests on folio 61*B* also another notation (which, however, he does not use) for the progression of powers of x, which he calls "die Cossische Progresʒ." We quote the following:

"Es mag aber die Cossische Progresʒ auch also verzeychnet werden:

$$\begin{array}{ccccc} 0 & 1 & 2 & 3 & 4 \\ 1 \cdot 1A & \cdot 1AA & \cdot 1AAA & \cdot 1AAAA & \cdot \text{ etc.} \end{array}$$

[1] Cantor, *op. cit.*, Vol. II (2. ed., 1913), p. 444.

Item auch also:

$$\begin{array}{ccccc} 0 & 1 & 2 & 3 & 4 \\ 1 \cdot 1B & \cdot \ 1BB & \cdot \ 1BBB & \cdot \ 1BBBB & \cdot \ \text{etc.} \end{array}$$

Item auch also:

$$\begin{array}{ccccc} 0 & 1 & 2 & 3 & 4 \\ 1 & \cdot \ 1C & \cdot \ 1CC & \cdot \ 1CCC & \cdot \ 1CCCC & \cdot \ \text{etc.} \end{array}$$

Vnd so fort an von andern Buchstaben."[1]

Fɪɢ. 63a.—This shows p. 134 of Stifel's edition of Rudolff's *Coss* (1553)

[1] Treutlein, *op. cit.*, Vol. II (1879), p. 34.

We see here introduced the idea of repeating a letter to designate powers, an idea carried out extensively by Harriot about seventy-five

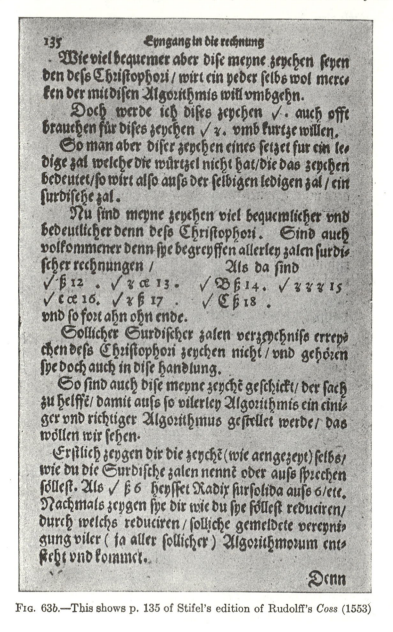

Fig. 63b.—This shows p. 135 of Stifel's edition of Rudolff's *Coss* (1553)

years later. The product of two quantities, of which each is represented by a letter, is designated by juxtaposition.

GERMAN: NICOLAUS COPERNICUS
(1566)

157. Copernicus died in 1543. The quotation from his *De revolutionibus orbium coelestium* (1566; 1st ed., 1543)[1] shows that the exposition is devoid of algebraic symbols and is almost wholly rhetorical. We find a curious mixture of modes of expressing numbers: Roman numerals, Hindu-Arabic numerals, and numbers written out in words. We quote from folio 12:

"Circulum autem communi Mathematicorum consensu in CCCLX. partes distribuimus. Dimetientem uero CXX. partibus asciscebant prisci. At posteriores, ut scrupulorum euitarent inuolutionem in multiplicationibus & diuisionibus numerorum circa ipsas lineas, quae ut plurimum incommensurabiles sunt longitudine, saepius etiam potentia, alij duodecies centena milia, alij uigesies, alij aliter rationalem constituerunt diametrum, ab eo tempore quo indicae numerorum figurae sunt usu receptae. Qui quidem numerus quemcunque alium, sine Graecum, sine Latinum singulari quadam promptitudine superat, & omni generi supputationum aptissime sese accommodat. Nos quoq, eam ob causam accepimus diametri 200000. partes tanquam sufficientes, que, possint errorem excludere patentem."

Copernicus does not seem to have been exposed to the early movements in the fields of algebra and symbolic trigonometry.

GERMAN: JOHANNES SCHEUBEL
(1545, 1551)

158. Scheubel was professor at the University of Tübingen, and was a follower of Stifel, though deviating somewhat from Stifel's notations. In Scheubel's arithmetic[2] of 1545 one finds the scratch method in division of numbers. The book is of interest because it does not use the + and − signs which the author used in his algebra; the + and − were at that time not supposed to belong to arithmetic proper, as distinguished from algebra.

---

[1] *Nicolai Coperrnici Torinensis de Revolvtionibus Orbium Coelestium, Libri VI. .... Item, de Libris Revolvtionvm Nicolai Copernici Narratio prima, per M. Georgium Ioachimum Rheticum ad D. Ioan. Schonerum scripta. Basileae* (date at the end of volume, M.D.LXVI).

[2] *De Nvmeris et Diversis Rationibvs seu Regulis computationum Opusculum, a Ioanne Scheubelio compositum* .... (1545).

Scheubel in 1550 brought out at Basel an edition of the first six books of Euclid which contains as an introduction an exposition of algebra,[1] covering seventy-six pages, which is applied to the working of examples illustrating geometric theorems in Euclid.

159. Scheubel begins with the explanation of the symbols for powers employed by Rudolff and Stifel, but unlike Stifel he retains a symbol for *numerus* or *dragma*. He explains these symbols, up to the twelfth power, and remarks that the list may be continued indefinitely. But there is no need, he says, of extending this unwieldy designation, since the ordinal natural numbers afford an easy nomenclature. Then he introduces an idea found in Chuquet, Grammateus, and others, but does it in a less happy manner than did his predecessors. But first let us quote from his text. After having explained the symbol for *dragma* and for x he says (p. 2): "The third of them ʒ, which, since it is produced by multiplication of the *radix* into itself, and indeed the first [multiplication], is called the *Prima* quantity and furthermore is noted by the syllable *Pri*. Even so the fourth ₵, since it is produced secondly by the multiplication of that same *radix* by the square, i.e., by the *Prima* quantity, is called the *Second* quantity, marked by the syllable *Se*. Thus the fifth sign ʒʒ, which springs thirdly from the multiplication of the *radix*, is called the *Tertia* quantity, noted by the syllable *Ter*. . . . ."[2] And so he introduces the series of symbols, *N., Ra., Pri., Se., Ter., Quar., Quin., Sex., Sep.* . . . . , which are abbreviations for the words *numerus, radix, prima quantitas* (because it arises from one multiplication), *secunda quantitas* (because it arises from two multiplications), and so on. This scheme gives rise to the oddity of designating x^n by the number $n-1$, such as we have not hitherto encountered. In Pacioli one finds the contrary relation, i.e., the designation of x^{n-1} by x^n (§ 136). Scheubel's notation does not coincide with that of Grammateus, who more judiciously had used *pri., se.,* etc., to designate x, x^2, etc. (§ 147). Scheubel's singular notation is illustrated by

[1] *Evclidis Megarensis, Philosophi et Mathematici excellentissimi, sex libri priores de Geometricis principijs, Graeci et Latini . . . . Algebrae porro regvlae, propter nvmerorum exempla, passim propositionibus adiecta, his libris praemissae sunt, eadenque demonstratae. Authore Ioanne Schevbelio, ... Basileae* (1550). I used the copy belonging to the Library of the University of Michigan.

[2] "Tertius de, ʒ. qui cū ex multiplicatione radicis in se producatur, et primo quidem: Prima quantitas, et Pri etiam syllaba notata, appelletur. Quartus uerò ₵ quia ex multiplicatione eiusdem radicis cum quadrato, hoc est, cum prima quantitate, secundò producitur: Se syllaba notata, Secunda quantitas dicitur. Sic character quintus, ʒʒ, quia ex multiplicatione radicis cum secunda quantitate tertio nascitur: Ter syllaba notata, Tertia etiam quantitias dicitur. . . . ."

Figure 64, where he shows the three rules for solving quadratic equations. The first rule deals with the solution of $4x^2+3x=217$, the second with $3x+175=4x^2$, the third with $3x^2+217=52x$. These different cases arose from the consideration of algebraic signs, it being desired that the terms be so written as to appear in the positive form. Only positive roots are found.

ALIVD EXEMPLVM.

PRIMI CANONIS.

SECVNDI CANONIS.

Pri.　　　ra.　　　N　　　　　ra.　　　N　　pri.

4　+　3　equales　217　　　3　+　175　æqu.　4

Hic, quia maximi characteris numerus non est unitas, diuisione, ut dictum est, ei succurri debet. Veniunt autem facta diuisione,

pri.　　　ra.　　　N　　　　　ra.　　　N　　pri.

1　+　$\frac{3}{4}$　æqu.　$\frac{217}{4}$　　　$\frac{3}{4}$　+　$\frac{175}{4}$　equ.　1

$\frac{3}{8}$ in se, $\frac{9}{64}$ + $\frac{217}{4}$　　　$\frac{3}{8}$ in se. $\frac{9}{64}$ + $\frac{175}{4}$

ueni. $\frac{3481}{64}$. Huius ra.　　　ueni. $\frac{2809}{64}$. Huius ra.

sunt $7\frac{3}{8}$ minus $\frac{3}{8}$　　　sunt $6\frac{5}{8}$ plus $\frac{3}{8}$

manent　7　　　　　　ueniunt　7

radicis ualor.　　　　　radicis ualor.

ALIVD TERTII CANONIS EXEMPLVM.

3 pri.　+　217 N　　　æquales　　　52 ra.

Et hic, quia maximi characteris numerus non est unitas, diuisione ei succurrendum erit. Veniunt autem hoc facto,

1 pri.　+　$\frac{217}{3}$ N　　　æquales　　　$5\frac{1}{3}$ N

$2\frac{6}{3}$ in se. $\frac{676}{9}$, minus $\frac{217}{3}$, manet $\frac{25}{9}$

Huius ra. qua. est $1\frac{2}{3}$ $\begin{cases} de \\ ad \end{cases}$ $8\frac{2}{3}$, & manent 7, uel proueniunt $10\frac{1}{3}$, Vterq radicis ualor, quod examinari potest.

FIG. 64.—Part of p. 28 in Scheubel's Introduction to his *Euclid*, printed at Basel in 1550.

Under proportion we quote one example (p. 41):

" 3 ra.+4 N. ualent 8 se.+4 pri.
quanti 8 ter.−4 ra.

$$Facit \frac{64\ sex.+32\ quin.-32\ ter.-16\ se.}{3\ ra.+4\ N.}$$ "

In modern notation:

$3x+4$ are worth $8x^3+4x^2$

how much $8x^4-4x$.

$$\text{Result } \frac{64x^7+32x^6-32x^4-16x^3}{3x+4} .$$

In the treatment of irrationals or *numeri surdi* Scheubel uses two notations, one of which is the abbreviation *Ra.* or *ra.* for *radix*, or "square root," *ra.cu.* for "cube root," *ra.ra.* for "fourth root." Confusion from the double use of *ra.* (to signify "root" and also to signify x) is avoided by the following implied understanding: If *ra.* is followed by a number, the square root of that number is meant; if *ra.* is preceded by a number, then *ra.* stands for x. Thus "8 *ra.*" means $8x$; "*ra.* 12" means $\sqrt{12}$.

Scheubel's second mode of indicating roots is by Rudolff's symbols for square, cube, and fourth roots. He makes the following statement (p. 35) which relates to the origin of $\sqrt{}$: "Many, however, are in the habit, as well they may, to note the desired roots by their points with a stroke ascending on the right side, and thus they prefix for the square root, where it is needed for any number, the sign $\sqrt{}$: for the cube root, $\wedge\!\!\wedge\!\!\sqrt{}$; and for the fourth root $\wedge\!\!\wedge\!\!\sqrt{}$."[1] Both systems of notation are used, sometimes even in the same example. Thus, he considers (p. 37) the addition of "*ra.* 15 *ad ra.* 17" (i.e., $\sqrt{15}+\sqrt{17}$) and gives the result "*ra.col.* 32+$\sqrt{}$1020" (i.e. $\sqrt{32+\sqrt{1,020}}$). The *ra.col.* (*radix collecti*) indicates the square root of the binomial. Scheubel uses also the *ra.re* (*radix residui*) and *radix binomij*. For example (p. 55), he writes "*ra.re.* $\sqrt{}$15−$\sqrt{}$12" for $\sqrt{\sqrt{15}-\sqrt{12}}$. Scheubel suggests a third notation for irrationals (p. 35), of which he makes no further use, namely, *radix se.* for "cube root," the abbreviation for *secundae quantitatis radix*.

The algebraic part of Scheubel's book of 1550 was reprinted in 1551 in Paris, under the title *Algebrae compendiosa facilisqve descriptio.*[2]

[1] "Solent tamen multi, et bene etiam, has desideratas radices, suis punctis cum lines quadam à dextro latere ascendente, notare, atque sic pro radice quidem quadrata, ubi haec in aliquo numero desideratur, notam $\sqrt{}$: pro cubica uerò, $\wedge\!\!\wedge\!\!\sqrt{}$: ac radicis radice deinde, $\wedge\!\!\wedge\!\!\sqrt{}$ praeponunt."

[2] Our information on the 1551 publication is drawn from H. Staigmüller, "Johannes Scheubel, ein deutscher Algebraiker des XVI. Jahrhunderts," *Abhandlungen zur Geschichte der Mathematik*, Vol. IX (Leipzig, 1899), p. 431–69; A. Witting and M. Gebhardt, *Beispiele zur Geschichte der Mathematik*, II. Teil

It is of importance as representing the first appearance in France of the symbols $+$ and $-$ and of some other German symbols in algebra.

Charles Hutton says of Scheubel's *Algebrae compendiosa* (1551): "The work is most beautifully printed, and is a very clear though succinct treatise; and both in the form and matter much resembles a modern printed book."[1]

MALTESE: WIL. KLEBITIUS
(1565)

160. Through the courtesy of Professor H. Bosmans, of Brussels, we are able to reproduce a page of a rare and curious little volume containing exercises on equations of the first degree in one unknown number, written by Wilhelm Klebitius and printed at Antwerp in 1565.[2] The symbolism follows Scheubel, particularly in the fancy form given to the plus sign. The unknown is represented by "1R."

The first problem in Figure 65 is as follows: Find a number whose double is as much below 30,000 as the number itself is below 20,000. In the solution of the second and third problems the notational peculiarity is that $\frac{1}{3}R. - \frac{1}{3}$ is taken to mean $\frac{1}{3}R. - \frac{1}{3}R.$, and $1R. - \frac{8}{8}$ to mean $1R. - \frac{8}{8}R.$

GERMAN: CHRISTOPHORUS CLAVIUS
(1608)

161. Though German, Christophorus Clavius spent the latter part of his life in Rome and was active in the reform of the calendar. His *Algebra*[3] marks the appearance in Italy of the German $+$ and $-$ signs, and of algebraic symbols used by Stifel. Clavius is one of the very first to use round parentheses to express aggregation. From his *Algebra* we quote (p. 15): "Pleriqve auctores pro signo $+$ ponunt literam *P*, vt significet plus: pro signo vero $-$ ponunt literam *M*, vt significet minus. Sed placet nobis vti nostris signis, vt à literis distinguantur, ne confusio oriatur." Translation: "Many authors put in place of the sign $+$ the letter *P*, which signifies "plus":

(Leipzig-Berlin, 1913), p. 25; Tropfke, *op. cit.*, Vol. I (1902), p. 195, 198; Charles Hutton, *Tracts on Mathematical and Philosophical Subjects*, Vol. II (London, 1812), p. 241–43; L. C. Karpinski, *Robert of Chester's* . . . . *Al-Khowarizmi*, p. 39–41.

[1] Charles Hutton, *op. cit.*, p. 242.

[2] The title is *Insvlae Melitensis, qvam alias Maltam vocant, Historia, quaestionib. aliquot Mathematicis reddita iucundior.* At the bottom of the last page: "*Avth. Wil. Kebitio.*"

[3] *Algebra Christophori Clavii Bambergensis e Societate Iesv.* (Romae, M.DC.VIII).

likewise, for the sign — they put the letter *M*, which signifies "minus." But we prefer to use our signs; as they are different from letters, no confusion arises."

In his arithmetic, Clavius has a distinct notation for "fractions of fractional numbers," but strangely he does not use it in the ordinary

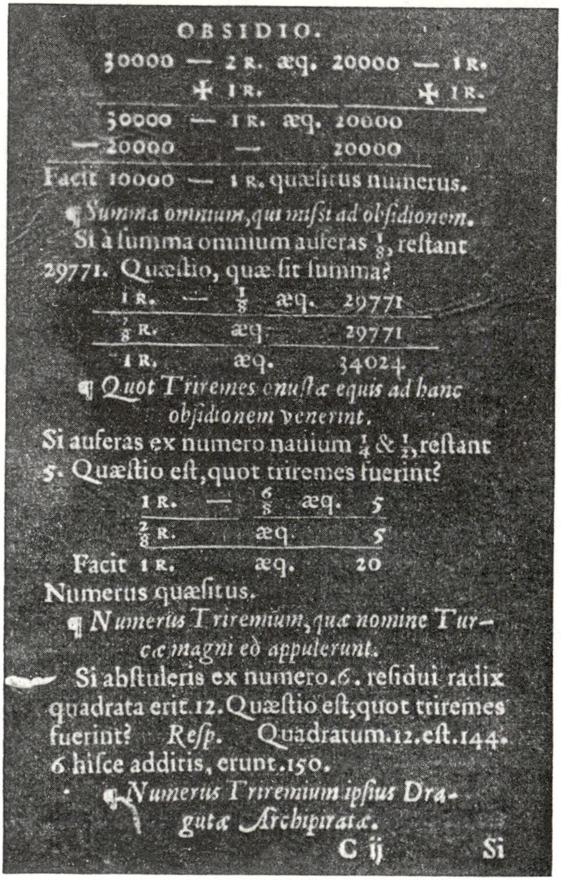

OBSIDIO.

$$30000 - 2\,\text{R. æq. } 20000 - 1\,\text{R.}$$
$$+ 1\,\text{R.} \qquad\qquad + 1\,\text{R.}$$

$$30000 - 1\,\text{R. æq. } 20000$$
$$-20000 \qquad - \qquad 20000$$

Facit 10000 — 1 R. quæsitus numerus.

¶ *Summa omnium, qui misi ad obsidionem.*

Si à summa omnium auferas $\frac{1}{3}$, restant 29771. Quæstio, quæ sit summa?

$$1\,\text{R.} \quad - \quad \tfrac{1}{3} \quad \text{æq.} \quad 29771$$
$$\tfrac{2}{3}\,\text{R.} \qquad \text{æq.} \qquad 29771$$
$$1\,\text{R,} \qquad \text{æq.} \qquad 34024$$

¶ *Quot Triremes onustæ equis ad hanc obsidionem venerint.*

Si auferas ex numero nauium $\frac{1}{4}$ & $\frac{1}{2}$, restant 5. Quæstio est, quot triremes fuerint?

$$1\,\text{R.} \quad - \quad \tfrac{6}{8} \quad \text{æq.} \quad 5$$
$$\tfrac{2}{8}\,\text{R.} \qquad \text{æq.} \qquad 5$$

Facit 1 R.   æq.   20

Numerus quæsitus.

¶ *Numerus Triremium, quæ nomine Tur—cæ magni eò appulerunt.*

Si abstuleris ex numero.6. residui radix quadrata erit.12. Quæstio est, quot triremes fuerint? *Resp.* Quadratum.12.est.144. 6.hisce additis, erunt.150.

¶ *Numerus Triremium ipsius Dra—gutæ Archipiratæ.*

C ij   Si

Fig. 65.—Page from W. Klebitius (1565)

multiplication of fractions. His $\frac{3}{5} \cdot \frac{4}{7} \cdot$ means $\frac{3}{5}$ of $\frac{4}{7}$. He says: "Vt praedicta minutia minutiae ita scribenda est $\frac{3}{5} \cdot \frac{4}{7} \cdot$ pronuntiaturque sic. Tres quintae quatuor septimarū vnius integri."[1] Similarly, $\frac{2}{3} \cdot \frac{3}{4} \cdot \frac{1}{6} \cdot \frac{1}{2} \cdot$ yields $\frac{6}{144}$. The distinctive feature in this notation is the

[1] *Epitome arithmeticae* (Rome, 1583), p. 68; see also p. 87.

omission of the fractional line after the first fraction.[1] The dot cannot be considered here as the symbol of multiplication. No matter what the operation may be, all numbers, fractional or integral, in the

C A P XXVIII. 159

Sit rurſus Binomium primum 72 + √ɀ 2880. Maius nomen 72. ſecabitur in duas partes producentes 720. quartam partem quadrati 2880. maioris nominis , hac ratione .

Semiſſis maioris nominis 72. eſt 36. a cuius quadrato 1296. detracta quarta pars prædicta 720. relinquit 576. cuius radix 24. addita ad ſemiſſem nominatam 36. & detracta ab eadem, facit partes quæſitas 60. & 12. Ergo radix Binomij eſt √ɀ 60 + √ɀ 12. quod

$$\begin{array}{r} √ɀ\ 60 + √ɀ\ 12 \\ √ɀ\ 60 + √ɀ\ 12 \\ \hline + √ɀ.\ 720 + 12 \\ 60 + √ɀ\ 720 \quad\ \\ \hline 72 + √ɀ\ 2880 \end{array}$$

hic probatum eſt per multiplicationem radicis in ſe quadratè .

Sit quoque elicienda radix ex hoc reſiduo ſexto √ɀ 60 — √ɀ 12. Maius nomen √ɀ 60. diſtribuetur in duas partes producétes 3. quartam partem quadrat: 12. minoris nominis, hoc pacto . Semiſſis maioris nominis √ɀ 60. eſt √ɀ 15. a cuius quadrato 15. detracta nominata pars quarta 3. relinquit 12. cuius radix √ɀ 12. addita ad ſemiſſem √ɀ 15. prædictam, & ab eadem ſublata facit partes √ɀ 15 + √ɀ 12. & √ɀ 15 — √ɀ 12. Ergo radix dicti Reſidui ſexti eſt √ɀ (√ɀ 15 + √ɀ 12) — √ɀ (√ɀ 15 — √ɀ 12) quod hic probatum eſt .

$$\begin{array}{r} √ɀ\ (√ɀ\ 15 + √ɀ\ 12) — √ɀ\ (√ɀ\ 15 — √ɀ\ 12) \\ √ɀ\ (√ɀ\ 15 + √ɀ\ 12) — √ɀ\ (√ɀ\ 15 — √ɀ\ 12) \\ \hline \end{array}$$

Quadrata partium. √ɀ 15 + √ɀ 12 & √ɀ 15 — √ɀ 12.
$$— √ɀ\ 3$$
$$— √ɀ\ 3$$

Summa . √ɀ 60 — √ɀ 12

Nam quadrata partium faciunt √ɀ 60. nimirum duplum √ɀ 15. Et ex vna parte √ɀ (√ɀ 15 + √ɀ 12) in alteram — √ɀ (√ɀ 15 — √ɀ 12) fit — √ɀ 3. quippe cum quadratum 12. ex quadrato 15. ſubductum relinquat 3. cui præponendum eſt ſignum √ɀ. cum ſigno —. propter Reſiduum . Duplum autem — √ɀ 3. facit — √ɀ 12.

Fig. 66.—A page in Clavius' *Algebra* (Rome, 1608). It shows one of the very earliest uses of round parentheses to express aggregation of terms.

arithmetic of Clavius are followed by a dot. The dot made the numbers stand out more conspicuously.

[1] In the edition of the arithmetic of Clavius that appeared at Cologne in 1601, p. 88, 126, none of the fractional lines are omitted in the foregoing passages.

As symbol of the unknown quantity Clavius uses[1] the German $\mathfrak{X}$. In case of additional unknowns, he adopts $1A$, $1B$, etc., but he refers to the notation $1q$, $2q$, etc., as having been used by Cardan, Nonius, and others, to represent unknowns. He writes: $3\mathfrak{X}+4A$, $4B-3A$ for $3x+4y$, $4z-3y$.

Clavius' *Astrolabium* (Rome, 1593) and his edition of the last nine books of *Euclid* (Rome, 1589) contain no algebraic symbolism and are rhetorical in exposition.

BELGIUM: SIMON STEVIN

(1585)

162. Stevin was influenced in his notation of powers by Bombelli, whose exponent placed in a circular arc became with Stevin an exponent inside of a circle. Stevin's systematic development of decimal fractions is published in 1585 in a Flemish booklet, *La thiende*,[2] and also in French in his *La disme*. In decimal fractions his exponents may be interpreted as having the base one-tenth. Page 16 (in Fig. 67) shows the notation of decimal fractions and the multiplication of 32.57 by 89.46, yielding the product 2913.7122. The translation is as follows:

"III. *Proposition, on multiplication:* Being given a decimal fraction to be multiplied, and the multiplier, to find their product.

"*Explanation of what is given:* Let the number to be multiplied be 32.57, and the multiplier 89.46. *Required*, to find their product. *Process:* One places the given numbers in order as shown here and multiplies according to the ordinary procedure in the multiplication of integral numbers, in this wise: [see the multiplication].

"Given the product (by the third problem of our *Arithmetic*) 29137122; now to know what this means, one adds the two last of the given signs, one (2) and the other (2), which are together (4). We say therefore that the sign of the last character of the product is (4), the which being known, all the others are marked according to their successive positions, in such a manner that 2913.7122 is the required product. *Proof:* The given number to be multiplied 32.57 (according to the third definition) is equal to $32\frac{5}{10}\frac{7}{100}$, together $32\frac{57}{100}$. And for the same reason the multiplier 89.46 becomes $89\frac{46}{100}$. Multiplying the said $32\frac{57}{100}$ by the same, gives a product (by the twelfth problem of our *Arithmetic*) $2913\frac{71222}{10000000}$; but this same value has also the said product 2913.7122; this is therefore the correct product, which we

[1] *Algebra*, p. 72.

[2] A facsimile edition of *La "thiende"* was brought out in 1924 at Anvers by H. Bosmans.

were to prove. But let us give also the reason why ② multiplied by ②, gives the product ④ (which is the sum of their numbers), also why ④ times ⑤ gives the product ⑨, and why ◎ times ③ gives ③, etc. We take $\frac{2}{10}$ and $\frac{3}{100}$ (which by the third definition of this Disme are .2 and .03; their product is $\frac{6}{1000}$ which, according to our third definition, is equal to .006. Multiplying, therefore, ① by ② gives the

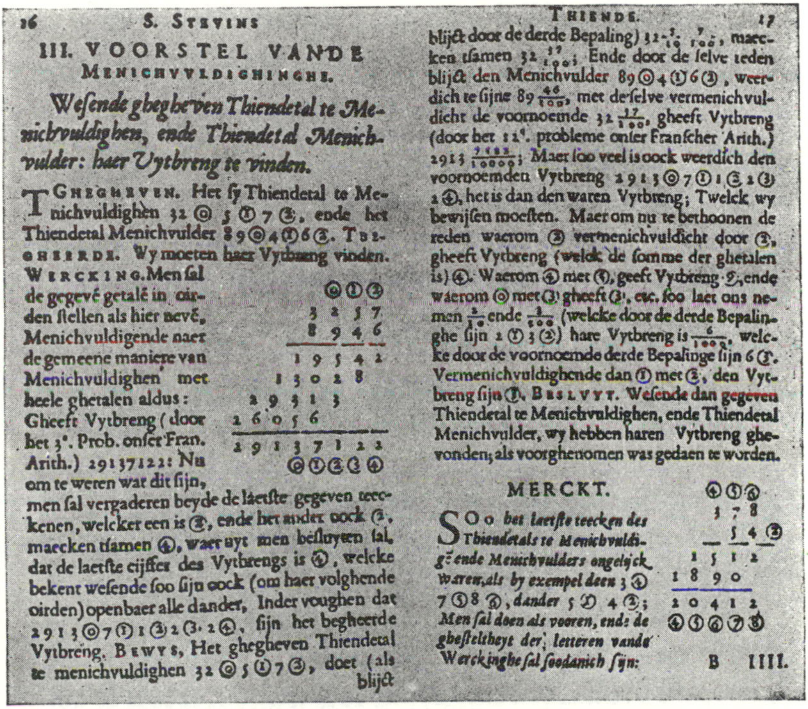

16 S. STEVINS

III. VOORSTEL VANDE MENICHVVLDIGHINGHE.

Wesende ghegheven Thiendetal te Menichvuldighen, ende Thiendetal Menichvulder: haer Vytbreng te vinden.

TGHEGHEVEN. Het sy Thiendetal te Menichvuldighen 32 ◎ 5 ① 7 ②, ende het Thiendetal Menichvulder 89 ◎ 4 ① 6 ②. TBEGHEERDE. Wy moeten haer Vytbreng vinden. WERCKING. Men sal de gegevē getalē in oirden stellen als hier nevē, Menichvuldigende naer de gemeene maniere van Menichvuldighen met heele ghetalen aldus: Gheeft Vytbreng (door het 3°. Prob. onser Fran. Arith.) 29137122: Nu om te weten wat dit sijn, men sal vergaderen beyde de laetste gegeven teecken, welcker een is ②, ende het ander oock ②, maecken tsamen ④, waer uyt men besluyten sal, dat de laetste cijffer des Vytbrengs is ④, welcke bekent wesende soo sijn oock (om haer volghende oirden) openbaer alle dander, Inder voughen dat 2913 ◎ 7 ① 1 ② 2 ③ 2 ④, sijn het begheerde Vytbreng. BEWYS. Het ghegheven Thiendetal te menichvuldighen 32 ◎ 5 ① 7 ②, doet (als blijckt

					◎	①	②
				3	2	5	7
					8	9	4	6
			1	9	5	4	2
		1	3	0	2	8	
	2	9	3	1	3		
2	6	0	5	6			
2	9	1	3	7	1	2	2
			◎	①	②	③	④

THIENDE. 17

blijckt door de derde Bepaling) $32\frac{5}{10}\frac{7}{100}$, maecken tsamen $32\frac{57}{100}$; Ende door de selve reden blijckt den Menichvulder 89 ◎ 4 ① 6 ②, weerdich te sijne $89\frac{46}{100}$, met deselve vermenichvuldicht de voornoemde $32\frac{57}{100}$, gheeft Vytbreng (door het 12°. probleme onser Franscher Arith.) $2913\frac{7122}{10000}$; Maer soo veel is oock weerdich den voornoemden Vytbreng 2913 ◎ 7 ① 1 ② 2 ③ 2 ④, het is dan den waren Vytbreng; Twelck wy bewijsen moesten. Maer om nu te bethoonen de reden waerom ② vermenichvuldicht door ②, gheeft Vytbreng (welck de somme der ghetalen is) ④. Waerom ④ met ⑤, geeft Vytbreng ⑨, ende waerom ◎ met ③ gheeft ③, etc. soo laet ons nemen $\frac{2}{10}$ ende $\frac{3}{100}$ (welcke door de derde Bepalinghe sijn 2 ① 3 ②) hare Vytbreng is $\frac{6}{1000}$, welcke door de voornoemde derde Bepalinge sijn 6 ③. Vermenichvuldighende dan ① met ②, den Vytbreng sijn ③. BESLVYT. Wesende dan gegeven Thiendetal te Menichvuldighen, ende Thiendetal Menichvulder, wy hebben haren Vytbreng ghevonden; als voorghenomen was gedaen te worden.

MERCKT.

SOo het laetste teecken des Thiendetals te Menichvuldighen ende Menichvulders ongelijck waren, als by exempel deen 3 ④ 7 ⑤ 8 ⑥, dander 5 ① 4 ②; Men sal doen als vooren, ende de ghesteltheyt der letteren vande Werckinghe sal soodanich sijn:

		④	⑤	⑥	
		3	7	8	
			5	4	②
	1	5	1	2	
1	8	9	0		
2	0	4	1	2	
④	⑤	⑥	⑦	⑧	

B IIII.

FIG. 67.—Two pages in S. Stevin's *Thiende* (1585). The same, in French, is found in *Les œuvres mathématiques de Simon Stevin* (ed. A. Girard; Leyden, 1634), p. 209.

product ③, a number made up of the sum of the numbers of the given signs. *Conclusion:* Being therefore given a decimal number as a multiplicand, and also a multiplier, we have found their product, as was to be done.

"*Note:* If the last sign of the numbers to be multiplied is not the same as the sign of the last number of the multiplier, if, for example, the one is 3④7⑤8⑥, and the other 5①4②, one proceeds as above and the disposition of the characters in the operation is as shown: [see process on p. 17]."

A translation of the *La disme* into English was brought out by Robert Norman at London in 1608 under the title, *Disme: The Art of*

QVESTION XX.

T Rouvons un ⊙ tel, que son quarré — 12 multiplié par la somme du double d icelui ⊙, & le quarré de — 2 & 4, le produict soit egal au quarré du produict de — 2 par icelui ⊙ requis.

CONSTRVCTION.

Soit le nombre requis   1 ① │ 4

Son quarré 1 ②, auquel ajousté — 12
  faict   1 ② — 12 │ 4

Qui multiplié par la somme du double du nombre requis, & le quarré de — 2 & 4, qui est
  par   2 ① + 8, faict 2 ③ + 8 ② — 24 ① — 96 │ 64

Egal au quarré du produict de — 2, par 1 ①
  premier en l'ordre, qui est à
          4 ②

  Lesquels reduicts, 1 ③ sera egale à — 2 ② + 12 ① +
48; Et 1 ① par le 71 probleme, vaudra 4.

Je di que 4 est le nombre requis. *Demonstration.* Le quarre de 4 est 16, qui avec — 12 faict 4, qui multiplié par 16 (16 pour la somme du double d'iceluy 4, & le quarre de — 2 & encore 4) faict 64, qui sont egales au quarré du produict de — 2, par le 4 trouvé, selon le requis; ce qu'il falloit demonstrer.

QVE-

Fig. 68.—From p. 98 of *L'arithmétique* in Stevin's *Œuvres mathématiques* (Leyden, 1634).

Tenths, or Decimall Arithmetike. Norman does not use circles, but round parentheses placed close together, the exponent is placed high, as in (²). The use of parentheses instead of circles was doubtless typographically more convenient.

Stevin uses the circles containing numerals also in algebra. Thus a circle with 1 inside means x, with 2 inside means x^2, and so on. In Stevin's *Œuvres* of 1634 the use of the circle is not always adhered to. Occasionally one finds, for x^4, for example,[1] the signs $(\overline{4})$ and (4).

The translation of Figure 68 is as follows: "To find a number such that if its square -12, is multiplied by the sum of double that number and the square of -2 or 4, the product shall be equal to the square of the product of -2 and the required number.

Solution

"Let the required number be . . . . x | 4

Its square x^2, to which is added -12 gives x^2-12 | 4

This multiplied by the sum of double the required number and the square of -2 or 4, i.e., by $2x+8$, gives $2x^3+8x^2-24x-96$ equal to the | 64
square of the product of -2 and x, i.e., equal to. . . . $4x^2$ Which reduced, $x^3 = -2x^2+12x+48$; and x, by the problem 71, becomes 4. I say that 4 is the required number.

"*Demonstration:* The square of 4 is 16, which added to -12 gives 4, which multiplied by 16 (16 being the sum of double itself 4, and the square of -2 or 4) gives 64, which is equal to the square of the product of -2 and 4, as required; which was to be demonstrated."

If more than one unknown occurs, Stevin marks[2] the first unknown "$1\odot$," the second "1 *secund.* $\odot$," and so on. In solving a Diophantine problem on the division of 80 into three parts, Stevin represents the first part by "$1\odot$," the second by "1 *secund.* $\odot$," the third by "$-\textcircled{1} -1$ *secund.* $\textcircled{1}+80$." The second plus $\frac{1}{6}$ the first $+6$ minus the binomial $\frac{1}{6}$ the second $+7$ yields him "$\frac{5}{6}$ *secund.* $\odot+\frac{1}{6}\odot-1$." The sum of the third and $\frac{1}{6}$ the second, $+7$, minus the binomial $\frac{1}{7}$ the third $+8$ yields him "$\frac{6}{7}\odot-\frac{29}{42}$ *secund.* $\odot+\frac{473}{7}$." By the conditions of the problem, the two results are equal, and he obtains "1 *Secund.* $\odot$ *Aequalem* $-\frac{111}{160}\odot+45$." In his *L'arithmétique*[3] one finds "12 *sec.* $\textcircled{4}+23\textcircled{1}M$ *sec.* $\textcircled{2}+10\textcircled{2}$," which means $12y^4+23xy^2+10x^2$, the M signifying here "multiplication" as it had with Stifel (§ 154). Stevin uses also D for "division."

163. For radicals Stevin uses symbols apparently suggested by

[1] *Les Œuvres mathématiques de Simon Stevin* (1634), p. 83, 85.

[2] Stevin, *Tomvs Qvintvs mathematicorvm Hypomnematvm de Miscellaneis* (Leiden, 1608), p. 516.

[3] Stevin, *Œuvres mathématiques* (Leyden, 1634), p. 60, 91, of "Le II. livre d'arith."

those of Christoff Rudolff, but not identical with them. Notice the shapes of the radicals in Figure 69. One stroke yields the usual square root symbol $\sqrt{}$, two strokes indicate the fourth root, three strokes the eighth root, etc. Cube root is marked by $\sqrt{}$ followed by a 3 inside a circle; $\sqrt{}\hspace{-0.1em}/$ followed by a 3 inside a circle means the cube root twice taken, i.e., the ninth root. Notice that $\sqrt{3}\,\text{)(}\,②$ means $\sqrt{3}$ times x^2, not $\sqrt{3x^2}$; the)(is a sign of separation of factors. In place of the u or v to express "universal" root, Stevin uses *bino* ("binomial") root.

Stevin says that $\tfrac{1}{2}$ placed within a circle means $x^{\frac{1}{2}}$, but he does not actually use this notation. His words are (p. 6 of *Œuvres* [*Arithmetic*]), "$\tfrac{1}{2}$ en un circle seroit le charactere de racine quarrée de ③, par ce que telle $\tfrac{1}{2}$ en circle multipliée en soy donne produict ③, et ainsi des autres." A notation for fractional exponents had been suggested much earlier by Oresme (§ 123).

LORRAINE: ALBERT GIRARD
(1629)

164. Girard[1] uses $+$ and $-$, but mentions $\div$ as another sign used for "minus." He uses $=$ for "difference entre les quantitez où il se treuve." He introduces two new symbols: ff, *plus que*; §, *moins que*. In further explanation he says: "Touchant les lettres de l'Alphabet au lieu des nombres: soit A & aussi B deux grandeurs: la somme est $A+B$, leur différence est $A=B$, (ou bien si A est majeur on dira que c'est $A-B$) leur produit est AB, mais divisant A par B viendra $\dfrac{A}{B}$ comme és fractions: les voyelles se posent pour les choses incognues." This use of the vowels to represent the unknowns is in line with the practice of Vieta.

The marks (2), (3), (4), ... , indicate the second, third, fourth, ... , powers. When placed before, or to the left, of a number, they signify the respective power of that number; when placed after a number, they signify the power of the unknown quantity. In this respect Girard follows the general plan found in Schöner's edition of the *Algebra* of Ramus. But

BRIEFVE COLLECTION DES CHARACTERES QV'ON VSERA EN CESTE ARITHMETIQVE.

Veu que la cognoissance des characteres est de grande consequence, par ce qu'on les vse en l'Arithmetique au lieu des mots, nous les ajousterons icy, (combien qu'au precedent chascun a esté amplement declaré

[Continued on page 159]

[1] *Invention nouvelle en l'Algebre, A Amsterdam* (M.DC.XXIX); réimpression par Dr. D. Bierens de Haan (Leiden, 1884), fol. *B*.

Girard adopts the practice of Stevin in using fractional exponents. Thus, "$(\frac{3}{2})49$" means $(\sqrt{49})^3 = 343$, while "$49(\frac{3}{2})$" means $49x^{\frac{3}{2}}$. He points out that 18(0) is the same as 18, that (1)18 is the same as 18(0).

We see in Girard an extension of the notations of Chuquet, Bombelli, and Stevin; the notations of Bombelli and Stevin are only variants of that of Chuquet.

The conflict between the notation of roots by the use of fractional exponents and by the use of radical signs had begun at the time of Girard. "Or pource que $\sqrt{}$ est en usage, on le pourra prendre au lieu de $(\frac{1}{2})$ à cause aussi de sa facilité, signifiant racine seconde, ou racine quarée; que si on veut poursuivre la progression on pourra au lieu de $\sqrt{}$ marquer $\sqrt{}$; & pour la racine cubique, ou tierce, ainsi $\sqrt{}$ ou bien $(\frac{1}{3})$, ou bié $\mathcal{C}$, ce qui peut estre au choix, mais pour en dire mon opinion les fractions sont plus expresses & plus propres à exprimer en perfection, & $\sqrt{}$ plus faciles et expedientes, comme $\sqrt{}$ 32 est à dire la racine de 32, & est 2. Quoy que ce soit l'un & l'autre sont facils

en la definition) par ordre tous ensemble côme s'ensuit.

Les characteres signifians quantitez, desquels l'explication se trouve es 14.15.16.17.18. definitions, sont tels.

⊙ Commencement de quantité qui est nombre Arith. ou radical quelconque.
① prime quantité.
② seconde quantité.
③ tierce quantité.
④ quarte quantité, &c.

Les characteres signifians postposées quantitez, desquels l'explication se trouve à la 28 definition, sont tels:

1 sec① Vne prime quantité secondement posée.
4 ter③ Quatre secondes quantitez tiercement posées, ou procedans de la prime quantité tiercement posée.
1 ① sec① Produict d'une prime quantité par une prime quantité secondement posée.
5 ④ ter② Produict de cinq quartes quantitez par une seconde quantité tiercement posée.

Les characteres signifians racine, desquels l'explication se trouve à la 29 & 30 definition sont tels:

$\sqrt{}$ Racine de quarré.
$\sqrt{}$ Racine de racine de quarré.
$\sqrt{}$ Racine de racine de racine de quarré.
$\sqrt{}$ Racine de racine de racine de racine de quarré.
$\sqrt{}$③ Racine de cube.
$\sqrt{}$③ Racine de racine de cube.
$\sqrt{}$④ Racine de quarte quantité.
$\sqrt{}$⑥ Racine de racine de quarte quantité, &c.

Le charactere signifiant la separation entre le signe de racine & la quantité, duquel l'explication se trouve à la 34. definition, est tel.

χ, Comme $\sqrt{}$ 3 χ ② n'est pas le mesme que $\sqrt{}$ 3 ②, comme dict est à ladicte 34. definition.

Les characteres signifians plus & moins, comme à la 36 definition, sont tels:

+ Plus.
— Moins.

Et pour expliquer la racine d'un multinomie (qu'aucuns appellent racine universelle) nous userons le vocable du multinomie, comme:

$\sqrt{}$ bino 2 + $\sqrt{}$ 3, c'est à dire racine quarrée de binomie, ou de la somme de 2 & $\sqrt{}$ 3.
$\sqrt{}$ trino $\sqrt{}$ 3 + $\sqrt{}$ 2 — $\sqrt{}$ 5, c'est à dire racine quarrée de trinomie, ou de la somme de $\sqrt{}$ 3 & $\sqrt{}$ 2 & — $\sqrt{}$ 5.
$\sqrt{}$ ③ bino $\sqrt{}$ 2 + $\sqrt{}$ 3, c'est à dire racine cubique de binomie $\sqrt{}$ 2 + $\sqrt{}$ 3.
$\sqrt{}$ bino 2 ② + 1 ①, c'est à dire racine quarrée de binomie 2 ② + 1 ①.
$\sqrt{}$ ③ bino 2 ② + 1 ①, c'est à dire racine cubique de binomie 2 ②, + 1 ①, &c.

FIG. 69.—From S. Stevin's *L'arithmétique* in *Œuvres mathématiques* (ed. A. Girard; Leyden, 1634), p. 19.

à comprendre, mais $\sqrt{}$ et $\mathcal{C}$ sont pris pour facilité." Girard appears to be the first to suggest placing the index of the root in the opening of the radical sign, as $\sqrt{}$. Sometimes he writes $\sqrt{}\sqrt{}$ for $\sqrt{}$.

The book contains other notations which are not specially explained. Thus the cube of $B+C$ is given in the form $B(B_q+C_q^3)+C(B_q^3+C_q)$.

We see here the use of round parentheses, which we encountered before in the *Algebra* of Clavius and, once, in Cardan. Notice also that C_q^3 means here $3C^2$.

Autre exemple	In Modern Symbols
"Soit 1(3) esgale à $-6(1)+20$	Let $x^3 = -6x+20$
Divisons tout par 1(1)	Divide all by x,
1(2) esgale à $-6+\dfrac{20}{1(1)}$."	$x^2 = -6+\dfrac{20}{x}$.

Again (fol. *F*3): "Soit 1(3) esgale à 12(1)-18 (impossible d'estre esgal)

    car le $\frac{1}{3}$ est 4   9 qui est $\frac{1}{2}$ de 18

    son cube 64   81 son quarré .

Et puis que 81 est plus que 64, l'equation est impossible & inepte."

Translation: "Let $x^3 = 12x-18$ (impossible to be equal)

    because the $\frac{1}{3}$ is 4   9 which is $\frac{1}{2}$ of 18

    its cube 64   81 its square

And since 81 is more than 64, the equation is impossible and inept."

A few times Girard uses parentheses also to indicate multiplication (see *op. cit.*, folios C_3^b, D_3^a, F_4^b).

GERMAN-SPANISH: MARCO AUREL
(1552)

165. Aurel states that his book is the first algebra published in Spain. He was a German, as appears from the title-page: *Libro primero de Arithmetica Algebratica ... por Marco Aurel, natural Aleman* (Valencia, 1552).[1] It is due to his German training that German algebraic symbols appear in this text published in Spain. There is hardly a trace in it of Italian symbolism. As seen in Figure 70, the plus $(+)$ and minus $(-)$ signs are used, also the German symbols for powers of the unknown, and the clumsy Rudolffian symbols for roots of different

[1] Aurel's algebra is briefly described by Julio Rey Pastor, *Los mathemáticos españoles del siglo XVI* (Oviedo, 1913), p. 36 n.; see *Bibliotheca mathematica*, Vol. IV (2d ser., 1890), p. 34.

orders. In place of the dot, used by Rudolff and Stifel, to express the root of a polynomial, Aurel employs the letter v, signifying universal root or *rayz vniuersal*. This v is found in Italian texts.

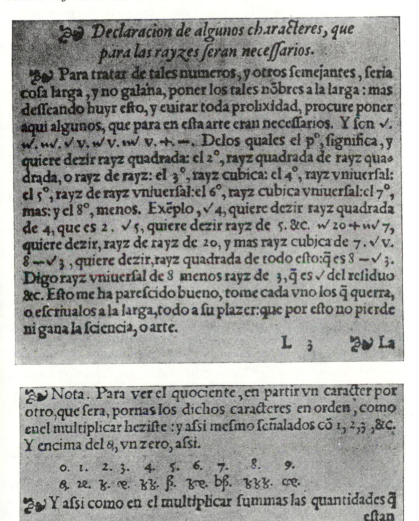

Fig. 70.—From Aurel's *Arithmetica algebratica* (1552). (Courtesy of the Library of the University of Michigan.) Above is part of fol. 43, showing the $+$ and $-$, and the radical signs of Rudolff, also the $\sqrt{}v$. Below is a part of fol. 73*B*, containing the German signs for the powers of the unknown and the sign for a given number.

166. Nuñez' *Libro de algebra* (1567)[1] bears in the Dedication the date December 1, 1564. The manuscript was first prepared in the Portuguese language some thirty years previous to Nuñez' preparation of this Spanish translation. The author draws entirely from Italian authors. He mentions Pacioli, Tartaglia, and Cardan.

The notation used by Nuñez is that of Pacioli and Tartaglia. He uses the terms *Numero, cosa, censo, cubo, censo de censo, relato primo, censo de cubo* or *cubo de censo, relato segundo, censo de censo de cēso, cubo de cubo, censo de relato primo*, and their respective abbreviations *co., ce., cu., ce.ce., re.p⁰, ce.cu.* or *cu.ce., re.seg⁰. ce.ce.ce., cu.cu., ce.re.p⁰.* He uses $\tilde{p}$ for *más* ("more"), and $\tilde{m}$ for *menos* ("les"). The only use made of the ✚ is in cross-multiplication, as shown in the following sentence (fol. 41): "... partiremos luego $\dfrac{12.}{1.co.}$ por $\dfrac{2.cu.\tilde{p}.8.}{1.ce.}$ como si fuessen puros quebrados, multiplicãdo en ✚, y verna por quociente $\dfrac{12.ce.}{2.ce.ce.\tilde{p}.8.co.}$ el qual quebrado abreuiado por numero y por dignidad verna a este quebrado $\dfrac{6.co.}{1.cu.\tilde{p}.4.}$." This expression, *multiplicando en* ✚, occurs often.

Square root is indicated by *R.*, cube root by *R.cu.*, fourth root by *R.R.*, eighth root by *R.R.R.* (fol. 207). Following Cardan, Nuñez uses *L.R.* and *R.V.* to indicate, respectively, the *ligatura* ("combination") of roots and the *Raiz vniuersal* ("universal root," i.e., root of a binomial or polynomial). This is explained in the following passage (fol. 45b): "... diziendo assi: *L.R.7p̃R.4.p̃.3.* que significa vna quantidad sorda compuesta de .3. y 2. que son 5. con la *R.7.* o diziendo assi: *L.R.3p̃2.co. Raiz vniuersal* es raiz de raiz ligada con numero o con otra raiz o dignidad. Como si dixessemos assi: *R.v. 22 p̃ R₉.*"

Singular notations are 2. *co.* ¼. for $2\frac{1}{4}x$ (fol. 32), and 2. *co.* ⅔ for $2\frac{2}{3}x$ (fol. 36b). Observe also that integers occurring in the running text are usually placed between dots, in the same way as was customary in manuscripts.

Although at this time our exponential notation was not yet invented and adopted, the notion of exponents of powers was quite well understood, as well as the addition of exponents to form the product

---

[1] *Libro de Algebra en arithmetica y Geometria. Compuesto por el Doctor Pedro Nuñez, Cosmographo Mayor del Rey de Portugal, y Cathedratico Jubilado en la Cathedra de Mathematicas en la Vniuersidad de Coymbra* (En Anvers, 1567).

of terms having the same base. To show this we quote from Nuñez the following (fol. 26b):

"... si queremos multiplicar .4. co. por .5. ce. diremos asi .4. por .5. hazen .20. y porque .1. denominaciõ de co. sũmado con .2. denominacion de censo hazen .3. que es denominaciõ de cubo. Diremos por tanto q .4. co. por .5. ce. hazen .20. cu. ... si multiplicamos .4. cu. por .8. ce.ce. diremos assi, la denominacion del cubo es .3. y la denominaciõ del censo de censo es .4. q̃ sũmadas hazẽ .7. q̃ sera la denominaciõ dela dignidad engẽdrada, y por que .4. por .8. hazen .32. diremos por tanto, que .4. cu. multiplicados por .8. ce.ce. hazen .32. dignida·les, que tienen .7. por denominacion, a que llaman relatos segundos."

Nuñez' division[1] of $12x^3+18x^2+27x+17$ by $4x+3$, yielding the quotient $3x^2+2\frac{1}{4}x+5\frac{1}{16}+\frac{1\frac{1}{16}}{4x+3}$, is as follows:

"Partidor .4.co.p̃.3 | $12.cu.p̃.18.ce.p̃.27.co.p̃.17.$
$12.cu.p̃.\ 9.ce.$

$9.ce.p̃.27.co.p̃.17.$
$9.ce.p̃.\ 6.co.\frac{9}{4}.$

$20.co.\frac{1}{4}.p̃.17.$
$20.co.\frac{1}{4}.p̃.15\frac{3}{16}.$

$1\frac{1}{16}$

3.$ce.p̃.2.co.\frac{1}{4}.p̃.5\frac{1}{16}.p̃.1\frac{1}{16}$
par .4.co.p̃.3."

Observe the "20.co.$\frac{1}{4}$" for $20\frac{1}{4}x$, the symbol for the unknown appearing between the integer and the fraction.

Cardan's solution of $x^3+3x=36$ is $\sqrt[3]{\sqrt{325}+18}-\sqrt[3]{\sqrt{325}-18}$, and is written by Nuñez as follows:

$$R.V.cu.R.325.p̃.18.m̃.R.V.cu,.R.325.m̃.18.$$

As in many other writers the V signifies *vniversal* and denotes, not the cube root of $\sqrt{325}$ alone, but of the binomial $\sqrt{325}+18$; in other words, the V takes the place of a parenthesis.

[1] See H. Bosmans, "Sur le 'Libro de algebra' de Pedro Nuñez," *Bibliotheca mathematica*, Vol. VIII (3d ser., 1908), p. 160–62; see also Tropfke, *op. cit.* (2d ed.), Vol. III, p. 136, 137.

167. Robert Recorde's arithmetic, the *Grovnd of Artes,* appeared in many editions. We indicate Recorde's singular notation for proportion:[1]

$$3 \diagdown\kern-1em\diagup 16s.$$
$$8 \diagup\kern-1em\diagdown 42s.\ 8d.$$ (direct) $3:8 = 16s.:42s.\ 8d.$

$$\tfrac{3}{4} \diagdown\kern-1em\diagup \tfrac{7}{13}$$
$$\tfrac{5}{12} \diagup\kern-1em\diagdown$$ (reverse) $\tfrac{5}{12}:\tfrac{3}{4} = \tfrac{7}{13}:x$

There is nothing in Recorde's notation to distinguish between the "rule of proportion direct" and the "rule of porportion reverse." The difference appears in the interpretation. In the foregoing "direct" proportion, you multiply 8 and 16, and divide the product by 3. In the "reverse" proportion, the processes of multiplication and division are interchanged. In the former case we have $8 \times 16 \div 3 = x$, in the second case we have $\tfrac{3}{4} \times \tfrac{7}{13} \div \tfrac{5}{12} = x$. In both cases the large strokes in $\mathbb{Z}$ serve as guides to the proper sequence of the numbers.

168. In Recorde's algebra, *The Whetstone of Witte* (London, 1557), the most original and historically important is the sign of equality (=), shown in Figure 71. Notice also the plus (+) and minus (−) signs which make here their first appearance in an English book.

In the designation of powers Recorde uses the symbols of Stifel and gives a table of powers occupying a page and ending with the eightieth power. The seventh power is denoted by $b\!\int\!{}_3$; for the eleventh, thirteenth, seventeenth powers, he writes in place of the letter b the letters c, d, E, respectively. The eightieth power is denoted by ${}_{3333}\!\int\!{}_3$, showing that the Hindu multiplicative method of combining the symbols was followed.

Figure 72 shows addition of fractions. The fractions to be added are separated by the word "to." Horizontal lines are drawn above and below the two fractions; above the upper line is written the new numerator and below the lower line is written the new denominator. In "Another Example of Addition," there are added the fractions $\dfrac{5x^6+3x^5}{6x^9}$ and $\dfrac{20x^3-6x^5}{6x^9}$

---

[1] *Op. cit.* (London, 1646), p. 175, 315. There was an edition in 1543 which was probably the first.

Square root Recorde indicates by $\sqrt{}$. or $\sqrt{3}$, cube root by $\mathcal{w}$. or $\mathcal{w}$.ȼ. Following Rudolff, he indicates the fourth root by

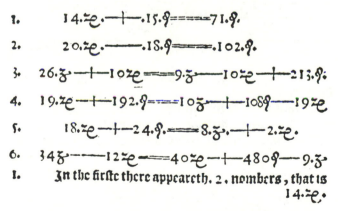

The Arte

as their workes doe extende) to diſtincte it onely into twoo partes. Whereof the firſte is, *when one number is equalle vnto one other.* And the ſeconde is, *when one number is compared as equalle vnto 2 other nombers.*

Alwaies willyng you to remēber, that you reduce your nombers, to their leaſte denominations, and ſmalleſte formes, before you procede any farther.

And again, if your *equation* be ſoche, that the greateſte denomination *Coßike,* be ioined to any parte of a compounde nomber, you ſhall tourne it ſo, that the nomber of the greateſte ſigne alone, maie ſtande as equalle to the reſte.

And this is all that neadeth to be taughte, concernyng this woorke.

Howbeit, for eaſie alteratiō of *equations.* I will propounde a fewe craples, bicauſe the extraction of their rootes, maie the more aptly bee wroughte. And to auoide the tediouſe repetition of theſe woordes: is equalle to: I will ſette as I doe often in woorke vſe, a paire of paralleles, or Gemowe lines of one lengthe, thus:======, bicauſe noe 2 thynges, can be moare equalle. And now marke theſe nombers.

Fig. 71.—From Robert Recorde's *Whetstone of Witte* (1557)

$\mathcal{w}$., but Recorde writes it also $\mathcal{w}$/₃₃. Instructive is the dialogue on these signs, carried on between master and scholar:

"*Scholar:* It were againste reason, to take reason for those signes, whiche be set voluntarily to signifie any thyng; although some tymes there bee a certaine apte conformitie in soche thynges. And in these

An other Example of Addition.

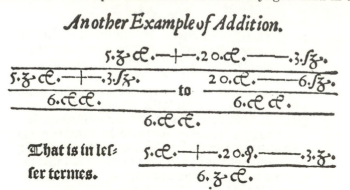

That is in lesser termes.

Here is noe multiplication, nor reduction to one common denominator: sith thei bee one all ready: nother can the nombers be reduced, to any other lesser: but the quantities onely be reduced as you see.

Scholar. I praie you let me proue.

An other Example.

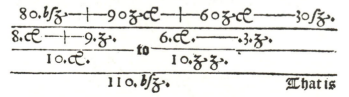

Master. Marke your worke well, before you reduce it.

Scholar. I see my faulte: I haue sette.2. nombers seuerally, with one signe *Cossike:* by reason I did not foresee, that.c.multiplied with.c.doeth make the like

Fig. 72.—Fractions in Recorde's *Whetstone of Witte* (1557)

figures, the number of their minomes, seameth disagreable to their order.

"*Master:* In that there is some reason to bee thewed: for as .$\surd$. declareth the multiplication of a nomber, ones by it self; so .$\sim\!\!\!\!\sim\!\!\!\!\surd$. representeth that multiplication *Cubike,* in whiche the roote is represented thrise. And .$\mathcal{W}$. standeth for .$\sqrt{}$.$\mathcal{N}$. that is .2. figures of *Square* multiplication: and is not expressed with .4. minomes. For so should it seme to expresse moare then .2. *Square* multiplications. But voluntarie signes, it is inoughe to knowe that this thei doe signifie.

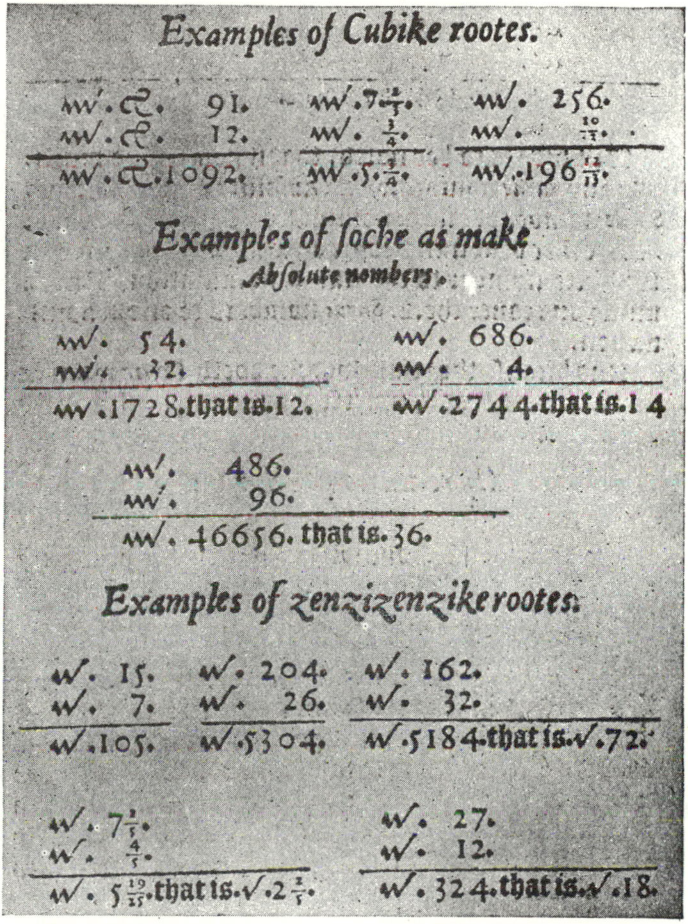

Fig. 73.—Radicals in Recorde's *Whetstone of Witte* (1557)

And if any manne can diuise other, moare easie or apter in use, that maie well be received."

Figure 73 shows the multiplication of radicals. The first two exercises are $\sqrt[3]{91} \times \sqrt[3]{12} = \sqrt[3]{1,092}$, $\sqrt[3]{7\frac{2}{3}} \times \sqrt[3]{\frac{3}{4}} = \sqrt[3]{5\frac{3}{4}}$. Under fourth roots one finds $\sqrt[4]{15} \times \sqrt[4]{7} = \sqrt[4]{105}$.

ENGLISH: JOHN DEE (1570)

169. John Dee wrote a Preface to Henry Billingsley's edition of *Euclid* (London, 1570). This Preface is a discussion of the mathematical sciences. The radical symbols shown in Figure 74 are those of Stifel. German influences predominated.

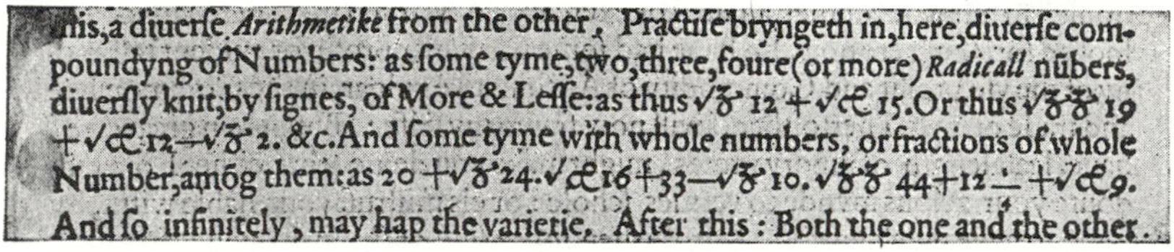
this, a diuerſe *Arithmetike* from the other. Practiſe bryngeth in, here, diuerſe compoundyng of Numbers: as ſome tyme, two, three, foure (or more) *Radicall* nūbers, diuerſly knit, by ſignes, of More & Leſſe: as thus √ȝ 12 + √ce 15. Or thus √ȝȝ 19 + √ce 12 − √ȝ 2. &c. And ſome tyme with whole numbers, or fractions of whole Number, amōg them: as 20 + √ȝ 24. √ce 16 + 33 − √ȝ 10. √ȝȝ 44 + 12 1/3 + √ce 9. And ſo infinitely, may hap the varietie. After this: Both the one and the other.

FIG. 74.—Radicals, John Dee's Preface to Billingsley's edition of *Euclid* (1570).

In Figure 75 Dee explains that if $a{:}b=c{:}d$, then also $a{:}a-b=c{:}c-d$. He illustrates this numerically by taking $9{:}6=12{:}8$. Notice Dee's use of the word "proportion" in the sense of "ratio." Attention is drawn to the mode of writing the two proportions 9.6:12.8 and 9.3:12.4, near the margin. Except for the use of a single colon (:),

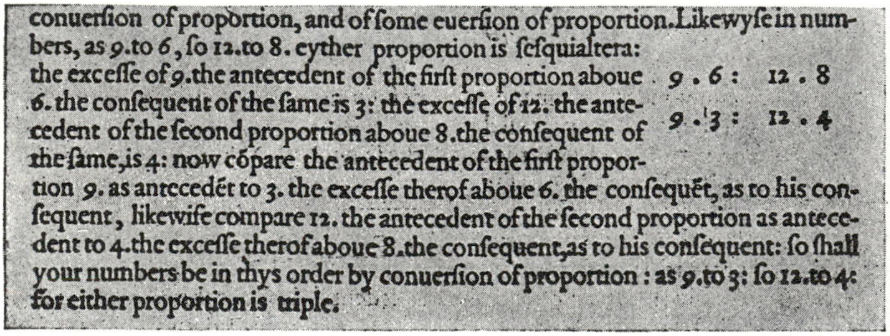
conuerſion of proportion, and of ſome euerſion of proportion. Likewyſe in numbers, as 9. to 6, ſo 12. to 8. eyther proportion is ſeſquialtera: the exceſſe of 9. the antecedent of the firſt proportion aboue 6. the conſequent of the ſame is 3: the exceſſe of 12. the antecedent of the ſecond proportion aboue 8. the conſequent of the ſame, is 4: now cōpare the antecedent of the firſt proportion 9. as antecedēt to 3. the exceſſe therof aboue 6. the conſequēt, as to his conſequent, likewiſe compare 12. the antecedent of the ſecond proportion as antecedent to 4. the exceſſe therof aboue 8. the conſequent, as to his conſequent: ſo ſhall your numbers be in thys order by conuerſion of proportion: as 9. to 3: ſo 12. to 4: for either proportion is triple.

9 . 6 : 12 . 8

9 . 3 : 12 . 4

FIG. 75.—Proportion in John Dee's Preface to Billingsley's edition of *Euclid* (1570).

in place of the double colon (: :), this is exactly the notation later used by Oughtred in his *Clavis mathematicae*. It is possible that Oughtred took the symbols from Dee. Dee's Preface also indicates the origin of these symbols. They are simply the rhetorical marks used in the text. See more particularly the second to the last line, "as 9. to 3: so 12. to 4:"

ENGLISH: LEONARD AND THOMAS DIGGES
(1579)

170. The *Stratioticos*[1] was brought out by Thomas Digges, the son of Leonard Digges. It seems that the original draft of the book was the work of Leonard; the enlargement of the manuscript and its preparation for print were due to Thomas.

The notation employed for powers is indicated by the following quotations (p. 33):

"In this Arte of Numbers Cossical, wae proceede from the Roote by Multiplication, to create all Squares, Cubes, Zenzizenzike, and Sur Solides, wyth all other that in this Science are used, the whyche by Example maye best bee explaned.

1	2	3	4	5	6	7	8	9	10	11	12
Roo.	*Sq.*	*Cu.*	*SqS.*	*S∫o.*	*SqC.*	*B∫S.*	*SSSq.*	*CC.*	*S∫S.*	*C∫S.*	*SSC.* "
2	4	8	16	32	64	128	256	512	1024	2048	4096

Again (p. 32):

". . . . Of these [Roote, Square, Cube] are all the rest composed. For the Square being four, againe squared, maketh his *Squared square* 16, with his Character ouer him. The nexte being not made by the Square or Cubike, Multiplication of any of the former, can not take his name from *Square* or *Cube*, and is therefore called a *Surd solide*, and is onely created by Multiplicatiõ of 2 the Roote, in 16 the SqS. making 32 with his cõuenient Character ouer him & for distinctiõ is tearmed $\overset{e}{y}$ first *Surd solide* . . . . the nexte being 128, is not made of square or Cubique Multiplication of any, but only by the Multiplication of the Squared Cube in his Roote, and therefore is tearmed the *B.S.solide*, or seconde S. solide. . . . .

"This I have rather added for custome sake, bycause in all parts of the world these *Characters* and names of *Sq.* and *Cu.* etc. are used, but bycause I find another kinde of *Character* by my Father deuised, farre more readie in *Multiplications, Diuisions*, and other *Cossical* operations, I will not doubt, hauing Reason on my side, to dissent from common custome in this poynt, and vse these Characters ensuing: [What follows is on page 35 and is reproduced here in Fig. 76]."

[1] *An Arithmeticall Militare Treatise, named Stratioticos: compendiously teaching the Science of Nũbers, as well in Fractions as Integers, and so much of the Rules and Aequations Algebraicall and Arte of Numbers Cossicall, as are requisite for the Profession of a Soldiour. Together with the Moderne Militare Discipline, Offices, Lawes and Dueties in euery wel gouerned Campe and Armie to be obserued: Long since attẽpted by Leonard Digges Gentleman, Augmented, digested, and lately finished, by Thomas Digges, his Sonne* . . . . (At London, 1579).

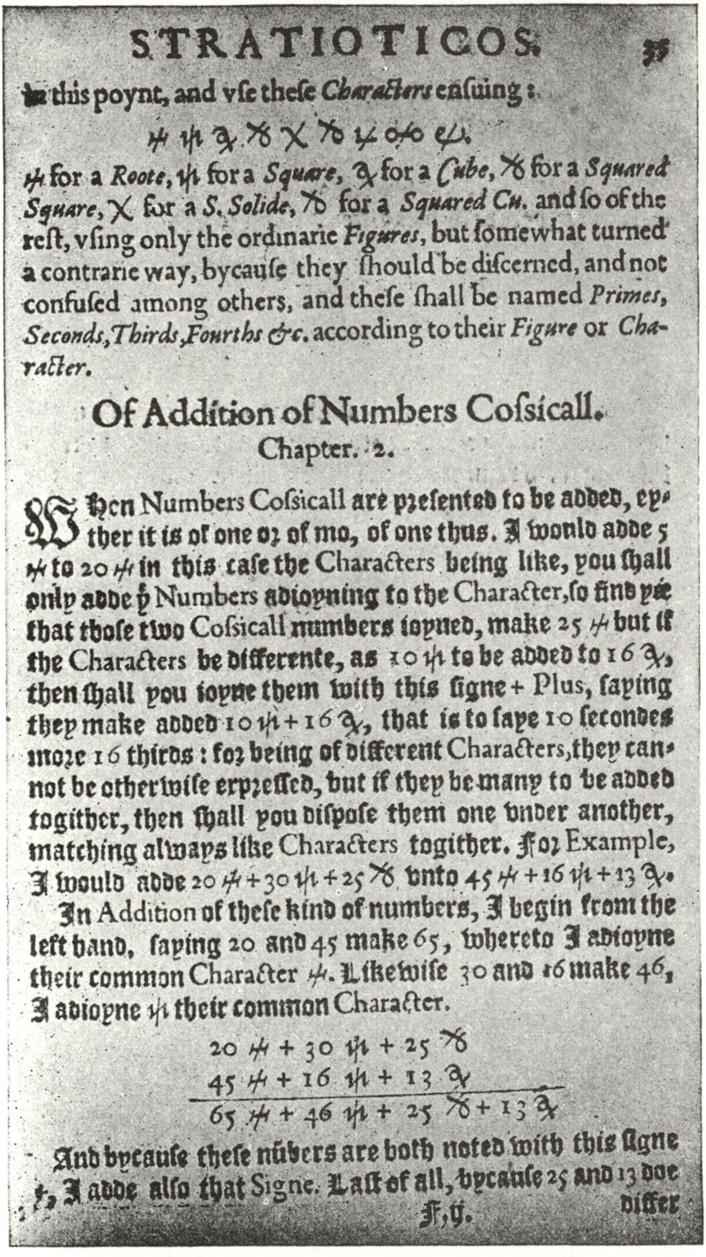

STRATIOTICOS.   35

in this poynt, and vſe theſe *Characters* enſuing :

$$\text{ᵷ ᷟ ᷟ ᷟ X ᷟ ᷟ ᷟ ᷟ}$$

ᵷ for a *Roote*, ᷟ for a *Square*, ᷟ for a *Cube*, ᷟ for a *Squared Square*, X for a *S. Solide*, ᷟ for a *Squared Cu.* and ſo of the reſt, vſing only the ordinarie *Figures*, but ſomewhat turned a contrarie way, bycauſe they ſhould be diſcerned, and not confuſed among others, and theſe ſhall be named *Primes*, *Seconds*, *Thirds*, *Fourths* &c. according to their *Figure* or *Cha-racter*.

Of Addition of Numbers Coſsicall.
Chapter. 2.

When Numbers Coſsicall are pꝛeſented to be added, ey-ther it is of one oꝛ of mo, of one thus. I woulo adde 5 ᵷ to 20 ᵷ in this caſe the Characters being like, you ſhall only adde y̆ Numbers adioyning to the Character, ſo find yee that thoſe two Coſsicall numbers ioyned, make 25 ᵷ but if the Characters be differente, as 10 ᷟ to be added to 16 ᷟ, then ſhall you ioyne them with this ſigne + Plus, ſaying they make added 10 ᷟ + 16 ᷟ, that is to ſaye 10 ſecondes moꝛe 16 thirds : foꝛ being of different Characters, they can-not be otherwiſe erpꝛeſſed, but if they be many to be added togither, then ſhall you diſpoſe them one vnder another, matching alwaps like Characters togither. Foꝛ Example, I woulo adde 20 ᵷ + 30 ᷟ + 25 ᷟ vnto 45 ᵷ + 16 ᷟ + 13 ᷟ.

In Addition of theſe kind of numbers, I begin from the left hand, ſaying 20 and 45 make 65, whereto I adioyne their common Character ᵷ. Likewiſe 30 and 16 make 46, I adioyne ᷟ their common Character.

$$20\ \text{ᵷ} + 30\ \text{ᷟ} + 25\ \text{ᷟ}$$
$$45\ \text{ᵷ} + 16\ \text{ᷟ} + 13\ \text{ᷟ}$$
$$\overline{65\ \text{ᵷ} + 46\ \text{ᷟ} + 25\ \text{ᷟ} + 13\ \text{ᷟ}}$$

And bycauſe theſe nũbers are both noted with this ſigne +, I adde alſo that Signe. Laſt of all, bycauſe 25 and 13 doe differ

ꝼ.ij.

FIG. 76.—Leonard and Thomas Digges, *Stratioticos* (1579), p. 35, showing the unknown and its powers to x^9.

As stated by the authors, the symbols are simply the numerals somewhat disfigured and crossed out by an extra stroke, to prevent confusion with the ordinary figures. The example at the bottom of page 35 is the addition of $20x+30x^2+25x^3$ and $45x+16x^2+13x^3$. It is noteworthy that in 1610 Cataldi in Italy devised a similar scheme for representing the powers of an unknown (§ 340).

The treatment of equations is shown on page 46, which is reproduced in Figure 77. Observe the symbol for zero in lines 4 and 7; this form is used only when the zero stands by itself.

A little later, on page 51, the authors, without explanation, begin to use a sign of equality. Previously the state of equality had been expressed in words, "equall to," "are." The sign of equality looks as if it were made up of two letters C in these positions $\mathcal{O}C$ and crossed by two horizontal lines. See Figure 78.

This sign of equality is more elaborate than that previously devised by Robert Recorde. The Digges sign requires four strokes of the pen; the Recorde sign demands only two, yet is perfectly clear. The Digges symbol appears again on five or more later pages of the *Stratioticos*. Perhaps the sign is the astronomical symbol for *Pisces* ("the Fishes"), with an extra horizontal line. The top equation on page 51 is $x^2=6x+27$.

ENGLISH: THOMAS MASTERSON
(1592)

171. The domination of German symbols over English authors of the sixteenth century is shown further by the *Arithmeticke* of Thomas Masterson (London, 1592). Stifel's symbols for powers are used. We reproduce (in Fig. 79) a page showing the symbols for radicals.

FRENCH: JACQUES PELETIER
(1554)

172. Jacques Peletier du Mans resided in Paris, Bordeaux, Beziers, Lyon, and Rome. He died in Paris. His algebra, *De occvlta Parte Nvmerorvm, Quam Algebram vocant, Libri duo* (Paris, 1554, and several other editions),[1] shows in the symbolism used both German and Italian influences: German in the designation of powers and roots, done in the manner of Stifel; Italian in the use of *p.* and *m.* for "plus" and "minus."

[1] All our information is drawn from H. Bosmans, "L'algèbre de Jacques Peletier du Mans," *Extrait de la revue des questions scientifiques* (Bruxelles: January, 1907), p. 1–61.

46   STRATIOTICOS.

Sometime it shall be requisite to take away some number from eyther part of the Æquation, as if I haue 6 ⅄ E-
quall to 12 ⅄ — 24, deducting from eyther part of the Æ-
quation 6 ⅄, there resteth ϙ Equall to 6 ⅄ — 24, and therefore of necessitie 6 ⅄ is equall to 24, for this Rule is gene-
rall. That if you bring an Æquation (by suche Deduction) to a ϙ on the one part, there must be some member in
the other conneced with the Signe Minus, the whiche is al-
wayes Equall to all the rest of that part of the Æquation.

Sometimes Reduction is made by adding togither all
suche parcels, as on the one side of the Æquation haue e-
qual Characters, as if 1 ✕ be Equal to 3 ꝗ + 16 ⅄ — 1 ⅄ —
10 ⅄. Héere by adding + 16 ⅄ to — 10 ⅄, there resulteth
+ 6 ⅄, so that I say 1 ✕ is equall to 3 ꝗ + 6 ⅄ — 1 ⅄, and
ϱ same diuided by 1 ⅄ maketh 1 ✗ Equal to 3 ⅄ + 6 — 1 ⅄.

Reduction of Fragments vvhich shall happen in Æquations to Integers.

Another kinde of Reduction there is of Fragmentes to
whole numbers, whiche cemneth in vse when an Æ-
quation is founde betwéene Fractions on the one or both
parts, as if $\frac{4\,⅄ + 2\,ꝗ}{2\,⅄}$ be Equall vnto $\frac{2\,ꝗ — 2\,⅄}{1\,⅄}$, by crosse
multiplication of the Denominator of the one in tha Nu-
merator of the other, I finde these two numbers produced
4 ꝗ + 2 ✗, and 6 ✗ — 4 ꝗ. Betwene these, the like Æquation remayneth, and the same first reduced by transporting
of Signes, maketh 4 ꝗ Equall to 6 ✗ — 2 ✗ — 4 ꝗ. Then
by Addition of 6 ✗ to — 2 ✗, there resulteth 4 ✗ — 4 ꝗ, e-
quall to 4 ꝗ. Againe, diuiding either part of the Æquation
by 4 ⅄, there resulteth 1 ⅄ Equall to 1 ⅄ — 1 ⅄. And last
of all, deducting 1 ⅄ from both partes of the Æquation, I
finde ϙ equall to 1 ⅄ — 2 ⅄, and therfore of necessity as was
declared.

FIG. 77.—Equations in Digges, *Stratioticos* (1579)

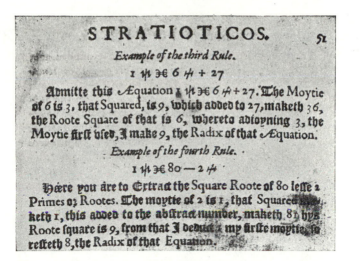

FIG. 78.—Sign of equality in Digges, *Stratioticos* (1579). This page exhibits also the solution of quadratic equations.

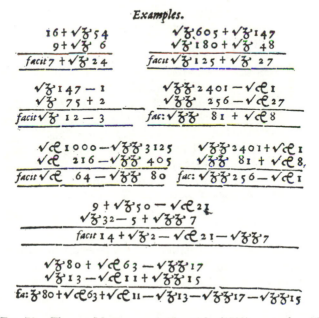

FIG. 79.—Thomas Masterson, *Arithmeticke* (1592), part of p. 45

Page 8 (reproduced in Fig. 80) is in translation: "[The arithmetical progression, according to the natural order of counting,]
furnishes us successive terms for showing the Radicand numbers
and their signs, as you see from the table given here [here appears the
table given in Fig. 80].

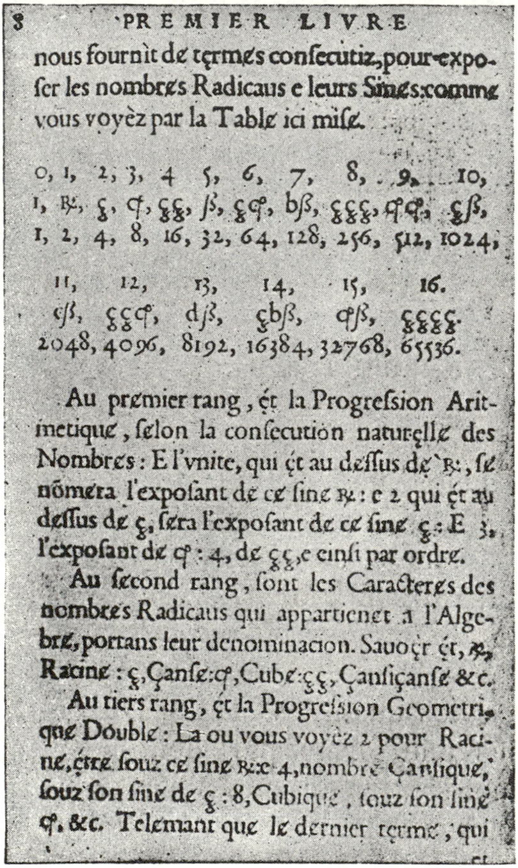

FIG. 80.—Designation of powers in J. Peletier's *Algebra* (1554)

"In the first line is the arithmetical progression, according to the
natural order of the numbers; and the one which is above the $\mathbb{R}$
numbers the exponent of this sign $\mathbb{R}$; the 2 which is above the $\mathfrak{z}$ is the
exponent of this sign $\mathfrak{z}$; and 3 is the exponent of *c*, 4 of $\mathfrak{zz}$, and so on.

"In the second line are the characters of the Radicand numbers which pertain to algebra, marking their denomination." Then are explained the names of the symbols, as given in French, viz., ℞ *racine*, ȝ *çanse*, ȼ *cube*, etc.

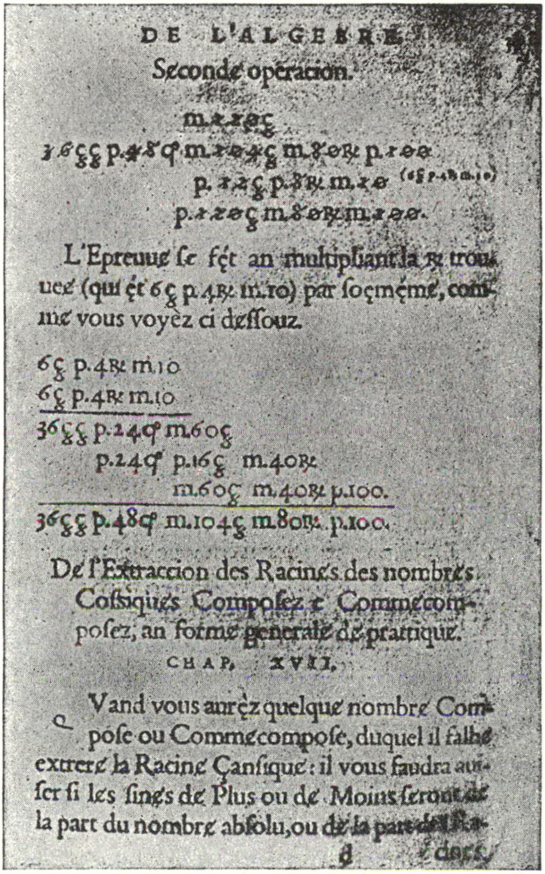

DE L'ALGEBRE.

Seconde opéracion.

m.120ç

36çç p.48ɋ m.104ç m.80℞ p.100

p.12ç p.8℞ m.10 (6ç p.4℞ m.10)

p.120ç m.80℞ m.100.

L'Epreuue se fét an multipliant la ℞ trouuee (qui ét 6ç p.4℞ m.10) par soçméme, comme vous voyèz ci dessouz.

6ç p.4℞ m.10

6ç p.4℞ m.10

36çç p.24ɋ m.60ç

p.24ɋ p.16ç m.40℞

m.60ç m.40℞ p.100.

36çç p.48ɋ m.104ç m.80℞ p.100.

De l'Extraccion des Racines des nombres Cossiques Composez e Commecomposez, an forme generale de pratique.

CHAP. XVII.

Quand vous aurèz quelque nombre Composé ou Commecomposé, duquel il falhe extrere la Racine Çansique: il vous faudra auiser si les sines de Plus ou de Moins seront de la part du nombre absolu, ou de la part des Racines.

d

FIG. 81.—Algebraic operations in Peletier's *Algebra* (1554)

Page 33 (shown in Fig. 81) begins with the extraction of a square root and a "proof" of the correctness of the work. The root extraction is, in modern symbols:

$$\begin{array}{l} \qquad\qquad\quad -120x^2 \\ 36x^4+48x^3-104x^2-80x+100 \\ \qquad\qquad\quad +\ 12x^2+\ 8x-10(6x^2+4x-10) \\ \qquad\qquad\quad +120x^2+80x-100\ . \end{array}$$

The "proof" is thus:

$$
\begin{array}{r}
6x^2 + 4x - 10 \\
6x^2 + 4x - 10 \\
\hline
36x^4 + 24x^3 - 60x^2 \\
+ 24x^3 + 16x^2 - 40x \\
- 60x^2 - 40x + 100 \\
\hline
36x^4 + 48x^3 - 104x^2 - 80x + 100
\end{array}
$$

Further on in this book Peletier gives:

$$\sqrt{3}\ 15\ p.\ \sqrt{3}8,\ \text{signifying}\ \sqrt{15} + \sqrt{8}\,.$$

$$\sqrt{3}\ .\ 15\ p.\ \sqrt{3}8,\ \text{signifying}\ \sqrt{15 + \sqrt{8}}\,.$$

FRENCH: JEAN BUTEON
(1559)

173. Deeply influenced by geometrical considerations was Jean Buteon,[1] in his *Logistica quae et Arithmetica vulgo dicitur* (Lugduni, 1559). In the part of the book on algebra he rejects the words *res*, *census*, etc., and introduces in their place the Latin words for "line," "square," "cube," using the symbols ρ, $\diamond$, $\square$. He employs also P and M, both as signs of operation and of quality. Calling the sides of an equation *continens* and *contentum*, respectively, he writes between them the sign [as long as the equation is not reduced to the simplest form and the *contentum*, therefore, not in its final form. Later the *contentum* is inclosed in the completed rectangle [ ]. Thus Buteon writes $3\rho\ M\ 7$ [8 and then draws the inferences, 3ρ [15], 1ρ [5]. Again he writes $\frac{1}{4}\diamond$ [100, hence $1\diamond$ [400], 1ρ [20]. In modern symbols: $3x - 7 = 8$, $3x = 15$, $x = 5$; $\frac{1}{4}x^2 = 100$, $x^2 = 400$, $x = 20$. Another example: $\frac{1}{8}\square\ P\ 2$ [218, $\frac{1}{8}\square$ [216, $1\square$ [1728], 1ρ [12]; in modern form $\frac{1}{8}x^3 + 2 = 218$, $\frac{1}{8}x^3 = 216$, $x^3 = 1,728$, $x = 12$.

When more than one unknown quantity arises, they are represented by the capitals A, B, C. Buteon gives examples involving only positive terms and then omits the P. In finding three numbers subject to the conditions $x + \frac{1}{2}y + \frac{1}{2}z = 17$, $y + \frac{1}{3}x + \frac{1}{3}z = 17$, $z + \frac{1}{4}x + \frac{1}{4}y = 17$, he writes:

$$1A\ ,\ \tfrac{1}{2}B\ ,\ \tfrac{1}{2}C\ [17$$
$$1B\ ,\ \tfrac{1}{3}A\ ,\ \tfrac{1}{3}C\ [17$$
$$1C\ ,\ \tfrac{1}{4}A\ ,\ \tfrac{1}{4}B\ [17$$

[1] Our information is drawn from G. Werthheim's article on Buteon, *Bibliotheca mathematica*, Vol. II (3d ser., 1901), p. 213–19.

and derives from them the next equations in the solution:

$$2A \cdot 1B \cdot 1C \text{ [34}$$
$$1A \cdot 3B \cdot 1C \text{ [51}$$
$$1A \cdot 1B \cdot 4C \text{ [68, etc.}$$

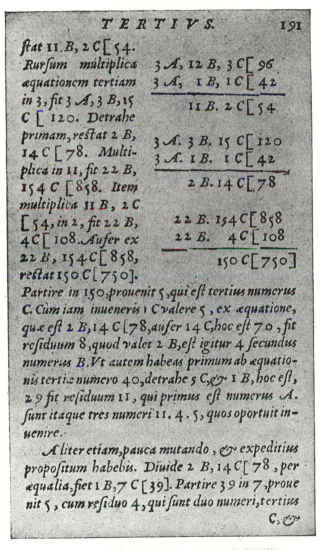

Fig. 82.—From J. Buteon, *Arithmetica* (1559)

In Figure 82 the equations are as follows:

$$
\begin{aligned}
3A + 12B + 3C &= 96 \\
3A + 1B + 1C &= 42 \\
\hline
11B + 2C &= 54
\end{aligned}
$$

$$
\begin{aligned}
3A + 3B + 15C &= 120 \\
3A + 1B + 1C &= 42 \\
\hline
2B + 14C &= 78
\end{aligned}
$$

$$
\begin{aligned}
22B + 154C &= 858 \\
22B + 4C &= 108 \\
\hline
150C &= 750
\end{aligned}
$$

FRENCH: GUILLAUME GOSSELIN
(1577)

174. A brief but very good elementary exposition of algebra was given by G. Gosselin in his *De arte magna*, published in Paris in 1577. Although the plus (+) and minus (−) signs must have been more or less familiar to Frenchmen through the *Algebra* of Scheubel, published in Paris in 1551 and 1552, nevertheless Gosselin does not use them. Like Peletier, Gosselin follows the Italians on this point, only Gosselin uses the capital letters P and M for "plus" and "minus," instead of the usual and more convenient small letters.[1] He defines his notation for powers by the following statement (chap. vi, fol. v):

$$L \cdot 2 \cdot Q \cdot 4 \cdot C \cdot 8 \cdot QQ \cdot 16 \cdot RP \cdot 32 \cdot QC \cdot 64 \cdot RS \cdot 128 \cdot CC \cdot 512 \,.$$

Here RP and RS signify, respectively, *relatum primum* and *relatum secundum*.

Accordingly,

"$12L \; M \; 1Q \; P \; 48$ aequalia $144 \; M \; 24L \; P \; 2Q$"

means

$$12x - x^2 + 48 = 144 - 24x + 2x^2 \,.$$

[1] Our information is drawn mainly from H. Bosmans' article on Gosselin, *Bibliotheca mathematica*, Vol. VII (1906–7), p. 44–66.

The translation of Figure 83 is as follows:

" . . . . Thus I multiply $4x-6x^2+7$ by $3x^2$ and there results $12x^3-18x^4+21x^2$ which I write below the straight line; then I multi-

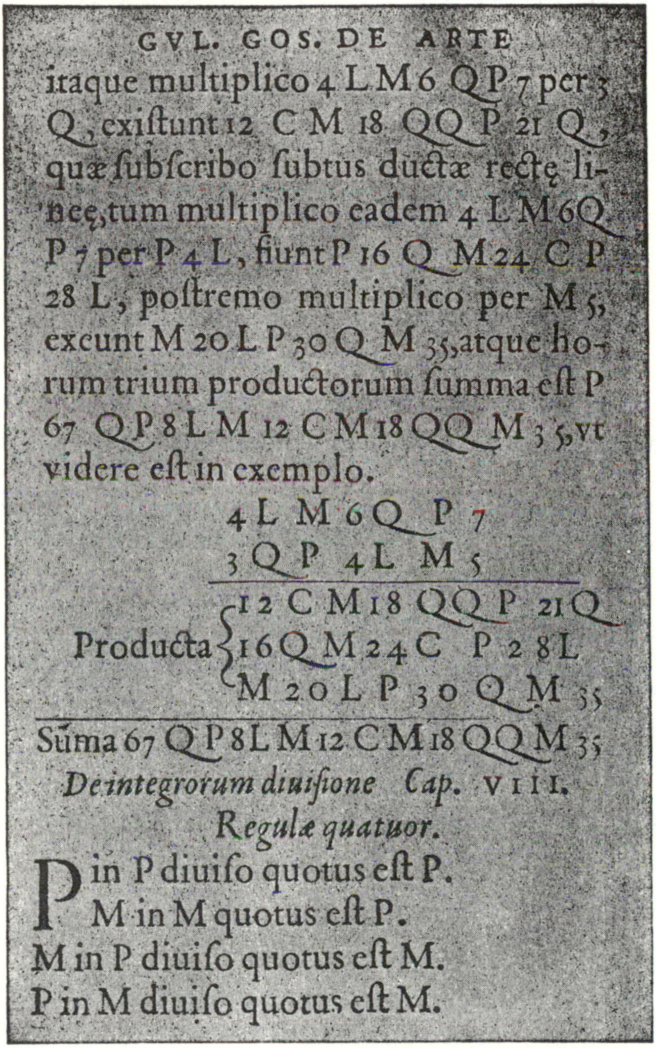

GVL. GOS. DE ARTE

itaque multiplico 4 L M 6 Q P 7 per 3 Q, exiſtunt 12 C M 18 QQ P 21 Q, quæ ſubſcribo ſubtus ductæ rectæ lineæ, tum multiplico eadem 4 L M 6 Q P 7 per P 4 L, fiunt P 16 Q M 24 C P 28 L, poſtremo multiplico per M 5, excunt M 20 L P 30 Q M 35, atque horum trium productorum ſumma eſt P 67 Q P 8 L M 12 C M 18 QQ M 35, vt videre eſt in exemplo.

$$4\ L\ M\ 6\ Q\ P\ 7$$
$$3\ Q\ P\ 4\ L\ M\ 5$$
$$\overline{}$$

Producta $\begin{cases} 12\ C\ M\ 18\ QQ\ P\ 21\ Q \\ 16\ Q\ M\ 24\ C\ \ \ P\ 2\ 8\ L \\ M\ 20\ L\ P\ 30\ Q\ M\ 35 \end{cases}$

Sũma 67 Q P 8 L M 12 C M 18 QQ M 35

De integrorum diuiſione   Cap. VIII.

Regulæ quatuor.

P in P diuiſo quotus eſt P.
  M in M quotus eſt P.
M in P diuiſo quotus eſt M.
P in M diuiſo quotus eſt M.

Fig. 83.—Fol. 45*v*° of Gosselin's *De arte magna* (1577)

ply the same $4x-6x^2+7$ by $+4x$, and there results $+16x^2-24x^3$ $+28x$; lastly I multiply by -5 and there results $-20x+30x^2-35$.

And the sum of these three products is $67x^2+8x-12x^3-18x^4-35$, as will be seen in the example.

$$4x - 6x^2 + 7$$
$$3x^2 + 4x - 5$$

$$\text{Products}\begin{cases} 12x^3-18x^4+21x^2 \\ 16x^2-24x^3+28x \\ -20x+30x^2-35 \end{cases}$$

Sum $\overline{67x^2+8x-12x^3-18x^4-35}$.

Oᴜ the Division of Integers, chapter viii

Four Rules

$+$ divided in $+$ the quotient is $+$
$-$ divided in $-$ the quotient is $+$
$-$ divided in $+$ the quotient is $-$
$+$ divided in $-$ the quotient is $-$ "

175. Proceeding to radicals we quote (fol. 47B): "Est autem laterum duplex genus simplicium et compositorum. Simplicia sunt L9, LC8, LL16, etc. Composita vero ut LV24 P L29, LV6 P L8." In translation: "There are moreover two kinds of radicals, simple and composite. The simple are like $\sqrt{9}$, $\sqrt[3]{8}$, $\sqrt[4]{16}$, etc. The composite are like $\sqrt{24+\sqrt{29}}$, $\sqrt{6+\sqrt{8}}$." First to be noticed is the difference between L9 and 9L. They mean, respectively, $\sqrt{9}$ and 9x. We have encountered somewhat similar conventions in Pacioli, with whom ℞ meant a power when used in the form, say, "℞ . 5ᵉ" (i.e., x^4), while ℞ meant a root when followed by a number, as in ℞ .200. (i.e., $\sqrt{200}$) (see § 135). Somewhat later the same principle of relative position occurs in Albert Girard, but with a different symbol, the circle. Gosselin's LV meant of course latus universale. Other examples of his notation of radicals are LVL10 P L5, for $\sqrt{\sqrt{10}+\sqrt{5}}$, and LVCL5 P LC10 for $\sqrt[3]{\sqrt{5}+\sqrt[3]{10}}$.

In the solution of simultaneous equations involving only positive terms Gosselin uses as the unknowns the capital letters A, B, C, ... (similar to the notation of Stifel and Buteon), and omits the sign P for "plus"; he does this in five problems involving positive terms, following here an idea of Buteo. In the problem 5, taken from Buteo, Gosselin finds four numbers, of which the first, together with half of the remaining, gives 17; the second with the third of the remaining gives 12; and the third with a fourth of the remaining gives 13; and the fourth with a sixth of the remaining gives 13. Gosselin lets A, B, C, D be the four numbers and then writes:

Modern Notation

" $1A\frac{1}{2}B\frac{1}{2}C\frac{1}{2}D$ aequalia 17 , $x+\frac{1}{2}y+\frac{1}{2}z+\frac{1}{2}w=17$,
$1B\frac{1}{3}A\frac{1}{3}C\frac{1}{3}D$ aequalia 12, etc. " $y+\frac{1}{3}x+\frac{1}{3}z+\frac{1}{3}w=12$.

He is able to effect the solution without introducing negative terms.

In another place Gosselin follows Italian and German writers in representing a second unknown quantity by q, the contraction of *quantitas*. He writes (fols. 84B, 85A) "1L P 2q M 20 aequalia sunt 1L P 30" (i.e., $1x+2y-20=1x+30$) and obtains "2q aequales 50, fit 1q 25" (i.e., $2y=50$, $y=25$).

FRENCH: FRANCIS VIETA
(1591 and Later)

176. Sometimes, Vieta's notation as it appears in his early publications is somewhat different from that in his collected works, edited by Fr. van Schooten in 1646. For example, our modern $\dfrac{3BD^2-3BA^2}{4}$ is printed in Vieta's *Zeteticorum libri v* (Tours, 1593) as

$$\text{" }\frac{B \text{ in } D \text{ quadratum } 3 - B \text{ in } A \text{ quadratum } 3}{4}\text{ "},$$

while in 1646 it is reprinted[1] in the form

$$\text{" }\frac{B \text{ in } Dq\ 3 - B \text{ in } Aq\ 3}{4}\text{ "}.$$

Further differences in notation are pointed out by J. Tropfke:[2]

Zeteticorum libri v (1593)

Fol. 3B: " $\dfrac{B \text{ in } A}{D} + \left\{ \dfrac{\begin{array}{c} B \text{ in } A \\ -B \text{ in } H \end{array}}{F} \right\}$ aequabuntur B ."

Modern: $\dfrac{Bx}{D} + \dfrac{Bx - B \cdot H}{F} = B$'.

Lib. II, 22:   " $l\,\dfrac{25}{3} - l\,\dfrac{5}{3}$."

[1] *Francisci Vietae Opera mathematica* (ed. Fr. à Schooten; Lvgdvni Batavorvm, 1646), p. 60. This difference in notation has been pointed out by H. Bosmans, in an article on Oughtred, in *Extrait des annales de la société scientifique de Bruxelles*, Vol. XXXV, fasc. 1 (2d part), p. 22.

[2] *Op. cit.*, Vol. III (2d ed., 1922), p. 139.

Lib. IV, 10: " B in $\left\{\begin{array}{l} D \text{ quadratum} \\ +B \text{ in } D \end{array}\right\}$."

Modern: $B(D^2+BD)$.

Lib. IV, 20: " D in $\left\{\begin{array}{l} B \text{ cubum } 2 \\ -D \text{ cubo} \end{array}\right\}$."

Modern: $D(2B^3-D^3)$.

Van Schooten edition of Vieta (1646)

P. 46: " $\dfrac{B \text{ in } A}{D}+\dfrac{B \text{ in } A-B \text{ in } H}{F}$ aequabitur B ."

P. 56: " $\sqrt{\dfrac{25}{3}}-\sqrt{\dfrac{5}{3}}$."

P. 70: " B in $\overline{D \text{ quad.}+B \text{ in } D}$."

P. 74: " $\overline{D}$ in $\overline{B \text{ cubum } 2-D \text{ cubo}}$."

Figure 84 exhibits defective typographical work. As in Stifel's *Arithmetica integra*, so here, the fractional line is drawn too short. In the translation of this passage we put the sign of multiplication ($\times$) in place of the word *in*: ". . . . Because what multiplication brings about above, the same is undone by division, as $\dfrac{B \times A}{B}$, i.e., A; and $\dfrac{B \times A^2}{B}$ is A^2.

Thus in additions, required, to $\dfrac{A^2}{B}$ to add Z. The sum is $\dfrac{A^2+Z \times B}{B}$; or required, to $\dfrac{A^2}{B}$ to add $\dfrac{Z^2}{G}$. The sum is $\dfrac{G \times A^2+B \times Z^2}{B \times G}$.

In subtraction, required, from $\dfrac{A^2}{B}$ to subtract Z. The remainder is $\dfrac{A^2-Z \times B}{B}$. Or required, from $\dfrac{A^2}{B}$ to subtract $\dfrac{Z^2}{G}$. The remainder is $\dfrac{A^2 \times G-Z^2 \times B}{B \times G}$."

Observe that Vieta uses the signs plus ($+$) and minus ($-$), which had appeared at Paris in the *Algebra* of Scheubel (1551). Outstanding in the foregoing illustrations from Vieta is the appearance of capital letters as the representatives of general magnitudes. Vieta was the first to do this systematically. Sometimes, Regiomontanus, Rudolff, Adam Riese, and Stifel in Germany, and Cardan in Italy, used letters at an earlier date, but Vieta extended this idea and first made it an essential part of algebra. Vieta's words,[1] as found in his *Isagoge*, are: "That this work may be aided by a certain artifice, given magnitudes are to be distinguished from the uncertain required ones by a symbolism, uniform and always readily seen, as is possible by designating the required quantities by letter A or by other vowel letters A, I, O, V, Y, and the given ones by the letters B, G, D or by other consonants."[2]

Vieta's use of letters representing known magnitudes as coefficients of letters representing unknown magnitudes is altogether new. In discussing Vieta's designation of unknown quantities by vowels,

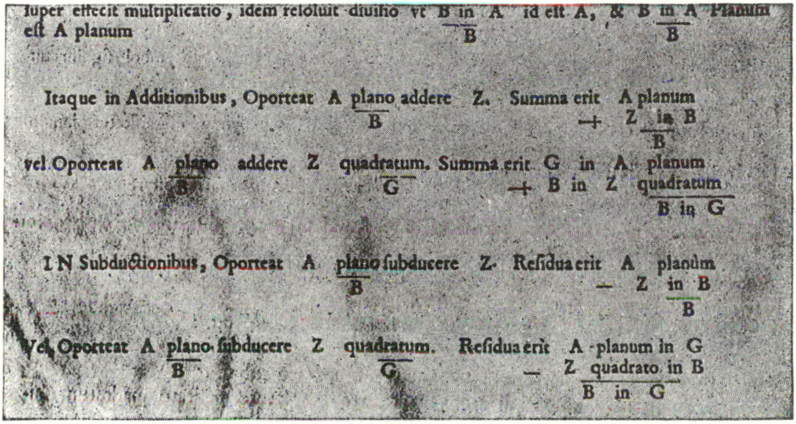

Fig. 84.—From Vieta's *In artem analyticam Isagoge* (1591). (I am indebted to Professor H. Bosmans for this photograph.)

C. Henry remarks: "Thus in a century which numbers fewer Orientalists of eminence than the century of Vieta, it may be difficult not to regard this choice as an indication of a renaissance of Semitic languages; every one knows that in Hebrew and in Arabic only the consonants are given and that the vowels must be recovered from them."[3]

177. Vieta uses $=$ for the expression of arithmetical difference. He says: "However when it is not stated which magnitude is the greater and which is the less, yet the subtraction must be carried out,

[1] Vieta, *Opera mathematica* (1646), p. 8.

[2] "Quod opus, ut arte aliqua juvetur, symbolo constanti et perpetuo ac bene conspicuo date magnitudines ab incertis quaesititiis distinguantur, ut r·'. magnitudines quaesititias elemento A aliave litera vocali, E, I, O, V, Y d· ·s elementis B, G, D, aliisve consonis designando."

[3] "Sur l'origine de quelques notations mathématiques," *Revue archéologique*, Vol. XXXVIII (N.S., 1879), p. 8.

the sign of difference is $=\!=$, i.e., an uncertain minus. Thus, given A^2 and B^2, the difference is $A^2=\!=B^2$, or $B^2=\!=A^2$."[1]

We illustrate Vieta's mode of writing equations in his *Isagoge:* "*B* in *A* quadratum plus *D* plano in *A* aequari *Z* solido," i.e., $BA^2+D^2A=Z^3$, where *A* is the unknown quantity and the consonants are the known magnitudes. In Vieta's *Ad Logisticen speciosam notae priores* one finds: "*A* cubus, $+A$ quadrato in *B* ter, $+A$ in *B* quadratum ter, $+B$ cubo," for $A^3+3A^2B+3AB^2+B^3$.[2]

We copy from Vieta's *De emendatione aequationum tractatus secundus* (1615),[3] as printed in 1646, the solution of the cubic $x^3+3B^2x=2Z^3$:

"Proponatur *A* cubus $+$ *B* plano 3 in *A*, aequari *Z* solido 2. Oportet facere quod propositum est. *E* quad. $+A$ in *E*, aequetur *B* plano. Vnde *B* planum ex hujus modi aequationis constitutione, intelligitur rectangulum sub duobus lateribus quorum minus est *E*, differentia à majore *A*. igitur $\dfrac{B \text{ planum} - E \text{ quad.}}{E}$ erit *A*. Quare

$$\frac{B \text{ plano-plano-planum} - E \text{ quad. in } B \text{ plano-planum } 3+E \text{ quad.}}{E \text{ cubo}} -$$

$$\frac{\text{quad. in } B \text{ planum } 3 - E \text{ cubo-cubo}}{E} + \frac{B \text{ pl. pl. } 3. - B \text{ pl. in Eq. } 3}{E} \text{ aequa-}$$

bitur *Z* solido 2 .

"Et omnibus per *E* cubum ductis et ex arte concinnatis, *E* cubi quad.$+Z$ solido 2 in *E* cubum, aequabitur *B* plani-cubo.[4]

"Quae aequatio est quadrati affirmate affecti, radicem habentis solidam. Facta itaque reductio est quae imperabatur.

"*Confectarium:* Itaque si *A* cubus $+$ *B* plano 3 in *A*, aequetur *Z* solido 2, & $\sqrt{B}$ plano-plano-plani $+$ *Z* solido-solido $-$ *Z* solido, aequetur *D* cubo. Ergo $\dfrac{B \text{ planum} - D \text{ quad.}}{D}$, sit *A* de qua quaeritur."

Translation: "Given $x^3+3B^2x=2Z^3$. To solve this, let $y^2+yx=B^2$. Since B^2 from the constitution of such an equation is understood to be a rectangle of which the less of the two sides is y, and the difference between it and the larger side is x. Therefore $\dfrac{B^2-y^2}{y}=x$. Whence

$$\frac{B^6-3B^4y^2+3B^2y^4-y^6}{y^3}+\frac{3B^4-3B^2y^2}{y}=2Z^3 .$$

[1] "Cum autem non proponitur utra magnitudo sit major vel minor, et tamen subductio facienda est, nota differentiae est $=\!=$ id est, minus incerto: ut propositis *A* quadrato et *B* plano, differentia erit *A* quadratum$=\!=B$ plano, vel *B* planum $A=\!=$quadrato" (Vieta, *Opera mathematica* [1646], p. 5).

[2] *Ibid.*, p. 17.   [3] *Ibid.*, p. 149.

[4] "*B* plani-cubo" should be "*B* cubo-cubo," and "*E* cubi quad." should be "*E* cubo-cubo."

All terms being multiplied by y^3, and properly ordered, one obtains $y^6 + 2Z^3y^3 = B^6$. As this equation is quadratic with a positive affected term, it has also a cube root. Thus the required reduction is effected.

"*Conclusion:* If therefore $x^3 + 3B^2x = 2Z^3$, and $\sqrt{B^6 + Z^6} - Z^3 = D^3$, then $\dfrac{B^2 - D^2}{D}$ is x, as required."

The value of x in $x^3 + 3B^2x = 2Z^3$ is written on page 150 of the 1646 edition thus:

" $\sqrt{C.\sqrt{B \text{ plano-plano-plani}} + Z \text{ solido-solido} + Z \text{ solido}} -$

$\sqrt{C.\sqrt{B \text{ plano-plano-plani}} + Z \text{ solido-solido}.} - Z \text{ solido} .$"

The combining of vinculum and radical sign shown here indicates the influence of Descartes upon Van Schooten, the editor of Vieta's collected works. As regards Vieta's own notations, it is evident that compactness was not secured by him to the same degree as by earlier writers. For powers he did not adopt either the Italian symbolism of Pacioli, Tartaglia, and Cardan or the German symbolism of Rudolff and Stifel. It must be emphasized that the radical sign, as found in the 1646 edition of his works, is a modification introduced by Van Schooten. Vieta himself rejected the radical sign and used, instead, the letter l (*latus*, "the side of a square") or the word *radix*. The l had been introduced by Ramus (§ 322); in the *Zeleticorum*, etc., of 1593 Vieta wrote l. 121 for $\sqrt{121}$. In the 1646 edition (p. 400) one finds $\sqrt{2 + \sqrt{2 + \sqrt{2 + \sqrt{2}}}}$, which is Van Schooten's revision of the text of Vieta; Vieta's own symbolism for this expression was, in 1593,[1]

"Radix binomiae 2

$+$Radix binomiae $\left\{ \begin{array}{l} 2 \\ +\text{radix binomiae} \end{array} \right.$ $\left\{ \begin{array}{l} 2 \\ +\text{radice 2 ,"} \end{array} \right.$

and in 1595,[2]

" R. bin. $2 + R$. bin. $2 + R$. bin. $2 + R$. 2. ,"

a notation employed also by his contemporary Adrian Van Roomen.

178. Vieta distinguished between number and magnitude even in his notation. In numerical equations the unknown number is no longer represented by a vowel; the unknown number and its powers are represented, respectively, by N (*numerus*), Q (*quadratus*), C (*cubus*), and

---

[1] *Variorum de rebus mathem. Responsorum liber VIII* (Tours, 1593), corollary to Caput XVIII, p. 12$v°$. This and the next reference are taken from Tropfke, *op. cit.*, Vol. II (1921), p. 152, 153.

[2] *Ad Problema quod omnibus mathematicis totius orbis construendum proposuit Adrianus Romanus, Francisci Vietae responsum* (Paris, 1595), Bl. *A IV°*.

combinations of them. Coefficients are now written to the left of the letters to which they belong.

Thus,[1] "Si $65C - 1QQ$, aequetur 1,481,544, fit $1N57$," i.e., if $65x^3 - x^4 = 1,481,544$, then $x = 57$. Again,[2] the "$B3$ in A quad." occurring in the regular text is displaced in the accompanying example by "$6Q$," where $B = 2$.

Figure 85 further illustrates the notation, as printed in 1646.

Vieta died in 1603. The *De emendatione aeqvationvm* was first printed in 1615 under the editorship of Vieta's English friend, Alexander Anderson, who found Vieta's manuscript incomplete and con-

THEOREMA I.

SI A cubus $+$ B in A quadr. 3 $+$ D plano in A, æquetur B cubo 2 $-$ D plano in B. A quad. $+$ B in A 2, æquabitur B quad. 2 $-$ D plano.

Quoniam enim A quadr. $+$ B in A 2, æquatur B quadr. 2 $-$ D plano. Ductis igitur omnibus in A. A cubus $+$ B in A quad. 2, æquabitur B quad. in A 2 $-$ D plano in A.

Et iisdem ductis in B. B in A quad. $+$ B quadr. in A 2, æquabitur B cubo 2 $-$ D plano in B. Iungatur ducta æqualia æqualibus. A cubus $+$ B in A quad. 3 $+$ B quad. in A 2, æquabitur B quad. in A 2 $-$ D plano in A $+$ B cubo 2 $-$ D plano in B.

Et deleta utrinque adfectione B quad. in A 2, & ad æqualitatis ordinationem, translata per antithesin D plani in A adfectione. A cubus $+$ B in A quadr. 3 $+$ D plano in A, æquabitur B cubo 2 $-$ D plano in B. Quod quidem ita se habet.

$1C + 30Q + 44N$, *æquatur* 1560.   *Igitur* $1Q + 20N$, *æquabitur* 156. *& fit* $1N6$.

FIG. 85.—From Vieta's *De emendatione aeqvationvm*, in *Opera mathematica* (1646), p. 154.

taining omissions which had to be supplied to make the tract intelligible. The question arises, Is the notation N, Q, C due to Vieta or to Anderson?[3] There is no valid evidence against the view that Vieta did use them. These letters were used before Vieta by Xylander in his edition of Diophantus (1575) and in Van Schooten's edition[4] of the *Ad problema, quod omnibus mathematicis totius orbis construendum proposuit Adrianus Romanus*. It will be noticed that the letter N stands here for x, while in some other writers it is used in the designation of absolute number as in Grammateus (1518), who writes our $12x^3 - 24$ thus: "12 ter. mi. $24N$." After Vieta N appears as a mark for absolute number in the *Sommaire de l'algebre* of Denis Henrion[5]

---

[1] Vieta, *Opera mathematica* (1646), p. 223.   [2] *Op. cit.*, p. 130.

[3] See Eneström, *Bibliotheca mathematica*, Vol. XIII (1912–13), p. 166, 167.

[4] Vieta, *Opera mathematica* (1646), p. 306, 307.

[5] Denis Henrion, *Les qvinze livres des elemens d'Evclide* (4th ed.; Paris, 1631), p. 675–788. First edition, Paris, 1615. (Courtesy of Library of University of Michigan.)

which was inserted in his French edition of Euclid. Henrion did not adopt Vieta's literal coefficients in equations and further showed his conservatism in having no sign of equality, in representing the powers of the unknown by R, q, c, qq, β, qc, $b\beta$, qqq, cc, $q\beta$, $c\beta$, qqc, etc., and in using the "scratch method" in division of algebraic polynomials, as found much earlier in Stifel.[1] The one novel feature in Henrion was his regular use of round parentheses to express aggregation.

ITALIAN: BONAVENTURA CAVALIERI
(1647)

179. Cavalieri's *Geometria indivisibilibvs* (Bologna, 1635 and 1653) is as rhetorical in its exposition as is the original text of Euclid's *Elements*. No use whatever is made of arithmetical or algebraic signs, not even of $+$ and $-$, or p and m.

An invasion of German algebraic symbolism into Italy had taken place in Clavius' *Algebra*, which was printed at Rome in 1608. That German and French symbolism had gained ground at the time of Cavalieri appears from his *Exercitationes geometriae sex* (1647), from which Figure 86 is taken. Plus signs of fancy shape appear, also Vieta's *in* to indicate "times." The figure shows the expansion of $(a+b)^n$ for $n = 2, 3, 4$. Observe that the numerical coefficients are written after the literal factors to which they belong.

ENGLISH: WILLIAM OUGHTRED
(1631, 1632, 1657)

180. William Oughtred placed unusual emphasis upon the use of mathematical symbols. His symbol for multiplication, his notation for proportion, and his sign for difference met with wide adoption in Continental Europe as well as Great Britain. He used as many as one hundred and fifty symbols, many of which were, of course, introduced by earlier writers. The most influential of his books was the *Clavis mathematicae*, the first edition[2] of which appeared in 1631, later Latin editions of which bear the dates of 1648, 1652, 1667, 1693.

---

[1] M. Stifel, *Arithmetica integra* (1544), fol. 239A.

[2] The first edition did not contain *Clavis mathematicae* as the leading words in the title. The exact title of the 1631 edition was: *Arithmeticae in|numeris et speci-|ebvs institvtio:|Qvae tvm logisticae, tvm analyti|cae, atqve adeo|totivs mathematicae, qvasi|clavis|est.|—Ad nobilissimvm spe|ctatissimumque iuvenem Dn. Gvilel|mvm Howard, Ordinis, qui dici|tur, Balnei Equitem, honoratissimi Dn.| Thomae, Comitis Arvndeliae & | Svrriae, Comitis Mareschal|li Angliae, &c. filium.—|Londini,|Apud Thomam Harpervm,| M. DC. xxxi.*

A second impression of the 1693 or fifth edition appeared in 1698. Two English editions of this book came out in 1647 and 1694.

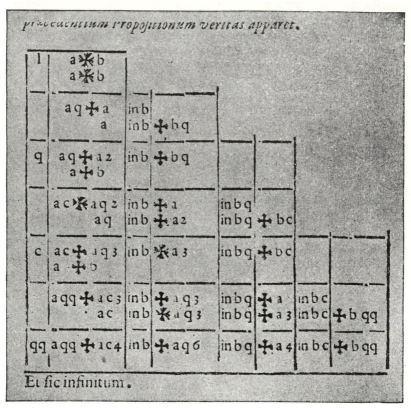

Fig. 86.—From B. Cavalieri's *Exercitationes* (1647), p. 268

We shall use the following abbreviations for the designation of tracts which were added to one or another of the different editions of the *Clavis mathematicae:*

Eq. = *De Aequationum affectarvm resolvtione in numeris*

Eu. = *Elementi decimi Euclidis declaratio*

So. = *De Solidis regularibus, tractatus*

An. = *De Anatocismo, sive usura composita*

Fa. = *Regula falsae positionis*

Ar. = *Theorematum in libris Archimedis de sphaera & cylindro declaratio*

Ho. = *Horologia scioterica in plano, Geometricè delineandi modus*

In 1632 there appeared, in London, Oughtred's *The Circles of Proportion*, which was brought out again in 1633 with an *Addition vnto the Vse of the Instrvment called the Circles of Proportion*.[1] Another edition bears the date 1660. In 1657 was published Oughtred's *Trigonometria*,[2] in Latin, and his *Trigonometrie*, an English translation.

We have arranged Oughtred's symbols, as found in his various works, in tabular form.[3] The texts referred to are placed at the head of the table, the symbols in the column at the extreme left. Each number in the table indicates the page of the text cited at the head of the column containing the symbol given on the left. Thus, the notation :: in geometrical proportion occurs on page 7 of the *Clavis* of 1648. The page assigned is not always the first on which the symbol occurs in that volume.

[1] In our tables this *Addition* is referred to as *Ad.*

[2] In our tables *Ca.* stands for *Canones sinuum tangentium*, etc., which is the title for the tables in the *Trigonometria*.

[3] These tables were first published, with notes, in the *University of California Publications in Mathematics*, Vol. I, No. 8 (1920), p. 171–86.

181.   OUGHTRED'S MATHEMATICAL SYMBOLS

Symbols	Meanings of Symbols	Clavis mathematica							Circ. of Prop. 1632, 1633	Trigono. (Latin), 1657	Opusc. Posth., 1677	Oughtr. Explic., 1682
		1631	1647	1648	1652	1667	1693	1694				
=	Equal to	38	34	53	30	15	16	73	20	3	3	29
0|56	Separatrix[2]	1	1	1	1	1	1	2	3	13	63	1
0.56	Separatrix	.....	.....	.....	.....	.....	.....	.....	.....	235	.....	.....
.|56	Separatrix[3]	.....	.....	.....	.....	.....	.....	17	.....	.....	.....	5
0,56	Separatrix	.....	.....	.....	.....	.....	.....	.....	.....	.....	221	.....
0|00005	.00005	.....	.....	3	3	3	3	.....	.....	.....	.....	2
$a.b$	Ratio $a:b$, or $a-b$	5	8	7	12	7	7	25	7	5	3	27
2.314	⌠ Separating[5]	.....	.....	.....	.....	.....	.....	.....	4	235	.....	.....
2,314	⌡ the mantissa	.....	Eq.136	158	150	113	113	175	.....	.....	.....	.....
2.314	— Characteristic	.....	Eq.167	158	150	150	150	207	.....	.....	.....	.....
:	Arith. proportion[6]	.....	.....	22	21	21	21	32	.....	.....	.....	.....
:	$a:b$, ratio[7]	.....	.....	An.162	.....	.....	.....	24	10	36	140	.....
R.S	Given ratio	21	28	33	32	25	25	49	19	.....	87	42
::	Geomet. proportion[8]	5	8	7	7	7	7	11	1	3	3	27
÷	Contin. proportion	13	18	16	16	16	16	25	.....	34	142	29
⋯	Contin. proportion	.....	.....	.....	.....	.....	.....	.....	.....	.....	.....	114
=	Geom.[9] proportion	.....	.....	.....	.....	.....	.....	.....	.....	.....	.....	85
: : (,[10]	(	45	57	107	104	52	53	149	96	.....	101	.....
:	(	40	58	99	92	56	92	119	35	32	101	75
: .	( )	.....	115	106	104	104	104	95	.....	.....	102	53
:	( )	.....	58	.....	95	95	63	122	.....	.....	.....	93
. :	( )	.....	65	58	57	57	63	95	.....	.....	.....	97
.	( )[11]	.....	.....	.....	.....	.....	An. 42	97	.....	.....	.....	116
. .	( )[11]	.....	.....	.....	.....	.....	.....	89	.....	.....	.....	101
(	(,[12]	.....	.....	.....	.....	.....	.....	.....	.....	.....	.....	81
∵	Therefore	.....	.....	.....	.....	.....	.....	.....	.....	.....	151	.....
+	Addition[13]	2	3	3	57	3	3	4	99	3	3	5
p^l	Addition	49	3	3	57	3	3	4	96	.....	112	5
mo	Addition	.....	.....	.....	.....	.....	.....	4	.....	.....	.....	.....
—	Subtraction	2	3	3	10	3	3	4	21	3	4	5
±	Plus or minus	51	57	106	56	53	17	140	.....	16	.....	97
mi	Subtraction	.....	3	66	57	3	3	4	96	.....	130	.....
e	Less[14]	.....	.....	.....	.....	.....	.....	4	.....	.....	.....	.....
2̄	Negative 2	.....	1	9	1	1	5	16	.....	.....	.....	8
×	Multiplication[15]	7	10	10	10	10	10	13	37	32	143	.....

182. OUGHTRED'S MATHEMATICAL SYMBOLS—*Cont.*

Symbols	Meanings of Symbols	Clavis mathematicae							Circ. of Prop.[1] 1632, 1633	Trigono. (Latin), 1657	Opusc. Posth., 1677	Oughtr. Explic., 1682
		1631	1647	1648	1652	1667	1693	1694				
$Hq\ bq$	$\times$ By juxtaposition	7	11	10	11	10	37	13	......	5	87	17
in	Multiplication[16]	7	10	10	10	10	10	......	96	......	219	59
$\frac{a}{b}$	Fraction, division	8	12	11	11	11	11	23	21	16	9	5
$a)b(c$	$b \div a = c$	10	14	13	14	14	13	21	......	......	99	50
$\frac{4}{5}]\frac{2}{3}[\frac{2}{3}$	$\frac{2}{3} + \frac{4}{5} = \frac{2}{3}$	......									156	......
Aq	AA	7	11	10	10	10	10	14	......	......	104	17
Ac	AAA	7	11	10	10	10	10	14	......	......	105	25
Aqq	$AAAA$	7	11	10	10	10	10	14	......	......	106	41
Aqc	$AAAAA$	7	11	10	10	10	10	55	......	......	......	67
Acc	$AAAAAA$	7	11	10	10	10	10	55	......	......	......	41
ABq	$\overline{AB}^2$ [17]	......	11	11	11	11	11	15	......	......	......	......
$\boxed{4} \cdots \boxed{10}$	4th .... 10th power	23	37	35	34	34	......	53	......	......	......	55
$|4| \cdots [10]$	4th .... 10th power	......	......	......	35	35	34	52	......	......	......	......
$a^2 \cdots a^7$	$a^2 \cdots a^7$	......	......	......	......	......	......	......	......	......	205	24
Q	*Quaesitum*	......	17	16	16	16	16	25	......	......	......	......
Q	Square[18]	38	33	31	57	30	30	47	28	5	100	75
$Q\,u$	Square	......	......	......	......	......	......	......	......	......	......	105
C	Cube	38	33	31	30	30	30	47	28	......	62	53
$C\,u$	Cube	......	136	128	123	123	123	175	......	......	......	......
QQ	4th power	45	33	61	30	30	30	47	......	......	210	......
QC	5th power	......	33	31	30	30	30	47	......	......	210	......
D	Diameter	......	......	187	*Eu.* 21	*Eu.* 21	*Eu.* 20	......	37	......	......	......
L, l	*Latus, radix*[19]	......	121	113	110	110	110	158	37	......	......	139
$\angle$	Angle	......	......	......	......	......	......	......	......	19	192	......
$\angle\angle$	Angles	......	......	......	......	......	......	......	......	16	......	......
P	Perimeter	......	......	......	......	......	......	......	37	......	......	......
P	$ZA - Aq$	41	......	......	......	......	......	......	......	......	......	......
R	Radius	......	120	111	109	109	109	154	37	32	211	......
β, R	Remainder	......	152	134	126	128	142	......	......	......	......	45
R	Rational	......	......	166	*Eu.* 1	*Eu.* 1	*Eu.* 1	......	......	......	......	......
Ω	*Superficies curva*	......	......	......	*Ar.* I	*Ar.* 1	*Ar.* 1	......	......	......	......	......
$\sqrt{}$	Root	......	33	31	30	30	30	47	......	......	102	......
$\sqrt{}$	Square root	......	53	48	47	47	47	70	......	......	134	......
$\sqrt{q}$	Square root	35	49	48	46	46	46	65	96	......	......	......
$\sqrt{b}$	*Latus binomii*	......	33	31	30	30	30	47	......	......	......	......

183.   OUGHTRED'S MATHEMATICAL SYMBOLS—*Cont.*

Symbols	Meanings of Symbols	Clavis mathematicae							Circ. of Prop.[1] 1632, 1633	Trigono. (Latin), 1657	Opusc. Posth., 1677	Oughtr. Explic., 1682
		1631	1647	1648	1652	1667	1693	1694				
√ r	Latus residui	......	34	31	30	30	30	47	......	......	......	......
√ u	Sq. rt. of polyno.[20]	......	55	53	53	52	52	96	......	......	......	......
√ qq	4th root	35	52	47	46	46	48	69	......	......	......	......
√ c	Cube root	35	52	49	46	46	46	69	......	......	......	......
√ qc	5th root	35	49	47	46	46	46	65	......	......	......	......
√ cc	6th root	37	52	49	48	48	48	69	......	......	......	......
√ ccc	9th root	......							......	......	......	......
√ cccc	12th root	37	52	49	48	48	48	69	......	......	......	......
√ qu	Square root	49	......									......
√[12] or √[12] ; rq, rc	12th root	37	52	50	49	49	49	69	......	......	......	......
r, ru	Square root	......							......	......	......	74,96
A, E	Nos., $A > E$	21	33	31	30	30	36	47	......	......	87	53
Z	$A+E$[21]	21	33	31	30	30	30	47	19	16	87	53
X	$A-E$	21	33	31	30	30	30	47	......	16	87	53
Z	A^2+E^2	41	33	31	30	30	30	47	......		98	54
X	A^2-E^2	41	33	31	30	30	30	47	......		99	54
Z	A^3+E^3	44	33	31	39	30	30	47				94
X	A^3-E^3	44	33	31	30	30	30	47				94
	$a+e$	......	......	167	Eu. 1	Eu. 1	Eu. I		......	......	......	......
	$a-e$	......	......		Eu. 1	Eu. 1	Eu. I		......	......	......	......
	a^2+b^2	......	......	167	Eu. 2	Eu. 2	Eu. I		......	......	......	......
	a^2-b^2	......	......	167	Eu. 2	Eu. 2	Eu. I		......	......	......	......
	Majus[22]	......	Ho. 17	166	145	Eu. 1	Eu. 1		......	......	......	......
	Minus	......	Ho. 17	166	Eu. 1	Eu. 1	Eu. 1		......	......	......	......
	Non majus	......	......	166	Eu. 1	Eu. 1	Eu. 1		......	......	......	......
	Non minus	......	......	166	Eu. 1	Eu. 1	Eu. 1		......	......	......	......
	Minus[23]	......	Ho. 30	......								
	Minus[23]	......	......	......	Ho. 31	Ho. 29	......					
∺	Major ratio	......	......	166	Eu. 1	Eu. 1	Eu. I		......	......	11	......
∷	Minor ratio	......	......	166	Eu. 1	Eu. 1	Eu. I		......	......	6	......
<	Less than[22]	......	......	......			......		......	......	4	......
>	Greater than	......	......	......			......		......	......	4	......
	Commensurabilia	......	......	166	Eu. 1	Eu. 1	Eu. I		......	......	......	......
	Incommensurabilia	......	......	166	Eu. 1	Eu. 1	Eu. I		......	......	......	......

184. OUGHTRED'S MATHEMATICAL SYMBOLS—*Cont.*

Symbols	Meanings of Symbols	Clavis mathematicae							Circ. of Prop.¹ 1632, 1633	Trigono. (Latin), 1657	Opusc. Posth., 1677	Oughtr. Explic., 1682
		1631	1647	1648	1652	1667	1693	1694				
	Commens. potentia	....	....	166	*Eu.* 1	*Eu.* 1	*Eu.* I	....			....
	Incommens. potentia	....	....	166	*Eu.* 1	*Eu.* 1	*Eu.* I	....			....
	Rationale	....	....	166	*Eu.* 1	*Eu.* 1	*Eu.* I	....			....
	Irrationale	....	....	166	*Eu.* 1	*Eu.* 1	*Eu.* I	....			....
	Medium	....	....	166	*Eu.* 1	*Eu.* 1	*Eu.* I	....			....
	Line, cut extr. and mean ratio	....	....	166	*Eu.* 1	*Eu.* 1	*Eu.* I	....			....
	Major ejus portio	....	....	166	*Eu.* 1	*Eu.* 1	*Eu.* I	....			....
	Minor ejus portio	....	....	166	*Eu.* 1	*Eu.* 1	*Eu.* I	....			....
sim	Simile	....	....	166	*Eu.* 1	*Eu.* 1	*Eu.* I	....		33	....
	Proxime majus	....	....	166	*Eu.* 1	*Eu.* 1	*Eu.* I	....			....
	Proxime minus	....	....	166	*Eu.* 1	*Eu.* 1	*Eu.* I	....			....
	Aequale vel minus	....	....	166	*Eu.* 1	*Eu.* 1	*Eu.* I	....			....
	Aequale vel majus	....	....	166	*Eu.* 1	*Eu.* 1	*Eu.* I	....			....
	Rectangulum	51	....	167	*Eu.* 2	*Eu.* 2	*Eu.* I	....		17	149	....
	Quadratum	....	....	167	*Eu.* 2	*Eu.* 2	*Eu.* I	....			....
	Triangulum	....	....	167	*Eu.* 2	*Eu.* 2	*Eu.* I	....			147	....
	Latus, radix	....	....	167	*Eu.* 2	*Eu.* 2	*Eu.* I	....			....
	Media proportion	....	....	167	*Eu.* 2	*Eu.* 2	*Eu.* I	....			....
	*Differentia*²⁴	....	....	....	*Eu.* 2	*Eu.* 2	*Eu.* I	....			....
||	Parallel	....	....	....	....	....	....	....			197	....
log	Logarithm	....	172	158	150	150	122	207	....	17	....	....
log:Q:	Log. of square	....	135	127	122	122	122	174	....	....	....
S	Sine²⁵	....	*Ho.* 29	....	....	....	....	....	96	5	172	....
t	Tangent	....	*Ho.* 29	....	....	....	....	....	96	3	174	....
se	Secant	....	....	....	....	....	....	....		14	....	....
sv	*Sinus versus*	76	107	99	98	98	98	140	....	....	....
s ver	*Sinus versus*²⁶	....	....	....	....	....	....		5	....	....
sin:com	Sine complement	....	....	....	....	....	....	*Ad.* 69	....	....	....
s co	Cosine	....	....	....	....	....	....		96	3	174	....
t co	Cotangent	....	....	....	....	....	....		96	3	....	....
se co	Cosecant	....	....	....	....	....	....		4	....	....
sin	Sine	....	....	....	....	*Ho.* 41	*Ho.* 41	*Ho.* 42	*Ad.* 69	35	....	37
tan	Tangent	....	....	....	....	....	....		*Ad.* 69	*Ca.* 3	....	....
sec	Secant	....	....	....	....	....	....		*Ad.* 41	....	....	....
sec:parall	Sum of secants	....	....	....	....	....	....		*Ad.* 41	....	....	....

185. OUGHTRED'S MATHEMATICAL SYMBOLS—*Cont.*

Columns 1631–1694 fall under the heading *Clavis Mathematicae*.

Symbols	Meanings of Symbols	1631	1647	1648	1652	1667	1693	1694	Circ. of Prop.[1] 1632, 1633	Trigono. (Latin), 1657	Opusc. Posth., 1677	Oughtr. Explic., 1682
tang	Tangent	.....	Ho. 29	.....	Ho. 41	Ho. 41	Ho. 42	.....	12	235	.....	.....
C	.01 of a degree	.....	.....	.....	.....	.....	.....	.....	.....	236	.....	.....
Cent	.01 of a degree	.....	.....	.....	.....	.....	.....	.....	.....	235	.....	.....
° ′ ″	Degr., min., sec.	.....	21	20	21	20	21	32	66	.....	.....	36
Ho. ′ ″	Hours, min., sec.	.....	.....	.....	.....	.....	.....	.....	67	.....	.....	.....
,	$180\ \overset{\circ}{=}$ angle	.....	.....	.....	.....	.....	.....	.....	.....	2	.....	.....
$\overline{V}$	Equal in no. of degr.	.....	.....	.....	.....	.....	.....	.....	.....	6	.....	.....
$\frac{\pi}{\delta}$	$\pi = 3.1416$	.....	72	69	66	66	66	99	.....	.....	.....	.....
———	Canceled[27]	68	100	94	90	90	90	131	.....	.....	.....	.....
M	Mean proportion	.....	.....	.....	Ar. 1	Ar. 1	Ar. 1	.....	.....	.....	.....	.....
m	Minus	⌐	.....	.....	.....	.....	.....	.....	.....	.....	.....	.....
(fraction glyph)	$\frac{3}{2}\times\frac{4}{3}=2,\ \frac{3}{2}\div\frac{4}{3}=\frac{9}{8}$[28]	20	32	30	29	29	29	45	.....	.....	.....	.....
Gr.	Degree	.....	20	19	Ho. 23	Ho. 23	19	29	.....	235	.....	.....
min.	Minute	.....	.....	.....	.....	.....	.....	.....	Ad. 19	.....	.....	.....
⌒	*Differentia*	.....	.....	.....	.....	.....	.....	.....	.....	.....	134	.....
⊢—⊣	*Aequalia tempore*	.....	.....	.....	.....	.....	.....	.....	.....	.....	68	.....
Lo	Logarithm	.....	.....	.....	.....	.....	.....	.....	.....	.....	Ca. 2	.....
|	Separatrix	.....	.....	.....	.....	.....	.....	.....	.....	.....	244	.....
D	*Differentia*	.....	.....	.....	.....	.....	.....	.....	19	.....	237	.....
Tri, tri	Triangle	.....	76	191	Eu. 26	70	69	.....	.....	.....	24	.....
M	Cent. minute of arc	.....	.....	.....	.....	.....	.....	.....	.....	.....	Ca. 2	.....
X	Multiplication[29]	.....	.....	.....	.....	.....	.....	.....	.....	5	101	16
Z cru	Z sum, X diff.	.....	.....	.....	.....	.....	.....	.....	.....	.....	17	.....
Z crur	of sides of	.....	.....	.....	.....	.....	.....	.....	.....	.....	16	.....
X cru	rectangle[30]	.....	.....	.....	.....	.....	.....	.....	.....	.....	17	.....
X crur	or triangle	.....	.....	.....	.....	.....	.....	.....	.....	.....	16	.....
A	Unknown	38	53	51	50	50	50	72	.....	.....	113	84
L	Altit. frust. of pyramid or cone	77	109	101	99	99	99	141	.....	.....	.....	.....
T	Altit. of part cut off	77	109	101	99	99	99	142	.....	.....	.....	.....
ι	First term	13	85, 18	80, 17	78, 16	78, 16	78, 16	116, 26	19	.....	.....	30, 116
ω	Last term	.....	85, 18	80, 17	78, 16	78, 16	78, 16	116, 26	.....	.....	.....	30, 116
T	No. of terms	.....	85	80	78	78	78	116	.....	.....	.....	11
X	Common differ.	.....	85	80	78	78	78	116	.....	.....	.....	116
Z	Sum of all terms	.....	85, 18	80, 17	78, 16	78, 16	78 , 16	116, 26	19	.....	.....	30, 116

The last five rows (First term, Last term, No. of terms, Common differ., Sum of all terms) are bracketed together and labeled "progressions" (o).

186. Historical notes[1] to the tables in §§ 181–85:

1. All the symbols, except "Log," which we saw in the 1660 edition of the *Circles of Proportion*, are given in the editions of 1632 and 1633.

2. In the first half of the seventeenth century the notation for decimal fractions engaged the attention of mathematicians in England as it did elsewhere (see §§ 276–89). In 1608 an English translation of Stevin's well-known tract was brought out, with some additions, in London by Robert Norton, under the title, *Disme: The Art of Tenths, or, Decimall Arithmetike* (§ 276). Stevin's notation is followed also by Henry Lyte in his *Art of Tens or Decimall Arithmetique* (London, 1619), and in *Johnsons Arithmetick* (2d ed.; London, 1633), where 3576.725 is

written 3576|725. William Purser in his *Compound Interest and Annuities* (London, 1634), p. 8, uses the colon (:) as the separator, as did Adrianus Metius in his *Geometriae practicae pars I et II* (Lvgd., 1625), p. 149, and Rich. Balam in his *Algebra* (London, 1653), p. 4. The decimal point or comma appears in John Napier's *Rabdologia* (Edinburgh, 1617). Oughtred's notation for decimals must have delayed the general adoption of the decimal point or comma.

3. This mixture of the old and the new decimal notation occurs in the *Key* of 1694 (*Notes*) and in Gilbert Clark's *Oughtredus explicatus*[2] only once; no reference is made to it in the table of errata of either book. On Oughtred's *Opuscula mathematica hactenus inedita*, the mixed notation 128,57 occurs on p. 193 fourteen times. Oughtred's regular notation 128|57 hardly ever occurs in this book. We have seen similar mixed notations in the *Miscellanies: or Mathematical Lucubrations, of Mr. Samuel Foster, Sometime publike Professor of Astronomie in Gresham Colledge, in London*, by John Twysden (London, 1659), p. 13 of the "Observationes eclipsium"; we find there 32.466, 31.008.

4. The dot (.), used to indicate ratio, is not, as claimed by some writers, used by Oughtred for division. Oughtred does not state in his book that the dot (.) signifies division. We quote from an early and a late edition of the *Clavis*. He says in the *Clavis* of 1694, p. 45, and in the one of 1648, p. 30, "to continue ratios is to multiply them *as if they were* fractions." Fractions, as well as divisions, are indicated by a horizontal line. Nor does the statement from the *Clavis* of 1694, p. 20, and the edition of 1648, p. 12, "In Division, as the Divisor is to Unity, so is the Dividend to the Quotient," prove that he looked upon ratio as an indicated division. It does not do so any more than the sentence from the *Clavis* of 1694, and the one of 1648, p. 7, "In Multiplication, as 1 is to either of the factors, so is the other to the Product," proves that he considered ratio an indicated multiplication. Oughtred says (*Clavis* of 1694, p. 19, and the one of 1631, p. 8): "If Two Numbers stand one above another with a Line drawn between them, 'tis as much

as to say, that the upper is to be divided by the under; as $\frac{12}{4}$ and $\frac{5}{12}$."

---

[1] N. 1 refers to the *Circles of Proportion*. The other notes apply to the superscripts found in the column, "Meanings of Symbols."

[2] This is not a book written by Oughtred, but merely a commentary on the *Clavis*. Nevertheless, it seemed desirable to refer to its notation, which helps to show the changes then in progress.

In further confirmation of our view we quote from Oughtred's letter to W. Robinson: "Division is wrought by setting the divisor under the dividend with a line between them."[1]

5. In Gilbert Clark's *Oughtredus explicatus* there is no mark whatever to separate the characteristic and mantissa. This is a step backward.

6. Oughtred's language (*Clavis* of 1652, p. 21) is: "Ut 7.4 : 12.9 vel 7.7 − 3 : 12.12 − 3. Arithmeticè proportionales sunt." As later in his work he does not use arithmetical proportion in symbolic analysis, it is not easy to decide whether the symbols just quoted were intended by Oughtred as part of his algebraic symbolism or merely as punctuation marks in ordinary writing. Oughtred's notation is adopted in the article "Caractere" of the *Encyclopédie méthodique* (*mathématiques*), Paris: Liège, 1784 (see § 249).

7. In the publications referred to in the table, of the years 1648 and 1694, the use of : to signify ratio has been found to occur only once in each copy; hence we are inclined to look upon this notation in these copies as printer's errors. We are able to show that the colon (:) was used to designate geometric ratio some years before 1657, by at least two authors, Vincent Wing the astronomer, and a schoolmaster who hides himself behind the initials "R.B." Wing wrote several works.

8. Oughtred's notation $A.B::C.D$, is the earliest serviceable symbolism for proportion. Before that proportions were either stated in words as was customary in rhetorical modes of exposition, or else was expressed by writing the terms of the proportion in a line with dashes or dots to separate them. This practice was inadequate for the needs of the new symbolic algebra. Hence Oughtred's notation met with ready acceptance (see §§ 248–59).

9. We have seen this notation only once in this book, namely, in the expression $R.S. = 3.2$.

10. Oughtred says (*Clavis* of 1694, p. 47), in connection with the radical sign, "If the Power be included between two Points at both ends, it signifies the universal Root of all that Quantity so included; which is sometimes also signified by b and r, as the $\sqrt{b}$ is the Binomial Root, the $\sqrt{r}$ the Residual Root." This notation is in no edition strictly adhered to; the second : is often omitted when all the terms to the end of the polynomial are affected by the radical sign or by the sign for a power. In later editions still greater tendency to a departure from the original notation is evident. Sometimes one dot takes the place of the two dots at the end; sometimes the two end dots are given, but the first two are omitted; in a few instances one dot at both ends is used, or one dot at the beginning and no symbol at the end; however, these cases are very rare and are perhaps only printer's errors We copy the following illustrations:

$Q : A - E$: est $Aq - 2AE + Eq$, for $(A-E)^2 = A^2 - 2AE + E^2$ (from *Clavis* of 1631, p. 45)

$\frac{1}{2}BCq \pm \sqrt{q} : \frac{1}{4}BCqq - CMqq. = \begin{matrix} BAq \\ CAq \end{matrix}\Big\}$,   for $\frac{1}{2}\overline{BC}^2 \pm \sqrt{(\frac{1}{4}\overline{BC}^4 - \overline{CM}^4)} = \overline{BA}^2$ or $\overline{CA}^2$
(from *Clavis* of 1648, p. 106)

$\sqrt{q} : BA + CA = BC + D$, for $\sqrt{(BA+CA)} = BC + D$ (from *Clavis* of 1631, p. 40)

$\frac{AB}{2} + \sqrt{q}\frac{ABq}{4} - \frac{C \times S}{R} : = A.$, for $\frac{\overline{AB}}{2} + \sqrt{\left(\frac{\overline{AB}^2}{4} - \frac{\overline{C \times S}}{R}\right)} = A.$ (from *Clavis* of 1652, p. 95)

[1] Rigaud, *Correspondence of Scientific Men of the Seventeenth Century*, Vol. I (1841), Letter VI, p. 8.

$Q.Hc+Ch$: for $(Hc+Ch)^2$ (from *Clavis* of 1652, p. 57)

$Q.A-X=$, for $(A-X)^2=$ (from *Clavis* of 1694, p. 97)

$\frac{B}{2}+r.u.\ \frac{Bq}{4}-CD.=A$, for $\frac{B}{2}+\sqrt{\left(\frac{B^2}{4}-CD\right)}=A$ (from   *Oughtredus   explicatus*

[1682], p. 101)

11. These notations to signify aggregation occur very seldom in the texts referred to and may be simply printer's errors.

12. Mathematical parentheses occur also on p. 75, 80, and 117 of G. Clark's *Oughtredus explicatus*.

13. In the *Clavis* of 1631, p. 2, it says, "Signum additionis siue affirmationis, est+plus" and "Signum subductionis, siue negationis est−minus." In the edition of 1694 it says simply, "The Sign of Addition is + more" and "The Sign of Subtraction is − less," thereby ignoring, in the definition, the double function played by these symbols.

14. In the errata following the Preface of the 1694 edition it says, for *"more* or *mo.* r. [ead] *plus* or *pl.*"; for *less* or *le.* r.[ead] *minus* or *mi.*"

15. Oughtred's *Clavis mathematicae* of 1631 is not the first appearance of $\times$ as a symbol for multiplication. In Edward Wright's translation of John Napier's *Descriptio*, entitled *A Description of the Admirable Table of Logarithms* (London, 1618), the letter "X" is given as the sign of multiplication in the part of the book called "An Appendix to the Logarithms, shewing the practise of the calculation of Triangles, etc."

The use of the letters x and X for multiplication is not uncommon during the seventeenth and beginning of the eighteenth centuries. We note the following instances: Vincent Wing, *Doctrina theorica* (London, 1656), p. 63; John Wallis, *Arithmetica infinitorum* (Oxford, 1655), p. 115, 172; *Moore's Arithmetick in two Books*, by Jonas Moore (London, 1660), p. 108; *Antoine Arnauld, Novveavx elemens de geometrie* (Paris, 1667), p. 6; Lord Brounker, *Philosophical Transactions*, Vol. II (London, 1668), p. 466; *Exercitatio geometrica, auctore Laurentio Lorenzinio, Vincentii Viviani discipulo* (Florence, 1721). John Wallis used the $\bowtie$ in his *Elenchus geometriae Hobbianae* (Oxoniae, 1655), p. 23.

16. *in* as a symbol of multiplication carries with it also a collective meaning; for example, the *Clavis* of 1652 has on p. 77, "Erit $\frac{1}{2}Z+\frac{1}{2}B$ *in* $\frac{1}{2}Z-\frac{1}{2}B=\frac{1}{4}Zq-\frac{1}{4}Bq$."

17. That is, the line AB squared.

18. These capital letters precede the expression to be raised to a power. Seldom are they used to indicate powers of monomials. From the *Clavis* of 1652, p. 65, we quote:

$$\text{"}Q:A+E:+Eq=2Q:\tfrac{1}{2}A+E:+2Q.\tfrac{1}{2}A\text{ ,"}$$

$$\text{i.e., } (A+E)^2+E^2=2(\tfrac{1}{2}A+E)^2+2\left(\frac{A}{2}\right)^2.$$

19. L and l stand for the same thing, "side" or "root," l being used generally when the coefficients of the unknown quantity are given in Hindu-Arabic numerals, so that all the letters in the equation, viz., l, q, c, qq, qc, etc., are small letters. The *Clavis* of 1694, p. 158, uses L in a place where the Latin editions use l.

20. The symbol $\sqrt{u}$ does not occur in the *Clavis* of 1631 and is not defined in the later editions. The following throws light upon its significance. In the 1631 edition, chap. xvi, sec. 8, p. 40, the author takes $\sqrt{qBA}+B=CA$, gets from it $\sqrt{qBA}=CA-B$, then squares both sides and solves for the unknown A. He passes

next to a radical involving two terms, and says: "Item $\sqrt{q}$ vniuers : $BA+CA$: $-$ $D=BC$: vel per transpositionem $\sqrt{q}$: $BA+CA=BC+D$"; he squares both sides and solves for A. In the later editions he writes "$\sqrt{u}$" in place of "$\sqrt{q}$ vniuers : "

21. The sum $Z=A+E$ and the difference $X=A-E$ are used later in imitation of Oughtred by Samuel Foster in his *Miscellanies* (London, 1659), "Of Projection," p. 8, and by Sir Jonas Moore in his *Arithmetick* (3d ed.; London, 1688), p. 404; John Wallis in his *Operum mathematicorum pars prima* (Oxford, 1657), p. 169, and other parts of his mathematical writings.

22. Harriot's symbols $>$ for "greater" and $<$ for "less" were far superior to the corresponding symbols used by Oughtred.

23. This notation for "less than" in the *Ho.* occurs only in the explanation of "Fig. *EE.*" In the text (chap. ix) the regular notation explained in *Eu.* is used.

24. The symbol ∞ so closely resembles the symbol $\backsim$ which was used by John Wallis in his *Operum mathematicorum pars prima* (Oxford, 1657), p. 208, 247, 334, 335, that the two symbols were probably intended to be one and the same. It is difficult to assign a good reascn why Wallis, who greatly admired Oughtred and was editor of the later Latin editions of his *Clavis mathematicae*, should purposely reject Oughtred's ∞ and intentionally introduce $\backsim$ as a substitute symbol.

25. Von Braunmühl, in his *Geschichte der Trigonometrie* (2. Teil; Leipzig, 1903), p. 42, 91, refers to Oughtred's *Trigonometria* of 1657 as containing the earliest use of abbreviations of trigonometric functions and points out that a half-century later the army of writers on trigonometry had hardly yet reached the standard set by Oughtred. This statement must be modified in several respects (see §§ 500–526).

26. This reference is to the English edition, the *Trigonometrie* of 1657. In the Latin edition there is printed on p. 5, by mistake, s instead of s *versus.* The table of errata makes reference to this misprint.

27. The horizontal line was printed beneath the expression that was being crossed out. Thus, on p. 68 of the *Clavis* of 1631 there is:

$$BGqq - \underline{BGq \times 2BK \times BD} + BKq \times BDq$$
$$= BGq \times BDq + BGq \times BKq - \underline{BGq \times 2BK \times BD} + BGq \times 4CAq.$$

28. This notation, says Oughtred, was used by ancient writers on music, who "are wont to connect the terms of ratios, either to be continued" as in $\frac{3}{2} \times \frac{4}{3} = 2$, "or diminish'd" as in $\frac{3}{2} \div \frac{4}{3} = \frac{9}{8}$.

29. See n. 15.

30. *Cru* and *crur* are abbreviations for *crurum*, side of a rectangle or right triangle. Hence $Z\ cru$ means the sum of the sides, $X\ cru$, the difference of the sides.

187. Oughtred's recognition of the importance of notation is voiced in the following passage:

". . . . Which Treatise being not written in the usuall synthetical manner, nor with verbous expressions, but in the inventive way of Analitice, and with symboles or notes of things instead of words, seemed unto many very hard; though indeed it was but their owne diffidence, being scared by the newness of the delivery; and not any difficulty in the thing it selfe. For this specious and symbolicall manner, neither racketh the memory with multiplicity of words, nor chargeth the phantasie with comparing and laying things together; but plainly presenteth to the eye the whole course and processe of every operation and argumentation."[1]

Again in his *Circles of Proportion* (1632), p. 20:

"This manner of setting downe theoremes, whether they be Proportions, or Equations, by Symboles or notes of words, is most excellent, artificiall, and doctrinall. Wherefore I earnestly exhort every one, that desireth though but to looke into these noble Sciences Mathematicall, to accustome themselves unto it: and indeede it is easie, being most agreeable to reason, yea even to sence. And out of this working may many singular consectaries be drawne: which without this would, it may be, for ever lye hid."

ENGLISH: THOMAS HARRIOT

(1631)

188. Thomas Harriot's *Artis analyticae praxis* (London, 1631) appeared as a posthumous publication. He used small letters in place of Vieta's capitals, indicated powers by the repetition of factors, and invented > and < for "greater" and "less."

Harriot used a very long sign of equality =. The following quotation shows his introduction of the now customary signs for "greater" and "smaller" (p. 10):

"Comparationis signa in sequentibus vsurpanda.

Aequalitatis === ut a === b. significet a aequalem ipi b.

Maioritatis ⊃ ut a ⊃ b. significet a maiorem quam b.

Minoritatis ⊂ ut a ⊂ b significet a minorem quam b."

Noteworthy is the notation for multiplication, consisting of a vertical line on the right of two expressions to be multiplied together of which one is written below the other; also the notation for complex fractions in which the principal fractional line is drawn double. Thus (p. 10):

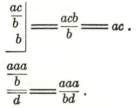

$$\left.\begin{array}{c}\dfrac{ac}{b}\\[4pt]b\end{array}\right| == \dfrac{acb}{b} == ac \,.$$

$$\dfrac{\dfrac{aaa}{b}}{d} == \dfrac{aaa}{bd}\,.$$

---

[1] William Oughtred, *The Key of the Mathematicks* (London, 1647), Preface.

Harriot places a dot between the numerical coefficient and the other factors of a term. Excepting only a very few cases which seem to be printer's errors, this notation is employed throughout. Thus (p. 60):

"Aequationis $aaa-3.baa+3.bba$══$+2.bbb$ est 2.b. radix
radici quaesititiae a. aequalis ."

Probably this dot was not intended as a sign of multiplication, but simply a means of separating a numeral from what follows, according to a custom of long standing in manuscripts and early printed books.

On the first twenty-six pages of his book, Harriot frequently writes all terms on one side of an equation. Thus (p. 26):

"Posito igitur $cdf=aaa$. est $aaa-cdf\ \overline{}\ $══ 0
$a+b$

Est autem ex genesi $aaa-cdf\ \overline{}\ $══ $aaaa+baaa-cdfa-bcdf$.
$a+b$

quae est aequatio originalis hic designata.
Ergo . . . . $aaaa+baaa-cdfa-bcdf$. ══ 0 ."

Sometimes Harriot writes underneath a given expression the result of carrying out the indicated operations, using a brace, but without using the regular sign of equality. This is seen in Figure 87. The first equation is $52=-3a+aaa$, where the vowel a represents the unknown. Then the value of a is given by Tartaglia's formula, as $\sqrt[3]{26+\sqrt{675}}+\sqrt[3]{26-\sqrt{675}}=4$. Notice that "$\sqrt{}3$.)" indicates that the cube root is taken of the binomial $26+\sqrt{675}$.

In Figure 88 is exhibited Harriot's use of signs of equality placed vertically and expressing the equality of a polynomial printed above a horizontal line with a polynomial printed below another horizontal line. This exhibition of the various algebraic steps is clever.

FRENCH: PIERRE HÉRIGONE
(1634, 1644)

189. A full recognition of the importance of notation and an almost reckless eagerness to introduce an exhaustive set of symbols is exhibited in the *Cursus mathematicus* of Pierre Hérigone, in six volumes, in Latin and French, published at Paris in 1634 and, in a second edition, in 1644. At the beginning of the first volume is given

In duabus antecedentibus æquationibus accidit interdum binomia cubica folutio-
nis radicalibus implicata explicari poffe per radices itidem binomias, quæ per fummam
vel differentiam conftituant tandem radicem fimplicem æquationis explicatoriam. Huius generis folutionum exempla funt quæ fequuntur.

Fig. 87.—From Thomas Harriot's *Artis analyticae praxis* (1631), p. 101

Lemma.

Si dari poffit radix aliqua æquationis radici *a.* æqualis, quæ radicibus *b. c. d.* inæ-
qualis fit, efto illa *f.* fiue alia quæcunque.

Pofito igitur $f = a.$ erit $ffff - bfff + bcff$
$- cfff + bdff$
$- dfff + cdff - bcdf$:
$+ ffff - bfff + bcff$
$- cfff + bdff$
$- dfff + cdff = + bcdf.$

Ergo . . . . $\dfrac{+2.ffff - 2.cfff + 2.cdff - 2.dfff}{\| \atop +2.bfff - 2.bcff + 2.bcdf - 2.bdff}$

Hoc eft . . . . . $\dfrac{+ffff - cfff + cdff - dfff}{\| \atop +bfff - bcff + bcdf - bdff}$

$\dfrac{+fff - cff + dcf - dff}{f} = \dfrac{+fff - cff + cdf - dff}{b}$

Ergo . . . . . $f = b.$ Quod eft contra Lemmatis hypothefin.

Non eft igitur $f = a.$ vt erat pofitum. Quod de alia quacunque ex fimili de-
ductione demonftrandum eft.

Fig. 88.—From Thomas Harriot's *Artis analyticae praxis* (1631), p. 65

an explanation of the symbols. As found in the 1644 edition, the list
is as follows:

+ plus

~ minus

·~: differentia

  ⊲ inter se, *entrélles*

⊲ *n* in, *en*

⊲ *ntr.* inter, *entre*

11 vel, *ou*

π, ad, *à*

5< pentagonum,   penta-
gone

6< hexagonum

√·4< latus quadrati

√·5< latus pentagoni

a2 *A* quadratum

a3 *A* cubus

a4 *A* quadrato-quadratū.
et sic infinitum.

= parallela

⊥ perpendicularis

·· est nota genitini, *signifie (de)*

; est nota numeri plural-

is, signifie le plurier

2|2 aequalis

3|2 maior

2|3 minor

⅓ tertia pars

¼ quarta pars

⅔ duae tertiae

a,b, 11 *ab* rectangulum quod sit
ductu *A* in *B*

· est punctum

— est recta linea

<, ∠ est angulus

⌐ est angulus rectus

⊙ est circulus

↷ ☾ est   pars   circumfer-
entiae circuli

⌒, ⌒ est segmentu circuli

△ est triangulum

□ est quadratum

▭ est rectangulum

◇ est parallelogrammum

◇ piped. est parallelepipedum

In this list the symbols that are strikingly new are those for equality
and inequality, the ~ as a minus sign, the — being made to represent
a straight line. Novel, also, is the expression of exponents in Hindu-
Arabic numerals and the placing of them to the right of the base, but
not in an elevated position. At the beginning of Volume VI is given a
notation for the aggregation of terms, in which the comma plays a
leading rôle:

"□· *a2~5a+6, a~4:* virgula, la virgule, dis-
tinguit multiplicatorem *a~4 à* multiplicădo
a2~5a+6.
Ergo ₒ □ 5+4+3, 7~3:~10, est 38."

Modern: The rectangle (a^2-5a+6) $(a-4)$,
Rectangle $(5+4+3)$ $(7-3)-10=38$.

"*hg* π *ga* 2|2 *hb* π *bd,* signifi. *HG* est *ad GA,* **vt**
HB ad *BD* ."

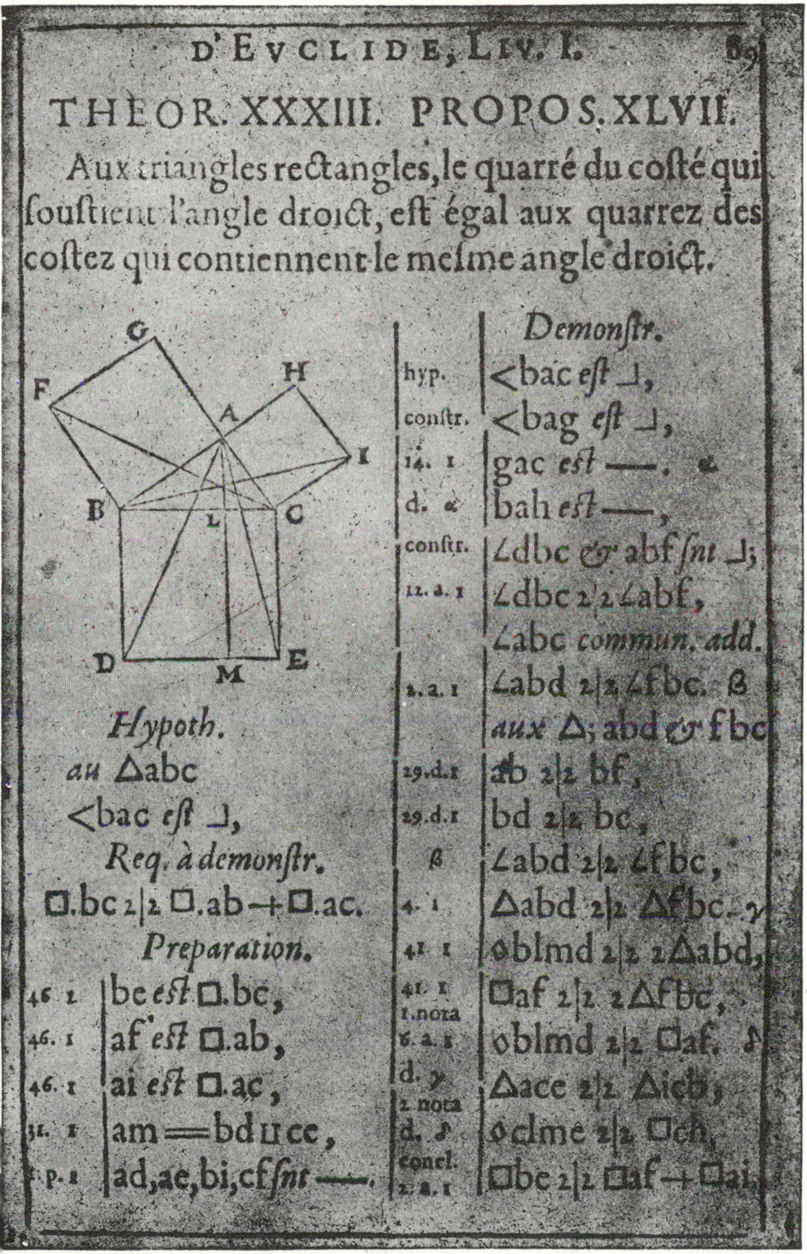

D'EVCLIDE, LIV. I.   89

THEOR. XXXIII. PROPOS. XLVII.

Aux triangles rectangles, le quarré du costé qui soustient l'angle droict, est égal aux quarrez des costez qui contiennent le mesme angle droict.

Hypoth.

au △abc

<bac est ⌐,

Req. à demonstr.

□.bc 2|2 □.ab + □.ac.

Preparation.

46.1	be est □.bc,
46.1	af est □.ab,
46.1	ai est □.ac,
31.1	am = bd ıı ce,
1.p.1	ad, ae, bi, cf snt ——.

Demonstr.

hyp.	<bac est ⌐,
constr.	<bag est ⌐,
14.1	gac est ——. &
d. &	bah est ——,
constr.	∠dbc & abf snt ⌐;
12.a.1	∠dbc 2|2 ∠abf,
	∠abc commun. add.
2.a.1	∠abd 2|2 ∠fbc. β
	aux △; abd & fbc
29.d.1	ab 2|2 bf,
29.d.1	bd 2|2 bc,
β	∠abd 2|2 ∠fbc,
4.1	△abd 2|2 △fbc. γ
41.1	oblmd 2|2 2△abd,
41.1 / 1.nota	□af 2|2 2△fbc,
6.a.1	oblmd 2|2 □af. δ
d.γ / 2.nota	△ace 2|2 △icb,
d.δ	oclme 2|2 □ch,
concl. / 2.a.1	□be 2|2 □af + □ai.

FIG. 89.—From P. Herigone, *Cursus mathematicus*, Vol. VI (1644); proof of the Pythagorean theorem.

Modern: $hg : ga = hb : bd$.

> "$\sqrt{}\cdot 16 + 9$ est 5, se pormoit de serire plus dis-
> tinctement ainsi ,
> $\sqrt{}\cdot(16+9)$ 11$\sqrt{}\cdot\overline{16+9}$, est 5:$\sqrt{}\cdot 9$, +4, sont
> 7:$\sqrt{}\cdot 9$, +$\sqrt{}\cdot 4$ sont 5: "

Modern: $\sqrt{}\cdot 16 + 9$ is 5, can be written more clearly thus,
$\sqrt{}\cdot(16+9)$ or $\sqrt{}\cdot\overline{16+9}$, is 5; $\sqrt{}\cdot 9$, +4, are 7;
$\sqrt{}\cdot 9$, +$\sqrt{}\cdot 4$ are 5 .

FRENCH: JAMES HUME
(1635, 1636)

190. The final development of the modern notation for positive integral exponents took place in mathematical works written in French. Hume was British by birth. His *Le traité d'algèbre* (Paris, 1635) contains exponents and radical indexes expressed in Roman numerals. In Figure 90 we see that in 1635 the plus (+) and minus (−) signs were firmly established in France. The idea of writing exponents without the bases, which had been long prevalent in the writings of Chuquet, Bombelli, Stevin, and others, still prevails in the 1635 publication of Hume. Expressing exponents in Roman symbols made it possible to write the exponent on the same line with the coefficient without confusion of one with the other. The third of the examples in Figure 90 exhibits the multiplication of $8x^2 + 3x$ by $10x$, yielding the product $80x^3 + 30x^2$.

The translation of part of Figure 91 is as follows: "*Example:* Let there be two numbers $\sqrt{9}$ and $\overset{\scriptscriptstyle ii}{\sqrt{}}8$, to reduce them it will be necessary to take the square of $\overset{\scriptscriptstyle ii}{\sqrt{}}8$, because of the II which is with 9, and the square of the square of $\sqrt{9}$ and you obtain $\sqrt{6561}$ and $\overset{\scriptscriptstyle v}{\sqrt{}}64$. . . . .

$$\overset{\scriptscriptstyle ii}{\sqrt{}}8 \text{ to } \overset{\scriptscriptstyle v}{\sqrt{}}64 \quad | \quad \overset{\scriptscriptstyle ii}{\sqrt{}}3 \text{ to } \overset{\scriptscriptstyle iv}{\sqrt{}}8 \text{ [should be } \overset{\scriptscriptstyle iv}{\sqrt{}}9]$$
$$\sqrt{9} \text{ to } \overset{\scriptscriptstyle v}{\sqrt{}}729 \quad | \quad \sqrt{2} \text{ to } \overset{\scriptscriptstyle v}{\sqrt{}}9 \text{ [should be } \overset{\scriptscriptstyle v}{\sqrt{}}8]$$
$$\overset{\scriptscriptstyle v}{\sqrt{}}3 \text{ to } \overset{\scriptscriptstyle vi}{\sqrt{}}9$$
$$\sqrt{2} \text{ to } \overset{\scriptscriptstyle iv}{\sqrt{}}32 \text{ ."}$$

The following year, Hume took an important step in his edition of *L'algèbre de Viète* (Paris, 1636), in which he wrote A^{iii} for A^3. Except for the use of the Roman numerals one has here the notation used by Descartes in 1637 in his *La géométrie* (see § 191).

FRENCH: RENÉ DESCARTES
(1637)

191. Figure 92 shows a page from the first edition of Descartes' *La géométrie*. Among the symbolic features of this book are: (1) the use of small letters, as had been emphasized by Thomas Harriot;

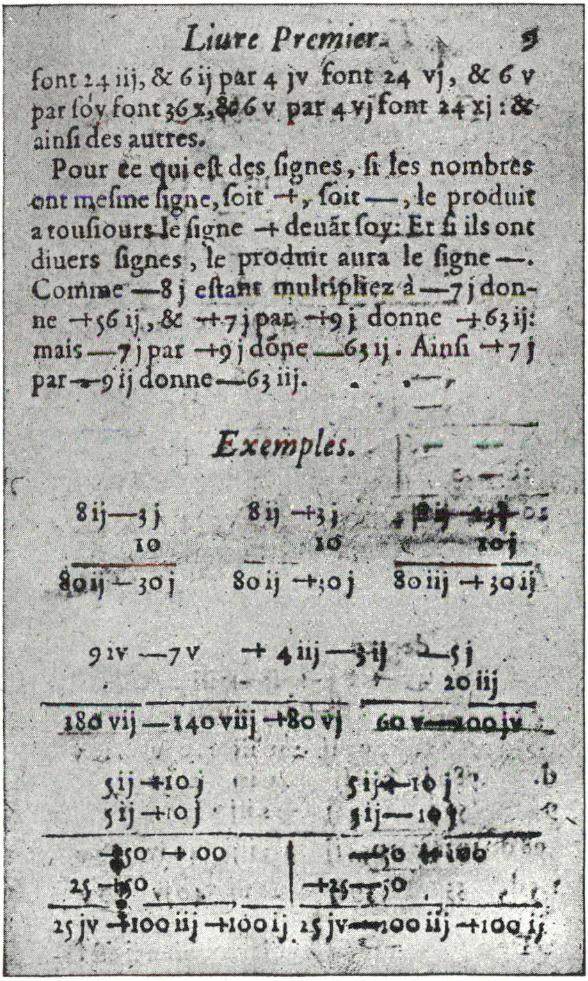

Liure Premier. 5

font 24 iij, & 6 ij par 4 jv font 24 vj, & 6 v par soy font 36 x, & 6 v par 4 vj font 24 xj : & ainsi des autres.

Pour ce qui est des signes, si les nombres ont mesme signe, soit +, soit —, le produit a tousiours le signe + deuãt soy: Et si ils ont diuers signes, le produit aura le signe —. Comme —8 j estant multipliez à —7 j donne +56 ij., & +7 j par +9 j donne +63 ij: mais —7 j par +9 j dõne —63 ij. Ainsi +7 j par —9 ij donne —63 iij.

Exemples.

8 ij — 3 j / 10 / 80 ij — 30 j

8 ij + 3 j / 10 / 80 ij + 30 j

8 ij + 3 j / 10 j / 80 iij + 30 ij

9 iv — 7 v / + 4 iij — 3 ij / 180 vij — 140 viij + 80 vj

— 5 j / 20 iij / 60 v — 100 iv

5 ij + 10 j / 5 ij + 10 j / + 50 + 100 / 25 + 50 / 25 jv + 100 iij + 100 ij

5 ij + 10 j / 5 ij — 10 j / — 50 + 100 / + 25 — 50 / 25 jv — 100 iij + 100 ij

FIG. 90.—Roman numerals for unknown numbers in James Hume, *Algèbre* (Paris, 1635).

(2) the writing of the positive integral exponents in Hindu-Arabic numerals and in the position relative to the base as is practiced today,

except that aa is sometimes written for a^2; (3) the use of a new sign of equality, probably intended to represent the first two letters in the word *aequalis*, but apparently was the astronomical sign, ♉ taurus,

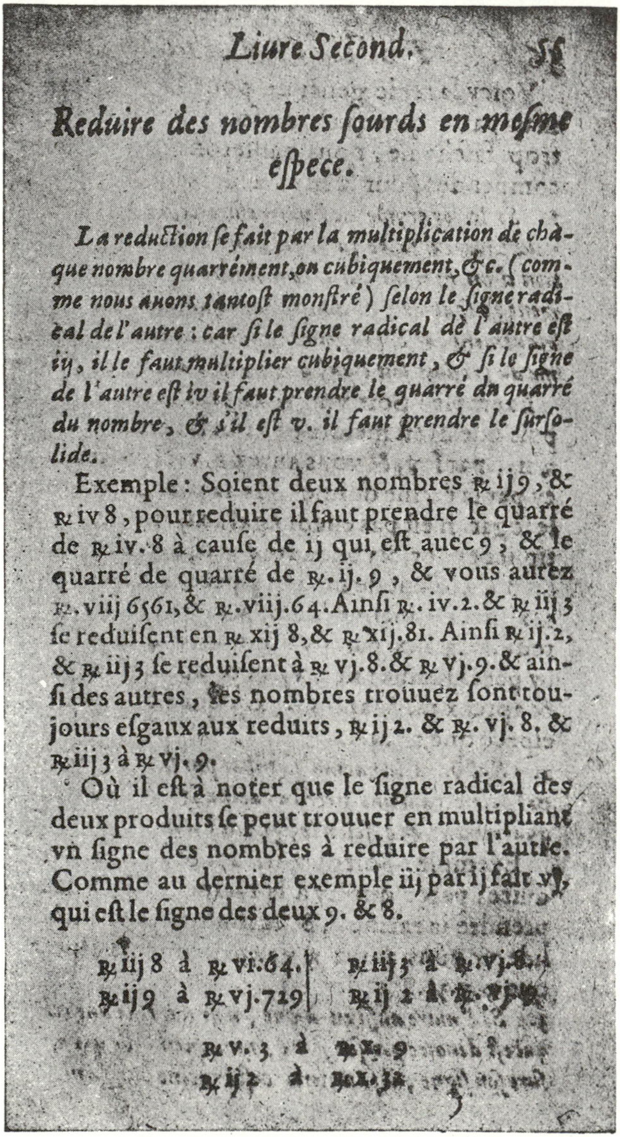

Fig. 91.—Radicals in James Hume, *Algèbre* (1635)

LIVRE PREMIER. 303

angle, iufques a O, en forte qu'N O foit efgale a N L, la toute O M eft z la ligne cherchée. Et elle s'exprime en cete forte

$$z \infty \tfrac{1}{2} a + \sqrt{\tfrac{1}{4} aa + bb}.$$

Que fi iay $yy \infty - ay + bb$, & qu'y foit la quantité qu'il faut trouuer, ie fais le mefme triangle rectangle N L M, & de fa baze M N i'ofte N P efgale a N L, & le refte P M eft y la racine cherchée. De façon que iay

$$y \infty - \tfrac{1}{2} a + \sqrt{\tfrac{1}{4} aa + bb}.$$

Et tout de mefme fi i'auois $x^4 \infty - ax^2 + b$. P M feroit x^2. & i'aurois

$$x \infty \sqrt{- \tfrac{1}{2} a + \sqrt{\tfrac{1}{4} aa + bb}}:$$

& ainfi des autres.

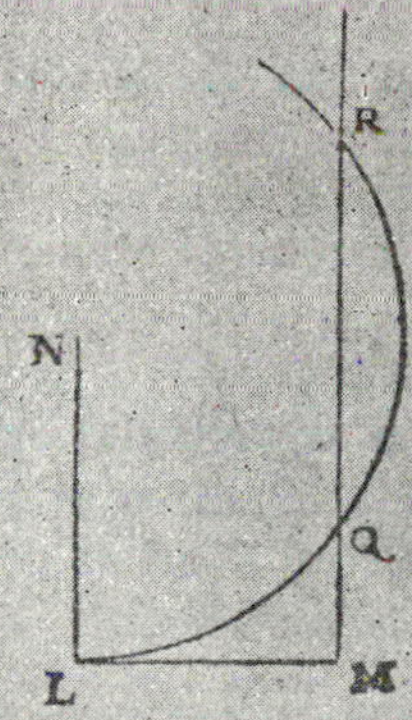

Enfin fi i'ay

$$z^2 \infty az - bb:$$

ie fais N L efgale à $\tfrac{1}{2} a$, & L M efgale à b côme deuāt, puis, au lieu de ioindre les poins M N, ie tire M Q R parallele a L N. & du centre N par L ayant defcrit vn cercle qui la couppe aux poins Q & R, la ligne cherchée z eft M Q, oubiē M R, car en ce cas elle s'exprime en deux façons, a fçauoir $z \infty \tfrac{1}{2} a + \sqrt{\tfrac{1}{4} aa - bb}$, & $z \infty \tfrac{1}{2} a - \sqrt{\tfrac{1}{4} aa - bb}$.

Et fi le cercle, qui ayant fon centre au point N, paffe par le point L, ne couppe ny ne touche la ligne droite M Q R, il n'y a aucune racine en l'Equation, de façon qu'on peut affurer que la conftruction du problefme propofé eft impoffible.

Au

FIG. 92.—A page from René Descartes, *La géométrie* (1637)

placed horizontally, with the opening facing to the left; (4) the uniting of the ˈvinculum with the German radical sign $\sqrt{}$, so as to give $\sqrt{}$, an adjustment generally used today.

The following is a quotation from Descartes' text (ed., Paris, 1886, p. 2): "Mais souvent on n'a pas besoin de tracer ainsi ces lignes sur le papier, et il suffit de les désigner par quelques lettres, chacune par une seule. Comme pour ajouter le ligne *BD* à *GH*, je nomme l'une *a* et l'autre *b*, et écris *a+b;* et *a−b* pour soustraire *b* de *a;* et *ab* pour les multiplier l'une par l'autre; et $\frac{a}{b}$ pour diviser *a* par *b;* et *aa* ou *a²* pour multiplier *a* par soi-même; et *a³* pour le multiplier encore une fois par *a*, et ainsi à l'infini."

The translation is as follows: "But often there is no need thus to trace the lines on paper, and it suffices to designate them by certain letters, each by a single one. Thus, in adding the line *BD* to *GH*, I designate one *a* and the other *b*, and write *a+b;* and *a−b* in sub-· tracting *b* from *a;* and *ab* in multiplying the one by the other; and $\frac{a}{b}$ in dividing *a* by *b;* and *aa* or *a²* in multiplying *a* by itself; and *a³* in multiplying it once more again by *a*, and thus to infinity."

ENGLISH: ISAAC BARROW
(1655, 1660)

192. An enthusiastic admirer of Oughtred's symbolic methods was Isaac Barrow,[1] who adopted Oughtred's symbols, with hardly any changes, in his Latin (1655) and his English (1660) editions of *Euclid*. Figures 93 and 94 show pages of Barrow's *Euclid*.

ENGLISH: RICHARD RAWLINSON
(1655–68)

193. Sometime in the interval 1655–68 Richard Rawlinson, of Oxford, prepared a pamphlet which contains a collection of litho-graphed symbols that are shown in Figure 95, prepared from a crude freehand reproduction of the original symbols. The chief interest lies in the designation of an angle of a triangle and its opposite side by the same letter—one a capital letter, the other letter small. This simple device was introduced by L. Euler, but was suggested many years earlier by Rawlinson, as here shown. Rawlinson designated spherical

---

[1] For additional information on his symbols, see §§ 456, 528.

triangles by conspicuously rounded letters and plane triangles by letters straight in part.

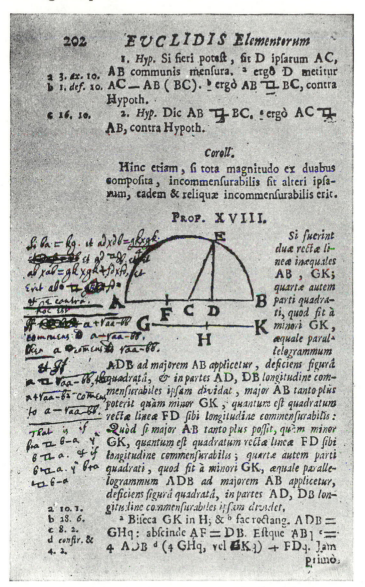

FIG. 93.—Latin edition (1655) of Barrow's *Euclid*. Notes by Isaac Newton. (Taken from *Isaac Newton: A Memorial Volume* [ed. W. J. Greenstreet; London, 1927], p. 168.)

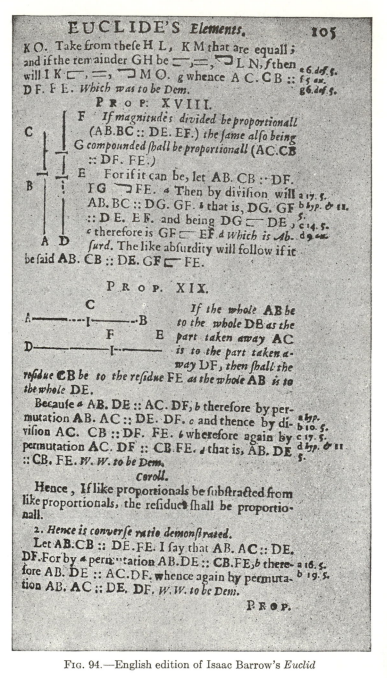

Fig. 94.—English edition of Isaac Barrow's *Euclid*

SWISS: JOHANN HEINRICH RAHN
(1659)

194. Rahn published in 1659 at Zurich his *Teutsche Algebra*, which was translated by Thomas Brancker and published in 1668 at London, with additions by John Pell. There were some changes in the symbols as indicated in the following comparison:

Meaning	German Edition, 1659		English Edition, 1668	
1. Multiplication.............	*	(p. 7)	Same	(p. 6)
2. $a+b$ times $a-b$..........	$\begin{matrix} a+b \\ a-b \end{matrix}\Big\|$	(p. 14)	Same	(p. 12)
3. Division.................	÷	(p. 8)	Same	(p. 7)
4. Cross-multiplication.......	*×	(p. 25)	*X	(p. 23)
5. Involution...............	Archimedean spiral (Fig. 96)	(p. 10)	Ligature of omicron and sigma (Fig. 97)	(p. 9)
6. Evolution................	Ligature of two epsilons (Fig.96)	(p. 11)	Same	(p. 9)
7. *Erfüll ein quadrat* Compleat the square $\Big\}$.....	$E\ \square$	(p 16)	$C\ \square$	(p. 14)
8. Sixth root...............	$\sqrt{}\ qc.\begin{cases} aaa=\sqrt{a} \\ aa=\sqrt{c}.a \end{cases}$	(p. 34)	*cubo-cubick* √ of $aaa=\sqrt{a}$ *cubo-cubick* √ of $aa=\sqrt{c}.a$	(p. 32)
9. Therefore...............	∴ (usually)	(p. 53)	∵ (usually)	(p. 37)
10. Impossible (absurd)	⊉	(p. 61)	ƆI	(p. 48)
11. Equation expressed in another way................	,	(p. 67)	Same	(p. 54)
12. Indeterminate, "liberty of assuming an equation".......	(*	(p. 89)	Same	(p. 77)
13. Nos. in outer column referring to steps numbered in middle column............	1, 2, 3, etc.	(p. 3)	1, 2, 3, etc.	(p. 3)
14. Nos in outer column *not* referring to numbers in middle column.................	1, 2, 3, etc.	(p. 3)	Ī, 2̄, 3̄, etc.	(p. 3)

REMARKS ON THESE SYMBOLS

No. 1.—Rahn's sign * for multiplication was used the same year as Brancker's translation, by N. Mercator, in his *Logarithmotechnia* (London, 1668), p. 28.

No. 4.—If the lowest common multiple of abc and ad is required, Rahn writes $\dfrac{abc}{ad}=\dfrac{bc}{d}$; then $\dfrac{abc}{ad}*×\dfrac{bc}{d}$ yields $abcd$ in each of the two cross-multiplications.

No. 8.—Rahn's and Brancker's modes of indicating the higher powers and roots differ in principle and represent two different procedures which had been competing for supremacy for several centuries. Rahn's $\sqrt{}\ qc.$ means the sixth root, $2×3=6$, and represents the Hindu idea. Brancker's *cubo-cubick* root means the "sixth root," $3+3=6$, and represents the Diophantine idea.

No. 9.—In both editions occur both ∴ and ∵, but ∴ prevails in the earlier edition; ∵ prevails in the later.

No. 10.—The symbols indicate that the operation is impossible or, in case of a root, that it is imaginary.

No. 11.—The use of the comma is illustrated thus: The marginal column (1668, p. 54) gives "6, 1," which means that the sixth equation "$Z=A$" and the first equation "$A=6$" yield $Z=6$.

No. 12.—For example, if in a right triangle h, b, c. we know only $b-c$, then one of the three sides, say c, is indeterminate.

Page 73 of Rahn's *Teutsche Algebra* (shown in Fig. 96) shows: (1) the first use of ÷ in print, as a sign of division; (2) the Archimedean spiral for involution; (3) the double epsilon for evolution; (4) the use of capital letters B, D, E, for given numbers, and small letters a, b, for unknown numbers; (5) the $*$ for multiplication; (6) the first use of $\therefore$ for "therefore"; (7) the three-column arrangement of which the left column contains the directions, the middle the numbers of

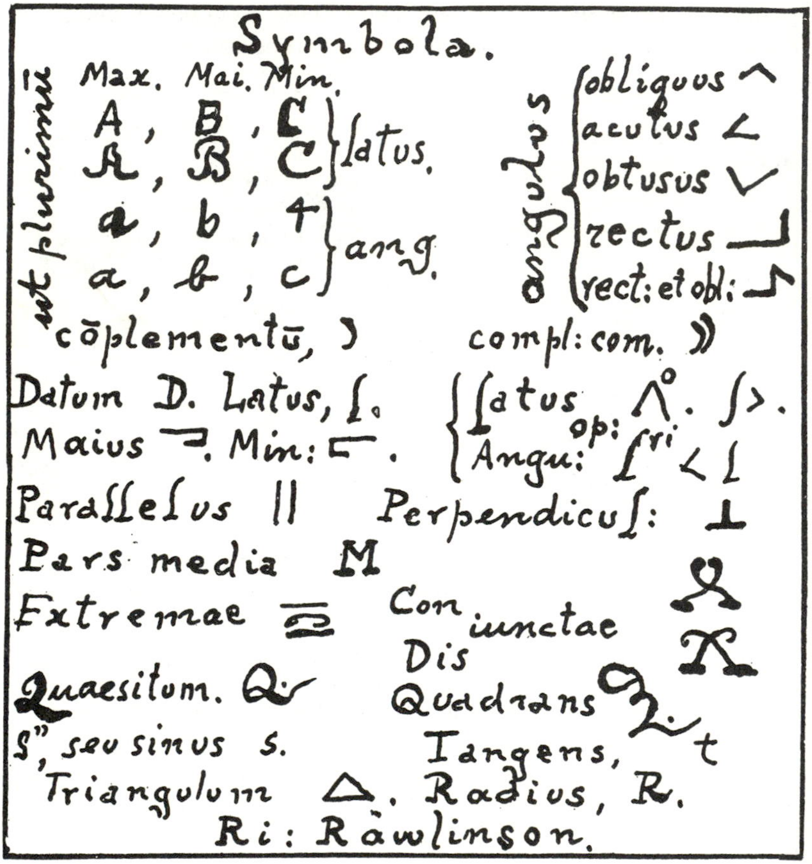

FIG. 95.—Freehand reproduction of Richard Rawlinson's symbols

the lines, the right the results of the operations. Thus, in line 3, we have "line 1, raised to the second power, gives $aa+2ab+bb=DD$."

ENGLISH: JOHN WALLIS
(1655, 1657, 1685)

195. Wallis used extensively symbols of Oughtred and Harriot, but of course he adopted the exponential notation of Descartes (1637). Wallis was a close student of the history of algebra, as is illustrated

by the exhibition of various notations of powers which Wallis gave in 1657. In Figure 98, on the left, are the names of powers. In the first column of symbols Wallis gives the German symbols as found in Stifel, which Wallis says sprang from the letters *r*, *z*, *c*, ∫, the first

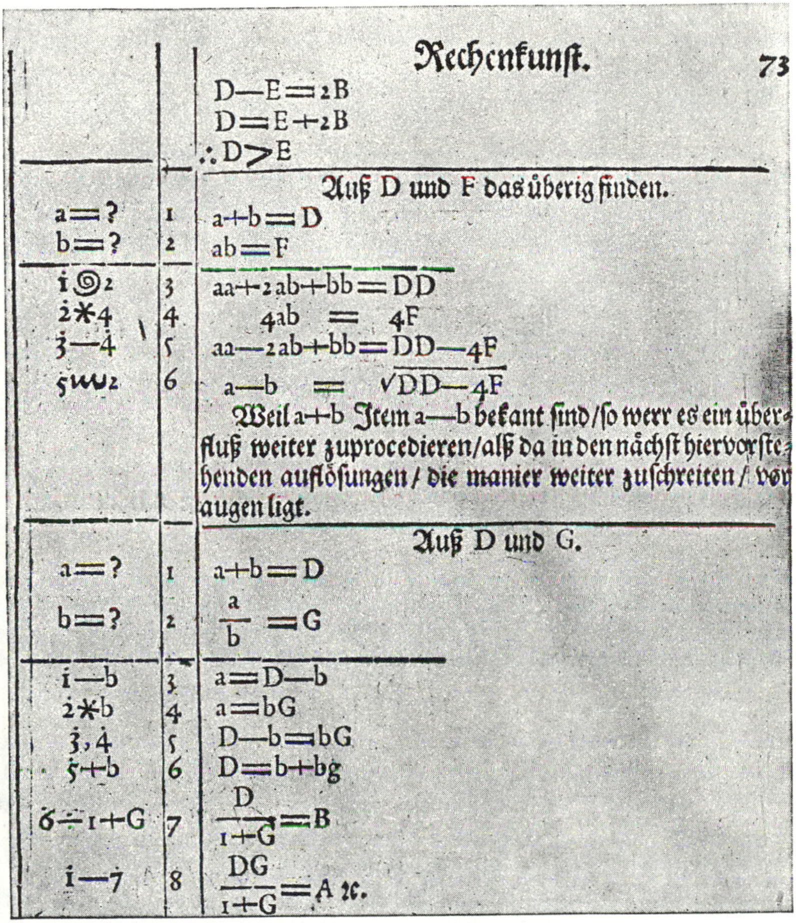

Rechenkunst. 73

D—E = 2B
D = E + 2B
∴ D > E

Auß D und F das übrig finden.

a = ?	1	a + b = D
b = ?	2	ab = F
1 ⑨ 2	3	aa + 2ab + bb = DD
2 * 4	4	4ab = 4F
3 — 4	5	aa — 2ab + bb = DD — 4F
5 ɯ 2	6	a — b = √DD — 4F

Weil a + b Item a — b bekant sind/so werr es ein überfluß weiter zuprocedieren/alß da in den nächst hiervorstehenden auflösungen/die manier weiter zuschreiten/vor augen ligt.

Auß D und G.

a = ?	1	a + b = D
b = ?	2	$\frac{a}{b}$ = G
1 — b	3	a = D — b
2 * b	4	a = bG
3, 4	5	D — b = bG
5 + b	6	D = b + bg
6 ÷ 1 + G	7	$\frac{D}{1+G}$ = B
1 — 7	8	$\frac{DG}{1+G}$ = A &c.

FIG. 96.—From Rahn, *Teutsche Algebra* (1659)

letters of the words *res*, *zensus*, *cubus*, *sursolidus*. In the second column are the letters *R*, *Q*, *C*, *S* and their combinations, Wallis remarking that for *R* some write *N*; these were used by Vieta in numerical equations. In the third column are Vieta's symbols in literal algebra, as abbreviated by Oughtred; in the fourth column Harriot's procedure is indicated; in the fifth column is Descartes' exponential notation.

In his *Arithmetica infinitorum*[1] he used the colon as a symbol for aggregation, as $\sqrt{:a^2+1}$ for $\sqrt{a^2+1}$, $\sqrt{:aD-a^2:}$ for $\sqrt{aD-a^2}$; Oughtred's notation for ratio and proportion, $\because$ for continued proportion. As the sign for multiplication one finds in this book X and X, both signs occurring sometimes on one and the same page (for instance, p. 172). In a table (p. 169) he puts □ for a given number: "Verbi gratiâ; si numerus hâc notâ □ designatus supponatur cognitus, reliqui omnes etiam cognoscentur." It is in this book and in his *De*

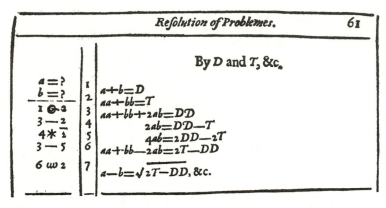

Fig. 97.—From Brancker's translation of Rahn (1668). The same arrangement of the solution as in 1659, but the omicron-sigma takes the place of the Archimedean spiral; the ordinal numbers in the outer column are not dotted, while the number in that column which does not refer to steps in the middle column carries a bar, $\bar{2}$. Step 5 means "line 4, multiplied by 2, gives $4ab = 2DD - 2T$."

sectionibus conicis that Wallis first introduces ∞ for infinity. He says (p. 70): "Cum enim primus terminus in serie Primanorum sit 0, primus terminus in serie reciproca erit ∞ vel infinitus : (sicut, in divisione, si diviso sit 0, quotiens erit infinitus)"; on pages 152, 153: " . . . . quippe $\frac{1}{\infty}$ (pars infinite parva) habenda erit pro nihilo," "$\infty \times \frac{1}{\infty}B=B$," "Nam ∞, ∞ +1 ∞ −1, perinde sunt"; on page 168: "Quamvis enim ∞ ×0 non aliquem determinate numerum designet. . . . ." An imitation of Oughtred is Wallis' "$\Pi\!\!\!\Pi : 1|\frac{3}{2}$," which occurs in his famous determination by interpolation of $\frac{4}{\pi}$ as the ratio of two infinite products. At this place he represents our $\frac{4}{\pi}$ by the symbol □.

[1] *Johannis Wallisii Arithmetica infinitorum* (Oxford, 1655).

He says also (p. 175): "Si igitur ut $\sqrt{}$:3×6: significat terminum medium inter 3 et 6 in progressione Geometrica aequabili 3, 6, 12, etc. (continue multiplicando 3×2×2 etc.) ita m :1|$\frac{3}{2}$: significet terminum medium inter 1 et $\frac{3}{2}$ in progressione Geometrica decrescente 1, $\frac{3}{2}$, $\frac{15}{8}$, etc. (continue multiplicando 1×$\frac{3}{2}$×$\frac{5}{4}$, etc.) erit $\square = \mathit{m}$:1|$\frac{3}{2}$: Et propterea circulus est ad quadratum diametri, ut 1 ad m :1|$\frac{3}{2}$." He uses this symbol again in his *Treatise of Algebra* (1685), pages 296, 362.

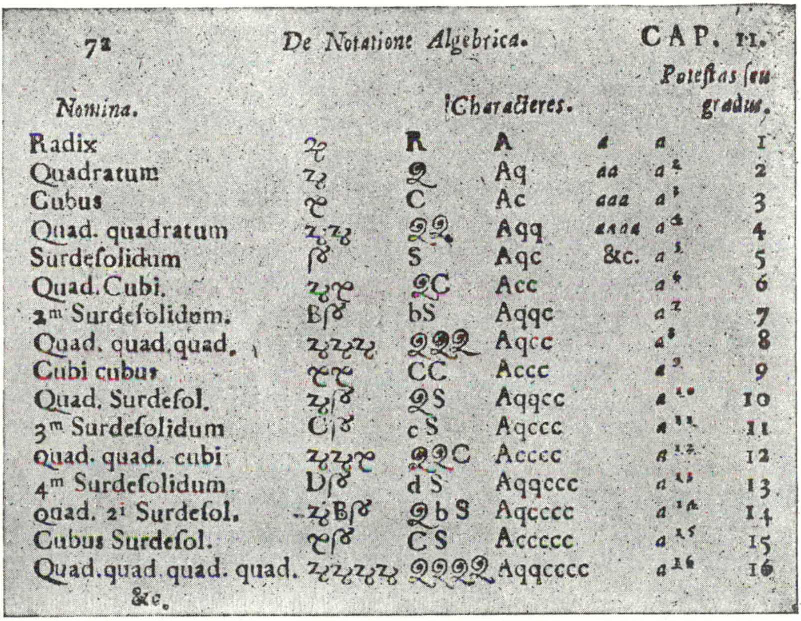

72	De Notatione Algebrica.				CAP. II.
Nomina.		*Characteres.*			*Potestas seu gradus.*
Radix	$\mathcal{R}$	R	A	*a*	*a*	1
Quadratum	$\mathcal{R}\mathcal{R}$	Q	Aq	*aa*	a^2	2
Cubus	$\mathcal{C}$	C	Ac	*aaa*	a^3	3
Quad. quadratum	$\mathcal{R}\mathcal{R}\mathcal{R}\mathcal{R}$	QQ	Aqq	*aaaa*	a^4	4
Surdesolidum	$\int^\delta$	S	Aqc	&c.	a^5	5
Quad. Cubi.	$\mathcal{R}\mathcal{C}$	QC	Acc		a^6	6
2m Surdesolidum.	B$\int^\delta$	bS	Aqqc		a^7	7
Quad. quad. quad.	$\mathcal{R}\mathcal{R}\mathcal{R}$	QQQ	Aqcc		a^8	8
Cubi cubus	$\mathcal{C}\mathcal{C}$	CC	Accc		a^9	9
Quad. Surdesol.	$\mathcal{R}\int^\delta$	QS	Aqqcc		a^{10}	10
3m Surdesolidum	C$\int^\delta$	cS	Aqccc		a^{11}	11
Quad. quad. cubi	$\mathcal{R}\mathcal{R}\mathcal{C}$	QQC	Accccc		a^{12}	12
4m Surdesolidum	D$\int^\delta$	dS	Aqqccc		a^{13}	13
Quad. 2i Surdesol.	$\mathcal{R}$B$\int^\delta$	QbS	Aqcccc		a^{14}	14
Cubus Surdesol.	$\mathcal{C}\int^\delta$	CS	Accccc		a^{15}	15
Quad. quad. quad. quad.	$\mathcal{R}\mathcal{R}\mathcal{R}\mathcal{R}$	QQQQ	Aqqcccc		a^{16}	16
&c.						

Fig. 98.—From John Wallis, *Operum mathematicorum pars prima* (Oxford, 1657), p. 72.

The absence of a special sign for division shows itself in such passages as (p. 135): "Ratio rationis hujus $\frac{1}{2\square}$ ad illam $\frac{1}{2}$, puta $\left.\frac{1}{2}\right)\frac{1}{2\square}\left(\frac{1}{\square}\right.$, erit. . . . . " He uses Oughtred's clumsy notation for decimal fractions, even though Napier had used the point or comma in 1617. On page 166 Wallis comes close to the modern radical notation; he writes "$\sqrt{}^6R$" for $\sqrt[6]{R}$. Yet on that very page he uses the old designation "$\sqrt{}qqR$" for $\sqrt[4]{R}$.

His notation for continued fractions is shown in the following quotation (p. 191):

"Esto igitur fractio ejusmodi continue fracta quaelibet, sic designata, $\dfrac{a}{a}\dfrac{b}{\beta}\dfrac{c}{\gamma}\dfrac{d}{\delta}\dfrac{e}{\epsilon}$, etc.,"

where

$$\frac{a}{a}\frac{b}{\beta} \equiv \frac{a\beta}{a\beta + b} .$$

The suggestion of the use of negative exponents, introduced later by Isaac Newton, is given in the following passage (p. 74): "Ubi autem series directae indices habent 1, 2, 3, etc. ut quae supra seriem Aequalium tot gradibus ascendunt; habebunt hae quidem (illis reciprocae) suos indices contrarios negativos $-1, -2, -3$, etc. tanquam tot gradibus infra seriem Aequalium descendentes."

In Wallis' *Mathesis universalis*,[1] the idea of positive and negative integral exponents is brought out in the explanation of the Hindu-Arabic notation. The same principle prevails in the sexagesimal notation, "hoc est, minuta prima, secunda, tertia, etc. ad dextram descendendo," while ascending on the left are units "quae vocantur Sexagena prima, secunda, tertia, etc. hoc modo.

$$\overset{\backslash\backslash\backslash\backslash}{4}\overset{}{9}, \quad \overset{\backslash\backslash\backslash}{3}\overset{}{6}, \quad \overset{\backslash\backslash}{2}\overset{}{5}, \quad \overset{\backslash}{1}\overset{}{5}, \quad \overset{\circ}{1}, \quad 1\overset{\prime}{5}, \quad 2\overset{\prime\prime}{5}, \quad 3\overset{\prime\prime\prime}{6}, \quad 4\overset{\prime\prime\prime\prime}{9} ."$$

That the consideration of sexagesimal integers of denominations of higher orders was still in vogue is somewhat surprising.

On page 157 he explains both the "scratch method" of dividing one number by another and the method of long division now current, except that, in the latter method, he writes the divisor underneath the dividend. On page 240: "$A, M, V \overset{\cdot\cdot}{-}$" for arithmetic proportion, i.e., to indicate $M - A = V - M$. On page 292, he introduces a general root d in this manner: "$\sqrt{d}R^d = R$." Page 335 contains the following interesting combination of symbols:

	in Modern Symbols
"Si $\overbrace{A \cdot B \cdot \underbrace{C : a \cdot \beta}} \cdot \gamma$	If $A:B = a:\beta$,
	and $B:C = \beta:\gamma$,
Erit $A \cdot C :: a \cdot \gamma$."	then $A:C = a:\gamma$.

196. In the *Treatise of Algebra*[2] (p. 46), Wallis uses the decimal point, placed at the lower terminus of the letters, thus: 3.14159,

[1] *Johannis Wallisii Mathesis universalis: sive, Arithmeticum opus integrum* (Oxford, 1657), p. 65–68.

[2] *Op. cit.* (London, 1685).

26535. . . . . , but on page 232 he uses the comma, "12,756," ",3936."
On page 67, describing Oughtred's *Clavis mathematicae*, Wallis says:
"He doth also (to very great advantage) make use of several Ligatures,
or Compendious Notes, to signify the *Summs, Differences*, and *Rectangles* of several Quantities. As for instance, Of two quantities A
(the Greater, and E (the Lesser,) the Sum he calls Z, the Difference
X, the Rectangle Æ. . . . ." On page 109 Wallis summarizes various
practices: "The Root of such Binomial or Residual is called a Root
universal; and thus marked $\sqrt{u}$, (Root universal,) or $\sqrt{b}$, (Root of a
Binomial,) or $\sqrt{r}$, (Root of a Residual,) or drawing a Line over the
whole Compound quantity; or including it (as Oughtred used to do)
within two colons; or by some other distinction, whereby it may appear, that the note of Radicality respects, not only the single quantity
next adjoining, but the whole Aggregate. As $\sqrt{b}:2+\sqrt{3}\cdot\sqrt{r}:2-$
$\sqrt{3}\cdot\sqrt{u}:2\pm\sqrt{3}\cdot\sqrt{2\pm\sqrt{3}}\cdot\sqrt{}:2\pm\sqrt{3}$; etc."

On page 227 Wallis uses Rahn's sign $\div$ for division; along with the
colon as the sign of aggregation it gives rise to oddities in notation
like the following: "$ll-2laa+a^4:\div bb$."

On page 260, in a geometric problem, he writes "$\square AE$" for the
square of the line AE; he uses $\stackrel{\frown}{\smile}$ for the absolute value of the
difference.

On page 317 his notation for infinite products and infinite series is
as follows:

"$1\times1\frac{1}{3}\times1\frac{1}{24}\times1\frac{1}{48}\times1\frac{1}{80}\times1\frac{1}{120}\times1\frac{1}{168}\times$ etc."

"$1+\frac{1}{8}A+\frac{1}{24}B+\frac{1}{48}C+\frac{1}{80}D+\frac{1}{120}E+\frac{1}{168}F+$ etc." ;

on page 322:

"$\sqrt{}:2-\sqrt{}:2+\sqrt{}:2+\sqrt{2}$" for $\sqrt{2-\sqrt{2+\sqrt{2+\sqrt{2}}}}$.

On page 332 he uses fractional exponents (Newton having introduced the modern notation for negative and fractional exponents in
1676) as follows:

"$\sqrt{}^5:c^5+c^4x-x^5:$ or $\overline{c^5+c^4x-x^5}|^{\frac{1}{5}}$."

The difficulties experienced by the typesetter in printing fractional
exponents are exhibited on page 346, where we find, for example,
"$d\frac{1}{2}\ x\frac{1}{2}$" for $d^{\frac{1}{2}}x^{\frac{1}{2}}$. On page 123, the factoring of 5940 is shown as
follows:

"11)5)3)3)3)2)2) 5940 (2970(1485(495(165(55(11(1 ."

In a letter to John Collins, Wallis expresses himself on the sign of multiplication: "In printing my things, I had rather you make use of Mr. Oughtred's note of multiplication, $\times$, than that of $*$; the other being the more simple. And if it be thought apt to be mistaken for X, it may [be] helped by making the upper and lower angles more obtuse $\bowtie$."[1] "I do not understand why the sign of multiplication $\times$ should more trouble the convenient placing of the fractions than the other signs $+ - = > ::$."[2]

Wallis, in presenting the history of algebra, stressed the work of Harriot and Oughtred. John Collins took some exception to Wallis' attitude, as is shown in the following illuminating letter. Collins says:[3] "You do not like those words of Vieta in his theorems, ex adjunctione plano solidi, plus quadrato quadrati, etc., and think Mr. Oughtred the first that abridged those expressions by symbols; but I dissent, and tell you 'twas done before by Cataldus, Geysius, and Camillus Gloriosus,[4] who in his first decade of exercises, (not the first tract,) printed at Naples in 1627, which was four years before the first edition of the Clavis, proposeth this equation just as I here give it you, viz., $1ccc + 16qcc + 41qqc - 2304cc - 18364qc - 133000qq - 54505c + 3728q + 8064N$ aequatur 4608, finds N or a root of it to be 24, and composeth the whole out of it for proof, just in Mr. Oughtred's symbols and method. Cataldus on Vieta came out fifteen years before, and I cannot quote that, as not having it by me. . . . . And as for Mr. Oughtred's method of symbols, this I say to it; it may be proper for you as a commentator to follow it, but divers I know, men of inferior rank that have good skill in algebra, that neither use nor approve it. . . . . Is not A^5 sooner wrote than Aqc? Let A be 2, the cube of 2 is 8, which squared is 64: one of the questions between Magnet Grisio and Gloriosus is whether $64 = A_{cc}$ or A_{qc}. The Cartesian method tells you it is A^6, and decides the doubt."

EXTRACT FROM ACTA ERUDITORUM[5]

197. "Monendum denique, nos in posterum in his Actis usuros esse Signis *Leibnitianis*, ubi cum *Algebraicis* res nobis fuerit, ne typothetis

[1] John Wallis to John Collins, July 21, 1668 (S. P. Rigaud, *Correspondence of Scientific Men of the Seventeenth Century*, Vol. II [Oxford, 1841], p. 492).

[2] Wallis to Collins, September 8, 1668 (*ibid.*, p. 494).

[3] Letter to John Wallis, about 1667 (*ibid.*, p. 477–80).

[4] "*Exercitationum Mathematicarum Decas prima*, Nap. 1627, and probably Cataldus' *Transformatio Geometrica*, Bonon. 1612."

[5] Taken from *Acta eruditorum* (Leipzig, 1708), p. 271.

taedia & molestias gratis creemus, utque ambiguitates evitemus. Loco igitur lineolae characteribus supraducendae parenthesin adhibebimus, immo in multiplicatione simplex comma, ex. gr. loco $\overline{Vaa+bb}$ scribemus V($aa+bb$) & pro $\overline{aa+bb}\times c$ ponemus $aa+bb$, c. Divisionem designabimus per duo puncta, nisi peculiaris quaedam circumstantia morem vulgarem adhiberi suaserit. Ita nobis erit $a:b=\dfrac{a}{b}$. Et hinc peculiaribus signis ad denotandam proportionem nobis non erit opus. Si enim fuerit ut a ad b ita c ad d, erit $a:b=c:d$. Quod potentias attinet, $\overline{aa+bb}^m$ designabimus per $(aa+bb)^m$: unde & $\sqrt[m]{\overline{aa+bb}}$ erit$=(aa+bb)^{1:m}$ & $\sqrt[m]{\overline{aa+bb}^n}=(aa+bb)^{n:m}$. Nulli vero dubitamus fore, ut Geometrae omnes Acta haec legentes Signorum Leibnitianorum praestantiam animadvertant, & nobiscum in eadem consentiant."

The translation is as follows: "We hereby issue the reminder that in the future we shall use in these *Acta* the Leibnizian signs, where, when algebraic matters concern us, we do not choose the typographically troublesome and unnecessarily repugnant, and that we avoid ambiguity. Hence we shall prefer the parenthesis to the characters consisting of lines drawn above, and in multiplication by all means simply the comma; for example, in place of $\sqrt{aa+bb}$ we write $\sqrt{}(aa+bb)$ and for $\overline{aa+bb}\times c$ we take $aa+bb$, c. Division we mark with two dots, unless indeed some peculiar circumstance directs adherence to the usual practice. Accordingly, we have $a:b=\dfrac{a}{b}$. And it is not necessary to denote proportion by any special sign. For, if a is to b as c is to d, we have $a:b=c:d$. As regards powers, $\overline{aa+bb}^m$, we designate them by $(aa+bb)^m$; whence also $\sqrt[m]{aa+bb}$ becomes $=(aa+bb)^{1:m}$ and $\sqrt[m]{\overline{aa+bb}^n}=(aa+bb)^{n:m}$. We do not doubt that all geometers who read the *Acta* will recognize the excellence of the Leibnizian symbols and will agree with us in this matter."

EXTRACT FROM MISCELLANEA BEROLINENSIA[1]

198. "*Monitum De Characteribus Algebraicis.*—Quoniam variant Geometrae in characterum usu, nova praesertim Analysi inventa; quae res legentibus non admodum provectis obscuritatem parit; ideo è re visum est exponere, quomodo Characteres adhibeantur Leibnitiano more, quem in his Miscellaneis secuturi sumus. *Literae*

[1] Taken from *Miscellanea Berolinensia* (1710), p. 155. Article due to G. W. Leibniz.

minusculae a, b, x, y solent significare magnitudines, vel quod idem est, numeros indeterminatos: Majusculae verō, ut A, B, X, Y puncta figurarum; ita ab significat factum ex a in b, sed AB rectam à puncto A ad punctum B ductam. Huic tamen observationi adeo alligati non sumus, ut non aliquando minusculas pro punctis, majusculas pro numeris vel magnitudinibus usurpemus, quod facile apparebit ex modo adhibendi. Solent etiam literae priores, ut a, b, pro quantitatibus cognitis vel saltem determinatis adhiberi, sed posteriores, ut x, y, pro incognitis vel saltem pro variantibus.

"Interdum pro literis adhibentur Numeri, sed qui idem significant quod literae, utiliter tamen usurpantur relationis exprimendae gratia. Exempli causa: Sint binae aequationes generales secundi gradus pro incognita, x; eas sic exprimere licebit: $10xx \rightarrow\!\!-\!\!\vdash\!\!\rightarrow 11x \rightarrow\!\!-\!\!\vdash\!\!\rightarrow 12 = 0$ & $20xx \rightarrow\!\!-\!\!\vdash\!\!\rightarrow 21x \rightarrow\!\!-\!\!\vdash\!\!\rightarrow 22 = 0$ ita in progressu calculi ex ipsa notatione apparet quantitatis cujusque relatio; nempe 21 (ex. gr.) per notam dextram, quae est 1 agnoscitur esse coëfficiens ipsius x simplicis, at per notam sinistram 2 agnoscitur esse ex. aeq. secunda: sed et servatur lex quaedam homogeneorum. Et ope harum duarum aequationum tollendo x, prodit aequatio, in qua similiter se habere oportet 10, 11, 12 et 12, 11, 10; item 20, 21, 22 et 22, 21, 20; et denique 10, 11, 12 se habent ut. 20, 21, 22. id est si pro 10, 11, 12 substituas 20, 21, 22 et vice versa manet eadem aequatio; idemque est in caeteris. Tales numeri tractantur ut literae, veri autem numeri, discriminis causa, parenthesibus includuntur vel aliter discernuntur. Ita in tali sensu 11.20. significat numeros indefinitos 11 et 20 in se invicem ductos, non vero significat 220 quasi essent Numeri veri. Sed hic usus ordinarius non est, rariusque adhibetur.

"*Signa, Additionis* nimirum et *Subtractionis*, sunt $\rightarrow\!\!-\!\!\vdash$ plus, $-$ minus, $\rightarrow\!\!-\!\!\vdash$ plus vel minus, $\overline{\rightarrow\!\!-\!\!\vdash}$ priori oppositum minus vel plus. At $(\underline{\,-\!\vdash\,})$ vel $(\overline{\,-\!\vdash\,})$ est nota ambiguitatis signorum, independens à priori; et $((\underline{\,-\!\vdash\,})$ vel $((\overline{\,-\!\vdash\,})$ alia independens ab utraque; Differt autem *Signum ambiguum a Differentia* quantitatum, quae etsi aliquando incerta, non tamen ambigua est. . . . . . Sed differentia inter a et b, significat $a-b$, si a sit majus, et $b-a$ si b sit majus, quod etiam appellari potest moles ipsius $a-b$, intelligendo (exempli causa) ipsius $-\!\!\vdash 2$ et ipsius $-\!\!-2$ molem esse eandem, nempe $-\!\!\vdash 2$; ita si $a-b$ vocemus c utique *mol. c*, seu moles ipsius c erit $-\!\!\vdash 2$, quae est quantitas affirmativa sive c sit affirmativa sive negativa, id est, sive sit c idem quod $-\!\!\vdash 2$, sive c sit idem quod $-\!\!-2$. Et quantitates duae diversae eandem molem habentes semper habent idem quadratum.

"*Multiplicationem* plerumque signifare contenti sumus per nudam appositionem: sic *ab* significat *a* multiplicari per *b*, Numeros multiplicantes solemus praefigere, sic 3*a* significat triplum ipsius *a* interdum tamen punctum vel comma interponimus inter multiplicans et multiplicandum, velut cum 3, 2 significat 3 multiplicari per 2, quod facit 6, si 3 et 2 sunt numeri veri; et *AB, CD* significat rectam *AB* duci in rectam *CD*, atque inde fieri rectangulum. Sed et commata interdum hoc loco adhibemus utiliter, velut *a*, *b*✚*c*, vel *AB, CD* ——┼–*EF*, id est, *a* duci in *b* ——┼–*c*, vel *AB* in *CD* ——┼–*EF;* sed de his mox, ubi de vinculis. Porro propria Nota Multiplicationis non solet esse necessaria, cum plerumque appositio, qualem diximus, sufficiat. Si támen utilis aliquando sit, adhibebitur potius ⌢ quam ⋈, quia hoc ambiguitatem parit, et ita *AB*⌢*CD* significat *AB* duci in *CD*.

"*Diviso* significatur interdum more vulgari per subscriptionem diuisoris sub ipso dividendo, intercedente linea, ita *a* dividi per *b*, significatur vulgo per $\frac{a}{b}$; plerumque tamen hoc evitare praestat, efficereque, ut in eadem linea permaneatur, quod sit interpositis duobus punctis; ita ut *a:b* significat *a* dividi per *b*. Quod si *a:b* rursus dividi debeat per *c*, poterimus scribere *a:b, :c*, vel (*a:b*):*c*. Etsi enim res hoc casu (sane simplici) facile aliter exprimi posset, fit enim *a:*(*bc*) vel *a:bc* non tamen semper divisio actu ipse facienda est, sed saepe tantum indicanda, et tunc praestat operationis dilatae processum per commata vel parentheses indicari. . . . . Et exponens interdum lineolis includitur hac modo ③(*AB* ——┼–*BC*) quo significatur cubus rectae *AB* ——┼–*BC*. . . . . a^{e+n} et utiliter interdum lineola subducitur, ne literae exponentiales aliis confundantur; posset etiam scribi $\boxed{e+n}$ a. . . . .

" . . . . ita $\sqrt[2]{}(a^3)$ vel $\sqrt{}$ ③ (a^3) rursus est *a*, ... sed $\sqrt[3]{}2$ vel $\sqrt{}$ ③ 2 significat radicem cubicam ex eodem numero, et $\sqrt[e]{}2$ vel $\sqrt{}$ $\boxed{e}$ 2 significat, radicem indeterminati gradus *e* ex 2 extrahendam. . . . .

"Pro vinculis vulgo solent adhiberi ductus linearum; sed quia lineis una super alia ductis, saepe nimium spatii occupatur, aliasque ob causas commodius plerumque adhibentur commata et parentheses. Sic *a*, $\overline{b\text{·}\!\!\!\dashv\text{·}\!\!<\!c}$ idem est quod *a*, *b*·┤·<*c* vel *a*(*b*·┤·<*c*); et $\overline{a\text{·}\!\!\dashv\text{·}\!\!—<\!b}$, $\overline{c\text{·}\!\!\dashv\text{·}\!\!—<\!d}$ idem quod *a*·┤·—<*b*, *c*·┤·—<*d* vel (*a*·┤·—<) (*c*·┤·—<), id est, ·┤·—<*a*·┤·—<*b* multiplicatum per *c*·┤·—<*d*. Et similiter vincula in vinculis exhibentur. Ita *a*, $\overline{bc\text{·}\dashv\!\!—<\overline{ef\text{·}\dashv\!\!—<g}}$ etiam sic exprimetur, *a*(*bc*·┤—<*e*(*f*·┤—<*g*)) Et *a*, $\overline{\overline{bc\text{·}\dashv\!\!—<\overline{ef\text{·}\dashv\!\!—<g}}\text{·}\dashv\!\!—<hlm}$, *n* potest etiam

sic exprimi: $-\!\!\!\!\downarrow\!\!\!\!-\!\!\!\prec(a(bc\!\!\succ\!\!-\!\!\!\downarrow\!\!\!\bullet e(f+g))+hlm)n$. Quod de vinculis multiplicationis, idem intelligi potest *de vinculis divisionis*, exempli gratia

$$\overline{\begin{array}{ccc} \overline{\dfrac{a}{\begin{array}{cc}\dfrac{b}{c}-\!\!\downarrow\!\!\prec & \overline{e}\\ & f-\!\!\downarrow\!\!\prec g\end{array}}}-\!\!\downarrow\!\!\prec\dfrac{\begin{array}{c}h\\ l\end{array}}{m}\\ n\end{array}}$$ sic scribetur in una linea

$$(a:((b:c)-\!\!\downarrow\!\!\prec(e:,f-\!\!\downarrow\!\!\prec g))-\!\!\downarrow\!\!\prec h:(l:m)):n$$

nihilque in his difficultatis, modo teneamus, quicquid parenthesin aliquam implet pro una quantitate haberi, ... Idemque igitur locum habet in vinculis extractionis radicalis.

Sic $\sqrt{\overline{a^4-\!\!\downarrow\!\!\prec\sqrt{e,\overline{f-\!\!\downarrow\!\!\prec}g}}}$ idem est quod $\sqrt{(a^4-\!\!\downarrow\!\!\prec\sqrt{(e(f-\!\!\downarrow\!\!\prec g)))}}$
vel $\sqrt{(a^4-\!\!\downarrow\!\!\prec\sqrt{(e,f-\!\!\downarrow\!\!\prec g))}}$.

Et pro $\dfrac{\sqrt{\overline{aa-\!\!\downarrow\!\!\prec b}\sqrt{\overline{cc-\!\!\downarrow\!\!\prec dd}}}}{e-\!\!\downarrow\!\!\prec\sqrt{f}\sqrt{gg-\!\!\downarrow\!\!\prec hh-\!\!\downarrow\!\!\prec kk}}$

scribi poterit $\sqrt{(aa-\!\!\downarrow\!\!\prec b}\sqrt{(cc-\!\!\downarrow\!\!\prec dd)):,}$
$\qquad\qquad e-\!\!\downarrow\!\!\prec\sqrt{(f}\sqrt{(gg-\!\!\downarrow\!\!\prec hh)-\!\!\downarrow\!\!\prec kk)}$ . . . .

itaque $a=b$ significat, a, esse equale ipsi b, et $a=\!\!\!\!=b$ significat a esse majus quam b, et a $=\!\!\!\!=b$ significat a esse minus quam b.

"Sed et *proportionalitas* vel analogia de quantitatibus enunciatur, id est, rationis identitas, quam possumus in Calculo exprimere per notam aequalitatis, ut non sit opus peculiaribus notis. Itaqua a
esse ad b, sic ut l ad m, sic exprimere poterimus $a:b=l:m$, id est $\dfrac{a}{b}=\dfrac{l}{m}$.
Nota continue proportionalium erit $\because$, ita ut $\because a.b.c$ etc. sint continuè proportionales. Interdum nota *Similitudinis* prodest, quae est ∞, item nota similitudinis aequalitatis simul, seu nota *congruitatis* $\cong$, Sic $DEF\infty PQR$ significabit Triangula haec duo esse similia; at $DEF\cong PQR$ significabit congruere inter se. Huic si tria inter se habeant eandem rationem quam tria alia inter se, poterimus hoc exprimere nota similitudinis, ut $a;b;\infty l;m;$ n quod significat esse a ad b, ut l ad m, et a ad c ut l ad n, et b ad c ut m ad n. . . . ."

The translation is as follows:

"*Recommendations on algebraic characters.*—Since geometers differ in the use of characters, especially those of the newly invented analysis, a situation which perplexes those followers who as yet are not very far advanced, it seems proper to explain the manner of using the characters in the Leibnizian procedure, which we have adopted in the

Miscellanies. The small letters a, b, x, y, signify magnitudes, or what is the same thing, indeterminate numbers. The capitals on the other hand, as A, B, X, Y, stand for points of figures. Thus ab signifies the result of a times b, but AB signifies the right line drawn from the point A to the point B. We are, however, not bound to this convention, for not infrequently we shall employ small letters for points, capitals for numbers or magnitudes, as will be easily evident from the mode of statement. It is customary, however, to employ the first letters a, b, for known or fixed quantities, and the last letters x, y, for the unknowns or variables.

"Sometimes numbers are introduced instead of letters, but they signify the same as letters; they are convenient for the expression of relations. For example, let there be two general equations of the second degree having the unknown x. It is allowable to express them thus: $10xx+11x+12=0$ and $20xx+21x+22=0$. Then, in the progress of the calculation the relation of any quantity appears from the notation itself; thus, for example, in 21 the right digit which is 1 is recognized as the coefficient of x, and the left digit 2 is recognized as belonging to the second equation; but also a certain law of homogeneity is obeyed. And eliminating x by means of these two equations, an equation is obtained in which one has similarity in 10, 11, 12 and 12, 11, 10; also in 20, 21, 22 and 22, 21, 20; and lastly in 10, 11, 12 and 20, 21, 22. That is, if for 10, 11, 12, you substitute 20, 21, 22 and vice versa, there remains the same equation, and so on. Such numbers are treated as if letters. But for the sake of distinction, they are included in parentheses or otherwise marked. Accordingly, $11 \cdot 20$. signifies the indefinite numbers 11 and 20 multiplied one into the other; it does not signify 220 as it would if they were really numbers. But this usage is uncommon and is rarely applied.

"The *signs of addition* and subtraction are commonly $+$ plus, $-$ minus, $\pm$ plus or minus, $\mp$ the opposite to the preceding, minus or plus. Moreover $(\pm)$ or $(\mp)$ is the mark of ambiguity of signs that are independent at the start; and $((\pm)$ or $((\mp)$ are other signs independent of both the preceding. Now the symbol of ambiguity differs from the difference of quantities which, although sometimes undetermined, is not ambiguous. . . . . But $a-b$ signifies the difference between a and b when a is the greater, $b-a$ when b is the greater, and this absolute value (moles) may however be called itself $a-b$, by understanding that the absolute value of $+2$ and -2, for example, is the same, namely, $+2$. Accordingly, if $a-b$ is called c, then *mol. c* or the absolute value of c is $+2$, which is an affirmative quantity whether

c itself is positive or negative; i.e., either c is the same as $+2$, or c is the same as -2. Two different quantities having the same absolute value have always the same square.

"*Multiplication* we are commonly content to indicate by simple apposition: thus, ab signifies a multiplied by b. The multiplier we are accustomed to place in front; thus $3a$ means the triple of a itself. Sometimes, however, we insert a point or a comma between multiplier and multiplicand; thus, for example, $3,2$ signifies that 3 is multiplied by 2, which makes 6, when 3 and 2 are really numbers; and AB,CD signifies the right line AB multiplied into the right line CD, producing a rectangle. But we also apply the comma advantageously in such a case, for example,[1] as $a,b+c$, or $AB,CD+EF$; i.e., a multiplied into $b+c$, or AB into $CD+EF$; we speak about this soon, under vinculums. Formerly no sign of multiplication was considered necessary for, as stated above, commonly mere apposition sufficed. If, however, at any time a sign seems desirable use $\frown$ rather than $\times$, because the latter leads to ambiguity; accordingly, $AB\frown CD$ signifies AB times CD.

"*Division* is commonly marked by writing the divisor beneath its dividend, with a line of separation between them. Thus a divided by b is ordinarily indicated by $\frac{a}{b}$; often, however, it is preferable to avoid this notation and to arrange the signs so that they are brought into one and the same line; this may be done by the interposition of two points; thus $a:b$ signifies a divided by b. If $a:b$ in turn is to be divided by c, we may write $a:b$, $:c$, or $(a:b):c$. However, this should be expressed more simply in another way, namely, $a:(bc)$ or $a:bc$, for the division cannot always be actually carried out, but can be only indicated, and then it becomes necessary to mark the delayed process of the operation by commas or parentheses. . . . . Exponents are frequently inclosed by lines in this manner $\boxed{3}$ $(AB+BC)$, which means the cube of the line $AB+BC$ . . . . ; the exponents of a^{l+n} may also be advantageously written between the lines, so that the literal exponents will not be confounded with other letters; thus it may be written $\boxed{l+n}\,a$. . . . . From $\sqrt[3]{}\,(a^3)$ or $\sqrt{}\,\boxed{3}\,(a^3)$ arises a . . . . ; but $\sqrt[3]{2}$ or $\sqrt{}\,\boxed{3}\,2$ means the cube root of the same number, and $\sqrt[e]{2}$ or $\sqrt{}\,\boxed{e}\,2$ signifies the extraction of a root of the indeterminate order e. . . . .

"For aggregation it is customary to resort to the drawing of

[1] A similar use of the comma to separate factors and at the same time express aggregation occurs earlier in Hérigone (see § 189).

lines, but because lines drawn one above others often occupy too much space, and for other reasons, it is often more convenient to introduce commas and parentheses. Thus $a, \overline{b+c}$ is the same as $a, b+c$ or $a(b+c)$; and $\overline{a+b}, \overline{c+d}$ is the same as $a+b, c+d$, or $(a+b)(c+d)$, i.e., $+a+b$ multiplied by $c+d$. And, similarly, vinculums are placed under vinculums. For example, $a, \overline{bc+e\overline{f+g}}$ is expressed also thus, $a(bc+e(f+g))$, and $a, \overline{\overline{bc+e\overline{f+g}}+hlm},n$ may be written also $+(a(bc+e(f+g))+hlm)n$. What relates to vinculums in multiplication applies to vinculums in division. For example,

$$\frac{\dfrac{a}{\dfrac{b}{c}+\dfrac{e}{f+g}}+\dfrac{h}{\dfrac{l}{m}}}{n}$$ may be written in one line thus:

$$(a: ((b:c)+(e:, f+g))+h:(l:m)):n \, ,$$

and there is no difficulty in this, as long as we observe that whatever fills up a given parenthesis be taken as one quantity. . . . . The same is true of vinculums in the extraction of roots. Thus $\sqrt{a^4+\sqrt{e, \overline{f+g}}}$ is the same as $\sqrt{(a^4+\sqrt{(e(f+g))})}$ or $\sqrt{(a^4+\sqrt{(e, f+g)})}$. And for $\dfrac{\sqrt{aa+b\sqrt{cc+dd}}}{e+\sqrt{f\sqrt{gg+hh}+kk}}$ one may write $\sqrt{(aa+b\sqrt{(cc+dd)})}:, e+\sqrt{(f\sqrt{(gg+hh)}+kk)}$. Again $a=b$ signifies that a is equal to b, and $a\hspace{-3pt}=\hspace{-3pt}\text{—}\,b$ signifies that a is greater than b, and $a\,\text{—}\hspace{-3pt}=\hspace{-3pt}b$ that a is less than b. Also proportionality or analogia of quantities, i.e., the identity of ratio, may be represented; we may express it in the calculus by the sign of equality, for there is no need of a special sign. Thus, we may indicate that a is to b as l is to m by $a:b=l:m$, i.e., $\dfrac{a}{b}=\dfrac{l}{m}$. The sign for continued proportion is $\div$, so that $\div a, b, c,$ and d are continued proportionals.

"There is adopted a sign for similitude; it is $\backsim$; also a sign for both similitude and equality, or a sign of congruence, $\backsimeq$ accordingly, $DEF \backsim PQR$ signifies that the two triangles are similar; but $DEF \backsimeq PRQ$ marks their congruence. Hence, if three quantities have to one another the same ratio that three others have to one another, we may mark this by a sign of similitude, as $a; b; c \backsim l; m; n$ means that a is to b as l is to m, and a is to c as l is to n, and b is to c as m is to n. . . . ."

In the second edition of the *Miscellanea Berolinensia*, of the year

1749, the typographical work is less faulty than in the first edition of 1710; some slight errors are corrected, but otherwise no alterations are made, except that Harriot's signs for "greater than" and "less than" are adopted in 1749 in place of the two horizontal lines of unequal length and thickness, given in 1710, as shown above.

199. *Conclusions.*—In a letter to Collins, John Wallis refers to a change in algebraic notation that occurred in England during his lifetime: "It is true, that as in other things so in mathematics, fashions will daily alter, and that which Mr. Oughtred designed by great letters may be now by others be designed by small; but a mathematician will, with the same ease and advantage, understand Ac, and a^3 or aaa."[1] This particular diversity is only a trifle as compared with what is shown in a general survey of algebra in Europe during the fifteenth, sixteenth, and seventeenth centuries. It is discouraging to behold the extreme slowness of the process of unification.

In the latter part of the fifteenth century $\tilde{p}$ and $\tilde{m}$ became symbols for "plus" and "minus" in France (§ 131) and Italy (§ 134). In Germany the Greek cross and the dash were introduced (§ 146). The two rival notations competed against each other on European territory for many years. The $\tilde{p}$ and $\tilde{m}$ never acquired a foothold in Germany. The German + and − gradually penetrated different parts of Europe. It is found in Scheubel's *Algebra* (§ 158), in Recorde's *Whetstone of Witte*, and in the *Algebra* of Clavius. In Spain the German signs occur in a book of 1552 (§ 204), only to be superseded by the $\tilde{p}$ and $\tilde{m}$ in later algebras of the sixteenth century. The struggle lasted about one hundred and thirty years, when the German signs won out everywhere except in Spain. Organized effort, in a few years, could have ended this more than a century competition.

If one takes a cross-section of the notations for radical expressions as they existed in algebra at the close of the sixteenth century, one finds four fundamental symbols for indicating roots, the letters R and l, the radical sign $\sqrt{}$ proper and the fractional exponent. The letters R and l were sometimes used as capitals and sometimes as small letters (§§ 135, 318–22). The student had to watch his step, for at times these letters were used to mark, not roots, but the unknown quantity x and, perhaps, also its powers (§ 136). When R stood for "root," it became necessary to show whether the root of one term or of several terms was meant. There sprang up at least seven different symbols for the aggregation of terms affected by the R, namely, one of Chuquet (§ 130), one of Pacioli (§ 135), two of Cardan (§ 141), the round paren-

[1] See Rigaud, *op. cit.*, Vol. II, p. 475.

thesis of Tartaglia (§ 351), the upright and inverted letter L of Bombelli (§ 144), and the r *bin.* and r *trinomia* of A. V. Roomen (§ 343). There were at least five ways of marking the orders of the root, those of Chuquet (§ 130), De la Roche (§ 132), Pacioli (§ 135), Ghaligai (§ 139), and Cardan (Fig. 46). With A. M. Visconti[1] the signs $R.ce\ cu.$ meant the "sixth root"; he used the multiplicative principle, while Pacioli used the additive one in the notation of radicals. Thus the letter R carried with it at least fifteen varieties of usage. In connection with the letter l, signifying *latus* or "root," there were at least four ways of designating the orders of the roots and the aggregation of terms affected (§§ 291, 322). A unique line symbolism for roots of different orders occurs in the manuscripts of John Napier (§ 323).

The radical signs for cube and fourth root had quite different shapes as used by Rudolff (§§ 148, 326) and Stifel (§ 153). Though clumsier than Stifel's, the signs of Rudolff retained their place in some books for over a century (§ 328). To designate the order of the roots, Stifel placed immediately after the radical sign the German abbreviations of the words *zensus, cubus, zensizensus, sursolidus,* etc. Stevin (§ 163) made the important innovation of numeral indices. He placed them within a circle. Thus he marked cube root by a radical sign followed by the numeral 3 coraled in a circle. To mark the root of an aggregation of terms, Rudolff (§§ 148, 348) introduced the dot placed after the radical sign; Stifel sometimes used two dots, one before the expression, the other after. Stevin (§§ 163, 343) and Digges (§§ 334, 343) had still different designations. Thus the radical sign carried with it seven somewhat different styles of representation. Stevin suggested also the possibility of fractional exponents (§ 163), the fraction being placed inside a circle and before the radicand.

Altogether there were at the close of the sixteenth century twenty-five or more varieties of symbols for the calculus of radicals with which the student had to be familiar, if he desired to survey the publications of his time.

Lambert Lincoln Jackson makes the following historical observations: "For a hundred years after the first printed arithmetic many writers began their works with the line-reckoning and the Roman numerals, and followed these by the Hindu arithmetic. The teaching of numeration was a formidable task, since the new notation was so unfamiliar to people generally."[2] In another place (p. 205) Jackson

---

[1] "Abbreviationes," *Practica numerorum, et mensurarum* (Brescia, 1581).

[2] *The Educational Significance of Sixteenth Century Arithmetic* (New York, 1906), p. 37, 38.

states: "Any phase of the growth of mathematical notation is an interesting study, but the chief educational lesson to be derived is that notation always grows too slowly. Older and inferior forms possess remarkable longevity, and the newer and superior forms appear feeble and backward. We have noted the state of transition in the sixteenth century from the Roman to the Hindu system of characters, the introduction of the symbols of operation, $+$, $-$, and the slow growth toward the decimal notation. The moral which this points for twentieth-century teachers is that they should not encourage history to repeat itself, but should assist in hastening new improvements."

The historian Tropfke expresses himself as follows: "How often has the question been put, what further achievements the patriarchs of Greek mathematics would have recorded, had they been in possession of our notation of numbers and symbols! Nothing stirs the historian as much as the contemplation of the gradual development of devices which the human mind has thought out, that he might approach the truth, enthroned in inaccessible sublimity and in its fullness always hidden from earth. Slowly, only very slowly, have these devices become what they are to man today. Numberless strokes of the file were necessary, many a chink, appearing suddenly, had to be mended, before the mathematician had at hand the sharp tool with which he could make a successful attack upon the problems confronting him. The history of algebraic language and writing presents no uniform picture. An assemblage of conscious and unconscious innovations, it too stands subject to the great world-law regulating living things, the principle of selection. Practical innovations make themselves felt, unsuitable ones sink into oblivion after a time. The force of habit is the greatest opponent of progress. How obstinate was the struggle, before the decimal division met with acceptance, before the proportional device was displaced by the equation, before the Indian numerals, the literal coefficients of Vieta, could initiate a world mathematics."[1]

Another phase is touched by Treutlein: "Nowhere more than in mathematics is intellectual content so intimately associated with the form in which it is presented, so that an improvement in the latter may well result in an improvement of the former. Particularly in arithmetic, a generalization and deepening of concept became possible only after the form of presentation had been altered. The history of our science supplies many examples in proof of this. If the Greeks had been in possession of our numeral notation, would their

[1] Tropfke, *Geschichte der Elementar-Mathematik*, Vol. II (Leipzig, 1921), p. 4, 5.

mathematics not present a different appearance? Would the binomial theorem have been possible without the generalized notation of powers? Indeed could the mathematics of the last three hundred years have assumed its degree of generality without Vieta's pervasive change of notation, without his introduction of general numbers? These instances, to which others from the history of modern mathematics could be added, show clearly the most intimate relation between substance and form."[1]

B. SPECIAL SURVEY OF THE USE OF NOTATIONS

SIGNS FOR ADDITION AND SUBTRACTION

200. *Early symbols.*—According to Hilprecht,[2] the early Babylonians had an ideogram, which he transliterates *LAL*, to signify "minus." In the hieratic papyrus of Ahmes and, more clearly in the hieroglyphic translation of it, a pair of legs walking forward is the sign of addition; away, the sign of subtraction.[3] In another Egyptian papyrus kept in the Museum of Fine Arts in Moscow,[4] a pair of legs walking forward has a different significance; there it means to square a number.

Figure 99, translated, is as follows (reading the figure from right to left):

"$\frac{2}{3}$ added and $\frac{1}{3}$ [of this sum] taken away, 10 remains.

Make $\frac{1}{10}$ of this 10: the result is 1, the remainder 9.

$\frac{2}{3}$ of it, namely, 6, added to it; the total is 15. $\frac{1}{3}$ of it is 5.

When 5 is taken away, the remainder is 10."

In the writing of unit fractions, juxtaposition meant addition, the unit fraction of greatest value being written first and the others in descending order of magnitude.

While in Diophantus addition was expressed merely by juxtaposition (§ 102), a sporadic use of a slanting line / for addition, also a semi-elliptical curve ⊋ for subtraction, and a combination of the two

---

[1] Treutlein, "Die deutsche Coss," *Abhandlungen z. Geschichte der Mathematik*, Vol. II (Leipzig, 1879), p. 27, 28.

[2] H. V. Hilprecht, *Babylonian Expedition: Mathematical etc. Tablets* (Philadelphia, 1906), p. 23.

[3] A. Eisenlohr, *op. cit.* (2d ed.), p. 46 (No. 28), 47, 237. See also the improved edition of the Ahmes papyrus, *The Rhind Mathematical Papyrus*, by T. Eric Peet (London, 1923), Plate J, No. 28; also p. 63.

[4] Peet, *op. cit.*, p. 20, 135: *Ancient Egypt* (1917), p. 101.

ր for the total result has been detected in Greek papyri.[1] Diophantus'
sign for subtraction is well known (§ 103). The Hindus had no mark
for addition (§ 106) except that, in the Bakhshali *Arithmetic, yu* is
used for this purpose (§ 109). The Hindus distinguished negative
quantities by a dot (§§ 106, 108), but the Bakhshali *Arithmetic* uses
the sign + for subtraction (§ 109). The Arab al-Qalasâdî in the fifteenth century indicated addition by juxtaposition and had a special
sign for subtraction (§ 124). The Frenchman Chuquet (1484), the
Italian Pacioli (1494), and the sixteenth-century mathematicians in
Italy used $\bar{p}$ or p: for plus and $\bar{m}$ or m: for "minus" (§§ 129, 134).

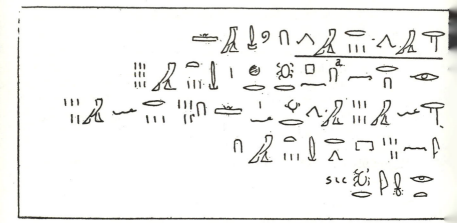

Fig. 99.—From the hieroglyphic translation of the Ahmes papyrus, Problem
28, showing a pair of legs walking forward, to indicate addition, and legs walking
away, to indicate subtraction. (Taken from T. E. Peet, *The Rhind Mathematical
Papyrus*, Plate J, No. 28.)

201. *Origin and meanings of the signs* + *and* −.—The modern
algebraic signs + and − came into use in Germany during the last
twenty years of the fifteenth century. They are first found in manuscripts. In the Dresden Library there is a volume of manuscripts,
C. 80. One of these manuscripts is an algebra in German, written in
the year 1481,[2] in which the minus sign makes its first appearance in

[1] H. Brugsch, *Numerorum apud veteres Aegyptios demoticorum doctrina. Ex
papyris* (Berlin, 1849), p. 31; see also G. Friedlein, *Zahlzeichen und das elementare
Rechnen* (Erlangen, 1869), p. 19 and Plate I.

[2] E. Wappler, *Abhandlungen zur Geschichte der Mathematik*, Vol. IX (1899), p.
539, n. 2; Wappler, *Zur Geschichte der deutschen Algebra im 15. Jahrhundert, Zwickauer Gymnasialprogramm von 1887*, p. 11–30 (quoted by Cantor, *op. cit.*, Vol. II [2d
ed., 1900], p. 243, and by Tropfke, *op. cit.*, Vol. II [2d ed., 1921], p. 13).

algebra (Fig. 100); it is called *minnes*. Sometimes the − is placed after the term affected. In one case −4 is designated "4 das ist −."
Addition is expressed by the word *vnd*.

In a Latin manuscript in the same collection of manuscripts, C. 80, in the Dresden Library, appear both symbols + and − as signs of operation (Fig. 101), but in some rare cases the + takes the place of *et* where the word does not mean addition but the general "and."[1] Repeatedly, however, is the word *et* used for addition.

It is of no little interest that J. Widman, who first used the + and − in print, studied these two manuscripts in the manuscript volume C. 80 of the Dresden Library and, in fact, annotated them. One of his marginal notes is shown in Figure 102. Widman lectured at the University of Leipzig, and a manuscript of notes

1. Ältestes Minuszeichen.
Dresd. C. 80.   Deutsche Algebra, fol. 368ʳ
(um 1486)

$$15 - 22x$$

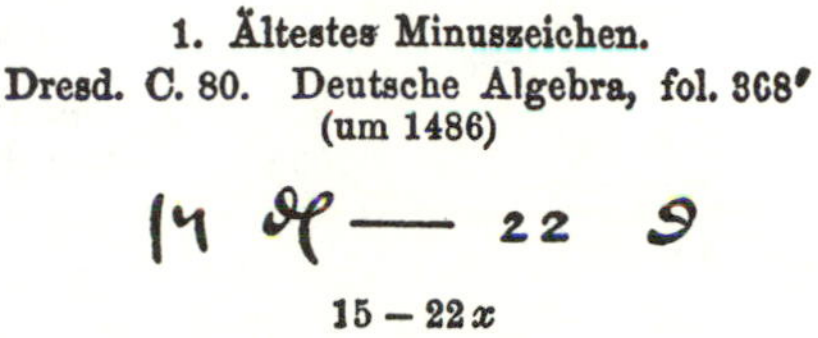

FIG. 100.—Minus sign in a German MS, C. 80, Dresden Library. (Taken from J. Tropfke, *op. cit.*, Vol. II [1921], p. 14.)

taken in 1486 by a pupil is preserved in the Leipzig Library (Codex Lips. 1470).[2] These notes show a marked resemblance to the two Dresden manuscripts.

The view that our + sign descended from one of the florescent forms for *et* in Latin manuscripts finds further support from works on

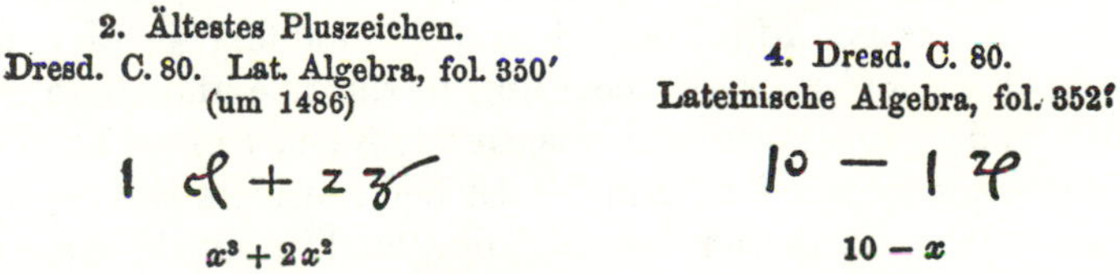

2. Ältestes Pluszeichen.
Dresd. C. 80. Lat. Algebra, fol. 350ʳ
(um 1486)

$$x^3 + 2x^2$$

4. Dresd. C. 80.
Lateinische Algebra, fol. 352ʳ

$$10 - x$$

FIG. 101.—Plus and minus signs in a Latin MS, C. 80, Dresden Library. (Taken from Tropfke, *op. cit.*, Vol. II [2d ed., 1921], p. 14.)

paleography. J. L. Walther[3] enumerates one hundred and two different abbreviations found in Latin manuscripts for the word *et;* one of these, from a manuscript dated 1417, looks very much like the modern

[1] Wappler, *Programm* (1887), p. 13, 15.

[2] Wappler, *Zeitschrift Math. u. Physik*, Vol. XLV (Hist. lit. Abt., 1900), p. 7–9.

[3] *Lexicon diplomaticvm abbreviationes syllabarvm et vocvm in diplomatibvs et codicibvs a secvlo VIII. ad XVI. . . . . . Studio Joannis Lvdolfi VValtheri . . . .* (Ulmae, 1756), p. 456–59.

$+$. The downward stroke is not quite at right angles to the horizontal stroke, thus +.

Concerning the origin of the minus sign $(-)$, we limit ourselves to the quotation of a recent summary of different hypotheses: "One knows nothing certain of the origin of the sign $-$; perhaps it is a simple bar used by merchants to separate the indication of the tare, for a long time called *minus*, from that of the total weight of merchandise; according to L. Rodet (*Actes Soc. philol. Alençon*, Vol. VIII [1879], p. 105) this sign was derived from an Egyptian hieratic sign. One has also sought the origin of our sign $-$ in the sign employed by Heron and Diophantus and which changed to $\top$ before it became $-$. Others still have advanced the view that the sign $-$ has its origin in the ὀβελός of the Alexandrian grammarians. None of these hypotheses is supported by plausible proof."[1]

Zusatz von WIDMANN.
5. Dresd. C. 80, fol. 349′
(um 1486)

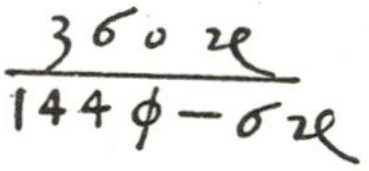

$$\dfrac{360\,x}{144 - 6x}$$

FIG. 102.—Widman's marginal note to MS C. 80, Dresden Library. (Taken from Tropfke.)

202. The sign $+$ first occurs in print in Widman's book in the question: "Als in diese exēpel 16 ellñ pro 9 fl $\frac{1}{3}$ vñ $\frac{1}{4}+\frac{1}{8}$ eynss fl wy kūmē 36 ellñ machss alsso Addir $\frac{1}{3}$ vñ $\frac{1}{4}$ vñ $\frac{1}{8}$ zu sāmen kumpt $\frac{4}{6}\frac{7}{6}$ eynss fl Nu secz vñ machss nach der regl vñ kūmē 22 fl $\frac{1}{80}$ eynsz fl dz ist gerad 3 hlr in gold."[2] In translation: "Thus in this example, 16 ells [are bought] for 9 florins [and] $\frac{1}{3}$ and $\frac{1}{4}+\frac{1}{8}$ of a florin, what will 36 ells cost? Proceed thus: Add $\frac{1}{3}$ and $\frac{1}{4}$ and $\frac{1}{8}$ obtaining $\frac{4}{6}\frac{7}{6}$ of a florin. Now put down and proceed according to the rule and there results 22 florin, and $\frac{1}{80}$ of a florin which is exactly 3 heller in gold." The $+$ in this passage stands for "and." Glaisher considers this $+$ a misprint for $v\bar{n}$ (the contraction for *vnnd*, our "and"), but there are other places in Widman where $+$ clearly means "and," as we shall see later. There is no need of considering this $+$ a misprint.

On the same leaf Widman gives a problem on figs. We quote from the 1498 edition (see also Fig. 54 from the 1526 edition):

[1] *Encyclopédie des scien. math.*, Tome I, Vol. I (1904), p. 31, 32, n. 145.

[2] Johann Widman, *Behēde vnd hubsche Rechenung auff allen Kauffmanschafft* (Leipzig, 1489), unnumbered p. 87. Our quotation is taken from J. W. L. Glaisher's article, "On the Early History of Signs $+$ and $-$ and on the Early German Arithmeticians," *Messenger of Mathematics*, Vol. LI (1921–22), p. 6. Extracts from Widman are given by De Morgan, *Transactions of the Cambridge Philosophical Society*, Vol. XI, p. 205, and by Boncompagni, *Bulletino*, Vol. IX, p. 205.

"*Veygen.*—Itm̄ Eyner Kaufft 13 lagel veygen vn̄ nympt ye 1 ct pro 4 fl ½ ort Vnd wigt itliche lagel als dan hye nochuolget. vn̄ ich wolt wissen was an der sum brecht

	4+ 5	Wiltu dass
	4−17	wyssen der
	3+36	dess gleichn̄
	4−19	Szo sum —
	3+44	mir die ct
	3+22	Vnd lb vn̄
Czentner	3−11 lb	was — ist
	3+50	dz ist mi⁹
	4−16	dz secz besū
	3+44	der vn̄ wer
	3+29	dē 4539
	3−12	lb (So du
	3+ 9	die ct zcu lb

gemacht hast Vnnd das + das ist mer dar zu addirest) vnd 75 min⁹ Nu solt du fur holcz abschlahn̄ albeg fur eyn lagel 24 lb vn̄ dz ist 13 mol 24· vn̄ macht 312 lb dar zu addir dz — dz ist 75 lb vnnd werden 387 Die subtrahir vonn 4539 Vnnd pleybn̄ 4152 lb Nu sprich 100 lb das ist 1 ct pro 4 fl ⅛ wie kummen 4152 lb vnd kūmen 171 fl 5 ss 4 hlr ⅘ Vn̄ ist recht gemacht."[1]

In free translation the problem reads: "*Figs.*—Also, a person buys 13 barrels of figs and receives 1 centner for 4 florins and ½ ort (4⅛ florins), and the weight of each barrel is as follows: 4 ct+5 lb, 4 ct−17 lb, 3 ct+36 lb, 4 ct−19 lb, 3 ct+44 lb, 3 ct+22 lb, 3 ct−11 lb, 3 ct+50 lb, 4 ct−16 lb, 3 ct+44 lb, 3 ct+29 lb, 3 ct−12 lb, 3 ct+9 lb; and I would know what they cost. To know this or the like, sum the ct and lb and what is −, that is minus, set aside, and they become 4539 lb (if you bring the centners to lb and thereto add the +, that is more) and 75 minus. Now you must subtract for the wood 24 lb for each barrel and 13 times 24 is 312 to which you add the −, that is 75 lb and it becomes 387 which subtract from 4,539 and there remains 4152 lb. Now say 100 lb that is 1 ct for 4⅛ fl, what do 4152 lb come to, and they come to 171 fl 5 ss 4⅘ hlr which is right."

Similar problems are given by Widman, relating to pepper and soap. The examination of these passages has led to divergent opinions on the original significance of the + and −. De Morgan suspected

[1] The passage is quoted and discussed by Eneström, *Bibliotheca mathematica*, Vol. IX (3d ser., 1908–9), p. 156, 157, 248; see also *ibid.*, Vol. VIII, p. 199.

that they were warehouse marks, expressing excess or deficiency in weights of barrels of goods.[1] M. W. Drobisch,[2] who was the first to point out the occurrence of the signs + and − in Widman, says that Widman uses them in passing, as if they were sufficiently known, merely remarking, "Was − ist das ist minus vnd das + das ist mer." C. I. Gerhardt,[3] like De Morgan, says that the + and − were derived from mercantile practice.

But Widman assigned the two symbols other significations as well. In problems which he solved by false position the error has the + or − sign prefixed.[4] The − was used also to separate the terms of a proportion. In "11630−198   4610−78" it separates the first and second and the third and fourth terms. The "78" is the computed term, the fractional value of the fourth term being omitted in the earlier editions of Widman's arithmetic. The sign + occurs in the heading "Regula augmenti + decrementi" where it stands for the Latin *et* ("and"), and is not used there as a mathematical symbol. In another place Widman gives the example, "Itm̄ eyner hat kaufft 6 eyer−2 ⅍ pro 4 ⅍ +1 ey" ("Again, someone has bought 6 eggs− 2 ⅍ for 4 ⅍+1 egg"), and asks for the cost of one egg. Here the − is simply a dash separating the words for the goods from the price. From this and other quotations Glaisher concludes that Widman used + and − "in all the ways in which they are used in algebra." But we have seen that Widman did not restrict the signs to that usage; the + was used for "and" when it did not mean addition; the − was used to indicate separation. In other words, Widman does not restrict the use of + and − to the technical meanings that they have in algebra.

203. In an anonymous manuscript,[5] probably written about the time when Widman's arithmetic appeared, use is made of symbolism in the presentation of algebraic rules, in part as follows:

   "*Conditiones circa + vel − in additione*

   $\left.\begin{array}{l} + \; et \; + \\ - \; et \; - \end{array}\right\rangle facit \left.\begin{array}{l} + \\ - \end{array}\right\rangle$ *addatur non sumendo respectum quis numerus sit superior.*

---

[1] De Morgan, *op. cit.*, Vol. XI, p. 206.

[2] *De Joannis Widmanni . . . . compendio* (Leipzig, 1840), p. 20 (quoted by Glaisher, *op. cit.*, p. 9).

[3] *Geschichte der Mathematik in Deutschland* (1877), p. 36: ". . . . dass diese Zeichen im kaufmännischen Verkehr üblich waren."

[4] Glaisher, *op. cit.*, p. 15.

[5] *Regulae Cosae vel Algebrae*, a Latin manuscript, written perhaps about 1450, but "surely before 1510," in the Vienna Library.

Si fuerit $\left\{\begin{array}{l} + \text{ et } - \\ - \text{ et } + \end{array}\right\rangle$ simpliciter subtrahatur minor numerus a majori et residuo sua ascribatur nota,"[1] and similarly for subtraction. This manuscript of thirty-three leaves is supposed to have been used by Henricus Grammateus (Heinrich Schreiber) in the preparation of his *Rechenbuch* of 1518 and by Christoff Rudolff in his *Coss* of 1525.

Grammateus[2] in 1518 restricts his use of + and − to technical algebra: "Vnd man braucht solche zaichen als + ist vnnd, − mynnder" ("And one uses such signs as + [which] is 'and,' − 'less' "). See Figure 56 for the reproduction of this passage from the edition of 1535. The two signs came to be used freely in all German algebras, particularly those of Grammateus, Rudolff (1525), Stifel (1544), and in Riese's manuscript algebra (1524). In a text by Eysenhut[3] the + is used once in the addition of fractions; both + and − are employed many times in the *regula falsi* explained at the end of the book.

Arithmetics, more particularly commercial arithmetics, which did not present the algebraic method of solving problems, did not usually make use of the + and − symbols. L. L. Jackson says: "Although the symbols + and − were in existence in the fifteenth century, and appeared for the first time in print in Widman (1489), as shown in the illustration (p. 53), they do not appear in the arithmetics as signs of operation until the latter part of the sixteenth century. In fact, they did not pass from algebra to general use in arithmetic until the nineteenth century."[4]

204. *Spread of the + and − symbols.*—In Italy the symbols $\bar{p}$ and $\tilde{m}$ served as convenient abbreviations for "plus" and "minus" at the end of the fifteenth century and during the sixteenth. In 1608 the German Clavius, residing in Rome, used the + and − in his algebra brought out in Rome (see Fig. 66). Camillo Gloriosi adopted them in his *Ad theorema geometricum* of 1613 and in his *Exercitationes mathematicae, decas I* (Naples, 1627) (§ 196). The + and − signs were used by B. Cavalieri (see Fig. 86) as if they were well known. The +

[1] C. I. Gerhardt, "Zur Geschichte der Algebra in Deutschland," *Monatsberichte der k. pr. Akademie d. Wissenschaften z. Berlin* (1870), p. 147.

[2] Henricus Grammateus, *Ayn New Kunstlich Buech* (Nürnberg: Widmung, 1518; publication probably in 1521). See Glaisher, *op. cit.*, p. 34.

[3] *Ein künstlich rechenbuch auff Zyffern / Lini vnd Wälschen Practica* (Augsburg, 1538). This reference is taken from Tropfke, *op. cit.*, Vol. I (2d ed., 1921), p. 58.

[4] *The Educational Significance of Sixteenth Century Arithmetic* (New York, 1906), p. 54.

and − were used in England in 1557 by Robert Recorde (Fig. 71) and
in Holland in 1637 by Gillis van der Hoecke (Fig. 60). In France and
Spain the German + and −, and the Italian $\bar{p}$ and $\tilde{m}$, came in sharp
competition. The German Scheubel in 1551 brought out at Paris an
algebra containing the + and − (§ 158); nevertheless, the $\bar{p}$ and $\tilde{m}$
(or the capital letters P, M) were retained by Peletier (Figs. 80, 81),
Buteo (Fig. 82), and Gosselin (Fig. 83). But the adoption of the German signs by Ramus and Vieta (Figs. 84, 85) brought final victory for
them in France. The Portuguese P. Nuñez (§ 166) used in his algebra
(published in the Spanish language) the Italian $\bar{p}$ and $\tilde{m}$. Before this,
Marco Aurel,[1] a German residing in Spain, brought out an algebra at
Valencia in 1552 which contained the + and − and the symbols for
powers and roots found in Christoff Rudolff (§ 165). But ten years
later the Spanish writer Pérez de Moya returned to the Italian sym-
bolism[2] with its $\bar{p}$ and $\tilde{m}$, and the use of n., co., ce, cu, for powers and
r, rr, rrr for roots. Moya explains: "These characters I am moved to
adopt, because others are not to be had in the printing office."[3] Of
English authors[4] we have found only one using the Italian signs for
"plus" and "minus," namely, the physician and mystic, Robert Fludd,
whose numerous writings were nearly all published on the Continent.
Fludd uses Ᵽ and ₥ for "plus" and "minus."

The + and −, and the $\bar{p}$ and $\tilde{m}$, were introduced in the latter part
of the fifteenth century, about the same time. They competed with
each other for more than a century, and $\bar{p}$ and $\tilde{m}$ finally lost out in the
early part of the seventeenth century.

205. *Shapes of the plus sign.*—The plus sign, as found in print, has
had three principal varieties of form: (1) the Greek cross +, as it is
found in Widman (1489); (2) the Latin cross, $\dagger$ more frequently
placed horizontally, −+ or +−; (3) the form ✠, or occasionally some
form still more fanciful, like the eight-pointed Maltese cross ✵, or a
cross having four rounded vases with tendrils drooping from their
edges.

The Greek cross, with the horizontal stroke sometimes a little

---

[1] *Libro primero de Arithmetica Algebratica . . . . por Marco Aurel, natural
Aleman* (Valencia, 1552).

[2] J. Rey Pastor, *Los mathemáticos españoles del siglo XVI* (Oviedo, 1913), p. 38.

[3] "Estos characteres me ha parecido poner, porque no auia otros en la imprenta" (*Ad theorema geometricvm, á nobilissimo viro propositum, Joannis Camilli
Gloriosi responsum* [Venetiis, 1613], p. 26).

[4] See C. Henry, *Revue archeologique*, N.S., Vol. XXXVII, p. 329, who quotes
from Fludd, *Utriusque cosmi . . . . Historia* (Oppenheim, 1617).

longer than the vertical one, was introduced by Widman and has been the prevailing form of plus sign ever since. It was the form commonly used by Grammateus, Rudolff, Stifel, Recorde, Digges, Clavius, Dee, Harriot, Oughtred, Rahn, Descartes, and most writers since their time.

206. The Latin cross, placed in a horizontal position, thus —|—, was used by Vieta[1] in 1591. The Latin cross was used by Romanus,[2] Hunt,[3] Hume,[4] Hérigone,[5] Mengoli,[6] Huygens,[7] Fermat,[8] by writers in the *Journal des* Sçavans,[9] Dechales,[10] Rolle,[11] Lamy,[12] L'Hospital,[13] Swedenborg,[14] Pardies,[15] Kresa,[16] Belidor,[17] De Moivre,[18] and Michelsen.[19] During the eighteenth century this form became less common and finally very rare.

Sometimes the Latin cross receives special ornaments in the form of a heavy dot at the end of each of the three shorter arms, or in the form of two or three prongs at each short arm, as in H. Vitalis.[20] A very ostentatious twelve-pointed cross, in which each of the four equal

---

[1] Vieta, *In artem analyticam isagoge* (Turonis, 1591).

[2] *Adriani Romani Canon triangvlorvm sphaericorum* . . . . (Mocvntiae, 1609).

[3] Nicolas Hunt, *The Hand-Maid to Arithmetick* (London, 1633), p. 130.

[4] James Hume, *Traité de l'algebre* (Paris, 1635), p. 4.

[5] P. Herigone, "Explicatis notarvm," *Cvrsvs mathematicvs*, Vol. I (Paris, 1634).

[6] Petro Mengoli, *Geometriae speciosae elementa* (Bologna, 1659), p. 33.

[7] *Christiani Hvgenii Holorogivm oscillatorivm* (Paris, 1673), p. 88.

[8] P. de Fermat, *Diophanti Alexandrini Arithmeticorum libri sex* (Toulouse, 1670), p. 30; see also Fermat, *Varia opera* (1679), p. 5.

[9] *Op. cit.* (Amsterdam, 1680), p. 160; *ibid.* (1693), p. 3, and other places.

[10] K. P. Claudii Francisci Milliet Dechales, *Mundus mathematicus*, Vol. I (Leyden, 1690), p. 577.

[11] M. Rolle, *Methode pour resoudre les egalitez de tous les degreez* (Paris, 1691) p. 15.

[12] Bernard Lamy, *Elemens des mathematiques* (3d ed.; Amsterdam, 1692), p. 61.

[13] L'Hospital, *Acta eruditorum* (1694), p. 194; *ibid.* (1695), p. 59; see also other places, for instance, *ibid.* (1711), Suppl., p. 40.

[14] Emanuel Swedenborg, *Daedalus Hyperborens* (Upsala, 1716), p. 5; reprinted in *Kungliga Vetenskaps Societetens i Upsala Tvåhundr aårsminne* (1910).

[15] *Œuvres du R. P. Pardies* (Lyon, 1695), p. 103.

[16] J. Kresa, *Analysis speciosa trigonometriae sphericae* (Prague, 1720), p. 57.

[17] B. F. de Belidor, *Nouveau cours de mathématique* (Paris, 1725), p. 10.

[18] A. de Moivre, *Miscellanea analytica* (London, 1730), p. 100.

[19] J. A. C. Michelsen, *Theorie der Gleichungen* (Berlin, 1791).

[20] "Algebra," *Lexicon mathematicum authore Hieronymo Vitali* (Rome, 1690).

arms has three prongs, is given by Carolo Renaldini.[1] In seventeenth-
and eighteenth-century books it is not an uncommon occurrence to
have two or three forms of plus signs in one and the same publication,
or to find the Latin cross in an upright or horizontal position, according to the crowded condition of a particular line in which the symbol
occurs.

207. The cross of the form ✠ was used in 1563 and earlier by the
Spaniard De Hortega,[2] also by Klebotius,[3] Romanus,[4] and Descartes.[5] It occurs not infrequently in the *Acta eruditorum*[6] of Leipzig,
and sometimes in the *Miscellanea Berolinensia*.[7] It was sometimes
used by Halley,[8] Weigel,[9] Swedenborg,[10] and Wolff.[11] Evidently this
symbol had a wide geographical distribution, but it never threatened
to assume supremacy over the less fanciful Greek cross.

A somewhat simpler form, ✝, consists of a Greek cross with four
uniformly heavy black arms, each terminating in a thin line drawn
across it. It is found, for example, in a work of Hindenburg,[12] and
renders the plus signs on a page unduly conspicuous.

Occasionally plus signs are found which make a "loud" display
on the printed page. Among these is the eight-pointed Maltese cross,

[1] *Caroli Renaldini Ars analytica mathematicvm* (Florence, 1665), p. 80, and
throughout the volume, while in the earlier edition (Anconnae, 1644) he uses both
the heavy cross and dagger form.

[2] Fray Juã de Hortega, *Tractado subtilissimo d'arismetica γ geometria* (Granada, 1563), leaf 51. Also (Seville, 1552), leaf 42.

[3] Guillaume Klebitius, *Insvlae Melitensis, quam alias Maltam vocant, Historia,
Quaestionib. aliquot Mathematicis reddita incundior* (Diest [Belgium], 1565). I
am indebted to Professor H. Bosmans for information relating to this book.

[4] Adr. Romanus, "Problema," *Ideae mathematicae pars prima* (Antwerp, 1593).

[5] René Descartes, *La géométrie* (1637), p. 325. This form of the plus sign is in-
frequent in this publication; the ordinary form (+) prevails.

[6] See, for instance, *op. cit.* (1682), p. 87; *ibid.* (1683), p. 204; *ibid.* (1691),
p. 179; *ibid.* (1694), p. 195; *ibid.* (1697), p. 131; *ibid.* (1698), p. 307; *ibid.* (1713),
p. 344.

[7] *Op. cit.*, p. 156. However, the Latin cross is used more frequently than the
form now under consideration. But in Vol. II (1723), the latter form is prevalent.

[8] E. Halley, *Philosophical Transactions*, Vol. XVII (London, 1692–94), p. 963;
ibid. (1700–1701), Vol. XXII, p. 625.

[9] *Erhardi Weigelii Philosophia Mathematica* (Jena, 1693), p. 135.

[10] E. Swedenborg, *op. cit.*, p. 32. The Latin cross is more prevalent in this
book.

[11] Christian Wolff, *Mathematisches Lexicon* (Leipzig, 1716), p. 14.

[12] Carl Friedrich Hindenburg, *Infinitinomii dignitatum . . . . leges ac Formulae*
(Göttingen, 1779).

of varying shape, found, for example, in James Gregory,[1] Corachan,[2] Wolff,[3] and Hindenburg.[4]

Sometimes the ordinary Greek cross has the horizontal stroke very much heavier or wider than the vertical, as is seen, for instance, in Fortunatus.[5] A form for plus —/— occurs in Johan Albert.[6]

208. *Varieties of minus signs.*—One of the curiosities in the history of mathematical notations is the fact that notwithstanding the extreme simplicity and convenience of the symbol − to indicate subtraction, a more complicated symbol of subtraction ÷ should have been proposed and been able to maintain itself with a considerable group of writers, during a period of four hundred years. As already shown, the first appearance in print of the symbols + and − for "plus" and "minus" is found in Widman's arithmetic. The sign − is one of the very simplest conceivable; therefore it is surprising that a modification of it should ever have been suggested.

Probably these printed signs have ancestors in handwritten documents, but the line of descent is usually difficult to trace with certainty (§ 201). The following quotation suggests another clue: "In the west-gothic writing before the ninth century one finds, as also Paoli remarks, that a short line has a dot placed above it ÷, to indicate *m*, in order to distinguish this mark from the simple line which signifies a contraction or the letter *N*. But from the ninth century down, this same west-gothic script always contains the dot over the line even when it is intended as a general mark."[7]

In print the writer has found the sign ÷ for "minus" only once. It occurs in the 1535 edition of the *Rechenbüchlin* of Grammateus (Fig. 56). He says: "Vnd mañ brauchet solche zeichen als + ist mehr / vnd ÷ / minder."[8] Strange to say, this minus sign does not occur in the first edition (1518) of that book. The corresponding passage of the earlier edition reads: "Vnd man braucht solche zaichen

[1] *Geometriae pars vniversalis* (Padua, 1668), p. 20, 71, 105, 108.

[2] Juan Bautista Corachan, *Arithmetica demonstrada* (Barcelona, 1719), p. 326.

[3] Christian Wolff, *Elementa matheseos universae*, Tomus I (Halle, 1713), p. 252.

[4] *Op. cit.*

[5] P. F. Fortunatus, *Elementa matheseos* (Brixia, 1750), p. 7.

[6] Johan Albert, *New Rechenbüchlein auff der federn* (Wittemberg, 1541); taken from Glaisher, *op. cit.*, p. 40, 61.

[7] Adriano Cappelli, *Lexicon abbreviaturam* (Leipzig, 1901), p. xx.

[8] Henricus Grammateus, *Eyn new Künstlich behend and gewiss Rechenbüchlin* (1535; 1st ed., 1518). For a facsimile page of the 1535 edition, see D. E. Smith, *Rara arithmetica* (1908), p. 125.

als + ist vnnd / − mynnder." Nor does Grammateus use ∸ in other parts of the 1535 edition; in his mathematical operations the minus sign is always −.

The use of the dash and two dots, thus ÷, for "minus," has been found by Glaisher to have been used in 1525, in an arithmetic of Adam Riese,[1] who explains: "Sagenn sie der warheit zuuil so bezeychenn sie mit dem zeychen + plus wu aber zu wenigk so beschreib sie mit dem zeychen ÷ minus genant."[2]

No reason is given for the change from − to ÷. Nor did Riese use ÷ to the exclusion of −. He uses ÷ in his algebra, *Die Coss*, of 1524, which he did not publish, but which was printed[3] in 1892, and also in his arithmetic, published in Leipzig in 1550. Apparently, he used − more frequently than ÷.

Probably the reason for using ÷ to designate − lay in the fact that − was assigned more than one signification. In Widman's arithmetic − was used for subtraction or "minus," also for separating terms in proportion,[4] and for connecting each amount of an article (wool, for instance) with the cost per pound (§ 202). The symbol − was also used as a rhetorical symbol or dash in the same manner as it is used at the present time. No doubt, the underlying motive in introducing ÷ in place of − was the avoidance of confusion. This explanation receives support from the German astronomer Regiomontanus,[5] who, in his correspondence with the court astronomer at Ferrara, Giovanni Bianchini, used − as a sign of equality; and used for subtraction a different symbol, namely, $\overline{i9}$ (possibly a florescent form of $\tilde{m}$). With him 1 $\overline{i9}$ r^e meant $1-x$.

Eleven years later, in 1546, Gall Splenlin, of Ulm, had published at Augsburg his *Arithmetica künstlicher Rechnung*, in which he uses ÷, saying: "Bedeut das zaichen + züuil, und das ÷ zü wenig."[6] Riese and Splenlin are the only arithmetical authors preceding the middle of the sixteenth century whom Glaisher mentions as using ÷ for subtraction or "minus."[7] Caspar Thierfeldern,[8] in his *Arithmetica*

[1] *Rechenung auff der linihen vnd federn in zal, masz, vnd gewicht* (Erfurt, 1525; 1st ed., 1522).

[2] This quotation is taken from Glaisher, *op. cit.*, p. 36.

[3] See Bruno Berlet, *Adam Riese* (Leipzig, Frankfurt am Main, 1892).

[4] Glaisher, *op. cit.*, p. 15.

[5] M. Curtze, *Abhandlungen zur Geschichte der mathematischen Wissenschaften*, Vol. XII (1902), p. 234; Karpinski, *Robert of Chester, etc.*, p. 37.

[6] See Glaisher, *op. cit.*, p. 43.

[7] *Ibid.*, Vol. LI, p. 1–148.   [8] See Jackson, *op. cit.*, p. 55, 220.

(Nuremberg, 1587), writes the equation (p. 110), "18 fl. $\div$ 85 gr. gleich 25 fl. $\div$ 232 gr."

With the beginning of the seventeenth century $\div$ for "minus" appears more frequently, but, as far as we have been able to ascertain only in German, Swiss, and Dutch books. A Dutch teacher, Jacob Vander Schuere, in his *Arithmetica* (Haarlem, 1600), defines $+$ and $-$, but lapses into using $\div$ in the solution of problems. A Swiss writer, Wilhelm Schey,[1] in 1600 and in 1602 uses both $\div$ and $\dot{\div}$ for "minus." He writes $9 \div 9$, $5 \div 12$, $6 \div 28$, where the first number signifies the weight in *centner* and the second indicates the excess or deficiency of the respective "pounds." In another place Schey writes "9 fl. $\dot{\div}$ 1 ort," which means "9 florins less 1 ort or quart." In 1601 Nicolaus Reymers,[2] an astronomer and mathematician, uses regularly $\div$ for "minus" or subtraction; he writes

$$\text{"XXVIII} \quad \text{XII} \quad \text{X} \quad \text{VI} \quad \text{III} \quad \text{I} \quad \text{O}$$
$$\text{1 gr.} \quad 65532+18 \quad \div 30 \quad \div 18 \quad +12 \quad \div 8\text{ "}$$
$$\text{for } x^{28} = 65,532x^{12} + 18x^{10} - 30x^6 - 18x^3 + 12x - 8 \,.$$

Peter Roth, of Nürnberg, uses $\dot{\div}$ in writing[3] $3x^2 - 26x$. Johannes Faulhaber[4] at Ulm in Württemberg used $\div$ frequently. With him the horizontal stroke was long and thin, the dots being very near to it. The year following, the symbol occurs in an arithmetic of Ludolf van Ceulen,[5] who says in one place: "Subtraheert $\sqrt{7}$ van, $\sqrt{13}$, rest $\sqrt{13}$, weynigher $\sqrt{7}$, daerom stelt $\sqrt{13}$ voren en $\sqrt{7}$ achter, met een sulck teecken $\div$ tusschen beyde, vvelck teecmin beduyt, comt alsoo de begeerde rest $\sqrt{13} \div \sqrt{7} -$." However, in some parts of the book $-$ is used for subtraction. Albert Girard[6] mentions $\div$ as the symbol for "minus," but uses $-$. Otto Wesellow[7] brought out a book in which

---

[1] *Arithmetica oder die Kunst zu rechnen* (Basel, 1600–1602). We quote from D. E. Smith, *op. cit.*, p. 427, and from Matthäus Sterner, *Geschichte der Rechenkunst* (München and Leipzig, 1891), p. 280, 291.

[2] *Nicolai Raimari Ursi Dithmarsi . . . . arithmetica analytica, vulgo Cosa, oder Algebra* (zu Frankfurt an der Oder, 1601). We take this quotation from Gerhardt, *Geschichte der Mathematik in Deutschland* (1877), p. 85.

[3] *Arithmetica philosophica* (1608). We quote from Treutlein, "Die deutsche Coss," *Abhandlungen zur Geschichte der Mathematik*, Vol. II (Leipzig, 1879), p. 28, 37, 103.

[4] *Numerus figuratus sive arithmetica analytica* (Ulm, 1614), p. 11, 16.

[5] *De arithmetische en geometrische Fondamenten* (1615), p. 52, 55, 56.

[6] *Invention nouvelle en l'algebre* (Amsterdam, 1629), no paging. A facsimile edition appeared at Leiden in 1884.

[7] *Flores arithmetici* (drüdde vnde veerde deel; Bremen, 1617), p. 523.

+ and ÷ stand for "plus" and "minus," respectively. These signs are used by Follinus,[1] by Stampioen (§ 508), by Daniel van Hovcke[2] who speaks of + as signifying "mer en ÷ min.," and by Johann Ardüser[3] in a geometry. It is interesting to observe that only thirteen years after the publication of Ardüser's book, another Swiss, J. H. Rahn, finding, perhaps, that there existed two signs for subtraction, but none for division, proceeded to use ÷ to designate division. This practice did not meet with adoption in Switzerland, but was seized upon with great avidity as the symbol for division in a far-off country, England. In 1670 ÷ was used for subtraction once by Huygens[4] in the *Philosophical Transactions.* Johann Hemelings[5] wrote ∺ for "minus" and indicated, in an example, $14\frac{1}{2}$ legions less 1250 men by "14 1/2 Legion ÷ 1250 Mann." The symbol is used by Tobias Beutel,[6] who writes "$81 \div 1R6561 \div 162. R. + 1.$ *zenss*" to represent our $81 - \sqrt{6561 - 162x + x^2}$. Kegel[7] explains how one can easily multiply by 41, by first multiplying by 6, then by 7, and finally subtracting the multiplicand; he writes "$7 \div 1$." In a set of seventeenth-century examination questions used at Nürnberg, reference is made to cossic operations involving quantities, "durch die Signa + und ÷ connectirt."[8]

The vitality of this redundant symbol of subtraction is shown by its continued existence during the eighteenth century. It was employed by Paricius,[9] of Regensburg. Schlesser[10] takes ∹ to represent

[1] Hermannus Follinus, *Algebra sive liber de rebus occultis* (Coloniae, 1622), p. 113, 185.

[2] *Cyffer-Boeck* . . . . (den tweeden Druck: Rotterdam, 1628), p. 129–33.

[3] *Geometriae theoricae et practicae. Oder von dem Feldmässen* (Zürich, 1646), fol. 75.

[4] In a reply to Slusius, *Philosophical Transactions*, Vol. V (London, 1670), p. 6144.

[5] *Arithmetisch-Poetisch-u. Historisch-Erquick Stund* (Hannover, 1660); *Selbstlehrendes Rechen-Buch* . . . . *durch Johannem Hemelingium* (Frankfurt, 1678). Quoted from Hugo Grosse, *Historische Rechenbücher des 16. and 17. Jahrhunderts* (Leipzig, 1901), p. 99, 112.

[6] *Geometrische Gallerie* (Leipzig, 1690), p. 46.

[7] Johann Michael Kegel, *New vermehrte arithmetica vulgaris et practica italica* (Frankfurt am Main, 1696). We quote from Sterner, *op. cit.*, p. 288.

[8] Fr. Unger, *Die˙ Methodik der praktischen Arithmetik in historischer Entwickelung* (Leipzig, 1888), p. 30.

[9] Georg Heinrich Paricius, *Praxis arithmetices* (1706). We quote from Sterner, *op. cit.*, p. 349.

[10] Christian Schlesser, *Arithmetisches Haupt-Schlüssel* . . . . *Die Coss—oder Algebra* (Dresden and Leipzig, 1720).

"minus oder weniger." It was employed in the *Philosophical Transactions* by the Dutch astronomer N. Cruquius;[1] ÷ is found in Hübsch[2] and Crusius.[3] It was used very frequently as the symbol for subtraction and "minus" in the *Maandelykse Mathematische Liefhebbery*, Purmerende (1754–69). It is found in a Dutch arithmetic by Bartjens[4] which passed through many editions. The vitality of the symbol is displayed still further by its regular appearance in a book by van Steyn,[5] who, however, uses — in 1778.[6] Halcke states, "÷ of — het teken van *substractio minus* of min.,"[7] but uses — nearly everywhere. Praalder, of Utrecht, uses ordinarily the minus sign —, but in one place[8] he introduces, for the sake of clearness, as he says, the use of ÷ to mark the subtraction of complicated expressions. Thus, he writes "$= \div \overline{9\frac{1}{2} + 2\sqrt{26}}$." The ÷ occurs in a Leipzig magazine,[9] in a Dresden work by Illing,[10] in a Berlin text by Schmeisser,[11] who uses it also in expressing arithmetical ratio, as in "$2 \div 6 \div 10$." In a part of Klügel's[12] mathematical dictionary, published in 1831, it is stated that ÷ is used as a symbol for division, "but in German arithmetics is employed also to designate subtraction." A later use of it for "minus," that we have noticed, is in a Norwegian arithmetic.[13] In fact, in Scandinavian

---

[1] *Op. cit.*, Vol. XXXIII (London, 1726), p. 5, 7.

[2] J. G. G. Hübsch, *Arithmetica portensis* (Leipzig, 1748).

[3] David Arnold Crusius, *Anweisung zur Rechen-Kunst* (Halle, 1746), p. 54.

[4] *De vernieuwde Cyfferinge van Mr. Willem Bartjens, ... vermeerdert—ende verbetert, door Mr. Jan van Dam. . . . . en van alle voorgaande Fauten gezuyvert door Klaas Bosch* (Amsterdam, 1771), p. 174–77.

[5] Gerard van Steyn, *Liefhebbery der Reekenkonst* (eerste deel; Amsterdam, 1768), p. 3, 11, etc.

[6] *Ibid.* (2ᵉ Deels, 2ᵉ Stuk, 1778), p. 16.

[7] *Mathematisch Zinnen-Confect . . . . door Paul Halcken . . . . Uyt het Hoogduytsch vertaald . . . . dor Jacob Oostwoud* (Tweede Druk, Te Purmerende, 1768), p. 5.

[8] *Mathematische Voorstellen . . . . door . . . . Ludolf van Keulen . . . . door Laurens Praalder* (Amsterdam, 1777), p. 137.

[9] J. A. Kritter, *Leipziger Magazin für reine and angewandte Mathematik* (herausgegeben von J. Bernoulli und C. F. Hindenburg, 1788), p. 147–61.

[10] Carl Christian Illing, *Arithmetisches Handbuch für Lehrer in den Schulen* (Dresden, 1793), p. 11, 132.

[11] Friedrich Schmeisser, *Lehrbuch der reinen Mathesis* (1. Theil, Berlin, 1817), p. 45, 201.

[12] G. S. Klügel, "Zeichen," *Mathematisches Wörterbuch*. This article was written by J. A. Grunert.

[13] G. C. Krogh, *Regnebog for Begyndere* (Bergen, 1869), p. 15.

countries the sign ÷ for "minus" is found occasionally in the twentieth century. For instance, in a Danish scientific publication of the year 1915, a chemist expresses a range of temperature in the words "fra+18° C. til ÷ 18° C."[1] In 1921 Ernst W. Selmer[2] wrote "0,72÷ 0,65 = 0,07." The difference in the dates that have been given, and the distances between the places of publication, make it certain that this symbol ÷ for "minus" had a much wider adoption in Germany, Switzerland, Holland, and Scandinavia than the number of our citations would indicate. But its use seems to have been confined to Teutonic peoples.

Several writers on mathematical history have incidentally called attention to one or two authors who used the symbol ÷ for "minus," but none of the historians revealed even a suspicion that this symbol had an almost continuous history extending over four centuries.

209. Sometimes the minus sign − appears broken up into two or three successive dashes or dots. In a book of 1610 and again of 1615, by Ludolph van Ceulen,[3] the minus sign occasionally takes the form − −. Richard Balam[4] uses three dots and says "3 · · · 7, 3 from 7"; he writes an arithmetical proportion in this manner: "2 · · · 4 = 3 · · · 5." Two or three dots are used in René Descartes' *Géométrie*, in the writings of Marin Mersenne,[5] and in many other seventeenth-century books, also in the *Journal des Sçavans* for the year 1686, printed in Amsterdam, where one finds (p. 482) "1 − − − R − − − 11" for $1 - \sqrt{-11}$, and in volumes of that *Journal* printed in the early part of the eighteenth century. Hérigone used ∼ for "minus" (§ 189), the − being pre-empted for *recta linea*.

From these observations it is evident that in the sixteenth and seventeenth centuries the forms of type for "minus" were not yet standardized. For this reason, several varieties were sometimes used on the same page.

This study emphasizes the difficulty experienced even in ordinary

[1] Johannes Boye Petersen, *Kgl. Danske Vidensk. Selskabs Skrifter, Nat. og. Math. Afd., 7. Raekke*, Vol. XII (Kopenhagen, 1915), p. 330; see also p. 221, 223, 226, 230, 238.

[2] *Skrifter utgit av Videnskapsselskapet i Kristiania* (1921)," Historisk-filosofisk Klasse" (2. Bind; Kristiania, 1922), article by Ernst W. Selmer, p. 11; see also p. 28, 29, 39, 47.

[3] *Circvlo et adscriptis liber. . . . . Omnia e vernacülo Latina fecit et annotationibus illustravit Willebrordus Snellius* (Leyden, 1610), p. 128.

[4] *Algebra* (London, 1653), p. 5.

[5] *Cogitata Physico-Mathematica* (Paris, 1644), Praefatio generalis, "De Rationibus atque Proportionibus," p. xii, xiii.

arithmetic and algebra in reaching a common world-language. Centuries slip past before any marked step toward uniformity is made. It appears, indeed, as if blind chance were an uncertain guide to lead us away from the Babel of languages. The only hope for rapid approach of uniformity in mathematical symbolism lies in international co-operation through representative committees.

210. *Symbols for "plus or minus."*—The $\pm$ to designate "plus or minus" was used by Albert Girard in his *Tables*[1] of 1626, but with the interpolation of *ou*, thus "$\overset{+}{\underset{-}{ou}}$." The $\pm$ was employed by Oughtred in his *Clavis mathematicae* (1631), by Wallis,[2] by Jones[3] in his *Synopsis*, and by others. There was considerable experimentation on suitable notations for cases of simultaneous double signs. For example, in the third book of his *Géométrie*, Descartes uses a dot where we would write $\pm$. Thus he writes the equation "$+y^6 \cdot 2py^4 \overset{+pp}{} 4ryy - qq \eqcirc 0$" and then comments on this: "Et pour les signes $+$ ou $-$ que iay omis, s'il y a eu $+p$ en la precedente Equation, il faut mettre en celle $-$ cy $+ 2p$, ou s'il ya eu $- p$, il faut mettre $- 2p;$ & au contraire s'il ya eu $+ r$, il faut mettre $- 4r$, ..." The symbolism which in the *Miscellanea Berolinensia* of 1710 is attributed to Leibniz is given in § 198.

A different notation is found in Isaac Newton's *Universal Arithmetick:* "I denoted the Signs of b and c as being indeterminate by the Note $\perp$, which I use indifferently for $+$ or $-$, and its opposite $\top$ for the contrary."[4] These signs appear to be the $+$ with half of the vertical stroke excised. William Jones, when discussing quadratic equations, says: "Therefore if $\vee$ be put for the Sign of any Term, and $\wedge$ for the contrary, all *Forms of Quadratics* with their *Solutions*, will be reduc'd to this one. If $xx \vee ax \vee b = 0$ then $\wedge \frac{1}{2}a \pm \overline{aa \wedge b}\big|^{\frac{1}{2}}$."[5] Later in the book (p. 189) Jones lets two horizontal dots represent any sign: "Suppose any *Equation* whatever, as $x^n \mathrel{..} ax^{n-1} \mathrel{..} bx^{n-2} \mathrel{..} cx^{n-3} \mathrel{..} dx^{n-4}$, etc. $\mathrel{..} A = 0$."

A symbol $\wr$ standing for $\pm$ was used in 1649 and again as late as 1695, by van Schooten[6] in his editions of Descartes' geometry, also

[1] See *Bibliotheca mathematica* (3d ser., 1900), Vol. I, p. 66.

[2] J. Wallis, *Operum mathematicorum pars prima* (Oxford, 1657), p. 250.

[3] William Jones, *Synopsis Palmariorum matheseos* (London, 1706), p. 14.

[4] *Op. cit.* (trans. Mr. Ralphson . . . . rev. by Mr. Cunn; London, 1728), p. 172; also *ibid.* (rev. by Mr. Cunn . . . . expl. by Theaker Wilder; London, 1769), p. 321.

[5] *Op. cit.*, p. 148.

[6] *Renati Descartes Geometria* (Leyden, 1649), Appendix, p. 330; *ibid.* (Frankfurt am Main, 1695), p. 295, 444, 445.

by De Witt.[1] Wallis[2] wrote ৪ for $+$ or $-$, and ৪ for the contrary. The sign ৪ was used in a restricted way, by James Bernoulli;[3] he says, "৪ significat $+$ in pr. e $-$ in post. hypoth.," i.e., the symbol stood for $+$ according to the first hypothesis, and for $-$, according to the second hypothesis. He used this same symbol in his *Ars conjectandi* (1713), page 264. Van Schooten wrote also ৪ for $\mp$. It should be added that ৪ appears also in the older printed Greek books as a ligature or combination of two Greek letters, the omicron o and the upsilon v. The ৪ appears also as an astronomical symbol for the constellation Taurus.

Da Cunha[4] introduced $\pm$' and $\pm$', or $\pm$' and $\mp$', to mean that the upper signs shall be taken simultaneously in both or the lower signs shall be taken simultaneously in both. Oliver, Wait, and Jones[5] denoted positive or negative N by $\pm N$.

211. The symbol $[a]$ was introduced by Kronecker[6] to represent 0 or $+1$ or -1, according as a was 0 or $+1$ or -1. The symbol "sgn" has been used by some recent writers, as, for instance, Peano,[7] Netto,[8] and Le Vavasseur, in a manner like this: "sgn $A = +1$" when $A > 0$, "sgn $A = -1$" when $A < 0$. That is, "sgn A" means the "sign of A." Similarly, Kowalewski[9] denotes by "sgn $\mathfrak{P}$" $+1$ when $\mathfrak{P}$ is an even, and -1 when $\mathfrak{P}$ is an odd, permutation.

The symbol $\sqrt{a^2}$ is sometimes taken in the sense[10] $\pm a$, but in equations involving $\sqrt{\ }$, the principal root $+a$ is understood.

212. *Certain other specialized uses of $+$ and $-$.*—The use of each of the signs $+$ and $-$ in a double sense—first, to signify addition and subtraction; second, to indicate that a number is positive and negative—has met with opposition from writers who disregarded the advantages resulting from this double use, as seen in $a-(-b)=a+b$,

[1] Johannis de Witt, *Elementa Cvrvarvm Linearvm. Edita Operâ Francisci à Schooten* (Amsterdam, 1683), p. 305.

[2] John Wallis, *Treatise of Algebra* (London, 1685), p. 210, 278.

[3] *Acta eruditorum* (1701), p. 214.

[4] J. A. da Cunha, *Principios mathematicos* (Lisbon, 1790), p. 126.

[5] *Treatise on Algebra* (2d ed.; Ithaca, 1887), p. 45.

[6] L. Kronecker, *Werke*, Vol. II (1897), p. 39.

[7] G. Peano, *Formulario mathematico*, Vol. V (Turin, 1908), p. 94.

[8] E. Netto and R. le Vavasseur, *Encyclopédie des scien. math.*, Tome I, Vol. II (1907), p. 184; see also A. Voss and J. Molk, *ibid.*, Tome II, Vol. I (1912), p. 257, n. 77.

[9] Gerhard Kowalewski, *Einführung in die Determinantentheorie* (Leipzig, 1909), p. 18.

[10] See, for instance, *Encyclopédie des scien. math.*, Tome II, Vol. I, p. 257, n. 77.

and who aimed at extreme logical simplicity in expounding the elements of algebra to young pupils. As a remedy, German writers proposed a number of new symbols which are set forth by Schmeisser as follows:

"The use of the signs $+$ and $-$, not only for opposite magnitudes . . . . but also for Addition and Subtraction, frequently prevents clearness in these matters, and has even given rise to errors. For that reason other signs have been proposed for the positive and negative. Wilkins (*Die Lehre von d. entgegengesetzt. Grössen etc.*, Brschw., 1800) puts down the positive without signs ($+a=a$) but places over the negative a dash, as in $-a=\bar{a}$. v. Winterfeld (*Anfangsgr. d. Rechenk.*, 2te Aufl. 1809) proposes for positive the sign ⊢ or Γ, for negative ⊣ or ⌐. As more scientific he considers the inversion of the letters and numerals, but unfortunately some of them as i, r, o, x, etc., and 0, 1, 8, etc., cannot be inverted, while others, by this process, give rise to other letters as b, d, p, q, etc. Better are the more recent proposals of Winterfeld, to use for processes of computation the signs of the waxing and waning moon, namely for Addition), for Subtraction (, for Multiplication)), for Division ((, but as he himself acknowledges, even these are not perfectly suitable. . . . . Since in our day one does not yet, for love of correctness, abandon the things that are customary though faulty, it is for the present probably better to stress the significance of the concepts of the positive and additive, and of the negative and subtractive, in instruction, by the retention of the usual signs, or, what is the same thing, to let the qualitative and quantitative significance of $+$ and $-$ be brought out sharply. This procedure has the advantage moreover of more fully exercising the understanding."[1]

Wolfgang Bolyai[2] in 1832 draws a distinction between $+$ and $-$, and ⊹ and ⊢; the latter meaning the (intrinsic) "positive" and "negative." If A signifies ⊢B, then $-A$ signifies ⊹B.

213. In more recent time other notations for positive and negative numbers have been adopted by certain writers. Thus, Spitz[3] uses $\leftarrow a$ and $\rightarrow a$ for positive a and negative a, respectively. Méray[4] prefers $\overrightarrow{a}$, $\overleftarrow{a}$; Padé,[5] a_ρ, a_n; Oliver, Wait, and Jones[6] employ an ele-

[1] Friedrich Schmeisser, *op. cit.*, p. 42, 43.

[2] *Tentamen* (2d ed., T. I.; Budapestini, 1897), p. xi.

[3] C. Spitz, *Lehrbuch der alg. Arithmetik* (Leipzig, 1874), p. 12.

[4] Charles Méray, *Leçons nouv. de l'analyse infin.*, Vol. I (Paris, 1894), p. 11.

[5] H. Padé, *Premières leçons d'algèbre élém.* (Paris, 1892), p. 5.

[6] *Op. cit.*, p. 5.

vated $+$ or $-$ (as in $+10$, -10) as signs of "quality"; this practice has been followed in developing the fundamental operations in algebra by a considerable number of writers; for instance, by Fisher and Schwatt,[1] and by Slaught and Lennes.[2] In elementary algebra the special symbolisms which have been suggested to represent "positive number" or "negative number" have never met with wide adoption. Stolz and Gmeiner[3] write a, $\bar{a}$, for positive a and negative a. The designation . . . . -3, -2, -1, 0, $+1$, $+2$, $+3$, ... , occurs in Huntington's *Continuum* (1917), page 20.

214. A still different application of the sign $+$ has been made in the theory of integral numbers, according to which Peano[4] lets $a+$ signify the integer immediately following a, so that $a+$ means the integer $(a+1)$. For the same purpose, Huntington[5] and Stolz and Gmeiner[6] place the $+$ in the position of exponents, so that $5^+=6$.

215. *Four unusual signs.*—The Englishman Philip Ronayne used in his *Treatise of Algebra* (London, 1727; 1st ed., 1717), page 4, two curious signs which he acknowledged were "not common," namely, the sign $-\ominus$ to denote that "some Quantity indefinitely Less than the Term that next precedes it, is to be added," and the sign $\ominus-$ that such a quantity is "to be subtracted," while the sign ϕ may mean "either $-\ominus$ or $\ominus-$ when it matters not which of them it is." We have not noticed these symbols in other texts.

How the progress of science may suggest new symbols in mathematics is illustrated by the composition of velocities as it occurs in Einstein's addition theorem.[7] Silberstein uses here $\#$ instead of $+$.

216. *Composition of ratios.*—A strange misapplication of the $+$ sign is sometimes found in connection with the "composition" of ratios. If the ratios $\dfrac{NP}{CN}$ and $\dfrac{AN}{CN}$ are multiplied together, the product

---

[1] G. E. Fisher and I. J. Schwatt, *Text-Book of Algebra* (Philadelphia, 1898), p. 23.

[2] H. E. Slaught and U. J. Lennes, *High School Algebra* (Boston, 1907), p. 48.

[3] Otto Stolz und J. A. Gmeiner, *Theoretische Arithmetik* (2d ed.; Leipzig, 1911), Vol. I, p. 116.

[4] G. Peano, *Arithmetices principia nova methodo exposita* (Turin, 1889); "Sul concetto di numero," *Rivista di matem.*, Vol. I, p. 91; *Formulaire de mathématiques*, Vol. II, § 2 (Turin, 1898), p. 1.

[5] E. V. Huntington, *Transactions of the American Mathematical Society*, Vol. VI (1905), p. 27.

[6] Op. cit., Vol. I, p. 14. In the first edition Peano's notation was used.

[7] C. E. Weatherburn, *Advanced Vector Analysis* (London, 1924), p. xvi.

$\dfrac{NP}{CN} \cdot \dfrac{AN}{CN}$, according to an old phraseology, was "compounded" of the first two ratios.[1] Using the term "proportion" as synonymous with "ratio," the expression "composition of proportions" was also used. As the word "composition" suggests addition, a curious notation, using $+$, was at one time employed. For example, Isaac Barrow[2] denoted the "compounded ratio" $\dfrac{NP}{CN} \cdot \dfrac{AN}{CN}$ in this manner, "$NP \cdot CN + AN \cdot CN$." That is, the sign of addition was used in place of a sign of multiplication, and the dot signified ratio as in Oughtred.

In another book[3] Barrow again multiplies equal ratios by equal ratios. In modern notation, the two equalities are

$$(PL+QO):QO=2BC:(BC-CP) \text{ and } QO:BC=BC:(BC+CP) .$$

Barrow writes the result of the multiplication thus:

$$PL+QO \cdot QO+QO \cdot BC=2BC \cdot BC-CP+BC \cdot BC+CP .$$

Here the $+$ sign occurs four times, the first and fourth times as a symbol of ordinary addition, while the second and third times it occurs in the "addition of equal ratios" which really means the multiplication of equal ratios. Barrow's final relation means, in modern notation,

$$\frac{PL+QO}{QO} \cdot \frac{QO}{BC}=\frac{2BC}{BC-CP} \cdot \frac{BC}{BC+CP} .$$

Wallis, in his *Treatise of Algebra* (London, 1685), page 84, comments on this subject as follows: "But now because *Euclide* gives to this the name of *Composition*, which word is known many times to impart an *Addition;* (as when we say the Line *ABC* is compounded of *AB* and *BC;*) some of our more ancient Writers have chanced to call it *Addition of Proportions;* and others, following them, have continued that form of speech, which abides in (in divers Writers) even to this day: And the Dissolution of this composition they call *Subduction of Proportion*. (Whereas that should rather have been called *Multiplication*, and this Division.)"

A similar procedure is found as late as 1824 in J. F. Lorenz' trans-

[1] See Euclid, *Elements*, Book VI, Definition 5. Consult also T. L. Heath, *The Thirteen Books of Euclid's "Elements,"* Vol. II (Cambridge, 1908), p. 132–35, 189, 190.

[2] *Lectiones opticae* (1669), Lect. VIII, § V, and other places.

[3] *Lectiones geometricae* (1674), Lect. XI, Appendix I, § V.

lation from the Greek of Euclid's *Elements* (ed. C. B. Mollweide; Halle, 1824), where on page 104 the Definition 5 of Book VI is given thus: "Of three or more magnitudes, A, B, C, D, which are so related to one another that the ratios of any two consecutive magnitudes $A:B$, $B:C$, $C:D$, are equal to one another, then the ratio of the first magnitude to the last is said to be *composed* of all these ratios so that

$$A:D = (A:B) + (B:C) + (C:D)" \quad \left(\text{in modern notation, } \frac{A}{D} = \frac{A}{B} \cdot \frac{B}{C} \cdot \frac{C}{D}\right).$$

SIGNS OF MULTIPLICATION

217. *Early symbols.*—In the early Babylonian tablets there is, according to Hilprecht,[1] an ideogram A-DU signifying "times" or multiplication. The process of multiplication or division was known to the Egyptians[2] as *wshtp*, "to incline the head"; it can hardly be regarded as being a mathematical symbol. Diophantus used no symbol for multiplication (§ 102). In the Bakhshālī manuscript multiplication is usually indicated by placing the numbers side by side (§ 109). In some manuscripts of Bhāskara and his commentators a dot is placed between factors, but without any explanation (§ 112). The more regular mark for product in Bhāskara is the abbreviation *bha*, from *bhavita*, placed after the factors (§ 112).

Stifel in his *Deutsche Arithmetica* (Nürnberg, 1545) used the capital letter M to designate multiplication, and D to designate division. These letters were again used for this purpose by S. Stevin[3] who expresses our $3xyz^2$ thus: 3 ① M *sec* ① M *ter* ②, where *sec* and *ter* mean the "second" and "third" unknown quantities.

The M appears again in an anonymous manuscript of 1638 explaining Descartes' *Géométrie* of 1637, which was first printed in 1896;[4] also once in the Introduction to a book by Bartholinus.[5]

Vieta indicated the product of A and B by writing "*A in B*" (Fig. 84). Mere juxtaposition signified multiplication in the Bakhshālī tract, in some fifteenth-century manuscripts, and in printed algebras designating $6x$ or $5x^2$; but $5\frac{1}{3}$ meant $5 + \frac{1}{3}$, not $5 \times \frac{1}{3}$.

[1] H. V. Hilprecht, *Babylonian Expedition*, Vol. XX, Part 1, *Mathematical etc. Tablets* (Philadelphia, 1906), p. 16, 23.

[2] T. Eric Peet, *The Rhind Mathematical Papyrus* (London, 1923), p. 13.

[3] *Œuvres mathematiques* (ed. Albert Girard; Leyden, 1634), Vol. I, p. 7.

[4] Printed in *Œuvres de Descartes* (éd. Adam et Tannery), Vol. X (Paris, 1908), p. 669, 670.

[5] Er. Bartholinus, *Renati des Cartes Principia matheseos universalis* (Leyden, 1651), p. 11. See J. Tropfke, *op. cit.*, Vol. II (2d ed., 1921), p. 21, 22.

218. *Early uses of the St. Andrew's cross, but not as a symbol of multiplication of two numbers.*—It is well known that the St. Andrew's cross (×) occurs as the symbol for multiplication in W. Oughtred's *Clavis mathematicae* (1631), and also (in the form of the letter *X*) in an anonymous Appendix which appeared in E. Wright's 1618 edition of John Napier's *Descriptio*. This Appendix is very probably from the pen of Oughtred. The question has arisen, Is this the earliest use of × to designate multiplication? It has been answered in the negative—incorrectly so, we think, as we shall endeavor to show.

In the *Encyclopédie des sciences mathématiques*, Tome I, Volume I (1904), page 40, note 158, we read concerning ×, "One finds it between factors of a product, placed one beneath the other, in the Commentary added by Oswald Schreckenfuchs to Ptolemy's *Almagest*, 1551."[1] As will be shown more fully later, this is not a correct interpretation of the symbolism. Not two, but four numbers are involved, two in a line and two others immediately beneath, thus:

$$315172 \diagdown \diagup 295448$$
$$395093 \diagup \diagdown 174715$$

The cross does not indicate the product of any two of these numbers, but each bar of the cross connects two numbers which are multiplied. One bar indicates the product of 315172 and 174715, the other bar the product of 395093 and 295448. Each bar is used as a symbol singly; the two bars are not considered here as one symbol.

Another reference to the use of × before the time of Oughtred is made by E. Zirkel,[2] of Heidelberg, in a brief note in which he protests against attributing the "invention" of × to Oughtred; he states that it had a period of development of over one hundred years. Zirkel does

[1] *Clavdii Ptolemaei Pelusiensis Alexandrini Omnia quae extant Opera* (Basileae, 1551), Lib. ii, "Annotationes."

[2] Emil Zirkel, *Zeitschr. f. math. u. naturw. Unterricht*, Vol. LII (1921), p. 96. An article on the sign ×, which we had not seen before the time of proofreading, when R. C. Archibald courteously sent it to us, is written by N. L. W. A. Gravelaar in *Wiskundig Tijdschrift*, Vol. VI (1909–10), p. 1–25. Gravelaar cites a few writers whom we do not mention. His claim that, before Oughtred, the sign × occurred as a sign of multiplication, must be rejected as not borne out by the facts. It is one thing to look upon × as two symbols, each indicating a separate operation, and quite another thing to look upon × as only one symbol indicating only one operation. This remark applies even to the case in § 229, where the four numbers involved are conveniently placed at the four ends of the cross, and each stroke connects two numbers to be subtracted one from the other.

not make his position clear, but if he does not mean that $\times$ was used before Oughtred as a sign of multiplication, his protest is pointless.

Our own studies have failed to bring to light a clear and conclusive case where, before Oughtred, $\times$ was used as a symbol of multiplication. In medieval manuscripts and early printed books $\times$ was used as a mathematical sign, or a combination of signs, in eleven or more different ways, as follows: (1) in solutions of problems by the process of two false positions, (2) in solving problems in compound proportion involving integers, (3) in solving problems in simple proportion involving fractions, (4) in the addition and subtraction of fractions, (5) in the division of fractions, (6) in checking results of computation by the processes of casting out the 9's, 7's, or 11's, (7) as part of a group of lines drawn as guides in the multiplication of one integer by another, (8) in reducing radicals of different orders to radicals of the same order, (9) in computing on lines, to mark the line indicating "thousands," (10) to take the place of the multiplication table above 5 times 5, and (11) in dealing with amicable numbers. We shall briefly discuss each of these in order.

219. *The process of two false positions.*—The use of $\times$ in this process is found in the *Liber abbaci* of Leonardo[1] of Pisa, written in 1202. We must begin by explaining Leonardo's use of a single line or bar. A line connecting two numbers indicates that the two numbers are to be multiplied together. In one place he solves the problem: If 100 *rotuli* are worth 40 *libras*, how many *libras* are 5 *rotuli* worth? On the margin of the sheet stands the following:

$$
\begin{array}{cc}
l. & \text{B.} \\
40 & 100 \\
& \text{———}\, 5 \text{ B.}
\end{array}
$$

The line connecting 40 and 5 indicates that the two numbers are to be multiplied together. Their product is divided by 100, but no symbolism is used to indicate the division. Leonardo uses single lines over a hundred times in the manner here indicated. In more complicated problems he uses two or more lines, but they do not necessarily

[1] Leonardo of Pisa, *Liber abbaci* (1202) (ed. B. Boncompagni; Roma, 1857), Vol. I, p. 84.

form crosses. In a problem involving five different denominations of money he gives the following diagram:[1]

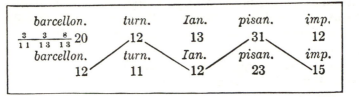

Here the answer 20+ is obtained by taking the product of the connected numbers and dividing it by the product of the unconnected numbers.

Leonardo uses a cross in solving, by double false position, the problem: If 100 *rotuli* cost 13 *libras*, find the cost of 1 *rotulus*. The answer is given in *solidi* and *denarii*, where 1 *libra* = 20 *solidi*, 1 *solidus* = 12 *denarii*. Leonardo assumes at random the tentative answers (the two false positions) of 3 *solidi* and 2 *solidi*. But 3 *solidi* would make this cost of 100 *rotuli* 15 *libra*, an error of +2 *libras*; 2 *solidi* would make the cost 10, an error of −3. By the underlying theory of two false positions, the errors in the answers (i.e., the errors $x-3$ and $x-2$ *solidi*) are proportional to the errors in the cost of 100 *rotuli* (i.e., +2 and −3 *libras*); this proportion yields $x=2$ *solidi* and $7\frac{1}{5}$ *denarii*. If the reader will follow out the numerical operations for determining our x he will understand the following arrangement of the work given by Leonardo (p. 319):

"Additum ex 13 *multiplicationibus*

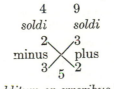

Additum ex erroribus."

Observe that Leonardo very skilfully obtains the answer by multiplying each pair of numbers connected by lines, thereby obtaining the products 4 and 9, which are added in this case, and then dividing 13 by 5 (the sum of the errors). The cross occurring here is not one symbol, but two symbols. Each line singly indicates a multiplication. It would be a mistake to conclude that the cross is used here as a symbol expressing multiplication.

[1] *Ibid.,* Vol. I, p. 127.

The use of two lines crossing each other, in double or single false position, is found in many authors of later centuries. For example, it occurs in MS 14908 in the Munich Library,[1] written in the interval 1455–64; it is used by the German Widman,[2] the Italian Pacioli,[3] the Englishman Tonstall,[4] the Italian Sfortunati,[5] the Englishman Recorde,[6] the German Splenlin,[7] the Italians Ghaligai[8] and Benedetti,[9] the Spaniard Hortega,[10] the Frenchman Trenchant,[11] the Dutchman Gemma Frisius,[12] the German Clavius,[13] the Italian Tartaglia,[14] the Dutchman Snell,[15] the Spaniard Zaragoza,[16] the Britishers Jeake[17] and

[1] See M. Curtze, *Zeitschrift f. Math. u. Physik*, Vol. XL (Leipzig, 1895). Supplement, *Abhandlungen z. Geschichte d. Mathematik*, p. 41.

[2] Johann Widman, *Behēde vnd hubsche Rechenung* (Leipzig, 1489). We have used J. W. L. Glaisher's article in *Messenger of Mathematics*, Vol. LI (1922), p. 16.

[3] L. Pacioli, *Summa de arithmetica, geometria, etc.* (1494). We have used the 1523 edition, printed at Toscolano, fol. 99[b], 100[a], 182.

[4] C. Tonstall, *De arte supputandi* (1522). We have used the Strassburg edition of 1544, p. 393.

[5] Giovanni Sfortunati da Siena, *Nvovo Lvme. Libro di Arithmetica* (1534), fol. 89–100.

[6] R. Recorde, *Grovnd of Artes* (1543[?]). We have used an edition issued between 1636 and 1646 (title-page missing), p. 374.

[7] Gall Splenlin, *Arithmetica künstlicher Rechnung* (1645). We have used J. W. L. Glaisher's article in *op. cit.*, Vol. LI (1922), p. 62.

[8] Francesco Ghaligai, *Pratica d'arithmetica* (Nuovamente Rivista ... ; Firenze, 1552), fol. 76.

[9] *Io. Baptistae Benedicti Diversarvm specvlationvm mathematicarum, et physicarum Liber* (Turin, 1585), p. 105.

[10] Juan de Hortega, *Tractado subtilissimo de arismetica y de geometria* (emendado por Lonçalo Busto, 1552), fol. 138, 215[b].

[11] Jan Trenchant, *L'arithmetiqve* (4th ed.; Lyon, 1578), p. 216.

[12] Gemma Frisius, *Arithmeticae Practicae methodvs facilis* (iam recens ab ipso authore emendata . . . . Parisiis, 1569), fol. 33.

[13] Christophori Clavii Bambergensis, *Opera mathematica* (Mogvntiae, 1612), Tomus secundus; "Numeratio," p. 58.

[14] *L'arithmetique de Nicolas Tartaglia Brescian* (traduit par Gvillavmo Gosselin de Caen ... Premier Partie; Paris, 1613), p. 105.

[15] *Willebrordi Snelli Doctrinae Triangvlorvm Canonicae liber qvatvor* (Leyden, 1627), p. 36.

[16] *Arithmetica Vniversal ... avthor* El M. R. P. Joseph Zaragoza (Valencia, 1669), p. 111.

[17] Samuel Jeake, ΛΟΓΙΣΤΙΚΗΛΟΓΙΑ *or Arithmetick* (London, 1696; Preface 1674), p. 501.

Wingate,[1] the Italian Guido Grandi,[2] the Frenchman Chalosse,[3] the Austrian Steinmeyer,[4] the Americans Adams[5] and Preston.[6] As a sample of a seventeenth-century procedure, we give Schott's solution[7] of $\frac{x}{2}-\frac{x}{6}-\frac{x}{8}=30$. He tries $x=24$ and $x=48$. He obtains errors -25 and -20. The work is arranged as follows:

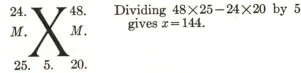

24.   48.   Dividing $48\times25-24\times20$ by 5
M.   M.   gives $x=144$.

25. 5. 20.

220. *Compound proportion with integers.*—We begin again with Leonardo of Pisa (1202)[8] who gives the problem: If 5 horses eat 6 quarts of barley in 9 days, for how many days will 16 quarts feed 10 horses? His numbers are arranged thus:

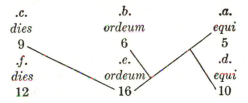

.c.   .b.   .a.
dies   ordeum   equi
9   6   5
.f.   .e.   .d.
dies   ordeum   equi
12   16   10

The answer is obtained by dividing $9\times16\times5$ by the product of the remaining known numbers. Answer 12.

Somewhat different applications of lines crossing each other are given by Nicolas Chuquet[9] and Luca Pacioli[10] in dealing with numbers in continued proportion.

[1] *Mr. Wingate's Arithmetick*, enlarged by John Kersey (11th ed.), with supplement by George Shelley (London, 1704), p. 128.

[2] Guido Grandi, *Instituzioni di arithmetia pratica* (Firenze, 1740), p. 104.

[3] *L'arithmetique par les fractions ...* par M. Chalosse (Paris, 1747), p. 158.

[4] *Tirocinium Arithmeticum a P. Philippo Steinmeyer* (Vienna and Freiburg, 1763), p. 475.

[5] Daniel Adams, *Scholar's Arithmetic* (10th ed.; Keene, N.H., 1816), p. 199.

[6] John Preston, *Lancaster's Theory of Education* (Albany, N.Y., 1817), p. 349.

[7] G. Schott, *Cursus mathematicus* (Würzburg, 1661), p. 36.

[8] *Op. cit.*, p. 132.

[9] Nicolas Chuquet, *Le Triparty en la Science des Nombres* (1484), edited by A. Marre, in *Bullettino Boncompagni*, Vol. XIII (1880), p. 700; reprint (Roma, 1881), p. 115.

[10] Luca Pacioli, *op. cit.*, fol. 93a.

Chuquet finds two mean proportionals between 8 and 27 by the scheme

where 12 and 18 are the two mean proportionals sought; i.e., 8, 12, 18, 27 are in continued proportion.

221. *Proportions involving fractions.*—Lines forming a cross ($\times$), together with two horizontal parallel lines, were extensively applied to the solution of proportions involving fractions, and constituted a most clever device for obtaining the required answer mechanically. If it is the purpose of mathematics to resolve complicated problems by a minimum mental effort, then this device takes high rank.

The very earliest arithmetic ever printed, namely, the anonymous booklet gotten out in 1478 at Treviso,[1] in Northern Italy, contains an interesting problem of two couriers starting from Rome and Venice, respectively, the Roman reaching Venice in 7 days, the Venetian arriving at Rome in 9 days. If Rome and Venice are 250 miles apart, in how many days did they meet, and how far did each travel before they met? They met in $3\frac{15}{16}$ days. The computation of the distance traveled by the courier from Rome calls for the solution of the proportion which we write $7:250 = \frac{63}{16} : x$.

The Treviso arithmetic gives the following arrangement:

$$
\begin{array}{c}
112 \\
\frac{7}{1} \diagdown\!\!\!\diagup \frac{250}{1} \text{------} \frac{63}{16}
\end{array}
$$

The connecting lines indicate what numbers shall be multiplied together; namely, 1, 250, and 63, also 7, 1, and 16. The product of the latter—namely, 112—is written above on the left. The author then finds $250\times63=15,750$ and divides this by 112, obtaining $140\frac{5}{8}$ miles.

These guiding lines served as Ariadne threads through the maze of a proportion involving fractions.

We proceed to show that this magical device was used again by Chuquet (1484), Widman (1489), and Pacioli (1494). Thus Chuquet[2]

---

[1] The Treviso arithmetic of 1478 is described and partly given in facsimile by Boncompagni in *Atti dell'Accademia Pontificia de' nuovi Lincei*, Tome XVI (1862–63; Roma, 1863), see p. 568.

[2] Chuquet, in Boncompagni, *Bullettino*, Vol. XIII, p. 636; reprint, p. (84).

uses the cross in the problem to find two numbers in the ratio of $\frac{2}{3}$ to $\frac{3}{4}$ and whose sum is 100. He writes $-\overset{3}{\underset{4}{\times}}\overset{2}{\underset{3}{-}}$; multiplying 3 by 3, and 2 by 4, he obtains two numbers in the proper ratio. As their sum is only 17, he multiplies each by $\frac{100}{17}$ and obtains $47\frac{1}{17}$ and $52\frac{16}{17}$.

Johann Widman[1] solves the proportion $9 : \frac{53}{8} = \frac{89}{8} : x$ in this manner: "Secz also $\overset{9}{\underset{1}{\times}}\overset{53\text{——}89}{\underset{8\text{——}8}{—}}$ machss nach der Regel vnd küpt 8 fl. 35s 9 helr $\frac{5}{12}$." It will be observed that the computer simply took the products of the numbers connected by lines. Thus $1\times53\times89 = 4,717$ gives the numerator of the fourth term; $9\times8\times8 = 576$ gives the denominator. The answer is 8 florins and a fraction.

Such settings of numbers are found in Luca Pacioli,[2] Ch. Rudolph,[3] G. Sfortunati,[4] O. Schreckenfuchs,[5] Hortega,[6] Tartaglia,[7] M. Steinmetz,[8] J. Trenchant,[9] Hermann Follinus,[10] J. Alsted,[11] P. Hérigone,[12] Chalosse,[13] J. Perez de Moya.[14] It is remarkable that in England neither Tonstall nor Recorde used this device. Recorde[15] and Leonard Digges[16]

---

[1] Johann Widman, op. cit.; see J. W. L. Glaisher, op. cit., p. 6.

[2] Luca Pacioli, op. cit. (1523), fol. 18, 27, 54, 58, 59, 64.

[3] Christoph Rudolph, Kunstliche Rechnung (1526). We have used one of the Augsburg editions, 1574 or 1588 (title-page missing), CVII.

[4] Giovanni Sfortunati da Siena, Nvovo Lvme. Libro di Arithmetica (1534), fol. 37.

[5] O. Schreckenfuchs, op. cit. (1551).

[6] Juan de Hortega, op. cit. (1552), fol. 92a.

[7] N. Tartaglia, General Trattato di Nvmeri (la prima parte, 1556), fol. 111b, 117a.

[8] Arithmeticae Praecepta .... M. Mavricio Steinmetz Gersbachio (Leipzig, 1568) (no paging).

[9] J. Trenchant, op. cit., p. 142.

[10] Hermannvs Follinvs, Algebra sive liber de rebvs occvltis (Cologne, 1622), p. 72.

[11] Johannis-Henrici Alstedii Encyclopaedia (Hernborn, 1630), Lib. XIV, Cossae libri III, p. 822.

[12] Pierre Herigone, Cvrsvs mathematici, Tomus VI (Paris, 1644), p. 320.

[13] L'Arithmetique par les fractions ... par M. Chalosse (Paris, 1747), p. 71.

[14] Juan Perez de Moya, Arithmetica (Madrid, 1784), p. 141. This text reads the same as the edition that appeared in Salamanca in 1562.

[15] Robert Recorde, op. cit., p. 175.

[16] (Leonard Digges), A Geometrical Practical Treatise named Pantometria (London, 1591).

use a slightly different and less suggestive scheme, namely, the capital letter Z for proportions involving either integers or fractions. Thus, $3:8=16:x$ is given by Recorde in the form 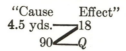 . This rather unusual notation is found much later in the *American Accomptant* of Chauncey Lee (Lansinburgh, 1797, p. 223) who writes,

"Cause   Effect"
4.5 yds. ——— 18
          90 ——— Q

and finds $Q=90\times18\div4.5=360$ dollars.

222. *Addition and subtraction of fractions.*—Perhaps even more popular than in the solution of proportion involving fractions was the use of guiding lines crossing each other in the addition and subtraction of fractions. Chuquet[1] represents the addition of $\frac{2}{3}$ and $\frac{4}{5}$ by the following scheme:

" 10   12 "
   —   —
    2 ╲   ╱ 4
       ╳
    3 ╱   ╲ 5
        ·15·

The lower horizontal line gives $3\times5=15$; we have also $2\times5=10$, $3\times4=12$; hence the sum $\frac{22}{15}=1\frac{7}{15}$.

The same line-process is found in Pacioli,[2] Rudolph,[3] Apianus.[4] In England, Tonstall and Recorde do not employ this intersecting line-system, but Edmund Wingate[5] avails himself of it, with only slight variations in the mode of using it. We find it also in Oronce Fine,[6] Feliciano,[7] Schreckenfuchs,[8] Hortega,[9] Baëza,[10] the Italian

---

[1] Nicolas Chuquet, *op. cit.*, Vol. XIII, p. 606; reprint p. (54).

[2] Luca Pacioli, *op. cit.* (1523), fol. 51, 52, 53.

[3] Christoph Rudolph, *op. cit.*, under addition and subtraction of fractions.

[4] Petrus Apianus, *Kauffmansz Rechnung* (Ingolstadt, 1527).

[5] E. Wingate, *op. cit.* (1704), p. 152.

[6] *Orontii Finei Delphinatis, liberalivm Disciplinarvm professoris Regii Protomathesis: Opus varium* (Paris, 1532), fol. 4b.

[7] Francesco Feliciano, *Libro de arithmetica e geometria* (1550).

[8] O. Schreckenfuchs, *op. cit.*, "Annot.," fol. 25b.

[9] Hortega, *op. cit.* (1552), fol. 55a, 63b.

[10] *Nvmerandi doctrina*, authore Lodoico Baëza (Paris, 1556), fol. 38b.

translation of Fine's works,[1] Gemma Frisius,[2] Eyçaguirre,[3] Clavius,[4] the French translation of Tartaglia,[5] Follinus,[6] Girard,[7] Hainlin,[8] Caramuel,[9] Jeake,[10] Corachan,[11] Chalosse,[12] De Moya,[13] and in slightly modified form in Crusoe.[14]

223. *Division of fractions.*—Less frequent than in the preceding processes is the use of lines in the multiplication or division of fractions, which called for only one of the two steps taken in solving a proportion involving fractions. Pietro Borgi (1488)[15] divides $\frac{3}{4}$ by $\frac{4}{5}$

thus: $\begin{array}{ccc} "4 & 3 & 15" \\ - & \times & - & — \\ 5 & 4 & 16 \end{array}$. In dividing $\frac{1}{2}$ by $\frac{1}{3}$, Pacioli[16] writes

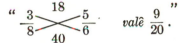

$$\begin{array}{cc} "2 & 3" \\ \frac{1}{3} & \times & \frac{1}{2} \end{array}$$

and obtains $\frac{3}{2}$ or $1\frac{1}{2}$.

Petrus Apianus (1527) uses the $\times$ in division. Juan de Hortega (1552)[17] divides $\frac{3}{8}$ by $\frac{5}{6}$, according to the following scheme:

$$\begin{array}{ccc} " & 18 & \\ \frac{3}{8} & \times & \frac{5}{6} \\ & 40 & \end{array} ." \quad valê \quad \frac{9}{20} ."$$

[1] *Opere di Orontio Fineo del Definato.* ... Tradotte da Cosimo Bartoli (Venice, 1587), fol. 31.

[2] *Arithmeticae Practicae methodvs facilis*, per Gemmam Frisium ... iam recèns ab ipso authore emendata ... (Paris, 1569), fol. 20.

[3] Sebastian Fernandez Eyçaguirre, *Libro de Arithmetica* (Brussels, 1608), p. 38.

[4] Chr. Clavius, *Opera omnia*, Tom. I (1611), Euclid, p. 383.

[5] *L'Arithmetique de Nicolas Tartaglia Brescian*, traduit ... par Gvillavmo Gosselin de Caen (Paris, 1613), p. 37.

[6] *Algebra sive Liber de Rebvs Occvltis*, ... Hermannvs Follinvs (Cologne, 1622), p. 40.

[7] Albert Girard, *Invention Nouvelle en L'Algebre* (Amsterdam, 1629).

[8] Johan. Jacob Hainlin, *Synopsis mathematica* (Tübingen, 1653), p. 32.

[9] *Joannis Caramvelis Mathesis Biceps Vetus et Nova* (Companiae, 1670), p. 20.

[10] Samuel Jeake, *op. cit.*, p. 51.

[11] Juan Bautista Corachan, *Arithmetica demonstrada* (Barcelona, 1719), p. 87.

[12] *L'Arithmetique par les fractions* ... par M. Chalosse (Paris, 1747), p. 8.

[13] J. P. de Moya, *op. cit.* (1784), p. 103.

[14] George E. Crusoe, *Y Mathematics?* ("Why Mathematics?") (Pittsburgh, Pa., 1921), p. 21.

[15] Pietro Borgi, *Arithmetica* (Venice, 1488), fol. 33B.

[16] L. Pacioli, *op. cit.* (1523), fol. 54a.

[17] Juan de Hortega, *op. cit.* (1552), fol. 66a.

We find this use of $\times$ in division in Sfortunati,[1] Blundeville,[2] Steinmetz,[3] Ludolf van Ceulen,[4] De Graaf,[5] Samuel Jeake,[6] and J. Perez de 'Moya.[7] De la Chapelle, in his list of symbols,[8] introduces $\times$ as a regular sign of division, *divisé par*, and $\times$ as a regular sign of multiplication, *multiplié par*. He employs the latter regularly in multiplication, but he uses the former only in the division of fractions, and he explains that in $\frac{6}{7}\times\frac{3}{4}=\frac{24}{21}$, "le sautoir $\times$ montre que 4 doit multiplier 6 & que 3 doit multiplier 7," thus really looking upon $\times$ as two symbols, one placed upon the other.

224. In the multiplication of fractions Apianus[9] in 1527 uses the parallel horizontal lines, thus, $\frac{1-3}{2-5}$. Likewise, Michael Stifel[10] uses two horizontal lines to indicate the steps. He says: "Multiplica numeratores inter se, et proveniet numerator productae summae. Multiplica etiam denominatores inter se, et proveniet denominator productae summae."

225. *Casting out the 9's, 7's, or 11's.*—Checking results by casting out the 9's was far more common in old arithmetics than by casting out the 7's or 11's. Two intersecting lines afforded a convenient grouping of the four results of an operation. Sometimes the lines appear in the form $\times$, at other times in the form $+$. Luca Pacioli[11] divides 97535399 by 9876, and obtains the quotient 9876 and remainder 23. Casting out the 7's (i.e., dividing a number by 7 and noting the residue), he obtains for 9876 the residue 6, for 97535399 the residue 3, for 23 the residue 2. He arranges these residues thus: "$\frac{6|2}{6|3}$ ".

Observe that multiplying the residues of the divisor and quotient, 6 times $6=36$, one obtains 1 as the residue of 36. Moreover, $3-2$ is also 1. This completes the check.

[1] Giovanni Sfortvnati da Siena, *Nvovo Lvme. Libro di Arithmetica* (1534), fol. 26.

[2] *Mr. Blundevil. His Exercises contayning eight Treatises* (London, 1636), p. 29.

[3] M. Mavricio Steinmetz Gersbachio, *Arithmeticae praecepta* (1568) (no paging).

[4] Ludolf van Ceulen, *De arithm.* (title-page gone) (1615), p. 13.

[5] Abraham de Graaf, *De Geheele Mathesis of Wiskonst* (Amsterdam, 1694), p. 14.

[6] Samuel Jeake, *op. cit.*, p. 58.   [7] Juan Perez de Moya, *op. cit.*, p. 117.

[8] De la Chapelle, *Institutions de géométrie* (4th éd.; Paris, 1765), Vol. I, p. 44, 118, 185.

[9] Petrus Apianus, *op. cit.* (1527).

[10] M. Stifel, *Arithmetica integra* (Nuremberg, 1544), fol. 6.

[11] Luca Pacioli, *op. cit.* (1523), fol. 35.

Nicolas Tartaglia[1] checks, by casting out the 7's, the division $912345 \div 1987 = 459$ and remainder 312.

Casting the 7's out of 912345 gives 0, out of 1987 gives 6, out of 459 gives 4, out of 312 gives 4. Tartaglia writes down " $\dfrac{4\,|\,4}{6\,|\,0}$ ".

Here 4 times $6 = 24$ yields the residue 3; 0 minus 4, or better 7 minus 4, yields 3 also. The result "checks."

Would it be reasonable to infer that the two perpendicular lines $+$ signified multiplication? We answer "No," for, in the first place, the authors do not state that they attached this meaning to the symbols and, in the second place, such a specialized interpretation does not apply to the other two residues in each example, which are to be *subtracted* one from the other. The more general interpretation, that the lines are used merely for the convenient grouping of the four residues, fits the case exactly.

Rudolph[2] checks the multiplication 5678 times $65 = 369070$ by casting out the 9's (i.e., dividing the sum of the digits by 9 and noting the residue); he finds the residue for the product to be 7, for the factors to be 2 and 8. He writes down

Here 8 times $2 = 16$, yielding the residue 7, written above. This residue is the same as the residue of the product; hence the check is complete. It has been argued that in cases like this Rudolph used $\times$ to indicate multiplication. This interpretation does not apply to other cases found in Rudolph's book (like the one which follows) and is wholly indefensible. We have previously seen that Rudolph used $\times$ in the addition and subtraction of fractions. Rudolph checks the proportion $9 : 11 = 48 : x$, where $x = 58\tfrac{6}{9}$, by casting out the 7's, 9's, and 11's as follows:

"(7)   (9)   (11)"

Take the check by 11's (i.e., division of a number by 11 and noting the residue). It is to be established that $9x = 48$ times 11, or that 9

[1] N. Tartaglia, *op. cit.* (1556), fol. 34B.

[2] Chr. Rudolph, *Kunstliche Rechnung* (Augsburg, 1574 or 1588 ed.) A VIII.

times 528 = 48 times 99. Begin by casting out the 11's of the factors 9 and 48; write down the residues 9 and 4. But the residues of 528 and 99 are both 0. Multiplying the residues 9 and 0, 4 and 0, we obtain in each case the product 0. This is shown in the figure. Note that here we do not take the product 9 times 4; hence $\times$ could not possibly indicate 9 times 4.

The use of $\times$ in casting out the 9's is found also in Recorde's *Grovnd of Artes* and in Clavius[1] who casts out the 9's and also the 7's.

Hortega[2] follows the Italian practice of using lines $+$, instead of $\times$, for the assignment of resting places for the four residues considered. Hunt[3] uses the Latin cross $\underline{+}$. The regular $\times$ is used by Regius (who also casts out the 7's),[4] Lucas,[5] Metius,[6] Alsted,[7] York,[8] Dechales,[9] Ayres,[10] and Workman.[11]

In the more recent centuries the use of a cross in the process of casting out the 9's has been abandoned almost universally; we have found it given, however, in an English mathematical dictionary[12] of 1814 and in a twentieth-century Portuguese cyclopedia.[13]

226. *Multiplication of integers.*—In Pacioli the square of 37 is found mentally with the aid of lines indicating the digits to be multiplied together, thus:

1369

[1] Chr. Clavius, *Opera omnia* (1612), Tom. I (1611), "Numeratio," p. 11.

[2] Juan de Hortega, *op. cit.*, fol. 42b.

[3] Nicolas Hunt, *Hand-Maid to Arithmetick* (London 1633).

[4] Hudalrich Regius, *Vtrivsqve Arithmetices Epitome* (Strasburg, 1536), fol. 57; *ibid.* (Freiburg-in-Breisgau, 1543), fol. 56.

[5] Lossius Lucas, *Arithmetices Erotemata Pverilia* (Lüneburg, 1569), fol. 8.

[6] *Adriani Metii Alcmariani Arithmeticae libri dvo:* Leyden, Arith. Liber I, p. 11.

[7] Johann Heinrich Alsted, *Methodus Admirandorum mathematicorum novem Libris* (Tertia editio; Herbon, 1641), p. 32.

[8] Tho. York, *Practical Treatise of Arithmetick* (London, 1687), p. 38.

[9] R. P. Claudii Francisci Milliet Dechales Camberiensis, *Mundus Mathematicus. Tomus Primus, Editio altera* (Leyden, 1690), p. 369.

[10] John Ayres, *Arithmetick made Easie*, by E. Hatton (London, 1730), p. 53.

[11] Benjamin Workman, *American Accountant* (Philadelphia, 1789), p. 25.

[12] Peter Barlow, *Math. & Phil. Dictionary* (London, 1814), art. "Multiplication."

[13] *Encyclopedia Portugueza* (Porto), art. "Nove."

From the lower 7 two lines radiate, indicating 7 times 7, and 7 times 3. Similarly for the lower 3. We have here a cross as part of the line-complex. In squaring 456 a similar scheme is followed; from each digit there radiate in this case three lines. The line-complex involves three vertical lines and three well-formed crosses ✕. The multiplication of 54 by 23 is explained in the manner of Pacioli by Mario Bettini[1] in 1642.

There are cases on record where the vertical lines are omitted, either as deemed superfluous or as the result of an imperfection in the typesetting. Thus an Italian writer, Unicorno,[2] writes:

It would be a rash procedure to claim that we have here a use of ✕ to indicate the product of two numbers; these *lines* indicate the product of 6 and 70, and of 50 and 8; the lines are not to be taken as *one* symbol; they do not mean 78 times 56. The capital letter X is used by F. Ghaligai in a similar manner in his *Algebra*. The same remarks apply to J. H. Alsted[3] who uses the X, but omits the vertical lines, in finding the square of 32.

A procedure resembling that of Pacioli, but with the lines marked as arrows, is found in a recent text by G. E. Crusoe.[4]

227. *Reducing radicals to radicals of the same order.*—Michael Stifel[5] in 1544 writes: "Vt volo reducere $\sqrt{z}$ 5 et $\sqrt{ơ}$ 4 ad idem signum, sic stabit exemplum ad regulam

[1] Mario Bettino, *Apiaria Vniversae philosophiae mathematicae* (Bologna, 1642), "Apiarivm vndecimvm," p. 37.

[2] S. Joseppo Vnicorno, *De l'arithmetica universale* (Venetia, 1598), fol. 20. Quoted from C. le Paige, "Sur l'origine de certains signes d'opération," *Annales de la société scientifique de Bruxelles* (16th year, 1891–92), Part II, p. 82.

[3] J. H. Alsted, *Methodus Admirandorum Mathematicorum Novem libris exhibens universam mathesin* (tertiam editio; Herbon, 1641), p. 70.

[4] George E. Crusoe, *op. cit.*, p. 6.

[5] Michael Stifel, *Arithmetica integra* (1544), fol. 114.

$\sqrt{zcf}125$ et $\sqrt{zcf}16$." Here $\sqrt{5}$ and $\sqrt[3]{4}$ are reduced to radicals of the same order by the use of the cross $\times$. The orders of the given radicals are two and three, respectively; these orders suggest the cube of 5 or 125 and the square of 4, or 16. The answer is $\sqrt[6]{125}$ and $\sqrt[6]{16}$.

Similar examples are given by Stifel in his edition of Rudolff's *Coss*,[1] Peletier,[2] and by De Billy.[3]

228. *To mark the place for "thousands."*—In old arithmetics explaining the computation upon lines (a modified abacus mode of computation), the line on which a dot signified "one thousand" was marked with a $\times$. The plan is as follows:

$\times$————————1000
                                 500
            ————————100
                                 50
            ————————50
                                 5
            ————————1

This notation was widely used in Continental and English texts.

229. *In place of multiplication table above 5×5.*—This old procedure is graphically given in Recorde's *Grovnd of Artes* (1543?). Required to multiply 7 by 8. Write the 7 and 8 at the cross as shown here; next, $10-8=2$, $10-7,=3$; write the 2 and 3 as shown:

Then, $2\times3=6$, write the 6; $7-2=5$, write the 5. The required product is 56. We find this process again in Oronce Fine,[4] Regius,[5]

[1] Michael Stifel, *Die Coss Christoffs Rudolffs* (Amsterdam, 1615), p. 136. (First edition, 1553.)

[2] *Jacobi Peletarii Cenomani, de occvlta parte nvmerorvm, qvam Algebram vocant, Libri duo* (Paris, 1560), fol. 52.

[3] Jacqves de Billy, *Abregé des Preceptes d'Algebre* (Reims, 1637), p. 22. See also the *Nova Geometriae Clavis*, authore P. Jacobo de Billy (Paris, 1643), p. 465.

[4] *Orontii Finei Delphinatis, liberalivm Disciplinarvm prefossoris* Regii Protomathesis: Opus uarium (Paris, 1532), fol. 4b.

[5] Hudalrich Regius, *Vtrivsqve arithmetices Epitome* (Strasburg, 1536), fol. 53; *ibid.* (Freiburg-in-Breisgau, 1543), fol. 56.

Stifel,[1] Boissiere,[2] Lucas,[3] the Italian translation of Oronce Fine,[4] the French translation of Tartaglia,[5] Alsted,[6] Bettini.[7] The French edition of Tartaglia gives an interesting extension of this process, which is exhibited in the product of 996 and 998, as follows:

$$\begin{array}{l} 996 \diagdown 4 \\ 998 \diagup 2 \\ \hline 994 \ 0 \ 0 \ 8 \end{array}$$

230. *Amicable numbers.*—N. Chuquet[8] shows graphically that 220 and 284 are amicable numbers (each the sum of the factors of the other) thus:

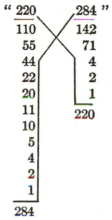

" 220   284 "
110   142
 55   71
 44   4
 22   2
 20   1
 11   220
 10
  5
  4
  2
  1
284

The old graphic aids to computation which we have described are interesting as indicating the emphasis that was placed by early arithmeticians upon devices that appealed to the eye and thereby contributed to economy of mental effort.

231. *The St. Andrew's cross used as a symbol of multiplication.*— As already pointed out, Oughtred was the first (§ 181) to use ✕ as the

[1] Michael Stifel, *Arithmetica integra* (Nuremberg, 1544), fol. 3.

[2] Claude de Boissiere, Daulphinois, *L'Art d'Arythmetique* (Paris, 1554), fol. 15b.

[3] Lossius Lucas, *Arithmetices Erotemata Pverilia* (Lüneburg, 1569), fol. 8.

[4] *Opere di Orontio Fineo.* ... Tradotte da Cosimo Bartoli (Bologna, 1587), "Della arismetica," libro primo, fol. 6, 7.

[5] *L'arithmetique de Nicolas Tartaglia* ... traduit ... par Gvillavmo Gosselin de Caen. (Paris, 1613), p. 14.

[6] *Johannis-Henrici Alstedii Encyclopaedia* (Herbon, 1630), Lib. XIV, p. 810.

[7] Mario Bettino, *Apiaria* (Bologna, 1642), p. 30, 31.

[8] N. Chuquet, *op. cit.*, Vol. XIII, p. 621; reprint, p. (69).

sign of multiplication of two numbers, as $a \times b$ (see also §§ 186, 288). The cross appears in Oughtred's *Clavis mathematicae* of 1631 and, in the form of the letter X, in E. Wright's edition of Napier's *Descriptio* (1618). Oughtred used a small symbol $\times$ for multiplication (much smaller than the signs $+$ and $-$). In this practice he was followed by some writers, for instance, by Joseph Moxon in his *Mathematical Dictionary* (London, 1701), p. 190. It seems that some objection had been made to the use of this sign $\times$, for Wallis writes in a letter of September 8, 1668: "I do not understand why the sign of multiplication $\times$ should more trouble the convenient placing of the fractions than the older signs $+ - = > :: .$"[1] It may be noted that Oughtred wrote the $\times$ small and placed it high, between the factors. This practice was followed strictly by Edward Wells.[2]

On the other hand, in A. M. Legendre's famous textbook *Géométrie* (1794) one finds (p. 121) a conspicuously large-sized symbol $\times$, for multiplication. The following combination of signs was suggested by Stringham:[3] Since $\times$ means "multiplied by," and / "divided by," the union of the two, viz., $\times/$, means "multiplied or divided by."

232. *Unsuccessful symbols for multiplication.*—In the seventeenth century a number of other designations of multiplication were proposed. Hérigone[4] used a rectangle to designate the product of two factors that were separated by a comma. Thus, "$\square 5+4+3, 7\sim3 : \sim10, est 38$" meant $(5+4+3) \cdot (7-3) - 10 = 38$. Jones, in his *Synopsis palmariorum* (1706), page 252, uses the $\sqsupset$, the Hebrew letter *mem*, to denote a rectangular area. A six-pointed star was used by Rahn and, after him, by Brancker, in his translation of Rahn's *Teutsche Algebra* (1659). "The Sign of Multiplication is [$*$] i.e., multiplied with." We encounter this use of $*$ in the *Philosophical Transactions*.[5]

Abraham de Graaf followed a practice, quite common among Dutch writers of the seventeenth and eighteenth centuries, of placing symbols on the right of an expression to signify direct operations (multiplication, involution), and placing the same symbols on the

[1] S. P. Rigaud, *Correspondence of Scientific Men of the Seventeenth Century* (Oxford, 1841), Vol. II, p. 494.

[2] Edward Wells, *The Young Gentleman's Arithmetic and Geometry* (2d ed.; London, 1723); "Arithmetic," p. 16, 41; "Geometry," p. 283, 291.

[3] Irving Stringham, *Uniplanar Algebra* (San Francisco, 1893), p. xiii.

[4] P. Herigone, *Cursus mathematici* (1644), Vol. VI, *explicatio notarum.* (First edition, 1642.)

[5] *Philosophical Transactions*, Vol. XVII, (1692–94), p. 680. See also §§ 194, 547.

left of an expression to signify inverse operations. Thus, Graaf[1] multiplies x^2+4 by $2\frac{1}{4}$ by using the following symbolism:

$$" \quad \frac{x \; x \; tot \; 4}{als \; \frac{9}{4} \; xx \; tot \; 9} \; 2\frac{1}{4} \quad "$$

In another place he uses this same device along with double commas, thus

$$" \quad \frac{\overline{a+b} \; , \; , \; -cc}{\overline{a+b} \; , \; , \; -ccd} \; d \quad "$$

to represent $(a+b)(-cc)(d) = (a+b)(-ccd)$.

Occasionally the comma was employed to mark multiplication, as in Hérigone (§ 189), F. Van Schooten,[2] who in 1657 gives $\dfrac{23,23,11,2}{3,3,3\sqrt{113,5}}$, where all the commas signify "times," as in Leibniz (§§ 197, 198, 547), in De Gua[3] who writes "3, 4, 5 . . . . &c. $\overline{n-m-2}$," in Petrus Horrebowius[4] who lets "A,B" stand for A times B, in Abraham de Graaf[5] who uses one or two commas, as in "$\overline{p-b},a$" for $(p-b)a$. The German Hübsch[6] designated multiplication by $\int$, as in $\frac{2}{5}\int\frac{1}{3}$.

233. *The dot for multiplication.*—The dot was introduced as a symbol for multiplication by G. W. Leibniz. On July 29, 1698, he wrote in a letter to John Bernoulli: "I do not like $\times$ as a symbol for multiplication, as it is easily confounded with x; . . . . often I simply relate two quantities by an interposed dot and indicate multiplication by $ZC \cdot LM$. Hence, in designating ratio I use not one point but two points, which I use at the same time for division." It has been stated that the dot was used as a symbol for multiplication before Leibniz, that Thomas Harriot, in his *Artis analyticae praxis* (1631), used the dot in the expressions "$aaa-3 \cdot bba = +2 \cdot ccc$." Similarly, in explaining cube root, Thomas Gibson[7] writes, in 1655, "$3 \cdot bb$," "$3 \cdot bcc$," but it

---

[1] Abraham de Graaf, *Algebra of Stelkonst* (Amsterdam, 1672), p. 8.

[2] *Francisci à Schooten. ... Exercitationum mathematicarum liber primus* (Leyden, 1657), p. 89.

[3] L'Abbe' de Gua, *Histoire de l'academie r. d. sciences, année 1741* (Paris, 1744), p. 81.

[4] *Petri Horrebowii. . . . . Operum mathematico-physicorum tomus primus* (Havniae, 1740), p. 4.

[5] Abraham de Graaf, *op. cit.* (1672), p. 87.

[6] J. G. G. Hübsch, *Arithmetica Portensis* (Leipzig, 1748). Taken from Wildermuth's article, "Rechnen," in K. A. Schmid's *Encyklopaedie des gesammten Erziehungs- und Unterrichtswesens* (1885).

[7] Tho. Gibson, *Syntaxis mathematica* (London, 1655), p. 36.

is doubtful whether either Harriot or Gibson meant these dots for multiplication. They are introduced without explanation. It is much more probable that these dots, which were placed after numerical coefficients, are survivals of the dots habitually used in old manuscripts and in early printed books to separate or mark off numbers appearing in the running text. Leibniz proposed the dot after he had used other symbols for over thirty years. In his first mathematical publication, the *De arte combinatoria*[1] of 1666, he used a capital letter C placed in the position ◌ for multiplication, and placed in the position ◌ for division. We have seen that in 1698 he advocated the point. In 1710 the Leibnizian symbols[2] were explained in the publication of the Berlin Academy (§ 198); multiplication is designated by apposition, and by a dot or comma (*punctum vel comma*), as in 3,2 or $a,b+c$ or $AB,CD+EF$. If at any time some additional symbol is desired, ◌ is declared to be preferable to ×.

The general adoption of the dot for multiplication in Europe in the eighteenth century is due largely to Christian Wolf. It was thus used by L. Euler; it was used by James Stirling in Great Britain, where the Oughtredian × was very popular.[3] Whitworth[4] stipulates, "The full point is used for the sign of multiplication."

234. *The St. Andrew's cross in notation for transfinite ordinal numbers.*—The notation $\omega \times 2$, with the multiplicand on the left, was chosen by G. Cantor in the place of 2ω (where ω is the first transfinite ordinal number), because in the case of three ordinal transfinite numbers, α, β, γ. the product $\alpha^{\beta} \cdot \alpha^{\gamma}$ is equal to $\alpha^{\beta+\gamma}$ when α^{β} is the multiplicand, but when α^{γ} is the multiplicand the product is $\alpha^{\gamma+\beta}$. In transfinite ordinals, $\beta+\gamma$ is not equal to $\gamma+\beta$.

SIGNS FOR DIVISION AND RATIO

235. *Early symbols.*—Hilprecht[5] states that the Babylonians had an ideogram *IGI-GAL* for the expression of division. Aside from their fractional notation (§ 104), the Greeks had no sign for division. Diophantus[6] separates the dividend from the divisor by the words ἐν

---

[1] G. W. Leibniz, *Opera omnia*, Vol. II (Geneva, 1768), p. 347.

[2] *Miscellanea Berolinensia* (Berlin), Vol. I (1710), p. 156.

[3] See also §§ 188, 287, 288; Vol. II, §§ 541, 547.

[4] W. A. Whitworth, *Choice and Chance* (Cambridge, 1886), p. 19.

[5] H. V. Hilprecht, *The Babylonian Expedition Mathematical, etc., Tablets from the Temple Library of Nippur* (Philadelphia, 1906), p. 22.

[6] Diophantus, *Arithmetica* (ed. P. Tannery; Leipzig, 1893), p. 286. See also G. H. F. Nesselmann, *Algebra der Griechen* (Berlin, 1842), p. 299.

μορίῳ or μορίου, as in the expression $\overset{\breve{v}}{\delta}\bar{\varsigma}$ λείψει $s\bar{s}$ $\overline{\kappa\delta}$ μορίου $\overset{\breve{v}}{\delta}\bar{a}\mu^{\delta}$ $\overline{\iota\beta}$ λείψεις$\bar{s}\bar{\varsigma}$, which means $(7x^2-24x)\div(x^2+12-7x)$. In the Bakhshālī arithmetic (§ 109) division is marked by the abbreviation *bhâ* from *bhâga*, "part." The Hindus often simply wrote the divisor beneath the dividend. Similarly, they designated fractions by writing the denominator beneath the numerator (§§ 106, 109, 113). The Arabic author[1] al-Ḥaṣṣâr, who belongs to the twelfth century, mentions the use of a fractional line in giving the direction: "Write the denominators below a [horizontal] line and over each of them the parts belonging to it; for example, if you are told to write three-fifths and a third of a fifth, write thus, $\dfrac{3\ \ 1}{5\ \ 3}$." In a second example, four-thirteenths and three-elevenths of a thirteenth is written $\dfrac{4\ \ 3}{13\ \ 11}$. This is the first appearance of the fractional line, known to us, unless indeed Leonardo of Pisa antedates al-Ḥaṣṣâr. That the latter was influenced in this matter by Arabic authors is highly probable. In his *Liber abbaci* (1202) he uses the fractional line (§ 122). Under the caption[2] "De diuisionibus integrorum numerorum" he says: "Cum super quemlibet numerum quedam uirgula protracta fuerit, et super ipsam quilibet alius numerus descriptus fuerit, superior numerus partem uel partes inferioris numeri affirmat; nam inferior denominatus, et superior denominans appellatur. Vt si super binarium protracta fuerit uirgula, et super ipsam unitas descripta sit ipsa unitas unam partem de duabus partibus unius integri affirmat, hoc est medietatem sic $\frac{1}{2}$." ("When above any number a line is drawn, and above that is written any other number, the superior number stands for the part or parts of the inferior number; the inferior is called the denominator, the superior the numerator. Thus, if above the two a line is drawn, and above that unity is written, this unity stands for one part of two parts of an integer, i.e., for a half, thus $\frac{1}{2}$.") With Leonardo, an indicated division and a fraction stand in close relation. Leonardo writes also $\cdot\dfrac{1\ \ 5\ \ 7}{2\ \ 6\ \ 10}$, which means, as he explains, seven-tenths, and five-sixths of one-tenth, and one-half of one-sixth of one-tenth.

236. One or two lunar signs, as in 8)24 or 8)24(, which are often employed in performing long and short division, may be looked upon as symbolisms for division. The arrangement 8)24 is found in Stifel's

[1] H. Suter, *Bibliotheca mathematica* (3d ser.), Vol. II (1901), p. 24.

[2] *Il Liber abbaci di Leonardo Pisano* (ed. B. Boncompagni; Roma, 1857), p. 23, 24.

Arithmetica integra (1544)[1], and in W. Oughtred's different editions of his *Clavis mathematicae.* In Oughtred's *Opuscula posthuma* one finds also $\frac{4}{3}]\frac{3}{2}[\frac{2}{3}$, (§ 182). Joseph Moxon[2] lets $D)A+B-C$ signify our $(A+B-C) \div D.$

Perhaps the earliest to suggest a special symbol for division other than the fractional line, and the arrangement 5)15 in the process of dividing, was Michael Stifel[3] in his *Deutsche Arithmetica* (1545). By the side of the symbols + and − he places the German capitals 𝔐 and 𝔇, to signify multiplication and division, respectively. Strange to say, he did not carry out his own suggestion; neither he nor seemingly any of his German followers used the 𝔐 and 𝔇 in arithmetic or algebraic manipulation. The letters M and D are found again in S. Stevin, who expressed our $\dfrac{5x^2}{y} \cdot z^2$ in this manner:[4]

$$5②D \text{ sec } ①M \text{ ter } ②,$$

where *sec* and *ter* signify the "second" and "third" unknown quantity.

The inverted letter Ↄ is used to indicate division by Gallimard,[5] as in

$$\text{"12 Ↄ } 4=3\text{"} \text{ and } \text{"}a^2b^2 \text{ Ↄ } a^2 \text{."}$$

In 1790 Da Cunha[6] uses the horizontal letter ᴆ as a mark for division.

237. *Rahn's notation.*—In 1659 the Swiss Johann Heinrich Rahn published an algebra[7] in which he introduced ÷ as a sign of division (§ 194). Many writers before him had used ÷ as a minus sign (§§ 164, 208). Rahn's book was translated into English by Thomas Brancker (a graduate of Exeter College, Oxford) and published, with additions from the pen of Joh. Pell, at London in 1668. Rahn's *Teutsche Algebra* was praised by Leibniz[8] as an "elegant algebra," nevertheless it did not enjoy popularity in Switzerland and the symbol ÷ for division

---

[1] Michael Stifel, *Arithmetica integra* (Nürnberg, 1544), fol. 317$V°$, 318$r°$. This reference is taken from J. Tropfke, *op. cit.,* Vol. II (2d ed., 1921), p. 28, n. 114.

[2] Joseph Moxon, *Mathematical Dictionary* (3d ed.; London, 1701), p. 190, 191.

[3] Michael Stifel, *Deutsche Arithmetica* (Nürnberg, 1545), fol. 74$v°$. We draw this information from J. Tropfke, *op. cit.,* Vol. II (2d ed., 1921), p. 21.

[4] S. Stevin, *Œuvres* (ed. A. Girard, 1634), Vol. I, p. 7, def. 28.

[5] J. E. Gallimard, *La Science du calcul numerique,* Vol. I (Paris, 1751), p. 4; *Methode ... d'arithmetique, d'algèbre et de géométrie* (Paris, 1753), p. 32.

[6] J. A. da Cunha, *Principios mathematicos* (1790), p. 214.

[7] J. H. Rahn, *Teutsche Algebra* (Zürich, 1659).

[8] *Leibnizens mathematische Schriften* (ed. C. I. Gerhardt), Vol. VII, p. 214.

was not adopted by his countrymen. In England, the course of events was different. The translation met with a favorable reception; Rahn's ÷ and some other symbols were adopted by later English writers, and came to be attributed, not to Rahn, but to John Pell. It so happened that Rahn had met Pell in Switzerland, and had received from him (as Rahn informs us) the device in the solution of equations of dividing the page into three columns and registering the successive steps in the solution. Pell and Brancker never claimed for themselves the introduction of the ÷ and the other symbols occurring in Rahn's book of 1569. But John Collins got the impression that not only the three-column arrangement of the page, but all the new algebraic symbols were due to Pell. In his extensive correspondence with John Wallis, Isaac Barrow, and others, Collins repeatedly spoke of ÷ as "Pell's symbol." There is no evidence to support this claim (§ 194).[1]

The sign ÷ as a symbol for division was adopted by John Wallis and other English writers. It came to be adopted regularly in Great Britain and the United States, but not on the European Continent. The only text not in the English language, known to us as using it, is one published in Buenos Aires;[2] where it is given also in the modified form $\cdot / \cdot$, as in $\frac{3}{4} \cdot / \cdot \frac{5}{8} = \frac{34}{20}$. In an American arithmetic,[3] the abbreviation ÷rs was introduced for "divisors," and ÷nds for "dividends," but this suggestion met with no favor on the part of other writers.

238. *Leibniz' notations.*—In the *Dissertatio de arte combinatoria* (1668)[4] G. W. Leibniz proposed for division the letter C, placed horizontally, thus ꙡ, but he himself abandoned this notation and introduced the colon. His article of 1684 in the *Acta eruditorum* contains for the first time in print the colon (:) as the symbol for division.[5] Leibniz says: ". . . . notetur, me divisionem hic designare hoc modo: $x:y$, quod idem est ac x divis. per y seu $\frac{x}{y}$." In a publication of the year 1710[6] we read: "According to common practice, the division

---

[1] F. Cajori, "Rahn's Algebraic Symbols," *Amer. Math. Monthly*, Vol. XXXI (1924), p. 65–71.

[2] Florentino Garcia, *El aritmético Argentino* (5th ed.; Buenos Aires, 1871), p. 102. The symbol ÷ and its modified form are found in the first edition of this book, which appeared in 1833.

[3] *The Columbian Arithmetician*, "by an American" (Haverhill [Mass.], 1811), p. 41.

[4] Leibniz, *Opera omnia*, Tom. II (Geneva, 1768), p. 347.

[5] See *Leibnizens mathematische Schriften* (ed. C. I. Gerhardt), Vol. V (1858), p. 223. See also M. Cantor, *Gesch. d. Mathematik*, Vol. III (2d ed.; Leipzig), p. 194.

[6] *Miscellanea Berolinensia* (Berlin, 1710), p. 156. See our § 198.

is sometimes indicated by writing the divisor beneath the dividend, with a line between them; thus a divided by b is commonly indicated by $\frac{a}{b}$; very often however it is desirable to avoid this and to continue on the same line, but with the interposition of two points; so that $a:b$ means a divided by b. But if, in the next place $a:b$ is to be divided by c, one may write $a:b,:c$, or $(a:b):c$. Frankly, however, in this case the relation can be easily expressed in a different manner, namely $a:(bc)$ or $a:bc$, for the division cannot always be actually carried out but often can only be indicated and then it becomes necessary to mark the course of the deferred operation by commas or parentheses."

In Germany, Christian Wolf was influential through his textbooks in spreading the use of the colon (:) for division and the dot (·) for multiplication. His influence extended outside Germany. A French translation of his text[1] uses the colon for division, as in "$(a-b):b$." He writes: "$a:mac = b:mbc$."

239. In Continental Europe the Leibnizian colon has been used for division and also for ratio. This symbolism has been adopted in the Latin countries with only few exceptions. In 1878 Balbontin[2] used in place of it the sign ÷ preferred by the English-speaking countries. Another Latin-American writer[3] used a slanting line in this manner, $\left(\frac{6}{7}\backslash 3\right) = \frac{6:3}{7} = \frac{2}{7}$ and also $12\backslash 3 = 4$. An author in Peru[4] indicates division by writing the dividend and divisor on the same line, but inclosing the former in a parenthesis. Accordingly, "$(20)5$" meant $20 \div 5$. Sometimes he uses brackets and writes the proportion $2:1\frac{1}{2} = 20:15$ in this manner: "$2:1[1]2::20:15$."

240. There are perhaps no symbols which are as completely observant of political boundaries as are ÷ and : as symbols for division. The former belongs to Great Britain, the British dominions, and the United States. The latter belongs to Continental Europe and the Latin-American countries. There are occasional authors whose prac-

[1] C. Wolf, *Cours de mathématique*, Tom. I (Paris, 1747), p. 110, 118.

[2] Juan Maria Balbontin, *Tratado elemental de aritmetica* (Mexico, 1878), p. 13.

[3] Felipe Senillosa, *Tratado elemental de arismética* (neuva ed.; Buenos Aires, 1844), p. 16. We quote from p. 47: "Este signo deque hemos hecho uso en la particion (\) no es usado generalmente; siendo el que se usa los dos punctos (:) ó la forma de quebrado. Pero un quebrado denota mas bien un cociente ó particion ejecutada que la operacion ó acto del partir; así hemos empleado este signo \ con analogia al del multiplicar que es éste: ×."

[4] Juan de Dios Salazar, *Lecciones de aritmetica* (Arequipa, 1827), p. v, 74, 89.

tices present exceptions to this general statement of boundaries, but their number is surprisingly small. Such statements would not apply to the symbolisms for the differential and integral calculus, not even for the eighteenth century. Such statements would not apply to trigonometric notations, or to the use of parentheses or to the designation of ratio and proportion, or to the signs used in geometry.

Many mathematical symbols approach somewhat to the position of world-symbols, and approximate to the rank of a mathematical world-language. To this general tendency the two signs of division $\div$ and : mark a striking exception. The only appearance of $\div$ signifying division that we have seen on the European Continent is in an occasional translation of an English text, such as Colin Maclaurin's *Treatise of Algebra* which was brought out in French at Paris in 1753. Similarly, the only appearance of : as a sign for division that we have seen in Great Britain is in a book of 1852 by T. P. Kirkman.[1] Saverien[2] argues against the use of more than one symbol to mark a given operation. "What is more useless and better calculated to disgust a beginner and embarrass even a geometer than the three expressions $\cdot$, :, $\div$, to mark division?"

241. *Relative position of divisor and dividend.*—In performing the operation of division, the divisor and quotient have been assigned various positions relative to the dividend. When the "scratch method" of division was practiced, the divisor was placed beneath the dividend and moved one step to the right every time a new figure of the quotient was to be obtained. In such cases the quotient was usually placed immediately to the right of the dividend, but sometimes, in early writers, it was placed above the dividend. In short division, the divisor was often placed to the left of the dividend, so that $a)b(c$ came to signify division.

A curious practice was followed in the Dutch journal, the *Maandelykse Mathematische Liefhebberye* (Vol. I [1759], p. 7), where $a)$——— signifies division by a, and ———$(a$ means multiplication by a. Thus:

$$" \quad x)\frac{xy=b-a+x}{ergo\ y=\dfrac{b-a+x}{x}} \quad ."$$

James Thomson called attention to the French practice of writing the divisor on the right. He remarks: "The French place the divisor

[1] T. P. Kirkman, *First Mnemonial Lessons in Geometry, Algebra and Trigonometry* (London, 1852).

[2] Alexandre Saverien, *Dictionnaire universel de mathematique et de physique* (Paris, 1753), "Caractere."

to the right of the dividend, and the quotient below it. . . . . This mode gives the work a more compact and neat appearance, and possesses the advantage of having the figures of the quotient near the divisor, by which means the practical difficulty of multiplying the divisor by a figure placed at a distance from it is removed. . . . . This method might, with much propriety, be adopted in preference to that which is employed in this country."[1]

The arrangement just described is given in Bézout's arithmetic,[2] in the division of 14464 by 8, as follows:

$$\text{``}14464 \overline{\left| \begin{array}{l} 8 \\ \overline{1808} \end{array} \right.} \text{''}$$

242. *Order of operations in terms containing both* ÷ *and* ×.—If an arithmetical or algebraical term contains ÷ and ×, there is at present no agreement as to which sign shall be used first. "It is best to avoid such expressions."[3] For instance, if in 24÷4×2 the signs are used as they occur in the order from left to right, the answer is 12; if the sign × is used first, the answer is 3.

Some authors follow the rule that the multiplications and divisions shall be taken in the order in which they occur.[4] Other textbook writers direct that multiplications in any order be performed first, then divisions as they occur from left to right.[5] The term $a \div b \times b$ is interpreted by Fisher and Schwatt[6] as $(a \div b) \times b$. An English committee[7] recommends the use of brackets to avoid ambiguity in such cases.

243. *Critical estimates of* : *and* ÷ *as symbols.*—D. André[8] expresses himself as follows: "The sign : is a survival of old mathematical notations; it is short and neat, but it has the fault of being symmetrical toward the right and toward the left, that is, of being a symmetrical sign of an operation that is asymmetrical. It is used less and less.

[1] James Thomson, *Treatise on Arithmetic* (18th ed.; Belfast, 1837).

[2] *Arithmétique de Bézout* ... par F. Peyrard (13th ed.; Paris, 1833).

[3] M. A. Bailey, *American Mental Arithmetic* (New York, 1892), p. 41.

[4] Hawkes, Luby, and Touton, *First Course of Algebra* (New York, 1910), p. 10.

[5] Slaught and Lennes, *High School Algebra, Elementary Course* (Boston, 1907), p. 212.

[6] G. E. Fisher and I. J. Schwatt, *Text-Book of Algebra* (Philadelphia, 1898), p. 85.

[7] "The Report of the Committee on the Teaching of Arithmetic in Public Schools," *Mathematical Gazette*, Vol. VIII (1917), p. 238. See also p. 296.

[8] Désiré André, *Des Notations mathématiques* (Paris, 1909), p. 58, 59.

. . . . When it is required to write the quotient of a divided by b, in the body of a statement in ordinary language, the expression $a:b$ really offers the typographical advantage of not requiring, as does $\frac{a}{b}$, a wider separation of the line in which the sign occurs from the two lines which comprehend it."

In 1923 the National Committee on Mathematical Requirements[1] voiced the following opinion: "Since neither $\div$ nor :, as signs of division, plays any part in business life, it seems proper to consider only the needs of algebra, and to make more use of the fractional form and (where the meaning is clear) of the symbol /, and to drop the symbol $\div$ in writing algebraic expressions."

244. *Notations for geometrical ratio.*—William Oughtred introduced in his *Clavis mathematicae* the dot as the symbol for ratio (§ 181). He wrote (§ 186) geometrical proportion thus, $a.b::c.d$. This notation for ratio and proportion was widely adopted not only in England, but also on the European Continent. Nevertheless, a new sign, the colon (:), made its appearance in England in 1651, only twenty years after the first publication of Oughtred's text. This colon is due to the astronomer Vincent Wing. In 1649 he published in London his *Urania practica*, which, however, exhibits no special symbolism for ratio. But his *Harmonicon coeleste* (London, 1651) contains many times Oughtred's notation $A.B::C.D$, and many times also the new notation $A:B::C:D$, the two notations being used interchangeably. Later there appeared from his pen, in London, three books in one volume, *Logistica astronomica* (1656), *Doctrina spherica* (1655), and *Doctrina theorica* (1655), each of which uses the notation $A:B:: C:D$.

A second author who used the colon nearly as early as Wing was a schoolmaster who hid himself behind the initials "R.B." In his book entitled *An Idea of Arithmetik*, at first designed for the use of "the Free Schoole at Thurlow in Suffolk . . . . by R.B., Schoolmaster there" (London, 1655), one finds $1.6::4.24$ and also $A:a::C:c$.

W. W. Beman pointed out in *L'Intermédiaire des mathématiciens*, Volume IX (1902), page 229, that Oughtred's Latin edition of his *Trigonometria* (1657) contains in the explanation of the use of the tables, near the end, the use of : for ratio. It is highly improbable that the colon occurring in those tables was inserted by Oughtred himself. In the *Trigonometria* proper, the colon does not occur, and Ought-

---

[1] *Report of the National Committee on Mathematical Requirements under the Auspices of the Mathematical Association of America, Inc.* (1923), p. 81.

red's regular notation for ratio and proportion $A.B::C.D$ is followed throughout. Moreover, in the English edition of Oughtred's trigonometry, printed in the same year (1657), but subsequent to the Latin edition, the passage of the Latin edition containing the : is recast, the new notation for ratio is abandoned, and Oughtred's notation is introduced. The : used to designate ratio (§ 181) in Oughtred's *Opuscula mathematica hactenus inedita* (1677) may have been introduced by the editor of the book.

It is worthy of note, also, that in a text entitled *Johnsons Arithmetik; In two Bookes* (2d ed.; London, 1633), the colon (:) is used to designate a fraction. Thus ¾ is written 3:4. If a fraction be considered as an indicated division, then we have here the use of : for division at a period fifty-one years before Leibniz first employed it for that purpose in print. However, dissociated from the idea of a fraction, division is not designated by any symbol in Johnson's text. In dividing 8976 by 15 he writes the quotient "598 6:15."

As shown more fully elsewhere (§ 258), the colon won its way as the regular symbol for geometrical ratio, both in England and the European Continent.

245. Oughtred's dot and Wing's colon did not prevent experimentation with other characters for geometric ratio, at a later date. But none of the new characters proposed became serious rivals of the colon. Richard Balam,[1] in 1653, used the colon as a decimal separatrix, and proceeded to express ratio by turning the colon around so that the two dots became horizontal; thus "3 . . 1" meant the geometrical ratio 1 to 3. This designation was used by John Kirkby[2] in 1735 for arithmetical ratio; he wrote arithmetical proportion "9 . . 6 = 6 . . 3." In the algebra of John Alexander,[3] of Bern, geometrical ratio is expressed by a dot, $a.b$, and also by $a \overset{\cdot\cdot}{\cdot} b$. Thomas York[4] in 1687 wrote a geometrical proportion "33600   7 : : 153600   32," using no sign at all between the terms of a ratio.

In the minds of some writers, a geometrical ratio was something more than an indicated division. The operation of division was associated with rational numbers. But a ratio may involve incommensu-

---

[1] Richard Balam, *Algebra: or The Doctrine of Composing, Inferring, and Resolving an Equation* (London, 1653), p. 4.

[2] John Kirkby, *Arithmetical Institutions* (London, 1735), p. 28.

[3] *Synopsis algebraica, opus posthumum Iohannis Alexandri, Bernatis-Helvetii. In usum scholae mathematicae apud Hospitium-Christi Londinense* (London, 1693), p. 16, 55. An English translation by Sam. Cobb appeared at London in 1709.

[4] Thomas York, *Practical Treatise of Arithmetik* (London, 1687), p. 146.

rable magnitudes which are expressible by two numbers, one or both of which are irrational. Hence ratio and division could not be marked by the same symbol. Oughtred's ratio *a.b* was not regarded by him as an indicated division, nor was it a fraction. In 1696 this matter was taken up by Samuel Jeake[1] in the following manner: "And so by some, to distinguish them [ratios] from Fractions, instead of the intervening Line, two Pricks are set; and so the Ratio Sesquialtera is thus expressed $\frac{3}{2}$." Jeake writes the geometrical proportion, "$\underset{1}{7} \cdot \underset{1}{9} :: 7 \cdot 9$."

Emanuel Swedenborg starts out, in his *Daedalus Hyperboreus* (Upsala, 1716), to designate geometric proportion by : :: :, but on page 126 he introduces $\because$ as a *signum analogicum* which is really used as a symbol for the ratio of quantities. On the European Continent one finds Hérigone[2] using the letter π to stand for "proportional" or ratio; he writes π where we write : . On the other hand, there are isolated cases where : was assigned a different usage; the Italian L. Perini[3] employs it as separatrix between the number of feet and of inches; his "11:4" means 11 feet 4 inches.

246. Discriminating between ratio and division, F. Schmeisser[4] in 1817 suggested for geometric ratio the symbol .., which (as previously pointed out) had been used by Richard Balam, and which was employed by Thomas Dilworth[5] in London, and in 1799 by Zachariah Jess,[6] of Wilmington, Delaware. Schmeisser comments as follows: "At one time ratio was indicated by a point, as in *a.b*, but as this signifies multiplication, Leibniz introduced two points, as in *a:b*, a designation indicating division and therefore equally inconvenient, and current only in Germany. For that reason have Mönnich, v. Winterfeld, Krause and other thoughtful mathematicians in more recent time adopted the more appropriate designation *a..b*." Schmeisser writes (p. 233) the geometric progression: "$\div 3..6..12 ..24..48..96 \ldots$"

[1] Samuel Jeake, ΛΟΓΙΣΤΙΚΗΛΟΓΊΑ, *or Arithmetick* (London, 1696), p. 410.

[2] Peter Herigone, *Cursus mathematicus*, Vol. I (Paris, 1834), p. 8.

[3] Lodovico Perini, *Geometria pratica* (Venezia, 1750), p. 109.

[4] Friedrich Schmeisser, *Lehrbuch der reinen Mathesis*, Erster Theil, "Die Arithmetik" (Berlin, 1817), Vorrede, p. 58.

[5] Thomas Dilworth, *The Schoolmaster's Assistant* (2d ed.; London, 1784). (First edition, about 1744.)

[6] Zachariah Jess, *System of Practical Surveying* (Wilmington, 1799), p. 173.

Similarly, A. E. Layng,[1] of the Stafford Grammar School in England, states: "The Algebraic method of expressing a ratio $\dfrac{A}{B}$ being a very convenient one, will also be found in the Examples, where it should be regarded as a symbol for the words *the ratio of A to B*, and not as implying the operation of division; it should not be used for *book-work*."

247. *Division in the algebra of complex numbers.*—As, in the algebra of complex numbers, multiplication is in general not commutative, one has two cases in division, one requiring the solution of $a = bx$, the other the solution of $a = yb$. The solution of $a = bx$ is designated by Peirce[2] $\dfrac{a}{b\times}$, by Schröder[3] $\dfrac{a}{b}$, by Study[4] and Cartan $\dfrac{a}{b}$.

The solution of $a = yb$ is designated by Peirce $\dfrac{a}{\times b}$ and by Schröder $a:b$, by Study and Cartan $\dfrac{a}{.b}$. The $\times$ and the . indicate in this notation the place of the unknown factor. Study and Cartan use also the notations of Peirce and Schröder.

SIGNS OF PROPORTION

248. *Arithmetical and geometrical progression.*—The notation $\div\!\cdot$ was used by W. Oughtred (§ 181) to indicate that the numbers following were in continued geometrical proportion. Thus, $\div\!\cdot$ 2, 6, 18, 54, 162 are in continued geometric proportion. During the seventeenth and eighteenth centuries this symbol found extensive application; beginning with the nineteenth century the need of it gradually passed away, except among the Spanish-American writers. Among the many English writers using $\div\!\cdot$ are John Wallis,[5] Richard Sault[6], Edward Cocker,[7] John Kersey,[8] William Whiston,[9] Alexander Mal-

[1] A. E. Layng, *Euclid's Elements of Geometry* (London, 1891), p. 219.

B. Peirce, *Linear Associative Algebra* (1870), p. 17; *Amer. Jour. of Math.*, Vol. IV (1881), p. 104.

[3] E. Schröder, *Formale Elemente der absoluten Algebra* (Progr. Bade, 1874).

[4] E. Study and E. Cartan, *Encyclopédie des scien. math.*, Tom. I, Vol. I (1908), p. 373.

[5] *Phil. Trans.*, Vol. V (London, 1670), p. 2203.

[6] Richard Sault, *A New Treatise of Algebra* (London [no date]).

[7] *Cocker's Artificial Arithmetick*, by Edward Cocker, perused and published by John Hawkes (London, 1684), p. 278.

[8] John Kersey, *Elements of Algebra* (London, 1674), Book IV, p. 177.

[9] A. Tacquet's edition of *W. Whiston's Elementa Euclidea geometriae* (Amsterdam, 1725), p. 124.

colm,[1] Sir Jonas Moore,[2] and John Wilson.[3] Colin Maclaurin indicates in his *Algebra* (1748) a geometric progression thus: "$\div\div 1:q:q^2$: $q^3:q^4:q^5$: etc." E. Bézout[4] and L. Despiau[5] write for arithmetical progression "$\div 1.3.5.7.9$," and "$\div\div 3:6:12$" for geometrical progression.

Symbols for arithmetic progression were less common than for geometric progression, and they were more varied. Oughtred had no symbol. Wallis[6] denotes an arithmetic progression A, B, C, $D \rightleftharpoons$, or by a, b, c, d, e, $f \div\div$. The sign $\div$, which we cited as occurring in Bézout and Despiau, is listed by Saverien[7] who writes "$\div 1.2.3.4.5$, etc." But Saverien gives also the six dots $:::$, which occur in Stone[8] and Wilson.[9] A still different designation, $\div\div$, for arithmetical progression is due to Kirkby[10] and Emerson,[11] another $\div\div\div$ to Clark,[12] again another $\div\div$ is found in Blassière.[13] Among French writers using $\div$ for arithmetic progression and $\div\div$ for geometric progression are Lamy,[14] De Belidor,[15] Suzanne,[16] and Fournier;[17] among Spanish-American

[1] Alexander Malcolm, *A New System of Arithmetick* (London, 1730), p. 115.

[2] Sir Jonas Moore, *Arithmetick in Four Books* (3d ed.; London, 1688), beginning of the Book IV.

[3] John Wilson, *Trigonometry* (Edinburgh, 1714), p. 24.

[4] E. Bézout, *Cours de mathématiques*, Tome I (2. éd.; Paris, 1797), "Arithmétique," p. 130, 165.

[5] *Select Amusements in Philosophy of Mathematics* . . . . translated from the French of M. L. Despiau, Formerly Professor of Mathematics and Philosophy at Paris. . . . . Recommended . . . . by Dr. Hutton (London, 1801), p. 19, 37, 43.

[6] John Wallis, *Operum mathematicorvm Pars Prima* (Oxford, 1657), p. 230, 236.

[7] A. Saverien, *Dictionnaire universel de mathematique et de physique* (Paris, 1753), art. "Caractere."

[8] E. Stone, *New Mathematical Dictionary* (London, 1726), art. "Characters."

[9] John Wilson, *Trigonometry* (Edinburgh, 1714).

[10] John Kirkby, *Arithmetical Institutions containing a compleat System of Arithmetic* (London, 1735), p. 36.

[11] W. Emerson, *Doctrine of Proportion* (1763), p. 27.

[12] Gilbert Clark, *Oughtredus explicatus* (London, 1682), p. 114.

[13] J. J. Blassière, *Institution du calcul numerique et litteral* (a La Haye, 1770), end of Part II.

[14] Bernard Lamy, *Elemens des mathematiques* (3d ed.; Amsterdam, 1692), p. 156.

[15] B. F. de Belidor, *Nouveau Cours de mathématique* (Paris, 1725), p. 71, 139.

[16] H. Suzanne, *De la Manière d' étudier' es Mathématiques* (2. éd.; Paris, 1810), p. 208.

[17] C. F. Fournier, *Éléments d'Arithmétique et d'Algèbre*, Vol. II (Nantes, 1822).

writers using these two symbols are Senillosa,[1] Izquierdo,[2] Liévano,[3] and Porfirio da Motta Pegado.[4] In German publications ÷ for arithmetical progression and ∺ for geometric progression occur less frequently than among the French. In the 1710 publication in the *Miscellanea Berolinensia*[5] ∺ is mentioned in a discourse on symbols (§ 198). The ∺ was used in 1716 by Emanuel Swedenborg.[6]

Emerson[7] designated harmonic progression by the symbol ∺ and harmonic proportion by ∴ .

249. *Arithmetical proportion* finds crude symbolic representation in the *Arithmetic* of Boethius as printed at Augsburg in 1488 (see Figure 103). Being, in importance, subordinate to geometrical proportion, the need of a symbolism was less apparent. But in the seventeenth century definite notations came into vogue. William Oughtred appears to have designed a symbolism. Oughtred's language (*Clavis* [1652], p. 21) is "Ut 7.4:12.9 vel 7.7−3:12.12−3. Arithmeticè proportionales sunt." As later in his work he does not use arithmetical proportion in symbolic analysis, it is not easy to decide whether the symbols just quoted were intended by Oughtred as part of his algebraic symbolism or merely as punctuation marks in ordinary writing. John Newton[8] says: "As 8,5:6,3. Here 8 exceeds 5, as much as 6 exceeds 3."

Wallis[9] says: "Et pariter 5,3; 11,9; 17,15; 19,17. sunt in eadem progressione arithmetica." In P. Chelucci's[10] *Institutiones analyticae*, arithmetical proportion is indicated thus: 6.8 ∵ 10.12. Oughtred's notation is followed in the article "Caractère" of the *Encyclopédie*

---

[1] Felipe Senillosa, *Tratado elemental de Arismetica* (Neuva ed.; Buenos Aires, 1844), p. 46.

[2] Gabriel Izquierdo, *Tratado de Aritmética* (Santiago [Chile], 1859), p. 167.

[3] Indalecio Liévano, *Tratado de Aritmetica* (2. éd.; Bogota, 1872), p. 147.

[4] Luiz Porfirio da Motta Pegado, *Tratade elementar de arithmetica* (2. éd.; Lisboa, 1875), p. 253.

[5] *Miscellanea Berolinensia* (Berolini, 1710), p. 159.

[6] Emanuel Swedberg, *Daedalus hyperboreus* (Upsala, 1716), p. 126. Facsimile reproduction in *Kungliga Vetenskaps Societetens i Upsala Tvåhundraårsminne* (Upsala, 1910).

[7] W. Emerson, *Doctrine of Proportion* (London, 1763), p. 2.

[8] John Newton, *Institutio mathematica or mathematical Institution* (London, 1654), p. 125.

[9] John Wallis, *op. cit.* (Oxford, 1657), p. 229.

[10] Paolino Chelucci, *Institutiones analyticae* (editio post tertiam Romanam prima in Germania; Vienna, 1761), p. 3. See also the first edition (Rome, 1738), p. 1–15.

méthodique (*Mathématiques*) (Paris: Liège, 1784). Lamy[1] says: "Proportion arithmétique, 5,7 ∵ 10,12.c'est à dire qu'il y a même différence entre 5 et 7, qu'entre 10 et 12."

In Arnauld's geometry[2] the same symbols are used for arithmetical progression as for geometrical progression, as in 7.3::13.9 and 6.2::12.4.

Samuel Jeake (1696)[3] speaks of "⦂ Three Pricks or Points, sometimes in disjunct proportion for the words *is as*."

A notation for arithmetical proportion, noticed in two English seventeenth-century texts, consists of five dots, thus ∶∙∶; Richard Balam[4] speaks of "arithmetical disjunct proportionals" and writes "2.4 ∶∙∶3.5"; Sir Jonas Moore[5] uses ∶∙∶ and speaks of "disjunct proportionals." Balam adds, "They may also be noted thus, 2...4 = 3...5." Similarly, John Kirkby[6] designated arithmetrical proportion in this manner, 9..6 = 6..3, the symbolism for arithmetical ratio being 8..2. L'Abbé Deidier (1739)[7] adopts 20.2.∴78.60. Before that Weigel[8] wrote "(o) 3| ∵ 4.7" and "(o) 2.| ∵ 3.5." Wolff (1710),[9] Panchaud,[10] Saverien,[11] L'Abbé Foucher,[12] Emerson,[13] place

[1] B. Lamy, *Elemens des mathematiques* (3. éd.; Amsterdam, 1692), p. 155.

[2] Antoine Arnauld, *Nouveaux elemens de geometrie* (Paris, 1667); also in the edition issued at The Hague in 1690.

[3] Samuel Jeake, ΛΟΓΙΣΤΙΚΗΛΟΓΊΑ or *Arithmetick* (London, 1696; Preface, 1674), p. 10–12.

[4] Richard Balam, *Algebra: or the Doctrine of Composing, Inferring, and Resolving an Equation* (London, 1653), p. 5.

[5] Sir Jonas Moore, *Moore's Arithmetick: In Four Books* (3d ed.; London, 1688), the beginning of Book IV.

[6] Rev. Mr. John Kirkby, *Arithmetical Institutions containing a compleat System of Arithmetic* (London, 1735), p. 27, 28.

[7] L'Abbé Deidier, *L'Arithmétiques des géomètres, ou nouveau élémens de mathématiques* (Paris, 1739), p. 219.

[8] *Erhardi Weigelii Specimina novarum inventionum* (Jenae, 1693), p. 9.

[9] Chr. v. Wolff, *Anfangsgründe aller math. Wissenschaften* (1710), Vol. I, p. 65. See J. Tropfke, *op. cit.*, Vol. III (2d ed., 1922), p. 12.

[10] Benjamin Panchaud, *Entretiens ou leçons mathématiques*, Premier Parti (Lausanne et Genève, 1743), p. vii.

[11] A. Saverien, *Dictionnaire universel* (Paris, 1753), art. "Proportion arithmetique."

[12] L'Abbé Foucher, *Géométrie métaphysique ou essai d'analyse* (Paris, 1758), p. 257

[13] W. Emerson, *The Doctrine of Proportion* (London, 1763), p. 27.

the three dots as did Chelucci and Deidier, viz., $a.b \cdot.\cdot c.d$. Cosalli[1]
writes the arithmetical proportion $a:b \cdot.\cdot c:d$. Later Wolff[2] wrote $a-b$
$=c-d$.

Blassière[3] prefers $2:7 \div 10:15$. Juan Gerard[4] transfers Oughtred's
signs for geometrical proportion to arithmetical proportion and
writes accordingly, $9.7::5.3$. In French, Spanish, and Latin-American texts Oughtred's notation, $8.6:5.3$, for arithmetical proportion
has persisted. Thus one finds it in Benito Bails,[5] in a French text for
the military,[6] in Fournier,[7] in Gabriel Izquierdo,[8] in Indalecio Liévano.[9]

250. *Geometrical proportion.*—A presentation of geometrical proportion that is not essentially rhetorical is found in the Hindu Bakhshālī arithmetic, where the proportion $10:1\frac{63}{60}=4:1\frac{63}{150}$ is written in
the form[10]

10	163	4	*pha* 163
1	60	1	150

It was shown previously (§ 124) that the Arab al-Qalasâdî (fifteenth
century) expresses the proportion $7.12=84:144$ in this manner:
$144 \therefore 84 \therefore 12 \therefore 7$. Regiomontanus in a letter writes our modern
$a:b:c$ in the form $a.b.c$, the dots being simply signs of separation.[11] In
the edition of the *Arithmetica* of Boethius, published at Augsburg in
1488, a crude representation of geometrical and arithmetical propor-

[1] *Scritti inediti del P. D. Pietro Cossali* . . . . pubblicati da B. Boncompagni
(Rome, 1857), p. 75.

[2] Chr. v. Wolff., *op. cit.* (1750), Vol. I, p. 73.

[3] J. J. Blassière, *Institution du calcul numerique et litteral* (a La Haye 1770),
the end of Part II.

[4] Juan Gerard, *Tratado completo de aritmética* (Madrid, 1798), p. 69.

[5] Benito Bails, *Principios de matematica de la real academia de San Fernando*
(2. ed.), Vol. I (Madrid, 1788), p. 135.

[6] *Cours de mathématiques, à l'usage des écoles impériales militaires* ... rédigé
par ordre de M. le Général de Division Bellavène ... (Paris, 1809), p. 52. Dedication signed by "Allaize, Billy, Puissant, Boudrot, Professeurs de mathématiques à
l'Ecole de Saint-Cyr."

[7] C. F. Fournier, *Eléments d'arithmétique et d'algèbre*, Tome II (Nantes, 1842),
p. 87.

[8] Gabriel Izquierdo, *op. cit.* (Santiago [Chile], 1859), p. 155.

[9] Indalecio Liévano, *Tratado aritmetica* (2d ed.; Bogota, 1872), p. 147.

[10] G. R. Kaye, *The Bakhshālī Manuscript*, Parts I and II (Calcutta, 1927),
p. 119.

[11] M. Curtze, *Abhandlungen z. Geschichte d. Mathematik*, Vol. XII (1902),
p. 253.

tion is given, as shown in Figure 103. The upper proportion on the left is geometrical, the lower one on the left is arithmetical. In the latter, the figure 8 plays no part; the 6, 9, and 12 are in arithmetical proportion. The two exhibitions on the right relate to harmonical and musical proportion.

Proportion as found in the earliest printed arithmetic (in Treviso,

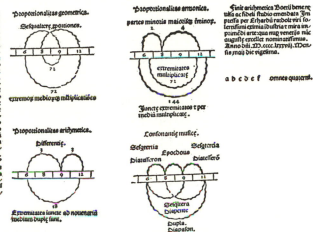

FIG. 103.—From the *Arithmetica* of Boethius, as printed in 1488, the last two pages. (Taken from D. E. Smith's *Rara arithmetica* [Boston, 1898], p. 28.)

1487) is shown in Figure 39. Stifel, in his edition of Rudolff's *Coss* (1553), uses vertical lines of separation, as in

$$\text{``100 } \mid \tfrac{1}{6}\, z \mid 100\ z \mid \text{Facit } \tfrac{1}{6}\, zz \ .\text{''}$$

Tartaglia[1] indicates a proportion thus:

$$\text{``Se £ 3// val } \beta\ 4 \text{ // che valeranno £ 28.''}$$

Chr. Clavius[2] writes:

$$\text{``9 . 126 . 5 . ? fiunt 70 .''}$$

---

[1] N. Tartaglia, *La prima parte del General Tratato di Nvmeri, etc.* (Venice 1556), fol. 129*B*.

[2] Chr. Clavius, *Epitome arithmeticae practicae* (Rome, 1583), p. 137.

This notation is found as late as 1699 in Corachan's arithmetic[1] in such statements as

"$A . B . C . D .$
$5 . 7 . 15 . 21 .$ "

Schwenter[2] marks the geometric proportion 68——51——85, then finds the product of the means $51 \times 85 = 4335$ and divides this by 68. In a work of Galileo,[3] in 1635, one finds:

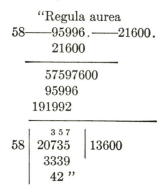

"Regula aurea
58——95996.——21600.
           21600
——————————————
    57597600
    95996
  191992

In other places in Galileo's book the three terms in the proportion are not separated by horizontal lines, but by dots or simply by spacing. Johan Stampioen,[4] in 1639, indicates our $a:b=b:c$ by the symbolism:

"$a,, b$ gel $: b ,, c .$"

Further illustrations are given in § 221.

These examples show that some mode of presenting to the eye the numbers involved in a geometric proportion, or in the application of the rule of three, had made itself felt soon after books on mathematics came to be manufactured. Sometimes the exposition was rhetorical, short words being available for the writing of proportion. As late as 1601 Philip Lansberg[5] wrote "ut 5 ad 10; ita 10 ad 20," meaning

[1] Ivan Bavtista Corachan, *Arithmetica demonstrada* (Valencia, 1699), p. 199.

[2] Daniel Schwenter, *Geometriae practicae novae et auctae tractatus* (Nürnberg, 1623), p. 89.

[3] *Systema Cosmicvm, aucthore Galilaeo Galilaei. . . . . Ex Italica lingua latine conversum* (Florence, 1635), p. 294.

[4] Johan Stampioen, *Algebra ofte nieuwe Stel-Regel* (The Hague, 1639), p. 343.

[5] Philip Lansberg, *Triangulorum geometriae libri quatuor* (Middelburg [Zeeland], 1663), p. 5.

5:10 = 10:20. Even later the Italian Cardinal Michelangelo Ricci[1] wrote "esto AC ad CB, ut 9 ad 6." If the fourth term was not given, but was to be computed from the first three, the place for the fourth term was frequently left vacant, or it was designated by a question mark.

251. *Oughtred's notation.*—As the symbolism of algebra was being developed and the science came to be used more extensively, the need for more precise symbolism became apparent. It has been shown (§ 181) that the earliest noted symbolism was introduced by Oughtred. In his *Clavis mathematicae* (London, 1631) he introduced the notation $5.10::6.12$ which he retained in the later editions of this text, as well as in his *Circles of Proportion* (1632, 1633, 1660), and in his *Trigonometria* (1657).

As previously stated (§ 169) the suggestion for this symbolism may have come to Oughtred from the reading of John Dee's Introduction to Billingley's *Euclid* (1570). Probably no mathematical symbol has been in such great demand in mathematics as the dot. It could be used, conveniently, in a dozen or more different meanings. But the avoidance of confusion necessitates the restriction of its use. Where then shall it be used, and where must other symbols be chosen? Oughtred used the dot to designate ratio. That made it impossible for him to follow John Napier in using the dot as the separatrix in decimal fractions. Oughtred could not employ two dots (:) for ratio, because the two dots were already pre-empted by him for the designation of aggregation, $:A+B:$ signifying $(A+B)$. Oughtred reserved the dot for the writing of ratio, and used four dots to separate the two equal ratios. The four dots were an unfortunate selection. The sign of equality (=) would have been far superior. But Oughtred adhered to his notation. Editions of his books containing it appeared repeatedly in the seventeenth century. Few symbols have met with more prompt adoption than those of Oughtred for proportion. Evidently the time was ripe for the introduction of a definite unambiguous symbolism. To be sure the adoption was not immediate. Nineteen years elapsed before another author used the notation $A.B::C.D$. In 1650 John Kersey brought out in London an edition of Edmund Wingate's *Arithmetique made easie*, in which this notation is used. After this date, the publications employing it became frequent, some of them being the productions of pupils of Oughtred. We have

[1] *Michaelis Angeli Riccii exercitatio geometrica de maximis et minimis* (London, 1668), p. 3.

seen it in Vincent Wing,[1] Seth Ward,[2] John Wallis,[3] in "R.B.," a schoolmaster in Suffolk,[4] Samuel Foster,[5] Sir Jonas Moore,[6] and Isaac Barrow.[7] John Wallis[8] sometimes uses a peculiar combination of processes, involving the simplification of terms, during the very act of writing proportion, as in "$\frac{8}{2}A = 4A \cdot \frac{6}{2}A = 3A :: \frac{4}{2}A = 2A \cdot \frac{2}{2}A :: 8 \cdot 6 ::$ 4.3." Here the dot signifies ratio.

The use of the dot, as introduced by Oughtred, did not become universal even in England. As early as 1651 the astronomer, Vincent Wing (§ 244), in his *Harmonicon Coeleste* (London), introduced the colon (:) as the symbol for ratio. This book uses, in fact, both notations for ratio. Many times one finds $A \cdot B :: C \cdot D$ and many times $A : B :: C : D$. It may be that the typesetter used whichever notation happened at the moment to strike his fancy. Later, Wing published three books (§ 244) in which the colon (:) is used regularly in writing ratios. In 1655 another writer, "R.B.," whom we have cited as using the symbols $A \cdot B :: C \cdot D$, employed in the same publication also $A : B :: C : D$. The colon was adopted in 1661 by Thomas Streete.[9]

That Oughtred himself at any time voluntarily used the colon as the sign for ratio does not appear. In the editions of his *Clavis* of 1648 and 1694, the use of : to signify ratio has been found to occur only once in each copy (§ 186); hence one is inclined to look upon this notation in these copies as printer's errors.

252. *Struggle in England between Oughtred's and Wing's notations, before 1700.*—During the second half of the seventeenth century there was in England competition between (.) and (:) as the symbols for the designation of the ratio (§§ 181, 251). At that time the dot maintained its ascendancy. Not only was it used by the two most influ-

---

[1] Vincent Wing, *Harmonicon coeleste* (London, 1651), p. 5.

[2] Seth Ward, *In Ismaelis Bullialdi astronomiae philolaicae fundamenta inquisitio brevis* (Oxford, 1653), p. 7.

[3] John Wallis, *Elenchus geometriae Hobbianae* (Oxford, 1655), p. 48; *Operum mathematicorum pars altera* (Oxford, 1656), the part on *Arithmetica infinitorum*, p. 181.

[4] *An Idea of Arithmetick, at first designed for the use of the Free Schoole at Thurlow in Suffolk. . . . .* By R. B., Schoolmaster there (London, 1655), p. 6.

[5] *Miscellanies: or mathematical Lucrubations of Mr. Samuel Foster . . . .* by John Twyden (London, 1659), p. 1.

[6] Jonas Moore, *Arithmetick in two Books* (London, 1660), p. 89; Moore's *Arithmetique in Four Books* (3d ed.; London, 1688), Book IV, p. 461.

[7] Isaac Barrow's edition of *Euclid's Data* (Cambridge, 1657), p. 2.

[8] John Wallis, *Adversus Marci Meibomii de Proportionibus Dialogum* (Oxford, 1657), "Dialogum," p. 54.

[9] Thomas Streete, *Astronomia Carolina* (1661). See J. Tropfke, *Geschichte der Elementar-Mathematik*, 3. Bd., 2. Aufl. (Berlin und Leipzig, 1922), p. 12.

ential English mathematicians before Newton, namely, John Wallis and Isaac Barrow, but also by David Gregory,[1] John Craig,[2] N. Mercator,[3] and Thomas Brancker.[4] I. Newton, in his letter to Oldenburg of October 24, 1676,[5] used the notation . :: . , but in Newton's *De analysi per aequationes terminorum infinitas,* the colon is employed to designate ratio, also in his *Quadratura curvarum.*

Among seventeenth-century English writers using the colon to mark ratio are James Gregory,[6] John Collins,[7] Christopher Wren,[8] William Leybourn,[9] William Sanders,[10] John Hawkins,[11] Joseph Raphson,[12] E. Wells,[13] and John Ward.[14]

253. *Struggle in England between Oughtred's and Wing's notations during 1700–1750.*—In the early part of the eighteenth century, the dot still held its place in many English books, but the colon gained in ascendancy, and in the latter part of the century won out. The single dot was used in John Alexander's *Algebra* (in which proportion is written in the form $a \, . \, b :: c \, . \, X$ and also in the form $a \overline{..} b : c \overline{..} X)$[15] and, in John Colson's translation of Agnesi (before 1760).[16] It was used

[1] David Gregory in *Phil. Trans.,* Vol. XIX (1695–97), p. 645.

[2] John Craig, *Methodus figurarum lineis rectis et curvis* (London, 1685). Also his *Tractatus mathematicus* (London, 1693), but in 1718 he often used : :: . in his *De Calculo Fluentium Libri Duo,* brought out in London.

[3] N. Mercator, *Logarithmotechnia* (London, 1668), p. 29.

[4] Th. Brancker, *Introduction to Algebra* (trans. of Rhonius; London, 1668), p. 37.

[5] John Collins, *Commercium epistolicum* (London, 1712), p. 182.

[6] James Gregory, *Vera circuli et hyperbolae quadratura* (Patavia, 1668), p. 33.

[7] J. Collins, *Mariners Plain Scale New Plain'd* (London, 1659).

[8] *Phil. Trans.,* Vol. III (London), p. 868.

[9] W. Leybourn, *The Line of Proportion* (London, 1673), p. 14.

[10] William Sanders, *Elementa geometriae* (Glasgow, 1686), p. 3.

[11] *Cocker's Decimal Arithmetick* . . . . perused by John Hawkins (London, 1695) (Preface dated 1684), p. 41.

[12] J. Raphson, *Analysis aequationum universalis* (London, 1697), p. 26.

[13] E. Wells, *Elementa arithmeticae numerosae et speciosae* (Oxford, 1698), p. 107.

[14] John Ward, *A Compendium of Algebra* (2d ed.; London, 1698), p. 62.

[15] *A Synopsis of Algebra.* Being the Posthumous Work of John Alexander, of Bern in Swisserland. To which is added an Appendix by Humfrey Ditton. . . . . Done from the Latin by Sam. Cobb M.A. (London, 1709), p. 16. The Latin edition appeared at London in 1693.

[16] Maria Gaetana Agnesi, *Analytical Institutions,* translated into English by the late Rev. John Colson. . . . . Now first printed . . . . under the inspection of Rev. John Hellins (London, 1801).

by John Wilson[1] and by the editors of Newton's *Universal arithmetick*.[2] In John Harris' *Lexicon technicum* (1704) the dot is used in some articles, the colon in others, but in Harris' translation[3] of G. Pardies' geometry the dot only is used. George Shelley[4] and Hatton[5] used the dot.

254. *Sporadic notations.*—Before the English notations . :: . and : :: : were introduced on the European Continent, a symbolism consisting of vertical lines, a modification of Tartaglia's mode of writing, was used by a few continental writers. It never attained popularity, yet maintained itself for about a century. René Descartes (1619–21)[6] appears to have been the first to introduce such a notation $a|b||c|d$. In a letter[7] of 1638 he replaces the middle double stroke by a single one. Slusius[8] uses single vertical lines in designating four numbers in geometrical proportion, $p|a|e|d-a$. With Slusius, two vertical strokes $||$ signify equality. Jaques de Billy[9] marks five quantities in continued proportion, thus $3-R5|R5-1|2|R5+1|3+R5$, where R means "square root." In reviewing publications of Huygens and others, the original notation of Descartes is used in the *Journal des Sçavans* (Amsterdam)[10] for the years 1701, 1713, 1716. Likewise, Picard,[11] De la Hire,[12] Abraham de Graaf,[13] and Parent[14] use the notation $a|b||xx|ab$.

[1] John Wilson, *Trigonometry* (Edinburgh, 1714), p. 24.

[2] I. Newton, *Arithmetica universalis* (ed. W. Whiston; Cambridge, 1707), p. 9; *Universal Arithmetick*, by Sir Isaac Newton, translated by Mr. Ralphson . . . . revised . . . . by Mr. Cunn (London, 1769), p. 17.

[3] *Plain Elements of Geometry and Plain Trigonometry* (London, 1701), p. 63.

[4] G. Shelley, *Wingate's Arithmetick* (London, 1704), p. 343.

[5] Edward Hatton, *An Intire System of Arithmetic* (London, 1721), p. 93.

[6] *Œuvres des Descartes* (éd. Adam et Tannery), Vol. X, p. 240.

[7] *Op. cit.*, Vol. II, p. 171.

[8] *Renati Francisci Slusii mesolabum seu duae mediae proportionales, etc.* (1668), p. 74. See also Slusius' reply to Huygens in *Philosophical Transactions* (London), Vols. III–IV (1668–69), p. 6123.

[9] Jaques de Billy, *Nova geometriae clavis* (Paris, 1643), p. 317.

[10] *Journal des Sçavans* (Amsterdam, année 1701), p. 376; *ibid.* (année 1713), p. 140, 387; *ibid.* (année 1716), p. 537.

[11] J. Picard in *Mémoires de l'Académie r. des sciences* (depuis 1666 jusqu'à 1699), Tome VI (Paris, 1730), p. 573.

[12] De la Hire, *Nouveaux elemens des sections coniques* (Paris, 1701), p. 184. J. Tropfke refers to the edition of 1679, p. 184.

[13] Abraham de Graaf, *De vervulling van der geometria en algebra* (Amsterdam, 1708), p. 97.

[14] A. Parent, *Essais et recherches de mathematique et de physique* (Paris, 1713), p. 224.

It is mentioned in the article "Caractere" in Diderot's *Encyclopédie* (1754). La Hire writes also "$aa\,||\,xx\,||\,ab$" for $a^2:x^2=x^2:ab$.

On a subject of such universal application in commercial as well as scientific publications as that of ratio and proportion, one may expect to encounter occasional sporadic attempts to alter the symbolism. Thus Hérigone[1] writes "$hg\ \pi\ ga\ 2\,|\,2\ hb\ \pi\ bd,\ signifi.$ *HG* est ad *GA*, *vt* *HB* ad *BD*," or, in modern notation, $hg:ga=hb:bd;$ here $2\,|\,2$ signifies equality, π signifies ratio. Again Peter Mengol,[2] of Bologna, writes "$a;r:a2;ar$" for $a:r=a^2:ar$. The London edition of the algebra of the Swiss J. Alexander[3] gives the signs . :: . but uses more often designations like $b\mathbin{\,\dot{-}\,}a:d\ \Big|\ \dfrac{ad}{b}$. Ade Mercastel,[4] of Rouen, writes $2\,,,3\,;;8\,,,12$. A close approach to the marginal symbolism of John Dee is that of the Spaniard Zaragoza[5] $4.3:12.9$. More profuse in the use of dots is J. Kresa[6] who writes $x\ldots r::r\ldots\dfrac{rr}{x}$, also $AE..EF::AD..DG$. The latter form is adopted by the Spaniard Cassany[7] who writes $128..119$ $::3876$; it is found in two American texts,[8] of 1797.

In greater conformity with pre-Oughtredian notations is van Schooten's notation[9] of 1657 when he simply separates the three given numbers by two horizontal dashes and leaves the place for the unknown number blank. Using Stevin's designation for decimal fractions, he writes "65————$95,753\,\text{③}-1$." Abraham de Graaf[10] is

---

[1] Pierre Herigone, *Cvrsvs mathematici* (Paris, 1644), Vol. VI, "Explicatio notarum." The first edition appeared in 1642.

[2] Pietro Mengoli, *Geometriae speciosae elementa* (Bologna, 1659), p. 8.

[3] *Synopsis algebraica*, Opus posthumum Johannis Alexandri, Bernatis-Helvetii (London, 1693), p. 135.

[4] Jean Baptiste Adrien de Mercastel, *Arithmétique démontrée* (Rouen, 1733), p. 99.

[5] Joseph Zaragoza, *Arithmetica vniversal* (Valencia, 1669), p. 48.

[6] Jacob Kresa, *Analysis speciosa trigonometriae sphericae* (Prague, 1720), p. 120, 121.

[7] Francisco Cassany, *Arithmetica Deseada* (Madrid, 1763), p. 102.

[8] *American Tutor's Assistant*. By sundry teachers in and near Philadelphia (3d ed.; Philadelphia, 1797), p. 57, 58, 62, 91–186. In the "explanation of characters," : :: : is given. The second text is Chauncey Lee's *American Accomptant* (Lansingburgh, 1797), where one finds (p. 63) $3..5::6..10$.

[9] Francis à Schooten, Leydensis, *Exercitationum mathematicarum liber primus* (Leyden, 1657), p. 19.

[10] Abraham de Graaf, *De Geheele mathesis of wiskonst* (Amsterdam, 1694), p. 16.

partial to the form $2-4=6-12$. Thomas York[1] uses three dashes 125—429—10—?, but later in his book writes "33600 7 :: 153600 32," the ratio being here indicated by a blank space. To distinguish ratios from fractions, Samuel Jeake[2] states that by some authors "instead of the intervening Line, two Pricks are set; and so the *Ratio sesquialtera* is thus expressed $\overset{3}{\underset{2}{:}}$." Accordingly, Jeake writes "$\overset{1}{\underset{7}{:}} \cdot \overset{1}{\underset{9}{:}}$ :: 9.7."

In practical works on computation with logarithms, and in some arithmetics a rhetorical and vertical arrangement of the terms of a proportion is found. Mark Forster[3] writes:

"As Sine of 40 deg.   9,8080675
To 1286   3,1092401
So is Radius   10,0000000
To the greatest Random 2000 3,3011726
Or, For Random at 36 deg."

As late as 1789 Benjamin Workman[4] writes "$\dfrac{\text{lb.} \quad \text{d.} \quad \text{lb.}}{1-7-112}$."

255. *Oughtred's notation on the European Continent.*—On the European Continent the dot as a symbol of geometrical ratio, and the four dots of proportion, . :: ., were, of course, introduced later than in England. They were used by Dulaurens,[5] Prestet,[6] Varignon,[7] Pardies,[8] De l'Hospital,[9] Jakob Bernoulli,[10] Johann Bernoulli,[11] Carré,[12] Her-

[1] Thomas York, *Practical Treatise of Arithmetick* (London, 1687), p. 132, 146.

[2] Samuel Jeake, ΛΟΓΙΣΤΙΚΗΛΟΓΊΑ, or *Arithmetick* (London, 1696 [Preface, 1674]), p. 411.

[3] Mark Forster, *Arithmetical Trigonometry* (London, 1690), p. 212.

[4] Benjamin Workman, *American Accountant* (Philadelphia, 1789), p. 62.

[5] Francisci Dulaurens, *Specimina mathematica* (Paris, 1667), p. 1.

[6] Jean Prestet, *Elemens des mathematiques* (Preface signed "J.P.") (Paris, 1675), p. 240. Also *Nouveaux elemens des mathematiques*, Vol. I (Paris, 1689), p. 355.

[7] P. Varignon in *Journal des Sçavans*, année 1687 (Amsterdam, 1688), p. 644. Also Varignon, *Eclaircissemens sur l'analyse des infiniment petits* (Paris, 1725), p. 16.

[8] *Œuvres du R. P. Ignace-Gaston Pardies* (Lyon, 1695), p. 121.

[9] De l'Hospital, *Analyse des infiniment petits* (Paris, 1696), p. 11.

[10] Jakob Bernoulli in *Acta eruditorum* (1687), p. 619 and many other places.

[11] Johann Bernoulli in *Histoire de l'académie r. des sciences*, année 1732 (Paris, 1735), p. 237.

[12] L. Carré, *Methode pour la Mesure des Surfaces* (Paris, 1700), p. 5.

mann,[1] and Rolle;[2] also by De Reaumur,[3] Saurin,[4] Parent,[5] Nicole,[6] Pitot,[7] Poleni,[8] De Mairan,[9] and Maupertuis.[10] By the middle of the eighteenth century, Oughtred's notation $A.B::C.D$ had disappeared from the volumes of the Paris Academy of Sciences, but we still find it in textbooks of Belidor,[11] Guido Grandi,[12] Diderot,[13] Gallimard,[14] De la Chapelle,[15] Fortunato,[16] L'Abbé Foucher,[17] and of Abbé Girault de Koudou.[18] This notation is rarely found in the writings of German authors. Erhard Weigel[19] used it in a philosophical work of 1693. Christian Wolf[20] used the notation "$DC.AD::EC.ME$" in 1707, and in 1710 "$3.12::5.20$" and also "$3:12=5:20$." Beguelin[21] used the dot for ratio in 1773. From our data it is evident that $A.B::C.D$ began

[1] J. Hermann in *Acta eruditorum* (1702), p. 502.

[2] M. Rolle in *Journal des Sçavans*, année 1702 (Amsterdam, 1703), p. 399.

[3] R. A. F. de Reaumur, *Histoire de l'académie r. des sciences*, année 1708 (Paris, 1730), "Mémoires," p. 209, but on p. 199 he used also the notation : :: :.

[4] J. Saurin, *op. cit.*, année 1708, "Mémoires," p. 26.

[5] Antoine Parent, *op. cit.*, année 1708, "Mémoires," p. 118.

[6] F. Nicole, *op. cit.*, année 1715 (Paris, 1741), p. 50.

[7] H. Pitot, *op. cit.*, année 1724 (Paris, 1726), "Mémoires," p. 109.

[8] Joannis Poleni, *Epistolarvm mathematicarvm Fascicvlvs* (Patavii, 1729).

[9] J. J. de Mairan, *Histoire de l'académie r. des sciences*, année 1740 (Paris, 1742), p. 7.

[10] P. L. Maupertuis, *op. cit.*, année 1731 (Paris, 1733), "Mémoires," p. 465.

[11] B. F. de Belidor, *Nouveau Cours de mathématique* (Paris, 1725), p. 481.

[12] Guido Grandi, *Elementi geometrici piani e solide de Euclide* (Florence, 1740).

[13] Denys Diderot, *Mémoires sur différens sujets de Mathématiques* (Paris, 1748), p. 16.

[14] J. E. Gallimard, *Géométrie élémentaire d'Euclide* (nouvelle éd.; Paris, 1749), p. 37.

[15] De la Chapelle, *Traité des sections coniques* (Paris, 1750), p. 150.

[16] F. Fortunato, *Elementa matheseos* (Brescia, 1750), p. 35.

[17] L'Abbé Foucher, *Géometrie métaphysique ou Essai d'analyse* (Paris, 1758), p. 257.

[18] L'Abbé Girault de Koudou, *Leçons analytiques du calcul des fluxions et des fluentes* (Paris, 1767), p. 35.

[19] *Erhardi Weigelii Philosophia mathematica* (Jenae, 1693), "Specimina novarum inventionum," p. 6, 181.

[20] C. Wolf in *Acta eruditorum* (1707), p. 313; Wolf, *Anfangsgründe aller mathematischen Wissenschaften* (1710), Band I, p. 65, but later Wolf adopted the notation of Leibniz, viz., $A:B=C:D$. See J. Tropfke, *Geschichte der Elementar-Mathematik*, Vol. III (2d ed.; Berlin und Leipzig, 1922), p. 13, 14.

[21] Nicolas de Beguelin in *Nouveaux mémoires de l'académie r. des sciences et belles-lettres*, année 1773 (Berlin, 1775), p. 211.

to be used in the Continent later than in England, and it was also later to disappear on the Continent.

256. An unusual departure in the notation for geometric proportion which involved an excellent idea was suggested by a Dutch author, Johan Stampioen,[1] as early as the year 1639. This was only eight years after Oughtred had proposed his . :: . Stampioen uses the designation $A,,B=C,,D$. We have noticed, nearly a century later, the use of two commas to represent ratio, in a French writer, Mercastel. But the striking feature with Stampioen is the use of Recorde's sign of equality in writing proportion. Stampioen anticipates Leibniz over half a century in using = to express the equality of two ratios. He is also the earliest writer that we have seen on the European Continent to adopt Recorde's symbol in writing ordinary equations. He was the earliest writer after Descartes to use the exponential form a^3. But his use of = did not find early followers. He was an opponent of Descartes whose influence in Holland at that time was great. The employment of = in writing proportion appears again with James Gregory[2] in 1668, but he found no followers in this practice in Great Britain.

257. *Slight modifications of Oughtred's notation.*—A slight modification of Oughtred's notation, in which commas took the place of the dots in designating geometrical ratios, thus $A,B::C,D$, is occasionally encountered both in England and on the Continent. Thus Sturm[3] writes "$3b,2b::2b,\frac{4bb}{3b}$ sive $\frac{4b}{3}$," Lamy[4] "$3,6::4,8$," as did also Ozanam,[5] De Moivre,[6] David Gregory,[7] L'Abbé Deidier,[8] Belidor,[9] who also uses the regular Oughtredian signs, Maria G. Agnesi,[10]

[1] Johan Stampioen d'Jonghe, *Algebra ofte Nieuwe Stel-Regel* ('s Graven-Haye, 1639).

[2] James Gregory, *Geometriae Pars Vniversalis* (Padua, 1668), p. 101.

[3] Christopher Sturm in *Acta eruditorum* (Leipzig, 1685), p. 260.

[4] R. P. Bernard Lamy, *Elemens des mathematiques*, troisième edition revue et augmentée sur l'imprismé à Paris (Amsterdam, 1692), p. 156.

[5] J. Ozanam, *Traité des lignes du premier genre* (Paris, 1687), p. 8; Ozanam, *Cours de mathématique*, Tome III (Paris, 1693), p. 139.

[6] A. de Moivre in *Philosophical Transactions*, Vol. XIX (London, 1698), p. 52; De Moivre, *Miscellanea analytica de seriebus* (London, 1730), p. 235.

[7] David Gregory, *Acta eruditorum* (1703), p. 456.

[8] L'Abbé Deidier, *La Mesure des Surfaces et des Solides* (Paris, 1740), p. 181.

[9] B. F. de Belidor, *op. cit.* (Paris, 1725), p. 70.

[10] Maria G. Agnesi, *Instituzioni analitiche*, Tome I (Milano, 1748), p. 76.

Nicolaas Ypey,[1] and Manfredi.[2] This use of the comma for ratio, rather than the Oughtredian dot, does not seem to be due to any special cause, other than the general tendency observable also in the notation for decimal fractions, for writers to use the dot and comma more or less interchangeably.

An odd designation occurs in an English edition of Ozanam,[3] namely, "$A.2.B.3::C.4.D.6$," where A,B,C,D are quantities in geometrical proportion and the numbers are thrown in by way of concrete illustration.

258. *The notation* : :: : *in Europe and America.*—The colon which replaced the dot as the symbol for ratio was slow in making its appearance on the Continent. It took this symbol about half a century to cross the British Channel. Introduced in England by Vincent Wing in 1651, its invasion of the Continent hardly began before the beginning of the eighteenth century. We find the notation $A:B::C:D$ used by Leibniz,[4] Johann Bernoulli,[5] De la Hire,[6] Parent,[7] Bomie,[8] Saulmon,[9] Swedenborg,[10] Lagny,[11] Senès,[12] Chevalier de Louville,[13] Clairaut,[14] Bouguer,[15] Nicole (1737, who in 1715 had used . :: .),[16] La

---

[1] Nicolaas Ypey, *Grondbeginselen der Keegelsneeden* (Amsterdam, 1769), p. 3.

[2] Gabriello Manfredi, *De Constructione Aequationum differentialium primi gradus* (1707), p. 123.

[3] J. Ozanam, *Cursus mathematicus*, translated "by several Hands" (London, 1712), Vol. I, p. 199.

[4] *Acta eruditorum* (1684), p. 472.

[5] Johanne (I) Bernoulli in *Journal des Sçavans*, année 1698 (Amsterdam, 1709), p. 576. See this notation used also in l'année 1791 (Amsterdam, 1702), p. 371.

[6] De la Hire in *Histoire de l'académie r. des sciences*, année 1708 (Paris, 1730), "Mémoires," p. 57.

[7] A. Parent in *op. cit.*, année 1712 (Paris, 1731), "Mémoires," p. 98.

[8] Bomie in *op. cit.*, p. 213.

[9] Saulmon in *op. cit.*, p. 283.

[10] Emanuel Swedberg, *Daedalus Hyperboreus* (Upsala, 1716).

[11] T. F. Lagny in *Histoire de l'académie r. des sciences*, année 1719 (Paris, 1721), "Mémoires," p. 139.

[12] Dominique de Senès in *op. cit.*, p. 363.

[13] De Louville in *op. cit.*, année 1724 (Paris, 1726), p. 67.

[14] Clairaut in *op. cit.*, année 1731 (Paris, 1733), "Mémoires," p. 484.

[15] Pierre Bougver in *op. cit.*, année 1733 (Paris, 1735), "Mémoires," p. 89.

[16] F. Nicole in *op. cit.*, année 1737 (Paris, 1740), "Mémoires," p. 64.

Caille,[1] D'Alembert,[2] Vicenti Riccati,[3] and Jean Bernoulli.[4] In the Latin edition of De la Caille's[5] *Lectiones* four notations are explained, namely, $3.12::2.8$, $3:12::2:8$, $3:12=2:8$, $3|12||2|8$, but the notation $3:12::2:8$ is the cne actually adopted.

The notation : :: : was commonly used in England and the United States until the beginning of the twentieth century, and even now in those countries has not fully surrendered its place to $:=:$. As late as 1921 : :: : retains its place in Edwards' *Trigonometry*,[6] and it occurs in even later publications. The : :: : gained full ascendancy in Spain and Portugal, and in the Latin-American countries. Thus it was used in Madrid by Juan Gerard,[7] in Lisbon by Joao Felix Pereira[8] and Luiz Porfirio da Motta Pegado,[9] in Rio de Janeiro in Brazil by Francisco Miguel Pires[10] and C. B. Ottoni,[11] at Lima in Peru by Maximo Vazquez[12] and Luis Monsante,[13] at Buenos Ayres by Florentino Garcia,[14] at Santiage de Chile by Gabriel Izquierdo,[15] at Bogota in Colombia by Indalecio Liévano,[16] at Mexico by Juan Maria Balbontin.[17]

[1] La Caille in *op. cit.*, année 1741 (Paris, 1744), p. 256.

[2] D'Alembert in *op. cit.*, année 1745 (Paris, 1749), p. 367.

[3] Vincenti Riccati, *Opusculorum ad res physicas et mathematicas pertinentium. Tomus primus* (Bologna, 1757), p. 5.

[4] Jean Bernoulli in *Nouveaux mémoires de l'académie r. des sciences et belles-lettres*, année 1771 (Berlin, 1773), p. 286.

[5] N. L. de la Caille, *Lectiones elementares mathematicae* . . . . in Latinum traductae et ad editionem Parisinam anni MDCCLIX denuo exactae a C [arolo] S [cherffer] e S. J. (Vienna, 1762), p. 76.

[6] R. W. K. Edwards, *An Elementary Text-Book of Trigonometry* (new ed.; London, 1921), p. 152.

[7] Juan Gerard, Presbitero, *Tratado completo de aritmética* (Madrid, 1798), p. 69.

[8] J. F. Pereira, *Rudimentos de arithmetica* (Quarta Edição; Lisbon, 1863), p. 129.

[9] Luiz Porfirio da Motta Pegado, *Tratado elementar de arithmetica* (Secunda edição; Lisbon, 1875), p. 235.

[10] Francisco Miguel Pires, *Tratado de Trigonometria Espherica* (Rio de Janeiro, 1866), p. 8.

[11] C. B. Ottoni, *Elementos de geometria e trigonometria rectilinea* (4th ed.; Rio de Janeiro, 1874), "Trigon.," p. 36.

[12] Maximo Vazquez, *Aritmetica practica* (7th ed.; Lima, 1875), p. 130.

[13] Luis Monsante, *Lecciones de aritmetica demostrada* (7th ed.; Lima, 1872), p. 171.

[14] Florentino Garcia, *El aritmética Argentino* (5th ed.; Buenos Aires, 1871), p. 41; first edition, 1833.

[15] Gabriel Izquierdo, *Tratado de aritmética* (Santiago, 1859), p. 157.

[16] Indalecio Liévano, *Tratado de aritmetica* (2d ed.; Bogota, 1872), p. 148.

[17] Juan Maria Balbontin, *Tratado elemental de aritmetica* (Mexico, 1878), p. 96.

259. *The notation of Leibniz.*—In the second half of the eighteenth century this notation, $A:B::C:D$, had gained complete ascendancy over $A.B::C.D$ in nearly all parts of Continental Europe, but at that very time it itself encountered a serious rival in the superior Leibnizian notation, $A:B=C:D$. If a proportion expresses the equality of ratios, why should the regular accepted equality sign not be thus extended in its application? This extension of the sign of equality = to writing proportions had already been made by Stampioen (§ 256). Leibniz introduced the colon (:) for ratio and for division in the *Acta eruditorum* of 1684, page 470 (§ 537). In 1693 Leibniz expressed his disapproval of the use of special symbols for ratio and proportion, for the simple reason that the signs for division and equality are quite sufficient. He[1] says: "Many indicate by $a \div b \div\!\!\div c \div d$ that the ratios a to b and c to d are alike. But I have always disapproved of the fact that special signs are used in ratio and proportion, on the ground that for ratio the sign of division suffices and likewise for proportion the sign of equality suffices. Accordingly, I write the ratio a to b thus: $a:b$ or $\frac{a}{b}$ just as is done in dividing a by b. I designate proportion, or the equality of two ratios by the equality of the two divisions or fractions. Thus when I express that the ratio a to b is the same as that of c to d, it is sufficient to write $a:b=c:d$ or $\frac{a}{b}=\frac{c}{d}$."

Cogent as these reasons are, more than a century passed before his symbolism for ratio and proportion triumphed over its rivals.

Leibniz's notation, $a:b=c:d$, is used in the *Acta eruditorum* of 1708, page 271. In that volume (p. 271) is laid the editorial policy that in algebra the Leibnizian symbols shall be used in the *Acta*. We quote the following relating to division and proportion (§ 197): "We shall designate division by two dots, unless circumstance should prompt adherence to the common practice. Thus, we shall have $a:b=\frac{a}{b}$. Hence with us there will be no need of special symbols for denoting proportion. For instance, if a is to b as c is to d, we have $a:b=c:d$."

The earliest influential textbook writer who adopted Leibniz' notation was Christian Wolf. As previously seen (§ 255) he sometimes

[1] G. W. Leibniz, *Matheseos universalis pars prior*, de Terminis incomplexis, No. 16; reprinted in *Gesammelte Werke* (C. I. Gerhardt), 3. Folge, II³, Band VII (Halle, 1863), p. 56.

wrote $a.b = c.d$. In 1710[1] he used both $3.12 :: 5.20$ and $3:12 = 5:20$, but from 1713[2] on, the Leibnizian notation is used exclusively.

One of the early appearances of $a:b = c:d$ in France is in Clairaut's algebra[3] and in Saverien's dictionary,[4] where Saverien argues that the equality of ratios is best indicated by $=$ and that $::$ is superfluous. It is found in the publications of the Paris Academy for the year 1765,[5] in connection with Euler who as early as 1727 had used it in the commentaries of the Petrograd Academy.

Benjamin Panchaud brought out a text in Switzerland in 1743,[6] using $: = :$. In the Netherlands[7] it appeared in 1763 and again in 1775.[8] A mixture of Oughtred's symbol for ratio and the $=$ is seen in Pieter Venema[9] who writes $. = .$.

In Vienna, Paulus Mako[10] used Leibniz' notation both for geometric and arithmetic proportion. The Italian Petro Giannini[11] used $: = :$ for geometric proportion, as does also Paul Frisi.[12] The first volume of *Acta Helvetia*[13] gives this symbolism. In Ireland, Joseph Fenn[14] used it about 1770. A French edition of Thomas Simpson's geometry[15] uses $: = :$. Nicolas Fuss[16] employed it in St. Petersburgh. In England,

[1] Chr. Wolf, *Anfangsgründe aller mathematischen Wissenschaften* (Magdeburg, 1710), Vol. I, p. 65. See J. Tropfke, *Geschichte der Elementar-Mathematik*, Vol. III (2d ed.; Berlin and Leipzig, 1922), p. 14.

[2] Chr. Wolf, *Elementa matheseos universae*, Vol. I (Halle, 1713), p. 31.

[3] A. C. Clairaut, *Elemens d'algebre* (Paris, 1746), p. 21.

[4] A. Saverien, *Dictionnaire universel de mathematique et physique* (Paris, 1753), arts. "Raisons semblables," "Caractere."

[5] *Histoire de l'académie r. des sciences*, année 1765 (Paris, 1768), p. 563; *Commentarii academiae scientiarum . . . .* ad annum 1727 (Petropoli, 1728), p. 14.

[6] Benjamin Panchaud, *Entretiens ou leçons mathématiques* (Lausanne, Genève, 1743), p. 226.

[7] A. R. Maudvit, *Inleiding tot de Keegel-Sneeden* (Shaage, 1763).

[8] J. A. Fas, *Inleiding tot de Kennisse en het Gebruyk der Oneindig Kleinen* (Leyden, 1775), p. 80.

[9] Pieter Venema, *Algebra ofte Stel-Konst*, Vierde Druk (Amsterdam, 1768), p. 118.

[10] Pavlvs Mako, *Compendiaria matheseos institutio* (editio altera; Vindobonae, 1766), p. 169, 170.

[11] Petro Giannini, *Opuscola mathematica* (Parma, 1773), p. 74.

[12] *Paulli Frisii Operum*, Tomus Secundus (Milan, 1783), p. 284.

[13] *Acta Helvetica, physico-mathematico-Botanico-Medica*, Vol. I (Basel, 1751), p. 87.

[14] Joseph Fenn, *The Complete Accountant* (Dublin, [n.d.]), p. 105, 128.

[15] Thomas Simpson, *Elémens de géométrie* (Paris, 1766).

[16] Nicolas Fuss, *Leçons de géométrie* (St. Pétersbourg, 1798), p. 112.

John Cole[1] adopted it in 1812, but a century passed after this date before it became popular there.

The Leibnizian notation was generally adopted in Europe during the nineteenth century.

In the United States the notation : :: : was the prevailing one during the nineteenth century. The Leibnizian signs appeared only in a few works, such as the geometries of William Chauvenet[2] and Benjamin Peirce.[3] It is in the twentieth century that the notation : = : came to be generally adopted in the United States.

A special symbol for variation sometimes encountered in English and American texts is $\propto$, introduced by Emerson.[4] "To the Common Algebraic Characters already receiv'd I add this $\propto$, which signifies a general Proportion; thus, $A \propto \dfrac{BC}{D}$, signifies that A is in a constant ratio to $\dfrac{BC}{D}$." The sign was adopted by Chrystal,[5] Castle,[6] and others.

SIGNS OF EQUALITY

260. *Early symbols.*—A symbol signifying "it gives" and ranking practically as a mark for equality is found in the linear equation of the Egyptian Ahmes papyrus (§ 23, Fig. 7). We have seen (§ 103) that Diophantus had a regular sign for equality, that the contraction *pha* answered that purpose in the Bakhshālī arithmetic (§ 109), that the Arab al-Qalasâdî used a sign (§ 124), that the dash was used for the expression of equality by Regiomontanus (§ 126), Pacioli (§ 138), and that sometimes Cardan (§ 140) left a blank space where we would place a sign of equality.

261. *Recorde's sign of equality.*—In the printed books before Recorde, equality was usually expressed rhetorically by such words as *aequales, aequantur, esgale, faciunt, ghelijck,* or *gleich,* and sometimes by the abbreviated form *aeq.* Prominent among the authors expressing equality in some such manner are Kepler, Galileo, Torricelli, Cavalieri, Pascal, Napier, Briggs, Gregory St. Vincent, Tacquet, and Fermat. Thus, about one hundred years after Recorde, some of

---

[1] John Cole, *Stereogoniometry* (London, 1812), p. 44, 265.

[2] William Chauvenet, *Treatise on Elementary Geometry* (Philadelphia, 1872), p. 69.

[3] Benjamin Peirce, *Elementary Treatise on Plane and Solid Geometry* (Boston, 1873), p. xvi.

[4] W. Emerson, *Doctrine of Fluxions* (3d ed.; London, 1768), p. 4.

[5] G. Chrystal, *Algebra*, Part I, p. 275.

[6] Frank Castle, *Practical Mathematics for Beginners* (London, 1905), p. 317.

the most noted mathematicians used no symbol whatever for the expression of equality. This is the more surprising if we remember that about a century before Recorde, Regiomontanus (§ 126) in his correspondence had sometimes used for equality a horizontal dash —, that the dash had been employed also by Pacioli (§ 138) and Ghaligai (§ 139). Equally surprising is the fact that apparently about the time of Recorde a mathematician at Bologna should independently originate the same symbol (Fig. 53) and use it in his manuscripts.

Recorde's =, after its début in 1557, did not again appear in print until 1618, or sixty-one years later. That some writers used symbols in their private manuscripts which they did not exhibit in their printed books is evident, not only from the practice of Regiomontanus, but also from that of John Napier who used Recorde's = in an algebraic manuscript which he did not publish and which was first printed in 1839.[1] In 1618 we find the = in an anonymous Appendix (very probably due to Oughtred) printed in Edward Wright's English translation of Napier's famous *Descriptio*. But it was in 1631 that it received more than general recognition in England by being adopted as the symbol for equality in three influential works, Thomas Harriot's *Artis analyticae praxis*, William Oughtred's *Clavis mathematicae*, and Richard Norwood's *Trigonometria*.

262. *Different meanings of* =.—As a source of real danger to Recorde's sign was the confusion of symbols which was threatened on the European Continent by the use of = to designate relations other than that of equality. In 1591 Francis Vieta in his *In artem analyticen isagoge* used = to designate arithmetical difference (§ 177). This designation was adopted by Girard (§ 164), by Sieur de Var-Lezard[2] in a translation of Vieta's *Isagoge* from the Latin into French, De Graaf,[3] and by Franciscus à Schooten[4] in his edition of Descartes' *Géométrie*. Descartes[5] in 1638 used = to designate *plus ou moins*, i.e., ±.

Another complication arose from the employment of = by Johann

---

[1] Johannis Napier, *De Arte Logistica* (Edinburgh, 1839), p. 160.

[2] I. L. Sieur de Var-Lezard, *Introduction en l'art analytic ov nouvelle algèbre de François Viète* (Paris, 1630), p. 36.

[3] Abraham de Graaf, *De beginselen van de Algebra of Stelkonst* (Amsterdam, 1672), p. 26.

[4] Renati Descartes, *Geometria* (ed. Franc. à Schooten; Francofvrti al Moenvm, 1695), p. 395.

[5] *Œuvres de Descartes* (éd. Adam et Tannery), Vol. II (Paris, 1898), p. 314, 426.

Caramuel[1] as the separatrix in decimal fractions; with him $102 = 857$ meant our 102.857. As late as 1706 G. H. Paricius[2] used the signs $=$, $:$, and $-$ as general signs to separate numbers occurring in the process of solving arithmetical problems. The confusion of algebraic language was further increased when Dulaurens[3] and Reyher[4] designated parallel lines by $=$. Thus the symbol $=$ acquired five different meanings among different continental writers. For this reason it was in danger of being discarded altogether in favor of some symbol which did not labor under such a handicap.

263. *Competing symbols.*—A still greater source of danger to our $=$ arose from competing symbols. Pretenders sprang up early on both the Continent and in England. In 1559, or two years after the appearance of Recorde's algebra, the French monk, J. Buteo,[5] published his *Logistica* in which there appear equations like "$1A$, $\frac{1}{3}B$, $\frac{1}{3}C$ [14" and "$3A.3B.15C[120$," which in modern notation are $x + \frac{1}{3}y + \frac{1}{3}z = 14$ and $3x + 3y + 15z = 120$. Buteo's [functions as a sign of equality. In 1571, a German writer, Wilhelm Holzmann, better known under the name of Xylander, brought out an edition of Diophantus' *Arithmetica*[6] in which two parallel vertical lines || were used for equality. He gives no clue to the origin of the symbol. Moritz Cantor[7] suggests that perhaps the Greek word ἴσοι ("equal") was abbreviated in the manuscript used by Xylander, by the writing of only the two letters ιι. Weight is given to this suggestion in a Parisian manuscript on Diophantus where a single ι denoted equality.[8] In 1613, the Italian writer Giovanni Camillo Glorioso used Xylander's two vertical lines for equality.[9] It was used again by the Cardinal Michaelangelo Ricci.[10] This character was adopted by a few Dutch and French

[1] Joannis Caramuelis, *Mathesis Biceps vetus et nova* (1670), p. 7.

[2] Georg Heinrich Paricius, *Praxis arithmetices* (Regensburg, 1706). Quoted by M. Sterner, *Geschichte der Rechenkunst* (München und Leipzig, 1891), p. 348.

[3] François Dulaurens, *Specimina mathematica* (Paris, 1667).

[4] Samuel Reyher, *Euclides* (Kiel, 1698).

[5] J. Buteo, *Logistica* (Leyden, 1559), p. 190, 191. See J. Tropfke, *op. cit.*, Vol. III (2d ed.; Leipzig, 1922), p. 136.

[6] See Nesselmann, *Algebra der Griechen* (1842), p. 279.

[7] M. Cantor, *Vorlesungen über Geschichte der Mathematik*, Vol. II (2d ed.; Leipzig, 1913), p. 552.

[8] M. Cantor, *op. cit.*, Vol. I (3d ed.; 1907), p. 472.

[9] *Joannis Camillo Gloriosi, Ad theorema geometricvm* (Venetiis, 1613), p. 26.

[10] Michaelis Angeli Riccii, *Exercitatio geometrica de maximis et minimis* (Londini, 1668), p. 9.

mathematicians during the hundred years that followed, especially in the writing of proportion. Thus, R. Descartes,[1] in his *Opuscules de 1619–1621*, made the statement, "ex progressione $1|2||4|8||16|32||$ habentur numeri perfecti 6, 28, 496." Pierre de Carcavi, of Lyons, in a letter to Descartes (Sept. 24, 1649), writes the equation "$+1296-3060a+2664a^2-1115a^3+239a^4-25a^5+a^6 || 0$," where "la lettre a est l'inconnuë en la maniere de Monsieur Vieta" and $||$ is the sign of equality.[2] De Monconys[3] used it in 1666; De Sluse[4] in 1668 writes our $be=a^2$ in this manner "$be || aa$." De la Hire (§ 254) in 1701 wrote the proportion $a:b=x^2:ab$ thus: "$a|b||xx|ab$." This symbolism is adopted by the Dutch Abraham de Graaf[5] in 1703, by the Frenchman Parent[6] in 1713, and by certain other writers in the *Journal des Sçavans*.[7] Though used by occasional writers for more than a century, this mark $||$ never gave promise of becoming a universal symbol for equality. A single vertical line was used for equality by S. Reyher in 1698. With him, "$A|B$" meant $A=B$. He attributes[8] this notation to the Dutch orientalist and astronomer Jacob Golius, saying: "Especially indebted am I to Mr. Golio for the clear algebraic mode of demonstration with the sign of equality, namely the rectilinear stroke standing vertically between two magnitudes of equal measure."

In England it was Leonard and Thomas Digges, father and son, who introduced new symbols, including a line complex ⚡ for equality (Fig. 78).[9]

The greatest oddity was produced by Hérigone in his *Cursus mathematicus* (Paris, 1644; 1st ed., 1634). It was the symbol "$2|2$." Based on the same idea is his "$3|2$" for "greater than," and his "$2|3$" for "less than." Thus, $a^2+ab=b^2$ is indicated in his symbolism by

[1] *Œuvres de Descartes*, Vol. X (1908), p. 241.

[2] *Op. cit.*, Vol. V (1903), p. 418.

[3] *Journal des voyages de Monsieur de Monconys* (Troisième partie; Lyon, 1666), p. 2. Quoted by Henry in *Revue archéologique* (N.S.), Vol. XXXVII (1879), p. 333.

[4] *Renati Francisci Slusii Mesolabum*, Leodii Eburonum (1668), p. 51.

[5] Abraham de Graaf, *De Vervulling van de Geometria en Algebra* (Amsterdam, 1708), p. 97.

[6] A. Parent, *Essais et recherches de mathématique et de physique* (Paris, 1713), p. 224.

[7] *Journal des Sçavans* (Amsterdam, for 1713), p. 140; *ibid.* (for 1715), p. 537; and other years.

[8] Samuel Reyher, *op. cit.*, Vorrede.

[9] Thomas Digges, *Stratioticos* (1590), p. 35.

"a2+ba2|2b2." Though clever and curious, this notation did not appeal. In some cases Hérigone used also ⊔ to express equality. If this sign is turned over, from top to bottom, we have the one used by F. Dulaurens[1] in 1667, namely, ⊓; with Dulaurens ⊓ signifies "majus," ⊓ signifies "minus"; Leibniz, in some of his correspondence and unpublished papers, used[2] ⊓ and also[3] = ; on one occasion he used the Cartesian[4] ∝ for identity. But in papers which he printed, only the sign = occurs for equality.

Different yet was the equality sign ₃ used by J. V. Andrea[5] in 1614.

The substitutes advanced by Xylander, Andrea, the two Digges, Dulaurens, and Hérigone at no time seriously threatened to bring about the rejection of Recorde's symbol. The real competitor was the mark ∝, prominently introduced by René Descartes in his *Géométrie* (Leyden, 1637), though first used by him at an earlier date.[6]

264. *Descartes' sign of equality.*—It has been stated that the sign was suggested by the appearance of the combined *ae* in the word *aequalis*, meaning "equal." The symbol has been described by Cantor[7] as the union of the two letters *ae*. Better, perhaps, is the description given by Wieleitner[8] who calls it a union of *oe* reversed; his minute examination of the symbol as it occurs in the 1637 edition of the *Géométrie* revealed that not all of the parts of the letter *e* in the combination *oe* are retained, that a more accurate way of describing that symbol is to say that it is made up of two letters *o*, that is, *oo* pressed against each other and the left part of the first excised. In some of the later appearances of the symbol, as given, for example, by van Schooten in 1659, the letter *e* in *oe*, reversed, remains intact. We incline to the opinion that Descartes' symbol for equality, as it appears in his *Géométrie* of 1637, is simply the astronomical symbol

[1] F. Dulaurens, *Specimina mathematica* (Paris, 1667).

[2] C. I. Gerhardt, *Leibnizens mathematische Schriften*, Vol. I, p. 100, 101, 155, 163, etc.

[3] *Op. cit.*, Vol. I, p. 29, 49, 115, etc.

[4] *Op. cit.*, Vol. V, p. 150.

[5] *Joannis Valentini Andreae, Collectaneorum Mathematicorum decades XI* (Tubingae, 1614). Taken from P. Treutlein, "Die deutsche Coss," *Abhandlungen zur Geschichte der Mathematik*, Vol. II (1879), p. 60.

[6] *Œuvres de Descartes* (éd. Ch. Adam et P. Tannery), Vol. X (Paris, 1908), p. 292, 299.

[7] M. Cantor, *op. cit.*, Vol. II (2d ed., 1913), p. 794.

[8] H. Wieleitner in *Zeitschr. für math. u. naturwiss. Unterricht*, Vol. XLVII (1916), p. 414.

for Taurus, placed sideways, with the opening turned to the left. This symbol occurs regularly in astronomical works and was therefore available in some of the printing offices.

Descartes does not mention Recorde's notation; his *Géométrie* is void of all bibliographical and historical references. But we know that he had seen Harriot's *Praxis*, where the symbol is employed regularly. In fact, Descartes himself[1] used the sign = for equality in a letter of 1640, where he wrote "$1C-6N=40$" for $x^3-6x=40$. Descartes does not give any reason for advancing his new symbol ∞. We surmise that Vieta's, Girard's, and De Var-Lezard's use of = to denote arithmetical "difference" operated against his adoption of Recorde's sign. Several forces conspired to add momentum to Descartes' symbol ∞. In the first place, the *Géométrie*, in which it first appeared in print, came to be recognized as a work of genius, giving to the world analytic geometry, and therefore challenging the attention of mathematicians. In the second place, in this book Descartes had perfected the exponential notation, a^n (n, a positive integer), which in itself marked a tremendous advance in symbolic algebra; Descartes' ∞ was likely to follow in the wake of the exponential notation. The ∞ was used by F. Debeaune[2] as early as October 10, 1638, in a letter to Roberval.

As Descartes had lived in Holland several years before the appearance of his *Géométrie*, it is not surprising that Dutch writers should be the first to adopt widely the new notation. Van Schooten used the Cartesian sign of equality in 1646.[3] He used it again in his translation of Descartes' *Géométrie* into Latin (1649), and also in the editions of 1659 and 1695. In 1657 van Schooten employed it in a third publication.[4] Still more influential was Christiaan Huygens[5] who used ∞ as early as 1646 and in his subsequent writings. He persisted in this usage, notwithstanding his familiarity with Recorde's symbol through the letters he received from Wallis and Brouncker, in which it occurs many times.[6] The Descartian sign occurs in the writings of Hudde and De Witt, printed in van Schooten's 1659 and later editions of Descartes' *Géométrie*. Thus, in Holland, the symbol was adopted by

---

[1] *Œuvres de Descartes*, Vol. III (1899), p. 190.

[2] *Ibid.*, Vol. V (1903), p. 519.

[3] Francisci à Schooten, *De organica conicarum sectionum* (Leyden, 1646), p. 91.

[4] Francisci à Schooten, *Exercitationvm mathematicarum liber primus* (Leyden, 1657), p. 251.

[5] *Œuvres complètes de Christiaan Huygens*, Tome I (La Haye, 1888), p. 26, 526.

[6] *Op. cit.*, Tome II, p. 296, 519; Tome IV, p. 47, 88.

the most influential mathematicians of the seventeenth century. It worked its way into more elementary textbooks. Jean Prestet[1] adopted it in his *Nouveaux Élémens*, published at Paris in 1689. This fact is the more remarkable, as in 1675 he[2] had used the sign =. It seems to indicate that soon after 1675 the sign ∞ was gaining over = in France. Ozanam used ∞ in his *Dictionaire mathematique* (Amsterdam, 1691), though in other books of about the same period he used ∼, as we see later. The Cartesian sign occurs in a French text by Bernard Lamy.[3]

In 1659 Descartes' equality symbol invaded England, appearing in the Latin passages of Samuel Foster's *Miscellanies*. Many of the Latin passages in that volume are given also in English translation. In the English version the sign = is used. Another London publication employing Descartes' sign of equality was the Latin translation of the algebra of the Swiss Johann Alexander.[4] Michael Rolle uses ∞ in his *Traité d'algèbre* of 1690, but changes to = in 1709.[5] In Holland, Descartes' equality sign was adopted in 1660 by Kinckhvysen,[6] in 1694 by De Graaf,[7] except in writing proportions, when he uses =. Bernard Nieuwentiit uses Descartes' symbol in his *Considerationes* of 1694 and 1696, but preferred = in his *Analysis infinitorum* of 1695. De la Hire[8] in 1701 used the Descartian character, as did also Jacob Bernoulli in his *Ars Conjectandi* (Basel, 1713). Descartes' sign of equality was widely used in France and Holland during the latter part of the seventeenth and the early part of the eighteenth centuries, but it never attained a substantial foothold in other countries.

265. *Variations in the form of Descartes' symbol.*—Certain variations of Descartes' symbol of equality, which appeared in a few texts, are probably due to the particular kind of symbols available or improvisable in certain printing establishments. Thus Johaan Cara-

[1] Jean Prestet, *Nouveaux Élémens des mathématiques*, Vol. I (Paris, 1689), p. 261.

[2] J. P. [restet] *Élémens des mathématiques* (Paris, 1675), p. 10.

[3] Bernard Lamy, *Elemens des mathematiques* (3d ed.; Amsterdam, 1692), p. 93.

[4] *Synopsis Algebraica, Opus posthumum Johannis Alexandri, Bernatis-Helvetii. In usum scholae mathematicae apud Hospitium-Christi Londinense* (Londini, 1693), p. 2.

[5] *Mem. de l'académie royale des sciences*, année 1709 (Paris), p. 321.

[6] Gerard Kinckhvysen, *De Grondt der Meet-Konst* (Te Haerlem, 1660), p. 4.

[7] Abraham de Graaf, *De Geheele Mathesis of Wiskonst* (Amsterdam, 1694), p. 45.

[8] De la Hire, *Nouveaux élémens des sections coniques* (Paris, 1701), p. 184.

muel[1] in 1670 employed the symbol Æ; the 1679 edition of Fermat's[2] works gives ∞ in the treatise *Ad locos planos et solidos isagoge*, but in Fermat's original manuscripts this character is not found.[3] On the margins of the pages of the 1679 edition occur also expressions of which "*DA{BE*" is an example, where *DA = BE*. J. Ozanam[4] employs ∽ in 1682 and again in 1693; he refers to ⇌ as used to mark equality, "mais nous le changerons en celuy-cy, ∾ ; que nous semble plus propre, et plus naturel." Andreas Spole[5] said in 1692: "∼ vel = est nota aequalitates." Wolff[6] gives the Cartesian symbol inverted, thus ∝ .

266. *Struggle for supremacy.*—In the seventeenth century, Recorde's = gained complete ascendancy in England. We have seen its great rival ∞ in only two books printed in England. After Harriot and Oughtred, Recorde's symbol was used by John Wallis, Isaac Barrow, and Isaac Newton. No doubt these great names helped the symbol on its way into Europe.

On the European Continent the sign = made no substantial headway until 1650 or 1660, or about a hundred years after the appearance of Recorde's algebra. When it did acquire a foothold there, it experienced sharp competition with other symbols for half a century before it fully established itself. The beginning of the eighteenth century may be designated roughly as the time when all competition of other symbols practically ceased. Descartes himself used = in a letter of September 30, 1640, to Mersenne. A Dutch algebra of 1639 and a tract of 1640, both by J. Stampioen,[7] and the *Teutsche Algebra* of the Swiss Johann Heinrich Rahn (1659), are the first continental textbooks that we have seen which use the symbol. Rahn says, p. 18: "Bey disem anlaasz habe ich das namhafte gleichzeichen = zum ersten gebraucht, bedeutet ist gleich, $2a = 4$ heisset $2a$ ist gleich 4." It was used by Bernhard Frenicle de Bessy, of magic-squares fame, in a

---

[1] J. Caramuel, *op. cit.*, p. 122.

[2] *Varia opera mathematica D. Petri de Fermat* (Tolosae, 1679), p. 3, 4, 5.

[3] *Œuvres de Fermat* (ed. P. Tannery et C. Henry), Vol. I (Paris, 1891), p. 91.

[4] *Journal des Sçavans* (de l'an 1682), p. 160; Jacques Ozanam, *Cours de Mathematiques*, Tome I (Paris, 1692), p. 27; also Tome III (Paris, 1693), p. 241.

[5] Andreas Spole, *Arithmetica vulgaris et specioza* (Upsaliae, 1692), p. 16. See G. Eneström in *L'Intermédiaire des mathématiciens*, Tome IV (1897), p. 60.

[6] Christian Wolff, *Mathematisches Lexicon* (Leipzig, 1716), "Signa," p. 1264.

[7] Johan Stampioen d'Jonghe, *Algebra ofte Nieuwe Stel-Regel* ('s Graven-Hage, 1639); *J. Stampioenii Wisk-Konstich ende Reden-maetich Bewijs* (s'Graven-Hage, 1640).

letter[1] to John Wallis of December 20, 1661, and by Huips[2] in the same year. Leibniz, who had read Barrow's *Euclid* of 1655, adopted the Recordean symbol, in his *De arte combinatoria of 1666* (§ 545), but then abandoned it for nearly twenty years. The earliest textbook brought out in Paris that we have seen using this sign is that of Arnauld[3] in 1667; the earliest in Leyden is that of C. F. M. Dechales[4] in 1674.

The sign = was used by Prestet,[5] Abbé Catelan and Tschirnhaus,[6] Hoste,[7] Ozanam,[8] Nieuwentijt,[9] Weigel,[10] De Lagny,[11] Carré,[12] L'Hospital,[13] Polynier,[14] Guisnée,[15] and Reyneau.[16]

This list constitutes an imposing array of names, yet the majority of writers of the seventeenth century on the Continent either used Descartes' notation for equality or none at all.

267. With the opening of the eighteenth century the sign = gained rapidly; James Bernoulli's *Ars Conjectandi* (1713), a posthumous publication, stands alone among mathematical works of prominence of that late date, using ∞. The dominating mathematical advance of the time was the invention of the differential and integral calculus. The fact that both Newton and Leibniz used Recorde's symbol led to its general adoption. Had Leibniz favored Descartes'

[1] *Œuvres complètes des Christiaan-Huygens* (La Haye), Tome IV (1891), p. 45.

[2] Frans van der Huips, *Algebra ofte een Noodige* (Amsterdam, 1661), p. 178. Reference supplied by L. C. Karpinski.

[3] Antoine Arnauld, *Nouveaux Elemens de Geometrie* (Paris, 1667; 2d ed., 1683).

[4] C. F. Dechales, *Cvrsvs sev Mvndvs Mathematicvs*, Tomvs tertivs (Lvgdvni, 1674), p. 666; Editio altera, 1690.

[5] J. P[restet], *op. cit.* (Paris, 1675), p. 10.

[6] *Acta eruditorum* (anno 1682), p. 87, 393.

[7] P. Hoste, *Recueil des traites de mathematiques*, Tome III (Paris, 1692), p. 93.

[8] Jacques Ozanam, *op. cit.*, Tome I (nouvelle éd.; Paris, 1692), p. 27. In various publications between the dates 1682 and 1693 Ozanam used as equality signs ∼, ∞, and =.

[9] Bernard Nieuwentijt, *Analysis infinitorum.*

[10] *Erhardi Weigelii Philosophia mathematica* (Jenae, 1693), p. 135.

[11] Thomas F. de Lagny, *Nouveaux élémens d'arithmétique, et d'algèbre* (Paris, 1697), p. 232.

[12] Louis Carré, *Methode pour la mesure des surfaces* (Paris, 1700), p. 4.

[13] Marquis de l'Hospital, *Analyse des Infiniment Petits* (Paris, 1696, 1715).

[14] Pierre Polynier, *Élémens des Mathématiques* (Paris, 1704), p. 3.

[15] Guisnée, *Application de l'algèbre à la géométrie* (Paris, 1705).

[16] Charles Reyneau, *Analyse demontrée*, Tome I (1708).

∞, then Germany and the rest of Europe would probably have joined France and the Netherlands in the use of it, and Recorde's symbol would probably have been superseded in England by that of Descartes at the time when the calculus notation of Leibniz displaced that of Newton in England. The final victory of = over ∞ seems mainly due to the influence of Leibniz during the critical period at the close of the seventeenth century.

The sign of equality = ranks among the very few mathematical symbols that have met with universal adoption. Recorde proposed no other algebraic symbol; but this one was so admirably chosen that it survived all competitors. Such universality stands out the more prominently when we remember that at the present time there is still considerable diversity of usage in the group of symbols for the differential and integral calculus, for trigonometry, vector analysis, in fact, for every branch of mathematics.

The difficulty of securing uniformity of notation is further illustrated by the performance of Peter van Musschenbroek,[1] of Leyden, an eighteenth-century author of a two-volume text on physics, widely known in its day. In some places he uses = for equality and in others for ratio; letting $S.$ $s.$ be distances, and $T.$ t .times, he says: "Erit $S.$ $s.$ :: $T.$ $t.$ exprimunt hoc Mathematici scribendo, est $S = T.$ sive Spatium est uti tempus, nam signum = non exprimit aequalitatem, sed rationem." In writing proportions, the ratio is indicated sometimes by a dot, and sometimes by a comma. In 1754, Musschenbroek had used ∞ for equality.[2]

268. *Variations in the form of Recorde's symbol.*—There has been considerable diversity in the form of the sign of equality. Recorde drew the two lines very long (Fig. 71) and close to each other, =====. This form is found in Thomas Harriot's algebra (1631), and occasionally in later works, as, for instance, in a paper of De Lagny[3] and in Schwab's edition of Euclid's *Data*.[4] Other writers draw the two lines very short, as does Weigel[5] in 1693. At Upsala, Emanuel

[1] Petro van Musschenbroek, *Introductio ad philosophiam naturalem*, Vol. I (Leyden, 1762), p. 75, 126.

[2] Petri van Musschenbroek, *Dissertatio physica experimentalis de magnete* (Vienna), p. 239.

[3] De Lagny in *Mémoires de l'académie r. d. sciences* (depuis 1666 jusqu'à 1699), Vol. II (Paris, 1733), p. 4.

[4] Johann Christoph Schwab, *Euclids Data* (Stuttgart, 1780), p. 7.

[5] *Erhardi Weigeli Philosophia mathematica* (Jena, 1693), p. 181.

Swedenborg[1] makes them very short and slanting upward, thus //. At times one encounters lines of moderate length, drawn far apart $=$, as in an article by Nicole[2] and in other articles, in the *Journal des Sçavans*. Frequently the type used in printing the symbol is the figure 1, placed horizontally, thus[3] $\rightleftarrows$ or[4] $\rightleftharpoons$.

In an American arithmetic[5] occurs, "$1+6,=7,\times 6=42,\div 2=21$."

Wolfgang Bolyai[6] in 1832 uses $\doteq$ to signify absolute equality; $\overline{\overline{}}$, equality in content; $A(=B$ or $B=)A$, to signify that each value of A is equal to some value of B; $A(=)B$, that each of the values of A is equal to some value of B, and vice versa.

To mark the equality of vectors, Bellavitis[7] used in 1832 and later the sign $\rightleftharpoons$.

Some recent authors have found it expedient to assign $=$ a more general meaning. For example, Stolz and Gmeiner[8] in their theoretical arithmetic write $a \circ b = c$ and read it "*a* mit *b* ist *c*," the $=$ signifying "is explained by" or "is associated with." The small circle placed between a and b means, in general, any relation or *Verknüpfung*.

De Morgan[9] used in one of his articles on logarithmic theory a double sign of equality $= \ =$ in expressions like $(be^{\beta\sqrt{-1}})^x = \ = ne^{\nu\sqrt{-1}}$, where β and ν are angles made by b and n, respectively, with the initial line. He uses this double sign to indicate "that every symbol shall express not merely the length and direction of a line, but also the quantity of revolution by which a line, setting out from the unit line, is supposed to attain that direction."

[1] Emanuel Swedberg, *Daedalus Hyperboreus* (Upsala, 1716), p. 39. See facsimile reproduction in *Kungliga Vetenskaps Societetens i Upsala Tvåhundraårsminne* (Upsala, 1910).

[2] François Nicole in *Journal des Sçavans*, Vol. LXXXIV (Amsterdam, 1728), p. 293. See also année 1690 (Amsterdam, 1691), p. 468; année 1693 (Amsterdam, 1694), p. 632.

[3] James Gregory, *Geometria Pars Vniversalis* (Padua, 1668); Emanuel Swedberg, *op. cit.*, p. 43.

[4] H. Vitalis, *Lexicon mathematicum* (Rome, 1690), art. "Algebra."

[5] *The Columbian Arithmetician*, "by an American" (Haverhill [Mass.], 1811), p. 149.

[6] Wolfgangi Bolyai de Bolya, *Tentamen* (2d ed.), Tom. I (Budapestini, 1897), p. xi.

[7] Guisto Bellavitis in *Annali del R. Lomb.-Ven.* (1832), Tom. II, p. 250–53.

[8] O. Stolz und J. A. Gmeiner, *Theoretische Arithmetik* (Leipzig), Vol. I (2d ed.; 1911), p. 7.

[9] A. de Morgan, *Trans. Cambridge Philos. Society*, Vol. VII (1842), p. 186.

269. *Variations in the manner of using it.*—A rather unusual use of equality signs is found in a work of Deidier[1] in 1740, viz.,

$$\frac{0+1+2=3}{2+2+2=6}=\frac{1}{2}\ ;\qquad \frac{0.\ 1.\ 4.=5}{4,\ 4,\ 4,=12}=\frac{1}{3}+\frac{1}{12}\ .$$

H. Vitalis[2] uses a modified symbol: "Nota $\backsimeq$ significat repetitam aequationem .... *vt* $10\rightleftharpoons 6.\ +4\backsimeq 8\ +2$." A discrimination between $=$ and ∞ is made by Gallimard[3] and a few other writers; "$=$, est égale à; ∞ qui signifie tout simplement, égal à , ou , qui est égal à."

A curious use, in the same expressions, of $=$, the comma, and the word *aequalis* is found in a Tacquet-Whiston[4] edition of Euclid, where one reads, for example, "erit $8\times 432=3456$ aequalis $8\times 400=3200$, $+8\times 30=240,+8\times 2=16$."

L. Gustave du Pasquier[5] in discussing general complex numbers employs the sign of double equality $\equiv$ to signify "equal by definition."

The relations between the coefficients of the powers of x in a series may be expressed by a formal equality involving the series as a whole, as in

$$1+n_{(1)}x+n_{(2)}x^2+\ \cdots\ \overline{\overline{f}}\ (1+x)\{1+(n+1)_{(1)}x+(n-1)_{(2)}x^2+\ \cdots\ ,$$

where the symbol $\overline{\overline{f}}$ indicates that the equality is only formal, not arithmetical.[6]

270. *Nearly equal.*—Among the many uses made in recent years of the sign $\sim$ is that of "nearly equal to," as in "$e\sim\frac{1}{4}$"; similarly, $e\cong\frac{1}{4}$ is allowed to stand for "equal or nearly equal to."[7] A. Eucken[8] lets $\simeq$ stand for the lower limit, as in "$J\simeq 45.10^{-40}$ (untere Grenze)," where J means a mean moment of inertia. Greenhill[9] denotes approximate

---

[1] L'Abbé Deidier, *La mesure des surfaces et des solides* (Paris, 1740), p. 9.

[2] H. Vitalis, *loc. cit.*

[3] J. E. Gallimard, *La Science du calcul numerique*, Vol. I (Paris, 1751), p. 3.

[4] Andrea Tacquet, *Elementa Euclidea geometriae* [after] Gulielmus Whiston (Amsterdam, 1725), p. 47.

[5] *Comptes Rendus du Congrès International des Mathématicians* (Strasbourg, 22–30 Septembre 1920), p. 164.

[6] Art. "Algebra" in *Encyclopaedia Britannica* (11th ed., 1910).

[7] A. Kratzer in *Zeitschrift für Physik*, Vol. XVI (1923), p. 356, 357.

[8] A. Eucken in *Zeitschrift der physikalischen Chemie*, Band C, p. 159.

[9] A. G. Greenhill, *Applications of Elliptic Functions* (London, 1892), p. 303, 340, 341.

equality by $\underset{\wedge\wedge\wedge}{\wedge\wedge\wedge}$. An early suggestion due to Fischer[1] was the sign $\asymp$ for "approximately equal to." This and three other symbols were proposed by Boon[2] who designed also four symbols for "greater than but approximately equal to" and four symbols for "less than but approximately equal to."

SIGNS OF COMMON FRACTIONS

271. *Early forms.*—In the Egyptian Ahmes papyrus unit fractions were indicated by writing a special mark over the denominator (§§ 22, 23). Unit fractions are not infrequently encountered among the Greeks (§ 41), the Hindus and Arabs, in Leonardo of Pisa (§ 122), and in writers of the later Middle Ages in Europe.[3] In the text *Triśatika*, written by the Hindu Śrīdhara, one finds examples like the following: "How much money is there when half a *kākini*, one-third of this and one-fifth of this are added together?

Statement
$$\begin{array}{|cc|} \hline 1 & 1 \\ 1 & 2 \\ \hline \end{array} \quad \begin{array}{|ccc|} \hline 1 & 1 & 1 \\ 1 & 2 & 3 \\ \hline \end{array} \quad \begin{array}{|cccc|} \hline 1 & 1 & 1 & 1 \\ 1 & 2 & 3 & 5 \\ \hline \end{array}$$
Answer. *Varātikas* 14."

This means $1\times\frac{1}{2}+1\times\frac{1}{2}\times\frac{1}{3}+1\times\frac{1}{2}\times\frac{1}{3}\times\frac{1}{5}=\frac{7}{10}$, and since 20 *varātikas* = 1 *kākini*, the answer is 14 *varātikas*.

John of Meurs (early fourteenth century)[4] gives $\frac{7}{8}$ as the sum of three unit fractions $\frac{1}{2}$, $\frac{1}{4}$, and $\frac{1}{8}$, but writes "$\frac{1}{2}\ \frac{1}{2}\ \frac{1}{8}$," which is an ascending continued fraction. He employs a slightly different notation for $\frac{1}{24}$, namely, "$\frac{1}{4}\ \frac{1}{8}\circ\frac{1}{2}\circ$."

Among Heron of Alexandria and some other Greek writers the numerator of any fraction was written with an accent attached, and was followed by the denominator marked with two accents (§ 41). In some old manuscripts of Diophantus the denominator is placed above the numerator (§ 104), and among the Byzantines the denominator is found in the position of a modern exponent;[5] $\vartheta^{\iota\alpha}$ signified accordingly $\frac{9}{11}$.

[1] Ernst Gottfried Fischer, *Lehrbuch der Elementar-Mathematik*, 4. Theil, *Anfangsgründe der Algebra* (Berlin und Leipzig, 1829), p. 147. Reference given by R. C. Archibald in *Mathematical Gazette*, Vol. VIII (London, 1917), p. 49.

[2] C. F. Boon, *Mathematical Gazette*, Vol. VII (London, 1914), p. 48.

[3] See G. Eneström in *Bibliotheca mathematica* (3d ser.), Vol. XIV (1913–14), p. 269, 270.

[4] Vienna Codex 4770, the *Quadripartitum numerorum*, described by L. C. Karpinski in *Bibliotheca mathematica* (3d ser.), Vol. XIII (1912–13), p. 109.

[5] F. Hultsch, *Metrologicorum scriptorum reliquiae*, Vol. I (Leipzig, 1864), p. 173–75.

The Hindus wrote the denominator beneath the numerator, but without a separating line (§§ 106, 109, 113, 235).

In the so-called arithmetic of John of Seville,[1] of the twelfth century (?), which is a Latin elaboration of the arithmetic of al-Khowârizmî, as also in a tract of Alnasavi (1030 A.D.),[2] the Indian mode of writing fractions is followed; in the case of a mixed number, the fractional part appears below the integral part. Alnasavi pursues this course consistently[3] by writing a zero when there is no integral part; for example, he writes $\frac{1}{11}$ thus: $\overset{\text{``o''}}{\underset{11}{|}}$.

272. *The fractional line* is referred to by the Arabic writer al-Ḥaṣṣâr (§§ 122, 235, Vol. II § 422), and was regularly used by Leonardo of Pisa (§§ 122, 235). The fractional line is absent in a twelfth-century Munich manuscript;[4] it was not used in the thirteenth-century writings of Jordanus Nemorarius,[5] nor in the *Gernardus algorithmus demonstratus*, edited by Joh. Schöner (Nürnberg, 1534), Part II, chapter i.[6] When numerator and denominator of a fraction are letters, Gernardus usually adopted the form *ab* (*a* numerator, *b* denominator), probably for graphic reasons. The fractional line is absent in the Bamberger arithmetic of 1483, but occurs in Widman (1489), and in a fifteenth-century manuscript at Vienna.[7] While the fractional line came into general use in the sixteenth century, instances of its omission occur as late as the seventeenth century.

273. Among the sixteenth- and seventeenth-century writers omitting the fractional line were Baëza[8] in an arithmetic published at Paris, Dibuadius[9] of Denmark, and Paolo Casati.[10] The line is

[1] Boncompagni, *Trattati d'aritmetica*, Vol. II, p. 16–72.

[2] H. Suter, *Bibliotheca mathematica* (3d ser.), Vol. VII (1906–7), p. 113–19.

[3] M. Cantor, *op. cit.*, Vol. I (3d ed.), p. 762.

[4] Munich MS Clm 13021. See *Abhandlungen über Geschichte der Mathematik*, Vol. VIII (1898), p. 12–13, 22–23, and the peculiar mode of operating with fractions.

[5] *Bibliotheca mathematica* (3d ser.), Vol. XIV, p. 47.

[6] *Ibid.*, p. 143.

[7] Codex Vindob. 3029, described by E. Rath in *Bibliotheca mathematica* (3d ser.), Vol. XIII (1912–13), p. 19. This manuscript, as well as Widman's arithmetic of 1489, and the anonymous arithmetic printed at Bamberg in 1483, had as their common source a manuscript known as Algorismus Ratisponensis.

[8] *Nvmerandi doctrina authore Lodoico Baeza* (Lvtetia, 1556), fol. 45.

[9] *C. Dibvadii in arithmeticam irrationalivm Evclidis* (Arnhemii, 1605).

[10] Paolo Casati, *Fabrica et Vso Del Compasso di Proportione* (Bologna, 1685) [Privilege, 1662], p. 33, 39, 43, 63, 125.

usually omitted in the writings of Marin Mersenne[1] of 1644 and 1647. It is frequently but not usually omitted by Tobias Beutel.[2]

In the middle of a fourteenth-century manuscript[3] one finds the notation $3\,\overline{5}$ for $\frac{3}{5}$, $4\,\overline{7}$ for $\frac{4}{7}$. A Latin manuscript,[4] Paris 7377A, which is a translation from the Arabic of Abu Kamil, contains the fractional line, as in $\frac{1}{9}$, but $\frac{1}{9}\frac{1}{9}$ is a continued fraction and stands for $\frac{1}{9}$ plus $\frac{1}{81}$, whereas $\frac{6}{9}\frac{1}{2}$ as well as $\frac{9}{8}\frac{1}{2}$ represent simply $\frac{1}{18}$. Similarly, Leonardo of Pisa,[5] who drew extensively from the Arabic of Abu Kamil, lets $\frac{5}{8}\frac{6}{8}$ stand for $\frac{5}{64}$, there being a difference in the order of reading. Leonardo read from right to left, as did the Arabs, while authors of Latin manuscripts of about the fourteenth century read as we do from left to right. In the case of a mixed number, like $3\frac{1}{5}$, Leonardo and the Arabs placed the integer to the right of the fraction.

274. *Special symbols for simple fractions* of frequent occurrence are found. The Ahmes papyrus has special signs for $\frac{1}{2}$ and $\frac{2}{3}$ (§ 22); there existed a hieratic symbol for $\frac{1}{4}$ (§ 18). Diophantus employed special signs for $\frac{1}{2}$ and $\frac{2}{3}$ (§ 104). A notation to indicate one-half, almost identical with one sometimes used during the Middle Ages in connection with Roman numerals, is found in the fifteenth century with the Arabic numerals. Says Cappelli: "I remark that for the designation of one-half there was used also in connection with the Arabic numerals, in the XV. century, a line between two points, as $4 \div$ for $4\frac{1}{2}$, or a small cross to the right of the number in place of an exponent, as $4^\dagger$, presumably a degeneration of 1/1, for in that century this form was used also, as 7 1/1 for $7\frac{1}{2}$. Toward the close of the XV. century one finds also often the modern form $\frac{1}{2}$."[6] The Roman designation of certain unit fractions are set forth in § 58. The peculiar designations employed in the Austrian cask measures are found in § 89. In a fifteenth-century manuscript we find: "Whan pou hayst write pat, for pat pat leues, write such a merke as is here ^{w} vpon his hede, pe quych

---

[1] Marin Mersenne, *Cogitata Physico-mathematica* (Paris, 1644), "Phaenomena ballistica"; *Novarvm observationvm Physico-mathematicarvm*, Tomvs III (Paris, 1647), p. 194 ff.

[2] Tobias Beutel, *Geometrische Gallerie* (Leipzig, 1690), p. 222, 224, 236, 239, 240, 242, 243, 246.

[3] *Bibliotheca mathematica* (3d ser.), Vol. VII, p. 308-9.

[4] L. C. Karpinski in *ibid.*, Vol. XII (1911-12), p. 53, 54.

[5] Leonardo of Pisa, *Liber abbaci* (ed. B. Boncompagni, 1857), p. 447. Noteworthy here is the use of *e* to designate the absence of a number.

[6] A. Cappelli, *Lexicon Abbreviaturarum* (Leipzig, 1901), p. L.

merke schal betoken halfe of þe odde þat was take away";[1] for example, half of 241 is 120^w. In a mathematical roll written apparently in the south of England at the time of Recorde, or earlier, the character $\sim$ stands for one-half, a dot · for one-fourth, and $\sim$ for three-fourths.[2] In some English archives[3] of the sixteenth and seventeenth centuries one finds one-half written in the form $\underline{\underline{1}}$. In the earliest arithmetic printed in America, the *Arte para aprendar todo el menor del arithmetica* of Pedro Paz (Mexico, 1623), the symbol $\underline{\circ}$ is used for $\frac{1}{2}$ a few times in the early part of the book. This symbol is taken from the *Arithmetica practica* of the noted Spanish writer, Juan Perez de Moya, 1562 (14th ed., 1784, p. 13), who uses $\underline{\circ}$ and also $\overset{\circ}{\underset{m}{}}$ for $\frac{1}{2}$ or *medio*.

This may be a convenient place to refer to the origin of the sign % for "per cent," which has been traced from the study of manuscripts by D. E. Smith.[4] He says that in an Italian manuscript an "unknown writer of about 1425 uses a symbol which, by natural stages, developed into our present %. Instead of writing ' per 100', ' P 100' or 'P cento,' as had commonly been done before him, he wrote 'P σ^{-o}' for 'P ℔,' just as the Italians wrote $\overset{o}{1}$, $\overset{o}{2}$, ... and 1°, 2°, ... for primo, secundo, etc. In the manuscripts which I have examined the evolution is easily traced, the σ^{-} becoming ⅌ about 1650, the original meaning having even then been lost. Of late the 'per' has been dropped; leaving only ⅌ or %." By analogy to %, which is now made up of two zeros, there has been introduced the sign %₀, having as many zeros as 1,000 and signifying *per mille*.[5] Cantor represents the fraction $(100+p)/100$ "by the sign 1, 0p, not to be justified mathematically but in practice extremely convenient."

275. *The solidus.*[6]—The ordinary mode of writing fractions $\frac{a}{b}$ is typographically objectionable as requiring three terraces of type. An effort to remove this objection was the introduction of the solidus, as in a/b, where all three fractional parts occur in the regular line of type. It was recommended by De Morgan in his article on "The Calculus

---

[1] R. Steele, *The Earliest Arithmetics in English* (London, 1922), p. 17, 19. The *p* in "pou," "pat," etc., appears to be our modern *th*.

[2] D. E. Smith in *American Mathematical Monthly*, Vol. XXIX (1922), p. 63.

[3] *Antiquaries Journal*, Vol. VI (London, 1926), p. 272.

[4] D. E. Smith, *Rara arithmetica* (1898), p. 439, 440.

[5] Moritz Cantor, *Politische Arithmetik* (2. Aufl.; Leipzig, 1903), p. 4.

[6] The word "solidus" in the time of the Roman emperors meant a gold coin (a "solid" piece of money); the sign / comes from the old form of the initial letter *s*, namely, ʃ, just as £ is the initial of *libra* ("pound"), and *d* of *denarius* ("penny").

of Functions," published in the *Encyclopaedia Metropolitana* (1845). But practically that notation occurs earlier in Spanish America. In the *Gazetas de Mexico* (1784), page 1, Manuel Antonio Valdes used a curved line resembling the sign of integration, thus $1/4$, $3/4$; Henri Cambuston[1] brought out in 1843, at Monterey, California, a small arithmetic employing a curved line in writing fractions. The straight solidus is employed, in 1852, by the Spaniard Antonio Serra Y Oliveres.[2] In England, De Morgan's suggestion was adopted by Stokes[3] in 1880. Cayley wrote Stokes, "I think the 'solidus' looks very well indeed . . . . ; it would give you a strong claim to be President of a Society for the Prevention of Cruelty to Printers." The solidus is used frequently by Stolz and Gmeiner.[4]

While De Morgan recommended the solidus in 1843, he used $a:b$ in his subsequent works, and as Glaisher remarks, "answers the purpose completely and it is free from the objection to $\div$ viz., that the pen must be twice removed from the paper in the course of writing it."[5] The colon was used frequently by Leibniz in writing fractions (§ 543, 552) and sometimes also by Karsten,[6] as in $1:3=\frac{1}{3}$; the $\div$ was used sometimes by Cayley.

G. Peano adopted the notation b/a whenever it seemed convenient.[7]

Alexander Macfarlane[8] adds that Stokes wished the solidus to take the place of the horizontal bar, and accordingly proposed that the terms immediately preceding and following be welded into one, the welding action to be arrested by a period. For example, m^2-n^2/m^2+n^2 was to mean $(m^2-n^2)/(m^2+n^2)$, and a/bcd to mean $\frac{a}{bcd}$, but $a/bc\cdot d$ to mean $\frac{a}{bc}d$. "This solidus notation for algebraic expressions oc-

[1] Henri Cambuston, *Definicion de las principales operaciones de arismetica* (1843), p. 26.

[2] Antonio Serra Y Oliveres, *Manuel de la Tipografia Española* (Madrid, 1852), p. 71.

[3] G. G. Stokes, *Math. and Phys. Papers*, Vol. I (Cambridge, 1880), p. vii. See also J. Larmor, *Memoirs and Scient. Corr. of G. G. Stokes*, Vol. I (1907), p. 397.

[4] O. Stolz and J. A. Gmeiner, *Theoretische Arithmetik* (2d ed.; Leipzig, 1911), p. 81.

[5] J. W. L. Glaisher, *Messenger of Mathematics*, Vol. II (1873), p. 109.

[6] W. J. G. Karsten, *Lehrbegrif der gesamten Mathematik*, Vol. I (Greifswald, 1767), p. 50, 51, 55.

[7] G. Peano, *Lezioni di analisi infinitesimale*, Vol. I (Torino, 1893), p. 2.

[8] Alexander Macfarlane, *Lectures on Ten British Physicists* (New York, 1919), p. 100, 101.

curring in the text has since been used in the *Encyclopaedia Britannica*, in Wiedemann's *Annalen* and quite generally in mathematical literature." It was recommended in 1915 by the Council of the London Mathematical Society to be used in the current text.

"The use of small fractions in the midst of letterpress," says Bryan,[1] "is often open to the objection that such fractions are difficult to read, and, moreover, very often do not come out clearly in printing. It is especially difficult to distinguish $\frac{1}{3}$ from $\frac{1}{8}$. . . . . For this reason it would be better to confine the use of these fractions to such common forms as $\frac{1}{4}$, $\frac{1}{2}$, $\frac{3}{4}$, $\frac{1}{3}$, and to use the notation 18/22 for other fractions."

SIGNS OF DECIMAL FRACTIONS

276. *Stevin's notation.*—The invention of decimal fractions is usually ascribed to the Belgian Simon Stevin, in his *La Disme*, published in 1585 (§ 162). But at an earlier date several other writers came so close to this invention, and at a later date other writers advanced the same ideas, more or less independently, that rival candidates for the honor of invention were bound to be advanced. The *La Disme* of Stevin marked a full grasp of the nature and importance of decimal fractions, but labored under the burden of a clumsy notation. The work did not produce any immediate effect. It was translated into English by R. Norton[2] in 1608, who slightly modified the notation by replacing the circles by round parentheses. The fraction .3759 is given by Norton in the form $3^{(1)}7^{(2)}5^{(3)}9^{(4)}$.

277. Among writers who adopted Stevin's decimal notation is Wilhelm von Kalcheim[3] who writes 693 ② for our 6.93. He applies it also to mark the decimal subdivisions of linear measure: "Die Zeichen sind diese: ⓪ ist ein ganzes oder eine ruthe: ① ist ein erstes / prime oder schuh: ② ist ein zweites / secunde oder Zoll: ③ ein drittes / korn oder gran: ④ ist ein viertes stipflin oder minuten: und so forthan." Before this J. H. Beyer writes[4] 8 7̌98 for 8.00798; also

---

[1] G. H. Bryan, *Mathematical Gazette*, Vol. VIII (1917), p. 220.

[2] *Disme: the Art of Tenths, or Decimall Arithmetike, ... invented by the excellent mathematician, Simon Stevin.* Published in English with some additions by Robert Norton, Gent. (London, 1608). See also A. de Morgan in *Companion to the British Almanac* (1851), p. 11.

[3] *Zusammenfassung etlicher geometrischen Aufgaben. . . . .* Durch Wilhelm von Kalcheim, genant Lohausen Obristen (Bremen, 1629), p. 117.

[4] Johann Hartmann Beyer, *Logistica decimalis, das ist die Kunstrechnung mit den zehntheiligen Brüchen* (Frankfurt a/M., 1603). We have not seen Beyer's

14.3761 for 14.00003761, 123.4.5.9.8.7.2. or 123.4.5.9.8.7.2

or 123. 459. 872 for 123.459872, 643 for 0.0643.

That Stevin's notation was not readily abandoned for a simpler one is evident from Ozanam's use[1] of a slight modification of it as late as 1691, in passages like "$\frac{6667}{10000}$ ég. à 6 6 6 7," and 3 9 8 for our 3.98.

278. *Other notations used before 1617.*—Early notations which one might be tempted to look upon as decimal notations appear in works whose authors had no real comprehension of decimal fractions and their importance. Thus Regiomontanus,[2] in dividing 85869387 by 60000, marks off the last four digits in the dividend and then divides by 6 as follows:

$$8\ 5\ 8\ 6\ |\ 9\ 3\ 8\ 7$$
$$1\ 4\ 3\ 1$$

In the same way, Pietro Borgi[3] in 1484 uses the stroke in dividing 123456 by 300, thus

"*per* 300
1 2 3 4 | 5 6
  4 1 1   ——— .
  4 1 1$\frac{156}{300}$."

Francesco Pellos (Pellizzati) in 1492, in an arithmetic published at Turin, used a point and came near the invention of decimal fractions.[4]

Christoff Rudolff[5] in his *Coss* of 1525 divides 652 by 10. His words are: "Zu exempel / ich teile 652 durch 10. stet also 65/2. ist 65 der quocient vnnd 2 das übrig. Kompt aber ein Zal durch 100 zů teilen / schneid ab die ersten zwo figuren / durch 1000 die ersten drey / also weiter für yede o ein figur." ("For example, I divide 652 by 10. It gives 65/2; thus, 65 is the quotient and 2 the remainder. If a number is to be divided by 100, cut off the first two figures, if by

___

book; our information is drawn from J. Tropfke, *Geschichte der Elementar-Mathematik*, Vol. I (2d ed.; Berlin and Leipzig, 1921), p. 143; S. Günther, *Geschichte der Mathematik*, Vol. I (Leipzig, 1908), p. 342.

[1] J. Ozanam, *L'Usage du Compas de Proportion* (a La Haye, 1691), p. 203, 211.

[2] *Abhandlungen zur Geschichte der Mathematik*, Vol. XII (1902), p. 202, 225.

[3] See G. Eneström in *Bibliotheca mathematica* (3d ser.), Vol. X (1909–10), p. 240.

[4] D. E. Smith, *Rara arithmetica* (1898), p. 50, 52.

[5] Quoted by J. Tropfke, *op. cit.*, Vol. I (2d ed., 1921), p. 140.

1,000 the first three, and so on for each 0 a figure.") This rule for division by 10,000, etc., is given also by P. Apian[1] in 1527.

In the *Exempel Büchlin* (Vienna, 1530), Rudolff performs a multiplication involving what we now would interpret as being decimal fractions.[2] Rudolff computes the values 375 $(1+\frac{5}{100})^n$ for $n=1$, 2, ... , 10. For $n=1$ he writes 393 | 75, which really denotes 393.75; for $n=3$ he writes 434 | 109375. The computation for $n=4$ is as follows:

$$
\begin{array}{l}
4\ 3\ 4\ |\ 1\ 0\ 9\ 3\ 7\ 5 \\
\ \ \ 2\ 1\ \ \ 7\ 0\ 5\ 4\ 6\ 8\ 7\ 5 \\
4\ 5\ 5\ |\ 8\ 1\ 4\ 8\ 4\ 3\ 7\ 5
\end{array}
$$

Here Rudolff uses the vertical stroke as we use the comma and, in passing, uses decimals without appreciating the importance and generality of his procedure.

F. Vieta fully comprehends decimal fractions and speaks of the advantages which they afford;[3] he approaches close to the modern notations, for, after having used (p. 15) for the fractional part smaller type than for the integral part, he separated the decimal from the integral part by a vertical stroke (p. 64, 65); from the vertical stroke to the actual comma there is no great change.

In 1592 Thomas Masterson made a close approach to decimal fractions by using a vertical bar as separatrix when dividing £337652643 by a million and reducing the result to shillings and pence. He wrote:[4]

$$
\text{facit}
\begin{cases}
l. & 3\ 3\ 7 \ |\ 6\ 5\ 2\ 6\ 4\ 3 \\
s. & \ \ \ 1\ 3 \ |\ 0\ 5\ 2\ 8\ 6\ 0 \\
d. & \ \ \text{---} \ |\ 6\ 3\ 4\ 3\ 2\ 0
\end{cases}
\text{"}
$$

John Kepler in his *Oesterreichisches Wein-Visier-Büchlein* (Lintz, MDCXVI), reprinted in Kepler's *Opera omnia* (ed. Ch. Frisch), Volume V (1864), page 547, says: "Fürs ander, weil ich kurtze Zahlen brauche, derohalben es offt Brüche geben wirdt, so mercke, dass alle Ziffer, welche nach dem Zeichen (◖) folgen, die gehören zu

---

[1] P. Apian, *Kauffmannsz Rechnung* (Ingolstadt, 1527), fol. ciijr°. Taken from J. Tropfke, *op. cit.*, Vol. I (2d ed., 1921), p. 141.

[2] See D. E. Smith, "Invention of the Decimal Fraction," *Teachers College Bulletin* (New York, 1910–11), p. 18; G. Eneström, *Bibliotheca mathematica* (3d ser.), Vol. X (1909–10), p. 243.

[3] F. Vieta, *Universalium inspectionum*, p. 7; Appendix to the *Canon mathematicus* (1st ed.; Paris, 1579). We copy this reference from the *Encyclopédie des scienc. math.*, Tome I, Vol. I (1904), p. 53, n. 180.

[4] A. de Morgan, *Companion to the British Almanac* (1851), p. 8.

dem Bruch, als der Zehler, der Nenner darzu wird nicht gesetzt, ist aber allezeit eine runde Zehnerzahl von so vil Nullen, als vil Ziffer nach dem Zeichen kommen. Wann kein Zeichen nicht ist, das ist eine gantze Zahl ohne Bruch, vnd wann also alle Ziffern nach dem Zeichen gehen, da heben sie bissweilen an von einer Nullen. Dise Art der Bruch-rechnung ist von Jost Bürgen zu der sinusrechnung erdacht, vnd ist darzu gut, dass ich den Bruch abkürtzen kan, wa er vnnötig lang werden wil, ohne sonderen Schaden der vberigen Zahlen; kan ihne auch etwa auff Erhaischung der Notdurfft erlengern. Item lesset sich also die gantze Zahl vnd der Bruch mit einander durch alle species Arithmeticae handlen wie nur eine Zahl. Als wann ich rechne 365 Gulden mit 6 per cento, wievil bringt es dess Jars Interesse? dass stehet nun also:

3(65

6 mal

——————

facit 21(90

vnd bringt 21 Gulden vnd 90 hundertheil, oder 9 zehentheil, das ist 54 kr."

Joost Bürgi[1] wrote 1414 for 141.4 and 001414 for 0.01414; on the title-page of his *Progress-Tabulen* (Prag, 1620) he wrote 230270̊022 for our 230270.022. This small circle is referred to often in his *Gründlicher Unterricht*, first published in 1856.[2]

279. *Did Pitiscus use the decimal point?*—If Bartholomaeus Pitiscus of Heidelberg made use of the decimal point, he was probably the first to do so. Recent writers[3] on the history of mathematics are

[1] See R. Wolf, *Viertelj. Naturf. Ges.* (Zürich), Vol. XXXIII (1888), p. 226.

[2] *Grunert's Archiv der Mathematik und Physik*, Vol. XXVI (1856), p. 316–34.

[3] A. von Braunmühl, *Geschichte der Trigonometrie*, Vol. I (Leipzig, 1900), p. 225.
M. Cantor, *Vorlesungen über Geschichte der Mathematik*, Vol. II (2d ed.; Leipzig, 1913), p. 604, 619.
G. Eneström in *Bibliotheca mathematica* (3d ser.), Vol. VI (Leipzig, 1905), p. 108, 109.
J. W. L. Glaisher in *Napier Tercentenary Memorial Volume* (London, 1913), p. 77.
N. L. W. A. Gravelaar in *Nieuw Archief voor Wiskunde* (2d ser.; Amsterdam), Vol. IV (1900), p. 73.
S. Günther, *Geschichte der Mathematik*, 1. Teil (Leipzig, 1908), p. 342.
L. C. Karpinski in *Science* (2d ser.), Vol. XLV (New York, 1917), p. 663–65.
D. E. Smith in *Teachers College Bulletin, Department of Mathematics* (New York, 1910–11), p. 19.
J. Tropfke, *Geschichte der Elementar-Mathematik*, Vol. I (2d ed.; Leipzig, 1921), p. 143.

divided on the question as to whether or not Pitiscus used the decimal point, the majority of them stating that he did use it. This disagreement arises from the fact that some writers, apparently not having access to the 1608 or 1612 edition of the *Trigonometria*[1] of Pitiscus, reason from insufficient data drawn from indirect sources, while others fail to carry conviction by stating their conclusions without citing the underlying data.

Two queries are involved in this discussion: (1) Did Pitiscus employ decimal fractions in his writings? (2) If he did employ them, did he use the dot as the separatrix between units and tenths?

Did Pitiscus employ decimal fractions? As we have seen, the need of considering this question arises from the fact that some early writers used a symbol of separation which we could interpret as separating units from tenths, but which they themselves did not so interpret. For instance,[2] Christoff Rudolff in his *Coss* of 1525 divides 652 by 10, "stet also 65|2. ist 65 der quocient vnnd 2 das übrig." The figure 2 looks like two-tenths, but in Rudolff's mind it is only a remainder. With him the vertical bar served to separate the 65 from this remainder; it was not a decimal separatrix, and he did not have the full concept of decimal fractions. Pitiscus, on the other hand, did have this concept, as we proceed to show. In computing the chord of an arc of 30° (the circle having 10^7 for its radius), Pitiscus makes the statement (p. 44): "All these chords are less than the radius and as it were certain parts of the radius, which parts are commonly written $\frac{5176381}{10000000000}$. But much more brief and necessary for the work, is this writing of it .05176381. For those numbers are altogether of the same value, as these two numbers 09. and $\frac{9}{10}$ are." In the original Latin the last part reads as follows: ".... quae partes vulgo sic scriberentur $\frac{5176381}{10000000000}$. Sed multò compendiosior et ad calculum accommodatior est ista scriptio .05176381. Omnino autem idem isti numeri valent, sicut hi duo numeri 09. et $\frac{9}{10}$ idem valent."

One has here two decimals. The first is written .05176381. The dot on the left is not separating units from tenths; it is only a rhetorical mark. The second decimal fraction he writes 09., and he omits the dot on the left. The zero plays here the rôle of decimal separatrix.

---

[1] I have used the edition of 1612 which bears the following title: *Bartholamei | Pitisci Grunbergensis | Silesij | Trigonometriae | Sioe. De dimensione Triangulor [um] Libri Qvinqve. Jtem | Problematvm variorv. [m] nempe | Geodaeticorum, | Altimetricorum, | Geographicorum, | Gnomonicorum, et | Astronomicorum: | Libri Decem. | Editio Tertia. | Cui recens accessit Pro | blematum Arckhitectonicorum Liber | unus | Francofurti. | Typis Nicolai Hofmanni: | Sumptibus Ionae Rosae.| M.DCXII.*

[2] Quoted from J. Tropfke, *op. cit.*, Vol. I (1921), p. 140.

The dots appearing here are simply the punctuation marks written after (sometimes also before) a number which appears in the running text of most medieval manuscripts and many early printed books on mathematics. For example, Clavius[1] wrote in 1606: "Deinde quia minor est $\frac{4}{7}$. quam $\frac{3}{5}$. erit per propos .8. minutarium libri 9. Euclid. minor proportio 4. ad 7. quam 3. ad 5."

Pitiscus makes extensive use of decimal fractions. In the first five books of his *Trigonometria* the decimal fractions are not preceded by integral values. The fractional numerals are preceded by a zero; thus on page 44 he writes 02679492 (our 0.2679492) and finds its square root which he writes 05176381 (our 0.5176381). Given an arc and its chord, he finds (p. 54) the chord of one-third that arc. This leads to the equation (in modern symbols) $3x - x^3 = .5176381$, the radius being unity. In the solution of this equation by approximation he obtains successively 01, 017, 0174 . . . . and finally 01743114. In computing, he squares and cubes each of these numbers. Of 017, the square is given as 00289, the cube as 0004913. This proves that Pitiscus understood operations with decimals. In squaring 017 appears the following:

$$
\begin{array}{r}
\text{`` }001\,.7 \\
2\ 7 \\
1\ 89 \\
\hline
002\ 89\,.4\text{ ''}
\end{array}
$$

What rôle do these dots play? If we put $a = \frac{1}{10}$, $b = \frac{7}{100}$, then $(a+b)^2 = a^2 + (2a+b)b$; $001 = a^2$, $027 = (2a+b)$, $00189 = (2a+b)b$, $00289 \equiv (a+b)^2$. The dot in 001.7 serves simply as a separator between the 001 and the digit 7, found in the second step of the approximation. Similarly, in 00289.4, the dot separates 00289 and the digit 4, found in the third step of the approximation. It is clear that the dots used by Pitiscus in the foreging approximation are not decimal points.

The part of Pitiscus' *Trigonometria* (1612) which bears the title "Problematvm variorvm . . . . libri vndecim" begins a new pagination. Decimal fractions are used extensively, but integral parts appear and a vertical bar is used as decimal separatrix, as (p. 12) where he says, "pro . . . . 13|00024. assumo 13. fractione scilicet $\frac{24}{100000}$ neglecta." ("For 13.00024 I assume 13, the fraction, namely, $\frac{24}{100,000}$ being neglected.") Here again he displays his understanding of decimals, and he uses the dot for other purposes than a decimal separatrix. The writer has carefully examined every appearance of

[1] *Christophori Clavius* . . . . *Geometria practica* (Mogvntiae, 1606), p. 343.

dots in the processes of arithmetical calculation, but has failed to find the dot used as a decimal separatrix. There are in the Pitiscus of 1612 three notations for decimal fractions, the three exhibited in 0522 (our .522), 5|269 (our 5.269), and the form (p. 9) of common fractions, $121\frac{418}{1000}$. In one case (p. 11) there occurs the tautological notation $29|\frac{95}{100}$ (our 29.95).

280. But it has been affirmed that Pitiscus used the decimal point in his trigonometric Table. Indeed, the dot does appear in the Table of 1612 hundreds of times. Is it used as a decimal point? Let us quote from Pitiscus (p. 34): "Therefore the radius for the making of these Tables is to be taken so much the more, that there may be no error in so many of the figures towards the left hand, as you will have placed in the Tables: And as for the superfluous numbers they are to be cut off from the right hand toward the left, after the ending of the calculation. So did Regiomontanus, when he would calculate the tables of sines to the radius of 6000000; he took the radius 60000000000. and after the computation was ended, he cut off from every sine so found, from the right hand toward the left four figures, so Rhaeticus when he would calculate a table of sines to the radius of 10000000000 took for the radius 1000000000000000 and after the calculation was done, he cut off from every sine found from the right hand toward the left five figures: But I, to find out the numbers in the beginning of the Table, took the radius of 100000 00000 00000 00000 00000. But in the Canon itself have taken the radius divers numbers for necessity sake: As hereafter in his place shall be declared."

On page 83 Pitiscus states that the radius assumed is unity followed by 5, 7, 8, 9, 10, 11, or 12 ciphers, according to need. In solving problems he takes, on page 134, the radius 10^7 and writes sin 61°46′ = 8810284 (the number in the table is 88102.838); on page 7 ("Probl. var.") he takes the radius 10^5 and writes sin 41°10′ = 65825 (the number in the Table is 66825.16). Many examples are worked, but in no operation are the trigonometric values taken from the Table written down as decimal fractions. In further illustration we copy the following numerical values from the Table of 1612 (which contains sines, tangents, and secants):

" sin 2″ = 97   sec   3″ = 100000.00001.06
  sin 3″ = 1.45   sec 2°30′ = 100095.2685.
  tan 3″ = 1.45   sec 3°30′ = 100186.869
              sin 89°59′59″ = 99999.99999.88
              tan 89°59′59″ = 20626480624.
  sin 30°31′ = 50778.90   sec 30°31′ = 116079.10"

To explain all these numbers the radius must be taken 10^{12}. The 100000.00001.06 is an integer. The dot on the right is placed between tens and hundreds. The dot on the left is placed between millions and tens of millions.

When a number in the Table contains two dots, the left one is always between millions and tens of millions. The right-hand dot is between tens and hundreds, except in the case of the secants of angles between 0°19′ and 2°31′ and in the case of sines of angles between 87°59′ and 89°40′; in these cases the right-hand dot is placed (probably through a printer's error) between hundreds and thousands (see sec. 2°30′). The tangent of 89°59′59″ (given above) is really 20626480624-0000000, when the radius is 10^{12}. All the figures below ten millions are omitted from the Table in this and similar cases of large functional values.

If a sine or tangent has one dot in the Table and the secant for the same angle has two dots, then the one dot for the sine or tangent lies between millions and tens of millions (see sin 3″, sec 3″).

If both the sine and secant of an angle have only one dot in the Table and $r = 10^{12}$, that dot lies between millions and tens of millions (see sin 30°31′ and sec 30°31′). If the sine or tangent of an angle has no dots whatever (like sin 2″), then the figures are located immediately below the place for tens of millions. For all angles above 2°30′ and below 88° the numbers in the Table contain each one and only one dot. If that dot were looked upon as a decimal point, correct results could be secured by the use of that part of the Table. It would imply that the radius is always to be taken 10^5. But this interpretation is invalid for any one of the following reasons: (1) Pitiscus does not always take the $r = 10^5$ (in his early examples he takes $r = 10^7$), and he explicitly says that the radius may be taken 10^5, 10^7, 10^8, 10^9, 10^{10}, 10^{11}, or 10^{12}, to suit the degrees of accuracy demanded in the solution. (2) In the numerous illustrative solutions of problems the numbers taken from the Table are always in integral form. (3) The two dots appearing in some numbers in the Table could not both be decimal points. (4) The numbers in the Table containing no dots could not be integers.

The dots were inserted to facilitate the selection of the trigonometric values for any given radius. For $r = 10^5$, only the figures lying to the left of the dot between millions and tens of millions were copied. For $r = 10^{10}$, the figures to the left of the dot between tens and hundreds were chosen, zeroes being supplied in cases like sin 30°31′, where there was only one dot, so as to yield sin 30°31′ = 5077890000.

For $r = 10^7$, the figures for 10^5 and the two following figures were copied from the Table, yielding, for example, sin 30°31′ = 5077890. Similarly for other cases.

In a Table[1] which Pitiscus brought out in 1613 one finds the sine of 2°52′30″ given as 5015.71617.47294, thus indicating a different place assignment of the dots from that of 1612. In our modern tables the natural sine of 2°52′30″ is given as .05015. This is in harmony with the statement of Pitiscus on the title-page that the Tables are computed "ad radium 1.00000.00000.00000." The observation to be stressed is that these numbers in the Table of Pitiscus (1613) are not decimal fractions, but integers.

Our conclusions, therefore, are that Pitiscus made extended use of decimal fractions, but that the honor of introducing the dot as the separatrix between units and tenths must be assigned to others.

J. Ginsburg has made a discovery of the occurrence of the dot in the position of a decimal separatrix, which he courteously permits to be noted here previous to the publication of his own account of it. He has found the dot in Clavius' *Astrolabe*, published in Rome in 1593, where it occurs in a table of sines and in the explanation of that table (p. 228). The table gives sin 16°12′ = 2789911 and sin 16°13′ = 2792704. Clavius places in a separate column 46.5 as a correction to be made for every second of arc between 16°12′ and 16°13′. He obtained this 46.5 by finding the difference 2793 "between the two sines 2789911.2792704," and dividing that difference by 60. He identifies 46.5 as signifying $46\frac{5}{10}$. This dot separates units and tenths. In his works, Clavius uses the dot regularly to separate any two successive numbers. The very sentence which contains 46.5 contains also the integers "2789911.2792704." The question arises, did Clavius in that sentence use both dots as general separators of two pairs of numbers, of which one pair happened to be the integers 46 and the five-tenths, or did Clavius consciously use the dot in 46.5 in a more restricted sense as a decimal separatrix? His use of the plural "duo hi numeri 46.5" goes rather against the latter interpretation. If a more general and more complete statement can be found in Clavius, these doubts may be removed. In his *Algebra* of 1608, Clavius writes all decimal fractions in the form of common fractions. Nevertheless, Clavius unquestionably deserves a place in the history of the introduction of the dot as a decimal separatrix.

More explicit in statement was John Napier who, in his *Rabdologia*

[1] B. Pitiscus, *Thesavrvs mathematicvs. sive Canon sinum* (Francofurti, 1613), p. 19.

of 1617, recommended the use of a "period or comma" and uses the comma in his division. Napier's *Constructio* (first printed in 1619) was written before 1617 (the year of his death). In section 5 he says: "Whatever is written after the period is a fraction," and he actually uses the period. In the Leyden edition of the *Constructio* (1620) one finds (p. 6) "25.803. idem quod $25\frac{803}{1000}$."

281. The point occurs in E. Wright's 1616 edition of Napier's *Descriptio*, but no evidence has been advanced, thus far, to show that the sign was intended as a separator of units and tenths, and not as a more general separator as in Pitiscus.

282. *The decimal comma and point of Napier.*—That John Napier in his *Rabdologia* of 1617 introduced the comma and point as separators of units and tenths, and demonstrated that the comma was intended to be used in this manner by performing a division, and properly placing the comma in the quotient, is admitted by all historians. But there are still historians inclined to the belief that he was not the first to use the point or comma as a separatrix between units and tenths. We copy from Napier the following: "Since there is the same facility in working with these fractions as with whole numbers, you will be able after completing the ordinary division, and adding a period or comma, as in the margin, to add to the dividend or to the remainder one cypher to obtain

```
                        6 4
                      1 3 6
                    3 1 6
                  1 1 8,0 0 0
              1 4 1
            4 0 2
          4 2 9
          8 6 1 0 9 4,0 0 0(1 9 9 3,2 7 3
          4 3 2
          3 8 8 8
            3 8 8 8
              1 2 9 6
          ─────────────────
                  8 6 4
                  3 0 2 4
                    1 2 9 6
```

tenths, two for hundredths, three for thousandths, or more afterwards as required: And with these you will be able to proceed with the working as above. For instance, in the preceding example, here repeated, to which we have added three cyphers, the quotient will become 1 9 9 3,2 7 3, which signifies 1 9 9 3 units and 2 7 3 thousandth parts or $\frac{273}{1000}$."[1]

Napier gives in the *Rabdologia* only three examples in which decimals occur, and even here he uses in the text the sexagesimal exponents for the decimals in the statement of the results.[2] Thus he writes 1994.9160 as 1994,9 $\overset{\prime}{1}\,\overset{\prime\prime}{6}\,\overset{\prime\prime\prime}{0}\overset{\prime\prime\prime\prime}{}$; in the edition brought out at Leyden in 1626, the circles used by S. Stevin in his notation of decimals are used in place of Napier's sexagesimal exponents.

Before 1617, Napier used the decimal point in his *Constructio*, where he explains the notation in sections 4, 5, and 47, but the *Constructio* was not published until 1619, as already stated above. In section 5 he says: "Whatever is written after the period is a fraction," and he actually uses the period. But in the passage we quoted from *Rabdologia* he speaks of a "period or comma" and actually uses a comma in his illustration. Thus, Napier vacillated between the period and the comma; mathematicians have been vacillating in this matter ever since.

In the 1620 edition[3] of the *Constructio*, brought out in Leyden, one reads: "Vt 10000000.04, valet idem, quod $10000000\frac{4}{100}$. Item 25.803. idem quod $25\frac{803}{1000}$. Item 9999998.0005021, idem valet quod $9999998\frac{5021}{10000000}$. & sic de caeteris."

283. *Seventeenth-century notations after 1617.*—The dot or comma attained no ascendancy over other notations during the seventeenth century.

In 1623 John Johnson (the *survaighour*)[4] published an *Arithmatick* which stresses decimal fractions and modifies the notation of Stevin by omitting the circles. Thus, £ 3. 2 2 9 1 6 is written

$$\pounds\,3\,\left|\overset{\text{1. 2. 3. 4. 5.}}{2\,2\,9\,1\,6}\right.,$$

while later in the text there occurs the symbolism 31 | 2500 and 54 | 2625, and also the more cautious "358 | 49411 fifths" for our 358.49411.

---

[1] John Napier, *Rabdologia* (Edinburgh, 1617), Book I, chap. iv. This passage is copied by W. R. Macdonald, in his translation of John Napier's *Constructio* (Edinburgh, 1889), p. 89.

[2] J. W. L. Glashier, "Logarithms and Computation," *Napier Tercentenary Memorial Volume* (ed. Cargill Gilston Knott; London, 1915), p. 78.

[3] *Mirifici logarithmorvm Canonis Constructio . . . . authore & Inventore Ioanne Nepero, Barone Merchistonii, etc.* (Scoto. Lvgdvni, M.DC.XX.), p. 6.

[4] From A. de Morgan in *Companion to the British Almanac* (1851), p. 12.

Henry Briggs[1] drew a horizontal line under the numerals in the decimal part which appeared in smaller type and in an elevated position; Briggs wrote 5_{9321} for our 5.9321. But in his Tables of 1624 he employs commas, not exclusively as a decimal separatrix, although one of the commas used for separation falls in the right place between units and tenths. He gives $-0,22724,3780$ as the logarithm of $\frac{16}{27}$.

A. Girard[2] in his *Invention nouvelle* of 1629 uses the comma on one occasion; he finds one root of a cubic equation to be $1\frac{532}{1000}$ and then explains that the three roots expressed in decimals are 1,532 and 347 and $-1,879$. The 347 is .347; did Girard consider the comma unnecessary when there was no integral part?

Bürgi's and Kepler's notation is found again in a work which appeared in Poland from the pen of Joach. Stegman;[3] he writes 39(063. It occurs again in a geometry written by the Swiss Joh. Ardüser.[4]

William Oughtred adopted the sign 2|5 in his *Clavis mathematicae* of 1631 and in his later publications.

In the second edition of Wingate's *Arithmetic* (1650; ed. John Kersey) the decimal point is used, thus: .25, .0025.

In 1651 Robert Jager[5] says that the common way of natural arithmetic being tedious and prolix, God in his mercy directed him to what he published; he writes upon decimals, in which 16|7249 is our 16.7249.

Richard Balam[6] used the colon and wrote 3:04 for our 3.04. This same symbolism was employed by Richard Rawlyns,[7] of Great Yarmouth, in England, and by H. Meissner[8] in Germany.

[1] Henry Briggs, *Arithmetica logarithmica* (London, 1624), Lectori. S.

[2] De Morgan, *Companion to the British Almanac* (1851), p. 12; *Invention nouvelle*, fol. *E*2.

[3] Joach. Stegman, *Institutionum mathematicarum libri II* (Rakow, 1630), Vol. I, cap. xxiv, "De logistica decimali." We take this reference from J. Tropfke, *op. cit.*, Vol. I (2d ed., 1921), p. 144.

[4] Joh. Ardüser, *Geometriae theoricae et practicae XII libri* (Zürich, 1646), fol. 306, 180*b*, 270*a*.

[5] Robert Jager, *Artificial Arithmetick in Decimals* (London, 1651). Our information is drawn from A. de Morgan in *Companion to the British Almanac* (1851), p. 13.

[6] Rich. Balam, *Algebra* (London, 1653), p. 4.

[7] Richard Rawlyns, *Practical Arithmetick* (London, 1656), p. 262.

[8] H. Meissner, *Geometria tyronica* (1696[?]). This reference is taken from J. Tropfke, *op. cit.*, Vol. I (2d ed., 1921), p. 144.

Sometimes one encounters a superposition of one notation upon another, as if one notation alone might not be understood. Thus F. van Schooten[1] writes 58,5 ① for 58.5, and 638,82 ② for 638.82. Tobias Beutel[2] writes $645._{1\,0\,0\,0}^{8\,7\,9}$. A. Tacquet[3] sometimes writes $25.8\ \overset{\text{i ii iii iv v}}{0\ 0\ 0\ 7\ 9}$, at other times omits the dot, or the Roman superscripts.

Samuel Foster[4] of Gresham College, London, writes $31.\underline{0\,0\,8}$; he does not rely upon the dot alone, but adds the horizontal line found in Briggs.

Johann Caramuel[5] of Lobkowitz in Bohemia used two horizontal parallel lines, like our sign of equality, as 22=3 for 22.3, also 92= 123,345 for 92.123345. In a Parisian text by Jean Prestet[6] $272097792\overset{\text{vi}}{}$ is given for 272.097792; this mode of writing had been sometimes used by Stevin about a century before Prestet, and in 1603 by Beyer.

William Molyneux[7] of Dublin had three notations; he frequently used the comma bent toward the right, as in 30ˌ24. N. Mercator[8] in his *Logarithmotechnia* and Dechales[9] in his course of mathematics used the notation as in 12[345.

284. The great variety of forms for separatrix is commented on by Samuel Jeake in 1696 as follows: "For distinguishing of the Decimal Fraction from Integers, it may truly be said, *Quot Homines, tot Sententiae;* every one fancying severally. For some call the Tenth Parts, the *Primes;* the Hundredth Parts, *Seconds;* the Thousandth Parts, *Thirds,* etc. and mark them with *Indices* equivalent over their heads. As to express 34 integers and $_{1\,0\,0\,0\,0}^{1\,4\,2\,6}$ Parts of an Unit, they do it thus, 34.1. 4. 2. 6. Or thus, $34.\overset{(1)}{1}.\overset{(2)}{4}.\overset{(3)}{2}.\overset{(4)}{6}$. Others thus, 34,1426'''''; or thus, 34,1426$^{(4)}$. And some thus, 34.1 . 4 . 2 . 6 . setting the Decimal Parts

---

[1] Francisci à Schooten, *Exercitationvm mathematicarum liber primus* (Leyden, 1657), p. 33, 48, 49.

[2] Tobias Beutel, *Geometrischer Lust-Garten* (Leipzig, 1690), p. 173.

[3] *Arithmeticae theoria et praxis, autore Andrea Tacqvet* (2d ed.; Antwerp, 1665), p. 181–88.

[4] Samuel Foster, *Miscellanies: or Mathematical Lvcvbrations* (London, 1659), p. 13.

[5] *Joannis Caramvels Mathesis Biceps. Vetus, et Nova* (Companiae, 1670), "Arithmetica," p. 191.

[6] Jean Prestet, *Nouveaux elemens des mathematiques,* Premier volume (Paris, 1689), p. 293.

[7] William Molyneux, *Treatise of Dioptricks* (London, 1692), p. 165.

[8] N. Mercator, *Logarithmotechnia* (1668), p. 19.

[9] A. de Morgan, *Companion to the British Almanac* (1851), p. 13.

at little more than ordinary distance one from the other. . . . . Others distinguish the Integers from the Decimal Parts only by placing a Cöma before the Decimal Parts thus, 34,1426; a good way, and very useful. Others draw a Line under the Decimals thus, 34<u>1426</u>, writing them in smaller Figures than the Integers. And others, though they use the Cöma in the work for the best way of distinguishing them, yet after the work is done, they use a Rectangular Line after the place of the Units, called *Separatrix*, a separating Line, because it separates the Decimal Parts from the Integers, thus 34|1426. And sometimes the Cöma is inverted thus, 34'1426, contrary to the true Cöma, and set at top. I sometimes use the one, and sometimes the other, as cometh to hand." The author generally uses the comma. This detailed statement from this seventeenth-century writer is remarkable for the omission of the point as a decimal separatrix.

285. *Eighteenth-century discard of clumsy notations.*—The chaos in notations for decimal fractions gradually gave way to a semblance of order. The situation reduced itself to trials of strength between the comma and the dot as separatrices. To be sure, one finds that over a century after the introduction of the decimal point there were authors who used besides the dot or comma the strokes or Roman numerals to indicate primes, seconds, thirds, etc. Thus, Chelucci[1] in 1738 writes $\overset{0}{5}.\overset{I}{8}\ \overset{II}{6}\overset{III}{4}\overset{IV}{2}$, also $\overset{I}{4}.\overset{IV}{2}\ \overset{}{5}$ for 4.2005, $\overset{II}{3}.\overset{V}{5}\ \overset{}{7}$ for 3.05007.

W. Whiston[2] of Cambridge used the semicolon a few times, as in 0;9985, though ordinarily he preferred the comma. O. Gherli[3] in Modena, Italy, states that some use the sign 35|345, but he himself uses the point. E. Wells[4] in 1713 begins with 75.25, but later in his arithmetic introduces Oughtred's |75. Joseph Raphson's translation into English of I. Newton's *Universal Arithmetick* (1728),[5] contains 732,|569 for our 732.569. L'Abbé Deidier[6] of Paris writes the

[1] Paolino Chelucci, *Institutiones analyticae . . . . auctore Paulino A. S. Josepho Lucensi* (Rome), p. 35, 37, 41, 283.

[2] Isaac Newton, *Arithmetica Vniversalis* (Cambridge, 1707), edited by G. W[histon], p. 34.

[3] O. Gherli, *Gli elementi . . . . delle mathematiche pure*, Vol. I (Modena, 1770), p. 60.

[4] Edward Wells, *Young gentleman's arithmetick* (London, 1713), p. 59, 105, 157.

[5] *Universal Arithmetick, or Treatise of Arithmetical Composition and Resolution . . . .* transl. by the late Mr. Joseph Ralphson, & revised and corrected by Mr. Cunn (2d ed.; London, 1728), p. 2.

[6] L'Abbé Deidier, *L'Arithmétique des Géomètres, ou nouveaux élémens de mathématiques* (Paris, 1739), p. 413.

decimal point and also the strokes for tenths, hundredths, etc. He says: "Pour ajouter ensemble $32.6'\ 3''\ 4'''$ et $8.5'\ 4''.3'''$ —

$$
\begin{array}{r}
32\ 6\ 3\ 4^{\mathrm{III}} \\
8\ 5\ 4\ 3^{\mathrm{III}} \\
\hline
41\ 1\ 7\ 7^{\mathrm{III}}\ "
\end{array}
$$

A somewhat unusual procedure is found in Sherwin's *Tables*[1] of 1741, where a number placed inside a parenthesis is used to designate the number of zeroes that precede the first significant figure in a decimal; thus, (4) 2677 means .00002677.

In the eighteenth century, trials of strength between the comma and the dot as the separatrix were complicated by the fact that Leibniz had proposed the dot as the symbol of multiplication, a proposal which was championed by the German textbook writer Christian Wolf and which met with favorable reception throughout the Continent. And yet Wolf[2] himself in 1713 used the dot also as separatrix, as "loco $5\frac{47}{10000}$ scribimus 5.0047." As a symbol for multiplication the dot was seldom used in England during the eighteenth century, Oughtred's $\times$ being generally preferred. For this reason, the dot as a separatrix enjoyed an advantage in England during the eighteenth century which it did not enjoy on the Continent. Of fifteen British books of that period, which we chose at random, nine used the dot and six the comma. In the nineteenth century hardly any British authors employed the comma as separatrix.

In Germany, France, and Spain the comma, during the eighteenth century, had the lead over the dot, as a separatrix. During that century the most determined continental stand in favor of the dot was made in Belgium[3] and Italy.[4] But in recent years the comma has finally won out in both countries.

[1] H. Sherwin, *Mathematical Tables* (3d ed.; rev. William Gardiner, London, 1741), p. 48.

[2] Christian Wolf, *Elementa matheseos universae*, Tomus I (Halle, 1713), p. 77.

[3] Désiré André, *Des Notations Mathématiques* (Paris, 1909), p. 19, 20.

[4] Among eighteenth-century writers in Italy using the dot are Paulino A. S. Josepho Lucensi who in his *Institutiones analyticae* (Rome, 1738) uses it in connection with an older symbolism, "3.05007"; G. M. della Torre, *Istituzioni arimmetiche* (Padua, 1768); Odoardo Gherli, *Elementi delle matematiche pure*, Modena, Tomo I (1770); Peter Ferroni, *Magnitudinum exponentialium logarithmorum et trigonometriae sublimis theoria* (Florence, 1782); F. A. Tortorella, *Arithmetica degl'idioti* (Naples, 1794).

286. *Nineteenth century: different positions for dot and comma.*—
In the nineteenth century the dot became, in England, the favorite
separatrix symbol. When the brilliant but erratic Randolph Churchill
critically spoke of the "damned little dots," he paid scant respect to
what was dear to British mathematicians. In that century the dot
came to serve in England in a double capacity, as the decimal symbol
and as a symbol for multiplication.

Nor did these two dots introduce confusion, because (if we may
use a situation suggested by Shakespeare) the symbols were placed in
Romeo and Juliet positions, the Juliet dot stood on high, above
Romeo's reach, her joy reduced to a decimal over his departure, while
Romeo below had his griefs multiplied and was "a thousand times the
worse" for want of her light. Thus, $2 \cdot 5$ means $2\frac{5}{10}$, while 2.5 equals
10. It is difficult to bring about a general agreement of this kind,
but it was achieved in Great Britain in the course of a little over half
a century. Charles Hutton[1] said in 1795: "I place the point near the
upper part of the figures, as was done also by Newton, a method which
prevents the separatrix from being confounded with mere marks of
punctuation." In the Latin edition[2] of Newton's *Arithmetica uni-
versalis* (1707) one finds, "Sic numerus 732'|569. denotat septingentas
triginta duas unitates, ... qui et sic 732,|569, vel sic 732˙569. vel
etiam sic ˉ732|569, nunnunquam scribitur . . . . 57104'2083 . . . .
0'064." The use of the comma prevails; it is usually placed high, but
not always. In Horsely's and Castillon's editions of Newton's *Arithmetica universalis* (1799) one finds in a few places the decimal notation 35'72; it is here not the point but the comma that is placed on
high. Probably as early as the time of Hutton the expression "decimal point" had come to be the synonym for "separatrix" and was
used even when the symbol was not a point. In most places in Horsley's and Castillon's editions of Newton's works, the comma 2,5 is
used, and only in rare instances the point 2.5. The sign $2 \cdot 5$ was used
in England by H. Clarke[3] as early as 1777, and by William Dickson[4]
in 1800. After the time of Hutton the $2 \cdot 5$ symbolism was adopted by
Peter Barlow (1814) and James Mitchell (1823) in their mathematical
dictionaries. Augustus de Morgan states in his *Arithmetic:* "The

---

[1] Ch. Hutton, *Mathematical and Philosophical Dictionary* (London, 1795),
art. "Decimal Fractions."

[2] I. Newton, *Arithmetica universalis* (ed. W. Whiston; Cambridge, 1707), p. 2.
See also p. 15, 16.

[3] H. Clarke, *Rationale of Circulating Numbers* (London, 1777).

[4] W. Dickson in *Philosophical Transactions*, Vol. VIII (London, 1800), p. 231.

student is recommended always to write the decimal point in a line with the top of the figures, or in the middle, as is done here, and never at the bottom. The reason is that it is usual in the higher branches of mathematics to use a point placed between two numbers or letters which are multiplied together."[1] A similar statement is made in 1852 by T. P. Kirkman.[2] Finally, the use of this notation in Todhunter's texts secured its general adoption in Great Britain.

The extension of the usefulness of the comma or point by assigning it different vertical positions was made in the arithmetic of Sir Jonas Moore[3] who used an elevated and inverted comma, 116'64. This notation never became popular, yet has maintained itself to the present time. Daniel Adams,[4] in New Hampshire, used it, also Juan de Dios Salazar[5] in Peru, Don Gabriel Ciscar[6] of Mexico, A. de la Rosa Toro[7] of Lima in Peru, and Federico Villareal[8] of Lima. The elevated and inverted comma occurs in many, but not all, the articles using decimal fractions in the *Enciclopedia-vniversal ilvstrada Evropeo-Americana* (Barcelona, 1924).

Somewhat wider distribution was enjoyed by the elevated but not inverted comma, as in 2'5. Attention has already been called to the occurrence of this symbolism, a few times, in Horsley's edition of Newton's *Arithmetica universalis*. It appeared also in W. Whiston's edition of the same work in 1707 (p. 15). Juan de Dios Salazar of Peru, who used the elevated inverted comma, also uses this. It is Spain and the Spanish-American countries which lead in the use of this notation. De La-Rosa Toro, who used the inverted comma, also used this. The 2'5 is found in Luis Monsante[9] of Lima; in Maximo

---

[1] A. de Morgan, *Elements of Arithmetic* (4th ed.; London, 1840), p. 72.

[2] T. P. Kirkman, *First Mnemonical Lessons in Geometry, Algebra and Trigonometry* (London, 1852), p. 5.

[3] *Moore's Arithmetick: In Four Books* (3d ed.; London, 1688), p. 369, 370, 465.

[4] Daniel Adams, *Arithmetic* (Keene, N.H., 1827), p. 132.

[5] Juan de Dios Salazar, *Lecciones de Aritmetica*, Teniente del Cosmògrafo major de esta Republica del Perŭ (Arequipa, 1827), p. 5, 74, 126, 131. This book has three different notations: 2,5; 2'5; 2'5.

[6] Don Gabriel Ciscar, *Curso de estudios elementales de Marina* (Mexico, 1825).

[7] Agustin de La-Rosa Toro, *Aritmetica Teorico-Practica* (tercera ed.; Lima, 1872), p. 157.

[8] D. Federico Villareal, *Calculo Binomial* (P. I. Lima [Peru], 1898), p. 416.

[9] Luis Monsante, *Lecciones de Aritmetica Demostrada* (7th ed.; Lima, 1872), p. 89.

Vazquez[1] of Lima; in Manuel Torres Torija[2] of Mexico; in D. J. Cortazár[3] of Madrid. And yet, the Spanish-speaking countries did not enjoy the monopoly of this symbolism. One finds the decimal comma placed in an elevated position, 2'5, by Louis Bertrand[4] of Geneva, Switzerland.

Other writers use an inverted wedge-shaped comma,[5] in a lower position, thus: $2_{\blacktriangle}5$. In Scandinavia and Denmark the dot and the comma have had a very close race, the comma being now in the lead. The practice is also widely prevalent, in those countries, of printing the decimal part of a number in smaller type than the integral part.[6] Thus one frequently finds there the notations $2_{,5}$ and $2._5$. To sum up, in books printed within thirty-five years we have found the decimal notations[7] $2{\cdot}5$, $2{}^{\cdot}5$, 2,5, 2'5, 2'5, $2_{\blacktriangle}5$, $2_{,5}$, $2._5$.

287. The earliest arithmetic printed on the American continent which described decimal fractions came from the pen of Greenwood,[8] professor at Harvard College. He gives as the mark of separation "a Comma, a Period, or the like," but actually uses a comma. The arithmetic of "George Fisher" (Mrs. Slack), brought out in England, and also her *The American Instructor* (Philadelphia, 1748) contain both the comma and the period. Dilworth's *The Schoolmaster's Assistant*, an English book republished in America (Philadelphia, 1733), used the period. In the United States the decimal point[9] has always had the

---

[1] Maximo Vazquez, *Aritmetica practica* (septiema ed.; Lima, 1875), p. 57.

[2] Manuel Torres Torija, *Nociones de Algebra Superior y elementos fundamentales de cálculo differencial é Integral* (México, 1894), p. 137.

[3] D. J. Cortazár, *Tratado de Aritmética* (42d ed.; Madrid, 1904).

[4] L. Bertrand, *Developpment nouveaux de la partie elementaire des mathematiques*, Vol. I (Geneva, 1778), p. 7.

[5] As in A. F. Vallin, *Aritmética para los niños* (41st ed.; Madrid, 1889), p. 66.

[6] Gustaf Haglund, *Samlying of Öfningsexempel till Lärabok i Algebra*, Fjerde Upplagan (Stockholm, 1884), p. 19; *Öfversigt af Kongl. Vetenskaps-Akademiens Förhandlingar*, Vol. LIX (1902; Stockholm, 1902, 1903), p. 183, 329; *Oversigt over det Kongelige Danske Videnskabernes Selskabs, Fordhandlinger* (1915; Kobenhavn, 1915), p. 33, 35, 481, 493, 545.

[7] An unusual use of the elevated comma is found in F. G. Gausz's *Fünfstellige vollständige Logar. u. Trig. Tafeln* (Halle a. S., 1906), p. 125; a table of squares of numbers proceeds from $N = 0'00$ to $N = 10'00$. If the square of 63 is wanted, take the form 6'3; its square is 39'6900. Hence $63^2 = 3969$.

[8] Isaac Greenwood, *Arithmetick Vulgar and Decimal* (Boston, 1729), p. 49. See facsimile of a page showing decimal notation in L. C. Karpinski, *History of Arithmetic* (Chicago, New York, 1925), p. 134.

[9] Of interest is Chauncey Lee's explanation in his *American Accomptant* (Lasingburgh, 1797), p. 54, that, in writing denominate numbers, he separates

lead over the comma, but during the latter part of the eighteenth and the first half of the nineteenth century the comma in the position of 2,5 was used quite extensively. During 1825–50 it was the influence of French texts which favored the comma. We have seen that Daniel Adams used 2'5 in 1827, but in 1807 he[1] had employed the ordinary 25,17 and ,375. Since about 1850 the dot has been used almost exclusively. Several times the English elevated dot was used in books printed in the United States. The notation 2·5 is found in Thomas Sarjeant's *Arithmetic*,[2] in F. Nichols' *Trigonometry*,[3] in American editions of Hutton's *Course of Mathematics* that appeared in the interval 1812–31, in Samuel Webber's *Mathematics*,[4] in William Griev's *Mechanics Calculator*, from the fifth Glasgow edition (Philadelphia, 1842), in *The Mathematical Diary* of R. Adrain[5] about 1825, in Thomas Sherwin's *Common School Algebra* (Boston, 1867; 1st ed., 1845), in George R. Perkins' *Practical Arithmetic* (New York, 1852). Sherwin writes: "To distinguish the sign of Multiplication from the period used as a decimal point, the latter is elevated by inverting the type, while the former is larger and placed down even with the lower extremities of the figures or letters between which it stands." In 1881 George Bruce Halsted[6] placed the decimal point halfway up and the multiplication point low.

It is difficult to assign definitely the reason why the notation 2·5 failed of general adoption in the United States. Perhaps it was due to mere chance. Men of influence, such as Benjamin Peirce, Elias Loomis, Charles Davies, and Edward Olney, did not happen to become interested in this detail. America had no one of the influence of De Morgan and Todhunter in England, to force the issue in favor of 2·5. As a result, 2·5 had for a while in America a double meaning, namely, 2 5/10 and 2 times 5. As long as the dot was seldom used to

---

the denominations "in a vulgar table" by two commas, but "in a decimal table" by the decimal point; he writes £ 175,, 15,, 9, and 1.41.

[1] Daniel Adams, *Scholar's Arithmetic* (4th ed.; Keene, N.H., 1807).

[2] Thomas Sarjeant, *Elementary Principles of Arithmetic* (Philadelphia, 1788), p. 80.

[3] F. Nichols, *Plane and Spherical Trigonometry* (Philadelphia, 1811), p. 33.

[4] Samuel Webber, *Mathematics*, Vol. I (Cambridge, 1801; also 1808, 2d ed.), p. 227.

[5] R. Adrain, *The Mathematical Diary*, No. 5, p. 101.

[6] George Bruce Halsted, *Elementary Treatise on Mensuration* (Boston, 1881).

express multiplication, no great inconvenience resulted, but about 1880 the need of a distinction arose. The decimal notation was at that time thoroughly established in this country, as 2.5, and the dot for multiplication was elevated to a central position. Thus with us $2 \cdot 5$ means 2 times 5.

Comparing our present practice with the British the situation is this: We write the decimal point low, they write it high; we place the multiplication dot halfway up, they place it low. Occasionally one finds the dot placed high to mark multiplication also in German books, as, for example, in Friedrich Meyer[1] who writes $2 \cdot 3 = 6$.

288. It is a notable circumstance that at the present time the modern British decimal notation is also the notation in use in Austria where one finds the decimal point placed high, but the custom does not seem to prevail through any influence emanating from England. In the eighteenth century P. Mako[2] everywhere used the comma, as in 3,784. F. S. Mozhnik[3] in 1839 uses the comma for decimal fractions, as in 3,1344, and writes the product "2 . 3..n." The *Sitzungsberichte der philosophisch-historischen Classe d. K. Akademie der Wissenschaften*, Erster Band (Wien, 1848), contains decimal fractions in many articles and tables, but always with the low dot or low comma as decimal separatrix; the low dot is used also for multiplication, as in "1.2.3...r."

But the latter part of the nineteenth century brought a change. The decimal point is placed high, as in $1 \cdot 63$, by I. Lemoch[4] of Lemberg. N. Fialkowski of Vienna in 1863 uses the elevated dot[5] and also in 1892.[6] The same practice is followed by A. Steinhauser of Vienna,[7] by Johann Spielmann[8] and Richard Supplantschitsch,[9] and by Karl

[1] Friedrich Meyer, *Dritter Cursus der Planimetrie* (Halle a/S., 1885), p. 5.

[2] P. Mako e S.I., *De . . . . aequationvm resolvtionibvs libri dvo* (Vienna, 1770), p. 135; *Compendiaria Matheseos Instilvtio. . . . .* Pavlvs Mako e S.I. in Coll. Reg. Theres Prof. Math. et Phys. Experim. (editio tertia; Vienna, 1771).

[3] Franz Seraphin Mozhnik, *Theorie der numerischen Gleichungen* (Wien, 1839), p. 27, 33.

[4] Ignaz Lemoch, *Lehrbuch der praktischen Geometrie*, 2. Theil, 2. Aufl. (Wien, 1857), p. 163.

[5] Nikolaus Fialkowski, *Das Decimalrechnen mit Rangziffern* (Wien, 1863), p. 2.

[6] N. Fialkowski, *Praktische Geometrie* (Wien, 1892), p. 48.

[7] Anton Steinhauser, *Lehrbuch der Mathematik. Algebra* (Wien, 1875), p. 111, 138.

[8] Johann Spielmann, *Močniks Lehrbuch der Geometrie* (Wien, 1910), p. 66.

[9] Richard Supplantschitsch, *Mathematisches Unterrichtswerk, Lehrbuch der Geometrie* (Wien, 1910), p. 91.

Rosenberg.[1] Karl Zahradníček[2] writes $0\dot{}35679.1\dot{}0765.1\dot{}9223.0\dot{}3358$, where the lower dots signify multiplication and the upper dots are decimal points. In the same way K. Wolletz[3] writes $(-0\dot{}0462)$. $0\dot{}0056$.

An isolated instance of the use of the elevated dot as decimal separatrix in Italy is found in G. Peano.[4]

In France the comma placed low is the ordinary decimal separatrix in mathematical texts. But the dot and also the comma are used in marking off digits of large numbers into periods. Thus, in a political and literary journal of Paris (1908)[5] one finds "2,251,000 drachmes," "Fr. 2.638.370 75," the francs and centimes being separated by a vacant place. One finds also "601,659 francs 05" for Fr. 601659. 05. It does not seem customary to separate the francs from centimes by a comma or dot.

That no general agreement in the notation for decimal fractions exists at the present time is evident from the publication of the International Mathematical Congress in Strasbourg (1920), where decimals are expressed by commas[6] as in 2,5 and also by dots[7] as in 2.5. In that volume a dot, placed at the lower border of a line, is used also to indicate multiplication.[8]

The opinion of an American committee of mathematicians is expressed in the following: "Owing to the frequent use of the letter x, it is preferable to use the dot (a raised period) for multiplication in the few cases in which any symbol is necessary. For example, in a case like $1 \cdot 2 \cdot 3 \ldots (x-1) \cdot x$, the center dot is preferable to the symbol $\times$; but in cases like $2a(x-a)$ no symbol is necessary. The committee recognizes that the period (as in $a.b$) is more nearly international than the center dot (as in $a \cdot b$); but inasmuch as the period will continue to be used in this country as a decimal point,

[1] Karl Rosenberg, *Lehrbuch der Physik* (Wien, 1913), p. 125.

[2] Karl Zahradníček, *Močniks Lehrbuch der Arithmetik und Algebra* (Wien, 1911), p. 141.

[3] K. Wolletz, *Arithmetik und Algebra* (Wien, 1917), p. 163.

[4] Giuseppe Peano, *Risoluzione graduale delle equazioni numeriche* (Torino, 1919), p. 8. Reprint from *Atti della r. Accad. delle Scienze di Torino*, Vol. LIV (1918–19).

[5] *Les Annales*, Vol. XXVI, No. 1309 (1908), p. 22, 94.

[6] *Comptes rendus du congrès international des mathématiques* (Strasbourg, 22–30 Septembre 1920; Toulouse, 1921), p. 253, 543, 575, 581.

[7] *Op. cit.*, p. 251.

[8] *Op. cit.*, p. 153, 252, 545.

it is likely to cause confusion, to elementary pupils at least, to attempt to use it as a symbol for multiplication."[1]

289. *Signs for repeating decimals.*—In the case of repeating decimals, perhaps the earliest writer to use a special notation for their designation was John Marsh,[2] who, "to avoid the Trouble for the future of writing down the Given Repetend or Circulate, whether Single or Compound, more than once," distinguishes each "by placing a Period over the first Figure, or over the first and last Figures of the given Repetend." Likewise, John Robertson[3] wrote $0,\dot{3}$ for $0,33\ldots$, $0,\dot{2}\dot{3}$ for $0,2323\ldots$, $0,\dot{7}8\dot{5}$ for $0,785785\ldots$. H. Clarke[4] adopted the signs $.\acute{6}$ for $.666\ldots$, $.\acute{6}4\acute{2}$ for $.642642\ldots$. A choice favoring the dot is shown by Nicolas Pike[5] who writes, $\dot{3}7\dot{9}$, and by Robert Pott[6] and James Pryde[7] who write $\cdot\dot{3}$, $\cdot\dot{4}\dot{5}$, $\cdot\dot{3}4\dot{5}67$. A return to accents is seen in the *Dictionary* of Davies and Peck[8] who place accents over the first, or over the first and last figure, of the repetend, thus: $.\acute{2}$, $.\acute{5}723\acute{}$, $2.4\acute{}18\acute{}$.

SIGNS OF POWERS

290. *General remarks.*—An ancient symbol for squaring a number occurs in a hieratic Egyptian papyrus of the late Middle Empire, now in the Museum of Fine Arts in Moscow.[9] In the part containing the computation of the volume of a frustrated pyramid of square base there occurs a hieratic term, containing a pair of walking legs Λ and signifying "make in going," that is, squaring the number. The Diophantine notation for powers is explained in § 101, the Hindu notation in §§ 106, 110, 112, the Arabic in § 116, that of Michael Psellus in § 117. The additive principle in marking powers is referred

[1] *The Reorganization of Mathematics in Secondary Schools*, by the National Committee on Mathematical Requirements, under the auspices of the Mathematical Association of America (1923), p. 81.

[2] John Marsh, *Decimal Arithmetic Made Perfect* (London, 1742), p. 5.

[3] John Robertson, *Philosophical Transactions* (London, 1768), No. 32, p. 207–13. See Tropfke, *op. cit.*, Vol. I (1921), p. 147.

[4] H. Clarke, *The Rationale of Circulating Numbers* (London, 1777), p. 15, 16.

[5] Nicolas Pike, *A New and Complete System of Arithmetic* (Newbury-port, 1788), p. 323.

[6] Robert Pott, *Elementary Arithmetic, etc.* (Cambridge, 1876), Sec. X, p. 8.

[7] James Pryde, *Algebra Theoretical and Practical* (Edinburgh, 1852), p. 278.

[8] C. Davies and W. G. Peck, *Mathematical Dictionary* (1855), art. "Circulating Decimal."

[9] See *Ancient Egypt* (1917), p. 100–102.

to in §§ 101, 111, 112, 124. The multiplicative principle in marking powers is elucidated in §§ 101, 111, 116, 135, 142.

Before proceeding further, it seems desirable to direct attention to certain Arabic words used in algebra and their translations into Latin. There arose a curious discrepancy in the choice of the principal unknown quantity; should it be what we call x, or should it be x^2? al-Khowârizmî and the older Arabs looked upon x^2 as the principal unknown, and called it *mâl* ("assets," "sum of money").[1] This viewpoint may have come to them from India. Accordingly, x (the Arabic *jidr*, "plant-root," "basis," "lowest part") must be the square root of *mâl* and is found from the equation to which the problem gives rise. By squaring x the sum of money could be ascertained.

Al-Khowârizmî also had a general term for the unknown, *shai* ("thing"); it was interpreted broadly and could stand for either *mâl* or *jidr* (x^2 or x). Later, John of Seville, Gerard of Cremona, Leonardo of Pisa, translated the Arabic *jidr* into the Latin *radix*, our $x;$ the Arabic *shai* into *res*. John of Seville says in his arithmetic:[2] "Quaeritur ergo, quae res cum. X. radicibus suis idem decies accepta radice sua efficiat 39." ("It is asked, therefore, what thing together with 10 of its roots or what is the same, ten times the root obtained from it, yields 39.") This statement yields the equation $x^2+10x=39$. Later *shai* was also translated as *causa*, a word which Leonardo of Pisa used occasionally for the designation of a second unknown quantity. The Latin *res* was translated into the Italian word *cosa*, and from that evolved the German word *coss* and the English adjective "cossic." We have seen that the abbreviations of the words *cosa* and *cubus*, viz., *co.* and *cu.*, came to be used as algebraic symbols. The words *numerus, dragma, denarius*, which were often used in connection with a given absolute number, experienced contractions sometimes employed as symbols. Plato of Tivoli,[3] in his translation from the Hebrew of the *Liber embadorum* of 1145, used a new term, *latus* ("side"), for the first power of the unknown, x, and the name *embadum* ("content") for the second power, x^2. The term *latus* was found mainly in early Latin writers drawing from Greek sources and was used later by Ramus (§ 322), Vieta (§ 327), and others.

291. *Double significance of "R" and "l."*—There came to exist considerable confusion on the meaning of terms and symbols, not only

[1] J. Ruska, *Sitzungsberichte Heidelberger Akad., Phil.-hist. Klasse* (1917), Vol. II, p. 61 f.; J. Tropfke, *op. cit.*, Vol. II (2d ed., 1921), p. 106.

[2] Tropfke, *op. cit.*, Vol. II (2d ed., 1921), p. 107.

[3] M. Curtze, *Bibliotheca mathematica* (3d ser.), Vol. I (1900), p. 322, n. 1.

because *res* (x) occasionally was used for x^2, but more particularly because both *radix* and *latus* had two distinct meanings, namely, x and $\sqrt{x}$. The determination whether x or $\sqrt{x}$ was meant in any particular case depended on certain niceties of designation which the unwary was in danger of overlooking (§ 137).

The letter l (*latus*) was used by Ramus and Vieta for the designation of roots. In some rare instances it also represented the first power of the unknown x. Thus, in Schoner's edition of Ramus[1] $5l$ meant $5x$, while $l5$ meant $\sqrt{5}$. Schoner marks the successive powers "*l.*, *q.*, *c.*, *bq.*, *ſ.*, *qc.*, *b.ſ.*, *tq.*, *cc.*" and named them *latus, quadratus, cubus, biquadratus,* and so on. Ramus, in his *Scholarvm mathematicorvm libri unus et triginti* (1569), uses the letter l only for square root, not for x or in the designation of powers of x; but he uses (p. 253) the words *latus, quadratus, latus cubi* for x, x^2, x^3.

This double use of l is explained by another pupil of Ramus, Bernardus Salignacus,[2] by the statement that if a number precedes the given sign it is the coefficient of the sign which stands for a power of the unknown, but if the number comes immediately after the l the root of that number is to be extracted. Accordingly, $2q$, $3c$, $5l$ stand respectively for $2x^2$, $3x^3$, $5x$; on the other hand, $l5$, $lc8$, $lbq16$ stand respectively for $\sqrt{5}$, $\sqrt[3]{8}$, $\sqrt[4]{16}$. The double use of the capital L is found in G. Gosselin (§§ 174, 175).

B. Pitiscus[3] writes our $3x-x^3$ thus, $3l-1c$, and its square $9q-6.bq+1qc$, while Willebrord Snellius[4] writes our $5x-5x^3+x^5$ in the form $5l- -5c+1\beta$. W. Oughtred[5] writes $x^5-15x^4+160x^3-1250x^2+6480x=170304782$ in the form $1qc-15qq+160c-1250q+6480l=170304782$.

Both *l.* and *R.* appear as characters designating the first power

---

[1] *Petri Rami Veromandui Philosophi . . . . arithmetica libri duo et geometriae septem et viginti. Dudum quidem a Lazaro Schonero . . . .* (Francofvrto ad moenvm, MDCXXVII). P. 139 begins: "De Nvmeris figvratis Lazari Schoneri liber." See p. 177.

[2] *Bernardi Salignaci Burdegalensis Algebrae libri duo* (Francofurti, 1580). See P. Treutlein in *Abhandl. zur Geschichte der Mathematik*, Vol. II (1879), p. 36.

[3] *Batholomaei Pitisci . . . . Trigonometriae editio tertia* (Francofurti, 1612), p. 60.

[4] Willebrord Snellius, *Doctrinae triangvlorvm cononicae liber qvatvor* (Leyden, 1627), p. 37.

[5] William Oughtred, *Clavis mathematicae*, under "De aequationum affectarum resolutione in numeris" (1647 and later editions).

in a work of J. J. Heinlin[1] at Tübingen in 1679. He lets N stand for *unitas, numerus absolutus,* ⅃, *l.,* ℞. for *latus vel radix; z., q.* for *quadratus, zensus; ce, c* for *cubus; zz, qq, bq* for *biquadratus.* But he utilizes the three signs ⅃, *l, R* also for indicating roots. He speaks[2] of "Latus cubicum, vel Radix cubica, cujus nota est *Lc. R.c.* ⅃. *ce.*"

John Wallis[3] in 1655 says "Est autem lateris *l*, numerus pyramidalis l^3+3l^2+2l" and in 1685 writes[4] "$ll-2laa+a^4 : \div bb$," where the l takes the place of the modern x and the colon is a sign of aggregation, indicating that all three terms are divided by b^2.

292. The use of ℞. (*Radix*) to signify root and also power is seen in Leonardo of Pisa (§ 122) and in Luca Pacioli (§§ 136, 137). The sign ℞ was allowed to stand for the first power of the unknown x by Peletier in his algebra, by K. Schott[5] in 1661, who proceeds to let $Q.$ stand for x^2, $C.$ for x^3, $Biqq$ or $qq.$ for x^4, $Ss.$ for x^5, $Cq.$ for x^6, $SsB.$ for x^7, $Triq.$ or qqq for x^8, $Cc.$ for x^9. One finds ℞ in W. Leybourn's publication of J. Billy's[6] *Algebra*, where powers are designated by the capital letters N, R, Q, QQ, S, QC, $S2$, $QQQ.$, and where $x^2=20-x$ is written "$1Q=20-1R.$"

Years later the use of $R.$ for x and of ꝗ. (an inverted capital letter E, rounded) for x^2 is given by Tobias Beutel[7] who writes "21 ꝗ, gleich 2100, 1 ꝗ. gleigh 100, 1R. gleich 10."

293. *Facsimilis of symbols in manuscripts.*—Some of the forms for radical signs and for x, x^2, x^3, x^4, and x^5, as found in early German manuscripts and in Widman's book, are tabulated by J. Tropfke, and we reproduce his table in Figure 104.

In the Munich manuscript *cosa* is translated *ding;* the symbols in Figure 104, C2a, seem to be modified d's. The symbols in C3 are signs for *res.* The manuscripts C3b, C6, C7, C9, H6 bear on the evolution of the German symbol for x. Paleographers incline to the view that it is a modification of the Italian *co*, the *o* being highly disfigured. In B are given the signs for dragma or *numerus.*

[1] *Joh. Jacobi Heinlini Synopsis mathematica universalis* (3d ed.; Tübingen, 1679), p. 66.

[2] *Ibid.,* p. 65.   [3] John Wallis, *Arithmetica infinitorum* (Oxford, 1655), p. 144.

[4] John Wallis, *Treatise of Algebra* (London, 1685), p. 227.

[5] P. Gasparis Schotti . . . . *Cursus mathematicus* (Herbipoli [Würtzburg], 1661), p. 530.

[6] *Abridgement of the Precepts of Algebra.* The Fourth Part. Written in French by James Billy and now translated into English. . . . . Published by Will. Leybourn (London, 1678), p. 194.

[7] Tobias Beutel, *Geometrische Galleri* (Leipzig, 1690), p. 165.

294. *Two general plans for marking powers.*—In the early development of algebraic symbolism, no signs were used for the powers of given numbers in an equation. As given numbers and coefficients were not represented by letters in equations before the time of Vieta, but were specifically given in numerals, their powers could be computed on the spot and no symbolism for powers of such numbers was needed. It was different with the unknown numbers, the determination of which constituted the purpose of establishing an equation. In consequence,

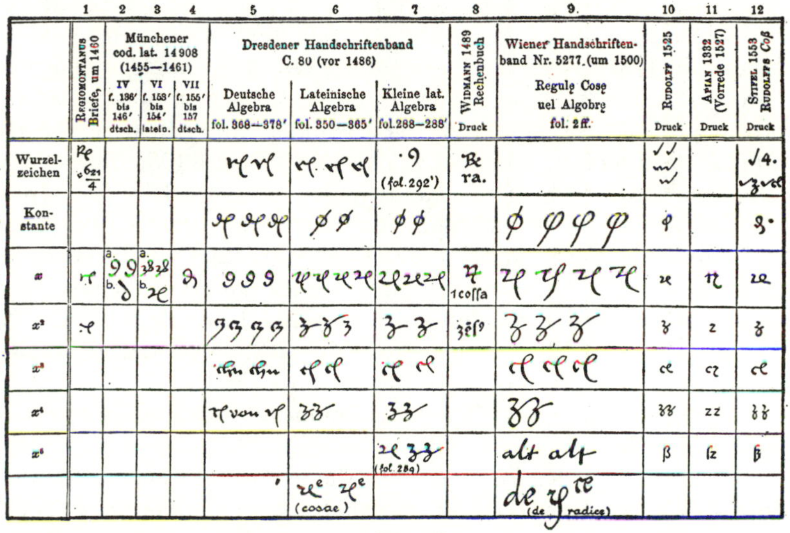

	1	2	3	4	5	6	7	8	9	10	11	12
	Regiomontanus Briefe, um 1460	Münchener cod. lat. 14908 (1455—1461) IV f.138' bis 146' dtsch.	VI f.153' bis 154' latein.	VII f.155' bis 157 dtsch.	Dresdener Handschriftenband C. 80 (vor 1486) Deutsche Algebra fol.368—378'	Lateinische Algebra fol.350—365'	Kleine lat. Algebra fol.288—288'	Widmann 1489 Rechenbuch Druck	Wiener Handschriften-band Nr.5277 (um 1500) Regule Cose uel Algobre fol.2ff.	Rudolff 1525 Druck	Apian 1532 (Vorrede 1527) Druck	Stifel 1553 Rudolff's Coß Druck
A.	Wurzel-zeichen						(fol. 292')					
B.	Konstante											
C.	a							1 coffa				
D.	x^2											
E.	x^3											
F.	x^4											
G.	x^6						(fol. 284)					
H.						(cosae)			(de radice)			

NB. Cod. Dresd. C. 80 fol. 289, 292' sind etwas spätere Eintragungen.

Fig. 104.—Signs found in German manuscripts and early German books. (Taken from J. Tropfke, *op. cit.*, Vol. II [2d ed., 1921], p. 112.)

one finds the occurrence of symbolic representation of the unknown and its powers during a period extending over a thousand years before the introduction of the literal coefficient and its powers.

For the representation of the unknown there existed two general plans. The first plan was to use some abbreviation of a name signifying unknown quantity and to use also abbreviations of the names signifying the square and the cube of the unknown. Often special symbols were used also for the fifth and higher powers whose orders were prime numbers. Other powers of the unknown, such as the fourth, sixth, eighth powers, were represented by combinations of those symbols. A good illustration is a symbolism of Luca Paciola, in which *co.* (*cosa*) represented x, *ce.* (*censo*) x^2, *cu.* (*cubo*) x^3, *p.r.* (*primo relato*) x^5; combinations of these yielded *ce.ce.* for x^4, *ce.cu.* for

x^6, etc. We have seen these symbols also in Tartaglia and Cardan, in the Portuguese Nuñez (§ 166), the Spanish Perez de Moya in 1652, and Antich Rocha[1] in 1564. We may add that outside of Italy Pacioli's symbols enjoyed their greatest popularity in Spain. To be sure, the German Marco Aurel wrote in 1552 a Spanish algebra (§ 165) which contained the symbols of Rudolff, but it was Perez de Moya and Antich Rocha who set the fashion, for the sixteenth century in Spain; the Italian symbols commanded some attention there even late in the eighteenth century, as is evident from the fourteenth unrevised impression of Perez de Moya's text which appeared at Madrid in 1784. The 1784 impression gives the symbols as shown in Figure 105, and also the explanation, first given in 1562, that the printing office does not have these symbols, for which reason the ordinary letters of the alphabet will be used.[2] Figure 105 is interesting, for it purports to show the handwritten forms used by De Moya. The symbols are not the German, but are probably derived from them. In a later book, the *Tratado de Mathematicas* (Alcala, 1573), De Moya gives on page 432 the German symbols for the powers of the unknown, all except the first power, for which he gives the crude imitation *Ze.* Antich Rocha, in his *Arithmetica*, folio 253, is partial to capital letters and gives the successive powers thus: *N, Co, Ce, Cu, Cce, R, CeCu, RR, Ccce, Ccu,* etc. The same fondness for capitals is shown in his *Mas* for "more" (§ 320).

We digress further to state that the earliest mathematical work published in America, the *Sumario compendioso* of Juan Diez Freyle[3]

---

[1] *Arithmetica* por Antich Rocha de Gerona compuesta, y de varios Auctores recopilada (Barcelona, 1564, also 1565).

[2] Juan Perez de Moya, *Aritmetica practica, y especulativa* (14th ed.; Madrid, 1784), p. 263: "Por los diez caractéres, que en el precedente capítulo se pusieron, uso estos. Por el qual dicen numero *n.* por la cosa, *co.* por el censo, *ce.* por cubo, *cu.* por censo, de censo, *cce.* por el primero relato, *R.* por el censo, y cubo, *ce.cu.* por segundo relato, *RR.* por censo de censo de censo, *cce.* por cubo de cubo, *ccu.* Esta figura *r.* quiere decir raíz quadrada. Esta figura *rr.* denota raíz quadrada de raíz quadrada. Estas *rrr.* denota raíz cúbica. De estos dos caractéres, *p. m.* notarás, que la *p.* quiere decir mas, y la *m.* menos, el uno es copulativo, el otro disyuntivo, sirven para sumar, y restar cantidades diferentes, como adelante mejor entenderás. Quando despues de *r.* se pone *u.* denota raíz quadrada universal: y asi *rru.* raíz de raíz quadrada universal: y de esta suerte *rrru.* raíz cúbica universal. Esta figura *ig.* quiere decir igual. Esta *q.* denota cantidad, y asi *qs.* cantidades: estos caractéres me ha parecido poner, porque no habia otros en la Imprenta; tú podrás usar, quando hagas demandas, de los que se pusieron en el segundo capítulo, porque son mas breves, en lo demás todos son de una condicion."

[3] Edition by D. E. Smith (Boston and London, 1921).

(City of Mexico, 1556) gives six pages to algebra. It contains the words *cosa*, *zenso*, or *censo*, but no abbreviations for them. The work does not use the signs $+$ or $-$, nor the $\tilde{p}$ and $\tilde{m}$. It is almost purely rhetorical.

The data which we have presented make it evident that in Perez de Moya, Antich Rocha, and P. Nuñez the symbols of Pacioli are used and that the higher powers are indicated by the combinations of symbols of the lower powers. This *general principle* underlies the notations of Diophantus, the Hindus, the Arabs, and most of the Germans and Italians before the seventeenth century. For convenience we shall call this the "Abbreviate Plan."

Cap. II. *En el qual se ponen algunos caractéres, que sirven por cantidades proporcionales.*

En este capítulo se ponen algunos caractéres, dando à cada uno el nombre y valor que le conviene. Los quales son inventados por causa de brevedad; y es de saber, que no es de necesidad, que estos, y no otros hayan de ser, porque cada uno puede usar de lo que quisiere, è inventar mucho mas, procediendo con la proporcion que le pareciere. Los caractéres son estos.

Fig. 105.—The written algebraic symbols for powers, as given in Perez de Moya's *Arithmetica* (Madrid, 1784), p. 260 (1st ed., 1562). The successive symbols are called *cosa es raíz, censo, cubo, censo de censo, primero relato, censo y cubo, segundo relato, censo de censo de censo, cubo de cubo.*

The second plan was not to use a symbol for the unknown quantity itself, but to limit one's self in some way to simply indicating by a numeral the power of the unknown quantity. As long as powers of only one unknown quantity appeared in an equation, the writing of the index of its power was sufficient. In marking the first, second, third, etc., powers, only the numerals for "one," "two," "three," etc., were written down. A good illustration of this procedure is Chuquet's 10^2 for $10x^2$, 10^1 for $10x$, and 10^0 for 10. We shall call this the "Index Plan." It was stressed by Chuquet, and passed through several stages of development in Bombelli, Stevin, and Girard. Then, after the introduction of special letters to designate one or more unknown quantities, and the use of literal coefficients, this notation was perfected by Hérigone and Hume; it finally culminated in the present-day form in the writings of Descartes, Wallis, and Newton.

295. Early symbolisms.—In elaborating the notations of powers according to the "Abbreviate Plan" cited in § 294, one or the other of two distinct principles was brought into play in combining the symbols of the lower powers to mark the higher powers. One was the additive principle of the Greeks in combining powers; the other was the multiplicative principle of the Hindus. Diophantus expressed the fifth power of the unknown by writing the symbols for x^2 and for x^3, one following the other; the indices 2 and 3 were *added*. Now, Bhāskara writes his symbols for x^2 and x^3 in the same way, but lets the two designate, not x^5, but x^6; the indices 2 and 3 are *multiplied*. This difference in designation prevailed through the Arabic period, the later Middle Ages in Europe down into the seventeenth century. It disappeared only when the notations of powers according to the "Abbreviate Plan" passed into disuse. References to the early symbolisms, mainly as exhibited in our accounts of individual authors, are as follows:

ABBREVIATE PLAN

ADDITIVE PRINCIPLE

Diophantus, and his editors Xylander, Bachet, Fermat (§ 101)
al-Karkhî, eleventh century (§ 116)
Leonardo of Pisa (§ 122)
Anonymous Arab (§ 124)
Dresden Codex C. 80 (§ 305, Fig. 104)
M. Stifel, (1545), *sum; sum; sum:* x^3 (§ 154)
F. Vieta (1591), and in later publications (§ 177)
C. Glorioso, 1527 (§ 196)
W. Oughtred, 1631 (§ 182)
Samuel Foster, 1659 (§ 306)

MULTIPLICATIVE PRINCIPLE

Bhāskara, twelfth century (§ 110–12)
Arabic writers, except al-Karkhî (§ 116)
L. Pacioli, 1494, *ce. cu.* for x^6 (§ 136)
H. Cardano, 1539, 1545 (§ 140)
N. Tartaglia, 1556–60 (§ 142)
Ch. Rudolff, 1525 (§ 148)
M. Stifel, 1544 (§ 151)
J. Scheubel, 1551, follows Stifel (§ 159)
A. Rocha, 1565, follows Pacioli (§ 294)
C. Clavius, 1608, follows Stifel (§ 161)
P. Nuñez, 1567, follows Pacioli and Cardan (§ 166)
R. Recorde, 1557, follows Stifel (§ 168)

L. and T. Digges, 1579 (§ 170)
A. M. Visconti, 1581 (§ 145)
Th. Masterson, 1592 (§ 171)
J. Peletier, 1554 (§ 172)
G. Gosselin, 1577 (§ 174)
L. Schoner, 1627 (§ 291)

NEW NOTATIONS ADOPTED

Ghaligai and G. del Sodo, 1521 (§ 139)
M. Stifel, 1553, repeating factors (§ 156)
J. Buteon, 1559 (§ 173)
J. Scheubel, *N, Ra, Pri, Se* (§ 159)
Th. Harriot, repeating factors (§ 188)
Johann Geysius, repeating factors (§ 196, 305)
John Newton, 1654 (§ 305)
Nathaniel Torporley (§ 305)
Joseph Raphson, 1702 (§ 305)
Samuel Foster, 1659, use of lines (§ 306)

INDEX PLAN

Psellus, nomenclature without signs (§ 117)
Neophytos, scholia (§§ 87, 88)
Nicole Oresme, notation for fractional powers (§ 123)
N. Chuquet, 1484, 12^3 for $12x^3$ (§ 131)
E. de la Roche, 1520 (§ 132)
R. Bombelli, 1572 (§ 144)
Grammateus, 1518, *pri, se., ter. quart.* (§ 147)
G. van der Hoecke, 1537, *pri, se,* 3^a (§ 150)
S. Stevin (§ 162)
A. Girard, 1629 (§ 164)
L. & T. Digges, 1579 (§ 170, Fig. 76)
P. Hérigone, 1634 (§ 189)
J. Hume, 1635, 1636 (§ 190)

296. *Notations applied only to an unknown quantity, the base being omitted.*—As early as the fourteenth century, Oresme had the exponential concept, but his notation stands in historical isolation and does not constitute a part of the course of evolution of our modern exponential symbolism. We have seen that the earliest important steps toward the modern notation were taken by the Frenchman Nicolas Chuquet, the Italian Rafael Bombelli, the Belgian Simon Stevin, the Englishmen L. and T. Digges. Attention remains to be called to a symbolism very similar to that of the Digges, which was contrived by Pietro Antonio Cataldi of Bologna, in an algebra of 1610

and a book on roots of 1613. Cataldi wrote the numeral exponents in their natural upright position,[1] and distinguished them by crossing them out. His "5 $\mathcal{Z}$ *via* 8 $\mathcal{A}$ *fa* 40 $\mathcal{7}$" means $5x^3 \cdot 8x^4 = 40x^7$. His sign for x is $\mathcal{1}$. He made only very limited use of this notation.

The drawback of Stevin's symbolism lay in the difficulty of writing and printing numerals and fractions within the circle. Apparently as a relief from this cumbrousness, we find that the Dutch writer, Adrianus Romanus, in his *Ideae Mathematicae pars prima* (Antwerp, 1593), uses in place of the circle two rounded parentheses and vinculums above and below; thus, with him $1(\overline{45})$ stands for x^{45}. He uses this notation in writing his famous equation of the forty-fifth degree. Franciscus van Schooten[2] in his early publications and when he quotes from Girard uses the notation of Stevin.

A notation more in line with Chuquet's was that of the Swiss Joost Bürgi who, in a manuscript now kept in the library of the observatory at Pulkowa, used Roman numerals for exponents and wrote[3]

$$\overset{\text{vi}}{8}+\overset{\text{v}}{12}-\overset{\text{iv}}{9}+\overset{\text{iii}}{10}+\overset{\text{ii}}{3}+\overset{\text{i}}{7}-\overset{0}{4} \text{ for } 8x^6+12x^5-9x^4+10x^3+3x^2+7x-4 \ .$$

In this notation Bürgi was followed by Nicolaus Reymers (1601) and J. Kepler.[4] Reymers[5] used also the cossic symbols, but chose R in place of $\mathcal{X}$; occasionally he used a symbolism as in $25\text{IIII}+20\text{II}-10\text{III}-8\text{I}$ for the modern $25x^4+20x^2-10x^3-8x$. We see that Cataldi, Romanus, Fr. van Schooten, Bürgi, Reymers, and Kepler belong in the list of those who followed the "Index Plan."

297. *Notations applied to any quantity, the base being designated.*— As long as literal coefficients were not used and numbers were not generally represented by letters, the notations of Chuquet, Bombelli,

[1] G. Wertheim, *Zeitschr. f. Math. u. Physik*, Vol. XLIV (1899), Hist.-Lit. Abteilung, p. 48.

[2] Francisci à Schooten, *De Organica conicarum sectionum . . . . Tractatus* (Leyden, 1646), p. 96; Schooten, *Renati Descartes Geometria* (Frankfurt a./M., 1695), p. 359.

[3] P. Treutlein in *Abhandlungen zur Geschichte der Mathematik*, Vol. II (Leipzig, 1879), p. 36, 104.

[4] In his "De Figurarum regularium" in *Opera omnia* (ed. Ch. Frisch), Vol. V (1864), p. 104, Kepler lets the radius AB of a circle be 1 and the side BC of a regular inscribed heptagon be R. He says: "In hac proportione continuitatem fingit, ut sicut est $AB1$ ad BC $1R$, sic sit $1R$ ad $1z$, et $1z$ as 1 $\mathcal{q}$, et 1 $\mathcal{q}$ ad $1zz$, et $1zz$ ad $1z$ $\mathcal{q}$ et sic perpetuo, quod nos commodius signabimus per apices six, 1, 1ᴵ, 1ᴵᴵ, 1ᴵᴵᴵ, 1ᴵⱽ, 1ⱽ, 1ⱽᴵ, 1ⱽᴵᴵ, etc."

[5] N. Raimarus Ursus, *Arithmetica analytica* (Frankfurt a. O., 1601), Bl. *C3v°*. See J. Tropfke, *op. cit.*, Vol. II (2d ed., 1921), p. 122.

Stevin, and others were quite adequate. There was no pressing need of indicating the powers of a given number, say the cube of twelve; they could be computed at once. Moreover, as only the unknown quantity was raised to powers which could not be computed on the spot, why should one go to the trouble of writing down the base? Was it not sufficient to put down the exponent and omit the base? Was it not easier to write $\overset{\text{v}}{16}$ than $16x^5$? But when through the innovations of Vieta and others, literal coefficients came to be employed, and when several unknowns or variables came to be used as in analytic geometry, then the omission of the base became a serious defect in the symbolism. It will not do to write $15x^2-16y^2$ as $\overset{\text{ii}}{15}-\overset{\text{ii}}{16}$. In watching the coming changes in notation, the reader will bear this problem in mind. Vieta's own notation of 1591 was clumsy: *D quadratum* or *D. quad.* stood for D^2, *D cubum* for D^3; *A quadr.* for x^2, *A* representing the unknown number.

In this connection perhaps the first writer to be mentioned is Luca Pacioli who in 1494 explained, as an alternative notation of powers, the use of *R* as a base, but in place of the exponent he employs an ordinal that is too large by unity (§ 136). Thus *R. 30*ᵃ stood for x^{29}. Evidently Pacioli did not have a grasp of the exponential concept.

An important step was taken by Romanus[1] who uses letters and writes bases as well as the exponents in expressions like

$$A(4)+B(4)+4A(3) \ in \ B+6A(2) \ in \ B(2)+4A \ in \ B(3)$$

which signifies

$$A^4+B^4+4A^3B+6A^2B^2+4AB^3 \ .$$

A similar suggestion came from the Frenchman, Pierre Hérigone, a mathematician who had a passion for new notations. He wrote our a^3 as *a3*, our $2b^4$ as *2b4*, and our $2ba^2$ as *2ba2*. The coefficient was placed *before* the letter, the exponent *after*.

In 1636 James Hume[2] brought out an edition of the algebra of Vieta, in which he introduced a superior notation, writing down the base and elevating the exponent to a position above the regular line and a little to the right. The exponent was expressed in Roman

[1] See H. Bosmans in *Annales Société scient. de Bruxelles*, Vol. XXX, Part II (1906), p. 15.

[2] James Hume, *L'Algèbre de Viète, d'une methode nouvelle claire et facile* (Paris 1636). See *Œuvres de Descartes* (ed. Charles Adam et P. Tannery), Vol. V, p. 504, 506–12.

numerals. Thus, he wrote A^{iii} for A^3. Except for the use of the Roman numerals, one has here our modern notation. Thus, this Scotsman, residing in Paris, had almost hit upon the exponential symbolism which has become universal through the writings of Descartes.

298. *Descartes' notation of 1637.*—Thus far had the notation advanced before Descartes published his *Géométrie* (1637) (§ 191). Hérigone and Hume almost hit upon the scheme of Descartes. The only difference was, in one case, the position of the exponent, and, in the other, the exponent written in Roman numerals. Descartes expressed the exponent in Arabic numerals and assigned it an elevated position. Where Hume would write $5a^{\mathrm{iv}}$ and Hérigone would write $5a4$, Descartes wrote $5a^4$. From the standpoint of the printer, Hérigone's notation was the simplest. But Descartes' elevated exponent offered certain advantages in interpretation which the judgment of subsequent centuries has sustained. Descartes used positive integral exponents only.

299. *Did Stampioen arrive at Descartes' notation independently?*— Was Descartes alone in adopting the notation $5a^4$ or did others hit upon this particular form independently? In 1639 this special form was suggested by a young Dutch writer, Johan Stampioen.[1] He makes no acknowledgment of indebtedness to Descartes. He makes it appear that he had been considering the two forms ^3a and a^3, and had found the latter preferable.[2] Evidently, the symbolism a^3 was adopted by Stampioen after the book had been written; in the body of his book[3] one finds aaa, $bbbb$, $fffff$, $gggggg$, but the exponential notation above noted, as described in his passage following the Preface, is not used. Stampioen uses the notation a^4 in some but not all parts of a controversial publication[4] of 1640, on the solution of cubic equations, and directed against Waessenaer, a personal friend of Descartes. In view of the fact that Stampioen does not state the originators of any of the notations which he uses, it is not improbable that his a^3 was taken from Descartes, even though Stampioen stands out as an opponent of Descartes.[5]

---

[1] Johan Stampioen d'Jonghe, *Algebra ofte Nieuwe Stel-Regel* (The Hague, 1639). See his statement following the Preface.

[2] Stampioen's own words are: "*aaa·* dit is a drievoudich in hem selfs gemen-nichvuldicht. men soude oock daer voor konnen stellen ^3a ofte better a^3."

[3] J. Stampioen, *op. cit.*, p. 343, 344, 348.

[4] *I. I. Stampioenii Wis-Konstigh ende Reden-Maetigh Bewijs* ('s Graven-Hage, 1640), unpaged Introduction and p. 52–55.

[5] *Œuvres de Descartes*, Vol. XII (1910), p. 32, 272–74.

300. *Notations used by Descartes before 1637.*—Descartes' indebtedness to his predecessors for the exponential notation has been noted. The new features in Descartes' notation, $5a^3$, $6ab^4$, were indeed very slight. What notations did Descartes himself employ before 1637?

In his *Opuscules de 1619–1621* he regularly uses German symbols as they are found in the algebra of Clavius; Descartes writes[1]

$$\text{`` } 36-3z-6 \, \mathcal{X} \, aequ. \, 1 \, \mathcal{C} \, \text{,''}$$

which means $36-3x^2-6x=x^3$. These *Opuscules* were printed by Foucher de Careil (Paris, 1859–60), but this printed edition contains corruptions in notation, due to the want of proper type. Thus the numeral 4 is made to stand for the German symbol $\mathcal{X}$; the small letter γ is made to stand for the radical sign $\sqrt{\ }$. The various deviations from the regular forms of the symbols are set forth in the standard edition of Descartes' works. Elsewhere (§ 264) we call attention that Descartes[2] in a letter of 1640 used the Recordian sign of equality and the symbols N and C of Xylander, in the expression "$1C-6N=40$." Writing to Mersenne, on May 3, 1638, Descartes[3] employed the notation of Vieta, "$Aq+Bq+A$ *in B bis*" for our a^2+b^2+2ab. In a posthumous document,[4] of which the date of composition is not known, Descartes used the sign of equality found in his *Géométrie* of 1637, and P. Hérigone's notation for powers of given letters, as $b3x$ for b^3x, $a3z$ for a^3z. Probably this document was written before 1637. Descartes[5] used once also the notation of Dounot (or Deidier, or Bar-le-Duc, as he signs himself in his books) in writing the equation $1C-9Q+13N$ *eq.* $\sqrt{288}-15$, but Descartes translates it into $y^3-9y^2+13y-12\sqrt{2}+15 \, \infty \, 0$.

301. *Use of Hérigone's notation after 1637.*—After 1637 there was during the seventeenth century still very great diversity in the exponential notation. Hérigone's symbolism found favor with some writers. It occurs in Florimond Debeaune's letter[6] of September 25, 1638, to Mersenne in terms like $2y4$, $y3$, $2l2$ for $2y^4$, y^3, and $2l^2$, re-

---

[1] *Ibid.*, Vol. X (1908), p. 249–51. See also E. de Jonquières in *Bibliotheca mathematica* (2d ser.), Vol. IV (1890), p. 52, also G. Eneström, *Bibliotheca mathematica* (3d ser.), Vol. VI (1905), p. 406.

[2] *Œuvres de Descartes*, Vol. III (1899), p. 190.

[3] *Ibid.*, Vol. II (1898), p. 125; also Vol. XII, p. 279.

[4] *Ibid.*, Vol. X (1908), p. 299.

[5] *Ibid.*, Vol. XII, p. 278.   [6] *Ibid.*, Vol. V (1903), p. 516.

spectively. G. Schott[1] gives it along with older notations. Pietro
Mengoli[2] uses it in expressions like $a4+4a3r+6a2r2+4ar3+r4$ for our
$a^4+4a^3r+6a^2r^2+4ar^3+r^4$. The Italian Cardinal Michelangelo Ricci[3]
writes "AC_2 in CB_3" for $\overline{AC}^2 . \overline{CB}^3$. In a letter[4] addressed to Ozanam
one finds $b4+c4\frown a4$ for $b^4+c^4=a^4$. Chr. Huygens[5] in a letter of June
8, 1684, wrote $a3+aab$ for a^3+a^2b. In the same year an article by
John Craig[6] in the *Philosophical Transactions* contains $a3y+a4$ for
a^3y+a^4, but a note to the "Benevole Lector" appears at the end apolo-
gizing for this notation. Dechales[7] used in 1674 and again in 1690
(along with older notations) the form $A4+4A3B+6A2B2+4AB3+
B4$. A Swedish author, Andreas Spole,[8] who in 1664–66 sojourned
in Paris, wrote in 1692 an arithmetic containing expressions $3a3+
3a2-2a-2$ for $3a^3+3a^2-2a-2$. Joseph Moxon[9] lets "$A-B.(2)$"
stand for our $(A-B)^2$, also "$A-B.(3)$" for our $(A-B)^3$. With the
eighteenth century this notation disappeared.

302. *Later use of Hume's notation of 1636.*—Hume's notation of
1636 was followed in 1638 by Jean de Beaugrand[10] who in an anonymous letter to Mersenne criticized Descartes and states that the equation $x^{IV}+4x^{III}-19x^{II}-106x-120$ has the roots $+5$, -2, -3, -4.
Beaugrand also refers to Vieta and used vowels for the unknowns, as
in "$A'''+3AAB+ADP$ *esgale a* ZSS." Again Beaugrand writes
"$E'''\circ -13E-12$" for $x^3-13x-12$, where the $\circ$ apparently designates the omission of the second term, as does $\divideontimes$ with Descartes.

303. *Other exponential notations suggested after 1637.*—At the time
of Descartes and the century following several other exponential
notations were suggested which seem odd to us and which serve to

[1] G. Schott, *Cursus mathematicus* (Würzburg, 1661), p. 576.

[2] *Ad Maiorem Dei Gloriam Geometriae speciosae Elementa,* . . . . Petri Mengoli
(Bologna, 1659), p. 20.

[3] *Michaelis Angeli Riccii Exercitatio geometrica* (Londini, 1668), p. 2. [Preface,
1666.]

[4] *Journal des Sçavans,* l'année 1680 (Amsterdam, 1682), p. 160.

[5] *Ibid.,* l'année 1684, Vol. II (2d ed.; Amsterdam, 1709), p. 254.

[6] *Philosophical Transactions,* Vol. XV–XVI (London, 1684–91), p. 189.

R. P. *Claudii Francisci Milliet Dechales Camberiensis Mundus mathematicus,*
Tomus tertius (Leyden, 1674), p. 664; Tomus primus (editio altera; Leyden,
1690), p. 635.

[8] Andreas Spole, *Arithmetica vulgaris et specioza* (Upsala, 1692). See G. Eneström in *L'Intermédiaire des mathématiciens,* Vol. IV (1897), p. 60.

[9] Joseph Moxon, *Mathematical Dictionary* (London, 1701), p. 190, **191**.

[10] *Œuvres de Descartes,* Vol. V (1903), p. 506, **507**.

indicate how the science might have been retarded in its progress under the handicap of cumbrous notations, had such wise leadership as that of Descartes, Wallis, and Newton not been available. Rich. Balam[1] in 1653 explains a device of his own, as follows: "(2) : 3 : , the Duplicat, or Square of 3, that is, 3×3; (4) : 2 : , the Quadruplicat of 2, that is, $2 \times 2 \times 2 \times 2 = 16$." The Dutch J. Stampioen[2] in 1639 wrote $\square A$ for A^2; as early as 1575 F. Maurolycus[3] used $\square$ to designate the square of a line. Similarly, an Austrian, Johannes Caramuel,[4] in 1670 gives "$\square 25$. est Quadratum Numeri 25. hoc est, 625." Huygens[5] wrote "1000(3)10" for $1,000 = 10^3$, and "1024(10)2" for $1024 = 2^{10}$. A Leibnizian symbolism[6] explained in 1710 indicates the cube of $AB + BC$ thus: $\boxed{3}$ $(AB + BC)$; in fact, before this time, in 1695 Leibniz[7] wrote $\boxed{m}$ $\overline{y + a}$ for $(y + a)^m$.

304. Descartes preferred the notation aa to a^2. Fr. van Schooten,[8] in 1646, followed Descartes even in writing qq, xx rather than q^2, x^2, but in his 1649 Latin edition of Descartes' geometry he wrote preferably x^2. The symbolism xx was used not only by Descartes, but also by Huygens, Rahn, Kersey, Wallis, Newton, Halley, Rolle, Euler— in fact, by most writers of the second half of the seventeenth and of the eighteenth centuries. Later, Gauss[9] was in the habit of writing xx, and he defended his practice by the statement that x^2 did not take up less space than xx, hence did not fulfil the main object of a symbol. The x^2 was preferred by Leibniz, Ozanam, David Gregory, and Pascal.

305. The reader should be reminded at this time that the representation of positive integral powers by the repetition of the factors was suggested very early (about 1480) in the Dresden Codex C. 80 under the heading *Algorithmus de additis et minutis* where $x^2 = z$ and $x^{10} = zzzzz$; it was elaborated more systematically in 1553 by M.

---

[1] Rich. Balam, *Algebra, or The Doctrine of Composing, Inferring, and Resolving an Equation* (London, 1653), p. 9.

[2] Johan Stampioen, *Algebra* (The Hague, 1639), p. 38.

[3] *D. Francisc' Mavrolyci Abbatis messanensis Opuscula Mathematica* (Venice, 1575) (Euclid, Book XIII), p. 107.

[4] *Joannis Caramvelis Mathesis Biceps. Vetus et Nova* (Companiae, 1670), p. 131, 132.

[5] *Christiani Hugenii Opera* . . . . quae collegit . . . . Guilielmus Jacobus's Gravesande (Leyden, 1751), p. 456.

[6] *Miscellanea Berolinensia* (Berlin, 1710), p. 157.

[7] *Acta eruditorum* (1695), p. 312.

[8] *Francisci à Schooten Leydensis de Organica conicarum sectionum* . . . . *Tractatus* (Leyden, 1646), p. 91 ff.

[9] M. Cantor, *op. cit.*, Vol. II (2d ed.), p. 794 n.

Stifel (§ 156). One sees in Stifel the exponential notation applied, not to the unknown but to several different quantities, all of them known. Stifel understood that a quantity with the exponent zero had the value 1. But this notation was merely a suggestion which Stifel himself did not use further. Later, in Alsted's *Encyclopaedia*,[1] published at Herborn in Prussia, there is given an explanation of the German symbols for *radix, zensus, cubus*, etc.; then the symbols from Stifel, just referred to, are reproduced, with the remark that they are preferred by some writers. The algebra proper in the *Encyclopaedia* is from the pen of Johann Geysius[2] who describes a similar notation $2a$, $4aa$, $8aaa$, ... , $512aaaaaaaaa$ and suggests also the use of Roman numerals as indices, as in $2l^{\text{I}}$ $4q^{\text{II}}$ $8c^{\text{III}}$ . . . . $512cc^{\text{IX}}$. Forty years after, Caramvel[3] ascribes to Geysius the notation aaa for the cube of a, etc.

In England the repetition of factors for the designation of powers was employed regularly in Thomas Harriot. In a manuscript preserved in the library of Sion College, Nathaniel Torporley (1573–1632) makes strictures on Harriot's book, but he uses Harriot's notation.[4] John Newton[5] in 1654 writes $aaaaa$. John Collins writes in the *Philosophical Transactions* of 1668 $aaa-3aa+4a=N$ to signify $x^3-3x^2+4x=N$. Harriot's mode of representation is found again in the *Transactions*[6] for 1684. Joseph Raphson[7] uses powers of g up to g^{10}, but in every instance he writes out each of the factors, after the manner of Harriot.

306. The following curious symbolism was designed in 1659 by Samuel Foster[8] of London:

┐	┘	┑	⌐┘	⌐|	⊐	⊐|	⊐⌐
q	c	qq	qc	cc	qqc	qcc	ccc
2	3	4	5	6	7	8	9

[1] *Johannis-Henrici Alstedii Encyclopaedia* (Herborn, 1630), Book XIV, "Arithmetica," p. 844.

[2] *Ibid.*, p. 865–74.

[3] *Joannis Caramvelis Mathesis Biceps* (Campaniae, 1670), p. 121.

[4] J. O. Halliwell, *A Collection of Letters Illustrative of the Progress of Science in England* (London, 1841), p. 109–16.

[5] John Newton, *Institutio Mathematica or a Mathematical Institution* (London, 1654), p. 85.

[6] *Philosophical Transactions*, Vol. XV–XVI (London, 1684–91), p. 247, 340.

Josepho Raphson, *Analysis Aequationum universalis* (London, 1702). [First edition, 1697.]

Samuel Foster, *Miscellanies, or Mathematical Lucubrations* (London, 1659), p. 10.

Foster did not make much use of it in his book. He writes the proportion

$$\text{`` } At\ AC\ .\ AR :: \overline{CD}|:\overline{RP}|\ \text{,''}$$

which means

$$AC:AR = \overline{CD}^2:\overline{RP}^2\ .$$

An altogether different and unique procedure is encountered in the *Maandelykse Mathematische Liefhebberye* (1754–69), where $\overset{m}{\nu}\!\!-\!\!-$ signifies extracting the mth root, and $-\!\!-\!\!-\nu\!\!\!/$ signifies raising to the mth power. Thus,

$$\text{`` }\begin{array}{c} \dfrac{y^n}{b} = \overset{m}{\nu}\overline{ay} \\ \hline \dfrac{y^{mn}}{b^m} = ay \end{array}\ \overset{m}{\Big\rfloor}\ \text{''}.$$

307. *Spread of Descartes' notation.*—Since Descartes' *Géométrie* appeared in Holland, it is not strange that the exponential notation met with prompter acceptance in Holland than elsewhere. We have already seen that J. Stampioen used this notation in 1639 and 1640. The great disciple of Descartes, Fr. van Schooten, used it in 1646, and in 1649 in his Latin edition of Descartes' geometry. In 1646 van Schooten indulges[1] in the unusual practice of raising some (but not all) of his coefficients to the height of exponents. He writes $x^3 - {}^3aax - {}^2a^3 \infty$ 0 to designate $x^3 - 3a^2x - 2a^3 = 0$. Van Schooten[2] does the same thing in 1657, when he writes ${}^2ax$ for $2ax$. Before this Marini Ghetaldi[3] in Italy wrote coefficients in a *low* position, as subscripts, as in the proportion,

$$\text{`` } ut\ AQ\ ad\ A_2\ in\ B\ ita\ \frac{x}{y}\ ad\ m_2 + n\ \text{,''}$$

which stands for $A^2:2AB = \dfrac{x}{y}:(2m+n)$. Before this Albert Girard[4] placed the coefficients where we now write our exponents. I quote: "Soit un binome conjoint $B+C$. Son Cube era $B(B_q + C_q^3) + C(B_q^3 + C_q)$." Here the cube of $B+C$ is given in the form corresponding

---

[1] Francisci à Schooten, *De organica conicarum sectionum . . . . tractatus* (Leyden, 1646), p. 105.

[2] Francisci à Schooten, *Exercitationum mathematicarum liber primus* (Leyden, 1657), p. 227, 274, 428, 467, 481, 483.

[3] Marini Ghetaldi, *De resolutione et compositione mathematica libri quinque. Opus posthumum* (Rome, 1630). Taken from E. Gelcich, *Abhandlungen zur Geschichte der Mathematik*, Vol. IV (1882), p. 198.

[4] A. Girard, *Invention nouvelle en l'algebre* (1629), "3 C."

to $B(B^2+3C^2)+C(3B^2+C^2)$. Much later, in 1679, we find in the collected works of P. Fermat[1] the coefficients in an elevated position: ^2D in A for $2DA$, ^2R in E for $2RE$.

The Cartesian notation was used by C. Huygens and P. Mersenne in 1646 in their correspondence with each other,[2] by J. Hudde[3] in 1658, and by other writers.

In England, J. Wallis[4] was one of the earliest writers to use Descartes' exponential symbolism. He used it in 1655, even though he himself had been trained in Oughtred's notation.

The Cartesian notation is found in the algebraic parts of Isaac Barrow's[5] geometric lectures of 1670 and in John Kersey's *Algebra*[6] of 1673. The adoption of Descartes' a^4 in strictly algebraic operations and the retention of the older A_q, A_c for A^2, A^3 in geometric analysis is of frequent occurrence in Barrow and in other writers. Seemingly, the impression prevailed that A^2 and A^3 suggest to the pupil the purely arithmetical process of multiplication, AA and AAA, but that the symbolisms A_q and A_c conveyed the idea of a geometric square and geometric cube. So we find in geometrical expositions the use of the latter notation long after it had disappeared from purely algebraic processes. We find it, for instance, in W. Whiston's edition of Tacquet's *Euclid*,[7] in Sir Isaac Newton's *Principia*[8] and *Opticks*,[9] in B. Robins' *Tracts*,[10] and in a text by K. F. Hauber.[11] In the *Philosophical Transactions* of London none of the pre-Cartesian notations for powers appear, except a few times in an article of 1714 from the pen of R. Cotes, and an occasional tendency to adhere to the primitive, but very

[1] P. Fermat, *Varia opera* (Toulouse, 1679), p. 5.

[2] C. Huygens, *Œuvres*, Vol. I (La Haye, 1888), p. 24.

[3] *Joh. Huddeni Epist. I de reductione aequationum* (Amsterdam, 1658); Matthiessen, *Grundzüge der Antiken u. Modernen Algebra* (Leipzig, 1878), p. 349.

[4] John Wallis, *Arithmetica infinitorum* (Oxford, 1655), p. 16 ff.

[5] Isaac Barrow, *Lectiones Geometriae* (London, 1670), Lecture XIII (W. Whewell's ed.), p. 309.

[6] John Kersey, *Algebra* (London, 1673), p. 11.

[7] See, for instance, *Elementa Euclidea geometriae auctore Andrea Tacquet, ...* Gulielmus Whiston (Amsterdam, 1725), p. 41.

[8] Sir Isaac Newton, *Principia* (1687), Book I, Lemma xi, Cas. 1, and in other places.

[9] Sir Isaac Newton, *Opticks* (3d ed.; London, 1721), p. 30.

[10] Benjamin Robins, *Mathematical Tracts* (ed. James Wilson, 1761), Vol. II, p. 65.

[11] Karl Friderich Hauber, *Archimeds zwey Bücher über Kugel und Cylinder* (Tübingen, 1798), p. 56 ff.

lucid method of repeating the factors, as aaa for a^3. The modern exponents did not appear in any of the numerous editions of William Oughtred's *Clavis mathematicae;* the last edition of that popular book was issued in 1694 and received a new impression in 1702. On February 5, 1666–67, J. Wallis[1] wrote to J. Collins, when a proposed new edition of Oughtred's *Clavis* was under discussion: "It is true, that as in other things so in mathematics, fashions will daily alter, and that which Mr. Oughtred designed by great letters may be now by others designed by small; but a mathematician will, with the same ease and advantage, understand A_c or aaa." As late as 1790 the Portuguese J. A. da Cunha[2] occasionally wrote A_q and A_c. J. Pell wrote r^2 and t^2 in a letter written in Amsterdam on August 7, 1645.[3] J. H. Rahn's *Teutsche Algebra*, printed in 1659 in Zurich, contains for positive integral powers two notations, one using the Cartesian exponents, a^3, x^4, the other consisting of writing an Archemidean spiral (Fig. 96) between the base and the exponent on the right. Thus $a \odot 3$ signifies a^3. This symbol is used to signify involution, a process which Rahn calls *involviren*. In the English translation, made by T. Brancker and published in 1668 in London, the Archimedean spiral is displaced by the omicron-sigma (Fig. 97), a symbol found among several English writers of textbooks, as, for instance, J. Ward,[4] E. Hatton,[5] Hammond,[6] C. Mason,[7] and by P. Ronayne[8]—all of whom use also Rahn's and Brancker's $\sim\!\!\!\sim\!\!\!/$ to signify evolution. The omicron-sigma is found in Birks;[9] it is mentioned by Saverien,[10] who objects to it as being superfluous.

Of interest is the following passage in Newton's *Arithmetick*,[11] which consists of lectures delivered by him at Cambridge in the period 1669–85 and first printed in 1707: "Thus $\sqrt{64}$ denotes 8; and $\sqrt{3:64}$

[1] Rigaud, *Correspondence of Scientific Men of the Seventeenth Century*, Vol. I (Oxford, 1841), p. 63.

[2] J. A. da Cunha, *Principios mathematicos* (1790), p. 158.

[3] J. O. Halliwell, *Progress of Science in England* (London, 1841), p. 89.

[4] John Ward, *The Young Mathematician's Guide* (London, 1707), p. 144.

[5] Edward Hatton, *Intire System of Arithmetic* (London, 1721), p. 287.

[6] Nathaniel Hammond, *Elements of Algebra* (London, 1742).

[7] C. Mason in the *Diarian Repository* (London, 1774), p. 187.

[8] Philip Ronayne, *Treatise of Algebra* (London, 1727), p. 3.

[9] Anthony and John Birks, *Arithmetical Collections* (London, 1766), p. viii.

[10] A. Saverien, *Dictionnaire universel de mathematique et de physique* (Paris, 1753), "Caractere."

[11] Newton's *Universal Arithmetick* (London, 1728), p. 7.

denotes 4. . . . . There are some that to denote the Square of the first Power, make use of q, and of c for the Cube, qq for the Biquadrate, and qc for the Quadrato-Cube, etc. . . . . Others make use of other sorts of Notes, but they are now almost out of Fashion."

In the eighteenth century in England, when parentheses were seldom used and the vinculum was at the zenith of its popularity, bars were drawn horizontally and allowed to bend into a vertical stroke[1] (or else were connected with a vertical stroke), as in $A \times \overline{B^{\frac{1}{n}}}\Big|^n$ or in $\overline{a+b}\Big|^{\frac{m}{n}}$.

In France the Cartesian exponential notation was not adopted as early as one might have expected. In J. de Billy's *Nova geometriae clavis* (Paris, 1643), there is no trace of that notation; the equation $x+x^2=20$ is written "$1R+1Q$ *aequatur* 20." In Fermat's edition[2] of Diophantus of 1670 one finds in the introduction $1QQ+4C+10Q+20N+1$ for $x^4+4x^3+10x^2+20x+1$. But in an edition of the works of Fermat, brought out in 1679, after his death, the algebraic notation of Vieta which he had followed was discarded in favor of the exponents of Descartes.[3] B. Pascal[4] made free use of positive integral exponents in several of his papers, particularly the *Potestatum numericarum summa* (1654).

In Italy, C. Renaldini[5] in 1665 uses both old and new exponential notations, with the latter predominating.

308. *Negative, fractional, and literal exponents.*—Negative and fractional exponential notations had been suggested by Oresme, Chuquet, Stevin, and others, but the modern symbolism for these is due to Wallis and Newton. Wallis[6] in 1656 used positive integral exponents and speaks of negative and fractional "indices," but he does not actually write a^{-1} for $\frac{1}{a}$, $a^{\frac{3}{2}}$ for $\sqrt{a^3}$. He speaks of the series

[1] See, for instance, A. Malcolm, *A New System of Arithmetick* (London, 1730), p. 143.

[2] *Diophanti Alexandrini arithmeticarum Libri Sex, cum commentariis G. B. Bacheti V. C. et observationibus D. P. de Fermat* (Tolosae, 1670), p. 27.

[3] See *Œuvres de Fermat* (éd. Paul Tannery et Charles Henry), Tome I (Paris, 1891), p. 91 n.

[4] *Œuvres de Pascal* (éd. Leon Brunschvicg et Pierre Boutroux), Vol. III (Paris, 1908), p. 349–58.

[5] Caroli Renaldinii, *Ars analytica* . . . . (Florence, 1665), p. 11, 80, 144.

[6] J. Wallis, *Arithmetica infinitorum* (1656), p. 80, Prop. CVI.

$\dfrac{1}{\sqrt{1}}$, $\dfrac{1}{\sqrt{2}}$, $\dfrac{1}{\sqrt{3}}$, etc., as having the "index $-\frac{1}{2}$." Our modern notation involving fractional and negative exponents was formally introduced a dozen years later by Newton[1] in a letter of June 13, 1676, to Oldenburg, then secretary of the Royal Society of London, which explains the use of negative and fractional exponents in the statement, "Since algebraists write a^2, a^3, a^4, etc., for aa, aaa, $aaaa$, etc., so I write $a^{\frac{1}{2}}$, $a^{\frac{3}{2}}$, $a^{\frac{5}{3}}$, for $\sqrt{a}$, $\sqrt{a^3}$, $\sqrt{c\,a^5}$; and I write a^{-1}, a^{-2}, a^{-3}, etc., for $\dfrac{1}{a}$, $\dfrac{1}{aa}$, $\dfrac{1}{aaa}$, etc." He exhibits the general exponents in his binomial formula first announced in that letter:

$$\overline{P+PQ}\Big|^{\frac{m}{n}}=P^{\frac{m}{n}}+\frac{m}{n}\ AQ+\frac{m-n}{2n}\ BQ+\frac{m-2n}{3n}\ CQ+\frac{m-3n}{4n}\ DQ+\ ,\ \text{etc.,}$$

where $A=P^{\frac{m}{n}}$, $B=\dfrac{m}{n}P^{\frac{m}{n}}Q$, etc., and where $\dfrac{m}{n}$ may represent any real and rational number. It should be observed that Newton wrote here literal exponents such as had been used a few times by Wallis,[2] in 1657, in expressions like $\sqrt{^dR^d}=R$, $AR^m\times AR^n=A^2R^{m+n}$, which arose in the treatment of geometric progression. Wallis gives also the division $AR^m)AR^{m+n}(R^n$. Newton[3] employs irrational exponents in his letter to Oldenburg of the date October 24, 1676, where he writes $\overline{x^{\sqrt{2}}+x^{\sqrt{7}}}\Big|^{\sqrt[3]{\frac{2}{3}}}=y$. Before Wallis and Newton, Vieta indicated general exponents a few times in a manner almost rhetorical;[4] his

$$A\ potestas+\frac{E\ potestate-A\ potesta}{E\ gradui+A\ gradu}\ in\ A\ gradum$$

is our

$$x^m+\frac{y^m-x^m}{y^n+x^n}\cdot x^n\ ,$$

the two distinct general powers being indicated by the words *potestas* and *gradus*. Johann Bernoulli[5] in 1691–92 still wrote $3\,\square\,\sqrt[3]{ax+xx}$ for

[1] *Isaaci Newtoni Opera* (ed. S. Horsely), Tom. IV (London, 1782), p. 215.

[2] J. Wallis, *Mathesis universalis* (Oxford, 1657), p. 292, 293, 294.

[3] See J. Collins, *Commercium epistolicum* (ed. J. B. Biot and F. Lefort; Paris, 1856), p. 145.

[4] Vieta, *Opera mathematica* (ed. Fr. van Schooten, 1634), p. 197.

[5] Iohannis I Bernoulli, *Lectiones de calculo differentialium* . . . . von Paul Schafheitlin. Separatabdruck aus den *Verhandlungen der Naturforschenden Gesellschaft in Basel*, Vol. XXXIV, 1922.

$3\sqrt[3]{(ax+x^2)^2}$, $4C\sqrt[4]{yx+xx}$ for $4\sqrt[4]{(yx+x^2)^3}$, $5QQ\sqrt[5]{ayx+x^3+zyx}$ for $5\sqrt[5]{(ayx+x^3+zyx)^4}$. But fractional, negative, and general exponents were freely used by D. Gregory[1] and were fully explained by W. Jones[2] and by C. Reyneau.[3] Reyneau remarks that this theory is not explained in works on algebra.

309. *Imaginary exponents.*—The further step of introducing imaginary exponents is taken by L. Euler in a letter to Johann Bernoulli,[4] of October 18, 1740, in which he announces the discovery of the formula $e^{+x\sqrt{-1}}+e^{-x\sqrt{-1}}=2\cos x$, and in a letter to C. Goldbach,[5] of December 9, 1741, in which he points out as a curiosity that the fraction $\dfrac{2^{+\sqrt{-1}}+2^{-\sqrt{-1}}}{2}$ is nearly equal to $1\frac{0}{3}$. The first appearance of imaginary exponents in print is in an article by Euler in the *Miscellanea Berolinensia* of 1743 and in Euler's *Introductio in analysin* (Lausannae, 1747), Volume I, page 104, where he gives the all-important formula $e^{+v\sqrt{-1}}=\cos v+\sqrt{-1}\sin v$.

310. At an earlier date occurred the introduction of variable exponents. In a letter of 1679, addressed to C. Huygens, G. W. Leibniz[6] discussed equations of the form $x^x-x=24, x^z+z^x=b, x^x+z^z=c$. On May 9, 1694, Johann Bernoulli[7] mentions expressions of this sort in a letter to Leibniz who, in 1695, again considered exponentials in the *Acta eruditorum*, as did also Johann Bernoulli in 1697.

311. Of interest is the following quotation from a discussion by T. P. Nunn, in the *Mathematical Gazette*, Volume VI (1912), page 255, from which, however, it must not be inferred that Wallis actually wrote down fractional and negative exponents: "Those who are acquainted with the work of John Wallis will remember that he invented negative and fractional indices in the course of an investigation into methods of evaluating areas, etc. He had

---

[1] David Gregory, *Exercitatio geometrica de dimensione figurarum* (Edinburgh, 1684), p. 4–6.

[2] William Jones, *Synopsis palmariorum matheseos* (London, 1706), p. 67, 115–19.

[3] Charles Reyneau, *Analyse demontrée* (Paris, 1708), Vol. I, Introduction.

[4] See G. Eneström, *Bibliotheca mathematica* (2d ser.), Vol. XI (1897), p. 49.

[5] P. H. Fuss, *Correspondance mathématique et physique* (Petersburg, 1843), Vol. I, p. 111.

[6] C. I. Gerhardt, *Briefwechsel von G. W. Leibniz mit Mathematikern* (2d ed.; Berlin, 1899), Vol. I, p. 568.

[7] Johann Bernoulli in *Leibnizens Mathematische Schriften* (ed. C. I. Gerhardt), Vol. III (1855), p. 140.

discovered that if the ordinates of a curve follow the law $y = kx^n$ its area follows the law $A = \dfrac{1}{n+1} \cdot kx^{n+1}$, n being (necessarily) a positive integer. This law is so remarkably simple and so powerful as a method that Wallis was prompted to inquire whether cases in which the ordinates follow such laws as $y = \dfrac{k}{x^n}$, $y = k \sqrt[r]{x}$ could not be brought within its scope. He found that this extension of the law would be possible if $\dfrac{k}{x^n}$ could be written kx^{-n}, and $k\sqrt[r]{x}$ as $kx^{\frac{1}{n}}$. From this, from numerous other historical instances, and from general psychological observations, I draw the conclusion that extensions of notation should be taught because and when they are needed for the attainment of some practical purpose, and that logical criticism should come after the suggestion of an extension to assure us of its validity."

312. *Notation for principal values.*—When in the early part of the nineteenth century the multiplicity of values of a^n came to be studied, where a and n may be negative or complex numbers, and when the need of defining the principal values became more insistent, new notations sprang into use in the exponential as well as the logarithmic theories. A. L. Cauchy[1] designated all the values that a^n may take, for given values of a and n $[a \neq 0]$, by the symbol $((a))^x$, so that $((a))^x = e^{xla} \cdot e^{2kx\pi i}$, where l means the tabular logarithm of $|a|$, $e = 2.718 \ldots$, $\pi = 3.141 \ldots$, $k = 0, \pm 1, \pm 2, \ldots$. This notation is adopted by O. Stolz and J. A. Gmeiner[2] in their *Theoretische Arithmetik*.

Other notations sprang up in the early part of the last century. Martin Ohm elaborated a general exponential theory as early as 1821 in a Latin thesis and later in his *System der Mathematik* (1822–33).[3] In a^x, when a and x may both be complex, $\log a$ has an infinite number of values. When, out of this infinite number some particular value of $\log a$, say $\propto$, is selected, he indicates this by writing $(a||\propto)$. With this understanding he can write $x \log a + y \log a = (x+y) \log a$, and consequently $a^x \cdot a^y = a^{x+y}$ is a *complete* equation, that is, an equation in which both sides have the same number of values, representing exactly the same expressions. Ohm did not introduce the particular value of a^x which is now called the "principal value."

[1] A. L. Cauchy, *Cours d'analyse* (Paris, 1821), chap. vii, § 1.

[2] O. Stolz and J. A. Gmeiner, *Theoretische Arithmetik* (Leipzig), Vol. II (1902), p. 371–77.

[3] Martin Ohm, *Versuch eines vollkommen consequenten Systems der Mathematik*, Vol. II (2d ed., 1829), p. 427. [First edition of Vol. II, 1823.]

Crelle[1] let $|u|^k$ indicate some fixed value of u^k, preferably a real value, if one exists, where k may be irrational or imaginary; the two vertical bars were used later by Weierstrass for the designation of absolute value (§ 492).

313. *Complicated exponents.*—When exponents themselves have exponents, and the latter exponents also have exponents of their own, then clumsy expressions occur, such as one finds in Johann I Bernoulli,[2] Goldbach,[3] Nikolaus II Bernoulli,[4] and Waring.

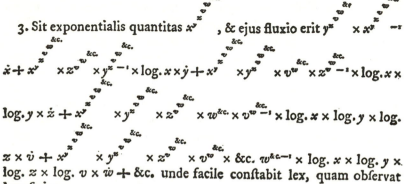

2. Sit data exponentialis quantitas $x^y \times v$, & per præcedentem methodum inveniri poteſt ejus fluxio $x^y \dot{v} + v y x^{y-1} \dot{x} + v \times x^y \times \log. x \times \dot{y}.$

3. Sit exponentialis quantitas $x^{y^{z^w}}$, & ejus fluxio erit ...

Fig. 106.—E. Waring's "repeated exponents." (From *Meditationes analyticae* [1785], p. 8.)

De Morgan[5] suggested a new notation for cases where exponents are complicated expressions. Using a solidus, he proposes $a \wedge \{(a+bx)/(c+ex)\}$, where the quantity within the braces is the exponent of a. He returned to this subject again in 1868 with the statement: "A convenient notation for repeated exponents is much wanted: not a working symbol, but a contrivance for preventing the symbol from wasting a line of text. The following would do perfectly well, $x|a|b|c|d$,

[1] A. L. Crelle in *Crelle's Journal*, Vol. VII (1831), p. 265, 266.

[2] Iohann I Bernoulli, *Acta eruditorum* (1697), p. 125–33.

[3] P. H. Fuss, *Correspondance math. et phys. ... du XVIIIᵉ siècle*, Vol. II (1843), p. 128.

[4] *Op. cit.*, p. 133.

[5] A. de Morgan, "Calculus of Functions," *Encyclopaedia Metropolitana*, Vol. II (1845), p. 388.

in which each *post* means all which follows is to be placed on the top of it. Thus:[1]

$$x|a|b|c|d = x^{a|b|c|d} = x^{a^{b|c|d}} = x^{a^{b^{c|d}}} = x^{a^{b^{c^d}}}.\text{''}$$

When the base and the successive exponents are all alike, say a, Woepcke[2] used the symbol $\overset{m}{\hat{a}}$ for $\left((a^a)^{a^{..}}\right)^a$, and $\overset{m}{\check{a}}$ for $_a\left(a^{..^{(a^a)}}\right)$, where m indicates the number of repetitions of a. He extended this notation to cases where a is real or imaginary, not zero, and m is a positive or negative integer, or zero. A few years later J. W. L. Glaisher suggested still another notation for complicated exponents, namely, $a\uparrow x^{\frac{1}{n}} + \frac{b}{x^n}\uparrow$, the arrows merely indicating that the quantity between them is to be raised so as to become the exponent of a. Glaisher prefers this to "$a\ Exp.\ u$" for a^u. Harkness and Morley[3] state, "It is usual to write $exp\ (z) = e^z$, when z is complex." The contraction "exp" was recommended by a British Committee (§ 725) in 1875, but was ignored in the suggestions of 1916, issued by the Council of the London Mathematical Society. G. H. Bryan stresses the usefulness of this symbol.[4]

Another notation was suggested by H. Schubert. If a^a is taken as an exponent of a, one obtains $a^{(a^a)}$ or a^{a^a}, and so on. Schubert designates the result by $(a; b)$, indicating that a has been thus written b times.[5] For the expression $(a; b)^{(a; c)}$ there has been adopted the sign $(a; b+c)$, so that $(a; b)^{(a; c)} = (a; c+1)^{(a; b-1)}$.

314. *D. F. Gregory*[6] in 1837 made use of the sign $(+)^r$, r an integer, to designate the repetition of the operation of multiplication. Also, $(+a^2)^{\frac{1}{2}} = +^{\frac{1}{2}}(a^2)^{\frac{1}{2}} = +^{\frac{1}{2}}a$, where the $+^{\frac{1}{2}}$ "will be different, according as we suppose the $+$ to be equivalent to the operation repeated an even or an odd number of times. In the former case it will be equal to $+$, in the latter to $-$. And generally, if we raise $+a$ to any power m, whether whole or fractional, we have $(+a)^m = +^m a^m$. . . . . So long as

[1] A. de Morgan, *Transactions of the Cambridge Philosophical Society*, Vol. XI, Part III (1869), p. 450.

[2] F. Woepcke in *Crelle's Journal*, Vol. XLII (1851), p. 83.

[3] J. Harkness and F. Morley, *Theory of Functions* (New York, 1893), p. 120.

[4] *Mathematical Gazette*, Vol. VIII (London, 1917), p. 172, 220.

[5] H. Schubert in *Encyclopédie d. scien. math.*, Tome I, Vol. I (1904), p. 61. L. Euler considered a^{a^a}, etc. See E. M. Lémeray, *Proc. Edinb. Math. Soc.*, Vol. XVI (1897), p. 13.

[6] *The Mathematical Writings of Duncan Farquharson Gregory* (ed. William Walton; Cambridge, 1865), p. 124–27, 145.

m is an integer, rm is an integer, and $+^{rm}a^m$ has only one value; but if m be a fraction of the form $\frac{p}{q}$, $+^{r\frac{p}{q}}$ will acquire different values, according as we assign different values to r .... $\sqrt{(-a)} \times \sqrt{(-a)} = \sqrt{(+a^2)} = \sqrt{(+)}\sqrt{(a^2)} = -a$; for in this case we know how the $+$ has been derived, namely from the product $- -= +$, or $-^2 = +$, which of course gives $+^{\frac{1}{2}} = -$, there being here nothing indeterminate about the $+$. It was in consequence of sometimes tacitly assuming the existence of $+$, and at another time neglecting it, that the errors in various trigonometrical expressions arose; and it was by the introduction of the factor $\cos 2r\pi + -^{\frac{1}{2}} \sin 2r\pi$ (which is equivalent to $+^r$) that Poinsot established the formulae in a more correct and general shape." Gregory finds "$\sin (+^{\frac{p}{q}}c) = +^{\frac{p}{q}} \sin c$."

A special notation for the positive integral powers of an imaginary root r of $x^{n-1} + x^{n-2} + \dots + x + 1 = 0$, n being an odd prime, is given by Gauss;[1] to simplify the typesetting he designates r, rr, r^3, etc., by the symbols [1], [2], [3], etc.

315. *Conclusions.*—There is perhaps no symbolism in ordinary algebra which has been as well chosen and is as elastic as the Cartesian exponents. Descartes wrote a^3, x^4; the extension of this to general exponents a^n was easy. Moreover, the introduction of fractional and negative numbers, as exponents, was readily accomplished. The irrational exponent, as in $a^{\sqrt{2}}$, found unchallenged admission. It was natural to try exponents in the form of pure imaginary or of complex numbers (L. Euler, 1740). In the nineteenth century valuable interpretations were found which constitute the general theory of b^n where b and n may both be complex. Our exponential notation has been an aid for the advancement of the science of algebra to a degree that could not have been possible under the old German or other early notations. Nowhere is the importance of a good notation for the rapid advancement of a mathematical science exhibited more forcibly than in the exponential symbolism of algebra.

SIGNS FOR ROOTS

316. *Early forms.*—Symbols for roots appear very early in the development of mathematics. The sign $\sqsubset$ for square root occurs in two Egyptian papyri, both found at Kahun. One was described by F. L.

---

[1] C. F. Gauss, *Disquisitiones arithmeticae* (Leipzig, 1801), Art. 342; *Werke*, Vol. I (1863), p. 420.

Griffith[1] and the other by H. Schack-Schackenburg.[2] For Hindu signs see §§ 107, 108, 112; for Arabic signs see § 124.

317. *General statement.*—The principal symbolisms for the designation of roots, which have been developed since the influx of Arabic learning into Europe in the twelfth century, fall under four groups having for their basic symbols, respectively, R (*radix*), l (*latus*), the sign $\sqrt{}$, and the fractional exponent.

318. *The sign R; first appearance.*—In a translation[3] from the Arabic into Latin of a commentary of the tenth book of the *Elements* of Euclid, the word *radix* is used for "square root." The sign R came to be used very extensively for "root," but occasionally it stood also for the first power of the unknown quantity, x. The word *radix* was used for x in translations from Arabic into Latin by John of Seville and Gerard of Cremona (§ 290). This double use of the sign R for x and also for square root is encountered in Leonardo of Pisa (§§ 122, 292)[4] and Luca Pacioli (§§ 135–37, 292).

Before Pacioli, the use of R to designate square root is also met in a correspondence that the German astronomer Regiomontanus (§ 126)[5] carried on with Giovanni Bianchini, who was court astronomer at Ferrara in Italy, and with Jacob von Speier, a court astronomer at Urbino (§ 126).

In German manuscripts referred to as the Dresden MSS C. 80, written about the year 1480, and known to have been in the hands of J. Widman, H. Grammateus, and Adam Riese, there is a sign consisting of a small letter, with a florescent stroke attached (Fig. 104). It has been interpreted by some writers as a letter r with an additional stroke. Certain it is that in Johann Widman's arithmetic of 1489 occurs the crossed capital letter R, and also the abbreviation *ra* (§ 293).

Before Widman, the Frenchman Chuquet had used R for "root"

[1] F. L. Griffith, *The Petrie Papyri, I. Kahun Papyri*, Plate VIII.

[2] H. Schack-Schackenburg in *Zeitschrift für aegyptische Sprache und Altertumskunde*, Vol. XXXVIII (1900), p. 136; also Plate IV. See also Vol. XL, p. 65.

[3] M. Curtze, *Anaritii in decem libros elementorum Euclidis commentarii* (Leipzig, 1899), p. 252–386.

[4] *Scritti di Leonardo Pisano* (ed. B. Boncompagni), Vol. II (Rome, 1862), "La practica geometriae," p. 209, 231. The word *radix*, meaning x, is found also in Vol. I, p. 407.

[5] M. Curtze, "Der Briefwechsel Riomantan's, etc.," *Abhandlungen zur Geschichte der mathematischen Wissenschaften*, Vol. XII (Leipzig, 1902), p. 234, 318.

in his manuscript, *Le Triparty* (§ 130). He[1] indicates R^2 16. as 4, "R^4. 16. *si est* . 2.," "R^5. 32. *si est* . 2.*"

319. *Sixteenth-century use of R.*—The different uses of R. made in Pacioli's *Summa* (1494, 1523) are fully set forth in §§ 134–38. In France, De la Roche followed Chuquet in the use of R. (§ 132). The symbol appears again in Italy in Ghaligai's algebra (1521), and in later editions (§ 139), while in Holland it appeared as early as 1537 in the arithmetic of Giel Van der Hoecke (§ 150) in expressions like "Item wildi aftrecken R $\frac{1}{5}$ van R $\frac{4}{5}$ resi R $\frac{1}{5}$"; i.e., $\sqrt{\frac{4}{5}} - \sqrt{\frac{1}{5}} = \sqrt{\frac{1}{5}}$. The employment of R in the calculus of radicals by Cardan is set forth in §§ 141, 199. A promiscuous adoption of different notations is found in the algebra of Johannes Scheubel (§§ 158, 159) of the University of Tübingen. He used Widman's abbreviation *ra*, also the sign $\sqrt{}$; he indicates cube root by *ra. cu.* or by $\mathcal{NV}$, fourth root by *ra. ra.* or by $\mathcal{NV}$. He suggests a notation of his own, of which he makes no further use, namely, *radix se.*, for cube root, which is the abbreviation of *radix secundae quantitatis*. As the sum "*ra. 15 ad ra. 17*" he gives "*ra. col.* $32 + \sqrt{1020}$," i.e., $\sqrt{15} + \sqrt{17} = \sqrt{32 + \sqrt{1020}}$. The *col., collecti*, signifies here aggregation.

Nicolo Tartaglia in 1556 used R extensively and also parentheses (§§ 142, 143). Francis Maurolycus[2] of Messina in 1575 wrote "*r. 18*" for $\sqrt{18}$, "*r. v. 6 m̃. r 7⅓*" for $\sqrt{6 - \sqrt{7\frac{1}{3}}}$. Bombelli's radical notation is explained in § 144. It thus appears that in Italy the R had no rival during the sixteenth century in the calculus of radicals. The only variation in the symbolism arose in the marking of the order of the radical and in the modes of designation of the aggregation of terms that were affected by R.

320. In Spain[3] the work of Marco Aurel (1552) (§ 204) employs the signs of Stifel, but Antich Rocha, adopting the Italian abbreviations in adjustment to the Spanish language, lets, in his *Arithmetica* of 1564, "15 *Mas ra. q.* 50 *Mas ra. q.* 27 *Mas ra. q.* 6" stand for $15 + \sqrt{50} + \sqrt{27} + \sqrt{6}$. A few years earlier, J. Perez de Moya, in his *Aritmetica practica y speculativa* (1562), indicates square root by *r*,

---

[1] *Le Triparty en la science des nombres par Maistre Nicolas Chuquet Parisien* ... par M. Aristide Marre in Boncompagni's *Bullettino*, Vol. XIII, p. 655; (reprint, Rome, 1881), p. 103.

[2] *D. Francisci Mavrolyci, Abbatis Messanensis, Opuscula mathematica* (Venice, 1575), p. 144.

[3] Our information on these Spanish authors is drawn partly from Julio Rey Pastor, *Los Matemáticos españoles de siglo XVI* (Oviedo, 1913), p. 42.

cube root by *rrr*, fourth root by *rr*, marks powers by *co.*, *ce.*, *cu.*, *c. ce.*, and "plus" by *p*, "minus" by *m*, "equal" by *eq.*

In Holland, Adrianus Romanus[1] used a small *r*, but instead of *v* wrote a dot to mark a root of a binomial or polynomial; he wrote

r bin. 2+*r bin.* 2+*r bin.* 2+*r* 2. to designate $\sqrt{2+\sqrt{2+\sqrt{2+\sqrt{2}}}}$.

In Tartaglia's arithmetic, as translated into French by Gosselin[2] of Caen, in 1613, one finds the familiar R *cu* to mark cube root. A modification was introduced by the Scotsman James Hume,[3] residing in Paris, who in his algebra of 1635 introduced Roman numerals to indicate the order of the root (§ 190). Two years later, the French text by Jacqves de Billy[4] used RQ, RC, RQC for $\sqrt{}$, $\sqrt[3]{}$, $\sqrt[6]{}$, respectively.

321. *Seventeenth-century use of R.*—During the seventeenth century, the symbol R lost ground steadily but at the close of the century it still survived; it was used, for instance, by Michael Rolle[5] who employed the signs $2+\overline{R.-121.}$ to represent $2+\sqrt{-121}$, and $R.$ *trin.* $\overline{6aabb-9a^4b-b^3}$ to represent $\sqrt{6a^2b^2-9a^4b-b^3}$. In 1690 H. Vitalis[6] takes R_2 to represent *secunda radix*, i.e., the radix next after the square root. Consequently, with him, as with Scheubel, 3. $R.$ 2* 8, meant $3\sqrt[3]{8}$, or 6.

The sign R or R, representing a radical, had its strongest foothold in Italy and Spain, and its weakest in England. With the close of the seventeenth century it practically passed away as a radical sign; the symbol $\sqrt{}$ gained general ascendancy. Elsewhere it will be pointed out in detail that some authors employed R to represent the unknown *x*. Perhaps its latest regular appearance as a radical sign is in the Spanish text of Perez de Moya (§ 320), the first edition of which appeared in 1562. The fourteenth edition was issued in 1784; it still gave *rrr* as signifying cube root, and *rr* as fourth root. Moya's book offers a most striking example of the persistence for centuries of old and clumsy notations, even when far superior notations are in general use.

---

[1] *Ideae Mathematicae Pars Prima,* . . . . *Adriano Romano Lovaniensi* (Antwerp, 1593), following the Preface.

[2] *L'Arithmetique de Nicolas Tartaglia Brescian,* traduit ... par Gvillavmo Gosselin de Caen, Premier Partie (Paris, 1613), p. 101.

[3] James Hume, *Traité de l'algebre* (Paris, 1635), p. 53.

[4] Jacqves de Billy, *Abregé des Preceptes d'Algebre* (Rheims, 1637), p. 21.

[5] *Journal des Sçavans* de l'An 1683 (Amsterdam, 1709), p. 97.

[6] *Lexicon mathematicum ... authore Hieronymo Vitali* (Rome, 1690), art. "Algebra."

322. *The sign l.*—The Latin word *latus* ("side of a square") was introduced into mathematics to signify root by the Roman surveyor Junius Nipsus,[1] of the second century A.D., and was used in that sense by Martianus Capella,[2] Gerbert,[3] and by Plato of Tivoli in 1145, in his translation from the Arabic of the *Liber embadorum* (§ 290). The symbol *l* (*latus*) to signify root was employed by Peter Ramus[4] with whom "*l* 27 *ad l* 12" gives "*l* 75," i.e., $\sqrt{27} + \sqrt{12} = \sqrt{75}$; "*ll* 32 *de ll* 162" gives "*ll* 2," i.e., $\sqrt[4]{32}$ from $\sqrt[4]{162} = \sqrt[4]{2}$. Again,[5] "8−*l* 20 *in* 2 *quotus est* (4−*l* 5." means $8 - \sqrt{20}$, divided by 2, gives the quotient $4 - \sqrt{5}$. Similarly,[6] "*lr.* *l*112−*l*76" meant $\sqrt{\sqrt{112} - \sqrt{76}}$; the *r* signifying here *residua*, or "remainder," and therefore *lr.* signified the square root of the binomial difference.

In the 1592 edition[7] of Ramus' arithmetic and algebra, edited by Lazarus Schoner, "*lc* 4" stands for $\sqrt[3]{4}$, and "*l bq* 5" for $\sqrt[4]{5}$, in place of Ramus' "*ll* 5." Also, $\sqrt{2}$. $\sqrt{3} = \sqrt{6}$, $\sqrt{6} \div \sqrt{2} = \sqrt{3}$ is expressed thus:[8]

"Esto multiplicandum *l₂* per *l₃* factus erit *l* 6.

$$l\,2.$$
$$l\,3.$$
$$\overline{}$$
$$l\,6.$$

$$l\,6 \Big(l\,3\,."$$
$$l\,2$$

It is to be noted that with Schoner the *l* received an extension of meaning, so that 5*l* and *l*5, respectively, represent $5x$ and $\sqrt{5}$, the *l* standing for the first power of the unknown quantity when it is not

[1] *Die Schriften der römischen Feldmesser* (ed. Blume, Lachmann, Rudorff; Berlin, 1848–52), Vol. I, p. 96.

[2] Martianus Capella, *De Nuptiis* (ed. Kopp; Frankfort, 1836), lib. VII, § 748.

[3] *Gerberti opera mathematica* (ed. Bubnow; Berlin, 1899), p. 83. See J. Tropfke, *op. cit.*, Vol. II (2d ed., 1921), p. 143.

[4] *P. Rami Scholarvm mathematicarvm libri unus et triginti* (Basel, 1569), Lib. XXIV, p. 276, 277.

[5] *Ibid.*, p. 179.

[6] *Ibid.*, p. 283.

[7] *Petri Rami ... Arithmetices libri duo, et algebrae totidem: à Lazaro Schonero* (Frankfurt, 1592), p. 272 ff.

[8] *Petri Rami ... Arithmeticae libri duo et geometriae septem et viginti, Dudum quidem, à Lazaro Schonero* (Frankfurt a/M., 1627), part entitled "De Nvmeri figvratis Lazari Schoneri liber," p. 178.

followed by a number (see also § 290). A similar change in meaning resulting from reversing the order of two symbols has been observed in Pacioli in connection with $\cancel{R}$ (§§ 136, 137) and in A. Girard in connection with the circle of Stevin (§ 164). The double use of the sign l, as found in Schoner, is explained more fully by another pupil of Ramus, namely, Bernardus Salignacus (§ 291).

Ramus' l was sometimes used by the great French algebraist Francis Vieta who seemed disinclined to adopt either $\cancel{R}$ or $\sqrt{\ }$ for indicating roots (§ 177).

This use of the letter l in the calculus of radicals never became popular. After the invention of logarithms, this letter was needed to mark logarithms. For that reason it is especially curious that Henry Briggs, who devoted the latter part of his life to the computation and the algorithm of logarithms, should have employed l in the sense assigned it by Ramus and Vieta. In 1624 Briggs used $l, l_{(3)}, ll$ for square, cube, and fourth root, respectively. "Sic $l_{(3)}$ 8 [i.e., $\sqrt[3]{8}$], latus cubicum Octonarii, id est 2. sic l bin $2 + l$ 3. [i.e., $\sqrt{2+\sqrt{3}}$] latus binomii $2+l3$." Again, "ll $85\frac{1}{3}$ [i.e., $\sqrt[4]{85\frac{1}{3}}$]. Latus $85\frac{1}{3}$ est $9^{237604307}$, et huius lateris latus est $3^{03934274}$. cui numero aequatur ll $85\frac{1}{3}$."[1]

323. *Napier's line symbolism.*—John Napier[2] prepared a manuscript on algebra which was not printed until 1839. He made use of Stifel's notation for radicals, but at the same time devised a new scheme of his own. "It is interesting to notice that although Napier invented an excellent notation of his own for expressing roots, he did not make use of it in his algebra, but retained the cumbrous, and in some cases ambiguous notation generally used in his day. His notation was derived from this figure

$$
\begin{array}{c|c|c}
1 & 2 & 3 \\
\hline
4 & 5 & 6 \\
\hline
7 & 8 & 9
\end{array}
$$

in the following way: ⊔ prefixed to a number means its square root, ⊐ its fourth root, □ its fifth root, ⌐ its ninth root, and so on, with extensions of obvious kinds for higher roots."[3]

[1] Henry Briggs, *Arithmetica logarithmica* (London, 1624), Introduction.

[2] *De Arte Logistica Joannis Naperi Merchistonii Baronis Libri qui supersunt* (Edinburgh, 1839), p. 84.

[3] J. E. A. Steggall, "De arte logistica," *Napier Tercentenary Memorial Volume* (ed. Cargill Gilston Knott; London, 1915), p. 160.

THE SIGN $\sqrt{}$

324. *Origin of* $\sqrt{}$.—This symbol originated in Germany. L. Euler guessed that it was a deformed letter r, the first letter in *radix*.[1] This opinion was held generally until recently. The more careful study of German manuscript algebras and the first printed algebras has convinced Germans that the old explanation is hardly tenable; they have accepted the a priori much less probable explanation of the evolution of the symbol from a dot. Four manuscript algebras have been available for the study of this and other questions.

The oldest of these is in the Dresden Library, in a volume of manuscripts which contains different algebraic treatises in Latin and one in German.[2] In one of the Latin manuscripts (see Fig. 104, $A7$), probably written about 1480, dots are used to signify root extraction. In one place it says: "In extraccione radicis quadrati alicuius numeri preponatur numero vnus punctus. In extraccione radicis quadrati radicis quadrati prepone numero duo puncta. In extraccione cubici radicis alicuius numeri prepone tria puncta. In extraccione cubici radicis alicuius radicis cubici prepone 4 puncta."[3] That is, one dot (·) placed before the radicand signifies square root; two dots (..) signify the square root of the square root; three dots (...) signify cube root; four dots (....), the cube root of the cube root or the ninth root. Evidently this notation is not a happy choice. If one dot meant square root and two dots meant square root of square root (i.e., $\sqrt{\sqrt{}}$), then three dots should mean square root of square root of square root, or eighth root. But such was not actually the case; the three dots were made to mean cube root, and four dots the ninth root. What was the origin of this dot-system? No satisfactory explanation has been found. It is important to note that this Dresden manuscript was once in the possession of Joh. Widman, and that Adam Riese, who in 1524 prepared a manuscript algebra of his own, closely followed the Dresden algebra.

325. The second document is the Vienna MS[4] No. 5277, Regule-

[1] L. Euler, *Institutiones calculi differentialis* (1775), p. 103, art. 119; J. Tropfke, *op. cit.*, Vol. II (2d ed., 1921), p. 150.

[2] M. Cantor, *Vorles. über Geschichte der Mathematik*, Vol. II (2. Aufl., 1900), p. 241.

[3] E. Wappler, *Zur Geschichte der deutschen Algebra im XV. Jahrhundert*, Zwickauer Gymnasialprogramm von 1887, p. 13. Quoted by J. Tropfke, *op. cit.*, Vol. II (1921), p. 146, and by M. Cantor, *op. cit.*, Vol. II (2. Aufl., 1900), p. 243.

[4] C. J. Gerhardt, *Monatsberichte Akad.* (Berlin, 1867), p. 46; *ibid.* (1870), p. 143–47; Cantor. *op. cit.*, Vol. II (2d ed., 1913), p. 240, 424.

Cose–uel Algobre–. It contains the passage: "Quum $\mathfrak{z}$ assimiletur radici de radice punctus deleatur de radice, $\mathfrak{z}$ in se ducatur et remanet adhuc inter se aequalia"; that is, "When $x^2 = \sqrt{x}$, erase the point before the x and multiply x^2 by itself, then things equal to each other are obtained." In another place one finds the statement, *per punctum intellige radicem*—"by a point understand a root." But no dot is actually used in the manuscript for the designation of a root.

The third manuscript is at the University of Göttingen, Codex Gotting. Philos. 30. It is a letter written in Latin by Initius Algebras,[1] probably before 1524. An elaboration of this manuscript was made in German by Andreas Alexander.[2] In it the radical sign is a heavy point with a stroke of the pen up and bending to the right, thus $\int$. It is followed by a symbol indicating the index of the root; $\int \mathfrak{z}$ indicates square root; $\int c^e$, cube root; $\int cc^e$, the ninth root, etc. Moreover, $\int cs \lfloor 8 + \int 22_{\mathfrak{z}}$ stands for $\sqrt{8 + \sqrt{22}}$, where cs (i.e., *communis*) signifies the root of the binomial which is designated as one quantity, by lines, vertical and horizontal. Such lines are found earlier in Chuquet (§ 130). The $\mathfrak{z}$, indicating the square root of the binomial, is placed as a subscript after the binomial. Calling these two lines a "gnomon," M. Curtze adds the following:

"This gnomon has here the signification, that what it embraces is not a length, but a power. Thus, the simple 8 is a length or simple number, while $\lfloor 8_{\mathfrak{z}}$ is a square consisting of eight areal units whose linear unit is $\int \mathfrak{z} \lfloor 8$. In the same way $\lfloor 8_{c^e}$ would be a cube, made up of 8 cubical units, of which $\int c^e \lfloor 8$ is its side, etc. A double point, with the tail attached to the last, signifies always the root of the root. For example, $.\int c^e \lfloor 88$ would mean the cube root of the cube root of 88. It is identical with $\int cc^e 88$, but is used only when the radicand is a so-called median [*Mediale*] in the Euclidean sense."[3]

326. The fourth manuscript is an algebra or *Coss* completed by Adam Riese[4] in 1524; it was not printed until 1892. Riese was familiar with the small Latin algebra in the Dresden collection, cited above;

[1] *Initius Algebras: Algebrae Arabis Arithmetici viri clarrisimi Liber ad Ylem geometram magistrum suum*. This was published by M. Curtze in *Abhandlungen zur Geschichte der mathematischen Wissenschaften*, Heft XIII (1902), p. 435–611. Matters of notation are explained by Curtze in his introduction, p. 443–48.

[2] G. Eneström, *Bibliotheca mathematica* (3d ser.), Vol. III (1902), p. 355–60.

[3] M. Curtze, *op. cit.*, p. 444.

[4] B. Berlet, *Adam Riese, sein Leben, seine Rechenbücher und seine Art zu rechnen; die Coss von Adam Riese* (Leipzig-Frankfurt a/M., 1892).

he refers also to Andreas Alexander.[1] For indicating a root, Riese does not use the dot, pure and simple, but the dot with a stroke attached to it, though the word *punct* ("point") occurs. Riese says: "Ist, so ʒ vergleicht wird $\sqrt{}$ vom radix, so mal den ʒ in sich multipliciren vnnd das punct vor dem Radix aussleschn."[2] This passage has the same interpretation as the Latin passage which we quoted from the Vienna manuscript.

We have now presented the main facts found in the four manuscripts. They show conclusively that the dot was associated as a symbol with root extraction. In the first manuscript, the dot actually appears as a sign for roots. The dot does not appear as a sign in the second manuscript, but is mentioned in the text. In the third and fourth manuscripts, the dot, pure and simple, does not occur for the designation of roots; the symbol is described by recent writers as a dot with a stroke or tail attached to it. The question arises whether our algebraic sign $\sqrt{}$ took its origin in the dot. Recent German writers favor that view, but the evidence is far from conclusive. Johannes Widman, the author of the *Rechnung* of 1489, was familiar with the first manuscript which we cited. Nevertheless he does not employ the dot to designate root, easy as the symbol is for the printer. He writes down $\cancel{R}$ and *ra*. Christoff Rudolff was familiar with the Vienna manuscript which uses the dot with a tail. In his *Coss* of 1525 he speaks of the *Punkt* in connection with root symbolism, but uses a mark with a very short heavy downward stroke (almost a point), followed by a straight line or stroke, slanting upward (see Fig. 59). As late as 1551, Scheubel,[3] in his printed *Algebra*, speaks of points. He says: "Solent tamen multi, et bene etiam, has desideratas radices, suis punctis cū linea quadam a dextro latere ascendente, notare. . . . ." ("Many are accustomed, and quite appropriately, to designate the desired roots by points, from the right side of which there ascends a kind of stroke.") It is possible that this use of "point" was technical, signifying "sign for root," just as at a later period the expression "decimal point" was used even when the symbol actually written down to mark a decimal fraction was a comma. It should be added that if Rudolff looked upon his radical sign as really a dot, he would have been less likely to have used the dot again for a second purpose in his radical symbolism, namely, for the purpose of designating that

[1] B. Berlet, *op. cit.*, p. 29, 33.

[2] C. I. Gerhardt, *op. cit.* (1870) p. 151.

[3] J. Scheubel, *Algebra compendiosa* (Paris, 1551), fol. 25*B*. Quoted from J. Tropfke, *op. cit.*, Vol. II (2d ed., 1921), p. 149.

the root extraction must be applied to two or more terms following the $\sqrt{}$; this use of the dot is shown in § 148. It is possible, perhaps probable, that the symbol in Rudolff and in the third and fourth manuscripts above referred to is not a point at all, but an r, the first letter in *radix*. That such was the understanding of the sixteenth-century Spanish writer, Perez de Moya (§ 204), is evident from his designations of the square root by r, the fourth root by rr, and the cube root by rrr. It is the notation found in the first manuscript which we cited, except that in Moya the r takes the place of the dot; it is the notation of Rudolff, except that the sign in Rudolff is not a regularly shaped r. In this connection a remark of H. Wieleitner is pertinent: "The dot appears at times in manuscripts as an abbreviation for the syllable *ra*. Whether the dot used in the Dresden manuscript represents this normal abbreviation for *radix* does not appear to have been specially examined."[1]

The history of our radical sign $\sqrt{}$, after the time of Rudolff, relates mainly to the symbolisms for indicating (1) the index of the root, (2) the aggregation of terms when the root of a binomial or polynomial is required. It took over a century to reach some sort of agreement on these points. The signs of Christoff Rudolff are explained more fully in § 148. Stifel's elaboration of the symbolism of Andreas Alexander as given in 1544 is found in §§ 153, 155. Moreover, he gave to the $\sqrt{}$ its modern form by making the heavy left-hand, downward stroke, longer than did Rudolff.

327. *Spread of* $\sqrt{}$.—The German symbol of $\sqrt{}$ for root found its way into France in 1551 through Scheubel's publication (§ 159); it found its way into Italy in 1608 through Clavius; it found its way into England through Recorde in 1557 (§ 168) and Dee in 1570 (§ 169); it found its way into Spain in 1552 through Marco Aurel (§§ 165, 204), but in later Spanish texts of that century it was superseded by the Italian $R\!\!\!/$. The German sign met a check in the early works of Vieta who favored Ramus' l, but in later editions of Vieta, brought out under the editorship of Fr. van Schooten, the sign $\sqrt{}$ displaced Vieta's earlier notations (§ 176, 177).

In Denmark Chris. Dibuadius[2] in 1605 gives three designations of square root, $\sqrt{}$, $\sqrt{}Q$, $\sqrt{}\mathring{3}$; also three designations of cube root, $\sqrt{}C$, $\sqrt{}c$, $\sqrt{}c^e$; and three designations of the fourth root, $\sqrt{}\sqrt{}$, $\sqrt{}QQ$, $\sqrt{}\mathring{3}\mathring{3}$.

Stifel's mode of indicating the order of roots met with greater

[1] H. Wieleitner, *Die Sieben Rechnungsarten* (Leipzig-Berlin, 1912), p. 49.

[2] *C. Dibvadii in Arithmeticam irrationalivm Euclidis* (Arnhem, 1605).

general favor than Rudolff's older and clumsier designation (§§ 153, 155).

328. *Rudolff's signs outside of Germany.*—The clumsy signs of Christoff Rudolff, in place of which Stifel had introduced in 1544 and 1553 better symbols of his own, found adoption in somewhat modified form among a few writers of later date. They occur in Aurel's *Spanish Arithmetica*, 1552 (§ 165). They are given in Recorde, *Whetstone of Witte* (1557) (§ 168), who, after introducing the first sign, $\sqrt{\ }$., proceeds: "The seconde signe is annexed with *Surde Cubes*, to expresse their rootes. As this $.\sqrt{\ }.16$ whiche signifieth the *Cubike roote* of .16. And $.\sqrt{\ }.20.$ betokeneth the *Cubike roote* of .20. And so forthe. But many tymes it hath the *Cossike* signe with it also: as $\sqrt{\ }.c^e$ 25 the *Cubike roote* of .25. And $\sqrt{\ }.c^e.32.$ the *Cubike roote* of .32. The thirde figure doeth represente a *zenzizenzike roote*. As $.\sqrt{\ }.12.$ is the *zenzizenzike roote* of .12. And $\sqrt{\ }.35.$ is the *zenzizenzike roote* of .35. And likewaies if it haue with it the *Cossike* signe $.\mathfrak{zz}.$ As $\sqrt{\mathfrak{zz}}24$ the *zenzizenzike roote* of .24. and so of other."

The Swiss Ardüser in 1627 employed Rudolff's signs for square root and cube root.[1] J. H. Rahn in 1659 used $\sqrt{\ }$ for evolution,[2] which may be a modified symbol of Rudolff; Rahn's sign is adopted by Thomas Brancker in his English translation of Rahn in 1668, also by Edward Hatton[3] in 1721, and by John Kirkby[4] in 1725. Ozanam[5] in 1702 writes $\sqrt[4]{5}+\sqrt[4]{2}$ and also $\sqrt{\ }5+\sqrt{\ }2$. Samuel Jeake[6] in 1696 gives modifications of Rudolff's signs, along with other signs, in an elaborate explanation of the "characters" of "Surdes"; $\sqrt{\ }$ means root, $\sqrt{\ }$: or V or V$\sqrt{\ }$ universal root, $\sqrt{\ }$ or $\sqrt{\ }\mathfrak{z}$ square root, $\sqrt{\ }$ or $\sqrt{\ }\phi$ cube root, $\sqrt{\ }$ or $\sqrt{\ }\mathfrak{zz}$ squared square root, $\sqrt{\ }$ or $\sqrt{\ }\rho$ sursolide root.

On the Continent, Johann Caramuel[7] in 1670 used $\sqrt{\ }$ for square root and repeated the symbol $\sqrt{\ }\sqrt{\ }$ for cube root: "$\sqrt{\ }\sqrt{\ }27.$ est Radix Cubica Numeri 27. hoc est, 3."

[1] Johann Ardüser, *Geometriae Theoricae et Practicae*, XII. *Bücher* (Zürich, 1627), fol. 81*A*.

[2] Johann Heinrich Rahn, *Teutsche Algebra* (Zürich, 1659).

[3] Edward Hatton, *An Intire System of Arithmetic* (London, 1721), p. 287.

[4] John Kirkby, *Arithmetical Institutions* (London, 1735), p. 7.

[5] J. Ozanam, *Nouveaux Elemens d'algebre* ... par M. Ozanam, I. Partie (Amsterdam, 1702), p. 82.

[6] Samuel Jeake, ΛΟΓΙΣΤΙΚΗΛΟΓΙΑ, *or Arithmetick* (London, 1696), p. 293.

[7] *Joannis Caramvelis Mathesis Biceps. Vetus, et Nova* (Campaniae, 1670), p. 132.

329. *Stevin's numeral root-indices.*—An innovation of considerable moment were Stevin's numeral indices which took the place of Stifel's letters to mark the orders of the roots. Beginning with Stifel the sign $\sqrt{}$ without any additional mark came to be interpreted as meaning specially square root. Stevin adopted this interpretation, but in the case of cube root he placed after the $\sqrt{}$ the numeral 3 inclosed in a circle (§§ 162, 163). Similarly for roots of higher order. Stevin's use of numerals met with general but not universal adoption. Among those still indicating the order of a root by the use of letters was Descartes who in 1637 indicated cube root by $\sqrt{}C$. But in a letter of 1640 he[1] used the 3 and, in fact, leaned toward one of Albert Girard's notations, when he wrote $\sqrt{}3).20 + \sqrt{}392$ for $\sqrt[3]{20 + \sqrt{392}}$. But very great diversity prevailed for a century as to the exact position of the numeral relative to the $\sqrt{}$. Stevin's $\sqrt{}$, followed by numeral indices placed within circles, was adopted by Stampioen,[2] and by van Schooten.[3]

A. Romanus displaced the circle of Stevin by two round parentheses, a procedure explained in England by Richard Sault[4] who gives $\sqrt{}\overline{(3)a+b}$ or $\overline{a+b}|^{3}$. Like Girard, Harriot writes $\sqrt{}3.)26 + \sqrt{}675$ for $\sqrt[3]{26 + \sqrt{675}}$ (see Fig. 87 in § 188). Substantially this notation was used by Descartes in a letter to Mersenne (September 30, 1640), where he represents the *racine cubique* by $\sqrt{}3)$, the *racine sursolide* by $\sqrt{}5)$, the *B sursolide* by $\sqrt{}7)$, and so on.[5] Oughtred sometimes used square brackets, thus $\sqrt{}[12]1000$ for $\sqrt[12]{1000}$ (§ 183).

330. A step in the right direction is taken by John Wallis[6] who in 1655 expresses the root indices in numerals without inclosing them in a circle as did Stevin, or in parentheses as did Romanus. However, Wallis' placing of them is still different from the modern; he writes $\sqrt{}^{3}R^2$ for our $\sqrt[3]{R^2}$. The placing of the index within the opening of the radical sign had been suggested by Albert Girard as early as 1629. Wallis' notation is found in the universal arithmetic of the Spaniard, Joseph Zaragoza,[7] who writes $\sqrt{}^{4}243 - \sqrt{}^{3}27$ for our $\sqrt[4]{243} - \sqrt[3]{27}$, and

---

[1] *Œuvres de Descartes*, Vol. X, p. 190.

[2] *Algebra ofte Nieuwe Stel-Regel ... door Johan Stampioen d'Jongle 's* Graven-Hage (1639), p. 11.

[3] Fr. van Schooten, *Geometria à Renato des Cartes* (1649), p. 328.

[4] Richard Sault, *A New Treatise of Algebra* (London, n.d.).

[5] *Œuvres de Descartes*, Vol. III (1899), p. 188.

[6] John Wallis, *Arithmetica infinitorum* (Oxford, 1655), p. 59, 87, 88.

[7] Joseph Zaragoza, *Arithmetica vniversal* (Valencia, 1669), p. 307.

$\sqrt[2]{}(7+\sqrt[2]{}13)$ for our $\sqrt{7+\sqrt{13}}$. Wallis employs this notation[1] again in his *Algebra* of 1685. It was he who first used general indices[2] in the expression $\sqrt[d]{}R^d=R$. The notation $\sqrt[4]{}19$ for $\sqrt[4]{19}$ crops out again[3] in 1697 in De Lagny's $\sqrt[3]{}54-\sqrt[3]{}16=\sqrt[3]{}2$; it is employed by Thomas Walter;[4] it is found in the *Maandelykse Mathematische Lief-hebberye* (1754–69), though the modern $\sqrt[4]{}$ is more frequent; it is given in Castillion's edition[5] of Newton's *Arithmetica universalis*.

331. The Girard plan of placing the index in the opening of the radical appears in M. Rolle's *Traité d'Algèbre* (Paris, 1690), in a letter of Leibniz[6] to Varignon of the year 1702, in the expression $\sqrt[2]{1+\sqrt{-3}}$, and in 1708 in (a review of) G. Manfred[7] with literal index, $\sqrt[m]{aa+bb}^n$. At this time the Leibnizian preference for $\sqrt{}(aa+bb)$ in place of $\sqrt{aa+bb}$ is made public;[8] a preference which was heeded in Germany and Switzerland more than in England and France. In Sir Isaac Newton's *Arithmetica universalis*[9] of 1707 (written by Newton sometime between 1673 and 1683, and published by Whiston without having secured the consent of Newton) the index numeral is placed after the radical, and low, as in $\sqrt{}3:64$ for $\sqrt[3]{64}$, so that the danger of confusion was greater than in most other notations.

During the eighteenth century the placing of the root index in the opening of the radical sign gradually came in vogue. In 1732 one finds $\sqrt[3]{25}$ in De la Loubère;[10] De Lagny[11] who in 1697 wrote $\sqrt[3]{}$, in 1733 wrote $\sqrt[3]{}$; Christian Wolff[12] in 1716 uses in one place the astro-

[1] John Wallis, *A Treatise of Algebra* (London, 1685), p. 107; *Opera*, Vol. II (1693), p. 118. But see also *Arithmetica infinitorum* (1656), Prop. 74.

[2] *Mathesis universalis* (1657), p. 292.

[3] T. F. de Lagny, *Nouveaux elemens d'arithmetique et d'algebre* (Paris, 1697), p. 333.

[4] Thomas Walter, *A new Mathematical Dictionary* (London, n.d., but published in 1762 or soon after), art. "Heterogeneous Surds."

[5] *Arithmetica universalis* . . . . *auctore Is. Newton* . . . . *cum commentario Johannis Castillionei* . . . . , Tomus primus (Amsterdam, 1761), p. 76.

[6] *Journal des Sçavans*, année 1702 (Amsterdam, 1703), p. 300.

[7] *Ibid.*, année 1708, p. 271.   [8] *Ibid.*

[9] *Isaac Newton, Arithmetica universalis* (London, 1707), p. 9; Tropfke, Vol. II, p. 154.

[10] Simon de la Loubère, *De la Résolution des équations* (Paris, 1732), p. 119.

[11] De Lagny in *Mémoires de l'académie r. des sciences*, Tome XI (Paris, 1733), p. 4.

[12] Christian Wolff, *Mathematisches Lexicon* (Leipzig, 1716), p. 1081.

nomical character representing Aries or the ram, for the radical sign, and writes the index of the root to the right; thus Υ^3 signifies cube root. Edward Hatton[1] in 1721 uses $\sqrt[3]{}$, $\sqrt[4]{}$, $\sqrt[5]{}$; De la Chapelle[2] in 1750 wrote $\sqrt[3]{b^3}$. Wolff[3] in 1716 and Hindenburg[4] in 1779 placed the index to the left of the radical sign, $^3\sqrt{Z}$; nevertheless, the notation $\sqrt[3]{}$ came to be adopted almost universally during the eighteenth century. Variations appear here and there. According to W. J. Greenstreet,[5] a curious use of the radical sign is to be found in Walkingame's *Tutor's Assistant* (20th ed., 1784). He employs the letter V for square root, but lets V^3 signify cube or third power, V^4 the fourth power. On the use of capital letters for mathematical signs, very often encountered in old books, as V, for $\sqrt{}$, $\succ$ for $>$, Greenstreet remarks that "authors in the eighteenth century complained of the meanness of the Cambridge University Press for using daggers set sideways instead of the usual $+$." In 1811, an anonymous arithmetician[6] of Massachusetts suggests $^2 4$ for $\sqrt[2]{4}$, $^3 8$ for $\sqrt[3]{8}$, $^m 8$ for $\sqrt[m]{8}$.

As late as 1847 one finds[7] the notaton $^3\sqrt{b}$, $^m\sqrt{abc}$, for the cube root and the mth root, the index appearing in front of the radical sign. This form was not adopted on account of the limitations of the printing office, for in an article in the same series, from the pen of De Morgan, the index is placed inside the opening of the radical sign.[8] In fact, the latter notation occurs also toward the end of Parker's book (p. 131).

In a new algorithm in logarithmic theory A. Bürja[9] proposed the sign $\sqrt[n]{a}$ to mark the nth root of the order N, of a, or the number of which the nth power of the order N is a.

[1] Edward Hatton, *op. cit.* (London, 1721), p. 287.

[2] De la Chapelle, *Traité des sections coniques* (Paris, 1750), p. 15.

[3] Christian Wolff, *Mathematisches Lexicon* (Leipzig, 1716), "Signa," p. 1265.

[4] Carl F. Hindenburg, *Infinitinomii dignitatum leges ac formulae* (Göttingen, 1779), p. 41.

[5] W. J. Greenstreet in *Mathematical Gazette*, Vol. XI (1823), p. 315.

[6] *The Columbian Arithmetician*, "by an American" (Havershall, Mass., 1811), p. 13.

[7] Parker, "Arithmetic and Algebra," *Library of Useful Knowledge* (London, 1847), p. 57.

[8] A. de Morgan, "Study and Difficulties of Mathematics," *ibid.*, *Mathematics*, Vol. I (London, 1847), p. 56.

[9] A. Bürja in *Nouveaux mémoires d. l'académie r. d. scienc. et bell.-lett.*, année 1778 et 1779 (Berlin, 1793), p. 322.

332. *Rudolff and Stifel's aggregation signs.*—Their dot symbolism for the aggregation of terms following the radical sign $\sqrt{}$ was used by Peletier in 1554 (§ 172). In Denmark, Chris. Dibuadius[1] in 1605 marks aggregation by one dot or two dots, as the case may demand. Thus $\sqrt{.5+\sqrt{3}.+\sqrt{2}}$ means $\sqrt{5+\sqrt{3}+\sqrt{2}}$; $\sqrt{.5+\sqrt{3}+\sqrt{2}}$ means $\sqrt{5+\sqrt{3}+\sqrt{2}}$.

W. Snell's translation[2] into Latin of Ludolf van Ceulen's book on the circle contains the expression

$$\sqrt{.2+\sqrt{.2-\sqrt{.2-\sqrt{.2+\sqrt{2\tfrac{1}{2}}-\sqrt{2\tfrac{1}{4}}}}}} ,$$

which is certainly neater than the modern

$$\sqrt{2+\sqrt{2-\sqrt{2-\sqrt{2+\sqrt{2\tfrac{1}{2}}-\sqrt{2\tfrac{1}{4}}}}}} .$$

The Swiss, Johann Ardüser,[3] in 1627, represents $\sqrt{(2-\sqrt{3})}$ by "$\sqrt{.2\div\sqrt{3}}$" and $\sqrt{[2+\sqrt{(2+\sqrt{2})}]}$ by "$\sqrt{.2+\sqrt{.2+\sqrt{2}}}$." This notation appears also in one of the manuscripts of René Descartes,[4] written before the publication of his *Géométrie* in 1637.

It is well known that Oughtred in England modified the German dot symbolism by introducing the colon in its place (§ 181). He had settled upon the dot for the expression of ratio, hence was driven to alter the German notation for aggregation. Oughtred's[5] double colons appear as in "$\sqrt{q:aq-eq:}$" for our $\sqrt{(a^2-e^2)}$.

We have noticed the use of the colon to express aggregation, in the manner of Oughtred, in the *Arithmetique made easie*, by Edmund Wingate (2d ed. by John Kersey; London, 1650), page 387; in John Wallis' *Operum mathematicorum pars altera* (Oxonii, 1656), page 186, as well as in the various parts of Wallis' *Treatise of Algebra* (London, 1685) (§ 196), and also in Jonas Moore's *Arithmetick in two Books* (London, 1660), Second Part, page 14. The 1630 edition of Wingate's book does not contain the part on algebra, nor the symbolism in question; these were probably added by John Kersey.

[1] *C. Dibvadii in arithmeticam irrationalivm Evclidis* (Arnhem, 1605), Introduction.

[2] *Lvdolphi à Cevlen de Circvlo Adscriptis Liber ... omnia é vernaculo Latina fecit ... Willebrordus Snellius* (Leyden, 1610), p. 5.

[3] Johann Ardüser, *Geometriae, Theoreticae practicae*, XII. Bücher (Zürich, 1627), p. 97, 98.

[4] *Œuvres de Descartes*, Vol. X (1908) p. 248.

[5] *Euclidis declaratio*, p. 9, in Oughtred's *Clavis* (1652).

333. *Descartes' union of radical sign and vinculum.*—René Descartes, in his *Géométrie* (1637), indicates the cube root by $\sqrt{}C.$ as in

$$\sqrt{C.\tfrac{1}{2}q+\sqrt{\tfrac{1}{4}qq-\tfrac{1}{27}p^3}} \text{ for our } \sqrt[3]{\tfrac{1}{2}q+\sqrt{\tfrac{1}{4}q^2-\tfrac{1}{27}p^2}}.$$

Here a noteworthy innovation is the union of the radical sign $\sqrt{}$ with the vinculum —— (§ 191). This union was adopted in 1640 by J. J. Stampioen,[1] but only as a redundant symbol. It is found in Fr. van Schooten's 1646 edition of the collected works of Vieta (§ 177), in van Schooten's conic sections,[2] as also in van Schooten's Latin edition of Descartes' geometry.[3] It occurs in J. H. Rahn's algebra (1659) and in Brancher's translation of 1668 (§ 194).

This combination of radical sign $\sqrt{}$ and vinculum is one which has met with great favor and has maintained a conspicuous place in mathematical books down to our own time. Before 1637, this combination of radical sign and vinculum had been suggested by Descartes (*Œuvres*, Vol. X, p. 292). Descartes also leaned once toward Girard's notation.

Great as were Descartes' services toward perfecting algebraic notation, he missed a splendid opportunity of rendering a still greater service. Before him Oresme and Stevin had advanced the concept of fractional as well as of integral exponents. If Descartes, instead of extending the application of the radical sign $\sqrt{}$ by adding to it the vinculum, had discarded the radical sign altogether and had introduced the notation for fractional as well as integral exponents, then it is conceivable that the further use of radical signs would have been discouraged and checked; it is conceivable that the unnecessary duplication in notation, as illustrated by $b^{\frac{3}{4}}$ and $\sqrt[4]{b^3}$, would have been avoided; it is conceivable that generations upon generations of pupils would have been saved the necessity of mastering the operations with two difficult notations when one alone (the exponential) would have answered all purposes. But Descartes missed this opportunity, as did later also I. Newton who introduced the notation of the fractional exponent, yet retained and used radicals.

[1] J. J. Stampioen, *Wis-Konstich ende Reden-Maetich Bewijs* (The Hague, 1640), p. 6.

[2] *Francisci à Schooten Leydensis de organica conicarum sectionum* (Leyden, 1646), p. 91.

[3] Francisci à Schooten, *Renati Descartes Geometria* (Frankfurt a/M., 1695), p. 3. [First edition, 1649.]

334. *Other signs of aggregation of terms.*—Leonard and Thomas Digges,[1] in a work of 1571, state that if "the side of the Pentagon. [is] 14, the containing circles semidiameter [is]

$$\sqrt{\tfrac{3}{8}}.V.98+\sqrt{\tfrac{3}{8}}1920\tfrac{4}{5}" \text{ i.e., } \sqrt{98+\sqrt{1920\tfrac{4}{5}}}.$$

In the edition of 1591[2] the area of such a pentagon is given as

$$\sqrt{\tfrac{3}{8}} \; vni. \; 60025+\sqrt{\tfrac{3}{8}}2882400500.$$

Vieta's peculiar notations for radicals of 1593 and 1595 are given in § 177. The *Algebra* of Herman Follinvs[3] of 1622 uses parentheses in connection with the radical sign, as in $\sqrt{\tfrac{3}{8}}(22+\sqrt{\tfrac{3}{8}}9)$, our $\sqrt{22+\sqrt{9}}$. Similarly, Albert Girard[4] writes $\sqrt{(2\tfrac{1}{2}+\sqrt{3\tfrac{1}{4}})}$, with the simplification of omitting in case of square root the letter marking the order of the root. But, as already noted, he does not confine himself to this notation. In one place[5] he suggests the modern designation $\sqrt[2]{}$, $\sqrt[3]{}$, $\sqrt[5]{}$.

Oughtred writes $\sqrt{u}$ or $\sqrt{b}$ for universal root (§ 183), but more commonly follows the colon notation (§ 181). Hérigone's notation of 1634 and 1644 is given in § 189. The Scotsman, James Gregory,[6] writes

$$\sqrt{C(b^2c^3az^3 \ldots)}, \; \sqrt{q}\left(\sqrt{q\tfrac{\tfrac{3}{8}^5}{l}+\frac{16\tfrac{3}{8}^2}{25}}\right).$$

In William Molyneux[7] one finds $\sqrt{CP^q-Px^q}$ for $\sqrt[3]{P^q-Px^q}$. Another mode of marking the root of a binomial is seen in a paper of James Bernoulli[8] who writes $\sqrt{,}\; ax-x^2$ for $\sqrt{ax-x^2}$. This is really the old idea of Stifel, with Hérigone's and Leibniz's comma taking the place of a dot.

The union of the radical sign and vinculum has maintained itself widely, even though it had been discouraged by Leibniz and others who aimed to simplify the printing by using, as far as possible, one-line symbols. In 1915 the Council of the London Mathematical So-

---

[1] *A Geometrical Practise, named Pantometria, framed by Leonard Digges,* . . . . *finished by Thomas Digges his sonne* (London, 1571) (pages unnumbered).

[2] *A Geometrical Practical Treatize named Pantometria* (London, 1591), p. 106.

[3] Hermann Follinvs, *Algebra sive liber de Rebvs occvltis* (Cologne, 1622), p. 157.

[4] Albert Girard, *Invention nouvelle en l'algebre* (Amsterdam, 1629).

[5] *Loc. cit.,* in "Caracteres de puissances et racines."

[6] James Gregory, *Geometriae pars vniversalis* (Patavii, 1668), p. 71, 108.

[7] William Molyneux, *A Treatise of Dioptricks* (London, 1692), p. 299.

[8] Jacob Bernoulli in *Acta eruditorum* (1697), p. 209.

ciety, in its *Suggestions for Notation and Printing*,[1] recommended that $\sqrt{2}$ or $2^{\frac{1}{2}}$ be adopted in place of $\sqrt{2}$, also $\sqrt{(ax^2+2bx+c)}$ or $(ax^2+2bx+c)^{\frac{1}{2}}$ in place of $\sqrt{ax^2+2bx+c}$. Bryan[2] would write $\sqrt{-1}$ rather than $\sqrt{-1}$.

335. *Redundancy in the use of aggregation signs.*—J. J. Stampioen marked aggregation of terms in three ways, any one of which would have been sufficient. Thus,[3] he indicates $\sqrt{b^3+6a^2b^2+9a^4b}$ in this manner, $\sqrt{\cdot(bbb+6aa\ bb+9aaaa\ b)}$; he used here the vinculum, the round parenthesis, and the dot to designate the aggregation of the three terms. In other places, he restricts himself to the use of dots, either a dot at the beginning and a dot at the end of the expression, or a dot at the beginning and a comma at the end, or he uses a dot and parentheses.

Another curious notation, indicating fright lest the aggregation of terms be overlooked by the reader, is found in John Kersey's symbolism of 1673,[4] $\sqrt{(2):\frac{1}{2}r-\sqrt{\frac{1}{4}rr-s}}$: for $\sqrt{\frac{1}{2}r-\sqrt{\frac{1}{4}r^2-s}}$. We observe here the superposition of two notations for aggregation, the Oughtredian colon placed before and after the binomial, and the vinculum. Either of these without the other would have been sufficient.

336. *Peculiar Dutch symbolism.*—A curious use of $\sqrt{}$ sprang up in Holland in the latter part of the seventeenth century and maintained itself there in a group of writers until the latter part of the eighteenth century. If $\sqrt{}$ is placed before a number it means "square root," if placed after it means "square." Thus, Abraham de Graaf[5] in 1694 indicates by $\sqrt{\dfrac{12\frac{1}{4}}{3\frac{1}{2}}}$ the square root of the fraction, by $\dfrac{15}{225}\sqrt{}$ the square of the fraction. This notation is used often in the mathematical journal, *Maandelykse Mathematische Liefhebberije*, published at Amsterdam from 1754 to 1769. As late as 1777 it is given by L. Praalder[6] of Utrecht, and even later (1783) by Pieter Venema.[7] We have here the

[1] *Mathematical Gazette*, Vol. VIII (1917), p. 172.

[2] *Op. cit.*, Vol. VIII, p. 220.

[3] *J. J. Stampioenii Wis-Konstigh Ende Reden-Maetigh Bewijs* (The Hague, 1640), p. 6.

[4] John Kersey, *Algebra* (London, 1673), p. 95.

[5] Abraham de Graaf, *De Geheele Mathesis* (Amsterdam, 1694), p. 65, 69.

[6] Laurens Praalder, *Mathematische Voorstellen* (Amsterdam, 1777), p. 14, 15 ff.

[7] Pieter Venema, *Algebra ofte Stel-Konst*, Vyfde Druk (Amsterdam, 1783), p. 168, 173.

same general idea that was introduced into other symbolisms, according to which the significance of the symbol depends upon its relative position to the number or algebraic expression affected. Thus with Pacioli $R_{\cdot}200$ meant $\sqrt{200}$, but $R3^a$ meant the second power (§§ 135, 136). With Stevin (§ 162, 163), ③20 meant 20^3, but 20③ meant $20x^3$. With L. Schoner $5l$ meant $5x$, but $l5$ meant $\sqrt{5}$ (§ 291). We may add that in the 1730 edition of Venema's algebra brought out in New York City radical expressions do not occur, as I am informed by Professor L. G. Simons, but a letter placed on the left of an equation means division of the members of the equation by it; when placed on the right, multiplication is meant. Thus (p. 100):

$$``\; b \left|\begin{array}{l} b\dot{x}=2a \\ \hline x=\dfrac{2a}{b} \end{array}\right. \;"$$

and (p. 112):

$$`` \dfrac{5-\dfrac{4500}{x}=\!=\!1\tfrac{1}{2}-\dfrac{1000}{x}}{5x-4500=1\tfrac{1}{2}x-1000}\; x\; . \;"$$

Similar is Prändel's use of $\sqrt{}$ as a marginal symbol, indicating that the square root of both sides of an equation is to be taken. His marginal symbols are shown in the following:[1]

$$`` \dfrac{V^2}{a}=\tfrac{1}{4}(a-p)$$

$$\times a \qquad V^2=\tfrac{1}{4}(a^2-ap)=\tfrac{1}{4}a(a-p)\; .$$

$$\sqrt{} \qquad V=\tfrac{1}{2}\sqrt{a(a-p)}\; . "$$

337. *Principal root-values.*—For the purpose of distinguishing between the principal value of a radical expression and the other values, G. Peano[2] indicated by $\sqrt[m]{}*a$ all the m values of the radical, reserving $\sqrt[m]{a}$ for the designation of its "principal value." This notation is adopted by O. Stolz and J. A. Gmeiner[3] in their *Theoretische Arithmetik* (see also § 312).

[1] J. G. Prändel, *Kugldreyeckslehre und höhere Mathematik* (München, 1793), p. 97.

[2] G. Peano, *Formulaire des mathématiques* (first published in *Rivista di Matematica*), Vol. I, p. 19.

[3] O. Stolz und J. A. Gmeiner, *Theoretische Arithmetik* (Leipzig), Vol. II (1902), p. 355.

338. *Recommendation of United States National Committee.*— "With respect to the root sign, $\sqrt{\ }$, the committee recognizes that convenience of writing assures its continued use in many cases instead of the fractional exponent. It is recommended, however, that in algebraic work involving complicated cases the fractional exponent be preferred. Attention is called to the fact that the symbol $\sqrt{a}$ (a representing a positive number) means only the positive square root and that the symbol $\sqrt[n]{a}$ means only the principal nth root, and similarly for $a^{\frac{1}{2}}$, $a^{\frac{1}{n}}$."[1]

SIGNS FOR UNKNOWN NUMBERS

339. *Early forms.*—Much has already been said on symbolisms used to represent numbers that are initially unknown in a problem, and which the algebraist endeavors to ascertain. In the Ahmes papyrus there are signs to indicate "heap" (§ 23); in Diophantus a Greek letter with an accent appears (§ 101); the Chinese had a positional mode of indicating one or more unknowns; in the Hindu Bakhshālī manuscript the use of a dot is invoked (§ 109). Brahmagupta and Bhāskara did not confine the symbolism for the unknown to a single sign, but used the names of colors to designate different unknowns (§§ 106, 108, 112, 114). The Arab Abu Kamil[2] (about 900 A.D.), modifying the Hindu practice of using the names of colors, designated the unknowns by different coins, while later al-Karkhî (following perhaps Greek sources)[3] called one unknown "thing," a second "measure" or "part," but had no contracted sign for them. Later still al-Qalasâdî used a sign for unknown (§ 124). An early European sign is found in Regiomontanus (§ 126), later European signs occur in Pacioli (§§ 134, 136), in Christoff Rudolff (§§ 148, 149, 151),[4] in Michael Stifel who used more than one notation (§§ 151, 152), in Simon Stevin (§ 162), in L. Schoner (§ 322), in F. Vieta (§§ 176–78), and in other writers (§§ 117, 138, 140, 148, 164, 173, 175, 176, 190, 198).

Luca Pacioli remarks[5] that the older textbooks usually speak of

---

[1] *Report of the National Committee on Mathematical Requirements under the Auspices of the Mathematical Association of America* (1923), p. 81.

[2] H. Suter, *Bibliotheca mathematica* (3d ser.), Vol. XI (1910–11), p. 100 ff.

[3] F. Woepcke, *Extrait du Fakhri* (Paris, 1853), p. 3, 12, 139–43. See M. Cantor, *op. cit.*, Vol. I (3d ed., 1907), p. 773.

[4] Q. Eneström, *Bibliotheca mathematica* (3d ser.), Vol. VIII (1907–8), p. 207.

[5] L. Pacioli, *Summa*, dist. VIII, tract 6, fol. 148*B*. See M. Cantor, *op. cit.*, Vol. II (2d ed., 1913), p. 322.

the first and the second *cosa* for the unknowns, that the newer writers prefer *cosa* for the unknown, and *quantita* for the others. Pacioli abbreviates those *co.* and $\wp\beta^a$.

Vieta's convention of letting vowels stand for unknowns and consonants for knowns (§§ 164, 176) was favored by Albert Girard, and also by W. Oughtred in parts of his *Algebra*, but not throughout. Near the beginning Oughtred used Q for the unknown (§ 182).

The use of N (*numerus*) for x in the treatment of numerical equations, and of Q, C, etc., for the second and third powers of x, is found in Xylander's edition of Diophantus of 1575 (§ 101), in Vieta's *De emendatione aequationum* of 1615 (§ 178), in Bachet's edition of Diophantus of 1621, in Camillo Glorioso in 1627 (§ 196). In numerical equations Oughtred uses l for x, but the small letters q, c, qq, qc, etc., for the higher powers of x (§ 181). Sometimes Oughtred employs also the corresponding capital letters. Descartes very often used, in his correspondence, notations different from his own, as perhaps more familiar to his correspondents than his own. Thus, as late as 1640, in a letter to Mersenne (September 30, 1640), Descartes[1] writes "$1C-6N=40$," which means $x^3-6x=40$. In the *Regulae ad directionem ingenii*, Descartes represents[2] by a, b, c, etc., known magnitudes and by A, B, C, etc., the unknowns; this is the exact opposite of the use of these letters found later in Rahn.

Crossed numerals representing powers of unknowns.—Interesting is the attitude of P. A. Cataldi of Bologna, who deplored the existence of many different notations in different countries for the unknown numbers and their powers, and the inconveniences resulting from such diversity. He points out also the difficulty of finding in the ordinary printing establishment the proper type for the representation of the different powers. He proposes[3] to remove both inconveniences by the use of numerals indicating the powers of the unknown and distinguishing them from ordinary numbers by crossing them out, so that Ø, ⨰, 2̸, 3̸, ... , would stand for x^0, x', x^2, x^3. . . . . Such crossed numerals, he argued, were convenient and would be found in printing offices since they are used in arithmetics giving the scratch method of dividing, called by the Italians the "a Galea" method. The reader will recall that Cataldi's notation closely resembles that of Leonard

---

[1] *Œuvres de Descartes*, Vol. III (1899), p. 190, 196, 197; also Vol. XII, p. 279.

[2] *Op. cit.*, Vol. X (1908), p. 455, 462.

[3] P. A. Cataldi, *Trattato dell'algebra proportionale* (Bologna, 1610), and in his later works. See G. Wertheim in *Bibliotheca mathematica* (3d ser.), Vol. II (1901), p. 146, 147.

and Thomas Digges in England (§ 170). These symbols failed of adoption by other mathematicians. We have seen that in 1627 Camillo Glorioso, in a work published at Naples,[1] wrote N for x, and q, c, qq, qc, cc, qqc, qcc, and ccc for x^2, x^3, ... , x^9, respectively (§ 196). In 1613 Glorioso had followed Stevin in representing an unknown quantity by 1⊙.

340. *Descartes' z, y, x.*—The use of z, y, x . . . . to represent unknowns is due to René Descartes, in his *La géométrie* (1637). Without comment, he introduces the use of the first letters of the alphabet to signify *known* quantities and the use of the last letters to signify *unknown* quantities. His own language is: "... l'autre, LN, est $\frac{1}{2}a$ la moitié de l'autre quantité connue, qui estoit multipliée par z, que ie suppose estre la ligne inconnue."[2] Again: "... ie considere ... Que le segment de la ligne AB, qui est entre les poins A et B, soit nommé x, et que BC soit nommé y; ... la proportion qui est entre les costés AB et BR est aussy donnée, et ie la pose comme de z a b; de façon qu' AB estant x, RB sera $\dfrac{bx}{z}$, et la toute CR sera $y + \dfrac{bx}{z}$. ..." Later he says: "et pour ce que CB et BA sont deux quantités indeterminées et inconnuës, ie les nomme, l'une y; et l'autre x. Mais, affin de trouver le rapport de l'une a l'autre, ie considere aussy les quantités connuës qui determinent la description de cete ligne courbe: comme GA que je nomme a, KL que je nomme b, et NL, parallele a GA, que ie nomme C."[3] As co-ordinates he uses later only x and y. In equations, in the third book of the *Géométrie*, x predominates. In manuscripts written in the interval 1629–40, the unknown z occurs only once.[4] In the other places x and y occur. In a paper on Cartesian ovals,[5] prepared before 1629, x alone occurs as unknown, y being used as a parameter. This is the earliest place in which Descartes used one of the last letters of the alphabet to represent an unknown. A little later he used x, y, z again as known quantities.[6]

Some historical writers have focused their attention upon the x, disregarding the y and z, and the other changes in notation made by

---

[1] Camillo Gloriosi, *Exercitationes mathematical, decas I* (Naples, 1627). Also *Ad theorema geometricvm, á nobilissimo viro propositum, Joannis Camilli Gloriosi* (Venice, 1613), p. 26. It is of interest that Glorioso succeeded Galileo in the mathematical chair at Padua.

[2] *Œuvres de Descartes*, Vol. VI (1902), p. 375.

[3] *Ibid.*, p. 394.

[4] *Ibid.*, Vol. X, p. 288–324.

[5] *Ibid.*, p. 310.   [6] *Ibid.*, p. 299.

Descartes; these writers have endeavored to connect this x with older symbols or with Arabic words. Thus, J. Tropfke,[1] P. Treutlein,[2] and M. Curtze[3] advanced the view that the symbol for the unknown used by early German writers, $\mathfrak{X}$, looked so much like an x that it could easily have been taken as such, and that Descartes actually did interpret and use it as an x. But Descartes' mode of introducing the knowns a, b, c, etc., and the unknowns z, y, x makes this hypothesis improbable. Moreover, G. Eneström has shown[4] that in a letter of March 26, 1619, addressed to Isaac Beeckman, Descartes used the symbol $\mathfrak{X}$ as a symbol in form distinct from x, hence later could not have mistaken it for an $\mathfrak{X}$. At one time, before 1637, Descartes[5] used x along the side of $\mathfrak{X}$; at that time x, y, z are still used by him as symbols for known quantities. German symbols, including the $\mathfrak{X}$ for x, as they are found in the algebra of Clavius, occur regularly in a manuscript[6] due to Descartes, the *Opuscules de 1619–1621*.

All these facts caused Tropfke in 1921 to abandon his old view[7] on the origin of x, but he now argues with force that the resemblance of x and $\mathfrak{X}$, and Descartes' familiarity with $\mathfrak{X}$, may account for the fact that in the latter part of Descartes' *Géométrie* the x occurs more frequently than z and y. Eneström, on the other hand, inclines to the view that the predominance of x over y and z is due to typographical reasons, type for x being more plentiful because of the more frequent occurrence of the letter x, to y and z, in the French and Latin languages.[8]

There is nothing to support the hypothesis on the origin of x due to Wertheim,[9] namely, that the Cartesian x is simply the notation of the Italian Cataldi who represented the first power of the unknown by a crossed "one," thus $\mathcal{I}$. Nor is there historical evidence

---

[1] J. Tropfke, *Geschichte der Elementar-Mathematik*, Vol. I (Leipzig, 1902), p. 150.

[2] P. Treutlein, "Die deutsche Coss," *Abhandl. z. Geschichte d. mathematischen Wiss.*, Vol. II (1879), p. 32.

[3] M. Curtze, *ibid.*, Vol. XIII (1902), p. 473.

[4] G. Eneström, *Bibliotheca mathematica* (3d ser.), Vol. VI (1905), p. 316, 317, 405, 406. See also his remarks in *ibid.* (1884) (Sp. 43); *ibid.* (1889), p. 91. The letter to Beeckman is reproduced in *Œuvres de Descartes*, Vol. X (1908), p. 155.

[5] *Œuvres de Descartes*, Vol. X (Paris, 1908), p. 299. See also Vol. III, Appendix II, No. 48*g*.

[6] *Ibid.*, Vol. X (1908), p. 234.

[7] J. Tropfke, *op. cit.*, Vol. II (2d ed., 1921), p. 44–46.

[8] G. Eneström, *Bibliotheca mathematica* (3d ser.), Vol. VI, p. 317.

[9] G. Eneström, *ibid.*

to support the statement found in Noah Webster's *Dictionary*, under the letter x, to the effect that "x was used as an abbreviation of Ar. *shei* a thing, something, which, in the Middle Ages, was used to designate the unknown, and was then prevailingly transcribed as *xei*."

341. *Spread of Descartes' signs.*—Descartes' x, y, and z notation did not meet with immediate adoption. J. H. Rahn, for example, says in his *Teutsche Algebra* (1659): "Descartes' way is to signify *known* quantities by the *former* letters of the alphabet, and *unknown* by the *latter* [z, y, x, etc.]. But I choose to signify the *unknown* quantities by *small* letters and the *known* by capitals." Accordingly, in a number of his *geometrical* problems, Rahn uses a and A, etc., but in the book as a whole he uses z, y, x freely.

As late as 1670 the learned bishop, Johann Caramuel, in his *Mathesis biceps* ... , Campagna (near Naples), page 123, gives an old notation. He states an old problem and gives the solution of it as found in Geysius; it illustrates the rhetorical exposition found in some books as late as the time of Wallis, Newton, and Leibniz. We quote: "Dicebat Augias Herculi: Meorum armentorum media pars est in tali loco octavi in tali, decima in tali, 20ma in tali 60ma in tali, & 50 . sunc hic. Et Geysius libr. 3 *Cossa* Cap. 4. haec pecora numeraturus sic scribit.

"Finge 1. *a*. partes $\frac{1}{2}a$, $\frac{1}{8}a$, $\frac{1}{10}a$, $\frac{1}{20}a$, $\frac{1}{60}a$ & additae (hoc est, in summam reductae) sunt $\frac{19}{24}a$ a quibus de 1. *a*. sublatis, restant $\frac{5}{24}$ a aequalia 50. Jam, quia sictus, est fractio, multiplicando reducatur, & 1. *a*. aequantur 240. Hic est numerus pecorum Augiae." ("Augias said to Hercules: 'Half of my cattle is in such a place, $\frac{1}{8}$ in such, $\frac{1}{10}$ in such, $\frac{1}{20}$ in such, $\frac{1}{60}$ in such, and here there are 50. And Geysius in Book 3, *Cossa*, Chap. 4, finds the number of the herd thus: Assume 1. *a*., the parts are $\frac{1}{2}a$, $\frac{1}{8}a$, $\frac{1}{10}a$, $\frac{1}{20}a$, $\frac{1}{60}a$, and these added [i.e., reduced to a sum] are $\frac{19}{24}a$ which subtracted from 1. *a*, leaves $\frac{5}{24}a$, equal to 50. Now, the fraction is removed by multiplication, and 1. *a* equal 240. This is the number of Augias' herd.' ")

Descartes' notation, x, y, z, is adopted by Gerard Kinckhuysen,[1] in his *Algebra* (1661). The earliest systematic use of three co-ordinates in analytical geometry is found in De la Hire, who in his *Nouveaux élémens des sections coniques* (Paris, 1679) employed (p. 27) x, y, v. A. Parent[2] used x, y, z; Euler[3], in 1728, t, x, y; Joh. Bernoulli,[4] in 1715,

---

[1] Gerard Kinckhuysen, *Algebra ofte Stel-Konst* (Haerlem, 1661), p. 6.

[2] A. Parent, *Essais et recherches de math. et de phys.*, Vol. I (Paris, 1705).

[3] Euler in *Comm. Aca. Petr.*, II, 2, p. 48 (year 1728, printed 1732).

[4] Leibniz and Bernoulli, *Commercium philosophicum et mathematicum*, Vol. II (1745), p. 345.

x, y, z in a letter (February 6, 1715) to Leibniz. H. Pitot[1] applied the three co-ordinates to the helix in 1724.

SIGNS OF AGGREGATION

342. *Introduction.*—In a rhetorical or syncopated algebra, the aggregation of terms could be indicated in words. Hence the need for symbols of aggregation was not urgent. Not until the fifteenth and sixteenth centuries did the convenience and need for such signs definitely present itself. Various devices were invoked: (1) the horizontal bar, placed below or above the expression affected; (2) the use of abbreviations of words signifying aggregation, as for instance u or v for *universalis* or *vniversalis*, which, however, did not always indicate clearly the exact range of terms affected; (3) the use of dots or commas placed before the expression affected, or at the close of such an expression, or (still more commonly) placed both before and after; (4) the use of parentheses (round parentheses or brackets or braces). Of these devices the parentheses were the slowest to find wide adoption in all countries, but now they have fairly won their place in competition with the horizontal bar or vinculum. Parentheses prevailed for typographical reasons. Other things being equal, there is a preference for symbols which proceed in orderly fashion as do the letters in ordinary printing, without the placing of signs in high or low positions that would break a line into two or more sublines. A vinculum at once necessitates two terraces of type, the setting of which calls for more time and greater technical skill. At the present time

---

[1] H. Pitot, *Mémoires de l'académie d. scien.*, année 1724 (Paris, 1726). Taken from H. Wieleitner, *Geschichte der Mathematik*, 2. Teil, 2. Hälfte (Berlin und Leipzig, 1921), p. 92.

To what extent the letter x has been incorporated in mathematical language is illustrated by the French expression *être fort en x*, which means "being strong in mathematics." In the same way, *tête à x* means "a mathematical head." The French give an amusing "demonstration" that old men who were *tête à x* never were pressed into military service so as to have been conscripts. For, if they were conscripts, they would now be *ex*-conscripts. Expressed in symbols we would have

$$\theta x = ex\text{-conscript.}$$

Dividing both sides by x gives

$$\theta = e\text{-conscript.}$$

Dividing now by e yields

$$\text{conscript} = \frac{\theta}{e}.$$

According to this, the conscript would be *la tête assurée* (i.e., θ over e, or, the *head assured* against casualty), which is absurd.

the introducing of typesetting machines and the great cost of type-setting by hand operate against a double or multiple line notation. The dots have not generally prevailed in the marking of aggregation for the reason, no doubt, that there was danger of confusion since dots are used in many other symbolisms—those for multiplication, division, ratio, decimal fractions, time-derivatives, marking a number into periods of two or three digits, etc.

343. *Aggregation expressed by letters.*—The expression of aggregation by the use of letters serving as abbreviations of words expressing aggregation is not quite as old as the use of horizontal bars, but it is more common in works of the sixteenth century. The need of marking the aggregation of terms arose most frequently in the treatment of radicals. Thus Pacioli, in his *Summa* of 1494 and 1523, employs v (*vniversale*) in marking the root of a binomial or polynomial (§ 135). This and two additional abbreviations occur in Cardan (§ 141). The German manuscript of Andreas Alexander (1524) contains the letters *cs* for *communis* (§ 325); Chr. Rudolff sometimes used the word "collect," as in "$\sqrt{}$ des collects $17+\sqrt{208}$" to designate $\sqrt{17+\sqrt{208}}$.[1] J. Scheubel adopted *Ra. col.* (§ 159). S. Stevin, Fr. Vieta, and A. Romanus wrote *bin.*, or *bino.*, or *binomia*, *trinom.*, or similar abbreviations (§ 320). The u or v is found again in Pedro Nuñez (who uses also L for "ligature"),[2] Leonard and Thomas Digges (§ 334), in J. R. Brasser[3] who in 1663 lets v signify "universal radix" and writes "$v\sqrt{}.8 \div \sqrt{45}$" to represent $\sqrt{8-\sqrt{45}}$. W. Oughtred sometimes wrote $\sqrt{u}$ or $\sqrt{b}$ (§§ 183, 334). In 1685 John Wallis[4] explains the notations $\sqrt{b}:2+\sqrt{3}$, $\sqrt{r}:2-\sqrt{3}$, $\sqrt{u}:2\pm\sqrt{3}$, $\sqrt{2\pm\sqrt{3}}$, $\sqrt{}:2\pm\sqrt{3}$, where b means "binomial," u "universal," r "residual," and sometimes uses redundant forms like $\sqrt{b}:\sqrt{5}+1:$.

344. *Aggregation expressed by horizontal bars or vinculums.*—The use of the horizontal bar to express the aggregation of terms goes back to the time of Nicolas Chuquet who in his manuscript (1484) under-lines the parts affected (§ 130). We have seen that the same idea is followed by the German Andreas Alexander (§ 325) in a manuscript of 1545, and by the Italian Raffaele Bombelli in the manuscript edition

[1] J. Tropfke, *op. cit.*, Vol. II (1921), p. 150.

[2] Pedro Nuñez, *Libra de algebra en arithmetica y geometria* (Anvers, 1567), fol. 52.

[3] J. R. Brasser, *Regula of Algebra* (Amsterdam, 1663), p. 27.

[4] John Wallis, *Treatise of Algebra* (London, 1685), p. 109, 110. The use of letters for aggregation practically disappeared in the seventeenth century.

of his algebra (about 1550) where he wrote $\sqrt[3]{2+\sqrt{-121}}$ in this manner:[1] $R^3[2.p.R[0\ \widetilde{m}\ 121]]$; parentheses were used and, in addition, vinculums were drawn underneath to indicate the range of the parentheses. The employment of a long horizontal brace in connection with the radical sign was introduced by Thomas Harriot[2] in 1631; he expresses aggregation thus: $\sqrt{ccc+\sqrt{ccccc-bbbbb}}$. This notation may, perhaps, have suggested to Descartes his new radical symbolism of 1637. Before that date, Descartes had used dots in the manner of Stifel and Van Ceulen. He wrote[3] $\sqrt{.2-\sqrt{2}}.$ for $\sqrt{2-\sqrt{2}}$. He attaches the vinculum to the radical sign $\sqrt{}$ and writes $\sqrt{a^2+b^2}$, $\sqrt{-\frac{1}{2}a+\sqrt{\frac{1}{4}aa+bb}}$, and in case of cube roots $\sqrt{C.\ \frac{1}{2}q+\sqrt{\frac{1}{4}gg-27p^3}}$. Descartes does not use parentheses in his *Géométrie*. Descartes uses the horizontal bar only in connection with the radical sign. Its general use for aggregation is due to Fr. van Schooten, who, in editing Vieta's collected works in 1646, discarded parentheses and placed a horizontal bar above the parts affected. Thus Van Schooten's "$B\ in$ $\overline{D\ quad.+B\ in\ D}$" means $B(D^2+BD)$. Vieta[4] himself in 1593 had written this expression differently, namely, in this manner:

$$``B\ in\ \begin{cases} D.\ quadratum \\ +B\ in\ D \end{cases}\text{''}\ .$$

B. Cavalieri in his *Geometria indivisibilibae* and in his *Exercitationes geometriae sex* (1647) uses the vinculum in this manner, $\overline{AB}$, to indicate that the two letters A and B are not to be taken separately, but conjointly, so as to represent a straight line, drawn from the point A to the point B.

Descartes' and Van Schooten's stressing the use of the vinculum led to its adoption by J. Prestet in his popular text, *Elemens des Mathématiques* (Paris, 1675). In an account of Rolle[5] the cube root is to be taken of $2+\overline{R.-121}$, i.e., of $2+\sqrt{-121}$. G. W. Leibniz[6] in a

[1] See E. Bertolotti in *Scientia*, Vol. XXXIII (1923), p. 391 n.

[2] Thomas Harriot, *Artis analyticae praxis* (London, 1631), p. 100.

[3] R. Descartes, *Œuvres* (éd. Ch. Adam et P. Tannery), Vol. X (Paris, 1908), p. 286 f., also p. 247, 248.

[4] See J. Tropfke, *op. cit.*, Vol. II (1921), p. 30.

[5] *Journal des Sçavans* de l'an 1683 (Amsterdam, 1709), p. 97.

[6] G. W. Leibniz' letter to D. Oldenburgh, Feb. 3, 1672–73, printed in J. Collin's *Commercium epistolicum* (1712).

letter of 1672 uses expressions like $\overline{a \backsim b} \backsim \overline{b \backsim c} \backsim \overline{b \backsim c} \backsim c \backsim \overline{c \backsim d}$, where $\backsim$ signifies "difference." Occasionally he uses the vinculum until about 1708, though usually he prefers round parentheses. In 1708 Leibniz' preference for round parentheses (§ 197) is indicated by a writer in the *Acta eruditorum*. Joh. (1) Bernoulli, in his *Lectiones de calculo differentialium*, uses vinculums but no parentheses.[1]

345. In England the notations of W. Oughtred, Thomas Harriot, John Wallis, and Isaac Barrow tended to retard the immediate introduction of the vinculum. But it was used freely by John Kersey (1673)[2] who wrote $\sqrt{(2)} : \overline{\frac{1}{2}r - \sqrt{\frac{1}{4}rr - s}}$: and by Newton, as, for instance, in his letter to D. Oldenburgh of June 13, 1676, where he gives the binomial formula as the expansion of $\overline{P+PQ}\Big|^{\frac{m}{n}}$. In his *De Analysi per Aequationes numero terminorum Infinitas*, Newton writes[3]

$\overline{\overline{\overline{y-4}\times y+5}\times y-12}\times y+17=0$ to represent $\{[(y-4)y-5]y-12\}y+17 = 0$. This notation was adopted by Edmund Halley,[4] David Gregory, and John Craig; it had a firm foothold in England at the close of the seventeenth century. During the eighteenth century it was the regular symbol of aggregation in England and France; it took the place very largely of the parentheses which are in vogue in our day. The vinculum appears to the exclusion of parentheses in the *Geometria organica* (1720) of Colin Maclaurin, in the *Elements of Algebra* of Nicholas Saunderson (Vol. I, 1741), in the *Treatise of Algebra* (2d ed.; London, 1756) of Maclaurin. Likewise, in Thomas Simpson's *Mathematical Dissertations* (1743) and in the 1769 London edition of Isaac Newton's *Universal Arithmetick* (translated by Ralphson and revised by Cunn), vinculums are used and parentheses do not occur. Some use of the vinculum was made nearly everywhere during the eighteenth century, especially in connection with the radical sign $\sqrt{}$, so as to produce $\sqrt{\overline{}}$. This last form has maintained its place down to the present time. However, there are eighteenth-century writers who avoid the vinculum altogether even in connection with the radical sign, and use

---

[1] The Johannis (1) Bernoullii *Lectiones de calculo differentialium*, which remained in manuscript until 1922, when it was published by Paul Schafheitlin in *Verhandlungen der Naturforschenden Gesellschaft in Basel*, Vol. XXXIV (1922).

[2] John Kersey, *Algebra* (London, 1673), p. 55.

[3] *Commercium epistolicum* (éd. Biot et Lefort; Paris, 1856), p. 63.

[4] *Philosophical Transactions* (London), Vol. XV–XVI (1684–91), p. 393; Vol. XIX (1695–97), p. 60, 645, 709.

parentheses exclusively. Among these are Poleni (1729),[1] Cramer (1750),[2] and Cossali (1797).[3]

346. There was considerable vacillation on the use of the vinculum in designating the square root of minus unity. Some authors wrote $\sqrt{-1}$; others wrote $\sqrt{-1}$ or $\sqrt{(-1)}$. For example, $\sqrt{-1}$ was the designation adopted by J. Wallis,[4] J. d'Alembert,[5] I. A. Segner,[6] C. A. Vandermonde,[7] A. Fontaine.[8] Odd in appearance is an expression of Euler,[9] $\sqrt{(2\sqrt{-1}-4)}$. But $\sqrt{(-1)}$ was preferred by Du Séjour[10] in 1768 and by Waring[11] in 1782; $\sqrt{-1}$ by Laplace[12] in 1810.

347. It is not surprising that, in times when a notation was passing out and another one taking its place, cases should arise where both are used, causing redundancy. For example, J. Stampioen in Holland sometimes expresses aggregation of a set of terms by three notations, any one of which would have been sufficient; he writes[13] in 1640, $\sqrt{.\overline{(aaa+6aab+9bba)}}$, where the dot, the parentheses, and the vinculum appear; John Craig[14] writes $\sqrt{2ay-y^2}$: and $\sqrt{}:\sqrt{6a^4-\frac{3}{2}a^2}$, where the colon is the old Oughtredian sign of aggregation, which is here superfluous, because of the vinculum. Tautology in notation is found in Edward Cocker[15] in expressions like $\sqrt{\overline{aa+bb}}$, $\sqrt{}:c+\frac{1}{4}bb-\frac{1}{2}b$, and

[1] Ioannis Poleni, *Epistolarvm mathematicarvm fascicvlvs* (Padua, 1729).

[2] Gabriel Cramer, *L'Analyse des lignes courbes algébriques* (Geneva, 1750).

[3] Pietro Cossali, *Origini ... dell'algebra*, Vol. I (Parma, 1797).

[4] John Wallis, *Treatise of Algebra* (London, 1685), p. 266.

[5] J. d'Alembert in *Histoire de l'académie r. des sciences*, année 1745 (Paris, 1749), p. 383.

[6] I. A. Segner, *Cursus mathematici*, Pars IV (Halle, 1763), p. 44.

[7] C. A. Vandermonde in *op. cit.*, année 1771 (Paris, 1774), p. 385.

[8] A. Fontaine, *ibid.*, année 1747 (Paris, 1752), p. 667.

[9] L. Euler in *Histoire de l'académie r. d. sciences et des belles lettres*, année 1749 (Berlin, 1751), p. 228.

[10] Du Séjour, *ibid.* (1768; Paris, 1770), p. 207.

[11] E. Waring, *Meditationes algebraicae* (Cambridge; 3d ed., 1782), p. xxxvl, etc.

[12] P. S. Laplace in *Mémoires d. l'académie r. d. sciences*, année 1817 (Paris, 1819), p. 153.

[13] *I. I. Stampionii Wis-Konstigh ende Reden-Maetigh Bewijs* (The Hague, 1640), p. 7.

[14] John Craig, *Philosophical Transactions*, Vol. XIX (London, 1695–97), p. 709.

[15] *Cockers Artificial Arithmetick. . . . . Composed by Edward Cocker. . . . . Perused, corrected and published by John Hawkins* (London, 1702) ["To the Reader," 1684], p. 368, 375.

a few times in John Wallis.[1] In the *Acta eruditorum* (1709), page 327, one finds $ny\sqrt{a} = \frac{2}{3}\sqrt{[(x-nna)^3]}$, where the [ ] makes, we believe, its first appearance in this journal, but does so as a redundant symbol.

348. *Aggregation expressed by dots.*—The denoting of aggregation by placing a dot before the expression affected is first encountered in Christoff Rudolff (§ 148). It is found next in the *Arithmetica integra* of M. Stifel, who sometimes places a dot also at the end. He writes[2]

$\sqrt{z}.12 + \sqrt{z}\ 6 + .\sqrt{z}.12 - \sqrt{z}\ 6$ for our $\sqrt{12 + \sqrt{6}} + \sqrt{12 - \sqrt{6}}$; also $\sqrt{z}.144 - 6 + \sqrt{z}.144 - 6$ for $\sqrt{144 - 6} + \sqrt{144 - 6}$. In 1605 C. Dibuadius[3] writes $\sqrt{.2 - \sqrt{.2 + \sqrt{.2 + \sqrt{.2 + \sqrt{.2 + \sqrt{2}}}}}}$ as the side of a regular polygon of 128 sides inscribed in a circle of unit radius, i.e.,

$$\sqrt{2 - \sqrt{2 + \sqrt{2 + \sqrt{2 + \sqrt{2 + \sqrt{2}}}}}}$$ (see also § 332). It must be admitted that this old notation is simpler than the modern. In Snell's translation[4] into Latin (1610) of Ludolph van Ceulen's work on the circle is given the same notation, $\sqrt{.2 + \sqrt{.2 - \sqrt{.2 - \sqrt{.2 + \sqrt{2\frac{1}{2}}}}}} - \sqrt{2\frac{1}{4}}$. In Snell's 1615 translation[5] into Latin of Ludolph's arithmetic and geometry is given the number $\sqrt{.2 - \sqrt{.2\frac{1}{2} + \sqrt{1\frac{1}{4}}}}$ which, when divided by $\sqrt{.2 + \sqrt{.2\frac{1}{2} + \sqrt{1\frac{1}{4}}}}$, gives the quotient $\sqrt{5} + 1 - \sqrt{.5} + \sqrt{20}$. The Swiss Joh. Ardüser[6] in 1646 writes $\sqrt{.2 \div \sqrt{.2 + \sqrt{.2 + \sqrt{.2 + \sqrt{2 + \sqrt{.2 + \sqrt{.2 + \sqrt{3}}}}}}}}$, etc., as the side of an inscribed polygon of 768 sides, where $\div$ means "minus."

The substitution of two dots (the colon) in the place of the single dot was effected by Oughtred in the 1631 and later editions of his *Clavis mathematicae*. With him this change became necessary when he adopted the single dot as the sign of ratio. He wrote ordinarily $\sqrt{q}:BC_q - BA_q:$ for $\sqrt{BC^2 - BA^2}$, placing colons before and after the terms to be aggregated (§ 181).[7]

[1] John Wallis, *Treatise of Algebra* (London, 1685), p. 133.

[2] M. Stifel, *Arithmetica integra* (Nürnberg, 1544), fol. 135$v°$. See J. Tropfke, *op. cit.*, Vol. III (Leipzig, 1922), p. 131.

[3] *C. Dibvadii in arithmeticam irrationalivm Evclidis decimo elementorum libro* (Arnhem, 1605).

[4] Willebrordus Snellius, *Lvdolphi à Cevlen de Circvlo et adscriptis liber ... è vernaculo Latina fecit ...* (Leyden, 1610), p. 1, 5.

[5] *Fvndamenta arithmetica et geometrica. ... Lvdolpho a Cevlen, ... in Latinum translata a Wil. Sn.* (Leyden, 1615), p. 27.

[6] Joh. Ardüser, *Geometriae theoricae et practicae XII libri* (Zürich, 1646), fol. 181b.

[7] W Oughtred, *Clavis mathematicae* (1652), p. 104.

Sometimes, when all the terms to the end of an expression are to be aggregated, the closing colon is omitted. In rare instances the opening colon is missing. A few times in the 1694 English edition, dots take the place of the colon. Oughtred's colons were widely used in England. As late as 1670 and 1693 John Wallis[1] writes $\sqrt{\,}:5-2\sqrt{3}:$. It occurs in Edward Cocker's[2] arithmetic of 1684, Jonas Moore's arithmetic[3] of 1688, where $C:A+E$ means the cube of $(A+E)$. James Bernoulli[4] gives in 1689 $\sqrt{\,}:a+\sqrt{\,}:a+\sqrt{\,}:a+\sqrt{\,}:a+\sqrt{\,}:a+$, etc. These methods of denoting aggregation practically disappeared at the beginning of the eighteenth century, but in more recent time they have been reintroduced. Thus, R. Carmichael[5] writes in his *Calculus of Operations*: "$D.\ uv=u.\ Dv+Du.\ v$." G. Peano has made the proposal to employ points as well as parentheses.[6] He lets $a.bc$ be identical with $a(bc)$, $a:bc.d$ with $a[(bc)d]$, $ab.cd:e.fg \therefore hk.l$ with $\{[(ab)(cd)][e(fg)]\}\ [(hk)l]$.

349. *Aggregation expressed by commas.*—An attempt on the part of Hérigone (§ 189) and Leibniz to give the comma the force of a symbol of aggregation, somewhat similar to Rudolff's, Stifel's, and van Ceulen's previous use of the dot and Oughtred's use of the colon, was not successful. In 1702 Leibniz[7] writes $c-b,\ l$ for $(c-b)l$, and $c-b,\ d-b,\ l$ for $(c-b)(d-b)l$. In 1709 a reviewer[8] in the *Acta eruditorum* represents $(m\div[m-1])x^{(m-1)\div m}$ by $(m,:m-1)x^{m-1,:m}$, a designation somewhat simpler than our modern form.

350. *Aggregation expressed by parenthesis* is found in rare instances as early as the sixteenth century. Parentheses present comparatively no special difficulties to the typesetter. Nevertheless, it took over two centuries before they met with general adoption as mathematical symbols. Perhaps the fact that they were used quite extensively as purely rhetorical symbols in ordinary writing helped to

[1] John Wallis in *Philosophical Transactions*, Vol. V (London, for the year 1670), p. 2203; *Treatise of Algebra* (London, 1685), p. 109; Latin ed. (1693), p. 120.

[2] Cocker's *Artificial Arithmetick* . . . . perused . . . . by John Hawkins (London, 1684), p. 405.

[3] Moore's *Arithmetick: in Four Books* (London, 1688; 3d ed.), Book IV, p. 425.

[4] *Positiones arithmeticae de seriebvs infinitis* . . . . *Jacobo Bernoulli* (Basel, 1689).

[5] R. Carmichael, *Der Operationscalcul*, deutsch von C. H. Schnuse (Braunschweig, 1857), p. 16.

[6] G. Peano, *Formulaire mathématique*, Édition de l'an 1902–3 (Turin, 1903), p. 4.

[7] G. W. Leibniz in *Acta eruditorum* (1702), p. 212.

[8] Reviewer in *ibid.* (1709). p. 230. See also p. 180.

retard their general adoption as mathematical symbols. John Wallis, for example, used parentheses very extensively as symbols containing parenthetical rhetorical statements, but made practically no use of them as symbols in algebra.

As a rhetorical sign to inclose an auxiliary or parenthetical statement parentheses are found in Newton's *De analysi per equationes numero terminorum infinitas*, as given by John Collins in the *Commercium epistolicum* (1712). In 1740 De Gua[1] wrote equations in the running text and inclosed them in parentheses; he wrote, for example, "... seroit $(\overline{7a-3x}\cdot dx = 3\sqrt{2ax-xx}\cdot dx)$ et où l'arc de cercle. ..."

English mathematicians adhered to the use of vinculums, and of colons placed before and after a polynomial, more tenaciously than did the French; while even the French were more disposed to stress their use than were Leibniz and Euler. It was Leibniz, the younger Bernoullis, and Euler who formed the habit of employing parentheses more freely and to resort to the vinculum less freely than did other mathematicians of their day. The straight line, as a sign of aggregation, is older than the parenthesis. We have seen that Chuquet, in his *Triparty* of 1484, underlined the terms that were to be taken together.

351. *Early occurrence of parentheses.*—Brackets[2] are found in the manuscript edition of R. Bombelli's *Algebra* (about 1550) in the expressions like $R^3[2\tilde{m}R[0\tilde{m}.121]]$ which stands for $\sqrt[3]{2-\sqrt{-121}}$. In the printed edition of 1572 an inverted capital letter L was employed to express *radix legata;* see the facsimile reproduction (Fig. 50). Michael Stifel does not use parentheses as signs of aggregation in his printed works, but in one of his handwritten marginal notes[3] occurs the following: ".... *faciant aggregatum* $(12-\sqrt{44})$ *quod sumptum cum* $(\sqrt{44}-2)$ *faciat* 10" (i.e., ".... One obtains the aggregate $(12-\sqrt{44})$, which added to $(\sqrt{44}-2)$ makes 10"). It is our opinion that these parentheses are punctuation marks, rather than mathematical symbols; signs of aggregation are not needed here. In the 1593 edition of F. Vieta's *Zetetica*, published in Turin, occur braces and brackets (§ 177) sometimes as open parentheses, at other times as closed ones. In Vieta's collected works, edited by Fr. van Schooten

---

[1] Jean Paul de Gua de Malves, *Usages de l'analyse de Descartes* (Paris, 1740), p. 302.

[2] See E. Bortolotti in *Scientia*, Vol. XXXIII (1923), p. 390.

[3] E. Hoppe, "Michael Stifels handschriftlicher Nachlass," *Mitteilungen Math. Gesellschaft Hamburg*, III (1900), p. 420. See J. Tropfke, *op. cit.*, Vol. II (2d ed., 1921), p. 28, n. 114.

in 1646, practically all parentheses are displaced by vinculums. However, in J. L. de Vaulezard's translation[1] into French of Vieta's *Zetetica* round parentheses are employed. Round parentheses are encountered in Tartaglia,[2] Cardan (but only once in his *Ars Magna*),[3] Clavius (see Fig. 66), Errard de Bar-le-Duc,[4] Follinus,[5] Girard,[6] Norwood,[7] Hume,[8] Stampioen, Henrion, Jacobo de Billy,[9] Renaldini[10] and Foster.[11] This is a fairly representative group of writers using parentheses, in a limited degree; there are in this group Italians, Germans, Dutch, French, English. And yet the mathematicians of none of the countries represented in this group adopted the general use of parentheses at that time. One reason for this failure lies in the fact that the vinculum, and some of the other devices for expressing aggregation, served their purpose very well. In those days when machine processes in printing were not in vogue, and when typesetting was done by hand, it was less essential than it is now that symbols should, in orderly fashion, follow each other in a line. If one or more vinculums were to be placed above a given polynomial, such a demand upon the printer was less serious in those days than it is at the present time.

[1] J. L. de Vaulezard's *Zététiques de F. Viète* (Paris, 1630), p. 218. Reference taken from the *Encyclopédie d. scien. math.*, Tom I, Vol. I, p. 28.

[2] N. Tartaglia, *General trattato di numeri e misure* (Venice), Vol. II (1556), fol. 167b, 169b, 170b, 174b, 177a, etc., in expressions like "$\cancel{R}v.(\cancel{R}$ 28 men R 10)" for $\sqrt{\sqrt{28}-\sqrt{10}}$; fol. 168b, "men (22 men $\cancel{R}6$" for $-(22-\sqrt{6})$, only the opening part being used. See G. Eneström in *Bibliotheca mathematica* (3d ser.), Vol. VII (1906–7), p. 296. Similarly, in *La Quarta Parte del general trattato* (1560), fol. 40B, he regularly omits the second part of the parenthesis when occurring on the margin, but in the running text both parts occur usually.

[3] H. Cardano, *Ars magna*, as printed in *Opera*, Vol. IV (1663), fol. 438.

[4] I. Errard de Bar-le-Duc, *La geometrie et practique generale d'icelle* (3d ed.; revué par D. H. P. E. M.; Paris, 1619), p. 216.

[5] Hermann Follinus, *Algebra sive liber de rebvs occvltis* (Cologne, 1622), p. 157.

[6] A. Girard, *Invention nouvelle en l'algebre* (Amsterdam, 1629), p. 17.

[7] R. Norwood, *Trigonometrie* (London, 1631), Book I, p. 30.

[8] Jac. Humius, *Traite de l'algebre* (Paris, 1635).

[9] Jacobo de Billy, *Novae geometriae clavis algebra* (Paris, 1643), p. 157; also in an *Abridgement of the Precepts of Algebra* (written in French by James de Billy; London, 1659), p. 346.

[10] Carlo Renaldini, *Opus algebricum* (1644; enlarged edition, 1665). Taken from Ch. Hutton, *Tracts on Mathematical and Philosophical Subjects*, Vol. II (1812), p. 297.

[11] Samuel Foster, *Miscellanies: or Mathematical Lucubrations* (London, 1659), p. 7.

And so it happened that in the second half of the seventeenth century, parentheses occur in algebra less frequently than during the first half of that century. However, voices in their favor are heard. The Dutch writer, J. J. Blassière,[1] explained in 1770 the three notations $(2a+5b)(3a-4b)$, $(2a+5b)\times(3a-4b)$, and $\overline{2a+5b}\times\overline{3a-4b}$, and remarked: "Mais comme la première manière de les enfermer entre des Parenthèses, est la moins sujette à erreur, nous nous en servirons dans la suite." E. Waring in 1762[2] uses the vinculum but no parentheses; in 1782[3] he employs parentheses and vinculums interchangeably. Before the eighteenth century parentheses hardly ever occur in the *Philosophical Transactions* of London, in the publications of the Paris Academy of Sciences, in the *Acta eruditorum* published in Leipzig. But with the beginning of the eighteenth century, parentheses do appear. In the *Acta eruditorum*, Carré[4] of Paris uses them in 1701, G. W. Leibniz[5] in 1702, a reviewer of Gabriele Manfredi[6] in 1708. Then comes in 1708 (§ 197) the statement of policy[7] in the *Acta eruditorum* in favor of the Leibnizian symbols, so that "in place of $\sqrt{aa+bb}$ we write $\sqrt{(aa+bb)}$ and for $\overline{aa+bb}\times c$ we write $aa+bb,c$ .... we shall designate $\overline{aa+bb}^m$ by $(aa+bb)^m$: whence $\overset{m}{\sqrt{}}\overline{aa+bb}$ will be $=(aa+bb)^{1:m}$ and $\overset{m}{\sqrt{}}\overline{aa+bb^n}=(aa+bb)^{n:m}$. Indeed, we do not doubt that all mathematicians reading these *Acta* recognize the preeminence of Mr. Leibniz' symbolism and agree with us in regard to it."

From now on round parentheses appear frequently in the *Acta eruditorum*. In 1709 square brackets make their appearance.[8] In the *Philosophical Transactions* of London[9] one of the first appearances of parentheses was in an article by the Frenchman P. L. Maupertuis in 1731, while in the *Histoire de l'académie royale des sciences* in Paris,[10]

[1] J. J. Blassière, *Institution du calcul numerique et litteral.* (a la Haye, 1770), 2. Partie, p. 27.

[2] E. Waring, *Miscellanea analytica* (Cambridge, 1762).

[3] E. Waring, *Meditationes algebraicae* (Cambridge; 3d ed., 1782).

[4] L. Carré in *Acta eruditorum* (1701), p. 281.

[5] G. W. Leibniz, *ibid.* (1702), p. 219.

[6] Gabriel Manfredi, *ibid.* (1708), p. 268.

[7] *Ibid.* (1708), p. 271.

[8] *Ibid.* (1709), p. 327.

[9] P. L. Maupertuis in *Philosophical Transactions*, for 1731–32, Vol. XXXVII (London), p. 245.

[10] Johann II Bernoulli, *Histoire de l'académie royale des sciences*, année 1732 (Paris, 1735), p. 240 ff.

Johann (John) Bernoulli of Bale first used parentheses and brackets in
the volume for the year 1732. In the volumes of the Petrograd
Academy, J. Hermann[1] uses parentheses, in the first volume, for the
year 1726; in the third volume, for the year 1728, L. Euler[2] and
Daniel Bernoulli used round parentheses and brackets.

352. The constant use of parentheses in the stream of articles from
the pen of Euler that appeared during the eighteenth century contributed vastly toward accustoming mathematicians to their use.
Some of his articles present an odd appearance from the fact that the
closing part of a round parenthesis is much larger than the opening
part,[3] as in $(1-\frac{z}{\pi})(1-\frac{z}{\pi-s})$. Daniel Bernoulli[4] in 1753 uses round
parentheses and brackets in the same expression while T. U. T.
Aepinus[5] and later Euler use two types of round parentheses of this
sort, $C(\beta+\gamma)(M-1)+AMつ$. In the publications of the Paris
Academy, parentheses are used by Johann Bernoulli (both round and
square ones),[6] A. C. Clairaut,[7] P. L. Maupertuis,[8] F. Nicole,[9] Ch. de
Montigny,[10] Le Marquis de Courtivron,[11] J. d'Alembert,[12] N. C. de
Condorcet,[13] J. Lagrange.[14] These illustrations show that about the
middle of the eighteenth century parentheses were making vigorous
inroads upon the territory previously occupied in France by vinculums almost exclusively.

[1] J. Hermann, *Commentarii academiae scientiarum imperialis Petropolitanae*, Tomus I ad annum 1726 (Petropoli, 1728), p. 15.

[2] *Ibid.*, Tomus III (1728; Petropoli, 1732), p. 114, 221.

[3] L. Euler in *Miscellanea Berolinensia*, Vol. VII (Berlin, 1743), p. 93, 95, 97, 139, 177.

[4] D. Bernoulli in *Histoire de l'académie r. des sciences et belles lettres*, année 1753 (Berlin, 1755), p. 175.

[5] Aepinus in *ibid.*, année 1751 (Berlin, 1753), p. 375; année 1757 (Berlin, 1759), p. 308–21.

[6] *Histoire de l'académie r. des sciences*, année 1732 (Paris, 1735), p. 240, 257.

[7] *Ibid.*, année 1732, p. 385, 387.

[8] *Ibid.*, année 1732, p. 444.

[9] *Ibid.*, année 1737 (Paris, 1740), "Mémoires," p. 64; also année 1741 (Paris, 1744), p. 36.

[10] *Ibid.*, année 1741, p. 282.

[11] *Ibid.*, année 1744 (Paris, 1748), p. 406.

[12] *Ibid.*, année 1745 (Paris, 1749), p. 369, 380.

[13] *Ibid.*, année 1769 (Paris, 1772), p. 211.

[14] *Ibid.*, année 1774 (Paris, 1778), p. 103.

353. *Terms in an aggregate placed in a vertical column.*—The employment of a brace to indicate the sum of coefficients or factors placed in a column was in vogue with Vieta (§ 176), Descartes, and many other writers. Descartes in 1637 used a single brace,[1] as in

$$x^4 - 2ax^3 + 2aa \atop - \ cc \Big\} xx - 2a^3x + a^4 \infty 0 \ ,$$

or a vertical bar[2] as in

$$x^4 + \tfrac{1}{2}aa \left| \begin{matrix} -a^3 \\ zz -acc \end{matrix} \right| \begin{matrix} + \tfrac{5}{16}a^4 \\ z -\tfrac{1}{4}aacc \end{matrix} \infty 0 \ .$$

Wallis[3] in 1685 puts the equation $aaa + baa + cca = ddd$, where a is the unknown, also in the form

$$\begin{matrix} 1 \\ aaa \end{matrix} \left| \begin{matrix} + \ b \\ aa \end{matrix} \right| \begin{matrix} +cc \\ a \end{matrix} \right| \begin{matrix} =ddd \\ 1 \end{matrix} \ .$$

Sometimes terms containing the same power of x were written in a column without indicating the common factor or the use of symbols of aggregation; thus, John Wallis[4] writes in 1685,

$$aaa + baa + bca = +bcd$$
$$+caa - bda$$
$$-daa - cda$$

Giovanni Poleni[5] writes in 1729,

$$y^6 + xxy^4 - 2ax^3yy + aax^4 = 0$$
$$-2axy^4 + aaxxyy$$
$$- \ aay^4$$

The use of braces for the combination of terms arranged in columns has passed away, except perhaps in recording the most unusual algebraic expressions. The tendency has been, whenever possible, to discourage symbolism spreading out vertically as well as horizontally. Modern printing encourages progression line by line.

354. *Marking binomial coefficients.*—In the writing of the factors in binomial coefficients and in factorial expressions much diversity of practice prevailed during the eighteenth century, on the matter of

[1] Descartes, *Œuvres* (éd. Adam et Tannery), Vol. VI, p. 450.

[2] *Ibid.*

[3] John Wallis, *Treatise of Algebra* (London, 1685), p. 160.

[4] John Wallis, *op. cit.*, p. 153.

[5] Joannis Poleni, *Epistolarvm mathematicarvm fascicvlvs* (Padua, 1729) (no pagination).

the priority of operations indicated by $+$ and $-$, over the operations of multiplication marked by $\cdot$ and $\times$. In $n \cdot n - 1 \cdot n - 2$ or $n \times n - 1 \times n - 2$, or $n, n-1, n-2$, it was understood very generally that the subtractions are performed first, the multiplications later, a practice contrary to that ordinarily followed at that time. In other words, these expressions meant $n(n-1)(n-2)$. Other writers used parentheses or vinculums, which removed all inconsistency and ambiguity. Nothing was explicitly set forth by early writers which would attach different meanings to nn and $n \cdot n$ or $n \times n$. And yet, $n \cdot n - 1 \cdot n - 2$ was not the same as $nn - 1n - 2$. Consecutive dots or crosses tacitly conveyed the idea that what lies between two of them must be aggregated as if it were inclosed in a parenthesis. Some looseness in notation occurs even before general binomial coefficients were introduced. Isaac Barrow[1] wrote "$L - M \times : R + S$" for $(L-M)(R+S)$, where the colon designated aggregation, but it was not clear that $L - M$, as well as $R + S$, were to be aggregated. In a manuscript of Leibniz[2] one finds the number of combinations of n things, taken k at a time, given in the form

$$\frac{n \frown n - 1 \frown n - 2, \text{ etc.}, \; n - k + 1}{1 \frown 2 \frown 3, \text{ etc.}, \frown k} \; .$$

This diversity in notation continued from the seventeenth down into the nineteenth century. Thus, Major Edward Thornycroft (1704)[3] writes $m \times m - 1 \times m - 2 \times m - 3$, etc. A writer[4] in the *Acta eruditorum* gives the expression $n, n-1$. Another writer[5] gives $\dfrac{(n, n-1, n-2)}{2, 3}$. Leibniz'[6] notation, as described in 1710 (§ 198), contains $e \cdot e - 1 \cdot e - 2$ for $e(e-1)(e-2)$. Johann Bernoulli[7] writes $n \cdot n - 1 \cdot n - 2$. This same notation is used by Jakob (James) Bernoulli[8] in a

[1] Isaac Barrow, *Lectiones mathematicae*, Lect. XXV, Probl. VII. See also Probl. VIII.

[2] D. Mahnke, *Bibliotheca mathematica* (3d ser.), Vol. XIII (1912–13), p. 35. See also *Leibnizens Mathematische Schriften*, Vol. VII (1863), p. 101.

[3] E. Thornycroft in *Philosophical Transactions*, Vol. XXIV (London, 1704–5), p. 1963.

[4] *Acta eruditorum* (Leipzig, 1708), p. 269.

[5] *Ibid.*, Suppl., Tome IV (1711), p. 160.

[6] *Miscellanea Berolinensia* (Berlin, 1710), p. 161.

[7] Johann Bernoulli in *Acta eruditorum* (1712), p. 276.

[8] Jakob Bernoulli, *Ars Conjectandi* (Basel, 1713), p. 99.

posthumous publication, by F. Nicole[1] who uses $x+n\cdot x+2n\cdot x+3n$, etc., by Stirling[2] in 1730, by Cramer[3] who writes in a letter to J. Stirling $a\cdot a+b\cdot a+2b$, by Nicolaus Bernoulli[4] in a letter to Stirling $r\cdot r+b\cdot r+2b\cdot$ ... by Daniel Bernoulli[5] $l-1\cdot l-2$, by Lambert[6] $4m-1\cdot$ $4m-2$, and by König[7] $n\cdot n-5\cdot n-6\cdot n-7$. Euler[8] in 1764 employs in the same article two notations: one, $n-5\cdot n-6\cdot n-7$; the other, $n(n-1)(n-2)$. Condorcet[9] has $n+2\times n+1$. Hindenburg[10] of Göttingen uses round parentheses and brackets, nevertheless he writes binomial factors thus, $m\cdot m-1\cdot m-2\ldots m-s+1$. Segner[11] and Ferroni[12] write $n\cdot n-1\cdot n-2$. Cossali[13] writes $4\times-2=-8$. As late as 1811 A. M. Legendre[14] has $n\cdot n-1\cdot n-2\ldots 1$.

On the other hand, F. Nicole,[15] who in 1717 avoided vinculums, writes in 1723, $x\cdot\overline{n+n}\cdot\overline{x+2n}$, etc. Stirling[16] in 1730 adopts $\overline{z-1}\cdot\overline{z-2}$. De Moivre[17] in 1730 likewise writes $\overline{m-p}\times\overline{m-q}\times\overline{m-s}$, etc. Similarly, Dodson,[18] $n\cdot\overline{n-1}\cdot\overline{n-2}$, and the Frenchman F. de Lalande,[19]

[1] Nicole in *Histoire de l'académie r. des sciences*, année 1717 (Paris, 1719), "Mémoires," p. 9.

[2] J. Stirling, *Methodus differentialis* (London, 1730), p. 9.

[3] Ch. Tweedie, *James Stirling* (Oxford, 1922), p. 121.   [4] *Op. cit.*, p. 144.

[5] Daniel I. Bernoulli, "Notationes de aequationibus," *Comment. Acad. Petrop.*, Tome V (1738), p. 72.

[6] J. H. Lambert, *Observationes in Acta Helvetica*, Vol. III.

[7] S. König, *Histoire de l'académie r. des sciences et des belles lettres*, année 1749 (Berlin, 1751), p. 189.

[8] L. Euler, *op. cit.*, année 1764 (Berlin, 1766), p. 195, 225.

[9] N. C. de Condorcet in *Histoire de l'académie r. des sciences*, année 1770 (Paris, 1773), p. 152.

[10] Carl Friedrich Hindenburg, *Infinitinomii dignatum leges . . . . ac formulae* (Göttingen, 1779), p. 30.

[11] J. A. de Segner, *Cursus mathematici*, pars II (Halle, 1768), p. 190.

[12] P. Ferroni, *Magnitudinum exponentialium . . . . theoria* (Florence, 1782), p. 29.

[13] Pietro Cossali, *Origine, trasporto in Italia ... dell'algebra*, Vol. I (Parma, 1797), p. 260.

[14] A. M. Legendre, *Exercices de calcul intégral*, Tome I (Paris, 1811), p. 277.

[15] *Histoire de l'académie r. des sciences*, année 1723 (Paris, 1753), "Mémoires," p. 21.

[16] James Stirling, *Methodus differentialis* (London, 1730), p. 6.

[17] Abraham de Moivre, *Miscellanea analytica de seriebus* (London, 1730), p. 4.

[18] James Dodson, *Mathematical Repository*, Vol. I (London, 1748), p. 238.

[19] F. de Lalande in *Histoire de l'académie r. des sciences*, année 1761 (Paris, 1763), p. 127.

$m \cdot (m+1) \cdot (m+2)$. In Lagrange[1] we encounter in 1772 the strictly modern form $(m+1)(m+2)(m+3), \ldots$, in Laplace[2] in 1778 the form $(i-1) \cdot (i-2) \ldots (i-r+1)$.

The omission of parentheses unnecessarily aggravates the interpretation of elementary algebraic expressions, such as are given by Kirkman,[3] viz., $-3=3\times-1$ for $-3=3\times(-1)$, $-m\times-n$ for $(-m)(-n)$.

355. *Special uses of parentheses.*—A use of round parentheses and brackets which is not strictly for the designation of aggregation is found in Cramer[4] and some of his followers. Cramer in 1750 writes two equations involving the variables x and y thus:

$$A \ldots x' - [-1]x^{n-1} + [1^2]x^{n-2} - [1^3]x^{n-3} + \&c. \ldots [1^n] = 0 ,$$
$$B \ldots (0)x^° + (1)x' + (2)x^2 + (3)x^3 + \&c. \ldots + (m)x^m = 0 ,$$

where $1, 1^2, 1^3, \ldots$, within the brackets of equation A do not mean powers of unity, but the coefficients of x, which are rational functions of y. The figures 0, 1, 2, 3, in B are likewise coefficients of x and functions of y. In the further use of this notation, (02) is made to represent the product of (0) and (2); (30) the product of (3) and (0), etc. Cramer's notation is used in Italy by Cossali[5] in 1799.

Special uses of parentheses occur in more recent time. Thus W. F. Sheppard[6] in 1912 writes

$$(n,r) \text{ for } n(n-1) \ldots (n-r+1)/r!$$
$$[n,r] \text{ for } n(n+1) \ldots (n+r-1)/r!$$
$$(n,2s+1) \text{ for } (n-s)(n-s+1) \ldots (n+s)/(2s+1)!$$

356. *A star to mark the absence of terms.*—We find it convenient to discuss this topic at this time. René Descartes, in *La Géométrie* (1637), arranges the terms of an algebraic equation according to the descending order of the powers of the unknown quantity x, y, or z. If any power of the unknown below the highest in the equation is

[1] J. Lagrange in *ibid.*, année 1772, Part I (Paris, 1775), "Mémoires," p. 523.

[2] P. S. Laplace in *ibid.*, année 1778 (Paris, 1781), p. 237.

[3] T. P. Kirkman, *First Mnemonical Lessons in Geometry, Algebra and Trigonometry* (London, 1852), p. 8, 9.

Gabriel Cramer, *Analyse des Lignes courbes algébriques* (Geneva, 1750), p. 660.

[5] Pietro Cossali, *op. cit.*, Vol. II (Parma, 1799), p. 41.

[6] W. F. Sheppard in *Fifth International Mathematical Congress*, Vol. II, p. 355.

lacking, that fact is indicated by a *, placed where the term would have been. Thus, Descartes writes $x^6 - a^4bx = 0$ in this manner:[1]

$$x^6 \ast \ast \ast \ast - - a^4bx \ast \infty 0 .$$

He does not explain why there was need of inserting these stars in the places of the missing terms. But such a need appears to have been felt by him and many other mathematicians of the seventeenth and eighteenth centuries. Not only were the stars retained in later editions of *La Géométrie*, but they were used by some but not all of the leading mathematicians, as well as by many compilers of textbooks. Kinckhuysen[2] writes "$x^5 \ast \ast \ast \ast - b \infty 0$." Prestet[3] in 1675 writes $a^3 \ast\ast + b^3$, and retains the * in 1689. The star is used by Baker,[4] Varignon,[5] John Bernoulli,[6] Alexander,[7] A. de Graaf,[8] E. Halley.[9] Fr. van Schooten used it not only in his various Latin editions of Descartes' *Geometry*, but also in 1646 in his *Conic Sections*,[10] where he writes $z^3 \infty \ast - pz + q$ for $z^3 = -pz + q$. In W. Whiston's[11] 1707 edition of I. Newton's *Universal Arithmetick* one reads $aa \ast - bb$ and the remark ". . . . locis vacuis substituitur nota *." Raphson's English 1728 edition of the same work also uses the *. Jones[12] uses * in 1706, Reyneau[13] in 1708; Simpson[14] employs it in 1737 and Waring[15] in 1762. De Lagny[16]

[1] René Descartes, *La géométrie* (Leyden, 1637); *Œuvres de Descartes* (éd. Adam et Tannery), Vol. VI (1903), p. 483.

[2] Gerard Kinckhuysen, *Algebra ofte Stel-Konst* (Haarlem, 1661), p. 59.

[3] *Elemens des mathematiques* (Paris, 1675), Epître, by J. P.[restet], p. 23. *Nouveaux elemens des Mathematiques*, par Jean Prestet (Paris, 1689), Vol. II, p. 450.

[4] Thomas Baker, *Geometrical Key* (London, 1684), p. 13.

[5] *Journal des Sçavans*, année 1687, Vol. XV (Amsterdam, 1688), p. 459. The star appears in many other places of this *Journal*.

[6] John Bernoulli in *Acta eruditorum* (1688), p. 324. The symbol appears often in this journal.

[7] John Alexander, *Synopsis Algebraica* ... (Londini, 1693), p. 203.

[8] Abraham de Graaf, *De Geheele Mathesis* (Amsterdam, 1694), p. 259.

[9] E. Halley in *Philosophical Transactions*, Vol. XIX (London, 1695–97), p. 61.

[10] Francisci à Schooten, *De organica conicarum sectionum* (Leyden, 1646), p. 91.

[11] *Arithmetica universalis* (Cambridge, 1707), p. 29.

[12] W. Jones, *Synopsis palmariorum matheseos* (London, 1706), p. 178.

[13] Charles Reyneau, *Analyse demontrée*, Vol. I (Paris, 1708), p. 13, 89.

[14] Thomas Simpson, *New Treatise of Fluxions* (London, 1737), p. 208.

[15] Edward Waring, *Miscellanea Analytica* (1762), p. 37.

[16] *Mémoires de l'académie r. d. sciences. Depuis 1666 jusqu'à 1699*, Vol. XI (Paris, 1733), p. 241, 243, 250.

employs it in 1733, De Gua[1] in 1741, MacLaurin[2] in his *Algebra*, and Fenn[3] in his *Arithmetic*. But with the close of the eighteenth century the feeling that this notation was necessary for the quick understanding of elementary algebraic polynomials passed away. In more advanced fields the star is sometimes encountered in more recent authors. Thus, in the treatment of elliptic functions, Weierstrass[4] used it to mark the absence of a term in an infinite series, as do also Greenhill[5] and Fricke.[6]

[1] *Histoire de l'académie r. d. sciences*, année 1741(Paris, 1744), p. 476.

[2] Colin Maclaurin, *Treatise of Algebra* (2d éd.; London, 1756), p. 277.

[3] Joseph Fenn, *Universal Arithmetic* (Dublin, 1772), p. 33.

[4] H. A. Schwarz, *Formeln und Lehrsätze . . . . nach Vorlesungen des Weierstrass* (Göttingen, 1885), p. 10, 11.

[5] A. G. Greenhill, *Elliptic Functions* (1892), p. 202, 204.

[6] R. Fricke, *Encyklopädie d. Math. Wissenschaften*, Vol. II[2] (Leipzig, 1913), p. 269.

SYMBOLS IN GEOMETRY
(ELEMENTARY PART)

A. ORDINARY ELEMENTARY GEOMETRY

357. The symbols sometimes used in geometry may be grouped roughly under three heads: (1) pictographs or pictures representing geometrical concepts, as △ representing a triangle; (2) ideographs designed especially for geometry, as ∼ for "similar"; (3) symbols of elementary algebra, like + and −.

Early use of pictographs.—The use of geometrical drawings goes back at least to the time of Ahmes, but the employment of pictographs in the place of words is first found in Heron's *Dioptra*. Heron (150 A.D.) wrote △ for triangle, $\underset{..}{\underline{ov}}$ for parallel and parallelogram, also $\underline{\rho'}$ for parallelogram, □ᐟ for rectangle, ○̆ for circle.[1] Similarly, Pappus (fourth century A.D.) writes ○ and ⊙ for circle, ▽ and △ for triangle, ∟ for right angle, $\underline{\underline{ol}}$ or = for parallel, □ for square.[2] But these were very exceptional uses not regularly adopted by the authors and occur in few manuscripts only. They were not generally known and are not encountered in other mathematical writers for about one thousand years. Paul Tannery calls attention to the use of the symbol □ in a medieval manuscript to represent, not a square foot, but a cubic foot; Tannery remarks that this is in accordance with the ancient practice of the Romans.[3] This use of the square is found in the *Triparty* of Chuquet (§ 132) and in the arithmetic of De la Roche.

358. Geometric figures were used in astrology to indicate roughly the relative positions of two heavenly bodies with respect to an observer. Thus ☌, ☍, □, △, ✳ designated,[4] respectively, conjunction,

[1] *Notices et extraits des manuscrits de la Bibliothèque impériale*, Vol. XIX, Part II (Paris, 1858), p. 173.

[2] *Pappi Alexandrini Collectionis quae supersunt* (ed. F. Hultsch), Vol. III, Tome I (Berlin, 1878), p. 126–31.

[3] Paul Tannery, *Mémoires scientifiques*, Vol. V (Toulouse and Paris, 1922). p. 73.

[4] Kepler says: "Quot sunt igitur aspectus? Vetus astrologia agnoscit tantum quinque: conjunctionem (☌), cum radii planetarum binorum in Terram descendentes in unam conjunguntur lineam; quod est veluti principium aspectuum omnium. 2) Oppositionem (☍), cum bini radii sunt ejusdem rectae partes, seu

opposition, at right angles, at 120°, at 60°. These signs are reproduced in Christian Wolff's *Mathematisches Lexicon* (Leipzig, 1716), page 188. The ✳, consisting of three bars crossing each other at 60°, was used by the Babylonians to indicate degrees. Many of their war carriages are pictured as possessing wheels with six spokes.[1]

359. In Plato of Tivoli's translation (middle of twelfth century) of the *Liber embadorum* by Savasorda who was a Hebrew scholar, at Barcelona, about 1100 A.D., one finds repeatedly the designations $\stackrel{\frown}{abc}$, $\stackrel{\frown}{ab}$ for arcs of circles.[2] In 1555 the Italian Fr. Maurolycus[3] employs △, □, also ✳ for hexagon and ∴ for pentagon, while in 1575 he also used ⊡. About half a century later, in 1623, Metius in the Netherlands exhibits a fondness for pictographs and adopts not only ◺, □, but a circle with a horizontal diameter and small drawings representing a sphere, a cube, a tetrahedron, and an octohedron. The last four were never considered seriously for general adoption, for the obvious reason that they were too difficult to draw. In 1634, in France, Hérigone's *Cursus mathematicus* (§ 189) exhibited an eruption of symbols, both pictographs and arbitrary signs. Here is the sign < for angle, the usual signs for triangle, square, rectangle, circle, also ⌐ for right angle, the Heronic = for parallel, ◇ for parallelogram, ⌒ for arc of circle, ⌓ for segment, — for straight line, ⊥ for perpendicular, 5< for pentagon, 6< for hexagon.

In England, William Oughtred introduced a vast array of characters into mathematics (§§ 181–85); over forty of them were used in symbolizing the tenth book of Euclid's *Elements* (§§ 183, 184), first printed in the 1648 edition of his *Clavis mathematicae*. Of these symbols only three were pictographs, namely, ⃞ for rectangle, □ for square, △ for triangle (§ 184). In the first edition of the *Clavis* (1631), the ⃞ alone occurs. In the *Trigonometria* (1657), he employed ∠ for angle and ⦞ for angles (§ 182), ‖ for parallel occurs in Oughtred's

---

cum duae quartae partes circuli a binis radiis interceptae sunt, id est unus semi-circulus. 3) Tetragonum seu quadratum (□), cum una quarta. 4) Trigonum seu trinum (△), cum una tertia seu duae sextae. 5) Hexagonum seu sextilem (✳), cum una sexta." See Kepler, *Opera omnia* (ed. Ch. Frisch), Vol. VI (1866), p. 490, quoted from "Epitomes astronomiae" (1618).

[1] C. Bezold, *Ninive und Babylon* (1903), p. 23, 54, 62, 124. See also J. Tropfke, *op. cit.*, Vol. I (2d ed., 1921), p. 38.

[2] See M. Curtze in *Bibliotheca mathematica* (3d ser.), Vol. I (1900), p. 327, 328.

[3] *Francisci Maurolyci Abbatis Messanensis Opuscula Mathematica* (Venice, 1575), p. 107, 134. See also Francisco Maurolyco in Boncompagni's *Bulletino*, Vol. IX, p. 67.

Opuscula mathematica hactenus inedita (1677), a posthumous work (§ 184).

Klügel[1] mentions a cube ▱ as a symbol attached to cubic measure, corresponding to the use of □ in square measure.

Euclid in his *Elements* uses lines as symbols for magnitudes, including numbers,[2] a symbolism which imposed great limitations upon arithmetic, for he does not add lines to squares, nor does he divide a line by another line.

360. *Signs for angles.*—We have already seen that Hérigone adopted < as the sign for angle in 1634. Unfortunately, in 1631, Harriot's *Artis analyticae praxis* utilized this very symbol for "less than." Harriot's > and < for "greater than" and "less than" were so well chosen, while the sign for "angle" could be easily modified so as to remove the ambiguity, that the change of the symbol for angle was eventually adopted. But < for angle persisted in its appearance, especially during the seventeenth and eighteenth centuries. We find it in W. Leybourn,[3] J. Kersey,[4] E. Hatton,[5] E. Stone,[6] J. Hodgson,[7] D'Alembert's *Encyclopédie*,[8] Hall and Steven's *Euclid*,[9] and Th. Reye.[10] John Caswell[11] used the sign ⋛ to express "equiangular."

A popular modified sign for angle was ∠, in which the lower stroke is horizontal and usually somewhat heavier. We have encountered this in Oughtred's *Trigonometria* (1657), Caswell,[12] Dulaurens,[13]

[1] G. S. Klügel, *Math. Wörterbuch*, 1. Theil (Leipzig, 1803), art. "Bruchzeichen."

[2] See, for instance, Euclid's *Elements*, Book V; see J. Gow, *History of Greek Mathematics* (1884), p. 106.

[3] William Leybourn, *Panorganon: or a Universal Instrument* (London, 1672), p. 75.

[4] John Kersey, *Algebra* (London, 1673), Book IV, p. 177.

[5] Edward Hatton, *An Intire System of Arithmetic* (London, 1721), p. 287.

[6] Edmund Stone, *New Mathematical Dictionary* (London, 1726; 2d ed., 1743), art. "Character."

[7] James Hodgson, *A System of Mathematics*, Vol. I (London, 1723), p. 10.

[8] *Encyclopédie ou Dictionnaire raissonné, etc.* (Diderot), Vol. VI (Lausanne et Berne, 1781), art. "Caractere."

[9] H. S. Hall and F. H. Stevens, *Euclid's Elements*, Parts I and II (London, 1889), p. 10.

[10] Theodor Reye, *Die Geometrie der Lage* (5th ed.; Leipzig, 1909), 1. Abteilung, p. 83.

[11] John Caswell, "Doctrine of Trigonometry," in Wallis' *Algebra* (1685).

[12] John Caswell, "Trigonometry," in *ibid.*

[13] Francisci Dulaurens, *Specimina mathematica duobus libris comprehensa* (Paris, 1667), "Symbols."

Jones,[1] Emerson,[2] Hutton,[3] Fuss,[4] Steenstra,[5] Klügel,[6] Playfair,[7] Kambly,[8] Wentworth,[9] Fiedler,[10] Casey,[11] Lieber and von Lühmann,[12] Byerly,[13] Müller,[14] Mehler,[15] C. Smith,[16] Beman and Smith,[17] Layng,[18] Hopkins,[19] Robbins,[20] the National Committee (in the U.S.A.).[21] The plural "angles" is designated by Caswell $\angle\,\angle$; by many others thus, $\measuredangle$. Caswell also writes $Z\angle\,\angle$ for the "sum of two angles," and $X\angle\,\angle$ for the "difference of two angles." From these quotations it is evident that the sign $\angle$ for angle enjoyed wide popularity in different countries. However, it had rivals.

361. Sometimes the same sign is inverted, thus $\reflectbox{\ensuremath{\angle}}$ as in John Ward.[22] Sometimes it is placed so as to appear $\wedge$, as in the *Ladies*

[1] William Jones, *Synopsis palmariorum matheseos* (London, 1706), p. 221.

[2] [W. Emerson], *Elements of Geometry* (London, 1763), p. 4.

[3] Charles Hutton, *Mathematical and Philosophical Dictionary* (1695), art. "Characters."

[4] Nicolas Fuss, *Leçons de géométrie* (St. Petersbourg, 1798), p. 38.

[5] Pibo Steenstra, *Grondbeginsels der Meetkunst* (Leyden, 1779), p. 101.

[6] G. S. Klügel, *Math. Wörterbuch*, fortgesetzt von C. B. Mollweide und J. A. Grunert, 5. Theil (Leipzig, 1831), art. "Zeichen."

[7] John Playfair, *Elements of Geometry* (Philadelphia, 1855), p. 114.

[8] L. Kambly, *Die Elementar-Mathematik*, 2. Theil: *Planimetrie*, 43. Aufl. (Breslau, 1876).

[9] G. A. Wentworth, *Elements of Geometry* (Boston, 1881; Preface, 1878).

[10] W. Fiedler, *Darstellende Geometrie*, 1. Theil (Leipzig, 1883), p. 7.

[11] John Casey, *Sequel to the First Six Books of the Elements of Euclid* (Dublin, 1886).

[12] H. Lieber und F. von Lühmann, *Geometrische Konstruktions-Aufgaben*, 8. Aufl. (Berlin, 1887), p. 1.

[13] W. E. Byerly's edition of Chauvenet's *Geometry* (Philadelphia, 1905), p. 44.

[14] G. Müller, *Zeichnende Geometrie* (Esslingen, 1889), p. 12.

[15] F. G. Mehler, *Hauptsätze der Elementar Mathematik*, 8. Aufl. (Berlin, 1894), p. 4.

[16] Charles Smith, *Geometrical Conics* (London, 1894).

[17] W. W. Beman and D. E. Smith, *Plane and Solid Geometry* (Boston, 1896), p. 10.

[18] A. E. Layng, *Euclid's Elements of Geometry* (London, 1890), p. 4.

[19] G. Irving Hopkins, *Inductive Plane Geometry* (Boston, 1902), p. 12.

[20] E. R. Robbins, *Plane and Solid Geometry* (New York, [1906]), p. 16.

[21] *Report by the National Committee on Mathematical Requirements*, under the auspices of the Mathematical Association of America, Inc. (1923), p. 77.

[22] John Ward, *The Young Mathematicians' Guide* (9th ed.; London, 1752), p. 301, 369.

Diary[1] and in the writings of Reyer,[2] Bolyai,[3] and Ottoni.[4] This position is widely used in connection with one or three letters marking an angle. Thus, the angle ABC is marked by L. N. M. Carnot[5] $\stackrel{\frown}{ABC}$ in his *Géométrie de position* (1803); in the *Penny Cyclopedia*(1839), article "Sign," there is given $A\stackrel{\frown}{\ }B$; Binet,[6] Möbius,[7] and Favaro[8] wrote $\stackrel{\frown}{ab}$ as the angle formed by two straight lines a and b; Favaro wrote also $P\stackrel{\frown}{D}C$. The notation $a\stackrel{\frown}{\ }b$ is used by Stolz and Gmeiner,[9] so that $a\stackrel{\frown}{\ }b = -b\stackrel{\frown}{\ }a$; Nixon[10] adopted $\stackrel{\frown}{A}$, also $\stackrel{\frown}{ABC}$; the designation $A\stackrel{\frown}{P}M$ is found in Enriques,[11] Borel,[12] and Durrell.[13]

362. Some authors, especially German, adopted the sign $\not\subset$ for angle. It is used by Spitz,[14] Fiedler,[15] Halsted,[16] Milinowski,[17] Meyer,[18]

[1] *Leybourne's Ladies Diary*, Vol. IV, p. 273.

[2] *Samuel Reyhers .... Euclides, dessen VI. erste Bücher auf sonderbare Art mit algebraischen Zeichen, also eingerichtet, sind, dass man derselben Beweise auch in anderen Sprachen gebrauchen kann* (Kiel, 1698).

[3] Wolfgangi Bolyai de Bolya, *Tentamen* (2d ed.), Tome II (Budapestini, 1904; 1st ed., 1832), p. 361.

[4] C. B. Ottoni, *Elementos de Geometria e Trigonometria* (4th ed.; Rio de Janeiro, 1874), p. 67.

[5] See Ch. Babbage, "On the Influence of Signs in Mathematical Reasoning," *Transactions Cambridge Philos. Society*, Vol. II (1827), p. 372.

[6] J. P. Binet in *Journal de l'école polyt.*, Vol. IX, Cahier 16 (Paris, 1813), p. 303.

[7] A. F. Möbius, *Gesammelte Werke*, Vol. I (Leipzig, 1885), "Barycyentrischer Calcul, 1827," p. 618.

[8] A. Favaro, *Leçons de Statique graphique*, trad. par Paul Terrier, 1. Partie (Paris, 1879), p. 51, 75.

[9] O. Stolz und J. A. Gmeiner, *Theoretische Arithmetik* (Leipzig), Vol. II (1902), p. 329, 330.

[10] R. C. J. Nixon, *Euclid Revised* (3d ed.; Oxford, 1899), p. 9.

[11] Federigo Enriques, *Questioni riguardanti la geometria elementare* (Bologna, 1900), p. 67.

[12] Emile Borel, *Algèbre* (2d cycle; Paris, 1913), p. 367.

[13] Clement V. Durell, *Modern Geometry; The Straight Line and Circle* (London, 1920), p. 7, 21, etc.

[14] Carl Spitz, *Lehrbuch der ebenen Geometrie* (Leipzig und Heidelberg, 1862), p. 11.

[15] W. Fiedler, *Darstellende Geometrie*, 1. Theil (Leipzig, 1883), p. 7.

[16] George Bruce Halsted, *Mensuration* (Boston, 1881), p. 28; *Elementary Synthetic Geometry* (New York, 1892), p. vii.

[17] A. Milinowski, *Elem.-Synth. Geom. der Kegelschnitte* (Leipzig, 1883), p. 3.

[18] Friedrich Meyer, *Dritter Cursus der Planimetrie* (Halle, a/S, 1885), p. 81.

Fialkowski,[1] Henrici and Treutlein,[2] Brückner,[3] Doehlemann,[4] Schur,[5] Bernhard,[6] Auerbach and Walsh,[7] Mangoldt.[8]

If our quotations are representative, then this notation for angle finds its adherents in Germany and the United States. A slight modification of this sign is found in Byrne[9] ◁.

Among sporadic representations of angles are the following: The capital letter[10] L, the capital letter[11] $\vee$, or that letter inverted,[12] $\wedge$, the inverted capital letter[13] $\forall$; the perpendicular lines[14] $\lrcorner$ or $\llcorner$, $\overline{pq}$ the angle made by the lines[15] p and q, (ab) the angle between the rays,[16] a and b, $\widehat{ab}$ the angle between the lines[17] a and b, or (u, v) the angle[18] formed by u and v.

363. Passing now to the designation of special angles we find $\angle$ used to designate an oblique angle.[19] The use of a pictograph for the designation of right angles was more frequent in former years than now and occurred mainly in English texts. The two perpendicular lines $\llcorner$ to designate "right angle" are found in Reyher;[20] he lets

[1] N. Fialkowski, *Praktische Geometrie* (Wien, 1892), p. 15.

[2] J. Henrici und P. Treutlein, *Lehrbuch der Elementar-Geometrie*, 1. Teil, 3. Aufl. (Leipzig, 1897), p. 11.

[3] Max Brückner, *Vielecke und Vielfache-Theorie und Geschichte* (Leipzig, 1900), p. 125.

[4] Karl Doehlemann, *Projektive Geometrie*, 3. Aufl. (Leipzig, 1905), p. 133.

[5] F. Schur, *Grundlagen der Geometrie* (Leipzig und Berlin, 1909), p. 79.

[6] Max Bernhard, *Darstellende Geometrie* (Stuttgart, 1909), p. 267.

[7] Matilda Auerbach and Charles B. Walsh, *Plane Geometry* (Philadelphia, [1920]), p. vii.

[8] Hans V. Mangoldt, *Einführung in die höhere Mathematik*, Vol. I (Leipzig, 1923), p. 190.

[9] Oliver Byrne, *Elements of Euclid* (London, 1847), p. xxviii.

[10] John Wilson, *Trigonometry* (Edinburgh, 1714), "Characters Explained."

[11] A. Saverien, *Dictionnaire de math. et phys.* (Paris, 1753), "Caractere."

[12] W. Bolyai, *Tentamen* (2d ed.), Vol. I (1897), p. xi.

[13] Joseph Fenn, *Euclid* (Dublin, 1769), p. 12; J. D. Blassière, *Principes de géométrie élémentaire* (The Hague, 1782), p. 16.

[14] H. N. Robinson, *Geometry* (New York, 1860), p. 18; *ibid.* (15th ed., New York), p. 14.

[15] Charlotte Angas Scott, *Modern Analytical Geometry* (London, 1894), p. 253.

[16] Heinrich Schröter, *Theorie der Kegelschnitte* (2d ed; Leipzig, 1876), p. 5.

[17] J. L. S. Hatton, *Principles of Projective Geometry* (Cambridge, 1913), p. 9.

[18] G. Peano, *Formulaire mathématique* (Turin, 1903), p. 266.

[19] W. N. Bush and John B. Clarke, *Elements of Geometry* (New York, [1905]).

[20] *Samuel Reyhers Euclides* (Kiel, 1698).

∧A∟ stand for "angle A is a right angle," a symbolism which could be employed in any language. The vertical bar stands for equality (§ 263). The same idea is involved in the signs $a \uparrow b$, i.e., "angle a is equal to angle b." The sign ∟ for right angle is found in Jones,[1] Hatton,[2] Saverien,[3] Fenn,[4] and Steenstra.[5] Kersey[6] uses the sign ⊥, Byrne[7] ◻. Mach[8] marks right angles ⊥. The Frenchman Hérigone[9] used the sign ⌐, the Englishman Dupius[10] ⌐ for right angle.

James Mills Peirce,[11] in an article on the notation of angles, uses "*Greek letters* to denote the *directions* of lines, without reference to their length. Thus if ρ denotes the axis in a system of polar co-ordinates, the polar angle will be $\breve{\rho}$." Accordingly, $\dfrac{\beta}{\alpha} = -\dfrac{\alpha}{\beta}$.

More common among more recent American and some English writers is the designation "rt. $\angle$" for right angle. It is found in G. A. Wentworth,[12] Byerly's *Chauvenet*,[13] Hall and Stevens,[14] Beman and Smith,[15] Hopkins,[16] Robbins,[17] and others.

Some writers use instead of pictographs of angles abbreviations of the word. Thus Legendre[18] sometimes writes "Angl. ACB";

[1] William Jones, *Synopsis palmariorum matheseos* (London, 1706), p. 221.

[2] Edward Hatton, *An Intire System of Arithmetik* (London, 1721), p. 287.

[3] A. Saverien, *Dictionnaire*, "Caractere."

[4] Joseph Fenn, *Euclid* (Dublin, 1769), p. 12.

[5] Pibo Steenstra, *Grondbeginsels der Meetkunst* (Leyden, 1779), p. 101.

[6] John Kersey, *Algebra* (London, 1673), Book IV, p. 177.

[7] Oliver Byrne, *The Elements of Euclid* (London, 1847), p. xxviii.

[8] E. Mach, *Space and Geometry* (trans. T. J. McCormack, 1906), p. 122.

[9] P. Herigone, *Cursus mathematicus* (Paris, 1634), Vol. I, "Explicatio notarum."

[10] N. F. Dupius, *Elementary Synthetic Geometry* (London, 1889), p. 19.

[11] J. D. Runkle's *Mathematical Monthly*, Vol. I, No. 5 (February, 1859), p. 168, 169.

[12] G. A. Wentworth, *Elements of Plane and Solid Geometry* (3d ed.; Boston, 1882), p. 14.

[13] W. E. Byerly's edition of *Chauvenet's Geometry* (1887).

[14] H. S. Hall and F. H. Stevens, *Euclid's Elements*, Parts I and II (London, 1889), p. 10.

[15] W. W. Beman and D. E. Smith, *Plane and Solid Geometry* (Boston, 1896), p. 10.

[16] G. I. Hopkins, *Inductive Plane Geometry* (Boston, 1902), p. 12.

[17] E. R. Robbins, *Plane and Solid Geometry* (New York, [1906]), p. 16.

[18] A. M. Legendre, *Éléments de Géométrie* (Paris, 1794), p. 42.

A. von Frank,[1] "*Wkl*," the abbreviation for *Winkel*, as in "*Wkl DOQ*."

The advent of non-Euclidean geometry brought Lobachevski's notation Π (ρ) for angle of parallelism.[2]

The sign $\underline{\vee}$ to signify equality of the angles, and $\underline{\perp}$ to signify the equality of the sides of a figure, are mentioned in the article "Caractere" by D'Alembert in Diderot' *Encyclopédie* of 1754 and of 1781[3] and in the Italian translation of the mathematical part (1800); also in Rees's *Cyclopaedia* (London, 1819), article "Characters," and in E. Stone's *New Mathematical Dictionary* (London, 1726), article "Characters," but Stone defines $\underline{\vee}$ as signifying "equiangular or similar." The symbol is given also by a Spanish writer as signifying *angulos iguales*.[4] The sign $=°$ to signify "equal number of degrees" is found in Palmer and Taylor's *Geometry*,[5] but failed to be recommended as a desirable symbol in elementary geometry by the National Committee on Mathematical Requirements (1923), in their *Report*, page 79.

Halsted suggested the sign $\not\subset$ for spherical angle and also the letter Ω to represent a "steregon," the unit of solid angle.[6]

364. *Signs for "perpendicular."*—The ordinary sign to indicate that one line is perpendicular to another, $\perp$, is given by Hérigone[7] in 1634 and 1644. Another Frenchman, Dulaurens,[8] used it in 1667. In 1673 Kersey[9] in England employed it. The inverted capital letter L was used for this purpose by Caswell,[10] Jones,[11] Wilson,[12] Saverien,[13]

[1] A. von Frank in *Archiv der Mathematik und Physik* von J. A. Grunert (2d ser.), Vol. XI (Leipzig, 1892), p. 198.

[2] George Bruce Halsted, *N. Lobatschewsky, Theory of Parallels* (Austin, 1891), p. 13.

[3] *Encyclopédie au Dictionnaire raisonné des sciences*, ... by Diderot, Vol. VI (Lausanne et Berne, 1781), art. "Caractere."

[4] Antonio Serra y Oliveres, *Manuel de la Tipografia Española* (Madrid, 1852), p. 70.

[5] C. I. Palmer and D. P. Taylor, *Plane Geometry* (1915), p. 16.

[6] G. B. Halsted, *Mensuration* (Boston, 1881), p. 28.

[7] Pierre Herigone, *Cursus mathematicus*, Vol. I (Paris, 1634), "Explicatio notarum."

[8] F. Dulaurens, *Specimina mathematica duobus libris comprehensa* (Paris, 1667), "Symbols."

[9] John Kersey, *Algebra* (London, 1673), Book IV, p. 177.

[10] J. Caswell's *Trigonometry* in J. Wallis' *Algebra* (1685).

[11] W. Jones, *op. cit.*, p. 253.

[12] J. Wilson, *Trigonometry* (Edinburgh, 1714), "Characters Explained."

[13] A. Saverien, *Dictionnaire*, "Caractere."

and Mauduit.[1] Emerson[2] has the vertical bar extremely short, ⊥. In the nineteenth century the symbol was adopted by all writers using pictographs in geometry. Sometimes ⌊ₛ was used for "perpendiculars." Thomas Baker[3] adopted the symbol ⊥ for perpendicular.

365. *Signs for triangle, square, rectangle, parallelogram.*—The signs △, □, ▭ or ▯, ▱ are among the most widely used pictographs. We have already referred to their occurrence down to the time of Hérigone and Oughtred (§ 184). The ▱ for parallelogram is of rare occurrence in geometries preceding the last quarter of the nineteenth century, while the △, □, and ▭ occur in van Schooten,[4] Dulaurens,[5] Kersey,[6] Jones,[7] and Saverien.[8] Some authors use only two of the three. A rather curious occurrence is the Hebrew letter "mem," ▭, to represent a rectangle; it is found in van Schooten,[9] Jones,[10] John Alexander,[11] John I Bernoulli,[12] Ronayne,[13] Klügel's *Wörterbuch*,[14] and De Graaf.[15] Newton,[16] in an early manuscript tract on fluxions (October, 1666), indicates the area or fluent of a curve by prefixing a rectangle to the ordinate (§ 622), thus □ $\frac{axx-x^3}{ab+xx}$, where x is the abscissa, and the fraction is the ordinate.

After about 1880 American and English school geometries came to employ less frequently the sign ▭ for rectangle and to introduce more often the sign ▱ for parallelogram. Among such authors are

[1] A. R. Mauduit, *Inleiding tot de Kleegel-Sneeden* (The Hague, 1763), "Symbols."

[2] [W. Emerson], *Elements of Geometry* (London, 1763).

[3] Thomas Baker, *Geometrical Key* (London, 1684), list of symbols.

[4] Fr. van Schooten, *Exercitationvm mathematicorvm liber primus* (Leyden, 1657).

[5] F. Dulaurens, *loc. cit.*, "Symbols."

[6] J. Kersey, *Algebra* (1673).

[7] W. Jones, *op. cit.*, p. 225, 238.

[8] Saverien, *loc. cit.*

[9] Franciscus van Schooten, *op. cit.* (Leyden, 1657), p. 67.

[10] W. Jones, *op. cit.*, p. 253.

[11] *Synopsis Algebraica, opus posthumum Iohannis Alexandri* (London, 1693), p. 67.

[12] John Bernoulli in *Acta eruditorum* (1689), p. 586; *ibid.* (1692), p. 31.

[13] Philip Ronayne, *Treatise of Algebra* (London, 1727), p. 3.

[14] J. G. Klügel, *Math. Wörterbuch*, 5. Theil (Leipzig, 1831), "Zeichen."

[15] Abraham de Graaf, *Algebra of Stelkonst* (Amsterdam, 1672), p. 81.

[16] S. P. Rigaud, *Historical Essay on . . . . Newton's Principia* (Oxford, 1838), Appendix, p. 23.

Halsted,[1] Wentworth,[2] Byerly,[3] in his edition of Chauvenet, Beman and Smith,[4] Layng,[5] Nixon,[6] Hopkins,[7] Robbins,[8] and Lyman.[9] Only seldom do both $\square$ and $\square$ appear in the same text. Halsted[10] denotes a parallelogram by $||g'm$.

Special symbols for right and oblique spherical triangles, as used by Jean Bernoulli in trigonometry, are given in Volume II, § 524.

366. *The square as an operator.*—The use of the sign $\square$ to mark the operation of squaring has a long history, but never became popular. Thus N. Tartaglia[11] in 1560 denotes the square on a line tc in the expression "il $\square$ de. tc." Cataldi[12] uses a black square to indicate the square of a number. Thus, he speaks of $8\frac{11}{4}$, "il suo $\blacksquare$ è 75 $\frac{1489}{1936}$." Stampioen[13] in 1640 likewise marks the square on BC by the "$\square$ BC." Caramvel[14] writes "$\square 25$. est Quadratum Numeri 25. hoc est, 625."

A. de Graaf[15] in 1672 indicates the square of a binomial thus: $\sqrt{a} \pm \sqrt{b}$, "zijn $\square$ is $a+b \pm 2\sqrt{ab}$." Johann I Bernoulli[16] wrote $3 \square \sqrt[3]{ax+xx}$ for $3\sqrt{(ax+x^2)^2}$. Jakob Bernoulli in 1690[17] designated

---

[1] G. B. Halsted, *Elem. Treatise on Mensuration* (Boston, 1881), p. 28.

[2] G. A. Wentworth, *Elements of Plane and Solid Geometry* (3d ed.; Boston, 1882), p. 14 (1st ed., 1878).

[3] W. E. Byerly's edition of *Chauvenet's Geometry* (1887), p. 44.

[4] W. W. Beman and D. E. Smith, *Plane and Solid Geometry* (Boston, 1896), p. 10.

[5] A. E. Layng, *Euclid's Elements of Geometry* (London, 1890), p. 4.

[6] R. C. J. Nixon, *Euclid Revised* (3d ed., Oxford, 1899), p. 6.

[7] G. J. Hopkins, *Inductive Plane Geometry* (Boston, 1902), p. 12.

[8] E. R. Robbins, *Plane and Solid Geometry* (New York, [1906]), p. 16.

[9] E. A. Lyman, *Plane and Solid Geometry* (New York, 1908), p. 18.

[10] G. B. Halsted, *Rational Geometry* (New York, 1904), p. viii.

[11] N. Tartaglia, *La Quinta parte del general trattato de nvmeri et misvre* (Venice, 1560), fols. 82AB and 83A.

[12] *Trattato del Modo Brevissimo di trouare la Radice quadra delli numeri,* . . . . Di Pietro Antonio Cataldi (Bologna, 1613), p. 111.

[13] J. Stampioen, *Wis-Konstich ende Reden-maetich Bewys* ('S Graven-Hage, 1640), p. 42.

[14] *Joannis Caramvelis mathesis biceps. vetus, et nova* (1670), p. 131.

[15] Abraham de Graaf, *Algebra of Stelkonst* (Amsterdam, 1672), p. 32.

[16] Johannis I Bernoulli, *Lectiones de calculo differentialium* . . . . von Paul Schafheitlin, Separatabdruck aus den *Verhandlungen der Naturforschenden Gesellschaft in Basel*, Vol. XXXIV (1922).

[17] Jakob Bernoulli in *Acta eruditorum* (1690), p. 223.

the square of $\frac{5}{6}$ by $\boxed{2}\frac{5}{6}$, but in his collected writings[1] it is given in the modern form $(\frac{5}{6})^2$. Sometimes a rectangle, or the Hebrew letter "mem," is used to signify the product of two polynomials.[2]

367. *Sign for circle.*—Although a small image of a circle to take the place of the word was used in Greek time by Heron and Pappus, the introduction of the symbol was slow. Hérigone used ⊙, but Oughtred did not. One finds ⊙ in John Kersey,[3] John Caswell,[4] John Ward,[5] P. Steenstra,[6] J. D. Blassière,[7] W. Bolyai,[8] and in the writers of the last half-century who introduced the sign ▭ for parallelogram. Occasionally the central dot is omitted and the symbol ◯ is used, as in the writings of Reyher[9] and Saverien. Others, Fenn for instance, give both ◯ and ⊙, the first to signify circumference, the second circle (area). Caswell[10] indicates the perimeter by Ȯ. Metius[11] in 1623 draws the circle and a horizontal diameter to signify *circulus*.

368. *Signs for parallel lines.*—Signs for parallel lines were used by Heron and Pappus (§ 701); Hérigone used horizontal lines = (§ 189) as did also Dulaurens[12] and Reyher,[13] but when Recorde's sign of equality won its way upon the Continent, vertical lines came to be used for parallelism. We find ‖ for "parallel" in Kersey,[14] Caswell, Jones,[15] Wilson,[16] Emerson,[17] Kambly,[18] and the writers of the last

---

[1] *Opera Jakob Bernoullis*, Vol. I, p. 430, 431; see G. Eneström, *Bibliotheca mathematica* (3d ser.), Vol. IX (1908–9), p. 207.

[2] See P. Herigone, *Cursus mathematici* (Paris, 1644), Vol. VI, p. 49.

[3] John Kersey, *Algebra* (London, 1673), Book IV, p. 177.

[4] John Caswell in Wallis' *Treatise of Algebra*, "Additions and Emendations," p. 166. For "circumference" Caswell used the small letter *c*.

[5] J. Ward, *The Young Mathematician's Guide* (9th ed.; London, 1752), p. 301, 369.

[6] P. Steenstra, *Grondbeginsels der Meetkunst* (Leyden, 1779), p. 281.

[7] J. D. Blassière, *Principes de géométrie élémentaire* (The Hague, 1723), p. 16.

[8] W. Bolyai, *Tentamen* (2d ed.), Vol. II (1904), p. 361 (1st ed., 1832).

[9] *Samuel Reyhers, Euclides* (Kiel, 1698), list of symbols.

[10] John Caswell in Wallis' *Treatise of Algebra* (1685), "Additions and Emendations," p. 166.

[11] Adriano Metio, *Praxis nova geometrica* (1623), p. 44.

[12] Fr. Dulaurens, *Specimina mathematica* (Paris, 1667), "Symbols."

[13] S. Reyher, *op. cit.* (1698), list of symbols.

[14] John Kersey, *Algebra* (London, 1673), Book IV, p. 177.

[15] W. Jones, *Synopsis palmariorum matheseos* (London, 1706).

[16] John Wilson, *Trigonometry* (Edinburgh, 1714), characters explained.

[17] [W. Emerson], *Elements of Geometry* (London, 1763), p. 4.

[18] L. Kambly, *Die Elementar-Mathematik*, 2. Theil, *Planimetrie*, 43. Aufl. (Breslau, 1876), p. 8.

fifty years who have been already quoted in connection with other pictographs. Before about 1875 it does not occur as often as do △, □, ⌑. Hall and Stevens[1] use "par[l] or ||" for parallel. Kambly[2] mentions also the symbols ╫ and ╪ for parallel.

A few other symbols are found to designate parallel. Thus John Bolyai in his *Science Absolute of Space* used |||. Karsten[3] used ╫; he says: "Man pflege wohl das Zeichen ╫ statt des Worts: *Parallel* der Kürze wegen zu gebrauchen." This use of that symbol occurs also in N. Fuss.[4] Thomas Baker[5] employed the sign ∽.

With Kambly ╫ signifies rectangle. Häseler[6] employs ╬ as "the sign of parallelism of two lines or surfaces."

369. *Sign for equal and parallel.*—╫is employed to indicate that two lines are equal and parallel in Klügel's *Wörterbuch;*[7] it is used by H. G. Grassmann,[8] Lorey,[9] Fiedler,[10] Henrici and Treutlein.[11]

370. *Signs for arcs of circles.*—As early a writer as Plato of Tivoli (§ 359) used $\overset{\frown}{ab}$ to mark the arc ab of a circle. Ever since that time it has occurred in geometric books, without being generally adopted. It is found in Hérigone,[12] in Reyher,[13] in Kambly,[14] in Lieber and Lühmann.[15] W. R. Hamilton[16] designated by ⌒LF the arc "from F to L." These

[1] H. S. Hall and F. H. Stevens, *Euclid's Elements*, Parts I and II (London, 1889), p. 10.

[2] L. Kambly, *op. cit.*, 2. Theil, *Planimetrie*, 43. Aufl. (Breslau, 1876), p. 8.

[3] W. J. G. Karsten, *Lehrbegrif der gesamten Mathematik*, 1. Theil (Greifswald, 1767), p. 254.

[4] Nicolas Fuss, *Leçons de géométrie* (St. Petersbourg, 1798), p. 13.

[5] Thomas Baker, *Geometrical Key* (London, 1684), list of symbols.

[6] J. F. Häseler, *Anfangsgründe der Arith., Alg., Geom. und Trig.* (Lemgo), *Elementar-Geometrie* (1777), p. 72.

[7] G. S. Klügel, *Mathematisches Wörterbuch*, fortgesetzt von C. B. Mollweide, J. A. Grunert, 5. Theil (Leipzig, 1831), "Zeichen."

[8] H. G. Grassmann, *Ausdehnungslehre von 1844* (Leipzig, 1878), p. 37; *Werke* by F. Engel (Leipzig, 1894), p. 67.

[9] Adolf Lorey, *Lehrbuch der ebenen Geometrie* (Gera und Leipzig, 1868), p. 52.

[10] Wilhelm Fiedler, *Darstellende Geometrie*, 1. Theil (Leipzig, 1883), p. 11.

[11] J. Henrici und P. Treutlein, *Lehrbuch der Elementar-Geometrie*, 1. Teil, 3. Aufl. (Leipzig, 1897), p. 37.

[12] P. Herigone, *op. cit.* (Paris, 1644), Vol. I, "Explicatio notarum."

[13] *Samuel Reyhers, Euclides* (Kiel, 1698), Vorrede.

[14] L. Kambly, *op. cit.* (1876).

[15] H. Lieber und F. von Lühmann, *Geometrische Konstructions-Aufgaben*, 8. Aufl. (Berlin, 1887), p. 1.

[16] W. R. Hamilton in *Cambridge & Dublin Math'l. Journal*, Vol. I (1846), p. 262.

references indicate the use of ⌢ to designate arc in different countries. In more recent years it has enjoyed some popularity in the United States, as is shown by its use by the following authors: Halsted,[1] Wells,[2] Nichols,[3] Hart and Feldman,[4] and Smith.[5] The National Committee on Mathematical Requirements, in its *Report* (1923), page 78, is of the opinion that "the value of the symbol ⌢ in place of the short word *arc* is doubtful."

In 1755 John Landen[6] used the sign $(P\acute{Q}R)$ for the circular arc which measures the angle $P\acute{Q}R$, the radius being unity.

371. *Other pictographs.*—We have already referred to Hérigone's use (§ 189) of $5<$ and $6<$ to represent pentagons and hexagons. Reyher actually draws a pentagon. Occasionally one finds a half-circle and a diameter ⌓ to designate a segment, and a half-circle without marking its center or drawing its diameter to designate an arc. Reyher in his *Euclid* draws ⌲ for trapezoid.

Pictographs of solids are very rare. We have mentioned (§ 359) those of Metius. Saverein[7] draws ■, ▲, ▬, ■ to stand, respectively, for cube, pyramid, parallelopiped, rectangular parallelopiped, but these signs hardly belong to the category of pictographs. Dulaurens[8] wrote ③ for cube and ④ for *aequi quadrimensum*. Joseph Fenn[9] draws a small figure of a parallelopiped to represent that solid, as Metius had done. Halsted[10] denotes symmetry by ⋅⊹⋅ .

Some authors of elementary geometries have used algebraic symbols and no pictographs (for instance, Isaac Barrow, Karsten, Tacquet, Leslie, Legendre, Playfair, Chauvenet, B. Peirce, Todhunter), but no author since the invention of symbolic algebra uses pictographs without at the same time availing himself of algebraic characters.

372. *Signs for similarity and congruence.*—The designation of "similar," "congruent," "equivalent," has brought great diversity of notation, and uniformity is not yet in sight.

Symbols for similarity and congruence were invented by Leibniz.

[1] G. B. Halsted, *Mensuration* (Boston, 1881).

[2] Webster Wells, *Elementary Geometry* (Boston, 1886), p. 4.

[3] E. H. Nichols, *Elements of Constructional Geometry* (New York, 1896).

[4] C. A. Hart and D. D. Feldman, *Plane Geometry* (New York, [1911]), p. viii.

[5] Eugene R. Smith, *Plane Geometry* (New York, 1909), p. 14.

[6] John Landen, *Mathematical Lucubrations* (London, 1755), Sec. III, p. 93.

[7] A. Saverein, *Dictionnaire*, "Caractere."

[8] F. Dulaurens, *op. cit.* (Paris, 1667), "Symbols."

[9] Joseph Fenn, *Euclid's Elements of Geometry* (Dublin, [ca. 1769]), p. 319.

[10] G. B. Halsted, *Rational Geometry* (New York, 1904), p. viii.

In Volume II, § 545, are cited symbols for "coincident" and "congruent" which occur in manuscripts of 1679 and were later abandoned by Leibniz. In the manuscript of his *Characteristica Geometrica* which was not published by him, he says: "similitudinem ita notabimus: $a \sim b$."[1] The sign is the letter S (first letter in *similis*) placed horizontally. Having no facsimile of the manuscript, we are dependent upon the editor of Leibniz' manuscripts for the information that the sign in question was $\sim$ and not $\backsim$. As the editor, C. I. Gerhardt, interchanged the two forms (as pointed out below) on another occasion, we do not feel certain that the reproduction is accurate in the present case. According to Gerhardt, Leibniz wrote in another manuscript $\simeq$ for congruent. Leibniz' own words are reported as follows: "ABC $\simeq CDA$. Nam $\sim$ mihi est signum similitudinis, et $=$ aequalitatis, unde congruentiae signum compono, quia quae simul et similia et aequalia sunt, ea congrua sunt."[2] In a third manuscript Leibniz wrote $|\simeq|$ for coincidence.

An anonymous article printed in the *Miscellanea Berolinensia* (Berlin, 1710), under the heading of "Monitum de characteribus algebraicis," page 159, attributed to Leibniz and reprinted in his collected mathematical works, describes the symbols of Leibniz; $\backsim$ for similar and $\backsimeq$ for congruent (§ 198). Note the change in form; in the manuscript of 1679 Leibniz is reported to have adopted the form $\sim$, in the printed article of 1710 the form given is $\backsim$. Both forms have persisted in mathematical writings down to the present day. As regards the editor Gerhardt, the disconcerting fact is that in 1863 he reproduces the $\backsim$ of 1710 in the form[3] $\sim$.

The Leibnizian symbol $\sim$ was early adopted by Christian von Wolf; in 1716 he gave $\sim$ for *Aehnlichkeit*,[4] and in 1717 he wrote "$= et$ $\sim$" for "equal and similar."[5] These publications of Wolf are the earliest in which the sign $\sim$ appears in print. In the eighteenth and early part of the nineteenth century, the Leibnizian symbols for "similar" and "congruent" were seldom used in Europe and not at all in England and America. In England $\sim$ or $\backsim$ usually expressed "difference," as defined by Oughtred. In the eighteenth century the signs for congruence occur much less frequently even than the signs

---

[1] Printed in *Leibnizens Math. Schriften* (ed. C. I. Gerhardt), Vol. V, p. 153.

[2] *Op. cit.*, p. 172.

[3] *Leibnizens Math. Schriften*, Vol. VII (1863), p. 222.

[4] Chr. Wolffen, *Math. Lexicon* (Leipzig, 1716), "Signa."

[5] Chr. V. Wolff, *Elementa Matheseos universalis* (Halle, 1717), Vol. I, § 236; see Tropfke, *op. cit.*, Vol. IV (2d ed., 1923), p. 20.

for similar. We have seen that Leibniz' signs for congruence did not use both lines occurring in the sign of equality =. Wolf was the first to use explicitly $\sim$ and = for congruence, but he did not combine the two into one symbolism. That combination appears in texts of the latter part of the eighteenth century. While the $\cong$ was more involved, since it contained one more line than the Leibnizian $\simeq$, it had the advantage of conveying more specifically the idea of congruence as the superposition of the ideas expressed by $\sim$ and =. The sign $\sim$ for "similar" occurs in Camus' geometry,[1] $\frown$ for "similar" in A. R. Mauduit's conic sections[2] and in Karsten,[3] $\sim$ in Blassière's geometry,[4] $\cong$ for congruence in Häseler's[5] and Reinhold's geometries,[6] $\frown$ for similar in Diderot's *Encyclopédie*,[7] and in Lorenz' geometry.[8] In Klügel's *Wörterbuch*[9] one reads, "$\frown$ with English and French authors means difference"; "with German authors $\frown$ is the sign of similarity"; "Leibniz and Wolf have first used it." The signs $\sim$ and $\cong$ are used by Mollweide;[10] $\sim$ by Steiner[11] and Koppe;[12] $\frown$ is used by Prestel,[13] $\cong$ by Spitz;[14] $\sim$ and $\cong$ are found in Lorey's geometry,[15] Kambly's

[1] C. E. L. Camus, *Élémens de géométrie* (nouvelle éd.; Paris, 1755).

[2] A. R. Mauduit, *op. cit.* (The Hague, 1763), "Symbols."

[3] W. J. G. Karsten, *Lehrbegrif der gesamten Mathematik*, 1. Theil (1767), p. 348.

[4] J. D. Blassière, *Principes de géometrié élémentaire* (The Hague, 1787), p. 16.

[5] J. F. Häseler, *op. cit.* (Lemgo, 1777), p. 37.

[6] C. L. Reinhold, *Arithmetica Forensis*, 1. Theil (Ossnabrück, 1785), p. 361.

[7] Diderot *Encyclopédie ou Dictionnaire raisoné des sciences* (1781; 1st ed., 1754), art. "Caractere" by D'Alembert. See also the Italian translation of the mathematical part of Diderot's *Encyclopédie*, the *Dizionario enciclopedico delle matematiche* (Padova, 1800), "Carattere."

[8] J. F. Lorenz, *Grundriss der Arithmetik und Geometrie* (Helmstädt, 1798), p. 9.

[9] G. S. Klügel, *Mathematisches Wörterbuch*, fortgesetzt von C. B. Mollweide, J. A. Grunert, 5. Theil (Leipzig, 1831), art. "Zeichen."

[10] Carl B. Mollweide, *Euklid's Elemente* (Halle, 1824).

[11] Jacob Steiner, *Geometrische Constructionen* (1833); *Ostwald's Klassiker*, No. 60, p. 6.

[12] Karl Koppe, *Planimetrie* (Essen, 1852), p. 27.

[13] M. A. F. Prestel, *Tabelarischer Grundriss der Experimental-physik* (Emden, 1856), No. 7.

[14] Carl Spitz, *Lehrbuch der ebenen Geometrie* (Leipzig und Heidelberg, 1862), p. 41.

[15] Adolf Lorey, *Lehrbuch der ebenen Geometrie* (Gera und Leipzig, 1868), p. 118.

Planimetrie,[1] and texts by Frischauf[2] and Max Simon.[3] Lorey's book contains also the sign ≅ a few times. Peano[4] uses ∽ for "similar" also in an arithmetical sense for classes. Perhaps the earliest use of ∼ and ≅ for "similar" and "congruent" in the United States are by G. A. Hill[5] and Halsted.[6] The sign ∽ for "similar" is adopted by Henrici and Treutlein,[7] ≅ by Fiedler,[8] ∼ by Fialkowski,[9] ∽ by Beman and Smith.[10] In the twentieth century the signs entered geometries in the United States with a rush: ≅ for "congruent" were used by Busch and Clarke;[11] ≅ by Meyers,[12] ≅ by Slaught and Lennes,[13] ∼ by Hart and Feldman;[14] ≅ by Shutts,[15] E. R. Smith,[16] Wells and Hart,[17] Long and Brenke;[18] ≅ by Auerbach and Walsh.[19]

That symbols often experience difficulty in crossing geographic or national boundaries is strikingly illustrated in the signs ∼ and ≅. The signs never acquired a foothold in Great Britain. To be sure, the symbol ∽ was adopted at one time by a member of the University

[1] L. Kambly, *Die Elementar-Mathematik*, 2. Theil, *Planimetrie*, 43. Aufl. (Breslau, 1876).

[2] J. Frischauf, *Absolute Geometrie* (Leipzig, 1876), p. 3.

[3] Max Simon, *Euclid* (1901), p. 45.

[4] G. Peano, *Formulaire de mathématiques* (Turin, 1894), p. 135.

[5] George A. Hill, *Geometry for Beginners* (Boston, 1880), p. 92, 177.

[6] George Bruce Halsted, *Mensuration* (Boston, 1881), p. 28, 83.

[7] J. Henrici und P. Treutlein, *Elementar-Geometrie* (Leipzig, 1882), p. 13, 40.

[8] W. Fiedler, *Darstellende Geometrie*, 1. Theil (Leipzig, 1883), p. 60.

[9] N. Fialkowski, *Praktische Geometrie* (Wien, 1892), p. 15.

[10] W. W. Beman and D. E. Smith, *Plane and Solid Geometry* (Boston, 1896), p. 20.

[11] W. N. Busch and John B. Clarke, *Elements of Geometry* (New York, 1905]).

[12] G. W. Meyers, *Second-Year Mathematics for Secondary Schools* (Chicago, 1910), p. 10.

[13] H. E. Slaught and N. J. Lennes, *Plane Geometry* (Boston, 1910).

[14] C. A. Hart and D. D. Feldman, *Plane Geometry* (New York, 1911), p. viii.

[15] G. C. Shutts, *Plane and Solid Geometry* [1912], p. 13.

[16] Eugene R. Smith, *Solid Geometry* (New York, 1913).

[17] W. Wells and W. W. Hart, *Plane and Solid Geometry* (Boston, [1915]), p. x.

[18] Edith Long and W. C. Brenke, *Plane Geometry* (New York, 1916), p. viii.

[19] Matilda Auerbach and Charles Burton Walsh, *Plane Geometry* (Philadelphia, [1920]), p. xi.

of Cambridge,[1] to express "is similar to" in an edition of Euclid. The book was set up in type, but later the sign was eliminated from all parts, except one. In a footnote the student is told that "in writing out the propositions in the Senate House, Cambridge, it will be advisable not to make use of this symbol, but merely to write the word short, thus, *is simil*." Moreover, in the Preface he is informed that "more competent judges than the editor" advised that the symbol be eliminated, and so it was, except in one or two instances where "it was too late to make the alteration," the sheets having already been printed. Of course, one reason for failure to adopt $\smile$ for "similar" in England lies in the fact that $\smile$ was used there for "difference."

373. When the sides of the triangle ABC and $A'B'C'$ are considered as being vectors, special symbols have been used by some authors to designate different kinds of similarity. Thus, Stolz and Gmeiner[2] employ $\stackrel{.}{\backsim}$ to mark that the similar triangles are uniformly similar (*einstimmig ähnlich*), that is, the equal angles of the two triangles are all measured clockwise, or all counter-clockwise; they employ $\stackrel{\sim}{\backsim}$ to mark that the two triangles are symmetrically similar, that is, of two numerically equal angles, one is measured clockwise and the other counter-clockwise.

The sign $\sim$ has been used also for "is [or are] measured by," by Alan Sanders;[3] the sign $\cong$ is used for "equals approximately," by Hudson and Lipka.[4] A. Pringsheim[5] uses the symbolism $a_\nu \cong ab_\nu$ to express that $\lim_{\nu=+\infty} \dfrac{a_\nu}{b_\nu} = a$.

374. The sign $\cong$ for congruence was not without rivals during the nineteenth century. Occasionally the sign $\equiv$, first introduced by Riemann[6] to express identity, or non-Gaussian arithmetical congruence of the type $(a+b)^2 = a^2 + 2ab + b^2$, is employed for the expression of geometrical congruence. One finds $\equiv$ for congruent in W. Bolyai,[7]

[1] *Elements of Euclid . . . . from the Text of Dr. Simson.* By a Member of the University of Cambridge (London, 1827), p. 104.

[2] O. Stolz und J. A. Gmeiner, *Theoretische Arithmetik* (Leipzig), Vol. II (1902), p. 332.

[3] Alan Sanders, *Plane and Solid Geometry* (New York, [1901]), p. 14.

[4] R. G. Hudson and J. Lipka, *Manual of Mathematics* (New York, 1917), p. 68.

[5] A. Pringsheim, *Mathematische Annalen*, Vol. XXXV (1890), p. 302; *Encyclopédie des scien. Math.*, Tom. I, Vol. I (1904), p. 201, 202.

[6] See L. Kronecker, *Vorlesungen über Zahlentheorie* (Leipzig, 1901), p. 86; G. F. B. Riemann, *Elliptische Funktionen* (Leipzig, 1899), p. 1, 6.

[7] W. Bolyai, *Tentamen* (2d ed.), Tom. I (Budapest, 1897), p. xi.

H. G. Grassmann,[1] Dupuis,[2] Budden,[3] Veronese,[4] Casey,[5] Halsted,[6] Baker,[7] Betz and Webb,[8] Young and Schwarz,[9] McDougall.[10] This sign ≡ for congruence finds its widest adoption in Great Britain at the present time. Jordan[11] employs it in analysis to express equivalence.

The idea of expressing similarity by the letter S placed in a horizontal position is extended by Callet, who uses ∽, ◖, ◖̲, to express "similar," "dissimilar," "similar or dissimilar."[12] Callet's notation for "dissimilar" did not meet with general adoption even in his own country.

The sign ≡ has also other uses in geometry. It is used in the Riemannian sense of "identical to," not "congruent," by Busch and Clarke,[13] Meyers,[14] E. R. Smith,[15] Wells and Hart.[16] The sign ≡ or ╳ is made to express "equivalent to" in the *Geometry* of Hopkins.[17]

The symbols ∼ and ⌣ for "similar" have encountered some competition with certain other symbols. Thus "similar" is marked ‖‖ in the geometries of Budden[18] and McDougall.

The relation "coincides with," which Leibniz had marked with |≃|, is expressed by ⧧ in White's *Geometry*.[19] Cremona[20] denotes by

[1] H. G. Grassmann in *Crelle's Journal*, Vol. XLII (1851), p. 193–203.

[2] N. F. Dupuis, *Elementary Synthetic Geometry* (London, 1899), p. 29.

[3] E. Budden, *Elementary Pure Geometry* (London, 1904), p. 22.

[4] Guiseppe Veronese, *Elementi di Geometria*, Part I (3d ed.; Verona, 1904), p. 11.

[5] J. Casey, *First Six Books of Euclid's Elements* (7th ed.; Dublin, 1902).

[6] G. B. Halsted, *Rational Geometry* (New York, 1904), p. vii.

[7] Alfred Baker, *Transactions of the Royal Society of Canada* (2d ser., 1906–7), Vol. XII, Sec. III, p. 120.

[8] W. Betz and H. E. Webb, *Plane Geometry* (Boston, [1912]), p. 71.

[9] John W. Young and A. J. Schwartz, *Plane Geometry* (New York, [1905]).

[10] A. H. McDougall, *The Ontario High School Geometry* (Toronto, 1914), p. 158.

[11] Camille Jordan, *Cours d'analyse*, Vol. II (1894), p. 614.

[12] François Callet, *Tables portatives de logarithmes* (Paris, 1795), p. 79. Taken from Désiré André, *Notations mathématiques* (Paris, 1909), p. 150.

[13] W. N. Busch and John B. Clarke, *Elements of Geometry* (New York, [1905]).

[14] G. W. Meyers, *Second-Year Mathematics for Secondary Schools* (Chicago, 1910), p. 119.

[15] Eugene R. Smith, *Solid Geometry* [1913].

[16] W. Wells and W. W. Hart, *Plane and Solid Geometry* (Boston, [1905]), p. x.

[17] Irving Hopkins, *Manual of Plane Geometry* (Boston, 1891), p. 10.

[18] E. Budden, *Elementary Pure Geometry* (London, 1904), p. 22.

[19] Emerson E. White, *Elements of Geometry* (New York City, 1895).

[20] Luigi Cremona, *Projective Geometry* (trans. Ch. Leudesdorf; 2d ed.; Oxford, 1893), p. 1.

$a.BC \equiv A'$ that the point common to the plane a and the straight line BC coincides with the point A'. Similarly, a German writer[1] of 1851 indicates by $a \equiv b$, $A \equiv B$ that the two points a and b or the two straights A and B coincide (*zusammenfallen*).

375. *The sign* $\Leftrightarrow$ *for equivalence.*—In many geometries congruent figures are marked by the ordinary sign of equality, $=$. To distinguish between congruence of figures, expressed by $=$, and mere equivalence of figures or equality of areas, a new symbol $\Leftrightarrow$ came to be used for "equivalent to" in the United States. The earliest appearance of that sign known to us is in a geometry brought out by Charles Davies[2] in 1851. He says that the sign "denotes equivalency and is read *is equivalent to*." The curved parts in the symbol, as used by Davies, are not semicircles, but semiellipses. The sign is given by Davies and Peck,[3] Benson,[4] Wells,[5] Wentworth,[6] McDonald,[7] Macnie,[8] Phillips and Fisher,[9] Milne,[10] McMahon,[11] Durell,[12] Hart and Feldman.[13] It occurs also in the trigonometry of Anderegg and Roe.[14] The signs $\Leftrightarrow$ and $=$ for equivalence and equality (i.e. congruence) are now giving way in the United States to $=$ and $\cong$ or $\simeq$.

We have not seen this symbol for equivalence in any European book. A symbol for equivalence, $\triangleq$, was employed by John Bolyai[15] in cases like $AB \triangleq CD$, which meant $\angle CAB = \angle ACD$. That the line BN is parallel and equal to CP he indicated by the sign "$BN || \triangleq CP$."

[1] *Crelle's Journal*, Vol. XLII (1851), p. 193–203.

[2] Charles Davies, *Elements of Geometry and Trigonometry from the Works of A. M. Legendre* (New York, 1851), p. 87.

[3] Charles Davies and W. G. Peck, *Mathematical Dictionary* (New York, 1856), art. "Equivalent."

[4] Lawrence S. Benson, *Geometry* (New York, 1867), p. 14.

[5] Webster Wells, *Elements of Geometry* (Boston, 1886), p. 4.

[6] G. A. Wentworth, *Text-Book of Geometry* (2d ed.; Boston, 1894; Preface, 1888), p. 16. The first edition did *not* use this symbol.

[7] J. W. Macdonald, *Principles of Plane Geometry* (Boston, 1894), p. 6.

[8] John Macnie, *Elements of Geometry* (ed. E. E. White; New York, 1895), p. 10.

[9] A. W. Phillips and Irving Fisher, *Elements of Geometry* (New York, 1896), p. 1.

[10] William J. Milne, *Plane and Solid Geometry* (New York, [1899]), p. 20.

[11] James McMahon, *Elementary Geometry (Plane)* (New York, [1903]), p. 139.

[12] Fletcher Durrell, *Plane and Solid Geometry* (New York, 1908), p. 8.

[13] C. A. Hart and D. D. Feldman, *Plane Geometry* (New York, [1911]), p. viii.

[14] F. Anderegg and E. D. Roe, *Trigonometry* (Boston, 1896), p. 3.

[15] W. Bolyai, *Tentamen* (2d ed.), Vol. II, Appendix by John Bolyai, list of symbols. See also G. B. Halsted's translation of that Appendix (1896).

376. *Lettering of geometric figures.*—Geometric figures are found in the old Egyptian mathematical treatise, the Ahmes papyrus (1550 B.C. or older), but they are not marked by signs other than numerals to indicate the dimensions of lines.

The designation of points, lines, and planes by a letter or by letters was in vogue among the Greeks and has been traced back[1] to Hippocrates of Chios (about 440 B.C.).

The Greek custom of lettering geometric figures did not find imitation in India, where numbers indicating size were written along the sides. However, the Greek practice was adopted by the Arabs, later still by Regiomontanus and other Europeans.[2] Gerbert[3] and his pupils sometimes lettered their figures and at other times attached Roman numerals to mark lengths and areas. The Greeks, as well as the Arabs, Leonardo of Pisa, and Regiomontanus usually observed the sequence of letters a, b, g, d, e, z, etc., omitting the letters c and f. We have here the Greek-Arabic succession of letters of the alphabet, instead of the Latin succession. Referring to Leonardo of Pisa's *Practica geometriae* (1220) in which Latin letters are used with geometric figures, Archibald says: "Further evidence that Leonardo's work was of Greek-Arabic extraction can be found in the fact that, in connection with the 113 figures, of the section *On Divisions*, of Leonardo's work, the lettering in only 58 contains the letters c or f; that is, the Greek-Arabic succession a b g d e z . . . . is used almost as frequently as the Latin a b c d e f g . . . . ; elimination of Latin letters added to a Greek succession in a figure, for the purpose of numerical examples (in which the work abounds), makes the balance equal."[4]

Occasionally one encounters books in which geometric figures are not lettered at all. Such a publication is Scheubel's edition of Euclid,[5] in which numerical values are sometimes written alongside of lines as in the Ahmes papyrus.

An oddity in the lettering of geometric figures is found in Ramus' use[6] of the vowels a, e, i, o, u, y and the employment of consonants only when more than six letters are needed in a drawing.

[1] M. Cantor, *op. cit.*, Vol. I (3d ed., 1907), p. 205.

[2] J. Tropfke, *op. cit.*, Vol. IV (2d ed., 1923), p. 14, 15.

[3] *Œuvres de Gerbert* (ed. A. Olleris; Paris, 1867), Figs. 1–100, following p. 475.

[4] R. C. Archibald, *Euclid's Book on Divisions of Figures* (Cambridge, 1915), p. 12.

[5] *Evclides Megarensis* . . . . *sex libri priores* . . . . authore Ioanne Schevbelio (Basel, [1550]).

[6] P. *Rami Scholarvm mathematicorvm libri vnus et triginta* (Basel, 1569).

In the designation of a group of points of equal rank or of the same property in a figure, resort was sometimes taken to the repetition of one and the same letter, as in the works of Gregory St. Vincent,[1] Blaise Pascal,[2] John Wallis,[3] and Johann Bernoulli.[4]

377. The next advancement was the introduction of indices attached to letters, which proved to be an important aid. An apparently unconscious use of indices is found in Simon Stevin,[5] who occasionally uses dotted letters $\dot{B}$, $\ddot{B}$ to indicate points of equal significance obtained in the construction of triangles. In a German translation[6] of Stevin made in 1628, the dots are placed beneath the letter B, $\underset{\cdot\cdot}{B}$. Similarly, Fr. van Schooten[7] in 1649 uses designations for points:

$$C, 2C, 3C;\ S, 2S, 3S;\ T, 2T, 3T;\ V, 2V, 3V\ .$$

This procedure is followed by Leibniz in a letter to Oldenburg[8] of August 27, 1676, in which he marks points in a geometric figure by $_1B, _2B, _3B, _1D, _2D, _3D$. The numerals are here much smaller than the letters, but are placed on the same level with the letters (see also § 549). This same notation is used by Leibniz in other essays[9] and again in a treatise of 1677 where he lets a figure move so that in its new position the points are marked with double indices like $1\textcircled{1}$ and $1 \cdot \overline{D}$. In 1679 he introduced a slight innovation by marking the points of the principal curve $3b, 6b, 9b\ .\ .\ .\ .$, generally yb, the curves of the entire curve $\overrightarrow{yb}$. The point $3b$ when moved yields the points $1\ 3b,\ 2\ 3b,\ 3\ 3b;$ the surface generated by $\overrightarrow{yb}$ is marked $\overrightarrow{zyb}$. Leibniz used indices also in his determinant notations (Vol. II, § 547).

---

[1] Gregory St. Vincent, *Opus geometricum* (Antwerp, 1647), p. 27, etc. See also Karl Bopp, "Die Kegelschnitte des Gregorius a St. Vincentio" in *Abhandlungen zur Gesch. d. math. Wissensch.*, Vol. XX (1907), p. 131, 132, etc.

[2] Blaise Pascal, "Lettre de Dettonville a Carcavi," *Œuvres complètes*, Vol. III (Paris, 1866), p. 364–85; *Œuvres* (ed. Faugere; Paris, 1882), Vol. III, p. 270–446.

[3] John Wallis, *Operum mathematicorum pars altera* (Oxford, 1656), p. 16–160.

[4] Johann Bernoulli, *Acta eruditorum* (1697), Table IV; *Opera omnia* (1742), Vol. I, p. 192.

[5] S. Stevin, *Œuvres* (éd. A. Girard; Leyden, 1634), Part II, "Cosmographie," p. 15.

[6] See J. Tropfke, *op. cit.*, Vol. II (2d ed., 1921), p. 46.

[7] F. van Schooten, *Geometria à Renato des Cartes* (1649), p. 112.

[8] J. Collins, *Commercium epistolicum* (ed. J. B. Biot and F. Lefort, 1856), p. 113.

[9] *Leibniz Mathematische Schriften*, Vol. V (1858), p. 99–113. See D. Mahnke in *Bibliotheca mathematica* (3d ser.), Vol. XIII (1912–13), p. 250.

I. Newton used dots and strokes for marking fluxions and fluents (§§ 567, 622). As will be seen, indices of various types occur repeatedly in specialized notations of later date. For example, L. Euler[1] used in 1748

$$x' \quad x'' \quad x'''$$
$$y' \quad y'' \quad y'''$$

as co-ordinates of points of equal significance. Cotes[2] used such strokes in marking successive arithmetical differences. Monge[3] employed strokes, $K^{\backprime}$, $K^{\backprime\backprime}$, $K^{\backprime\backprime\backprime}$, and also $\backprime K'$, $\backprime\backprime K'$.

378. The introduction of different kinds of type received increased attention in the nineteenth century. Wolfgang Bolyai[4] used Latin and Greek letters to signify quantities, and German letters to signify points and lines. Thus, $\widetilde{\mathfrak{ab}}$ signifies a line $\mathfrak{ab}$ infinite on both sides; $\mathfrak{a}\widetilde{\mathfrak{b}}$ a line starting at the point $\mathfrak{a}$ and infinite on the side $\mathfrak{b}$; $\widetilde{\mathfrak{a}}\mathfrak{b}$ a line starting at $\mathfrak{b}$ and infinite on the side $\mathfrak{a}$; $\widetilde{\mathfrak{P}}$ a plane P extending to infinity in all directions.

379. A remarkable symbolism, made up of capital letters, lines, and dots, was devised by L. N. M. Carnot.[5] With him,

$A, B, C, \ldots .$ marked points

$\overline{AB}$, $\widehat{AB}$ marked the segment AB and the circular arc AB

$\overline{BCD}$ marked that the points B, C, D are collinear, C being placed between B and D

$\overline{AB} \cdot \overline{CD}$ is the point of intersection of the indefinite lines AB, CD

$\widehat{ABCD}$ marked four points on a circular arc, in the order indicated

$\widehat{AB} \cdot \widehat{CD}$ is the point of intersection of the two arcs AB and CD

$F \, \overline{AB} \cdot \overline{CD}$ is the straight line which passes through the points F and $\overline{AB} \cdot \overline{CD}$

---

[1] L. Euler in *Histoire de l'Academie r. d. sciences et d. belles lettres*, année **1748** (Berlin, 1750), p. 175.

[2] Roger Cotes, *Harmonia mensurarum* (Cambridge, 1722), "Aestimatio errorum," p. 25.

[3] G. Monge, *Miscellanea Taurinensia* (1770/73). See H. Wieleitner, *Geschichte der Mathematik*, II. Teil, II. Hälfte (1921), p. 51.

[4] Wolfgangi Bolyai de Bolya, *Tentamen* (2d ed.), Tom. I (Budapestini, 1897), p. xi.

[5] L. N. M. Carnot, *De la Corrélation des figures de géométrie* (Paris, an IX = 1801), p. 40–43.

$= \mid =$	signifies equipollence, or identity of two objects
$\overset{\wedge}{ABC}$	marks the angle formed by the straight lines, AB, BC, B being the vertex
$\overline{AB\ \overset{\wedge}{CD}}$	is the angle formed by the two lines AB and CD
$\triangle ABC$	the triangle having the vertices A, B, C
$\varDelta\ ABC$	is a right triangle
$\overline{\overline{ABC}}$	is the area of the triangle ABC

A criticism passed upon Carnot's notation is that it loses its clearness in complicated constructions.

Reye[1] in 1866 proposed the plan of using capital letters, A, B, C, for points; the small letters a, b, c, ... , for lines; a, β, γ, ... , for planes. This notation has been adopted by Favaro and others.[2] Besides, Favaro adopts the signs suggested by H. G. Grassmann,[3] AB for a straight line terminating in the points A and B, Aa the plane passing through A and a, aa the point common to a and a; ABC the plane passing through the points A, B, C; $a\beta\gamma$ the point common to the planes a, β, γ, and so on. This notation is adopted also by Cremona,[4] and some other writers.

The National Committee on Mathematical Requirements (1923) recommends (*Report*, p. 78) the following practice in the lettering of geometric figures: "Capitals represent the vertices, corresponding small letters represent opposite sides, corresponding small Greek letters represent angles, and the primed letters represent the corresponding parts of a congruent or similar triangle. This permits speaking of a (alpha) instead of 'angle A' and of 'small a' instead of BC."

380. *Sign for spherical excess.*—John Caswell writes the spherical excess $c = A + B + C - 180°$ thus: "$E = \angle \angle \angle - 2\lrcorner$." Letting π stand for the periphery of a great circle, G for the surface of the sphere, R for the radius of the sphere, he writes the area $\triangle$ of a spherical triangle thus:[5]

$$2\pi\triangle = EG = 2R\pi E,$$
$$\triangle = RE.$$

[1] Reye, *Geometrie der Lage* (Hannover, 1866), p. 7.

[2] Antonio Favaro, *Leçons de Statique graphique* (trad. par Paul Terrier), 1. Partie (Paris, 1879), p. 2.

[3] H. Grassmann, *Ausdehnungslehre* (Leipzig, Berlin, 1862).

[4] Luigi Cremona, *Projective Geometry* (trans. Charles Leudesdorf; Oxford, 1885), chap. i.

[5] John Wallis, *Treatise of Algebra* (London, 1685), Appendix on "Trigonometry" by John Caswell, p. 15.

The letter E for spherical excess has retained its place in some books[1] to the present time. Legendre,[2] in his *Éléments de géométrie* (1794, and in later editions), represents the spherical excess by the letter S. In a German translation of this work, Crelle[3] used for this excess the sign ε. Chauvenet[4] used the letter K in his *Trigonometry*.

381. *Symbols in the statement of theorems.*—The use of symbols in the statement of geometric theorems is seldom found in print, but is sometimes resorted to in handwriting and in school exercises. It occurs, however, in William Jones's *Synopsis palmariorum*, a book which compresses much in very small space. There one finds, for instance, "An $\angle$ in a Segment $>$, $=$, $<$ Semicircle is Acute, Right, Obtuse."[5]

To Julius Worpitzky (1835–95), professor at the Friedrich Werder Gymnasium in Berlin, is due the symbolism $S.S.S.$ to recall that two triangles are congruent if their three sides are equal, respectively; and the abbreviations $S.W.S.$, $W.S.W.$ for the other congruence theorems.[6] Occasionally such abbreviations have been used in America, the letter a ("angle") taking the place of the letter W (*Winkel*), so that *asa* and *sas* are the abbreviations sometimes used. The National Committee on Mathematical Requirements, in its *Report* of 1923, page 79, discourages the use of these abbreviations.

382. *Signs for incommensurables.*—We have seen (§§ 183, 184) that Oughtred had a full set of ideographs for the symbolic representation of Euclid's tenth book on incommensurables. A different set of signs was employed by J. F. Lorenz[7] in his edition of Euclid's *Elements;* he used the Latin letter C turned over, as in $A \supset B$, to indicate that A and B are commensurable; while $A \cup B$ signified that A and B are incommensurable; $A \frown B$ signified that the lines A and B are commensurable only in power, i.e., A^2 and B^2 are commensurable, while A and B were not; $A \smile B$, that the lines are incommensurable even in power, i.e., A and B are incommensurable, so are A^2 and B^2.

[1] W. Chauvenet, *Elementary Geometry* (Philadelphia, 1872), p. 264; A. W. Phillips and I. Fisher, *Elements of Geometry* (New York, [1896]), p. 404.

[2] A. M. Legendre, *Éléments de géométrie* (Paris, 1794), p. 319, n. xi.

[3] A. L. Crelle's translation of Legendre's *Géométrie* (Berlin, 1822; 2d ed., 1833). Taken from J. Tropfke, *op. cit.*, Vol. V (1923), p. 160.

[4] William Chauvenet, *Treatise on Plane and Spherical Trigonometry* (Philadelphia, 1884), p. 229.

[5] William Jones, *Synopsis palmariorum matheseos* (London, 1706), p. 231.

[6] J. Tropfke, *op. cit.*, Vol. IV (2d ed., 1923), p. 18.

[7] Johann Friederich Lorenz, *Euklid's Elemente* (ed. C. B. Mollweide; Halle, 1824), p. xxxii, 194.

383. *Unusual ideographs in elementary geometry.*—For "is measured by" there is found in Hart and Feldman's *Geometry*[1] and in that of Auerbach and Walsh[2] the sign $\propto$, in Shutt's *Geometry*[3] the sign $\mathrel{\perp}$. Veronese[4] employs $\equiv|\equiv$ to mark "not equal" line segments.

A horizontal line drawn underneath an equation is used by Kambly[5] to indicate *folglich* or "therefore"; thus:

$$\angle r + q = 2R$$
$$\angle s + q = 2R$$

$$\angle r + q = s + q$$

$$\angle r = s$$

384. *Algebraic symbols in elementary geometry.*—The use of algebraic symbols in the solution of geometric problems began at the very time when the symbols themselves were introduced. In fact, it was very largely geometrical problems which for their solution created a need of algebraic symbols. The use of algebraic symbolism in applied geometry is seen in the writings of Pacioli, Tartaglia, Cardan, Bombelli, Widman, Rudolff, Stifel, Stevin, Vieta, and writers since the sixteenth century.

It is noteworthy that printed works which contained pictographs had also algebraic symbols, but the converse was not always true. Thus, Barrow's *Euclid* contained algebraic symbols in superabundance, but no pictographs.

The case was different in works containing a systematic development of geometric theory. The geometric works of Euclid, Archimedes, and Apollonius of Perga did not employ algebraic symbolism; they were purely rhetorical in the form of exposition. Not until the seventeenth century, in the writings of Hérigone in France, and Oughtred, Wallis, and Barrow in England, was there a formal translation of the geometric classics of antiquity into the language of syncopated or symbolic algebra. There were those who deplored this procedure; we proceed to outline the struggle between symbolists and rhetoricians.

[1] C. A. Hart and Daniel D. Feldman, *Plane Geometry* (New York, [1911]), p. viii.

[2] M. Auerbach and C. B. Walsh, *Plane Geometry* (Philadelphia, [1920]), p. xi.

[3] George C. Shutt, *Plane and Solid Geometry* [1912], p. 13.

[4] Giuseppe Veronese, *Elementi di geometria*, Part I (3d ed., Verona), p. 12.

[5] Ludwig Kambly, *Die Elementar-Mathematik*, 2. Theil: *Planimetrie* (Breslau, 1876), p. 8, 1. Theil: *Arithmetik und Algebra*, 38. Aufl. (Breslau, 1906), p. 7.

PAST STRUGGLES BETWEEN SYMBOLISTS AND RHETORI-
CIANS IN ELEMENTARY GEOMETRY

385. For many centuries there has been a conflict between individual judgments, on the use of mathematical symbols. On the one
side are those who, in geometry for instance, would employ hardly
any mathematical symbols; on the other side are those who insist on
the use of ideographs and pictographs almost to the exclusion of
ordinary writing. The real merits or defects of the two extreme views
cannot be ascertained by a priori argument; they rest upon experience
and must therefore be sought in the study of the history of our science.

The first printed edition of Euclid's *Elements* and the earliest
translations of Arabic algebras into Latin contained little or no mathematical symbolism.[1] During the Renaissance the need of symbolism
disclosed itself more strongly in algebra than in geometry. During the
sixteenth century European algebra developed symbolisms for the
writing of equations, but the arguments and explanations of the
various steps in a solution were written in the ordinary form of verbal
expression.

The seventeenth century witnessed new departures; the symbolic
language of mathematics displaced verbal writing to a much greater
extent than formerly. The movement is exhibited in the writings of
three men: Pierre Hérigone[2] in France, William Oughtred[3] in England, and J. H. Rahn[4] in Switzerland. Hérigone used in his *Cursus
mathematicus* of 1634 a large array of new symbols of his own design.
He says in his Preface: "I have invented a new method of making
demonstrations, brief and intelligible, without the use of any lan-

[1] Erhard Ratdolt's print of *Campanus' Euclid* (Venice, 1482). Al-Khowârizmî's algebra was translated into Latin by Gerard of Cremona in the twelfth century. It was probably this translation that was printed in Libri's *Histoire des sciences mathématique en Italie*, Vol. I (Paris, 1838), p. 253–97. Another translation
into Latin, made by Robert of Chester, was edited by L. C. Karpinski (New York,
1915). Regarding Latin translations of Al-Khowârizmî, see also G. Eneström,
Bibliotheca mathematica (3d ser.), Vol. V (1904), p. 404; A. A. Björnbo, *ibid.* (3d
ser.), Vol. VII (1905), p. 239–48; Karpinski, *Bibliotheca mathematica* (3d ser.),
Vol. XI, p. 125.

[2] Pierre Herigone, *op. cit.*, Vol. I–VI (Paris, 1634; 2d ed., 1644).

[3] William Oughtred, *Clavis mathematicae* (London, 1631, and later editions);
also Oughtred's *Circles of Proportion* (1632), *Trigonometrie* (1657), and minor
works.

[4] J. H. Rahn, *Teutsche Algebra* (Zürich, 1659), Thomas Brancker, *An Introduction to Algebra* (trans. out of the High-Dutch; London, 1668).

guage." In England, William Oughtred used over one hundred and fifty mathematical symbols, many of his own invention. In geometry Oughtred showed an even greater tendency to introduce extensive symbolisms than did Hérigone. Oughtred translated the tenth book of Euclid's *Elements* into language largely ideographic, using for the purpose about forty new symbols.[1] Some of his readers complained of the excessive brevity and compactness of the exposition, but Oughtred never relented. He found in John Wallis an enthusiastic disciple. At the time of Wallis, representatives of the two schools of mathematical exposition came into open conflict. In treating the "Conic Sections"[2] no one before Wallis had employed such an amount of symbolism. The philosopher Thomas Hobbes protests emphatically: "And for . . . . your Conic Sections, it is so covered over with the scab of symbols, that I had not the patience to examine whether it be well or ill demonstrated."[3] Again Hobbes says: "Symbols are poor unhandsome, though necessary scaffolds of demonstration";[4] he explains further: "Symbols, though they shorten the writing, yet they do not make the reader understand it sooner than if it were written in words. For the conception of the lines and figures . . . . must proceed from words either spoken or thought upon. So that there is a double labour of the mind, one to reduce your symbols to words, which are also symbols, another to attend to the ideas which they signify. Besides, if you but consider how none of the ancients ever used any of them in their published demonstrations of geometry, nor in their books of arithmetic . . . . you will not, I think, for the future be so much in love with them. . . . ."[5] Whether there is really a double translation, such as Hobbes claims, and also a double labor of interpretation, is a matter to be determined by experience.

386. Meanwhile the *Algebra* of Rahn appeared in 1659 in Zurich and was translated by Brancker into English and published with additions by John Pell, at London, in 1668. The work contained some new symbols and also Pell's division of the page into three columns. He marked the successive steps in the solution so that all steps in the process are made evident through the aid of symbols, hardly a word

[1] Printed in Oughtred's *Clavis mathematicae* (3d ed., 1648, and in the editions of 1652, 1667, 1693). See our §§ 183, 184, 185.

[2] John Wallis, *Operum mathematicorum*, Pars altera (Oxford), *De sectionibus conicis* (1655).

[3] Sir William Molesworth, *The English Works of Thomas Hobbes*, Vol. VII (London, 1845), p. 316.

[4] *Ibid.*, p. 248.   [5] *Ibid.*, p. 329.

of verbal explanation being necessary. In Switzerland the three-column arrangement of the page did not receive enthusiastic reception. In Great Britain it was adopted in a few texts: John Ward's *Young Mathematician's Guide*, parts of John Wallis' *Treatise of Algebra*, and John Kirkby's *Arithmetical Institutions*. But this almost complete repression of verbal explanation did not become widely and permanently popular. In the great mathematical works of the seventeenth century—the *Géométrie* of Descartes; the writings of Pascal, Fermat, Leibniz; the *Principia* of Sir Isaac Newton—symbolism was used in moderation. The struggles in elementary geometry were more intense. The notations of Oughtred also met with a most friendly reception from Isaac Barrow, the great teacher of Sir Isaac Newton, who followed Oughtred even more closely than did Wallis. In 1655, Barrow brought out an edition of Euclid in Latin and in 1660 an English edition. He had in mind two main objects: first, to reduce the whole of the *Elements* into a portable volume and, second, to gratify those readers who prefer "symbolical" to "verbal reasoning." During the next half-century Barrow's texts were tried out. In 1713, John Keill of Oxford edited the *Elements* of Euclid, in the Preface of which he criticized Barrow, saying: "Barrow's Demonstrations are so very short, and are involved in so many notes and symbols, that they are rendered obscure and difficult to one not versed in Geometry. There, many propositions, which appear conspicuous in reading Euclid himself, are made knotty, and scarcely intelligible to learners, by his Algebraical way of demonstration. . . . . The *Elements* of all Sciences ought to be handled after the most simple Method, and not to be involved in Symbols, Notes, or obscure Principles, taken elsewhere." Keill abstains altogether from the use of symbols. His exposition is quite rhetorical.

William Whiston, who was Newton's successor in the Lucasian professorship at Cambridge, brought out a school *Euclid*, an edition of Tacquet's *Euclid* which contains only a limited amount of symbolism. A more liberal amount of sign language is found in the geometry of William Emerson.

Robert Simson's edition of Euclid appeared in 1756. It was a carefully edited book and attained a wide reputation. Ambitious to present Euclid unmodified, he was careful to avoid all mathematical signs. The sight of this book would have delighted Hobbes. No scab of symbols here!

That a reaction to Simson's *Euclid* would follow was easy to see. In 1795 John Playfair, of Edinburgh, brought out a school edition of Euclid which contains a limited number of symbols. It passed through many editions in Great Britain and America. D. Cresswell, of Cambridge, England, expressed himself as follows: "In the demonstrations of the propositions recourse has been made to symbols. But these symbols are merely the representatives of certain words and phrases, which may be substituted for them at pleasure, so as to render the language employed strictly comfortable to that of ancient Geometry. The consequent diminution of the bulk of the whole book is the least advantage which results from this use of symbols. For the demonstrations themselves are sooner read and more easily comprehended by means of these useful abbreviations; which will, in a short time, become familiar to the reader, if he is not beforehand perfectly well acquainted with them."[1] About the same time, Wright[2] made free use of symbols and declared: "Those who object to the introduction of Symbols in Geometry are requested to inspect Barrow's *Euclid*, Emerson's *Geometry*, etc., where they will discover many more than are here made use of." "The difficulty," says Babbage,[3] "which many students experience in understanding the propositions relating to ratios as delivered in the fifth book of Euclid, arises entirely from this cause [tedious description] and the facility of comprehending their algebraic demonstrations forms a striking contrast with the prolixity of the geometrical proofs."

In 1831 R. Blakelock, of Cambridge, edited Simson's text in the symbolical form. Oliver Byrne's *Euclid* in symbols and colored diagrams was not taken seriously, but was regarded a curiosity.[4] The Senate House examinations discouraged the use of symbols. Later De Morgan wrote: "Those who introduce algebraical symbols into elementary geometry, destroy the peculiar character of the latter to

---

[1] *A Supplement to the Elements of Euclid*, Second Edition . . . . by D. Cresswell, formerly Fellow of Trinity College (Cambridge, 1825), Preface. Cresswell uses algebraic symbols and pictographs.

[2] J. M. F. Wright, *Self-Examination in Euclid* (Cambridge, 1829), p. x.

[3] Charles Babbage, "On the Influence of Signs in Mathematical Reasoning," *Transactions Cambridge Philos. Society*, Vol. II (1827), p. 330.

[4] Oliver Byrne, *The Elements of Euclid in which coloured diagrams and symbols are used* (London, 1847). J. Tropfke, *op. cit.*, Vol. IV (1923), p. 29, refers to a German edition of Euclid by Heinrich Hoffmann, *Teutscher Euclides* (Jena, 1653), as using color. The device of using color in geometry goes back to Heron (*Opera*, Vol. IV [ed. J. L. Heiberg; Leipzig, 1912], p. 20) who says: "And as a surface one can imagine every shadow and every color, for which reason the Pythagoreans called surfaces 'colors.'" Martianus Capella (*De nuptiis* [ed. Kopp, 1836], No. 708) speaks of surfaces as being "ut est color in corpore."

every student who has any mechanical associations connected with those symbols; that is, to every student who has previously used them in ordinary algebra. Geometrical reasons, and arithmetical process, have each its own office; to mix the two in elementary instruction, is injurious to the proper acquisition of both."[1]

The same idea is embodied in Todhunter's edition of Euclid which does not contain even a plus or minus sign, nor a symbolism for proportion.

The viewpoint of the opposition is expressed by a writer in the *London Quarterly Journal* of 1864: "The amount of relief which has been obtained by the simple expedient of applying to the elements of geometry algebraic notation can be told only by those who remember to have painfully pored over the old editions of Simson's *Euclid*. The practical effect of this is to make a complicated train of reasoning at once intelligible to the eye, though the mind could not take it in without effort."

English geometries of the latter part of the nineteenth century and of the present time contain a moderate amount of symbolism. The extremes as represented by Oughtred and Barrow, on the one hand, and by Robert Simson, on the other, are avoided. Thus a conflict in England lasting two hundred and fifty years has ended as a draw. It is a stupendous object-lesson to mathematicians on mathematical symbolism. It is the victory of the golden mean.

387. The movements on the Continent were along the same lines, but were less spectacular than in England. In France, about a century after Hérigone, Clairaut[2] used in his geometry no algebraic signs and no pictographs. Bézout[3] and Legendre[4] employed only a moderate amount of algebraic signs. In Germany, Karsten[5] and Segner[6] made only moderate use of symbols in geometry, but Reyher[7] and Lorenz[8]

[1] A. de Morgan, *Trigonometry and Double Algebra* (1849), p. 92 n.

[2] A. C. Clairaut, *Élémens de géométrie* (Paris, 1753; 1st ed., 1741).

[3] E. Bézout, *Cours de Mathématiques*, Tom. I (Paris: n. éd., 1797), *Élémens de géométrie*.

[4] A. M. Legendre, *Éléments de Géométrie* (Paris, 1794).

[5] W. J. G. Karsten, *Lehrbegrif der gesamten Mathematik*, I. Theil (Greifswald, 1767), p. 205–484.

[6] I. A. de Segner, *Cursus mathematici Pars I: Elementa arithmeticae, geometriae et calculi geometrici* (editio nova; Halle, 1767).

[7] *Samuel Reyhers . . . . Euclides* (Kiel, 1698).

[8] J. F. Lorenz, *Euklid's Elemente*, auf's neue herausgegeben von C. B. Mollweide (5th ed., Halle, 1824; 1st ed., 1781; 2d ed., 1798).

used extensive notations; Lorenz brought out a very compact edition of all books of Euclid's *Elements*.

Our data for the eighteenth and nineteenth centuries have been drawn mainly from the field of elementary mathematics. A glance at the higher mathematics indicates that the great mathematicians of the eighteenth century, Euler, Lagrange, Laplace, used symbolism freely, but expressed much of their reasoning in ordinary language. In the nineteenth century, one finds in the field of logic all gradations from no symbolism to nothing but symbolism. The well-known opposition of Steiner to Plücker touches the question of sign language.

The experience of the past certainly points to conservatism in the use of symbols in elementary instruction. In our second volume we indicate more fully that the same conclusion applies to higher fields. Individual workers who in elementary fields proposed to express practically everything in ideographic form have been overruled. It is a question to be settled not by any one individual, but by large groups or by representatives of large groups. The problem requires a consensus of opinion, the wisdom of many minds. That widsom discloses itself in the history of the science. The judgment of the past calls for moderation.

The conclusion reached here may be stated in terms of two schoolboy definitions for salt. One definition is, "Salt is what, if you spill a cupful into the soup, spoils the soup." The other definition is, "Salt is what spoils your soup when you don't have any in it."

ALPHABETICAL INDEX

(Numbers refer to paragraphs)

Abacus, 39, 75, 119

Abu Kamil, 273; unknown quantity, 339

Acta eruditorum, extracts from, 197

Adam, Charles, 217, 254, 300, 344

Adams, D., 219, 286, 287

Addition, signs for: general survey of, 200–216; Ahmes papyrus, 200; Al-Qalasâdî, 124; Bakhshālī MS, 109; Diophantus, 102; Greek papyri, 200; Hindus, 106; Leibniz, 198; *et* in Regiomontanus, 126

Additive principle in notation for powers, 116, 124, 295; in Pacioli, 135; in Gloriosus, 196

Additive principles: in Babylonia, 1; in Crete, 32; in Egypt, 19, 49; in Rome, 46, 49; in Mexico, 49; among Aztecs, 66

Adrain, R., 287

Aepinus, F. V. T., parentheses, 352

Aggregation of terms: general survey of, 342–56; by use of dots, 348; Oughtred, 181, 183, 186, 251; Romanus, 320; Rudolff, 148; Stifel, 148, 153; Wallis, 196. By use of comma, 189, 238; *communis* radix, 325; *Ra. col.* in Scheubel, 159; aggregation of terms, in radical expressions, 199, 319, 332, 334; redundancy of symbols, 335; signs used by Bombelli, 144, 145; Clavius, 161; Leibniz, 198, 354; Macfarlane, 275; Oughtred, 181, 183, 251; Pacioli. *See* Parentheses, Vinculum

Agnesi, M. G., 253, 257

Agrippa von Nettesheim, 97

Ahmes papyrus, 23, 260; addition and subtraction, 200; equality, 260; general drawings, 357, 376; unknown quantity, 339; fractions, 22, 23, 271, 274

Akhmim papyrus, 42

Aladern, J., 92

Alahdab, 118

Al-Battani, 82

Albert, Johann, 207

Alexander, Andreas, 325, 326; aggregation, 343, 344

Alexander, John, 245, 253, 254; equality, 264; use of star, 356

Algebraic symbols in geometry, 384

Algebras, Initius, 325

Al-Ḥaṣṣâr, 118, 235, 272; continued fractions, 118

Ali Aben Ragel, 96

Al-Kalsadi. *See* Al-Qalasâdî

Al-Karkhî, survey of his signs, 116, 339

Al-Khowârizmî, survey of his signs, 115; 271, 290, 385

Allaize, 249

Alligation, symbols for solving problems in, 133

Al-Madjrītī, 81

Alnasavi, 271

Alphabetic numerals, 28, 29, 30, 36, 38, 45, 46, 87; for fractions, 58, 59; in India, 76; in Rome, 60, 61

Al-Qalasâdî: survey of his signs, 124; 118, 200, 250; equality, 124, 260; unknown, 339

Alsted, J. H., 221, 225, 229, 305

Amicable numbers, 218, 230

Anatolius, 117

Anderegg, F., and E. D. Rowe: equivalence, 375

André, D., 95, 243, 285

Andrea, J. V., 263

Angle: general survey of, 360–63; sign for, in Hérigone, 189, 359; oblique angle, 363; right angle, 363; spherical angle, 363; solid angle, 363; equal angles, 363

Anianus, 127

Apian, P., 148, 222, 223, 224, 278

Apollonius of Perga, 384

Arabic numerals. *See* Hindu-Arabic numerals

Arabs, early, 45; Al-Khowârizmî, 81, 115, 271, 290, 385; Al-Qalasâdî, 118, 124, 200, 250, 260, 339; Al-Madjrītī, 81; Alnasavi, 271; Al-Karkhî, 116, 339; Ali Aben Ragel, 96

Arc of circle, 370
Archibald, R. C., 218, 270, 376
Archimedes, 41, 384
Ardüser, Johann, 208, 283; aggregation, 348; radical sign, 328, 332
Arithmetical progression, 248; arithmetical proportion, 249, 255
Arnauld, Antoine, 249; equality, 266
Āryabhaṭa, 76
Astronomical signs, relative position of planets, 358
Athelard of Bath, 81, 82
Attic signs, 33, 34, 35, 84
Auerbach, M., and C. B. Walsh: angle, 361; congruent in geometry, 372; is measured by, 383
Aurel, Marco, 165, 204, 327
Ayres, John, 225
Aztecs, 66

Babbage, Charles, 386; quoted, 386
Babylonians, 1–15; ideogram for multiplication, 217; ideogram for division, 235
Bachet, C. G., 101, 339
Baëza, L., 222
Bailey, M. A., 242
Baillet, J., 41
Bails, Benito, 249
Baker, Alfred, congruence in geometry, 374
Baker, Th.: parallel, 368; perpendicular, 364; use of star, 356
Bakhshālī MS: survey of signs, 109; 106, 200, 217, 235, 250, 260; equality, 260, 109; unknown quantity, 339
Balam, R., 186, 209, 245, 246, 283, 303; arithmetical proportion, 249
Balbontin, J. M., 239, 258
Ball, W. W. R., 96
Ballantine, J. P., 90
Bamberg arithmetic, 272
Barlaam, 40
Bar-le-Duc, I. Errard de. See Deidier, Dounot
Barlow, Peter, 225, 286
Barreme, N., 91
Barrow, Isaac: survey of his signs, 192; 216, 237, 371, 384, 386; aggregation, 345, 354; equality, 266; geometrical proportion, 251, 252; powers, 307
Bartholinus, E., 217
Bartjens, William, 52, 208

Barton, G. A., 9, 10, 12
Beaugrand, Jean de, 302
Beeckman, I., 340
Beguelin, Nic. de, 255
Belidor, Bernard Forest de, 206, 248, 255, 257
Bellavitis, G., 268
Beman, W. W., 244. See also Beman and Smith
Beman, W. W., and D. E., Smith: angle, 360; right angle, 363; parallelogram, 365; similar, 372
Benedetti, J. B., 219
Benson, L. S., equivalence, 375
Berlet, Bruno, 148, 326
Bernhard, Max, angle, 362
Bernoulli, Daniel (b. 1700), parentheses, 351, 352
Bernoulli, Jakob I (James), 210, 255; aggregation, 348, 354; equality, 264, 267; radical signs, 334; □ as operator, 366
Bernoulli, Johann I (John), 233, 255, 258, 309, 310, 341; aggregation, 344; lettering figures, 376; "mem," 365; use of star, 356; radical expressions, 308; □ as operator, 366
Bernoulli, Johann II (b. 1710), 258; parentheses, 351, 352
Bernoulli, Johann III (b. 1744), 208, 365
Bernoulli, Nicolaus (b. 1687), exponents, 313
Bertrand, Louis, 286
Bettini, Mario, 96, 226, 229
Betz, W., and H. E. Webb, congruent in geometry, 374
Beutel, Tobias, 208, 273, 283, 292
Beyer, J. H., 277, 283
Bezold, C., 358
Bézout, E., 91, 241, 248; geometry, 387
Bhāskara: survey of his signs, 110–14; 109, 217, 295; unknown quantity, 339
Bianchini, G., 126, 138, 208, 318
Biart, L., 66
Biernatzki, 71
Billingsley's Euclid, 169, 251
Billy, J. de, 227, 249, 253, 254, 292; aggregation, 351; exponents, 307; use of R, 320
Binet, J., angle, 361
Biot, 71

Birks, John, omicron-sigma, 307

Björnbo, A. A., 385

Blakelock, R., 386

Blassière, J. J., 248, 249; circle, 367; parentheses, 351; similar, 372

Blundeville, Th., 91, 223

Bobynin, V. V., 22

Boethius, 59; apices, 81; proportions, 249, 250

Boëza, L., 273

Boissiere, Claude de, 229

Bolognetti, Pompeo, 145

Bolyai, John, 368

Bolyai, Wolfgang, 212, 268; angle, 361, 362; circle, 367; congruent in geometry, 374; different kinds of type, 378; equivalence, 376

Bombelli, Rafaele: survey of his signs, 144, 145; 162, 164, 190, 384; aggregation, 344; use of R, 319, 199

Bomie, 258

Boncompagni, B., 91, 129, 131, 132, 219, 271, 273, 359

Boon, C. F., 270

Borel, E., angle, 361

Borgi (or Borghi) Pietro, survey of his signs, 133; 223, 278

Bortolotti, E., 47, 138, 145, 344, 351

Bosch, Klaas, 52, 208

Bosmans, H., 160, 162, 172, 176, 297

Boudrot, 249

Bouguer, P., 258

Bourke, J. G., 65

Braces, 353

Brackets, 347; in Bombelli, 351, 352

Brahmagupta: survey of his signs, 106–8; 76, 80, 112, 114; unknown quantities, 339

Brancker, Thomas, 194, 237, 252, 307, 386; radical sign, 328, 333, unknown quantity, 341

Brandis, 88

Brasch, F. E., 125

Brasser, J. R., 343

Briggs, H., 261; decimal fractions, 283; use of l for root, 322

Brito Rebello, J. I., 56

Bronkhorst, J. (Noviomagus), 97

Brouncker, W., 264

Brown, Richard, 35

Brückner, Mac, angle, 362

Brugsch, H., 16, 18, 200

Bryan, G. H., 334, 275

Bubnov, N., 75

Budden, E., similar, 374

Bühler, G., 80

Bürgi, Joost, 278, 283; powers, 296

Burja, Abel, radical sign, 331

Bush, W. N., and John B. Clarke: angle, 363; congruent in geometry, 372; ≡, 374

Buteon, Jean: survey of his signs, 173; 132, 204, 263; equality, 263

Byerly, W. E.: angle, 360; parallelogram, 365; right angle, 363

Byrne, O.: angle, 362; edition of Euclid, 386; right angle, 363

Cajori, F., 75, 92

Calculus, differential and integral, 365, 377

Calderón (Span. sign), 92

Callet, Fr., 95; similar, 374

Cambuston, H., 275

Campanus' *Euclid*, 385

Camus, C. E. L., 372

Cantor, Moritz, 27, 28, 31–34, 36, 38, 46, 47, 69, 71, 74, 76, 81, 91, 96, 97, 100, 116, 118, 136, 144, 201, 238, 263, 264, 271, 304, 324, 339; per mille, 274; lettering of figures, 376

Capella, Martianus, 322, 386

Cappelli, A., 48, 51, 93, 94, 208, 274

Caramuel, J., 91, 92; decimal separatrix, 262, 283; equality, 265; powers, 303, 305, 366; radical signs, 328; unknowns, 341

Cardano (Cardan), Hieronimo: survey of his signs, 140, 141; 152, 161, 176, 166, 177, 384; aggregation, 343, 351; equality, 140, 260; use of R, 199, 319; use of round parentheses once, 351

Carlos le-Maur, 96

Carmichael, Robert, calculus, 348

Carnot, L. N. M.: angle, 705; geometric notation, 379

Carra de Vaux, 75

Carré, L., 255, 266; parentheses, 351

Cartan, E., 247

Casati, P., 273

Casey, John: angle, 360; congruent in geometry, 374

Cassany, F., 254

Castillon, G. F., 286; radical signs, 330

Casting out the 9's, 7's, 11's, 218, 225

Castle, F., variation, 259

Caswell, John: circle, 367; equiangular, 360; parallel, 368; perimeter, 367; perpendicular, 364; spherical excess, 380

Cataldi, P. A., 148, 170, 196, 296; □ as an operator, 366; unknown quantities, 339

Catelan, Abbé, 266

Cauchy, A. L.: negative numerals, 90; principal values of a^n, 312

Cavalieri, B.: survey of his signs, 179; 204, 261; aggregation, 344

Cayley, A., 275

Census, word for x^2, 116, 134

Chace, A. B., 23

Chalosse, 219, 221

Chapelle, De la, 223, 255; radical sign, 331

Chauvenet, W., 259; algebraic symbols, 371; right angle, 363; spherical excess, 380

Chelucci, Paolino, 249, 285

China, 69, 119, 120; unknown quantity, 339

Chrystal, G., variation, 259

Chu Shih-Chieh, survey of his signs, 119, 120

Chuquet, N.: survey of his signs, 129–32; 117, 145, 164, 190, 200, 219, 220, 222, 230, 296, 308; aggregation, 344, 350; use of square, 132, 357; use of R, 199 318

Churchill, Randolph, 286

Cifrão (Portuguese sign), 94

Circle: arcs of, 359; pictograph for, 357, 359, 367, 371; as a numeral, 21

Ciscar, G., 286

Clairaut, A. C., 258; his geometry, 387; parentheses, 352

Clark, Gilbert, 186, 248

Clarke, H., 286, 289

Clavius, C.: survey of his signs, 161; 91, 179, 204, 205, 219, 222, 250, 279, 300; aggregation, 351; decimal point, 280; plus and minus, 199; powers, 300; radical sign, 327

Cobb, Sam, 253

Cocker, Edward: aggregation, 347, 348

Codex Vigilanus, 80

Coefficients: letters as coefficients, 176–78, written after the literal part, 179; written above the line, 307

Cole, John, 259

Colebrooke, H. Th., 76, 80, 91, 106, 107, 108, 110, 112–14

Collins, John, 195, 199, 237, 196, 252, 305, 307, 308, 344; aggregation, 344, 350

Colomera y Rodríguez, 92

Colon: for aggregation, 332; separatrix, 245; sign for ratio, 244, 251, 258, 259

Color: used in marking unknowns, 107, 108, 112, 114; colored diagrams, 386; colored quipu, 62, 64

Colson, John, 253; negative numerals, 90.

Comma: for aggregation, 334, 342, 349; for multiplication, 232, 233; for ratio, 256, 257; decimal fractions, 278, 282, 283, 284, 285

Condorcet, N. C. de, parentheses, 352, 354

Congruent, signs in geometry, 372–75

Continued fractions, 118, 273; in John of Meurs, 271; in Wallis, 196

Copernicus, N., 157

Corachan, J. B., 207, 250

Cortazár, J., 286

Cosa, 290, 318; in Buteon, 173; in Chuquet, 131; in De la Roche, 132; in Pacioli, 134, 136, 339; in Rudolff, 149

Cossali, P., 249; aggregation, 345, 354, 355

Cotes, Roger, 307

Courtivron, le Marquis de, parentheses, 352

Craig, John, 252, 253, 301; aggregation, 345, 347

Cramer, G., aggregation, 345, 354, 355

Crelle, A. L., spherical excess, 380

Cremona, L., coincides with, 374

Cresswell, D., 386

Creszfeldt, M. C., 91

Cretan numerals, 32

Crocker, E., 248

Cruquius, N., 208

Crusius, D. A., 96, 208

Crusoe, G. E., 222, 226

Cube of a number, Babylonians, 15

Cuentos("millions"), abbreviation for, 92

Cuneiform symbols, 1–15

Cunha, J. A. da. See Da Cunha, J. A.

Cunn, Samuel, aggregation, 210, 345

Curtze, M., 81, 85, 91, 123, 126, 138 219, 250, 290, 318, 325. 340, 359

Cushing, F. H., 65

Dacia, Petrus de, 91

Da Cunha, J. A., 210, 236, 307

Dagomari, P., 91

D'Alembert, J., 258; angle, 360, 363; imaginary $\sqrt{-1}$, 346; parentheses, 352; similar, 372

Dash. See Line

Dasypodius, C., 53

Datta, B., 75

Davies, Charles, 287; equivalence, 375

Davies, Charles, and W. G. Peck: equivalence, 375, repeating decimals, 289

Davila, M., 92

Debeaune, F., 264, 301

De Bessy, Frenicle, 266

Dechales, G. F. M., 206, 225; decimals, 283; equality, 266; powers, 201

Decimal fractions: survey of, 276–89; 186, 351; in Leibniz, 537; in Stevin, 162; in Wallis, 196; repeating decimals, 289

Decimal scale: Babylonian, 3; Egyptian, 16; in general, 58; North American Indians, 67

Decimal separatrix: colon, 245; comma, 282, 284, 286; point, 287, 288; point in Austria, 288

Dee, John: survey of his signs, 169; 205, 251, 254; radical sign, 327

De Graaf, A. See Graaf, Abraham de

Degrees, minutes, and seconds, 55; in Regiomontanus, 126, 127

De Gua. See Gua, De

Deidier, L'Abbé, 249, 257, 269, 285, 300, 351

De Lagny. See Lagny, T. F. de

Delahire, 254, 258, 264

De la Loubere, 331

Delambre, 87

De la Roche, E.: survey of his signs, 132; 319, radical notation, 199; use of square, 132, 357

Del Sodo, Giovanni, 139

De Moivre, A., 206, 207, 257; aggregation, 354

De Montigny. See Montigny, De

De Morgan, Augustus, 202, 276, 278, 283; algebraic symbols in geometry,

386; complicated exponents, 313; decimals, 286, 287; equality, 268; radical signs, 331; solidus, 275

Demotic numerals, 16, 18

Descartes, René: survey of his signs, 191; 177, 192, 196, 205, 207, 209, 210, 217, 256, 386; aggregation, 344, 353; equality, 264, 265, 300; exponential notation, 294, 298–300, 302–4, 315; geometrical proportion, 254; plus or minus, 262; radical sign, 329, 332, 333; unknown quantities, 339, 340; use of a star, 356

Despiau, L., 248

Determinants, suffix notation in Leibniz, 198

De Witt, James, 210, 264

Dibuadius, Christophorus, 273, 327, 332; aggregation, 348

Dickson, W., 286

Diderot, Denys, 255; *Encyclopédie*, 254

Didier. See Bar-le-Duc

Diez de la Calle, Juan, 92

Diez freyle, Juan, 290

Difference (arithmetical): $=$ symbol for, 164, 177, 262; in Leibniz, 198, 344; in Oughtred, 184, 372

Digges, Leonard and Thomas: survey of their signs, 170; 205, 221, 339; aggregation, 343; equality, 263; powers, 296; radical signs, 199, 334

Dilworth, Th., 91, 246, 287

Diophantus: survey of his signs, 101–5; 41, 87, 111, 117, 121, 124, 135, 200, 201, 217, 235; equality, 260, 104, 263; fractions, 274; powers of unknown. 295, 308, 339

Distributive, ideogram of Babylonians, 15

Division. signs for: survey of, 235–47; Babylonians, 15; Egyptians, 26; Bakhshālī, 109; Diophantus, 104; Leonardo of Pisa, 235. 122; Leibniz, 197, 198; Oughtred, 186; Wallis, 196; complex numbers, 247; critical estimate, 243; order of operations involving $\div$ and $\times$, 242; relative position of dividend and divisor, 241; scratch method, 196; $\div$, 237, 240; :, 238, 240; $\mathfrak{D}$ 154, 162, 236

Dixon, R. B., 65

Dodson, James, 354

Doehlmann, Karl, angle, 362

Dot: aggregation, 181, 183, 251, 348; as radical sign, 324–26; as separatrix in decimal fractions, 279, 283–85;

demand for, 251; for ratio, 244; geometrical ratii, 251–53; in complex numbers, 247; multiplication, in Bhaskara, 112, 217; in later writers, 188, 233, 287, 288; negative number, 107; to represent zero, 109

Dounot (Deidier, or Bar-le-Duc), 300, 351

Drachm or dragma, 149, 151, 158, 293

Drobisch, M. W., 202

Ducange, 87

Dulaurens, F., 255; angle, 360; equality, 263; majus, 263; parallel, 368; perpendicular, 364; pictographs, 365; solids, 371

Dumesnil, G., 96

Duodecimal scale, 3; among Romans, 58, 59

Du Pasquier, L. Gustave, 269

Dupuis, N. F., 95; congruent in geometry, 374; right angle, 363

Durell, Clement V., angle, 361

Durell, Fletcher, 375

Du Séjour. See Séjour, Du

Edwards, R. W. K., 258

Eells, W. C., 67

Egypt, 16; multiplication, 217; square-root sign, 100

Egyptian numerals, 16–23

Einstein, A., 215

Eisenlohr, A., 23

El-Hassar. See al-Ḥaṣṣâr

Emerson, W., 248, 249; angle, 360; geometry of, 386; parallel, 368; perpendicular, 364; variation, 259

Eneström, G., 91, 135, 136, 139, 150, 141, 271, 278, 325, 339, 340, 351, 385

Enriques, F., angle, 361

Equal and parallel, 369

Equality: survey of, 260–70; Ahmes papyrus, 260; al-Qalasâdî, 124; Bakhshâlî MS, 109; Buteon, 173; in Bolognetti, 145; Cardan, 140; dash in Regiomontanus, 126; dash in Ghaligai, 139; dash in Pacioli, 138; Descartes, 191, 300, 363; Digges, 170, 263; Diophantus, 104; Harriot, 188; Hérigone, 189; in proportion, 251, 256; Recorde, 167

Equivalence, 375

Eratosthenes, 41

Etruian signs, 46, 49

Eucken, A., 270

Euclid's Elements, 158, 166, 169, 179, 216, 318, 384, 385; Newton's annotation, 192; Barrow's editions, 192; Billingsley's edition, 251; Elements (Book X), 318, 332; lines for magnitudes, 359

Euler, L., 387; aggregation, 350, 352, 354; imaginary exponents, 309; indices in lettering, 377; lettering of triangle, 194; origin of $\sqrt{}$, 324; powers, 304; imaginary $\sqrt{-1}$, 346

Eutocius, 41

Evans, A. J., 32

Exponents: survey of, 296–315; 129, 131; Bombelli, 144, 162; Chuquet, 131; Descartes, 191; Leibniz, 198; Nuñez, 165; Stevin, 162; general exponents in Wallis, 195; fractional, 123, 129, 131, 162, 164, 196; negative, 131, 195, 308, 311; placed before the base, 198; placed on line in Hérigone, 189; Roman numerals placed above the line in Hume, 190

Eyçaguirre, S. F., 222

Eysenhut, 203

Factoring, notation for process in Wallis, 196

Fakhri, 339

Falkenstein, K., 127

False positions, 202, 218, 219

Favaro, A.: angle, 361; use of different letters, 379

Faye, P. L., 65

Feliciano, F., 222

Fenn, Joseph, 259; angle, 362; circle, 367; right angle, 363; solids, 371; use of star, 356

Fermat, P., 101, 206, 261, 386; coefficients, 307; equality, 265; powers, 307

Ferroni, P., aggregation, 354

Fialkowski, N., 288; angle, 362; similar, 372

Fiedler, W.: angle, 360, 362; congruent in geometry, 372; equal and parallel, 369

Fine, O., 222, 229

Fischer, E. G., 270

"Fisher, George" (Mrs. Slack), 287

Fisher, G. E., and I. J. Schwatt, 213, 242

Fludd, Robert, 204

Follinus, H., 208, 221, 222; aggregation, 351; radical signs, 334

Fontaine, A., imaginary $\sqrt{-1}$, 346

Ford, W. B., negative numerals, 90

Fortunatus, F., 207, 255

Foster, Mark, 254

Foster, S., 186, 251; decimals, 186, 283; equality, 264; parentheses, 351; powers, 306

Foucher, L'Abbé, 249, 255

Fournier, C. F., 248, 249

Fractions: common fractions (survey of), 271–75; Babylonian, 12, 13, 14, 15; addition and subtraction of, 222; Bakhshālī MS, 109; complex fractions in Stevin, 188; Diophantus, 104; division of, 224; duodecimal, 58, 59, 61; Egyptian, 18, 22, 23, 24; fractional line, 122, 235, 272, 273, 391; fraction not a ratio, 245; Greeks, 41, 42, 104; Hindus, 106, 113, 235; juxtaposition means addition, 217; in Austrian cask measure, 89; in Recorde, 167; Leibniz, 197, 198; multiplication, 224; Romans, 58, 59; special symbol for simple fractions, 274; : to denote fractions, 244; unit fractions, 22, 41, 42. *See* Decimal fractions

Frank, A. von, angle, 363

Frank, Sebastian, 55

Frenicle de Bessy, 266

Fricke, R., use of a star, 356

Friedlein, G., 33, 46, 58, 59, 60, 74, 97, 200

Frisch, Chr., 278, 296

Frischauf, J., similar and congruent, 372

Frisi, P., 259

Fuss, N., 259; angle, 360; parallel, 368

Galileo G., 250, 261, 339

Gallimard, J. E., 236, 239, 255; equality, 269

Ganesa, 110

Gangad'hara, 91

Garcia, Florentino, 237, 258

Gardiner, W., 285, 367

Garner, J. L., 66

Gauss, K. F., powers, 304, 314

Gausz, F. G., 286

Gebhardt, M., 159

Gelcich, E., 307

Geminus, 41

Gemna Frisius, 91, 219, 222

Geometrical progression, 248; proportion, 249, 250

Geometrical (pictograph) symbols, 189; in Rich. Rawlinson, 194

Geometry: survey of symbols, 357–87; symbols in statement of theorems, 381

Gerard, Juan, 249, 258

Gerard of Cremona, 118, 290, 318, 385

Gerbert (Pope Silvester II), 61, 322, 376

Gerhardt, C. I., 121, 147, 149, 202, 203, 310, 325, 326, 372

Gernardus, 272

Geysius, J., 196, 305, 341

Ghaligai, Fr.: survey of his signs, 139; 126, 138, 219, 226; equality, 139, 260; the letter R, 199, 319

Gherli, O., 285

Ghetaldi, M., 307

Giannini, P., 259

Gibson, Thomas, 233

Ginsburg, J., 79, 280

Girana Tarragones, H., 55

Girard, A.: survey of his signs, 164; 162, 163, 208, 210, 217, 296; aggregation, 351; coefficients, 307; decimal fractions, 283; difference, 262; powers, 322; radicals, 329–31, 334

Girault de Koudou (or Keroudou), Abbé, 255

Glaisher, J. W. L., 202, 208, 275, 282; complicated exponents, 313

Glorioso, C., 196, 204, 263; unknowns, 339

Gobar numerals, 86, 87, 88, 129

Goldbach, C., 309, 379; figurate numbers, 381; exponents, 313

Golius, Jacob, 263

Gonzalez Davila, Gil, 92

Gonzalo de las Casas, J., 92

Gordon, Cosmo, 134

Gosselin, G.: survey of his signs, 174; 204, 320; use of R, 320; use of capital L, 174, 175, 290

Gow, J., 26, 41, 359

Graaf, Abraham de, 223, 232, 254; equality, 263, 264; "mem," 365; radicals, 336; use of star, 356; = for difference, 262; $\square$ as an operator, 366

Grammateus (Heinrich Schreiber): survey of his signs, 147; 150, 203, 205, 208, 318
Grandi, Guido, 255
Grassmann, H. G.: congruent in geometry, 374; equal and parallel, 369
Greater or less, Oughtred, 183, 186. *See* Inequality
Greek cross, 205
Greek numerals, 33–44, 92; algebra in Planudes, 121
Green, F. W., 16
Greenhill, A. G., use of star, 356; approximately equal, 270
Greenstreet, W. J., 331
Greenwood, I., 287
Gregory, David, 252, 257, 304, 308; aggregation, 345
Gregory, Duncan F., 314
Gregory, James, 207, 252, 256, 268; radical signs, 334
Gregory St. Vincent, 261
Griev, W., 287
Griffith, F. L., 16, 18
Grisio, M., 196
Grosse, H., 208
Grotefend, 1, 47
Grunert, J. A., 47, 208
Gua, De, 232; aggregation, 350; use of a star, 356
Guisnée, 266
Günther, S., 152, 277

Haan, Bierens de, 91, 164
Häseler, J. F., 368; geometric congruence, 372
Haglund, G., 287
Halcke, P., 208
Hall, H. S., and F. H. Stevens: angle, 360; right angle, 363; parallel, 368
Halley, E., 207, 304; aggregation, 345; use of star, 356
Halliwell, J. O., 91, 305
Halma, N., 43, 44
Halsted, G. B., 287, 375; angle, 362, 363; arcs, 370; pictographs, 365; similar and congruent, 372, 374; symmetry, 371
Hamilton, W. R., arcs, 370
Hammond, Nathaniel, 307
Hankel, H., 33, 59
Harmonic progression, 248

Harrington, M. R., 65
Harriot, Th.: survey of his signs, 188; 156, 192, 196, 205, 217, 233; aggregation, 344, 345; equality, 261, 266, 268; greater or less, 188, 360; repetition of factors, 305; radicals, 329
Harris, John, 253
Harsdörffer, P., 96
Hart, C. A., and D. D. Feldman: arcs, 370; equivalent, 375; is measured by, 383; similar, 372
Hartwell, R., 91
Hatton, Edward, 253, 307; angle, 360, 363; radical signs, 328, 331
Hatton, J. L. S., angle, 362
Hauber, K. F., 307
Hawkes, John, 248
Hawkes, Luby and Touton, 242
Hawkins, John, 252
Heath, Sir Thomas, 101, 103, 104, 105, 116, 216
Hebrew numerals, 29–31, 36
Heiberg, J. L., 41, 43, 44, 84, 88, 386
Heilbronner, J. C., 40, 47, 97
Heinlin, J. J., 291
Hemelings, J., 208
Henrici, J., and P. Treutlein: angle, 361; equal and parallel, 369; similar, 372
Henrion, D., 178
Henry, C., 176, 204, 263
Hérigone, P.: survey of his signs, 189; 198, 206, 209, 221, 232, 245, 385, 387; angle, 189, 360; arc of circle, 370; circle, 367; equality, 263; greater than, 263; perpendicular, 364; pictographs, 189, 359, 365; powers, 297, 298, 301; radical signs, 189, 334; ratio, 254; right angle, 363
Hermann, J., 255; parentheses, 351
Herodianic signs, 33, 38
Heron of Alexandria, 41, 103, 201, 271; circle, 367; parallel, 368; pictograph, 357; colored surfaces, 386
Hieratic numerals, 16, 18, 23, 24, 25, 36, 201
Hieroglyphic numerals, 16, 17, 18, 22; problem in Ahmes papyrus translated into hieroglyphic writing, 200
Hill, George A., similar and congruent, 372
Hill, G. F., 74, 80–82, 89
Hilprecht, H. V., 6, 10, 15, 200, 217, 235
Hincks, E., 4, 5

Hindenburg, C. F., 207, 208; aggregation, 354; radical signs, 331

Hindu algebra, 107, 200; division and fractions, 235

Hindu-Arabic numerals: survey of, 74–99; 54; al-Qalasâdî, 124; Al-Khowârizmî, 115; Chuquet, 129; first occurrences, 79, 80; forms, 81–88, 128; Hindu-Arabic notation, 196; local value, 78; shape of figure *five*, 56, 127; shape of zero in Digges, 170; shape of figure *one* in Treviso arithmetic, 86

Hipparchus, 44

Hippocrates of Chios, lettering figures, 376

Hire, De la, 254, 258, 264, 341

Hobbes, Thomas, controversy with Wallis, 385

Hodgson, James, angle, 360

Hoecke, van der, survey of his signs, 150; 147, 204, 319

Hoernle, A. F. R., 109

Hoffmann, H., 386

Holzmann, W. *See* Xylander

Hopkins, G. Irving: angle, 360; right angle, 363; parallelogram, 365; ≡, 374

Hoppe, E., 351

Horrebowius, P., 232

Horsley, S., 286

Hortega, Juan de, 207, 219, 221, 222, 223, 225

Hospital, L', 206, 255, 266

Hoste, P., 266

Hostus, M., 97

Houël, G. J., 95

Hübsch, J. G. G., 208, 232

Hudde, J., 264, 307

Huguetan, Gilles, 132

Huips, Frans van der, 266

Hultsch, Fr., 41, 59, 272, 357

Humbert, G., 51

Humboldt, Alex. von, 49, 87, 88

Hume, James: survey of his signs, 190; 206, 297, 298, 302; parentheses, 351; use of R, 320

Hunt, N., 206, 225

Huntington, E. V., 213, 214

Hutton, Ch., 91, 107, 159, 286, 351; angle, 360

Huygens, Chr., 206, 208, 254, equality, 264; powers, 301, 303, 304, 307, 310

Hypatia, 117

Hypsicles, 44

Ibn Albanna, 118

Ibn Almunⱪim, 118

Ibn Khaldûn, 118

Identity, Riemann's sign, 374

Ideographs, 385; unusual ones in elementary geometry, 383

Iḫwān aṣ-ṣafā, 83

Illing, C. C., 208

Imaginary $\sqrt{-1}$ or $\sqrt{(-1)}$ or $\sqrt{-1}$, 346

Incommensurable: survey of, 382; sign for, in Oughtred, 183, 184; sign for, in J. F. Lorenz, 382

Inequality (greater or less): in Harriot, 188; in Hérigone, 189; in Oughtred, 182. *See* Greater or less

Infinity, Wallis' sign ∞, 196

Isidorus of Seville, 80

Izquierdo, G., 248, 249, 258

Jackson, L. L., 208; quoted, 199

Jacobs, H. von, 96

Jager, R., 283

Japanese numerals, 71, 73

Jeake, S., 219, 223, 245, 249, 254, 284, radical signs, 328

Jenkinson, H., 74

Jess, Zachariah, 246

John of Seville, 271, 290, 318

Johnson, John, 283

Johnson's *Arithmetic*, 186, 244

Jones, William, 210, 308; angle, 360, 363; parallel, 368; perpendicular, 364; pictographs in statement of geometric theorems, 381; use of a star, 356

Jordan, C., use of ≡, 374

Juxtaposition, indicating addition, 102; indicating multiplication, 122, 217

Kästner, A. G., 40

Kalcheim, Wilhelm von, 277

Kambly, L.: angle, 360; arc of circle, 370; horizontal line for "therefore," 383; parallel, 368; similar and congruent, 372

Karabacek, 80

Karpinski, L. C., 42, 48, 74, 79–81, 87, 92, 115, 116, 122, 159, 208, 266, 271, 273, 287, 385

Karsten, W. J. G.: algebraic symbols, 371; parallel, 368; similar, 372; signs in geometry, 387; division, 275

Kaye, G. R., 75, 76, 77, 80, 109, 250

Kegel, J. M., 208

Keill, J.: edition of Euclid, 386

Kepler, J., 261, 278, 283; astronomical signs, 358; powers, 296

Kersey, John, 248, 251, 304, 307; aggregation, 345; angle, 360; circle, 367; parallel, 368; perpendicular, 364; pictographs, 365; radical signs, 332, 335; right angle, 363

Kinckhuysen, G., 264, 341; use of star, 356

Kirkby, John, 245, 248, 386; arithmetical proportion, 249; sign for evolution, 328

Kirkman, T. P., 240, 286; aggregation, 354

Klebitius, Wil., 160, 207

Klügel, G. S., 47, 208; angle, 360; "mem," 365; pictographs, 359, 360; similar, 372

Knots records: in Peru, 62–64; in China, 69

Knott, C. G., 282

Köbel, J., 55

König, J. S., aggregation, 354

Köppe, K., similar and congruent, 372

Kosegarten, 88

Kowalewski, G., 211

Kratzer, A., 270, 271

Krause, 246

Kresa, J., 206, 254

Kritter, J. A., 208

Krogh, G. C., 208

Kronecker, L., 374; [a], 211

Kubitschek, 34

La Caille, Nicolas Louis de, 258

Lagny, T. F., de, 258, 266, 268; radical signs, 330, 331; use of a star, 356

Lagrange, J., 387; parentheses, 352, 354

Lalande, F. de, 95; aggregation, 354

Lampridius, Aelius, 51

Lamy, B., 206, 248, 249, 255, 257, 264

Landen, John, circular arc, 370

Lansberg, Philip, 250

Laplace, P. S., 99, 387; aggregation, 354; imaginary $\sqrt{-1}$, 346

Latin cross, 205, 206

Latus ("side"), 290; survey of, 322; use of l for x, 186, 322; use of L for powers and roots, 174, 175

Layng, A. E., 246; angle, 360; parallelogram, 365

Lee, Chauncey, 221, 254, 287

Leechman, J. D., 65

Legendre, A. M., 231; aggregation, 354; angle, 363; algebraic signs, 371; geometry, 387

Leibniz, G. W., 197, 198, 233, 237, 341, 386; aggregation, 344, 349–51, 354; dot for multiplication, 285; equality, 263, 266, 267: fractions, 275; geometrical proportion, 255, 258, 259; geometric congruence, 372; lettering figures, 377; powers, 303, 304; quotations from, 197, 198, 259; radical sign, 331; signs for division, 238, 244, 246; variable exponents, 310

Lemoch, I., 288

Lemos, M., 91

Lenormant, F., 5

Leonardo of Pisa: survey of his signs, 122; 91, 134, 219, 220, 235; fractions, 271, 273; letters for numbers, 351; lettering figures, 376; radix, 290, 292, 318

Lepsius, R., 5

Leslie, John, 371

Less than, 183. See Inequality

Lettering of geometric figures, 376

Letters: use of, for aggregation, 342, 343; capital, as coefficients by Vieta, 176; Cardan, 141; Descartes, 191; Leibniz, 198; Rudolff, 148; small, by Harriot, 188; lettering figures, 376

Leudesdorf, Ch., 379

Leupold, J., 96

Le Vavasseur, R., 211

Leybourn, William, 252, 292; angle, 360

Libri, 91, 116, 385

Lieber, H., and F. von Lühmann: angle, 360; arc of circle, 370

Lietzmann, W., 96

Liévano, I., 248, 249, 258

Line: fractional line, 235, 239; as sign of equality, 126, 138, 139; as sign of division, 235; as sign of aggregation. See Vinculum

Lipka, J., equal approximate, 373

Lobachevski, angle of parallelism, 363

Local value (principle of): Babylonians, 5, 78; Hebrews, 31; Hindus, 78, 88; Maya, 68, 78; Neophytos, 88; Turks, 84

Locke, L. L., 62, 63

Long, Edith, and W. C. Brenke, congruent in geometry, 372

Loomis, Elias, 287

Lorenz, J. F., 216; geometry, 387; incommensurables, 382; similar, 372

Lorey, Adolf: equal and parallel, 369; similar and congruent, 372

Loubère, De la, 331

Louville, Chevalier de, 258

Lucas, Edouard, 96

Lucas, Lossius, 225, 229

Ludolph van Ceulen, 148, 208, 209, 223; aggregation, 344, 348, 349; radical sign, 332

Lumholtz, K., 65

Lutz, H. F., 12, 13

Lyman, E. A., parallelogram, 365

Lyte, H., 186

Maandelykse Mathematische Liefhebberye, 330, 336

Macdonald, J. W., equivalent, 375

Macdonald, W. R., 282

McDougall, A. H.: congruent in geometry, 374; similar, 374

Macfarlane, A., 275

Mach, E., right angle, 363

Maclaurin, Colin, 240, 248; aggregation, 345; use of a star, 356

McMahon, James, equivalent, 375

Macnie, J., equivalent, 375

Mairan, Jean Jaques d'Orton de, 255

Mako, Paulus, 259, 288

Mal, Arabic for x^2, 116, 290

Malcolm, A., 248

Manfredi, Gabriele, 257, 331; parentheses, 351

Mangoldt, Hans von, angle, 362

Marini, 49

Marquardt, J., 51

Marre, Aristide, 129

Marsh, John, 289

Martin, T. H., 97

Mason, C., 307

Masterson, Thomas, 171, 278

Mauduit, A. R., 259; perpendicular, 364; similar, 372

Maupertuis, P. L., 255; aggregation, 351, 352

Maurolicus, Fr., 303; pictographs in geometry, 359; use of R, 319

Maya, 68, 5

Mehler, F. G., angle, 360

Meibomius, M., 251

Meissner, Bruno, 14, 15

Meissner, H., 283

"Mem," Hebrew letter for rectangle, 365, 366

Mengoli, Petro, 206, 254, 301

Menher, 148

Méray, Ch., 213

Mercastel, J. B. A. de, on ratio, 254, 256

Mercator, N., 252; use of star, 194; decimals, 283

Mersenne, M., 209, 266, 273, 301, 302, 339

Metius, Adrian, 186, 225; circle, 367; pictographs, 359, 371

Meurs, John of, 271

Meyer, Friedrich, angle, 362

Meyer, H., 287

Meyers, G. W., congruent in geometry, 372; $\equiv$, 374

Michelsen, J. A. C., 206

Mikami, Yoshio, 119, 120

Milinowski, A., angle, 362

Milne, W. J., equivalent, 375

Minus sign: survey of, 208–16; in Bombelli, 144; Buteon, 173; Cavalieri, 179; Cardan, 140; Clavius, 161; Diophantus, 103; Gosselin, 174; Hérigone, 189; Pacioli, 134; Peletier, 172; Recorde, 167; Regiomontanus, 126, 208; sign $\bar{m}$, 131, 132, 134, 142, 172–74, 200; sign $\sim$, 189; sign $\div$, 164, 208, 508; Tartaglia, 142, 143; Vieta, 176; not used in early arithmetics, 158

Mitchell, J., 286

Möbius, A. F., angle, 361

Mohammed, 45

Molesworth, W., 385

Molk, J., 211

Möller, G., 16, 18, 21

Mollweide, C. B., 47, 216; similar and congruent, 372

Molyneux, W., 283; radical signs, 334

Mommsen, Th., 46, 51

Monconys, De, 263

Monge, G., lettering, 377
Mönich, 246
Monsante, L., 258, 286
Montigny, Ch. de, parentheses, 352
Moore, Jonas, 186, 248; aggregation, 348; arithmetical proportion, 249; decimals, 286; geometrical proportion, 251; radicals, 332
Moraes Silva, Antonio de, 94
Morley, S. G., 68
Moxon, J., 231, 236, 303
Moya, Pérez de: 92, 204, 221, 223, 294; use of R, 320, 321, 326
Mozhnik, F. S., 288
Müller, A., 96
Müller, C., 41
Müller, G., angle, 360
Müller, O., 49
Multiplication: survey of signs, 217–34; Bakhshālī MS, 109; Cavalieri, 179; comma in Hérigone, 189; in Leibniz, 197, 198, 232, 536; cross-multiplication marked by X, 141, 165; Diophantus, 102; dot, 112, 188, 233, 287, 288; Hindus, 107, 112; order of operations involving ÷ and $\times$, 242; Wallis, 196; of integers, 226, 229; Stevin, 162; Stifel, 154; $\times$, 186, 195, 288; in Oughtred, 180, 186, 231; in Leibniz, 197, 198; in Vieta, 176–78, 186; $\frown$ in Leibniz, 198; star used by Rahn, 194; juxtaposition, 122, 217
Multiplicative principles, in numeral system: Aztecs, 66; Babylonians, 1; Cretans, 32; Egyptians, 19, 21; Romans, 50, 51, 55; Al-Kharkhī, 116; in algebraic notation, 101, 111, 116, 135, 142
Musschenbroek, van, 267

Nagl, A., 34, 54, 85, 89
Nallino, C. A., 82
Napier, John, 196, 218, 231, 251, 261; decimal point, 195, 282; line symbolism for roots, 323, 199
National Committee on Mathematical Requirements (in U.S.), 243, 288; angle, 360; radical signs, 338
Nau, F., 79
Negative number, sign for: Bakshālī, 109; Hindu, 106
Nemorarius, Jordanus, 272
Neomagus. See Noviomagus
Neophytos, 87, 88, 129, 295

Nesselmann, G. H. F., 31, 41, 50, 101, 235
Netto, E., 211
Newton, John, 249, 305
Newton, Sir Isaac, 196, 252, 253, 386; aggregation, 345; decimals, 285, 286; equality, 266, 267; exponential notation, 294, 303, 304, 307, 308, 377; radical sign, 331, 333; ratio and proportion, 253; annotations of Euclid, 192
Nichols, E. H., arcs, 370
Nichols, F., 287
Nicole, F., 255, 258, 268; parentheses, 352, 354
Nieuwentijt, B., 264, 266
Nipsus, Junius, 322
Nixon, R. C. J.: angle, 361; parallelogram, 365
Nonius. See Nuñez
Nordenskiöld, E., 64
Norman, Robert, 162
Norton, R., 186, 276
Norwood, 4, 261; aggregation, 351
Notation, on its importance: Oughtred, 187; Wallis, 199; L. L. Jackson, 199; Tropfke, 199; Treutlein, 199; Babbage, 386
Noviomagus (Bronkhorst, Jan), 97
Numbers, absolute, signs for, Hindus, 107
Numerals: alphabetic, 28, 29–31; Arabic (early), 45; Arabic (later), 86; Aztec, 66; Babylonian, 1–15; Brahmi, 77; Chinese, 69–73; Cretan, 32; Egyptian, 16–25; Fanciful hypotheses, 96; forms of, 85, 86; freak forms, 89; Gobar numerals, 86; Greek, 32–44, 87; grouping of, 91–94; Hindu-Arabic, 74, 127; Kharoshthi, 77; Phoenicians and Syrians, 27, 28; relative size, 95; Roman, 46, 47; Tamul, 88; North American Indians, 67; Peru, 62–64; negative, 90
Nuñez, Pedro: survey of his signs, 166, 204; aggregation, 343; several unknowns, 161
Nunn, T. P., quoted, 311

Ocreatus, N., 82
Octonary scale, 67
Ohm, Martin, 312; principal values of a^n, 312
Oldenburgh, H., 252, 308, 344, 377; aggregation, 344, 345

Oliver, Wait, and Jones (joint authors), 210, 213

Olleris, A., 61,

Olney, E., 287

Omicron-sigma, for involution, 307

Oppert, J., 5

Oresme, N., survey of his signs, 123; 129, 308, 333

Ottoni, C. B., 258; angle, 361

Oughtred, William: survey of his signs, 180–87; 91, 148, 169, 192, 196, 205, 210, 218, 231, 236, 244, 248, 382, 385; aggregation, 343, 345, 347–49; arithmetical proportion, 249; cross for multiplication, 285; decimals, 283; equality, 261, 266; geometrical proportion, 251–53, 255, 256; greater or less, 183; pictographs, 359; powers, 291; radical signs, 329, 332, 334; unknown quantity, 339

Ozanam, J., 257, 264; equality, 264, 265, 266, 277; powers, 301, 304; radical sign, 328

Pacioli, Luca: survey of his signs, 134–38; 91, 117, 126, 132, 145, 166, 177, 200, 219, 220, 221, 222, 223, 225, 226, 294, 297, 359, 384; aggregation, 343; equality, 138, 260; powers, 297, 322; radix, 292, 297, 318, 199; unknown, 339

Pade, H., 213

Palmer, C. I., and D. P. Taylor, equal number of degrees, 363

Panchaud, B., 249, 259

Paolo of Pisa, 91

Pappus, 55; circle, 367; pictographs, 357

Parallel lines, 359, 368

Parallelogram, pictograph for, 357, 359, 365

Pardies, G., 206, 253, 255

Parent, Antoine, 254, 255, 258; equality, 263; unknowns, 341

Parentheses: survey of, 342–52; braces, 188, 351; brackets, 347, 351; round, in Clavius, 161; Girard, 164; Hérigone, 189; Leibniz, 197, 238; marking index of root, 329; Oughtred, 181, 186. See Aggregation

Paricius, G. H., 208, 262

Parker, 331

Pascal, B., 261, 304, 307; lettering figures, 376

Pasquier, L. Gustave du, 269

Pastor, Julio Rey, 165, 204

Paz, P., 274

Peano, G., 214, 275, 288; aggregation, 348; angle, 362; principal values of roots, 337; "sgn," 211; use of ∽, 372

Peet, T. E., 23, 200, 217

Peirce, B., 247, 259, 287; algebraic symbols, 371

Peise, 14

Peletier, Jacques: survey of his signs, 172; 174, 204, 227, 292; aggregation in radicals, 332

Pell, John, 194, 237, 307, 386

Pellizzati. See Pellos, Fr.

Pellos, Fr., 278

Penny, sign for, 275

Per cent, 274

Pereira, J. F., 258

Perini, L., 245

Perkins, G. R., 287

Perny, Paul, 69

Perpendicular, sign for, 359, 364

Peruvian knots, 62–64, 69; Peru MSS, 92

Peruzzi, house of, 54

Peurbach, G., 91, 125

Phillips, A. W., and Irving Fisher: equivalent, 375; spherical excess, 380

Phoenicians, 27, 36

Pi (π): for "proportional," 245; $\frac{4}{\pi}$ and □, 196

Picard, J., 254

Piccard, 96

Pictographs, 357–71, 384, 385

Pihan, A. P., 25, 30, 73

Pike, Nicolas, 91, 289

Pires, F. M., 258

Pitiscus, B., 279–81

Pitot, H., 255, 341

Planudes, Maximus, survey of his signs, 121; 87

Plato, 7

Plato of Tivoli, 290, 322; arcs of circles, 359, 370

Playfair, John: angle, 360; algebraic symbols, 371; edition of Euclid, 386

Pliny, 50

Plücker, J., 387

"Plus or minus," 210, 196; Leibniz, 198; Descartes, 262, 210

Plus signs: general survey of, 201–16; 186, 199, in Bakhshālī, 109; Bombelli, 144; Cavalieri, 179; Cardan, 140; Clavius, 161; letter e, 139; not used in early arithmetics, 158, Recorde, 167; Scheubel, 158; shapes of, 265; sign $\tilde{p}$, 131, 132, 134, 139, 142, 172–74, 200; spread of $+$ and $-$, 204, Tartaglia, 143, Vieta, 176, Widmann, 146

Poebel, Arno, 15

Poinsot, L., 314

Polemi, G., aggregation, 345, 353

Polynier, P., 266

Porfirio da Motta Pegado, L., 248, 258

Pott, A. F., 66

Potts, Robert, 289

Pound, sign for, 275

Powers: survey of, 290–315; Arabic signs, 116; Bombelli, 144; Cardan, 140; complicated exponents, 313; Digges, 170; expressed by V, 331; fractional, 123, 129; in geometry, 307; Ghaligai, 139; general remarks, 315; Girard, 164; Grammateus, 147; Hindu signs, 107, 110, 112; Hume, 190; irrational, 308; negative and literal, 131, 195, 308, 311; Nuñez, 165; Pacioli, 134, 135; principal values, 312; Peletier, 172; Psellus, 117; repetition of factors, 305; Recorde, 167; Rudolff, 148; square in Egyptian papyrus, 100; Schoner, 322; Stifel, 151; Tartaglia, 142, 143; variable exponents, 310; Vieta, 176, 177; Van der Hoecke, 148; Wallis, 291; fifth and seventh, 135; aa for a^2, 304

Powers: additive principle in marking, 101, 111, 112, 117, 124; multiplicative principle in marking, 101, 111, 116, 135, 142

Powers, S., 65

Praalder, L., 208, 336

Prändel, J. G., 336

Prestel, M. A. F., similar, 372

Prestet, J., 255, 264; aggregation, 344; decimals, 283; equality, 266; use of star, 356

Preston, J., 219

Principal values, 211, 312, 337

Principle of local value. See Local value

Pringsheim, Alfred, limit, 373

Priscian, 53

Progression. See Arithmetical progression, Geometrical progression

Proportion: survey of, 248–59; al-Qalasâdî, 124; arithmetical proportion, 186, 249, 255; continued proportion, 254; compound proportion, 218, 220; geometrical proportion, 244, 249, 250, 254–58; Grammateus, 147; in earliest printed arithmetics, 128; Oughtred, 181; proportion involving fractions, 221; Recorde, 166: Tartaglia, 142; Wallis, 196; variation, 259

Pryde, James, 289

Psellus, Michael, survey of his signs, 117

Ptolemy, 41, 43, 44, 87, 125, 218

Puissant, 249

Purbach, G., 91, 125

Purser, W., 186

Quadratic equations, 26

Quaternary scale, 67

Quinary scale, 67

Quibell, J. E., 16

Quipu of Peru and North America, 62–65

Radicals: Leibniz, 198; Wallis, 196; reduced to same order, 218, 227; radical sign $\sqrt{}$, survey of, 199, 324–38; radical $\sqrt{}$, with literal index, 330, 331

Radix, 290, 291, 292; R for x, 296, 307, 318; R for root, survey of, 318–21; R for powers, in Pacioli, 136. See Roots

Raether, 96

Rahn, J. H.: survey of his signs, 194; 205, 208, 232, 237, 385, 386; Archimedian spiral, 307; equality, 266; powers, 304, 307; radical signs, 328, 333; unknowns, 341; * for multiplication, 194

Ralphson. See Raphson

Rama-Crishna Deva, 91

Ramus, P., 164, 177, 204, 290, 291; lettering figures, 376; use of l, 322

Raphson, J., 210, 252, 285, 305; aggregation, 345; use of a star, 356

Ratdolt, Erhard, 385

Rath, E., 272

Ratio: arithmetical, 245; "composition of ratios," 216; geometric (survey of), 244, 252; Hérigone, 189; Oughtred, 181, 186, 251, 252; not a division, 245, 246; of infinite products, 196; sporadic signs, 245, 246

Rawlinson, H., 5

Rawlinson, Rich., survey of his signs, 193

Rawlyns, R., 283

Reaumur, R. A. F. de, 255

Recio, M., 92

Recorde, R.: survey of his signs, 167–68; 145, 204, 205, 219, 221, 222, 225, 229, 256, 274; equality, 260–70; plus and minus, 199; radical sign, 327, 328

Rectangle: "mem," 365; pictograph for, 357, 359, 368

Rees's *Cyclopaedia*, 363

Regiomontanus: survey of his signs, 125–27; 134, 138, 176, 208, 250; decimal fractions, 278, 280; equality, 126, 260, 261; lettering figures, 376; R for "radix," 318; unknown, 339

Regius, Hudalrich, 225, 229

Regula falsi. See False positions

Reinhold, C. L., geometric congruence, 372

Renaldini, C., 206, 307; aggregation, 351

Res ("thing"), 134, 290, 293

Reye, Theodor: angle, 360; use of different letters, 379

Reyher, S., 262, 263; angle, 361; arc of circle, 370; circle, 367; geometry, 387; parallel, 368; right angle, 363; trapezoid, 371

Reymers, Nicolaus, 208, 296

Reyneau, Ch., 266, 308; use of a star, 356

Rhabdas, Nicolas, 42

Rhind papyrus. *See* Ahmes papyrus

Riccati, Vincente, 258

Ricci, M. A., 250, 263, 301

Richman, J. B., 92

Riemann, G. F. B., ≡, 374

Riese, Adam, 59, 148, 176, 208; radicals, 326

Rigaud, S. P., 199, 231, 196, 365

Robbins, E. R.: angle, 360; right angle, 363; parallelogram, 365

Robert of Chester, 385. *See* Karpinski

Robertson, John, 289

Roberval, G. P., 264

Robins, Benjamin, 307

Robinson, H. N., angle, 362

Roby, H. J., 46

Rocha, Antich, 320, 294

Roche, De la. *See* De la Roche

Roder, Christian, 126

Rodet, L., 201

Rolle, Michael, 82, 206, 255; equality, 264, 304; aggregation, 344; radical sign, 331; use of R, 321

Roman numerals, 46–61, 92, 93

Romanus, A., 206, 207; aggregation, 343; powers, 296, 297; radical signs, 329, 330, 199; use of R, 320

Ronayne, Philip, 215, 307; "mem" for rectangle, 365

Roomen, Adriaen van. *See* Romanus

Roots: survey of, 316–38; al-Qalasâdî, 124; Hindus, 107, 108; Leonardo of Pisa, 122; Nuñez, 165; principal values, 337; Recorde, 168; spread of $\sqrt{}$ symbol, 327; sign $\sqrt{}$ in Rudolff, 148, 155, in Stifel, 153, 155, in Scheubel, 159, in Stevin, 163, in Girard, 164, in Peletier, 172, in Vieta, 177, in Hérigone, 189, in Descartes, 191, V *bino*, 163; sign R, survey of, 318–21; in Regiomontanus, 126, in Chuquet, 130, 131, in De la Roche, 132, in Pacioli, 135, in Tartaglia, 142, in Cardan, 141, in Bombelli, 144, in Bolognetti, 145, in Scheubel, 159, in Van der Hoecke, 150; *radix relata*, 135, 142; *Ra. col.* in Scheubel, 159; $R\sqrt{}$., 135, 141, 165; *radix distincta*, 141; *radix legata*, 144; R for x, 137, 160, 318; R to mark powers, 136; R to mark both power and root in same passage in Pacioli, 137; L as radical in Gosselin, 175; $\sqrt{}$ and dot for square of binomial, 189

Rosen, F., 115

Rosenberg, Karl, 288

Roth, Peter, 208

Rudolff, Chr.: survey of his signs, 148, 149; 168, 177, 203, 204, 205, 221, 222, 225, 227; aggregation, 148; Coss of 1525, 151, 153, 728; Stifel's edition, 155; decimal fractions, 278, 279; freak numerals, 89, 91, 158; geometrical proportion, 250; radical sign, 165, 199, 326, 328; unknown quantity, 339

Ruska, Julius, 45, 83, 97, 290

Ryland, J., 16

Sacrobosco, J. de, 82, 91, 127

Saez, Liciniano, 52, 92

St. Andrew's cross, 218; in complex numbers, 247

St. Vincent, Gregory, 261; lettering figures, 376

Salazar, Juan de Dios, 239, 286

Salignacus, B., 291, 322

Sanders, Alan, is measured by, 373

Sanders, W., 252

Sarjeant, Th., 287

Saulmon, 258

Sault, Richard, 248, 329

Saunderson, Nicholas, aggregation, 345

Saurin, Abbé, 255

Savérien, A., 240, 248, 249, 259; angle, 362; circle, 367; omicron-sigma, 307; perpendicular, 364; pictographs, 365; right angle, 363; solids, 371

Scales: quinary, in Egypt, 16; duodecimal, in Babylonia, 3, in Egypt, 16; vigesimal, in Egypt, 16, Maya, 68; sexagesimal, in Babylonia, 5, 8, in Egypt, 16. See Decimal scale

Schack, H., 26

Schafheitlin, P., 308, 344, 366

Scherffer, C., 258

Scheubel, Johann: survey of his signs, 158, 159; 160, 174, 176, 204, 319; aggregation, 343; plus and minus, 199, 768; radical sign, 326, 327; use of ℞, 319

Schey, W., 208

Schlesser C., 208

Schmeisser, F., 208, 212, 246

Schmid, K. A., 91

Schnuse, C. H., 348

Schöne, H., 103

Schöner, Joh., 272

Schoner, L., 291; unknown, 339; use of ., 322

Schooten, Van. See Van Schooten

Schott, G., 219; powers, 301

Schott, K., 91, 292

Schreiber, Heinrich. See Grammateus

Schrekenfuchs, O., 218, 221, 222

Schröder, E., 247

Schrön, L., 95

Schröter, Heinrich, angle, 362

Schubert, H., complicated exponents, 313

Schur, F., angle, 362

Schwab, J. C., 268

Schwarz, H. A., 356

Schwenter, D., 250

Scott, Charlotte A., angle, 362

Scratch method of multiplication and division, 128, 133, 195, 241

Sebokht, S., 79

Segner, J. A. de, 91; aggregation, 354; geometry, 387; imaginary $\sqrt{-1}$, 346

Séjour, Du, imaginary $\sqrt{(-1)}$, 346

Selling, E., 90

Selmer, E. W., 208

Senès, D. de, 258

Senillosa, F., 239, 248

Senkereh, tablets of, 5

Serra y Oliveres, A., 275; angles, 363

Sethe, Kurt, 16, 17, 18, 21, 22

Sexagesimal system: in Babylonia, 5, 78; Egypt, 16; Greece, 43, 87; Western Europe, 44; Wallis, 196; sexagesimal fractions, 12; degrees, minutes, and seconds, 55, 126

Sfortunati, G., 219, 221, 223

Sgn, 211

Shai, Arabic for "thing," 290

Shelley, George, 253

Sheppard, W. F., parentheses, 355

Sherman, C. P., 96

Sherwin, H., 285

Sherwin, Thomas, 287

Shutts, G. C., congruent in geometry, 372

Sieur de Var. Lezard, I. L., 262

Sign ∞, 53, 196

Sign ⊙, 375

Sign ∾ or ഗ, 41, 372, 373

Sign ≡, 374

Sigüenza, y Góngora, 51

Silberstein, L., 215

Similar, survey of signs, 372-74

Simon, Max, similar and congruent, 372

Simpson, Th., 259; aggregation, 345; use of a star, 356

Simson, Robert, Euclid, 372, 386

Slack, Mrs., 287

Slaught, H. E., and Lennes, N. J. (joint authors), 213; congruent in geometry, 372

Slusius, R. F., 254, 263; equality, 263

Smith, C., angle, 360

Smith, D. E., 47, 74, 80, 81, 147, 154, 208, 274, 278. See also Beman and Smith

Smith, Eugene R.: arcs, 370; congruent in geometry, 372; ⊨, 374

Smith, George, 5

Snell, W., 219; aggregation, 348; radical signs, 332

Solidus, 275, 313

Solomon's ring, 96

Spain, *calderón* in MSS, 92, 93

Speier, Jacob von, 126, 318

Spenlin, Gall, 208, 219

Spherical excess, 380

Spielmann, I., 288

Spier, L., 65

Spitz, C., 213; angle, 362; similar, 372

Spole, Andreas, 265, 301

Square: Babylonians, 15; □ to mark cubes, in Chuquet, 131; to mark cube roots, in De la Roche, 132; □ for given number, in Wallis, 196; pictograph, 357, 359, 365; as an operator, 366

Square root: Babylonian, 15; al-Qalasâdî's sign, 124; Egyptian sign, 100

Śrīdhara, 112; fractions, 271

Staigmüller, H., 159

Stampioen, J., 250, 256, 259; aggregation, 347, 351; equality, 266; exponents, 299, 303, 307; radicals, 329; radical signs, 333, 335; □ as an operator, 366

Star: to mark absence of terms, 356; for multiplication, 194, 195, 232; in Babylonian angular division, 358

Steele, Robert, 82, 274

Steenstra, P.: angle, 360; right angle, 363; circle, 367

Stegall, J. E. A., 323

Stegman, J., 283

Steiner, Jacob: similar, 372; and Plücker, 387

Steinhauser, A., 288

Steinmetz, M., 221, 223

Steinmeyer, P., 219

Sterner, M., 96, 262

Stevin, S.: survey of his signs, 162, 163; 123, 164, 190, 217, 236, 254, 728; aggregation, 343; decimal fractions, 276, 282, 283; powers, 296, 308; lettering figures, 377; radicals, 199, 329, 330, 333; unknowns, 339, 340

Steyn, G. van, 208

Stifel, M.: survey of his signs, 151–56; 59, 148, 158, 161, 167, 169, 170, 171, 172, 175, 176, 177, 192, 205, 217, 224, 227, 229, 236, 384; aggregation, 344, 348, 349, 351; geometric proportion, 250; multiplication of fractions, 152; repetition of factors, 305; radical sign, 199, 325, 327, 328, 329, 334; unknowns, 339

Stirling, James, 233, 354

Stokes, G. G., 275

Stolz, O., and Gmeiner, J. A. (joint authors), 213, 214, 268; angle, 361; principal values, 312, 337; solidus, 275; uniformly similar, 373

Stone, E., angle, 360, 363

Streete, Th., 251

Stringham, I., multiplied or divided by, 231

Study, E., 247

Sturm, Christoph, 257

Subtraction, principle of: in al-Qalasâdî, 124; in Babylonia, 10; in India, 49; in Rome, 48, 49

Subtraction: survey of, 200–216; Diophantus, 103; Hindus, 106, 108, 109, 114, 200; Greek papyri, 200

Sun-Tsu, 72

Supplantschitsch, R., 288

Surd, sign for, Hindu, 107, 108

Suter, H., 81, 235, 271, 339

Suzanne, H., 248

Swedenborg, Em., 206, 207, 245, 248, 258; equality, 268

Symbolism, on the use of, 39, 40; by Stifel, 152. *See* Sign

Symbolists versus rhetoricians, 385

Symbols: value of, 118; by Oughtred, 187

Symmetrically similar triangles, 373

Symmetry, symbol for, 371

Syncopated notations, 105

Syrians, 28, 36

Tacquet, A., 261, 269, 283, 307; algebraic symbols, 371

Tamul numerals, 88

Tannery, P., 42, 88, 101, 103, 104, 117, 121, 217, 235, 254, 300, 344, 357

Tartaglia, N.: survey of his signs, 142, 143; 145, 166, 177, 219, 221, 222, 225, 229, 384; geometrical proportion, 250, 254; parentheses, 351; □ as an operator, 366; use of ℞, 199, 319

Ternary scale, 67
Terquem, 132
Terrier, Paul, 379
Texeda, Gaspard de, 92
Theon of Alexandria, 87
Thierfeldern, C., 208
Thing. *See* Cosa
Thompson, Herbert, 42
Thomson, James, 241
Thornycroft, E., 354
Thousands, Spanish and Portuguese signs for, 92, 93, 94
Thureau-Dangin, François, 12
Todhunter, I., 286, 287; algebraic symbols, 715; edition of Euclid, 386
Tonstall, C., 91, 219, 221, 222
Torija, Manuel Torres, 286
Toro, A. de la Rosa, 286
Torporley, N., 305
Torricelli, E., 261
Touraeff, B., 100
Transfinite ordinal number, 234
Trenchant, J., 219, 221
Treutlein, P., 96, 147, 148, 151, 154, 156, 263, 296, 340; quoted, 199
Treviso arithmetic, 86, 221
Triangles, pictograph for, 357, 359, 365; *s a s* and *a s a*, 381
Tropfke, J., 91, 136, 140, 149, 151, 159, 176, 201, 203, 217, 255, 263, 277, 289, 293, 296, 324, 340, 343, 344, 348, 353, 359, 376, 386; quoted, 199
Tschirnhaus, E. W. von, 266
Tweedie, Ch., 354
Twysden, John, 186

Unger, F., 81, 91, 208
Ungnad, A., 14
Unicorno, J., 226
Unknown number: survey of, 399–41; Ahmes papyrus, 16; al-Qalasâdî, 124; Bakhshālī MS, 109; Cataldi, 340; Chinese, 120; Digges, 170; Diophantus, 101; Hindus, 107, 108, 112, 114; Leibniz, 198; more than one unknown, 136, 138, 140, 148, 152, 161, 173, 175, 217, 339; Pacioli, 134; Psellus, 117; Regiomontanus, 126; represented by vowels, 164, 176; Roman numerals in Hume, 190; Oughtred, 182, 186; Rudolff, 148, 149, 151; Schoner, 322; Stevin, 162, 217; Stifel, 151, 152; Vieta, 176–78; fifth power of, 117

Valdes, M. A., 275, 372
Vallin, A. F., 286
Van Ceulen. *See* Ludolph van Ceulen
Van Dam, Jan, 52, 208
Van der Hovcke, Daniel, 208
Van der Hoecke, Gielis, survey of his signs, 150; 147, 204, 319
Vandermonde, C. A., imaginary $\sqrt{-1}$, 346
Van der Schuere, Jacob, 208
Van Musschenbroek, P., 267
Van Schooten, Fr., Jr., 176, 177, 210, 232; aggregation, 344, 351; decimals, 283; difference, 262; equality, 264; geometrical proportion, 254; lettering of figures, 377; pictographs, 365; powers, 296, 304, 307, 308; radical sign, 327, 329, 333; use of a star, 356; comma for multiplication, 232
Van Steyn, G., 208
Variation, 259
Varignon, P., 255; radical sign, 331; use of star, 356
Vaulezard, J. L. de, 351
Vazquez, M., 258, 286
Vectors, 373
Venema, P., 259; radical signs, 336
Veronese, G.: congruent in geometry, 374; not equal, 383
Vieta, Francis: survey of his signs, 176–78; 188, 196, 204, 206, 262, 384; aggregation, 343, 344, 351, 353; decimal fractions, 278; general exponents, 308; indicating multiplication, 217; *l* for *latus*, 290, 322, 327; letters for coefficients, 199, 360; powers, 297, 307; radical sign $\sqrt{}$, 327, 333; use of vowels for unknowns, 164, 176, 339
Vigesimal scale: Aztecs, 66; Maya, 68; North American Indians, 67
Villareal, F., 286
Vinculum: survey of, 342–46; in Bombelli, 145; Chuquet, 130; Hérigone, 189; joined to radical by Descartes, 191, 333; Leibniz, 197; Vieta, 177
Visconti, A. M., 145, 199
Vitalis, H.; 206, 268, 269; use of R, 321
Vnicorno, J. *See* Unicorno, J.
Voizot, P., 96
Voss, A., 211

Wachter, G., 96

Waessenaer, 297

Walkingame, 331

Wallis, John: survey of his signs. 195, 196; 28, 132, 139, 186, 192, 196, 210, 231, 237, 248, 264, 384, 385, 386; aggregation, 345, 347, 348, 353; arithmetical proportion, 249; equality, 266; general exponents, 308, 311; general root indices, 330; geometrical proportion, 251, 252; imaginaries, 346; lettering figures, 376; parentheses, 350; quoted, 195, 199, 307; radical signs, 330, 332; sexagesimals, 44

Walter, Thomas, 330

Walther, J. L., 48, 201

Wappler, E., 50, 82, 201, 324

Ward, John, 252, 307, 386; angle, 361; circle, 367

Ward, Seth, 251

Waring, E.: aggregation, 351; complicated exponents, 313; imaginary $\sqrt{-1}$, 346; use of a star, 356

Weatherburn, C. E., 215

Webber, S., 287

Webster, Noah, 340

Weidler, 96

Weierstrass, K., use of star, 356

Weigel, E., 207, 249, 255; equality, 266, 268

Wells, E., 231, 252, 285

Wells, Webster: arcs, 370; equivalent, 375

Wells, W., and W. W. Hart: congruent in geometry, 372; ≡, 374

Wentworth, G. A.: angle, 360; equivalent, 375; parallelogram, 365; right angle, 363

Wertheim, G., 148, 173, 296, 339

Wersellow, Otto, 208

Whiston, W., 248, 269, 285, 286, 307; edition of Tacquet's *Euclid*, 386; radical signs, 331; use of a star, 356

White, E. E., coincides with, 374

Whitworth, W. A., dot for multiplication, 233

Widman, Johann, 146, 201, 202, 205, 208, 219, 221, 272, 384; radical sign, 293, 318

Wieleitner, H., 264, 326, 341

Wilczynski, E. J., 524

Wildermuth, 91

Wilkens, 212

Wilson, John, 248, 253; angle, 362; perpendicular, 364; parallel, 368

Wing, V., 44, 244, 251, 252, 253, 258

Wingate, E., 222, 251, 283; radical signs, 332

Winterfeld, von, 212, 246

Witting, A., 159

Woepcke, F., 74, 118, 124, 339

Wolf, R., 278

Wolff (also Wolf), Chr., 207, 233, 238; arithmetical proportion, 249; astronomical signs, 358; dot for multiplication, 285; equality, 265; geometrical proportion, 255, 259; radical sign, 331; similar, 372

Wolletz, K., 288

Workman, Benjamin, 225, 254

Worpitzky, J., notation for equal triangles, 381

Wren, Chr., 252

Wright, Edward, 231, 261; decimal fractions, 281

Wright, J. M. F., quoted, 386

Xylander (Wilhelm Holzmann), 101, 121, 263, 178; unknowns, 339

York, Thomas, 225, 245, 254

Young, J. W., and A. J. Schwartz, congruent in geometry, 374

Ypey, Nicolaas, 257

Zahradníček, K., 288

Zaragoza, J., 219, 254; radical signs, 330

Zero: symbol for, 5, 11, 68, 84, 109; forms of, in Hindu-Arabic numerals, 81, 82, 83; omicron, 87

Zirkel, E., 218

Volume II

NOTATIONS MAINLY IN HIGHER MATHEMATICS

PREFACE TO THE SECOND VOLUME

The larger part of this volume deals with the history of notations in higher mathematics. The manuscript for the parts comprising the two volumes of this History was completed in August, 1925, but since then occasional alterations and additions have been made whenever new material or new researches came to my notice.

Some parts of this History appeared as separate articles in scientific and educational journals, but later the articles were revised and enlarged.

I am indebted to Professor R. C. Archibald and to Professor L. C. Karpinski for aid in the arduous task of reading the proofs of this volume.

FLORIAN CAJORI

UNIVERSITY OF CALIFORNIA

TABLE OF CONTENTS

INTRODUCTION TO THE SECOND VOLUME

PARAGRAPHS

I. TOPICAL SURVEY OF SYMBOLS IN ARITHMETIC AND ALGEBRA
    (ADVANCED PART) 388–510

    Letters Representing Magnitudes 388–94
        Greek Period 388
        Middle Ages 389
        Renaissance 390
        Vieta in 1591 391
        Descartes in 1637 392
        Different Alphabets 393
        Astronomical Signs. 394
    The Letters π and e 395–401
        Early Signs for 3.1415.. 395
        First Occurrence of Sign π 396
        Euler's Use of π 397
        Spread of Jones's Notation 398
        Signs for the Base of Natural Logarithms . 399
        The Letter e 400
        B. Peirce's Signs for 3.141 .... and 2.718 401
    The Evolution of the Dollar Mark 402–5
        Different Hypotheses 402
        Evidence in Manuscripts and Early Printed Books . 403
        Modern Dollar Mark in Print 404
        Conclusion 405
    Signs in the Theory of Numbers 406–20
        Divisors of Numbers, Residues 407
        Congruence of Numbers 408
        Prime and Relatively Prime Numbers 409
        Sums of Numbers 410
        Partition of Numbers 411
        Figurate Numbers 412
        Diophantine Expressions 413
        Number Fields 414
        Perfect Numbers 415
        Mersenne's Numbers 416
        Fermat's Numbers 417
        Cotes's Numbers 418

PARAGRAPHS

Bernoulli's Numbers 419
Euler's Numbers 420
Signs for Infinity and Transfinite Numbers 421
Signs for Continued Fractions and Infinite Series. . . . 422–38
  Continued Fractions 422
  Tiered Fractions 434
  Infinite Series 435
Signs in the Theory of Combinations 439–58
  Binomial Formula 439
  Product of Terms of Arithmetical Progression 440
  Vandermonde's Symbols 441
  Combinatorial School of Hindenburg 443
  Kramp on Combinatorial Notations 445
  Signs of Argand and Ampère 446
  Thomas Jarrett 447
  Factorial n 448
  Subfactorial N 450
  Continued Products 451
  Permutations and Combinations 452
  Substitutions 453
  Groups 454
  Invariants and Covariants 456
  Dual Arithmetic 457
  Chessboard Problem 458
Determinant Notations 459–68
  Seventeenth Century 459
  Eighteenth Century 460
  Early Nineteenth Century 461
  Modern Notations 462
  Compressed Notations 463
  Jacobian 464
  Hessian 465
  Cubic Determinants 466
  Infinite Determinants 467
  Matrix Notations 468
Signs for Logarithms 469–82
  Abbreviation for "Logarithm" 469
  Different Meanings of log x, lx, and Lx 470
  Power of a Logarithm 472
  Iterated Logarithms 473
  Marking the Characteristic 474
  Marking the Last Digit 478
  Sporadic Notations 479
  Complex Numbers 480

PARAGRAPHS

Exponentiation 481
Dual Logarithms 482
Signs of Theoretical Arithmetic 483–94
Signs for "Greater" or "Less" 483
Sporadic Symbols for "Greater" or "Less" 484
Improvised Type 485
Modern Modifications 486
Absolute Difference 487
Other Meanings of $\smile$ and $\sim$ 489
A Few Other Sporadic Symbols 491
Signs for Absolute Value 492
Zeroes of Different Origin 493
General Combinations between Magnitudes or Numbers 494
Symbolism for Imaginaries and Vector Analysis 495–510
Symbols for the Square Root of Minus One 495
De Morgan's Comments on $\sqrt{-1}$ 501
Notation for a Vector 502
Length of a Vector 504
Equality of Vectors 505
Products of Vectors 506
Certain Operators 507
Rival Vector Systems 508
Attempts at Unification 509
Tensors 510

II. SYMBOLS IN MODERN ANALYSIS 511–700
Trigonometric Notations 511–37
Origin of the Modern Symbols for Degrees, Minutes, and
    Seconds 511
Signs for Radians 515
Marking Triangles 516
Early Abbreviations of Trigonometric Lines 517
Great Britain during 1602–18 518
European Continent during 1622–32 519
Great Britain during 1624–57 520
Seventeenth-Century English and Continental Practices
    are Independent 521
England during 1657–1700 522
The Eighteenth Century 524
Trigonometric Symbols of the Eighteenth Century . . 525
Trigonometric Symbols of the Nineteenth Century . . 526
Less Common Trigonometric Functions 527
Quaternion Trigonometry 528
Hyperbolic Functions 529
Parabolic Functions 531

PARAGRAPHS

Inverse Trigonometric Functions. 532
John Herschel's Notation for Inverse Functions . . . 533
Martin Ohm's Notation for Inverse Functions . . . 534
Persistance of Rival Notations for Inverse Functions . 535
Inverse Hyperbolic Functions 536
Powers of Trigonometric Functions 537
Survey of Mathematical Symbols Used by Leibniz . . . 538–65
Introduction 538
Tables of Symbols 542
Remarks on Tables 563–65
Differential and Integral Calculus 566–639
1. Introduction 566
2. Symbols for Fluxions, Differentials, and Derivatives . 567
   a) Total Differentiation during the Seventeenth and Eighteenth Centuries. Newton, Leibniz, Landen, Fontaine, Lagrange (1797), Pasquich, Grüson, Arbogast, Kramp 567–78
   b) Criticisms of Eighteenth-Century Notations. Woodhouse, Lacroix, Lagrange 579
   c) Total Differentiation during the Nineteenth Century. Barlow, Mitchell, Herschel, Peacock, Babbage, Crelle, Cauchy (1823, 1829), M. Ohm, Cauchy and Moigno (1840), B. Peirce, Carr, Peacock, Fourier 582
   d) Partial Differentials and Partial Derivatives. Euler, Karsten, Fontaine, Monge, Condorcet, Legendre, Lagrange (1788), Lacroix, Da Cunha, L'Huilier, Lagrange (1797), Arbogast, Lagrange (1801), Crelle, Barlow, Cauchy, M. Ohm, W. R. Hamilton, W. Bolyai, Cauchy and Moigno, C. G. J. Jacobi, Hesse, B. Peirce, Strauch, Duhamel, Carr, Méray, Muir, Mansion . . . 593
3. Symbols for Integrals, Leibniz 620
4. Early Use of Leibnizian Notation in Great Britain. . 621
5. Symbols for Fluents: Later Notations in Integral Calculus. Newton, Reyneau, Crelle, Euler, Fourier, Volterra, Peano, E. H. Moore, Cauchy's Residual Calculus 622
6. Calculus Notations in the United States. 630
7. Symbols for Passing to the Limit. L'Huilier, Weierstrass Oliver, Riemann, Leathem, Dirichlet, Pringsheim, Scheffer, Peano, W. H. Young 631
8. The Sign $\frac{0}{0}$. 638
9. Concluding Observations 639

PARAGRAPHS

Finite Differences 640, 641
  Early Notations 640
  Later Notations 641
Symbols in Theory of Functions 642–66
A. Symbols for Functions in General 642
B. Symbols for Some Special Functions 647
    Symmetric Functions 647
    Gamma and Beta Functions 649
    Elliptic Functions 651
    Theta Functions 656
    Zeta Functions 659
    Power Series 661
    Laplace, Lamé, and Bessel Functions 662
    Logarithm-Integral, Cosine-Integral, etc. 665
Symbols in Mathematical Logic. 667–99
  Some Early Symbols 667
  The Sign for "Therefore" 668
  The Sign for "Because" 669
  The Program of Leibniz 670
  Signs of
    H. Lambert 671
    G. J. von Holland 672
    G. F. Castillon 673
    J. D. Gergonne 674
    Bolyai 675
    Bentham 676
    A. de Morgan 677
    G. Boole 678
    W. S. Jevons 679
    Macfarlane 680
    C. S. Peirce 681
    Ladd-Franklin and Mitchell 682
    R. G. Grassmann 684
    E. Schroeder 685
    J. H. MacColl 686
    G. Frege 687
    G. Peano 688
    A. N. Whitehead 692
    E. H. Moore 693
    Whitehead and Russell 695
    P. Poretsky 696
    L. Wittgenstein 697
    Remarks by Rignano and Jourdain 698
  A Question 699

                                                              PARAGRAPHS
III. SYMBOLS IN GEOMETRY (ADVANCED PART) 700–711
   1. Recent Geometry of Triangle and Circle, etc.   700
      Geometrographie   701
      Signs for Polyhedra.   702
      Geometry of Graphics   703
   2. Projective and Analytical Geometry   704
      Signs for Projectivity and Perspectivity   705
      Signs for Harmonic and Anharmonic Ratios . . .   706
      Descriptive Geometry   707
      Analytical Geometry   708
      Plücker's Equations   709
      The Twenty-seven Lines on a Cubic Surface . . .   710
      The Pascal Hexagram   711

IV. THE TEACHINGS OF HISTORY 712–50
  A. The Teachings of History as Interpreted by Various
      Writers. Individual Judgments 712–25
     Review of D. André   712
     Quotations from A. de Morgan   713
              J. W. L. Glaisher   714
              D. E. Smith   715
              A. Savérien   716
              C. Maclaurin   717
              Ch. Babbage   718
              E. Mach   719
              B. Branford   720
              A. N. Whitehead   721
              H. F. Baker   722
              H. Burckhardt   723
              P. G. Tait   724
              O. S. Adams.   724
              A British Committee   725
  B. Empirical Generalizations on the Growth of Mathe-
      matical Notations 727–33
     Forms of Symbols   727
     Invention of Symbols   727
     Nature of Symbols   728
     Potency of Symbols   729
     Selection and Spread of Symbols.   730
     State of Flux   731
     Defects in Symbolism   732
     Individualism a Failure   733

                                                     PARAGRAPHS
C. Co-operation in Some Other Fields of Scientific Endeavor   734
   Electric Units   734
   Star Chart and Catalogue.   735

D. Group Action Attempted in Mathematics   736
   In Vector Analysis   737
   In Potential and Elasticity   738
   In Actuarial Science   739

E. Agreements To Be Reached by International Committees
   the Only Hope for Uniformity of Notations   . . .   740

ALPHABETICAL INDEX

ILLUSTRATIONS

FIGURES   PARAGRAPHS

107. B. PEIRCE'S SIGNS FOR 3.141 . . . . AND 2.718 400

108. FROM J. M. PEIRCE'S TABLES, 1871 401

109. PILLAR DOLLAR OF 1661 402

110. FORMS THAT ARE NOT DOLLAR SYMBOLS 402

111. SYMBOLS FOR THE SPANISH DOLLAR OR PESO TRACED FROM MANU-
     SCRIPT LETTERS, CONTRACTS, AND ACCOUNT-BOOKS . . . 403

112. THE MODERN DOLLAR MARK IN THE MAKING 403

113. DOLLAR MARKS IN L'HOMMEDIEU'S DIARY, 1776 403

114. FROM CHAUNCEY LEE'S AMERICAN ACCOMPTANT, 1797 . . . 404

115. FROM CHAUNCEY LEE'S AMERICAN ACCOMPTANT, PAGE 142 . . 404

116. MULTIPLICATION TABLE FOR SEXAGESIMAL FRACTIONS . . . 513

117. MARKING THE GIVEN AND REQUIRED PARTS OF A TRIANGLE, 1618 518

118. THE GIVEN AND REQUIRED PARTS OF ANOTHER TRIANGLE, 1618 518

119. ILLUSTRATING GIRARD'S NOTATION IN TRIGONOMETRY 519

120. A PAGE OF ISAAC NEWTON'S NOTEBOOK SHOWING TRIGONOMETRIC
     SYMBOLS 522

121. LEIBNIZ' FIGURE IN MSS DATED OCTOBER 26, 1675 570

122. LEIBNIZ' FIGURE IN MSS DATED OCTOBER 29, 1675 570

123. FROM ARBOGAST'S *Calcul des Dérivations* (1880), PAGE xxi . . 578

124. MANUSCRIPT OF LEIBNIZ, DATED OCTOBER 29, 1675, IN WHICH
     HIS SIGN OF INTEGRATION FIRST APPEARS 620

125. G. FREGE'S NOTATION AS FOUND IN HIS *Grundgesetze* (1893), VOL-
     UME I, PAGE 70 687

126. BOW'S NOTATION 702

INTRODUCTION TO THE SECOND VOLUME

It has been the endeavor to present in the two volumes of this *History* a fairly complete list of the symbols of mathematics down to the beginning of the nineteenth century, and a fairly representative selection of the symbols occurring in recent literature in pure mathematics. That we have not succeeded in gathering all the symbols of modern mathematics is quite evident. Anyone hunting, for even an hour, in the jungle of modern mathematical literature is quite certain to bag symbolisms not mentioned in this *History*. The task of making a complete collection of signs occurring in mathematical writings from antiquity down to the present time transcends the endurance of a single investigator. If such a history were completed on the plan of the present work, it would greatly surpass this in volume. At the present time the designing of new symbols is proceeding with a speed that is truly alarming.

Diversity of notation is bound unnecessarily to retard the spread of a knowledge of the new results that are being reached in mathematics. What is the remedy? It is hoped that the material here presented will afford a strong induction, facilitating the passage from the realm of conjecture as to what constitutes a wise course of procedure to the realm of greater certainty. If the contemplation of the mistakes in past procedure will afford a more intense conviction of the need of some form of organized effort to secure uniformity, then this *History* will not have been written in vain.

ADDENDA

PAGE 28, line 3, *add the following:* In the Commonplace Book of
Samuel B. Beach, B.A., Yale, 1805, now kept in the Yale University Library, there is given under the year 1804, "the annual
expence about $700," for the upkeep of the lighthouse in New
Haven. The dollar mark occurs there in the conventional way
now current. Prof. D. E. Smith found the symbol $ very nearly
in the present form in Daboll's *Schoolmaster's Assistant*, 4th edition, 1799, p. 20. Dr. J. M. Armstrong of St. Paul, Minn., writes
that in the *Medical Repository*, New York (a quarterly publication), Vol. III, No. 3, November and December, 1805, and
January, 1806, p. 312, the $ is used as it is today.

PAGE 29, line 6, *add the following:* Since this volume was printed,
important additional and confirmatory material appeared in our
article "New Data on the Origin and Spread of the Dollar
Mark" in the *Scientific Monthly*, September, 1929, p. 212–216.

PAGE 145, line 1, *for* Gioseppe *read* Giuseppe Moleti

PAGE 323, lines 8 and 9, *for* in G. Cramer . . . found earlier *read*
in Claude Rabuel's *Commentaires sur la Géométrie de M. Descartes*,
Lyons, 1730, and in G. Cramer . . . found also.

In the alphabetical index *insert* Mahnke, D., 542, 543, 563.

I

TOPICAL SURVEY OF SYMBOLS IN ARITHMETIC AND ALGEBRA (ADVANCED PART)

LETTERS REPRESENTING MAGNITUDES

388. *Greek period.*—The representation of general numbers by letters goes back to Greek antiquity. Aristotle uses frequently single capital letters, or two letters, for the designation of magnitude or number. For example, he says: "If A is what moves, B what is being moved, and Γ the distance over which it was moved, and Δ the time during which it was moved, then the same force A, in the same time could move the half of B twice as far as Γ, or also in half the time Δ exactly as far as Γ."[1] In other places[2] he speaks of the "$B\Gamma$ any force," "the time EZ." In another place he explains how much time and trouble may be saved by a general symbolism.[3]

Euclid[4] in his *Elements* represented general numbers by segments of lines, and these segments are marked by one letter,[5] or by two letters placed at the ends of the segment,[6] much the same way as in Aristotle. Euclid used the language of *line* and *surface* instead of numbers and their products. In printed editions of the *Elements* it became quite customary to render the subject more concrete by writing illustrative numerical values alongside the letters. For example, Clavius in 1612 writes (Book VII, Prop. 5, scholium) "A, $\frac{2}{3}$ D $\frac{3}{4}$," and again (Book VIII, Prop. 4), "A, 6.B, 5.C, 4.D, 3." In Robert Simson's translation of Euclid and in others, the order of the English Alphabet is substituted for that of the Greek, thus A B Γ Δ E Z H Θ, etc., in Euclid are $A\ B\ C\ D\ E\ F\ G\ H$, etc., in Simson and others.[7]

---

[1] Aristotle *Physics* vii. 5.

[2] *Ibid.* viii. 10.

[3] Aristotle *Analytica posteriora* i. 5, p. 74 *a* 17. Reference taken from Gow, *History of Greek Mathematics* (Cambridge, 1884), p. 105, n. 3.

[4] Euclid's *Elements*, Book 7.

[5] Euclid's *Elements*, Book 7, Prop. 3 (ed. J. L. Heiberg), Vol. II (1884), p. 194–98.

[6] Euclid's *Elements*, Book 7, Prop. 1 (ed. Heiberg), Vol. II, p. 188–90.

[7] A. de Morgan in *Companion to the British Almanac*, for 1849, p. 5.

According to Pappus,[1] it was Apollonius of Perga who, like Archimedes, divided numbers into groups or myriads and spoke of double, triple myriads, and so on, and finally of the "κ fold" myriad. This general expression of a myriad of as high an order as we may wish marks a decided advance in notation. Whether it was really due to Apollonius, or whether it was invented by Pappus, for the more elegant explanation of the Apollonian system, cannot now be determined. But Apollonius made use of general letters, in the manner observed in Euclid, as did also Pappus, to an even greater extent.[2] The small Greek letters being used to represent numbers, Pappus employed the Greek capitals to represent general numbers.[3] Thus, as Cantor says, "The possibility presents itself to distinguish as many general magnitudes as there are capital letters."[4]

It is of some interest that Cicero,[5] in his correspondence, used letters for the designation of quantities. We have already seen that Diophantus used Greek letters for marking different powers of the unknown and that he had a special mark $\mu^{\overset{o}{}}$ for given numbers. We have seen also a symbol $rû$ for known quantities, and $yâ$ and other symbols for unknown quantities (Vol. I, § 106).

389. *Middle Ages.*—The Indian practice of using the initial letters of words as abbreviations for quantities was adopted by the Arabs of the West and again by the translators from the Arabic into Latin. As examples, of Latin words we cite *radix, res, census,* for the unknown and its square; the word *dragma* for absolute number.

In Leonardo of Pisa's *Liber abbaci* (1202),[6] the general representation of given numbers by small letters is not uncommon. He and other writers of the Middle Ages follow the practice of Euclid. He uses letters in establishing the correctness of the rules for proving operations by casting out the 9's. The proof begins thus: "To show the foundation of this proof, let *.a.b.* and *.b.g.* be two given numbers which we wish to add, and let *.a.g.* be the number joint from them

[1] Pappus *Collectio*, Book II (ed. Hultsch), Vol. I (1876), p. 2–29. See M. Cantor, *Vorlesungen über Geschichte der Mathematik*, Vol. I (3d ed., 1907), p. 347.

[2] See G. H. F. Nesselmann, *Die Algebra der Griechen* (Berlin, 1842), p. 128–30.

[3] *Pappus* (ed. Hultsch), Vol. I (1876), p. 8.

[4] M. Cantor, *op. cit.*, Vol. I (3d ed., 1907), p. 455.

[5] *Epistolae ad atticum*, Lib. II, epistola 3.

[6] *Scritti di Leonardo Pisano matematico* ... da Bald. Boncompagni (Rome, 1857–62), Vol. I, p. 394, 397, 441. See G. Eneström, *Bibliotheca mathematica* (3d ser.); Vol. XIII (1912–13), p. 181.

. . . ."[1] Observe the use of dots to bring into prominence letters occurring in the running text, a practice very common in manuscripts
of that time. In another place Leonardo proposes a problem: *a*
horses eat *b* oats in *c* days, *d* horses eat *e* oats in *f* days; if the same
amount of feed is eaten in the two cases, then the first product
.a.b.c. is equal to the second *.d.e.f.*, etc.[2] Still more frequent representation of numbers by letters occurs in Jordanus Nemorarius'
Arithmetica. Jordanus died 1237; the arithmetic was brought out in
print in 1496 and 1514 in Paris by Faber Stapulensis. Letters are
used instead of special particular numbers.[3] But Jordanus Nemorarius
was not able to profit by this generality on account of the fact that he
had no signs of operation—no sign of equality, no symbols for subtraction, multiplication, or division. He marked addition by juxtaposition. He represented the results of an operation upon two letters
by a new letter.[4] This procedure was adopted to such an extent that
the letters became as much an impediment to rapid progress on a
train of reasoning as the legs of a centipede are in a marathon race.
Letters are used occasionally in the arithmetic (1321) of the Jewish
writer Levi ben Gerson[5] of Avignon who, like Nemorarius, has no
signs of operation. Gerson uses letters in treating of permutations
and combinations. A similar cumbrousness of procedure with letters
is found in the printed editions (1483) of the *Algorismus de integris*
of the Italian Prosdocimo de' Beldomandi.[6] Less extensive, but
more skilful, use of letters is made in the thirteenth century by
Meister Gernardus[7] and in the fourteenth century by the Frenchman
N. Oresme,[8] who even prefixed numerical coefficients to the letters
in passages like ".4.*a*. excedunt. .3.*a*. in sesquitertia."

390. *Renaissance.*—The employment of letters to represent the
various powers of unknown quantities by Chuquet in his manuscript
Triparty, by Pacioli, and by the early algebraists of the sixteenth

[1] *Scritti di Leonardo Pisano*, Vol. I, p. 20, ll. 9–28.

[2] *Op. cit.*, Vol. I, p. 132, 133.

[3] See M. Curtze in *Zeitschrift f. Mathematik und Physik*, Vol. XXXVI (1891),
histor.-liter. Abt., p. 1–3.

[4] P. Treutlein in *Abhandlungen zur Geschichte der Mathematik*, Vol. II (1879),
p. 132–33.

[5] Levi ben Gerson, *Die Praxis des Rechners* (trans. G. Lange; Frankfurt a. M.,
1909); J. Carlebach, *Lewi ben Gerson als Mathematiker* (Berlin, 1910).

[6] M. Cantor, *op. cit.*, Vol. II (2d ed., 1913), p. 206.

[7] G. Eneström, *Bibliotheca mathematica* (3d ser.), Vol. XIV (1914), p. 99 ff.

[8] N. Oresme, *Algorismus proportionum* (ed. M. Curtze; Berlin, 1868), p. 22.

century has been explained elsewhere. The manuscript algebra of Adam Riese, found in 1855 in the Library at Marienberg, contains some use of letters to represent given general numbers.[1] Parts of his *Coss* were written in 1524, other parts in the interval 1544–59. Gram mateus in his *Rechenbuch* of 1518 uses in one place[2] letters as the terms of a proportion: "Wie sich hadt a zum b also hat sich c zum d. Auch wie sich hadt a $\widetilde{zu}$ c also had sich b zum d." One finds isolated cases indicating the employment of letters for given numbers in other writers, for instance, Chr. Rudolff (1525) who writes $\sqrt{c}+\sqrt{d}$, Cardan (1570)[3] who explains that $R\dfrac{a}{b}$ is equal to $\dfrac{Ra}{Rb}$, that is, that $\sqrt{\dfrac{a}{b}}=\dfrac{\sqrt{a}}{\sqrt{b}}$. During the close of the fifteenth and early part of the sixteenth century the development of symbols of operation in algebra proceeded rapidly, but quantities supposed to be given were, as a rule, represented by actual numbers; numerical coefficients were employed with few exceptions. A reader who goes over the explanations of quadratic and cubic equations in works of Tartaglia, Cardan, Stifel, is impressed by the fact. As yet literal coefficients, as we write them in $ax^2+bx=c$ and $x^3+ax=b$, were absent from algebra. In consequence there could not be a general treatment of the solution of a cubic. In its place there was given a considerable number of special cases, illustrated by equations having particular numerical coefficients appropriately chosen for each case. Thus, Cardan,[4] on August 4, 1539, discusses the irreducible case of the cubic, not by considering $x^3=ax+b$, when $\left(\dfrac{a}{3}\right)^3>\left(\dfrac{b}{2}\right)^2$, but by taking $x^3=9x+10$, where $27>25$.

391. *Vieta in 1591.*—The extremely important step of introducing the systematic use of letters to denote general quantities and general numbers as coefficients in equations is due to the great French algebraist F. Vieta, in his work *In artem analyticam isagoge* (Tours, 1591). He uses capital letters which are primarily representatives of lines and surfaces as they were with the Greek geometricians, rather than pure numbers. Owing to this conception, he stresses the idea of homogeneity of the terms in an equation. However, he does not confine himself to three dimensions; the geometric limitation is aban-

---

[1] Bruno Berlet, *Adam Riese, sein Leben, seine Rechenbücher und seine Art zu rechnen. Die Cos von Adam Riese* (Leipzig, 1892), p. 35–62.

[2] Grammateus, *Rechenbuch* (1518), Bl. CIII; J. Tropfke, *Geschichte der Elementar-Mathematik*, Vol. II (2d ed., 1921), p. 42.

[3] J. Tropfke, *op. cit.*, Vol. II (2d ed., 1921), p. 42.

[4] M. Cantor, *op. cit.*, Vol. II (2d ed., 1913), p. 489.

doned, and he proceeds as high as ninth powers—*solido-solido-solidum*. The homogeneity is illustrated in expressions like "*A planum* $+Z$ *in B*," the A is designated *planum*, a "surface," so that the first term may be of the same dimension as is the second term, Z times B. If a letter B represents geometrically a length, the product of two B's represents geometrically a square, the product of three B's represents a cube.

Vieta uses capital vowels for the designation of unknown quantities, and the consonants for the designation of known quantities. His own words are in translation: "As one needs, in order that one may be aided by a particular device, some unvarying, fixed and clear symbol, the given magnitudes shall be distinguished from the unknown magnitudes, perhaps in this way that one designate the required magnitudes with the letter A or with another vowel E, I, O, U, Y, the given ones with the letters B, G, D or other consonants."[1]

392. *Descartes in 1637.*—A geometric interpretation different from that of Vieta was introduced by René Descartes in his *La géométrie* (1637). If b and c are lengths, then bc is not interpreted as an area, but as a length, satisfying the proportion $bc:b=c:1$. Similarly, $\frac{b}{c}$ is a line satisfying the proportion $\frac{b}{c}:1=b:c$.

With Descartes, if b represents a given number it is always a positive number; a negative number would be marked $-b$. It was J. Hudde[2] who first generalized this procedure and let a letter B stand for a number, positive *or* negative.

393. *Different alphabets.*—While the Greeks, of course, used Greek letters for the representation of magnitudes, the use of Latin letters became common during the Middle Ages.[3] With the development of other scripts, their use in mathematics was sometimes invoked. In 1795 J. G. Prändel expressed himself on the use of Latin type in algebraic language as follows:

"Why Latin and Greek letters are chosen for algebraic calculation, while German letters are neglected, seems, in books composed in

[1] Vieta, *Isagoge* (Tours, 1591), fol. 7.

[2] J. Hudde, *De reductione aequationum* (1657), published at the end of the first volume of F. Van Schooten's second Latin edition of René Descartes' *Géométrie* (Amsterdam, 1659), p. 439. See G. Eneström in *Encyclopédie des scien. math.*, Tom. I, Vol. II (1907), p. 1, n. 2; also *Bibliotheca mathematica* (3d ser.), Vol. IV (1903), p. 208; *The Geometry of Descartes*, by Smith & Latham, Open Court (Chicago, 1925), p. 301.

[3] See, for instance, Gerbert in *Œuvres de Gerbert*, par A. Olleris (Paris, 1867), p. 429–45.

our language, due to the fact that thereby algebraic quantities can be instantaneously distinguished from the intermixed writing. In Latin, French and English works on algebra the want of such a convenience was met partly by the use of capital letters and partly by the use of italicized letters. After our German language received such development that German literature flourishes in other lands fully as well as the Latin, French and English, the proposal to use German letters in Latin or French books on algebra could not be recounted as a singular suggestion."[1]

"The use of Greek letters in algebraic calculation, which has found wide acceptance among recent mathematicians, cannot in itself encumber the operations in the least. But the uncouthness of the Greek language, which is in part revealed in the shape of its alphabetic characters, gives to algebraic expressions a certain mystic appearance."[2]

Charles Babbage[3] at one time suggested the rule that all letters that denote quantity should be printed in italics, but all those which indicate operations should be printed in roman characters.

The detailed use of letters and of subscripts and superscripts of letters will be treated under the separate topics of algebra and geometry.[4]

That even highly trained mathematicians may be attracted or repelled by the kind of symbols used is illustrated by the experience of Weierstrass who followed Sylvester's papers on the theory of algebraic forms until Sylvester began to employ Hebrew characters which caused him to quit reading.[5]

394. *Astronomical signs.*—We insert here a brief reference to astronomical signs; they sometimes occur as mathematical symbols. The twelve zodiacal constellations are divisions of the strip of the celestial sphere, called the "zodiac"; they belong to great antiquity.[6] The symbols representing these constellations are as follows:

[1] Johann Georg Prändel's *Algebra* (München, 1795), p. 4.   [2] *Ibid.*, p. 20.

[3] Charles Babbage, art. "Notation," in *Edinburgh Encyclopedia* (Philadelphia, 1832).

[4] Consult Vol. I, §§ 141, 148, 176, 188, 191, 198, 342, 343; Vol. II, 395–401, 443, 444, 561, 565, 681, 732.

[5] E. Lampe in *Naturwissenschaftliche Rundschau*, Vol. XII (1897), p. 361; quoted by R. E. Moritz, *Memorabilia mathematica* (1914), p. 180.

[6] Arthur Berry, *Short History of Astronomy* (New York, 1910), p. 13, 14; W. W. Bryan, *History of Astronomy* (London, 1907), p. 3, 4; R. Wolf, *Geschichte der Astronomie* (München, 1877), p. 188–91; Gustave Schlegel, *Uranographie chinoise*, Vol. I (Leyden, 1875), Book V, "Des zodiaques et des planètes."

♈	Aries, the Ram	♎	Libra, the Balance
♉	Taurus, the Bull	♏	Scorpio, the Scorpion
♊	Gemini, the Twins	♐	Sagitarius, the Archer
♋	Cancer, the Crab	♑	Capricornus, the Goat
♌	Leo, the Lion	♒	Aquarius, the Water-**Bearer**
♍	Virgo, the Maid	♓	Pisces, the Fishes

The signs for the planets, sun, moon, etc., are as follows:

☉	Sun	♃	Jupiter
☽	Moon	♄	Saturn
⊕	Earth	☊	Ascending node
☿	Mercury	☋	Descending node
♀	Venus	☌	Conjunction
♂	Mars	☍	Opposition

According to Letronne,[1] the signs for the five planets and the sun and moon occur in two manuscripts of the tenth century; these signs, except that for the moon, are not found in antiquity. The early forms of the signs differ somewhat from those given in printed books. The signs for ascending and descending nodes of the moon's orbit occur in a Greek manuscript of the fourteenth century.[2] Some forms bear resemblance to the Hindu-Arabic numerals. Particularly, those for Jupiter and Saturn look like the four and five, respectively. In the twelfth century there were marked variations in the forms of the Hindu-Arabic numerals and also in the forms of the signs for the planets, sun, and moon. It is believed by some[3] that these astronomical signs and numeral signs (being used often by the same persons) mutually influenced each other, with regard to their forms, before the introduction of printing. Hence the resemblances.

Several of the astronomical signs appear as mathematical symbols. Apparently, the sign for Pisces was chosen by L. and T. Digges as their sign of equality, but they added an additional horizontal stroke, as a cross-line. Such strokes were applied by them also to their symbols for powers of the unknown (Vol. I, § 170). The sign for Taurus, placed horizontally, with the open end to the left, was, we believe, the sign of

[1] Letronne, *Revue archéologique* (1st ser.), Vol. III (Paris, 1846), p. 153, 253–63.

[2] P. Tannery, *Mémoires scientifiques*, Vol. IV (1920), p. 356, 359. Tannery gives facsimile reproductions.

[3] G. Horn–D'Arturo, "Numeri Arabici e simboli celesti," *Pubblicazioni dell'Osservatorio astronomico della R. Università di Bologna*, Vol. I (Roma, 1925), p. 187–204.

equality in the 1637 edition of Descartes' *Géométrie* (§ 191). The sign for Aries, placed horizontally, serves with Kästner[1] for "greater than" and "smaller than." The sign for earth is employed extensively in the modern logical exposition of algebra (§ 494). Extensive use of astronomical signs occurs in Leibniz' letters[2] to Jacob Bernoulli; for instance, $\int \mathfrak{Q} \mathring{\varphi} \, \mathfrak{D} \, dx$, where each astronomical sign stands for a certain analytic expression (§ 560). Kästner employed the signs for Sun, Moon, Mars, Venus, Jupiter, in the marking of equations, in the place of our modern Roman or Hindu-Arabic numerals.[3] Cauchy sometimes let the sign for Taurus stand for certain algebraic expressions.[4]

THE LETTERS π AND e

395. *Early signs for* 3.1415. . . . ' —John Wallis,[5] in his *Arithmetica infinitorum* (1655), lets the square $\square$ or, in some cases, the Hebrew letter "mem" which closely resembles a square, stand for $4/3,14149$ . . . . ; he expresses $\square$ as the ratio of continued products and also, as William Brounker had done before him, in the form of a continued fraction.

Perhaps the earliest use of a single letter to represent the ratio of the length of a circle to its diameter occurs in a work of J. Chr. Sturm,[6] professor at the University of Altdorf in Bavaria, who in 1689 used the letter e in a statement, "si diameter alicuius circuli ponatur a, circumferentiam appellari posse ea (quaecumque enim inter eas fuerit ratio, illius nomen potest designari littera e)." Sturm's letter failed of general adoption.

Before Sturm the ratio of the length of a circle to its diameter was represented in the fractional form by the use of two letters. Thus,

---

[1] A. G. Kästner, *Anfangsgründe der Arithmetik, Geometrie* . . . . (Göttingen, 1758), p. 89, 385.

[2] C. I. Gerhardt, *Leibnizens Mathematische Schriften*, Vol. III (Halle, 1855), p. 100.

[3] A. G. Kästner, *Anfangsgründe der Analysis endlicher Grössen* (Göttingen, 1760), p. 55, 269, 336, 358, 414, 417, and other places.

[4] A. L. Cauchy, *Comptes rendus*, Vol. XXIV (1847); *Œuvres complètes* (1st ser.), Vol. X, p. 282.

[5] John Wallis, *Arithmetica infinitorum* (Oxford, 1655), p. 175, 179, 182.

[6] J. Christoph Sturm, *Mathesis enucleata* (Nürnberg, 1689), p. 81. This reference is taken from A. Krazer's note in *Euleri opera omnia* (1st ser.), Vol. VIII, p. 134.

William Oughtred[1] designated the ratio (§ 185) by $\frac{\pi}{\delta}$. He does not define π and δ separately, but no doubt π stood for periphery and δ for diameter. The radius he represents by R. We quote from page 66 of the 1652 edition: "Si in circulo sit $7.22::\delta.\pi::113.355$: erit $\delta.\pi::2R.P$: periph. Et $\pi.\delta::\frac{1}{2}P.R$: semidiam. $\delta.\pi::Rq$. Circul. Et $\pi.\delta::\frac{1}{4}Pq$. Circul." Oughtred's notation was adopted by Isaac Barrow[2] and by David Gregory.[3] John Wallis[4] in 1685 represented by π the "periphery" described by the center of gravity in a revolution. In 1698 De Moivre[5] designated the ratio of the length of the circle to the *radius* by $\frac{c}{r}$.

396. *First occurrence of the sign* π.—The modern notation for $3.14159\ldots$. was introduced in 1706. It was in that year that William Jones[6] made himself noted, without being aware that he was doing anything noteworthy, through his designation of the ratio of the length of the circle to its diameter by the letter π. He took this step without ostentation. No lengthy introduction prepares the reader for the bringing upon the stage of mathematical history this distinguished visitor from the field of Greek letters. It simply came, unheralded, in the following prosaic statement (p. 263):

"There are various other ways of finding the *Lengths* or *Areas* of particular *Curve Lines*, or *Planes*, which may very much facilitate the Practice; as for instance, in the *Circle*, the Diameter is to the Circumference as 1 to $\overline{\frac{16}{5}-\frac{4}{239}}-\frac{1}{3}\overline{\frac{16}{5^3}-\frac{4}{239^3}}-$, &c. $=3.14159$, &c. $=\pi$.

This *series* (among others for the same purpose, and drawn from the same Principle) I received from the Excellent Analyst, and my much esteem'd Friend Mr. *John Machin;* and by means thereof, *Van*

---

[1] W. Oughtred, *Clavis mathematicae* (1652), p. 66. This symbolism is given in the editions of this book of 1647, 1648, 1652, 1667, 1693, 1694. It is used also in the Appendix to the *Clavis*, on "Archimedis de Sphaera et Cylindro declaratio." This Appendix appeared in the editions of 1652, 1667, 1693.

[2] W. Whewell, *The Mathematical Works of Isaac Barrow* (Cambridge, 1860), p. 380, Lecture XXIV.

[3] David Gregory, *Philosophical Transactions*, Vol. XIX (London, 1697), p. 652 except that he writes $\frac{\pi}{\rho}$, ρ being the radius.

[4] John Wallis, *Treatise of Algebra* (1685), "Additions and Emendations," p. 170.

[5] De Moivre, *Philosophical Transactions*, Vol. XIX (1698), p. 56.

[6] William Jones, *Synopsis palmariorum matheseos* (London, 1706), p. 263.

Ceulen's Number, or that in Art. 64.38. may be Examin'd with all desirable Ease and Dispatch." Then he writes "$d = c \div \pi$" and "$c = d \times \pi$."

This was not the first appearance of the letter π in Jones's book of 1706. But in earlier passages the meanings were different. On page 241 it was used in lettering a geometric figure where it represented a point. On page 243 one finds "Periphery (π)," as previously found in Wallis.

Nor did the appearance of $\pi = 3.14159 \ldots$ on the stage attract general attention. Many mathematicians continued in the old way. In 1721 P. Varignon[1] wrote the ratio $\delta.\pi$, using for ratio the dot of Oughtred.

397. *Euler's use of* π.—In 1734 Euler[2] employed p instead of π and g instead of $\dfrac{\pi}{2}$. In a letter of April 16, 1738, from Stirling to Euler, as well as in Euler's reply, the letter p is used.[3] But in 1736 he[4] designated that ratio by the sign $1 : \pi$ and thus either consciously adopted the notation of Jones or independently fell upon it. Euler says: "Si enim est $m = \frac{1}{2}$ terminus respondens inuenitur $\dfrac{\pi}{2}$ denotante $1 : \pi$ rationem diametri ad peripheriam." But the letter is not restricted to this use in his *Mechanica*, and the definition of π is repeated when it is taken for $3.1415 \ldots$ He represented $3.1415 \ldots$ again by π in 1737[5] (in a paper printed in 1744), in 1743,[6] in 1746,[7] and in 1748.[8] Euler and Goldbach used $\pi = 3.1415 \ldots$ repeatedly in their correspondence in 1739. Johann Bernoulli used in 1739, in his correspondence with Euler, the letter c (*circumferentia*), but in a letter of 1740

[1] Pierre Varignon, *Histoire de l'Académie r. des sciences*, année 1721 (Paris, 1723), *Mémoires*, p. 48.

[2] Euler in "De summis serierum reciprocarum," *Comm. Acad. Petr.*, Vol. VII (1734–35), p. 123 ff. See von Braunmühl, *Vorlesungen über Geschichte der Trigonometrie*, Vol. II (Leipzig, 1903), p. 110.

[3] Charles Tweedie's *James Stirling* (Oxford, 1922), p. 179, 180, 185, 188.

[4] L. Euler, *Mechanica sive motus scientia analytice exposita*, Vol. I (Petrograd, 1736), p. 119, 123; Vol. II, p. 70, 80.

[5] L. Euler in *Comm. Acad. Petr. ad annum 1737*, IX (1744), p. 165. See A. von Braunmühl, *op. cit.*, Vol. II, p. 110. Euler says: "Posito π pro peripheria circuli, cuius diameter est 1, ... "

[6] L. Euler in *Miscellanea Berolinensia*, Vol. VII (1743), p. 10, 91, 136.

[7] L. Euler in *Histoire de l'académie r. des sciences, et de belles lettres*, année 1745 (Berlin, 1746), p. 44.

[8] L. Euler, *op. cit.*, année 1748 (Berlin, 1750), p. 84.

he began to use π. Likewise, Nikolaus Bernoulli employed π in his letters to Euler of 1742.[1] Particularly favorable for wider adoption was the appearance of π for 3.1415 .... in Euler's *Introductio in analysin infinitorum* (1748). In most of his later publications, Euler clung to π as his designation of 3.1415. . . . .

398. *Spread of Jones's notation.*—In 1741, $\pi = 3.14159$ .... is used in Sherwin's *Tables*.[2] Nevertheless, mathematicians in general were slow in following suit. In 1748 Diderot[3] wrote, "Soit le rapport du diametre à la circumférence $= \dfrac{1}{c}$ ... ." J. A. Segner varied in his practice; in 1751[4] he let π stand for the ratio, but in 1767 he[5] represented 3.14159 .... by $\delta : \pi$, as did Oughtred more than a century earlier. Says Segner: "Si ratio diametri ad peripheriam circuli, quam dedimus, vel alia verae satis propinqua, $\delta : \pi$, et sit diameter circuli data d, erit eiusdem circuli peripheria $= \dfrac{\pi}{\delta} . d$." Again, later, he lets π be "dimidium peripheriae" of the circle.[6] Even more vacillating was Kästner, who in his *Anfangsgründe* of 1758 lets $1 : P$ stand for the ratio of diameter to circumference,[7] and π for the circumference. He uses P in this sense in his plane geometry, and the early part of his solid geometry. Then all of a sudden he writes (p. 323) the ratio in the form $1 : \pi$ and continues this notation over nine consecutive pages. Further on (p. 353) in his trigonometry he puts $\cos u = \pi$ and $\sin u = p$; he writes, on page 367, $\cos A = \pi$, and on page 389, $\cos AP = \pi$. It cannot be said that in 1758 Kästner had settled upon any one fixed use of the letter π. In 1760 his practice had not changed;[8] he lets π be coefficient of the $(n+1)$th term of an equation; later he puts π equal to the algebraic irrational $\sqrt{a}$,then $\pi = \sqrt[a]{-1}$, then π is an angle APM

---

[1] See Paul H. von Fuss, *Correspondance mathématique et physique de quelques célèbres géomètres du XVIII siècle* (1843). Also F. Rudio, *Archimedes, Huygens, Lambert, Legendre* (Leipzig, 1892), p. 53.

[2] H. Sherwin's *Mathematical Tables* (3d éd.; revised by William Gardiner, London, 1741), p. 44.

[3] Denys Diderot, *Mémoires sur différens sujets de mathematiques* (Paris, 1748), p. 27.

[4] J. A. Segner, *Histoire de l'académie*, année 1751 (Berlin, 1753), p. 271.

[5] J. A. de Segner, *Cursus Mathematici*, Pars I (2d ed.; Halae, 1767), p. 309.

[6] Segner, *op. cit.*, Pars IV (Halae, 1763), p. 3.

[7] A. G. Kästner, *Anfangsgründe der Arithmetik, Geometrie und Trigonometrie* (Göttingen, 1758), p. 267, 268.

[8] A. G. Kästner, *Anfangsgründe der Analysis endlicher Grössen* (Göttingen, 1760), p. 107, 117, 211, 228, 254, 296, 326, 327, 413, 432.

and $\pi = R$, then $\pi = \tan MpH$, then π is a coefficient in the cubic $z^3 + \pi z + p = 0$. After that $\pi = 3.14159 \ldots$, then π is a general exponent of the variable x, and is $= 0$ in a particular case and $= 3$ in another, then again π is the coefficient of a term in an equation, and an exponent of x. Evidently π was still serving him in the rôle of a general-utility symbol. But in 1771, at last, Kästner[1] regularly reserved $\pi = 3.14159 \ldots$

Nicolas de Beguelin[2] in 1751 adopted the notation $\pi = 3.14159 \ldots$ as did also Daniel Bernoulli[3] in 1753, G. W. Krafft[4] in 1753, Daviet de Foncenex[5] in 1759.

Another noted German writer of textbooks of the eighteenth century, W. J. G. Karsten, uses π in the first volume of his *Lehrbegrif*[6] to represent a polygon, and uses no letter for $3.14 \ldots$ But in the second volume he is definite: "Wenn man hinführo ein für allemahl die Zahl 3, 1415926 u.s.f. $= \pi$ setzt, $\ldots$ so ist $p = 2r\pi = \pi d$." One finds π for $3.14159 \ldots$ in publications of C. A. Vandermonde[7] in 1770, and Laplace[8] in 1782. About the middle of the eighteenth century the letter π was used frequently by French mathematicians in mechanics and astronomy for other designations than $3.141 \ldots$, but in the latter part of that century $3.141 \ldots$ came to be generally designated by π. Unusual is the procedure of Wessel,[9] who writes $\pi = 360°$, and of L. N. M. Carnot, who, in his *Géométrie de position* (1803), page 138, takes the radius to be unity and *a-fourth* of the length of the circle to be π, so that "sin $(\pi \pm a) = +\cos a$." Another unusual procedure is that of D. Lardner,[10] who lets π be the "approxi-

[1] A. G. Kästner, *Dissertationes mathematicae et physicae* (Altenbvrgi, 1771), p. 41, 66, 67.

[2] Beguelin, *op. cit.*, année 1751 (Berlin, 1753), p. 444.

[3] Daniel Bernoulli, *Histoire de l'académie*, année 1753 (Berlin, 1755), p. 156.

[4] Georg Wolffgang Krafft, *Institutiones Geometriae Sublimioris* (Tübingen, 1753), p. 122.

[5] Daviet de Foncenex in *Miscellanea philosophico-mathematica Taurinensis*, Vol. I (1759), p. 130.

[6] W. J. G. Karsten, *Lehrbegrif der gesamten Mathematik. 1. Theil* (Greifswald, 1767), p. 304, 412.

[7] Vandermonde in *Histoire de l'Académie des Sciences*, année 1770 (Paris, 1773), p. 494.

[8] Laplace in *op. cit.*, année 1782 (Paris, 1785), p. 15.

[9] Caspar Wessel, *Essai sur la représentation analytique de la direction* (Copenhague, 1897), p. 15. This edition is a translation from the Danish (1799).

[10] Dionysius Lardner, *The First Six Books of the Elements of Euclid* (London, 1828), p. 278.

mate ratio of the circumference of a circle to its diameter," but does not state which approximate value it represents. The Italian, Pietro Ferroni,[1] in 1782 wrote the capital letters P for 3.14159 . . and Π for 6.283 . . Perhaps the earliest elementary French schoolbook to contain π in regular use was A. M. Legendre's *Éléments de géométrie* (1794), page 121.

399. *Signs for the base of natural logarithms.*—The need of a symbol to represent the base of the natural system of logarithms presented itself early in the development of the calculus. Leibniz[2] used the letter b in letters to Huygens of October 3/13, 1690, and January 27, 1691. In the latter he considers $t = \int \frac{d\nu}{1-\nu^2}$ and writes $b^t = \frac{1+\nu}{1-\nu}$, "$b$ estant une grandeur constante, dont le logarithme est 1, et le logarithme de 1 estant 0." A reviewer[3] of G. Cheyne's *Fluxionum methodus inversa* writes in 1703, "$\int dx : x = lx$ et $X^x = a^y$. (seu cum $la = 1$) $x\ lx = y$," thus suggesting the letter a.

400. *The letter e.*—The introduction of the letter e to represent the base of the natural system of logarithms is due to L. Euler. According to G. Eneström, it occurs in a manuscript written in 1727 or 1728, but which was not published until 1862.[4] Euler used e again in 1736 in his *Mechanica*,[5] Volume I, page 68, and in other places, as well as in articles of the years[6] 1747 and 1751. Daniel Bernoulli[7] used e in this sense in 1760, J. A. Segner[8] in 1763, Condorcet[9] in 1771, Lambert[10] in

[1] Pietro Ferroni, *Magnitudinum exponentialium . . . . theoria Florence* (1782), p. 228, 252.

[2] C. I. Gerhardt, *Leibnizens Mathematische Schriften*, Vol. II (Berlin, 1850), p. 53, 76.

[3] *Acta eruditorum* (Leipzig, 1703), p. 451.

[4] Euler's art., "Meditatio in experimenta explosione tormentorum nuper instituta," in the *Opera posthuma* (1862), Vol. II, p. 800–804. See G. Eneström, *Bibliotheca mathematica* (3d ser.), Vol. XIV (1913–14), p. 81.

[5] L. Euler, *Mechanica sive motus scientia analytice exposita* (St. Petersburg, 1736), p. 251, 256; also in *Comm. Acad. Petr.*, Vol. VII (1740) p. 146.

[6] L. Euler in *Histoire de l'Academie r. d. sciences et d. belles lettres de Berlin*, année 1745 (Berlin, 1746), p. 185; année 1751 (Berlin, 1753), p. 270.

[7] Daniel Bernoulli in *Histoire de l'Académie r. d. sciences*, année 1760 (Paris, 1766), p. 12.

[8] J. A. Segner, *Cursus mathematici*, Paris IV (Halae, 1763), p. 60.

[9] N. C. de Condorcet, *Histoire de l'académie*, année 1771 (Paris, 1774), p. 283.

[10] J. H. Lambert in *Histoire de l'Académie r. d. sciences et d. belles lettres*, année 1764 (Berlin, 1766), p. 188; année 1764 (Berlin, 1766), p. 412.

1764, J. A. Fas[1] in 1775. On the other hand, D'Alembert[2] in 1747 and in 1764 used the letter c for 2.718 . . . . , as did also the astronomer Daniel Melandri[3] of Upsala in 1787. The letter e for 2.718 is found in Abbé Sauri,[4] in E. Bézout,[5] in C. Kramp.[6] In Italy, P. Frisi,[7]

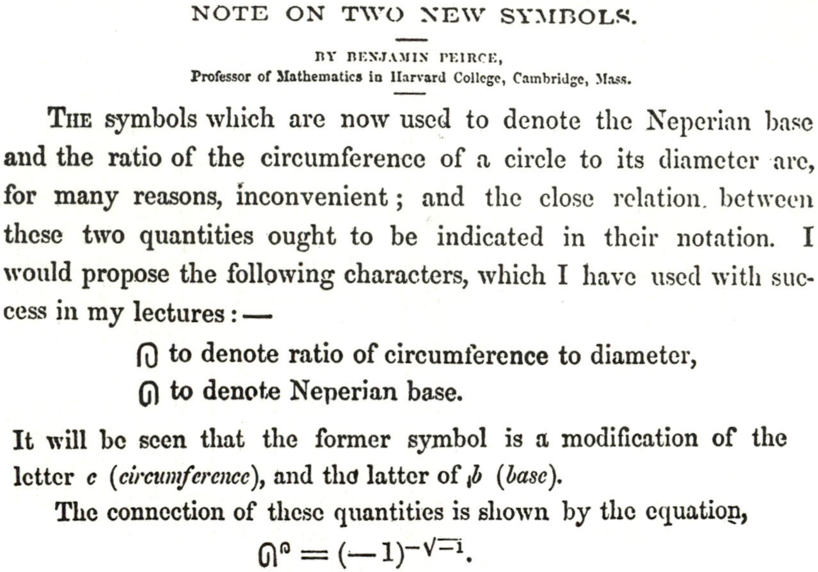

NOTE ON TWO NEW SYMBOLS.

BY BENJAMIN PEIRCE,
Professor of Mathematics in Harvard College, Cambridge, Mass.

THE symbols which are now used to denote the Neperian base and the ratio of the circumference of a circle to its diameter are, for many reasons, inconvenient; and the close relation between these two quantities ought to be indicated in their notation. I would propose the following characters, which I have used with success in my lectures : —

⋂ to denote ratio of circumference to diameter,

⋒ to denote Neperian base.

It will be seen that the former symbol is a modification of the letter c (*circumference*), and the latter of b (*base*).

The connection of these quantities is shown by the equation,

$$\text{⋒}^{\text{⋂}} = (-1)^{-\sqrt{-1}}.$$

FIG. 107.—B. Peirce's signs for 3.141 . . . . and 2.718 . . . .

in 1782, and Pietro Ferroni,[8] in the same year, used C for 2.718 . . . . , but Paoli[9] adopted the e. A. de Morgan[10] in 1842 used the epsilon ϵ for 2.718 . . . . and E for $e^{\sqrt{-1}}$.

---

[1] J. A. Fas, *Inleiding tot de Kennisse en het gebruyk der Oneindig Kleinen* (Leyden, 1775), p. 71.

[2] D'Alembert in *Histoire de l'académie*, année 1747 (Berlin, 1748), p. 228; année 1764 (Berlin, 1766), p. 412.

[3] Daniel Melandri in *Nova Acta Helvetica physico-mathematica*, Vol. I (Basel, 1787), p. 102.

[4] L'Abbé Sauri, *Cours de mathématiques*, Tome III (Paris, 1774), p. 35.

[5] E. Bézout, *Cours de mathématiques*, Tome I (2d ed.; Paris, 1797), p. 124.

[6] C. Kramp, *Eléments d'arithmétique* (Cologne, 1808), p. 28.

[7] Paulii Frisii, *Operum tomus primus* (Mediolani, 1782), p. 195.

[8] Pietro Ferroni, *Magnitudinum exponentialium logarithmorum et Trigonometriae sublimis theoria* (Florence, 1782), p. 64.

[9] Pietro Paoli, *Elementi d'algebra*, Tomo I (Pisa, 1794), p. 216.

[10] A. de Morgan, "On the Foundations in Algebra," *Transactions of the Cambridge Philos. Society*, Vol. VII (Cambridge, 1842), p. 185.

The use of the letter *c* in place of *e*, found in the writings of a few French and Italian mathematicians, occurs again in the *Analytic Mechanics* of Benjamin Peirce.[1]

401. *B. Peirce's signs for 3.141 . . . . and 2.718 . . . .* ˙—An extraordinary innovation in notation for π and *e* was suggested in 1859 by Benjamin Peirce. He made the statement[2] shown in Figure 107.

His sons, Charles Saunders Peirce and James Mills Peirce, used this notation in their articles; the latter placed the symbols shown in Figure 108 on the title-page of his *Three and Four Place Tables* (Boston, 1871). But Peirce's other pupils, Joseph Winlock, Chauncey

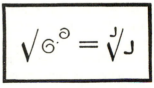

FIG. 108.—From J. M. Peirce's *Tables* (1871)

Wright, and Truman Henry Safford, used the symbol π in the first volume of the *Mathematical Monthly*.

THE EVOLUTION OF THE DOLLAR MARK

402. *Different hypotheses.*—There are few mathematical symbols the origin of which has given rise to more unrestrained speculation and less real scientific study than has our dollar mark, $. About a dozen different theories have been advanced by men of imaginative minds, but not one of these would-be historians permitted himself to be hampered by the underlying facts. These speculators have dwelt with special fondness upon monogrammatic forms, some of which, it must be admitted, maintain considerable antecedent probability. Breathes there an American with soul so dead that he has not been thrilled with patriotic fervor over the "U.S. theory" which ascribes the origin of the $ mark to the superposition of the letters *U* and *S?* This view of its origin is the more pleasing because it makes the symbol a strictly American product, without foreign parentage, apparently as much the result of a conscious effort or an act of invention as is the sewing machine or the cotton gin. If such were the case, surely some traces of the time and place of invention should be traceable; there ought to be the usual rival claimants. As a matter of fact, no one has ever advanced real evidence in the form of old manuscripts, or connected the symbol with a particular place or individual. Nor

[1] Benjamin Peirce, *Analytic Mechanics* (New York, 1855), p. 52.

[2] J. D. Runkle's *Mathematical Monthly*, Vol. I, No. 5 (February, 1859), p. 167, 168, "Note on Two New Symbols."

have our own somewhat extensive researches yielded evidence in
support of the "U.S. theory." The theory that the $ is an entwined
U and *S*, where *U S* may mean "United States" or one "Uncle Sam,"
was quoted in 1876 from an old newspaper clipping in the *Notes and
Queries* (London);[1] it is given in cyclopedic references. In the absence
of even a trace of evidence from old manuscripts, this explanation
must give way to others which, as we shall find, rest upon a strong
basis of fact. Possibly these statements suffice for some minds. However, knowing that traditional theories are dear to the heart of man,
an additional coup de grâce will not be superfluous. The earliest
high official of the United States government to use the dollar mark
was Robert Morris, the great financier of the Revolution. Letters in
his own handwriting, as well as those penned by his secretary, which
we have seen,[2] give the dollar mark with only one downward stroke,
thus, $. To assume that the symbol is made up of the letters *U* and *S*
is to assert that Robert Morris and his secretary did not know what
the real dollar symbol was; the letter *U* would demand two downward
strokes, connected below. As a matter of fact, the "U.S. theory" has
seldom been entertained seriously. Perhaps in derision of this fanciful
view, another writer declares "surely the stars and stripes is the
obvious explanation."[3]

Minds influenced less by patriotic motives than by ecclesiastical
and antiquarian predilections have contributed other explanations of
our puzzle. Thus the monogrammatic form of *I H S* (often erroneous-
ly interpreted as *Jesus, Hominum Salvator*) has been suggested.[4] The
combination of *H S* or *I I S*, which were abbreviations used by the
Romans for a coin called *sestertius*, have been advocated.[5] We should
expect the supporters of these hypotheses to endeavor to establish an
unbroken line of descent from symbols used at the time of Nero to the
symbols used in the time of Washington. But sober genealogical
inquiries of this sort were never made or, if made, they brought dis-
aster to the hypotheses.

A suggestion worthy of serious attention is the Portuguese symbol,
cifrão, used in designating "thousands," as in 13$786 (Vol. I, § 94).
Somehow this symbol is supposed to have received the new meaning of

---

[1] *Notes and Queries* (5th ser.), Vol. VI (London, 1876), p. 386; Vol. VII, p. 98.

[2] Letter of 1792, in Harper Memorial Library, University of Chicago; Robt.
Morris' Private Letter Book, in MSS Div. of Library of Congress.

[3] *Notes and Queries* (5th ser.), Vol. VI, p. 434.

[4] *Standard Dictionary*, art. "Dollar."

[5] M. Townsend, *U.S., an Index, etc.* (Boston, 1890), p. 420.

"dollar" and to have been transferred from its old position in thousands' place to the new position in front of the number affected. The burden of proof that the two transformations actually took place lies with the advocates of this theory. But such a proof has never been attempted in print. The present writer has examined many books and many manuscripts from which support might be expected for such a theory, if true, but nowhere has he found the slightest evidence. The names of monetary units used in Brazil at the beginning of the nineteenth century were *reis, veintein, tuston, pataca, patacon, cruzado*, and none of these was represented by the symbol $.

An interesting hypothesis is advanced by the noted historian, T. F. Medina, of Santiago de Chile. He suggests that perhaps the dollar mark was derived from the stamp of the mint of Potosi in Bolivia. This stamp was the monogrammatic *p* and *s*. Against the validity of this explanation goes the fact that forms of *p* and *s* were used as abbreviations of the *peso* before the time of the establishment of the mint at Potosi.

All the flights of fancy were eclipsed by those who carried the $ back to the "Pillars of Hercules." These pillars were strikingly impressed upon the "pillar dollar," the Spanish silver coin widely used in the Spanish-American colonies of the seventeenth and eighteenth centuries.[1] The "Pillars of Hercules" was the ancient name of the opposite promontories at the Straits of Gibraltar. The Mexican "globe dollar" of Charles III exhibited between the pillars two globes representing the old and new worlds as subject to Spain. A Spanish banner or a scroll around the Pillars of Hercules was claimed to be the origin of the dollar mark.[2] The theory supposes that the mark stamped on the coins was copied into commercial documents. No embarrassments were experienced from the fact that no manuscripts are known which show in writing the imitation of the pillars and scroll. On the contrary, the imaginative historian mounted his Pegasus and pranced into antiquity for revelations still more startling. "The device of the two pillars was stamped upon the coins" of the people who "built Tyre and Carthage"; the Hebrews had "traditions of the pillars of Jachin and Boaz in Solomon's Temple," "still further back in the remote ages we find the earliest known origin of the symbol in connection with the Deity. It was a type of reverence with the first people of the human race who worshipped the sun and the plains of

[1] *Notes and Queries* (5th ser.), Vol. VII (London, Feb. 24, 1877); *New American Cyclopedia*, Vol. VI (1859), art. "Dollar."

[2] M. Townsend, *op. cit.*, p. 420.

central Asia." The author of this romance facetiously remarks, "from thence the descent of the symbol to our own time is obvious."[1] Strange to say, the ingenious author forgot to state that this connection of the dollar mark with ancient deities accounts for the modern phrase, "the almighty dollar."

FIG. 109.—"Pillar dollar" of 1661, showing the "Pillars of Hercules." (From *Century Dictionary*, under "Pillar.")

Most sober-minded thinkers have been inclined to connect the dollar symbol with the figure 8. We have seen four varieties of this theory. The Spanish dollars were, as a rule, equivalent to eight smaller monetary units, universally known in Spain as *reales* or *reals*. The "pillar dollar" shows an 8 between the two pillars. The Spanish dollar was often called a "piece of eight." What guess could be more natural than that the 8 between two pillars suggested the abbreviation 8, which changed into $? So attractive is this explanation that those who advanced it did not consider it worth while to proceed to the prosaic task of finding out whether such symbols were actually employed in financial accounts by merchants of English and Latin America. Other varieties of theorizing claimed a union of P and 8 ("piece of eight")[2] or of R and 8 ("eight *reales*")[3] or of |8| (the vertical

---

[1] *American Historical Record*, Vol. III (Philadelphia), p. 407–8; *Baltimore American* (June 3, 1874).

[2] M. Townsend, *op. cit.*, p. 420; *Scribner's Magazine*, Vol. XLII (1907), p. 515.

[3] M. Townsend, *op. cit.*, p. 420.

lines being marks of separation)¹ or of 8/.² The "*P*8 theory" has been
given in Webster's *Unabridged Dictionary*, not In its first edition, but
in the editions since the fourth (1859) or fifth (1864). It is claimed that
this widely accepted theory rests on manuscript evidence.³ One writer
who examined old tobacco account-books in Virginia reproduces
lithographically the fancifully shaped letter *p* used to represent the
"piece of eight" in the early years. This part of his article is valuable.
But when it comes to the substantiation of the theory that $ is a
combination of *P* and 8, and that the $ had a purely local evolution in

Fig. 110.—Forms that are not dollar symbols

the tobacco districts of Virginia, his facts do not bear out his theory.
He quotes only one instance of manuscript evidence, and the reasoning in connection with that involves evident confusion of thought.⁴
To us the "*P*8 theory" seemed at one time the most promising working
hypothesis, but we were obliged to abandon it, because all evidence
pointed in a different direction. We sent inquiries to recent advocates
of this theory and to many writers of the present day on early American and Spanish-American history, but failed to get the slightest
manuscript evidence in its favor. None of the custodians of manuscript records was able to point out facts in support of this view. We
ourselves found some evidence from which a superficial observer
might draw wrong inferences. A few manuscripts, particularly one of
the year 1696 from Mexico (Oaxaca), now kept in the Ayer Collection of the Newberry Library in Chicago, give abbreviations for the
Spanish word *pesos* (the Spanish name for Spanish dollars) which
consist of the letter *p* with a mark over it that looks like a horizontal
figure 8. This is shown in Figure 110. Is it an 8? Paleographic study
goes against this conclusion; the mark signifies *os*, the last two letters
in *pesos*. This is evident from several considerations. The fact that
in the same manuscript exactly the same symbol occurs in *vez^{os}*,
the contraction for *vezinos*, or "neighbors," may suffice; an 8 is mean-
ingless here.

¹ *Notes and Queries* (5th ser.), Vol. VII (London), p. 317.

² *Scribner's Magazine*, Vol. XLII (1907), p. 515.

³ *American Historical Record*, Vol. III, p. 271.   ⁴ *Ibid.*, Vol. III, p. 271.

We have now described the various hypotheses.[1] The reader may have been amused at the widely different conclusions reached. One author gives to the $ "a pedigree as long as chronology itself." Others allow it only about 125 years. One traces it back to the worshipers of the sun in Central Asia, another attributes it to a bookkeeper in a Virginia tobacco district. Nearly every one of the dozen theories seemed so simple to its advocate as to be self-evident.

403. *Evidence in manuscripts and early printed books.*—The history of the dollar mark is difficult to trace. The vast majority of old documents give monetary names written out in full. This is the case also in printed books. Of nine Spanish commercial arithmetics of the seventeenth and eighteenth centuries, five gave no abbreviations whatever for the *peso* (also called *piastre, peso de* 8 *reales,* "piece of eight," "Spanish dollar"). In fact, some did not mention the *peso* at all. The reason for the omission of *peso* is that the part of Spain called Castile had monetary units called *reales, ducados, maravedises,* etc.; the word *peso* was used mainly in Spanish America and those towns of Spain that were in closest touch with the Spanish colonies. After the conquest of Mexico and Peru, early in the sixteenth century, Spanish-American mints, established in the various points in the Spanish possessions, poured forth the Spanish dollar in such profusion that it became a universal coin, reaching before the close of the century even the Philippines and China. In the seventeenth century the Spanish "piece of eight" was known in Virginia, and much was done to promote the influx of Spanish money into that colony. The United States dollar, adopted in 1785, was avowedly modeled on the average weight of the Spanish-dollar coins in circulation. Thomas Jefferson speaks of the dollar as "a known coin, and most familiar of all to the minds of the people."[2] No United States dollars were actually coined before the year 1794.[3] We proceed to unfold our data and to show the evolution of the dollar mark by stages so easy and natural that the conclusion is irresistible. There are no important "missing links." To enable the critical reader to verify our data, we give the sources of our evidence. No man's *ipse dixit* is a law in the world of scientific research.

We begin with information extracted from early Spanish printed books, consisting of abbreviations used for *peso* or *pesos.*

---

[1] Other possible lines of research on the origin of $ were suggested by Professor D. E. Smith in his *Rara Arithmetica* (1908), p. 470, 471, 491.

[2] D. K. Watson, *History of American Coinage* (1899), p. 15.

[3] Gordon, *Congressional Currency,* p. 118.

Ivan Vasquez de Serna[1]......1620......*Pes.*, *pes de* 8 *real*
Francisco Cassany[2]..........1763......*p*, also *ps.*
Benito Bails[3]...............1790......*pe*, seldom *p*
Manuel Antonio Valdes[4]......1808......*ps.*

Here we have the printed abbreviations *Pes.*, *ps*, *pe*, *p*. More interesting and convincing are the abbreviations found in manuscripts which record commercial transactions. We can give only a small part of the number actually seen. In our selection we are not discriminating against symbols which might suggest a conclusion different from our own. As a matter of fact, such discrimination would be difficult to make, for the reason that all the abbreviations for the *peso*, or "piece of eight," or *piastre* that we have examined point unmistakably to only one conclusion. We say this after having seen many hundreds of these symbols in manuscripts, antedating 1800, and written in Mexico, the Philippines, San Felipe de Puerto, New Orleans, and the colonies of the United States. It was a remarkable coincidence that all three names by which the Spanish dollar was best known, namely, the *peso*, *piastre*, and "piece of eight," began with the letter *p* and all three were pluralized by the use of the letter *s*. Hence *p* and *ps* admirably answered as abbreviations of any of these names. The symbols in Figure 111 show that the usual abbreviations was *ps* or *p*, the letter *p* taking sometimes a florescent form and the *s* in *ps* being as a rule raised above the *p*. The *p* and the *s* are often connected, showing that they were written in these instances by one uninterrupted motion of the pen. As seen in Figure 111, the same manuscript sometimes shows widely different shapes. The capital *P* is a rare occurrence. We have seen it used at the beginning of sentences and a few times written in ledgers at the top of columns of figures. In the sixteenth century the *ps* had above it a mark indicating the omission of part of the word, thus, *p̃ s*. Sometimes the contraction of the word *pesos* was *pss.* or *pos*. Not infrequently two or more different abbreviations are found in one and the same manuscript. The body of the text may contain the word written out in full, or

[1] Ivan Vasquez de Serna, *Reducciones de oro* (Cadiz, 1620), p. 263 ff. (In the Hispanic Museum, New York City.)

[2] Don Fr. Cassany, *Arithmetica deseada* (Madrid, 1763). (In the Library of Congress.)

[3] Don Benito Bails, *Arismetica* (Madrid, 1790). (In the Library of the American Philosophical Society, Philadelphia.)

[4] Don. M. A. Valdes, *Gazetas de Mexico* (1808). (In the Newberry Library, Chicago.)

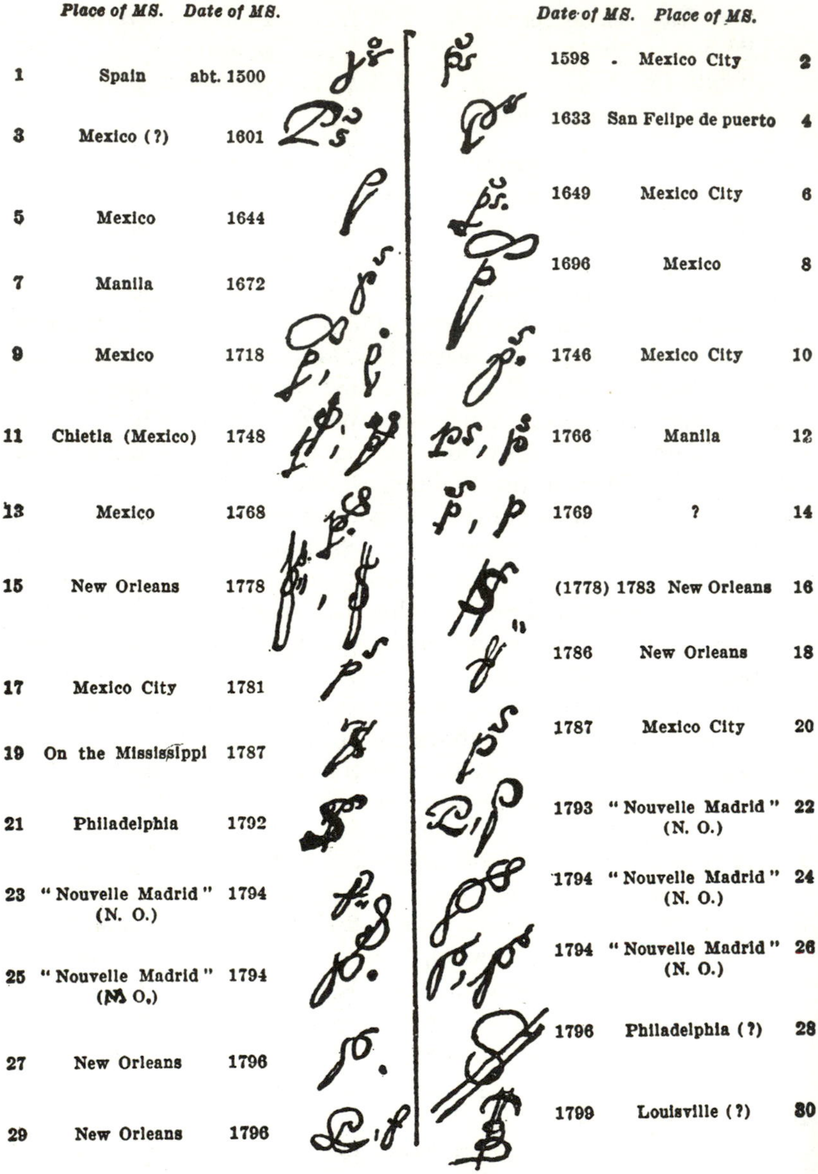

Place of MS.	Date of MS.			Date of MS.	Place of MS.	
1	Spain	abt. 1500		1598	Mexico City	2
3	Mexico (?)	1601		1633	San Felipe de puerto	4
5	Mexico	1644		1649	Mexico City	6
7	Manila	1672		1696	Mexico	8
9	Mexico	1718		1746	Mexico City	10
11	Chietla (Mexico)	1748		1766	Manila	12
13	Mexico	1768		1769	?	14
15	New Orleans	1778		(1778) 1783	New Orleans	16
				1786	New Orleans	18
17	Mexico City	1781		1787	Mexico City	20
19	On the Mississippi	1787				
21	Philadelphia	1792		1793	"Nouvelle Madrid" (N. O.)	22
23	"Nouvelle Madrid" (N. O.)	1794		1794	"Nouvelle Madrid" (N. O.)	24
25	"Nouvelle Madrid" (N. O.)	1794		1794	"Nouvelle Madrid" (N. O.)	26
27	New Orleans	1796		1796	Philadelphia (?)	28
29	New Orleans	1796		1799	Louisville (?)	30

FIG. 111.—Symbols for the Spanish dollar or *peso*, traced from MS letters, contracts, and account-books. No. 1: The historian, Dr. Cayetano Coll y Toste, of Porto Rico, says that this was the written symbol "during the time of Christopher Columbus." Nos. 2, 3, 6, 9, 10, 11, 13, 14, 17, 20 are traced from MSS owned by W. W. Blake, Avenida 16 de Septiembre 13, Mexico City. Nos. 15, 16,

18, 19 are from the Draper Collection in Wis. Hist. Libr., Madison; Nos. 15, 16 in Clark MSS, Vol. XLVIII J, p. 37, 38; Nos. 18, 19 in Clark MSS, Vol. I, p. 136, 143. Nos. 4, 5, 7, 8, 12 are from the Ayer Collection, Newberry Libr., Chicago. No. 21 from letter of Robert Morris to the Hon. Jeremiah Wadsworth, Esq., Hartford, Conn., in Harper Mem. Libr., University of Chicago. Nos. 22, 23, 24, 25, 26, 27, 28, 29, 30 are from MSS in Chicago Hist. Soc. Libr.; No. 22 in the Menard Collection, Vol. LXIV; Nos. 23, 24 in the Menard Collection, Vol. LX, p. 187; Nos. 25, 30 in Autogr. Letters, Vol. LXI; No. 26 in the Menard Collection, Vol. LXII; Nos. 27, 29 in the Menard Collection, Vol. LXIII; No. 28 in Autogr. Letters, Vol. LXXI, p. 76. The "N.O." in the figure, following "Nouvelle Madrid," should be "Mo."

contracted to *pss* or *pos*, while the margin or the head of a column of figures may exhibit *ps* or simply *p*. These were the abbreviations used by the Spanish-Americans from the sixteenth century down to about 1820 or 1830. The transition from the *ps* to our modern dollar mark was not made by the Spaniards; it was made by the English-speaking people who came in contact with the Spaniards. At the time when Mexico achieved its independence (1821), the $ was not yet in vogue there. In a Mexican book of 1834 on statistics[1] both the *ps* and the $ are used. Our $ was introduced into Hawaii by American missionaries in a translation of Warren Colburn's *Mental Arithmetic* in 1835.[2]

The transition from the florescent p^s to our dollar mark is seen in Figure 112. Apparently it is a change introduced unconsciously, in the effort to simplify the complicated motion of the pen called for in the florescent p^s. No manuscript on this point is so interesting and convincing as the two contemporaneous copies, made by the same hand, of a letter written in 1778 by Oliver Pollock, then "commercial agent of the United States at New Orleans." Pollock rendered great service to the United States, being to the west what Robert Morris was to the east. Pollock's letter is addressed to George Roger Clark, who was then heading an expedition for the capture of the Illinois country. Both copies of that letter show the $ in the body of the letter, while in the summary of accounts, at the close, the $ and the florescent p^s are both used. These documents show indeed "the modern dollar mark in the making." In the copy from which our photograph is taken, Figure 112, the 8613 dollars is indicated by the regular $, while in the other copy it is represented by the fancy p^s. Carefully examining the two symbols in our photograph, we see that the p^s is made by one continuous motion of the pen, in this order: Down on the left—up on the right—the loop of the *p*—the *s* above. On the other

[1] J. A. Escudero, *Noticias estadisticas del Estado de Chihuahua* (Mexico, 1834).

[2] Copy in the Newberry Library, Chicago.

hand, the $ symbol is made by two motions: One motion down and up for the *p*, the other motion the curve for the *s*, one symbol being superimposed upon the other.

Mr. Augustus H. Fiske, of Cambridge, Massachusetts, has pointed out to the present writer that the modern dollar mark occurs in a diary of Ezra l'Hommedieu for the year 1776. L'Hommedieu was a native of Southold, Long Island, and a Yale graduate. He was a member of the New York Provincial Assembly, which, on July 10, 1776, styled itself the Convention of the Representatives of the State

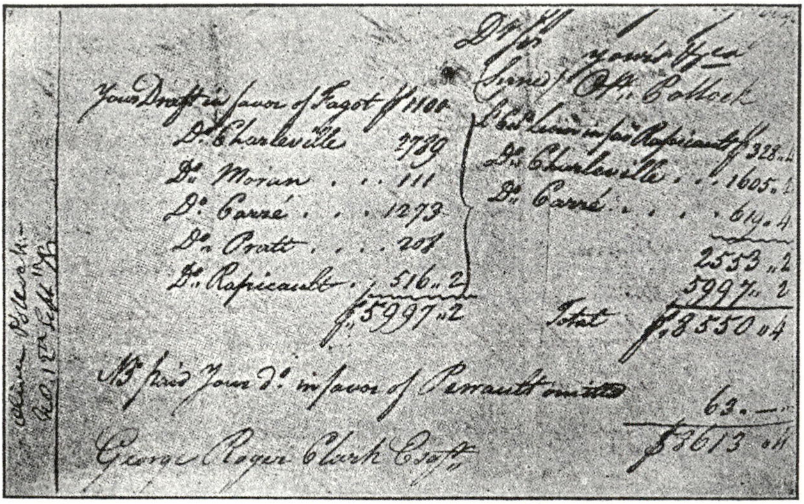

Fig. 112.—The modern dollar mark in the making. (From copy of letter by Oliver Pollock at New Orleans to George Roger Clark, 1778, Wis. Hist. Libr., Madison, Draper Collection, Vol. XXXVIII J, p. 37.)

of New York. The first date in the diary is June 10, 1776; the last is December 5, 1776. Before August 21, 1776, most sums of money are expressed in pounds and shillings. When dollars are mentioned, the word "dollar" is written out in full. On August 21 occurs the first dollar symbol (see tracing 1 in Fig. 113). Under date of August 28 the treasurer is to advance $10 for removing military stores from New York (tracing 2). On October 2 a loan of $100,000 is obtained from the Continental Congress (tracing 3); on October 3 and 4 the same sum is referred to in a similar way (tracings 4 and 5). On October 4 the treasurer is to pay 6412\frac{2}{3}$ bounty money to the rangers (tracing 6). The $ signs now appear more frequently. Their shapes are shown in the remaining tracings. We see in this diary the gradual substitution of the conventional sign $ for the spelled word. The first eleven tracings have the S crossed by only one line; the last three have the double lines.

The origin of the dollar mark is simplicity itself. It is an evolution from p^s. When the p was made by one long stroke only, as in Figure 111, Nos. 12, 14, 17, 20, then the mark took the form $, as used by Robert Morris (Fig. 111, No. 21). Before 1800 the regular mark $ was seldom used. In all our researches we have encountered it in eighteenth-century manuscripts not more than thirty or forty times. None

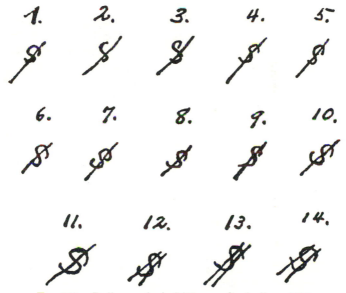

Fig. 113.—Dollar marks in L'Hommedieu's diary, 1776

of these antedates L'Hommedieu's diary of 1776. But the dollar money was then very familiar. In 1778 theater prices in printed advertisements in Philadelphia ran, "Box, one dollar." An original manuscript document of 1780 gives thirty-four signatures of subscribers, headed by the signature of George Washington. The subscribers agree to pay the sum annexed to their respective names, "in the promotion of support of a dancing assembly to be held in Morristown this present winter. The sums are given in dollars, but not one of the signers used the $ symbol; they wrote "Dollars," or "Doll," or "D."[1]

[1] *American Historical and Literary Curiosities* (Philadelphia, 1861), Plates 52, 22.

It is interesting to observe that Spanish-Americans placed the *ps* after the numerals, thus 65*ps*, while the English colonists, being accustomed to write £ before the number of pounds, usually wrote the $ to the left of the numerals, thus $65. It follows after the numerals in some letters written by Joseph Montfort Street, at Prairie du Chien in Wisconsin and Rock Island, Illinois, in 1832 and 1836.[1] The dollar mark $ is usually written after the number, as in 85$, in many Latin-American arithmetics; for instance, in those of Gabriel Izquierdo,[2] Florentino Garcia,[3] Luis Monsante,[4] Agustin de La-Rosa Toro,[5] and Maximo Vazquez.[6] In the Argentine Republic the $ is still frequently written to the right of the numerals, like this, 65$.

404. *Modern dollar mark in print.*—It has been said by various writers, including myself, that the first appearance of the dollar mark in print is in Chauncey Lee's *American Accomptant*, printed at Lansinburgh, New York, in 1797. The statement is inaccurate. Lee's sign for "dimes" more nearly resembles our dollar mark than does his sign for "dollars." We may premise that Lee assumes in his arithmetic the attitude of a reformer. Impressed by the importance of decimal fractions, he declares that "vulgar fractions are a very unimportant, if not useless part of Arithmetic,"[7] and in his Introduction he proceeds to propose a decimal system of weights and measures. In his table "Of Federal Money" (p. 56) he introduces, without comment, new signs for mills, cents, dimes, dollars, and eagles, as shown in Figures

---

[1] See correspondence of Joseph Montfort Street in the Iowa State Historical Department, in one volume. On p. 31 is a letter from Street to the Hon. Lewis Cass, secretary of the Department of War, Oct. 4, 1832, in which occurs "800 $"; on p. 57 is a letter from Street to David Lowry, Nov., 1836, in which one finds "27000 $," but also "$ 300,000."

[2] Gabriel Izquierdo, *Tratado de aritmética* (Santiago, 1859), p. 219 ff.

[3] Florentino Garcia, *El Aritmético Argentino*, Quinta edicion (Buenos Aires, 1871), p. 24 ff. It is noteworthy that the dollar mark in its modern form occurs in the first edition of this text, 1833. The author, Garcia y Coates, probably followed English texts, for the first edition contains the sign ÷ for division; the author was teaching in the "Academia española e inglesa" in Buenos Aires, and the general arrangement of the original text reminds one of arithmetics in the English language.

[4] Luis Monsante, *Lecciones de aritmetica demostrada*, Sétima edicion (Lima, 1872), p. 119 ff.

[5] Agustin de La-Rosa Toro, *Aritmetica Teorico-Practica*, Tercera edition (Lima, 1872), p. 126.

[6] *Aritmetica practica ... por* M. Maximo Vazquez ... Séptima edicion (Lima, 1875), p. 113.

[7] Ch. Lee, *American Accomptant* (1797), p. xiv.

114 and 115. It will be seen that his sign for "dollars" involves four strokes, the "dollar" being the fourth denomination of units. Two of the four strokes are curved, and they regularly inclose a space. They do not suggest the letter S in p^s from which our dollar mark descended. The probability is that Lee was familiar with the dollar mark as it was found in handwritten business documents of his day, and that he modified it in the manner shown in his book, in order to arrive at a unified plan for constructing signs for mills, cents, dimes, and dollars. It appears, therefore, that Lee's publication marks a

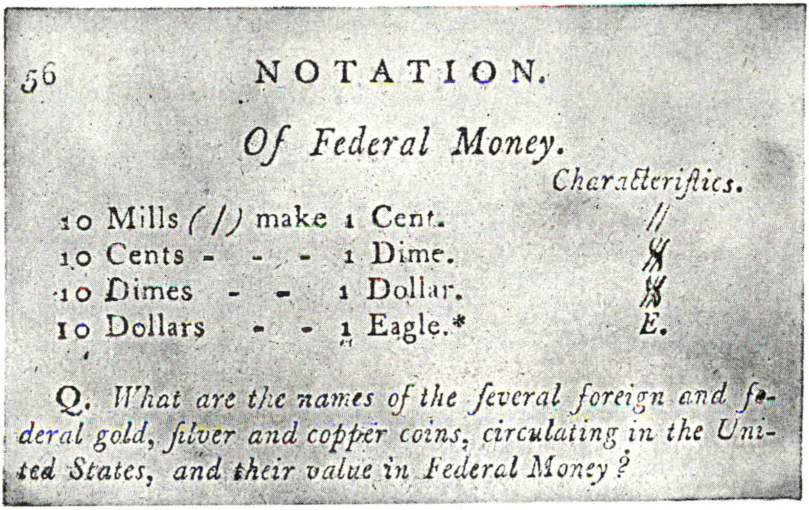

56 NOTATION.

Of Federal Money.

			Characteriſtics.
10 Mills (//)	make	1 Cent.	//
10 Cents	- - -	1 Dime.	
10 Dimes	- -	1 Dollar.	
10 Dollars	- -	1 Eagle.*	E.

Q. *What are the names of the ſeveral foreign and federal gold, ſilver and copper coins, circulating in the United States, and their value in Federal Money?*

FIG. 114.—From Chauncey Lee's *American Accomptant* (1797), showing the signs proposed by him for mills, cents, dimes, dollars, and eagles. (Courtesy of the Library of the University of Michigan.)

side excursion and does not constitute a part of the actual path of descent of our dollar mark. Erroneous, therefore, is the view of a writer[1] claiming that Lee's symbol for the dollar constitutes the primitive and true origin of the dollar mark.

After 1800 the symbol began to be used, in print, and also more frequently in writing. On September 29, 1802, William A. Washington wrote a letter on the disposal of part of the land above the Potomac belonging to the estate of George Washington. In this letter there is mention of "$20," "$30," and "$40" per acre.[2] In fact, the ledger

[1] See *Bankers Magazine*, Vol. LXII (1908), p. 857.

[2] Letter in Harper Memorial Library, University of Chicago.

kept by George Washington himself, now preserved in the Omaha
Public Library, contains the $ frequently. The earliest date in the
ledger is January 1, 1799.

The dollar mark occurs a few times in Daniel Adams' *Scholar's
Arithmetic; or Federal Accountant*, Keene, New Hampshire (4th ed.,
1807) (p. 87, 88). The more common designation in this text is
"Dolls." or "D." Adams gives also "D.," "d.," "c.," "m." for dollars, dimes, cents, mills, respectively. With him, the dollar mark
has the modern form except that the two strokes are not vertical on
the page, but slanting like a solidus. The same form is found in an
anonymous publication, *"The Columbian Arithmetician.* By an American" (Haverhill, Mass., 1811).

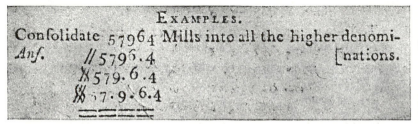

FIG. 115.—From Chauncey Lee's *American Accomptant,* p. 142

In newspapers the dollar mark rarely occurs during the first de-
cennium of the nineteenth century. In the *Boston Patriot* of September 1, 1810, one finds "$1 12," (the cents being separated from
the dollars by a blank space), but we have failed to find the mark in
the numbers of the *Columbian Centinel*, published in Boston, for the
period from August to December, 1801, and for August 12, 1809.

Samuel Webber's *Arithmetic* (Cambridge, Mass., 1812) contains
the dollar mark but not before page 125. In the first treatment of
"Federal Money," page 95, the abbreviations "E.," "D.," "d.,"
"c.," "m." are used; in the second treatment, page 125, our dollar
mark is introduced. The shape of the mark in this book is peculiar.
The S is unusually broad and is heavy; the two slanting parallel
lines are faint, and are close to each other, so short as hardly to project beyond the curves of the S, either above or below. Most of the
business problems deal with English money.

In Jacob Willetts' *Scholars' Arithmetic* (2d ed., Poughkeepsie,
1817), the dollar mark appears in its modern form with the two strokes
straight up and down. Willetts uses also the abbreviations "Dol."
and "D."

405. *Conclusion.*—It has been established that the $ is the lineal descendant of the Spanish abbreviation p^s for *pesos*, that the change from the florescent p^s to $ was made about 1775 by English-Americans who came into business relations with Spanish-Americans, and that the earliest printed $ dates back to the opening of the nineteenth century.

SIGNS IN THE THEORY OF NUMBERS

406. In the article on the theory of numbers in the *Encyclopédie des sciences mathématiques*, Tome I, Volume III (1906), page 3, which was originally written by P. Bachmann and later revised by E. Maillet, the great multiplicity and duplication of notations is deplored in the following statement: "Il n'y a malheureusement pas d'entente au sujet des notations relatives aux fonctions arithmétiques qui interviennent dans la théorie des nombres." In the present article we cannot do more than enumerate what seem to be the more important symbols introduced; the preparation of an exhaustive list would seem very onerous and also of comparatively little additional value.

407. *Divisors of numbers; residues.*—The notation $\int n$ for the sum of the divisors of n was introduced by Euler;[1] when n is prime and k an integer, he wrote $\int n^k = 1 + n + n^2 + \ldots + n^k = \dfrac{n^{k+1}-1}{n-1}$. Barlow[2] employed the sign ᔆ to save repetition of the words "divisible by," and the sign ᔆ to express "of the form of." Cunningham[3] lets "$\sigma(N)$ denote the sum of the sub-factors of N (including 1, but excluding N). It was found that, with most numbers, $\sigma^n N = 1$, when the operation (σ) is repeated often enough." Here $\sigma^2(n)$ means $\sigma\{\sigma(n)\}$. Dickson writes $s(n)$ in place of Cunningham's $\sigma(n)$; Dickson[4] lets $\sigma(n)$ represent the sum of the divisors (including 1 and n) of n; he lets also $\sigma_k(n)$ represent the sum of the kth powers of the divisors of n.

The "symbol of Legendre" is $\left(\dfrac{N}{c}\right)$; it represents the residue, $+1$

[1] L. Euler, "De numeris amicabilibus," *Opuscula varii argumenti*, Vol. II (Berlin, 1750), p. 23–107; *Commentationes arithmeticae*, Vol. I (Petrograd, 1849), p. 102, 103; *Opera omnia* (1st ser.), Vol. VI, p. 21. See L. E. Dickson, *History of the Theory of Numbers*, Vol. I (1919), p. 42.

[2] Peter Barlow, "Theory of Numbers" in *Encyclopaedia Metropolitana*, *Pure Sciences*, Vol. I (1845), p. 648.

[3] Allan Cunningham, *Proc. London Math. Soc.*, Vol. XXXV (1902–3), p. 40; L. E. Dickson, *op. cit.*, Vol. I (1919), p. 48.

[4] L. E. Dickson, *op. cit.*, Vol. I, p. 53.

or -1, when $N^{\frac{c-1}{2}}$ is divided by c. In Legendre's words: "Comme les quantités analogues à $N^{\frac{c-1}{2}}$ se recontreront fréquemment dans le cours de nos recherches, nous emploierons la caractère abrégé $\left(\dfrac{N}{c}\right)$ pour exprimer le reste que donne $N^{\frac{c-1}{2}}$ divisé par c; reste qui, suivant ce qu'on vient de voir, ne peut être que $+1$ ou -1."[1]

Legendre's notation was extended by Jacobi.[2] If $p = ff'f'' \ldots .$, where $f, f', f'' \ldots .$ are uneven prime numbers, then Jacobi defines the symbol $\left(\dfrac{x}{p}\right)$ by the equation $\left(\dfrac{x}{p}\right) = \left(\dfrac{x}{f}\right)\left(\dfrac{x}{f'}\right)\left(\dfrac{x}{f''}\right) \ldots . = \pm 1$. In Jacobi's words: "Ist nemlich, um diese Verallgemeinerung für die quadratischen Reste anzudeuten, p irgend eine ungerade Zahl $= ff'f''$ $\ldots .$, wo $f, f', f'' \ldots .$ gleiche oder verschiedene Primzahlen bedeuten, so dehne ich die schöne Legendre'sche Bezeichnung auf zusammengesetzte Zahlen p in der Art aus, dass ich mit $\left(\dfrac{x}{p}\right)$, wenn x zu p Primzahl ist, das Product $\left(\dfrac{x}{f}\right)\left(\dfrac{x}{f'}\right)\left(\dfrac{x}{f''}\right) \ldots .$ bezeichne." Legendre's and Jacobi's symbols have been written[3] also $(N|c)$ and $(x|p)$. A sign similar to that of Legendre was introduced by Dirichlet[4] in connection with biquadratic residues; according as $k^{\frac{c-1}{4}} \equiv +1$ or -1 (mod. c), he wrote $\left(\dfrac{k}{c}\right)_4 = +1$ or -1, respectively. Dickson[5] writes $(k/c)_4$.

Another symbol analogous to that of Legendre was introduced by Dirichlet[6] in the treatment of complex numbers; he designated by $\left[\dfrac{k}{m}\right]$ the number $+1$, or -1, according as k is, or is not, quadratic residue of m, such that one has $k^{\frac{1}{2}(p-1)} \equiv \left[\dfrac{k}{m}\right]$ (mod. m), k and m being

[1] A. M. Legendre, *Essai sur la théorie des nombres* (Paris, 1798), p. 186.

[2] C. G. J. Jacobi, *Crelle's Journal*, Vol. XXX (1846), p. 172; *Werke*, Vol. VI (Berlin, 1891), p. 262. Reprinted from *Monatsberichten der Königl. Akademie d. Wissenschaften zu Berlin vom Jahr 1837*. See L. E. Dickson, *op. cit.*, Vol. III, p. 84.

[3] G. B. Mathews, *Theory of Numbers* (Cambridge, 1892), p. 33, 42.

[4] G. L. Dirichlet, *Abhand. d. K. Preussisch. Akad. d. Wissensch.* (1833), p. 101–21; *Werke*, Vol. I (1889), p. 230.

[5] L. E. Dickson, *op. cit.*, Vol. II, p. 370.

[6] G. L. Dirichlet, *Crelle's Journal*, Vol. XXIV (1842), p. 307; *Werke*, Vol. I (1889), p. 551.

complex integers and p the norm of m. Eisenstein[1] employed the sign $\left[\dfrac{n}{m}\right]$ in biquadratic residues of complex numbers to represent the complex unit to which the power $n^{\frac{1}{4}(p-1)}$ (mod. m) is congruent, where m is any primary prime number, n a primary two-term prime number different from m, p the norm of m.

Jacobi[2] lets $A^{(x)}$ be the excess of the number of divisors of the form $4m+1$, of x, over the number of divisors of the form $4m+3$, of x. His statement is "Sit $B^{(x)}$ numerus factorum ipsius x, qui formam $4m+1$ habent, $C^{(x)}$ numerus factorum, qui formam $4m+3$ habent, facile patet, fore $A^{(x)}=B^{(x)}-C^{(x)}$." Glaisher[3] represents this excess, for a number n, by $E(n)$. Dickson[4] lets $E_r(n)$ or $E'_r(n)$ be the excess of the sum of the rth powers of those divisors of n which (whose complementary divisors) are of the form $4m+1$ over the sum of the rth powers of those divisors which (whose complementary divisors) are of the form $4m+3$. He also lets $\Delta'_r(n)$ be the sum of the rth powers of those divisors of n whose complementary divisors are odd. Bachmann[5] lets $t(n)$, and Landau[6] lets $T(n)$, be the number of divisors of the positive integer n; Landau writes $\mathcal{T}(n)=\Sigma T(n)$, $C=$ Euler's constant $0.57721\ldots$ and $R(x)=\mathcal{T}(x)-x\ \log\ x-(2C-1)x$, $x>0$. Dickson lets the sign $T(n)=\mathcal{T}(1)+\mathcal{T}(2)+\ldots+\mathcal{T}(n)$, where $\mathcal{T}(n)$ is the number of divisors[7] of n. Bachmann[8] designates by $\tilde{\omega}(n)$ the number of distinct prime divisors of n. Landau and Dickson[9] designated by $Og(T)$ a function whose quotient by $g(T)$ remains numerically less than a fixed finite value for all sufficiently large values of T. Before them Bachmann[10] used the form $O(n)$.

The constant C, referred to above as "Euler's constant," was in-

[1] G. Eisenstein, *Crelle's Journal*, Vol. XXX (1846), p. 192.

[2] C. G. J. Jacobi, *Werke*, Vol. I (1881), p. 162, 163; Dickson, *op. cit.*, Vol. I, p. 281.

[3] J. W. L. Glaisher, *Proc. London Math. Soc.*; Vol. XV (1883), p. 104–12.

[4] L. E. Dickson, *op. cit.*, Vol. I, p. 296.

[5] P. Bachmann, *Encyklopädie d. math. Wissensch.*, Vol. I (Leipzig, 1898–1904), p. 648.

[6] Edmund Landau, *Nachrichten von der K. Gesellschaft d. Wissensch. zu Göttingen, Math.-Phys. Klasse* (1920), p. 13.

[7] L. E. Dickson, *op. cit.*, Vol. I, p. 279.

[8] P. Bachmann, *op. cit.*, p. 648.

[9] L. E. Dickson, *op. cit.*, Vol. I, p. 305; Landau, *Handbuch der* . . . . *Primzahlen*, Vol. I (1909), p. 31, 59.

[10] P. Bachmann, *Analytische Zahlentheorie* (Leipzig, 1894), p. 401.

troduced by Euler[1] who represented it by the letter C. We mention it here, even though strictly it is not part of the theory of numbers. Mertens[2] designated it by a German capital letter $\mathfrak{C}$. It is known also as "Mascheroni's constant."[3] Mascheroni[4] in 1790 designated it by the letter A. This designation has been retained by Ernst Pascal.[5] Gauss[6] wrote $\Psi_0 = -0{,}5772156$. W. Shanks[7] adopted the designation "E. or Eul. constant." The letter E for "Eulerian constant" was adopted by Glaisher[8] in 1871 and by Adams[9] in 1878. Unfortunately, E is in danger of being confused with E_{2m}, sometimes used to designate Eulerian numbers.

In formulas for the number of different classes of quadratic forms of negative determinant, Kronecker[10] introduced the following notation which has been adopted by some later writers: n is any positive integer; m is any positive odd integer; r is any positive integer of the form $8k-1$; s is any positive integer of the form $8k+1$; $G(n)$ is the number of non-equivalent classes of quadratic forms of determinant $-n$; $F(n)$ is the number of classes of determinant $-n$, in which at least one of the two outer coefficients is odd; $X(n)$ is the sum of all odd divisors of n; $\Phi(n)$ is the sum of all divisors of n; $\Psi(n)$ is the sum of the divisors of n which are $>\sqrt{n}$, minus the sum of those which are $<\sqrt{n}$; $\Phi'(n)$ is the sum of the divisors of n of the form $8k\pm1$, minus the sum of the divisors of the form $8k\pm3$; $\Psi'(n)$ is the sum both of the divisors of the form $8k\pm1$ which are $>\sqrt{n}$ and of the divisors of the form $8k\pm3$ which are $<\sqrt{n}$, minus the sum of both the divisors of the form $8k\pm1$ which are $<\sqrt{n}$ and of the divisors of the form $8k\pm3$ which are $>\sqrt{n}$; $\phi(n)$ is the number of divisors of n of the form $4k+1$, minus the number of those of the form $4k-1$; $\psi(n)$ is the number of divisors of n of the form $3k+1$, minus the number of those of the form $3k-1$; $\phi'(n)$ is half the number of solutions in integers of $n=x^2+64y^2$;

[1] L. Euler, *Commentarii academiae Petropolitanae ad annum 1736*, Tom. VIII, p. 14–16; M. Cantor, *op. cit.*, Vol. III (2d ed.), p. 665; Vol. IV (1908), p. 277.

[2] F. Mertens, *Crelle's Journal*, Vol. LXXVII (1874), p. 290.

[3] Louis Saalschütz, *Bernoullische Zahlen* (Berlin, 1893), p. 193.

[4] L. Mascheroni, *Adnotationes ad calculum integralem Euleri* (Pavia, 1790–92), Vol. I, p. 11, 60. See also *Euleri Opera omnia* (1st ser.), Vol. XII, p. 431.

[5] Ernst Pascal, *Repertorium d. höheren Mathematik*, Vol. I (1900), p. 477.

[6] C. F. Gauss, *Werke*, Vol. III (Göttingen, 1866), p. 154.

[7] W. Shanks, *Proc. Royl. Soc. of London*, Vol. XV (1867), p. 431.

[8] J. W. L. Glaisher, *Proc. Roy. Soc. of London*, Vol. XIX (1871), p. 515.

[9] J. C. Adams, *Proc. Roy. Soc. of London*, Vol. XXVII (1878), p. 89.

[10] L. Kronecker in *Crelle's Journal*, Vol. LVII (1860), p. 248.

$\psi'(n)$ is half the number of solutions in integers of $n=x^2+3.64y^2$, in which positive, negative, and zero values of x and y are counted for both equations.

Some of the various new notations employed are indicated by the following quotation from Dickson: "Let $\chi_k(x)$ be the sum of the kth powers of odd divisors of x; $\chi_k'(x)$ that for the odd divisors $>\sqrt{x}$; $X_k'(x)$ the excess of the latter sum over the sum of the kth powers of the odd divisors $<\sqrt{x}$ of x; $\chi_k''(x)$ the excess of the sum of the kth powers of the divisors $8s\pm1>\sqrt{x}$ of x over the sum of the kth powers of the divisors $8s\pm3<\sqrt{x}$ of x."[1]

Lerch[2] of Freiburg in Switzerland wrote in 1905, $q(a)=(a^{p-1}-1)/p$, where p is an odd prime, and a any integer not a multiple of p. Similarly, Dickson[3] lets the sign q_u stand for the quotient $(u^{p-1}-1)/p$, or for $(u^{\phi(p)}-1)/p$.

Möbius[4] defined the function b_m to be zero if m is divisible by a square >1, but to be $(-1)^k$ if m is a product of k distinct primes >1, while $b_1=1$. Mertens[5] writes μn, Dickson,[6] $\mu(n)$, for the b_m of Möbius. This function is sometimes named after Mertens.

Dirichlet[7] used the sign $\left[\dfrac{n}{s}\right]$, when n and s are integers and $s\leqq n$, to designate the largest integer contained in $\dfrac{n}{s}$. Mertens[8] and later authors wrote $[x]$ for the largest integer $\leqq x$. Stolz and Gmeiner[9] write $\left[\dfrac{a}{b}\right]$ or $[a:b]$. Dirichlet[10] denoted by $N(a+bi)$ the norm a^2+b^2 of the complex number $a+bi$, a symbolism used by H. J. S. Smith[11] and others.

[1] L. E. Dickson, *op. cit.*, Vol. I, p. 305.

[2] M. Lerch, *Mathematische Annalen*, Vol. LX (1905), p. 471.

[3] L. E. Dickson, *op. cit.*, Vol. I, p. 105, 109; see also Vol. II, p. 768.

[4] A. F. Möbius in *Crelle's Journal*, Vol. IX (1832), p. 111; Möbius, *Werke*, Vol. IV, p. 598.

[5] F. Mertens, *Crelle's Journal*, Vol. LXXVII (1874), p. 289.

[6] L. E. Dickson, *op. cit.*, Vol. I, p. 441.

[7] G. L. Dirichlet, *Abhand. d. K. Preussisch. Akad. d. Wissensch. von 1849*, p. 69–83; *Werke*, Vol. II (1897), p. 52.

[8] F. Mertens, *op. cit.*, Vol. LXXVII (1874), p. 290.

[9] Stolz und J. A. Gmeiner, *Theoretische Arithmetik*, Vol. I (2d ed.; Leipzig, 1911), p. 29.

[10] G. L. Dirichlet, *Crelle's Journal*, Vol. XXIV (1842), p. 295.

[11] H. J. S. Smith, "Report on the Theory of Numbers," *Report British Association* (London, 1860), p. 254.

The designation by $\overset{\cdot}{n}$ of a multiple of the integer n is indicated in the following quotation from a recent edition of an old text: "Pour exprimer un multiple d'un nombre nous mettrons un point audessus de ce nombre: ainsi ... $\overline{327}$ signifie un multiple de 327 ..., $\overline{a,c}$ ... signifie un multiple commun aux deux nombres a, et c."[1]

408. *Congruence of numbers.*—The sign $\equiv$ to express congruence of integral numbers is due to C. F. Gauss (1801). His own words are: "Numerorum congruentiam hoc signo, $\equiv$, in posterum denotabimus, modulum ubi opus erit in clausulis adiungentes, $-16 \equiv 9$ (mod. 5), $-7 \equiv 15$ (modo 11)."[2] Gauss adds in a footnote: "Hoc signum propter magnam analogiam quae inter aequalitatem atque congruentiam invenitur adoptavimus. Ob eandem caussam ill. Le Gendre in comment. infra saepius laudanda ipsum aequalitatis signum pro congruentia retinuit, quod nos ne ambiguitas oriatur imitari dubitavimus."

The objection which Gauss expressed to Legendre's double use of the sign $=$ is found also in Babbage[3] who holds that Legendre violated the doctrine of *one notation for one thing* by letting $=$ mean: (1) ordinary equality, (2) that the two numbers between which the $=$ is placed will leave the same remainder when each is divided by the same given number. Babbage refers to Peter Barlow as using in his *Theory of Numbers* (London, 1811) the symbol $f\!f$ placed in a horizontal position (see § 407). Thus, says Babbage, Legendre used the symbolism $a^n = -1$, Gauss the symbolism $a^n \equiv -1$ (mod. p), and Barlow the symbolism $a^n \gtrless p$. Babbage argues that "we ought not to multiply the number of mathematical symbols without necessity." A more recent writer expresses appreciation of Gauss's symbol: "The invention of the symbol $\equiv$ by Gauss affords a striking example of the advantage which may be derived from appropriate notation, and marks an epoch in the development of the science of arithmetic."[4]

Among the earliest writers to adopt Gauss's symbol was C. Kramp of Strassbourg; he says: "J'ai adopté de même la signe de *congruence*, proposé par cet auteur, et composé de trois traits parallèles, au lieu de

[1] Claude-Gaspar Bachet, *Problèmes plaisants et délectables* (3d ed., par A. Labosne; Paris, 1874), p. 13.

[2] C. F. Gauss, *Disquisitiones arithmeticae* (Leipzig, 1801), art. 2; *Werke*, Vol. I (Göttingen, 1863), p. 10.

[3] Charles Babbage, art. "Notation" in the *Edinburgh Cyclopædia* (Philadelphia, 1832).

[4] G. B. Mathews, *Theory of Numbers* (Cambridge, 1892), Part I, sec. 29.

deux. Ce signe ṁ'a paru essentiel pour toute cette partie de l'analyse, qui admet les seules solutions en nombres entiers, tant positifs que négatifs."[1] The sign $\not\equiv$ is sometimes used for "incongruent."[2]

409. *Prime and relatively prime numbers.*—Peano[3] designates a prime by N_p. Euler[4] lets πD stand for the number of positive integers not exceeding D which are relatively prime to D ("denotet character πD multitudinem istam numerorum ipso D minorum, et qui cum eo nullum habeant divisorem communem"). Writing n for D, Euler's function πD was designated $\phi(n)$ by Gauss,[5] and $T(n)$ (totient of n) by Sylvester.[6] Gauss's notation has been widely used; it is found in Dedekind's edition of Dirichlet's *Vorlesungen über Zahlentheorie*[7] and in Wertheim's *Zahlentheorie*.[8] Jordan[9] generalized Euler's πD function, and represented by $[n, k]$ the number of different sets of k (equal or distinct) positive integers $\leqq n$, whose greatest common divisor is prime to n. In place of Jordan's $[n, k]$ Story[10] employed the symbol $\mathcal{T}^k(n)$, some other writers $\phi_k(n)$, and Dickson[11] $J_k(n)$.

Meissel designates by $\psi(n)$ the nth prime number,[12] so that, for instance, $\psi(4) = 5$, and by *rev* (*reversio*) the function which is the opposite[13] of ψ, so that "*rev* $\psi(x) = \psi$ *rev* $(x) = x$," and by E *rev* $(m) = \vartheta(m)$ the number of primes in the natural series from 1 to m inclusive,[14]

---

[1] Cf. Kramp, "Notations," *Élémens d'Arithmétique Universelle* (Cologne, 1808).

[2] See, for instance, L. E. Dickson, *Algebras and Their Arithmetics* (Chicago, 1923), p. 38.

[3] G. Peano, *Formulaire mathématique*, Tom. IV (1903), p. 68.

[4] *Acta Acad. Petrop.*, 4 II (or 8), for the year 1755 (Petrograd, 1780), p. 18; *Commentationes arithmeticae*, Vol. II (Petrograd, 1849), p. 127; L. E. Dickson, *op. cit.*, Vol. I, p. 61, 113.

[5] C. F. Gauss, *Disquisitiones arithmeticae* (Leipzig, 1801), No. 38. See also article by P. Bachmann and E. Maillet in *Encyclopédie des sciences mathématiques*, Tome I, Vol. III (1906), p. 3.

[6] J. J. Sylvester, *Philosophical Magazine* (5th ser.), Vol. XV (1883), p. 254.

[7] P. G. L. Dirichlet, *Vorlesungen über Zahlentheorie*, herausgegeben von R. Dedekind (3d ed.; Braunschweig, 1879), p. 19.

[8] G. Wertheim, *Anfangsgründe der Zahlentheorie* (Braunschweig, 1902), p. 42.

[9] C. Jordan, *Traité des substitutions* (Paris, 1870), p. 95–97.

[10] W. E. Story, *Johns Hopkins University Circulars*, Vol. I (1881), p. 132.

[11] L. E. Dickson, *op. cit.*, Vol. I, p. 147, 148.

[12] E. Meissel, *Crelle's Journal*, Vol. XLVIII, p. 310.

[13] E. Meissel, *ibid.*, p. 307.

[14] E. Meissel, *ibid.*, p. 313.

which Dickson[1] represents by $\theta(m)$. Landau[2] writes $\pi(x)$ for the number of primes $\leqq x$ in the series 1, 2, ... , $[x]$, where $[x]$ is the largest integer $\leqq x$. He also lets $f(x)$ be a simpler function of x, such that $\dfrac{\pi(x)-f(x)}{\pi(x)}\to 0$ when $x\to\infty$. R. D. Carmichael[3] uses $H\{y\}$ to represent the index of the highest power of the prime p dividing y, while Stridsberg[4] uses H_m to denote the index of the highest power of the prime p which divides $m!$

410. *Sums of numbers.*—Leibniz[5] used the long letter $\int$ not only as a sign of integration in his calculus, but also as the sign for the sum of integers. For example, he marked the sum of triangular numbers thus:

$$"1+3+\ 6+10+\text{etc.}=\int x$$

$$1+4+10+20+\text{etc.}=\int\int x$$

$$1+5+15+25+\text{etc.}=\int^3 x\ ."$$

This practice has been followed by some elementary writers; but certain modifications were introduced, as, for example, by De la Caille[6] who lets $\int$ stand for the sum of certain numbers, $\int^2$ for the sum of their squares, $\int^3$ for the sum of their cubes, etc. Bachmann[7] lets $\int_m(k)$ stand for the sum of the mth powers of the divisors of k, m being any given odd number. A. Thacker[8] of Cambridge used the notation $\phi(z)=1^n+2^n+\ \ldots\ldots+z^n$, where z is an integer and n a positive integer. Dickson[9] lets $\phi_k(n)$ be the sum of the kth powers of

[1] L. E. Dickson, *op. cit.*, Vol. I, p. 429.

[2] Edmund Landau, *Handbuch der Lehre von der Verteilung der Primzahlen* (Leipzig und Berlin), Vol. I (1909), p. 4.

[3] R. D. Carmichael, *Bull. Amer. Math. Soc.*, Vol. XV (1909), p. 217.

[4] E. Stridsberg, *Arkiv för Matematik, Astr., Fysik.*, Vol. VI (1911), No. 34. See L. E. Dickson, *op. cit.*, Vol. I, p. 264.

[5] G. W. Leibniz, "Historia et origo calculi differentialis" in *Leibnizens Mathematische Schriften* (ed. C. I. Gerhardt), Vol. V (1858), p. 397.

[6] D. de la Caille, *Lectiones elementares mathematicae . . . . in Latinum traductae . . . . a C(arolo) S.(cherffer)e S. J.* (Vienna, 1762), p. 107.

[7] *Encyklopädie d. Math. Wiss.*, Vol. 1 (1900–1904), p. 638.

[8] A. Thacker, *Crelle's Journal*, Vol. XL (1850), p. 89.

[9] L. E. Dickson, *op. cit.*, Vol. I, p. 140.

the integers $\leqq n$ and prime to n. Sylvester[1] in 1866 wrote $S_{j,i}$ to express the sum of all the products of j distinct numbers chosen from $1, 2, \ldots, i$ numbers. L. E. Dickson[2] writes $S_{n,m}$ for Sylvester's $S_{j,i}$, and also S_n for $1^n + 2^n + \ldots + (p-1)^n$. The sign S_n has been variously used for the sum of the nth powers of all the roots of an algebraic equation.[3] The use of Σ for "sum" is given in § 407.

The sign $F(a, N)$ stands for a homogeneous symmetric polynomial, a and N being integers.[4] Dickson[5] introduces the symbol $\underset{=}{n}$ in the following quotations: "The separation of two sets of numbers by the symbol $\underset{=}{n}$ shall denote that they have the same sum of kth powers for $k = 1, \ldots, n$. Chr. Goldbach noted that $a + \beta + \delta, a + \gamma + \delta, \beta + \gamma + \delta, \delta \underset{=}{2} a + \delta, \beta + \delta, \gamma + \delta, a + \beta + \gamma + \delta$."

411. *Partition of numbers.*—Euler's "De partitione numerorum"[6] contains no symbolism other than that of algebra. He starts out with the product $(1 + x^\alpha z)(1 + x^\beta z)(1 + x^\gamma z)(1 + x^\delta z)(1 + x^\epsilon z)$, etc., and indicates its product by

$$1 + Pz + Qz^2 + Rz^3 + Sz^4 +, \text{ etc.}$$

and considers the term $N x^n z^m$ "whose coefficient indicates in what various ways the number n may be the sum of m different terms of the series $a, \beta, \gamma, \delta, \epsilon, \zeta$, etc."

Chrystal[7] uses a quadripartite symbol to denote the number of partitions. "Thus, $P(n|p|q)$ means the number of partitions of n into p parts, the greatest of which is q"; "$P(n| \ast | \not> q)$ means the number of partitions of n into any number of parts, no one of which is to exceed q; $Pu(n| \not> p| \ast)$, the number of partitions of n into p or any less number of unequal parts unrestricted in magnitude." "$P(n| \ast 1, 2, 2^2, 2^3, \ldots)$ the number of partitions of n into any number of parts, each part being a number in the series $1, 2, 2^2, 2^3, \ldots$"

C. G. J. Jacobi represents the number of partitions, without repetition, of a number n, by $N(n = 1 \cdot x_1 + 2 \cdot x_2 + 3 \cdot x_3)$, $x_i \geqq 0$. Sylvester,[8]

[1] J. J. Sylvester, *Giornale di matematiche*, Vol. IV (Napoli, 1866), p. 344.

[2] L. E. Dickson, *op. cit.*, Vol. I, p. 95, 96.

[3] See, for instance, M. W. Drobisch, *Höhere numerische Gleichungen* (Leipzig, 1834), p. 133, 134.

[4] L. E. Dickson, *op. cit.*, Vol. I, p. 84.

[5] L. E. Dickson, *op. cit.*, Vol. II, p. 705.

[6] L. Euler, *Introductio in analysin infinitorum*, Vol. I, Editio Nova (Lugduni 1797), chap. xvi, p. 253.

[7] G. Chrystal, *Algebra*, Part II (Edinburgh, 1889), p. 527, 528.

[8] J. J. Sylvester, *Collected Mathematical Papers*, Vol. II (Cambridge, 1908), "On the Partitions of Numbers," p. 128.

using the terms "denumerant" and "denumeration," writes $\mathbb{C}$U and $\mathbb{C}$(U, V). "$\mathbb{C}$U in its explicit form $\dfrac{n;}{a,b,c;\ \ldots\ l;}$" "Herschel's symbol $r_{n,\ldots}$ as a linear function of the nth powers of the rth roots of unity, ... will be replaced by $\dfrac{n;}{r;}$." The coefficient[1] of t^n in the product of the series generated by $\dfrac{1}{1-t^a}\cdot\dfrac{1}{1-t^b}\cdot$, etc., is represented as $\dfrac{n;}{a,b,c,\ \ldots\ l;}$, the meaning of which is then extended and modified by Sylvester.[2] "A system of equations[3] in x, y, z, . . . u, may be denoted by $S(x, y, z, \ldots, u)$ or . . . . by S alone"; "$R_x S$ is the equation which expresses that the coefficient cluster and primary of S balance about the axis Ox." "If S' is what S becomes when we write in S, $fx+g$, or more generally ϕx, in place of x, we may denote S' symbolically by $\dfrac{\phi x}{x} S$," where $\dfrac{\phi x}{x}$ is an operative symbol. "$\dfrac{n-3;}{1,\ 2,\ 3,\ \ldots\ ,\ r;}$ expresses[4] the r-ary partibility of n when repetitions are allowed; $\dfrac{n-r\ \frac{r+1}{2};}{1, 2, 3\ \ldots\ r;}$ the same when repetitions are excluded."

The sign $p(n)$ is used to express the total number of ways in which a given positive integer n may be broken up into positive integral summands, counting as identical two partitions which are distinguished only by the order of the summands.[5]

When the order of summands is taken into account, then the resulting "compositions" are often marked by $c(n)$. G. H. Hardy and J. E. Littlewood[6] use g_k as a positive integer depending only on k, such that every positive integer is the sum of g_k or fewer positive kth powers; they also use $G(k)$ to represent a number such that every sufficiently large number can be represented as the sum of not more than $G(k)$ positive kth powers. We have $G_2=g_2=4$; $G_3\leq8$, $g_3=9$.

412. *Figurative numbers.*—An old symbolism for square numbers is the geometric square $\square$, for triangular numbers the triangle $\triangle$, and so on. Hence the names "polygonal" numbers and "pyramidal" numbers and the general name "figurate" numbers. L. Euler and C.

[1] *Op. cit.*, p. 132.   [2] *Op. cit.*, p. 133.   [3] *Op. cit.*, p. 138.   [4] *Op. cit.*, p. 155.

[5] A. J. Kempner, *American Mathematical Monthly*, Vol. XXX (1923), p. 357.

[6] G. H. Hardy and J. E. Littlewood, *Nachrichten von der K. Gesellschaft d. Wissensch. zu Göttingen, Math.-Phys. Klasse* (1920), p. 34. See also A. J. Kempner, *op. cit.*, p. 363, 364.

Goldbach used the geometric signs □ and △ in connection with number theory in their correspondence,[1] about 1748. Occasionally they write also ▭ to denote the product of two unequal factors. On September 7, 1748, and again later, Goldbach[2] lets □ be an even square and ⊡ an odd square; he lets also ⊟ or ⊟ stand for a square which may be even or odd. Goldbach writes "4⊟+⊡+⊡+⊡ = $8m+5\pm2$," where, in case + is taken, ⊟ is an odd square, and in case − is taken ⊟ is an even square.

Legendre[3] designates by "pol. x" a polygonal number of the order $m+2$, the side of the polygon being x, so that "pol. $2=m+2$, pol. $3=3m+3$" and pol. x, of the order $m+2$, is $\frac{m}{2}(x^2-x)+x$. For triangular numbers, $m=1$; for square numbers, $m=2$, etc.

In an arithmetical progression with the first term 1 and the common difference $m-2$, the sum of r terms is the rth m-gonal number, designated by $P_r^{(m)}$ in the *Encyclopédie des sciences mathématiques*, Tome I, Volume 1, page 30, but is designated p_m^r by L. E. Dickson,[4] who lets P_r^m stand for the pyramidal number $p_m'+p_m^2+\ldots+p_m^r$. Dickson says further: "We shall often write $\triangle_r$ or $\triangle(r)$ for the rth triangular number $r(r+1)/2$, $\triangle$ or $\triangle'$ for any triangular number, □ for any square, ②, ③, or ④ for a sum of two, three, or four squares. The rth *figurate* number of the order n is the binomial coefficient $f_n^r=\binom{r+n-1}{n}=\frac{(r+n-1)(r+n-2)\ldots r}{1.2\ldots n}$."[5]

413. *Diophantine expressions.*—In the solution of $ax+by=c$ and $ax^2+bx+c=$□, L. Euler wrote (v,a) for $av+1$, (v,a,b) for $(ab+1)v+b$, etc., while C. F. Gauss[6] employed the notations $A=[a]$, $B=[a,\beta]=\beta A+1$, $C=[a,\beta,\gamma]=\gamma B+A$, $D=[a,\beta,\gamma,\delta]=\delta C+B$, etc.

Gauss[7] employed the notation $\begin{pmatrix}a,&a',&a''\\b,&b',&b''\end{pmatrix}$ for $axx+a'x'x'+a''x''x''$

[1] P. H. Fuss, *Correspondance mathématique et physique ... du XVIIIème siècle* (St. Péterbourg), Vol. I (1843), p. 451–63, 550, 602, 630.

[2] P. H. Fuss, *op. cit.*, Vol. I, p. 476, 492.

[3] A. M. Legendre, *Théorie des nombres* (3d ed.), Vol. II (Paris, 1830), p. 340, 349.

[4] L. E. Dickson, *op. cit.*, Vol. II (1920), p. 1, 2.

[5] L. E. Dickson, *op. cit.*, Vol. II, p. 6, 7.

[6] C. F. Gauss, *Disquisitiones Arithmeticae* (1801), § 27; *Werke*, Vol. I (1863), p. 20. See L. E. Dickson, *op. cit.*, Vol. II, p. 49, 357.

[7] C. F. Gauss, *Disquisitiones Arithmeticae* (1801), arts. 266–85; *Werke*, Vol. I (1863), p. 300. L. E. Dickson, *op. cit.*, Vol. III, p. 206.

$+2bx'x''+2b'xx''+2b''xx'$. G. Eisenstein[1] considered the cubic form

$$f = (a, b, c, d) = ax^3 + 3bx^2y + 3cxy^2 + dy^3$$

with integral coefficients and let $A = b^2 - ac$, $B = bc - ad$, $C = c^2 - bd$. He called $Ax^2 + Bxy + Cy^2 = F$ the determining form of the cubic form f. The sign (a, b, c) is used by Dickson[2] for the binary quadratic form with integral coefficients $ax^2 + 2bxy + cy^2$.

Kronecker denoted by $K(D)$ the number of primitive classes of discriminant $D = b^2 - 4ac$. He put $H(D) = \sum_{h=1}^{\infty} \left(\dfrac{D}{h}\right)\dfrac{1}{h}$ and $\left(\dfrac{D}{h}\right) = \left(\dfrac{2^g}{D}\right)\left(\dfrac{D}{h'}\right)$ where $h = 2^g h'$, h' uneven, and $\left(\dfrac{2^g}{D}\right)$ and $\left(\dfrac{D}{h'}\right)$ are the Jacobi-Legendre signs.[3] Cresse[4] explains the more current notation thus: "$h(D)$ denotes the number of properly primitive, and $h'(D)$ the number of improperly primitive, classes of Gauss forms (a, b, c) of determinant $D = b^2 - ac$. Referring to Gauss' forms, $F(D)$, $G(D)$, $E(D)$, though printed in italics, will have the meaning which L. Kronecker assigned to them when printed in roman type. The class-number symbol $H(D)$ is defined as $G(D) - F(D)$. By $K(D)$ or $C(D)$ we denote the number of classes of primitive Kronecker forms of discriminant $D = b^2 - 4ac$."

$Val(\omega)$ is called the "valence of ω" and is a function of ω in a field S; it was used by Dedekind[5] in 1887. In the congruential theory of forms, L. E. Dickson uses E. H. Moore's notation, $GF[p^n]$, to represent[6] the Galois field of order p^n, the letter F being used here as in many other English articles to signify "field."

414. *Number fields.*—The German designation for number field or domain, namely, *Zahlkörper* or simply *Körper*, has given rise to the general notation[7] *Körper K*, used by Dedekind. For special number

---

[1] G. Eisenstein, *Crelle's Journal*, Vol. XXVII (1844), p. 89; L. E. Dickson, *op. cit.*, Vol. III, p. 253.

[2] L. E. Dickson, *op. cit.*, Vol. III, p. 2.

[3] L. Kronecker, *Sitzungsberichte Akad. d. Wissensch.* (Berlin, 1885), Vol. II, p. 768–80; L. E. Dickson, *op. cit.*, Vol. III, p. 138.

[4] G. H. Cresse in Dickson's *History*, Vol. III, p. 93.

[5] R. Dedekind, *Crelle's Journal*, Vol. LXXXIII (1877), p. 275. See L. E. Dickson, *op. cit.*, Vol. III, p. 125.

[6] L. E. Dickson, *op. cit.*, Vol. I, p. 249; Vol. III, p. 293.

[7] See P. G. L. Dirichlet, *Zahlentheorie*, herausgegeben von R. Dedekind (3d ed.; Braunschweig, 1879), p. 465 n.

fields Dedekind[1] used *Körper R* for the field of rational numbers, *Körper J* for the field of all complex numbers $\omega = x + yi$, where x and y are real and rational. He lets $N(\omega)$ stand for the norm, (modulus) of ω and its conjugate ω', *Körper* Ω signifies with him a finite field[2] of degree n, but Ω has been employed in the theory of equations as a more general symbol for field. In 1873, G. Cantor[3] used (Ω) to signify the *Inbegriff* of all numbers Ω which are rational functions with integral coefficients of the given series of linearly independent real numbers $\omega_1, \omega_2, \ldots$. In his *Was sind und was sollen die Zahlen* (1888), Dedekind designates by $A \ni S$ that a system A is a part of a system $S;$ he indicates by $\mathfrak{M}(A, B, C \ldots)$ a system made up of systems $A, B, C \ldots$ and by $\mathfrak{G}(A, B, C \ldots)$ the system composed of all elements common to $A, B, C \ldots;$ he marks by $\phi(s)$ an *Abbildung* of a system S whose elements are s.

The introduction of the symbol $GF[q^n]$ to represent the Galois field of order q^n is due to E. H. Moore[4] of Chicago, in 1893.

415. *Perfect numbers*.—Perfect numbers have been called also "Euclidean numbers," and have been represented by the symbol E_p where $E_p = 2^{p-1}(2^p - 1)$, when the second factor is prime. Some confusion is likely to arise between this and the E which sometimes is made to represent "Euler's constant," or "Euler's numbers." The notation N_{prf} (*nombre parfait*) was suggested by Nassò[5] in 1900 and adopted by Peano.[6] The notation P_m for a multiply perfect number n of multiplicity m (i.e., one, the sum of whose divisors, including n and 1, is the mth multiple of n) is used by Dickson.[7]

416. *Mersenne numbers*, named after Marin Mersenne, are marked M_q by Cunningham;[8] Mersenne in 1644 asserted that numbers $M_q = 2^q - 1$, q being prime and not greater than 257, are prime only when $q = 1, 2, 3, 5, 7, 13, 17, 19, 31, 67, 127, 257$.

[1] *Op. cit.*, p. 435, 436.

[2] *Op. cit.*, p. 473.

[3] G. Cantor, *Crelle's Journal*, Vol. LXXVII (1874), p. 261.

[4] E. H. Moore in *Mathematical Papers Read at the International Congress, at Chicago, 1893* (New York, 1896), p. 211. See also E. Galois, *Œuvres* (ed. E. Picard; Paris, 1897), p. 15–23.

[5] M. Nassò, *Rivista di matematica* (1900), p. 52.

[6] G. Peano, *Formulaire mathématique*, Vol. IV (1903), p. 144.

[7] L. E. Dickson, *op. cit.*, Vol. I (1919), p. 33.

[8] Allan Cunningham in *Mathematical Questions and Solutions from the Educational Times*, N.S., Vol. XIX (London, 1911), p. 81. See also L. E. Dickson, *op. cit.*, Vol. I (1919), p. 30.

417. *Fermat numbers*, $2^{2^n}+1$, are represented by F_n in Dickson's *History*.[1]

418. *Cotes's numbers* are positive fractional numbers represented by W. W. Johnson[2] by the sign ^nA_r; the numbers occur in Cotes's *Harmonia Mensurarum* as coefficients of a series. Cotes gave the values of the numbers for $n = 1, 2, \ldots, 10$. The values for $n = 1$ are embodied in the "trapezoidal rule," and for $n = 2$ in the "parabolic rule." A property of these numbers is that $^nA_0 + {}^nA_1 + \ldots + {}^nA_n = 1$.

419. *Bernoulli's numbers* are given in Jakob (James) Bernoulli's *Ars conjectandi* (1713), page 97. They occur in the formulas for the sums of the even powers of the first n integers, being the coefficients of n in the formulas. Bernoulli himself computed only the first five numbers: $\frac{1}{6}, -\frac{1}{30}, \frac{1}{42}, -\frac{1}{30}, \frac{5}{66}$.

We quote the first four formulas in Bernoulli:

$$\text{``}\int n \;\backsim\; \tfrac{1}{2}nn + \tfrac{1}{2}n\,.$$

$$\int nn \backsim \tfrac{1}{3}n^3 + \tfrac{1}{2}nn + \tfrac{1}{6}n\,.$$

$$\int n^3 \backsim \tfrac{1}{4}n^4 + \tfrac{1}{2}n^3 + \tfrac{1}{4}nn\,.$$

$$\int n^4 \backsim \tfrac{1}{5}n^5 + \tfrac{1}{2}n^4 + \tfrac{1}{3}n^3 \ast - \tfrac{1}{30}n\,.\text{''}$$

Bernoulli obtains by inspection the sum for n^c. Then he states, "Literae capitales A, B, C, D etc., ordine denotant coefficientes ultimorum terminorum pro $\int nn$, $\int n^4$, $\int n^6$, $\int n^8$, etc. nempe $A \backsim \frac{1}{6}$, $B \backsim -\frac{1}{30}$, $C \backsim \frac{1}{42}$, $D \backsim -\frac{1}{30}$." ("The capital letters A, B, C, D, etc., denote by their order the coefficients of the last terms for $\int nn$, $\int n^4$, $\int n^6$, $\int n^8$, etc., namely, $A = \frac{1}{6}$, $B = -\frac{1}{30}$, $C = \frac{1}{42}$, $D = -\frac{1}{30}$.")

Euler[3] in 1755 used the German type of the capital letters A, B, C, D, $\ldots$, to represent the absolute values of the Bernoullian coefficients, so that $A = \frac{1}{6}$, $B = \frac{1}{30}$, etc. Euler introduced the name "Bernoullian numbers." In a paper of 1769, on the summation of series, Euler[4] considers numbers representing the products of the Ber-

[1] L. E. Dickson, *op. cit.*, Vol. I, p. 375.

[2] Roger Cotes, *Harmonia Mensurarum* (Cambridge, 1722), "De methodo differentiali Newtoniana"; W. W. Johnson's article on the Cotesian numbers is in the *Quarterly Journal of Pure and Applied Mathematics* (1914), p. 52–65.

[3] L. Euler, *Institutiones calculi differentialis* (1755), Vol. II, § 122, p. 420. See also *Euleri Opera omnia* (1st ser.), Vol. X, p. 321; Vol. XII, p. 431.

[4] *Nova Comment. Petrop. XIV* (pro 1769), p. 129–67; M. Cantor, *op. cit.*, Vol. IV (1908), p. 262.

noullian numbers $\frac{1}{6}$, $\frac{1}{30}$, $\frac{1}{42}$, $\frac{1}{30}$, etc., by the numbers 6, 10, 14, 18, etc., respectively, and marks these products by the German forms of the capital letters A, B, C, D, etc. In a paper for the year 1781, Euler[1] designates the Bernoullian numbers in the same manner as he did in 1755. Euler's notation of 1755 was used by L. Mascheroni,[2] and by Gauss.[3] Von Staudt[4] employed the notation $B^{(n)}$ for the nth number; this notation occurs earlier in Klügel's *Wörterbuch.*[5]

According to G. Peano,[6] it was Euler who designated the Bernoullian numbers by the signs B_1, B_3, B_5, ... , but Peano does not give the reference. L. Saalschütz[7] says that formerly they were marked B_1, B_3, B_5, ... , as, B_2, B_4, B_6, ... , had the value zero. Scherk[8] and Grunert[9] write the first, second, ... , nth Bernoullian numbers thus: $\overset{1}{B}$, $\overset{3}{B}$, ... , $\overset{2n-1}{B}$. De Morgan[10] writes B_1, B_3, B_5, ... Binet[11] adopts the signs B_1, $-B_3$, B_5, $-B_7$, ... Others, for instance, Pascal,[12] represent these numbers by the even subscripts, B_2, B_4, B_6, ... The notation B_1, B_2, B_3, ... , which is now in wide use, was employed by Binet,[13] Ohm,[14] Raabe,[15] Stern,[16] Hermite,[17] and Adams.[18]

[1] *Acta Petrop. IV* (1781), II, p. 45; M. Cantor, *op. cit.,* Vol. IV, p. 277.

[2] L. Mascheroni, *Adnotationes ad Calculum Integralum Euleri* (Paris, 1790); *Euleri Opera omnia* (1st ser.), Vol. XII, p. 431.

[3] C. F. Gauss, *Werke,* Vol. III (Göttingen, 1866), p. 152.

[4] K. G. C. von Staudt, *Crelle's Journal,* Vol. XXI (1840), p. 373.

[5] G. S. Klügel's *Mathematisches Wörterbuch,* completed by J. A. Grunert, Vol. IV (Leipzig, 1823), "Summirung der Reihen," p. 636.

[6] G. Peano, *Formulaire mathématique,* Tome IV (Turin, 1903), § 79, p. 248.

[7] Louis Saalschütz, *Bernoullische Zahlen* (Berlin, 1893), p. 4.

[8] H. F. Scherk, *Crelle's Journal,* Vol. IV (1829), p. 299.

[9] J. A. Grunert, Supplement zu G. S. Klügel's *Wörterbuch,* Vol. I (Leipzig, 1833), "Bernoullische Zahlen."

[10] A. de Morgan, "Numbers of Bernoulli," *Penny Cyclopaedia.*

[11] J. Binet, *Comptes rendus,* Vol. XXXII (Paris, 1851), p. 920.

[12] Ernst Pascal, *Repertorium d. höheren Mathematik* (ed. A. Schepp), Vol. I (Leipzig, 1900), p. 474.

[13] J. Binet, *Journal de l'école polytechnique* (Paris,) Vol. XVI, Cahier 27 (1839), p. 240.

[14] Martin Ohm, *Crelle's Journal,* Vol. XX (1840), p. 11.

[15] J. L. Raabe, *ibid.,* Vol. XLII (1851), p. 350–51.

[16] M. A. Stern, *ibid.,* Vol. LXXXIV (1878), p. 267.

[17] C. Hermite, *ibid.,* Vol. LXXXI (1876), p. 222.

[18] J. C. Adams, *Proc. Roy. Soc. of London,* Vol. XXVII (1878), p. 88.

420. *Euler's numbers,*[1] E_{2m}, are coefficients occurring, as Ernst Pascal states, in the series sec $x = \sum_{0}^{\infty} E_{2m} \dfrac{x^{2m}}{2m!}$; the name was given to these numbers by Scherk.[2] Chrystal[3] marks Euler's numbers E_1, E_2, ...; thus $E_1 = 1$, $E_2 = 5$, $E_3 = 61$, $E_4 = 1385$, etc.

SIGNS FOR INFINITY AND TRANSFINITE NUMBERS

421. The sign ∞ to signify infinite number was introduced by John Wallis in 1655 in his *De sectionibus conicis* in this manner: *Esto enim* ∞ *nota numeri infiniti*[4] (see also § 196). The conjecture[5] has been made that Wallis, who was a classical scholar, adopted this sign from the late Roman symbol ∞ for 1,000. Nieuwentijt[6] uses the letter m to represent *quantitas infinita.*

Wallis' symbol for infinity came to be used at the beginning of the eighteenth century in the Calculus. Thus in the *Acta eruditorum* for 1708 (p. 344) one encounters the oddity "$dy = \infty$, seu infinito." The same publication[7] contains "∞^n potestas indeterminate numeri infiniti" in a review of Cheyne's *Philosophical Principles of Natural Religion* (London, 1705); Cheyne[8] used this symbol freely. The symbol came to be used extensively, for instance, by Johann Bernoulli[9] in dealing with tautochrones. Sometimes, particularly in

[1] L. Euler, *Institutiones calculi differentialis* (1755), p. 542.

[2] H. F. Scherk, *Vier mathematische Abhandlungen* (Berlin, 1825). The first *Abhandlung* concerns the Eulerian and Bernoullian numbers.

[3] G. Chrystal, *Algebra*, Part II (Edinburgh, 1889), p. 318.

[4] John Wallis, *Opera mathematica*, Vol. I (Oxford, 1695), p. 297, 405; *De sectionibus conicis*, Pars I, Prop. 1; also *Arithmetica infinitorum*, Prop. 91.

[5] W. Wattenbach, *Anleitung zur latein. Paläographie* (2d ed., 1872), Appendix, p. 41.

[6] Bernhardi Nieuwentiitii, *Analysis infinitorum seu curvilineorum proprietates ex polygonorum natura deductae* (Amstelaedami, 1695). This reference is taken from H. Weissenborn, *Die Principien der höheren Analysis* (Halle, 1856), p. 124.

[7] *Acta eruditorum* (Leipzig, 1710), p. 462.

[8] See also George Cheyne, *Philosophical Principles of Religion*, Part II (London, 1716), p. 20, 21.

[9] J. Bernoulli, *Histoire de l'academie r. d. scien.*, année 1730 (Paris, 1732), *Mémoires*, p. 78; *Opera omnia*, Vol. III (1742), p. 182, 183.

writings of Euler,[1] and some later writers,[2] the symbol is not a closed figure, but simply ∞. Another variation in form, due, apparently, to the exigencies of the composing-room, is seen in a book of Bangma,[3] containing "Sec $270° = 0 - 0$." B. Fontenelle, in his *Éléments de la géométrie de l'infini* (Paris, 1727), raises ∞ to fractional powers, as, for example (p. 43), "$\therefore 1 \cdot \infty^{\frac{1}{4}}, \infty^{\frac{2}{4}}, \infty^{\frac{3}{4}}, \infty$." In modern geometry expressions $\infty, \infty^2, \ldots, \infty^n$ are used, where the exponents signify the number of dimensions of the space under consideration.

In the theory of functions as developed by Weierstrass[4] and his followers, the symbol ∞ is used in more than one sense. First, it is used to represent an *actual* infinity. One says that a function $f(x)$, when $\dfrac{1}{f(a)} = 0$, is infinite at $x = a$, or for $x = a$, and one writes $f(a) = \infty$. In this case one does not use the $+$ or $-$ signs before ∞; that is, one does not write $+\infty$ or $-\infty$. In writing $f(\infty) = b$ one means the same as in writing $\left[f\left(\dfrac{1}{y} \right) \right]_{y=0} = b$; in writing $f(\infty) = \infty$ one means the same as in writing $\dfrac{1}{\left[f\left(\dfrac{1}{y} \right) \right]_{y=0}} = 0$. Second, in considering the *limit of a function* $f(x)$ the $+\infty$ and $-\infty$ may arise as *virtual infinities*. For example, one has $\lim\limits_{x = +\infty} f(x) = -\infty$, when, however large a positive constant P be taken, one can take a positive number Q, such that $f(x) < -P$ when $x > Q$.

In the theory of transfinite number, Georg Cantor[5] represents by ω an ordinal number of a rank next superior to any of the integers 1, 2, 3, ... Previously,[6] Cantor had used the sign ∞, but he discarded

[1] L. Euler in *Histoire de l'academie r. d. sciences et belles lettres*, année 1757 (Berlin, 1759), p. 311; année 1761 (Berlin, 1768), p. 203. Euler, *Institut. calculi diff.* (1755), Vol. I, p. 511, 745; Euler in *Novi Comment. acad. scient. imper. Petropolitanae*, Vol. V, for the years 1754, 1755 (Petropoli, 1760), p. 209.

[2] Matthias Hauser, *Anfangsgründe der Mathematik*, 1. Theil (Wien, 1778), p. 122.

[3] O. S. Bangma, *Verhandeling over de . . . . Driehoeks-meting* (Amsterdam, 1808), p. 26.

[4] K. Weierstrass, *Abh. Akad.* (Berlin, 1876), *Math.*, p. 12; *Funktionenlehre* (Berlin, 1886), p. 2; *Werke*, Vol. II, p. 78. See also A. Pringheim and J. Molk in *Encyclopédie des scien. math.*, Tom. II, Vol. I, p. 20, 31, 32 (1909).

[5] Georg Cantor, *Grundlagen einer allgemeinen Mannichfaltigkeitslehre* (Leipzig, 1883), p. 33.

[6] Georg Cantor, *Mathematische Annalen*, Vol. XVII (1880), p. 357.

that in 1882, in favor of ω, "because the sign ∞ is used frequently for the designation of undetermined infinities."[1] Cantor's first ordinal of the second class of numbers (II) is therefore ω; he[2] marks the first ordinal of the third class (III) by Ω; Bertrand Russell[3] marks the first of the $(\nu+2)$th class by ω_ν, so that ω_1 is the same as Ω.

Cantor designates by the sign aleph-zero $\aleph_0$, which is the Hebrew letter aleph with a subscript zero attached, the smallest transfinite *cardinal* number,[4] and by $\aleph_1$ the next superior cardinal number. Peano,[5] in 1895, indicated cardinal numbers by the abbreviation Nc.

Earlier than this the aleph had been used as a symbol for 60 in the Hebrew numeral notation, and it was used as a mathematical symbol for certain fundamental functions by H. Wronski,[6] in his *Philosophie des mathématiques* (1811).

The cardinal number of the aggregate constituting all numbers in the linear continuum[7] is marked $\mathfrak{c}$. The cardinal number of the aggregate F of all functions[8] of a real variable (known to be greater than $\mathfrak{c}$) is marked $\mathfrak{f}$.

The different alephs of well-ordered transfinite aggregates are marked $\aleph_0 \ \aleph_1 \ldots . \ \aleph_\nu, \ldots ., \ \aleph_\omega, \ldots ., \ \aleph_a$, and Schoenflies[9] writes the initial numbers of each class of ordinals, $\Omega_0, \ \Omega_1, \ldots ., \ \Omega_\nu, \ldots .,$ $\Omega_\omega, \ldots ., \ \Omega_d$, so that $\Omega_0 = \omega$ and $\Omega_1 = \Omega$, which latter is G. Cantor's designation for the first number in his third class, and is sometimes[10] written also ω_1. G. Cantor[11] marked a derived aggregate of μ by μ'; Peano marked it Du in 1895[12] and δu in 1903.

Other symbols used by G. Cantor[13] in his theory of aggregates are

[1] *Op. cit.*, Vol. XXI (1883), p. 577.   [2] *Op. cit.*, Vol. XXI, p. 582.

[3] B. Russell, *Principles of Mathematics*, Vol. I (1903), p. 322.

[4] G. Cantor, *Mathematische Annalen*, Vol. XLVI (1895), p. 492.

[5] G. Peano, *Formulaire de mathématiques*, Vol. I (1895), p. 66; Vol. IV (1903), p. 128.

[6] Gergonne's *Annales de mathématiques* (Nismes), Vol. III (1812, 1813), p. 53.

[7] A. Schoenflies, *Entwickelung der Mengenlehre und ihrer Anwendungen* (1913), p. 54; E. V. Huntington, *The Continuum* (2d ed., 1917), p. 80.

[8] A. Schoenflies, *op. cit.*, p. 60.   [9] A. Schoenflies, *op. cit.*, p. 124.

[10] E. V. Huntington, *op. cit.* (2d ed., 1917), p. 72.

[11] G. Cantor, *Mathematische Annalen*, Vol. V, p. 123.

[12] G. Peano, *Formulaire mathématique*, Vol. I (1895), p. 69; Vol. IV (1903), p. 121.

[13] G. Cantor, *Mathematische Annalen*, Vol. XLVI (1895), p. 481–512; see also P. E. B. Jourdain's translation of this article, and the one in Vol. XLIX (1897), p. 207–46, in a book, *Contributions to the Founding of the Theory of Transfinite Numbers* (Chicago and London, 1915).

$M = \{m\}$, where M suggests *Menge* ("aggregate"); $(M, N, P, \ldots)$ the uniting of the aggregates $M, N, P, \ldots$, which have no common elements into a single aggregate; $\overline{\overline{M}}$ the general symbol for *Mächtigkeit* ("power") or "cardinal number" of M; when two aggregates M and N are "equivalent," Cantor writes $M \backsim N$. For two aggregates M and N, the cardinal numbers are designated also by the signs $\mathfrak{a} = \overline{\overline{M}}$, $\mathfrak{b} = \overline{\overline{N}}$. The "covering-aggregate (*Belegungsmenge*) of N with M" is denoted by $(N|M)$, and $\mathfrak{a}^{\mathfrak{b}} = \overline{\overline{(N|M)}}$.

If in a simply ordered aggregate $M = \{m\}$, the element m_1 has a rank lower than m_2 in the given order of precedence, this relation is expressed $m_1 \prec m_2$, $m_2 \succ m_1$. If the infinite aggregates are multiply ordered F. Riesz[1] used the symbolism $a \overline{i} b$, $a < i\, b$, $a\, i > b$ to designate that in the ith order, a and b have equal rank, or a precedes b, or a follows b. If two ordered aggregates M and N are "similar," they are marked by Cantor $M \simeq N$; the "ordinal type" of M is marked $\overline{M}$. If α is the ordinal type, the corresponding cardinal number is $\bar{\alpha}$. If in an ordered aggregate M all the relations of precedence of its elements are inverted, the resulting ordered aggregate is denoted by $*M$ and its ordinal type $*\alpha$, provided $\alpha = \overline{M}$. Cantor[2] designates a linear continuous series having both a first and a last element, a series of type θ.

When two fundamental series $\{a_\nu\}$ and $\{a'_\nu\}$ of numbers of the first or second number-class are *zusammengehörig* ("coherent")[3] they are marked $\{a_\nu\} \| \{a'_\nu\}$. The "ϵ-numbers of the second number-class" are roots of the equation $\omega^\xi = \xi$. The symbol $E(\gamma)$ represents the limit of the fundamental series $\{\gamma_\nu\}$, and is an ϵ number.

G. Cantor[4] in 1880 marked the least common multiple of aggregates M and N by $\mathfrak{M}\{M, N\}$, but E. Zermelo[5] and A. Schoenflies[6] adopted the sign $\mathfrak{S}(M, N)$ suggested by the first letter in *Summe* ("sum"). The *Durchschnitt*, or all the elements common to M and N, are marked $[M, N]$ by Zermelo and $\mathfrak{D}(M, N)$ by Cantor and Schoenflies.[7] The sign C_ν is used to represent the continuum[8] of space of ν

---

[1] F. Riesz, *Mathematische Annalen*, Vol. LXI (1903), p. 407.

[2] G. Cantor, *Mathematische Annalen*, Vol. XLVI (1895), p. 511.

[3] G. Cantor, *ibid.*, Vol. XLIX (1897), p. 222.

[4] G. Cantor, *ibid.*, Vol. XVII (1880), p. 355.

[5] E. Zermelo, *Mathematische Annalen*, Vol. LXV (1908), p. 265.

[6] A. Schoenflies, *Entwickelung der Mengenlehre und ihrer Anwendungen* (Leipzig und Berlin, 1913), p. 10.

[7] A. Schoenflies, *op. cit.*, p. 11.   [8] A. Schoenflies, *op. cit.*, p. 11, 56.

dimensions. If $X_a = \Sigma x_a$ are ordered aggregates, their product yields an ordered product-aggregate which F. Hausdorff[1] represents by $\overset{A}{\underset{a}{\Pi}} X_a$ and Schoenflies[2] by $*\Pi X a;$ if such a product is a *Vollprodukt* ("complete product"), Hausdorff (writing Ma for Xa) marks it $((\overset{A}{\underset{a}{\Pi}} Ma))$ and its type by $((\overset{a}{\underset{a}{\Pi}} \mu a))$. He designates the "maximal product" by T.

In the theory of the equivalence of aggregates E. Zermelo[3] writes $\phi \in MN$, to express the *Abbildung* of M upon N, where M and N are aggregates having no elements in common and ϕ is a subaggregate such that each element of $M+N$ appears as element in one and only one element $\{m, n\}$ of ϕ.

Peano[4] designates by $a \, \underset{\cdot}{!} \, b$ the aggregate of all the couples formed by an object of a class a with an object of class b, while the sign $a ; b$ designates the couple whose two elements are the classes a and b. J. Rey Pastor[5] designates the existence of a one-to-one correspondence between infinite aggregates A and B, by the symbol $A \barwedge B$.

SIGNS FOR CONTINUED FRACTIONS AND INFINITE SERIES

422. *Continued fractions.*—We previously mentioned (§ 118) a notation for continued fractions, due to al Ḥaṣṣâr, about the beginning of the thirteenth century, who designates[6] by $\frac{3}{8}\frac{3}{8}\frac{2}{9}$ the ascending fraction

$$2 + \frac{3 + \frac{3}{5}}{8}$$
$$\overline{\qquad 9 \qquad} \; .$$

A Latin translation of the algebra of the famous Arabic algebraist and poet, Abu Kamil,[7] contains a symbolism for ascending continued

[1] F. Hausdorff, *Mathematische Annalen*, Vol. LXV (1908), p. 462.

[2] A. Schoenflies, *op. cit.*, p. 75–77.

[3] E. Zermelo, *Mathematische Annalen*, Vol. LXV (1908), p. 267, 268.

[4] G. Peano, *Formulaire mathématique*, Tome IV (1903), p. 125.

[5] *Revista matematica Hispano-Americana*, Vol. I (Madrid, 1919), p. 28.

[6] H. Suter in *Bibliotheca Mathematica* (3d ser.), Vol. II (1901), p. 28.

[7] Paris MS 7377 A, described by L. C. Karpinski, *Bibliotheca mathematica* (3d ser.), Vol. XII (1911–12). Our reference is to p. 53. See also Vol. X (1909–10), p. 18, 19, for H. Suter's translation into German of a translation into Italian of a Hebrew edition of Abu Kamil's book, *On the Pentagon and Decagon*, where $\frac{2}{3}\frac{2}{3}$ signifies $\frac{2}{3}$ plus $\frac{2}{3}$ of $\frac{1}{3}$, or $\frac{5}{3}\frac{2}{3}$.

fractions, which differs from that of al-Ḥaṣṣâr in proceeding from left to right. The fraction $\frac{1}{8}\frac{1}{9}$ stands for $\frac{1}{8}$ plus $\frac{1}{8}\frac{1}{1}$; the fraction $\frac{0}{9}\frac{1}{2}$ or $\frac{0}{8}\frac{1}{2}$ stands for $\frac{1}{2}$ of $\frac{1}{8}$. Abu Kamil himself presumably proceeded from right to left. The Arabic notation is employed also by Leonardo[1] of Pisa in his *Liber abbaci* of 1202 and 1228, and later still by the Arabic author al-Qualasâdî.[2] Al-Ḥaṣṣâr, it will be observed, also used the fractional line for ordinary fractions, as did also Leonardo of Pisa.

423. Leonardo[3] gives expressions like $\frac{7}{10}\frac{1}{10}$ 8, which must be read from right to left and which stands in our notation for $8 + \dfrac{1 + \frac{7}{10}}{10}$

or $8\frac{17}{100}$. He gives also $\frac{1}{5}\frac{0}{12}\frac{0}{20}$ as equal to $\frac{1}{2}\frac{0}{6}\frac{0}{10}\frac{0}{10}$; the value of these ascending continued fractions is $\frac{1}{1200}$.

Leonardo gives also two other notations, one of which he describes as follows: "Et si in uirga fuerint plures rupti, et ipsa uirga termina-uerit in circulo, tunc fractiones eius, aliter quam dictum sit denota-bunt, ut in hac $\cdot\frac{2}{3}\frac{4}{5}\frac{6}{7}\frac{8}{9}$○ cuius uirge fractiones denotant, octo nonas unius integri, et sex septimas de octononis, et quatuor quintas sex septimarum de octo nonis et duas tertias quattuor quintarum sex septimarum de octo nonis unius integri.[4] ("And if on the line there should be many fractions and the line itself terminated in a circle, then its fractions would denote other than what has been stated, as in this $\cdot\frac{2}{3}\frac{4}{5}\frac{6}{7}\frac{8}{9}$○, the line of which denotes the fractions eight-ninths of a unit, and six-sevenths of eight-ninths, and four-fifths of six-sevenths of eight-ninths, and two-thirds of four-fifths of six-sevenths of eight-ninths of an integral unit.") We have here an ascending continued fraction

$$\frac{8 + \dfrac{48 + \dfrac{192 + \dfrac{384}{3}}{5}}{7}}{9} \ .$$

Leonardo's third notation is described by him as follows: "Item si uirgule protraherunter super uirgam in hunc modum $\dfrac{1\ 1\ 1\ 5}{5\ 4\ 3\ 9}$, denotant fractiones eius quinque nonas et tertiam et quartam et quintam unius none."[5] ("Also if a short line be drawn above the

[1] Leonardo of Pisa, *Scritti* (pub. by B. Boncompagni), Vol. I (Rome, 1857), p. 24. See also *Encyclopédie des scienc. math.*, Tome I, Vol. I (1904), p. 318, n. 330.

[2] M. Cantor, *Vorlesungen über Geschichte der Mathematik*, Vol. I (3d ed., 1907), p. 813.

[3] Leonardo of Pisa, *op. cit.*, Vol. I, p. 85.   [4] *Op. cit.*, Vol. I (1857), p. 24.

[5] *Op. cit.*, Vol. I (1857), p. 24, also p. 91, 92, 97. See G. Eneström, *Bibliotheca mathematica* (3d ser.), Vol. XII (1911–12), p. 62.

fractional line in this manner $\dfrac{1\ 1\ 1\ 5}{5\ 4\ 3\ 9}$, they denote the following fractions, five-ninths, and one-third, one-fourth and one-fifth of one-ninth.") The foregoing fraction signifies, therefore, $\dfrac{5+\frac{1}{3}+\frac{1}{4}+\frac{1}{5}}{9}$; it is a complex fraction, but hardly of the "continued" type. Not to be confounded with these is Leonardo's notation $\bigcirc \frac{8\ 6\ 4\ 2}{9\ 7\ 5\ 3}$ which does not represent a continued fraction either, but signifies the multiplication[1] of the fractions, thus $\frac{2}{3}\cdot\frac{4}{5}\cdot\frac{6}{7}\cdot\frac{8}{9}$, and resembles a mode of writing, $\frac{3}{7}-\frac{2}{5}$, for $\frac{3}{7}\cdot\frac{2}{5}$, sometimes found in al-Ḥaṣṣâr.[2]

424. Pietro Antonio Cataldi in 1606[3] discusses the ascending continued fraction as "vna quātità scritta, ò proposta in forma di rotto di rotto" ("a quantity written or proposed in the form of a fraction of a fraction"). He explains an example and tells how mathematicians are accustomed to write it: "Sogliono i Pratici scriuere cosi $\frac{3}{8}\cdot\frac{2}{5}\cdot\frac{6}{8}\cdot\frac{4}{5}\cdot\frac{5}{8}$." The meaning of this appears from a simpler example explained on the previous page (142). He says: "poniamo 3. quarti, & $\frac{1}{2}$. quarto, cioè, $\frac{3}{4}.\&\,\frac{1}{2}$." or $\frac{7}{8}$. One has here a notation for an ascending continued fraction. Seven years later Cataldi introduced, for the first time, a notation for descending continued fractions, and he chose practically the symbolism that he had used for the ascending ones. In 1613 he explained: "Notisi, che nõ si potendo cõmodamẽte nella stampa formare i rotti, & rotti di rotti come andariano cioè così

$$4.\&\dfrac{2}{8}.\&\dfrac{2}{8}.\&\dfrac{2}{8}$$

come ci siamo sforzati di fare in questo, noi da qui inãzi gli formaremo tutti à q̄sta similitudine

$$4.\&\dfrac{2}{8}.\ \&\dfrac{2}{8}.\ \&\dfrac{2}{8}.$$

facendo vn punto all '8. denominatore de ciascum rotto, à significare, che il sequente rotto è rotto d'esso denominatore."[4] ("Observe that

---

[1] Leonardo of Pisa, *op. cit.*, Vol. I, p. 24.

[2] H. Suter, *Bibliotheca mathematica* (3d ser.), Vol. II, p. 27.

[3] Pietro Antonio Cataldi, *Seconda parte della pratica aritmetica* (Bologna, 1606), p. 143.

[4] *Trattato del Modo Brevissimo di trouare la Radice quadra delli numeri, et Regole da approssimarsi di continuo al vero nelle Radici de' numeri non quadrati* ... Di Pietro Antonio Cataldi Lettore delle Scienze Mathematiche nello Studio di Bologna (in Bologna, 1613), p. 70, 71.

since in printing when proceeding hurriedly, one cannot conveniently form fractions, and fractions of fractions, in this form

$$4.\ \&\ \frac{2}{8}.\ \&\ \frac{2}{8}.\ \&\ \frac{2}{8}$$

as here we are forcing ourselves to do it, we can easily denote all of them by adopting this imitation

$$4.\ \&\ \frac{2}{8}.\ \&\ \frac{2}{8}.\ \&\ \frac{2}{8}.$$

placing a point after the denominator 8. in each fraction, to signify that the fraction following is a fraction of the denominator.") But, in his simplified form of writing, as it appears further on in his book, the point is not placed after the numbers in the denominator, but is placed higher up, on a level with the "&." This is seen in Cataldi's approximation to $\sqrt{18}$ (p. 71):

"Di 18. la R̄ sia $4.\&\frac{2}{8}.\&\frac{2}{8}.\&\frac{2}{8}.\&\frac{2}{8}.\&\frac{2}{8}.\&\frac{2}{8}$

ò vogliamo dire $4.\&\frac{2}{8}.\&\frac{2}{8}.\&\frac{2}{8}.\&\frac{2}{8}.\&\frac{2}{8}.\&\frac{1}{4}$

cioè $4.\&\frac{2}{8}.\&\frac{2}{8}.\&\frac{2}{8}.\&\frac{2}{8}.\&\frac{8}{33}$

ò vogliamo dire $4.\&\frac{2}{8}.\&\frac{2}{8}.\&\frac{2}{8}.\&\frac{1}{4}\frac{4}{33}$

che è $4.\&\frac{2}{8}.\&\frac{2}{8}.\&\frac{2}{8}.\&\frac{33}{136}$

cioè $4.\&\frac{2}{8}.\&\frac{2}{8}.\&\frac{2}{8}.\&\frac{272}{1121}$

ò vogliamo dire $4.\&\frac{2}{8}.\&\frac{1}{4}\frac{136}{1121}$

che è $4.\&\frac{2}{8}.\&\frac{1121}{4620}$

cioè $4.\&\frac{9240}{38081}$."

Except for the "&" in place of "+," this notation conforms closely with one of the modern notations.

425. We come next to Wallis' representation of Brounker's expression[1] for $\frac{4}{\pi}$, namely,

" $\square = 1\frac{1}{2}\ \frac{9}{2}\ \frac{25}{2}\ \frac{49}{2}\ \frac{81}{2}$ & c.," and Wallis' " $\frac{a}{a}\ \frac{b}{\beta}\ \frac{c}{\gamma}\ \frac{d}{\delta}\ \frac{e}{\epsilon}$ &c. "

[1] John Wallis, *Arithmetica infinitorum* (Oxford, 1655), p. 182.

The second expression is Wallis' general representation[1] of a continued fraction. The omission of the signs of addition makes this notation inferior to that of Cataldi who used the "&." But Wallis adheres to his notation in his collected works[2] of 1695, as well as in his letter[3] to Leibniz of April 6, 1697. Leibniz,[4] on the other hand, used an improved form,

$$\text{“ } a+\frac{1}{b}+\frac{1}{c}+\frac{1}{d}+\frac{1}{e}+\text{etc.”}$$

in his letter to John Bernoulli of December 28, 1696.

A modern symbolism is found also in C. Huygens[5] who expressed the ratio $2640858:77708431$ in the form

$$\text{“}29+\tfrac{1}{2}+\tfrac{1}{2}+\tfrac{1}{1}+\tfrac{1}{5}+\tfrac{1}{1}+\tfrac{1}{4}\text{ etc.”}$$

The symbolism of Leibniz and Huygens came to be the regular form adopted by eighteenth-century writers and many writers of more recent date; as, for example, by J. A. Serret[6] in his *Higher Algebra*. Among eighteenth-century writers we cite especially L. Euler and J. Lagrange.

426. L. Euler[7] writes

$$\text{“ } a+\frac{a}{b}+\frac{\beta}{c}+\frac{\gamma}{d}+\text{etc.”}$$

In some of his articles the continued fractions written in this notation made extraordinary demands upon space, indeed as much as

[1] *Op. cit.*, p. 191.

[2] John Wallis, *Opera mathematica*, Vol. I (1695), p. 469, 474.

[3] C. I. Gerhardt, *Leibnizens Mathematische Schriften*, Vol. IV (Halle, 1859), p. 17.

[4] *Op. cit.*, Vol. III (Halle, 1855), p. 351, 352.

[5] Christian Huygens, *Descriptio automati planetarii* (The Hague, 1698); *Opuscula posthuma* (Amsterdam, 1728), Vol. II, p. 174–79. See also S. Günther, in *Bullettino Boncompagni*, Vol. VII (Rome, 1874), p. 247, 248.

[6] J. A. Serret, *Cours d'algèbre supérieure* (Paris, 1854), p. 491; German edition by G. Wertheim, Vol. I (2d ed.; Leipzig, 1878), p. 8.

[7] L. Euler, *Introductio in analysin infinitorum* (1748); ed. 1922, Vol. I, p. 362.

13 cm. or four-fifths of a large page.[1] In 1762 he devised a new notation in his *Specimen Algorithmi Singularis;*[2] a fraction

$$a + \frac{1}{b} + \frac{1}{c}$$

is represented by the symbol $\frac{(a, b, c)}{(b, c)}$; an infinite continued fraction by $\frac{(a, b, c, d, e \text{ etc.})}{(b, c, d, e \text{ etc.})}$. This symbolism offers superior advantages in computation, as, for example: $(a, b, c, d, e) = e(a, b, c, d) + (a, b, c)$, also $(a, b, c, d, e) = (e, d, c, b, a)$.

427. E. Stone[3] writes

$$\text{``}\frac{314159}{100000} \text{ will be} = 3 + \cfrac{1}{7 + \cfrac{1}{15 + \cfrac{1}{1 + \cfrac{1}{15 + \cfrac{1}{1 + \cfrac{1}{7 + \frac{1}{4}}}}}} \text{."}$$

This extension of the fractional lines to the same limit on the right is found earlier in Leibniz (§ 562) and is sometimes encountered later, as, for instance, in the book on continued fractions by O. Perron (1913).[4]

428. In the nineteenth century the need of more compact notations asserted itself and there was a return to Cataldi's practice of placing all the partial fractions on the same level. Sir John W. Herschel[5] writes the continued fraction

$$\frac{c_1}{a_1} + \cfrac{c_2}{a_2} + \cfrac{c_3}{a_3} + \ldots \frac{c_x}{a_x}$$

in the form

$$\frac{c_1}{a_1} + \frac{c_2}{a_2} + \frac{c_3}{a_3} + \cdots \frac{c_x}{a_x},$$

[1] See L. Euler in *Novi comment. academ. imper. Petropolitanae*, Vol. V, for the years 1754, 1755 (Petropoli, 1760), p. 225, 226, 231.

[2] *N. Comm. Petr. IX, pro annis 1762 et 1763* (Petropoli, 1764), p. 53–69. See M. Cantor, *op. cit.*, Vol. IV (1908), p. 155.

[3] E. Stone, *New Mathematical Dictionary* (2d ed.; London, 1743), art. "Ratio."

[4] Oskar Perron, *Lehre von den Kettenbrüchen* (Leipzig und Berlin, 1913), p. 3.

[5] J. F. W. Herschel, *Collection of Examples of . . . . Calculus of Finite Differences* (Cambridge, 1820), p. 148.

but states that this is "after the example" of Heinrich Bürmann of Mannheim, a worker in C. F. Hindenburg's school of combinatory analysis. The notation of Bürmann and Herschel has been used in England by Hall and Knight,[1] C. Smith,[2] G. Chrystal,[3] L. Ince,[4] and is still widely used there. Chrystal (Part II, p. 402) also uses symbolisms, such as:

$$\sqrt{13} = 3 + \frac{1}{1+} \; \frac{1}{1+} \; \frac{1}{1+} \; \frac{1}{1+} \; \frac{1}{6+} \; \frac{1}{1+} \cdots ,$$
$$\qquad\qquad * \qquad\qquad\qquad\qquad *$$

where the $*$ $*$ indicate the beginning and end of the cycle of partial quotients.

429. Möbius[5] says: "Der Raum-Ersparniss willen mögen nun die Kettenbrüche von der besagten Form, wie

$$\frac{1}{a}, \quad \frac{1}{a-\dfrac{1}{b}}, \quad \frac{1}{a-\dfrac{1}{b-\dfrac{1}{c}}}, \quad \frac{1}{a-\dfrac{1}{b-\dfrac{1}{c-\dfrac{1}{d}}}}$$

durch (a), (a, b), (a, b, c), (a, b, c, d), u.s.w. ausgedrückt werden." Möbius[6] represents the function

$$\frac{1}{(a, \ldots e)(b, \ldots e)(c, d, e)(d, e)(e)}$$

by the symbolism $[a, b, c, d, e]$. M. Stern[7] designated the continued fraction

$$a + \frac{b_1}{a_1+} \frac{b_2}{a_2+}$$
$$\cdot$$
$$\cdot$$
$$\cdot + \frac{b_m}{a_m}$$

[1] H. S. Hall and S. R. Knight, *Elementary Algebra* (ed. F. L. Sevenoak; New York, 1895), p. 369.

[2] C. Smith, *Elementary Algebra* (ed. Irving Stringham; New York, 1895), p. 530.

[3] G. Chrystal, *Algebra*, Part II (Edinburgh, 1889), p. 396, 397.

[4] Linsay Ince in *University of Edinburgh Mathematical Dept.* (1914), Paper No. 7, p. 2, 3.

[5] A. F. Möbius, *Crelle's Journal*, Vol. VI (1830), p. 217; *Werke*, Vol. IV (Leipzig, 1887), p. 507.

[6] Möbius, *Crelle's Journal*, Vol. VI, p. 220.

[7] M. Stern, *Theorie der Kettenbrüche und ihre Anwendung* (Berlin, 1834), p. 4, 22, 33; reprinted from *Crelle's Journal*, Vol. X, XI.

by $F(a, a_m)$; he lets a_1, a_m stand symbolically for the denominator of the equivalent ordinary fraction, and a, a_m for its numerator, so that

$F(a, a_m) = \dfrac{a, a_m}{a_1, a_m}$. Stern employs also the fuller form $F(a, a_m) = F(a +$

$b_1 : a_1 + b_2 : a_2$, etc.). For the special, infinite continued fraction $F(1:1+$ $1:2+9:2+25;2$, etc.) he suggests the form

$$\underset{0-\infty}{x} \ F[1:1+(2x+1)^2:2] .$$

430. J. H. T. Müller[1] devised the notation

$$b_0 + \frac{a_1}{b_1} + \frac{a_2}{b_2} + \cdots + \frac{a_n}{b_n},$$

where each $+$ sign may be replaced by a $-$ sign when parts are negative. Müller also wrote

$$b_0 + \left(\frac{a}{b}\right)_{(1+2+3+\cdots+n)}.$$

Müller's first symbolism found quite wide acceptance on the European continent, as is born out by statements of Chrystal[2] and Perron.[3] Perron remarks that sometimes the dots were omitted, and the continued fraction written in the form

$$b_0 + \frac{a_1}{b_1} + \frac{a_2}{b_2} + \cdots + \frac{a_n}{b_n} .$$

431. When all the numerators of the partial fractions are unity, G. Lejeune Dirichlet[4] wrote down simply the denominators b_r, in the following form:

$$(b_0, b_1, \ldots, b_{n-1}, b_n).$$

E. Heine[5] writes the continued fraction

$$\lambda_0 - \frac{\mu_1}{\lambda_1} - \frac{\mu_2}{\lambda_2} - \cdots \frac{\mu_n}{\lambda_n}$$

[1] J. H. T. Müller, *Lehrbuch der Mathematik. Erster Theil, Die gesammte Arithmetik enthaltend* (Halle, 1838), p. 384.

[2] G. Chrystal, *Algebra*, Part II (Edinburgh, 1889), p. 397.

[3] O. Perron, *op. cit.*, p. 3.

[4] G. Lejeune Dirichlet, *Abhandl. Akad. Berlin* (1854), *Math.*, p. 99; *Werke*, Vol. II (Berlin, 1897), p. 141.

[5] E. Heine, *Theorie der Kugelfunctionen* (Berlin, 1878), p. 261.

in the form

$$\begin{vmatrix} & \mu_1 & \mu_2 \ldots \ldots \mu_{n-1} & \mu_n \\ \lambda_0 & \lambda_1 & \lambda_2 \ldots \ldots \lambda_{n-1} & \lambda_n \end{vmatrix} .$$

This notation is followed by Pascal[1] in his *Repertorium*.

432. The modern notation, widely used on the European continent,

$$b_0 \pm \frac{a_1]}{|b_1} \pm \frac{a_2]}{|b_2} \pm \ldots \ldots \pm \frac{a_n]}{|b_n} ,$$

is due to Alfred Pringsheim,[2] who represents this continued faction also by the symbol

$$\left[b_0; \pm \frac{a_\nu}{b_\nu} \right]_1^n ;$$

by

$$\left[b_m; + \frac{a_\nu}{b_\nu} \right]_{m+1}^n$$

he represents the continued fraction whose first term is b_m and the denominator of the first partial fraction, b_{m+1}. Also in place of

$$\left[0; - \frac{a_\nu}{b_\nu} \right]_{m+1}^n$$

he writes simply

$$\left[- \frac{a_\nu}{b_\nu} \right]_{m+1}^n .$$

G. Peano[3] represents the continued fraction

$$\frac{1}{a_1} + \frac{1}{a_2} + \text{etc. to } a_n ,$$

by the symbolism $F_c(a_1, 1 \ldots \ldots n)$.

In the continued fraction

$$a_1 + \frac{b_2}{a_2} + \frac{b_3}{a_3} + \ldots \ldots \frac{b_n}{a_n} ,$$

---

[1] E. Pascal, *Repertorium der höheren Mathematik* (ed. P. Epstein and H. E. Timerding), Vol. I (2d ed., 1910), p. 444.

[2] Alfred Pringsheim, *Encyklopädie der mathematischen Wissenschaften*, 1. Band, 1. Theil (Leipzig, 1898–1904), p. 119. These notations occur also in the French edition, the *Encyclopédie des scien. math.*, Tome I, Vol. I (1904), p. 283, 284.

[3] G. Peano, *Formulaire mathématique*, Vol. IV (1903), p. 197.

the expression $p_n = a_n p_{n-1} + b_n p_{n-2}$, where $p_0 = 1$, $p_1 = a_1$, is called by Chrystal[1] a "continuant of the nth order" and is marked

$$p_n = K \begin{pmatrix} b_2, & \ldots & , b_n \\ a_1, & a_2, & \ldots & , a_n \end{pmatrix} ,$$

a symbol attributed[2] to Thomas Muir.

433. A notation for ascending continued fractions corresponding to Bürmann and Herschel's notation for descending ones is the following,

$$\frac{b_1 +}{a_1} \frac{b_2 +}{a_2} \frac{b_3 +}{a_3} \ldots \frac{+ b_n}{a_n} ,$$

used, among others, by Weld.[3] A. Pringsheim and J. Molk[4] write an ascending continued fraction

$$K^{(n)} = \cfrac{a_1 + \cfrac{a_2 + \cdots \cdot + \cfrac{a_n}{b_n}}{b_2}}{b_1}$$

in the form

$$K^{(n)} = \left\{ \frac{a_\nu}{b_\nu} \right\}_1^n .$$

434. *Tiered fractions.*—G. de Longchamps[5] indicates by

$$A_n = \left| \cfrac{a_1}{\cfrac{a_2}{\cfrac{\vdots}{\cfrac{a_n}{a_{n+1}}}}} \right|$$

tiered fractions (*fractions étagées*) involving the numbers $a_1, a_2, \ldots,$ a_{n+1}, the bars between the numbers signifying division. If the bars be assigned different lengths, A_n has a definite arithmetical value. Thus,

$$\cfrac{a_1}{\cfrac{a_2}{\cfrac{a_3}{a_4}}} = \frac{a_1 a_3 a_4}{a_2} \; ; \qquad \cfrac{\cfrac{a_1}{a_2}}{\cfrac{a_3}{a_4}} = \frac{a_1}{a_2 a_3 a_4} .$$

[1] G. Chrystal, *Algebra*, Part II (1889), p. 466.   [2] O. Perron, *op. cit.*, p. 6.

[3] Laenas G. Weld, *Theory of Determinants* (2d ed.; New York, 1896), p. 186.

[4] A. Pringsheim and J. Molk, *Encyclopédie des scien. math.*, Tome I, Vol. I (1904), p. 317.

[5] Gobierre de Longchamps, *Giornale di matem.* (1st ser.), Vol. XV (1877), p. 299.

A_n may have $n!$ different values, arising from the different relative lengths which may be assigned to the fractional lines.

435. *Infinite series.*[1]—The early writers in the field of infinite series or infinite products indicated such expressions by writing down the first few terms or factors and appending "&c." Thus Wallis in 1656 wrote: "1, 6, 12, 18, 24, &c.,"[2] and again "$1 \times \frac{3}{2} \times \frac{5}{4} \times \frac{7}{6}$, &c.,"[3] and similarly for infinite continued fractions.[4] He uses the "&c." also for finite expressions, as in "$0 + a^3 + b^3 + c^3$ &c. cujus terminus ultimus l^3, numerus terminorum $l+1$"[5] (whose last term [is] l^3, the number of terms $l+1$).

Nicholas Mercator writes "$ps = 1 - a + aa - a^3 + a^4$, &c.,"[6] and also "$\frac{a}{b} + \frac{ca}{bb} + \frac{cca}{b^3} + \frac{c^3a}{b^4}$, & deinceps continuando progressionem in infinitum."[7] James Gregory, in the same year, says: "Si fuerint quantitates continuè proportionales, A, B, C, D, E, F, &c. numero infinitae."[8] He places the sign $+$ before "&c." when the algebraic sum of the terms is expressed, as in "erit primus terminus $+\frac{1}{2}$ secundi $+\frac{1}{3}$ tertii $+\frac{1}{4}$ quarti $+\frac{1}{5}$ quinti $+$&c. in infinitum $= =$ spatio Hyperbolico S B KH."[9] In other passages he leaves off the "in infinitum."[10] Brounker[11] in 1668 writes

"$\frac{1}{1 \times 2} + \frac{1}{3 \times 4} + \frac{1}{5 \times 6} + \frac{1}{7 \times 8} + \frac{1}{9 \times 10}$, &c. in infinitum."

G. W. Leibniz[12] gives the quadrature of the circle by the series $\frac{1}{1} - \frac{1}{3} + \frac{1}{5}$, and ends with "$+\frac{1}{17}$ &c." Cotes[13] writes

$$"v* - \tfrac{1}{6}v^3 + \tfrac{5}{24}v^4 - \tfrac{23}{66}v^5, \text{ &c.}"$$

Wolff[14] says "$\frac{1}{2} + \frac{1}{4} + \frac{1}{8} + \frac{1}{16} + \frac{1}{32}$ und so weiter unendlich fort."

---

[1] See also § 408.

[2] John Wallis, *Arithmetica infinitorum* (Oxford, 1655), p. 26.

[3] *Op. cit.*, p. 175.   [4] *Op. cit.*, p. 191.   [5] *Op. cit.*, p. 145.

[6] *Logarithmo-technia* .... *auctore Nicolao Mercatore* (London, 1668), Propositio XVII, p. 32.

[7] *Op. cit.*, p. 25.

[8] James Gregory, *Exercitationes geometricae;* in the part on Mercator's quadrature of the hyperbola, p. 9.

[9] James Gregory, *op. cit.*, p. 11.   [10] James Gregory, *op. cit.*, p. 13.

[11] L. Viscount Brounker, *Philosophical Transactions* (London), Vol. II, p. 645–49; abridged edition by John Lowthrop, Vol. I (London, 1705), p. 10.

[12] *Philosoph. Transactions*, abridged, Vol. I (London, 1705), p. 16. The practice of Leibniz is exhibited also in articles in the *Acta eruditorum* (1691), p. 179; (1693), p. 178; (1694), p. 369; (1695), p. 312; (1701), p. 215.

[13] Roger Cotes, *Harmonia mensurarum* (Cambrigiae, 1722), p. 6.

[14] Christian Wolff, *Mathematisches Lexicon* (Leipzig, 1716), p. 176.

436. T. Watkins[1] uses dots in place of "&c." to represent the tail end of a series, but he does not write the $+$ or the $-$ after the last term written down. In a letter to Nikolaus I. Bernoulli, May 14, 1743, Euler[2] writes $1+x+x^2+x^3+\ldots+x^\infty$, and also $1-3+5-7+9-\ldots\pm(2\infty+1)$.

E. Waring[3] writes series thus, "$a+\beta+\gamma+\delta+\epsilon+\zeta+$etc."

The indefinite continuance of terms is designated by Schultz[4] in this manner, "$1+3+5+7+\ldots\sim$," the $\sim$ being probably intended for Wallis' ∞. Owing no doubt to the lack of proper type in printing offices Wallis' symbol was given often forms which stood also for "difference" or "similar." Thus the sign ∞ stands for "infinity" in a publication of F. A. Prym[5] in 1863.

The use of dots was resorted to in 1793 by Prändel[6] when he writes "$\Box$ Circ. $=4-\frac{2}{3}-\frac{1}{10}-\frac{1}{28}-\frac{1}{288}\ldots$" L'Huilier[7] ends an infinite series of positive terms with "$+\ldots$"

L'Abbé de Gua[8] writes a finite expression, marking the terms omitted, with dots and also with "&c.," "3, 4, 5 . . . . &c $\overline{n-m+2}$," the commas indicating here multiplication. F. Nicole[9] writes a procession of factors, using dots, but omitting the "&c." and also the bar: "$n\times n-1\times\ldots n-7$." The same course is pursued by Condorcet[10]. On the other hand, C. A. Vandermonde[11] used the "&c." as in "$a+b+c+$&c." Paulus Mako[12] of Austria designates that the series is infinite, by writing "$b, bm, bm^2, bm^3, \ldots bm^\infty$." C. F. Hindenburg[13] uses dots between, say, the fourth term and the nth

[1] Thomas Watkins, *Philosophical Transactions*, Vol. XXIX (1714–16), p. 144.

[2] L. Euler, *Opera posthuma*, I (Petropoli, 1862), p. 538. See G. Eneström in *Bibliotheca mathematica* (3d ser.), Vol. XI (1910–11), p. 272.

[3] E. Waring, *Meditationes algebraicae* (Cantibrigiae; 3d ed., 1782), p. 75.

[4] Johann Schultz, *Versuch einer genauen Theorie des Unendlichen* (Königsberg, 1788), p. xxvii.

[5] F. A. Prym, *Theoria nova functionum ultraellipticarum* (Berlin, 1863), p. 1, 3, 4.

[6] Johann Georg Prändel, *Kugldreyeckslehre u. höhere Mathematik* (München, 1793), p. 37.

[7] Simon l'Huilier, *Principiorum Calculi Differentialis et Integralis Expositio* (Tubingae, 1795), p. 27.

[8] De Gua in *Histoire de l'académie r. d. sciences*, année 1741 (Paris, 1744), p. 81.

[9] F. Nicole, *op. cit.*, année 1743 (Paris, 1746), p. 226.

[10] Le Marquis de Condorcet, *op. cit.*, année 1769 (Paris, 1772), p. 211.

[11] C. A. Vandermonde, *op. cit.*, année 1771 (Paris, 1774), p. 370.

[12] *Compendiaria matheseos institutio* . . . . *Paulus Mako* (3d ed.; Vienna, 1771), p. 210.

[13] C. F. Hindenburg, *Infinitinomii dignitatum* . . . . *historia leges ac formulae* . . . . *auctore Carolo Friderico Hindenburg* (Göttingen, 1779), p. 5, 6, 41.

term, the $+$ or the $-$ sign being prefixed to the last or nth term of the polynomial. However, at times he uses "&c." in place of the dots to designate the end of a polynomial, "$cC+dD+eE+$&c." E. G. Fischer[1] writes a finite expression $y=ax+bx^2+cx^3+\ldots+px^r$ and, in the case of an infinite series of positive terms, he ends with "$+$etc."

437. M. Stern[2] writes "$1-\frac{1}{3}+\frac{1}{5}-\frac{1}{7}$ etc. $=\frac{\pi}{4}$." Enno Heeren Dirksen[3] of Berlin indicates in 1829 by "$+$etc." that the sum of the terms of an infinite series is intended, but a few years later[4] he marks an infinite progression thus: "1, 2, 3, 4 in inf." Martin Ohm[5] says: "Jede unendliche Reihe $(P)\ldots a+bx+cx^2+dx^3\ldots$ in inf. kann durch das kombinatorische Aggregat $S[P_a\cdot x^a]$ ausgedrückt werden, wenn man nur statt a nach und nach 0, 1, 2, 3, 4, in inf. gesetzt denkt, und wenn P_a den Koeffizienten von x^a vorstellt." L. Seidel[6] writes "$f(x, n) f(x, 1)+f(x, 2)\ldots$ in inf."; A. N. Whitehead[7] in 1911 ends with "$+$etc. ....." A notation for infinite series now frequently used is to place dots before and after the nth term, as in[8] $a_0+a_1(1/x)+a_2(1/x^2)+\ldots+a_n(1/x)^n+\ldots$

There are recent examples of rather involved notations to indicate the number of each term in a finite series, like the following of O. Stolz and J. A. Gmeiner.[9]

$$\overset{1}{\frac{1}{b}}+\overset{2}{\frac{1}{b}}+\ldots+\overset{a}{\frac{1}{b}}=\frac{a}{b}.$$

[1] Ernst Gottfried Fischer, *Theorie der Dimensionszeichen* (Halle, 1794), p. 27, 54.

[2] M. Stern, *Theorie der Kettenbrüche* (Berlin, 1834), p. 34. Reprint from *Crelle's Journal*, Vol. X, XI.

[2] *Crelle's Journal* (Berlin, 1829), p. 170.

[4] *Abhandlungen der K. P. Akademie d. Wissenschaften* (Berlin, 1832), Th. I, p. 77–107.

[5] Martin Ohm, *Versuch eines vollkommen consequenten Systems der Mathematik*, Vol. II (Berlin, 1829), p. 264, 265.

[6] *Abhandlungen d. Math.-Phys. Classe d. k. Bayerischen Akademie der Wissenschaften*, Vol. V (München, 1850), p. 384.

[7] A. N. Whitehead, *An Introduction to Mathematics* (New York and London), p. 212.

[8] W. B. Ford, *Studies on Divergent Series and Summability* (New York, 1916), p. 28.

[9] O. Stolz und J. A. Gmeiner, *Theoretische Arithmetik* (Leipzig), Vol. I (2d. ed., 1911), p. 104, 105, 109, 114, 121.

438. The sign Σ for summation is due to L. Euler[1] (1755), who says, "summam indicabimus signo Σ." This symbol was used by Lagrange,[2] but otherwise received little attention during the eighteenth century. A widely used notation for the sum of a series is the capital letter S; it is given, for instance, in G. S. Klügel's *Wörterbuch*.[3] The Σ to express "sum" occurs in 1829 in Fourier's *Theory of Heat*,[4] published in 1822, and in C. G. J. Jacobi's[5] elliptic functions of 1829.

Cauchy[6] used three indices m, n, r, as in $\sum_{m}^{n} r$ fr. Alfred Pringsheim[7]

marks the sum of an infinite series thus, $\sum_{0}^{\infty} \nu \, a_\nu$.

Jahnke[8] adopts in one place, as a substitute for Σa_i, the simpler sign $\tilde{a}$, which he borrows from geodesists, to designate the sum $a_1 + a_2 + a_3 + \ldots$. Additional symbols for sum are given in § 410.

SIGNS IN THE THEORY OF COMBINATIONS

439. *Binomial formula.*—The binomial formula, as Newton wrote it in his letter to Oldenburg of June 13, 1676, took this form, $(P+PQ)^{\frac{m}{n}}$ $= P^{\frac{m}{n}} + \frac{m}{n} AQ + \frac{m-n}{2n} BQ + \frac{m-2n}{3n} CQ + \text{etc.}$,[9] where A stands for the first term $P^{\frac{m}{n}}$, B for the second term, and so on. Wallis,[10] in his *Algebra* of 1693, gives Newton's form of 1676 for the binomial formula. Leib-

---

[1] L. Euler, *Institutiones calculi differentialis* (St. Petersburg, 1755), Cap. I, § 26, p. 27.

[2] J. Lagrange, *Œuvres*, Vol. III, p. 451.

[3] G. S. Klügel's *Mathematisches Wörterbuch*, completed by J. A. Grunert, Part V (Leipzig, 1831), "Umformung der Reihen," p. 348; Part IV (1823), "Summirbare Reihe," p. 577.

[4] Joseph Fourier, *La théorie analytique de la chaleur* (Paris, 1822; Eng. trans. by A. Freeman, 1878), chap. iii, sec. 6, p. 208, and other places.

[5] C. G. J. Jacobi, *Fundamenta nova theoriae functionum ellipticarum* (1829); *Gesammelte Werke*, Vol. I (Berlin, 1881), p. 94.

[6] See G. Peano, *Formulaire mathématique*, Vol. IV (Turin, 1903), p. 132.

[7] See, for instance, *Encyklopädie der Math. Wissensch.*, Vol. I, Part I (Leipzig, 1898–1904), p. 77.

[8] E. Jahnke, *Archiv der Mathematik und Physik* (3d ser.), Vol. XXV (1916), p. 317.

[9] J. Collins, *Commercium Epistolicum* (London, 1712; ed. J. B. Biot and F. Lefort, Paris, 1856), p. 103.

[10] John Wallis, *De algebra tractatus* (1693), p. 376.

niz[1] wrote in 1695, "$\overline{m\,|\,y+a} \overset{(5)}{=} y^m + \dfrac{m}{1} y^{\frac{m-1}{\cdot}} a^1 + \dfrac{m \cdot m - 1}{1 \cdot 2} y^{\frac{m-2}{\cdot}} a^2$, etc.,"
where the (5) simply marks the equation as being equation No. 5.
In Newton's *Quadratura curvarum*,[2] 1704, occurs the following passage: "Quo tempore quantitas x fluendo evadit $x+0$, quantitas x^n evadet $\overline{x+0}\,|^n$; id est per methodum serierum infinitarum, $x^n + n0x^{n-1}$
$+ \dfrac{nn-n}{2} 00x^{n-2} +$ etc." In 1706 William Jones[3] gives the form

$$\overline{a+x}\,|^n = a^n + \frac{n-0}{1} a^{n-1}x + \frac{n-0}{1} \times \frac{n-1}{2} a^{n-2}x^2$$
$$+ \frac{n-0}{1} \times \frac{n-1}{2} \times \frac{n-2}{3} a^{n-3}x^3 + \text{etc.,}$$

but Jones proceeds to develop the form given by Newton, except that
Jones writes $a+ag$ where Newton has $P+PQ$. In 1747 Jones[4]
adopted the abbreviations "$n' = n \cdot \dfrac{n-1}{2}$; $n'' = n' \cdot \dfrac{n-2}{3}$; $n''' = n'' \cdot \dfrac{n-3}{4}$;
$n^{\mathrm{iv}} = n''' \cdot \dfrac{n-4}{5}$." Another notation due to Kramp will be noted further
on (§ 445). Cotes[5] wrote "$\overline{1+\mathcal{Q}}\,|^{\frac{m}{n}} = 1 + \dfrac{m}{n} \mathcal{Q} + \dfrac{m}{n} \times \dfrac{m-n}{2n} \mathcal{Q}\mathcal{Q} + \dfrac{m}{n} \times \dfrac{m-n}{2n}$
$\times \dfrac{m-2n}{3n} \mathcal{Q}^3 +$ etc." Euler designated the binomial coefficient
$\dfrac{n(n-1) \ldots (n-\rho+1)}{1 \cdot 2 \cdot 3 \ldots \rho}$ by $\left(\dfrac{n}{\rho}\right)$ in a paper written in 1778 but not
published until 1806,[6] and by $\left[\dfrac{m}{\rho}\right]$ in a paper of 1781, published in
1784.[7] Rothe[8] in 1820 denoted the pth binomial coefficient in the

[1] *Leibnizens Mathematische Schriften* (ed. C. I. Gerhardt), Vol. V (1858), p. 323.

[2] Isaac Newton, *Opera* (ed. S. Horsley), Vol. I (London, 1779), p. 336.

[3] William Jones, *Synopsis Palmariorum Matheseos* (London, 1706), p. 169, 170; the coefficients taking the form used by John Wallis in his *Algebra* of 1693, p. 358.

[4] W. Jones, *Philosophical Transactions for the Year 1747* (London, 1748), p. 563; abridged edition by John Martin, Vol. X (London, 1756), p. 17.

[5] Roger Cotes, *Harmonia mensurarum*, the tract *De methodo differentiali Newtoniana* (Cambridge, 1722), p. 30.

[6] L. Euler in *Nova acta acad. Petrop.*, Vol. XV (1806), *Math.*, p. 33

[7] L. Euler in *Acta Acad. Petrop.*, Vol. V (1784), pars prior, § 18, p. 89. See Cantor, *op. cit.*, Vol. IV, p. 206.

[8] H. A. Rothe, *Theorie der kombinatorischen Integrale* (Nürnberg, 1820). Taken from J. Tropfke, *op. cit.*, Vol. VI (2d ed., 1924), p. 44.

expansion of $(a+b)^n$ by n , and Ohm[1] in 1829 denoted the nth coefficient in $(a+b)^m$ by m_n. The notation $\binom{m}{p}$ which has become the more
common was introduced in 1827 by von Ettingshausen,[2] and was
used in 1851 by Raabe.[3] Stolz and Gmeiner[4] employ for binomial
coefficients $\binom{m}{r}$ and also m_r, thus following the symbolism of Rothe
and Ohm. Saalschütz[5] lets $(k)_h$ stand for $\binom{k}{h}$.

The quantic $ax^3+3bx^2y+3cxy^2+dy^3$, in which the terms are
affected with the binomial coefficients arising in the expansion of
$(x+y)^3$, is denoted by Cayley[6] by the abbreviation $(a, b, c, d \wr x, y)$.
When the terms are not affected by bionomial coefficients, Cayley put
the arrow-head on the parenthesis, writing for instance $(a, b, c, d \wr x, y)$
to denote $ax^3+bx^2y+cxy^2+dy^3$. Faà de Bruno[7] designated by $((x^p))F$
the coefficient of x^p in the development of F, which is a function of x.

An imitation of Newton's original mode of writing the binomial
formula is seen in Stirling's notation for series: "I designate the initial
terms of a series by the initial letters of the alphabet, A, B, C, D, etc.
A is the first, B the second, C the third, and *sic porro*. I denote any
term of this kind by the letter T, and the remaining terms in their
order of succession by the same letter affixed with the Roman numerals I, II, III, IV, V, VI, VII, etc., for the sake of distinction. Thus,
if T is the tenth, then T' is the eleventh, T'' the twelfth, T''' the
thirteenth, and so on. And whichever term of this kind is defined as
T, the ones that follow are generally defined by T' T'' T''' T^{iv}, etc.
The distance of the term T from any given term I denote by the indeterminate quantity $\mathfrak{z}$."[8] Thus, following the notation of Newton
("more Newtoniana"), Stirling puts, in the series $1, \frac{1}{2}x, \frac{3}{8}x^2, \frac{5}{16}x^3$, etc.,

$$A = 1, \; B = \tfrac{1}{2}Ax, \; C = \tfrac{3}{4}Bx, \; \ldots \ldots, \; T' = \frac{\mathfrak{z}+\frac{1}{2}}{\mathfrak{z}+1} \; Tx.$$

[1] Martin Ohm, *op. cit.*, Vol. II (1829), p. 79.

[2] Andreas von Ettingshausen, *Vorlesungen über höhere Mathematik*, Vol. I
(Vienna, 1827), p. 38. See E. Netto in Cantor, *op. cit.*, Vol. IV, p. 206; J. Tropfke,
op. cit., Vol. VI (1924), p. 44.

[3] J. L. Raabe in *Journal f. reine u. angewandte Mathematik*, Vol. XLII (1851),
p. 350. See *Encyclopédie des sciences Math.*, Tome I, Vol. I (1904), p. 67, n. 20, 21.

[4] O. Stolz and J. A. Gmeiner, *Theoretische Arithmetik*, Vol. I (Leipzig, 1902),
p. 187.

[5] Louis Saalschütz, *Bernoullische Zahlen* (Berlin, 1893), p. 3.

[6] See G. Salmon, *Modern Higher Algebra* (3d ed.; Dublin, 1876), p. 92.

[7] Faà de Bruno, *Théorie des formes binaires* (Turin, 1876), p. 161.

[8] James Stirling, *Methodus differentialis* (London, 1730), p. 3.

440. *Product of terms in an arithmetical progression.*—Machin,[1] in a paper on Kepler's problem, adopts the notation $\frac{a}{n+a}\Big(^m$ which he says, "denotes by its Index m on the Right-hand, that it is a Composite Quantity, consisting of so many Factors as there are Units in the Number m; and the Index a above, on the Left, denotes the common Difference of the Factors, decreasing in an Arithmetical Progression, if it be positive; or increasing, if it be negative; and so signifies, in the common Notation, the composite Number or Quantity, $\overline{n+a}\cdot\overline{n+a-a}\cdot\overline{n+a-2a}\cdot\overline{n+a-3a}\cdot$ and so on." Further on we shall encounter other notations for such a product, for instance, those of Kramp, Ampère, and Jarrett (§§ 445–47).

441. *Vandermonde's symbols.*—In 1770 C. A. Vandermonde, in an article on the resolution of equations,[2] adopted the following contractions:

"(A) pour $a+b+c+$&c
(A^2) pour $a^2+b^2+c^2+$&c
(AB) pour $ab+ac+$&$+bc+$&c$+$
(A^2B) pour $a^2b+b^2a+c^2a+$&c$+a^2c+b^2c+c^2b+$&c$+$&c$+$.

Et en géneral par $(A^\alpha B^\beta C^\gamma \ldots)r'$ indiquerai la somme de tous les termes différens qui résuteroient de celui-là, au moyen de toutes les substitutions possibles des petites lettres a, b, c, d, e, &c dans un ordre quel conque aux grandes, A, B, C, &c." He writes also $(A^\alpha B^\beta C^\gamma D^\delta E^\epsilon \ldots) = \{\alpha\beta\gamma\delta\epsilon \ldots\}$, or if several Greek letters are equal, he writes $\{\alpha^n\beta^p\gamma^q \ldots\}$.

442. Laplace[3] represented the resultant of three equations by the symbolism $(^1a^2\cdot b^3\cdot c)$, the indices being placed to the left of a letter and above.

443. *Combinatorial school of Hindenburg.*—The notation used by members of the combinatorial school in Germany is often so prolix and involved that a complete account of their symbolism transcends the limits of our space. The generalization of the binomial theorem so as to involve any power of any polynomial, be the number of terms

[1] John Machin, *Philosophical Transactions* (London), No. 147, p. 205 (Jan., etc., 1738); abridged by John Martyn, Vol. VIII, Part I, p. 78.

[2] C. A. Vandermonde in *Histoire de l'académie r. d. sciences*, année 1771 (Paris, 1774), p. 370, 371.

[3] P. S. Laplace, *Histoire de l'académie, r. d. sciences* (Paris), année 1772, p. 267, 294, 2. partie; Laplace, *Œuvres*, Vol. VIII, p. 365–406. See also Th. Muir, *The Theory of Determinants in Historical Order of Development*, Part I (2d ed., 1906), p. 30.

in the polynomial finite or infinite, was one of the problems considered. Carl Friedrich Hindenburg, in a treatise of 1778,[1] represents the binomial coefficients by capital German letters, thus

$$^m\mathfrak{A} = \frac{m}{1}, \qquad ^m\mathfrak{B} = \frac{m(m-1)}{1\cdot 2}, \qquad ^m\mathfrak{C} = \frac{m(m-1)(m-2)}{1\cdot 2\cdot 3}, \ldots$$

This notation is followed by G. S. Klügel in his *Wörterbuch*.[2] Netto remarks that Hindenburg adheres to the practice which had been at least in part abandoned by Leibniz as inconvenient—the practice of using the alphabetical arrangement of the letters for the designation of order and for enumeration. Hindenburg employs the upright Latin capitals A, B, C to mark combinations, the slanting Latin capitals *A, B, C* for permutations. A superscript stroke placed to the left of a letter, $'A$ $'A$, means a combination or permutation *without* repetitions, while a stroke placed to the right, A' A', means *with* repetitions. To mark coefficients that are not binomial, but polynomial, he uses *small* German letters in place of the *capital* German letters given above.

In a book of 1779, Hindenburg uses small and capital letters in Roman, Greek, and German type. He employs superscripts placed not only before or after a letter, but also above the letter, the superscripts either inclosed in a parenthesis or not. He[3] calls ^mb, ^mc, ^md, *signa indefinita*, used in the manner illustrated in the following expressions:

$$A^m = a^m$$

$$B^m = (A+b)^m = A^m + mA^{m-1}b + \frac{m\cdot m - 1}{1\cdot 2} A^{m-2}b^2 + \&c = a^m + {}^mb.$$

$$C^m = (B+c)^m = \ldots = B^m + {}^mc = a^m + {}^mb + {}^mc.$$

Further on (p. 17) Hindenburg puts

$$a[a+b+c+d \ldots +\omega] = 'A,$$
$$a['A+'B+'C+'D \ldots +'\Omega] = ''A,$$
$$a[''A+''B+''C+''D \ldots +''\Omega] = '''A, \ldots,$$
$$d[''D \ldots +''\Omega] = '''D, \text{ etc.},$$
$$^nD = d[^{n-1}D \ldots +^{n-1}\Omega].$$

[1] C. F. Hindenburg, *Methodus nova et facilis serierum infinitarum*, etc. (Göttingen, 1778). Our knowledge of this publication is drawn from E. Netto in Cantor, *op. cit.*, Vol. IV, p. 205–7.

[2] G. S. Klügel, *Mathematisches Wörterbuch*, 1. Theil (Leipzig, 1803), art "Binomial—Coefficienten."

[3] C. F. Hindenburg, *Infinitinomii dignitatum .... Historia, leges ac formulae* (Göttingen, 1779), p. 5, 17, 18, 20.

He introduces (p. 21) $\int$ for *summam partium* (the parts chosen·in a certain manner), which is to be distinguished from the sign of integration $\int$, but like $\int$ has *vim transitivam*, i.e., is distributive in addition. Later still (p. 159), writing $B = b+c+d+$&c, $C = c+d+e+$ &c, ... , he puts $\overset{2}{B} = bB+cC+dD+$&c, $\overset{2}{C} = cC+dD+eE+$&c, and so on; also (p. 161) $\overset{(2)}{B} = bC+cD+dE$ . . . . , $\overset{(2)}{C} = cD+dE$ . . . . , $\overset{(3)}{C} = c\overset{(2)}{D}+d\overset{(2)}{E}$ . . . . , and so on.

444. In 1796 and 1800 Hindenburg published collections of papers on combinatorial analysis. The first collection[1] opens with an article by J. N. Tetens of Copenhagen who lets $|n|$ stand for the coefficient of the nth term of a polynomial, also lets the coefficient of the nth term (*terminus generalis*) in the expansion of $(a+bx+cx^2+ +|n|x^{n-1}+ +)^m$ be indicated by $T(a+bx+ +|n|x^{n-1})^m$ or simply by $T(a+ +|n|)^m$. In a footnote Hindenburg compares the symbols of Tetens with symbols of his own; thus, $T(a+ +|n|)^m = p^m \mathfrak{X}^n$, etc., where in Hindenburg's notation of that time, $p^m \mathfrak{X}^n$ is the nth coefficient of the power p^m.

In 1800 Hindenburg[2] refers to Heinrich Bürmann who advanced a purely combinatorial notation in his *Developpment des fonctions combinatoires*. Hindenburg states that Bürmann's short and expressive signs cannot be explained at this time, but states that the simplest *monogrammata* of Bürmann from which the others are formed by additions and changes are $\sqcap$ for series, $\llcorner$ for combinations, $\lrcorner$ for discerption. In 1803 Hindenburg and Bürmann brought out a joint publication, *Ueber combinatorishe Analysis*, and in 1807 Bürmann published at Mannheim a *Pangraphie*, or system of universal notation. We have not seen these two publications. As we point out elsewhere (§ 428), J. W. F. Herschel adopted a few of Bürmann's symbols.

445. *Kramp on combinatorial notations.*—Kramp[3] of Cologne in 1808 expressed himself on matters of notation as follows: "If one designates by Q any polynomial, ordered according to the power of the variable x, such that $ax^n+bx^{n+r}+cx^{n+2r}+$, etc., the notation Qℸ1, Qℸ2, Qℸ3, etc., will be very convenient for designating the coefficients of that polynomial, namely, a, b, c, etc. The coefficient Qℸ m is accordingly the one preceding the power of x whose exponent is $n+(m-1)r$.

---

[1] *Sammlung combinatorisch-analytischer Abhandlungen*, herausgeg. von Carl Friedrich Hindenburg, Erste Sammlung (Leipzig, 1796), p. 6, 7.

[2] *Op. cit.*, Zweyte Sammlung (1800), p. xiii.

[3] Christian Kramp, *Élémens d'arithmétique universelle* (Cologne, 1808), "Notations."

The polynomial Q is therefore $Q\ ^{\mathsf{k}}1x^n + Q\ ^{\mathsf{k}}2 \cdot x^{n+r} + Q\ ^{\mathsf{k}}3 \cdot x^{n+2r} +$, etc. The power of the polynomial Q^h, according to this same notation, becomes $Q^{h}\ ^{\mathsf{k}}1 \cdot x^{nh} + Q^{h}\ ^{\mathsf{k}}2 \cdot x^{nh+r} + Q^{h}\ ^{\mathsf{k}}3 \cdot x^{nh+2r} +$, etc. And if by S one understands any other polynomial, ordered according to the powers of that same variable, such that $S\ ^{\mathsf{k}}1 \cdot x^l + S\ ^{\mathsf{k}}2 \cdot x^{l+r} + S\ ^{\mathsf{k}}3 \cdot x^{l+2r} +$, etc., the product of these two polynomials $Q^h S$ will be identical with $(Q^h S)\ ^{\mathsf{k}} \cdot x^{nh+l} + (Q^h S)\ ^{\mathsf{k}}2 x^{nh+l+r} + (Q^h S)\ ^{\mathsf{k}}3 \cdot x^{nh+l+2r} +$, etc.

"Professor Hindenburg appears to be the first mathematician who has felt the indispensable need of this *local* notation in the present state of Analysis; it is moreover generally adopted today by the mathematicians of his nation. Above all the polynomial coefficients $Q^{h}\ ^{\mathsf{k}}1$, $Q^{h}\ ^{\mathsf{k}}2$, $Q^{h}\ ^{\mathsf{k}}3$, etc. recur without ceasing in all that vast part of Analysis which has for its object the development in series of any function whether explicit or implicit. . . . .

"In the eighteenth chapter and in most of those which follow, I have conveniently employed the German letters $\mathfrak{a}$, $\mathfrak{b}$, $\mathfrak{c}$, $\mathfrak{d}$, etc., to denote the binomial factors of the power to the exponent n. One has therefore, $\mathfrak{a} = n$; $\mathfrak{b} = \dfrac{n(n-1)}{1 \cdot 2}$ and so on.

"I employ the Latin letter D, placed before the polynomial function Q, and I have named *first derivative (première dérivée)*, what results from that function when one multiplies all its terms by their respective exponents and one then divides them by x. In the case that $Q = a + bx + cx^2 + dx^3 +$ etc. one will therefore have, $DQ = b + 2cx + 3dx^2 + 4ex^3$ etc. One has in the same way for a *second derivative (seconde dérivée)*, $D^2Q = 2c + 6dx + 12ex^2 + 20fx^3 +$ etc. . . . .

"The capital German letters are in general signs of functions. I employ particularly to this end the letters $\mathfrak{F}$ and $\mathfrak{P}$. Thus $\mathfrak{F}(x)$ designates any function whatever of the variable; $\mathfrak{P}(x)$ designates another.

"The derivatives (*dérivées*) of Arbogast designate therefore the simple coefficients of the series; mine are veritable polynomial functions, developed in series, of which only the first terms are the derivatives of Arbogast. The notation of these last is identical with the local notation $\mathfrak{F}(a+x)\ ^{\mathsf{k}}1$, $\mathfrak{F}(a+x)\ ^{\mathsf{k}}2$, $\mathfrak{F}(a+x)\ ^{\mathsf{k}}3$ etc. for the first derivatives; and for the second, with $(\mathfrak{F}X)\ ^{\mathsf{k}}1$, $(\mathfrak{F}X)\ ^{\mathsf{k}}2$, $(\mathfrak{F}X)\ ^{\mathsf{k}}3$ etc. always in the case that the exponents of the powers of x in the function X are the terms of the progression of the natural numbers $0, 1, 2, 3$, etc. . . . .

"My researches on the calculus of derivatives go back to the year 1795. They appear for the first time in the work published in 1796 by Professor Hindenburg, under the title *Der polynomische Lehrsatz,* where this mathematician did me the honor of joining my essay on combinatorial analysis with his and also with those of Messrs. Tetens, Pfaff and Klügel. . . . .

"For the designation of the product whose factors form among themselves an arithmetical progression, such as $a(a+r)(a+2r)$ . . . . $(a+nr-r)$, I have retained the notation $a^{n\,|\,r}$ already proposed in my *analyse des refractions;* I have given it the name *facultés.* Arbogast substituted for it the choicer and more French designation of *factorielles."*

Klügel[1] uses Kramp's notation $a^{m,r}$ for $a(a+r)$ . . . . $(a+mr-r)$ and calls it a *Facultät.*

446. *Signs of Argand and Ampère.*—Independently of German writers, a few symbols were introduced by French writers. J. R. Argand,[2] in treating a "Problème de combinaisons," designates by (m, n) the "ensemble de toutes les manières de faire avec m choses n parts ... et par $Z(m, n)$ le *nombre* de ces manières." A few years later A. M. Ampère[3] lets $\overset{1}{[x]}=x$, $\overset{2}{[x]}=\overset{1}{[x]}(x+p)$, $\overset{3}{[x]}=\overset{2}{[x]}(x+2p)$ . . . . , $\overset{m+1}{[x]}=\overset{m}{[x]}(p+mp)$. This notation resembles that of Vandermonde; Ampère refers to the work of Vandermonde and Kramp. Ampère's notation is used by Lentheric[4] of Montpellier.

Crelle[5] adopted the signs when m is an integer: $(u, +x)^m = u(u+x)(u+2x)(u+3x)$ . . . . $(u+[m-1]x)$, $(u,+x)^{-y}=\dfrac{1}{(u-yx, +x)^y}$.

Schellbach[6] added to these symbols the following:

$$a_0^n, +k = a_0\, a_k\, a_{2k}\, a_{3k} \ldots\ldots a_{nk-k}\ ,$$
$$f^n(x, +y) = f(x)\, f(x+y)\, f(x+2y) \ldots\ldots f(x+ny-y),$$
$$(1-a_n, +1)^{-n} = \frac{1}{(1-a_0)(1-a_1)(1-a_2) \ldots\ldots (1-a_{n-1})}\ .$$

Schellbach gives the symbolism $n!a_\delta$ to mark the occurrence of n quantities $a_0, a_1, a_2, \ldots\ldots, a_{n-1}$, where δ takes successively the values

[1] G. S. Klügel, *Mathematisches Wörterbuch*, 1. Theil (Leipzig, 1803), art. "Facultät."

[2] J. R. Argand in J. D. Gergonne's *Annales de mathématiques pures et appliquées* (Nismes), Vol. VI (1815 and 1816), p. 21.

[3] *Op. cit.*, Vol. XV (1824–25), p. 370.

[4] *Op. cit.*, Vol. XVI (1825–26), p. 120.

[5] A. L. Crelle, in *Crelle's Journal*, Vol. VII (1831), p. 270, 271.

[6] Karl Heinrich Schellbach, *Crelle's Journal*, Vol. XII (1834), p. 74, 75. Schellbach gave a discussion of mathematical notation in this article.

0, 1, ... , $n-1$. Accordingly, $f(n!x_\delta)$ stands for a function of x_0, x_1, ... , x_{n-1}. He writes $(m, n!\, a_\delta)$ = the combinations, without repetition, of the elements $a_0, a_1, \ldots, a_{n-1}$, taken m at a time. And $[m, n!, a_\delta]$ = the combinations with repetitions. He lets also $n|a_\sigma = a_0 + a_1 + \ldots + a_{n-1}$, where σ takes successively the values 0, 1, ... , $n-1$. Also[1]

$$(a+b)^3 = 4 \mid [3 - \sigma \downarrow 1 + \sigma! \; a + 1 + \delta] \; [\sigma \uparrow 4 - \sigma! \; \delta - a] \left(\frac{b - \sigma, +1}{1, +1}\right)^3, \text{ and}$$

similarly for $(a+b)^n$, where the arrows indicate (p. 154) the direction of the combination.

447. *Thomas Jarrett.*—An extensive study of algebraic notations was made by Thomas Jarrett (1805–82), of Catharine Hall, at Cambridge in England. He[2] published an article in 1830, but gave a much fuller treatment of this subject in an essay[3] of 1831. He remarks that the demonstration of the legitimacy of the separation of the symbols of operation and quantity, with certain limitations, belongs to Servois. Jarrett refers to Arbogast, J. F. W. Herschel, Hindenburg, Lacroix, Laplace, Schweins, and Wronski. Jarrett points out that the following notations used by him are not original with him: First, $E_x\phi(x)$ for $\phi(x + Dx)$ is partly due to Arbogast; second, $d_x^n u$ for $\dfrac{d^n u}{dx^n}$ is due to Lacroix, although not used by him, being merely pointed out in a single line (*Calcul Diff.*, Vol. II, p. 527); third, $(u)_{x=a}$ for the value assumed by u, when x is put equal to a, belongs to Schweins.

The principal symbols introduced by Jarrett are as follows:[4]

$\overset{n}{S}_m a_m$, the sum of n terms, of which the mth is a_m (p. 1).

$\overset{n;\,r}{S}_m a_m$, the rth term must be omitted.

$\overset{r}{\underset{m}{S}} \overset{s}{\underset{n}{S}} a_{m,\,n}$, the sum of r terms of which the mth is $\overset{s}{\underset{m}{S}} a_{m,\,n}$ (p. 5).

$\overset{m,\,n}{S}_{r,\,+s}(^s a_r)$, the sum of every term that can be formed with the following conditions: each term in the product of m quantities in which r has the values of the successive

---

[1] Schellbach, *op. cit.*, p. 154, 156.

[2] Thomas Jarret in *Transactions of the Cambridge Philosophical Society*, Vol. III (Cambridge, 1830), p. 67.

[3] *An Essay on Algebraic Development containing the Principal Expansions in Common Algebra, in the Differential and Integral Calculus and in the Calculus of finite Differences.* . . . By the Rev. Thomas Jarrett, M.A., Fellow of Catharine Hall, and Professor of Arabic in the University of Cambridge (1831).

[4] Thomas Jarrett, *op. cit.* (1831), "Index to the Symbols."

natural numbers, while s has any m values such that their sum shall be n, zero being admissible as a value of s, and repetitions of the same value of that letter being allowed in the same term (p. 78).

$\varpi_a^m \phi^{(a)},$ coefficient of x^m in the development of $\phi(\overset{\infty}{S}_m a_{m-1} x^{m-1})$ (p. 79).

$\overset{n,r}{S}_m a_m,$ the sum of the series formed by giving to m every integral value from n to r both inclusive; zero being also taken as a value if n is either zero or negative (p. 136).

$\overset{n}{P}_m a_m,$ the product of n factors, of which the mth is a_m (p. 12).

$\overset{n,r}{P}_m a_m,$ the rth factor must be omitted.

$\lfloor a \atop n,m = a(a+m)(a+2m) \ldots (a+\overline{n-1}\cdot m)$ (p. 15).

$\lfloor a \atop n = a(a-1)(a-2) \ldots (a-n+1).$

$\lfloor a = a(a-1)(a-2) \ldots 2\cdot 1.$

$\overset{n}{\underset{m}{\{}} a_m + b_n \underset{m+1}{\{} \ldots \underset{n+1}{\{} c \underset{n+1}{\}} \ldots \}$ denotes the result of the combination of the symbols $\underset{1}{\{} a_1 + b_1 \underset{2}{\{} a_2 + b_2 \ldots \underset{n}{\{} a_n + b_n \underset{n+1}{\{} c \underset{n+1}{\}} \ldots \underset{1}{\}};$

the brackets being omitted after the expansion, if they are then without signification (p. 19).

Theorem. $\overset{n}{\underset{m}{\{}} a_m + b_m \underset{m+1}{\{} \ldots \underset{n+1}{\{} c = \overset{n}{S}_m a_m \cdot \overset{m-1}{P}_r b_r + c \cdot \overset{m}{\underset{r}{P}} b_r.$

$\overset{m,n}{C}_r a_r,$ the sum of every possible combination (without repetition of any letter in the same combination) that can be formed by taking m at a time of n quantities of which the rth is a_r.

$\overset{m,n;s}{C}_r \quad a_r,$ a_s is to be everywhere omitted.

$\overset{\overline{m,n-m}}{C}_{r,s}(a_r\cdot b_s),$ n quantities of which the rth is a_r, and n others of which the sth is b_s, every possible combination being formed of the first series, by taking them m at a time, each combination thus formed being multiplied by $n-m$ quantities of the second series, so taken that in each of the combinations the whole of the natural numbers from 1 to n shall appear as indices: thus (p. 22).

$$\overset{2,3}{C}_{r,s}(a_r \cdot b_s) = a_1a_2b_3b_4b_5 + a_1a_3b_2b_4b_5 + a_1a_4b_2b_3b_5 + a_1a_5b_2b_3b_4 +$$
$$a_2a_3b_1b_4b_5 + a_2a_4b_1b_3b_5 + a_2a_5b_1b_3b_4 + a_3a_4b_1b_2b_5 +$$
$$a_3a_5b_1b_2b_4 + a_4a_5b_1b_2b_3 .$$

$(\phi+\psi)_n u$ means $\{(\phi+\psi)(\phi+\psi) \ldots (\phi+\psi)\}u$, ($n$ round parentheses) (p. 41).

$\overset{n}{\underset{r}{\{}}(\phi_r+\psi_r)\underset{r+1}{\{}$ u means $\{(\phi_1+\psi_1)(\phi_2+\psi_2) \ldots (\phi_n+\psi_n)\}u$.

$(u)_{x=a}, \phi_{x=a}(u)$, denote respectively the values of u, and $\phi_x(u)$, when x is put equal to a; this substitution, in the latter case, not being made until after the operation indicated by ϕ_x has been performed (p. 45).

$E_x^{(u)}$ — means that in u, any function of x, $x+h$ is substituted for x.

$D_x^{(u)}$ — means the excess of the new value of u above the original value.

$E_{x,y}^{(u)}$ — expresses either $E_x \cdot E_y(u)$ or $E_y \cdot E_x(u)$ (p. 58).

$\mathcal{C}_{2m-1} = \overset{2m}{\underset{+r}{A}}\left(\dfrac{-1}{r+1}\right)$, where the right member is called the $(2m-1)$th number of Bernoulli; $\mathcal{C}_1 = \dfrac{1}{2\lfloor 3} = ,083$ (p. 89).

A turn through 180° of Jarrett's sign for n-factorial yields ⌐, a symbol introduced by Milne[1] in the treatment of annuities. He lets a stand for the expectation of life by an individual A, $_t a$ the probability of his surviving t years, and $⌐_t a$ the expectation of life after the expiration of t years. This sign was used similarly by Jones[2] who lets $a_{(m)}⌐_n$ be the present value of £1 per annum, to be entered upon after n years, m being the present age.

448. *Factorial* "n."—The frequency of the occurrence of n-factorial in algebra and general analysis gives this expression sufficient importance to justify a separate treatment, even at the risk of some slight repetition of statements. In 1751 Euler[3] represented the product $1.2.3. \ldots . m$ by the capital letter M. "Ce nombre de cas 1.2.3.4. $\ldots . m$ etant posé pour abréger $= M$. ..." Probably this was not intended as a general representation of such products, but was introduced simply as a temporary expedient. The very special relation

[1] Joshua Milne, *Annuities and Assurances*, Vol. I (London, 1815), p. 57, 58.

[2] David Jones, *Value of Annuities*, Vol. I (London, 1843), p. 209.

[3] L. Euler, "Calcul de la Probabilité dans le jeu de Recontre," *Histoire de l'académie r. d. sciences et des belles lettres de Berlin*, année 1751 (Berlin, 1753), p. 259, 265.

m to M would go against the use of M as the product of, for example,
1.2.3. . . . . r, or of 1.2.3.4.5.6. A little-known suggestion came in
1774 from J. B. Basedow[1] who used a star, thus $5^* = 5.4.3.2.1$. Other
abbreviations were used in 1772 by A. T. Vandermonde: "Je repré-
sente par $[\overset{n}{p}]$ le produit de n facteurs ... $p \cdot (p-1) \cdot (p-2) \cdot (p-3)$ ...
ou le produit de n termes consécutifs d'une suite dont les premières
différences sont 1, et les secondes différences sont zéro."[2] He writes
$[\overset{n}{p}] = p \cdot (p-1) \cdot (p-2) \cdot (p-3)$ . . . . $(p-n+1)$; he finds $[\overset{o}{p}] = 1$, $[\overset{-n}{p}]$

$$= \frac{1}{(p+1) \cdot (p+2) \cdot (p+3) \cdot (p+4) \ldots (p+n)}, \quad 2^4 \left[\frac{11}{2} \right]^4 = 11.9.7.5. ,$$

$[\overset{-5}{0}] = \dfrac{1}{1.2.3.4.5.}$, $\frac{1}{2}\pi = [\tfrac{1}{2}]^{\frac{1}{2}} [-\tfrac{1}{2}]^{-\frac{1}{2}} = \dfrac{2.2.4.4.6.6. \ldots}{1.3.3.5.5.7. \ldots}$. The special case
when $p = n = \mathrm{a}$ positive integer would yield the product $n(n-1)$
. . . . 3.2.1., but Vandermonde was operating with expressions in
form more general than this (see also § 441).

A sign for n-factorial arises as a special case of a more general
notation also in Christian Kramp of Strasbourg who in his *Élémens
d'arithmétique universelle* (1808) and in special articles[3] lets $a^{m|r}$ stand
for $a(a+r)(a+2r)$ . . . . $[a+(m-1)r]$, and uses the special forms
$a^{m|0} = a^m$, $a^{-m|-r} = \dfrac{1}{(a+r)^{m|r}}$, $1^{m|1} = 1.2.3 \ldots . m$ "ou a cette autre
forme plus simple $m!$" In 1808 Kramp said: "Je me sers de la nota-
tion trés simple $n!$ pour désigner le produit de nombres décroissans
depuis n jusqu'à l'unité, savoir $n(n-1)(n-2)$ . . . . 3.2.1. L'emploi
continuel de l'analyse combinatoire que je fais dans la plupart de mes
démonstrations, a rendu cette notation indispensable."[4] In a footnote
to Kramp's article, the editor, J. S. Gergonne, compares the notations
of Vandermonde and Kramp. "Vandermonde fait $a \cdot (a-1) \cdot (a-2)$
. . . . $(a-m+1) = [a]^m$, d'où il suit qu'en rapprochant les deux nota-
tions, ou a $[a]^m = (a-m+1)^{m|1} = a^{m|-1}$, ..., $1.2.3.4 \ldots . m = 1^{m|1}$
$= m!$" Kramp's notation $1^{m|1}$ found its way into Portugal where
Stockler[5] used it in 1824.

[1] Johann Bernhard Basedow, *Bewiesene Grundsätze der reinen Mathematik*,
Vol. I (Leipzig, 1774), p. 259.

[2] A. T. Vandermonde, *Histoire de l'académie r. d. sciences*, année 1772, Part I
(Paris, 1775), *Mém.*, p. 490, 491.

[3] J. D. Gergonne, *Annales de Mathématiques*, Vol. III (1812 et 1813), p. 1.

[4] C. Kramp, *Élémens d'arithmétique universelle* (Cologne, 1808), "Notations."
See also p. 219.

[5] Francisco de Borja Garção Stockler, *Methodo inverso dos Limites* (Lisbon,
1824), p. 35.

A new designation for n-factorial was introduced by Legendre. In 1808 he wrote in an article, "Dans la première qui traite des intégrales de la forme $\int \frac{x^{p-1}\,dx}{\sqrt[n]{(1-x^n)^{n-q}}}$, prises depuis $x=0$ jusqu'à $x=1$ désignées par Euler par le symbole $\left(\dfrac{p}{q}\right)$, on peut regarder comme choses nouvelles."[1] Legendre designated the integral by the capital Greek letter gamma, Γ; and continued this usage in his integral calculus[2] of 1811. As the value of this definite integral is n-factorial, the symbol $\Gamma(n+1)$ came to stand for that value.

449. Occasionally the small letter pi was used to represent n-factorial, as, for instance, in Ruffini's theory of equations[3] of 1799, where one finds "$\pi=1.2.3.4. \ldots . m$," or "un' Equazione del grado $1.2.3. \ldots . m$, che chiamo $\pi. \ldots .$" More frequent is the use of the capital letter Π; thus Gauss[4] employed for n-factorial the notation $\Pi(n)$; Jacobi[5] Π_4 for 4! H. Weber[6] in 1893 wrote $\Pi(m)=1.2.3. \ldots . m$. There are a few obsolete notations; for instance, Henry Warburton,[7] of Cambridge, England, in 1847, represents n-factorial by $1^{n|1}$, as a special case of $S^{n|1} \equiv s(s+1).(s+2) \ldots . (s+[n-1])$, which resembles Kramp's $a^{m|r}$.

De Morgan[8] (who ordinarily uses no contracted symbol) employed in 1838 the designation $[n]$, as a special case, apparently, of Vandermonde's[9] $[p]^m \equiv p(p-1)(p-2) \ldots . (p-[m-1])$. Carmichael[10] in 1855 and 1857, when hard pressed for space in writing formulas, uses the

[1] *Mémoires de l'institut national, sciences math. et phys.*, année 1808 (Paris, 1809), p. 14, 15.

[2] A. M. Legendre, *Exercices de calcul intégral*, Vol. I (Paris, 1811), p. 277.

[3] Paolo Ruffini, *Teoria generale delle equazioni* (Bologna, 1799), p. 97, 244, 254.

[4] C. F. Gauss, *Commentationes Societatis regiae scientiarum Gottingensis recentiores*, Vol. II (1811–13), *Math. mém.*, n° 1, p. 26; *Werke*, Vol. III (Göttingen, 1866), p. 146.

[5] C. G. J. Jacobi, *Crelle's Journal*, Vol. XII (1834), p. 264.

[6] H. Weber, *Mathematische Annalen*, Vol. XLIII (1893), p. 535.

[7] H. Warburton, *Cambridge Philosophical Transactions*, Vol. VIII (1849), p. 477.

[8] A. de Morgan, *Essay on Probabilities*, *Cabinet Cyclopoedia*, p. 15 (London, 1838).

[9] A. T. Vandermonde in *Mémoires de l'Académie des Sciences* (1772), première partie, p. 490.

[10] Robert Carmichael, *Treatise on the Calculus of Operations* (London, 1855); German edition, *Operations Calcul* (Braunschweig, 1857), p. 30–55.

designation $\bar{n}$. Weierstrass[1] in 1841 and later Schlömilch[2] chose the sign m' for our $m!$.

Relating to the origin of the notation $\lfloor n$ for "n-factorial," nothing has been given in histories, except the statement that it has been in use in England. The notation $\lfloor n$ was suggested in 1827 by Jarrett who had just graduated from St. Catherine's College in Cambridge, England, with the degree of B.A. It occurs in a paper "On Algebraic Notation" that was printed in 1830[3] (see § 447). The passage in question reads as follows: "A factorial of this kind consisting of m factors, of which n is the first, and of which the common difference is $\pm r$, may be denoted by $\dfrac{\lfloor n}{m \pm r}$; the particular case in which the common difference is -1, we may represent by $\dfrac{\lfloor n}{m}$ and if, in this case, $m \equiv n$, the index subscript may be omitted:
Thus

$$\frac{\lfloor n}{m \pm r} = n(n \pm r)(n \pm 2r) \ldots (n \pm \overline{m-1}r)$$

$$\frac{\lfloor n}{m} = n(n-1)(n-r) \ldots (n - \overline{m-1})$$

$$\lfloor n = n(n-1)(n-2) \ldots 1 \text{ ."}$$

For a quarter of a century the notation $\lfloor n$ was neglected. In 1846, Rev. Harvey Goodwin used it freely in an article, "On the Geometrical Representation of Roots of Algebraic Equations," that was printed in 1849.[4] In 1847 Goodwin published his *Elementary Course in Mathematics*, a popular educational manual which reached several editions, but, to our surprise, he did not make use of any contracted notation for factorial n in this text. In fact, the symbol $\lfloor n$ made no substantial headway in England until it was adopted by Todhunter about 1860, and was used in his popular texts. In his *History of Probability*[5] he says: "I have used no symbols which are not common to all mathematical literature, except $\lfloor n$ which is an abbreviation for the product $1.2. \ldots . n$, frequently used but not

---

[1] K. Weierstrass, *Mathematische Werke*, Vol. I (Berlin, 1894), p. 14, 50.

[2] O. Schlömilch, *Zeitschrift für Math. u. Physik*, Vol. II (1857), p. 139.

[3] *Transactions of the Cambridge Philosophical Society*, Vol. III, p. 67 (Cambridge, 1830).

[4] *Cambridge Philosophical Transactions*, Vol. VIII (1849), p. 343.

[5] Isaac Todhunter, *History of the Mathematical Theory of Probability*, p. viii and lx (Cambridge and London, 1865).

universally employed; some such symbol is much required, and I do not know of any which is preferable to this, and I have accordingly introduced it in all my publications." In 1861, $\lfloor n$ was used by Henry M. Jeffery.[1] Some use of it has been made in Sweden,[2] though less often than of $n!$.

In the United States $\lfloor n$ was probably introduced through Todhunter's texts. In the first volume (1874) of J. E. Hendricks' *Analyst* (Des Moines, Iowa), both the notation $\lfloor n$ and $n!$ are used by different writers. The latter notation, though simpler, was used in elementary texts of this country less frequently than the first. The notation $\lfloor n$ was adopted by such prominent textbook writers as Joseph Ficklin (*Complete Algebra*, copyright 1874), Charles Davies (revised Bourdon's *Elements of Algebra* [1877]), Edward Olney (1881), and about the same time by George A. Wentworth, Webster Wells, E. A. Bowser, and others. Thus it became firmly rooted in this country.[3] Among the French and German authors $\lfloor n$ has met with no favor whatever. In 1915 the Council of the London Mathematical Society,[4] in its "Suggestions for Notation and Printing," recommended the adoption of $n!$ for $\lfloor n$, $(n!)^2$ for $(\lfloor n)^2$, $2^n \cdot n!$ for $2^n \lfloor n$.

The notation $n!$ is used after Kramp in Gergonne's *Annales* by J. B. Durrande[5] in 1816, and by F. Sarrus[6] in 1819. Durrande remarks, "There is ground for surprise that a notation so simple and consequently so useful has not yet been universally adopted." It found wide adoption in Germany, where it is read "n-Fakultät."[7] Some texts in the English language suggest the reading "n-admiration"[8] (the exclamation point [!] being a note of admiration), but most texts prefer "factorial n," or "n-factorial." In Germany, Ohm,[9] whose books enjoyed popularity for many years, used the notation $n!$

[1] *Quarterly Journal of Mathematics*, Vol. IV, p. 364 (1861).

[2] *Encyclopédie des sciences mathématiques pures et appliquées*, Tome I, Vol. I, p. 65 (1904).

[3] In a few publications Jarrett's factorial symbol is given in the modified form $n\rfloor$. See, for example, Thomas Craig's *Treatise on Linear Differential Equations*, Vol. I, p. 463 (New York, 1889), and Webster's *New International Dictionary of the English Language* (Springfield, 1919), under the word "Factorial."

[4] *Mathematical Gazette*, Vol. VIII (London, 1917), p. 172.

[5] Gergonne, *Annales de mathématiques*, Vol. VII (1816 and 1817), p. 334.

[6] *Op. cit.*, Vol. XII (1821 and 1822), p. 36.

[7] E. Pascal, *Repertorium*, Vol. I, p. 43 (Leipzig und Berlin, 1910).

[8] W. E. Byerly, *Elements of the Differential Calculus*, p. 120 (Boston, 1880).

M. Ohm, *System der Mathematik*, Vol. II, p. 17 (Berlin, 1829).

about 1829. It was used in 1847 by Eisenstein, then Privat Docent in Berlin, in an article quoted by Sylvester[1] in 1857. Chrystal's *Algebra* (1889) came out for $n!$, though in the nineteenth century it was much less frequent in England than $\lfloor n$.

In the United States $n!$ was used by W. P. G. Bartlett[2] as early as the year 1858, and by Simon Newcomb[3] in 1859. It was adopted mainly by a group of men who had studied at Harvard, Pliny Earl Chase[4] (later professor of physics at Haverford), James Edward Oliver[5] (later professor of mathematics at Cornell), and C. S. Peirce[6] the logician. Afterward $n!$ was used in Newcomb's *Algebra* (1884), I. Stringham's edition of Charles Smith's *Algebra* (1897), M. Merriam and R. S. Woodward's *Higher Mathematics* (1898), Oliver, Wait, and Jones's *Algebra* (1887), and in others.

In the present century the notation $n!$ has gained almost complete ascendancy over its rivals. It is far more convenient to the printer.

Remarkable is the fact that many writings, both advanced and elementary, do not use any contracted notation for n-factorial; the expanded notation $1, 2, 3 \ldots n$ is adhered to. The facts are that a short mode of designation is not so imperative here as it is for "square root," "cube root," or "the nth power." We have seen that Harvey Goodwin of Caius College, Cambridge, made liberal use of $\lfloor n$ in a research article, but avoided it in his *Elementary Course*. Instinctively he shrunk from the introduction of it in elementary instruction. We have here the issue relating to the early and profuse use of symbolism in mathematics: Is it desirable? In the case of n-factorial some writers of elementary books of recognized standing avoid it. More than this, it has been avoided by many writers in the field of advanced mathematics, such as J. J. Sylvester, A. Cayley, H. Laurent, E. Picard, E. Carvallo, E. Borel, G. B. Airy, G. Salmon, W. Chauvenet, Faà de Bruno, P. Appell, C. Jordan, J. Bertrand, W. Fiedler, A. Clebsch.

[1] G. Eisenstein, *Mathematische Abhandlungen* (Berlin, 1847); J. J. Sylvester, *Quarterly Journal of Mathematics*, Vol. I, p. 201 (London, 1857).

[2] J. D. Runkle's *Mathematical Monthly*, Vol. I, No. 3, p. 84–87 (Cambridge, Mass., 1858).

[3] J. D. Runkle, *op. cit.*, Vol. I (1859), p. 331, 396.

[4] *Trans. American Philosoph. Society*, Vol. XIII, p. 25–33, N.S. (Philadelphia, 1869). Chase's paper is dated Sept. 18, 1863.

[5] *Op. cit.*, p. 69–72. Oliver's paper is dated May 6, 1864.

[6] *Memoirs American Academy of Arts and Sciences*, Vol. IX, p. 317, N.S. (Cambridge and Boston, 1867).

Of course, I am not prepared to say that these writers never used $n!$ or $\lfloor n$; I claim only that they usually avoided those symbols. These considerations are a part of the general question of the desirability of the use of symbols in mathematics to the extent advocated by the school of G. Peano in Italy and of A. N. Whitehead and B. Russell in England. The feeling against such a "scab of symbols" seems to be strong and widespread. If the adoption of only one symbol, like our $n!$ or $\lfloor n$, were involved, the issue would seem trivial, but when dozens of symbols are offered, a more serious situation arises. Certain types of symbols are indispensable; others possess only questionable value. Rich meaning is conveyed instantaneously by $\frac{dy}{dx}$, $\int y\,dx$, but $\lfloor n$ and $n!$ serve no other purpose than to save a bit of space. Writers who accept *in toto* the program of expressing all theorems and all reasoning by a severely contracted symbolism must frame notations for matters that can more conveniently be expressed by ordinary words or in less specialized symbolism. We know that intellectual food is sometimes more easily digested, if not taken in the most condensed form. It will be asked, To what extent can specialized notations be adopted with profit? To this question we reply, *only experience can tell*. It is one of the functions of the history of mathematics to record such experiences. Some light, therefore, may be expected from the study of the history of mathematics, as to what constitutes the most profitable and efficient course to pursue in the future. The history of mathematics can reduce to a minimum the amount of future experimentation. Hence algebraic notations deserve more careful historic treatment than they have hitherto received.

450. *Subfactorial "n"* was introduced in 1878 by W. Allen Whitworth[1] and represented by the sign $\lfloor\!\lfloor$, in imitation of the sign $\lfloor$ for factorial n. The value $\lfloor\!\lfloor n$ is obtained by multiplying $\lfloor\!\lfloor n-1$ by n and adding $(-1)^n$. Taking $\lfloor\!\lfloor 0 = 1$, Whitworth gets $\lfloor\!\lfloor 1 = 1 - 1 = 0$, $\lfloor\!\lfloor 2 = 0\times2+1 = 1$, $\lfloor\!\lfloor 3 = 3\times1-1 = 2$, $\lfloor\!\lfloor 4 = 4\times2+1 = 9$. Since $\frac{1}{e} = 1 - \frac{1}{\lfloor 1} + \frac{1}{\lfloor 2} - \frac{1}{\lfloor 3} +$, etc., Whitworth obtains,[2] when n is even, $\frac{\lfloor\!\lfloor n}{\lfloor n} > \frac{1}{e}$; when n is odd, $\frac{\lfloor\!\lfloor n}{\lfloor n} < \frac{1}{e}$; limit $\frac{\lfloor\!\lfloor n}{\lfloor n} = \frac{1}{e}$. Subfactorials chiefly occur in connection

[1] W. Allen Whitworth, *Messenger of Mathematics*, Vol. VII (1878), p. 145 See also Whitworth, *Choice and Chance* (Cambridge, 1886), Preface, p. xxxiii.

[2] See also *Choice and Chance* (1886), p. 108.

with permutations. Chrystal[1] uses in place of Whitworth's $\underline{||}$ the sign $n_]$ ·

451. *Continued products.*—An infinite product occurs in the writings of Vieta,[2] but without any symbolism to mark that the number of factors is infinite. As reproduced in his collected works of 1646, four factors are written down as follows:

$$``\sqrt{\tfrac{1}{2}+\sqrt{\tfrac{1}{2}}}\,,\text{ in }\sqrt{\tfrac{1}{2}+\sqrt{\tfrac{1}{2}+\sqrt{\tfrac{1}{2}}}}\,,\text{ in }\sqrt{\tfrac{1}{2}+\sqrt{\tfrac{1}{2}+\sqrt{\tfrac{1}{2}+\sqrt{\tfrac{1}{2}}}}}\,,\text{ in}$$

$$\sqrt{\tfrac{1}{2}+\sqrt{\tfrac{1}{2}+\sqrt{\tfrac{1}{2}+\sqrt{\tfrac{1}{2}+\sqrt{\tfrac{1}{2}}}}}}\,.\text{''}$$

The factors are separated by a comma and the preposition "in." The extension of the product to one that is infinite is expressed in the words, "in infinitum observata uniformi methodo." Continued products appear in Wallis' *Arithmetica infinitorum* (1655), in passages like the following (p. 179):

"invenietur

$$\square \begin{cases} \text{minor quam } \dfrac{3\times3\times5\times5\times7\times7\times9\times\ 9\times11\times11\times13\times13}{2\times4\times4\times6\times6\times8\times8\times10\times10\times12\times12\times14}\times\sqrt{1\tfrac{1}{13}} \\[2mm] \text{major quam } \dfrac{3\times3\times5\times5\times7\times7\times9\times\ 9\times11\times11\times13\times13}{2\times4\times4\times6\times6\times8\times8\times10\times10\times12\times12\times14}\times\sqrt{1\tfrac{1}{14}}\,, \end{cases}$$

where $\square$ stands for our $\dfrac{4}{\pi}$. Later (p. 180) Wallis makes the products in the numerator and denominator infinite by adding "&c." as follows:

"Dicimus, fractionem illam $\dfrac{3\times3\times5\times5\times7\times7\times\&c.}{2\times4\times4\times6\times6\times8\times\&c.}$ in infinitum continuatam, esse ipsissimum quaesitum numerum $\square$ praecise."

Gauss in 1812 introduced capital Π in the designation of continued products in his paper on the hypergeometric series[3] bearing the title "Disquisitiones generales circa seriem infinitam: $1+\dfrac{\alpha\beta}{1\cdot\gamma}\,x+$ $\dfrac{\alpha(\alpha+1)\beta(\beta+1)}{1\cdot2\cdot\gamma(\gamma+1)}\,xx+\dfrac{\alpha(\alpha+1)(\alpha+2)\beta(\beta+1)(\beta+2)}{1\cdot2\cdot3\cdot\gamma(\gamma+1)(\gamma+2)}\ x^3+$ etc.," which series he represents for brevity by the symbolism $F(\alpha,\beta,\gamma,x)$. The passage in which Π first appears is as follows:

"Introducamus abhinc sequentem notationem:

$$\Pi(k,\ z)=\frac{1\cdot2\cdot3\ldots\ldots k}{(z+1)(z+2)(z+3)\ldots\ldots(z+k)}\ k^z$$

[1] G. Chrystal, *Algebra*, Part II (1889), p. 25.

[2] F. Vieta, *Opera mathematica* (ed. Fr. à Schooten; Leyden, 1646), p. 400.

[3] C. F. Gauss, *Werke*, Vol. III (Göttingen, 1866), p. 123–61; see p. 144.

ubi k natura sua subintelligitur designare integrum positivum." As previously indicated (§ 449), Gauss later arrives at $\Pi z = 1.2.3. \ldots z$. He obtains further (p. 148), $\Pi(-\frac{1}{2}) = \sqrt{\pi}$. He represents[1] the function $\dfrac{d \log \Pi z}{dz}$ by ψz. Jacobi[2] uses Π for continued product. Jordan[3] uses the notation $\Pi_r(c^{\pi p^r} - \overline{K}^\pi)^e$ where the multiplication involves the factors resulting from the different values of r. Chrystal[4] represents the product $(1+u_1)(1+u_2) \ldots (1+u_n)$ by $\overset{n}{\Pi}(1+u_n)$, or "simply by P_n." A. Pringsheim[5] writes $\overset{n}{\underset{0}{\Pi}}(1+u_v)$ to indicate the $(n+1)$ factors from $(1+u_0)$ to $(1+u_n)$.

452. *Permutations and combinations.*—A. T. Vandermonde[6] gives $[n]^p$ as the number of arrangements of n distinct elements, taken p at a time. We have seen (§ 439) that Euler marked the number of combinations of n distinct things taken p at a time by $\left(\dfrac{n}{p}\right)$ in 1778 and by $\left[\dfrac{n}{p}\right]$ in 1781, and that nineteenth-century writers changed this to $\binom{n}{p}$. In English textbooks on algebra special symbols for such expressions were slow in appearing. Usually authors preferred to present the actual indicated products and quotients of products. Todhunter[7] uses no special symbolism; Peacock[8] introduces C_r for the combinations of n things taken r at a time, only in deriving the binomial formula and after he had given a whole chapter on permutations and combinations without the use of special notations. However, in 1869 and earlier, Goodwin[9] of Cambridge used $_nP_r$ for the number of permutations of n things taken r at a time, so that $_nP_n$ was equal to Jarret's $\lfloor n$ or Kramp's $n!$ Potts[10] begins his treatment by letting the number of

[1] C. F. Gauss, *Werke*, Vol. III, p. 201.

[2] C. G. J. Jacobi, *Gesammelte Werke*, Vol. I (Berlin, 1881), p. 95.

[3] C. Jordan, *Traité des substitutions* (Paris, 1870), p. 430.

[4] G. Chrystal, *Algebra*, Part II (1889), p. 135.

[5] *Encyklopädie d. Math. Wissensch.*, Vol. I, Part I (Leipzig, 1898–1904), p. 113.

[6] *Histoire de l'acad. d. sciences* (Paris, 1772), I, *Mém.*, p. 490.

[7] Isaac Todhunter, *Algebra* (new ed.; London, 1881), p. 286–97.

[8] George Peacock, *Treatise of Algebra* (Cambridge, 1830), chap. ix, p. 200–253. C_r is introduced on p. 255.

[9] Harvey Goodwin, *Elementary Course of Mathematics* (3d ed.; Cambridge, 1869), p. 75, 76

[10] Robert Potts, *Elementary Algebra*, Sec. XI (London, 1880), p. 1–8.

combinations of n different things taken r at a time be denoted by ^nC_r, the number of variations by ^nV_r, the number of permutations by ^nP_r. Whitworth[1] uses P_r^n, C_r^n; also R_r^n for the number of ways of selecting r things out of n when repetitions are allowed; $C_{m,n}$ for the number of orders in which m gains and n losses occur; $J_{m, n}$ for the number of orders in which losses never exceed gains. Chrystal[2] writes $_nP_r$, $_nC_r$; also $_nH_r$ for the number of r combinations of n letters when each letter may be repeated any number of times up to r.

In Germany symbolism appeared in textbooks somewhat earlier. Martin Ohm[3] lets P stand for permutations; $\overset{6}{P}$ $(aabccd)$ for the permutations of the six elements a, a, b, c, c, d; V for variations without repetition; V' for variations with repetition; $\overset{n}{\underset{(a, b, c, \ldots)}{V'}}$ for the variations of elements $a, b, c, \ldots$, with repetition, and taken n at a time; $\overset{n}{\underset{(a, b, c, d, \ldots)}{C}}$ for the combinations of $a, b, c, d, \ldots$, taken n at a time; C' for combinations with repetition; $Ns.$ for $numerus$ $specierum$; $Ns.\overset{m}{P'}(a^\alpha, b^\beta, c^\gamma, \ldots)$ for the number of permutations of m elements when a is repeated α times, b is repeated β times, etc.; $Ns.\overset{n}{\underset{(a, b, c \ldots)}{C}}$ for the number of combinations without repetition, of m elements taken n at a time. A little later (p. 43) Ohm introduces the symbolisms $^m\overset{n}{V'}$, $^m\overset{n}{C'}$, where m gives the sum of all the n elements appearing in each assemblage. Thus $\overset{7}{\underset{(0, 1, 2, 3, \ldots)}{\overset{4}{C'}}}$ means the combinations, with repetition, which can arise from the elements $0, 1, 2, 3, \ldots$, when each assemblage contains four elements. There are eleven combinations, viz., 0007, 0016, 0025, 0034, 0115, etc., the sum of the elements in each being 7.

Similar symbolisms are used by Wiegand[4] in 1853 and M. A. Stern[5] in 1860. Thus, Stern represents by $'\overset{m}{C}(1, 2 \ldots n)$ the combinations of the mth class (m elements taken each time), formed from

[1] W. A. Whitworth, *Choice and Chance* (Cambridge, 1886), p. 121.

[2] G. Chrystal, *Algebra*, Part II (Edinburgh, 1899), p. 10.

[3] Martin Ohm, *op. cit.*, Vol. II (Berlin, 1829), p. 33–38.

[4] August Wiegand, *Algebraische Analysis* (2d ed.; Halle, 1853), p. 2–8.

[5] M. A. Stern, *Lehrbuch der algebraischen Analysis* (Leipzig und Heidelberg, 1860), p. 19.

the elements 1, 2, ... , n, such that the sum of the elements in each combination is r; while $^rC(1m, 2m, 3m, ... , nm)$ means that each element may be repeated without restriction. E. Netto[1] in 1901 lets $P(a, b, ... , g)$, or simply P_n, be the number of permutations of n distinct elements; $P_n[a, b, ... , g]$ the assemblage of all the permutations themselves; $I^{(n)}_\kappa$ the number of permutations of n elements which have κ inversions (p. 94). He lets also (p. 119)

$$C^{(k)}(\smallint m; a_1, a_2, ... , a_n) \qquad \Gamma^{(k)}(\smallint m; a_1, a_2, ... , a_n)$$
$$C^{(k)}[\smallint m; a_1, a_2, ... , a_n] \qquad \Gamma^{(k)}[\smallint m; a_1, a_2, ... , a_n]$$
$$V^{(k)}(\smallint m; a_1, a_2, ... , a_n) \qquad \Phi^{(k)}(\smallint m; a_1, a_2, ... , a_n)$$
$$V^{(k)}[\smallint m; a_1, a_2, ... , a_n] \qquad \Phi^{(k)}[\smallint m; a_1, a_2, ... , a_n]$$

stand for the combinations and variations of the numbers $a_1, a_2, ... , a_n$, taken (k) at a time, and yielding in each assemblage the prescribed sum m; where the four symbols on the left allow no repetition, while the four symbols on the right allow unrestricted repetition, and where the round parentheses indicate the number of assemblages, and the brackets the collection of all assemblages themselves connected by the $+$ sign.

In the *Encyclopédie* (1904)[2] the number of arrangements (permutations) of n things taken p at a time is marked by A^p_n, the number of combinations by C^p_n and by $\binom{n}{p}$. Moreover,

$$\Gamma^p_n \equiv n(n+1) . . . . (n+p-1) \div p! .$$

453. *Substitutions.*—Ruffini,[3] in his theory of equations of 1799, introduces a symbolism for functions designed for the consideration of their behavior when subjected to certain substitutions. He writes:

$f(x)$ for any function of x.

$f(x)(y)$ for any function of x and y.

$f(x, y)$ for a function which retains its value when x and y are permuted, as $x^3+y^3+z^3=f(x, y, z)$.

$f(x, y)(z)$ for a function like x^2+y^2+z.

$f((x)(y), (z)(u))$ for a function like $\dfrac{x}{y}+\dfrac{z}{u}$ which remains the same in value when x and z interchange simultaneously when y and u interchange.

---

[1] Eugen Netto, *Lehrbuch der Combinatorik* (Leipzig, 1901), p. 2, 3.

[2] *Encyclopédie d. scien. math.*, Tom. I, Vol. I (1904), p. 67, article by E. Netto, revised by H. Vogt.

[3] Paolo Ruffini, *Teoria generale delle equazioni* (Bologna, 1799), p. 1–6.

$f((x, y), (z, u))$ for a function like $xy+zu$, which remains invariant when x and y, or z and u, are interchanged, or when both x and y are changed into z and u.

If $Y=f(r)(s)(t)(u) \ldots (z)$, say $= r^3x^3 - 4r^2st + rstu - 2s^2tx + 4ux^2$,

then, $r|Y$ represents the aggregate of terms in Y which contain r,

$r\overset{|s}{|Y}$ represents the aggregate of all the terms which contain r and do not contain s.

Cauchy[1] in 1815 represented a substitution in this manner:

$$\begin{pmatrix} 1 & 2 & 3 & . & . & . & . & n \\ a & \beta & \gamma & . & . & . & . & \zeta \end{pmatrix}$$

where every number of the upper line is to be replaced by the letter of the lower line immediately below.

Jordan[2] represents a substitution, S, by the notation $S = (ab \ldots k)(a'b' \ldots k')(a'' \ldots k'') \ldots$ where each parenthesis exhibits a cycle of substitutions. If A and B are two substitutions, then Jordan lets AB represent the result of the substitution A, followed by the substitution B, and the meanings of $A^2, A^3, \ldots, A^{-1}$ are evident.

E. Netto[3] writes substitutions in the form similar to that of Cauchy,

$$\begin{pmatrix} x_1, & x_2, & x_3, & \ldots, & x_n \\ x_{i_1}, & x_{i_2}, & x_{i_3}, & \ldots, & x_{i_n} \end{pmatrix}.$$

Netto uses also the form of cycles employed by Jordan. If A and B are two substitutions, and B^{-1} is the inverse of A, then the substitution $B^{-1}AB$ signifies the transformation[4] of A by B. Some writers—for instance, W. Burnside[5]—indicate the inverse substitution of s by s_{-1}. If s, t, v are substitutions,[6] then in their product stv the order of

[1] A. L. Cauchy, *Journal de l'école polyt.*, Tome X, Cah. 17, p. 56. Reference taken from Thomas Muir, *The Theory of Determinants in the Historical Order of Development*, Vol. I (London; 2d ed., 1906), p. 101.

[2] Camille Jordan, *Traité des substitutions* (Paris, 1870), p. 21.

[3] Eugen Netto, *Substitutionstheorie* (Leipzig, 1882); English translation by F. N. Cole (Ann Arbor, 1892), p. 19.

[4] C. Jordan, *op. cit.*, p. 23.

[5] W. Burnside, *Theory of Groups of Finite Order* (Cambridge, 1897), p. 4.

[6] L. E. Dickson, *Theory of Algebraic Equations* (New York, 1903), p. 11. See also Harold Hilton, *Introduction to the Theory of Groups of Finite Order* (Oxford, 1908), p. 12, 13.

applying the factors was from right to left with Cayley and Serret, but in the more modern use the order is from left to right, as in Jordan's work.

By the symbolism $\begin{bmatrix} f(z) \\ z \end{bmatrix}$ or simply $f(z)$, there is represented in J. A. Serret's *Algebra*[1] the substitution resulting from the displacing of each index z by the function $f(z)$, where z is given in order n values, which it is assumed agree, except in order, with the simultaneous values of $f(z)$. Similarly,[2] $\begin{pmatrix} az+b \\ z \end{pmatrix}$ represents the $n(n-1)$ linear substitutions, where z takes in succession all the indices $0, 1, 2, \ldots,$ $(n-1)$ of the n variables $x_0, x_1, \ldots, x_{n-1}$, and the values of $az+b$ are taken with respect to modulus of the prime n between the limits, 0 and $n-1$.

454. *Groups.*—The word "group" in a technical sense was first used by E. Galois in 1830. He[3] first introduced the mark ∞ into group theory by the statement that in the group $x_k, x_{ak+bl}, \dfrac{"k}{l}$ peut avoir les $\dfrac{}{l} \quad \dfrac{}{ck+dl}$ $p+1$ valeurs $\infty, 0, 1, 2, \ldots, p-1$. Ainsi, en convenant que k peut être infini, on peut écrire simplement $x_k, x_{\frac{ak+b}{ck+d}}$." In another place[4] Galois speaks of substitutions "de la forme $(a_{k_1, k_2}, a_{mk_1+nk_2})$." Again he represents a substitution group by writing down the substitutions in it,[5] as "le groupe *abcd, badc, cdab, dcba.*"

Jordan[6] represents the group formed by substitutions $A, B, C,$ $\ldots,$ by the symbol $(A, B, C, \ldots)$, but frequently finds it convenient to represent a group by a single letter, such as $F, G, H,$ or L.

Certain groups which have received special attention have been represented by special symbols. Thus, in groups of movement, the group formed by rotations about a single line is the cyclic group[7] C_a, the group formed by rotations through π about three mutually perpendicular intersecting lines is the quadratic group D, rotations still differently defined yield the tetrahedral group T, the octahedral group O, the

---

[1] J. A. Serret, *Handbuch der höheren Algebra* (trans. by G. Wertheim), Vol. II (2d ed.; Leipzig, 1879), p. 321.

[2] J. A. Serret, *op. cit.*, Vol. II, p. 353.

[3] E. Galois, *Œuvres* (ed. E. Picard; Paris, 1897), p. 28.

[4] *Op. cit.*, p. 54.

[5] *Op. cit.*, p. 45.

[6] C. Jordan, *op. cit.*, p. 22, 395.

[7] See Harold Hilton, *op. cit.* (Oxford, 1908), p. 113–15.

icosahedral group E. The consideration of rotary inversions leads to cyclic groups c_m, generated by an m-al rotary inversion about a point O and a line l, and to groups marked D_m, d_m, and so on. Hilton's symbolism is further indicated in such phrases[1] as "suppose that $\{a\}$ is a normal cyclic subgroup of the Group $G \equiv \{a, b\}$ generated by the elements a and b."

Let S_1, S_2, S_3, ..., be a given set of operations, and G_1, G_2, ..., a set of groups, then the symbol $\{S_1, S_2, S_3, ..., G_1, G_2, ... \}$ is used by W. Burnside to denote the group that arises by combining in every possible way the given operations and the operations of the given groups.[2]

The notation $\dfrac{G}{G'}$ was introduced by Jordan[3] in 1873 and then used by Hölder[4] and others to represent a "factor-group" of G, or the "quotient" of G by G'.

The symbol $\vartheta(P)$ has been introduced by Frobenius[5] to denote $(p-1)(p^2-1) \ .... \ (p^\kappa-1)$, where κ is the greatest of a series of exponents of p.

455. In the theory of continuous groups single letters T or U are used to designate transformations, T^{-1}, U^{-1} being inverse transformations. The notation TU was employed by Sophus Lie[6] to represent the product of T and U, the transformation T being followed by the transformation U. If (x) and (x') are points such that $(x') = T(x)$, then $(x) = T^{-1}(x')$. In the relation $(y') = TUT^{-1}(x')$ one has the symbolic notation for "the transformation of U by T." The index of a subgroup H of a group G is often represented by the symbol (G, H). G. Frobenius[7] uses this same symbol in a more general sense.

The need of symbolism in the study of groups made itself felt in applications to crystallography. A. Schoenflies[8] marks a revolution about an axis a through an angle a by $\mathfrak{A}(a)$ or simply by $\mathfrak{A}$. For

[1] *Op. cit.*, p. 169.

[2] W. Burnside, *Theory of Groups of Finite Order* (Cambridge, 1897), p. 27.

[3] C. Jordan, *Bulletin Soc. math.* (1st ser.), Vol. I (1873), p. 46.

[4] O. Hölder, *Mathematische Annalen*, Vol. XXXIV (1889), p. 31.

[5] G. Frobenius, *Sitzungsberichte d. k. p. Akademie der Wissensch.* (Berlin, 1895), p. 1028.

[6] S. Lie and F. Engel, *Theorie der Transformationsgruppen*, Vol. I (Leipzig, 1888), p. 223. See also *Encyclopédie des scien. math.*, Tom. II, Vol. IV, p. 162–63.

[7] G. Frobenius, *Crelle's Journal*, Vol. CI (1887), p. 273.

[8] Arthur Schoenflies, *Krystallsysteme und Krystallstructur* (Leipzig, 1891), p. 32, 36.

brevity, $\mathfrak{A}(na)$ is written $\mathfrak{A}^n$. A reflection (*Spiegelung*) is marked by $\mathfrak{S}$, an operation of the second kind (involving a reflection and a rotation) having the axis a and the angle a is marked $\overline{\mathfrak{A}}(a)$ or $\overline{\mathfrak{A}}$. The cyclic group of rotation is marked C_n, n indicating the number of rotations;[1] the Dieder (regular n gon) group by D_n, which is equal to $\left\{ \mathfrak{A}\left(\dfrac{2\pi}{n}\right), \mathfrak{U} \right\}$ where $\mathfrak{U}$ indicates an inversion (*Umklappung*). In common with other writers, Schoenflies indicates the tetrahedral group by T and the octahedral by O. In the treatment of these groups and the corresponding classes of crystals, he[2] uses h_p or l_p as a p-fold axis of symmetry, thereby following the notation of Bravais,[3] except that the latter uses capital letters for the designation of lines. Schoenflies[4] marks translation groups by Γ_r and space group by Γ. On pages 555, 556, he gives a table of space groups (32 in all), with their individual symbols, for the various classes of crystals.

456. *Invariants and covariants.*—In the treatment of this subject Cayley[5] let $\xi_1 = \delta_{x_1}$, $\eta_1 = \delta_{y_1}$, where δ_{x_1} and δ_{y_1} are symbols of differentiation relative to x_1 and y_1, and similarly for the more general x_p, y_p, and he writes $\xi_1\eta_2 - \xi_1\eta_1 = \overline{12}$, etc.; the result of operating r times with $\overline{12}$ he marks $\overline{12}^r$. For the Caleyan notation Aronhold, Clebsch,[6] and Gordan[7] advanced one of their own based on a symbolical representation of the coefficients in a quantic. A binary quantic $a_0x^n + na_1x^{n-1} + \ldots + a_n$ is represented by $(a_1x + a_2y)^n$. The quantic of the nth degree in three variables x_1, x_2, x_3 is symbolically written $(a_1x_1 + a_2x_2 + a_3x_3)^n$ or a_x^n, where a_1, a_2, a_3 are umbral symbols not regarded as having any meaning separately.

457. *Dual arithmetic.*—Perhaps the following quotations will convey an idea of the sign language introduced by Byrne in his *Dual*

[1] *Op. cit.*, p. 58, 61.   [2] *Op. cit.*, p. 72, 73.

[3] A. Bravais, "Mémoire sur les polyèdres de forme symétrique," *Journal de math. par Liouville*, Vol. XIV (Paris, 1849), p. 141–80. We may add here that Bravais' "symbole de la symétrie du polyèdre" (24 symbols in all) are tabulated on p. 179 of his article.

[4] *Op. cit.*, p. 301, 359, 362.

[5] A. Cayley, *Cambridge and Dublin Mathematical Journal*, Vol. I (1846), p. 104–22; *Collected Mathematical Papers*, Vol. I, p. 96, 97. See also § 462.

[6] R. F. A. Clebsch, *Theorie der Binären Formen.*

[7] P. Gordan, *Vorlesungen über Invariantentheorie.* See also G. Salmon, *Modern Higher Algebra* (3d ed.; Dublin, 1876), p. 266, 267; Faà de Bruno, *Théorie des formes binaires* (Turin, 1876), p. 293; E. B. Elliott, *Algebra of Quantics* (Oxford, 1913), p. 68.

Arithmetic. He says: ".... the continued product of $1 \cdot 2345678 \times$ $(1 \cdot 01)^4 (1 \cdot 001)^5 (1 \cdot 0001)^6$ which may be written $1 \cdot 2345678 \downarrow 0, 4, 5, 6, =$ $1 \cdot 291907$. The arrow $\downarrow$ divides the coefficient $1 \cdot 2345678$ and the powers of $1 \cdot 1, 1 \cdot 01, 1 \cdot 001, 1 \cdot 0001$, &c; 0, immediately follows the arrow because no power of $1 \cdot 1$ is employed; 6, is in the fourth place after $\downarrow$, and shows that this power operated upon periods of four figures each; 5 being in the third place after $\downarrow$ shows by its position, that its influence is over periods of three figures each; and 4 occupies the second place after $\downarrow$ ....."[1] A few years later Byrne[2] wrote: "Since any number may be represented in the form $N = 2^n 10^m \downarrow u_1, u_2, u_3, u_4, u_5$, etc., we may omit the bases 2 and 10 with as much advantage in perspicuity as we omitted the bases $1 \cdot 1, 1 \cdot 01, 1 \cdot 001$, &c. and write the above expression in the form $N = {}^m \downarrow {}^n u_1, u_2, u_3, u_4$, etc." "Any number N may be written as a continued product of the form $10^m \times (1 - \cdot 1)^{v_1} (1 - \cdot 01)^{v_2} (1 - .001)^{v_3} (1 - .0001)^{v_4}$, etc." or $10^m (\cdot 9)^{v_1} (\cdot 99)^{v_2} (\cdot 999)^{v_3} (\cdot 9999)^{v_4}$, etc." "In analogy with the notation used in the descending branch of dual arithmetic, this continued product may be written thus ${}'v_1 {}'v_2 {}'v_3 {}'v_4 {}'v_5 {}'v_6 \downarrow^m$ where any of the digits v_1, v_2, v_3, etc. as well as m may be positive or negative."[3]

"As in the ascending branch, the power of $10, m$, may be taken off the arrow and digits placed to the right when m is a $+$ whole number. Thus ${}'v_1 {}'v_2 {}'v_3 {}'v_4 \uparrow t'_1 \; t'_2 \; t'_3$ etc. represents the continued product $(\cdot 9)^{v_1}$ $(\cdot 99)^{v_2} (\cdot 999)^{v_3} (\cdot 9999)^{v_4} (9)^{t_1} (99)^{t_2} (999)^{t_3}$." Dual signs of addition $\updownarrow$ and subtraction $\rightarrow$ are introduced for ascending branches; the reversal of the arrows gives the symbols for descending branches.

458. *Chessboard problem.*—In the study of different non-linear arrangements of eight men on a chessboard, T. B. Sprague[4] lets each numeral in 61528374 indicate a row, and the position of each numeral (counted from left to right) indicates a column. He lets also i denote "inversion" so that $i(61528374) = 47382516$, r "reversion" (each number subtracted from 9) so that $r(61528374) = 38471625$, p "perversion" (interchanging columns and rows so that the 6 in first column becomes 1 in the sixth column, the 1 in the second column becomes 2 in the first column, the 5 in the third column becomes 3 in the fifth column, etc., and then arranging in the order of the new columns) so that $p(61528374) = 24683175$. It is found that $i^2 = r^2 = p^2 = 1$, $ir = ri$, $ip = pr$, $rp = pi$, $irp = rip = ipi$, $rpr = pir = pri$.

[1] Oliver Byrne, *Dual Arithmetic* (London, 1863), p. 9.

[2] Oliver Byrne, *op. cit.*, Part II (London, 1867), p. v.

[3] O. Byrne, *op. cit.*, Part II (1867), p. x.

[4] T. B. Sprague, *Proceedings Edinburgh Math. Soc.*, Vol. VIII (1890), p. 32.

DETERMINANT NOTATIONS

459. *Seventeenth century.*—The earliest determinant notations go back to the originator of determinants in Europe, Leibniz,[1] who, in a letter of April 28, 1693, to Marquis de l'Hospital, writes down the three equations

$$10+11x+12y=0 ,$$
$$20+21x+22y=0 ,$$
$$30+31x+32y=0 .$$

In this topographic notation the coefficients 10, 11, 12, etc., are to be interpreted as indices like the subscripts in our modern coefficient notation a_{10}, a_{11}, a_{12}, etc. Eliminating y and then x, Leibniz arrives at the relation

$$\begin{matrix} 1_0 & 2_1 & 3_2 \\ 1_1 & 2_2 & 3_0 \\ 1_2 & 2_0 & 3_1 \end{matrix} \quad aeqv. \quad \begin{matrix} 1_0 & 2_2 & 3_1 \\ 1_1 & 2_0 & 3_2 \\ 1_2 & 2_1 & 3_0 \end{matrix} ,$$

where $1_0\,2_1\,3_2$ represents a product of the three coefficients and, generally, each of the six rows represents such a product (§ 547).

460. *Eighteenth century.*—Independently of Leibniz, determinants were reinvented by Gabriel Cramer[2] who writes the coefficients of linear equations Z^1, Y^2, X^3, ... the letters indicating the column, the superscripts the rows, in which the coefficients occur. Cramer says: "Let there be several unknowns $z, y, x, v, \ldots$, and also the equations

$$A^1=Z^1z+Y^1y+X^1x+V^1v+ \ldots ,$$
$$A^2=Z^2z+Y^2y+X^2x+V^2v+ \ldots ,$$
$$A^3=Z^3z+Y^3y+X^3x+V^3v+ \ldots ,$$
$$A^4=Z^4z+Y^4y+X^4x+V^4v+ \ldots ,$$

. . . .

where the letters A^1, A^2, A^3, A^4, ... , do not as usual denote the powers of A, but the left side of the first, second, third, fourth, ... , equation, assumed to be known. Likewise $Z^1, Z^2, \ldots$, are the coefficients of z, $Y^1, Y^2, \ldots$, those of y, $X^1, X^2, \ldots$, those of x, V^1, V^2, ... , those of v, ... , in the first, second, ... , equation. This no-

---

[1] *Leibnizens Mathematische Schriften* (ed. C. I. Gerhardt, 3F.), Vol. II (1850), p. 229, 234, 239, 240, 241, 245, 261; Vol. V, p. 348. See also Vol. I, p. 120, 161; Vol. II, p. 7, 8; *Bibliotheca mathematica* (3d ser.), Vol. XIII (1912–13), p. 255; S. Günther, *Lehrbuch der Determinantentheorie* (Erlangen, 1877), p. 2, 3.

[2] Gabriel Cramer, *Introduction à l'analyse des lignes courbes algébriques* (Geneva, 1750), p. 657–59.

tation being adopted, one has, if only one equation with one unknown z is presented, $z = \dfrac{A^1}{Z^1}$. When there are two equations and two unknowns x and y, one finds

$$z = \frac{A^1 Y^2 - A^2 Y^1}{Z^1 Y^2 - Z^2 Y^1} \, ,$$

and

$$y = \frac{Z^1 A^2 - Z^2 A^1}{Z^1 Y^2 - Z^2 Y^1} \, . \text{''}$$

His method of solving simultaneous linear equations attracted the attention of Bézout,[1] Vandermonde,[2] and Laplace. Bézout lets (ab') stand for $ab' - a'b$; he similarly denotes a determinant of the third order $(ab'c'')$. An example of Vandermonde's symbols is seen in the two equations

$$\begin{aligned} {}^1_1\xi_1 + {}^1_2\xi_2 + {}^1_3 &= 0 \, , \\ {}^2_1\xi \; + {}^2_2\xi_2 + {}^2_3 &= 0 \, , \end{aligned}$$

in which the designation of coefficients by numerical superscripts and subscripts resembles the device of Leibniz. Furthermore, Vandermonde in 1772 adopts the novel notation

$$\frac{a \mid \beta}{a \mid b} \text{ for } a^\alpha \cdot b^\beta - b^\alpha \cdot a^\beta \, ,$$

and

$$\frac{a \mid \beta \mid \gamma}{a \mid b \mid c} \text{ for } a^\alpha \cdot \frac{\beta \mid \gamma}{b \mid c} + b^\alpha \cdot \frac{\beta \mid \gamma}{c \mid a} + c^\alpha \cdot \frac{\beta \mid \gamma}{a \mid b} \, ,$$

and finds

$$\frac{a \mid \beta}{a \mid b} = - \frac{a \mid \beta}{b \mid a} \, .$$

"Le symbole $++$ sert ici de charactéristique."

Laplace[3] expressed the resultant of three equations by $({}^1a \cdot {}^2b \cdot {}^3c)$, those indices being placed to the left of a letter and above.

461. *Early nineteenth century.*—The first use of the term "determinant" occurs in the following passage of K. F. Gauss: "Nume-

[1] E. Bézout, *Théorie générale des équations algébriques* (Paris, 1779), p. 211, 214.

[2] A. T. Vandermonde, *Histoire de l'acad. des scienc. Paris*, année 1772 (Paris, 1775), p. 516.

[3] P. S. Laplace, *Histoire de l'acad. des scienc. Paris*, année 1772 (2d partie), p. 267, 294; *Œuvres*, Vol. VIII, p. 365–406.

rum $bb-ac$, a cuius indole proprietates formae (a, b, c) imprimis pendere in sequentibus docebimus, *determinantem* huius formae uocabimus."[1]

The notation of Binet[2] is seen in the following quotation from his discussion of the multiplication theorem: "Désignons par $S(y', z'')$ une somme de résultantes, telle que $(y', z'')+(y'_{\prime\prime}, z'_{\prime\prime})+(y'_{\prime\prime\prime}, z'_{\prime\prime\prime})+$ etc. c'est-à-dire, $y'_1z''_1-z'_1y''_1+y'_{\prime\prime}z''_{\prime\prime}-z'_{\prime\prime}y''_{\prime\prime}+y'_{\prime\prime\prime}z''_{\prime\prime\prime}-z'_{\prime\prime\prime}y''_{\prime\prime\prime}+$etc.; et continuons d'employer la charactéristique Σ pour les intégrales relatives aux accens supérieurs des lettres. L'expression $\Sigma[S(y,z')\cdot S(v, \zeta')]$ devient. ..."

Cauchy[3] represents alternating functions by the symbolism $S(\pm a_1^1 a_2^2 a_3^3 \ldots a_n^n)$ or, writing $a_{r\cdot s}$ for a_r^s, the symbolism

$$S(\pm a_{1\cdot 1} a_{2\cdot 2} a_{3\cdot 3} \ldots a_{n\cdot n}) .$$

Cauchy represents by $(a_{1\cdot P}^{(p)})$ "un système symétrique de l'ordre P," embracing $P\times P$ determinants which may be arranged in a square as follows:[4]

$$\begin{cases} a_{1\cdot 1}^{(p)}\, a_{1\cdot 2}^{(p)} \ldots a_{1\cdot P}^{(p)} \\ a_{2\cdot 1}^{(p)}\, a_{2\cdot 2}^{(p)} \ldots a_{2\cdot P}^{(p)} \\ \cdot \quad \cdot \quad \cdot \quad \cdot \quad \cdot \\ a_{P\cdot 1}^{(p)}\, a_{P\cdot 2}^{(p)} \ldots a_{P\cdot P}^{(p)} , \end{cases}$$

where $P=n(n-1) \ldots (n-p+1)/1\cdot 2\cdot 3 \ldots p$. Of these determinants, $a_{1\cdot 1}^{(p)}$, for example, would be in our modern notation:

$$\begin{vmatrix} a_{1\cdot 1}\, a_{1\cdot 2} \ldots a_{1\cdot p} \\ a_{2\cdot 1}\, a_{2\cdot 2} \ldots a_{2\cdot p} \\ \cdot \quad \cdot \quad \cdot \quad \cdot \quad \cdot \\ a_{p\cdot 1}\, a_{p\cdot 2} \ldots a_{p\cdot p} \end{vmatrix} .$$

[1] C. F. Gauss, *Disquisitiones arithmeticae* (1801), XV, p. 2; Thomas Muir, *The Theory of Determinants in the Historical Order of Development* (London; 2d ed., 1906), p. 64.

[2] Jacques P. M. Binet, *Jour. de l'école polyt.*, Tome IX, Cahier 16 (1813), p. 288; T. Muir, *op. cit.*, Vol. I (1906), p. 84.

[3] A. L. Cauchy, *Jour. de l'école polyt.*, Tome X, Cahier 17 (1815), p. 52; *Œuvres*, IIᵉ Série, Vol. I, p. 113, 114, 155, 156.

[4] Th. Muir, *op. cit.*, Vol. I (1906), p. 111, 112.

Desnanot[1] in 1819, instead of writing indices in the position of exponents, places them above the letters affected, as in $\overset{k}{a}$, where k is the index of a. He writes for $\overset{k\ l}{a\ b} - \overset{k\ l}{b\ a}$ the abbreviation $(\overset{k\ l}{a\ b})$.

By the notation $P\left(\underset{n}{\overset{n}{a}};\ \underset{h}{s},\ \underset{h}{\overset{1}{a}}\right)$ Scherk[2] represents

$$
\begin{vmatrix}
 & \overset{2}{a} & \overset{3}{a} & \ldots\ldots & \overset{n}{a} \\
\underset{1}{s} & \underset{1}{a} & \underset{1}{a} & & \underset{1}{a} \\
 & \overset{2}{a} & \overset{3}{a} & \ldots\ldots & \overset{n}{a} \\
\underset{2}{s} & \underset{2}{a} & \underset{2}{a} & & \underset{2}{a} \\
 & . & . & . . . . . & . \\
 & \overset{2}{a} & \overset{3}{a} & \ldots\ldots & \overset{n}{a} \\
\underset{n}{s} & \underset{n}{a} & \underset{n}{a} & & \underset{n}{a}
\end{vmatrix}
$$

The notation of Schweins[3] is a modification of that of Laplace. Schweins uses $\|\ )$ where Laplace uses $(\ )$. Schweins writes $\left\|\ \overset{a_1}{A_1}\right) = \overset{a_1}{A_1}$, and in general

$$
\left\|\ \begin{matrix} a_1 & \ldots\ldots & a_n \\ A_1 & \ldots\ldots & A_n \end{matrix}\right) = \sum (-1)^x \begin{vmatrix} a_1 & \ldots\ldots & a_{n-x-1}\ a_{n-x+1} & \ldots\ldots & a_n \\ A_1 & \ldots\ldots\ldots\ldots\ldots\ldots\ldots & A_{n-1} \end{vmatrix}\Big) \\
\cdot A_{n,}^{a_{n-x}} \qquad\qquad x = 0, 1, \ldots\ldots, n-1 \ \cdot
$$

Reiss[4] adopts in 1829 the notation $(abc, \overline{123})$ for $a^1b^2c^3 - a^1b^3c^2 - a^2b^1c^3 + a^2b^3c^1 + a^3b^1c^2 - a^3b^2c^1$, the line above the 123 is to indicate permutation. He uses the same notation[5] in 1838.

[1] P. Desnanot, *Complément de la théorie des équations du premier degré* (Paris, 1819); Th. Muir, *op. cit.*, Vol. I (1906), p. 138.

[2] H. F. Scherk, *Mathematische Abhandlungen* (Berlin, 1825), 2. Abhandlung; Th. Muir, *op. cit.*, Vol. I (1906), p. 151, 152.

[3] Ferd. Schweins, *Theorie der Differenzen und Differentiale* (Heidelberg, 1825); *Theorie der Producte mit Versetzungen*, p. 317–431; Th. Muir, *op. cit.*, Vol. I (1906), p. 160.

[4] M. Reiss in *Correspondance math. et phys.*, Vol V (1829), p. 201–15; Th. Muir, *op. cit.*, Vol. I (1906), p. 179. .

[5] M. Reiss, *Correspondance math. et phys.*, Vol. X (1838), p. 229–90; Th. Muir, *op. cit.*, Vol. I (1906), p. 221.

Jacobi[1] chose the representation $\Sigma \pm a_{r_0,\,s_0} a_{r_1,\,s_1} \ldots\ldots a_{r_m,\,s_m}$, and in 1835 introduced also another:

$$a \left\{ \begin{array}{l} r_0,\ r_1,\ r_2,\ \ldots\ldots r_m \\ s_0,\ s_1,\ s_2,\ \ldots\ldots s_m \end{array} \right\} \ .$$

Sometimes he used[2] $(a_{0,\,0}\, a_{1,\,1} \ldots\ldots a_{m-1,\,m-1}\, a_{k,\,i})$, a notation resembling that of Bézout.

V. A. Lebesgue,[3] in his notation for minors of a determinant D, lets $[g,\, i]$ represent the determinant left over after the suppression of the row g and the column i, and he lets $\begin{bmatrix} g,\ i \\ h\ k \end{bmatrix}$ represent the determinant resulting from the suppression of the rows g and h, and the columns i and k.

In 1839, J. J. Sylvester[4] denotes by the name "zeta-ic multiplication," an operation yielding such products as the following:

$$\zeta(a_1 - b_1)(a_1 - c_1) = a_2 - a_1 b_1 - a_1 c_1 + b_1 c_1 \ .$$

For difference products he writes:

$$(b-a)(c-a)(c-b) = PD(a\ b\ c)$$
$$abc(b-a)(c-a)(c-b) = PD(0\ a\ b\ c) \ .$$

Combining the two notations, he represents the determinant of the system

$$\begin{array}{ccc} a_1 & a_2 & a_3 \\ b_1 & b_2 & b_3 \\ c_1 & c_2 & c_3 \end{array}$$

by $\zeta abc PD(abc)$ or $\zeta PD(0abc)$, and calls it "zeta-ic product of differences." Muir states (Vol. I, p. 230): "Now Sylvester's ζPD notation being unequal to the representation of the determinant $|a_1 b_2 c_4 d_5|$ in which the index-numbers do not proceed by common difference 1, he would seem to have been compelled to give a periodic character to the arguments of the bases in order to remove the difficulty. At any rate, the difficulty *is* removed; for the number of terms

[1] C. G. J. Jacobi in *Crelle's Journal*, Vol. XV (1836), p. 115; *Werke*, Vol. III, p. 295–320; Th. Muir, *op. cit.*, Vol. I (1906), p. 214.

[2] C. G. J. Jacobi, *Werke*, Vol. III, p. 588, 589.

[3] V. A. Lebesgue in *Liouville's Journal de Math.*, Vol. II (1837), p. 344; Th. Muir, *op. cit.*, Vol. I (1906), p. 220.

[4] J. J. Sylvester in *Philosophical Magazine*, Vol. XVI (1840), p. 37–43; *Collected Math. Papers*, Vol. I, p. 47–53; Th. Muir, *op. cit.*, Vol. I (1906), p. 228, 230.

in the period being 5 the index-numbers 4 and 5 become changeable into -1 and 0 and . . . . . Sylvester's form of the result (in zeta-i products) thus is $\zeta\{S_2(abcd) \cdot \zeta PD(0abcd)\} = \zeta_{-2}(0abcd)$."

In 1840 Cauchy[1] obtained as a result of elimination "la fonction alternée formée avec les quantités que présente le tableau

$$\begin{cases} a, & 0, & A, & 0, & 0, \\ b, & a, & B, & A, & 0, \\ c, & b, & C, & B, & A, \\ d, & c, & 0, & C, & B, \\ 0, & d, & 0, & 0, & C. \end{cases}$$

462. *Modern notations.*—A notation which has rightly enjoyed great popularity because of its objective presentation of the elements composing a determinant, in convenient arrangement for study, was given in 1841 by Cayley. He says: "Let the symbols

$$|\,\alpha\,|, \quad \begin{vmatrix} \alpha, & \beta \\ \alpha', & \beta' \end{vmatrix}, \quad \begin{vmatrix} \alpha\,, & \beta\,, & \gamma \\ \alpha', & \beta', & \gamma' \\ \alpha'', & \beta'', & \gamma'' \end{vmatrix}, \text{ etc.,}$$

denote the quantities

$$\alpha, \ \alpha\beta' - \alpha'\beta, \ \alpha\beta'\gamma'' - \alpha\beta''\gamma' + \alpha'\beta''\gamma - \alpha'\beta\gamma'' + \alpha''\beta\gamma' - \alpha''\beta'\gamma \text{ , etc.,}$$

the law of whose formation is tolerably well known. . . . ."[2] "Here then," says Muir, "we have for the first time in the notation of determinants, the pair of upright lines so familiar in all the later work. The introduction of them marks an epoch in the history, so important to the mathematician is this apparently trivial matter of notation. By means of them every determinant became representable, no matter how heterogeneous or complicated its elements might be; and the most disguised member of the family could be exhibited in its true lineaments. While the common characteristic of previous notations is their ability to represent the determinant of such a system as

$$\begin{array}{ccc} a_1 \ a_2 \ a_3 & \quad a_{1 \cdot 1} \ a_{1 \cdot 2} \ a_{1 \cdot 3} \\ b_1 \ b_2 \ b_3 & \text{or} \ a_{2 \cdot 1} \ a_{2 \cdot 2} \ a_{2 \cdot 3} \\ c_1 \ c_2 \ c_3 & \quad a_{3 \cdot 1} \ a_{3 \cdot 2} \ a_{3 \cdot 3} \end{array}$$

[1] A. L. Cauchy, *Exercices d'analyse et de phys. math.*, Vol. I, p. 385–422 *Œuvres complètes* (2d ser.), Vol. XI, p. 471; Th. Muir, *op. cit.*, Vol. I (1906), p. 242

[2] A. Cayley, *Cambridge Math. Journal*, Vol. II (1841), p. 267–71; *Collected Math. Papers*, Vol. I, p. 1–4; Th. Muir, *op. cit.*, Vol. II (1911), p. 5, 6.

and failure to represent in the case of systems like

$$\begin{array}{ccc} a\,b\,c & a\,b\,c & 4\,5\,6 \\ c\,a\,b & 1\,a\,b & 3\,2\,7 \\ b\,c\,a\,, & 0\,1\,a\,, & 8\,1\,0: \end{array}$$

Cayley's notation is equally suitable for all."

The first occurrence of Cayley's vertical-line notation for determinants and double vertical-line notation for matrices in *Crelle's Journal*[1] is in his "Mémoire sur les hyperdéterminants"; in *Liouville's Journal*, there appeared in 1845 articles by Cayley in which [ ] and { } are used in place of the vertical lines.[2] The notation { } was adopted by O. Terquem[3] in 1848, and by F. Joachimsthal[4] in 1849, who prefixes "dét," thus: "dét. { }." E. Catalan[5] wrote "dét. $(A, B, C \ldots)$," where $A, B, C, \ldots$, are the terms along the principal diagonal. The only objection to Cayley's notation is its lack of compactness. For that reason, compressed forms are used frequently when objective presentation of the elements is not essential. In 1843 Cayley said: "Representing the determinants

$$\begin{vmatrix} x_1 & y_1 & z_1 \\ x_2 & y_2 & z_2 \\ x_3 & y_3 & z_3 \end{vmatrix}, \text{ etc.,}$$

by the abbreviated notation $\overline{123}$, etc.; the following equation is identically true:

$$\overline{345}\cdot\overline{126} - \overline{346}\cdot\overline{125} + \overline{356}\cdot\overline{124} - \overline{456}\cdot\overline{123} = 0 \text{ ."}[6]$$

A little later Cayley wrote: "Consider the series of terms

$$\begin{array}{cccc} x_1 & x_2 & \ldots & x_n \\ A_1 & A_2 & \ldots & A_n \\ \cdot & \cdot & \cdot & \cdot \\ K_1 & K_2 & \ldots & K_n \end{array}$$

[1] A. Cayley, *Crelle's Journal*, Vol. XXX (1846), p. 2.

[2] A. Cayley, *Liouville's Journal*, Vol. X (1845), p. 105 gives [ ], p. 383 gives { }.

[3] *Nouvelles Annales de Math.*, Vol. VII (1848), p. 420; Th. Muir, *op. cit.*, Vol. II, p. 36.

[4] F. Joachimsthal, *Crelle's Journal*, Vol. XL (1849), p. 21–47.

[5] E. Catalan, *Bulletin de l'acad. roy. de Belgique*, Vol. XIII (1846), p. 534–55; Th. Muir, *op. cit.*, Vol. II, p. 37.

[6] A. Cayley, *Cambridge Math. Journal*, Vol. IV (1845), p. 18–20; *Collected Math. Papers*, Vol. I, p. 43–45; Th. Muir, *op. cit.*, Vol. II, p. 10.

the number of quantities $A, \ldots, K$ being equal to q ($q < n$). Suppose $q+1$ vertical rows selected, and the quantities contained in them formed into a determinant, this may be done in

$$\frac{n(n-1) \ldots (q+2)}{1 \cdot 2 \ldots (n-q-1)}$$

different ways. The system of determinants so obtained will be represented by the notation

$$\left\|\begin{array}{cccc} x_1 & x_2 \ldots & x_n \\ A_1 & A_2 \ldots & A_n \\ \cdot & \cdot \ \cdot \ \cdot & \cdot \\ K_1 & K_2 \ldots & K_n \end{array}\right\| \ ,"1$$

If this form is equated to zero, it signifies that each of these determinants shall be equated to zero, thus yielding a system of equations.

A function U, from which he considers three determinants to be derived,[2] is

$$\begin{aligned} & x(\alpha\xi + \beta\eta + \ldots) \\ & + y(\alpha'\xi + \beta'\eta + \ldots), \\ & + \ldots \ldots \ldots \end{aligned}$$

there being n lines and n terms in each line. Cayley would have represented this later, in 1855, by the notation

$$\begin{aligned} & (\alpha \ \beta \ \ldots \ \{\!\!\{ \xi, \eta \ \}\!\!\} x, y, \ldots) \\ & \left|\begin{array}{c} \alpha' \beta' \ \ldots \\ \ldots \ldots \end{array}\right| \end{aligned}$$

and called a "bipartite." A still later notation, as Muir[3] points out, is

$$\begin{array}{cc} \zeta & \eta \ \ldots \\ \hline \begin{array}{cc} \alpha & \beta \ \ldots \\ \alpha' & \beta' \ \ldots \\ \cdot \ \cdot & \cdot \ \cdot \ \cdot \end{array} & \begin{array}{c} x \\ y \\ \cdot \end{array} \end{array}$$

from which each term of the final expansion is very readily obtained by multiplying an element, β' say, of the square array, by the two elements (y, η) which lie in the same row and column with it but out-

[1] A. Cayley, *Cambridge Math. Journal*, Vol. IV, p. 119–20; *Collected Math. Papers*, Vol. I, p. 55–62; Th. Muir, *op. cit.*, Vol. II, p. 15.

[2] A. Cayley, *Trans. Cambridge Philos. Soc.*, Vol. VIII (1843), p. 1–16; *Collected Math. Papers*, Vol. I, p. 63–79.

[3] Thomas Muir, *op. cit.*, Vol. II, p. 18, 19.

side the array. The three determinants which are viewed as "derivational functions" of this function U are denoted by Cayley KU, FU, $\mathcal{T}U$, symbols which "possess properties which it is the object of this section to investigate," KU being what later came to be called the "discriminant" of U.

In the second part of this paper of 1843, Cayley deals (as Muir explains)[1] with a class of functions obtainable from the use of m sets of n indices in the way in which a determinant is obtainable from only two sets. The general symbol used for such a function is

$$\left\{ \begin{array}{cccc} A_{\rho_1} & \sigma_1 & \tau_1 & \ldots \\ \rho_2 & \sigma_2 & \tau_2 & \ldots \\ \ldots & \ldots & \ldots & \ldots \\ \rho_n & \sigma_n & \tau_n & \ldots \end{array} \right\}$$

which stands for the sum of all the different terms of the form $\pm r \cdot \pm_s \cdot \pm_t \ldots A_{\rho_{r_1}\sigma_{s_1}\tau_{t_1}} \ldots \times \ldots \times A_{\rho_{r_n}\sigma_{s_n}\tau_{t_n}} \ldots$ where $r_1, r_2, \ldots, r_n$; $s_1, s_2, \ldots, s_n$; $t_1, t_2, \ldots, t_n$; $\ldots$ denote any permutation, the same or different, of the series $1, 2, \ldots, n$, and where $\pm_r$ denotes $+$ or $-$, according as the number of inversions in $r_1, r_2, \ldots, r_n$ is even or odd. Using a $\dagger$ to denote that the ρ's are unpermutable, he obtains

$$\left\{ \begin{array}{ccc} A\rho_1\,\sigma_1 & \ldots \\ \ldots & \ldots \\ \rho_n\,\sigma_n & \ldots \end{array} \right\} = 1.2 \ldots m \left\{ \begin{array}{ccc} \dagger \\ A\rho_1\,\sigma_1 & \ldots \\ \ldots & \ldots \\ \rho_n\,\sigma_n & \ldots \end{array} \right\}$$

when the number of columns, m, is even.

In 1845 Cayley[2] gave the notation which may be illustrated by

$$\left\| \begin{array}{cccc} a_1 & a_2 & a_3 & a_4 \\ b_1 & b_2 & b_3 & b_4 \end{array} \right\| = \left\| \begin{array}{cccc} x_1 & x_2 & x_3 & x_4 \\ y_1 & y_2 & y_3 & y_4 \end{array} \right\|,$$

which represents six equations,

$$\left| \begin{array}{cc} a_1 & a_2 \\ b_1 & b_2 \end{array} \right| = \left| \begin{array}{cc} x_1 & x_2 \\ y_1 & y_2 \end{array} \right|, \qquad \left| \begin{array}{cc} a_1 & a_3 \\ b_1 & b_3 \end{array} \right| = \left| \begin{array}{cc} x_1 & x_3 \\ y_1 & y_3 \end{array} \right|, \text{ etc.}$$

This is a more general mode of using the double vertical lines which had been first introduced by Cayley in 1843.

[1] Th. Muir, *op. cit.*, Vol. II, p. 63, 64; *Collected Math. Papers*, Vol. I, p. 76.

[2] A. Cayley, *Cambridge Math. Journal*, Vol. IV (1845), p. 193–209; *Collected Math. Papers*, Vol. I, p. 80–94; Th. Muir, *op. cit.*, Vol. II, p. 33.

In 1851, J. J. Sylvester[1] introduced what he calls "a most powerful, because natural, method of notation." He says: "My method consists in expressing the same quantities biliterally as below:

$$
\begin{array}{llll}
a_1a_1 & a_1a_2 & \ldots & a_1a_n \\
a_2a_1 & a_2a_2 & \ldots & a_ra_n \\
\cdot & \cdot & \cdot \cdot \cdot \cdot \\
a_na_1 & a_na_2 & \ldots & a_na_n
\end{array}
$$

where, of course, whenever desirable, instead of $a_1, a_2, \ldots a_n$, and $a_1, a_2, \ldots, a_n$, we may write simply $a, b, \ldots, l$ and $a, \beta, \ldots, \lambda$, respectively. Each quantity is now represented by two letters; the letters themselves, taken separately, being symbols neither of quantity nor of operation, but mere umbrae or ideal elements of quantitative symbols. We have now a means of representing the determinant above given in compact form; for this purpose we need but to write one set of umbrae over the other as follows:

$$
\begin{pmatrix}
a_1 & a_2 \ldots a_n \\
a_1 & a_2 \ldots a_n
\end{pmatrix}.
$$

If we now wish to obtain the algebraic value of this determinant, it is only necessary to take $a_1, a_2, \ldots, a_n$ in all its $1.2.3 \ldots n$ different positions, and we shall have

$$
\left\{
\begin{array}{ll}
a_1 & a_2 \ldots a_n \\
a_1 & a_2 \ldots a_n
\end{array}
\right\}
= \sum \pm \{a_1a_{\theta_1} \times a_2a_{\theta_2} \times \ldots \times a_na_{\theta_n}\} ,
$$

in which expression $\theta_1, \theta_2, \ldots, \theta_n$ represents some order of the numbers $1, 2, \ldots, n$, and the positive or negative sign is to be taken according to the well-known dichotomous law." An extension to "compound" determinants is also indicated. Since $\left\{\begin{array}{ll} a & b \\ a & \beta \end{array}\right\}$ denotes

$aa \cdot b\beta - a\beta \cdot ba$, Sylvester lets $\left\{\begin{array}{ll} \overline{ab} & \overline{cd} \\ a\beta & \gamma\delta \end{array}\right\}$ denote

$$
\frac{ab}{a\beta} \times \frac{cd}{\gamma\delta} - \frac{ab}{\gamma\delta} \times \frac{cd}{a\beta} ,
$$

"that is,

$$
\left\{
\begin{array}{l}
(aa \times b\beta) \\
-(a\beta \times ba)
\end{array}
\right\}
\times
\left\{
\begin{array}{l}
(c\gamma \times d\delta) \\
-(c\delta \times d\gamma)
\end{array}
\right\}
-
\left\{
\begin{array}{l}
(a\gamma \times b\delta) \\
-(a\delta \times b\gamma)
\end{array}
\right\}
\times
\left\{
\begin{array}{l}
(ca \times d\beta) \\
-(c\beta \times da)
\end{array}
\right\} ,
$$

[1] J. J. Sylvester, *Phil. Magazine* (4), Vol. I (1851), p. 295–305; *Collected Math. Papers*, Vol. I, p. 241; Th. Muir, *op. cit.*, Vol. II, p. 58, 59.

and in general the compound determinant

$$\left\{ \begin{array}{ccc} \overline{a_1 \quad b_1 \ldots l_1} & \overline{a_2 \quad b_2 \ldots l_2} \ldots & \overline{a_r \quad b_r \quad l_r} \\ a_1 \quad \beta_1 \ldots \lambda_1 & a_2 \quad \beta_2 \ldots \lambda_2 \ldots & a_r \quad \beta_r \quad \lambda_r \end{array} \right\}$$

will denote

$$\sum \pm \left\{ \begin{array}{cc} a_1 & b_1 \ldots l_1 \\ a_{\theta_1} & \beta_{\theta_1} \ldots \lambda_{\theta_1} \end{array} \right\} \times \left\{ \begin{array}{cc} a_2 & b_2 \ldots l_2 \\ a_{\theta_2} & \beta_{\theta_2} \ldots \lambda_{\theta_2} \end{array} \right\} \times$$
$$\ldots \times \left\{ \begin{array}{cc} a_r & b_r \ldots l_r \\ a_{\theta_r} & \beta_{\theta_r} \ldots \lambda_{\theta_r} \end{array} \right\},$$

where, as before, we have the disjunctive equation $\theta_1, \theta_2, \ldots, \theta_r = 1, 2, \ldots, r$."

Sylvester gives a slightly different version of his umbral notation in 1852: "If we combine each of the n letters $a, b, \ldots, l$ with each of the other $n, a, \beta, \ldots, \lambda$, we obtain n^2 combinations which may be used to denote the terms of a determinant of n lines and columns, as thus:

$$\begin{array}{l} a\alpha, a\beta \ldots a\lambda \\ b\alpha, b\beta \ldots b\lambda \\ \cdot \quad \cdot \quad \cdot \quad \cdot \\ l\alpha \quad l\beta \ldots l\lambda \end{array} \cdot$$

It must be well understood that the single letters of either set are mere umbrae, or shadows of quantities, and only acquire a real signification when one letter of one set is combined with one of the other set. Instead of the inconvenient form above written, we may denote the determinant more simply by the matrix

$$\begin{array}{l} a, \quad b, \quad c \ldots l \\ a, \quad \beta, \quad \gamma \ldots \lambda \end{array} ."[1]$$

In place of Sylvester's umbral notation $\left\{ \begin{array}{c} a_1 a_2 \ldots a_n \\ b_1 b_2 \ldots b_n \end{array} \right\}$ Bruno[2] writes $\sum (\pm a_1^{\phi_1} a_2^{\phi_2} \ldots a_p^{\phi_p} \ldots a_n^{\phi_n} \}$.

Following Sylvester in matters of notation, Reiss[3] indicates the expansion of a five-line determinant in terms of the minors formed

---

[1] J. J. Sylvester, *Cambridge and Dublin Mathematical Journal*, Vol. VII (Cambridge, 1852), p. 75, 76; *Collected Mathematical Papers*, Vol. I (1904), p. 305.

[2] F. Faà de Bruno, *Liouville Journal de Math.* (1), Vol. XVII (1852), p. 190–92; Th. Muir, *op. cit.*, Vol. II, p. 72.

[3] M. Reiss, *Beiträge zur Theorie der Determinanten* (Leipzig, 1867); Th. Muir, *op. cit.*, Vol. III (1920), p. 21, 22.

from the second and fourth rows and the minors formed from the first, third, and fifth rows, by

$$si\begin{pmatrix} 2 & 4 & 1 & 3 & 5 \\ 1 & 2 & 3 & 4 & 5 \end{pmatrix} \overset{\beta_1\beta_2}{\underset{1\,2\,3\,4\,5}{\sum}} \epsilon \begin{pmatrix} 2 & 4 \\ \beta_1 & \beta_2 \end{pmatrix} \begin{pmatrix} 1 & 3 & 5 \\ \beta_3 & \beta_4 & \beta_5 \end{pmatrix},$$

where, firstly, $\begin{pmatrix} 2 & 4 \\ \beta_1 & \beta_2 \end{pmatrix} \begin{pmatrix} 1 & 3 & 5 \\ \beta_3 & \beta_4 & \beta_5 \end{pmatrix}$ is the typical product of two complementary minors; secondly, the accessories to the Σ imply that for β_1, β_2 are to be taken any two integers from 1, 2, 3, 4, 5; and thirdly, the remaining items are connected with the determination of sign.

A somewhat uncommon symbol is $\overline{r}\lvert s$, used by Dodgson[1] to designate the element in the (r, s)th place.

A somewhat clumsy notation due to E. Schering[2] is illustrated by the following: "A member of the determinant can accordingly always be represented by

$$\overset{\nu=n}{\underset{\nu=1}{\Pi}} E_{\eta_\nu} k_\nu \times \mathcal{J} \underset{m,\mu}{\Pi} (\eta_m - \eta_\mu)(k_m - k_\mu) \times \mathcal{J} \underset{a,a}{\Pi} (h_a - h_a)(a - a)$$
$$\times \mathcal{J} \underset{b,\beta}{\Pi} (k_b - k_\beta)(b - \beta) ,"$$

where E is any element denuded of its compound suffix, and $\mathcal{J}$ is a function-symbol corresponding to Scherk's ϕ, and being used such that $\mathcal{J}(x)$ is equal to $+1$ or 0 or -1 according as x is greater than, equal to, or less than 0.

Cayley's notation of a matrix[3] is exemplified by

$$\begin{vmatrix} a, & \beta, & \gamma & \cdots \\ a', & \beta', & \gamma' & \cdots \\ a'', & \beta'', & \gamma'' & \cdots \\ \cdot & \cdot & \cdot & \cdots \end{vmatrix},$$

a matrix being defined as a system of quantities arranged in the form of a square, but otherwise quite independent. Cayley says that the equations

$$\left.\begin{aligned} \xi &= ax & +\beta y & +\gamma z & \cdots \\ \eta &= a'x & +\beta'y & +\gamma'z & \cdots \\ \zeta &= a''x & +\beta''y & +\gamma''z & \cdots \end{aligned}\right\}$$

[1] C. L. Dodgson, *Elementary Theory of Determinants* (London, 1867); Th. Muir, *op. cit.*, Vol. III, p. 24.

[2] E. Schering, *Abhandlungen d. K. Gesellsch. d. Wissensch.* (Göttingen), Vol. XXII (1907); Th. Muir, *op. cit.*, Vol. III, p. 71.

[3] A. Cayley, *Crelle's Journal*, Vol. 50 (1855), p. 282–85; *Collected Math. Papers*, Vol. II, p. 185–88; Muir, *op. cit.*, Vol. II, p. 85.

may be written[1]

$$(\xi, \eta, \zeta \ldots) = \begin{vmatrix} a, & \beta, & \gamma, & \ldots \\ a', & \beta', & \gamma', & \ldots \\ a'', & \beta'', & \gamma'', & \ldots \\ . & . & . & . \end{vmatrix}^{\frown} (x, y, z \ldots)$$

and the solution of the equations in the matrix form

$$(x, y, z \ldots) = \left(\begin{vmatrix} a, & \beta, & \gamma & \ldots \\ a', & \beta', & \gamma' & \ldots \\ a'', & \beta'', & \gamma'' & \ldots \\ . & . & . & . \end{vmatrix}^{\overset{\frown}{-1}} \right) (\xi, \eta, \zeta \ldots)$$

where

$$\begin{vmatrix} a, & \beta, & \gamma & \ldots \\ a', & \beta', & \gamma' & \ldots \\ a'', & \beta'', & \gamma'' & \ldots \\ . & . & . & . \end{vmatrix}^{-1}$$

stands for the inverse matrix.

463. *Compressed notations.*—H. J. S. Smith,[2] when the elements are double suffixed, places between two upright lines a single element with variable suffixes and then appends an indication of the extent of the variability. For example, he writes

$$\sum \left(\pm \frac{\partial x_1}{\partial y_1} \frac{\partial x_2}{\partial y_2} \ldots \frac{\partial x_n}{\partial y_n} \right)$$

in the form

$$\left| \frac{dx_a}{dy_\beta} \right|, \qquad \left. \begin{array}{l} a = 1, 2, \ldots, n \\ \beta = 1, 2, \ldots, n \end{array} \right\} .$$

In 1866 L. Kronecker[3] made similar use of $|b_{ik}|$; he used also

$$| a_{h1}, a_{h2}, \ldots, a_{hn} |, \quad (h = 1, 2, \ldots, n) .$$

[1] In the *Collected Math. Papers*, Vol. II, p. 185, parentheses are placed outside of the vertical lines, on the first row, which were not given in 1855.

[2] H. J. S. Smith, *Report British Assoc.*, Vol. XXXII (1862), p. 503–26; *Collected Math. Papers*, Vol. I (1894), p. 230.

[3] L. Kronecker in *Crelle's Journal*, Vol. LXVIII (1868), p. 276; *Monatsberichte d. Akademie* (Berlin, 1866), p. 600; *Werke*, Vol. I (1895), p. 149, 150.

Salmon[1] ordinarily uses Cayley's two vertical bars, but often "for brevity" writes $(a_1, b_2, c_3 \ldots)$ where $a_1, b_2, c_3, \ldots$, are elements along the principal diagonal, which resembles designations used by Bézout and Jacobi.

E. H. Moore[2] makes the remark of fundamental import that a determinant of order t is uniquely defined by the unique definition of its t^2 elements in the form a_{uv}, where the suffixes uv run independently over any (the same) set of t distinct marks *of any description whatever*. Accordingly, in determinants of special forms, it is convenient to introduce in place of the ordinary $1, 2, \ldots, t$ some other set of t marks. Thus, if one uses the set of t bipartite marks

$$gj \qquad \begin{pmatrix} g = 1, \ldots, m \\ j = 1, \ldots, n \end{pmatrix}$$

and denotes a_{uv} by a_{fihk}, the determinant $A = |a_{fihk}|$ of order mn, where throughout

$$a_{fihk} = b_{fh}^{(i)} \cdot c_{ik}^{(h)} \qquad \begin{pmatrix} f, h = 1, \ldots, m \\ i, k = 1, \ldots, n \end{pmatrix},$$

is the product of the n determinants $B^{(i)}$ of order m and the m determinants $C^{(h)}$ of the order n:

$$A = B^{(1)} \ldots B^{(n)} \cdot C^{(1)} \ldots C^{(m)}, \text{ where } B^{(i)} = \left| b_{fh}^{(i)} \right|, C^{(h)} = \left| c_{ik}^{(h)} \right|.$$

Kronecker[3] introduced a symbol in his development of determinants which has become known as "Kronecker's symbol," viz., δ_{ik}, where $\begin{matrix} i = 1, 2, \ldots, m \\ k = 1, 2, \ldots, n \end{matrix}$, and $\delta_{ik} = 0$ when $i \gtrless k$, $\delta_{kk} = 1$. The usual notation is now δ_s^r; Murnaghan writes $[_s^r]$. A generalization of the ordinary "Kronecker symbol" was written by Murnaghan in 1924[4] in the form $\delta \begin{matrix} r_1, \ldots, r_m \\ s_1, \ldots, s_m \end{matrix}$ ($m \leq n$, in space of n dimensions), and in 1925[5] in the form $\begin{bmatrix} r_1 \, r_2 \ldots r_m \\ s_1 \, s_2 \ldots s_m \end{bmatrix}$, and appears in the outer multiplication of tensors.

[1] George Salmon, *Modern Higher Algebra* (Dublin, 1859, 3d. ed., Dublin, 1876), p. 1.

[2] E. H. Moore, *Annals of Mathematics* (2d ser.), Vol. I (1900), p. 179, 180.

[3] Leopold Kronecker, *Vorlesungen über die Theorie der Determinanten*, bearbeitet von Kurt Hensel, Vol. I (Leipzig, 1903), p. 316, 328, 349.

[4] F. D. Murnaghan, *International Mathematical Congress* (Toronto, 1924); Abstracts, p. 7.

[5] F. D. Murnaghan, *American Math. Monthly*, Vol. XXXII (1925), p. 234.

Sheppard[1] follows Kronecker in denoting by $|d_{qr}|$ and $|d_{rq}|$ the determinants whose elements in the qth column and rth row are respectively d_{qr} and d_{rq}. The notation of summation

$$\begin{pmatrix} p=1, 2, \ldots, m; \\ q=1, 2, \ldots, m \end{pmatrix} \sum_{s=1}^{m} D_{ps}d_{qs} = \begin{cases} D \text{ if } q=p \\ 0 \text{ if } q \neq p \end{cases},$$

where D_{ps} (cofactor of d_{ps} in the determinant D) is simplified by dropping the sign of summation and using Greek letters;[2] thus

$$\begin{pmatrix} p=1, 2, \ldots, m; \\ q=1, 2, \ldots, m \end{pmatrix} d^{p\sigma}d_{q\sigma} = \begin{cases} 1 \text{ if } q=p \\ 0 \text{ if } q \neq p \end{cases},$$

where $d^{p\sigma} \equiv$ (cofactor of $d_{p\sigma}$ in D) $\div D$. This, in turn,[3] is condensed into $|_q^p$; also $d^{\lambda\sigma}d_{\mu\sigma} = |_\mu^\lambda$ ('unit $\lambda\mu$'). The symbol adopted by A. Einstein is δ_μ^λ, while J. E. Wright, in his *Invariants of Quadratic Differential Forms*, uses $\eta_{\lambda\mu}$. Says Sheppard: "We can treat the statement $d^{\lambda\sigma}d_{\mu\sigma} = |_\mu^\lambda$ as a set of equations giving the values of $d^{\lambda\sigma}$ in terms of those of $d_{\lambda\sigma}$. If, for instance, $m=20$, the set $d_{\lambda\sigma}$ contains 400 elements,"[4] and this expression is a condensed statement of the 400 equations (each with 20 terms on one side) which give the 400 values of $d_{\lambda\sigma}$.

464. *The Jacobian.*—The functional determinant of Jacobi,[5] called the "Jacobian," is presented by Jacobi, himself, in the following manner: "Propositis variabilium $x, x_1, \ldots, x_n$ functionibus totidem $f, f_1, \ldots, f_n$, formentur omnium differentialia partialia omnium variabilium respectu sumta, unde prodeunt $(n+1)^2$ quantitates, $\frac{\partial f_i}{\partial x_k}$. Determinans $\sum \pm \frac{\partial f}{\partial x} \cdot \frac{\partial f_1}{\partial x_1} \ldots \frac{\partial f_n}{\partial x_n}$, voco Determinans functionale." It was represented by Donkin[6] in the form $\frac{\partial(f_1, f_2, \ldots, f_m)}{\partial(z_1, z_2, \ldots, z_m)}$, by Gordan[7] $\begin{pmatrix} f_1, \ldots, f_m \\ z_1, \ldots, z_m \end{pmatrix}$, by L. O. Hesse $(f_1, f_2, \ldots, f_m)$, and by Weld[8] $J(f_1, f_2, \ldots, f_m)$.

[1] W. F. Sheppard, *From Determinant to Tensor* (Oxford, 1923), p. 39.

[2] *Op. cit.*, p. 41, 47, 49.   [3] *Op. cit.*, p. 50.   [4] *Op. cit.*, p. 57.

[5] C. G. J. Jacobi, *Crelle's Journal*, Vol. XXII (1841), p. 327, 328; *Werke*, Vol. III, p. 395. See also *Encyclopédie des scien. math.*, Tome I, Vol. II (1907), p. 169.

[6] W. F. Donkin, *Philosophical Transactions* (London), Vol. CXLIV (1854), p. 72.

[7] P. Gordan, *Vorlesungen über Invariantentheorie* (Leipzig, 1885), Vol. I, p. 121.

[8] L. G. Weld, *Theory of Determinants* (New York, 1896), p. 195.

465. *Hessian.*—If u be a function of $x_1, x_2, \ldots, x_n$ and $y_1, \ldots, y_n$ its differential coefficients with respect to these variables, the Jacobian of $y_1, \ldots, y_n$ is called the Hessian of u and is denoted[1] by $H(u)$ so that $H(u) = |u_{ik}|$, where u_{ik} is $\dfrac{dy_i}{dx_k}$.

466. *Cubic determinants.*—The set of elements in the leading diagonal may be written $a_{111}, a_{222}, \ldots, a_{nnn}$. The cubic determinant may be denoted by[2] $\Sigma \pm a_{111}a_{222} \ldots a_{nnn}$, or by $|a_{ijk}|$ (i, j, k = 1, 2, $\ldots, n$). "The multiple-suffix notation is not very attractive," says Muir.[3]

467. *Infinite determinants.*—Kötterlitzsch[4] solved in 1870 a system of an infinity of linear equations which he writes

$$\begin{matrix} 0 \\ m \\ \infty \end{matrix} \left| \begin{matrix} \infty \\ \underset{0}{\overset{}{\Sigma^p}}f_0(m, p)x_p = \phi m \end{matrix} \right. .$$

The determinant of this system he represents by

$$R = \sum \pm f_0(0, 0)f_0(1, 1)f_0(2, 2) \ldots f_0(p, p) .$$

468. *Matrix notations.*—Cayley introduced two double vertical lines to denote a matrix in 1843 and 1845, as already shown. It was used by H. J. S. Smith.[5] Whitworth[6] uses in one place triple vertical lines to indicate that three determinant equations may be independently formed from the matrix.

The notation[7] $[a](x)$ for the product of the matrix $[a_{ik}]$ by the complex quantity (x) has been used by Peano and Bôcher; V. Volterra employs the equation between matrices, $\left[\dfrac{dx_{ik}}{dx}\right] = [a_{ik}][x_{ik}]$, and notations even more condensed, such as the fundamental differential operation Dx defined by the formula

$$D_x[x_{ik}] = [x_{ik}]^{-1}\left[\dfrac{dx_{ik}}{dx}\right] .$$

[1] R. F. Scott, *Treatise on the Theory of Determinants* (Cambridge, 1880), p. 142, 143.

[2] R. B. Scott, *op. cit.*, p. 89, 90; M. Noether, *Mathematische Annalen*, Vol. XVI (1880), p. 551–55.

[3] Th. Muir, *op. cit.*, Vol. III, p. 392.

[4] Th. Kötterlitzsch in *Zeitschrift f. Math. u. Physik* (Leipzig), Vol. XV (1870), p. 231.

[5] H. J. S. Smith, *Mathematical Papers*, Vol. I, p. 374.

[6] W. A. Whitworth, *Trilinear Coordinates* (Cambridge, 1866), p. xxv.

[7] See E. Vessiot in *Encyclopédie des scien. math.*, Tom. II, Vol. III (1910), p. 129.

The notation for fractional matrices,

$$\left\| \begin{array}{ccc} \dfrac{a_{11} \quad a_{12} \quad a_{13}}{a_{21} \quad a_{22} \quad a_{23}} \\ \hline a_{31} \quad a_{32} \quad a_{33} \end{array} \right\|,$$

has been used by E. H. Moore and M. Bôcher[1] to indicate that one may consider two matrices as equal whenever the elements of one matrix can be obtained from the elements of the other matrix by multiplying each element of the first, or each element of the second, by the same quantity not zero.

Besides the double vertical lines of Cayley, there are several other notations for matrices. Round parentheses have been used by many, for instance, by M. Bôcher.[2] G. Kowalewski[3] finds it convenient sometimes to use the double vertical lines, sometimes the round parentheses, and also often a single brace, placed to the left of the elements, thus:

$$\left\{ \begin{array}{l} x_{11}, \; x_{12}, \ldots, \; x_{1n}, \\ \cdot \quad \cdot \quad \cdot \quad \cdot \quad \cdot \quad \cdot \\ x_{m1}, \quad x_{m2}, \cdot \qquad , \; x_{mn} \, . \end{array} \right.$$

A variety of new notations for matrices are found in the work of Cullis.[4] A rectangular matrix is indicated as follows:

$$A = [a]_m^n = \begin{bmatrix} a_{11} & a_{12} \ldots a_{1n} \\ a_{21} & a_{22} \ldots a_{2n} \\ a_{m1} & a_{m2} \ldots a_{mn} \end{bmatrix}, \qquad [abc]_{1234} = \begin{bmatrix} a_1 & b_1 & c_1 \\ a_2 & b_2 & c_2 \\ a_3 & b_3 & c_3 \\ a_4 & b_4 & c_4 \end{bmatrix}$$

A', conjugate to A, is denoted thus:

$$A' = \underline{\bar{a}}\,_n^m = \begin{bmatrix} a_{11} & a_{21} \ldots a_{m1} \\ a_{12} & a_{22} \ldots a_{m2} \\ \cdot & \cdot \quad \cdot \quad \cdot \quad \cdot \\ a_{1n} & a_{2n} \ldots a_{mn} \end{bmatrix}.$$

[1] M. Bôcher, *Introduction to Higher Algebra* (New York, 1907), p. 86. See "Tribune publique," 19, No. 430, *Encyclopédie des scien. math.*, Tom. II, Vol. **V** (1912).

[2] Maxine Bôcher, *Introduction to Higher Algebra* (New York, 1919), p. 20.

[3] G. Kowalewski, *Determinantentheorie* (1909), p. 51, 55, 66, 214.

[4] C. E. Cullis, *Matrices and Determinoïds* (Cambridge), Vol. I (1913), p. 1–12.

Accordingly, $[a]_m^n$, $\overline{a}_n^m$ are conjugate (abbreviations). More general double-suffix notations are:

$$[a_{pq}]_m^n \; , \; \overline{a_{pq}}_n^m \; , \; [a_{p1}]_m^n = \begin{bmatrix} a_{p_11} & a_{p_12} & \ldots\ldots & a_{p_1n} \\ a_{p_21} & a_{p_2} & \ldots\ldots & a_{p_2n} \\ \cdot & \cdot & \cdot\;\;\cdot\;\;\cdot\;\;\cdot & \cdot \\ a_{p_m1} & a_{p_m2} & \ldots\ldots & a_{p_mn} \end{bmatrix} ,$$

and its conjugate

$$A' = \overline{a_{p1}}_n^m = \begin{bmatrix} a_{p_11} & a_{p_21} & \ldots\ldots & a_{p_m1} \\ a_{p_12} & a_{p_22} & \ldots\ldots & a_{p_m2} \\ \cdot & \cdot & \cdot\;\;\cdot\;\;\cdot\;\;\cdot & \cdot \\ a_{p_1n} & a_{p_2n} & \ldots\ldots & a_{p_mn} \end{bmatrix} .$$

Most general double-suffix notation:

$$A = \begin{bmatrix} uv & \ldots\ldots & w \\ & a & \\ pq & \ldots\ldots & r \end{bmatrix} = \begin{bmatrix} a_{pu} & a_{pv} & \ldots\ldots & a_{pw} \\ a_{qu} & a_{qv} & \ldots\ldots & a_{qw} \\ \cdot & \cdot & \cdot\;\;\cdot\;\;\cdot\;\;\cdot & \cdot \\ a_{ru} & a_{rv} & \ldots\ldots & a_{rw} \end{bmatrix} ,$$

$$A' = \begin{bmatrix} \overline{pq \ldots\ldots r} \\ a \\ \underline{uv \ldots\ldots w} \end{bmatrix} = \begin{bmatrix} a_{pu} & a_{qu} & \ldots\ldots & a_{ru} \\ a_{pv} & a_{qv} & \ldots\ldots & a_{rv} \\ \cdot & \cdot & \cdot\;\;\cdot\;\;\cdot\;\;\cdot & \cdot \\ a_{pw} & a_{qw} & \ldots\ldots & a_{rw} \end{bmatrix} .$$

Double-suffix notation for augmented matrices:

$$[a,b]_m^{n,\,r} = \begin{bmatrix} a_{11} & a_{12} & \ldots\ldots & a_{1n} & b_{11} & b_{12} & \ldots\ldots & b_{1r} \\ a_{21} & a_{22} & \ldots\ldots & a_{2n} & b_{21} & b_{22} & \ldots\ldots & b_{2r} \\ \cdot & \cdot & \cdot\;\;\cdot\;\;\cdot\;\;\cdot & \cdot & \cdot & \cdot & \cdot\;\;\cdot\;\;\cdot\;\;\cdot & \cdot \\ a_{m1} & a_{m2} & \ldots\ldots & a_{mn} & b_{m1} & b_{m2} & \ldots\ldots & b_{mr} \end{bmatrix} ,$$

$$\begin{bmatrix} a \\ b \end{bmatrix}_{m,\,r}^n = \begin{bmatrix} a_{11} & a_{12} & \ldots\ldots & a_{1n} \\ a_{21} & a_{22} & \ldots\ldots & a_{2n} \\ \cdot & \cdot & \cdot\;\;\cdot\;\;\cdot\;\;\cdot & \cdot \\ a_{m1} & a_{m2} & \ldots\ldots & a_{mn} \\ b_{11} & b_{12} & \ldots\ldots & b_{1n} \\ b_{21} & b_{22} & \ldots\ldots & b_{2n} \\ \cdot & \cdot & \cdot\;\;\cdot\;\;\cdot\;\;\cdot & \cdot \\ b_{r1} & b_{r2} & \ldots\ldots & b_{rn} \end{bmatrix} .$$

The conjugate matrices of $[a, b]_m^{n,\,r}$ and $\begin{bmatrix} a \\ b \end{bmatrix}_{m,\,r}^n$ are denoted

$$\overline{\underset{b}{a}}_{n,\,r}^m \qquad \text{and} \qquad \overline{a, b}_n^{m,\,r} .$$

Most general single-suffix notation for matrix and its conjugate:

$$[ab \ldots k]_{\alpha\beta \ldots k} \, , \quad \begin{bmatrix} a \\ b \\ \vdots \\ k \end{bmatrix}_{\alpha\beta \ldots k}$$

Corresponding to the notations for a matrix of orders m and n, Cullis uses symbols to represent determinoids of orders m and n,

$$(a)^n_m, \quad (a_{pq})^n_m, \quad (a_{p1})^n_m, \quad (a_{1q})^n_m, \quad \begin{pmatrix} uv \ldots w \\ a \\ pq \ldots r \end{pmatrix},$$

$$(abc \ldots k)_{123 \ldots m}, \quad (abc \ldots k)_{\alpha\beta\gamma \ldots k} \, .$$

The determinoid of any matrix A is defined as the algebraic sum of all its complete derived products, when each product is provided with a positive or negative sign in accordance with a fixed *rule of signs*.

For "matrix product" Noether[1] employed the symbol $(|)$, as in $(S^{\rho}|p_{\rho})$; Weitzenböck[2] changed the notation for the series S^{ρ} by writing $(a'_{\rho}|p_{\rho})$.

SIGNS FOR LOGARITHMS

469. *Abbreviations for "logarithm."*—The use by some authors of l to represent *latus* and signifying either "root" or "unknown quantity," did not interfere with its introduction as an abbreviation for "logarithm." Two or three abbreviations occur early, namely, "l" and "log.," also "lg." John Napier coined the word "logarithm," but did not use any abbreviation for it in his writings. Kepler,[3] in 1624, used the contraction "Log." So did H. Briggs[4] and W. Oughtred.[5] Oughtred wrote also "Log: Q:" for "Logarithm of the square of." Ursinus[6] employs "L."

[1] Emmy Noether, *Journal f. reine u. angewandte Mathematik*, Vol. CXXXIX (1910), p. 123, 124.

[2] R. Weitzenböck, *Encyklopädie d. math. Wissensch.*, Band III₃, Heft 6 (1922), p. 16.

[3] J. Kepler, *Chilias logarithmorum* (Marpurgi, 1624).

[4] Henry Briggs, *Logarithmicall Arithmetike* (London, 1631), p. 7.

[5] W. Oughtred, *Key of the Mathematicks* (London, 1647), p. 135, 172, and in later editions of his *Clavis mathematicae* (see § 184).

[6] B. Ursinus, *Magnus canon logarithmicus* (Cologne, 1624).

The Italian B. Cavalieri writes in 1632[1] "log." and in 1643[2] "l." Molyneux[3] of Dublin uses "Log. comp." for the complement of the logarithm. When the logarithmic function received the attention of eighteenth-century writers on the calculus, the contractions "l" and "L" became more frequent. W. Jones[4] wrote "L" for "logarithm" and further on "L" for the "log. sine," and "l" for arithmetical complement of "Log. sine." One finds "l" for logarithm in the *Acta eruditorum*,[5] in the textbook of Christian Wolf[6] (who uses also "Log."), in Brook Taylor,[7] in Johann Bernoulli,[8] in Denys Diderot[9] (who employs also "log."). Euler[10] uses "l" in 1728 and 1748. In Sherwin's *Tables* of 1741[11] one finds "Log." and "L.," in P. Varignon[12] "l *x*." Robert Shirtcliffe[13] writes "$L:A+L:B=L:A\times B$" for $\log A + \log B = \log AB$.

W. Gardiner, in the Introduction to his *Tables* (London, 1742), not only uses "L.," but suggests "L'" to represent the arithmetical complement in common logarithms, so that "$L.\frac{1}{a}=L'a$." Less happy was the suggestion "*cd* log *a*" for the complement of log *a*, made by J. A. Grunert.[14]

[1] Bonaventura Cavalieri, *Directorivm generale Vranometricvm* (Bologna, 1632). p. 169.

[2] Bonaventura Cavalieri, *Trigonometria* (Bologna, 1643).

[3] William Molyneux, *A Treatise of Dioptricks* (London, 1692), p. 92.

[4] W. Jones, *Synopsis palmariorum matheseos* (1706), p. 176, 279.

[5] *Acta eruditorum* (Leipzig, 1703), p. 451.

[6] Christian Wolf, *Elementa matheseos universae*, Vol. I (Halle, 1713), p. 231, 236, 508.

[7] Brook Taylor, *Methodus incrementorum*. See also *Phil. Trans.*, Vol. XXX (1717–19), p. 621.

[8] Johann Bernoulli in *Histoire de l'académie r. d. sciences*, année 1730 (Paris. 1732), *Mém.*, p. 78.

[9] *Memoires sur différens sujets de mathématiques par M. Diderot* (Paris, 1748). p. 42, 43.

[10] L. Euler's letter to J. Bernoulli, Dec. 10, 1728 (see G. Eneström in *Bibliotheca mathematica* [3d ser.], Vol. IV [1903], p. 351, 353); Euler, *Introductio in analysin infinitorum* (1748), Lib. I, § 104.

[11] *Sherwin's Mathematical Tables* (3d ed.; rev. William Gardiner; London, 1741), p. 16.

[12] P. Varignon, *Eclaircissemens sur l'analyse des infiniment petits* (Paris, 1725), p. 116.

[13] Robert Shirtcliffe, *The Theory and Practice of Gauging* (London, 1740), p. 31.

[14] J. A. Grunert, *Lehrbuch der Mathematik*, Vol. III (Brandenburg, 1832), p. 69. See J. Tropfke, *op. cit.*, Vol. II (2d ed., 1921), p. 210.

The capital letter "L" for logarithm occurs in Thomas Watkins,[1] in Count de Fagnano,[2] also in J. A. Fas,[3] C. F. Fournier.[4]

The sign "$|x$" for log x sometimes occurs;[5] the vertical stroke is probably a modified letter "l."

470. *Different meanings of "log x," "l x," and "L x."*—In more recent time the meanings of "log," "l," and "L" have been differentiated; Leibniz[6] employed "l" when natural logarithms are implied. Cauchy[7] proposed that "l" stand for natural logarithms, and "L" for the logarithms in any system whose base, $b > 1$. Stolz and Gmeiner[8] let "log" stand for the common logarithm of a positive real number, "l" for its natural logarithm, and "L" for the logarithms of a, where a is real or imaginary, though different from zero. Peano[9] writes "log x" for a logarithm to the base e, and "Log x" for one to the base 10. André[10] states that the logarithm to the base e is written "LP," "$\mathscr{L}P$," or "log P."

471. Another notation was proposed by Crelle,[11] in which the base is written above and to the left of the logarithm. This notation is sometimes encountered in more recent books. Thus, Stringham[12] denotes a logarithm to the base b by "^{b}log"; he denotes also a natural logarithm by "ln" and a logarithm to the complex modulul k, by "log $_k$." Stolz and Gmeiner[13] signify by "^{a} log. b" the "logarithm of b to the base a."

[1] Thomas Watkins in *Philosoph. Transactions*, Vol. XXIX (1714–16), p. 123.

[2] *Opere Mathematiche del Marchese Giulio Carlo De' Toschi di Fagnano*, Vol. III (1912), p. 30.

[3] J. A. Fas, *Inleiding tot de Kennisse en het Gebruyk der Oneindig Kleinen* (Leyden, 1775), p. 29.

[4] C. F. Fournier, *Éléments d'arithmétique et d'algèbre*, Tome II (Nantes, 1842), p. 147.

[5] *Mathematical Gazette*, Vol. XII (1924), p. 61.

[6] *Leibnizens Mathematische Schriften* (ed. C. I. Gerhardt), Vol. I, p. 117.

[7] A. L. Cauchy, *Cours d'analyse* (Paris, 1821), p. 52, 137, 171.

[8] O. Stolz und J. A. Gmeiner, *Theoretische Arithmetik*, Vol. II (Leipzig, 1902), p. 212, 365.

[9] G. Peano, *Lezioni di analisi infinitesimale*, Vol. I (Torino, 1893), p. 33.

[10] Désiré André, *Des Notations mathématiques* (Paris, 1909), p. 64.

[11] A. L. Crelle, *Sammlung Math. Aufsätze 1* (Berlin, 1821), p. 207. See *Encyclopédie des scien. math.*, Tome I, Vol. I (1904), p. 59.

[12] Irving Stringham, *Uniplanar Algebra* (San Francisco, 1893), p. xiii.

[13] O. Stolz und J. A. Gmeiner, *Theoretische Arithmetik*, Vol. II (Leipzig, 1902), p. 211.

Crelle[1] in 1831 and Martin Ohm[2] in 1846 write the base above the "log," thus, "$\overset{c}{\log} a$." This symbolism is found in many texts; for instance, in Kambly's *Elementar-Mathematik*.[3] A. Steinhauser[4] wrote "log. nat. z" and "log. brigg. z."

472. *The power of a logarithm.*—The power $(\log x)^m$ is indicated by Reyneau[5] by "$\mathrm{l}^m x$," while log (log x) is designated by "l. l x." Spence[6] lets "$\mathrm{L}^2 (x)$" stand for $(\log x)^2$. But more commonly parentheses are used for marking powers; for instance, Carmichael[7] uses the notation $(\log a)^l . (\log b)^m . (\log c)^n$ for powers. Bonnet[8] denotes by "$(\mathrm{ll}\ 3)^a$" the expression $[\log (\log 3)]^a$.

Pringsheim[9] marks by "$\mathrm{lg}_k^\rho n$" or by, say, "$\log_2^3 n$," the expression $\{\mathrm{lg}(\mathrm{lg}\ n)\}^3$, by "$\mathrm{L}_\kappa (x)$" the expression $\mathrm{lg}_0 x . \mathrm{lg}_1 x . \ldots . \mathrm{lg}_\kappa x$.

473. *Iterated logarithms* have received some consideration in the previous paragraph. They will occur again when we consider logarithmic notations which diverge widely from those ordinarily used. We note here the symbolism used by Pringsheim and Molk in their joint *Encyclopédie* article:

$$\text{``} {}^2\log_b a = \log_b (\log_b a) , \ldots ,$$
$$ {}^{k+1}\log_b a = \log_b ({}^k\log_b a) .\text{''}[10]$$

474. *Marking the characteristic.*—In the earliest logarithmic tables (except those of Bürgi) no fractions at all were used, the numbers being integers. When the point or comma first entered, it was not as a separator of integers from fractions, but as a mode of marking the integers into groups for more convenient and rapid reading. Thus,

[1] A. L. Crelle in *Crelle's Journal*, Vol. VII (1831), p. 327.

[2] M. Ohm, *Der Geist der Differential- and Integral-Rechnung* (Erlangen, 1846), p. 4.

[3] Ludwig Kambly, *Elementar-Mathematik*, 1. Teil, "Algebra" (Breslau, 1906), p. 57.

[4] Anton Steinhauser, *Lehrbuch der Mathematik*, "Algebra" (Vienna, 1875), p. 278, 283.

[5] Charles Reyneau, *Analyse demontrée*, Vol. II (Paris, 1708), p. 802.

[6] William Spence, *Theory of . . . . Logarithmic Transcendents* (London, 1809), p. viii.

[7] R. Carmichael, *Der Operationscalcul* (trans. C. H. Schnuse; Braunschweig, 1857), p. 87.

[8] Ossian Bonnet, *Journal de mathématiques*, Vol. VIII (Paris, 1845), p. 74.

[9] Alfred Pringsheim, *Mathematische Annalen*, Vol. XXXV (1890), p. 317, 333.

[10] A. Pringsheim and J. Molk, *Encyclopédie des scien. math.*, Tom. I, Vol. I (1907), p. 195.

in Napier's *Descriptio* of 1614 one finds, in the tables, the sine of 30°30′ given as "5, 0 7 5, 3 8 4" and its logarithm as "6, 7 8 1, 8 2 7." Here both the natural sine and its logarithm are integers, and the commas serve to facilitate the reading. As late as 1636, in Gunter's logarithms, the comma is used not as a separator of integers and fractions, but as a separator of the last five figures; the logarithm of 20 is given as "1 3 0 1, 0 2 9 9 9."

The use of the characteristic is found in the logarithmic tables of 1617 and 1624, published by Briggs,[1] who separated the characteristic from the mantissa by a comma. He placed also a comma after every fifth digit in the fourteen decimal places. Thus, Briggs writes log 1 6 5 0 3 thus: "4, 2 1 7 5 6, 2 8 9 9 6, 6 9 4 3." In 1624 Briggs writes the decimal "2⁵⁄," and its logarithm "0, 3 9 7." In the tables of H. Gellibrand,[2] published by Vlacq, the comma is used in the same manner. The omission of the characteristic in the tables of numbers occurs first in Sherwin (1705).[3]

475. There long existed, in fact there still exists, divergence in the mode of separating the mantissa and characteristic. Beutel[4] in 1690 writes the logarithm of "160 . $\frac{88}{100}$" as "2.2065203." Metzburg[5] uses the dot (sometimes the comma) as logarithmic separatrix, and the comma regularly as separatrix in ordinary decimal fractions.

Vega[6] uses regularly the comma as decimal separatrix and also as logarithmic separatrix, the dot being reserved for multiplication. André expresses himself on this matter as follows: "In the logarithmic tables of Lalande the characteristic is followed, not by a comma, but by a point, only the point is placed on the principal line as the comma would be placed. In a book on elementary Arithmetic, the numeral for units is separated from the one for tenths by a kind of hyphen: the number 7, 2 3 1 8 is written 7- 2 3 1 8. . . . . In a certain five-place table of logarithms the mantissa is divided into two parts, separated by an interval in which is placed a point: The mantissa 1 5 4 7 3 appears under the form 15 . 4 7 3. These diverse employments of the

[1] Henry Briggs, *Logarithmorum Chilias prima* (London, 1617); *Arithmetica Logarithmica sive logarithmorum chiliades triginta* (London, 1624).

[2] Henry Gellibrand, *Trigonometria Britannica* (Goudae, 1633).

[3] H. Sherwin, *Mathematical Tables* (London, 1705).

[4] Tobias Beutel, *Geometrischer Lust Garten* (Leipzig, 1690), p. 240, 241.

[5] Des Freyherrn von Metzburg, *Anleitung zur Mathematik*, 1. Theil (Wien, 1798), p. 88, 274; 3. Theil, p. 41.

[6] Georg Freyherr von Vega, *Vorlesungen über die Mathematik*, Band I, 3. Aufl. (Wien, 1802), p. 96, 364.

point constitute so many faults: the point should not enter a numerical expression other than as a contracted sign of multiplication."[1]

476. The logarithms of fractions less than one gave rise to a diversity of notation which has prevailed down to the present time.

Richard Norwood[2] in 1631 (writing "s" when he means "log sine") gives "s, D 6 7 deg. 2 3′ 9,9 6 5 2 4 8 0," where D is an angle of the triangle $A D B$. Here we have the characteristic increased by 10 and separated from the mantissa by a comma. Similarly, in John Ward,[3] "the Sine of the Angle C 49°−30′ 9,8810455."

The writing of the negative sign over the characteristic, to indicate that the characteristic is negative, is found in Oughtred's *Clavis mathematicae* (except the 1631 edition which does not consider logarithms). We quote from the 1652 edition (p. 122): "Ut numeri 436, Log: est 2,6394865 at numeri 4 3 6 0 0 , est 4, 6 3 9 4 8 6 5. et numeri 4|36, Log: est 0, 6 3 9 4 8 6 5. Denique numeri 0|0 0 4 3 6. Log: est $\bar{3}$, 6 3 9 4 8 6 5." Observe that the characteristic is separated from the mantissa by a comma, and that a different sign is used for decimal fractions. In his *Circles of Proportion* (London, 1632), page 4, Oughtred gives log 2 as "0. 3 0 1 0 3."

Ozanam[4] in 1692 gives the logarithm of $\frac{3}{4}$ as "−0. 1 2 4 9 3 8 8," the fractional part being here negative and the minus sign placed in front of the logarithm. Likewise, Benito Bails[5] gives "log. $\frac{17}{187}=$ −1, 0 4 1 3 9 3."

477. Oughtred's practice of placing the minus sign of a negative characteristic above the characteristic has found favor among British authors of recent time, such as H. S. Hall and S. R. Knight,[6] E. M. Langley,[7] C. G. Knott,[8] and F. Castle.[9] Langley directs computors "to reject the superfluous 10." The notation "$\bar{2}$. 7 9 5 8 0 0 " is found also in the *Report of the British Association* (1896), page 75. However, it has been far from universal even in England. A. de

[1] Désiré André, *Des Notations mathématiques* (Paris, 1909), p. 20.

[2] Richard Norwood, *Trigonometrie* (London, 1631), Book I, p. 20.

[3] *Posthumous Works of John Ward* (London, 1730), p. 74.

[4] J. Ozanam, *Cours de mathematiques*, Vol. II (new ed.; Paris, 1692), p. 65.

[5] *Principios de matematica de la real academia de San Fernando por Don Benito Bails* (2d. ed.), Vol. I (Madrid, 1788), p. 191.

[6] H. S. Hall and S. R. Knight, *Elementary Algebra*, ed. F. L. Sevenoak (New York, 1895), p. 339.

[7] Edward M. Langley, *A Treatise on Computation* (London, 1895), p. 111, 123.

[8] C. G. Knott, *Logarithmical Tables* (Edinburgh, 1914).

[9] Frank Castle, *Practical Mathematics for Beginners* (London, 1905), p. 127.

Morgan states that "9 is written for $\bar{1}$, that this result may be too great by 10, as are all the logarithms in the table of tangents"; there occurs also "19˙ 8 2 3 9 0 8 8 − 2 0."[1]

On the European continent this notation has been used to a limited extent. Fournier[2] in 1842 wrote "log. 0, 3 4 1=$\bar{1}$. 5 3 7 5 4." A. Fontaine[3] gives "$\dot{2}$ 3 0 2 5 8 5 0 9 2 9 9 4, etc. = 1 10," a designation which did not meet with favor. Some French writers, for example Bézout,[4] and many German, Austrian, and Dutch writers, place the minus sign of a negative characteristic before the logarithm. Segner[5] in 1767 said, "Numeri 0, 5 7 9 3 2 logarithmus est = − 1, 7 6 2 9 1 8 5, in quo signum − ad solam characteristicam refertur," but in the trigonometrical part he writes "log. sin. C = 9, 6 1 6 3 3 8 2."

C. Scherffer[6] writes "Log. sin. 36° = 9, 7 6 9 2 1 8 7" and "Log. $\frac{1}{2}$ = −1, 6 9 8 9 7 0 0." Blassière[7] says: "... pour l'index −1 on écrira ,9; pour −2, on substitue ,8; pour −3 on écrit ,7; et ainsi de suite." "1 0,6 = , 9, 7 7 8 1 5 1 3." The trend of European practice is shown further by the following quotation from Prändel:[8] "Some place the negative characteristic after the mantissa;[9] for example, 0. 8 7 5 0 6 1 3 − 1, and 0. 9 0 3 0 9 0 0 − 3. Others, as Prof. Böhm, write the minus sign over the prefixed characteristic, as in $\bar{3}$. 9 0 3 0 9 0 0. Our designation − 3, + 9 0 3 0 9 0 0 would be the best, except that in addition and subtraction the numbers do not appear in a convenient position relative to the other logarithms. We shall select a middle path and designate such hypothetical logarithms which have a negative characteristic and a positive mantissa thus, −1. 8 7 5 0 6 1 3 +),

[1] *Library of Useful Knowledge, Mathematics*, Vol. I (London, 1847): A. de Morgan, *Examples of the Processes of Arithmetic and Algebra*, p. 54, 55.

[2] C. F. Fournier, *Éléments d'arithmétique et d'algèbre*, Tome II (Nantes, 1842).

[3] *Mémoires donnés à l'académie royale des sciences*, non imprimés dans leur temps. Par M. Fontaine, de cette Académie (Paris, 1764), p. 3.

[4] E. Bézout, *Cours le mathématiques*, Tome I (2d ed.; Paris, 1797), p. 181, 188. He uses also negative mantissas.

[5] I. A. de Segner, *Cursus mathematici*, Part I (Halae, 1767), p. 135, 356.

[6] *Institutiones geometriae sphericae, ... a Carolo Scherffer* (Vienna, 1776), p. 66.

[7] J. J. Blassière, *Institution du Calcul numerique et litteral*, 2. partie (A la Haye, 1770), p. 383.

[8] Johann Georg Prändel, *Algebra* (München, 1795), p. 507, 508.

[9] This notation is found in the following books: J. V. Voigt's *Grundlehren der reinen Mathematik* (Jena, 1791), p. 192; J. F. Lorenz', *Grundlagen der allgemeinen Grössenberechnung* (2d ed.; Helmstadt, 1800), p. 60; A. Steinhauser's *Lehrbuch der Mathematik, Algebra* (Vienna, 1875), p. 278, 283.

$-3.9030900+$). Certainly these can be placed more conveniently one below the other.

"Other mathematicians understand by hypothetical logarithms the complement of the negative logarithms with respect to 10.000000. For example the hypothetical logarithm of -2.7482944 would be 7.2517056. But computation with this sort of logarithms becomes in the first place too intricate, and in the second place too mechanical."

478. *Marking the last digit.*—Gauss[1] states that von Prasse in his logarithmic *Tafeln* of 1810 placed in different type the last figure in the mantissa, whenever that figure represented an increase in the value of the mantissa. Babbage denoted the increase by a point subscript which the reader scarcely notices, but Lud. Schrön (1860) used a bar subscript which catches the eye at once "and is confusing."[2] Many writers, for instance Chrystal,[3] place a bar above the last digit, to show that the digit has been increased by a unit. F. G. Gausz[4] explains that $\bar{5}$ in the last digit means that it was raised from 4, that $\dot{5}$ means that the remaining decimal has been simply dropped, that a star ($*$) prefixed to the last three digits of the mantissa means that the first "difference" given on the line next below should be taken.

479. *Sporadic notations.*—A new algorithm for logarithms was worked out in 1778 by Abel Burja[5], who writes $\frac{a}{b}=m$ where b is the base, a the power (*dignité*), and m the logarithm, so that, expressed in the ordinary notation, $a=b^m$. He finds that $\frac{a}{b}-\frac{c}{d}=\frac{a}{b}-\frac{b}{b}=\frac{c^{\frac{b}{d}}}{b}=\frac{a:c^{\frac{b}{d}}}{b}$.

He calls *proportion logarithmique* four quantities of which the second is the same power of the first that the fourth is of the third; he designates it by the sign $\bowtie$ as in $2, 8 \bowtie 4, 64$. Burja proceeds:

"$a^{\bar{n}}$ ou a^n sera la n-ième puissance de a

$a^{\overset{n}{:}}$, la n-ième bipuissance de a ...

[1] K. F. Gauss, *Werke*, Vol. III (1866), p. 242.

[2] J. W. L. Glaisher's "Report on Mathematical Tables" in *Report of British Association* (1873), p. 58.

[3] G. Chrystal, *Algebra*, Part II (Edinburgh, 1889), p. 218.

[4] F. G. Gausz, *Logarithm. und trigonom. Tafeln* (Halle, a. S., 1906), "Erläuterungen."

[5] Abel Burja in *Nouveaux mémoires de l'académie r. d. sciences et des belles lettres*, année 1778 et 1779 (Berlin, 1793), p. 301, 321.

en general,

$a^{\overset{n}{N}}$ sera la n-ième puissance de l'ordre N de a."

The need of the novel symbolism suggested by Burja was not recognized by mathematicians in general. His signs were not used except by F. Murhard[1] of Göttingen who refers to them in 1798. By the sign "log $\frac{a}{b}$" Robert Grassmann[2] marked the quantity c which yields the relation $b^c = a$. The Germans considered several other notations. In the relation $2^3 = 8$, Rothe in 1811 and Martin Ohm[3] in 1823 suggested "8 ? 2 = 3," while Bischoff[4] in 1853 wrote $\wedge$ for "log," and Köpp[5] in 1860 proposed $\overbrace{8}\kern-1em\diagup = 3$. Köpp's sign is favored by Draenert. F. J. Studnička[6] of Prague opposes it as *hässlich* and favors the use of "l" for a natural logarithm, and ℓ_n (derived, as he says, from "log," "lg," ℓ,) for Briggian logarithms, and "$\ell_c{}^b$" for logarithms to any base c as the equivalent of Schlömilch's "$\log^c b$." E. Bardey opposes the introduction of any new logarithmic symbol. Paugger[7] simply inverts the radical sign $\sqrt{\ }$, and writes $\overset{m}{V}p = a$ and $\underset{a}{\diagup}p = m$ (identical with $\log_a p = m$). Later on in his *Operationslehre* (p. 81) Paugger uses a modification of the Greek letter λ, as shown in $a^m = b$, $\overset{m}{V}b = a$, $\underset{a}{\lambda}b = m$. The editor of the *Zeitschrift*, J. C. V. Hoffmann,[8] prefers in the case $a^n = p$, the symbolisms $n = \overbrace{8}\kern-1em\diagup$, and for antilogarithm $\diagdown\!\!\!_a = p$. J. Worpitzky and W. Erler[9] propose a modified l, namely, $\underset{a}{2}$, so that log (bc) to the base a would be written $\underset{a}{2}$."

Landen[10] marked by log $(RQ':PQ')$ the hyperbolic logarithm of $\frac{RQ'}{PQ'}$, or the measure of the ratio of RQ' to PQ.

[1] Friederich Murhard, *System der Elemente der allgemeinen Grössenlehre* (Lemgo, 1798), p. 260.

[2] Robert Grassmann, *Zahlenlehre oder Arithmetik* (Stettin, 1872), p. 45.

[3] See Draenert in *Zeitschr. f. math. u. naturwiss. Unterricht*, Vol. VIII (Leipzig, 1877), p. 266.

[4] Anton Bischoff, *Lehrbuch der Algebra* (Regensburg, 1853), p. 265.

[5] Köpp, *Ausführung gewöhnlicher Zifferrechnungen mittelst Logarithmen* (Osterprogramm des Realgymnasiums zu Eisenach, 1860); also in his *Trigonometrie* (1863) and *Arithmetik* (1864).

[6] *Zeitschr. f. math. u. naturw. Unterricht*, Vol. VIII (1877), p. 403.

[7] *Op. cit.*, Vol. VIII, p. 268, 269.

[8] *Op. cit.*, Vol. VIII, p. 270.   [9] *Op. cit.*, Vol. VIII, p. 404.

[10] John Landen, *Mathematical Lucubrations* (London, 1755), Sec. III, p. 93.

Schellbach[1] of Berlin adopted for the operation of *logarithmirung*, in place of the incomplete formula, log $a=c$, the equation $\overset{a}{\underset{b}{\times}} = c$; he expresses the theorem relating to change of modulus thus,

$$\overset{``\ \ a}{\underset{b}{\times}} \ \ \overset{b}{\underset{k}{\times}} = \overset{a\ \ "}{\underset{k}{\times}}.$$

As the notation here proposed is not always convenient, he suggests an alternative notation. Just as one has $a \times b$ and $a.b$, $\frac{a}{b}$ and $a:b$, so one may choose $\overset{a}{\underset{b}{\times}}$ and $a \vdots b$. He writes $(a+b) \vdots c$ for log $(a+b)$ to the modulus c, also $(a \vdots b) \vdots c$ for $\log_c \left(\log \frac{a}{b} \right)$.

480. *Complex numbers.*—Martin Ohm, in his treatment of logarithms of complex numbers, lets "log" represent the infinitely many logarithms of a complex number, "L" the tabular logarithm of the modulus. He states that when the concept of the general power a^x is given the concept of the general logarithm $b ? a$, if by it is meant every expression x such that one has $a^x = b$ or $e^{x \log a} = b$."[2] If x has an infinite number of values for each value of log a, one sees the reason for the appearance in $a ? b$ of two independent arbitrary constants, as is seen also in the investigations due to J. P. W. Stein, John Graves, and W. R. Hamilton.[3] Ohm says that, since b and a are taken completely general, and a^x has an infinity of values, $b ? a$ is wholly undetermined, unless it is stated for what value of log a the power a^x is to be taken. He adopts the special notation $b?(a||a)$, which means the logarithm of·b to the base a, when log $a = a$. He shows that $b?(a||a) = \frac{\log b}{a}$ is a complete equation. If $a = e = 2.718 \ldots .$, then Ohm's logarithms reduce to those previously developed by Euler.

De Morgan[4] in developing the general theory of logarithms used "log" for the numerical or tabular value of the logarithm, and let λx

[1] K. H. Schellbach, "Ueber die Zeichen der Mathematik," *Crelle's Journal*, Vol. XII (1834), p. 70–72.

[2] Martin Ohm, *System der Mathematik*, Vol. II (2d ed., 1829), p. 438, 415.

[3] See F. Cajori, *American Mathematical Monthly*, Vol. XX (1913), p. 173–78.

[4] A. de Morgan, *Trans. Cambridge Philos. Soc.*, Vol. VII (1842), p. 186.

stand for "any legitimate solution of $e^{\lambda x} = x$." Later[1] he called λA the "logometer of A" and used also the inverted letter $\curlyvee$, $\curlyvee A$ to mean the line whose logometer is A.

Peano[2] lets log x represent "la valeur principale du logarithme," and log* x, "la classe des solutions de l'équation $e^y = x$."

481. *Exponentiation* is the operation of finding x in $a^x = P$. Says André[3] "It is J. Bourget[4] who, for the first time, pointed out this lacuna and indicated the means of filling it up. He gave the operation of finding the exponent the name exponentiation and represented it by the sign $\div$ which he called 'exponented by.' The equality $64 \div 4 = 3$ is read; 64 exponented by 4 is equal to 3."

482. *Dual logarithms.*—A dual number of the ascending scale is always a continued product of powers of the numbers $1\cdot1$, $1\cdot01$, $1\cdot001$, $1\cdot0001$, etc., taken in order, the powers of the numbers alone being expressed. To distinguish these numbers from ordinary numbers the sign $\downarrow$ precedes them. Thus $\downarrow$ 6, 9, 7, $6 = (1\cdot1)^6 \times (1\cdot01)^9 \times (1\cdot001)^7 \times (1\cdot0001)^6$, ... , $\downarrow 6 = (1\cdot1)^6$, $\downarrow 0$, $6 = (1\cdot01)^6$, $\downarrow 0$, 0, $6 = (1\cdot001)^6$. When all but the last digit of a dual number are zeroes, the dual number is called a dual logarithm. Byrne[5] gives also tables of "Dual logarithms and dual numbers of the descending branch of Dual Arithmetic from '0 '0 '0 '1 '0 '0 '0 '0 $\uparrow$ to '3 '6 '9 '9 '0 '0 '0 '0 $\uparrow$ with corresponding natural numbers."

SIGNS IN THEORETICAL ARITHMETIC

483. *Signs for "greater" and "less."*—Harriot's symbols $>$ for greater and $<$ for less (§ 188) were far superior to the corresponding symbols ▭ and ▭ used by Oughtred. While Harriot's symbols are symmetric to a horizontal axis and asymmetric only to a vertical, Oughtred's symbols are asymmetric to both axes and therefore harder to remember. Indeed, some confusion in their use occurred in Oughtred's own works, as is shown in the table (§ 183). The first deviation from his original forms is in "Fig. EE" in the Appendix, called the *Horologio*, to his *Clavis*, where in the edition of 1647 there stands ▭ for $<$, and in the 1652 and 1657 editions there stands ▭ for $<$. In the text of the *Horologio* in all three editions, Oughtred's regular nota-

---

[1] A. de Morgan, *Trigonometry and Double Algebra* (London, 1849), p. 130.

[2] G. Peano, *Formulaire mathématique*, Vol. IV (Turin, 1903), p. 229.

[3] Désiré André, *op. cit.*, p. 63.

[4] J. Bourget, *Journal de Mathématiques élémentaires*, Vol. II, p. 12.

[5] Oliver Byrne, *Tables of Dual Logarithms* (London, 1867), p. 7–9. See also Byrne's *Dual Arithmetic* and his *Young Dual Arithmetician*.

tion is adhered to. Isaac Barrow used ⌐ for "majus" and ⌐ for "minus" in his *Euclidis Data* (Cambridge, 1657), page 1, and also in his *Euclid's Elements* (London, 1660), Preface, as do also John Kersey,[1] Richard Sault,[2] and Roger Cotes.[3] In one place John Wallis[4] writes ⌐ for >, ⌐ for <.

Seth Ward, another pupil of Oughtred, writes in his *In Ismaelis Bullialdi astronomiae philolaicae fundamenta inquisitio brevis* (Oxoniae, 1653), page 1, ⌐ for "majus" and ⌐ for "minus." For further notices of discrepancy in the use of these symbols, see *Bibliotheca mathematica*, Volume XII[3] (1911–12), page 64. Harriot's > and < easily won out over Oughtred's notation. Wallis follows Harriot almost exclusively; so do Gibson[5] and Brancker.[6] Richard Rawlinson of Oxford used ⌐ for greater and ⌐ for less (§ 193). This notation is used also by Thomas Baker[7] in 1684, while E. Cocker[8] prefers ⌐ for ⌐ . In the arithmetic of S. Jeake,[9] who gives " ⌐ greater, ⌐ . next greater, ⌐ . lesser, ⌐ . next lesser, ⌐ not greater, ⌐ . not lesser, ⌐ . equal or less, ⌐ . equal or greater," there is close adherence to Oughtred's original symbols.

Ronayne[10] writes in his *Algebra* ⌐ for "greater than," and ⌐ for "less than." As late as 1808, S. Webber[11] says: ". . . . . we write a ⌐ b, or $a \gtrdot b$; . . . . a ⌐ b, or $a < b$." In Isaac Newton's *De Analysi per Aequationes*, as printed in the *Commercium Epistolicum* of 1712, page 20, there occurs x ⌐ $\frac{1}{2}$, probably for $x < \frac{1}{2}$; apparently, Newton used here the symbolism of his teacher, I. Barrow, but in Newton's *Opuscula* (Castillion's ed., 1744) and in Lefort's *Commercium Epistolicum* (1856), page 74, the symbol is interpreted as meaning $x > \frac{1}{2}$. Eneström[12]

---

[1] John Kersey, *Elements of Algebra* (London, 1674), Book IV, p. 177.

[2] Richard Sault, *A New Treatise of Algebra* (London, n.d.).

[3] Roger Cotes, *Harmonia mensurarum* (Cambridge, 1722), p. 115.

[4] John Wallis, *Algebra* (1685), p. 127.

[5] Thomas Gibson, *Syntaxis mathematica* (London, 1655), p. 246.

[6] Thomas Brancker, *Introduction to Algebra* (trans. of Rahn's *Algebra;* London, 1668), p. 76.

[7] Thomas Baker, *Clavis geometrica* (London, 1684), fol. *d* 2 *a*.

[8] Edward Crocker, *Artificial Arithmetick* (London, 1684), p. 278.

[9] Samuel Jeake, Sr., ΛΟΓΙΣΤΙΚΗΛΟΓΓΊΑ *or Arithmetick* (London, 1696), p. 12

[10] Philip Ronayne, *Treatise of Algebra* (London, 1727), p. 3.

[11] Samuel Webber, *Mathematics*, Vol. I (Cambridge, Mass., 1808; 2d ed.), p. 233.

[12] G. Eneström, *Bibliotheca mathematica* (3d ser.), Vol. XII (1911–12), p. 74.

argues that Newton followed his teacher Barrow in the use of ⊐ and actually took $x<\frac{1}{2}$, as is demanded by the reasoning.

In E. Stone's *New Mathematical Dictionary* (London, 1726), article "Characters," one finds ⊏ or ⊐ for "greater" and ⊐ or ⊏ for "less." In the Italian translation (1800) of the mathematical part of Diderot's *Encyclopédie*, article "Carattere," the symbols are further modified, so that ⊏ and ⊓ stand for "greater than," ⊐ for "less than"; and the remark is added, "but today they are no longer used."

Brook Taylor[1] employed ⊐ and ⊏ for "greater" and "less," respectively, while E. Hatton[2] in 1721 used ⊏ and ⊐, and also > and <. The original symbols of Oughtred are used in Colin Maclaurin's *Algebra*.[3] It is curious that as late as 1821, in an edition of Thomas Simpson's *Elements of Geometry* (London), pages 40, 42, one finds ⊏ for > and ⊐ for <.

The inferiority of Oughtred's symbols and the superiority of Harriot's symbols for "greater" and "less" are shown nowhere so strongly as in the confusion which arose in the use of the former and the lack of confusion in employing the latter. The burden cast upon the memory by Oughtred's symbols was even greater than that of double asymmetry; there was difficulty in remembering the distinction between the symbol ⊏ and the symbol ⊏ . It is not strange that Oughtred's greatest admirers—John Wallis and Isaac Borrow—differed not only from Oughtred, but also from each other, in the use of these symbols. Perhaps nowhere is there another such a fine example of symbols ill chosen and symbols well chosen. Yet even in the case of Harriot's symbolism, there is on record at least one strange instance of perversion. John Frend[4] defined < as "greater than" and > as "less than."

484. *Sporadic symbols for "greater" or "less."*—A symbol constructed on a similar plan to Oughtred's was employed by Leibniz[5] in 1710, namely, "$a =$ significat a esse majus quam b, et $a =$ significat a esse minus quam b." Leibniz borrowed these signs from his teacher Erhard Weigel,[6] who used them in 1693. In the 1749 edition of the *Miscellanea Berolinensia* from which we now quote, these inequality

---

[1] Brook Taylor, *Phil. Trans.*, Vol. XXX (1717–19), p. 961.

[2] Edward Hatton, *Intire System of Arithmetic* (London, 1721), p. 287.

[3] Colin Maclaurin, *A Treatise of Algebra* (3d ed.; London, 1771).

[4] John Frend, *Principles of Algebra* (London, 1796), p. 3.

[5] *Miscellanea Berolinensia* (Berlin, 1710), p. 158.

[6] *Erhardi Weigelii Philosophia mathematica* (Jenae, 1693), p. 135.

symbolisms are displaced by $>$ and $<$. Other symbolisms which did not meet with favor were Albert Girard's[1] ff for *plus que* and § for *moins que*, and Samuel Reyhers'[2] $\dashv$ for $>$, $\vdash$ for $<$, $\sqsupseteq$ for twofold greater ratio or *duplicate* ratio, $\sqsupseteq$ for threefold greater ratio or *triplicate* ratio. Hérigone (§ 263) denoted "greater than" by 3|2 and "less than" by 2|3. F. Dulaurens[3] in 1667 adopted $\sqcap$ for "equal," $\sqcap$ for "greater," and $\sqcap$ for "less." John Bonnycastle[4] introduced in his *Algebra* $\diagup\!\!\!\diagdown$ for $>$ and $\diagdown\!\!\!\diagup$ for $<$.

John Wallis[5] used $\overline{\Rightarrow}$ for "equal or greater than," and again later[6] the signs $\geqslant$ ("equal or greater"), $\lesssim$ ("equal or less"), and $\gtreqqless$ ("equal, greater, or less"). The first two of these symbols have retained their places in some modern texts, for instance in one of A. R. Forsyth.[7] The third symbol was used by J. F. Lorenz[8] in his *Euklid's Elemente*. The signs $>$ and $<$ slowly worked their way into Continental Europe. The symbols $\geqq$, $\leqq$, which are less compact than those of Wallis, have been attributed to the Frenchman P. Bouguer.[9] In a letter written by Goldbach[10] to Euler, soon after 1734, one reads: "Der Hr. Bouguer brauchet, vor das signum, so Ew. schreiben $\prec$, dieses $\geqq$, welches zwar nicht compendiös, aber sehr expressif ist." This statement is of interest also as indicating that Euler had used $<$ crossed by a stroke, to indicate "not less than." The signs $>$ and $<$ were used by Rahn[11] (§ 194) and by Brancker[12] in his translation of Rahn's book into English (§ 194), also by Johann, Bernoulli,[13] and C. Wolf.[14]

[1] Albert Girard, *Invention nouvelle en l'algebre* (Amsterdam, 1629).

[2] Samuel Reyhers, *Euclides* (Kiel, 1698), Vorrede.

[3] Francisci Dulaurens, *Specimina mathematica duobus libris comprehensa* (Paris, 1667), in list of symbols.

[4] John Bonnycastle, *Introduction to Algebra* (London, 1824), p. 2.

[5] John Wallis, *De sectionibus conicis* (Oxford, 1655), p. 70.

[6] John Wallis in *Philosoph. Trans.*, Vol. V (London, 1670), p. 4011.

[7] A. R. Forsyth, *Theory of Functions of a Complex Variable* (Cambridge, 1893), p. 86.

[8] J. F. Lorenz, *Euklid's Elemente* (ed. C. B. Mollweide; Halle, 1824), p. xxxi.

[9] *Encyclopédie des scien. mathém.*, Tome I, Vol. I (1904), p. 23, n. 123.

[10] P. H. Fuss, *Correspondance mathématique et physique* (St. Peterburg, 1843), Vol. I, p. 304.

[11] Johann Heinrich Rahn, *Teutsche Algebra* (Zürich, 1659), p. 72.

[12] Thomas Brancker, *Introduction to Algebra* (trans. out of the High-Dutch into English; London, 1668), p. 59.

[13] Johann Bernoulli, *Acta eruditorum* (1687), p. 617.

[14] *Elementa matheseos universae*, Tomus I .... autore Christiano Wolfo (Halle, 1713), p. 30.

485. *Improvised type.*—How typesetters frequently improvised signs by the use of forms primarily intended for other purposes is illustrated in the inequality signs appearing in 1743 in papers of Euler.[1] The radical sign $\sqrt{}$ is turned into the position $\searrow$ for "greater than" and into $\nwarrow$ for "less than." A. G. Kästner[2] in 1758 has the astronomical sign for Aries or Ram to designate "greater than," as shown in 8⅋5. Friedrich Schmeisser[3] uses the letter Z, with the slanting stroke much heavier than the rest, as the sign for inequality (*inaequalitas*). Sometimes the sign $\angle$ is used for "less," as in parts of an article by L. Seidel,[4] notwithstanding the fact that this symbol is widely used for "angle."

486. *Modern modifications.*—A notation, the need of which has not been felt by later writers, was introduced in 1832 by E. H. Dirksen[5] of Berlin. He wrote "$n \cdot > Q$" for an infinite series whose terms remain greater than Q "by no assignable positive quantity," and "$n \cdot < Q$" when the terms remain less than Q "by no assignable positive quantity."

Another unusual notation was adopted by W. Bolyai[6] who in 1832 distinguished between $>$, $<$ and $\succ$, $\prec$. The latter were applied to the absolute values, as in "$-5 < $ et $\succ -2$."

Additional symbols are given in the *Algebra* of Oliver, Wait, and Jones:[7] $<$, $>$, $\leqq$, $\geqq$, mean "less than," "greater than," "smaller than," "larger than," respectively. Here "larger" and "smaller" take account of the size of the two numbers only; "greater" and "less" have a broader significance, meaning in some cases "higher" and "lower," "later" and "earlier." $=$ means that two numbers are equally large, as "$+1600 =- 1600$."

W. E. Byerly[8] used the sign $\prec$ for "equal to or less than" (see also § 682).

[1] L. Euler, *Miscellanea Berolinensia*, Vol. VII (Berlin, 1743), p. 170, 178.

[2] A. G. Kästner, *Anfangsgründe der Arithmetik, Geometrie . . . . Trigonometrie* (Göttingen, 1758), p. 30.

[3] Friedrich Schmeisser, *Lehrbuch der reinen Mathesis*, Erster Theil, *Die Arithmetik* (Berlin, 1817), p. 7.

[4] L. Seidel in *Abhandlungen der Math.-Phys. Classe d. k. Bayerisch. Akad. d. Wissensch.*, Vol. V (München, 1850), p. 386, 387.

[5] E. H. Dirksen, *Abhandlungen der K. P. Akademie d. Wissensch.* (Berlin, 1832), Th. I, p. 82.

[6] Wolfgangi Bolyai de Bolya, *Tentamen* (2d ed.), Tome I (Budapestini, 1897), p. xi.

[7] Oliver, Wait, and Jones, *Treatise on Algebra* (2d ed.; Ithaca, 1887), p. 5, 7.

[8] W. E. Byerly, *Elements of the Integral Calculus* (Boston, 1882), p. 12.

In connection with the theory of limits, Harriot's $>$ and $<$ have been given a curved form by Paul du Bois-Reymond[1] and by A. Pringsheim and J. Molk,[2] thus when $\lim \dfrac{a_\nu}{b_\nu}$ is zero, they write $a_\nu \prec b_\nu$; when the limit is $+\infty$, $a_\nu \succ b_\nu$.

The sign $\ll$ was introduced by H. Poincaré and by É. Borel[3] in comparing series like

$$u_0 + u_1 z + u_2 z^2 + \cdots \cdot \ll M\left[1 + \frac{z}{R^1} + \left(\frac{z}{R^1}\right)^2 + \cdots \cdot\right],$$

where the second series has positive coefficients; the modulus of each coefficient of the first series is less than the corresponding coefficient of the second series. The signs $\ll$ and $\gg$ are also used for "much less than" and "much greater than."[4]

487. *Signs for absolute difference.*—Oughtred's symbol ∞ for arithmetical or absolute "difference" (§ 184) came to be used widely, for instance, by Kersey,[5] Sault,[6] Jeake,[7] Hatton,[8] Donn,[9] and Cocker.[10] Hérigone[11] lets $\cdot\sim$: stand for "differentia," and $\sim$ for a minus sign.

There has been great confusion between ∞ and $\sim$, both in the designation of absolute difference in arithmetic and of similarity in geometry. In place of Oughtred's ∞, Wallis used[12] $\sim$; he used also[13] $\pm$ for plus or absolute difference. In an account of Oughtred's algebra, Wallis[14] gives ∞ for absolute difference; he says, "$a \infty b$ with me serves indifferently for $a - b$, or $b - a$, according as a or b is the great-

---

[1] *Annali di matematica pura ed applicata*, Serie II, Vol. IV (1870–71), p. 339.

[2] A. Pringsheim and Molk in *Encyclopédie des scien. math.*, Tom. I, Vol. I (1904), p. 202.

[3] Émile Borel, *Leçons sur les séries divergentes* (Paris, 1901), p. 142.

[4] *Zeitschrift für Physik*, Vol. XIII (1923), p. 366 (the sign was used by H. A. Kramers and W. Pauli); Torsten Heurlinger, *Untersuchungen über die Struktur der Bandenspectra* (Lund, 1918), p. 39.

[5] John Kersey, *Algebra* (London, 1673), p. 3.

[6] Richard Sault, *A New Treatise of Algebra* (London, [n.d.]).

[7] Samuel Jeake, ΛΟΓΙΣΤΙΚΗΛΟΓΙΑ or *Arithmetick* (London, 1696), p. 10–12.

[8] Edward Hatton, *An Intire System of Arithmetick* (London, 1721), p. 287.

[9] Benjamin Donn, *Mathematical Essays* (London, 1758), p. 280.

[10] Edward Cocker, *Artificial Arithmetick*, perused by John Hawkins (London, 1684), p. 275.

[11] Pierre Herigone, *Cursus mathematicus* (Paris, 1644), Vol. I, "Explicatio notarum" [1st ed., 1634].

[12] John Wallis, *Operum mathematicorum pars prima* (Oxford, 1657), p. 208.

[13] John Wallis, *op. cit.*, p. 334, 335.

[14] John Wallis, *Treatise of Algebra* (London, 1685), p. 127.

er." Probably Wallis' change in 1657 from Oughtred's $\backsim$ to $\sim$ was purely accidental. Jones[1] is faithful to Oughtred's $\backsim$ and uses also $\underline{\backsim}$, but Barrow,[2] who usually adheres to Oughtred's symbols, adopts the symbol $-\cdot\cdot$: which he defines as "The Difference, or Excess; Also, that the Quantities which follow, are to be subtracted, the Signs not being changed."

In England and the United States the sign $\sim$ or $\backsim$ for "absolute difference" has been used down to the present time. Thus one finds $\backsim$ in texts by Clarke,[3] J. Thomson,[4] Bridge.[5] The sign $\sim$ is found in the writings of Morton[6] who, from $A:B::C:D$ derives $A\sim B:B::C\sim D:D$ and explains that $\sim$ means "the difference of the magnitudes which are represented by them, without supposing the first to be the greater, as is the case when we write $A- -B$ by itself." It is found also in Wright,[7] Goodwin,[8] Hall and Knight,[9] and C. Smith.[10]

The algebra of Alexander[11] uses the sign $\curlyvee$ for "absolute difference." W. P. G. Bartlett[12] (Cambridge, Mass.) stated in 1859, "I have used a German notation $\backsim$, to denote the difference between $\dfrac{y}{p}$ and $\dfrac{y}{y_0}$, because p may fall on the other side of y_0."

488. On the European continent the signs $\sim$ and $\backsim$ came to be employed for the expression of absolute difference by a few authors outside of Germany. In Germany these symbols were widely used for another purpose, namely, to express geometric similarity. Praalder[13] uses for arithmetic difference an $\rightharpoondown$, looking like a letter f placed

[1] William Jones, *Synopsis palmariorum matheseos* (London, 1706), p. 248, 249.

[2] *Euclid's Elements . . . . Corrected by J. Barrow* (London, 1751), list of symbols.

[3] H. Clarke, *The Rationale of Circulation Numbers* (London, 1777), p. 207.

[4] James Thomson, *Treatise on Arithmetic* (18th ed.; Belfast, 1837), p. 101.

[5] B. Bridge, *The Three Conic Sections* (2d London ed.; New Haven, 1836), p. 9.

[6] [Pierce Morton], *Geometry, Plane Solid and Spherical* (London, 1830), p. 41.

[7] J. M. F. Wright, *Self-Examination in Euclid* (Cambridge, England, 1829).

[8] Harvey Goodwin, *Elementary Course in Mathematics, designed Principally for the Students of the University of Cambridge* (3d ed.; Cambridge 1869), p. 3.

[9] H. S. Hall and S. R. Knight, *Elementary Algebra* (ed. F. L. Sevenoak; New York, 1895), p. 374.

[10] C. Smith, *Elementary Algebra* (ed. Irving Stringham; New York, 1895), p. 18.

[11] *Synopsis algebraica, opus posthumum Johannis Alexandri, Bernatis-Helvetii* (London, 1693), p. 6.

[12] J. D. Runkle, *Mathematical Monthly*, Vol. I, No. 5 (Feb., 1859), p. 181.

[13] L. Praalder in his edition of the *Mathematische Voorstellen* of Ludolf van Keulen (Amsterdam, 1777), p. 205.

horizontally. The $\backsim$ for difference is found in Saverien,[1] Cagnoli,[2] and Da Cunha.[3] A Spaniard[4] writes $l\backsim L$, $1\underset{\sim}{+}B$, $45°\underset{\sim}{+}F$.

The use of = for absolute difference was noted in §§ 164, 177, 262.

489. *Other meanings of* $\backsim$ *and* $\sim$.—Argand[5] employed "$\sim a$" for our $a\sqrt{-1}$, and "$\backsim a$" for our $-a\sqrt{-1}$; Hudson and Lipka[6] use $\cong$ for "equals approximately," which is in line with the older usage of $\sim$ for *annähernd gleich* by Günther,[7] M. Cantor, and others. With Peano[8] $\sim$ signifies "not"; $\sim\epsilon$, "is not"; and $\sim=$, "is not equal."

490. The sign $\sim$ is used extensively to express equivalence. G. Eisenstein[9] said in 1844, "Sind f and f' aequivalent, welches ich so bezeichne: $f\sim f'$." The sign is used in this sense by A. Wiegand.[10] The notation $a_\nu \sim b_\nu$ was used by P. du Bois–Reymond[11] in the case of $\underset{\nu=+\infty}{\lim}\dfrac{a_\nu}{b_\nu}=a$, and by A. Pringsheim and J. Molk[12] when the series $\dfrac{a_1}{b_1},\dfrac{a_2}{b_2},\ldots,\dfrac{a_\nu}{b_\nu},\ldots$ diverges so as to have a superior limit and a distinct inferior limit, both finite and positive. André[13] attributes the use of $\sim$ for equivalence to Kronecker and his school. With G. H. Hardy and J. E.

[1] A. Saverien, *Dictionnaire universel de mathematique et de physique* (Paris 1753), art. "Caractere."

[2] Antonio Cagnoli, *Traité de Trigonométrie* ... Traduit de l'Italien par M. Chompré (Paris, 1786), p. 4.

[3] J. A. da Cunha, *Principios mathematicos* (Lisboa, 1790), p. 49.

[4] Dionisio Alcalá-Galciano, Capitan de Navío de la Real Armada, *Memoria sobre las observaciones de Latitud y Longitud en el Mar* (1796), p. 46.

[5] J. R. Argand, *Essai sur une manière de représenter les quantités imaginaires* (Paris, 1806). A. S. Hardy's translation (New York, 1881), p. 35.

[6] R. G. Hudson and J. Lipka, *A Manual of Mathematics* (New York, 1917), p. 68.

[7] S. Günther, "Die quadratischen Irrationalitäten der Alten," *Abhandlungen zur Geschichte der Mathematik*, Vol. IV (Leipzig, 1882), p. 11. Günther gives a reference to M. Cantor, "Gräko-indische Studien," *Zeitschr. f. Math. u. Phys.*, *Hist. Lit. Abth.*, Band XXII, S. 1 ff.

[8] G. Peano, *Lezioni di analsisi infinitesimale*, Vol. I (Torino, 1893), p. 10.

[9] G. Eisenstein in *Crelle's Journal*, Vol. XXVII (1844), p. 91.

[10] August Wiegand, *Algebraische Analysis und Anfangsgründe der Differential-Rechnung* (2d ed.; Halle, 1853), p. 44.

[11] P. du Bois–Reymond in *Annali di matematica pura ed applicata*, serie II, Vol. IV (1870–71), p. 339.

[12] A. Pringsheim and J. Molk in *Encyclopédie des scien. math.*, Tom. I, Vol. I (1904), p. 201, 202.

[13] Désire André, *Notations mathématiques* (Paris, 1909), p. 98; L. Kronecker, *Vorlesungen über Zahlentheorie*, Vol. I (Leipzig, 1901), p. 86.

Littlewood[1] "$f(n) \sim g(n)$ means that $\frac{f}{g} \to 1$ when $n \to \infty$." Bromwich[2] in presenting Poincaré's theory of asymptotic series places $\sim$ between the function and the series under consideration, to denote that "the series is asymptotic to the function." He writes also $a_n \sim kb_n$ when $a_n/b_n \to k > 0$. Similarly with W. B. Ford[3], $f(x) \sim a_0 + a_1/x + a_2/x^2 + \cdots\cdot$ represents symbolically the relation $\lim\limits_{x = +\infty} x^n[f(x) - (a_0 + a_1/x + a_2/x^2 + \cdots\cdot + a_n/x^n)] = 0; n = 0, 1, 2, 3, \ldots$.

491. *A few sporadic symbols.*—Kirkman[4] lets ($\pm$) denote "that the signs have to be properly affixed, after the permutations are written out."

Simpson[5] uses the symbolism, "If n, n, n, $\overset{|}{n}$, $\overset{||}{n}$, $\overset{|||}{n}$, &c. be a series of Terms in a decreasing Arithmetical Progression, whose common difference is $\dot n$. . . . . " Chauncey Lee[6] explains that ", $= = =$, the double line, drawn under a row of figures, shews that the operation is finished, and the answer stands over it."

492. *Signs for absolute value.*—There has been a real need in analysis for a convenient symbolism for "absolute value" of a given number, or "absolute number," and the two vertical bars introduced in 1841 by Weierstrass,[7] as in $|z|$, have met with wide adoption; it occurs in an article on the power series

$$F(x) = \sum_{\nu=-\infty}^{\nu=+\infty} A_\nu x^\nu \, ,$$

where x is a complex variable: "Ist dann r irgend eine bestimmte, innerhalb des Convergenzbezirks der Reihe liegende positive Grösse, so hat der absolute Betrag von $F(x)$, wenn man der Veränderlichen x

[1] G. H. Hardy and J. E. Littlewood, *Nachrichten von der K. Gesellschaft d. Wiss. zu Göttingen, Math.-Phys. Klasse* (1920), p. 34.

[2] T. J. I'A. Bromwich, *Introduction to the Theory of Infinite Series* (1908), p. 330. See also p. 11.

[3] W. B. Ford, *Studies on Divergent Series and Summability*, Michigan Science Series, Vol. II (New York, 1916), p. 22.

[4] T. P. Kirkman, *First Mnemonical Lessons in Geometry, Algebra and Trigonometry* (London, 1852), p. 166.

[5] Thomas Simpson, *Essays on . . . . Mathematics* (London, 1740), p. 87.

[6] Chauncey Lee, *The American Accomptant* (Lansingburgh, 1797), p. 64.

[7] K. Weierstrass, *Mathematische Werke*, Vol. I (Berlin, 1894), p. 67.

alle diejenigen Werthe beilegt, für welche $|x|=r$ ist, eine obere Grenze, die mit g bezeichnet werde: und es gilt der Satz: $|A_\mu|\leqq gr^{-\mu}$ für jeden ganzzahligen Werth von μ." This article was not printed at the time. Weierstrass used $|\ |$ again in an article read before the Berlin Academy of Sciences on December 12, 1859.[1] Weierstrass also employed the notation $|\ |$ for determinants,[2] and avoided confusion by the use of such phrases as "Die Determinante $|\omega_{a_y}|$."

André[3] states that the mark $\bar{z}$ has also been used to designate absolute value. The abbreviation[4] "mod z" from Argand's and Cauchy's word[5] "module" is not very convenient; nor is it desirable in view of the fact that "mod" has been widely adopted, since the time of Gauss, in the theory of numbers.

. 493. *Zeroes of different origin.*—Riabouchinski[6] gives a new extension of the notion of number leading to transfinite numbers differing from those of G. Cantor; he represents the ordinary symbolism $\lim\limits_{n\to\infty}\dfrac{a}{n}=$ o by the symbol $\dfrac{a}{J}=\pm$ o. As a is taken arbitrarily, he assumes the existence of an infinity of zeroes of different origin; if a is positive, the corresponding zero is marked $+$o or o, if negative the zero is marked $-$o. Let $\dfrac{1}{J}$ be ȯ, and J the inverse of passage to the limit, then $J\cdot\dfrac{5}{J}=J\cdot$ o$=5$. When the origin of a zero is not defined, he writes $J\cdot\pm$o$=a$, where a is arbitrary. He has $a\cdot$o$=\pm$o, $J=\dfrac{1}{\text{ȯ}}$, $J\cdot a=\dfrac{a}{\text{ȯ}}, a\cdot$o$=\dfrac{ab}{J}=ab\cdot$ȯ$=\pm$o, $1+J=J+1$. He designates by $\pm\overline{}$ the operation of the return to relative values or the inverse of the passage to absolute values. Accordingly, he lets $\overline{-1}$ or j represent the result of an impossible return to a relative value. Then, $|j|=-1$, $j=\pm\overline{-1}$.

[1] K. Weierstrass, *op. cit.*, Vol. I, p. 252.

[2] K. Weierstrass, *op. cit.*, Vol. IV, p. 524.

[3] D. André, *op. cit.*, p. 86.

[4] See G. Peano, *Formulaire mathématique*, Vol. IV (1903), p. 93.

[5] A. L. Cauchy, *Exercices de mathématiques* (Paris, 1829), Tome IV, p. 47; *Œuvres* (2d ser.), Vol. IX, p. 95.

[6] D. Riabouchinski, "Sur le calcul des valeurs absolues," *Comptes rendus du congrès international des mathématiciens* (Strasbourg, 22–30 Septembre 1920; Toulouse, 1921), p. 231–42.

494. *General combinations between magnitudes or numbers.*—In the establishment of a combination (*Verknüpfung*) between magnitudes or numbers, symbols like $a \frown b$ and $a \smile b$ were used by H. G. Grassmann[1] in 1844, where $\frown$ indicated an affirmative proposition or thesis, and $\smile$ a corresponding negative proposition or lysis. A second thesis was marked $\frown$. The two symbols for thesis and lysis were easily confused, one with the other, so that his brother R. Grassmann[2] used a small circle in place of the $\frown$; he used also a large circle with a dot at its center $\odot$, to mark a second relationship, as in $(aoe)b = ab \odot ae$. Likewise Stolz[3] used the small circle for the thesis and H. G. Grassmann's $\smile$ for the lysis. Hankel[4] in 1867 employed for thesis and lysis the rather cumbrous signs Θ (a, b) and $\lambda(a, b)$. R. Bettazzi[5] writes $S(a, b)$ and the inverse $D(a, b)$. The symbol $aob = c$ is used by Stolz and Gmeiner[6] to designate "*a* mit *b* ist *c*"; the small circle indicates the bringing of a and b into any relationship explained by c; the $=$ expresses here "is explained by." When two theses[7] are considered, the second is marked $\odot$; a second lysis is marked $\smile$. When $xob = a$, $boy = a$ are both single valued, they are marked respectively $x = a \smile b$, $y = a \smile b$. The small circle is used also by Huntington.[8] He says: "A system composed of a class K and a rule of combination o we shall speak of as a system (K, o)." When different classes come under consideration, different letters K, C represent them, respectively. When different rules of combination are to be indicated, a dot or small $+$ or $<$ is placed within the circle. Thus Huntington speaks of the "system $(K, C, \oplus, \odot, \oslash)$." Huntington also writes $A^{[n]}$ to represent $AoAo \ldots \ldots oA$, where n is a positive element. Dickson[9] marks by $\oplus$ the operation of addition, by $\odot$ that of multiplication, and by $\bigcirc$ the special operation of scalar multiplication.

[1] H. G. Grassmann, *Die lineale Ausdehnungslehre* (1844; 2d. ed., 1878), §§ 5 f.

[2] Robert Grassmann, *Die Grössenlehre* (Stettin, 1872), p. 27, 37.

[3] Otto Stolz, *Allgemeine Arithmetik* (Leipzig, 1885), p. 26, 28.

[4] H. Hankel, *Theorie der complexen Zahlensysteme* (Leipzig, 1867), p. 18–34.

[5] R. Bettazzi, *Theoria delle grandezze* (Pisa, 1890), p. 7, 11.

[6] O. Stolz und J. A. Gmeiner, *Theoretische Arithmetik* (Leipzig), Vol. I (2d ed., 1911), p. 7, 51–58.

[7] O. Stolz und J. A. Gmeiner, *op. cit.*, p. 63, 66.

[8] J. W. A. Young's *Monographs* (New York, 1911), p. 164, 186–99. See also E. V. Huntington in *Trans. Amer. Math. Soc.*, Vol. VI (1905), p. 209–29.

[9] L. E. Dickson, *Algebras and Their Arithmetics* (Chicago, 1923), p. 9.

SYMBOLISMS FOR IMAGINARIES AND VECTOR ANALYSIS

495. *Symbols for the square root of minus one.*—Nicolas Chuquet in his manuscript work of 1484, *Le Triparty*, solved $4+x^2=3x$ and obtained $x=\frac{3}{2}\pm\sqrt{2\frac{1}{4}-4}$ (expressed here in modern symbols). As $2\frac{1}{4}-4$ is negative, he pronounced the root impossible. In his own words: "Et pourtant que .2.¼. qui est la multiplicacion du moyen est moindre que le precedent Il senβ que ceste raiβ est impossible."[1] Chuquet did not admit imaginary roots as valid, and in the solution of the equation quoted, does not use symbolism for square root. In general, he indicates square root as in "$\mathcal{R}^2.21$" for $\sqrt{21}$.

The earliest mathematician seriously to consider imaginaries and to introduce them in the expression for the roots of equations was H. Cardan. He uses[2] the designation "$\mathcal{R}.\tilde{m}.$," that is, *radix minus*, to express the square root of a negative number. The marking of the imaginary roots in the collected works of 1663 is not quite the same as in the *Ars magna* of 1545. In 1545[3] he gives the following multiplication:

$$5p:\mathcal{R}m:15 \qquad\qquad 5+\sqrt{-15}$$
$$5m:\mathcal{R}m:15 \qquad\qquad 5-\sqrt{-15}$$

$$25m:m:15 \ \tilde{q}d \ est \ 40 \qquad\qquad 25-(-15)=40$$

In 1663 this passage is given thus:

$$5.\tilde{p}.\mathcal{R}.\tilde{m}.15.$$
$$5.\tilde{m}.\mathcal{R}.m.15.$$

$$25m.m.15.quad.est \ 40.$$

It must be noted that Cardan doubted the validity of arithmetical operations performed upon imaginary quantities.

496. Apparently it was Rafaele Bombelli who undertook to "demonstrate as valid in every case the formula of Scipio del Ferro" ("dimostrare valida in ogni caso la formula di Scipione Dal Ferro").[4] Bombelli's algebra was issued in manuscript edition[5] about 1550 and

[1] *Bullettino Boncompagni*, Vol. XIII (1880), p. 805.

[2] H. Cardan, *Ars magna* (Nuremberg, 1545), chap. xxxvii; (nouv. éd.; Bâle, 1570), p. 130; *Opera* (Lyon, 1663), p. 287.

[3] J. Tropfke, *op. cit.*, Vol. II (2d ed., 1921), p. 18; Vol. III (2d ed., 1922), p. 135.

[4] See E. Bortolotti in *Scientia*, Vol. XXXIII (1923), p. 389.

[5] A copy of the manuscript edition has only recently been found by E. Bortolotti in the *Biblioteca Comunale of Bologna* (Codex B, 1569).

in printed form in 1572 and 1579. In the manuscript edition Bombelli writes the equation $x^2+20=8x$ in the form $|{\overset{2}{\smile}}p.20$ à $8{\smile}$, and writes the root $4+\sqrt{-4}$ in the form $4.p.R[0\tilde{m}.4]$ and the root $4-\sqrt{-4}$ in the form $4.\tilde{m}R[0\tilde{m}.4]$. Similarly, he writes the cubic $x^3=15x+4$ in the form $|{\overset{3}{\smile}}.$ à $15{\smile}.p.4{\overset{0}{\smile}}$ and its roots $R^3[2.p.R[0\tilde{m}.121]].p$ $.R^3[2\tilde{m}R[0\tilde{m}.121]]$.

Further on, in his manuscript edition, Bombelli introduces a new phraseology and a new symbolism for the imaginary.[1] He says: "As one cannot call it either plus, or minus, I shall call it 'più di meno' when it is added, and 'meno di meno' when it is subtracted." This "più de meno" appears as the contraction for *più radice di meno;* the "meno di meno" for *meno radice di meno.* The "di meno" was abbreviated *d.m.* and stands for our $\sqrt{-1}$. Bombelli wrote "*p.dm.2· via.p.dm.* 1 *egual.* $\tilde{m}$ 2," which means $(+2i)(+i)=-2$; he wrote "1 *diviso da p. dm.* 1 *egual. m. dm.* 1," which means $1\div i=-i$.

That Bombelli was far in advance of his time on the algorism of imaginaries is strikingly evident, when one recalls that as late as the eighteenth century gross errors were committed. Even in Euler's *Elements of Algebra*[2] one finds $\sqrt{-2}.\sqrt{-3}=\sqrt{6}$, $\sqrt{-1}.\sqrt{-4}=2$, $\sqrt{+3}\div\sqrt{-3}=\sqrt{-1}$, $1\div\sqrt{-1}=\sqrt{-1}$. Euler was blind when he prepared this book, and he is probably not responsible for printers' errors.

497. Albert Girard in 1629 found the roots of the equation $x^4=4x-3$, using the symbolism $\sqrt{-2}$, in the following passage: "Si 1 (4) est esgale à 4 (1) −3, alors les quatre factions seront 0, 0, 4, 3, partant les quatre solutions seront 1, 1, $-1+\sqrt{-2}, -1-\sqrt{-2}$."[3]

He speaks of three kinds of roots of equations, namely, those "qui sont plus que rien; d'autres moins que rien; & d'autres envelopées, comme celles qui ont des $\sqrt{-}$, comme des $\sqrt{-3}$, ou autres nombres semblables." In Wallis[4] one finds the signs $\sqrt{-3}$, $\sqrt{-e}$, also in Isaac Newton[5] $\sqrt{-2}$. Before Euler the sign $\sqrt{-1}$, as distinct from $\sqrt{-a}$, seldom, if ever, occurs.

[1] See E. Bortolotti, *op. cit.*, p. 391, 394.

[2] Leonhard Euler, *Vollständige Anleitung zur Algebra*, Erster Theil (St. Petersburg, 1770), §§ 148, 149, p. 87, 88.

[3] Albert Girard, *Invention nouvelle en l'Algebre* (Amsterdam, 1629), fol. *F A* and *F B*.

[4] John Wallis, *Treatise of Algebra* (1685), p. 180.

[5] Sir Isaac Newton, *Universal Arithmetick* (London, 1728), p. 193.

Kästner[1] was among the first to represent a pure imaginary by a letter; for $\sqrt[\alpha]{-1}$ he wrote π where a is an even integer. He speaks of "den unmöglichen Factor $b+f\sqrt[\alpha]{-1}$ für den ich $b+f\pi$ schreiben will." Before this he had used π for $\sqrt{a}$, where a was positive.[2]

Caspar Wessel[3] in an *Essay* presented to the Danish Academy in 1797 designates "by $+1$ positive rectilinear unity, by $+\epsilon$ another unity, perpendicular to the first and having the same origin," and then writes "$\sqrt{-1}=\epsilon$"; "cos $v+\epsilon$ sin v."

498. It was Euler who first used the letter i for $\sqrt{-1}$. He gave it in a memoir presented in 1777 to the Academy at St. Petersburg, and entitled "De formulis differentialibus etc.," but it was not published until 1794 after the death of Euler.[4]

As far as is now known, the symbol i for $\sqrt{-1}$ did not again appear in print for seven years, until 1801. In that year Gauss[5] began to make systematic use of it; the example of Gauss was followed in 1808 by Kramp.[6]

499. Argand,[7] in his *Essai* (Paris, 1806), proposes the suppression of $\sqrt{-1}$ and the substitution for it of particular signs similar to the signs of operation $+$ and $-$. He would write $\sim$ for $+\sqrt{-1}$ and $\sim$ for $-\sqrt{-1}$, both indicating a rotation through 90°, the former

[1] A. G. Kästner, *Anfangsgründe der Analysis endlicher Grössen* (Göttingen, 1760), p. 133.

[2] A. G. Kästner, *op. cit.*, p. 117.

[3] Caspar Wessel, *Essai sur la représentation analytique de la direction* (Copenhague, 1897), p. 9, 10, 12. This edition is a translation from the Danish; the original publication appeared in 1799 in Vol. V of the *Nye Samling af det Kongelige Danske Videnskabernes Selskabs Skrifter*.

[4] The article was published in his *Institutiones calculi integralis* (2d ed.), Vol. IV (St. Petersburg, 1794), p. 184. See W. W. Beman in *Bull. Amer. Math. Soc.*, Vol. IV (1897–98), p. 274. See also *Encyclopédie des scien. math.*, Tom. I, Vol. I (1904), p. 343, n. 60.

[5] K. F. Gauss, *Disquisitiones arithmeticae* (Leipzig, 1801), No. 337; translated by A. Ch. M. Poullet-Delisle, *Recherches arithmétiques* (Paris, 1807); Gauss, *Werke*, Vol. I (Göttingen, 1870), p. 414.

[6] C. Kramp, *Élémens d'arithmétique universelle* (Cologne, 1808), "Notations."

[7] J. R. Argand, *Essai sur une manière de représenter les quantités imaginaires dans les constructions géométriques* (Paris, 1806; published without the author's name); a second edition was brought out by G. J. Hoüel (Paris, 1874); an English translation by A. S. Hardy appeared in New York in 1881. We quote from Hardy's edition, p. 35, 45.

in the positive direction, the latter in the negative. De Moivre's formula appears in Argand's *Essai* in the garb

$$\cos na \sim \sin na = (\cos a \sim \sin a)^n .$$

In 1813 Argand[1] interpreted $\sqrt{-1}^{\sqrt{-1}}$ as a sign of a unit vector perpendicular to the two vectors 1 and $\sqrt{-1}$. J. F. Français at first entertained the same view, but later rejected it as unsound.

W. R. Hamilton[2] in 1837 states that one may write a "couple" (a_1, a_2) thus: $(a_1, a_2) = a_1 + a_2 \sqrt{-1}$. Accordingly, $(0, 1)$ represents $\sqrt{-1}$. Similarly, Burkhardt[3] wrote the *Zahlenpaar* (a, b) for $a + bi$.

In manuscripts of J. Bolyai[4] are used the signs $+$, $-$, $+$, $\div$ for $+$, $-$, $\sqrt{-1}$, $-\sqrt{-1}$, but without geometrical interpretation.

Français[5] and Cauchy[6] adopted a_a as the notation for a vector. According to this, the notation for $\sqrt{-1}$ would be $1_{\frac{\pi}{2}}$. However, in 1847 Cauchy began to use i in a memoir on a new theory of imaginaries[7] where he speaks of " ... le signe symbolique $\sqrt{-1}$, auquel les géomètres allemands substituent la lettre i," and continues: "Mais il est évident que la théorie des imaginaires deviendrait beaucoup plus claire encore et beaucoup facile à saisir, qu'elle pourrait être mise à la portée de toutes les intelligences, si l'on parvenait à réduire les expressions imaginaires, et la lettre i elle-même, a n'être plus que des quantités réelles." Using in this memoir the letter i as a *lettre symbolique*, substituted for the variable x in the formulas considered, he obtains an *équation imaginaire*, in which the symbolic letter i should be considered as a real quantity, but indeterminate. The symbolic letter i, substituted for x in an integral function $f(x)$, indicates the value received, not by $f(x)$, but by the rest resulting from the alge-

---

[1] J. R. Argand, *Annales Math. pures appl.*, Vol. IV (1813–14), p. 146. See also *Encyclopédie des scien. math.*, Tom. I, Vol. I (1904), p. 358, n. 88.

[2] W. R. Hamilton, *Trans. Irish Acad.*, Vol. XVII (1837), p. 417, 418; *Lectures on Quaternions* (Dublin, 1853), Preface, p. 10.

[3] H. Burkhardt, *Einführung in die Theorie der analytischen Funktionen* (Leipzig, 1903), p. 3.

[4] See P. Stäckel, *Math. Naturw. Ber. Ungarn*, Vol. XVI (1899), p. 281. See also *Encyclopédie des scien. math.*, Tom. I, Vol. I (1904), p. 346, n. 63.

[5] J. F. Français in *Annalen Math. pures appl.*, Vol. IV (1813–14), p. 61.

[6] A. L. Cauchy in *Comptes rendus Acad. sci.* (Paris), Vol. XXIX (1849), p. 251; *Œuvres* (1st ser.), Vol. XI (Paris, 1899), p. 153.

[7] A. L. Cauchy, *Comptes rendus*, Vol. XXIV (1847), p. 1120 = *Œuvres complètes* (1st ser.), Vol. X, p. 313, 317, 318.

braic division of $f(x)$ by x^2+1. He obtains in a particular case $i^2+1=0$, and equations like $(a+bx)(c+dx)=ac+bdx^2+(ad+bc)x$ which gives, when one writes i for x $(a+bi)(c+di)=ac-bd+(ad+bc)i$. Cauchy[1] used the letter i for $\sqrt{-1}$ in papers of 1851 and later.

That the symbol i for $\sqrt{-1}$ was somewhat slow in securing general adoption is illustrated by the fact that as alert a writer as A. de Morgan used k for $\sqrt{-1}$ in his *Differential and Integral Calculus* of 1842 (p. 118). J. M. Peirce's form is shown in Figure 108 of § 401. A. C. Dixon[2] writes ι for $\sqrt{-1}$.

500. With the development of electromagnetism physicists and electricians, C. P. Steinmetz[3] for instance, let the letter i stand for strength of an electric current. In presenting the mathematical theory of electromagnetism they avoided confusion resulting from the double use of i by letting j represent $\sqrt{-1}$. For example, in a recent *Manual*,[4] one finds the notation $x = a + jb$, where $j = \sqrt{-1}$ and x is a vector. On the other hand, Bryan[5] introduces j for $\log -1$, in aeroplane calculations of the products of quantities, some positive and some negative, by the use of logarithmic tables; he writes $\log -2 = j0\cdot30103$, $\log -0\cdot02 = j\bar{2}\cdot30103$. An even number of j's yields a positive product. For $\sqrt{-1}$ Bryan uses i or ι.

501. *De Morgan's comments on* $\sqrt{-1}$.—The great rôle that $\sqrt{-1}$ has played in the evolution of algebraic theory is evident from the statements of Augustus de Morgan. In 1849 he spoke of "the introduction of the unexplained symbol $\sqrt{-1}$,"[6] and goes on to say: "The use, which ought to have been called *experimental*, of the *symbol* $\sqrt{-1}$, under the name of an *impossible* quantity, shewed that; come how it might, the intelligible results (when such things occurred) of the experiment were always true, and otherwise demonstrable. I am now going to try some new experiments." Consider also the following passage from De Morgan: "As soon as the idea of acquiring symbols

[1] See E. B. Jourdain, *Bibliotheca mathematica* (3d ser., Vol. VI (1905–6), p. 191, 192 n.; A. Brill and M. Noether, "Entwicklung der Theorie der algebr. Funktionen," *Jahresbericht d. d. Mathematiker Vereinigung*, Vol. III (1894), p. 194.

[2] A. C. Dixon, *Elliptic Functions* (London, 1894), p. 5.

[3] C. P. Steinmetz, *Alternating Current Phenomena* (New York, 1908), p. 34 ff.

[4] R. G. Hudson and J. Lipka, *A Manual of Mathematics* (New York, 1917), p. 68.

[5] G. H. Bryan, *Mathematical Gazette*, Vol. VIII (1917), p. 220.

[6] Augustus de Morgan, *Trigonometry and Double Algebra* (London, 1849), p. 41

and laws of combination, without giving meaning, has become familiar, the student has the notion of what I will call a *symbolic calculus;* which, with certain symbols and certain laws of combination, is *symbolic algebra:* an art, not a science; and an apparently useless art, except as it may afterwards furnish the grammar of a science. The proficient in a symbolic calculus would naturally demand a supply of meaning."[1] And again: "The first who used algebraical symbols in a general sense, Vieta, concluded that subtraction was a defect, and that expressions containing it should be in every possible manner avoided. *Vitium negationis,* was his phrase. Nothing could make a more easy pillow for the mind, than the rejection of all which could give any trouble; . . . . The next and second step, ... consisted in treating the results of algebra as necessarily true, and as representing some relation or other, however inconsistent they might be with the suppositions from which they were deduced. So soon as it was shewn that a particular result had no existence as a quantity, it was permitted, by definition, to have an existence of another kind, into which no particular inquiry was made, because the rules under which it was found that the new symbols would give true results, did not differ from those previously applied to the old ones. . . . . When the interpretation of the abstract negative quantity shewed that a part at least of the difficulty admitted a rational solution, the remaining part, namely that of the square root of a negative quantity, was received, and its results admitted, with increased confidence."[2]

502. *Notation for vector.*—Argand[3] adopted for a directed line or vector the notation $\overline{AB}$. Möbius[4] used the designation AB, where A is the origin and B the terminal point of the vector. This notation has met with wide adoption. It was used in Italy by G. Bellavitis,[5] in France by J. Tannery,[6] by E. Rouché and Ch. de Comberousse,[7]

---

[1] A. de Morgan, *op. cit.,* p. 92.

[2] A. de Morgan, *op. cit.,* p. 98, 99.

[3] J. R. Argand, *Essai sur une manière de représenter les quantités imaginaires dans les constructions géométriques* (Paris, 1806).

[4] A. F. Möbius, *Der barycentrische Calcul* (Leipzig, 1827); *Werke,* Vol. I (Leipzig, 1885), p. 25.

[5] G. Bellavitis, "Metodo delle Equipollenze," *Annali delle scienze del regno Lombardo-Veneto* (Padova), Vol. VII (1837), p. 243–61.

[6] J. Tannery, "Deux leçons de cinématique," *Ann. École Normal* (3d ser.), Vol. III (1886), p. 43–80.

[7] E. Rouché et Ch. de Comberousse, *Traité de géométrie* (7th ed.), Vol. I (Paris, 1900), p. 216.

and P. Appell.[1] On the other hand, W. R. Hamilton, P. G. Tait, and J. W. Gibbs designated vectors by small Greek letters. D. F. Gregory used the symbol $a(.)$ to "represent a straight line, as the result of transferring a point through a given space in a constant direction."[2] We have already called attention to the designation a_a of Français and Cauchy, which is found in some elementary books of the present century; Frank Castle[3] lets $A\,a$ designate a vector. Frequently an arrow is used in marking a vector, thus $\overrightarrow{R}$, where R represents the length of the vector. W. Voigt[4] in 1896 distinguished between "polar vectors" and "axial vectors" (representing an axis of rotation). P. Langevin[5] distinguishes between them by writing polar vectors $\overrightarrow{a}$, axial vectors $\overset{\smile}{a}$. Some authors, in Europe and America, represent a vector by bold-faced type a or b which is satisfactory for print, but inconvenient for writing.

Cauchy[6] in 1853 wrote a radius vector $\bar{r}$ and its projections upon the co-ordinate axes $\bar{x}$, $\bar{y}$, $\bar{z}$, so that $\bar{r}=\bar{x}+\bar{y}+\bar{z}$. Schell[7] in 1879 wrote $[AB]$; some others preferred (AB).

503. Wessel[8] in 1797 represented a vector as a sum $x+\eta y+ez$, with the condition that $\eta^2=-1$, $e^2=-1$. H. G. Grassmann[9] represents a point having x, y, z for its co-ordinates by $p=v_1e_1+v_2e_2+v_3e_3$, where the e_1, e_2, e_3 are unit lengths along the co-ordinate axes. W. R. Hamilton[10] wrote $\rho=ix+jy+kz$, where i, j, k are unit vectors per-

---

[1] P. Appell, *Traité de mécanique rationnelle* (3d ed.), Vol. I (Paris, 1909), p. 3.

[2] D. F. Gregory, *Cambridge Math. Jour.*, Vol. II (1842), p. 1; *Mathematical Writings* (Cambridge, 1865), p. 152.

[3] Frank Castle, *Practical Mathematics for Beginners* (London, 1905), p. 289. We take these references from *Encyclopédie des scien. math.*, Tom. IV, Vol. II (1912), p. 4.

[4] W. Voigt, *Compendium der theoretischen Physik*, Vol. II (Leipzig, 1896), p. 418–801.

[5] *Encyclopédie des scien. math.*, Tom. IV, Vol. II (1912), p. 25.

[6] A. L. Cauchy, *Comptes rendus Acad. sc.*, Vol. XXXVI (Paris, 1853), p. 75; *Œuvres* (1st ser.), Vol. XI (Paris, 1899), p. 443.

[7] W. Schell, *Theorie der Bewegung der Kräfte* (2d ed.), Vol. I (Leipzig, 1879).

[8] C. Wessel, "Om directionens analytiske Betegning," *Nye Samling af det Kongelige Danske Videns-Kabernes Selskabs Skrifter* (2), Vol. V (1799), p. 469–518. French translation (Copenhagen, 1897), p. 26. See *Encyclopédie des scien. math.*, Tom. IV, Vol. II (1912), p. 12.

[9] H. G. Grassmann, *Ausdehnungslehre von 1844* (2d ed.; Leipzig, 1878), p. 128, § 92.

[10] W. R. Hamilton, *Lectures on Quaternions* (Dublin, 1853), p. 59 (§ 65), p. 105 (§ 101).

pendicular to one another, such that $i.j = -j.i = k$, $j.k = -k.j = i$, $k.i = -ik = j$.

Stringham[1] denoted $\cos \beta + i \sin \beta$ by "cis β," a notation used also by Harkness and Morley.[2] Study[3] represented vectorial quantities by a biplane, or two non-perpendicular planes, the initial plane ϕ, and the final plane ϕ'. The biplane is represented by $\Re_\phi^{\phi'}$. Similarly, Study defines a "motor" by two non-perpendicular straight lines ($\mathfrak{X}$ the initial line, and $\mathfrak{Y}$ the final), and represents the motor by the symbol $\mathfrak{M}_{\mathfrak{X}}^{\mathfrak{Y}}$.

504. *Length of vector.*—The length of a vector was marked by Bellavitis AB (the same designation as for vector), by H. G. Grassmann[4] $\sqrt{R^2}$, by W. R. Hamilton[5] TR (i.e., tensor of the vector R), by R. Gans[6] $|R|$, the Weierstrassian symbol for absolute value. Later Gans discarded $|R|$ because $|R|$ is a function of R and functional symbols should precede or follow the entity affected, but $|R|$ does both.

505. *Equality of vectors.*—L. N. M. Carnot[7] employed $\doteq$ as *signe d'équipollence*, practically a sign of identity. "Si les droites $\overline{AB}$, $\overline{CD}$ concurrent au point E, j'écrirai $\overline{AB} \cdot \overline{CD} \doteq E$," where $\overline{AB} \cdot \overline{CD}$ means the point of intersection of the two lines.

To express the equality of vectors the sign $=$ has been used extensively. Möbius[8] had a few followers in employing $\equiv$; that symbol was used by H. G. Grassmann[9] in 1851. Bellavitis[10] adopted the astronomical sign *libra* $\libra$, H. G. Grassmann[11] in 1844 wrote $\doteqdot$, which was also chosen by Voigt[12] for the expression of complete equality. Hamil-

[1] Irving Stringham, *Uniplanar Algebra* (San Francisco, 1893), p. xiii, p. 101.

[2] J. Harkness and F. Morley, *Theory of Analytic Functions* (London, 1898), p. 18, 22, 48, 52, 170.

[3] E. Study, *Geometrie der Dynamen* (Leipzig, 1903), p. 30, 51; *Encyclopédie des scien. math.*, Tom. IV, Vol. II (1912), p. 55, 59.

[4] H. Grassmann, *Werke*, Vol. I¹ (Leipzig, 1894), p. 345; Vol. I² (1896), p. 118.

[5] W. R. Hamilton, *Cambr. and Dublin Math. Jour.*, Vol. I (1846), p. 2.

[6] R. Gans, *Einführung in die Vector Analysis* (Leipzig, 1905), p. 5.

[7] L. N. M. Carnot, *Géométrie de position* (Paris, 1803), p. 83, 84.

[8] A. F. Möbius, *Der baryc. Calcul* (1827), § 15; *Werke*, Vol. I, p. 39.

[9] *Crelle's Journal*, Vol. XLII, p. 193–203; Hermann Grassmann's *Gesammelte math. und physik. Werke*, Band II (ed. F. Engel; Leipzig, 1904), p. 90.

[10] G. Bellavitis, *op. cit.*, p. 243–61.

[11] Hermann Grassmann, *Gesammelte . . . . Werke*, Band I; *Ausdehnungslehre* (von 1844), p. 67.

[12] W. Voigt, *Nachricht. K. Gesellsch. d. Wissensch. z. Göttingen* (1904), p. 495–513.

ton[1] in 1845 wrote, ". . . . the symbolic equation, $D-C=B-A$, may denote that the point D is ordinarily related (in space) to the point C as B is to A, and may in that view be also expressed by writing the *ordinal analogy*, $D..C::B..A$; which admits of *inversion* and *alternation*." Peano[2] employed the symbolism $a-b=c-d$.

506. *Products of vectors.*—Proceeding to the different products of vectors, one observes that there has been and is great variety of notations. W. R. Hamilton assigned to two vectors only one product $\rho\rho'$, which is a quaternion and is the sum of two parts, the scalar of the quaternion $S\rho\rho'$, and the vector part of the quaternion $V\rho\rho'$.

H. G. Grassmann developed several products, of which the "internal product" or scalar product (the same as Hamilton's $S\rho\rho'$) and the "external product" or vector product (the same as Hamilton's $V\rho\rho'$) occupy a central place in vector analysis. Grassmann[3] represented in 1846 the scalar product by $a\times b$; in 1862[4] the scalar product by $[u|v]$ and in 1844 and 1862 the vector product by $[uv]$.

Résal[5] wrote for the scalar product $a\times b$, as did also Peano[6] in 1899, and later Burali-Forti and Marcolongo.[7] At an earlier date, 1888, Peano had written the scalar product $u|v$. Burali-Forti and Marcolongo give the vector product in the form $u\wedge v$. The simple form uv for the scalar product is employed by Somov,[8] Heaviside,[9] Föppl,[10] Ferraris.[11] Heaviside at one time followed Hamilton in writing the vector product, Vuv. Gibbs[12] represented the scalar product, called also "dot product," by $u.v$, and the vector product by $u\times v$; Gibbs's $u.v = -Suv$ of Hamilton.

[1] W. R. Hamilton, *Cambr. and Dublin Math. Jour.*, Vol. I (1846), p. 47.

[2] G. Peano, *Formulaire mathématique*, Vol. IV (1903), p. 253, 254.

[3] H. G. Grassmann, *Geometrische Analyse* (Leipzig, 1847); *Werke*, Vol. I[1] (Leipzig, 1894), p. 345.

[4] H. Grassmann, *Werke*, Vol I[1], p. 345; Vol. I[2], p. 56, 112.

[5] H. Résal, *Traité de cinématique pure* (Paris, 1862).

[6] G. Peano, *Formulaire de mathématiques*, Vol. II (Turin, 1899), p. 156.

[7] C. Burali-Forti et R. Marcolongo, *Elementi di calcolo vettoriale* (Bologna, 1909), p. 31. See *Encyclopédie des scien. math.*, Tom. IV, Vol. II, p. 14, 22.

[8] P. Somov, *Vector Analysis and Its Applications* (Russian) (St. Petersburg, 1907).

[9] O. Heaviside, *Electrical Papers*, Vol. II (London, 1892), p. 5.

[10] Föppl, *Geometrie der Wirbelfelder* (1897).

[11] G. Ferraris, *Lezioni di elettratecnica*, Kap. I (1899).

[12] J. W. Gibbs's *Vector Analysis*, by E. B. Wilson (New York, 1902), p. 50.

H. A. Lorentz[1] writes the scalar product $(u.v)$ and the vector product $[u.v]$; Henrici and Turner[2] adopted for the scalar product (u,v); Heun[3] has for the scalar product $\overline{u}\,\overline{v}$, and $\overline{uv}$ for the vector product; Timerding and Lévy, in the *Encyclopédie*,[4] use $u|v$ for the scalar product, and $u\times v$ for the vector product. Macfarlane at one time suggested cos ab and sin ab to mark scalar and vector products, to which Heaviside objected, because these products are more primitive and simpler than the trigonometric functions. Gibbs considered also a third product,[5] the *dyad*, which was a more extended definition and which he wrote uv without dot or cross.

507. *Certain operators.*—W. R. Hamilton[6] introduced the operator $\nabla = i\dfrac{d}{dx} + j\dfrac{d}{dy} + k\dfrac{d}{dz}$, a symbolic operator which was greatly developed in the later editions of P. G. Tait's *Treatise on Quaternions*. The sign ∇ was called "nabla" by Heaviside, "atled" by others, which is "delta," Δ, reversed. It is often read "del." The vector $-\nabla\phi$ is sometimes called the "gradient"[7] of the scalar ϕ and is represented by Maxwell and Riemann-Weber, as grad ϕ. The form $-S\nabla a$ is the quaternion representation of what Abraham and Langevin call div $\overrightarrow{a} = \dfrac{\partial a_x}{\partial x} + \dfrac{\partial a_y}{\partial y} + \dfrac{\partial a_z}{\partial z}$, and what Gibbs marked $\nabla.a$, and Heaviside ∇a. The mark $V\nabla a$, used by Tait, is Gibbs's $\nabla\times a$, Heaviside's "curl a," Wiechert's "Quirl a," Lorentz' "Rot a," Voigt's "Vort a," Abraham and Langevin's[8] "Rot $\overrightarrow{a}$"; Clifford's "divergence" of u, marked *div u* or *dv u*, is the $-S\nabla u$ of Hamilton[9] and the $\nabla.x$ of Gibbs.[10]

[1] See *Encyclopédie des scien. math.*, Tom. IV, Vol. II, p. 14, 22.

[2] O. Henrici and G. C. Turner, *Vectors and Rotors* (London, 1903).

[3] K. Heun, *Lehrbuch der Mechanik*, Vol. I (Leipzig, 1906), p. 14.

[4] *Encyclopédie des scien. math.*, Tom. IV, Vol. II (1912), p. 14, 22.

[5] See the Gibbs-Wilson *Vector-Analysis* (1901), chap. v.

[6] W. R. Hamilton, *Lectures on Quaternions* (1853), p. 610. The rounded letter ∂, to indicate that the derivatives in this operator are partial, is found in A. McAulay's *Octonions*, Cambridge, 1898, p. 85, and in C. J. Joly's *Quaternions*, London, 1905, p. 75.

[7] See, for instance, the *Encyclopédie des scien. math.*, Tom. IV, Vol. V, p. 15, 17, 18. Fuller bibliographical references are given here. C. Burali-Forti in *L'Enseignement mathématique*, Vol. XI (1909), p. 41.

[8] See, for instance, the *Encyclopédie des scien. math.*, Tom IV, Vol. V, p. 15, 17, 18, where fuller bibliographical references are given.

[9] C. Burali-Forti et R. Marcolongo, *op. cit.*, Vol. XI, p. 44.

[10] For additional historical details about ∇ see J. A. Schouten, *Grundlagen der Vector- und Affinoranalysis* (1914), p. 204–13.

508. *Rival vector systems.*—W. R. Hamilton and H. G. Grassmann evolved their algebras about the same time (1843, 1844). Hamilton's quaternions received the attention of Tait and some other mathematicians in Great Britain; there it was applied to physical problems. Grassmann's *Ausdehnungslehre* was obscurely described and needed interpretation, such as it received from Victor Schlegel in the years 1872–75; even after that it spread very slowly and found little use in applied mathematics.

In America quaternions engaged the attention of mathematicians at Harvard University and Yale University. Gibbs of Yale proceeded to modify the quaternion system. Like Arthur Cayley and William Thomson (Lord Kelvin) in Great Britain, he felt that quaternions were not as serviceable a tool of research as had been claimed. Gibbs rejected the quaternion concept, introduced two products of vectors in the place of Hamilton's one product, and built up a system called "vector analysis." Gibbs's was the first important abandonment of Hamilton's great creation. In 1881 Gibbs[1] began to issue his *Elements of Vector Analysis, arranged for the Use of Students of Physics* (New Haven, 1881–84) (lithographed issue) in which he introduced his notation $a.\beta$, $a \times \beta$, $a \times \beta . \gamma$, $(a.\beta)\gamma$, $a \times [\beta \times \gamma]$, using small Greek letters for vectors and small Latin letters for scalars. Tait in the Preface to the third edition of his *Elementary Treatise on Quaternions* opened a controversy by the statement that Gibbs "must be ranked as one of the retarders of quaternion progress, in virtue of his pamphlet on *Vector Analysis;* a sort of hermaphrodite monster, compounded of the notations of Hamilton and Grassmann." Gibbs made reply,[2] but the controversy converted neither side to the views of the other. In 1901 there was published E. B. Wilson's textbook on *Vector Analysis,* founded on the lectures of Gibbs.[3] It contained Gibbs's notation $a.b$ for the scalar product and $a \times b$ for the vector product.

It is unfortunate that no uniform notation has yet been reached. If the question were simply one involving the products of two vectors, it would not matter very much which notation was adopted. But the issue is more complicated. In the first place, what definitions for the product are the most useful in the applications of vector analysis? What concepts should be included in a minimum system? These matters cannot be considered here. In the second place, in the products of

---

[1] See *Scientific Papers of J. Willard Gibbs,* Vol. II (1906), p. 17–90.

[2] *Nature,* Vol. XLIV (London, 1891), p. 79–82.

[3] *Vector Analysis—Founded on the Lectures of J. Willard Gibbs* . . . . by Edwin Bidwell Wilson (New York, 1901).

three or more vectors, what notation will indicate in the simplest manner the kinds of products to be taken? What notation is not only free from ambiguity, but also least likely to be misunderstood? There has been much discussion of these topics. There have been those who hold that W. R. Hamilton's quaternions offered, all things considered, the best product and the best notation. Others preferred the American notation introduced by Gibbs. German and Italian mathematicians have different designations of their own.

509. *Attempts at unification.*—The disagreements of mathematicians on the value of quaternions and on the best substitute for quaternions led in 1895 to the first step toward the organization of an International Association for Promoting the Study of Quaternions and Allied Systems of Mathematics, which had among its presidents Charles Jasper Joly, professor of astronomy at Dublin, and Alexander Macfarlane, at one time professor at the University of Texas. Valuable discussions[1] were carried on, but no agreements on theoretical points and on notations were reached by this body. The death of Joly and Macfarlane, and the outbreak of the great war stopped the activities of the Association.

A second attempt at unification was started in 1903 when L. Prandtl of Göttingen sent out a circular suggesting certain rules for a uniform notation. In the main, he favored Gibbs's notation. In September, 1903, he read a paper on this subject at the Naturforscherversammlung at Kassel, and Felix Klein appointed a committee to investigate and report. Articles appeared in the *Jahresbericht*[2] by L. Prandtl, R. Mehmke, and A. Macfarlane. Professor Klein described the results as follows: "Its members could not fully agree and although each had the generous impulse of departing from his original position and take a step toward the position of the others, the outcome was practically this, that about three new notations were originated!"[3]

A third attempt at unification was planned for the fourth International Congress of Mathematicians that was held at Rome in 1908. But the Congress limited itself to the appointment of a Commission which should report at the International Congress to be held at

[1] See A. Macfarlane, *International Association for Promoting the Study etc.* (April, 1905), p. 22.

[2] *Jahresbericht d. d. Mathematiker Vereinigung*, Vol. XIII (1904), p. 36, 217, 229; Vol. XIV (1905), p. 211.

[3] F. Klein, *Elementar-mathematik vom höheren Standpunkte aus*, Teil I (Leipzig, 1908), p. 157.

Cambridge in 1912. The Commission as a whole was somewhat inactive, and the desired unification was not reached.

As a preparatory movement for the Congress at Rome in 1908, Burali-Forti and Marcolongo published a preliminary study,[1] partly historical, setting forth what had been done and what they thought should be done. Their notations for the scalar and vector products, $a \times b$, $a \wedge b$, were a departure from the notations of previous authors, except that Grassmann in one of his papers had used $a \times b$ for the scalar product. The critics of the Italian proposal complained that these "unifiers" introduced still greater diversity. Henri Fehr of Geneva invited the leaders in the advancement of vector analysis to publish their views in *L'Enseignement mathématique*, published in Paris.[2] H. E. Timerding[3] of Strasbourg thought that characters not usually found in printing establishments should be avoided; he would use only very small characters $\times$ for the internal product, and $\wedge$ for the external product. He disapproves of "mod a" for the length of a vector, considers $|a|$ confusing, and argues in favor of introducing the notion of a *bivector* in the elements of vector analysis. Timerding goes contrary to Combebiac of Bourges in emphasizing the need of a universal notation: "The present discord in vectorial terminology is deplorable and its consequences in the development of this science are very grave, for it renders extremely difficult the study of the various researches in vector analysis and the numerous applications." E. B. Wilson[4] of Boston stated that in vector analysis absolute uniformity was undesirable, because of the very great variety of its applications. Peano[5] of Turin favored the Italian symbols for the internal and external products, but like Timerding would use small symbols (small symbols tend to *unite*, large ones to *separate* the entities); Peano would discard $|a|$ and modify mod a by writing either ma or the Hamiltonian Ta ("tensor a"); parentheses, being used in arithmetic, should be avoided in denoting the internal and external products. In the

---

[1] C. Burali-Forti and R. Marcolongo in *Rendiconti del Circolo Matematico di Palermo*, Vol. XXIII, p. 324 (1907); Vol. XXIV, p. 65, 318 (1907); Vol. XXV, p. 352 (1908); Vol. XXVI, p. 369 (1908).

[2] A condensed account of the Italian recommendations, with the symbols placed in tabular arrangement, was published in *L'Enseignement mathématique*, Vol. XI (1909), p. 41.

[3] H. E. Timerding, *L'Enseignement mathématique*, Vol. XI (1909), p. 129–34.

[4] E. B. Wilson, *L'Enseignement mathématique*, Vol. XI (1909), p. 211. See also *Bull. Amer. Math. Soc.* (2d ser.), Vol. XVI (1910), p. 415–36.

[5] G. Peano, *op. cit.*, Vol. XI (1909), p. 216.

function $f(x)$, the parenthesis is useless; it would be well to return to the Lagrangian fx. Knott[1] of Edinburgh preferred the complete product of two vectors, as given by W. R. Hamilton; he says that the ∇ of Hamilton and Tait is only partially represented in the system of Gibbs by ∇, $\nabla.$, $\nabla\times$ and in other systems by *grad*, *div.*, *rot*. Knott maintains that Hamilton accomplished with four symbols S, V, ∇, ϕ what in the minimum system proposed by the Italians requires nine or eleven symbols $\times$, $\wedge$, *grad.*, *rot*, *div*, Δ, Δ', a, *grad* a, $\dfrac{du}{dP}$, $\dfrac{d\,\boldsymbol{u}}{dP}$.

Knott and Macfarlane[2] did not consider the Italian program superior to that of Gibbs.

Thus, a deadlock was reached on the subject of vectorial notation at the time when the great war broke out; since then international endeavors at unification have not yet been resumed.[3]

510. *Tensors*.—Recent developments of electrodynamics and of the theory of relativity have led to a new calculus of directed quantity, the calculus of tensors. The tensor of the Hamiltonian quaternions was simply a numerical factor which stretched the unit vector so that it attained the proper length. Hamilton marked the tensor of the vector q by writing Tq. He says: "I employ . . . . the two capital letters T and U as characteristics of the two operations which I call *taking the tensor*, and *taking the versor* respectively, ... which satisfy the two following general equations, or symbolical identities (in the present system of symbols):

$$q = Tq \times Uq \; ; \qquad q = Uq \times Tq \; ."[4]$$

Other notations have been suggested by Irving Stringham,[5] such as *tsr* for "tensor of" and *vsr* for "versor of," *amp* for "amplitude of." The recent use of the word "tensor" is different; the "tensor" is itself a directed quantity of a general type which becomes the ordinary vector in special cases. Elaborate notations which, for lack of space,

[1] C. G. Knott, *op. cit.*, Vol. XII (1910), p. 39.

[2] A. Macfarlane, *op. cit.*, Vol. XII (1910), p. 45.

[3] In 1912 a very full and detailed statement of the different notations which had been proposed up to that time was made by James Byrnie Shaw in the publication entitled *International Association for the Study of Quaternions and Allied Systems of Mathematics* (Lancaster, Pa.; June 12, 1912), p. 18–33. More recent lists of symbols are found in Shaw's *Vector Calculus* (New York, 1922), p. 12, 127, 136, 165, 166, 179–81, 248–52. See also a historical and critical article by C. Burali-Forti and R. Marcolongo in *Isis*, Vol. II (1914), p. 174–82.

[4] W. R. Hamilton, *Lectures on Quaternions* (Dublin, 1853), p. 88.

[5] Irving Stringham, *Uniplanar Algebra* (San Francisco, 1893), p. xiii.

must be omitted here were proposed in 1914 by J. A. Schouten.[1] For example, he introduces ⌐ and ∟ as signs of multiplication of vectors with each other, and ⌐ and ∟ as signs of multiplication of affinors (including tensors) with each other, and uses fifty or more multiples of these signs for multiplications, their inverses and derivations. In the multiplication of geometric magnitudes of still higher order (Septoren) he employs the sign —o and combinations of it with others. These symbols have not been adopted by other writers.

A tensor calculus was elaborated in 1901 chiefly by G. Ricci and T. Levi-Civita, and later by A. Einstein, H. Weyl, and others. There are various notations. The distance between two world-points very near to each other, in the theory of relativity, as expressed by Einstein in 1914[2] is $ds^2 = \sum_{\mu\nu} g_{\mu\nu} dx_\mu dy_\nu$, where $g_{\mu\nu}$ is a symmetric tensor of second rank (having two subscripts μ and ν) which embraces sixteen products $A_\mu B_\nu$ of two covariant vectors (A_μ) and (B_ν), where $x = x_1$, $x_2 = y$, $x_3 = z$, $x_4 = idt$. He refers to Minkowski and Laue and represents, as Laue[3] had done, a symmetric tensor by the black-faced letter **p** and its components by $\mathbf{p}_{xx}$, $\mathbf{p}_{xy}$, $\mathbf{p}_{xz}$, $\mathbf{p}_{yx}$, $\mathbf{p}_{yy}$, $\mathbf{p}_{yz}$, $\mathbf{p}_{zx}$, $\mathbf{p}_{zy}$, $\mathbf{p}_{zz}$. A world-tensor has sixteen components. Laue denotes the symmetric tensor by the black-faced letter **t**. Representing vectors by German letters, either capital or small, Laue marks the vector products of a vector and tensor by [q**p**] or [𝔄**t**].

In writing the expression for ds^2 Einstein in other places dropped the Σ, for the sake of brevity, the summation being understood in such a case.[4] Pauli uses the symbolism $x = x^1$, $y = x^2$, $z = x^3$, $u = x^4$, writing briefly x^i, and expressed the formula for the square of the distance thus, $ds^2 = g_{ik} dx^i dx^k$, where $g_{ik} = g_{ki}$; it being understood that i and k, independently of each other, assume the values 1, 2, 3, 4. Pauli[5] takes generally the magnitudes $a_{iklm....}{}^{rst}....$, in which the indices assume independently the values 1, 2, 3, 4, and calls them "tensor components," if they satisfy certain conditions of co-ordinate

[1] J. A. Schouten, *Grundlagen der Vektor- und Affinoranalysis* (Leipzig, 1914), p. 64, 87, 93, 95, 105, 111.

[2] A. Einstein, *Sitzungsberichte der Königl. P. Akademie der Wissensch.*, Vol. XXVIII (1914), p. 1033, 1036.

[3] M. Laue, *Das Relativitätsprinzip* (Braunschweig, 1911), p. 192, 193.

[4] Einstein explains the dropping of Σ in *Annalen der Physik*, Vol. XLIX (1916), p. 781. See H. A. Lorentz and Others, *The Principle of Relativity, a Collection of Original Memoirs* (London, 1923), p. 122.

[5] W. Pauli, Jr., in *Encyclopädie* (Leipzig, 1921), Band V2, Heft 4, Art. V 19, p. 569.

transformation.[1] The rank of the tensor is determined by the number of indices of its components. The tensors of the first rank are also called "vectors." If

$$g_{ik} = \delta_i^k = \begin{cases} 0 \text{ for } i \neq k \\ 1 \text{ for } i = k \end{cases}$$

as in the case of orthogonal co-ordinates, then these g_{ik} are tensor components which assume the same value in all co-ordinate systems. Einstein[2] had used the notation d_μ^ν (§ 463); Eddington[3] used g_σ^ν.

A tensor of the second rank is what Gibbs and Wilson had called a "dyadic."[4] C. E. Weatherburn[5] writes $r' = r \cdot (ai + bj + ck)$, where the expression in brackets is the operator called "dyadic," and each term of the expanded r' is called a "dyad." Weatherburn uses also the capital Greek letters Φ, Ψ, Ω to indicate dyadics, and these letters with the suffix c, as in Φ_c, etc., to indicate their conjugates. Gibbs and Wilson marked a dyad by juxtaposition ab; Heaviside by $a \cdot b$; Ignatowsky by $\mathfrak{A}$; $\mathfrak{B}$. These citations from various authors indicate wide divergence, both in nomenclature and notation in the tensor calculus.

In the consideration of the metrical continuum for Riemann's space, three-indices symbols have been introduced by E. B. Christoffel,[6] such that

$$\begin{bmatrix} gh \\ k \end{bmatrix} = \tfrac{1}{2} \left[\frac{\partial w_{gk}}{\partial x_h} + \frac{\partial w_{hk}}{\partial x_g} - \frac{\partial w_{gh}}{\partial x_k} \right], \quad \begin{Bmatrix} il \\ r \end{Bmatrix} = \sum_k \begin{bmatrix} il \\ k \end{bmatrix} \frac{E_{rk}}{E}.$$

Weyl[7] denotes Christoffel's three-index symbol by $\begin{bmatrix} ik \\ r \end{bmatrix}$ and $\begin{Bmatrix} ik \\ r \end{Bmatrix}$;

Eddington[8] by $[\mu\nu, \lambda]$ and $\{\mu\nu, \lambda\}$; Birkhoff[9] by $\Gamma_{i,jk}$ and Γ_{jk}^i.

[1] W. Pauli, Jr., op. cit., p. 572.

[2] H. A. Lorentz and Others, The Principle of Relativity, a Collection of Original Memoirs (London, 1923), p. 128.

[3] A. S. Eddington, Report on the Relativity Theory of Gravitation (1918), p. 34.

[4] R. Weitzenböck, Encyclopädie d. Math. Wissensch., Band CXI, 3, Heft 6 (1922), p. 25.

[5] C. E. Weatherburn, Advanced Vector Analysis (London, 1924), p. 81, 82.

[6] E. B. Christoffel, Journal f. reine u. angew. Mathematik, Vol. LXX (1869), p. 48, 49.

[7] Hermann Weyl, Raum-Zeit-Materie (Berlin, 1921), p. 119.

[8] A. S. Eddington, op. cit. (1918), p. 36.

[9] G. D. Birkhoff, Relativity and Modern Physics (Cambridge, 1923), p. 114.

II

SYMBOLS IN MODERN ANALYSIS

TRIGONOMETRIC NOTATIONS

511. *Origin of the modern symbols for degrees, minutes, and seconds.*
—Signs resembling those now in use are found in the *Syntaxis (Almagest)* of Ptolemy, where the Babylonian sexagesimal fractions are used in astronomical calculations. The units were sometimes called μοῖραι and frequently denoted by the abbreviation μ°. The *first sixtieths* or minutes were marked with one accent, the *second sixtieths* with two accents. Thus,[1] μοιρῶν μζ μβ′ μ″ stood for 47°42′40″; μ°β stood for 2°. From these facts it would seem that our signs °, ′, ″ for degrees, minutes, and seconds were of Greek origin. But it is difficult to uphold this view, especially for the sign ° for "degrees." Such a line of descent has not been established. However, Edward O. von Lippmann[2] holds to the view that ° is the Greek letter omicron taken from the word μοῖραι.

Medieval manuscripts and early printed books contain abbreviations of Latin words in place of the signs ° ′ ″. J. L. Walther[3] gives in his *Lexicon* abbreviations for *gradus* as they occur in early Latin manuscripts. The abbreviations are simply florescent forms of *gu, gdu, gdus.*

In Athelard of Bath's translation into Latin (twelfth century) of certain Arabic astronomical tables, the names *signa, gradus, minutae, secundae,* etc., are abbreviated. The contractions are not always the same, but the more common ones are *Sig., Gr., Min., Sec.*[4] In the Alfonsine tables,[5] published in 1252, one finds marked by $49^{\ddot{s}}\ 32^{i}\ 15^{\ddot{m}}\ 4^{\ddot{s}}$ the sexagesimals $49 \times 60 + 32 + 15 \times \dfrac{1}{60} + 4 \times \dfrac{1}{60^2}$.

[1] Sir Thomas Heath, *A History of Greek Mathematics*, Vol. I (Oxford, 1921), p. 45.

[2] E. O. von Lippmann, *Entstehung und Ausbreitung der Alchemie* (Berlin, 1919), p. 353.

[3] *Lexicon diplomaticum abbreviationes* . . . . *Joannis Lvdolfi Waltheri* . . . . (Ulmae, 1756), p. 138, 139.

[4] See H. Suter, *Die astronomischen Tafeln des Muḥammed ibn Mūsā Al-Khowārizmī* in der Bearbeitung des Maslama ibn Aḥmed Al-Madjrītī und der latein. Uebersetzung des Athelard von Bath (Copenhagen, 1914), p. xxiv.

[5] A. Wegener in *Bibliotheca mathematica* (3d ser.), Vol. VI (1905), p. 179.

512. A different designation is found in a manuscript in one of the Munich libraries,[1] which gives a table (Fig. 116) to facilitate the formation of the product of two sexagesimal fractions. In the left-hand column are given the symbols for "degrees," "minutes," "seconds," "thirds," etc., down to "ninths." The same symbols appear at the head of the respective columns. Using this table as one would a multiplication table, one finds, for example, that "thirds" times "seconds" give "fifths," where the symbol for "fifths" is in the row marked on the left by the symbol for "thirds" and in the column marked at the head by the symbol for "seconds."[2]

The notation of Regiomontanus is explained in § 126. The 1515 edition of Ptolemy's *Almagest*,[3] brought out in Cologne, contains the signs š, ǧ, m̃ for *signa, gradus, minutae*.

Oronce Fine[4] in 1535 used the marks *grad*, m̃, 2̃, 3̃. Sebastianus Munsterus[5] in 1536 had *S.*, *G.*, m̃ for *signa, gradus, minutae*, respectively. Regius[6] in 1536 refers to the Alfonsine tables and gives the signs *T*, *s*, *g*, *m*, *s*, *t*, *qr.*, and, in an example, writes *T*, *S*, *gr*, m̃i, 2ā, 3ā, 4ā, 5ā. Here 1*T* (*tota revolutio*) equals 12*S* (*signa*), 1*S* equals 30 *grad.*, 1 *gr.* equals 60 m̃i., etc. In 1541 Peter Apianus[7] wrote ǧ., m̃.

[1] M. Curtze in *Abhandlungen zur Geschichte der Mathematik*, Vol. VIII (1898), p. 21. Algorithmus-manuscripts Clm. 13021, fol. 27–29 of the Munich Staats-bibliothek.

[2] The multiplication of 2°45′ by 3°10′30″, yielding the product 8°43′52″30‴, is given as follows: "Sint quoque verbi gratia 2 gradus et 45 minuta, quae per 3 gradus et 10 minuta ac 30 secunda ducantur. Resolvatur itaque per 60 unusquisque ad suam ultimam differentiam, id est minuta et secunda scilicet 165 et 11430, quae si inter se ducantur, erunt 1885950. Quae omnia si per 60 dividantur ad secunda reducentur, eruntque 31432 secunda, et remanebunt 30 tertia. Quae secunda eadem divisione 523 minuta remanentibus 52 secundis ostendunt, et hi itidem divisa 8 gradus indicant 43 minutis supereminentibus hoc modo: $\begin{matrix} 8 \\ 43 \\ 52 \\ 30 \end{matrix}$."

[3] *Almagestū, Cl. Ptolemei Pheludiensis Alexandrini Astronomo' principis* . . . . *Colonie'sis Germani* (1515), fol. 78.

[4] Orontius Finaeus, *Arithmetica practica* (Paris, 1535), p. 46. We quote from Tropfke, *op. cit.*, Vol. I (2d ed., 1921), p. 43.

[5] Sebastianvs Munstervs, *Organum Uranicum* (Basel, 1536), p. 49.

[6] Hudalrich Regius, *Vtrivsqve arithmetices epitome* (Strasbourg, 1536), fol. 81*B*, 82*B*.

[7] Peter Apianus, *Instrvmentvm sinvvm, sev primi mobilis* (Nürnberg, 1541), in the tables.

513. In the 1566 edition[1] of the great work of N. Copernicus, the *De revolutionibus*, there is given on folio 15 a table of natural sines, arranged in three columns. The first column contains the degrees and minutes (named *circūferentiae pt., sec.*). The second column gives the sines of the angles to a radius of 100000 parts (named *Semisses dup. circūferen.*, i.e., "half the chord of double the arc"). The third

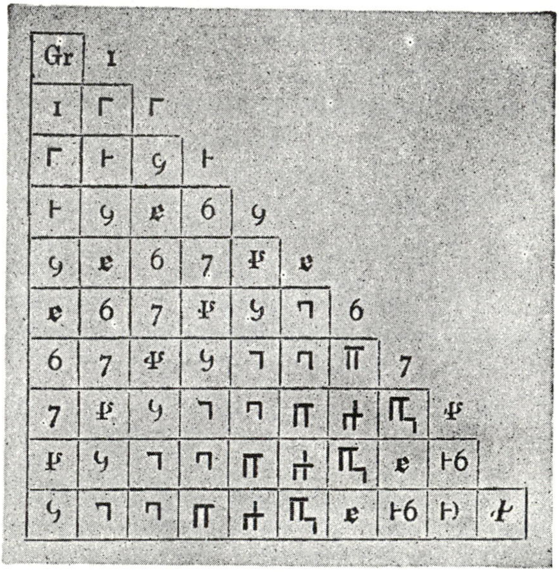

Fig. 116.—Multiplication table for sexagesimal fractions. MS in Munich Staatsbibliothek.

column contains the differences in the values of sines for angles differing by 10 minutes and has the heading *Differētiae* or "Differences" in the 1566 edition, but in the original edition of 1543 this column had the heading *Unius gradus partes*.[2] The sine of 30 *pt.* is given in the table as 50000. The *pt.* at the head of the first column stands for *partes*, in the text also called *gradus;* the *sec.* stands for *scrupulus*. Further on in the edition of 1566 the abbreviation *scr.* takes the place of *sec.*

Among the writers immediately following Copernicus, the Italian

[1] Nicolaus Copernicus, *De revolutionibus orbium celestium*, Liber VI (Basel, 1566).

[2] See the edition of 1873, brought out in Nürnberg, p. 44.

Gioseppe[1] uses *Gra., Mi.;* Schöner[2] writes *S., gr., m̃.* and also *S., G., M.;* Nunez[3] has *Gr., m̃.*, while Lansberge[4] has no sign for degrees and writes "23 28′ 30″." In the works of Clavius,[5] one still finds *G., M., S.*

In the foregoing publications, we have looked in vain for evidence which would connect the modern signs for circular measure directly with the Greek signs of Ptolemy. The modern signs came into use in the sixteenth century. No connection with ancient symbols has been established. We proceed to show that our modern symbols appear to be an outgrowth of the exponential concept.

514. Gemma Frisius[6] in 1540 wrote

"Integr. *Mi.*   2.   3.   4.
          36.   30.   24   50   15 "

for our modern $36°30'24''50'''15^{iv}$, but in the revised 1569 edition of his book published in Paris there is an Appendix on astronomical fractions, due to J. Peletier (dated 1558), where one finds

"Integra, *Mi* vel $\tilde{1}, \tilde{2}, \tilde{3}, \tilde{4}, \tilde{5}, \tilde{6}, \tilde{7}, \tilde{8}$, etc."
°   1, 2, 3, 4, 5, 6, 7, 8.

This is the first modern appearance that I have found of ° for *integra* or "degrees." It is explained that the denomination of the product of two such denominate numbers is obtained by combining the denominations of the factors; minutes times seconds give thirds, because $1+2=3$. The denomination ° for integers or degrees is necessary to impart generality to this mode or procedure. "Integers when multiplied by seconds make seconds, when multiplied by thirds make thirds" (fol. 62, 76). It is possible that Peletier is the originator of the ° for degrees. But nowhere in this book have I been able to find the modern angular notation ° ′ ″ used in writing angles. The ° is used only in multiplication. The angle of 12 minutes and 20 seconds is written "*S*. 0. *g̃*. 0 *m̃*. 12. $\tilde{2}$. 20" (fol. 76).

[1] *L'Efemeridi Dim. Gioseppe, Moleto matematico. per anni XVIII* (Venice, 1563), p. 32.

[2] *Opera mathematica Joannis Schoneri Carolostadii in vnvm volvmen congesta* (Nürnberg, 1551).

[3] *Petri Nonii Salaciensis Opera* (Basel [1556]), p. 41.

[4] *Philippi Lansbergi Geometriae liber quatuor* (Leyden, 1591), p. 18.

[5] *Ch. Clavii Bambergensis Opera Mathematica* (Mayence, 1611), Vol. II, p. 20.

[6] *Gemma Frisius, Arithmeticae practicae methodus facilis* (Strasbourg, 1559), p. 57v°. From Tropfke, *op. cit.*, Vol. I, p. 43.

Twelve years later one finds in a book of Johann Caramuel[1] the signs ° ′ ″ ‴ iv used in designating angles. In 1571 Erasmus Reinhold[2] gave an elaborate explanation of sexagesimal fractions as applied to angular measure and wrote "°63 ′13 ″53," and also "62° 54′ 18″." The positions of the ° ′ ″ are slightly different in the two examples. This notation was adopted by Tycho Brahe[3] who in his comments of 1573 on his *Nova Stella* writes 75° 5′, etc.

As pointed out by Tropfke,[4] the notation ° ′ ″ was used with only minute variations by L. Schoner (1586), Paul Reesen (1587), Raymarus Ursus (1588), Barth. Pitiscus (1600), Herwart von Hohenburg (1610), Peter Crüger (1612), Albert Girard (1626). The present writer has found this notation also in Rhaeticus (1596),[5] Kepler (1604),[6] and Oughtred.[7] But it did not become universal. In later years many authors designated degrees by *Grad.* or *Gr.*, or *G.*; minutes by *Min.* or *M.;* seconds by *Sec.* or *S.* Thus in Horsley's edition[8] of Isaac Newton's works one finds "79 *gr.* 47′ 48″."

When sexagesimals ceased to be used, except in time and angular measure, subdivisions of seconds into thirds became very uncommon. In the eighteenth century the expression found in H. Scherwin's *Tables* of 1741 (p. 24), "natural Sine of 1° 48′ 28″ 12‴," is decidedly uncommon. The unit *S* (*signa*) occurs as late as the eighteenth century.[9]

Some early writers place the signs ′, ″, and also the ° when it is used, above the numbers, and not in the now usual position where exponents are placed. They would write 7° 50′ in this manner, 7 50. Among such writers are Schoner, Reesen, Ursus, and Wright (1616).

---

[1] *Joannis Caramvelis, Mathesis Bicepts Vetus, et Nova* (Companiae, 1670), p. 61.

[2] E. Reinhold, *Prvtenicae tabulae coelestium motuum* (Tübingen, 1571), fol. 15.

[3] Tycho Brahe, *Opera omnia* (ed. I. L. E. Dreyer), Vol. I (Copenhagen, 1913), p. 137.

[4] Tropfke, *op. cit.*, Vol. I (2d ed., 1921), p. 43.

[5] *Opus Palatinvm triangvlis a Georgio Joachimo Rhetico coeptvm* (1696), p. 3 ff.

[6] J. Kepler, *Ad vitellionem paralipomena quibus astronomiae pars optica traditur* (Frankfurt, 1604), p. 103, 139, 237.

[7] William Oughtred in *Clavis mathematicae* (1631); anonymous "Appendix" to E. Wright's 1618 translation of John Napier's *Descriptio*.

[8] S. Horsley, *Newtoni Opera*, Vol. I (1779), p. 154.

[9] Jean Bernoulli, *Recueil pour des Astronomes*, Tome II (Berlin, 1772), p. 264.

A novel mark is adopted by Georg Vega,[1] he lets 45$^\lor$ stand for 45° but occasionally uses the full circle or zero.

The weight of evidence at hand favors the conclusion that the sign ° for degree is the numeral zero used as an exponent in the theory of sexagesimal fractions and that it is not the Greek letter omicron.

To prevent confusion[2] between circular measure and time measure, both of which involve the words "minutes" and "seconds," and both of which are used side by side by astronomers and navigators, it became the common practice to mark minutes and seconds of arc by ' and ", and minutes and seconds of time by ^{m} and ^{s} (see also § 36).

515. *Sign for radians.*—The word "radian" was first used in print in 1873 by James Thomson, a brother of Lord Kelvin.[3] It has been quite customary among many authors to omit the word "radian" in the use of circular measure and to write, for example, $\frac{\pi}{2}$ or π, when $\frac{\pi}{2}$ radians or π radians are meant. It has been proposed also to use the Greek letter ρ and to write 2ρ, $\frac{3}{5}\pi\rho$ for two radians and three-fifths of π radians.[4] Others[5] have used the capital letter R in a raised position, as $\frac{\pi}{4}^R$ for $\frac{\pi}{4}$ radians, or else the small letter[6] r as in 1^r for one radian, or again the small r in parentheses[7] as in $2^{(r)}$ for two radians.

The expression "circular measure" of an angle has led to the suggestion that radians be indicated by the small letter ^{c}, i.e., 2^c means two radians. However, "when the radian is the unit angle, it is customary to use Greek letters to denote the number of radians, and the symbol ^{c} is then often omitted. When capital English letters are used, it is usually understood that the angle is measured in degrees."[8]

[1] Georg Vega, *Vorlesungen über Mathematik*, Vol. II (Vienna, 1784), p. 5, 8. 16, 23, 27, 31, 185.

[2] D. André, *op. cit.*, p. 32.

[3] Cajori, *History of Mathematics* (2d ed., 1919), p. 484.

[4] G. B. Halsted, *Mensuration* (Boston, 1881), p. 83; T. U. Taylor and C. Puryear, *Plane and Spherical Trigonometry* (Boston, 1902), p. 92.

[5] A. G. Hall and F. G. Frink, *Plane Trigonometry* (New York, 1909), p. 73.

[6] G. N. Bauer and W. E. Brooke, *Plane and Spherical Trigonometry* (Boston, 1907), p. 7.

[7] P. R. Rider and A. Davis, *Plane Trigonometry* (New York, 1923), p. 29.

[8] W. E. Paterson, *Elementary Trigonometry* (Oxford, 1911), p. 29.

Evidently there is, as yet, no approach to uniformity in the designation of radians.

In geodesy a symbol is used for a radian expressed in seconds of arc (i.e., for $180\times60\times60\div\pi$); Albrecht[1] uses $\dfrac{1}{\sin 1''}$, Helmert[2] and W. D. Lambert[3] use ρ''.

516. *Marking triangles.*—One of the devices introduced at an early period to aid the eye when one enters upon the solution of a plane or spherical triangle was the use of strokes to indicate the given parts and the parts to be computed. Perhaps the earliest author to use strokes was Philipp van Lansberge, a native of Gand. He published a trigonometry with tables, *Triangulorum geometricorum libri quator*, the dedicatory epistle of which bears the date 1591.[4] If in a spherical triangle a side is given, two parallel strokes cross it;[5] if an angle is given, the two parallel strokes appear in the angular space close to the vertex. A required part is unmarked or else marked with a single stroke.

A slight modification of these marks is found in a manuscript[6] prepared in 1599 by Melchior Jöstel, of Wittenberg, who uses one stroke for each of the given parts and a small circle for each of the required parts. The stroke is drawn so as to cross a side or bisect an angle. If an angle is required, the small circle is drawn with its center approximately at the vertex.

The use of strokes is found again in the *Trigonometria* of Bartholomaeus Pitiscus, which was published in Augsburg in 1600; later editions appeared in 1608 and 1612. An English translation was made by R. Handson. Glaisher states[7] that Pitiscus used in the 1600 edi-

[1] Theodor Albrecht, *Formeln und Hülfstafeln für geographische Ortsbestimmungen* (Leipzig, 1879; 4th ed., 1908).

[2] Helmert, *Höhere Geodäsie.*

[3] W. D. Lambert, *Effect of Variations in the Assumed Figure of the Earth on the Mapping of a Large Area*, Spec. Pub. No. 100 (U.S. Coast and Geodetic Survey, 1924), p. 35.

[4] The book is printed in P. Lansbergii, *Opera omnia* (Middelburgi Zelandiae, 1663), p. 1–88. About Lansberge see also Ad. Quetelet, *Histoire des sciences mathématiques et physiques* (Bruxelles, 1864), p. 168–79; Delambre, *Histoire de l'astronomie moderne*, Tome II (Paris, 1821), p. 40.

[5] *P. Lansbergii opera omnia*, p. 77, 85–88.

[6] Printed in *Tychonis Brahe Dani Scripta Astronomica* (ed. I. L. E. Dreyer), Vol. I (Hauniae, 1913), p. 300.

[7] J. W. L. Glaisher, *Quarterly Journal of Pure and Applied Mathematics*, Vol. XLVI (London, 1915), p. 167.

tion (p. 47) a stroke to indicate the given parts of a triangle and a small circle to indicate the required parts. This notation is the same as that of Jöstel. Oughtred owned a copy of Pitiscus as we know from John Aubrey who refers to Oughtred's book: "I myselfe have his Pitiscus, imbellished with his excellent marginall notes, which I esteeme as a great rarity." There is no doubt that Oughtred received the suggestion of using strokes and little circles from Pitiscus.

In the 1618 edition of Edward Wright's translation, from Latin into English, of John Napier's *Mirifici Logarithmorum Canonis Descriptio*, there is an anonymous Appendix bearing the title, "An Appendix to the Logarithmes," probably from the pen of William Oughtred. It is historically a remarkable document, for it is here that one first finds the $\times$ in the form of the letter X to designate multiplication, the earliest table of natural logarithms, the first appearance of the radix method of computing logarithms, the first printed appearance of $=$ as a sign of equality after the time of Recorde, and the earliest use in England of abbreviations for the trigonometric functions.[1]

This "Appendix" marks the known parts of a triangle with one stroke and the unknown parts with a small circle. These signs are also found in Oughtred's *Circles of Proportion* (1632), in Richard Delamain's *Grammelogia* (1630), in John Well's *Sciographia* (1635), in the second (1636) and later editions of Edmund Gunter's *Works*, as well as in John Seller's *Practical Navigation* (1671[?]). Glaisher, from whom we take these data, remarks that among the books that he has seen in which the stroke and the little circle are used, it is only in the "Appendix" and in Oughtred's *Circles of Proportion* that the little circle is placed just inside the triangle and not with its center in the side.

That there should appear variations in the mode of marking the known and unknown parts of a triangle is to be expected. In Albert Girard's *Tables des Sinus, etc.* (1626), a given side of a spherical triangle is marked by two strokes drawn across it, and a given angle by a small arc of a circle following the angle. A required side is not marked, but a required angle is marked by a point in its interior.

In Briggs's *Trigonometria Britannica* (1633), one finds the use of a dotted line to denote the required parts, the full stroke being used for

<hr/>

[1] The "Appendix" in question is reprinted in different parts of an article by J. W. L. Glaisher in the *Quarterly Journal of Pure and Applied Mathematics* (London, 1915), p. 125–97. The "Appendix" will be found in proper sequence on p. 140–41, 148–52, 194–96, 141–43, 192–94.

the given parts, as in the previously named publications. This modified designation is employed also in Vlacq's *Trigonometria artificialis* (1633, 1665, 1675), in Gellibrand's *Institution Trigonometricall* (1635), in John Newton's *Trigonometria Britannica* (1658), and in other publications.[1]

We shall not trace this notation through the later centuries. While it does not appear regularly in textbooks of more recent time, some symbols for marking the given parts and the required parts have been and are still widely used in elementary teaching.

517. *Early abbreviations of trigonometric lines.*—An early need for contracted writing in trigonometry is seen in Francisco Maurolyco, of Messina, in Italy, who, in a manuscript[2] dated 1555 writes "sinus 2us arcus." In his printed edition of the *Menelai Sphaericorum libri tres* (1558),[3] he writes "sinus 1^m arcus" for "sinus rectus primus" (our sine), and also "sinus 2^m arcus" for "sinus rectus secundus" (our cosine).

Perhaps the first use of abbreviations for the trigonometric lines goes back to the physician and mathematician, Thomas Finck, a native of Flensburg in Schleswig-Holstein. We premise that we owe to him the invention in 1583 of the words "tangent" and "secant."[4] These terms did not meet with the approval of Vieta, because of the confusion likely to arise with the same names in geometry. Vieta[5] called the trigonometric tangent *Prosinus* and the trigonometric secant *Transsinuosa*. But Vieta's objection was overlooked or ignored. The trigonometric names "tangent" and "secant" were adopted by Tycho Brahe in a manuscript of 1591, by G. A. Magini in 1592, by Thomas Blundeville in 1594, and by B. Pitiscus in 1600. The need for abbreviations arose in connection with the more involved relations occurring in spherical trigonometry. In Finck's Liber XIV, which relates to spherical triangles, one finds proportions and abbreviations as follows: "sin.," "tan.," "sec.," "sin. com.," "tan. com.," and "sec. com." The last three are our cosine, cotangent, and cosecant. He gives the proportions:

[1] J. W. L. Glaisher, *op. cit.*, p. 166, 167.

[2] B. Boncompagni *Bullettino*, Vol. IX (1876), p. 76, also p. 74.

[3] A. von Braunmühl, *Bibliotheca mathematica* (3d ser.), Vol. I (1900), p. 64, 65.

[4] *Thomae Finkii Flenspurgensis Geometriae Rotundi Libri XIIII* (Bâle, 1583). All our information relating to this book is drawn from J. W. L. Glaisher's article in *Quarterly Journal of Pure and Applied Mathematics*, Vol. XLVI (1915), p. 172.

[5] Fr. Vieta, *Opera mathematica* (Leyden, 1646), p. 417, in the *Responsorum*, Liber VIII. The *Responsorum* was the first published in 1593.

(P. 360)	Rad.	sin. ai.	sin. a.	sin. ie.
(P. 364)	Sin. yu	sin. ys	sin. iu	sin. it
(P. 378)	Rad.	tan. com. a.	tan. com. i	sin. com. ia.
(P. 381)	Rad.	sin. com. ie.	sec. ia.	sec. ea.

It will be observed that the four terms of each proportion are written one after the other, without any symbols appearing between the terms. For instance, the first proportion yields sin $ie =$ sin $ai \cdot$ sin a. An angle of a spherical triangle is designated by a single letter, namely the letter at the vertex; a side by the letters at its extremities. Accordingly, the triangle referred to in the first proportion is marked $a\ i\ e$.

Glaisher remarks that Finck is not at all uniform in his manner of appreviating the names of the trigonometric lines, for one finds in his book also "sin. an.," "sin. ang.," "tang.," "sin. comp.," "sin. compl.," "sin. com. an.," "tan. comp.," etc. Finck used also "sin. sec." for *sinus secundus*, which is his name for the versed sine. The reader will observe that Finck used abbreviations which (as we shall see) ultimately became more widely adopted than those of Oughtred and Norwood in England, who are noted for the great emphasis which they laid upon trigonometric symbolism.

Similar contractions in treating spherical triangles are found not long after in Philipp van Lansberge's *Triangulorum geometricorum libri quator* (1591), previously referred to. In stating proportions he uses the abbreviations[1] "sin.," "sec.," "tang.," "sin. cŏp.," "tang. comp.," "tang. compl.," "sec. comp.," but gives the names in full whenever there is sufficient room in the line for them. For instance, he writes in a case of a right spherical triangle (p. 87), "ut radius ad sinum basis ita sec. comp. lat. ad sec. comp. ang."

A few years later, one finds a few abbreviations unrelated to those of Finck, in a manuscript of Jöstel, of Wittenberg, whom we have mentioned earlier. He was a friend of Longomontanus and Kepler. In 1599 he writes "S." for *Sinus*,[2] as does also Praetorius of Altdorf.

Later still one encounters quite different symbols in the *Canon triangulorum* of Adriaen van Roomen (Adrianus Romanus), of the Netherlands. He designates the sine by "S.," the tangent (which he called, as did Vieta, the *Prosinus*) by "P.", the secant or *Transsinuosa* by "T." The co-functions are indicated by writing $\mathcal{c}$. after the function. A rectangle or the product of two factors is indicated by

[1] Ph. van Lansberge, *op. cit.*, p. 85–88.

[2] *Tycho Brahe Dani Opera Omnia* (ed. J. L. E. Dreyer; Hauniae, 1913), Vol. I, p. 297 f.; A. von Braunmühl, *Vorlesungen über Geschichte der Trigonometrie,* 1. Teil (Leipzig, 1900), p. 230 n.

"§," followed by "sub." Thus Romanus writes our equation 1:(sin b cosec A) = (cot c+cot b cos A):cot C in the following ponderous manner:[1]

"Ut quadratum radii ad § sub S. AB & T. ¢ A,

Ita § sub P. ¢ $\left\{ \begin{array}{ll} AC \text{ \& Radio} & \text{ad § sub Radio \&} \\ AB \text{ \& S. } \, ¢ \, A & \text{Prosinus } \, ¢ \, C \end{array} \right\}$ $\begin{array}{l} \text{efficiente} \\ \text{coefficiente.''} \end{array}$

During the next quarter-century the leading experiments on trigonometric symbolism were carried on in Great Britain, where the invention of logarithms gave a tremendous impetus to trigonometric computation.

518. *Great Britain during 1602–18.*—We begin with Nathaniel Torporley's trigonometry, as found in his *Diclides Coelometricae seu valvae astronomicae universales* (London, 1602). Torporley, who had been at Christ Church, Oxford, is said to have been amanuensis to Vieta in France and to have conveyed ideas of Vieta to Thomas Harriot. De Morgan[2] describes Torporley's book as follows: "Torporley has given two tables of double entry, which Delambre says are the most obscure and incommodious that ever were made. The first is neither one nor the other; a and b being the arguments, and c the tabulated result, it amounts to tan c=tan a×sin b, the double entry being contrived like that of the common multiplication table. Of the second table, as the book is scarce, I subjoin half a dozen instances.

30° (24°)

G	G	M	M
21		27	
	8		33
12		54	

54° (70°)

G	G	M	M
27		55	
	26		5
1		50	

70° (54°)

G	G	M	M
39		14	
	30		46
8		28	

90° (19°)

G	G	M	M
71		58	
	18		2
53		56	

19° (90°)

G	G	M	M
9		30	
	9		30
0		0	

147° (59°)

G	G	M	M
88		3	
	58		57
29		6	

[1] *Adriani Romani Canon triangvlorum sphaericum* (Mayence, 1609), p. 121. See also A. von Braunmühl in *Bibliotheca mathematica* (3d ser.), Vol. I (1890), p. 65, who refers to p. 230 of the *Canon triangvlorum*.

[2] A. de Morgan, "On the Invention of Circular Parts," *Philosophical Magazine*, Vol. XXII (1843), p. 352.

As far as the formulae for right-angled [spherical] triangles are concerned, this table applies as follows. The sine of the angle on the left multiplied by the sine of the upper angle in the square compartment, gives the sine of the second angle in that compartment. Thus Torporley means to say that

$$\sin 24° \times \sin 21°27' = \sin 8°33' .$$

Those who like such questions may find out the meaning of the other parts of the table. . . . . Torporley was an astrologer."

The evidence at our command does not indicate that English mathematicians of the early part of the seventeenth century had seen the trigonometric work of Finck, Lansberge, and Romanus on the Continent. Some English authors were familiar with the very influential work of Pitiscus, but he did not use abbreviations for the trigonometric lines. The earliest efforts in the way of improved notation made in

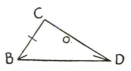

Fig. 117.—Marking the given and required parts of a triangle (1618)

England occur in that brief but remarkable anonymous "Appendix" in the 1618 edition of Edward Wright's translation of Napier's *Descriptio*. We quote the following from this "Appendix":

"THE CALCULATION OF A PLAINE OBLIQUE ANGLED TRIANGLE

"1. Either the foure ingredient parts are opposite two to two: as

$$sB + BC = sD + DC .$$

"2. Or the two sides being given with the angle comprehended within them, either of the other two angles is sought.

The angle B is 37°. 42' .
The side BC is 39 .
& the side BD is 85 ."[1]

Parts of the triangle are marked in the manner previously noted. The equation by which the unknown side DC is to be computed involves the Law of Sines. It is clear that the four terms are intended to represent *logarithmic* numbers; that is, sB means really log sin B; BC stands for $\log (BC)$, etc. If such were not the intention, the two signs of addition $+$ would have to be replaced each by the sign

[1] See *Quarterly Journal of Pure and Applied Mathematics*, Vol. XLVI (1915), p. 150.

of multiplication, ×. The abbreviations of trigonometric lines are "s" for "sine" and "t" for "tangent," "s*" (in one place "s*") for "cosine," and "t*" for "cotangent." Note that the author writes "s," not "s."; there are no dots after the abbreviations. This "Appendix" is the earliest book which contains the abbreviations of the words "sine" and "cosine" used as constituents of formulae other than proportions. In the writing of equations, the sign of equality (=) is used, perhaps for the first time in print sine 1557 when Robert Recorde introduced it in his *Whetstone of Witte.*

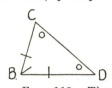

Fig. 118.—The given and the required parts of another triangle (1618)

Proportions were presented before 1631 in rhetorical form or else usually by writing the four terms in the same line, with a blank space or perhaps a dot or dash, to separate adjacent terms. These proportions would hardly take the rank of "formulae" in the modern sense. Another peculiarity which we shall observe again in later writers is found in this "Appendix." Sometimes the letter s in "s*" (cosine) is omitted and the reader is expected to supply it. Thus the equation which in modern form is

$$\log \cos \overline{BA} + \log \cos \overline{CA} = \log \cos \overline{BC}$$

is given in the "Appendix" in extremely compact form as

$$s_* BA + {}^* CA = {}_* BC \ ,$$

the *logarithmic* numbers being here understood.

As regards the lettering of a triangle, the "Appendix" of 1618 contains the following: "It will bee convenient in every calculation, to have in your view a triangle, described according to the present occasion: and if it bee a right angled triangle, to denote it with the Letters $A.B.C$: so that A may bee always the right angle; B the angle at the Base $B.A$ and C the angle at the Cathetus CA."[1] The "cathetus" is the vertical leg of the right triangle. J. W. L. Glaisher makes the following observation: "It is evident that A is taken to be the right angle in the right-angled triangle ABC in order that BA may represent the base and CA the cathetus, the letters indicating the words."[2] An oblique triangle is marked BCD with BD as the base, so that the perpendicular (or cathetus) drawn from the vertex to the base could be marked CA. The fact that this same unique mode of

[1] *Quarterly Journal of Mathematics*, Vol. XLVI (1915), p. 162.

[2] *Op. cit.*, p. 162.

lettering is employed by Oughtred in the books which bear his name
is one of several arguments in support of the view that Oughtred is
the author of the "Appendix."

519. *European Continent during 1622–32.*—Meanwhile, some attention to symbolism continued to be given on the Continent. The
Danish astronomer, C. S. Longomontanus,[1] in 1622 used the notation "S.R." for *sinus rectus* (sine), "S.T." for *sinus totus* (i.e., sin 90°
or radius), "T." and "Tang." for *tangens* (tangent), "Sec." for *secans*
(secant), "T. cōpl." for "cotangent," "Sec. Compl." for "cosecant."

In 1626 a work was published at the Hague by Albert Girard of
Lorraine with the title *Tables de sinus, tangentes et sécantes selon le
raid de 10000 parties,* of which a translation into Dutch appeared at

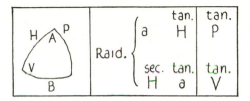

the Hague in 1629.[2] Girard uses in his formulae for right-angled
spherical triangles *H, P, B* to represent the hypothenuse, perpendicular, and base, respectively, *A* to represent the angle at the vertex subtended by the base, *V* the angle at the base subtended by the perpendicular; a small letter *a* denotes the complement of capital *A*,
i.e., $a = 90° - A$. By a single letter is meant its sine; if a tangent or
secant is intended, the abbreviation for the function is given. Thus
Girard writes the formula shown in Figure 119, which means in
modern notation

$$\tan P = \sin (90° - A) \tan H ,$$
$$\tan V = \sec H \tan (90° - A) .$$

When Girard uses brackets he means that the sine of the quantity
included shall be taken. Thus, remembering that $b = 90° - B$, etc.,

[1] Christian Longomontanus, *Astronomiae Danicae* (Amsterdam, 1622), Pars
prior, p. 12 ff.

[2] Our information is drawn from J. W. L. Glaisher, *op. cit.*, Vol. XLVI (1915),
p. 170–72; M. Cantor, *Vorlesungen über Geschichte der Mathematik,* Vol. II (2d
ed.; Leipzig, 1913), p. 708, 709; Vol. III (2d ed., 1901), p. 559; A. von Braunmühl,
Vorlesungen über Geschichte der Trigonometrie, Vol. I (Leipzig, 1900), p. 237.

one sees that $(A+b)-(d)$ signifies sin $(A+90°-B)-$sin $(90°-D)$, but in three cases he expresses the sine by writing the abbreviation "sin." above the angle or side, so that $\overset{\text{sin.}}{CD}$ stands for sin CD.

In further illustration we cite the following proportion from the 1629 edition, folio $K_9V°$, which applies to the spherical triangle ABD having AC as a perpendicular upon BD:

$$\overset{\text{sec.}}{BAC} \quad \overset{\text{sec.}}{DAC} \quad \overset{\text{tan.}}{AB} \quad \overset{\text{tan.}}{AD} .$$

In an Appendix of the 1629 edition of the *Tables*, "Tan. B" occurs twice, the "Tan." being written before the B, not above it.

The Dutch mathematician, Willebrord Snellius,[1] did not place much emphasis on notation, but in one place of his trigonometry of 1627 he introduces "sin.," "tang.," "sec.," "sin. com.," "tan. com.," "sin. vers.," "tan. com.," "sec. com." Here "comp." means "complement"; hence "sin. com." is our "cosine," etc.

A follower of Girard was J. J. Stampioen, who in his edition of van Schooten's *Tables of Sines*[2] gave a brief treatment of spherical trigonometry. Like Girard, he uses *equations* rather than *proportions*, but stresses symbolism more than Girard. Stampioen writes one of his formulae for an oblique spherical triangle ABC thus,

$$\frac{(:BC) \div (AB \dotplus AC) \text{ in } \square R}{\mid AB, \text{ in } AC \mid} = (:A) ,$$

$$\text{—Sin—}$$

which means

$$\frac{\text{vers } BC - \text{vers } (AB-AC)}{\text{sin } AB \cdot \text{sin } AC} = \text{vers } A .$$

520. *Great Britain during 1624–57.*—More general was the movement toward a special symbolism in England. Contractions similar to those which we found in the 1618 "Appendix" came to be embodied in regular texts in the course of the next dozen years. In 1624 "sin" for "sine" and "tan" for "tangent" were placed on the drawing representing Gunter's scale, but Gunter did not otherwise use them in that

[1] *Willebrordi Snelli a Royen R. F. Doctrinae triangvlorvm canonicae liber qvatvor* (Leyden, 1627), p. 69.

[2] *Tabulae Sinuum . . . . door Fr. van Schooten gecorrigeert . . . . door J. J. Stampioen* (Rotterdam, 1632). See A. von Braunmühl, *Bibliotheca mathematica* (3d ser.), Vol. I (1900), p. 67.

edition of the book;[1] in the 1636 edition, a Latin passage is interpolated which contains *Sin., Tan., Cosi.* Noteworthy is the abbreviation *cosi* to represent the new word "cosine." To him we owe the manufacture of this word out of "sine of the complement," or "complemental sine,"[2] which he introduced in his *Canon triangulorum* (London, 1620), along with the word "cotangent"; but at that time he did not use abbreviations. Glaisher[3] says the two words "cosine" and "cotangent" did not find ready acceptance, that they were not used by Briggs, Speidell, Norwood, nor by Oughtred, but that they occur in Edmund Wingate's *Use of Logarithmes* (London, 1633), in Wells' *Sciographia* (London, 1635), in John Newton's works, and in John Seller's *Practical Navigation* (1671[?]).

More generous in the use of symbols was a teacher of mathematicians in London, John Speidell,[4] who published in 1627 a treatise on spherical triangles in which he uses "Si.," "Si. Co.," "T.," "T. Co.," "Se.," "Se. Co.," for "sine," "cosine," "tangent," "cotangent," "secant," "cosecant." As in Thomas Finck, so here, there are given abbreviations for six trigonometric lines; as in Finck, the abbreviations have a number of variants. With Speidell there is no regularity in the printing as regards capitals and small letters, so that one finds "Co." and "co.," "T. Co." and "T. co.," "Se. Co." and "se. co." Moreover, he generally leaves out the symbol "Si.," so that "Co.," the abbreviation for "complement," stands for "cosine." He writes,

$$Si.\ AB - Si.\ AD - T.\ BC - T.\ DE$$
$$90. - Co. \overset{d.}{60}. \overset{m.}{27} - Se.\ Co. \overset{d.}{42} \cdot \overset{m.}{34} \cdot \tfrac{1}{3}.$$

which means

$$Sin\ AB : sin\ AD = tan\ BC : tan\ DE\ ,$$
$$sin\ 90° : cos\ 60°27' = cosec\ 42°34\tfrac{1}{3}' : cos\ x\ .$$

Conspicuous among writers on our subject was Richard Norwood, a "Reader of Mathematicks in London," who brought out in 1631 a widely used book, his *Trigonometrie.* Norwood explains (Book I, p. 20): "In the examples, *s* stands for *sine: t* for *tangent: ∫c* for *sine*

---

[1] E. Gunter, *Description and Use of the Sector, The Crosse-staffe and other Instruments* (London, 1624), Book II, p. 31.

[2] *Quarterly Journal of Mathematics*, Vol. XLVI (1915), p. 147, 187, 190.

[3] *Op. cit.*, p. 190.

[4] *A Breefe Treatise of Sphaericall Triangles.* By John Speidell, Professor and teacher of the Mathematickes in Queene-streete (London, 1627), p. 15, 29, 37.

complement: tc for *tangent complement:* ∫ec: for *secant.*" It is to be observed that these abbreviations are not followed by a period. As a sample, we give his arrangement of the work (p. 21) in solving a plane triangle *ABD*, right-angled at *B*, the side *A* and the angles being given and *DB* required:

$$\text{"s, D 67 deg. 23'.} \quad \text{comp:ar:} \quad 0,0347520$$
$$AB\ 768\ \text{paces.} \qquad\qquad 2,8853612$$
$$s,\ A\ 22\text{--}37. \qquad\qquad 9,5849685$$
$$DB\ 320\ \text{paces.} \qquad\qquad 2,5050817\text{"}$$

By "comp: ar:" is meant the "arithmetical complement." Further on in his book, Norwood uses his abbreviations not only in examples, but also in the running text. In formulae like (Book II, p. 13)

$$\text{sc } A + \text{Rad} = \text{tc } AG + \text{tc } AD \ ,$$

which applies to a spherical triangle *ADG* having *DG* as a quadrant, each term must be taken as representing logarithmic values.

Oughtred, who is probably the author of the "Appendix" of 1618, uses the notation "s" and "s co" in an undated letter to W. Robinson.[1] In this letter, Oughtred uses such formulae as

$$\frac{\text{sco. } PS \times \text{sco. } PZ}{\text{sco. } ZS} = \text{s. arcus prosthaph.}$$

and

$$2s \cdot \tfrac{1}{2} \left\{ {Z \atop X} \right\} + \text{sco. } ZS + 0.3010300(-2R)$$

$$-\text{s. } PS - \text{s } PZ(+2R) = \text{sco. } \angle P \ .$$

It is noticeable that in this letter he writes "s" and "sco" for natural and logarithmic sines and cosines indifferently, and passes from natural to the logarithmic, or vice versa, without remark.

In his *Clavis mathematicae* of 1631 (1st ed.), Oughtred does not use trigonometric lines, except the *sinus versus* which on page 76 is abbreviated "*sv.*" But the following year he brought out his *Circles of Proportion* which contain "sin" and "s" for "sine," "tan" and "t" for "tangent," "sec" and "se" for "secant," "s co" for "cosine," "tco" for "cotangent," the dot being omitted here as it was in Norwood and in the "Appendix" of 1618. The *Circles of Proportion* were brought out again at London in 1633 and at Oxford in 1660. In 1657 Oughtred published at London his *Trigonometria* in Latin and

[1] S. P. Rigaud, *Correspondence of Scientific Men of the 17th Century* (Oxford, 1841), Vol. I, p. 8–10.

also his *Trigonometrie* in English. Contractions for the trigonometric lines occur in all these works, sometimes in the briefest possible form "s," "t," "sco," "tco," at other times in the less severe form[1] of "*sin*," "*tan*," "*sec*," "*tang*." In his texts on trigonometry he uses also "s ver" for *sinus versus* ("versed sine"). In both the Latin and English editions, the abbreviations are sometimes printed in roman and sometimes in italic characters.

521. *Seventeenth-century English and continental practices are independent.*—Attention to chronological arrangement takes us to France where Pierre Hérigone brought out his six-volume *Cursus mathematicus* in 1634 and again in 1644. The entire work is replete with symbols (§ 189), some of which relate to trigonometry. It will be remembered that "equality" is expressed by 2|2, "minus" by $\sim$, "ratio" by π, "square" by $\square$, "rectangle" by $\square$, "multiplication" by ($\prime$). Small letters are used in the text and capitals in the accompanying figures. If in a plane triangle abc a perpendicular ad is drawn from the vertex a to the side bc, then cos $b = $ sin $\lfloor bad$. With these preliminaries the reader will recognize in the following the Law of Cosines:

$$2 \square\ ab,\ bc\ \pi\ \square\ bc + \square\ ba \sim \square\ ac\ 2|2\ \text{rad.}\ \pi\ \sin < bad\ .$$

Hérigone obtains a compact statement for the Law of Tangents in the plane triangle fgh by assuming

$$a\ 2|2\ hf + hg,\ b\ 2|2\ hf - hg,\ c\ 2|2\ \tfrac{1}{2}, f + g\ ,$$

and considering the special case when

$$\tfrac{1}{2}\ (f - g) = 8°43'\ .$$

His statement is:

$$a\ \pi\ \text{tangent}\ .c\ \ 2|2\ \ b\ \pi\ \text{tangent}.8g\ 43'\ .$$

We have seen that on the Continent abbreviations for the trigonometric lines were introduced earlier than in England by Thomas Finck and a few others. In England abbreviations occur as early as 1618 but did not become current until about 1631 or 1632 in the writings of Norwood and Oughtred. But they did not become universal in England at that time; they are absent from the writings of Briggs, Gellibrand, Blundeville,[2] Bainbridge (except occasionally when the

---

[1] For the exact references on the occurrence of each abbreviation in the various texts, see §§ 184, 185.

[2] *Mr. Blundevil His Exercises* . . . . (7th ed.; London, 1636), p. 99.

printer is crowded for space and resorts to "Tan." and "Sin."),[1] Wingate and Wells.[2] But these writers belong to the first half of the seventeenth century. Later English writers who abstained from the use of abbreviations in trigonometric names were Collins[3] and Forster.[4]

In the latter part of the seventeenth century trigonometric symbolism reached a wider adoption in England than on the Continent. The example set by Finck, Romanus, Girard, and Snellius did not find imitators on the Continent. Thus in the *Trigonometria* of B. Cavalieri (1643) no contractions appear. In A. Tacquet's *Opera mathematica* (Antwerp, 1669) one finds in the part entitled *Geometria practica* the names of the trigonometric lines written out in full. In Kaspar Schott's *Cursus mathematicus* (Würtzburg, 1661), the same practice is observed, except that in one place (p. 165) "sin." is used. At that time another difference between the English writers and those on the Continent was that, in abbreviating, continental writers placed a dot after each symbol (as in "sin."), while the English usually omitted the dot (as in "sin").

522. *England during 1657–1700.*—In England the abbreviations came to be placed also on Gunter's scales and on slide rules. To be sure, in the earliest printed explanation of a slide rule found in Richard Delamain's *Grammelogia* (London, 1630–33[?]), the designations on the slide rule (which was circular) were selected without reference to the names of the trigonometric lines represented. Thus, the "Circle of Tangents" was marked S on the fixed circle and Y on the movable circle; the "Circle of Sines" was marked D on the fixed circle and TT on the movable circle. Seth Partridge, in his *Instrument called the Double Scale of Proportion* (London, 1662), gives no drawings and uses no abbreviations of trigonometric lines. But in Wil. Hunt's *Mathematical Companion* (London, 1697) there is a description of a slide rule having a "Line of Sines" marked S, a "Line of Artificial Sines" marked SS, a "Line of Tangents" marked T.

Figure 120 shows a page[5] in Isaac Newton's handwriting, from Newton's notebook, now in the Pierpont Morgan Library, New York. This page was probably written about the time when Newton entered Cambridge as a student.

---

[1] *Cl. V. Johannis Bainbrigii Astronomiae . . . . canicvlaria* (Oxford, 1648), p. 55.

[2] *Quarterly Journal of Mathematics*, Vol. XLVI (1915), p. 160, 166.

[3] John Collins, *Description and Uses of the General Quadrant* (London, 1658); also *The Sector on a Quadrant* (London, 1658).

[4] Mark Forster, *Arithmetical Trigonometry* (London, 1690).

[5] Taken from *Isaac Newton* (ed. W. J. Greenstreet; London, 1927), p. 29.

During the latter part of the seventeenth century and during the eighteenth century the extreme contractions "s," "t," etc., continued in England along with the use of the moderate contractions "sin.," "tan.," etc. Occasionally slight variations appear. Seth Ward, in his

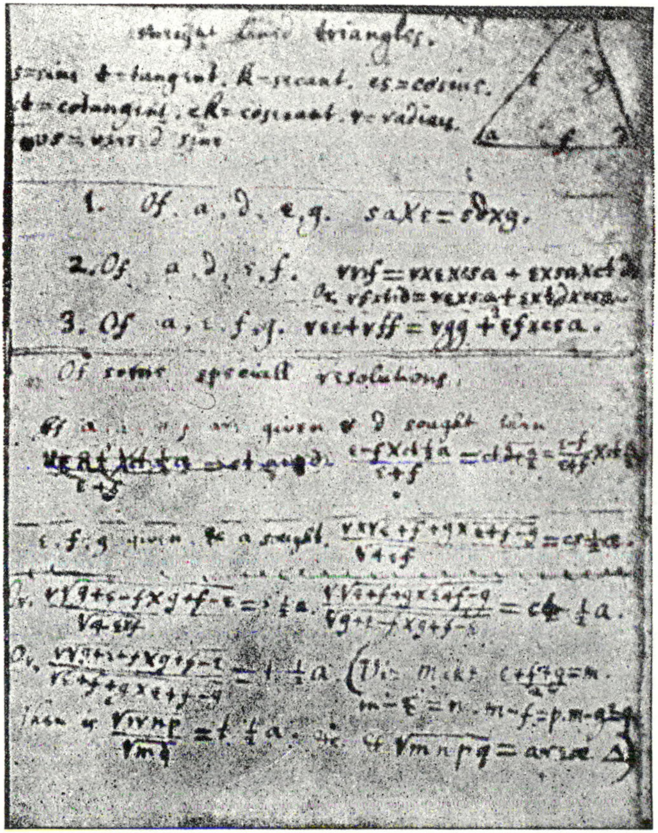

FIG. 120.—A page of Isaac Newton's notebook, showing trigonometric symbols and formulae.

Idea trigonometriae demonstratae (Oxford, 1654), used the following notation:

s. *Sinus*	⟩ *Complementum*
s' *Sinus Complementi*	Z *Summa*
t *Tangens*	X *Differentia*
𝒯 *Tangens Complementi*	C, *Crura*

The notation given here for "sine complement" is a close imitation of the s^* found in the anonymous (Oughtred's?) "Appendix" to E. Wright's translation of John Napier's *Descriptio* of the year 1618. Seth Ward was a pupil of Oughtred who claimed Seth Ward's exposition of trigonometry as virtually his own.

In the same year, 1654, John Newton published at London his *Institutio mathematica*. His idea seems to have been not to use symbolism except when space limitations demand it. On page 209, when dealing with formulae in spherical trigonometry, he introduces abruptly and without explanation the contractions "s," "t," "cs," "ct" for "sine," "tangent," "cosine," "cotangent." From that page we quote the following:

$$\text{``} 4 \begin{Bmatrix} \text{Ra.} + \text{cs } ABF \\ \text{t}FB + \text{ct } BC \end{Bmatrix} \quad \text{is equall to} \quad \begin{Bmatrix} \text{ct } AB + \text{t}FB \\ \text{cs } FBC + \text{Ra.} \end{Bmatrix}\text{''}$$

"1. cs AB+cs FC is equall to cs BC+cs AF.
2. s AF+ct C is equall to s FC+ct A. "

The capital letters are used in lettering the triangle and the small letters in abbreviating the four trigonometric lines. The two are easily distinguished from each other. It is to be noted that in these formulae the "s," "cs," "t," "ct" mean the *logarithmic* lines, but on the same page they are used also to represent the natural lines, in the proportion "As cs AF, to cs FC; so is cs AB, to cs BC."

In his *Trigonometria Britannica* (London, 1658), John Newton uses the abbreviations "s," "t," "ct" in the portions of the work written by himself, but not in those translated from Briggs or Gellibrand. On pages 4 and 5 he uses "s" for "sine" in proportions, such as "Rad. DE :: s $E.DL$," and on page 66 he uses (in similar proportions) "t" and "ct" for "tangent" and "cotangent."

In the British Museum there is a copy of a pamphlet of a dozen engraved pages, prepared sometime between 1654 and 1668, by Richard Rawlinson, of Oxford, which contains "s" for "sine," "t" for "tangent," ᵓ for "complement." But Rawlinson introduces another notation which is indeed a novelty, namely, the designation of the sides of a triangle by the capital letters A, B, C and the opposite angles by the small letters a, b, c, respectively—a notation usually ascribed to the Swiss mathematician Leonhard Euler who used a similar scheme for the first time in the *Histoire de l'académie de Berlin* (année 1753), p. 231. Rawlinson antedates Euler nearly a century, but for some reason (perhaps his obscurity as a mathematician and his lack of assertive personality), his extremely useful and important suggestion, so very simple, failed to attract the notice of mathematicians. In his notation, A was the largest and C the smallest side of a triangle. It is interesting to see how he distinguishes between plane and spherical triangles, by writing the letters in different script. Each letter for spherical triangles was curved in its parts; each letter for plane triangles had a conspicuous straight line as a part of itself. Mathematicians as a group have never felt the need of a notation bringing out this distinction.

John Seller[1] in his *Practical Navigation* (London, 1671[?]), adopts the abbreviations for the trigonometrical lines due to Norwood. That is, he designates the cosine "sc," not "cs"; and the cotangent "tc," not "ct."

Vincent Wing[2] adopted "s.," "t.," "sec.," "cs.," "ct."; William Leybourn[3] preferred "S.," "T.," "SC.," and "TC." for "sine," "tangent," "cosine," "cotangent"; and Jonas Moore,[4] "S.," "Cos.," "T.," "Cot."

John Taylor, in his *Thesaurarium mathematicae* (London, 1687), uses "S," "Sc," "T," "Tc," "Se," "Sec," as abbreviations for "sine," "cosine," "tangent," "cotangent," "secant," "cosecant."[5] Samuel Jeake,[6] in 1696, writes "s.," "cos.," "t.," "cot.," "sec.," "cosec."

Some new symbols were introduced by John Wallis and John Caswell in a tract on trigonometry published in the 1685 and 1693 editions of Wallis' *Algebra*. Caswell's abbreviations for the trigonometric functions, as they appear in his proportions, are "S." for "sine," "Σ." for "cosine," "T" for "tangent," "τ." for "cotangent," "$\int$" for "secant," "σ" for "cosecant," "V" for "versed sine," "v" for "$1+\Sigma$." Observe that the co-functions are designated by Greek letters. These abbreviations were introduced by John Wallis in his tract on "Angular Sections."[7] Half the perimeter of a triangle is represented by ζ. Caswell uses also ζ to represent half the sum of the sides of a triangle.

[1] See *Quarterly Journal of Mathematics*, Vol. XLVI (1915), p. 159.

[2] *Astronomia Britannica, authore Vincentio Wing, Mathem.* (London, 1669), p. 16, 48.

[3] William Leybourn, *Panorganon: or, a Universal Instrument* (London, 1672), p. 75.

[4] Jonas Moore, *Mathematical Compendium* (London, 1674). See A. von Braunmühl, *op. cit.*, Vol. II (1903), p. 45.

[5] *Quarterly Journal*, Vol. XLVI, p. 164.

[6] Samuel Jeake, ΛΟΓΙΣΤΙΚΗΛΟΓΙΑ, *or Arithmetick* (London, 1696,) p. 11.

[7] John Wallis, *Opera* (Oxford), Vol. II (1693), p. 591–92.

The expression

$$\text{``4 } mn \cdot Z + B : \times : Z - B \,.\colon Rq \cdot \Sigma^q \tfrac{1}{2} \text{ Ang.''}$$

means

$$4 \, mn : (Z+B)(Z-B) = R^2 : \cos^2 \tfrac{1}{2} \text{ Angle },$$

where m, n, B are the sides of a plane triangle, $Z = m + n$, R is the radius of the circle defining the trigonometric functions, and the "angle" is the one opposite the side B. Another quotation:

$$\text{``T } 60° + \text{T } 15° = \mathcal{f} \ 60° = 2 \text{ Rad. and } \textit{th.} \text{ 2T } 60° + \text{T } 15° =$$
$$\text{T } 60° + 2R = \text{T } 60° + \mathcal{f} \ 60° = \text{(by 1st Theorem) T } 75° \text{ ,''}$$

which signifies

$$\tan 60° + \tan 15° = \sec 60° = 2 \ R, \text{ and therefore}$$
$$2 \tan 60° + \tan 15° = \tan 60° + 2R = \tan 60° + \sec. 60° =$$
$$\text{(by 1st Theorem) } \tan 75° \text{ .}$$

Caswell's notation was adopted by Samuel Heynes,[1] J. Wilson, J. Ward, and some other writers of the eighteenth century.

523. An English writer, Thomas Urquhart,[2] published a book in which he aimed to introduce a new universal language into trigonometry. He introduced many strange-sounding words: "that the knowledge of severall things representatively confined within a narrow compasse, might the more easily be retained in memory susceptible of their impression." This is a case of misdirected effort, where a man in the seclusion of his chamber invents a complete artificial language, with the expectation that the masses outside will accept it with the eagerness that they would gold coins dropped to them. Urquhart marks the first case, in the solution of a plane right triangle, *Uale* or *Vale;* the first two vowels, *u* or *v*, and *a* give notice of the data, the third vowel of what is demanded. A reference to the table of abbreviations yields the information that *u* means "the subtendent side," *a* means a given angle, and *e* means the required side. Hence the problem is: "Given an [acute] angle of a right triangle and the hypothenuse, to find a side." The mode of solution is indicated by the relation "Rad-U-Sapy ☞ Yr." in which "Sapy" means artificial sine of the angle opposite the side required, "U" means the logarithm of the given hypothenuse, "Yr." the logarithm of the side required. The author had one admiring friend whose recommendation follows the Preface to the book.

[1] Sam. Heynes, *A Treatise of Trigonometry* (2d ed.; London, 1702), p. 6 ff.

[2] *The Trissotetras: or, a most Exquisite Table for resolving all manner of Triangles. . . . . .* By Sir Thomas Urquhart of Cromartie, Knight (London, 1645), p. 60 ff.

We close with a remark by von Braunmühl on the seventeenth-century period under consideration: "Much the same as in Germany during the last third of the seventeenth century were conditions in other countries on the Continent: Trigonometric developments were brought together and expounded more or less completely, to make them accessible to a wider circle of readers; but in this the abbreviated notations and the beginnings in the use of formulas, which we have repeatedly encountered in England, received little attention and application."[1]

524. *The eighteenth century.*—During the eighteenth century the use of abbreviations for the trigonometric lines became general on the European Continent, but the symbolism differed from that most popular in England in not being so intensely specialized as to be represented by a single letter. As a rule, the Continent used three letters (as in "sin.," "cos.," "tan.," etc.) in their formulae, while the English, between 1700 and about 1750 or 1760, used only one or two letters (as in s, cs, t). There are of course eighteenth-century authors in England, who use the three-letter abbreviations, just as there are writers on the mainland who employ only one or two letters. But, in general, there existed the difference which we have mentioned. Then again, as a rule, the continental writers used dots after the abbreviations, while the English did not. The trend of eighteenth-century practice is established in the following tables.

But we first give symbols proposed by Johann III Bernoulli (Jean Bernoulli, 1744–1807)[2], which do not lend themselves to tabular exhibition. In an effort to produce a symbolic trigonometry for writing succinctly the fundamental theorems, he introduced for right spherical triangles the symbols: $\lrcorner$ *l'angle droit,* $\smallsmile$ *l'hypothénuse,* $\wedge$ *un angle,* $\frown$ *un côté,* $\measuredangle$ *l'autre angle,* $\frown$ *l'autre côté,* $\bowtie$ *l'angle adjacent,* $\measuredangle$ *l'angle opposé,* $\in$ *le côté adjacent,* $\ni$ *le côté opposé.* For oblique spherical triangles he introduced the symbols: $\boxslash$ *un triangle sphérique rectangle,* $\triangle$ *un triangle sphérique obliquangle,* $\boxbslash$ *tout triangle sphérique rectangle,* $\boxbslash$ *tout triangle sphérique obliquangle,* $\sqcap$ *de même espece,* $\dashv$ *de différente espece,* $\vee$ *aigu,* $\smallsmile$ *obtus,* SS *segmens,* $\wedge\!\!\wedge$ *angles à la base,* $\wedge$ *angles au sommet.* He writes the formulae: Dans $\boxslash$ le R: $\smallint$in. $\smallsmile$ :: $\smallint$in. $\wedge$: $\smallint$in. $\ni$; Dans $\boxbslash$ les $\smallint$inus des SS :: les cotgts des $\wedge\!\!\wedge$.

[1] A. von Braunmühl, *Vorlesungen über Geschichte der Trigonometrie*, Vol. II (Leipzig, 1903), p. 50.

[2] Jean Bernoulli, *Recueil pour les Astronomes*, Tome III (Berlin, 1776), p. 133, 158, and Planche V. For this reference I am indebted to R. C. Archibald, of Brown University.

525. *Trigonometric symbols of the eighteenth century.*—

No.	Date	Author	Sine	Cosine	Tangent	Cotangent	Secant	Cosecant	Versed Sine
1	1704	W. Leybourn	s	cs	t	ct			
2	1706	W. Jones	s	s	t	ل	ſ	ſ	ν
3	1753	W. Jones	s	s·	t	t	f	f	ν
4	1714	J. Wilson	S, s	Σ, σ	T, t	τ			VS
5	1714	E. Wells	s	σ	t				
6	1720	J. Kresa	S.	S. 2.	T.	T. 2.	Sec.	S. 2.	
7	1726	J. Keill	S	Cos.	T.	Cot.			V.S.
8	1727	Ph. Ronayne	S	Σ	T, t	τ	ſ	σ	ν
9	1727	F. C. Maier	S, s	C, c	T, t				
10	1729	L. Euler	ſ	coſ, cos		ſc	ſ		
11	1730	J. Ward	S	Σ	τ	τc			
12	1737	Th. Simpson	Sine	Co-sine	Tangent	Co-tangent	Secant	Co-sec.	
13	1750	Th. Simpson	Sin.	Co-ſ.	Tang.	Cot., Co-t.	sec.	Co-seca.	
14	1748	L. Euler	sin.	cos.	tang.	cot.	sec.	cosec.	
15	1753	L. Euler	sin, sn	cos, cs	tang, tag, tng, tg				
16	1755	C. E. L. Camus	S.	co-S.	T	co-T	Séc.	co-Séc.	
17	1758	A. G. Kästner	sin	cos, Cos, cosin	tang, Tang	cot	sec, Sec	cosec	
18	{1754 1763}	D'Alembert	Sin, sin, sin	Cos, cos, cos	tang.	cot.			
19	1762	E. Waring	s, S, σ, sin.		Tan.		Sec S	Sec Com S	
20	1765	A. R. Mauduit	s	c	t	τ	Sec S		ν
21	1767	J. A. Segner	sin.	c	tan.	cot.			
22	1768	D. F. Rivard	sin.	cos.	tang.	cot.	séc.		
23	1770	P. Steenstra	S.	Cos.	T.	CoT., Cot.			
24	1770	S. Klügel	sin	cos	tang	cot	sec	cosec	
25	1772	C. Scherffer	sin.	cos.	tang.	cot.	sec.	cosec	
26	1772	G. de Koudon	sin.	cos.	tang.		sec.	coséc.	sinv.
27	1772	O. Gherli	Sen	Cos.	Tang.	Cot.	Sec.	Cosec.	
28	1774	J. Lagrange	sin	cos	tang, tang.				

No.	Date	Author	Sine	Cosine	Tangent	Cotangent	Secant	Cosecant	Versed Sine
29	1774	Sauri	sin.	co-sin.	tang,	co-tang,	séc.	co-séc.	...
30	1778	L. Bertrand	sin.	cos.	tang,	cot.	sec.	cosec.	...
31	1782	P. Frisius	sin.	cos.	tang,	cot.	sec.	cosec.	...
32	1782	P. Ferroni	Sin.	Cosin.	Tang.	Cotang.	Sec.	Cosec.	...
33	1784	G. Vega	sin	cos	tang	cot	sec	cosec	sinvers
34	1786	J. P. de Gua	$\{\Sigma, \Pi, \Gamma$ σ, π, γ	S, P, G s, p, g	$I\Sigma, I\Pi, I\Gamma$ $I\sigma, I\pi, I\gamma$	IS, IP, IG Is, Ip, Ig	$\Gamma\Sigma, \Gamma\Pi, \Gamma\Gamma$ $\Gamma\sigma, \Gamma\pi, \Gamma\gamma$	$\Gamma S, \Gamma P, \Gamma G$ $\Gamma s, \Gamma p, \Gamma g$	...
35	1786	J. B. J. Delambre	sin.	cos.	tang,	cot.	séc.	coséc.	...
36	1786	A. Cagnoli	sin.	cos.	tang,	cot.	sec.	coséc.	...
37	1787	D. Bernoulli	sin.	cos.	tang,	...	séc.	...	...
38	1787	J. J. Ebert	Sin.	Cos.	Tang., T.	Cot.	...	...	...
39	1788	B. Bails	sen.	cos., cosen.	tang,	cotang.	secante	cosec	...
40	1790	J. A. Da Cunha	sen	cos	tang	cot.	see	cosec	...
41	1794	P. Paoli	sen.	cos.	tang,	cot.	sec.	cosec.	...
42	1795	S. l'Huilier	sin.	cos.	tang,	cot.	sec.	...	...
43	1795	Ch. Hutton	sin., s.	cosin., s'	tan., t.	cotan., t'	...	...	...
44	1797	E. Bézout	sin.	cos.	tang,	cot.	séc.	...	...
45	1799	von Metzburg	Sin., S.	Cos.	T.	Cot.	Sec.	Cosec.	Sin. vers.
46	1800	J. F. Lorenz	sin	cos	tang	cot	...	...	...

REFERENCES TO EIGHTEENTH-CENTURY WRITERS CITED IN THE TABLE

1. William Leybourn, *Janua Mathematica*, Part II (London, 1704), p. 70.
2. William Jones, *Synopsis palmariorum matheseos* (London, 1706), p. 240.
3. William Jones, *Philosophical Transactions* (London, 1753), p. 560.
4. John Wilson, *Trigonometry* (Edinburgh, 1714), p. 41, 43.
5. Edward Wells, *The Young Gentleman's Trigonometry* (London, 1714), p. 15, 17.
6. Jakob Kresa, *Analysis speciosa Trigonometriae* (Pragae, 1720), p. 288.
7. John Keill, *Elements of Trigonometry* (Dublin, 1726), last page.
8. Philip Ronayne, *Treatise of Algebra* (London, 1727), Book II.
9. Friedrich Christian Maier, *Trigonometria* in *Commentarii Academiae Scient. Petropolitanae ad annum 1727*, Vol. II, p. 12–30.
10. L. Euler, *Commentarii Academiae Scient. Petropolitanae, ad annum 1729*, Pet. 1735.
11. John Ward, *Posthumous Works* (London, 1730), p. 437.
12. Thomas Simpson, *Treatise of Fluxions* (London, 1737), p. 176, 192.
13. Thomas Simpson, *Doctrine and Application of Fluxions* (London, 1750), p. 278–88.
14. L. Euler, *Introductio in analysin infinitorum* (Berlin, 1748), Vol. I, p. 93, 94, 103, 104.
15. Leonard Euler, *Histoire de l'Académie de Berlin* (Berlin, 1753), p. 223–57, 258–93. Euler exerted himself to arithmetize trigonometry and to use sin z, cos z, and tan z as functions of z. His efforts along this line are especially displayed in this article, "Svbsidium calcvli sinvvm," of 1754, where, conscious of the importance of his innovation, he states: "Simili autem modo mihi equidem angulorum sinus tangentesque primus in calculum ita transtulisse videor, ut instar reliquarum quantitatum tractari, cunctaeque operationes sine vllo impedimento peragi possent. . . . Tamen haec ipsa notandi ratio post modum vniversae analysi tanta attulit adiumenta, vt nouum fere campum patefecisse videatur, in quo Geometrae non sine notabili elaborauerint fructu. Ac si quidem ipsus Analysis praestantiam spectamus, eam praecipue soli idoneo quantitates signis denotandi modo tribuendam esse deprehendimus, quo minus erit mirandum, si commoda sinuum in algorithmum introductio tantum lucri attulerit." ("Similarly I believe that I have first introduced the sines and cosines of angles into the calculus, so that they can be treated like other quantities and can be dealt with in extended operations without any hindrance . . . so that this very mode of designation brings to general analysis great aid, that almost a new field is thereby seen to be opened up, in which geometers have worked not without great results. And if we behold the excellence of analysis we recognize that it must be attributed to the mode of denoting quantities by signs, hence it is less to be marveled if the skilful introduction of the sine into the algorithm has brought such great advantage.")
16. C. E. L. Camus, *Cours de mathématique*, 2. Partie (nouv. éd.; Paris, 1755), p. 402.
17. A. G. Kästner, *Anfangsgründe der Arithmetik, Geometrie . . . . Trigonometrie* (Göttingen, 1758), p. 339–47.
18. D'Alembert, *Recherches sur differens points importans du Systeme du Monde* (Paris), Vol. I (1754), p. 130, 131; Vol. II (1754), p. 35; Vol. III (1756), p. 11, 158.
19. E. Waring, *Miscellanea analytica* (Cambridge, 1762), p. 66, 91, 141.
20. Antoine Remi Mauduit, *Principes d'Astronomie Spherique* (Paris, 1765).

21. J. A. Segner, *Cursus mathematici*, Pars I (Halle, 1767), p. 336–40.

22. Dominique-François Rivard, *Éléments de mathématiques* (6th ed.; Paris, 1768), Seconde Partié, p. 288–90.

23. P. Steenstra, *Verhandeling over de Klootsche Drichoeks Meeting* (Holland, 1770), p. 50 ff.

24. Simon Klügel, *Analytische Trigonometrie* (Halle, 1770); A. von Braunmühl, *op. cit.*, Vol. II, p. 123, 135; G. S. Klügel, *Mathematisches Wörterbuch*, Vol. I (Leipzig, 1803), "Differentialformeln," p. 873.

25. Carolo Scheffler, *Institutionum analyticarum pars secunda* (Vindobonae, 1772), p. 134, 159, 163, 178.

26. Girault de Koudon, *Leçons analytiques du Calcul des Fluxions* (Paris, 1772), p. 35–40.

27. O. Gherli, *Gli elementi teorico-pratici delle matematiche pure* (Modena, 1772), Vol. III, p. 228, 272.

28. J. Lagrange, *Nouveaux mémoires de l'académie r.d. sciences et belles-lettres*, anno 1774 (Berlin, 1776), p. 279–87.

29. L'Abbé Sauri, *Cours complet de mathématiques*, Vol. I (Paris, 1774), p. 437, 445.

30. Louis Bertrand, *Developpement nouveau de la partie elementaire des mathematiques*, Vol. II (Geneve, 1778), p. 407–32.

31. Paul Frisius, *Operum tomus Primus* (Mediolani, 1782), p. 303.

32. Pietro Ferroni, *Magnitudinum exponentialium logarithmorum nova methodo pertractata* (Florentiae, 1782), p. 537.

33. Georg Vega, *Vorlesungen über die Mathematik* (Wien), Band II (1781), p. 180–84.

34. J. P. de Gua, *Mémoires de l'académie de Paris* (1786), p. 291–343. In a spherical triangle $G \, S \, P$, he writes cos $S = S$, cos $P = P$, cos $G = G$, sin $S = \Sigma$, etc.; if the sides are s, p, g, then sin $s = \sigma$, etc.

35. J. B. J. Delambre, in *Mémoires de l'Institut National de Sciences* (*Math et Phys.*), Vol. V, p. 395.

36. A. Cagnoli, *Traité de Trigonométrie* (Vérone, 1786), p. 3.

37. Daniel Bernoulli, *Nova Acta Helvetica phys.-math.* (Basel, 1787). Vol. I, p. 232.

38. Johann Jacob Ebert, *Unterweisung in den philos. u math. Wissenschaften* (Leipzig, 1787).

39. Benito Bails, *Principios de matematica de real academia* (Madrid, 1788; 2d ed.). Vol. I, p. 328.

40. J. A. da Cunha, *Principios Mathematicos* (Lisboa, 1790). p. 206.

41. Pietro Paoli, *Elementi d'Algebra*, Vol. II (Pisa, 1794), p. 94, 95, 125, 129.

42. Simon l'Huilier, *Principiorum calculi differentialis et integralis expositio* (Tubingae, 1795), p. 263.

43. Charles Hutton, *Mathematical and Philosophical Dictionary* (1795), art. "trigonometry."

44. E. Bézout, *Cours de Mathématiques*, Tome I (Paris, 1797), p. 200.

45. Freyherr von Metzburg, *Anleitung zur Mathematik*, Vol. III (Wien, 1799), p. 8.

46. Johann Friedrich Lorenz, *Grundlehren der allgemeinen Grössenberechnung* (Helmstadt, 1800), p. 306, 307.

526. Trigonometric symbols of the nineteenth century.—

No.	Date	Author	Sine	Cosine	Tangent	Cotangent	Secant	Cosecant
1...	1803	S. F. Lacroix	sin	cos	tang	cot	séc	coséc
2...	1811	S. D. Poisson	sin.	cos.	tan.	......	......	......
3...	1812	J. Cole	sin.	cos.	tan.	cot. cotan.	sec.	cosec.
4...	1813	A. L. Crelle	sin	cos	tang	cot	sec	cosec
5...	1814	P. Barlow	sin.	cos.	tan.	cot.	sec.	cosec.
6...	1814	C. Kramp	Sin.	Cos.	Tang.	Cot.	......	......
7...	1817	A. M. Legendre	sin	cos	tang	cot	séc	coséc
8...	1823	J. Mitchell	sin.	cos.	tan.	cotan.	sec.	cosec.
9...	1823	G. U. A. Vieth	sin.	cos.	tang.	cot.	sec.	......
10...	1826	N. H. Abel......	sin	cos	tang	......	......	......
11...	1827	J. Steiner	......	......	tg	......	......	......
12...	1827	C. G. J. Jacobi	sin	cos	tg, tang.	cotg	......	......
13...	1829	M. Ohm	Sin	Cos	Tg	Cotg	Sec	Cosec
13a..	1830	W. Bolyai	ⓢx	ⓢx	ⓣx	ⓣx	ⓢx	ⓙx
14...	1836	J. Day	sin	cos	tan	cot	sec	cosec
15...	1840	A. Cauchy	sin	cos	tang	......	......	......
16...	1861	B. Peirce	sin.	cos.	tang.	cotan.	sec.	cosec.
17...	1862	E. Loomis	sin.	cos.	tang.	cot.	sec.	cosec.
18...	1866	F. M. Pires	Sen	Cos	tg	Cot	......	......
19...	1875	J. Cortazar	sen	cos	tg	cot	......	......
20...	1875	I. Todhunter	sin	cos	tan	cot	sec	cosec
21...	1880	J. A. Serret	sin	cos	tang	......	......	......
22...	1881	Oliver, Wait, and Jones	sin	cos	tan	cot	sec	csc
23...	1886	A. Schönflies	......	......	tg	ctg	......	......
24...	1890	W. E. Byerly	sin	cos	tan	ctn	sec	csc
25...	1893	O. Stolz	sin	cos	tan	cot	sec	cosec
26...	1894	E. A. Bowser	sin	cos	tan	cot	sec	cosec
27...	1895	C. L. Dodgson	⌒	⌒	......	......	......	......
28...	1897	G. A. Wentworth	sin	cos	tan	cot	sec	csc
29...	1903	G. Peano	sin, s	cos , c	tng, t	/tng	/cos	/sin
30...	1903	Weber, Wellstein	sin	cos	tg	cotg	sec	cosec
31...	1911	E. W. Hobson	sin	cos	tan	cot	sec	cosec
32...	1913	Kenyon, Ingold	sin	cos	tan	ctn	sec	csc
33...	1917	L. O. de Toledo	sen	cos	tg	ctg	sec	cosec
34...	1911	{A. Pringsheim J. Molk}	sin	cos	tg	cot	séc	coséc
35...	1921	H. Rothe	sin	cos	tg	cot	sec	cosec
36...	1921	G. Scheffers......	sin	cos	tg	ctg	......	......

REFERENCES TO NINETEENTH-CENTURY WRITERS CITED IN THE TABLES

1. S. F. Lacroix, *Traité élémentaire de trigonométrie* (Paris, 1803), p. 9–17.
2. S. D. Poisson, *Traité de mécanique* (Paris, 1811), Vol. I, p. 23.
3. John Cole, *Stereogoniometry* (London, 1812).
4. August Leopold Crelle, *Rechnung mit veränderlichen Grössen*. Band I (Göttingen, 1813), p. 516, 517.
5. Peter Barlow, *Dictionary of Pure and Mixed Mathematics* (London, 1814), art. "Trigonometrical Definitions."
6. C. Kramp, *Annales de mathématiques pures et appliquées* (Nismes, 1814 and 1815), Vol. V, p. 11–13.
7. A. M. Legendre, *Éléments de géométrie* (11th ed.; Paris, 1817), p. 341.
8. James Mitchell, *Cyclopedia of the Mathematical and Physical Sciences* (1823), art. "Fluxion."

9. G. U. A. Vieth, *Kurze Anleitung zur Differentialrechnung* (Leipzig, 1823), p. 40–42.

10. N. H. Abel, *Journal für die r. u. a. Mathematik* (A. L. Crelle; Berlin), Vol. I (1826), p. 318–37; Vol. II (1827), p. 28.

11. J. Steiner, *Journal für die r. u. a. Mathematik* (A. L. Crelle; Berlin), Vol. II (1827), p. 59.

12. C. G. J. Jacobi, *Journal für die r. u. a Mathematik* (Berlin), Vol. II (1827), p. 5, 188.

13. Martin Ohm, *System der Mathematik*, 3. Theil (Berlin, 1829), p. 21, 79, 162.

13a. Wolfgang Bolyai, *Az arithmetica eleje* (Maros-Vásárhelyt, 1830). See B. Boncompagni, *Bullettino*, Vol. I (1868), p. 287.

14. Jeremiah Day, *Plane Trigonometry* (New Haven, 1836; 4th ed.), p. 54.

15. A. Cauchy, *Exercices d'Analyse*, Tome I (Paris, 1840).

16. Benjamin Peirce, *Plane and Spherical Trigonometry* (rev. ed.; Boston and Cambridge, 1861), p. 7.

17. Elias Loomis, *Trigonometry* (New York, 1862), p. 52.

18. Francisco Miguel Pires, *Tratado de Trigonometria Espherica* (Rio de Janeiro, 1866).

19. Juan Cortazar, *Tratado de Trigonometria* (15th ed.; Madrid, 1875), p. 5.

20. I. Todhunter, *Differential Calculus* (7th ed.; London, 1875), p. 49.

21. J. A. Serret, *Traité de Trigonométrie* (6th ed.; Paris, 1880), p. 6.

22. Oliver, Wait, and Jones, *Treatise on Trigonometry* (Ithaca, 1881).

23. A. Schönflies, *Geometrie der Bewegung* (Leipzig, 1886), p. 71.

24. W. E. Byerly, *Elements of the Integral Calculus* (Boston, 1890), p. 21, 22, 28, 30, 65.

25. Otto Stolz, *Grundzüge der Differential and Integralrechnung* (Leipzig, 1893), p. 38–42.

26. Edward A. Bowser, *Trigonometry* (Boston, 1894), p. 12.

27. Charles L. Dodgson, *Curiosa mathematica* (4th ed.; London, 1895), p. xix.

28. G. A. Wentworth, *Plane and Spherical Trigonometry* (Boston, 1897).

29. G. Peano, *Formulaire mathématique*, Vol. IV (1903), p. 228, 229.

30. Weber, Wellstein, *Encyklopädie der Elementar-Mathematik* (Leipzig, 1903).

31. E. W. Hobson, *Treatise on Plane Trigonometry* (3d ed.; Cambridge, 1911), p. 18, 32.

32. A. M. Kenyon and L. Ingold, *Trigonometry* (New York, 1913), p. 8.

33. L. O. de Toledo, *Tradado de Trigonometria* (tercera ed.; Madrid, 1917), p. 37.

34. A. Pringsheim, J. Molk, *Encyclopédie des sciences math.*, Tome II, Vol. II (1911), p. 69.

35. Hermann Rothe, *Vorlesungen über höhere Mathematik* (Wien, 1921), p. 107.

36. Georg Scheffers, *Lehrbuch der Mathematik* (Berlin und Leipzig), 5. Aufl. (1921), p. 498.

527. *Less common trigonometric functions*.—The frequent occurrence in practice of certain simple trigonometric expressions has led to suggestions of other functions. Thus, in navigation $\frac{1}{2}(1 - \cos A)$ has led to symbolisms such as[1] "$\frac{1}{2}$ sinver A" and "hav A" (i.e., haversine or half the versed sine of A). Another symbol "covers A"[2] or "cvs A"[3] stands for $1 - \sin A$. W. Bolyai[4] marked the versed cosine of x by

[1] A. von Braunmühl, *Vorlesungen über Geschichte der Trigonometrie*, 2. Teil (Leipzig, 1903), p. 231; P. R. Rider and A. Davis, *Plane Trigonometry* (New York, 1923), p. 42. The haversine function first appears in the tables of logarithmic versines of José de Mendoza y Rios (Madrid, 1801, also 1805, 1809), and later in a treatise on navigation of James Inman (1821). See J. D. White in *Nautical Magazine* (February and July, 1926).

[2] William Jones, op. cit., in 1706, represented the coversed sine by v, and A. R. Mauduit, *op. cit.*, in 1765, by u. The designation "covers A" is found, for example, in G. A. Wentworth, *Trigonometry* (2d ed.; Boston, 1903), p. 5; H. H. Ludlow, *Trigonometry* (3d ed., 1891), p. 33; A. M. Kenyon and L. Ingold, *Trigonometry* (New York, 1913), p. 8, 9.

[3] F. Anderegg and E. D. Rowe, *Trigonometry* (Boston, 1896), p. 10.

[4] Wolfgang Bolyai, *Az arithmetica eleje* (Maros-Vásárhelyt, 1830). See B. Boncompagni, *Bullettino*, Vol. I (1868), p. 287.

⊚x, and the versed sine of x by ⊙. A third, "exsec A," i.e., "external secant of A," signifies[1] sec $A - 1$.

528. *Quaternion trigonometry.*—I. Stringham[2] outlined a quaternion trigonometry by defining the cosine and sine of the angle between the vectors a and β in a right triangle whose sides are the vectors a, β, δ, in this manner, cq $\dfrac{\beta}{a} = \dfrac{a}{\beta}$, Sq $\dfrac{\beta}{a} = \dfrac{\delta}{\beta}$, where "cq" signifies "quaternion cosine" and "sq" signifies "quaternion sine." He obtains sq $\dfrac{\beta}{a}$+cq $\dfrac{\beta}{a} = 1$, which yields in scalar trigonometry T^2sq $\dfrac{\beta}{a}$+T^2 cq $\dfrac{\beta}{a} = 1$, "T" meaning "tensor."

529. *Hyperbolic functions* were first introduced in 1757 by Vicenzo Riccati[3] who used the notation Shx, Chx for hyperbolic sine and cosine. He writes our $\cosh^2 x - \sinh^2 x = 1$ in this manner, "$\overline{\text{Ch}.\mu}^2 - \overline{\text{Sh}.\mu}^2 = r^2$," where μ is the arc and r the *sinus totus*. Hyperbolic functions were further developed in 1768 by J. H. Lambert[4] who writes "sin h$(k-y)$" and "cos hk" for the hyperbolic sine and cosine. This use of "h" after the ordinary trigonometric contractions has retained its place in many books to the present time. Seventeen years after Riccati, L'Abbé Saurin[5] represented the *sinus hyperbolique* by "s.h.," the *co-sinus hyperbolique* by "c.h.," the *tangente hyperbolique* by "t.h.," the *co-tangente hyperbolique* by "cot.h.," and wrote

$$\text{``c.h. } n. \ x = \frac{(\text{c.h. } x + \text{s.h. } x)^n + (\text{c.h. } x - \text{s.h. } x)^n}{2 \cdot r^{n-1}} . \text{''}$$

Somewhat later Frullani[6] writes

$$\text{``} \int_0^\infty \text{Cos h sin h } x \text{ dh} = \pi . \text{''}$$

[1] R. G. Hudson and J. Lipka, *A Manual of Mathematics* (New York, 1917), p. 68; A. M. Kenyon and L. Ingold, *Trigonometry* (1913), p. 5.

[2] I. Stringham in *Johns Hopkins University Circulars*, I (1880), p. 35. See von Braunmühl, *op. cit.*, Vol. II, p. 246.

[3] *Institutiones analyticae a Vincentio Riccato . . . . et Hieronymo Saladino . . . .* Tomus Secundus (Bologna, 1767), p. 152.

[4] *Histoire de l'académie de Berlin* (1768), Vol. XXIV, p. 327; reference taken from S. Günther, *Lehre von den Hyperbelfunktionen* (Halle a/S., 1881).

[5] L'Abbé Sauri, *Cours complet de Mathématiques*, Tome IV (Paris, 1774), p. 222, 223.

[6] Frullani in *Mem. di mat. e di fis. della società italiana delle scienze*, Tomo XX, p. 66. See S. Günther, *op. cit.*, p. 36.

J. Houël[1] in 1864 uses "Sh," "Ch," "Th," and introduces "Amh u" to represent what he called the "hyperbolic amplitude of u" in analogy with the amplitude of an elliptic function. J. A. Serret[2] gives in 1857 the symbolism "cos. h. x," "sin. h. x," "tang. h. x."

For hyperbolic functions, E. W. Hobson writes cosh $u = \cos \iota u$, sin $hu = -\iota \sin \iota u$, tan h $u = -\iota \tan \iota u$, and similarly, "coth," "sech," "cosech."

A new notation as well as a new nomenclature was introduced in 1829 by C. Gudermann[3] in an article on "Potential-Funktion." He uses a capital initial letter in the ordinary trigonometric function for the designation of the hyperbolic functions. Moreover, the type used is German. Regarding this choice he himself says:[4] "Finally the choice of the initial syllables printed with German type: $\mathfrak{Cos}$, $\mathfrak{Sin}$, $\mathfrak{Tang}$, $\mathfrak{Cot}$, $\mathfrak{Arc}$. $\mathfrak{Sin}$, $\mathfrak{Arc}$. $\mathfrak{Cos}$, $\mathfrak{Arc}$. $\mathfrak{Tang}$, $\mathfrak{Arc}$. $\mathfrak{Cot}$, and the sign $\mathfrak{L}$ may not meet with favor outside of Germany. But one can take instead the initial syllables written in Latin capitals and for $\mathfrak{L}$ take L and thus by the use of capital letters express the difference from the cyclic functions." Also (p. 290): "To carry out a consideration relating to the construction of the tables, I set up the formula $\mathfrak{Cos}\ \phi \cdot \cos k = 1$ which expresses a relation between the two arcs ϕ and k which I designate by $\phi = \mathfrak{L}k$ and conversely by $k = \mathfrak{l}\phi$. I take $\mathfrak{L}k$ as the Längen-Zahl for the arc k, $\mathfrak{l}\phi$ as the Longitudinal-Zahl for the arc ϕ."

Gudermann's notation has been widely used in Germany. It was used in expository articles by Wilhelm Matzka,[5] J. A. Grunert,[6] Siegmund Günther,[7] and L. A. Sohncke,[8] and as late as 1908 by H. E. Timerding[9] who writes in German type $\mathfrak{Sin}$, $\mathfrak{Cos}$, $\mathfrak{Tang}$, $\mathfrak{Sec}$, as does

[1] J. Houël in *Nouvelles Annales de Mathématiques* (2d ser.), Vol. III (1864), p. 426, 432.

[2] J. A. Serret, *Traité de Trigonométrie* (2d ed.; Paris, 1857), p. 217.

[3] Gudermann in *Crelle's Journal*, Vol. IV (1829), p. 288.

[4] C. Gudermann, "Ueber die Potenzial-Functionen," in *op. cit.*, Vol. IV (1829), p. 295.

[5] W. Matzka in *Archiv der Math. u. Physik*, Vol. XXXVII (1861), p. 408.

[6] J. A. Grunert, *op. cit.*, Vol. XXXVIII (1862), p. 49.

[7] S. Günther, *Lehre von Hyperbelfunktionen* (Halle a/S., 1881).

[8] L. A. Sohncke, *Sammlung von Aufgaben aus der Differen. u. Integ. Rechnung* (Halle, 1885), p. 9.

[9] H. E. Timerding, *Geometrie der Kräfte* (Leipzig, 1908), p. 330. Consult this for quaternion notations (p. 18).

also H. Rothe[1] in 1921; and in 1912 by Hans von Mangoldt[2] who, however, uses Latin type. On the other hand, J. Frischauf[3] prefers small letters of German black-faced type.

530. The German notation and nomenclature made itself felt in America. Benjamin Peirce[4] in 1846 called the hyperbolic functions "potential functions" and designated them by the symbols for the ordinary trigonometric functions, but with initial capital letters; for example, Sin. $B = \frac{1}{2}(e^B - e^{-B})$. It is found also in articles that appeared in J. D. Runkle's *Mathematical Monthly*, for instance, Volume I, page 11.

However, the Gudermannian notation finally gave way at Harvard to that prevalent in France; in James Mills Peirce's *Mathematical Tables* (Boston, 1881), one finds "Sh u," "Ch u."

In England, W. K. Clifford[5] proposed for the hyperbolic sine, cosine, and tangent the signs "hs," "hc," and "ht," or else "sh," "ch," "th." In the United States, Clifford's first suggestion was adopted by W. B. Smith.[6] Prefixing the syllable "hy" was suggested by G. M. Minchin[7] in 1902. He would write "hy sin x," "hy tg x" in order to facilitate enunciation. This suggestion was adopted by D. A. Murray[8] in his *Calculus* of 1908.

531. *Parabolic functions.*—James Booth, in his *Theory of Elliptic Integrals* (1851), and in an article in the *Philosophical Transactions* for the year 1852,[9] develops the trigonometry of the parabola. For the parabola we have tan ω = tan ϕ sec χ + tan χ sec ϕ, ω, ϕ, χ being parabolic arcs. If we make the imaginary transformations tan ω = i sin ω', tan ϕ = i sin ϕ', tan χ = i sin χ', sec ϕ = cos ϕ', sec χ = cos χ', then the foregoing formula becomes sin ω' = sin ϕ' cos χ' + sin χ' cos ϕ', which is the ordinary expression for the sine of the sum of two circular arcs. In the trigonometry of the circle $\omega = \phi + \chi$; in the trigonometry of the parabola ω is such a function of ϕ and χ as will render

[1] Hermann Rothe, *Vorlesungen über höhere Mathematik* (Wien, 1921), p. 107.

[2] Hans von Mangoldt, *Einführung in die höhere Mathematik*, Vol. II (Leipzig, 1912), p. 518; Vol. III (1914), p. 154, 155.

[3] Johannes Frischauf, *Absolute Geometrie* (Leipzig, 1876), p. 54.

[4] Benjamin Peirce, *Curves, Functions, and Forces*, Vol. II (1846), p. 27, 28.

[5] W. K. Clifford, *Elements of Dynamic*, Part I (London, 1878), p. 89.

[6] W. B. Smith, *Infinitesimal Analysis*, Vol. I (New York, 1897), p. 87, 88.

[7] G. M. Minchin in *Nature*, Vol. LXV (London, 1902), p. 531.

[8] D. A. Murray, *Differential and Integral Calculus* (London, 1908), p. 414.

[9] James Booth, "On the Geometrical Properties of Elliptic Integrals," *Philosophical Transactions* (London, 1852), p. 385.

tan $[(\phi, \chi)] =$ tan ϕ sec $\chi +$ tan χ sec ϕ. Let this function (ϕ, χ) be designated $\phi \perp \chi$, so that tan $(\phi \perp \chi) =$ tan ϕ sec $\chi +$ tan χ sec ϕ; let it be designated $(\phi \top \chi)$ when tan $(\phi \top \chi) =$ tan ϕ sec $\chi -$ tan χ sec ϕ. Thus when tan ϕ is changed into i sin ϕ, sec ϕ into cos ϕ, and cot ϕ into $-i$ cosec ϕ, $\perp$ must be changed into $+$, $\top$, into $-$.

532. *Inverse trigonometric functions.*—Daniel Bernoulli was the first to use a symbolism for inverse trigonometric functions. In 1729 he used "A S." to represent "arcsine."[1] Euler[2] in 1736 introduced "A t" for "arctangent," in the definition: "expressio A t nobis denotet arcum circuli, cuius tangens est t existente radio $=1$." Later, in the same publication, he expressed[3] the arcsine simply by "A": "arcus cuius sinus est $\frac{x}{a}$ existente toto sinu $=1$ notetur per A $\frac{x}{a}$." To remove the ambiguity of using "A" for two different arc functions, he introduced[4] the sign "A t" for "arctangent," saying, "At.$\frac{q}{B}$ est arcus circuli cuius tangens est $\frac{q}{B}$ existente sinu toto $=1$." He used the sign "A t" in 1736 also in another place.[5] In 1737 Euler[6] put down "A sin $\frac{b}{c}$" and explained this as meaning the arc of a unit-circle whose sine is $\frac{b}{c}$. In 1744 he[7] uses "A $\overline{\text{tagt}}$" for *arcus, cujus tangens* $=t$. Lambert[8] fourteen years later says, "erit $\frac{x}{m} = \frac{x}{o}$ arcus sinui b respondens." Carl Scherffer[9] in Vienna gives "arc. tang." or "arc. tangent"; J. Lagrange[10] in the same year writes "arc. sin $\frac{1}{1+a}$." In

[1] Daniel Bernoulli in *Comment. acad. sc. Petrop.*, Vol. II (1727; printed 1728), p. 304–42. Taken from G. Eneström, *Bibliotheca mathematica* (3d ser)., Vol. XIV, p. 78. See also Vol. VI (1905), p. 319–21.

[2] L. Euler, *Mechanica sive motus scientia* (Petropoli, 1736), Vol. I, p. 184–86.

[3] L. Euler, *op. cit.*, Vol. II, p. 138.   [4] L. Euler, *op. cit.*, Vol. II, p. 303.

[5] L. Euler, *Comment. acad. sc. Petrop.*, Vol. VIII (1736; printed in 1741), p. 84, 85.

[6] L. Euler in *Commentarii academiae Petropolitanae ad annum 1737*, Vol. IX, p. 209; M. Cantor, *op. cit.*, Vol. III (2d ed.), p. 560.

[7] *Nova acta eruditorum* (1744), p. 325; Cantor, *op. cit.*, Vol. III (2d ed.), p. 560.

[8] *Acta Helvetica Physico-Math.-Botan.*, Vol. III (Basel, 1758), p. 141.

[9] Carolo Scherffer, *Institutionum analyticarum pars secunda* (Vienna, 1772), p. 144, 200.

[10] J. Lagrange in *Nouveaux·mémoires de l'académie r. d. sciences et belles-lettres*, année 1772 (Berlin, 1774), p. 277.

1776 Lambert[1] uses "arc. sin." The notation was adopted by French writers. Marquis de Condorcet[2] wrote arc $(\cos. = (1+P)^{-\frac{1}{2}})$ for our arc $\cos (1+P)^{-\frac{1}{2}}$. Bérard gave "Arc.$\left(\text{Tang.} = \dfrac{dy}{dx}\right)$." In Italy, Pietro Paoli[3] adopts "Arc. sen." for "arcsin."

533. *John Herschel's notation for inverse functions*, $\sin^{-1} x$, $\tan^{-1} x$, etc., was published by him in the *Philosophical Transactions of London*, for the year 1813. He says (p. 10): "This notation $\cos.^{-1} e$ must not be understood to signify $\dfrac{1}{\cos. e}$, but what is usually written thus, arc $(\cos. = e)$." He admits that some authors use $\cos.^m A$ for $(\cos. A)^m$, but he justifies his own notation by pointing out that since $d^2 x$, $\Delta^3 x$, $\Sigma^2 x$ mean ddx, $\Delta\Delta\Delta x$, $\Sigma\Sigma x$, we ought to write $\sin.^2 x$ for sin. sin. x, $\log.^3 x$ for log. log. log. x. Just as we write $d^{-n}\, V = \int^n V$, we may write similarly $\sin.^{-1}\, x = \text{arc } (\sin. = x)$, $\log.^{-1}\, x. = c^x$. Some years later Herschel explained that in 1813 he used $f^n(x)$, $f^{-n}(x)$, $\sin.^{-1} x$, etc., "as he then supposed for the first time. The work of a German Analyst, Burmann, has, however, within these few months come to his knowledge, in which the same is explained at a considerably earlier date. He [Burmann], however, does not seem to have noticed the convenience of applying this idea to the inverse functions $\tan^{-1}$, etc., nor does he appear at all aware of the inverse calculus of functions to which it gives rise." Herschel adds, "The symmetry of this notation and above all the new and most extensive views it opens of the nature of analytical operations seem to authorize its universal adoption."[4] Thus was initiated a notation which has kept its place in many English and American books to the present time.

534. *Martin Ohm's notation for inverse functions.*—In Germany, Ohm introduced a new notation for the inverse trigonometric functions which found no favor. He says: "What is here designated by $\dfrac{1}{\text{Sin}}\, y$, $\dfrac{1}{\cos}\, y$, etc., is usually indicated by arc. Sin y, arc. Cos y, etc., also by ang. $(\text{Sin} = y)$, ang. $(\text{Cos} = y)$, etc. Here this new notation is recommended because it admits a much more convenient mechanism

[1] Lambert in *Nouveaux mémoires de l'académie r. d. sciences et belles-lettres*, année 1776 (Berlin, 1779), p. 12.

[2] N. C. de Condorcet, *Histoire de l'académie r. d. sciences*, année 1769 (Paris, 1772), p. 255.

[3] Pietro Paoli, *Elementi d'algebra* (Pisa, 1794), Vol. II, p. 21.

[4] John F. W. Herschel, *Collection of Examples on Calculus of Finite Differences*, Cambridge, 1820, p. 5, 6.

for the calculus, inasmuch as, analogous to the usual forms of algebra, there follows from $\frac{1}{Tg}$ $y = x$ immediately $y = Tgx$; from $\frac{1}{Cos}$ $z = u$, immediately $z = Cos\ u$; etc."[1]

535. *Persistence of rival notations for inverse functions.*—The eighteenth-century notations, involving the word "arc" or its equivalent, were only mild contractions of the words which defined the meanings to be conveyed to the reader, and they maintained their place in most works of Continental Europe during the nineteenth century. C. Gudermann[2] in 1829 used the notation "arc (tang = z)," etc. In France, J. Houël employed the easier form "arcsin." The only marked inroad of Herschel's notation upon the Continent is found with G. Peano who in 1893[3] adopts the Eulerian $\overline{cos}\ x$, for arc cos x, and later[4] writes $\sin^{-1}$, $\cos^{-1}$, $\tan^{-1}$.

In Mexico the continental and English notations clashed. F. D. Covarrubias[5] writes "arco (sec = x)" and lets $\cos^{-1} x$ stand for $\frac{1}{\cos x}$. On the other hand, Manuel Torres Torija[6] uses "$x = \tan^{-1} y$" and says that it means the same as "$x = $ arc (tan = y)."

In England, Herschel's notation gained ground rapidly, but not instantaneously. We find, for instance, in 1838, "arc $\left(\tan.\ \frac{x}{c}\right)$" in a treatise by John West.[7]

In the United States the British influence greatly predominated during the nineteenth century. A rare exception is the occurrence of the continental symbols "arc tan," "arc csc," in a book of E. P. Seaver.[8] The use of Herschel's notation underwent a slight change in

[1] M. Ohm, *System der Mathematik*, Vol. II (Berlin, 1829), p. 372.

[2] C. Gudermann, *Journal für r. u. a. Mathematik*, Vol. IV (1829), p. 287.

[3] G. Peano, *Lezioni di analisi infinitesimale*, Vol. I (Turin, 1893), p. 43.

[4] G. Peano, *Formulaire mathématique, Edition de l'an 1902–'03* (Turin, 1903), p. 228, 229.

[5] F. D. Covarrubias, *Elementos de análisis trascendente ó cálculo infinitessimal* (2d ed.; Mexico, 1890), p. 48, 49.

[6] Manuel Torres Torija, *Nociones de Álgebra Superior y elementos fundamentales de cálculo differencial é integral* (México, 1894), p. 181.

[7] Rev. John West, *Mathematical Treatises* (ed. John Leslie; Edinburgh, 1838), p. 237.

[8] Edwin P. Seaver, *Formulas of Plane and Spherical Trigonometry* (Boston and Cambridge, 1871), p. 43.

Benjamin Peirce's books, to remove the chief objection to them, Peirce wrote:

$$\text{``cos}^{[-1]} x\text{,''} \quad \text{``tan}^{[-1]} x \text{.''}[1]$$

In the present century the continental notation of Europe has found entrance in the United States through several texts, as, for instance, those of Love,[2] Wilczynski and Slaught,[3] Kenyon and Ingold.[4]

536. *Inverse hyperbolic functions.*—Houël[5] marked the inverse hyperbolic sine, cosine, and tangent thus: "Arg Sh," "Arg Ch," "Arg Th." Similar notations have been used by other writers. The arcsin tanh u is called by Cayley the Gudermannian of u and is written[6] "gd u." When z is a complex variable, the symbolism "Arg sh z" was used by A. Pringsheim, G. Faber, and J. Molk[7] in their joint article in the *Encyclopédie* to indicate the infinitely fold function

$$\text{Arg sh } z = \frac{\text{Arc sin } iz}{i} ,$$ but the authors point out that in one respect the

notation is improper, because the function which is the inverse of "sh$z$" is not defined by the aid of the arc of the curve.

537. *Powers of trigonometric functions.*—Three principal notations have been used to denote, say, the square of sin x, namely, $(\sin x)^2$, $\sin x^2$, $\sin^2 x$. The prevailing notation at present is $\sin^2 x$, though the first is least likely to be misinterpreted. In the case of $\sin^2 x$ two interpretations suggest themselves: first, $\sin x \cdot \sin x$; second,[8] $\sin (\sin x)$. As functions of the last type do not ordinarily present themselves, the danger of misinterpretation is very much less than in case of $\log^2 x$, where $\log x \cdot \log x$ and $\log (\log x)$ are of frequent occurrence in analysis. In his *Introductio in analysin* (1748), Euler[9] writes $(\cos . z)^n$, but in an article of 1754 he adopts[10] $\sin . \psi^3$ for $(\sin \psi)^3$ and writes the

---

[1] B. Peirce, *Curves, Functions and Forces*, Vol. I (new ed.; Boston, 1852), p. 203.

[2] C. E. Love, *Differential and Integral Calculus* (New York, 1910), p. 132.

[3] E. J. Wilczynski and H. E. Slaught, *Plane Trigonometry* (1914), p. 217.

[4] A. M. Kenyon and L. Ingold, *Trigonometry* (1913), p. 102, 105.

[5] J. Houël, *Cours de calcul infinitesimal*, Vol. I (Paris, 1878), p. 199.

[6] G. Chrystal, *Algebra*, Part II (1889), p. 288.

[7] *Encyclopédie des sciences math.*, Tome II, Vol. II (1911), p. 80.

[8] See G. Peano, *Formulaire mathématique*, Vol. IV (1903), p. 229

[9] L. Euler, *Introductio in analysin infinitorum* (Lausannae, 1748), Vol. I, p. 98, 99.

[10] L. Euler in *Novi Comment. acad. scient. imper. Petropolitanae*, for the years 1754, 1755 (Petropoli, 1760), p. 172.

formula "$4 \sin \cdot \psi^3 = -\sin \cdot 3\,\psi + 3 \sin \cdot \psi$." William Jones in 1710 wrote "cs², $\frac{1}{2} < v$" for $\left(\cos \dfrac{v}{2} \right)^2$, and "s², $\frac{1}{2}v$" for $\left(\sin \dfrac{v}{2} \right)^2$.

The parentheses as in $(\sin x)^n$ were preferred by Karsten,[1] Scherffer,[2] Frisius,[3] Abel (in some passages),[4] Ohm.[5] It passed into disuse during the nineteenth century. Paoli[6] writes "$\overline{\text{sen.}a}^m$." The designation $\sin x^2$ for $(\sin x)^2$ is found in the writings of Lagrange,[7] Lorenz,[8] Lacroix,[9] Vieth,[10] Stolz;[11] it was recommended by Gauss.[12] The notation $\sin^n x$ for $(\sin x)^n$ has been widely used and is now the prevailing one. It is found, for example, in Cagnoli,[13] De Morgan,[14] Serret,[15] Todhunter,[16] Hobson,[17] Toledo,[18] Rothe.[19]

[1] W. J. G. Karsten, *Mathesis theoretica* (Rostock, 1760), p. 511.

[2] Carolo Scherffer, *Institutionum analyticarum pars secunda* (Vienna, 1772), p. 144.

[3] Paulli Frisii, *Operum tomus primus* (Milan, 1782), p. 303.

[4] N. H. Abel in *Crelle's Journal*, Vol. I (Berlin, 1826), p. 318–37; Vol. II (1827), p. 26.

[5] Martin Ohm, *System der Mathematik*, 3. Theil (Berlin, 1829), p. 21.

[6] Pietro Paoli, *Elementi d'algebra*, Vol. II (Pisa, 1794), p. 94, 95, 125, 129.

[7] J. Lagrange, *Méchanique analitique* (Paris, 1788), p. 272, 301.

[8] J. F. Lorenz, *Grundlehren der allgemeinen Grössenberechnung* (Helmstädt, 1800), p. 306, 307.

[9] S. F. Lacroix, *Traité élémentaire de trigonometrie* (Paris, 1803), p. 9–17.

[10] G. N. A. Vieth, *Kurze Anleitung zur Differenzialrechnung* (Leipzig, 1823), p. 40–42.

[11] Otto Stolz, *Grundzüge der Differential und Integralrechnung* (Leipzig, 1893), p. 38–42.

[12] See Grunert, *Archiv der Mathematik*, Vol. XXXVIII, p. 366.

[13] Antonio Cagnoli, *Traité de Trigonométrie* (trad. par Chompré; Paris, 1786), p. 20.

[14] A. de Morgan, *Trigonometry and Double Algebra* (London, 1849), p. 35.

[15] J. A. Serret, *Traité de Trigonométrie* (2d ed.; Paris, 1857), p. 12.

[16] I. Todhunter, *Plane Trigonometry* (6th ed.; London, 1876), p. 19.

[17] E. W. Hobson, *Treatise on Plane Trigonometry* (Cambridge, 1911), p. 19.

[18] L. O. de Toledo, *Tradado de Trigonometria* (tercera ed.; Madrid, 1917), p. 64.

[19] Hermann Rothe, *Vorlesungen über höhere Mathematik* (Wien, 1921), p. 261.

SURVEY OF MATHEMATICAL SYMBOLS USED BY LEIBNIZ

538. *Introduction.*—The leading rôle played by Leibniz in the development of mathematical notations induces us here to present a complete survey of all the mathematical symbols used by him.

Among the seventeenth-century mathematicians active in the development of modern notations a prominent rôle was played by five men—Oughtred, Hérigone, Descartes, Leibniz, and Newton. The scientific standing of these men varied greatly. Three of them—Descartes, Leibniz, and Newton—are generally proclaimed men of genius. The other two—Oughtred and Hérigone—were noted textbook writers. Newton and Descartes did not devote themselves persistently to work on notations. Descartes made only minute changes in the algebraic notations of some of his predecessors, but these changes were generally adopted, along with his great creation in the science of mathematics—his analytic geometry. In the case of Newton, mathematical notation was a minor topic of attention. His notation for fractional and negative exponents was foreshadowed by his predecessors and constituted a fairly obvious extension of the symbols used by Descartes and Wallis. His dot symbolism for fluxions suggested itself to him as a young man, but he made no use of it in his *Analysis* and his *Principia*. Nor did he experiment with it and with rectangles denoting fluents, for the purpose of improving his notation. In an unsigned article, where he[1] speaks of himself in the third person, he declared: "Mr. Newton doth not place his Method in Forms of Symbols, nor confine himself to any particular Sort of Symbols for Fluents and Fluxions."

Unlike Newton and Descartes, Leibniz made a prolonged study of matters of notation; like Oughtred and Hérigone, Leibniz used many new symbols, but there was an important difference of method. Oughtred adopted a large array of symbols in the early editions of his *Clavis*, without having experimented with rival symbols to ascertain their relative merits. For example, his notation for powers of numbers was a modification of that of Vieta; it was strikingly inferior to that of Descartes. Yet Oughtred, in later editions of his *Clavis*, hardly ever modified his own original signs even by a stroke. The same remark applies to Hérigone. The great difference in procedure between Leibniz and his two predecessors was that for about a third of a century Leibniz experimented with different symbols, corre-

---

[1] *Philosophical Transactions*, Vol. XXIX, for the years 1714–16 (London, 1717), p. 204.

sponded with mathematicians on the subject, and endeavored to ascertain their preferences. It would seem that on this topic he conferred with every mathematician of his acquaintance who was willing to listen sympathetically—with Jakob Bernoulli,[1] Johann Bernoulli,[2] Wallis,[3] Guido Grando,[4] Oldenburgh,[5] Hermann,[6] Huygens,[7] L'Hospital,[8] Tschirnhausen,[9] Zendrini,[10] and Wolf.[11]

539. Perhaps no mathematician has seen more clearly than Leibniz the importance of good notations in mathematics. His symbols for the differential and integral calculus were so well chosen indeed that one is tempted to ask in the words of Goethe's Faust, "War es ein Gott, der diese Zeichen schrieb?" But this excellence was no divine inspiration; it was the result of patient and painstaking procedure. He withheld these symbols from print for about ten years. Before committing himself to print he explained his dx and dy to Oldenburgh[12] and Tschirnhausen.[13] He stressed the importance of a notation which would identify the variable and its differential.[14] The dx indicated the variable x; dy, the variable y; dz, the variable z. This symbolism was immensely superior to the a chosen by Fermat as the small increment of x, and the a and e adopted by Barrows as the increments of x and y. As regards the integral calculus (§ 620, Fig. 124), Johann Bernoulli had been active in this field and was looked upon as the creator of the integral calculus,[15] notwithstanding Leibniz' publication of 1686. At one time Leibniz and Johann Bernoulli discussed in their letters both the name and the principal symbol of the integral calculus. Leibniz favored the name *calculus summatorius* and the long letter $\int$ as the symbol. Bernoulli favored the name *calculus integralis* and the capital letter I as the sign of integration.[16]

---

[1] *Leibnizens gesammelte Werke. Dritte Folge Mathematik* (*Leibnizens Mathematische Schriften* herausgegeben von C. I. Gerhardt), Vol. III (Halle, 1855), p. 10–110.

[2] *Op. cit.*, Vol. III, p. 133–973.

[3] *Op. cit.*, Vol. IV, p. 5–82.

[4] *Op. cit.*, Vol. IV, p. 209–26.

[5] *Op. cit.*, Vol. I, p. 11–168.

[6] *Op. cit.*, Vol. IV, p. 259–413.

[7] *Op. cit.*, Vol. II, p. 11–208.

[8] *Op. cit.*, Vol. II, p. 216–343.

[9] *Op. cit.*, Vol. IV, p. 429–539.

[10] *Op. cit.*, Vol. IV, p. 230–51.

[11] *Op. cit.*, Vol. VII, p. 14–188 (Wolf).

[12] *Op. cit.*, Vol. I, p. 154.

[13] C. I. Gerhardt, *Der Briefwechsel von G. W. Leibniz mit Mathematikern*, Band I (Berlin, 1899), p. 375.

[14] *Leibnizens Math. Schriften*, Vol. V, p. 231.

[15] *Op. cit.*, Vol. III, p. 115, 116.

[16] *Op. cit.*, Vol. III, p. 115, 168, 170–72, 177, 262, 272, 273; Vol. V, p. 320.

The word "integral" had been used in print first by Jakob Bernoulli,[1] although Johann claimed for himself the introduction of the term.[2] Leibniz and Johann Bernoulli finally reached a happy compromise, adopting Bernoulli's name "integral calculus," and Leibniz' symbol of integration. For many years Leibniz frequently attached a vinculum to his symbol dx for differential and also to his sign of integration. In both cases he later dropped the vinculum, it having been found by experience to be superfluous.

It is not generally known that Leibniz at one time invented a special sign[3] for the differential coefficient, $\dfrac{dz}{dx}$, namely, $d\dddot{z}$, the letter d being broken. This sign was free from the objection of requiring three terraces of type as is the case in ordinary fractions. He submitted this symbol to Johann Bernoulli who in his reply[4] rather favored its adoption, even to the exclusion of his own capital letter D which he had been using for this purpose. Leibniz himself, doubtless for good reasons, never brought his symbol into print and, in fact, did not again urge its use in his letters. This episode affords a fine example of masterful self-control.

540. In algebra and geometry an improved symbolism on which all could agree was greatly needed. Leibniz engaged in extensive experimentation. At different times he used four different symbols for equality, three for proportion, three for coincidence of geometric figures, two for similarity, four for congruence, five for multiplication, three for division, two for logarithms, and about half a dozen for the aggregation of terms. Besides this he had several different signs for powers and roots, two signs for "greater," and two for "less."

How Leibniz dealt with the problem of notations in his correspondence is illustrated by the following extract from a letter (February 22, 1696) to Johann Bernoulli:[5] "As regards signs, I see it clearly that it is to the interest of the Republic of Letters and especially of students, that learned men should reach agreement on signs. Accordingly I wish to get your opinion, whether you approve of marking by the sign $\int$ the sum, just as the sign d is displayed for differences; also whether you approve of my designation of ratio as if it were a division,

[1] Jakob Bernoulli in *Acta eruditorum* (1690), p. 218; Jakob Bernoulli, *Opera*, Vol. I, p. 423. He says, "Ergo et horum Integralia aequantur."

[2] *Leibnizens Math. Schriften*, Vol. III, p. 115 n., 163, 172.

[3] *Op. cit.*, Vol. III, p. 526.

[4] *Op. cit.*, Vol. III, p. 531.

[5] *Op. cit.*, Vol. III, p. 262.

by two dots, for example, that $a:b$ be the same as $\dfrac{a}{b}$; it is very easily typed, the spacing of the lines is not disturbed. And, as regards proportion, there are some who exhibit such a relation by $a:b::c:d$; since this really amounts to an equality of quotients, it is sufficient to write, as is my custom, $a:b=c:d$ or $\dfrac{a}{b}=\dfrac{c}{d}$. Perhaps it will be well to examine other symbols, concerning which more on another occasion." Bernoulli[1] expresses his assent to these suggestions, although, through force of habit, he continues with his old notations which were somewhat at variance with what Leibniz had proposed. "It is difficult to become adjusted," Bernoulli candidly remarks. Nor is he altogether pleased with Leibniz' colon for division, "since those who are accustomed to the ordinary mark for division $\left[\dfrac{a}{b}\right]$ can hardly distinguish at a glance between dividend and divisor."

In another letter to Johann Bernoulli, Leibniz says:[2] "In marking operations I diminish the task of type-setting; I use to express aggregation, instead of bars of raised vinculums the direct or inverted commas; thereby the line of type is not broken, nor the spacing disturbed and yet (if I am not deceived) everything is indicated accurately. However, I desire first to ascertain your opinion." In a later communication (July 29, 1698) Leibniz objects to the St. Andrews cross $\times$ which he himself had previously used as a sign of multiplication, because it is easily confused with the letter x. Johann Bernoulli approves of Leibniz' suggestions but frankly adds:[3] "Meanwhile I prefer to follow the accepted practice, rather than begin with the definitions of new signs, which can be done more conveniently when one is writing a whole treatise." Leibniz proceeds[4] next to propose a symbolism for "function," which was forcing its way as a new concept into mathematics. Apparently Leibniz was not satisfied with his symbols for functions, for he did not have them printed, and mathematicians of the eighteenth century invented other designs.

Tschirnhausen expressed the opinion that new terminology and new symbols render the science less comprehensible. He praises Vieta for having used only the letters of the alphabet instead of introducing other characters resembling monsters.[5] But the enthusiasm

[1] *Op. cit.*, Vol. III, p. 273. [3] *Op. cit.*, Vol. III, p. 531.

[2] *Op. cit.*, Vol. III, p. 276. [4] *Op. cit.*, Vol. III, p. 537.

[5] C. I. Gerhardt, *Briefwechsel von G. W. Leibniz mit Mathematikern*, Vol. I, p. xvii, 358, 388.

of Leibniz was not so easily crushed; he replies:[1] "I perform the calculus by certain new signs of wonderful convenience, concerning which, when I recently wrote you, you responded that your mode of exposition was the more customary and ordinarily intelligible way and that you avoided, as far as possible, novelty in matters of definition, since it effected indeed nothing but to render the sciences difficult." But the same objection might have been made by the arithmeticians of previous years when the Arabic in the place of the Roman characters were introduced, or by the old algebraists, when Vieta introduced letters in place of numbers. "In signs one observes an advantage in discovery which is greatest when they express the exact nature of a thing briefly and, as it were, picture it; then indeed the labor of thought is wonderfully diminished."

541. During the last fifteen or twenty years of his life, Leibniz reached definite conclusions as to the superiority of certain algebraic symbols over others. In making his decisions he was guided in part by the principle of economy, according to which there should not be unnecessary duplication of symbols. Accordingly, he discarded the four dots of Oughtred in writing proportion. He urged the avoidance of signs so closely resembling each other as to give rise to doubt or confusion. He came to realize the fundamental importance of adopting symbolisms which could be set up in a line like ordinary type,[2] without need of widening the spaces between lines to make room for symbols with sprawling parts. Considerations of this sort led him to discard the vinculum so freely used by him in his earlier practice, not only in the ordinary form for the aggregation of terms in a polynomial, but also in conjunction with the radical sign, the powers of polynomials, the differential dx, and the integral sign. In his later practice the vinculum was displaced by parentheses or by commas or dots. The same principle led him to advocate the printing of fractions in the running text by the use of the colon,[3] a device preferable to the modern solidus. I mention these details to illustrate how painstaking and careful he was. And what was the result of all these efforts, all his experimentation? Simply this, that no other mathematician has advanced as many symbols which have retained their place to the present time as has Leibniz. Of Oughtred's many symbols, the cross in multiplication is the only one which is still universally known and widely used. Of Hérigone's signs all have passed into innocuous

[1] *Op. cit.*, p. 375, 380, 403.

[2] *Leibnizens Mathematische Schriften*, Vol. III, p. 276.

[3] *Op. cit.*, Vol. III, p. 276.

desuetude. Among Leibniz' symbols which at the present time enjoy universal, or well-nigh universal, recognition and wide adoption are his dx, dy, his sign of integration, his colon for division, his dot for multiplication, his geometric signs for similar and congruence, his use of the Recordian sign of equality in writing proportions, his double-suffix notation for determinants.

Leibniz expressed his mathematical creed in a letter to L'Hospital (April 28, 1693) as follows:[1] "One of the secrets of analysis consists in the characteristic, that is, in the art of skilful employment of the available signs, and you will observe, Sir, by the small enclosure [on determinants] that Vieta and Descartes have not known all the mysteries." His ideals for mathematical symbolisms was part of his broader scheme on mathematical logic which has commanded the lively admiration of modern logicians. With exuberant optimism Leibniz prophesied the triumphal success of researches in this field in the famous affirmation:[2] "I dare say that this is the last effort of the human mind, and, when this project shall have been carried out, all that men will have to do will be to be happy, since they will have an instrument that will serve to exalt the intellect not less than the telescope serves to perfect their vision."

542. *Tables of symbols used by Leibniz in his manuscripts and in the papers he published. Abbreviations of titles:*

Ae = Acta eruditorum.

B = Briefwechsel von Gottfried Wilhelm Leibniz mit Mathematikern, herausgegeben von C. I. Gerhardt, Band I (Berlin, 1899).

Bm, XIII = *Bibliotheca mathematica* (3d ser.), Vol. XIII (ed. by G. Eneström).

Cat. crit. = Catalogue critique des Manuscrits de Leibniz, Fasc. II (Poitiers, 1914–24).

E = Entdeckung der höheren Analysis, von C. I. Gerhardt (Halle, 1855).

Ph., VII = C. I. Gerhardt, *Philosophische Schriften von Leibniz,* Vol. VII (Berlin, 1890).

Roman numerals followed by Arabic numerals indicate the volume and page of *Leibnizens gesammelte Werke. Dritte Folge, Mathematik* (*Leibnizens Mathematische Schriften,* herausgegeben von C. I. Gerhardt [1849–63]); thus "V, 196" means Volume V, page 196.

Mah. = Dietrich Mahnke, "Neue Einblicke in die Entdeckungsgeschichte der höheren Analysis," *Abhandlungen der Preuss. Akademie*

---

[1] *Op. cit.,* Vol. II, p. 240.

[2] Quoted by A. Padoa in *La Logique deductive* (Paris, 1912), p. 21.

der Wissenschaften zu Berlin (Jahrgang 1925), Phys.-math. Kl., No. 1 (Berlin, 1926).

$O = Opuscules$ *et Fragments inédits de Leibniz,* par L. Couturat (Paris, 1903).

P means that the article referred to was printed by Leibniz in the year indicated, or else was printed in the *Commercium epistolicum* of John Collins before the death of Leibniz. The references given in these tables are in most cases only a few out of the total number that might be given.

543. *Differential and derivative.*

Sign	Meaning	Year	Reference
b	Differential, $\xi b = x$	1673	$\begin{cases} Cat.\ crit.,\ \text{No. 575; Mah., 45,} \\ \quad \text{Table I} \end{cases}$
ξ	Number of differentials		
a, l	Differentials of x, y	1675	E, 123, 125
z	Differential of x	1675	E, 132
$\dfrac{x}{d}$	Differential of x	1675	E, 126; B, XIV
dx	Differential of x	1675	E, 134, 150–55; B, XIV; I, 128; II, 60, 108, 118; III, 262, 453; V, 106
$\partial, \bar{d}x$	Differential of x	1684 . . . P	V, 220, 257, 310, 315· VII, 222
		1675	E, 59, 135; I, 128; II, 44; IV, 479, 506; VII, 355
		1692P	V, 256
$\dfrac{dx}{dy}$	Derivative	1675	E, 137, 138; II, 19b
$dx:dy$	Derivative	1692 . . .	II, 118; III, 288
		1684P	V, 224
$\dfrac{d\bar{y}}{dx}$	Derivative	1676	E, 140, 146; B, 202, 224, 230; I, 159
ddv	Second differential	1684P	V, 221
		1694	II, 196; III, 167
$dd\bar{y}$	Second differential	1690	II, 43
$ddd y$	Third differential	1697	IV, 25
$d\bar{y}^3$	$d(y^3)$	1676	I, 154; B, 229
$\dfrac{d\bar{x}^e}{dx}$	$= ex^{e-1}$	1676	E, 141
$d.x^v$	$= x^v.,\dfrac{v}{x}dx+dv+dv.\log.x$	1695P	V, 325*
d^3n	Third differential	1695	III, 167; IV, 25
$d^m n$	mth differential	1695?	II, 294, III, 167, 221; IV, 25, 211
$\dfrac{m-1}{d}\cdot n$	$d^{m-1}n$	1695?	II, 294
$d^{-1}x$	$\int x$	1695	II, 274
$d^{1:2}x$	$x.\sqrt[2]{dx:x}$ fract. differential	1695	II, 302; III, 228
$d^{\frac{1}{2}}\overline{xy}$	Fractional differential	1695 . . .	II, 301; IV, 25
$d^{1:2}\overline{xy}$	Fractional differential	1695	II, 301
$dx.dx$	(dx)	?	E, 151
$\overline{dx}^2$	$(dx)^2$	1676	E, 141
$\mathbf{d}_\delta$	$dz:dx$	1698	III, 526, 531
δm	$\dfrac{\partial m}{\partial x}$	1694	II, 261
ϑm	$\dfrac{\partial m}{\partial y}$	1694	II, 261

* Misprint

544. *Signs for integral, sum.*—

Sign	Meaning	Year	Reference
omn. w	Sum of all the w's	1675	E, 58, 120, 121; B, XII, XIII
omn. $\overline{\text{omn. } w}$	Sum of the sum of all the w's	1675	E, 58, 120, 121; B, XII, XIII
$\int$	"omn." i.e., sum, integral	1675	E, 59, 125, 126, 132–38, 150–55; B, XII, XIII; II, 43, 108; III, 104, 116, 167, 262; IV, 506; V, 397
$\int$	. . . .	1686 . . . . P	Ae (1686), 297–99; (1693), 178
$\int \overline{x\,}l$	Integral sign and vinculum	1675	E, 59, 125, 126, 132–38, 140, 141, 152, 153; B, XII, XIII
		1676	I, 128; II, 119, 257, 275; IV, 506
		1686 . . . . P	V, 231, 313; III, 455
$\int$	Integral	1691P	Ae (1691), 178, 181; (1692), 276; V, 131, 275; VI, 144
$\int\int$	Integral of an integral	1675	E, 125
$\int \overline{\int \overline{l\,\frac{l}{a}}}$	Integral of an integral with vinculum	1675	E, 59, 125
$\int x \int y$	"Jam $\int x = \frac{x^2}{2}$"	1675	E, 137
$\int dv \int d\psi$	. . . .	1675	E, 138
$d \int x$ aequ. x	$d(\int x) = x$	?	E, 153
$\int^n$	nth sum or integral	1695	II, 301; III, 221

545. *Equality.*—

Sign	Meaning	Year	Reference
$=$	Equal	$\begin{cases} 1666P \\ 1675 \end{cases}$	V, 15 E, 132–47
$\sqcap$	Equal	1675	B, XII, XIII, 151–60; E, 118, 121–31
		1676	I, 101, 154, 160, 163; II, 11; IV, 477; V, 92, 154; VII, 141, 153, 251
$aeq.$ $\Big\}$	Equal	1679	I, 31; IV, 485; VII, 47, 88; E, 150–55
$aequ.$		1684P	V, 220
$=$	Equal	1675	I, 117; II, 60, 76, 116–19, 196, 222, 239; III, 167; IV, 337; V, 112, 241; VII, 55, 266; E, 132
$=$	Equal	1684 . . . . P	V, 125, 222, 256, 279; VI, 156; VII, 222
$\underset{=}{(n)}$	nth equation as in "$nG\overset{(7)}{=}\int \overline{g}$"	?	VI, 427; VII, 359, 364–68, 370, 371, 385, 389
		1694P	V, 305, 315, 323
$\overset{(n)}{\sqcap}$	nth equation	1675	E, 117, 118
	. . . .	Equality as in $g|7$ or $b+c|2+3$	?	VII, 83
∞	Equal	1675?	$Ph.$, VII, 214
∞	Equal	1675?	$Ph.$, VII, 228

546. *Plus, minus.—*

Sign	Meaning	Year	Reference
+	+	1666 .... P	V, 15
‾	‾	1666 .... P	V, 15
±	Plus or minus	1673	O, 100, 126; I, 117
±, ±		1684P	I, 27
±, ∓	The upper signs simultane-	1684P	V, 222; VII, 56
+, + }	ously, or the lower simul-	?	VII, 145
∓, ∓	taneously	?	VII, 139, 153
(±), (±)		1684P	V, 222
±, (±), ((±)) }	Three independent ambigui-	1684P	V, 222
±, ∓, (∓) }	ties in sign	?	V, 169, 170
(+)	Logical combination	?	V, 198
⊕	Logical combination	?	VII, 261, 285
−1 or I	"Minus one" (sign of quality)	?	V, 370

547. *Multiplication.—*

Sign	Meaning	Year	Reference
....	Apposition	1710 .... P	VII, 219
⌒	Multiplication	1666P	V, 15, 20, 24, 25
		?	VII, 33, 45, 49, 54, 100, 220
		1672	I, 29, 156
		1675	E, 128, 138, 139
		1677	B, 242
×	Multiplication	1676	I, 117; B, 242
, ...	Multiplication	1674	Cat. crit., No. 775; I, 100; VI, 228
		1710P	VII, 219
•	Multiplication	1676	I, 128; II, 196, 239; III, 36, 160, 341; IV, 211; V, 111, 112; VII, 54, 103, 175
⋅	Multiplication	1684, 1691P	V, 129, 222; VII, 219
;	Multiplication and separa-		
	tion as in 2³; 1.2.3	1713	III, 986
✱	Multiplication	1705	IV, 211

548. *Division.—*

Sign	Meaning	Year	Reference
$\frac{a}{b}$	Fraction, division	1666 .... P	V, 48, 49, 220
		1673	I, 46, 116; IV, 211; VII, 54, 170
∪	Division	1666P	V, 15, 20
		1675	E, 134; III, 526; VII, 91, 103
:	Division, ratio	1676	I, 90, 128; IV, 211, 400; VII, 170
		1684 .... P	V, 131, 223, 228

549. *Proportion.—*

Sign	Meaning	Year	Reference
: :: :	Proportion	1676	I, 146; II, 44, 97, 121; V, 241; VII, 309, 310, 356, 364, 378; *E*, 150
		1684*P*	V, 225
: ▬ :	Proportion	1691	II, 113, 118, 119, 196; III, 80, 84, 262, 288, 526; IV, 211; VII, 390, 17 (Wolf)
		1694*P*	V, 315
		1710*P*	VII, 222
$\frac{a}{b} = \frac{c}{d}$	Proportion	1696	III, 262; IV, 211
		1710*P*	VII, 222
∺	Continued proportion	1710*P*	VII, 222

Decimal fractions.—

Sign	Meaning	Year	Reference
3.14159, etc.	. . . .	1682*P* ?	V, 119, 120 VII, 361

550. *Greater or less.—*

Sign	Meaning	Year	Reference
⊓	Greater	1679	V, 153; VII, 55, 81
⊐	Less	1679	V, 153, 198; VII, 55, 81
⫤	Greater	1710*P*	*Miscell. Berolinensia*
⫣	Less	1710*P*	*Ibid.*
>	Greater	1749*P*	2d ed. of *ibid.*
<	Less	1749*P*	2d ed. of *ibid.*
⊐	Greater	1863*P*	VII, 222
⊏	Less	1863*P*	VII, 222

In reprint of article of 1710

551. *Aggregation of terms.—*

Sign	Meaning	Year	Reference
‾‾‾‾‾‾‾	Vinculum placed above expression	1675	E, 121, 124
		1676	I, 128; VI, 301; VII, 55, 80
		1684 .... P	V, 130, 131, 220, 325; VI, 144
_____	Vinculum placed beneath expression	?	VII, 83
$\boxed{n}\,e+fz^{h}$	Vinculum and nth power	1676	I, 116, 128
		1695	III, 175
		1694 .... P	V, 313, 323
$\sqrt{}$	Vinculum and radical sign	1675	I, 76, 120, 156; II, **43**; V, 408; VII, 55
		1694 .... P	V, 313, 319, 360
⌢	Aggregating terms beneath the brace	1678P	V, 117
⌣	Aggregating terms above the brace	1678P	V, 117
$\sqrt{}$( )	Parentheses and radical sign	1677 .... ?	I, 161; VII, 84, 90, 95, 97, 147
$\int$( )	Parentheses and sign of integration	1679 .... ?	II, 31, 61, 66, 195
		1690	II, 43
$\boxed{n}$( )	Parentheses and nth power	1701	III, 662
$\boxed{f}$( )	Parentheses and fractional power	1701	III, 663
(( ))	Parentheses within parentheses	1696	III, 276
( )	Parentheses	1715	IV, 400
$\sqrt{()}$	Redundant notation	1708?	IV, 356
$A\frown A-1,,\smile 2$	$A(A-1)\div 2$	1666P	V, 20
$1-h+m,\ :h$	$\dfrac{1-h+m}{h}$	1676	I, 128
$m,\ m-n$	$m(m-n)$	1676	I, 100
$,e-,\,f{:}g,,$	$(e-(f\div g))$	1696	III, 276
$a+,b{:}c,,$	$\left(a+\dfrac{b}{c}\right)$	1696	III, 276
$a+b,:\,,c+d$	$\dfrac{a+b}{c+d}$	?	VII, 55
$x.x-1.x-2$	$x(x-1)(x-2)$	1695	III, 160; VII, 101, **103**
$\sqrt{},\ 2bx-xx$	$\sqrt{}\,2bx-x^2$	1696	III, 288
$eh+d,:h=e+d(:h)$ ..	$(eh+d)\div h=e+\dfrac{d}{h}$	1705	IV, 337
$\sqrt{},,a+b,\,:,c+d,,:,\ l+m$	$\sqrt{\dfrac{a+b}{c+d}}\div(l+m)$	?	VII, 55
$\sqrt{}((a+b):(c+d)):(l+m)$	$\sqrt{\dfrac{a+b}{c+d}}\div(l+m)$	?	VII, 55
$\sqrt{},,x:,2b-x$	Redundant notation	1696	III, 288

552. *Powers and roots.—*

Sign	Meaning	Year	Reference
z^2, x^4	z^2, x^4	1675	E, 118
x^3, z^9	x^3, z^9	1676 .... P	I, 89; V, 129
aa, xx	a^2, x^2	?	E, 151
		1684 .... P	V, 222, 130
x^{a-b}	x^{a-b}	1684P	V, 222
y^{-1}	$\dfrac{1}{y}$	1684P	V, 222
$x^{-1:2}$	$x^{-\frac{1}{2}}$	1687?	V, 241
$x^{\frac{n}{\cdot}}$, $x^{\frac{n-1}{\cdot}}$	x^n, x^{n-1}	?	V, 106
$v^{\frac{1:n-1}{\cdot}}$	$\dfrac{1}{v^{n-1}}$	1676?	I, 128
$b^{\frac{t}{\cdot}}$..............	b^t, t variable	1691	II, 76
$x^y + y^z$	$x^y + y^z$	1676?	I, 160
$\boxed{2}\,r^2 + z^2$..........	$(r^2 + z^2)^2$	1676	I, 116
$\boxed{3}\,(AB + BC)$	$(AB + BC)^3$	1710P	VII, 220
		1701	III, 662, **663**
$\overline{\overline{w-e:f}\ \boxed{1:h}}$	$\left(\dfrac{w-e}{f}\right)^{\frac{1}{h}}$	1676	I, 128
$\overline{1+y}\,|^{\frac{2}{3}}$	$(1+y)^{\frac{2}{3}}$	1677	B, 242
$\boxed{m}\,\overline{y+a}$	$(y+a)^m$	1695P	V, 323
$\boxed{n}\!\sqrt[m]{y}$	$y^{\frac{n}{m}}$	?	VII, 184
$\boxed{p^e}\,y$	y^e	1697	IV, 25
$\overline{a+b}^3$	$(a+b)^3$	?	VII, 55
$\overset{2}{\sqrt{\ \ }}$, $\sqrt[3]{\ \ }$	$\sqrt{\ \ }$, $\sqrt[3]{\ \ }$	1676 ... ?	I, 156; II, 90, 119, 275
$\overset{e}{\sqrt{\ }}\,\mathfrak{D}$	eth root of polynomial $\mathfrak{D}$	1705	III, 104
$\dfrac{20}{1+20}\,\Big\}\,\boxed{3},\,\dfrac{5}{20}$	$\left(\dfrac{20}{1+20}\right)^3\cdot\dfrac{5}{20}$	?	VII, 136
$\sqrt{\boxed{3}}$	$\sqrt[3]{\ }$	1710P	*Miscell. Berolinensia*
$\sqrt{\text{③}}$	$\sqrt[3]{\ }$	1863	VII, 220, 221
$\sqrt{(3)}$	$\sqrt[3]{\ }$	?	VII, 139, 145

553. *Pure imaginary.—*

Sign	Meaning	Year	Reference
$\sqrt{-\frac{3}{4}}$	$\sqrt{-\frac{3}{4}}$	1675	I, 76
$\sqrt{-3}$..............	$\sqrt{-3}$	1693	II, 236
$\sqrt{-1}$..............	$\sqrt{-1}$	1675	I, 87
		1695 .... P	V, 326
$\sqrt{-1}$..............	$\sqrt{-1}$	1674?	II, 11
$\overset{2}{\sqrt{}}-1$, $\overset{4}{\sqrt{}}-1$........	$\sqrt{-1}$, $\sqrt[4]{-1}$	1703	III, 80, 905
$\overset{16}{\sqrt{}}-1$..............	$\sqrt[16]{-1}$	1703	III, 80
$\sqrt{(-\sqrt{-1})}$........	$\sqrt{(-1\sqrt{(-1)})}$	1702P	V, 360

554. *Logarithm.*—

Sign	Meaning	Year	Reference
log................	Logarithm to any base b	1690? 1691 .... P	II, 53 V, 130, 324
l................	Logarithm to any base b	1694 1691P	III, 153 V, 129

555. *Trigonometry.*—

Sign	Meaning	Year	Reference
t................	Tangent	1691	V, 129
s................	Sine	1691	V, 129; VII, 57
v................	Versed sine	1691 ...	V, 129; VII, 57
a................	Arc	1691	V, 129
r................	Radius	?	VII, 57

556. *Infinite series.*—

Sign	Meaning	Year	Reference
etc. in infinitum......	Series is infinite	1682 .... P	V, 121
etc.....	Series is infinite	1675	II, 63, 275; VII, 161, 128
etc...............	Series is infinite	1691 .... P	V, 129, 323
etc...............	Continued fraction is infinite	1696	III, 351; VII, 47, 48

557. *Similar, Congruent.*—

Sign	Meaning	Year	Reference
ꝏ	Coincident	1679	V, 150, 165
ꝝ	Congruent	1679	V, 150, 151, 154, 158, 165; II, 22
$E\,\overset{\frown}{.A\,.C}\,ꝝ\,F\,\overset{\frown}{.A\,.C}$....	I.e., $E\,.A\,ꝝ\,F\,.A,\,E\,.C\,ꝝ\,F\,.C,$ $A\,.C\,ꝝ\,A\,.C$	1679	V, 159, 160
⌀	Similar	1679	V, 153, 154, 172, 185; VII, 57, 222, 277
≌	Congruent	....	V, 172, 173, 185; VII, 222
≌	Alg. identity	1714?	V, 406
|≌|	Coincident	?	V, 185
≊	Coincident	?	V, 173
∽	Similar	1710P	*Miscell. Berolinensia*
⌒	Congruent	1710P	*Miscell. Berolinensia*
ꝏ	Congruent	?	VII, 263
∞	Coincident	?	VII, 57, 263
∞	Infinite	?	VII, 88
$\overset{\infty}{r;s;v}$	Relation between radius (r), sine (s), and versed sine (v)	?	VII, 57
$\underset{\infty}{x;a}$................		1697	III, 453
$a;b;c \propto l;m;n$	$a:b=l:m, a:c=l:n, b:c=m:n$	1710P	VII, 222
$a;b;c::l;m;n$........	"Same relations" as when $a^2+ab=c^2,\ l^2+lm=n^2$	1710P	VII, 222

558. *Function symbol.*—

Sign	Meaning	Year	Reference
ξ	Function of x, ξ being the Greek letter corresponding to x	1698?	III, 537
$\overline{x}\lfloor 1 \rfloor$	Function of x	1698?	III, 537
$\overline{x}\lfloor 2 \rfloor$	A second function of x	1698?	III, 537
$\overline{x;y}\lfloor 1 \rfloor$	Function of x and y	1698?	III, 537
$\overline{x;y}\lfloor 2 \rfloor$	Second function of x and y	1698?	III, 537
$\overline{x}\lfloor r.1 \rfloor$	A rational function of x	1698?	III, 537
$\overline{x}\lfloor r.2 \rfloor$	Second rational function of x	1698?	III, 537
$\overline{x}\lfloor ri.1 \rfloor$	Rational integral function of x	1698?	III, 537
$a b$	Symmetric function $ab+ac+ad+ \ldots$	?	
$a \, bc$	Symmetric function $abc+abd+ \ldots$	?	VII, 178, 191; *Bm.*, XIII, 37, 38
a^2	Symmetric function $a^2+b^2+ \ldots$	?	
a^3	Symmetric function $a^3+b^3+ \ldots$	?	

559. *Index numbers, determinant notation.*—

Sign	Meaning	Year	Reference
$x=10+11v+12v^2$, etc..... $y=20+21v+22v^2$, etc..... $z=30+31v+32v^2$, etc.....	These suppositional or fictitious numbers mark rows and columns	1693	II, 239; VII, 161
10.22	$10.22+12.20$	?	VII, 164
$11.21.30$	$11.21.30+11.20.31+$ $10.21.31$ Hence $xy=$ $10.20+10.21v+10.22v^2+$ $13.20v^3+ \ldots$	?	VII, 164
$\left. \begin{array}{l} 1_0\,2_1\,3_2 \\ 1_1\,2_2\,3_0 \\ 1_2\,2_0\,3_1 \end{array} \right\}$	....	1678	VII, 7, 8; *Bm*, XIII, 255
$1_0\,2_1\,3_2$	....	1684	*Bm*, XIII, 255
$\left. \begin{array}{l} 10 \cdot 21 \cdot 32 \\ 11 \cdot 22 \cdot 30 \\ 12 \cdot 20 \cdot 31 \end{array} \right\}$	....	1693	II, 239; *Bm*, XIII, 255
$\left. \begin{array}{l} 0=100+110x+101y+111xy+120xx+102yy \\ 0=200+210x+201y+211xy+220xx+202yy \end{array} \right\}$		1705	IV, 269
$\left. \begin{array}{l} 122.22+133.33+123.23+12.2+13.3+\text{⑩} \; aequ. \; 0 \\ 222.22+233.33+223.23+22.2+23.3+\text{㉓} \; aequ. \; 0 \end{array} \right\}$		1675	*Bm*, XIII, 254

560. *Astronomical symbols.*—

Sign	Meaning	Year	Reference
$\odot$ ....⎫		1694	III, 142; VII, 47
♀, ☽, ♃, ♂		1705	III, 100–108
$\odot$, ☽ .... ⎬	For algebraic expression or equation	1679	V, 174; VII, 31, 128, 147, 383
☽, $\odot$, ☿, ♂, ♄, ♃, ♀ .... ⎭		?	VII, 390
$\odot$, ☽	To mark places in operations	1703P	VII, 224, 225, 236
$+\odot+$ ☽ $aequ.$ ♀ (22)	Here (22) means equation No. 22	1679	V, 170

561. *Marking points, lines, surfaces.—*

Sign	Meaning	Year	Reference
$1B, 2B, 3B\ldots\ldots\ldots$	Points on same line	1677	*Bm*, XIII, 252
$1\,\text{⊕}, {}_1D\ldots\ldots\ldots$	Double indices marking a surface	1677	*Bm*, XIII, 252
$\underline{X}, \underline{Y}\ldots\ldots\ldots$	Segments of lines	?	V, 185
$Y, (Y),$ etc$\ldots\ldots\ldots$	Different positions of a moving point	1679	II, 23, 97
$X, (X)\ldots\ldots\ldots$	Different positions of a moving point	1679	II, 23, 97
$D_2M_3, M_3L, \text{⊙}M_2\ldots$	Curves	1689	VI, 189
$2G, 3G\ldots\ldots\ldots$	Bodies represented by small circles	1690P	VI, 196
$\overline{Y}\ldots\ldots\ldots$	All points of a *locus*	1679	V, 166
$\underline{y}\,b\ldots\ldots\ldots$	Line traced by moving point		
$\overline{xy}b\ldots\ldots\ldots$	Surface traced by moving curve	1679	V, 148, 149
$\overline{xyz}b\ldots\ldots\ldots$	Solid traced by moving surface		
${}_1C, {}_2C, {}_3C\ldots\ldots$	Points on a curve	1681	IV, 493; V, 100
${}_1H_1l, {}_2H_1d_2l\ldots\ldots$	Curves	1684P	V, 130
${}_1F, {}_2F, {}_3F\ldots\ldots$	Positions of moving point	1695	III, 219
$[D], (D), D\ldots\ldots$	Curve D (D) $[D]$	1688	V, 238
${}_1N_1\xi, {}_3N_3\xi\ldots\ldots$	Line marked by points	1691 .... P	V, 245, 296
${}_1A, {}_2A, {}_3B\ldots\ldots$	Line marked by points	1696	III, 240; VI, 223, 248, 261, 263, 469; VII, 58
$B(B), F(F), L(L)\ldots$	Curves marked by points	1704	III, 741
${}_1L_2L, {}_1B_2B\ldots\ldots$	Curves marked by points	1706	III, 793
$\ldots$ anguli ${}_2T$ ⊙_3M	Angle, vertex at ⊙	?	VI, 263
${}_1{}_3b, {}_2{}_3b, {}_3{}_3b\ldots\ldots$	Point $3b$ on a line takes these positions when line is moved	1679	*Bm*, XIII (1912), 252
$AY\frown A\ldots\ldots\ldots$	AY and then along the arc to A	1714	V, 401

562. *Continued fraction.—*

Sign	Meaning	Year	Reference
$a + \dfrac{1}{b+\dfrac{1}{c+\dfrac{1}{d+\dfrac{1}{e}+\text{etc.}}}}\ldots\ldots$	Continued fraction	1696	III, 351, 352; IV, 24; VII, 47
$\cfrac{1}{1+\cfrac{1}{1+\cfrac{1}{1+\cfrac{1}{1+\cfrac{1}{1+\cfrac{1}{1+}}}}}}$ *Aequ.* ⊙ $\ldots\ldots\ldots\ldots\ldots$ etc.		?	VII, 47

REMARKS ON TABLES

563. *Plus and minus.—*About 1674, in an article entitled "De la
Methode de l'universalité," Leibniz used an involved symbolism
which cannot be readily reproduced and explained in our tables
In one of his manuscripts (O 100–104) occurs $\overline{\mp}a - \overline{\mp}y$, which means
$+a-y$, or $-a+y$; $\overline{\mp}AB\overline{\pm}BC$, which means $+AB-BC$, or $-AB+$

BC, or $+AB+BC$; $(\overline{\mp})\overline{\mp}AB(\overline{\mp})\overline{\mp}BC$, which means $+AB+BC$, or $-AB+BC$, or $+AB-BC$, or $-AB-BC$; $\underline{\pm+}a\underline{+\pm}b$, which means the same as $\overset{+}{\cdot} a \dotplus b$, namely, $+a+b$, or $\underline{-a+b}$, or $+a+b$, or $+a-b$. He gives also some other forms, still more complicated, but he does not use these signs again in later manuscripts.

Multiplication.—Leibniz used both the comma and the dot for multiplication, or for multiplication and aggregation combined. Mahnke pointed out in *Isis* (Vol. IX [1927], p. 289–290) that at first Leibniz used the point and comma, not as signs for multiplication, but simply as signs of separation, and that in this practice he followed Pascal. In his *Dissertatio de arte combinatoria* of 1666 Leibniz (V, p. 40, 44, 51, 60) writes $2.2.2\frown$ for our $2\times2\times2$. These dots are simply signs of separation, for Leibniz writes also $4.18.28.15+$ to mark the sum $4+18+28+15$. Mahnke says that as late as 1673, in unpublished manuscripts, Leibniz writes $3\frown1.+10\frown2.=3\frown3.+2\frown7$, though by that time he used the dot frequently for multiplication. Since the time of Leibniz the dot gained the ascendancy over the comma as the sign of multiplication, while parentheses now generally serve the purpose of aggregation of terms.

564. *Greater and less.*—Reprints of the article "Monitum" in the *Miscellanea Berolinensia* of 1710 contain symbols for "greater" and "less" different from those of 1710. These changes are made without calling the reader's attention to them. Thus in the second edition of the volume of 1710, brought out at Berlin in 1749, the inequality symbols of Harriot are substituted; C. I. Gerhardt's reprint of the article of 1710 brought out in 1863 contains symbols which resemble those of Oughtred, but are not identical with them. We add that Leibniz' signs for "greater" and "less," as given in 1710, closely resemble the signs used by his teacher, E. Weigel.[1] Of the two unequal parallel lines, Weigel's shorter line is drawn the heavier, while Leibniz' longer line is drawn the heavier.

565. *Similar and congruent.*—The first appearance in print of the signs $\backsim$ and $\underline{\backsim}$ for "similar" and "congruent," respectively, was in 1710 in the *Miscellanea Berolinensia* in the anonymous article "Monitum," which is attributed to Leibniz. This form of the symbols was retained in the reprint of the volume of 1710, which appeared in 1749. But in 1863 C. I. Gerhardt reprinted the article in his edition of the mathematical works of Leibniz, and he changed the symbols to $\sim$

---

[1] *Erhardi Weigeli* . . . . *Philosophia mathematica Theologica* . . . . (Jena, 1693), *Specimina novarum inventionum. I Doctrinalia*, p. 135.

and $\simeq$, respectively, without offering any explanation of the reason for the changes. Both forms for "similar," viz., $\backsim$ and $\sim$, have retained their places in some of our modern textbooks.

Marking points, lines, surfaces.—Eneström, in *Bibliotheca mathematica* (3d ser.), Vol. XI (1910–11), p. 170, 171, shows that notations like $1C$, $2C$, $3C$ were used, before Leibniz, in Fr. van Schooten's Latin edition of Descartes' *Géométrie* (1649), where (p. 112) three points of a curve are marked in a figure C, $2C$, $3C$, while other points in the figure are marked $2S$, $3S$, $2T$, $3T$, $2V$, $3V$. In fact, such notations were used by Descartes himself, in the first (1637) edition of his *Géométrie*, pages 406 and 408. These designations occur also in the *Acta eruditorum* (1682), Plate IV; *ibid.* (1684), page 322; and in Newton's *Principia* (1687), pages 481–82.

DIFFERENTIAL AND INTEGRAL CALCULUS

1. INTRODUCTION

566. Without a well-developed notation the differential and integral calculus could not perform its great function in modern mathematics. The history of the growth of the calculus notations is not only interesting, but it may serve as a guide in the invention of fresh notations in the future. The study of the probable causes of the success or failure of past notations may enable us to predict with greater certainty the fate of new symbols which may seem to be required, as the subject gains further development.

Toward the latter part of the eighteenth century the notations of the differential and integral calculus then in vogue no longer satisfied all the needs of this advancing science, either from the standpoint of the new fundamental conceptions that came to be advanced or the more involved analytic developments that followed. Some new symbols were proposed soon after 1750, but the half-century of great unrest and much experimentation in matters of calculus notations was initiated by the publication of J. L. Lagrange's *Théorie des fonctions analytiques*, in 1797.

The fourth volume of Cantor's *Vorlesungen über Geschichte der Mathematik*, dealing with the period 1759–99, pays comparatively little attention to these notations. Nor has anyone yet published a detailed history of calculus notations for the nineteenth century. The present exposition aims to supply a want in this field. Except for certain details, the writer will touch lightly upon the early history of these notations, for that is given quite fully in our histories of mathematics and in the biographies of Newton and Leibniz.

2. Symbols for Fluxions, Differentials, and Derivatives

a) Total Differentiation During the Seventeenth and Eighteenth Centuries

567. *I. Newton.*—Newton's earliest use of dots, "pricked letters," to indicate velocities or fluxions is found on a leaf dated May 20, 1665; no facsimile reproduction of it has ever been made.[1] The earliest printed account of Newton's fluxional notation appeared from his pen in the Latin edition of Wallis' *Algebra*,[2] where one finds not only the fluxionary symbols $\dot{x}, \ddot{x}, \ldots$, but also the fluxions of a fraction and radical, thus $\dfrac{\overset{\cdot}{y}\overset{\cdot}{y}}{\underset{\cdot}{b-x}}$, $\sqrt{\overline{\overset{\cdot}{aa-xx}}}$.

Newton explained his notation again in his *Quadratura curvarum* (London, 1704), where he gave $\overset{\shortmid\shortmid}{x}, \overset{\shortmid}{x}, x, \dot{x}, \ddot{x}, \dddot{x}$, each of these terms being the fluxion of the one preceding, and the fluent of the one that follows. The $\overset{\shortmid}{x}$ and $\overset{\shortmid\shortmid}{x}$ are fluent notations. His notation for the fluxions of fractions and radicals did not meet with much favor because of the

---

[1] S. P. Rigaud, *Historical Essay on the First Publication of Sir Isaac Newton's "Principia"* (Oxford, 1838), Appendix, p. 23. Consult the remarks on this passage made by G. Eneström in *Bibliotheca mathematica* (3d ser.), Band XI (Leipzig, 1910–11), p. 276, and Band XII (1911–12), p. 268, and by A. Witting in Band XII, p. 56–60. See also *A Catalogue of the Portsmouth Collection of Books and Papers, written by or belonging to Isaac Newton* (Cambridge, 1888), p. xviii, 2. The use of dots before the time of Newton, in England, for the specialized purposes of indicating comparatively small quantities, is evident by reference to Leonard Digges and Nicolas Mercator. Digges said, in his Booke named *Tectonicon* (London, 1592), p. 15, where he explains his Table of Timber Measure, "Supposing the square of your Timber were 7. Inches, and that ye desired to know what measure of length of the Ruler would make a foot square: Seeke in the left margine seven Inches: and with him in that order toward the right hand, ye shal find 2. foote, 11. Inches, and $\frac{2}{7}$ · of an Inch. Note because the fraction $\dfrac{\cdot 2}{7}$ hath a prick by him, it betokeneth some small quantitie lesse then $\frac{2}{7}$ · of an Inch. If it had 2. prickes or points thus $:\dfrac{2}{7}$ it should signifie som litle quantitie more. Neither maketh it matter whether ye observe this pricking or no, the quantitie is so little to be added or pulled away." It is of some interest that in 1668, before Newton's notation had appeared in print, N. Mercator used the dot over the letter *I*, thus $\dot{I}$, to signify an infinitesimal, and over numbers, $\overset{\cdot}{6}4$ for example, to serve as a reminder that 64 was the coefficient of a power of an infinitesimal. See *Philosophical Transactions* (London), Vol. III, p. 759 ff. Reprinted in Maseres, *Scriptores logarithmici*, Vol. I, p. 231. See also *Nature*, Vol. CIX (1922), p. 477.

[2] J. Wallis, *De algebra tractatus* (Oxoniae, 1693), p. 390–96.

typographical difficulties. In another place[1] he writes "$\ddot{z}:\dot{x}$", which in modern notation would be $\dfrac{dz}{dt}:\dfrac{dx}{dt}$, or the derivative $\dfrac{dz}{dx}$.

English writers of the eighteenth century used the Newtonian dots almost exclusively, but sometimes with slight variations. Thus H. Ditton[2] and J. Clarke extended the use of two dots separated by a vinculum to denote the first fluxion. They explain: "l:$\overline{x+a}^{\,m}$ denotes the Log. of $\overline{x+a}^{\,m}$, and l:$\overset{\cdot\,\cdot}{\overline{x+a}}^{\,m}$ the Fluxion of that Log. Also l:$\overline{x+a}^{\,m}$ denotes the Log. of the Quantity x added to the m Power of a, and l:$\overset{\cdot\,\cdot}{x+a}^{\,m}$ the Fluxion of the Log. of the same."

568. The Newtonian symbols are found also on the European Continent. Thus the Frenchman, Alexis Fontaine,[3] uses $\dot{x}$, $\dot{y}$, $\dot{s}$ without defining the notation except by $\overset{|}{x}-x=\dot{x}$, $\overset{|}{y}-y=\dot{y}$, where x, y and $\overset{|}{x}$, $\overset{|}{y}$ are points on a curve, not far apart, and $\dot{s}$ is the arc from (x, y) to $(\overset{|}{x}, \overset{|}{y})$. The symbols are not used as velocities, but appear to be small constant or variable increments or variations, which may be used by the side of the Leibnizian differential "d," thus $d\dot{s}=\dfrac{y\,d\overset{|}{y}}{\dot{s}}$. In deriving the curve of quickest descent, he ends with the statement: "*Fluxion* $\dfrac{\dot{y}}{x^{\frac{1}{2}}\dot{s}}=0$, dont la *Fluente* est $\dfrac{\dot{y}}{x^{\frac{1}{2}}\dot{s}}=\dfrac{1}{a^{\frac{1}{2}}}$, que l'on trouvera être l'équation de la cycloide."

In a paper dated 1748 Fontaine[4] treats differential equations and uses $\dot{x}$, $\dot{y}$ in place of dx, dy. Differentiating $n(1-n)p+2nx+(2-5n)y=0$, he obtains $2n\dot{x}+(2-5n)\dot{y}=0$; eliminating n, he arrives at "l'équation aux premières différences $3p\dot{y}^2+10x\dot{y}^2-10y\dot{x}\dot{y}-2p\dot{x}\dot{y}-4x\dot{x}\dot{y}+4y\dot{x}^2=0$." The Newtonian dot was used also by a few other continental writers of the eighteenth century and were found regularly in the republications on the Continent of the works of Newton and of a

---

[1] Isaaci Newtoni, *Opera* (ed. S. Horsley; London), Vol. I (1779), p. 410.

[2] Humphrey Ditton, *An Institution of Fluxions* (2d ed. by John Clarke; London, 1726), p. 117.

[3] *Mémoires donnés à l'académie royale des sciences, non imprimés dans leur temps.* Par M. Fontaine, de cette Académie (Paris, 1764), p. 3. Fontaine uses the dot also in articles in *Histoire de l'académie r. des sciences*, année 1734 (Paris, 1736), *Mémoires*, p. 372; année 1767 (Paris, 1770), p. 588, 596. For further data see F. Cajori, "Spread of Calculus Notations," *Bull. Amer. Math. Soc.*, Vol. XXVII (1921), p. 455.

[4] Fontaine, *op. cit.*, p. 85.

few other Englishmen, among whom are John Muller, David Gregory, and Colin Maclaurin.[1]

After the middle of the eighteenth century, when feelings of resentment ran high between the adherents of Leibniz and their opponents, it is extraordinary that a mathematical journal should have been published on the European Continent for fifteen years, which often uses the Newtonian notation, but never uses the Leibnizian. The journal in question is a monthly, published at Amsterdam from 1754 to 1769. It appeared under the title, *Maandelykse Mathematische Liefhebbery* ("Monthly Mathematical Recreations"). It contains problems on maxima and minima, solved with the aid of fluxions dressed in the familiar garb of the Newtonian dots. The first twelve volumes of the journal give altogether forty-eight or more such problems; the last few volumes contain no fluxions. Among the contributors using fluxions were G. Centen, J. Schoen, J. Kok, J. T. Kooyman, F. Kooyman, A. Vryer, J. Bouman, D. Bocx. The two most noted were P. Halcke and Jakob Oostwoud. The latter was a teacher of mathematics in Oost-Zaandam near Amsterdam, who in 1766 became a member of the Hamburg Mathematical Society.[2] Later he collected and published the problems solved in his journal in three separate volumes.[3]

Among other Dutch adherents of the Newtonian notation may be cited P. Steenstra[4] and A. B. Strabbe.[5] Strabbe's text on the calculus uses the notation of Newton exclusively and develops the subject in the manner of the English mathematicians of the eighteenth century.

569. *Quotation from Newton.*—This quotation is of interest because it is Newton's own statement regarding his invention and use

[1] It should be stated, however, that in the Introduction and annotations to Newton's *Principia*, brought out by Thomas le Seur and François Jacquier at Cologne, in 1760, the Leibnizian notation is employed by the editors.

[2] Bierens de Haan in *Festschrift, herausgegeben von der Mathematischen Gesellschaft in Hamburg, anlässlich ihres 200 jährigen Jubelfestes 1890*, Erster Teil (Leipzig, 1890), p. 79.

[3] Bierens de Haan, *op. cit.*, p. 80, where the titles of the three publications are given in full. See also De Haan's *Bouwstoffen voor de Geschiedenis der Wis-en Natuurkundige Wetenschappen in de Nederlanden* (1878), p. 76–85, reprinted from *Verslagen en Mededeelingen der Kon. Akademie van Wetenschappen*, Afd. Natuurk., 2° Reeks, Deel VIII, IX, X, en XII.

[4] P. Steenstra, *Verhandeling over de Klootsche Driehoeks-Meeting* (Amsterdam, 1770), p. 167.

[5] Arnoldus Bastiaan Strabbe, *Erste Beginselen der Fluxie-rekening* (Amsterdam, 1799).

of fluxional notations. It occurs in a review of John Collins' *Commercium epistolicum* bearing on the controversy between Newton and Leibniz on the invention of the calculus. In that review, which appeared in the *Philosophical Transactions*,[1] Newton refers to himself in the third person: "Mr. Newton doth not place his Method in Forms of Symbols, nor confine himself to any particular Sort of Symbols for Fluents and Fluxions. Where he puts the Areas of Curves for Fluents, he frequently puts the Ordinates for Fluxions, and denotes the Fluxions by the Symbols of the Ordinates, as in his *Analysis*. Where he puts Lines for Fluents, he puts any Symbols for the Velocities of the Points which describe the Lines, that is, for the first Fluxions; and any other Symbols for the Increase of those Velocities, that is, for the second Fluxions, as is frequently done in his *Principia philosophiae*. And where he puts the Letters x, y, z for Fluents, he denotes their Fluxions, either by other Letters as p, q, r; or by the same Letters in other Forms as X, Y, Z or $\dot{x}$, $\dot{y}$, $\dot{z}$; or by any Lines as DE, FG, HI, considered as their Exponents. And this is evident by his Book of *Quadratures*, where he represents Fluxions by prickt Letters in the first Proposition, by Ordinates of Curves in the last Proposition, and by other Symbols, in explaining the Method and illustrating it with Examples, in the Introduction. Mr. Leibnitz hath no Symbols of Fluxions in his Method, and therefore Mr. Newton's Symbols of Fluxions are the oldest in the kind. Mr. Leibnitz began to use the Symbols of Moments or Differences dx, dy, dz, in the Year 1677. Mr. Newton represented Moments by the Rectangles under the Fluxions and the Moment o, when he wrote his *Analysis*, which was at least Forty Six Years ago. Mr. Leibnitz has used the symbols $\int x$, $\int y$, $\int z$ for the Sums of Ordinates ever since the Year 1686; Mr. Newton represented the same thing in his *Analysis*, by inscribing the Ordinate in a Square or Rectangle. All Mr. Newton's Symbols are the oldest in their several Kinds by many Years.

"And whereas it has been represented that the use of the Letter o is vulgar, and destroys the Advantages of the Differential Method: on the contrary, the Method of Fluxions, as used by Mr. Newton, has all the Advantages of the Differential, and some others. It is more elegant, because in his Calculus there is but one infinitely little Quantity represented by a Symbol, the Symbol o. We have no Ideas of infinitely little Quantities, and therefore Mr. Newton intro-

---

[1] *Philosophical Transactions*, Vol. XXIX, for the years 1714, 1715, 1716 (London, 1717), p. 294–307.

duced Fluxions into his Method, that it might proceed by finite Quantities as much as possible. It is more Natural and Geometrical, because founded upon the *primae quantitatum nascentium rationes*, which have a being in Geometry, whilst *Indivisibles*, upon which the Differential Method is founded have no Being either in Geometry or in Nature. There are *rationes primae quantitatum nascentium*, but not *quantitates primae nascentes*. Nature generates Quantities by continual Flux or Increase; and the ancient Geometers admitted such a Generation of Areas and Solids, when they drew one Line into another by local Motion to generate an Area, ...

"In the Year 1692, Mr. Newton at the Request of Dr. Wallis, sent to him a Copy of the first Proposition of the Book of Quadratures, with Examples thereof in the first, second and third Fluxions: as you may see in the second Volume of the Doctor's Works, p. 391, 392, 393 and 396. And therefore he had not then forgotten the Method of Second Fluxions."

570. *G. W. Leibniz.*—In the years 1672–76 Leibniz resided most of the time in Paris and, being influenced strongly by Cartesian geometry, he entered upon the study of quadratures. He divided figures by ordinates into elements and on October 26, 1675, letting $\sqcap$ express equality, wrote down the theorem:[1] "Differentiarum momenta ex perpendiculari ad axem aequantur complemento summae terminorum, sive Momenta Terminorum aequantur complemento summae summarum, sive omn. $\overline{xw} \sqcap$ ult. x, $\overline{\text{omn. } w}$,, $-$ omn. $\overline{\text{omn. } w}$. Sit $xw \sqcap a_\delta$,

fiet $w \sqcap \dfrac{a_\delta}{x}$, fiet omn. $\overline{\overline{a_\delta}} \sqcap$ ult. x omn. $\dfrac{a_\delta}{x} -$ omn. omn. $\dfrac{a_\delta}{x}$; ergo omn.

$\dfrac{a_\delta}{x} \sqcap$ ult. x, omn. $\dfrac{a_\delta}{x^2} -$ omn. omn. $\dfrac{a_\delta}{x^2}$, quo valore in aequ. praecedenti

inserto fiet: omn. $a_\delta \sqcap$ ult. x^2 omn. $\dfrac{a_\delta}{x^2} -$ ult. x, omn. omn. $\overline{\overline{\dfrac{a_\delta}{x^2}}} -$

$\overline{\text{omn. ult. } x. \text{ omn. } \dfrac{a_\delta}{x^2}} -$ omn. omn. $\dfrac{a_\delta}{x^2}$. Et ita iri potest infinitum."

("The moments of the differences about a perpendicular to the axis are equal to the complement of the sum of the terms, or, the moments of the terms are equal to the complement of the sum of the sums, that

[1] C. I. Gerhardt, *Der Briefwechsel von G. W. Leibniz mit Mathematikern*, Vol. I (Berlin, 1899), p. xii, xiii; C. I. Gerhardt, Die *Entdeckung der höheren Analysis* (Halle, 1855), p. 120. See also the English translation, J. M. Child, *The Early Mathematical Manuscripts of Leibniz* (Chicago and London, 1920), p. 70–80. See our §§ 538–65.

is, omn. $\overline{xw}$ ⊓, ... , and this can proceed in this manner indefinitely.") Referring to Leibniz' figure (Fig. 121), the top horizontal line is the axis, the w's are small parts of the horizontal lines, omn. w is a full horizontal line, x is a variable height having the altitude of the rectangle as its "ultimate" value.

The last equation given above forcibly exhibits the necessity of a simplified notation.

Three days later, October 29, 1675, Leibniz constructs what he called the *triangulum characteristicum*, which had been used before him by Pascal and Barrow, and considers the question whether expressions like "omn. l" and "omn. p," where l and p designate lines, can be multiplied one into the other. Such products can be formed

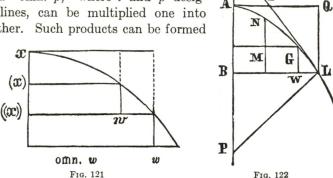

Fig. 121.   Fig. 122.

Fig. 121.—Leibniz' figure in MS dated Oct. 26, 1675. Note also the use of parentheses in designating different x's. (Taken from C. I. Gerhardt, *Die Entdeckung der höheren Analysis* [1855], p. 121.)

Fig. 122.—From the manuscript of Leibniz, Oct. 29, 1675, as reproduced by C. I. Gerhardt, *Entdeckung*, etc., p. 123.

if "omn. l" is a line and "omn. p" is a plane figure; the product is the volume of a solid. In a figure like Figure 122 he lets BL ⊓ y, WL ⊓ l, BP ⊓ p, TB ⊓ t, AB ⊓ x, GW ⊓ a, y ⊓ omn. l, and obtains $\frac{l}{a}$ ⊓ $\frac{p}{\text{omn. } l ⊓ y}$, consequently p ⊓ $\frac{\overline{\text{omn. } l}}{a}$ l, i.e., omn. l is to be multiplied into l; moreover, omn. p ⊓ $\frac{y^2}{2}$ ⊓ $\frac{\overline{\text{omn. } l \text{ ⊡}}}{2}$; "ergo habemus theorema, quod mihi videtur admirabile et novo huic calculo magni adjumenti loco futurum, nempe quod sit $\frac{\overline{\text{omn.} l \text{ ⊡}}}{2}$ ⊓ omn. $\overline{\text{omn. } l \frac{l}{a}}$ qualiscunque sit l." ("Hence we have a theorem that to me seems admirable, and one that will be of great service to this new calculus, namely. . . . .")

A little further on, in the manuscript dated October 29, 1675, he says: "Utile erit scribi $\int$ pro omn. ut $\int l$ pro omn. l, id est summa ipsorum l. Itaque fiet $\dfrac{\int \overline{l}^2}{2} \sqcap \int \overline{\int \overline{l} \dfrac{l}{a}}$ et $\int \overline{xl} \sqcap x \int \overline{l} - \int \int \overline{l}$." ("It will be useful to write $\int$ for omn., as $\int l$ for omn. l, that is, the sum of these l's. Thus, one obtains. . . . .") Here, then, is the origin of the fundamental symbol in the integral calculus.[1] Still further on Leibniz remarks, "Satis haec nova et nobilia, cum novum genus calculi inducant." ("These are sufficiently new and notable, since they will lead to a new calculus.")

It is in the same manuscript dated October 29, 1675, that he remarks,[2] "Si sit $\int l \sqcap ya$. Ponemus $l \sqcap \dfrac{ya}{d}$ nempe ut $\int$ augebit, ita d minuet dimensiones. $\int$ autem significat summam, d differentiam." ("Suppose that $\int l = ya$. Let $l = \dfrac{ya}{d}$; then just as $\int$ will increase, so d will diminish the dimensions. But $\int$ means sum, and d a difference.") Ten days later, November 11, 1675,[3] he remarks, "Idem est dx et $\dfrac{x}{d}$, id est differentia inter duas x proximas." ("dx and $\dfrac{x}{d}$ are the same, that is, the difference between two proximate x's.")[4] Further on in

---

[1] Why did Leibniz choose the long letter s (summa), rather than the letter o of the word omnia which he had been using? Was it because the long s stood out in sharper contrast to the other letters and could be more easily distinguished from them? Sixteen years earlier (in 1659) Pietro Mengoli of Bologna used o in summation, in his Geometriae speciosae elementa, p. 53, where he says: "massam ex omnibus abscissis, quàm significamus charactere O.a; et massam ex omnibus residuis O.r; et massam ex omnibus abscissis secundis O.a2."

[2] C. I. Gerhardt, Entdeckung (1855), p. 126.

[3] C. I. Gerhardt, Briefwechsel, p. xiv; C. I. Gerhardt, Entdeckung, p. 134. See also our § 532.

[4] Leibniz' representation of the differential of y successively by ω, l, $\dfrac{y}{d}$, and finally by dy goes against the conjecture that he might have received a suggestion for the notation dy, dx, from Antoine Arnauld who, in his Nouveaux Elémens de Géométrie (Paris, 1667), adopts $d'x$, $d'y$ as the designation of a small, aliquot part of x and y. "Could this French $d'x$ have been hovering before Leibniz when he made his choice of notation?" is the question which naturally arose in the mind of Karl Bopp in Abhandlungen zur Geschichte der mathematischen Wissenschaften, Vol. XIV (Leipzig, 1902), p. 252.

the same manuscript he enters upon the inquiry: "Videndum an $dx\,dy$ idem sit quod $d\,\overline{xy}$, et an $\dfrac{dx}{dy}$ idem quod d$\dfrac{x}{y}$." ("Let us see whether $dx\,dy$ is the same as $d\,\overline{xy}$, and whether $\dfrac{dx}{dy}$ is the same as $d\,\dfrac{x}{y}$.") And he concludes that neither pair of expressions is the same. Thus on November 11, 1675, he introduced the symbol dx and dy as the differentials of x and y, and also the derivative form $\dfrac{dx}{dy}$.

571. The first appearance of dx in print was in an article which Leibniz contributed to the *Acta eruditorum*, in 1684.[1] Therein occur the expressions "dw ad dx," "dx ad dy," and also "$dx:dy$," but not the form "$\dfrac{dx}{dy}$." For another symbol for the first derivative, at one time considered by Leibniz, see § 539. Leibniz used the second derivative[2] written in the form $ddx:d\overline{y}^2$. In his *Historia et origo calculi differentialis*, written by Leibniz not long before his death, he lets[3] $dx=1$, and writes $d(x)^3=3xx+3x+1$. In his paper, printed in 1684, dx appears to be finite when he says: "We now call any line selected at random dx" ("Iam recta aliqua pro arbitrio assumta vocetur dx").[4]

Leibniz writes $(dx)^2$ in the form $dx\,dx$; Child remarks,[5] "Leibniz does not give us an opportunity of seeing how he would have written the equivalent of $dx\,dx\,dx$; whether as dx^3 or $\overline{dx}^3$ or $(dx).^3$" In the case of a differential of a radical, Leibniz uses a comma after the d, thus:[6] $d, \sqrt[v]{x^a},$ aeq. $\dfrac{a}{b}\,dx\sqrt[v]{x^{a-b}}$. In imitation of Leibniz' $dd\,v$ for the second differential, Johann Bernoulli[7] denotes the fourth differential by $dddd\,z$. Not until 1695, or twenty years after Leibniz' first use of d,

---

[1] See also *Leibnizens Mathematische Schriften*, Vol. V (Halle, 1858), p. 220–26.

[2] *Acta eruditorum* (1693), p. 179; *Leibnizens Mathematische Schriften*, Vol. V (1858), p. 287.

[3] *Op. cit.*, Vol. V (Halle, 1858), p. 406.

[4] For the various explanations of the infinitesimal given by Leibniz at different times of his career see G. Vivanti, "Il Concetto d'Infinitesimo," *Giornale di Matematiche di Battaglini*, Vol. XXXVIII and XXXIX.

[5] J. M. Child, *Early Mathematical Manuscripts of Leibniz* (Chicago and London, 1920), p. 55.

[6] *Acta eruditorum* (1684), p. 469. *Leibnizens Math. Schriften*, Vol. V (Halle, 1858), p. 222.

[7] *Acta eruditorum* (1694), p. 439; Joh. Bernoulli, *Opera*, Vol. I, p. 125–28.

did he suggest the use of numbers in the writing of the higher differentials; in a letter[1] of October 20/30, 1695, addressed to Johann Bernoulli, he writes the equation showing the relationship between the two symbols, $d^m = \int^n$ when $n = -m$.

In the same year he published a reply to B. Nieuwentijt's[2] attacks upon the calculus, in which he writes ddx seu $d^2 \cdot x$, $dddx$ sive $d^3 \cdot x$.

De l'Hospital,[3] who was the earliest to write a textbook on the calculus, followed Leibniz' original practice and did not use numerals in the designation of the higher differentials; he wrote $dd\,y$, $ddd\,y$, $dddd\,y$. As late as 1764 we find these very notations in the writings of Fontaine.[4] William Hales,[5] an adherent of Newton's fluxions, in 1804, cites the symbols dx, $dd\,x$, $ddd\,x$, $dddd\,x$ and then declares them to be *minùs elegantèr* than the fluxionary symbols.

572. Minute adjustments of notation arose in the eighteenth century. Thus, Euler wrote $d \cdot x\,dy\,dz$ for $d(x\,dy\,dz)$. The same practice is observed by Lacroix; but $dx\,dy\,dz$ always meant the product of the three differentials. Bézout[6] points out that dx^2 means $(dx)^2$, and $d(x^2)$ means the "différentielle de x^2." Lacroix explains[7] that d^2y^2 is the same thing as $(d^2y)^2$ and, in general, d^ny^m indicates $(d^ny)^m$, that $d \cdot x^2$ means the differential of x^2 and $d \cdot uv$ of uv; the point following the d signifying that the operation applies to all that immediately follows.

No substantial changes in the notation for total differentiation were made until after the middle of the eighteenth century. A provisional, temporary notation Δ for differential coefficient or *différences des fonctions* was used in 1706 by Johann Bernoulli.[8] Previously he had used[9] the corresponding Latin letter D. The changes that came

[1] *Leibnizens Math. Schriften*, Vol. III (1855), p. 221.

[2] *Acta eruditorum* (1695), p. 310; *Leibnizens Math. Schriften*, Vol. V, p. 321.

[3] De l'Hospital, *Analyse des infiniment petits* (2d ed.; Paris, 1715), p. 55. The date of the first edition was 1696.

[4] A. Fontaine, *Mémoires donnés à l'académie r. des sciences, non imprimés dans leur temps* (1764), p. 73.

[5] Fr. Maseres, *Scriptores logarithmici*, Vol. V (London, 1804), p. 131.

[6] Etienne Bézout, *Cours de Mathématiques à l'usage du corps de l'artillerie*, Tome III (Paris, 1797), p. 18.

[7] S. F. Lacroix, *Traité du Calcul différentiel et du Calcul intégral*, Tome I (2d. ed.; Paris, 1810), p. 208; also his *Traité élémentaire de Calcul différentiel et de Calcul intégral* (Paris, 1802), p. 9, 18.

[8] G. Eneström, *Bibliotheca mathematica* (2d ser.), Vol. X (1896), p. 21.

[9] *Leibnizens Math. Schriften* (ed. C. I. Gerhardt), Vol. III, p. 531.

were due in most cases to new conceptions of the fundamental operations, conceptions which seemed to demand new symbols.

573. *J. Landen.*—One of the first in the field was John Landen, who published at London in 1758 his *Discourse concerning Residual Analysis* and in 1764 his *Residual Analysis*. These publications mark an attempt at arithmetization of the calculus, but the process was so complicated as to be prohibitive. In his *Mathematical Lucrubations* of 1755, Landen still follows Newtonian procedure and notation, but his *Discourse* of 1758 marks a departure. Still proceeding geometrically, he introduces (p. 12) two new symbols, "$[x|y]$ being put for the quotient of $y-u$ divided by $x-v$," i.e., for $\Delta y : \Delta x$. Then he, in fact, proceeds to the limit, "by taking v equal to x, and writing $[x \perp y]$ for the value of $[x|y]$ in the particular case when v is so taken." This notation $[x \perp y]$ for the first derivative labors under the disadvantage of not being compact and requiring many distinct strokes of the pen, but typographically it enjoys an advantage over $\dfrac{dy}{dx}$ in not requiring a fractional line and in calling for type of normal height. In considering maxima and minima, on page 24, Landen invents a new designation, "writing $[x \perp\!\!\perp y]$ for the value of the quotient of $[x \perp y]-[v \perp u]$ divided by $x-v$, when v is therein taken equal to x." Landen had no followers in the use of his symbols.

574. *A. Fontaine.*—In France, Alexis Fontaine introduced a notation for partial differentiation which we shall notice later; he takes $\dfrac{du}{dx}$ to represent a partial derivative and $\dfrac{1}{dx}\,du$ a complete derivative.[1] This form of the complete derivative is seldom found later. We have seen a trace of it in Da Cunha,[2] who in 1790 writes the differential

$$d\left(\frac{1}{dx}\,d\,\frac{dy}{dx}\right),$$

and in De Montferrier,[3] who in 1835 writes the derivative

$$\frac{1}{d\Phi\dot{x}} \cdot d\left[\frac{dF\dot{x}}{d\Phi\dot{x}}\right].$$

[1] A. Fontaine, *Mémoires donnés à l'académie r. des sciences, non imprimés dans leur temps* (Paris, 1764), in "Table" of the *Mémoires*.

[2] J. A. da Cunha, *Principios mathematicos* (Lisbon, 1790), p. 244.

[3] A. S. Montferrier, *Dictionnaire des sciences mathématiques* (Paris, 1835), art. "Différence," p. 460.

Here the point placed above the x is to indicate that, after the differentiations, the x is assigned the value which makes $\Phi\,x = 0$.

575. *J. L. Lagrange.*—A strikingly new treatment of the fundamental conceptions of the calculus is exhibited in J. L. Lagrange's *Théorie des fonctions analytiques* (1797). Not satisfied with Leibniz' infinitely small quantities, nor with Euler's presentation of dx as 0, nor with Newton's prime and ultimate ratios which are ratios of quantities at the instant when they cease to be quantities, Lagrange proceeded to search for a new foundation for the calculus in the processes of ordinary algebra. Before this time the derivative was seldom used on the European Continent; the differential held almost complete sway. It was Lagrange who, avoiding infinitesimals, brought the derivative into a supreme position. Likewise, he stressed the notion of a function. On page 2 he writes, "Ainsi fx désignera une fonction de x." On page 14 he proceeds to a new notation for the first, second, third, etc., derivatives of fx, with respect to x. He says: "... pour plus de simplicité et d'uniformité, on dénote par $f'x$ la première fonction dérivée de fx, par $f''x$ la première fonction dérivée de $f'x$, par $f'''x$ la première fonction dérivée de $f''x$, et ainsi de suite." "Nous appellerons la fonction fx, *fonction primitive*, par rapport aux fonctions $f'x$, $f''x$, etc., qui en dérivent, et nous appellerons celles-ci, fonctions dérivées, par rapport à celle-là. Nous nommerons de plus la première fonction dérivée $f'x$, *fonction prime;* la seconde dérivée $f''x$, *fonction seconde;* la troisième fonction dérivée $f'''x$, *fonction tierce*, et ainsi de suite. De la même manière, si y est supposé une fonction de x, nous dénoterons ses fonctions dérivées par y', y'', y''', etc."[1]

The same notation for the total derivatives of fx was used by Lagrange in his *Leçons sur le calcul des fonctions* which appeared at Paris in 1801 and again in 1806.[2]

[1] For typographical reasons the Council of the London Mathematical Society in 1915 expressed preference for x', r', θ'' over $\dot{x}$, $\dot{r}$, $\ddot{\theta}$. G. H. Bryan objects to this, saying: "It is quite useless to recommend the substitution of dashes for dots in the fluxional notation for velocities and accelerations, because dashes are so often used for other purposes. For example θ'' might be an angle of θ seconds. The only rational plan of avoiding the printer's difficulties is to place the superior dots *after* the letters instead of over them; thus $x'y'' - y'x''$ is all that is necessary" (*Mathematical Gazette*, Vol. VIII [1917], p. 172, 221). This last notation is actually found in a few publications at the opening of the nineteenth century (§ 630).

[2] Reprints are not always faithful in reproducing notations. Thus, in the 1806 edition of the *Leçons*, p. 10, 12, one finds fx, $f'x$, $f''x$, $f'''x$, while in the *Œuvres*, Vol. X, which purports to be a reprint of the 1806 edition, on p. 15, 17, one finds the corresponding parts given as $f(x)$, $f'(x)$, $f''(x)$, $f'''(x)$.

It should be noticed that the 1797 edition of Lagrange's *Théorie des fonctions* is not the first occurrence of the accent as the mark for derivatives. In 1770 Ψ' occurs for $\frac{d\Psi}{dx}$ in his *Nouvelle méthode pour résoudre les équations littérales*, and in 1759 the notation is found in a part of a memoir by François Daviet de Foncenex in the *Miscellanea Taurinensia*, believed to have been written for Foncenex by Lagrange himself.[1]

576. *J. Pasquich.*—Another writer who at this time attempted a reform of the calculus was Johann Pasquich, of Ofen (Budapest) in Hungary, in a paper entitled "Exponential Rechnung."[2] Postulating that every function can be expressed in the form $y = Ax^a + Bx^b + \ldots$, he calls the function $\epsilon y = aAx^a + bBx^b + \ldots$ the *exponential* of y; "ϵy" is really the limit of $\frac{x\Delta y}{\Delta x}$.

J. P. Grüson.—A change similar to that of Pasquich was suggested by Johann Philipp Grüson, of Berlin, in his *"Calcul d'exposition."*[3] He reaches the same limit which had been used by Pasquich. Grüson indicated it by an inverted and rounded E; thus $ɘF$ represented the limit of $\frac{x\Delta F}{\Delta x}$. This notation is mentioned by S. F. Lacroix in his *Traité du Calcul différentiel et du Calcul intégral*.

The notations of Pasquich and Grüson found no acceptance among the mathematicians of their time and have not been drawn upon since for the designation originally assigned. The notation of Lagrange, on the other hand, was adopted by many and has been found a useful adjunct of other calculus notations.

577. *L. F. A. Arbogast.*—Still another symbolism, which has been extensively used in recent years, was made public by L. F. A. Arbogast in his *De Calcul des Dérivations* (Strasbourg, 1800). This new calculus was offered as comprising the theory of series, and included the differential calculus as a special case. An outline of it was presented in the form of a memoir in 1789 to the Academy of Sciences at Paris, but was not published at the time. It was mentioned in the

[1] P. E. B. Jourdain in *Proceed. 5th Internat. Congress* (Cambridge, 1913), Vol. II, p. 540.

[2] J. Pasquich. "Anfangsgründe einer neuen Exponentialrechnung," *Archiv der reinen und angew. Math.*, Vol. II (1798), p. 385–424. Our information is drawn from M. Cantor, *op., cit.* Vol. IV, p. 667, 668.

[3] J. P. Grüson, "Le Calcul d'exposition," *Mém. académie* (Berlin, 1798), Pub. 1801, p. 151–216; (1799 and 1800) Pub. 1803, p. 157–88.

1797 edition of Lagrange's *Théorie des fonctions analytique* as well as in the first volume of S. F. Lacroix's *Traité du Calcul différentiel et du Calcul intégral*. Arbogast expresses himself in his Preface as follows:

"To form the algorithm of derivations, it became necessary to introduce new signs; I have given this subject particular attention, being persuaded that the secret of the power of analysis consists in the happy choice and use of signs, simple and characteristic of the things which they are to represent. In this regard I have set myself the following rules: (1) To make the notations as much as possible analogous to the received notations; (2) Not to introduce notations which are not needed and which I can replace without confusion by those already in use; (3) To select very simple ones, yet such that will exhibit all the varieties which the different operations require."

His principal symbol is the D, as the sign for "derivation." This symbol had been previously used by Johann Bernoulli (§§ 528, 560). During the latter part of the eighteenth century the D had been used by several authors to represent a finite difference. Arbogast lets $F(a+x)$ be any function (*une fonction quelconque*) of the binomial $a+x$; one knows, he says, that one can develop that function in a series proceeding according to the powers of x, viz., $a+bx+\dfrac{c}{1 \cdot 2} x^2 + \ldots$,

where $a = Fa$. He designates by D the operation upon Fa that yields b, so that $b = DFa$, $c = DDFa$, etc. While Arbogast's symbol D for our derivative has maintained its place in many books to the present time, a large variety of satellites to it, which Arbogast introduced, are now only of antiquarian interest. Placing a dot (p. 2) after the D gives him $D \cdot Fa$ to represent $DFa \cdot D \cdot a$ for cases where $D \cdot a$ is not 1.

He writes (p. 33): "$\underset{c}{D}{}^m$ au lieu de $\dfrac{D^m}{1 \cdot 2 \cdot 3 \ldots m}$." On page 308 he states: "Nous désignerons même à l'avenir les coëficiens différentiels $\dfrac{d\phi x}{dx}$, $\dfrac{d^2\phi x}{1 \cdot 2 \cdot dx^2}$, $\ldots$, dx étant invariable, par $\partial \phi x$, $\underset{c}{\partial^2} \phi x$, $\ldots$, ce qui est la même chose que les dérivées $D\phi x$, $\underset{c}{D^2}\phi x$, $\ldots$." Some of his signs are exhibited in Figure 123.

On page 330 Arbogast explains d^{-1}, d^{-2}, $\ldots$, and D^{-1}, D^{-2}, $\ldots$, as meaning, respectively, *différentielles inverses* and *dérivées inverses*. As an incidental rather than a systematic notation, the d^{-1} and D^{-1} have maintained themselves to the present day. This notation was used, for instance, by Georg Tralles[1] in 1804, by A. R. Forsyth in 1903.[2]

578. *C. Kramp.*—Arbogast's symbol for derivatives was adopted by C. Kramp, professor of mathematics and physics at Cologne, in his *Éléments d'Arithmétique universelle* (Cologne, 1808), pages 265,

D , signe des dérivations . . . . n.º 2.

D., D. , différence entre D sans point et avec point . . . . n.ᵒˢ 3 et 9.

Ɒⁿ , Ɒⁿ. , ... n.º 39.

Dᵐˑⁿ , Ɒᵐˑⁿ , Ɒˡˑᵐˑⁿˑ , ⱰˡⁿˑƉˑᵐˑˑƉˑⁿⁿˑ , ... n.º 111.

ⱰᵐˑƉˡⁿˑ , ... n.º 130.

ˡⁿⱰᵒˑ , ˡⁿƉᵒˑƉᵐˑ , ... n.º 153.

ⱰᵐˑƉˡⁿˑƉⁿʳˑ , ⱰᵐˑƉˡⁿˑƉⁿʳˑƉⁿˢˑˑˑ, ... n.ᵒˢ 161 et 166.

S , Sⁿ , Sˑ¹ , S¹ˑ , au lieu de D⁻¹ , D⁻ⁿ , Dˑ⁻¹ , D⁻¹ˑ , ... n.ᵒˢ 36 et 391.

Dⁿˑ1ᵐˑʳˑ , Dⁿˑ1ᵐˑ⁻ʳˑ , ... n.º 422.

d , signe ordinaire de la différentiation , ... n.ᵒˢ 1 et 352.

dᶜⁿ , ... n.º 352.

d¹ˑ , dˑ¹ , d„¹ , ... n.º 371.

∫ = d⁻¹ , signe ordinaire de l'intégration ; ∫ⁿ = d⁻ⁿ , ... n.º 384.

∂ , ∂ⁿ , ∂. , ∂ⁿˑ , ... n.º 352.

∂¹ˑ , ∂ˑ¹ , ∂„¹ , ∂¹ˑˑ , ∂ˑ¹ˑ , ∂„¹ˑ , ... n.º 376.

∫ , ∫ⁿ , ∫ˑ , ∫ ⁿˑ , au lieu de ∂⁻¹ , ∂⁻ⁿ , ∂⁻¹ˑ , ∂⁻ⁿˑ , ... n.º 384.

Δ, Σ , signes ordinaires des opérations qui donnent la différence et l'intégrale-somme.

Δ¹ˑ , Δˑ¹ , ... n.º 409.

E , Eⁿ , signes des états variés , ... n.º 442.

E¹ˑ , Eˑ¹ , E„¹ , ... n.º 444.

6

FIG. 123.—From Arbogast's *Calcul des Dérivations* (1800), p. xxi

271. He takes $X = Ax^a + Bx^b + Cx^c +$, etc., and then designates by DX "ce que devient cette fonction, lorsqu'on multiplie tous ses termes par leurs exposans respectifs, et qu'ensuite on divise par x." Similarly for D^2X, etc. With great faith in the combinatorial analysis then flourishing in Germany under the leadership of Hindenburg,

[1] G. Tralles, *Abhandlungen Berliner Akademie*, Math. Klasse (1804–11), p. 214 ff.

[2] A. R. Forsyth, *Treatise on Differential Equations* (2d ed., 1888), p. 43; (3d ed., 1903), p. 55.

Kramp says (p. x): "Arbogast is obliged to assume the theorem of Taylor, as well as the ordinary operations of the differential calculus, as perfectly [*parfaitement*] known and demonstrated. My derivatives on the contrary are perfectly independent of all notion of the infinitely little or of the limit, of the differential or of differences; and is equally disinclined [*et bien loin*] to be established on the theorem of Taylor, this theorem does not appear but as a simple corollary of a proposition much more general." Under the heading "Notations" he refers to the notation dx and $dy:dx$ as follows: "Later researches have convinced me of the absolute inutility of this constant factor or divisor dx, as well as the notion of the infinitely small from which it has always been considered inseparable. In supposing it equal to unity one banishes all idea of the infinite and one causes all this part of analysis to re-enter the domain of ordinary algebra." The notation D is used in connection with the coefficients arising in his general polynomial development.

b) CRITICISMS OF EIGHTEENTH-CENTURY NOTATIONS

579. *R. Woodhouse.*—A critical comparison of the Newtonian and Leibnizian notations was made in 1803 by Robert Woodhouse,[1] of Cambridge in England. He states: "In the simplest cases, perhaps, there is not much exercise of choice, and $\dot{x}$, $\ddot{x}$, $\dddot{x}$, $\dot{x}^2$, $\dot{x}^3$, $\dot{x}^n$ are as neat as dx, d^2x, d^4x, dx^2, dx^3, dx^n. Take a case from Waring, p. 299, *Meditationes Analyticae*,

$$P\left(\frac{\overset{n}{\dot{V}}}{\dot{x}^n}\right) + Q\left(\frac{\overset{n}{\dot{V}}}{\dot{x}^{n-1}\,\dot{y}}\right) + \text{etc.}; \text{this by differential}$$

notation is, $P \cdot \dfrac{d^nV}{dx^n} + Q \cdot \dfrac{d^nV}{dx^{n-1}\,dy} + \text{etc.}$

in which it appears to me, that $\dot{x}^n$ is not so clearly distinguished from $\overset{n}{\dot{x}}$ as dx^n is from d^nx.

"Again, $\dddot{xy}$ or $(xy)^{\cdots}$ is not so convenient as, $d^3(xy)$.

"Again, suppose x, x' x'', etc., to represent successive values of x, then according to the fluxionary notation, the first, second, etc., fluxions are $\dot{x}$, $\ddot{x'}$, $\dddot{x''}$; by differentials, dx, d^2x', d^3x''. Which notation has here the advantage, must be determined by inspection; and if the advantage is asserted to be with the fluxionary, it is impossible to state in words any irrefragable arguments to the contrary.

[1] R. Woodhouse, *Principles of Analytical Calculation* (Cambridge, 1803), p. xxvii.

"I put down a few more instances, $P\overset{r}{\dot{y}}+H\overset{r-1}{\dot{y}}+h\overset{r-2}{\dot{y}}$, (Waring, p. 206, *Meditat. Analyt.*) $Pd^r y+Hd^{r-1}y+hd^{r-2}y$; in the fluxionary notation the position of the symbols r, $r-1$, $r-2$, introduces ambiguity, since they may be mistaken for indices of P, H, h;

$$\left(\frac{\dot{x}}{\dot{z}}\right)^{\cdot} \qquad (fx)^{\cdot}, \qquad \frac{\ddot{u}}{\dot{x}^2\,\dot{y}} \, ,$$

$$d\left(\frac{dx}{dy}\right) \qquad dfx, \qquad \frac{d^3u}{dx^2\,dy} \, , \cdots$$

"The advantage on the side of the differential notation is not, it may be said, very manifest in these examples, and perhaps I had adhered to the notation most familiar to me, had not stronger reasons than what are contained in the preceding cases presented themselves, for adopting letters instead of dots as the significant symbols of operations. These reasons in a few words are: first, in the fluxionary notation, there is no simple mode of expressing the fluxionary or differential coefficients, that affect the terms of an expanded expression; thus, the form for the binomial without putting down the numeral coefficients, by fluxionary notation is,

$$(x+i)^m=x^m+\frac{\overline{\dot{x^m}}}{\dot{x}}\,i+\frac{\overline{\ddot{x^m}}}{1\,\cdot\,2\dot{x}^2}\,i^2+\text{etc.,}$$

an awkward mode of expression certainly; and even in the differential notation, the coefficients cannot be expressed, except by fractions, thus,

$$(x+i)^m=x^m+\frac{d(x^m)}{dx}\,i+\frac{d^2(x^m)}{1\,\cdot\,2\,\cdot\,dx^2}\,i^2+\text{etc.,}$$

but then, by a slight alteration of this notation, a very commodious one is obtained; thus, using the small capital D to denote the differential coefficient, that is, putting Dx^n for $\frac{d(x^n)}{dx}$, etc., we have,

$$(x+i)^m=x^m+Dx^m\,\cdot\,i+\frac{D^2x^m}{1\,\cdot\,2}\,\cdot\,i^2+\text{etc.}\,\ldots\ldots$$

We have the elements of a very clear and symmetrical notation, when
Δ denotes the entire difference,
d the differential, or part of the entire difference,
D the differential coefficient,
δ the variation."

580. *S. F. Lacroix.*—Calculus notations are discussed by Lacroix,[1] who says in part: "Euler[2] stresses . . . . the defects of the English notation, which becomes difficult to write and also to perceive when the number of points exceeds three, and one does not see how it is possible to indicate it in figures without danger of confusion with exponents; moreover, we know that it is easy to omit a point in writing or that the point may fail to show in the print. Many of these objections apply to the notation of Lagrange, which has besides the great inconvenience of depriving analysts of the right to represent by the same letter, differently marked with accents, such quantities as have analogous signification. The use of d is not subject to these difficulties; this character is most conspicuous, especially if, considering it as a sign of operation, we write it in Roman type, and reserve the Italic letter for the designation of magnitudes, a practice which Euler did not follow: finally the numerals applied to the character d show very clearly the number of times the differentiation has been repeated." Lacroix states that eight symbols have been proposed for

---

[1] S. F. Lacroix, *Traité du Calcul différentiel et du Calcul intégral* (2d. ed.), Tome I (Paris, 1810), p. 246.

[2] L. Euler says in his *Institutiones calculi differentialis* (Petrograd, 1755), p. 100, 101: "This mode of symbolizing, as a matter of fact, cannot be disapproved, when the number of points is small, since it can be recognized instantly by counting; however, if many points are to be written, it carries with it the greatest confusion and very many inconveniences. The tenth differential or fluxion is represented in this exceedingly inconvenient manner $\overset{\vdots}{\overset{\vdots}{y}}$, while by our mode of symbolizing, $d^{10}y$, it is most easily recognized. However, if a case arises in which much higher orders of differentials, perhaps even indefinite ones, are to be expressed, then the English notation is plainly unsuitable." On page 102 Euler says: "As in the differential calculus the letter d does not stand for a quantity, but for a sign [of operation], in order to avoid confusion in calculations, where many constant quantities occur, the letter d should not be employed for their designation; likewise one should avoid the introduction of the letter l as representing a quantity in calculation, when logarithms occur at the same time. It is to be desired that the letters d and l be written as slightly modified characters so as not to be confounded with the ordinary alphabetic forms which designate quantities. Just as, for example, in place of the letter r which first stood for radix [root], there has now passed into common usage this distorted form of it $\sqrt{\ }$." A concrete example of the confusion which may arise is given in *L'Enseignement mathématique*, Vol. I (Paris, 1899), p. 108, where Poincaré tells of an examination in mathematical physics in which the candidate became flustered and integrated the partial-differential equation $\dfrac{d^2z}{dt^2} = a^2 \dfrac{d^2z}{dx^2}$, by dividing by d^2z, then multiplying by dx^2, thereby obtaining $\dfrac{dx^2}{dt^2} = a^2$ and $\dfrac{dx}{dt} = \pm a$.

the (total) first derivative, the $\dfrac{dy}{dx}$ of Leibniz, the $\dfrac{\dot{y}}{\dot{x}}$ of Newton, the $[x \perp y]$ of Landen, the $f'(x)$ and $\dfrac{y'}{x'}$ of Lagrange, the $\varepsilon\, y$ of Pasquich, the $\dfrac{3y}{x}$ of Grüson, and the Dy of Arbogast. Lacroix takes the question of notations seriously (p. 248): "It is a principle avowed by all, that current notations should not be changed except when they are in manifest conflict with the concepts they are intended to represent, or when one can greatly shorten them, or finally when in modifying them, one renders them fit to develop new relations which cannot be exhibited without them. The signs of the differential calculus do not come under any of these cases: the enrichments which Mr. Lagrange has given to Analysis in his Théorie des fonctions, in his Résolution des équations numériques, in his Leçons sur le Calcul des fonctions may be represented with equal simplicity and elegance by the usual characters, as one can see in the first edition of this Treatise." Lacroix adds: "Before multiplying the signs in analysis that are already so very numerous that one may well stop to consider the embarrassment coming to those, who in studying and endeavouring to encompass the whole, must continually bring together formulas and operations that are analogous, yet are expressed in different characters." One should strive to "defend those [symbols] in which the Mécanique analytique and the Mécanique céleste are written."

581. *J. L. Lagrange.*—And what was Lagrange's attitude toward his own new symbolism?[1] In the autumn of his busy years his open-mindedness is splendidly displayed in his attitude taken in the Preface to the second edition of his *Mécanique Analitique* (the spelling of the title in more recent editions is *Mécanique analytique*), where he retains the Leibnizian notation. He says:

"The ordinary notation of the differential calculus has been retained, because it answers to the system of the infinitely small, adopted in this treatise. When one has well comprehended the spirit of this system and has convinced himself of the exactitude of its results by the geometrical method of prime and ultimate ratios, or by the analytical method of derived functions, one may employ the infinitely little as an instrument that is safe and convenient for abbreviating and simplifying demonstrations. It is in this manner that one shortens the demonstrations of the ancients by the method of indivisibles."

---

[1] J. L. Lagrange, *Mécanique Analytique* in *Œuvres*, Tome XI (Paris), p. xiv.

Nor had Lagrange any prejudice against the Newtonian dots. He used them in his *Leçons sur le calcul des fonctions* (1806), page 442, for indicating "dérivées par rapport à i," and in the second edition of his *Mécanique Analitique* to indicate time-derivatives. In the opinion of Thomson and Tait, "Lagrange has combined the two notations with admirable skill and taste."[1]

c) TOTAL DIFFERENTIATION DURING THE NINETEENTH CENTURY

582. *Barlow and Mitchell.*—A slight advantage of Lagrange's strokes over the Newtonian dots was that the former were not placed directly above the letter affected, but in the place where exponents are written. It is worthy of observation that at the opening of the nineteenth century a few English writers attempted to improve Newton's notation, in the case of fluxions of higher orders, by placing only one dot above the letter and then indicating the order by a number placed where exponents usually stand. Thus, Peter Barlow[2] writes $\dot{x}^n$ for the nth fluxion, but $\dot{x}^2$ for $(\dot{x})^2$. James Mitchell,[3] to denote the fluxions of a radical or fraction, places it in a parenthesis and writes the dot in the place for exponents. Sometimes, however, Mitchell and other authors write the fluxions of compound expressions by placing the letter F or f before them. This practice is, of course, objectionable, on account of the danger of misinterpreting the F or f as meaning, not "fluxion," but "function" or "fluent." We must remark also that this use of F or f was not new at this time; but is found much earlier in the writings of George Cheyne[4] in London (who also employed Φ to designate "fluxion of") and of A. Fontaine[5] in France.

583. *Herschel, Peacock, and Babbage.*—The adoption of the Leibnizian notation in England, advocated in 1802 by Woodhouse, was finally secured through the efforts of J. F. W. Herschel, Peacock, and Babbage at Cambridge. In his autobiography, Babbage says:[6]

[1] Thomson and Tait, *Treatise on Natural Philosophy* (new ed.; Cambridge, 1879), p. 158.

[2] P. Barlow, *Dictionary of Pure and Mixed Mathematics* (London, 1814), art. "Differential."

[3] James Mitchell, *Dictionary of the Mathematical and Physical Sciences* (London, 1823), art. "Fluxion."

[4] *Fluentium methodus inversa a Georgio Cheynaeo, M.D.* (Londini, 1703), p. 1 of the part against A. Pitcairne.

[5] Fontaine in *Histoire de l'académie r. des sciences*, année 1767 (Paris 1770), p. 589 ff.

[6] Ch. Babbage, *Passages from the Life of a Philosopher* (1864), p. 39. Quoted in *Mathematical Gazette*, Vol. XIII, p. 163.

"The progress of the notation of Leibniz at Cambridge was slow. . . . . It is always difficult to think and reason in a new language, and this difficulty discouraged all but men of energetic minds. I saw, however, that, by making it [the tutors'] interest to do so, the change might be accomplished. I therefore proposed to make a large collection of examples of the differential and integral calculus. . . . . I foresaw that if such a publication existed, all those tutors who did not approve of the change of the Newtonian notation would yet, in order to save their own time and trouble, go to the collection . . . . to find problems to set. . . . . I communicated to Peacock and Herschel my view, and proposed that they should each contribute a portion [to the *Examples*, 1820]. . . . . In a very few years the change was completely established." Says Glaisher:[1]

"Peacock was Moderator in 1817, and he ventured to introduce the symbol of differentiation into the examination, his colleague, however, retaining the old fluxional notation. The old system made its appearance once more in 1818, but in 1819 Peacock was again Moderator, and, having a colleague who shared his views, the change was fully accomplished. The introduction of the notation and language of the differential calculus into the Senate House examination forms an important landmark in the history of Cambridge mathematics."

584. *A. L. Crelle.*—Influenced by Lagrange's *Théorie des fonctions analytiques*, Crelle bases his calculus[2] upon Taylor's theorem. His various notations may be displayed by the different ways of writing that theorem, as shown in Article 48, page 144:

"Die Anwendung der verschiedenen . . . . vorgeschlagenen Bezeichnungen auf die allgemeine Entwicklung von $f(X+k)$ würde in der Zusammenstellung folgende sein:

$$f(X+k) = |X+k| = fX + kdfX + \frac{k^2}{2} d^2fX \ldots + \frac{k^n}{1 \cdot 2 \ldots n} d^nfX$$

$$= |X| + kd|X| + \frac{k^2}{2} d^2|X| \ldots + \frac{k^n}{1 \cdot 2 \ldots n} d^n|X|, \text{ oder wenn } fX = Y$$

$$\text{heisst, } = Y + kdY + \frac{k^2}{2} d^2Y \ldots + \frac{k^n}{1 \cdot 2 \ldots n} d^nY. \text{ Ferner:}$$

$$f(X+k) = fX + DfX + \tfrac{1}{2}D^2fX \ldots + \frac{1}{1 \cdot 2 \ldots n} D^n|X|, \ldots "$$

[1] J. W. L. Glaisher's "Presidential Address," *Proceedings London Mathematical Society*, Vol. XVIII (London, 1886–87), p. 18.

[2] August Leopold Crelle, *Rechnung mit veränderlichen Grössen*, Band I (Göttingen, 1813).

The variables x, y, z are given in small capitals, which gives the page an odd appearance. It will be noticed that with Crelle d does not signify "differential"; it signifies *Abgeleitete Grösse*, the "derived function" of Lagrange, or the "derivation" of Arbogast. Crelle's notation in partial differentiation and in integration will be given later.

585. *A. L. Cauchy.*—That Cauchy should be influenced greatly by Lagrange is to be expected. If we examine his published lessons of 1823 on the infinitesimal calculus,[1] and his lessons of 1829 on the differential calculus, we find that he availed himself of the Leibnizian dx, dy, $\frac{dy}{dx}$ and also of the Lagrangian F' and y' for the first derivatives. In this respect he set a standard which has prevailed widely down to our own day.

586. *M. Ohm.*—Returning to Germany, we find another attempt to rebuild the calculus notations in the works of Martin Ohm.[2] His symbolism of 1829 is made plain by his mode of writing Taylor's theorem:

$$f(x+h) = f(x) + \partial f(x) \cdot h + \partial^2 f(x) \cdot \frac{h^2}{2!} + \ldots .$$

One notices here the influence of Arbogast, though Ohm uses ∂ in place of D. Ohm writes also

$$\partial(A + B \cdot x^m)_x = mBx^{m-1} , \quad \partial(a^x)_a = x \cdot a^{x-1} , \quad \partial(a^x)_x = a^x \cdot \log a .$$

If y_x is a function of x, then $(\partial^\alpha y_x)_a$ is what becomes of $\partial^\alpha y_x$ when one writes a in place of x. He introduces also the Leibnizian dx, dy and writes

$$\frac{dy}{dx} = \partial y_x , \qquad d^2 y = \partial^2 y_x \cdot dx^2 , \qquad \frac{d^2 y}{dx^2} = \partial^2 y_x .$$

Ohm's notation found some following in Germany. For instance, F. Wolff[3] in 1845 writes f_x to designate a function of x, and $y - y' = \partial f_{x'}(x - x')$ as the equation of a line through the point x', y'; Wolff uses also the differential notation dx, dy, $d^2 y$, and the derivative $\frac{dx}{ds}$.

[1] Cauchy, *Résumé des leçons données à l'école royale polytechnique sur le calcul infinitésimal* (Paris, 1823); also his *Leçons sur le calcul différentiel* (Paris, 1829). Both are reprinted in his *Œuvres* (2d ser.), Vol. IV.

[2] Martin Ohm, *Versuch eines vollkommen consequenten Systems der Mathematik*, Vol. III (Berlin, 1829), p. 53, 67, 80, 94, 102, 103, 107.

[3] F. Wolff, *Lehrbuch der Geometrie*, Dritter Theil (Berlin, 1845), p. 178, 202, 278.

587. *Cauchy and Moigno.*—Meanwhile, some new designations introduced by Cauchy in France were clamoring for the right of way. Abbé Moigno[1] in the Introduction (p. xxi) to his calculus expresses regret that he could not altogether disregard custom and banish from his book "the vague and inconvenient notations $\dfrac{dy}{dx}$, $\dfrac{dy}{dx}\,dx$, $\dfrac{dz}{dx}$, $\dfrac{dz}{dy}$, $\dfrac{dz}{dx}\,dx$, $\dfrac{dz}{dy}\dfrac{dy}{dx}\,dx$, to substitute in their place the more compact notations $y'_x = D_x y$, $d_x y$, $z'_x = D_x z$, $z'_y = D_y z$, $d_x z$, $d_y z$, etc." On page 13 he lets $d_x z$ be a differential and z'_x the derivative, where $z = F(y)$ and $y = f(x)$, so that the $d_x z = z'_x dx$ and therefore $\dfrac{d_x z}{dx} = \dfrac{d_y z}{dy} \cdot \dfrac{d_x y}{dx}$, but "it is the custom to suppress" the indices x and y and to write $\dfrac{dz}{dx} = \dfrac{dz}{dy} \cdot \dfrac{dy}{dx}$.

588. *B. Peirce.*—Arbogast's D for derivative was adopted by Benjamin Peirce[2] in his *Curves, Functions, and Forces*, the first volume of which was published in Massachusetts in 1841. The calculus notation used in the United States before 1824 was almost exclusively Newtonian, from 1824 to 1841 almost entirely Leibnizian. In B. Peirce's book, "Differential coefficients are denoted by D, D', etc.; thus $Df \cdot x = \dfrac{df \cdot x}{dx}$, $D^2 \cdot f \cdot x = D \cdot D \cdot f \cdot x$, or, since dx is independent of x, $D^2 \cdot f \cdot x = \dfrac{d \cdot Df \cdot x}{dx} = \dfrac{d^2 f \cdot x}{dx^2}$." Again, "differentials are denoted by the letters δ, δ' etc., d, d', etc."; Peirce writes $e^x = D^n_. e^x$, and $D\ \mathrm{tang}^{[-1]} x = Dy = \dfrac{dy}{dx} = (1+x^2)^{-1}$. In Volume II (1846), page 15, he lets "also $d^n_c \cdot f \cdot x_0$ be the first differential coefficient of $f \cdot x_0$ which does not vanish," where $f \cdot x_0 \neq 0$ or ∞. Here the d_c meant "differential coefficient."

Since the time of B. Peirce the D has held its place in a few of America's best texts, but even in those it has not been used to the complete exclusion of the Leibnizian symbols. Some authors introduce the Dy, in preference to the $\dfrac{dy}{dx}$, to guard the student against the misconception of regarding the derivative as a fraction. Later in

[1] L'Abbé Moigno, *Leçons de calcul Différentiel et de calcul Intégral, rédigées d'après les méthodes et les ouvrages publiés ou inédits de M. A.-L. Cauchy*, Tome I (Paris, 1840).

[2] Benjamin Peirce in his *Elementary Treatise on Curves, Functions, and Forces* (new ed.; Boston and Cambridge), Vol. I (1852), p. 179, 182, 193, 203. (First edition, 1841).

the book they begin to avail themselves of the flexibility afforded by the $\frac{dy}{dx}$, which allows an easy passage from derivative to differential, or differential to derivative, by the application of simple algebraic rules.

589. *G. S. Carr*[1] introduced "experimentally" a radical departure from the usual notations; he lets y_x stand for $\frac{dy}{dx}$, y_{2x} for $\frac{d^2y}{dx^2}$, y_{3x} for $\frac{d^3y}{dx^3}$, and so on. This "shorter notation" has not met with general favor, for obvious reasons.

590. The more minute examination of the continuity of functions gave rise to the consideration of right-hand and of left-hand derivatives and to corresponding notations. Peano[2] uses D for derivative, writing $D(f, u, x)$ for the derivative of f in the class u, for the value x of the variable, and $D(f, x+Q_0, x)$ for the *dérivée à droite*, and $D(f, x-Q_0, x)$ for the *dérivée à gauche*.

591. *G. Peacock.*—A curious extension of the notion of a derivative $\frac{d^n y}{dx^n}$ to the case where n is not a positive integer, say $\frac{d^{\frac{1}{2}}y}{dx^{\frac{1}{2}}}$, was considered by Leibniz (§ 532), Euler, Fourier, Cauchy, and Liouville. The notion was first applied to differentials. Euler[3] wrote $\frac{d^{\frac{1}{2}}x}{\sqrt{dx}} = 2\sqrt{\frac{x}{\pi}}$, while according to the definition of Leibniz one has $\frac{d^{\frac{1}{2}}x}{\sqrt{dx}} = \sqrt{x}$. The definition of Liouville[4] agrees with that of Leibniz. The subject received the particular attention of George Peacock,[5] for the purpose of illustrating his "principle of the permanence of equivalent forms." Following Liouville, Peacock takes, for example,

$$\frac{d^n(e^{mx})}{dx^n} = m^n e^{mx}$$

[1] G. S. Carr, *Synopsis . . . . of Pure Mathematics* (London, 1886), p. 258, § 1405.

[2] G. Peano, *Formulaire mathématique*, Tome IV (Turin, 1903), p. 168.

[3] L. Euler, *Comment. Academiae Petropol. ad annos 1730 et 1731*, Vol. V, p. 36–57. See M. Cantor, *Vorlesungen*, Vol. III (2d ed.), p. 655.

[4] See G. Eneström, *Bibliotheca mathematica* (3d ser.), Vol. XIV (1913–14), p. 80, 81; C. W. Borchardt, *Monatsberichte d. Akad. d. Wissensch. zu Berlin* (1868), p. 623–25.

[5] George Peacock, "Report on the Recent Progress and Present State of Certain Branches of Analysis," *British Association Report for 1833* (London, 1834), p. 211–21, 241–49. Peacock gives Bibliography.

and then considers the propriety of assuming this relation as a definition of the operation when n is not necessarily a positive integer. This speculation did not lead to results commensurate with those obtained from the generalization of n in the exponential notation a^n, but it illustrates how a suitable notation is capable of suggesting generalizations which a purely rhetorical exposition would not readily do. In England this generalized differentiation demanded the attention also of D. F. Gregory, P. Kelland, and especially of Oliver Heaviside who made important use of it in electromagnetic theory.[1]

592. *J. Fourier.*—Fourier, in his *Traité de la chaleur* (1822), showed in connection with the solution of partial differential equations the advantage which may accrue from the separation of the symbols of operation from those of quantity. Thus, in chapter ix, Section IV: "If the proposed differential equation is

$$\frac{d^2v}{dt^2} + \frac{d^4v}{dx^4} + 2\frac{d^4v}{dx^2dy^2} + \frac{d^4v}{dy^4} = 0 \ ,$$

we may denote by D_Φ the function $\frac{d^2\Phi}{dx^2} + \frac{d^2\Phi}{dy^2}$, so that DD_Φ or D_Φ^2 can be formed by raising the binomial $\left(\frac{d^2}{dx^2} + \frac{d^2}{dy^2}\right)$ to the second degree, and regarding the exponents as orders of differentiation."

This symbolic method was developed further in England by Greatheed[2] and D. F. Gregory,[3] both of Cambridge; by Boole,[4] of Queen's University, Ireland; Carmichael[5] of Dublin; and by many others in Great Britain and on the Continent.

d) PARTIAL DIFFERENTIALS AND PARTIAL DERIVATIVES

593. *L. Euler.*—Partial derivatives appear in the writings of Newton, Leibniz, and the Bernoullis, but as a rule without any special symbolism. To be sure, in 1694, Leibniz,[6] in a letter to De l'Hospital, wrote "δm" for the partial derivative $\frac{\partial m}{\partial x}$ and "ϑm" for $\frac{\partial m}{\partial y}$, and De

---

[1] *Proceedings of the Royal Society of London*, Part I (Feb. 2, 1893); Part II (June 15, 1893); Part III (June 21, 1894). See also O. Heaviside, *Electromagnetic Theory*, Vol. II (London, 1899), p. 434, 435.

[2] S. S. Greatheed in *Philosophical Magazine* (Sept., 1837).

[3] D. F. Gregory in *Cambridge Mathematical Journal*, Vol. I, p. 123.

[4] George Boole, *Differential Equations* (4th ed.; London, 1877), chaps. xvi, xvii.

[5] Robert Carmichael, *Calculus of Operations* (London, 1855).

[6] *Leibnizens Mathematische Schriften*, Vol. II (Berlin, 1850), p. 261, 270.

l'Hospital uses "δm" in his reply of March 2, 1695 (see § 532). In 1728 L. Euler[1] used new letters, P, Q, R, to designate the partial derivatives with respect to x, y, z, respectively. Similarly, G. Monge[2] used the small letters p and q as partial derivatives of z with respect to x and y, respectively, and r, s, t as partial derivatives of the second order. In more recent time Monge's abbreviations have been employed by writers on differential equations and differential geometry. Euler indicated in 1755, in his *Institutiones calculi differentialis*, Volume I, page 195, the partial derivative of P with respect to y by the use of a parenthesis. Thus, $\left(\dfrac{dP}{dy}\right)$ signifies that "P is so differentiated that y alone is treated as a variable, and that differential is divided by dy."[3]

The notation $\left(\dfrac{dy}{dx}\right)$ for partial derivatives has been widely used, notwithstanding the objections which have been raised to it. If one writes

$$\left(\frac{dp}{dx}\right)^2 + \left(\frac{dp}{dy}\right)^2 + \left(\frac{dp}{dz}\right)^2 = 1 \text{ or } a\left(\frac{dy}{dx}\right),$$

the reader may be in doubt whether the parentheses are introduced to indicate the presence of partial derivatives or merely to render algebraic service in the expression of the square or the product. These objections to the contrary notwithstanding, this Eulerian notation long held its place in competition with other notations for partial derivatives. Some idea of its wide distribution both in time and localities will be conveyed by the following references: It was used by L'Abbé Sauri,[4] of Paris; Gherli,[5] in Modena; Waring,[6] of Cam-

[1] *Commentarii Academiae Scientiarum imperialis, Petropolitanae*, Tome III (1728) (ed. 1732), p. 115, 116; *Encyclopédie des sciences mathématiques*, Tome II, Vol. I (Paris and Leipzig, 1912), p. 284, n. 168.

[2] Gaspard Monge, *Application de l'analyse à la géométrie* (edited after 4th ed. of 1809 by J. Liouville; Paris, 1850), p. 53 and 71.

[3] "Brevitatis gratia autem hoc saltem capite quantitates r et q ita commode denotari solent, ut r indicetur per $\left(\dfrac{dP}{dy}\right)$, quae scriptura designatur P ita differentiari, ut sola y tanquam variabilis tractetur, atque differentiale istud per dy dividatur." Just above that passage Euler wrote "posito ergo x constante erit $dP = rdy$. . . . . Deinde posito y constante erit $dQ = qdx$."

[4] L'Abbé Sauri, *Cours Complet de mathématiques*, Tome V (Paris, 1774), p. 83.

[5] O. Gherli, *Gli Elementi teorico-pratici delle matematiche pure*, Tomo VII (Modena, 1777), p. 2.

[6] E. Waring, *Meditationes analyticae* (Cambridge, 1785), p. 166.

bridge; Laplace,[1] in Paris; by Monge,[2] Legendre,[3] Lagrange,[4] in 1759, 1792; by De Montferrier,[5] in Paris; Price,[6] at Oxford; Strong,[7] in New York; Resal,[8] at Paris; Newcomb,[9] at Washington; and Perry,[10] of London, who in 1902 expressed his preference for the Eulerian notation.

In 1776 Euler[11] uses $\dfrac{\partial^\lambda}{p} \cdot V$ to indicate the λth derivative, partial with respect to the variable p; he uses a corresponding notation $\dfrac{s}{x} \cdot V$ for partial integration. A similar notation is adopted by F. Servois[12] in 1814.

594. *W. J. G. Karsten.*—The need of some special designation must have been felt by many workers. Apparently unaware of Euler's notation, Karsten[13] in 1760, taking V as a function of x, y, z, u, etc., denotes the partial increment with respect to x by $\overset{x}{\Delta} V$, the partial increment of $\overset{x}{\Delta} V$ with respect to y by $\overset{y\,x}{\Delta\Delta} V$; correspondingly, he writes the partial differentials $\overset{x}{d} V$ and $\overset{y\,x}{dd} V$; he designates also the partial

---

[1] P. S. Laplace, *Traité de mécanique céleste*, Tome I (Paris), Vol. VII (1798), p. 7. *Histoire de l'académie r. des sciences*, année 1772 (Paris, 1775), *Mémoires*, p. 346.

[2] G. Monge in *Histoire de l'académie r. des sciences*, année 1784 (Paris, 1787), p. 119.

[3] A. M. Legendre, in *op. cit.*, année 1784, p. 372.

[4] J. L. Lagrange in *Mémoires de l'académie r. des sciences et belles-lettres*, année 1792 et 1793 (Berlin, 1798), p. 302; *Miscellanea philosophico-mathematica societatis privatae Taurinensis*, Tome I, "Propagation du son," p. 18.

[5] A. S. de Montferrier, *Dictionnaire des sciences mathématiques*, Tome I (Paris, 1835), art. "Différence."

[6] B. Price, *Treatise on Infinitesimal Calculus*, Vol. I (Oxford, 1852), p. 118.

[7] Theodore Strong, *Treatise of the Differential and Integral Calculus* (New York, 1869), p. 44.

[8] H. Resal, *Traité élémentaire de mécanique céleste* (2d ed.; Paris, 1884), p. 8.

[9] Simon Newcomb, *Elements of the Differential and Integral Calculus* (New York, 1887), p. 54.

[10] John Perry in *Nature*, Vol. LXVI (1902), p. 271.

[11] L. Euler in *Nova acta academiae i. scientiarum Petropolitanae*, Tome IV, ad annum 1786 (Petrograd, 1789), p. 18.

[12] F. Servois in *Gergonne's Annales de mathématiques*, Tome V (Nismes, 1814; 1815), p. 94, 95.

[13] W. J. G. Karsten, *Mathesis Theoretica Elementaris atque sublimior* (Rostochii, 1760), p. 775, 781, 782, 807, 808.

differential with respect to y of the partial integral with respect to x, $d^{y}\,^{x}\!\int P dx$. Similarly, for partial derivatives he writes $\dfrac{\overset{y}{dP}}{dy}=\dfrac{\overset{x}{dQ}}{dx}$.

This notation did not attract much attention. It involved, in the last illustration, four terraces of type, which was objectionable. Moreover, German writers in Karsten's time were not in the limelight, and not likely to secure a following outside of their own country.

595. *A. Fontaine.*—Attention must be paid to the notation in partial differentiation used by Alexis Fontaine,[1] of Paris, who presented certain memoirs before the French Academy in 1739, but which were not published until 1764. The notation for partial differentials used by him is found in textbooks even of our own time.

In the Table of Contents of the *Memoirs* contained in the volume, one finds μ given as a function of x, y, z, u, etc., and then

$$d\mu = \frac{d\mu}{dx}\,dx + \frac{d\mu}{dy}\,dy + \frac{d\mu}{dz}\,dz + \frac{d\mu}{du}\,du + \text{etc} \ .$$

$$\frac{dd\mu}{dx} = \frac{dd\mu}{dx^2}\,dx + \frac{dd\mu}{dx\,dy}\,dy + \frac{dd\mu}{dx\,dz}\,dz + \frac{dd\mu}{dx\,du}\,du + \text{etc.} \ ,$$

and similarly expressions for $\dfrac{dd\mu}{dy}$, $\dfrac{dd\mu}{dz}$, $\dfrac{dd\mu}{du}$. Evidently, the $d\mu$ in the left member of the first equation is the total differential of the first order, the $\dfrac{dd\mu}{dx}$ in the left member of the second equation is the total differential of the partial derivative $\dfrac{d\mu}{dx}$. The partial differential with x as the sole variable is here indicated by $\dfrac{du}{dx}\,dx$; with y as the sole variable, by $\dfrac{d\mu}{dy}\,dy$; and similarly for z, u, etc. This notation lacks compactness, but has the great merit of being easily understood and remembered.

Fontaine's equations in the Table of Contents are followed by the following explanation:

"Cette expression-ci $\dfrac{1}{dx}$ · $d\mu$ est donc bien différente de celle-cu $\dfrac{d\mu}{dx}$.

---

[1] *Mémoires donnés à l'académie royale des sciences, non imprimés dans leur temps.* Par M. Fontaine, de cette académie (Paris, 1764). The part in the book containing definitions, notations, and fundamental theorems of the *calcul intégra* on p. 24 is dated Nov. 19, 1738.

La première signifie la différence de μ divisée par dx; la seconde signifie le coëfficient de dx dans la différence de μ."

In other words, here the $\dfrac{d\mu}{dx}$ is our modern $\dfrac{\partial\mu}{\partial x}$. As the only derivatives occurring in Fontaine's equations are partial derivatives, no confusion could arise in the interpretation of derivatives; $\dfrac{dd\mu}{dx\,dy}\,dy$ meant the partial differential (with respect to y as the only variable) of the partial derivative $\dfrac{\partial\mu}{dx}$, while $\dfrac{dd\mu}{dx\,dy}$ indicated $\dfrac{\partial^2\mu}{\partial x\,\partial y}$. This interpretation is confirmed by statements on page 25.

The wide spread of Fontaine's notation for partial differentials may be displayed as follows: It was used by D'Alembert,[1] Lagrange,[2] Poisson,[3] Gauss,[4] Moigno,[5] Duhamel,[6] Hermite,[7] and Davies and Peck.[8]

A notation more along the line of Karsten's is found in Abbé Girault[9] de Koudou's calculus, which is a curious mixture of English phraseology and German symbols. We quote: "Cette expression $\overset{x}{d(X)}$ signifie Fluxion de X dans la supposition de x seule variable," where $X = x + y + z - u + a$.

596. *Monge, Condorcet, and Legendre.*—About 1770 G. Monge[10] wrote $\dfrac{\delta V}{dx}$ and $\dfrac{dV}{dy}$ to represent the partial derivatives with respect to

[1] D'Alembert in *Mémoires de l'académie*, Tome XIX, année 1763 (Berlin, 1763), p. 264.

[2] J. L. Lagrange, *Méchanique analitique* (1788), p. 137.

[3] S. D. Poisson in *Mémoires de l'académie r. des sciences de l'Institut de France*, année 1816, Tome I (Paris, 1818), p. 21.

[4] Gauss, *Werke*, Vol. V (Göttingen, 1867), p. 57. Gauss says: "Spectari itaque potererit z tamquam functio indeterminatarum x, y, cuius differentialia partialia secundum morem suetum, sed omissis vinculis, per $\dfrac{dz}{dx}\cdot dx$, $\dfrac{dz}{dy}\cdot dy$ denotabimus."

[5] L'Abbé Moigno, *Leçons de calcul différentiel et de calcul intégral* (Paris, 1840), Tome I, p. 20.

[6] J. M. C. Duhamel, *Éléments de calcul infinitésimal* (3d ed.; Paris, 1874), p. 273.

[7] Ch. Hermite, *Cours d'analyse*, 1. Partie (Paris, 1873), p. 80.

[8] Charles Davies and William G. Peck, *Mathematical Dictionary* (New York and Chicago, 1876), art. "Partial Differential."

[9] *Leçons analytiques du Calcul des Fluxions et des Fluentes, ou Calcul Différentiel et Intégrale*. Par M. L'Abbé Girault de Koudou (Paris, MDCCLXXII).

[10] Monge in *Miscellanea Taurinensia*, Tome V, Part II (1770–73), p. 94.

x and y, respectively. In an article of the year 1770 Marquis de Condorcet[1] indicated a partial differential of z with respect to x by dz, and a partial differential of z with respect to y by ∂z, or else he indicated by dz a total differential and by ∂z a partial. "Jusqu'ici l'une ou l'autre hypothèse donne le même résultat, mais lorsqu'il en sera autrement, j'aurai soin d'avertir de celle dont il sera question." In an article[2] of 1772 he alters the meaning of d and ∂ so as to signify differentiation, with respect to y and x, respectively.

The use of the rounded letter ∂ in the notation for partial differentiation occurs again in 1786 in an article by A. M. Legendre.[3] He introduces the modern notation $\dfrac{\partial v}{\partial x}$ for the partial derivative and says: "Pour éviter toute ambiguïté, je représenterai par $\dfrac{\partial v}{\partial x}$ le coéfficient de dx dans la différence de v et par $\dfrac{dv}{dx}$ la différence complète de v divisée par dx."

How did it happen that Legendre's happy suggestion remained unheeded? There were perhaps two reasons. One was that he himself failed to use his notation in later articles. The year following he not only abandoned his own notation, but also that of Euler which he had used in 1784, and employed no symbolism whatever to distinguish between total and partial differentiation. Perhaps a second reason lies in the fact that at that time there was a great scramble for the use of the letters D and d in different fields of mathematics. The straight d had been used by Leibniz and his followers, but writers on finite differences[4] laid claim to it, as well as to the capital D.

Thereupon Arbogast gave D the new meaning of "derivation." It looked indeed as if the different mathematical architects engaged in erecting a proud mathematical structure found themselves con-

---

[1] Condorcet in *Histoire de l'académie r. des sciences*, année 1770 (Paris, 1773), p. 152.

[2] Condorcet in *op. cit.*, année 1772, Part I (Paris, 1775), *Mémoires*, p. 14.

[3] Legendre in *op. cit.*, année 1786 (Paris, 1788), p. 8. The article is translated by Stäckel in *Ostwald's Klassiker der exakten Wissenschaften*, No. 78 (Leipzig, 1896), p. 64–65. See also *Encyclopédie des sciences mathématiques*, Tome II, Vol. 1, p. 284, n. 168.

[4] Among the writers who used the D in finite differences were L'Abbé Sauri, *Cours complet de mathématiques*, Tome III (Paris, 1774), p. 304; Georg Vega, *Vorlesungen über die Mathematik*, Vol. II (Wien, 1784), p. 395; J. A. da Cunha, *Principios mathematicos* (1790), p. 269. The small letter d was used less frequently in finite differences; we cite Charles Hutton who in his *Mathematical and Philosophical Dictionary* (London, 1895), art. "Differential Method," employs the d.

fronted with the curse of having their sign language confounded so that they could the less readily understand each other's speech. At this juncture certain writers on the calculus concluded that the interests of their science could be best promoted by discarding the straight-letter d and introducing the rounded ∂. Accordingly they wrote ∂y for the total differential, $\dfrac{\partial y}{\partial x}$ for the total derivative, and $\left(\dfrac{\partial y}{\partial x}\right)$ for the partial derivative. G. S. Klügel refers to this movement when he says in 1803:[1] "It is necessary to distinguish between the symbol for differential and that for a [finite] quantity by a special form of the letter. In France[2] and in the more recent memoirs of the Petersburg Academy, writers[3] have begun to designate the differential by the curved ∂." So Klügel himself adopts this symbolism from now on for his dictionary. Euler's *Institutiones calculi integralis*, which in its first edition of 1768–70 used the straight d, appeared in the third edition of 1824 with the round ∂, both for the total differential and the total derivative. This same notation is found in J. A. Grunert's calculus[4] of 1837. However, the movement failed of general adoption; for some years both d and ∂ were used by different writers for total differentiation.

In view of these countercurrents it is not surprising that Legendre's genial notation for partial differentiation failed of adoption.

597. *J. L. Lagrange.*—Lagrange, in his *Méchanique analitique* of 1788, used no special signs for partial differentiation; he even discarded Euler's parentheses. The reader was expected to tell from the context whether a derivative was total or partial. We have seen that this attitude was taken also by Legendre in 1787.

598. *S. F. Lacroix.*—Lagrange's procedure of 1788 was adopted by Lacroix whose three-volume treatise on the calculus was a standard work of reference in the early part of the nineteenth century. Lacroix says:[5] "First, let me observe that it [the partial derivative]

---

[1] G. S. Klügel, *Mathematisches Wörterbuch*, 1. Abth., 1. Theil (Leipzig, 1803), art. "Differentiale."

[2] In France, Le Marquis de Condorcet had used the ∂ in writing total derivatives in his *Probabilité des décisions*, in 1785.

[3] The *Nova Acta Petropolitana*, the first volume of which is for the year 1785, contain the rounded ∂.

[4] Johann August Grunert, *Elemente der Differential und Integralrechnung* (Leipzig, 1837).

[5] S. F. Lacroix, *Traité du calcul différentiel et du calcul integral*, Tome I (2d ed.; Paris, 1810), p. 243.

should be disencumbered of the parentheses which Euler employed. Really, $\frac{dz}{dx}$, $\frac{dz}{dy}$ are as clear as $\left(\frac{dz}{dx}\right)$, $\left(\frac{dz}{dy}\right)$ when one knows beforehand that z is a function of two independent variables x and y, which the statement of the question, or the meaning of which it is susceptible, always indicates. . . . . . Messrs. Lagrange and Legendre have long ago suppressed the parentheses in their calculations and I have felt that, without inconvenience, I may follow the example which they have set. Fontaine who was the first to apply the notation of Leibniz to partial differentials, proposed to designate by $\frac{1}{dx}\,d\mu$ the ratio of dx to the total differential of $d\mu$." In a more complicated function of several variables, of say x, y, z, where perhaps z is a function of x and y, Lacroix admits that some special notations should be used such as had been proposed by Lagrange (in 1797 and 1801).

Lacroix' attitude toward partial derivatives has had a very large following everywhere, almost down to the present; no special notation was employed for the ordinary partial derivative. This was the attitude of Poisson,[1] Cauchy (in 1821 and 1822),[2] Serret,[3] Duhamel,[4] Hermite,[5] and others in France. This was the practice of Boole,[6] Todhunter,[7] Lord Rayleigh,[8] Thomson and Tait,[9] in Great Britain, it being remarked by W. H. Young,[10] as late as 1910, that in English writings the ordinary $\frac{df}{dx}$ was sometimes used to represent the partial derivative. In Sweden a similar course was followed, as is exemplified in papers of A. V. Bäcklund.[11]

[1] S. P. Poisson, *Traité de mécanique* (2d ed.), Tome I (Paris, 1833), p. 36.

[2] A. Cauchy, *Mémoires de l'académie r. des sciences*, Tome V, années 1821 et 1822 (Paris, 1826), p. 47.

[3] *Notes de M. J. A. Serret sur le calcul différentiel et le calcul intégral par Lacroix* (no date, but after 1861). But in J. A. Serret's *Cours de calcul différentiel et intégral* (2d ed.; Paris, 1879), p. 29, etc., the Legendre-Jacobi notation is used.

[4] J. M. C. Duhamel, *Éléments de calcul infinitésimal* (3d ed.; Paris, 1874), p. 252.

[5] Ch. Hermite, *Cours d'Analyse*, 1. Partie (Paris, 1873), p. 78, 79.

[6] George Boole, *Differential Equations*.

[7] I. Todhunter, *Treatise on the Differential Calculus* (7th ed.; London, 1875)

[8] Baron Rayleigh, *The Theory of Sound* (London, 1877).

[9] Thomson and Tait, *Treatise on Natural Philosophy* (1879), p. 160.

[10] W. H. Young, *Fundamental Theorems of the Differential Calculus* (Cambridge University Press, 1910), p. 20.

[11] A. V. Bäcklund in *Lunds Univ. Årsskrift*, Tom. IX (Lund. 1872), p. 4.

This failure to observe a notational distinction between partial and total derivatives caused the less confusion, because the derivatives were usually obtained from expressions involving differentials, and in the case of differentials some distinctive symbolism was employed—it being either Fontaine's designation or some later notation.

599. *J. A. da Cunha.*—Portugal possessed an active mathematician, Da Cunha,[1] who in 1790, in a wonderfully compact, one-volume treatise of mathematics, used symbolism freely. With him, "d^x denotes the fluxion taken with respect to x, $\int^x$ the fluent taken relative to x." When $M = \dfrac{d^y O}{dy}$, then $\dfrac{d^x M}{dx} = \dfrac{{}^\backprime d^2 O}{dy\,dx^\backprime}$, the single, inverted commas in the second derivative being used as a reminder that each of the two differentiations is partial.

600. *S. Lhuilier.*—A notation for partial differentials different from Fontaine's and resembling Karsten's is given in Simon Lhuilier's book on the exposition of the principles of the calculus,[2] in 1795. When P is a function of x and y, he writes the first three partial differentials with respect to x thus: ${}^x d'P$, ${}^x d''P$, ${}^x d'''P$; in general, he indicates N partial differentiations with respect to x, followed by M partial differentiations with respect to y, by the differential symbol ${}^y d^M {}^x d^N P$. We shall see that Cauchy's notation of 1823 differs from Lhuilier's simply in the positions of the y and x and in the use of small letters in the place of M and N.

601. *J. L. Lagrange.*—We return to the great central figure in the field of pure mathematics at the close of the eighteenth century—Lagrange. The new foundations which he endeavored to establish for the calculus in his *Théorie des fonctions analytiques* (1797) carried with it also new notations, for partial differentiation. He writes (p. 92) $f', f'', f''', \ldots$, to designate the first, second, third, etc., functions (derivatives) of $f(x, y)$, relative to x alone; and the symbols $f_1, f_{11}, f_{111}, \ldots$, to designate the first, second, third, $\ldots$, derivatives, relative to y alone. Accordingly, he writes (p. 93) also $f_1'(x, y)$, $f_1''(x, y)$, $f_{11}'(x, y)$, $f_{11}''(x, y)$, etc., for our modern $\dfrac{\partial^2 f}{\partial y\, \partial x}$, $\dfrac{\partial^3 f}{\partial y\, \partial x^2}$, $\dfrac{\partial^3 f}{\partial y^2\, \partial x}$, $\dfrac{\partial^4 f}{\partial x^2 \partial y^2}$, etc. In the case of a function $F(x, y, z)$ of three variables he

---

[1] J. A. da Cunha, *Principios mathematicos* (Lisboa, 1790), p. 263.

[2] S. Lhuilier, *Principiorum calculi differentialis et integralis expositio elementaris* (Tübingen, 1795), p. 238.

represents (*Œuvres*, Vol. IX, p. 158), by $F'(x)$, $F'(y)$, $F'(z)$ the *primes fonctions* of $F(x, y, z)$ taken with respect to x, y, z, respectively, as independent variables. The danger of this notation is that $F'(x)$ might be taken to be the primitive F' function of the single variable x. Lagrange also (*ibid.*, p. 177) denotes by u', u_1, and $_1u$ the prime functions (first derivatives) of u with respect to x, y, z, respectively. Furthermore (*ibid.*, p. 273), u'', u_{11}, u_1' stand for our $\dfrac{\partial^2 u}{\partial x^2}$, $\dfrac{\partial^2 u}{\partial y^2}$, $\dfrac{\partial^2 u}{\partial x\,\partial y}$, respectively. In 1841 C. G. J. Jacobi declared that "the notation of Lagrange fails, if in a function of three or more variables one attempts to write down differentials of higher than the first order."[1] But Stäckel[2] remarked in 1896 that the Lagrangian defects can be readily removed by the use of several indices, as, for example, by writing $f_{\alpha\beta\gamma}$ for our $\dfrac{\partial^{\alpha+\beta+\gamma} f}{\partial x^\alpha \partial y^\beta \partial z^\gamma}$.

The historical data which we have gathered show that as early as 1760 Karsten had evolved a precise notation even for higher partial derivatives. It was capable of improvement by a more compact placing of the symbols. But this notation remained unnoticed.

That Lagrange was not altogether satisfied with his notation of 1797 appears from his *Résolution des équations numériques* (1798), page 210, where he introduces $\left(\dfrac{Z'}{a'}\right)$, $\left(\dfrac{Z'}{b'}\right)$, $\left(\dfrac{Z''}{a'^2}\right)$, $\left(\dfrac{Z''}{a'b'}\right)$ to designate the partial derivatives $\dfrac{\partial Z}{\partial a}$, $\dfrac{\partial Z}{\partial b}$, $\dfrac{\partial^2 Z}{\partial a^2}$, $\dfrac{\partial^2 Z}{\partial a\,\partial b}$, and then says: "Cette notation est plus nette et plus expressive que celle que j'ai employée dans la *Théorie des fonctions*, en plaçant les accens différemment, suivant les différentes variables auxquelles ils se rapportent." Lagrange adds the further comment: "In substituting the former in place of this, the algorithm of derived functions conserved all the advantages of the differential calculus, and will have this additional one of disencumbering the formulas of that multitude of d's which lengthen and disfigure them to a considerable extent, and which continually recall to mind the false notion of the infinitely little."

It will be noticed that Lagrange's notation for partial derivatives, as given in 1798 in his *Résolution des Équations numérique*, and again in his *Leçons sur le calcul des fonctions* (Paris, 1806), page 347,[3] closely

[1] C. G. J. Jacobi, "De determinantibus functionalibus," *Journal für d. r. & a. Mathematik*, Vol. XXII, p. 321; Jacobi, *Ges. Werke*, Vol. III, p. 397.

[2] P. Stäckel in *Ostwald's Klassiker der exakten Wiss.*, No. 78, p. 65.

[3] J. Lagrange, *Œuvres*, Vol. X.

resembles the notation previously used by Waring at Cambridge. The only difference is that Waring employs the fluxional dot, placed above the letter, as in $\left(\dfrac{\dot{Z}}{\dot{a}}\right)$, while Lagrange uses a stroke in the position assigned to an exponent, as in $\left(\dfrac{Z'}{a'}\right)$. It will not escape the reader that in ordinary derivatives as well, the Lagrangian f', f'', f''' bears close resemblance to the Newtonian $\dot{x}$, $\ddot{x}$, $\dddot{x}$, the stroke again replacing the dot. We shall point out that certain English and American writers shifted the Newtonian dots to the position where exponents are placed (§§ 579, 630). By so doing the notations of Newton and Lagrange were brought still closer together.

602. *L. F. A. Arbogast.*—In his *Calcul des Dérivations* (1800), page 89, Arbogast designates by $D^m \cdot \varphi(a, a)$ m total derivations, while $D^{m,n} \cdot \varphi(a, a)$ designates partial derivations, viz., m derivations with respect to a only, followed by n derivations with respect to a only. Accordingly, $D^{o,n} \cdot \varphi(a, a)$ or simply $D^{\cdot n}\varphi(a, a)$ indicates n partial derivations with respect to a. Also $D^{m,n}_c \cdot \varphi(a, a)$ means that $D^{m,n} \cdot \varphi(a, a)$ is to be divided by $(m!\, n!)$. He uses the Leibnizian d as the sign for the ordinary differential. The symbols d',, d,′, are partial differentials (p. 319). He writes d,′$+d'$,, $d^n = (d$,′$+d'$,$)^n$, $d^n(x, y, z) = (d$,,′$+d$,′$+d'$,$)^n$ $\times xyz$.

603. *J. L. Lagrange.*—In close alignment with the Arbogast symbols is the new notation for partial derivatives which Lagrange gave in his *Leçons sur le calcul des fonctions* (Paris, 1801; 2d ed., 1806), pages 329, 330, 331. Let $f(x, y)$ be a function of x and y, then f',(x, y) means the partial derivative with respect to x, f,′(x, y) the partial derivative with respect to y, f',′(x, y) the partial derivative with respect to x, followed by the partial derivative with respect to y. Higher partial derivatives are designated by $f^{m,n}(x, y)$.

Robert Woodhouse referred to these signs in his *Principles of Analytical Calculation* (Cambridge, 1803), page xxx, where he compares the English and continental notations and gives his reasons for adopting the continental. He says: "Many theorems are conveniently demonstrated by separating the scale of operation, thus, when applied to a rectangle xy, Δ, D, d, may be separated into parts, as Δ',, Δ,′, D',, D,^{d}, d',, d,′, where Δ',, d',, etc. mean respectively the entire difference and differential of xy, relatively to x only; Δ,′, d,′, relatively to y only; for instance, $d(xy) = x\,dy + y\,dx = d$,′$(xy) + d'$,$(xy) = (d'$,$+d'$,$)xy$: The advantage resulting from this change of notation, cannot be here distinctly pointed out, but it would be difficult, I apprehend, to introduce a similar and corresponding change in the fluxionary notation: and this is an additional reason, why I have made a deviation, for which the English reader may feel a propensity to blame me."

604. *A. L. Crelle.*—In this restless period in the development of the calculus, there were many minds, and divided councils. We have seen that new notations were suggested by Crelle,[1] who in partial derivatives has a notation resembling Euler's of 1776; Crelle states (p. 72): "When $z = |xy|$, then we let $\frac{d}{x}z$ or $\frac{d}{x}|xy|$, or $\frac{d}{x}fxy$ stand for the first partial derivation according to x of the magnitude $|xy|$, while $\frac{d}{y}z$ or $\frac{d}{y}|xy|$ or $\frac{d}{y}fxy$ stands for the partial derivation according to y, of the magnitude $|xy|$." Similarly, $\frac{d^2}{x^2}z$ or $\frac{d^2}{x^2}fxy$ or $\frac{d^2}{x^2}|xy|$ indicates the second partial derivative with respect to x, while $\frac{d^2}{xy}z$ represents the partial derivative with respect to x, followed by that with respect to y. It is to be noted that with Crelle d does not signify "differential," but *abgeleitete Grösse*, which Lagrange called the "derived function" and Arbogast the "derivation." The partial is distinguished from the total derivative, by having the independent variable with respect to which the differentiation takes place written below the d. Thus dz means the total derivative, $\frac{d}{x}z$ means the partial. On page 82 he introduces also the D for the total derivative.

605. *P. Barlow.*—A quaint notation was given in England in 1814 by Peter Barlow.[2] If u is any function of x, y, z, etc., let δ, d, D, etc., be the characteristics of their differentiation relatively to each of the variables, then the partial differences are shown here in the second and third lines,

x,	y,	z,	etc.,
$\dfrac{\delta u}{dx}$,	$\dfrac{du}{dy}$,	$\dfrac{Du}{dz}$,	etc.,
$\dfrac{\delta\delta u}{dx^2}$,	$\dfrac{d\delta u}{dx\,dy}$,	$\dfrac{D\delta u}{dx\,dz}$,	etc.

[1] A. L. Crelle, *Rechnung mit veränderlichen Grössen*, Vol. I (Göttingen, 1813).
[2] P. Barlow, *Dictionary of Mathematics* (1814), art. "Partial Differences."

The association of x, y, z, etc., with δ, d, D, etc., respectively, makes unnecessarily heavy demands upon the memory of the reader.

606. *A. L. Cauchy.*—Cauchy used a variety of different notations for partial differentiation at different times of his long career. In 1823, in his lessons on the infinitesimal calculus,[1] he represents partial differentials by $d_x u$, $d_y u$, $d_z u$, and the partial derivatives by $\dfrac{d_x u}{dx}$, $\dfrac{d_y u}{dy}$, $\dfrac{d_z u}{dz}$, but remarks that the latter are usually written, for brevity, $\dfrac{du}{dx}$, $\dfrac{du}{dy}$, $\dfrac{du}{dz}$. He writes also the more general partial derivative $\dfrac{d_x^l\, d_y^m\, d_z^n\, \ldots\, u}{dx^l\, dy^m\, dz^n\, \ldots}$, but adds that the letters x, y, z at the base of the d's are ordinarily omitted, and the notation $\dfrac{d^{l+m+n\,\ldots}\, u}{dx^l\, dy^m\, dz^n\, \ldots}$ is used. The same notation, and similar remarks on partial derivatives and partial differentials, are found in his lessons on the differential calculus[2] of 1829.

In England, John Hind[3] gives the general form of partial derivatives which we last quoted for Cauchy, and then remarks: "Another kind of notation attended with some conveniences has of late been partially adopted. In this, the differential coefficients are no longer expressed in a fractional form, but are denoted by the letter d with the principal variables suffixed: thus, $\dfrac{du}{dx}$ and $\dfrac{du}{dy}$ are equivalent to $d_x u$ and $d_y u$," and so that the foregoing partial derivative of the order $l+m+n+$etc. is written $d_x^l d_y^m d_z^n$ etc. u. This notation which Hind describes is that of Cauchy for partial differentials.

De Morgan, when considering partial processes, lets $\dfrac{du}{dx}$ stand for a partial derivative, and $\dfrac{d\cdot u}{dx}$ for a total derivative, the two being "totally distinct."[4]

[1] Cauchy, *Résumé des leçons données à l'école royale polytechnique sur le calcul infinitésimal* (Paris, 1823); *Œuvres* (2d ser.), Vol. IV, p. 50, 79.

[2] Cauchy, *Leçons sur le calcul différentiel* (Paris, 1829); *Œuvres* (2d ser.), Vol. IV, p. 513, 527.

[3] John Hind, *Principles of the Differential Calculus* (2d ed.; Cambridge, 1831), p. 372.

[4] Augustus De Morgan, *Differential and Integral Calculus* (London, 1842), p. 88–91.

607. *M. Ohm.*—The much-praised and much-criticized Martin Ohm,[1] in Germany, advanced notations as follows:

If f is a function of $x, y, z, \ldots$, then the partial derivatives with respect to x are indicated by $\partial f_x, \partial^2 f_x, \partial^3 f_x, \ldots , \partial^a f_x$. Similarly for partial derivatives with respect to y or z, etc. More generally, $\partial^{a,b,c} f_{x,y,z}$ means the ath partial derivative with respect to x, followed by the bth partial derivative with respect to y, and then by the cth partial derivative with respect to z. Ohm writes the form $\partial^{o,a,b} f_{x,y,z}$ and also $\partial^{a,b} f_{y,z}$; the latter notation is shorter but does not display to the eye which of the variables is constant. Ohm's is perhaps the fullest notation for partial differentiation developed up to that time. But Ohm was not content to consider simply functions which are expressed directly and explicitly in terms of their variables; he studied also functions $f_{x,y,z}, \ldots$, in which some of the variables $x, y, z, \ldots$, appear implicitly, as well as explicitly. For instance (p. 129), let $f = 3x^2 + 4xy - 5y^2 z^2 + 4xyz$, where $y = a + bx$ and $z = px^3$. He introduces the notation $f_x = 3x^2 + 4xy - 5y^2 z^2 + 4xyz$, where f_x is considered as the explicit function of x, and y and z are to be taken constant or independent of x; he represents by $f_{(x)}$ the foregoing expression when account is taken of $y = a + bx$ and $z = px^3$. Hence $\partial f_x = 6x + 4y + 4yz$, while $\partial f_{(x)} = 4a + (8b + 6)x - 30a^2 p^2 x^5 - 70abp^2 x^6 - 40b^2 p^2 x^7 + 16apx^3 + 20bpx^4$. The two partial derivatives are wholly different; the former is an incomplete partial derivative, the latter is a complete partial derivative. He states (p. 276) that in 1825 he brought out in Berlin his *Lehre vom Grössten und Kleinsten*, and introduced there a special notation for the complete and incomplete partial differentials which correspond to the complete and incomplete partial derivatives $f_{(x)}$ and f_x. In 1825 he marked the incomplete partial differentials $\dfrac{df}{dv} \cdot dv$; the complete partial differentials $\dfrac{df}{dv} \cdot dv$. Similarly for higher orders. Ohm states in 1829 (p. 276) that he has found his notation $\dfrac{df}{dx} \cdot dv$ adopted by other writers.

In the 1829 publication Ohm considers a great many cases of this type and gives a diffuse presentation of partial differentiation covering over one hundred and sixty pages. He compares his notations with those of Euler and Fontaine. $\dfrac{df}{dv}$ meant with Euler a complete

[1] Martin Ohm, *System der Mathematik*, Vol. III (1829), p. 118, 119.

derivative; with Fontaine, a partial derivative. As previously seen, Fontaine once marked a complete derivative $\frac{1}{dx} \cdot d\mu$, but generally he dealt with complete differentials rather than complete derivatives.

If f is a function of x and y, then the three notations are exhibited by Ohm as follows (p. 183):

$$df = df_x + df_y \qquad \text{(according to M. Ohm)},$$

$$df = \left(\frac{df}{dx}\right) \cdot dx + \left(\frac{df}{dy}\right) \cdot dy \qquad \text{(according to L. Euler)},$$

$$df = \frac{df}{dx} \cdot dx + \frac{df}{dy} \cdot dy \qquad \text{(according to A. Fontaine)},$$

where df means the complete differential. Ohm warns the reader (p. 182) that in Fontaine's notation for differentials the dx's and dy's cannot be canceled. We see in Ohm's intellectual *ménage* a large progeny of symbols, but it was a group with no one to bid them welcome.

608. *W. R. Hamilton.*—About the same time that Ohm was devising suitable symbols in Germany, William Rowan Hamilton in Great Britain felt the need of a suitable symbolism. At first he used parentheses[1] as Euler had done; they are found in an article which was written in 1824 and printed in 1828. In the table of contents of this article, probably prepared when it was printed, the partial derivatives appear in the garb $\frac{\delta i}{\delta x}$. This delta notation is used by Hamilton[2] in an article printed in 1830, and another printed in 1834,[3] where the expressions are called "partial differential coefficients." On account of the close resemblance between δ and ∂, this notation is almost identical with that proposed by Legendre forty years earlier.

609. *W. Bolyai.*—Quaint notations are found in Wolfgang Bolyai's[4] *Tentamen* (1832). He designates by $\odot x$ a function of x, by $^n_o|\odot x$ the nth differential, by $^n_o|\odot x, y$ or $^{|n}_{ox}\odot x, y$ the nth partial differential with respect to x.

[1] W. R. Hamilton in *Transactions of the Royal Irish Academy*, Vol. XV (Dublin, 1828), p. 152.

[2] *Op. cit.*, Vol. XVI (1830), p. 2.

[3] *Philosophical Transactions of the Royal Society* (London, 1834), p. 249.

[4] Wolfgangi Bolyai de Bolya, *Tentamen* (2d ed., 1897), p. 207, 640.

610. *Cauchy and Moigno.*—Cauchy slightly modified his former symbols. In 1840 L'Abbé Moigno[1] published a book on the calculus "d'après les méthodes et les ouvrages publiés ou inédits de M. A.-L. Cauchy." For higher partial differentials and derivatives he gives the notations $d_x^2 d_y d_z u$, $d_{x^2yz}^4$, $\dfrac{d^4 u}{dx^2\, dy\, dz}$. That these notations do not always distinguish the partial derivative from the total is evident from what is given on page 144 as partial derivatives: $\dfrac{dz}{dx}$, $\dfrac{dz}{dy}$, $\dfrac{d^2z}{dx^2}$, etc. That Moigno was not satisfied with his notations is evident from our previous quotations from his Introduction.

In the same year (1840) Cauchy himself brought out in Paris his *Exercices d'analyse et de physique mathématique*, where (p. 5) he lets D_x, D_y, D_z, D_t stand for the partial derivatives of a function of the independent variables x, y, z, t; D_t^2 stands for the second partial derivative.

More fully developed is his notation[2] of 1844. Taking s as a function of many variables, x, y, z, ... , u, v, w, Cauchy designates:

Partial increments . . . . $\Delta_1 s$, $\Delta_{11}s$, $\Delta_{111}s$, ...

    Différentielles partielles . . . . $d_1 s$, $d_{11}s$, $d_{111}s$, ...

If of the second order . . . . $d_1 d_{11}s$, $d_1 d_{111}s$ . . . .

When the variables are specified . . . . . . $\Delta_x s$, $\Delta_y s$, $\Delta_z s$, ...

    Différentielles partielles . . . . $d_x s$, $d_y s$, $d_z s$, ...

When of the second order . . . . $d_x d_y s$, $d_x d_z s$, ...

    Dérivées partielles . . . . $D_x D_y s$, $D_x D_z s$, ...

Etc.

The $D_x s$ as a partial derivative is distinguished from a total derivative by the x, provided that s is a function of more than one variable; otherwise the $D_x s$ and Ds mean the same. One objection to this symbolism is that the variables considered as constant during the differentiation are not exhibited to the eye.

611. *C. G. J. Jacobi.*—We come now to the researches of C. G. J. Jacobi who is usually credited with the invention of the notation of partial derivatives which has become popular in recent years. Had he instituted a careful historical survey of the notations which had been proposed by mathematicians before him, he would have found much of

[1] L'Abbé Moigno, *Leçons de calcul différentiel et de calcul intégral*, Tome I (Paris, 1840), p. 118–21.

[2] A. L. Cauchy, *Exercices d'analyse et de physique mathématique* (Paris, 1844), p. 12–17.

value—notations of Euler, Fontaine, Karsten, Legendre, Lagrange of 1797 and of 1801, Da Cunha, Arbogast, Crelle, Barlow, Ohm, Hamilton, and Cauchy. But Jacobi made no such survey. He happened to know the notations of Euler and of Lagrange of 1797. Not satisfied with these he devised one of his own. As is so common in the history of science, his invention happened to have been made by others before him. Nevertheless, the mathematical public, uncritical on matters of priority, is crediting Jacobi with devices first proposed by Legendre and Hamilton.

It was in 1841 that Jacobi published his paper[1] "De determinantibus functionalibus," in which he introduces d and ∂ for total and partial derivatives (*differentialia partialia*), respectively. He says: "To distinguish the partial derivatives from the ordinary, where all variable quantities are regarded as functions of a single one, it has been the custom of Euler and others to enclose the partial derivatives within parentheses. As an accumulation of parentheses for reading and writing is rather onerous, I have preferred to designate ordinary differentials by the character d and the partial differential by ∂. Adopting this convention, error is excluded. If we have a function f of x and y, I shall write accordingly,

$$df = \frac{\partial f}{\partial x}\, dx + \frac{\partial f}{\partial y}\, dy \ .$$

". . . . This distinction may be used also in the designation of integrations, so that the expressions

$$\int f(x,\, y)dx \ , \qquad \int f(x,\, y)\partial x$$

have different meaning; in the former y and therefore also $f(x,\, y)$ are considered functions of x, in the latter one carries out the integration according to x only and one regards y during the integration as a constant."

Jacobi proceeds to the consideration of a function of many variables. "Let f be a function of $x, x_1, \ldots , x_n$. Assume any n functions $w_1, w_2, \ldots , w_n$ of these variables and think of f as a function of the variables $x, w_1, w_2, \ldots , w_n$. Then, if $x_1, x_2, \ldots , x_n$ remain constant, the $w_1, w_2, \ldots , w_n$ are no longer constants when x varies, and also, when $w_1, w_2, \ldots , w_n$ remain constant, $x_1, x_2, \ldots , x_n$ do not remain constant. The expression $\dfrac{\partial f}{\partial x}$ will designate entirely different values according as these or other magnitudes are constant during the differentiation." To simplify matters, Jacobi stipulates: "If I say that f is a function of the variables x, x_1, ... , x_n, I wish it to be understood that, if this function is differentiated partially, the differentiation shall be so carried out that always only one of these varies, while all the others remain constant."

Jacobi proceeds: "If the formulas are to be free from every ambiguity, then also there should be indicated by the notation, not only according to which variable the differentiation takes place, but also the entire system of independent variables whose function is to be differentiated partially, in order that by the notation itself one may recognize also the magnitudes which remain constant during the differentiation. This is all the more necessary, as it is not possible to avoid the occurrence in the same computation and even in one and the same formula, of partial derivatives which relate to different systems of independent variables, as, for example, in the expression given above, $\dfrac{\partial f}{\partial x}+\dfrac{\partial f}{\partial u}\cdot\dfrac{\partial u}{\partial x}$, in which f is taken as a function of x and u, u on the other hand is a function of x and y. The $\dfrac{\partial f}{\partial x}$ [derived from $f(x, y)$] changed into this expression when u was introduced in the place of y as an independent variable. If however we write down, besides the dependent variable, also the independent variables occurring in the partial differentiations, then the above expression can be represented by the following formula which is free of every ambiguity:

$$\frac{\partial f(x, y)}{\partial x} = \frac{\partial f(x, u)}{\partial x}+\frac{\partial f(x, u)}{\partial u}\cdot\frac{\partial u(x, y)}{\partial x} \; ,,$$

By way of criticism of this notation, P. Stäckel[1] adds in his notes on Jacobi's paper: "This notation is ambiguous too, for the symbol is used in two different meanings, in as much as $f(x, y)$ is another function of x and y than is $f(x, u)$ of x and u." Jacobi proceeds to remark that one can conceive of complicated relations in which it would be exceedingly onerous completely to define partial derivatives by the notation alone; a formula written in only one line might have to be expanded so as to fill a page. To avoid such differences it is best in complicated cases to dispense with the designation in the formula of all the independent variables. To distinguish two systems of partial derivatives which belong to different systems of variables, one system

[1] *Ostwald's Klassiker*, No. 78, p. 65.

of derivatives may be inclosed in parentheses, as was the practice with Euler.

The notation $\frac{\partial u}{\partial x}$ advocated by Jacobi did not meet with immediate adoption. It took half a century for it to secure a generally recognized place in mathematical writing. In 1844 Jacobi used it in Volume XXIX of *Crelle's Journal*. When in 1857 Arthur Cayley[1] abstracted Jacobi's paper, Cayley paid no heed to the new notation and wrote all derivatives in the form $\frac{du}{dx}$.

G. A. Osborne refers in the Preface of his *Differential and Integral Calculus* (Boston, 1895) to this notation as having "recently come into such general use."

612. *O. Hesse.*—A few years after Jacobi had published the article containing the $\frac{\partial z}{\partial x}$ notation, Otto Hesse advanced another notation which has been found useful in analytical geometry. In his paper on points of inflection of cubic curves[2] he designates $\frac{\partial^2 u}{\partial x_\chi \partial x_\lambda}$ of the function u for brevity by $u_{\chi,\lambda}$. In accordance with this plan Salmon[3] writes the first partial derivatives of U with respect to x, y, and z, respectively, by U_1, U_2, U_3 and the second partial derivatives by U_{11}, U_{22}, U_{33}, U_{23}, U_{31}, U_{12}; or, as it is expressed in Fiedler's edition, U_i is the first partial derivative of U with respect to x_i.

613. *B. Peirce.*—Arbogast's and Cauchy's symbolism was in part adopted in the United States by Benjamin Peirce,[4] of Harvard, who introduced it thus: "Denoting by $D \cdot_x \cdot$, $D \cdot_u \cdot$, the differential coefficients taken on the supposition that x, u are respectively the independent variables, we have, at once, $D \cdot _x y = \frac{dy}{dx} = \frac{D \cdot_u y}{D \cdot_u x}$." Accordingly, D signifies the total derivative, $D \cdot_x y$ the partial derivative, with respect to x.

From an anonymous article in the *Mathematical Monthly* (probably from the pen of its editor, J. D. Runkle)[5] we cull the following comments:

[1] A. Cayley, "Report on Recent Progress in Theoretical Dynamics," *British Association Report*, 1857 (London, 1858), p. 24.

[2] Otto Hesse in *Crelle's Journal* (1843), Vol. XXVIII, p. 99; *Gesammelte Werke* (München, 1897), p. 124.

[3] George Salmon, *Higher Plane Curves* (3d ed.; Dublin, 1879), p. 49, 50.

[4] B. Peirce, *Curves, Functions, and Forces*, Vol. I (2d ed.; Boston, 1852), p. 225.

[5] *Mathematical Monthly* (ed. J. D. Runkle), Vol. I (Cambridge, 1859), p. 328.

"The student will observe, that $\frac{du}{dx}$ denotes the derivative of the quantity u; but the symbol, as separated from the quantity, and simply denoting the operation, is $\frac{d}{dx}$. Thus, $\frac{d}{dx} f(x)$ tells us to find the derivative of $f(x)$. The inconvenience of the use of the symbol $\frac{d}{dx}$, in this and like cases has led to the adoption of D in its place. If we wish to indicate at the same time the particular variable, as x, in reference to which the derivative is to be taken, then the symbol D_x is used. . . . . In the case of the general function, as $f(x)$, the notation $f'(x)$, $f''(x)$, $f'''(x)$, etc. to denote the successive derivatives, was used by Lagrange, and is most convenient. So far as we know, Prof. Peirce is the only author in this country who has used D; and we have made these remarks for the benefit of those students who meet with this notation only in the *Monthly*. . . . . $D x^n$ is better than $\frac{d}{dx} x^n$."

614. *G. W. Strauch.*—A special symbolism to distinguish between two partial derivatives with respect to x, one recognizing only explicit forms of x, the other observing also the implicit forms, was given by the Swiss, G. W. Strauch.[1] He assumes $U = \varphi(x, y, z)$ and $z = \psi(x, y)$. Following in part Cauchy's notation, Strauch uses the symbol $\frac{d_x U}{dx}$ to represent the "incomplete" partial differential coefficient with respect to x which is obtained by differentiating U only with respect to the explicit x; he uses the sybmol $\dfrac{d_x U}{\overline{dx}}$ to represent the "complete" partial differential coefficient with respect to x which is obtained by differentiating U with respect to the implicit as well as the explicit x. He has the following relation between the partial derivatives,

$$\frac{d_x U}{\overline{dx}} = \frac{d_x U}{dx} + \frac{d_z U}{dz} \cdot \frac{d_x z}{dx} \cdot$$

It will be remembered that the double horizontal line used here had been employed for the same purpose in 1825 by Ohm (§ 607).

615. *J. M. C. Duhamel.*—The D_x, D_y, etc., suggested by Cauchy for partial differentiation was used in many quarters. For example,

[1] G. W. Strauch, *Praktische Anwendungen für die Integration der totalen und partialen Differentialgleichungen* (Braunschweig, 1865), p. 7–13.

Ernest Pfannenstiel,[1] in 1882, wrote $D_x^2 z$, $D_{x,y}^2 z$, etc., to represent partial derivatives in a paper on partial-differential equations.

Duhamel[2] in ordinary cases used no distinctive device for partial differentiation, but when many variables and derivatives of any orders were involved, he designated "par $F_{x,y,x\ldots}^{m+n+p\cdots}$ (x, y) ou par $D_{x,y,x\ldots}^{m+n+p\cdots} u$ le résultat de m dérivations partielles effectuées par rapport à x sur la fonction $u = F(x, y)$, suivies de n dérivations partielles du résultat par rapport à y, lesquelles seront elles-mêmes suivies de p dérivations par rapport à x; et ainsi de suite." For partial differentials he wrote similarly, $d_{x,y,x\ldots}^{m+n+p\cdots} u$. It will be observed that, strictly speaking, there is nothing in this notation to distinguish sharply between partial and total operations; cases are conceivable in which each of the $m+n+p \ldots$ differentiations might be total.

616. *G. S. Carr*[3] proposed "experimentally" a notation corresponding to that for total derivation; he would write $\dfrac{\partial^5 u}{\partial x^2 \partial y^3}$ in the contracted form $u_{2x\,3y}$.

617. *Ch. Meray.*—An obvious extension of the notation used by Duhamel was employed by Ch. Méray.[4] Representing by $p, q, \ldots$, the orders of the partial derivatives with respect to the independent variables $x, y, \ldots$, respectively, and by the sum $p+q+ \ldots$ the total order of the derivatives, he suggests two forms, either

$$f_{x,y,\ldots}^{(p,\,q,\,\ldots)} (x, y, \ldots) \qquad \text{or} \qquad D_{x,y,\ldots}^{(p,\,q,\,\ldots)} f(x, y, \ldots) \ .$$

Applying this notation to partial differentials, Méray writes,

$$d_{x,y,\ldots}^{(p,\,q,\,\ldots)} f = f_{x,y,\ldots}^{(p,\,q,\,\ldots)} (x, y, \ldots) \, dx^p \, dy^q \ldots$$

Méray's notation is neither elegant nor typographically desirable. It has failed of general adoption. In the *Encyclopédie des sciences mathématiques*[5] J. Molk prefers an extension of the Legendre-Jacobi notation; thus $\dfrac{\partial^k u}{\partial x^p \, \partial y^q \ldots}$, where $k = p+q+ \ldots$

[1] *Societät der Wissenschaften zu Upsala* (Sept. 2, 1882).

[2] J. M. C. Duhamel, *Eléments de calcul Infinitésimal*, Tome I (3d ed.; Paris, 1874), p. 267.

[3] G. S. Carr, *Synopsis . . . . of the Pure Mathematics* (London, 1886), p. 266, 267.

[4] Ch. Méray, *Leçons nouvelles sur l'analyse infinitésimale*, Vol. I (Paris, 1894), p. 123.

[5] Tome II, Vol. I, p. 296.

618. *T. Muir.*—Thomas Muir stated[1] in 1901 that he found it very useful in lectures given in 1869 to write $\Phi\overline{x, y, z}$ in place of $\Phi(x, y, z)$ and to indicate the number of times the function had to be differentiated with respect to any one of the variables by writing that number above the vinculum; thus, $\Phi\overset{1\ \ 3\ \ 2}{\overline{x, y, z}}$ meant the result of differentiating once with respect to x, thrice with respect to y, and twice with respect to z. Applying this to Jacobi's example, in which $z = \Phi\overline{x, y}$, we should have satisfactorily $\dfrac{\partial z}{\partial x} = \Phi\overset{1}{\overline{x, y}}$; but when there is given $z = \Phi\overline{x, u}$ and $u = \Psi\overline{x, y}$, then arises the serious situation that we are not certain of the meaning of $\dfrac{\partial z}{\partial x}$, as it would stand for $\Phi\overset{1}{\overline{x, u}}$ or for $\Phi\overset{1}{\overline{x, u}} + \Phi\overset{1}{\overline{x, u}} \cdot \Psi\overset{1}{\overline{x, y}}$, according as u or y was to be considered constant. In such cases he declares the notation $\dfrac{\partial z}{\partial x}$ inadequate.

Thomas Muir explained his notation also in *Nature*, Volume LXVI (1902), page 271, but was criticized by John Perry who, taking a unit quantity of mere fluid, points out that v, p, t, E, Φ, are all known if any two (except in certain cases) are known, so that any one may be expressed as a function of any other two. Perry prefers the notation $\left(\dfrac{dE}{dv}\right)_p$, which is the Eulerian symbol to which p is affixed; according to Muir's suggestion, Perry says, one must let $E = f\overline{v, p}$ and write $f\overset{1}{\overline{v, p}}$. To this remark, Muir replies (p. 520) that he would write $E\overset{1}{\overline{v, p}}$, or, if a vinculum seems "curious," he would write $E(\overset{1}{v}, p)$. To these proposals A. B. Basset remarks (p. 577): "Dr. Muir's symbols may be very suitable for manuscripts or the blackboard, but the expense of printing them would be prohibitive. No book in which such symbols were used to any extent could possibly pay. On the other hand, the symbol $(dE/dv)_p$ can always be introduced into a paragraph of letter-press without using a justification or a vinculum; and this very much lessens the expense of printing."

In the case of $u = f(x, y)$ and $y = \phi(x, z)$, some authors, W. F. Os-

[1] Thomas Muir in *Proceedings of the Royal Society of Edinburgh*, Vol. XXIV (1902–3), p. 162.

good[1] for instance, write $\dfrac{\partial u}{\partial x}\Big]_y$ to indicate that y is held fast, and $\dfrac{\partial u}{\partial x}\Big]_z$ to indicate that z is held fast.

619. *P. Mansion.*—Muir and Perry were not alone in their rejection of the notation used by Jacobi in 1841. Paul Mansion,[2] for instance, in 1873 prefers to use the straight letter d for partial derivatives; thus, $\dfrac{dz}{dx_1}$, $\dfrac{dz}{dx_2}$, where z is an explicit or implicit function of the independent variables x_1, x_2, ... He uses also $\dfrac{\delta\varphi}{\delta x_1}$, $\dfrac{\delta\varphi}{\delta x_2}$ to designate "the derivatives of an *explicit* function $\varphi(x_1, x_2, \ldots.)$ of $x_1, x_2, \ldots.$ with respect to the letter X_1, to the letter X_2, etc., without considering whether $x_1, x_2, \ldots.$ are independent of each other or not."

3. Symbols for Integrals

620. *G. W. Leibniz.*—In a manuscript[3] dated October 29, 1675, as previously noted (§§ 544, 570), Leibniz introduces the symbol $\int l$ for *omn. l*, that is, for the sum of the l's. As a facsimile reproduction was published by C. I. Gerhardt in 1899, we have definite information as to the exact form of the $\int$. It was the long form of the letter s, frequently used at the time of Leibniz (see Fig. 124). Not until eleven years later did Leibniz use the symbol in print. In 1686 Leibniz used it eight times in an article in the *Acta eruditorum* (p. 297, 299), but in that paper it did not quite take the form found in his manuscript and in later printing; the lower part was amputated and the sign appeared thus, $\int$. It was simply the small letter s, as printed at that time. It resembled the modern type form for "f." He writes the cycloidal formula thus: $y = \sqrt{2x - xx} + \int dx : \sqrt{2x - xx}$. This form of the symbol appears in the *Acta eruditorum* for 1701 (p. 280), where Louis Carré, of Paris, writes $a \int dx = ax$, and again in 1704 (p. 313), in a reprint of John Craig's article from the *Philosophical Transactions of London*, No. 284, viz., $\int z\, dy$. We have noticed this form also in works of Manfredi[4] at Bologna in 1707 and of Wolf[5] at Halle in 1713.

---

[1] W. F. Osgood, *Differential and Integral Calculus* (New York, 1907), p. 306.

[2] *Théorie des équations aux dérivées partielles du premier ordre*, par M. Paul Mansion, professeur à l'université de Gand (1873), Tome XXV.

[3] C. I. Gerhardt, *Entdeckung der höheren Analysis* (Halle, 1855), p. 125.

[4] *De Constructione aequationum differentialium primi gradus.* Authore Gabriele Manfredio (Bononiae, 1707), p. 127.

[5] *Elementa matheseos universae*, tomus I . . . . Autore Christiano Wolfo (Halae, 1713), p. 474.

In *Leibnizens Mathematische Schriften* (ed. C. I. Gerhardt), Volume V (1858), page 231, the notation of the article of 1686 is not reproduced with precision; the sign of integration assumes in the reprint the regular form $\int$, namely, a slender, elongated form of the letter. John Bernoulli,[1] who used the term "integral" (first employed by Jacob Bernoulli, see § 539), proposed in his correspondence with

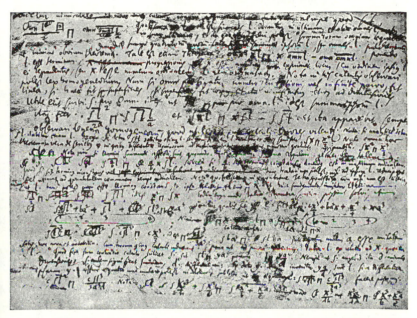

Fig. 124.—Facsimile of manuscript of Leibniz, dated Oct. 29, 1675, in which his sign of integration first appears. (Taken from C. I. Gerhardt's *Briefwechsel von G. W. Leibniz mit Mathematikern* [1899].)

Leibniz the letter I as the symbol of integration but finally adopted $\int$ in deference to Leibniz. We cannot agree with Klügel,[2] that I would have been more appropriate (*schicklicher*); this shorter symbol possessed less adaptability to the nineteenth-century need of indicating symbolically the limits of integration. In articles published in the *Acta eruditorum* of 1694 and 1695, Leibniz places a comma after the

[1] Johann Bernoulli to Leibniz, April 7, 1696, *Leibnizens Mathematische Schriften* (ed. C. I. Gerhardt), Vol. III (Halle, 1855), p. 262, 273.

[2] G. S. Klügel, *Mathematisches Wörterbuch*, 1. Abth., 2. Theil (Leipzig, 1805), art. "Integral," p. 746.

$\int$, thus, $\int$, $xx\,dx$, but Johann Bernoulli omits the comma in the volume for 1698 (p. 53). Some eighteenth- and early nineteenth-century writers placed a dot after the sign, thus, $\int.$, or the colon, as in $\int:$;

E. Waring[1] wrote $\int .\frac{\dot{x}}{x}(\log. x)$. In 1778 Euler[2] wrote $\int p\,dx \int q\,dx \int \tau\,dx$

...., which did not indicate a product like $\int p\,dx \cdot \int q\,dx \cdot \int \tau\,dx$ ....,

but was to be interpreted so that $\int$ applies to all that follows—a relation which some authors (Gherli, for instance, as we shall see) carefully indicate by the use of vinculums.

As regards the differential in the expression to be integrated, the usual practice has been to write it down, thus, $\int y\,dx$, but this practice has not been universal. Leibniz' conception of an integral as a summation would seem to require the use of the differential. In his printed paper of 1686 he adopts the form $\int p\,dy$, although in his manuscript of October 29, 1675, he had written $\int x^2 = \frac{x^3}{3}$. If integration is conceived as the converse of differentiation, then no serious objection can be raised to the omission of the differential. And so we find a few authors who prefer to omit the dx, as, for instance, Benjamin Peirce,[3] who writes $(n+1)\int .ax^n = ax^{n+1}$. On October 11–16, 1858, William R. Hamilton wrote to P. G. Tait: "And perhaps you may have adopted, even publicly—as Airy has done, using the (to me) uncouth notation $\int_\theta (\quad)$ for $\int (\quad)d\theta$—the system which rejects differentials."[4]

4. Early Use of Leibnizian Notation in Great Britain

621. Some years before Newton permitted his notation for fluxions and fluents to see the light of day in printed form, his friend, John Craig, used the notation of Leibniz, dp, dx, dy, in a book, the *Methodus figurarum*, published in 1685 in London. Craig employed the Leib-

---

[1] Edward Waring, *Meditationes analyticae* (1786), p. 23.

[2] L. Euler, *Institutiones Calculi Integralis*, Vol. IV, Suppl. IX, art. 37.

[3] B. Peirce, *Curves, Functions, and Forces*, Vol. II (Boston, 1846), p. 69.

[4] C. G. Knott, *Life and Scientific Work of Peter Guthrie Tait* (Cambridge, 1911), p. 121. For this reference and also some other data in this article I am indebted to Professor J. H. M. Wedderburn.

nizian notation again in 1693 in another booklet, the *Tractatus mathematicus*, as well as in articles printed in the *Philosophical Transactions of London* for the years 1701, 1703, 1704, 1708; his article of 1703 contains the sign of integration $\int$. In the *Transactions* for 1704–5, an article by Johann Bernoulli makes extensive use of the Leibnizian signs. In 1716 the English physician and philosophical writer, George Cheyne, brought out in London the *Philosophical Principles of Religion*, Part II, which contains a chapter from the pen of Craig on a discussion of zero and infinity, dated September 23, 1713. In this chapter Craig uses Leibniz' symbols for differentiation and integration. But in 1718 Craig made a complete change. In that year he issued a book, *De calculo fluentium*, in which he switches over to the exclusive use of the Newtonian notation. Evidently this change is a result of the controversy then raging between the supporters of Newton and Leibniz.

In Volume XXIII of the *Philosophical Transactions of London*, for the years 1702 and 1703, De Moivre uses dots for fluxions, but in integration he uses the Leibnizian sign $\int$. Thus, on page 1125, De Moivre writes "adeoq; $\dot{q} = \frac{3}{2}dv^2\dot{v} - \frac{3}{2}dv^2\dot{y}$, igitur $q = \frac{1}{2}dv^3 - \int \frac{3}{2}dv^2\dot{y}$. Ergo ad hoc perventum est ut fluentum quantitatem inveniamus cujus fluxio est $\frac{3}{2}dv^2\dot{y}$." Evidently his partisanship favoring Newton did not at this time prevent his resorting to the convenience of using the Leibnizian symbol of integration.

Even John Keill, who fought so prominently and unskilfully on the side of Newton, employed a similar mixed notation. In a paper dated November 24, 1713, and printed in the *Philosophical Transactions* for 1714–16, he adopts the symbolism $\int \phi\dot{x}$. A mixture of continental and British symbols is found in the writings of Waring and much later in Olinthus Gregory's *Treatise of Mechanics* (3d ed.; London, 1815) where on page 158 there is given $\int dy\sqrt{(\dot{x}^2 + \dot{y}^2)}$. Similarly, Playfair[1] wrote $\int \dot{x} \sin v$. John Brinkley[2] used the $\int$ with a horizontal bar added, as in $f\dot{p}s$. In times still more recent the Newtonian dot has been used to advantage, by the side of Leibnizian symbols, in the treatment of mechanics and the algebras of vectors.

Returning to the eighteenth century, we find Benjamin Robins[3]

[1] *The Works of John Playfair, Esq.*, Vol. III (Edinburgh, 1822), p. 16.

[2] *Transactions of the Royal Irish Academy*, Vol. XII (1815), p. 85; see also Vol. XIII (1818), p. 31.

[3] Benjamin Robins, *Remarks on Mr. Euler's Treatise on Motion, Dr. Smith's Compleat System of Opticks, and Dr. Jurin's Essay upon Distinct and Indistinct Vision* (London, 1739).

writing the Leibnizian signs in a review of a publication of L. Euler.
Joseph Fenn, an Irish writer who had studied on the Continent, and
was at one time professor of philosophy in the University of Nantes,
issued at Dublin, soon after 1768, a *History of Mathematics*. In it
Fenn had occasion to use the calculus, and he employs the Leibnizian
notation. In a "Plan of the Instructions" given in the Drawing
School established by the Dublin Society (p. lxxxix of the foregoing
volume) he discusses the tides and uses the calculus and the notation
of Leibniz. He uses them again in the second volume of the *Instructions* given in the Drawing School established by the Dublin Society
(Dublin, 1772). He is friendly to Newton, uses the terms "fluxion"
and "fluent," but never uses Newton's notation. He is perhaps the
last eighteenth-century writer in Great Britain who used the symbolism of Leibniz in differentiation and integration. The Leibnizian
notation was the earliest calculus notation in England which appeared
in print, but in the latter part of the eighteenth century it vanished
almost completely from British soil.

5. Symbols for Fluents; Later Notations in the Integral Calculus

622. *I. Newton.*—In his *Quadratura curvarum*,[1] of 1704, Newton
gave the symbol $\overset{\text{\tiny I}}{x}$ as the integral of x, $\overset{\text{\tiny II}}{x}$ as the integral of $\overset{\text{\tiny I}}{x}$. In fact, in
the succession, $\overset{\text{\tiny II}}{x}$, $\overset{\text{\tiny I}}{x}$, x, $\dot{x}$, $\ddot{x}$, $\dddot{x}$, each term was the fluxion (fluent) of
the preceding (succeeding) term. Another notation for integral was
the inclosure of the term in a rectangle,[2] as is explained in Newton's

De Analysi per equationes numero terminorum infinitas, where $\boxed{\dfrac{aa}{64x}}$

stands for $\displaystyle\int \frac{aa \cdot dx}{64x}$.

That Newton's notation for integration was defective is readily
seen. The $\overset{\text{\tiny I}}{x}$ was in danger of being mistaken for an abscissa in a series
of abscissas x, x', x''; the rectangle was inconvenient in preparing a
manuscript and well-nigh impossible for printing, when of frequent
occurrence. As a consequence, Newton's signs of integration were
never popular, not even in England. They were used by Brook
Taylor, in his *Methodus incrementorum* (London, 1715); he writes
(p. 2): "$\overset{\text{\tiny II}}{x}$ designat fluentem secondam ipsuis x" and (p. 38) "$\dot{p} = -r\dot{s}$,
adeoque $p = -\boxed{r\dot{s}}$."

[1] I. Newton, *Quadratura curvarum* (London, 1704); *Opera*, Vol. I, p. 338.

[2] I. Newton, *Opera* (Horsley's ed.), Vol. I (1779), p. 272.

In Newton's *Principia* (1687), Book II, Lemma II, fluents are represented simply by capital letters and their fluxions by the corresponding small letters. Newton says: "If the moments of any quantities A, B, C, etc., increasing or decreasing, by a perpetual flux, or the velocities of the mutations which are proportional to them, be called a, b, c, etc., the moment or mutation of the generated rectangle AB will be $aB+bA$." Here a velocity or fluxion is indicated by the same symbol as a moment. With Newton a "fluxion" was always a velocity, not an infinitely small quantity; a "moment" was usually, if not always, an infinitely small quantity. Evidently, this notation was intended only as provisional. Maclaurin does not use any regular sign of integration. He says[1] simply: "$\dot{y}z+\dot{z}y$, the fluent of which is yz." Nor have we been able to find any symbol of integration in Thomas Simpson's *Treatise of Fluxions* (London 1737, 1750), in Edmund Stone's *Integral Calculus*,[2] in William Hale's *Analysis fluxionum* (1804), in John Rowe's *Doctrine of Fluxions* (4th ed.; London, 1809), in S. Vince's *Principles of Fluxions* (Philadelphia, 1812). In John Clarke's edition of Humphry Ditton's text,[3] the letter F. is used for "fluent." Dominated by Wallis' concept of infinity, the authors state,

"$F.\dfrac{x^{\,n-1}\ \dot{x}}{x^m}$, will be Finite, Infinite, or more than Infinite, according as n is $>$, $=$, or $<$ than m." We have seen (§ 582) that the letter F was used also for "fluxion."

623. *Ch. Reyneau and others.*—Perhaps no mathematical symbol has encountered so little competition with other symbols as has $\int$; the sign $\int$ can hardly be called a competitor, it being simply another form of the same letter. The $\int$ was used in France by Reyneau[4] in 1708 and by L'Abbé Sauri in 1774; in Italy by Frisi,[5] by Gherli,[6] who, in the case of multiple integrals, takes pains to indicate by vinculums

[1] C. Maclaurin, *Treatise of Fluxions*, Book II (1742), p. 600.

[2] We have examined the French translation by Rondet, under the title *Analyse des infiniment petits comprenant le Calcul Intégral*, par M. Stone (Paris, 1735).

[3] *An Institution of Fluxions* .... by Humphry Ditton (2d ed., John Clarke; London, 1726), p. 159, 160.

[4] Ch. Reyneau, *Usage de L'Analyse*, Tome II (Paris, 1708), p. 734.

[5] *Paulli Frisii Operum Tomus primus* (Milan, 1782), p. 303.

[6] O. Gherli, *Gli elementi teorico-pratici delle matematiche pure*, Tomo VI (Modena, 1775), p. 1. 334.

the scope of each integration, thus, $\overline{\overline{\int \cdot dx \int \cdot dx \int \cdot y \, dx}}$. In Boscovich's[1] treatise of 1796 there is used part the time a small s, as in $s \, a \, x^m \, dx = \dfrac{a x^{m+1}}{m+1}$, then $\int y \, dx$, and finally $\int \dfrac{c y^2 \, dx}{2r}$, this large letter being used, apparently, because there was plenty of space in front of the fraction. In the books which we have quoted the use of $\int$ rather than $\int$ was probably due to the greater plenty of the former type in the respective printing offices.

Not infrequently the two symbols were used in the same publication, the $\int$ to indicate some specialized integral. Thus, Fourier said in an article of 1811, first published in 1824: ". . . . And taking the integral from $x=0$ to $x=\pi$, one has, on representing these integrations by the sign $\int$, $\int \varphi \sin \cdot i \, x \, dx = \ldots$." Further on he writes "$a_0 = \frac{1}{2} \int x \, dx$, ou $\dfrac{\pi^2}{4}$."[2] It will be seen that later Fourier suggested the notation for definite integrals now in general use. L. M. H. Navier,[3] in an article on fluid motion, remarks: "Le signe $\int$ désigne une intégration effectuée, dans toute l'entendu de la surface du fluide. ..." In the third edition of Lagrange's *Mécanique analytique*[4] we read: "Nous dénoterons ces intégrales totales, c'est-à-dire relatives à l'étendue de toute la masse, par la charactéristique majuscule $\int$, en conservant la charactéristique ordinaire $\int$ pour désigner les intégrales partielles ou indéfinies." Finally, we refer to C. Jordan, who, in his *Cours d'Analyse*, Volume I (Paris, 1893), page 37, uses $\int_E$ to mark "l'intégrale de la fonction f dans le domaine E," and says that it is generally designated thus.

[1] Ruggero Guiseppe Boscovich, *Elementi delle matematiche pure, Edizione terza Italiana* (Venezia, 1796), p. 477, 479, 484.

[2] *Mémoires de l'académie r. des sciences de L'Institut de France*, Tome IV, années 1819 et 1820 (Paris, 1824), p. 303, 309.

[3] *Ibid.*, année 1823, Tome VI (Paris, 1827), p. 412.

[4] K. L. Lagrange, *Œuvres*, Vol. XI, p. 85.

624. *A. L. Crelle.*—We have encountered only one writer on the European Continent who deliberately rejected Leibniz' sign $\int$ or $\backslash$, namely, A. L. Crelle,[1] of Cassel, the founder of *Crelle's Journal.* In 1813 he rejected the symbol as foreign to the nature of the subject. He argued that since in differentiation (or the finding of dy, d^2y) d appears in the position of a multiplier, the symbol for the inverse operation, namely, integration, should be a d placed in the position of a divisor, thus, $\frac{1}{d}$, $\frac{1}{d^2}$, ... Like Leibniz, Crelle was influenced in his selection by his conception of the operation called "integration." He looked upon it as the inverse of differentiation, and chose his symbol accordingly; Leibniz saw this inverse relation; nevertheless he looked upon integration primarily as a summation and accordingly chose the $\int$ as the first letter in *summa.* Crelle found no following whatever in his use of $\frac{1}{d}$.

625. *L. Euler.*—Limits of integration were at first indicated only in words. Euler was the first to use a symbol in his integral calculus,[2] of which the following is an illustration:

$$Q = \int \frac{x^{p-1}\,\partial x}{\sqrt[n]{(1-x^n)^{n-q}}} \begin{bmatrix} ab\ x^n = \frac{1}{2} \\ ad\ x\ = 1 \end{bmatrix}.$$

This notation, with the omission of the ab and ad, was used in 1819 or 1820 by F. Sarrus,[3] of Strasbourg, and H. G. Schmidten.[4]

626. *J. Fourier.*—Our modern notation for definite integrals constituted an important enrichment of the notation for integration. It was introduced by Joseph Fourier, who was an early representative of that galaxy of French mathematical physicists of the early part of the nineteenth century. In Fourier's epoch-making work of 1822 on *La Théorie analytique de la chaleur*[5] he says: "Nous désignons en génér-

---

[1] A. L. Crelle, *Darstellung der Rechnung mit veränderlichen Grössen*, Band I Göttingen, 1813), p. 88, 89.

[2] L. Euler, *Institutiones calculi integralis* (3d ed.), Vol. IV (1845), Suppl. V, p. 324. (First edition, 1768–70.)

[3] Gergonne, *Annales de mathématique*, Vol. XII (1819 and 1820), p. 36.

[4] *Op. cit.*, p. 211.

[5] *Œuvres de Fourier* (ed. G. Darboux), Tome I (Paris, 1888), p. 231; see also p. 216.

al par le signe $\int_a^b$ intégrale qui commence lorsque la variable équi-
valent à a, et qui est complète lorsque la variable équivant à b; et
nous écrivons l'équation (n) sous la forme suivante

$$\frac{\pi}{2}\,\varphi(x) = \tfrac{1}{2}\int_0^\pi \varphi(x)dx + \text{etc.}"$$

But Fourier had used this notation somewhat earlier in the *Mémoires*
of the French Academy for 1819–20, in an article of which the early
part of his book of 1822 is a reprint. The notation was adopted im-
mediately by G. A. A. Plana,[1] of Turin, who writes $\int_0^1 a^u\,du = \dfrac{a-1}{Log.\,a}$;
by A. Fresnel[2] in one of his memorable papers on the undulatory
theory of light; and by Cauchy.[3] This instantaneous display to the
eye of the limits of integration was declared by S. D. Poisson[4] to be
a *notation très-commode*.[5]

It was F. Sarrus[6] who first used the signs $|F(x)|_a^x$ or $|_a^x F(x)$ to indi-
cate the process of substituting the limits a and x in the general integral $F(x)$. This designation was used later by Moigno and Cauchy.

627. In Germany, Ohm advanced another notation for definite
integrals, viz., $\int_{x \div a} \varphi\,.\,dx$ or $(\partial^{-1}\varphi_x)_{x \div a}$, where x is the upper limit
and a the lower. He adhered to this notation in 1830 and 1846,

---

[1] J. D. Gergonne, *Annales de mathématiques*, Vol. XII (Nismes, 1819 and 1820),
p. 148.

[2] *Mémoires de l'Académie r. des sciences de l'Institut de France*, Tome V,
années 1821 et 1822 (Paris, 1826), p. 434.

[3] Cauchy, *Résumé des leçons données à l'école royale polytechnique sur la calcul
infinitésimal* (Paris, 1823), or *Œuvres* (2d ser.), Vol. IV, p. 188.

[4] *Mémoires de l'académie des sciences de L'Institut de France*, Tome VI, année
1823 (Paris, 1827), p. 574.

[5] Referring to Fourier's notation, P. E. B. Jourdain said: "Like all advances
in notation designed to aid, not logical subtlety, but rather the power possessed
by mathematicians of dealing rapidly and perspicuously with a mass of complicated data, this improvement has its root in the conscious or unconscious
striving after mental economy. This economical function naturally seems propor-
tionally greater if we regard mathematics as a means, and not primarily as a
subject of contemplation. It is from a mentally economical standpoint that we
must consider Fourier's notation" (Philip E. B. Jourdain, "The Influence of
Fourier's Theory of the Conduction of Heat on the Development of Pure Mathematics," *Scientia*, Vol. XXII [1917], p. 252).

[6] F. Sarrus in Gergonne's *Annales*, Vol. XIV (1823), p. 197

claiming for it greater convenience over that of Fourier in complicated expressions.[1]

Slight modification of the notation for definite integration was found desirable by workers in the theory of functions of a complex variable. For example, Forsyth[2] writes $\int_{B}$ where the integration is taken round the whole boundary B. Integration around a circle[3] is sometimes indicated by the sign $\oint$.

V. Volterra and G. Peano.—What C. Jordan[4] calls *l'intégrale par excès et par défaut* is represented by Vito Volterra[5] thus, $\int_{x_0}^{\overline{x_1}}$ and $\int_{\underline{x_0}}^{x_1}$.

G. Peano,[6] who uses $\int$ as the symbol for integration, designated by $\int (f, a \vdash b)$ the integral of f, extended over the interval from a to b, and by $\int^1 (f, a \vdash b)$ the *intégrale par excès*, and by $\int_1 (f, a \vdash b)$ the *intégrale par défaut*.

628. *E. H. Moore.*—In 1901, in treating improper definite integrals, E. H. Moore[7] adopted the notation for the (existent) narrow broad Ξ—integral:

$$\int_{a(\Xi)}^{b} F(x)dx = \underset{D_I = 0}{\overset{I|I(\Xi)}{L}} \int_{a}^{b} F_1(x)dx \ ,$$

$$\int_{a\{\Xi\}}^{b} F(x)dx = \underset{D_I = 0}{\overset{I|I\{\Xi\}}{L}} \int_{a}^{b} F_1(x)dx \ ,$$

where Ξ is a point-set of points ξ, I is an interval-set, D_I is the length of I, $I(\Xi)$ is an interval-set which incloses Ξ narrowly (i.e.,

[1] Martin Ohm, *Lehre vom Grössten und Kleinsten* (Berlin, 1825); *Versuch eines vollkommen consequenten Systems der Mathematik*, Vol. IV (Berlin, 1830), p. 136–37; *Geist der Differential- und Integral-Rechnung* (Erlangen, 1846), p. 51 ff.

[2] A. R. Forsyth, *Theory of Functions of a Complex Variable* (Cambridge, 1893), p. 27.

[3] H. A. Kramers in *Zeitschrift für Physik*, Vol. XIII (1923), p. 346.

[4] C. Jordan, *Cours d'Analyse* (3d ed.), Vol. I (1909), p. 34, 35.

[5] V. Volterra, *Giornale di matematiche* (Battaglini), Vol. XIX (1881), p. 340.

[6] G. Peano, *Formulaire mathématique*, Vol. IV (Turin, 1903), p. 178.

[7] E. H. Moore, *Transactions of the American Mathematical Society*, Vol. II (1901), p. 304, 305.

every interval I contains or incloses at least one point ξ of Ξ), $I\{\Xi\}$ is an interval-set which incloses Ξ broadly (i.e., not necessarily narrowly), $F_1(x)$ has the value 0 or $F(x)$ according as the point x of the interval ab lies or does not lie on I.

629. *Residual calculus.*—In various papers Cauchy developed and applied a *calcul des résidus* which bears a certain analogy to the infinitesimal calculus. If $f(x) = \infty$ for $x = x_1$, and $f(x) = (x - x_1)f(x)$, then $f(x) = \dfrac{f(x)}{x - x_1}$ when $x = x_1$, and $f(x_1)$ is called the residue of $f(x)$ with respect to x_1. Cauchy represents the operation of finding the residue by a special symbol. He says:[1] "Nous indiquerons cette extraction à l'aide de la lettre initiale $\mathcal{L}$, qui sera considérée comme une nouvelle caractéristique, et, pour exprimer le résidu intégral de $f(x)$, nous placerons la lettre $\mathcal{L}$ devant la fonction entourée de doubles parenthèses, ainsi qu'il suit: $\mathcal{L}((f(x)))$." Accordingly,

$\mathcal{L}\dfrac{f(x)}{((F(x)))}$ stands for the sum of residues with respect to the roots of $F(x) = 0$ only, while $\mathcal{L}\dfrac{((f(x)))}{F(x)}$ stands for the sum of the residues of $f(x)$ relative to $\dfrac{1}{f(x)} = 0$ only. Furthermore,[2]

$$\underset{x_0}{\overset{X}{\mathcal{L}}}\,\underset{y_0}{\overset{Y}{}}\,((f(z)))\ ,$$

where $z = x + y\sqrt{-1}$ represents the residue of $f(z)$ taken between the limits $x = x_0$ and $x = X$, $y = y_0$ and $y = Y$. Laurent[3] employs the notation $\mathcal{L}_c f(z)$, where $f(c) = \infty$, or simply $\mathcal{L}f(z)$ when no ambiguity arises. B. Peirce[4] followed in the main Cauchy's notation. D. F. Gregory[5] changes the fundamental symbol to the inverted numeral 3, viz., $\mathcal{E}$, and distinguishes between the integral residues and the partial residues by suffixing the root to the partial symbol. Thus $\mathcal{E}_a f(x)$, $\mathcal{E}_b f(x)$ are partial, $\mathcal{E}f(x)$ is integral.

[1] A. L. Cauchy, *Exercices de mathématiques* (Paris, 1826); *Œuvres complètes* (2d ser.), Vol. VI, p. 26.

[2] Cauchy, *op. cit.*, Vol. VI, p. 256.

[3] H. Laurent, *Traité d'analyse* (Paris), Vol. III (1888), p. 243.

[4] Benjamin Peirce, *Curves, Functions, and Forces*, Vol. II (Boston, 1846), p. 43–59.

[5] D. F. Gregory, *Mathematical Writings* (Cambridge, 1865), p. 73–86; *Cambridge Mathematical Journal*, Vol. I (1839), p. 145.

6. Calculus Notations in the United States

630. The early influence was predominantly English. Before 1766 occasional studies of conic sections and fluxions were undertaken at Yale[1] under the leadership of President Clap. Jared Mansfield published at New Haven, Connecticut, about 1800, and again in 1802, a volume of *Essays, Mathematical and Physical*, which includes a part on "Fluxionary Analysis." In a footnote on page 207 he states that the fluxions of x, y, z are usually denoted "with a point over them, but we have here denoted these by a point somewhat removed to the right hand," thus, $x\cdot$. Similar variations occur in Robert Woodhouse's *Principles of Analytical Calculation* (Cambridge [England], 1803), page xxvii, where we read, "Again $\overset{...}{xy}$ or $(xy)^{...}$ is not so convenient as $d^3(xy)$," and in a book of Prändel[2] where the English manner of denoting a fluxion is given as $x\cdot$ and $y\cdot$. This displacement of the dot is found in articles contributed by Elizur Wright, of Tallmadge, Ohio, to the *American Journal of Science and Arts* for the years 1828, 1833, 1834.

In 1801 Samuel Webber, then professor of mathematics and later president of Harvard College, published his *Mathematics Compiled from the Best Authors*. In the second volume he touched upon fluxions and used the Newtonian dots. This notation occurred also in the *Transactions of the American Philosophical Society*, of Philadelphia, in an article by Joseph Clay who wrote in 1802 on the "Figure of the Earth." It is found also in the second volume of Charles Hutton's *Course of Mathematics*, American editions of which appeared in 1812, 1816(?), 1818, 1822, 1828, and 1831.

An American edition of S. Vince's *Principles of Fluxions* appeared in Philadelphia in 1812. That early attention was given to the study of fluxions at Harvard College is shown by the fact that in the interval 1796–1817 there were deposited in the college library twenty-one mathematical theses which indicate by their titles the use of fluxions.[3] These were written by members of Junior and Senior classes. The last thesis referring in the title to fluxions is for the year 1832.

At West Point, during the first few years of its existence, neither fluxions nor calculus received much attention. As late as 1816 it is stated in the West Point curriculum that fluxions were "to be taught

[1] William L. Kingsley, *Yale College; a Sketch of Its History*, Vol. II, p. 497, 498.

[2] J. G. Prändel, *Kugldreyeckslehre und höhere Mathematik* (München, 1793), p. 197.

[3] Justin Winsor, *Bibliographical Contributions*, No. 32. Issued by the Library of Harvard University (Cambridge, 1888).

at the option of professor and student." In 1817, Claude Crozet, trained at the Polytechnic School in Paris, became teacher of engineering. A few times, at least, he used in print the Newtonian notation, as, for instance, in the solution, written in French, of a problem which he published in the *Portico*, of Baltimore, in 1817. Robert Adrain, later professor in Columbia College and also at the University of Pennsylvania, used the English notation in his earlier writings, for example, in the third volume of the *Portico*, but in Nash's *Ladies and Gentlemen's Diary*, No. II, published in New York in 1820, he employs the dx.

Perhaps the latest regular use of the dot notation in the United States is found in 1836 in a text of B. Bridge.[1]

In the *American Journal of Science and Arts* the earliest paper giving the Leibnizian symbols was prepared by A. M. Fisher, of Yale College. It is dated August, 1818, and was printed in Volume V (1822). The earliest articles in the *Memoirs of the American Academy of Arts and Sciences* which contain the "*d*-istic" signs bear the date 1818 and were from the pen of F. T. Schubert and Nathaniel Bowditch.[2] In these *Memoirs*, Theodore Strong, of Rutgers College, began to use the Leibnizian symbolism in 1829. But the publication which placed the Leibnizian calculus and notation within the reach of all and which marks the time of the real beginning of their general use in the United States was the translation from the French of Bézout's *First Principles of the Differential and Integral Calculus*, made by John Farrar, of Harvard University, in 1824. Between 1823 and 1831 five mathematical theses were prepared at Harvard, which contained in their titles the words "Differential Calculus" or "Integral Calculus."[3] After 1824 French influences dominated all instruction in higher analysis at Harvard.

7. Symbols for Passing to the Limit

631. *S. Lhuilier.*—The abbreviation "lim." for "limit" does not seem to have been used before Simon Lhuilier,[4] who writes "lim. $q:Q$"

---

[1] B. Bridge, *Treatise on . . . . the three Conic Sections.* From the second London edition (New Haven, 1836), p. 75.

[2] *Memoirs of the American Academy of Arts and Sciences*, Vol. IV, Part I (Cambridge, Mass., 1818), p. 5, 47.

[3] Justin Winsor, *Bibliographical Contributions*, No. 32. Issued by the Library of Harvard University (Cambridge, 1888).

[4] Lhuilier, *Exposition élémentaire des principes des calculs supérieurs* (Berlin, 1786), p. 24, 31.

and "Lim. $\dfrac{\Delta y}{\Delta x}$." It was also used by Garçaõ Stockler, of Lisbon, in 1794; by L. N. M. Carnot[1] (who used also L); and by Brinkley,[2] of Dublin.

A. L. *Cauchy* wrote "lim." and pointed out[3] that "lim. (sin. x)" has a unique value 0, while "lim. $\left(\left(\dfrac{1}{x}\right)\right)$" admits of two values and "lim. $\left(\left(\text{sin.}\ \dfrac{1}{x}\right)\right)$" of an infinity of values, the double parentheses being used to designate all the values that the inclosed function may take, as x approaches zero. In limit and remainder theorems Cauchy designated[4] by θ an undetermined value lying between 0 and 1 for representing a value $x+\theta h$ situated between x and $x+h$. Some writers employ the form ϑ of the Greek letter. The period in "lim." was gradually dropped, and "lim" came to be the recognized form.

632. *E. H. Dirksen*[5] in Berlin used the abbreviation *Gr.* of the German word *Grenze*, in place of the usual contraction of the Latin *limes*. Dirksen writes "$\overset{m=\infty}{Gr.}\ a_m = (+, -, E, \infty, 0)$" to indicate that the limit of a certain infinite series is "positiv und negativ, endlich, unendlich und Null."

633. *The Bolyais.*—About the same time[6] Wolfgang and Johann Bolyai wrote "$x \longleftarrow a$" to denote x approaching the limit a ("x tendere ad limitem a").

634. *K. Weierstrass.*—In the nineteenth century many writers felt the need of a symbolism indicating also the limit which the independent variable approaches. The notation which has been used widely in Europe is "$\text{Lim}_{x=a}$," to express "the limit as x approaches a." It is found in papers of Weierstrass, who in 1841[7] wrote "lim"

---

[1] *Œuvres mathématique du Citoyen Carnot* (à Basle, 1797), p. 168, 191–98.

[2] John Brinkley in *Trans. Royal Irish Academy*, Vol. XIII (Dublin, 1818), p. 30.

[3] A. L. Cauchy, *Cours d'analyse* (Paris, 1821), p. 13, 14.

[4] Cauchy, *Exercices de mathématiques* (Paris, 1826); *Œuvres complètes* (2d ser.), Vol. VI, p. 24.

[5] E. H. Dirksen, *Abhandlungen d. K. P. Akademie d. Wissensch.* (Berlin, 1832), Th. I, p. 77–107.

[6] Wolfgangi Bolyai de Bolya, *Tentamen* (2d ed.), Tome II (Budapestini, 1904), p. xix, 361. The first edition appeared in 1832.

[7] K. Weierstrass, *Mathematische Werke*, Band I (Berlin, 1894), p. 60.

and in 1854[1] "Lim. $p_n = \infty$." W. R. Hamilton in 1853[2] used "$\lim_{n=\infty} \{\ \}$."
$\quad n=\infty$

Familiar among admirers of Weierstrassian rigor are the expressions "ϵ method of proof" and "ϵ-definition," Weierstrass having begun in his early papers[3] to use ϵ in his arithmetized treatment of limits. The epsilon was similarly used by Cauchy[4] in 1821 and later, but sometimes he wrote[5] δ instead. Cauchy's δ is sometimes associated with Weierstrass' ϵ in phrases like "ϵ and δ methods" of demonstration. Klein[6] chose η in place of ϵ.

635. *J. E. Oliver.*—In the United States the symbol $\doteq$ has been used extensively. It seems to have been introduced independently of European symbolisms. Wolfgang Bolyai[7] used $\doteq$ to express "absolute equality." We have found one Austrian author, A. Steinhauser, using it for "nearly equal," before it was employed in the United States. Letting "Num. log." signify the antilogarithm, Steinhauser[8] writes "Num. log. $0 \cdot 015682 = 1 \cdot 03676 | 8 \doteq 1 \cdot 03677$, wobei das Zeichen $\doteq$ soviel als nahezu gleich bedeutet." In America the symbol $\doteq$ is due to James E. Oliver, of Cornell University,[9] and appeared in print in 1880 in W. E. Byerly's[10] *Calculus.* This symbol was used when it became desirable to indicate that the variable was not allowed to reach its limit, though in recent years it has been used, also when no such restriction seemed intended. Oliver[11] himself used it in print in 1888. He says: "Using $\doteq$ in the sense 'Approaches in value to, as a limit,' $U \doteq_{\Delta x} C$ would naturally mean 'U approaches C as a

[1] *Crelle's Journal,* Vol LI (1856), p. 27; *Mathematische Werke,* Band 1, p. 155.

[2] W. R. Hamilton, *Lectures on Quaternions* (Dublin, 1853), p. 569.

[3] K. Weierstrass, *op. cit.,* Vol. I, p. 56.

[4] Cauchy, *Cours d'analyse* (1821), p. 49, 50, 61, etc.

[5] Cauchy, *Résumé des leçons sur le calcul infinitésimal* in *Œuvres* (2d ser.), Vol. IV, p. 149.

[6] Felix Klein, *Anwendung der Differential- und Integralrechnung auf Geometrie* (Leipzig, 1907), p. 60.

[7] Wolfgangi Bolyai de Bolya, *Tentamen* (2d ed.), Vol. 1 (Budapest, 1897), p. xi, 214.

[8] Anton Steinhauser, *Lehrbuch der Mathematik,* "Algebra" (Wien, 1875). p. 292.

[9] See D. A. Murray, *An Elementary Course in the Integral Calculus* (New York, 1898), p. 4 n.

[10] W. E. Byerly, *Elements of the Differential Calculus* (Boston, 1880), p. 7.

[11] *Annals of Mathematics* (Charlottesville, Va.), Vol. IV (1888), p. 187, 188. The symbol is used also in Oliver, Wait, and Jones's *Algebra* (Ithaca, 1887), p. 129, 161.

limit, when Δx approaches its limit (viz. 0).' " Again he says: "Between its meaning as first used by me, 'Is nearly equal to,' and Byerly's modification, 'Approaches as a limit,' and perhaps yet other useful shades of interpretation, the context will sufficiently determine." To mark the degree of approximation, Oliver writes $\overset{1}{=}$, $\overset{2}{=}$, . . . , $\overset{n}{=}$, so that "$U \overset{n}{=}_h V$" would mean that $U - V$ or $(U-V)/\sqrt{UV}$ is of at least the nth order of smallness or negligibility, h being of the first order. James Pierpont[1] writes $a_n \overset{.}{=} l$ for lim $a_n = l$.

W. H. Echols,[2] of the University of Virginia, advocated the use of the symbol £ for the term "limit" and the symbol (=) in preference to $\overset{.}{=}$. E. W. Hobson[3] uses the sign $\backsim$ to indicate approach to a limit.

636. *J. G. Leathem.*—In England an arrow has been used in recent years in place of =. In 1905 J. G. Leathem,[4] of St. John's College, Cambridge, introduced $\longrightarrow$ to indicate continuous passage to a limit, and he suggested later (1912) that a dotted arrow might appropriately represent a saltatory approach to a limit.[5] The full arrow is meeting with general adoption nearly everywhere. G. H. Hardy[6] writes $\lim_{n \to \infty} (1/n) = 0$ and introduces the $\longrightarrow$ with the remark, "I have followed Mr. J. G. Leathem and Mr. T. J. I'A. Bromwich in always writing $\lim_{n \to \infty}$, $\lim_{x \to \infty}$, $\lim_{x \to a}$, and not $\lim_{n = \infty}$, $\lim_{x = \infty}$, $\lim_{x = a}$. This

[1] James Pierpont, *Theory of Functions of Real Variables* (New York), Vol. I (1905), p. 25.

[2] W. H. Echols, *Elementary Text-Book on the Differential and Integral Calculus* (New York, 1902), Preface, also p. 5. We quote from the Preface: ". . . . The use of the 'English pound' mark for the symbol of 'passing to the limit' is so suggestive and characteristic that its convenience has induced me to employ it in the text, particularly as it has been frequently used for this purpose here and there in the mathematical journals. The use of the 'parenthetical equality' sign (=) to mean 'converging to' has appeared more convenient . . . . than the dotted equality $\overset{.}{=}$, which has sometimes been used in American texts."

[3] E. W. Hobson, *Theory of Functions of a Real Variable* (2d ed.; Cambridge, 1921).

[4] J. G. Leathem, *Volume and Surface Integrals Used in Physics* (Cambridge, 1905).

[5] The arrow was used by Riemann in a lecture of 1856–57 in a manner which suggests limits. He says (*Gesammelte Math. Werke, Nachträge* [Leipzig, 1902], p. 67): "Ist a ein Verzweigungspunkt der Lösung einer linearen Differentialgleichung zweiter Ordnung und geht, während x sich im positiven Sinn um a bewegt, $\mathfrak{z}_1$ über in $\mathfrak{z}_3$ und $\mathfrak{z}_2$ in $\mathfrak{z}_4$, was kurz durch $\mathfrak{z}_1 \to \mathfrak{z}_3$ und $\mathfrak{z}_2 \to \mathfrak{z}_4$ angedeutet werden soll, so ist $\mathfrak{z}_3 = t\mathfrak{z}_1 + u\mathfrak{z}_2$, $\mathfrak{z}_4 = r\mathfrak{z}_1 + s\mathfrak{z}_2$. Ist ϵ irgend eine Konstante, so ist $\mathfrak{z}_1 + \epsilon\mathfrak{z}_2 \to \mathfrak{z}_3 + \epsilon\mathfrak{z}_4$."

[6] G. H. Hardy, *Course of Pure Mathematics* (1908), p. 116; see also the Preface.

change seems to me one of considerable importance, especially when '∞' is the 'limiting value.' I believe that to write $n = \infty$, $x = \infty$ (as if anything ever were 'equal to infinity'), however convenient it may be at a later stage, is in the early stages of mathematical training to go out of one's way to encourage incoherence and confusion of thought concerning the fundamental ideas of analysis." Sometimes the "lim" is omitted,[1] as in $\dfrac{\Delta u}{\Delta x} \longrightarrow u'$. F. S. Carey,[2] following a suggestion made by James Mercer, uses for monotone sequences an arrow with a single barb, either upper or lower; thus, $\longrightarrow$ "tends up to" the limit, $\longrightarrow$ "tends down to" the limit. Leathem[3] used sloped arrows for this purpose.

637. *G. L. Dirichlet.*—G. Lejeune Dirichlet[4] in 1837 introduced a symbolism which we give in his own words: "Es wird im Folgenden nöthig sein, diese beiden Werthe von $\varphi(\beta)$ zu unterscheiden, und wir werden sie durch $\varphi(\beta-0)$ und $\varphi(\beta+0)$ bezeichnen." We have here the first suggestion of our present notation, "lim $x = a+0$," when x converges to a by taking only values greater than a, and "lim $x = a-0$," when x converges to a by taking values all less than a.

A. Pringsheim.—The upper and the lower limit of y as a function of x was designated by M. Pasch,[5] of Giessen, as "$\displaystyle\lim_{v=\infty} \sup y$" and "$\displaystyle\lim_{v=\infty} \inf y$." For the same purpose A. Pringsheim, of Munich, introduced the notations $\displaystyle\overline{\lim_{v=+\infty}} \, a_\nu = A$ and $\displaystyle\underline{\lim_{v=+\infty}} \, a_\nu = a$ to denote, respectively, the upper and lower limits of the variable a_ν,[6] and $\displaystyle\overline{\underline{\lim_{\nu=+\infty}}} \, a_\nu$ to indicate that one may take at pleasure either the upper or the lower limit.

Paul du Bois-Reymond,[7] of the University of Freiburg in Baden,

---

[1] See, for example, Maurice Laboureur, *Cours de calcul algébrique différentiel et intégral* (Paris et Liège, 1913), p. 102.

[2] F. S. Carey, *Infinitesimal Calculus* (London, 1919), Preface.

[3] J. G. Leathem, *Mathematical Theory of Limits* (London, 1925), Preface.

[4] *Repertorium der Physik*, Vol. I (Berlin, 1837), p. 170.

[5] M. Pasch, *Mathematische Annalen*, Vol. XXX (1887), p. 134.

[6] *Sitzungsberichte Akad. München, Math.-phys. Classe*, Vol. XXVIII (1898), p. 62. See also *Encyclopédie des scienc. math.*, Tome I, Vol. I, p. 189, n. 224; p. 232, n. 45.

[7] Paul du Bois-Reymond in *Annali di mathematica pura ed applicata*. Diretti da F. Brioschi e L. Cremona (Milano, 2d ser.), Tomo IV (1870–71), p. 339.

introduced the notations $f(x) \succ \varphi(x)$, $f(x) \sim \varphi(x)$, $f(x) \prec \varphi(x)$, as equivalents to the formulae

$$\lim \frac{f(x)}{\varphi(x)} = \infty \ , \qquad \lim \frac{f(x)}{\varphi(x)} \text{ is finite,} \qquad \lim \frac{f(x)}{\varphi(x)} = 0 \ .$$

These symbols were used by A. Pringsheim[1] who in 1890 added $f_1(n) \cong af_2(n)$ for $\lim \frac{f_1(n)}{f_2(n)} = a$. The sign $\simeq$ is sometimes used to mark the lower limit,[2] as in "$J \simeq 45.10^{-40}$."

L. Scheffer.—The four values (viz., Du Bois-Reymond's *Unbestimmtheitsgrenzen*) which the difference-quotient $\frac{\Delta y}{\Delta x}$ approaches, as Δx approaches zero (y being a single-valued function of x), were designated by Dini[3] by the symbols Λ, λ, Λ', λ'; the first two being the right-hand upper and lower limits, the last two the left-hand upper and lower limits. Ludwig Scheffer[4] used in 1884 for these four limits, respectively, the symbols D^+, D_+, D^-, D_-, which were modified by M. Pasch[5] in 1887 to D^+, $_+D$, D^-, $_-D$.

G. Peano.—The countless values when x approaches zero, of $\lim \left(\left(\sin \frac{1}{x} \right) \right)$, cited by Cauchy in 1821, constitute a class, for the designation of which G. Peano[6] introduced the sign "Lm," in place of the ordinary "lim." By Lm x Peano means "les valeurs limites de x" or "la classe limite de x." The need of this new abbreviation is an indication that, since the time when Cauchy began to publish, a new concept, the idea of classes, has become prominent in mathematics. By Lm (f, u, x) Peano[7] designates "les valeurs limites de la fonction f, lorsque la variable, en variant dans la classe u, tend vers x."

Peano[8] designates an interval of integration by the use of a capital letter I placed horizontally and in an elevated position; thus $a \rightharpoondown b$ means the interval from a to b. He writes also $S^1(f, a \rightharpoondown b)$ for *l'intégrale*

[1] A. Pringsheim in *Mathematische Annalen*, Vol. XXXV (1890), p. 302.

[2] A. Eucken, *Zeitschrift der physikalischen Chemie*, Band C, p. 164.

[3] Dini, *Fondamenti per la teorica delle funzioni di variabili reali* (1878), art. 145, p. 190.

[4] L. Scheffer, *Acta mathematica*, Vol. V (Stockholm, 1884), p. 52, 53.

[5] M. Pasch, *Mathematische Annalen*, Vol. XXX (1887), p. 135.

[6] G. Peano, *Formulaire mathématique*, Tome IV (Turin, 1903), p. 148.

[7] *Op. cit.*, p. 164.

[8] G. Peano, *op. cit.*, p. 178, 179.

par excès or the inferior limit of the sums s' of f in the interval from a to b, and $S_1(f, a \vdash b)$ for *l'intégrale par défaut* or the superior limit of the sums s_1 of f in the interval a to b. Peano remarks that his notation $S(f, a \vdash b)$ has the advantage of being applicable also when the integral does not extend over an interval, but over any class. In an earlier publication[1] Peano used the sign $\vdash$ to mark an interval with the right end open and $\dashv$ one with the left end open. More common notations are[2] $(a\ \beta)$ for the interval from a to β; also $a \leqq \xi \leqq \beta$, $a < \xi \leqq \beta$, $a \leqq \xi < \beta$, $a < \xi < \beta$, the first and last intervals being closed and open, respectively, the others partially open, and a, β finite.

W. H. Young.—The consideration of the plurality of limits has brought forth other special notations. Thus W. H. Young[3] states in 1910: "We use the notation $\underset{x=a}{\operatorname{Llt}} f(x)$ to denote the set of all the limits of $f(x)$ at a point a, while, if it is known that there is an unique limit, we write $\underset{x=a}{\operatorname{Lt}} f(x)$ for that limit. . . . . If there are two independent variables x and y, the limits at $(a,\ b)$ . . . . are called *double limits*." He considers "*repeated limits* of $f(x, y)$ first with respect to x and then with respect to y, and written $\underset{y=b\ x=a}{\operatorname{Llt} \operatorname{Lt}} f(x, y)$." Similarly he considers repeated limits, first with respect to y and then with respect to x.

638. *The sign* $\frac{0}{0}$.—The indeterminate case, zero divided by zero, is treated by L'Hospital in his *Analyse des infiniments petits* (1696), but he does not use the symbol. Johann Bernoulli[4] discusses this indeterminedness in 1704, and uses such symbolism as $\frac{a0}{0}$, $\frac{0a}{0}$, $\frac{0m}{0n}$, but not $\frac{0}{0}$ in its nakedness. However, in 1730 he[5] did write $\frac{0}{0}$. G. Cramer[6] used $\frac{0}{0}$ in a letter to James Stirling, dated February 22, 1732. The symbol occurs repeatedly in D'Alembert's article "Différential" in

[1] G. Peano, *Analisi Infinitesimale*, Vol. I (Turin, 1893), p. 9, 10. See also *Formulario mathematico*, Vol. V (1908), p. 118.

[2] J. Harkness and F. Morley, *Theory of Analytic Functions* (London, 1898), p. 113–15.

[3] W. H. Young, *The Fundamental Theorems of the Differential Calculus* (Cambridge University Press, 1910), p. 3, 4.

[4] Johann Bernoulli in *Acta eruditorum* (1704), p. 376, 379, 380.

[5] Johann Bernoulli, *Opera omnia*, Vol. III, p. 183; *Mémoires de l'acad. r. d. sciences de Paris*, année 1730, p. 78.

[6] Charles Tweedie, *James Stirling; A Sketch of His Life and Works* (Oxford, 1922), p. 127.

Diderot's *Encyclopédie* of 1754. It became the battleground of contending philosophic thought on the calculus. A. T. Bledsoe[1] expresses himself with regard to it as follows: "We encounter $\frac{0}{0}$, the most formidable of all the symbols or enigmas in the differential calculus. . . . . Even Duhamel shrinks from a contact with it, although its adoption seems absolutely necessary to perfect the method of limits. . . . . This symbol is repudiated by Carnot and Lagrange. It is adopted by Euler and D'Alembert; but they do not proceed far before it breaks down under them. It is, nevertheless, one of the strongholds and defences of the method of limits, which cannot be surrendered or abandoned without serious and irreparable loss to the cause. . . . . This singular crusade of mathematicians against one poor symbol $\frac{0}{0}$, while all other symbols of indetermination are spared, is certainly a curious fact."

The sign $\frac{0}{0}$ was introduced quite frequently in textbooks on algebra of the first half of the nineteenth century.[2]

8. Concluding Observations

639. In considering the history of the calculus, the view advanced by Moritz Cantor presses upon the mind of the reader with compelling force. Cantor says: "We have felt that we must place the main emphasis upon the notation. This is in accordance with the opinion which we have expressed repeatedly that, even before Newton and Leibniz, the consideration of infinitesimals had proceeded so far that a suitable notation was essential before any marked progress could be made."[3]

Our survey of calculus notations indicates that this need was met, but met conservatively. There was no attempt to represent all reasoning in the calculus by specialized shorthand signs so as to introduce a distinct sign language which would exclude those of ordinary written or printed words. There was no attempt to restrict the exposition of

[1] Albert Taylor Bledsoe, *The Philosophy of Mathematics* (Philadelphia, 1886 [copyright, 1867]), p. 215, 222.

[2] For example, in C. F. Fournier, *Eléments d'arithmétique et d'algèbre*, Vol. II (Nantes, 1842), p. 217, 270.

[3] M. Cantor, *Vorlesungen über Geschichte der Mathematik*, Vol. III (2d ed.; Leipzig, 1901), p. 167.

theory and application of the calculus to ideographs. Quite the contrary. Symbols were not generally introduced, until their need had become imperative. Witness, for instance, the great hesitancy in the general acceptance of a special notation for the partial derivative.

It is evident that a sign, to be successful, must possess several qualifications: It must suggest clearly and definitely the concept and operation which it is intended to represent; it must possess adaptability to new developments in the science; it must be brief, easy to write, and easy to print. The number of desirable mathematical symbols that are available is small. The letters of the alphabet, the dot, the comma, the stroke, the bar, constitute the main source of supply. In the survey made in this paper, it was noticed that the forms of the fourth letter of the alphabet, d, D, and ∂, were in heavy demand. This arose from the fact that the words "difference," "differential," "derivative," and "derivation" all began with that letter. A whole century passed before any general agreement was reached among mathematicians of different countries on the specific use which should be assigned to each form.

The query naturally arises, Could international committees have expedited the agreement? Perhaps the International Association for the Promotion of Vector Analysis will afford an indication of what may be expected from such agencies.

An interesting feature in our survey is the vitality exhibited by the notation $\frac{dy}{dx}$ for derivatives. Typographically not specially desirable, the symbol nevertheless commands at the present time a wider adoption than any of its many rivals. Foremost among the reasons for this is the flexibility of the symbol, the ease with which one passes from the derivative to the differential by the application of simple algebraic processes, the intuitional suggestion to the mind of the reader of the characteristic right triangle which has dy and dx as the two perpendicular sides. These symbols readily call to mind ideas which reach to the very heart of geometric and mechanical applications of the calculus.

For integration the symbol $\int$ has had practically no rival. It easily adapted itself to the need of marking the limits of integration in definite integrals. When one considers the contributions that Leibniz has made to the notation of the calculus, and of mathematics in general, one beholds in him the most successful and influential builder of symbolic notation that the science ever has had.

FINITE DIFFERENCES

640. *Early notations.*—Brook Taylor explains his notation in the Introduction to his *Method of Increments*,[1] in a passage of which the following is a translation: "Indeterminate quantities I consider here as continually augmented by increments or diminished by decrements. The integral indeterminates I designate by the letters z, x, v, etc., while their increments, or the parts then added, I mark by the same letters dotted below, $\underset{.}{z}$, $\underset{.}{x}$, $\underset{.}{v}$, etc. I represent the increments of the increments, or the second increments of the integrals themselves, by the same letters twice dotted, $\underset{..}{z}$, $\underset{..}{x}$, $\underset{..}{v}$, etc. I designate the third increments by the letters triply dotted, $\underset{...}{z}$, $\underset{...}{x}$, $\underset{...}{v}$, etc., and so on. Indeed, for greater generality I sometimes write characters standing for the number of points: Thus, if n is 3, I designate $\underset{...}{x}$ by $\underset{n}{x}$ or $\overset{3}{x}$; if $n=0$, I designate the Integral itself x by $\underset{n}{x}$ or $\overset{0}{x}$; if n is -1, I designate the quantity of which the first increment is x by $\underset{n}{x}$ or $\underset{-1}{x}$; and so on. Often in this Treatise I mark the successive values of one and the same variable quantity by the same letter marked by lines; namely, by designating the present value by a simple letter, the preceding ones by superscribed grave accents, and the subsequent ones by strokes written below. Thus for example, $\overset{\backslash\backslash}{x}$, $\overset{\backslash}{x}$, x, $\underset{\backslash}{x}$, $\underset{\backslash\backslash}{x}$ are five successive values of the same quantity, of which x is the present value, $\overset{\backslash}{x}$ and $\overset{\backslash\backslash}{x}$ are preceding values, and $x_\backslash$ and $x_{\backslash\backslash}$ are subsequent values." Taylor uses at the same time the Newtonian symbols for fluxions and fluents. Two years later Taylor introduced a symbol for summation. He says:[2] "When Occasion requires that a variable Quantity, suppose x, is to be look'd upon as an Increment, I denote its Integral by the Letter included between two Hooks [ ]. Also the Integral of the Integral [x] or the second Integral of x, I denote by putting the Number 2 over the first of the Hooks, as $\overset{2}{[}x]$, ... so that it is $\overset{2}{[}x]=\overset{3}{[}x]$, $[x]=\overset{2}{[}x]$, $x=[x]$."

It was not long before alterations to Taylor's notation were made in England. In E. Stone's *New Mathematical Dictionary* (London, 1726), article "Series," preference is given to the modified symbols

[1] Brook Taylor, *Methodus Incrementorum directa et inversa* (London, 1715), p. 1.

[2] Brook Taylor in *Philosophical Transactions*, No. 353 (London, 1817), p. 676; *ibid.* (3d abridged ed., Henry Jones), Vol. IV (London, 1749), p. 130.

introduced by Samuel Cunn, a textbook writer and, with Raphson, a translator into English of Newton's *Arithmetica universalis*. Letting q be a constant increment, Cunn marks $Q+q$ by Q', $Q+2q$ by Q'', $Q+3q = Q'''$, and so on; also $Q-q$ by $\dot{Q}$, $Q-2q$ by $\ddot{Q}$, etc. Stone remarks that Taylor and others chose to denote the increments by the same letters with the integrals and only for the sake of distinction printed them beneath. Comparing the two notations, we have $q = \dot{Q}$, $n + \underset{\prime}{n} = \dot{n}$.

Taylor's notation, with some modifications, was employed by W. Emerson.[1] Any letters s, t, v, x, y, z, etc., are put for "integral quantities," the different values of any such quantity are denoted by the same letter with numeral subscripts. Emerson continues: "If Z be an integral, then z, z_1, z_2, z_3, etc. are the present value, and the first, second, third, etc. successive values thereof; and the preceding values are denoted with figures and negative signs; thus z_{-1}, z_{-2}, z_{-3}, z_{-4}, are the first, second, third, fourth preceding values. . . . . The increments are denoted with the same letters and points under them, thus $\underset{\cdot}{x}$ is the increment of x, and $\underset{\cdot}{z}$ is the increment of z. Also $\underset{1\cdot}{x}$ is the increment of $\underset{1}{x}$; and $\underset{n\cdot}{x}$ of $\underset{n}{x}$, etc. . . . . If x be any increment, then $[x]$ denotes the integral of x, and $\overset{2}{[x]}$ denotes the integral of $[x]$. . . . ."

Taylor's dot notation for differences prevailed with Edward Waring[2] who, using it along side of the fluxional notation, writes, for instance, "incrementum logarithmi quantitatis (x), cujus fluxio est $\dfrac{\dot{x}}{x}$, erit log. $(x+\underset{\prime}{x}) -$ log. $(x) = $ log. $\dfrac{x+\underset{\prime}{x}}{x}$, cujus fluxio erit $- \dfrac{x\dot{x}}{x(x+\underset{\prime}{x})}$ . . . . ."

On the European Continent, Leibniz used the sign d both for finite and infinitesimal differences; he employed $\int$ for the sum of terms in a series as well as for integration. For example, he says:[3] "Supposing that $1+1+1+1+1+$ etc. $= x$ or that x represents the natural numbers, for which $dx = 1$, then $1+3+6+10+$etc. $= \int x$, $1+4+10+20+$ etc. $= \int\int x$."

[1] William Emerson, *Method of Increments* (London, 1763), p. 2.

[2] Edward Waring, *Meditationes analyticae* (2d ed.; Cambridge, 1785), p. 307.

[3] *Historia et origo calculi differentialis* in *Leibnizens Mathematische Schriften*, Vol. V (1858), p. 397.

Wide acceptance has been given to the symbolism on finite differences introduced by L. Euler.[1] In place of Leibniz' small Latin letter d, he used the corresponding Greek capital letter Δ which Johann Bernoulli had used previously for differential coefficient (§§ 539, 572, 596). In *Institutiones calculi differentialis*, Euler begins in the first chapter by letting ω stand for the increment of a variable x, but soon after enters upon a more elaborate notation. Letting y, y^I, y^{II}, y^{III}, etc., be the values of y corresponding to the x-values x, $x+\omega$, $x+2\omega$, $x+3\omega$, etc., he puts $y^I-y=\Delta y$, $y^{II}-y^I=\Delta y^I$, $y^{III}-y^{II}=\Delta y^{II}$, etc. Thus Δy expresses the increment which the function y assumes, when in place of x one writes $x+\omega$. Proceeding further, he lets $\Delta\Delta y=\Delta y^I-\Delta y$, $\Delta\Delta y^I=\Delta y^{II}-\Delta y^I$, etc., and calls $\Delta\Delta y$ the "differentia secunda ipsius y; $\Delta\Delta y^I$ differentia secunda ipsius y^I," etc., and $\Delta^3 y$, $\Delta^3 y^I$ the "differentiae tertiae," $\Delta^4 y$, $\Delta^4 y^I$, "differentiae quartae," and so on. He represents the values of y corresponding to x, $x-\omega$, $x-2\omega$, etc., by y, y_I, y_{II}, etc., so that $y=\Delta y_I+y_I$, etc., and $y=\Delta y_I+\Delta y_{II}+\Delta y_{III}$, etc. It is here that he introduces also the letter Σ for "sum"; he says: "Quemadmodum ad differentiam denotandam vsi sumus signo Δ, ita summam indicabimus signo Σ." The general features of Euler's notations were adopted by Lagrange and Laplace in some of their papers. Lagrange[2] in 1772 assigns to the variables x, y, z . . . . the *accroissements* ξ, ψ, ζ, ... and to the function u the *accroissement* Δu, and the differences of higher order $\Delta^2 u$, $\Delta^3 u$, ... , $\Delta^\lambda u$. He supposes that λ may be negative and writes $\Delta^{-1}=\Sigma$, $\Delta^{-2}=\Sigma^2$, ... In 1792 Lagrange[3] lets T_0, T_1, T_2, ... , T_n, T_{n+1}, ... be the successive terms of a series D_1, D_2, D_3, the successive differences of the terms, i.e., $D_1=T_1-T_0$, $D_2=T_2-2T_1+T_0$, etc. He writes $D_m=(T_1-T_0)_m$, and by the analogy existing between exponents and the indices, he expands the binomial $(T_1-T_0)_m$. By making m negative the differences change into sums, which he represents by S_1, S_2, S_3, ... so that $S_1=D_{-1}$, $S_2=D_{-2}$, ... When interpolating between successive terms of T_0, T_1, T_2, ... he marks the terms interpolated $T_{\frac{1}{2}}$, $T_{\frac{3}{2}}$, $T_{\frac{5}{2}}$, ... When dealing simultaneously with two series, he employs for the second series the corresponding small letters t_0, t_1, t_2, ... , d_0, d_1, d_2, ... Lagrange considers also double and triple series, in which the

[1] Leonhard Euler, *Institutiones calculi differentialis* (Petersburgh, 1755), p. 3–7, 27.

[2] *Œuvres de Lagrange*, Vol. III (Paris, 1869), p. 443, 450, 451; *Nouveaux mémoires de l'acad. r. d. scienc. et bell.-lett. de Berlin*, année 1772.

[3] *Œuvres de Lagrange*, Vol. V (Paris, 1870), p. 667, 668, 673, 679; *Nouveaux mémoires de l'acad. r. d. scienc. et bell.-lett. de Berlin*, années 1792 et 1793.

terms vary in two or more different ways, and introduces symbols

$$T_{0,0} \quad T_{1,0} \quad T_{2,0} \ldots$$
$$T_{0,1} \quad T_{1,1} \quad T_{2,1} \ldots$$
$$T_{0,2} \quad T_{1,2} \quad T_{2,2} \ldots$$
$$\cdot \quad \cdot \quad \cdot \quad \cdot \quad \cdot$$

and, correspondingly, $D_{0,0}, D_{1,0}, \ldots, D_{0,1}, D_{1,1}, \ldots, D_{0,2}, D_{1,2}, \ldots$, or generally $D_{m,n}$ for the successive differences.

Laplace[1] lets the variables $t, t_1, t_2, \ldots$ increase respectively by $a, a_1, a_2, \ldots$, and lets u' be the value of a function of these variables after they have received their increment. He writes $u' - u = \Delta u$, and uses also the symbols $\Delta^2 u$, $\Delta^i u$, Σ, and Σ^i.

641. *Later notations.*—The symbols d, D, Δ, $\int$, Σ were used by Arbogast,[2] Français,[3] and others in the development of functions in series and in treatment of recurrent series. Kramp's substitution of the German letters $\mathfrak{D}$ and $\mathfrak{G}$ for the Greek Δ and Σ did not meet with wide acceptance.[4] These symbols were used sometimes as symbols of operation (d and D in differentiation, Δ for finite differences, $\int$ for integration, Σ for finite summation), and sometimes as if they were veritable algebraic quantities. Servois[5] studied the laws of the calculus with symbols and showed that Δ, D, Σ, $\int$ possessed the *propriété commutative*, *propriété distributive*, expressions which have secured permanent adoption. Later authors have used a great variety of symbolism in the calculus of operations, of which the more important are due to Boole,[6] Carmichael,[7] and Casorati.[8]

Euler's notation found entrance into England at the time of the

---

[1] *Œuvres de Laplace*, Vol. IX (Paris, 1893), p. 316, 318; *Mémoires de l'acad. r. d. scienc. de Paris*, année 1777 (Paris, 1780).

[2] L. F. A. Arbogast, *Du calcul des dérivations* (Strasbourg, an VIII [1800]), p. 1–159, 230. See A. Cayley in *Philosophical Transactions* (London), Vol. CLI (1861), p. 37. See also *Encyclopédie des scien. math.*, Tom. II, Vol. V, p. 4, 5 (1912).

[3] J. F. Français in *Annales math. pures appl.* (Gergonne), Vol. III (1812–13), p. 244–72. He uses the rounded letter ∂ in place of d.

[4] C. Kramp, *Élémens d'arithmétique universelle* (Cologne, 1808), "Notations."

[5] F. J. Servois, *Annales math. pures appl.* (Gergonne), Vol. V (1814–15), p. 93.

[6] G. Boole, *Philosophical Transactions*, Vol. CXXXIV (London, 1844), p. 225; *Math. Analysis of Logic* (Cambridge, 1847), p. 15–19. *Treatise on Differential Equations* (Cambridge, 1859), p. 371–401.

[7] R. Carmichael, *Treatise on the Calculus of Operations* (London, 1855).

[8] F. Casorati in *Annali di mat. pura ed appl.* (2d ser.), Vol. X (1880–82), p. 10, etc.

introduction of the Leibnizian notation of the calculus, in the early part of the nineteenth century. Euler's symbols are used by J. F. W. Herschel[1] and later by George Boole.[2] Boole says (p. 16, 17): "In addition to the symbol Δ, we shall introduce a symbol E to denote the operation of giving to x in a proposed subject function the increment unity;—its definition being $E_{u_x} = u_{x+1}$." "The two symbols are connected by the equation $E = 1 + \Delta$."

SYMBOLS IN THE THEORY OF FUNCTIONS

A. SYMBOLS FOR FUNCTIONS IN GENERAL

642. In 1694 Johann Bernoulli[3] represented by the letter n any function whatever of z ("posito n esse quantitatem quomodocunque formatam ex indeterminatis et constantibus") and then proceeded to consider the integral of ndz. In an article[4] of the following year, Jakob (James) Bernoulli proposed a problem in which he lets p and q be any two functions whatever of x ("... p et q quantitates utcunque datas per x denotant"). In 1697 Johann Bernoulli[5] let the capital letter X and also the corresponding Greek letter ξ represent functions of x ("per ξ intelligo quantitatem datam per x et constantes"; "X quantitati itidem ex x et constantibus compositae"). He writes also "ξX = quantitati pure dependenti ab x et constantibus." In a letter to Leibniz he said:[6] "For denoting any function of a variable quantity x, I rather prefer to use the capital letter having the same name X or the Greek ξ, for it appears at once of what variable it is a function; this relieves the memory." To this Leibniz replied:[7] "You do excellently in letting the mark of a function indicate of what letter it is the

[1] J. F. W. Herschel, *Calculus of Finite Differences* (Cambridge, 1820), p. 1, 4.

[2] George Boole, *Calculus of Finite Differences* (3d ed.; London, 1880), p. 2, 4, 5. [Preface to 1st ed. is dated April 18, 1860.]

[3] Johann Bernoulli in *Acta eruditorum* (Leipzig, 1694), p. 437. I am indebted for this reference and a few others to G. Eneström's article in *Bibliotheca mathematica* (2d ser.), Vol. V (1891), p. 89.

[4] Jakob Bernoulli in *Acta eruditorum* (1695), p. 553.

[5] Johann Bernoulli, in *Acta eruditorum* (1697), p. 115.

[6] *Leibnizens Mathematische Schriften*, Vol. III (1856), p. 531. Letter dated Aug. 16/26, 1698.

[7] Letter of Leibniz to Johann Bernoulli, n.d., in *Leibnizens Mathem. Schriften*, Vol. III (1856), p. 537. It should be stated that Leibniz' letter, as printed in the *Commercium Philosophicum et Mathematicum* of G. W. Leibniz and Johann Bernoulli, Vol. I (1745), p. 399, has some slight variations in the symbolism; thus, instead of $\overline{x}\lfloor r.1\rfloor$ we find $\overline{x}\lfloor r:1\rfloor$. Here, and in the other cases given in the letter, a colon is printed where in Gerhardt's edition there is simply a dot.

function, for example, that ξ be a function of x. If there are many functions of that same x, they may be distinguished by numbers. Occasionally I mark the sign of relationship in this manner $\overline{x}\lfloor 1 \rfloor$, $\overline{x}\lfloor 2 \rfloor$ etc., that is, any expression in x whatever; again, if one is formed from many, as from x and y, I write $\overline{x;y}\lfloor 1 \rfloor$, $\overline{x;y}\lfloor 2 \rfloor$. And when the form is rational, I add r, as in $\overline{x}\lfloor r.1 \rfloor$ and $\overline{x}\lfloor r.2 \rfloor$ or $\overline{x;y}\lfloor r.1 \rfloor$, $\overline{x;y}\lfloor r.2 \rfloor$. If the form is rational integral, I write $\overline{x}\lfloor ri.1 \rfloor$, $\overline{x}\lfloor ri.2 \rfloor$. But in the case of only one function, or only a few of them, the Greek letters suffice, or some such, as you are using." This notation to indicate functions Leibniz never had printed, nor did he refer to it again in his letters to Johann Bernoulli or to other correspondents. On the other hand, the notation proposed by Johann Bernoulli was used by his brother Jakob Bernoulli[1] in 1701, who without use of the word "function" defined B, F, G, C as functions, respectively of b, f, g, c by equations such as $adF=hdf$ or $F=\sqrt{(aa+ff)}$, where a is a constant. This mode of representation was employed extensively during the eighteenth century and not infrequently has been found convenient also in more recent times.

643. The use of a special symbol to stand for a function of a given variable is found again in a memoir[2] of Johann Bernoulli of 1718, where the Greek φ is employed for this purpose. He stated: "en prenant aussi φ pour la caractéristique de ces fonctions." He also writes (p. 246) the differential equation $dy:dz=(\varphi x \pm c):a$, and further on (p. 250), $dy:dx=(\varphi z \pm c):a$. It is to be observed that neither Johann Bernoulli nor his successors Clairaut and D'Alembert in their earlier writings inclose the variable or argument in parentheses.[3] The use of parentheses for this purpose occurs in Euler[4] in 1734, who says, "Si $f\left(\dfrac{x}{a}+c\right)$ denotet functionem quamcunque ipsius $\dfrac{x}{a}+c$." This is the first appearance of f for "function." About the same time Clairaut designates a function of x by Πx, Φx, or Δx, without using parentheses.[5] In

[1] Jakob Bernoulli in *Analysin magni problematis isoperimetrici* (Basel, 1701); Jacobi Bernoulli, *Opera*, Vol. II (Geneva, 1744), p. 907, 909.

[2] *Mémoires d. l'acad. d. sciences de Paris* (1718), p. 108; *Opera* (Lausanne et Genève, 1742), Vol. II, p. 243, 246, 250.

[3] G. Eneström, *Bibliotheca mathematica* (3d ser.), Vol. VI (1905), p. 111.

[4] L. Euler in *Comment. Petropol. ad annos 1734–1735*, Vol. VII (1840), p. 186, 187, second paging; reference taken from J. Tropfke, *op. cit.*, Vol. II (2d ed., 1921), p. 35.

[5] A. C. Clairaut in *Histoire d. l'acad. d. sciences*, année 1734 (Paris, 1736), *Mém.*, p. 197.

1747 D'Alembert[1] marked functions of the two variables u and s by the signs $\Delta u,s$, $\Gamma u,s$. In 1754 he says:[2] "Soit $\varphi(z)$ une fonction de la variable z, $dz\Delta(z)$ la différence de cette fonction; $dz\Gamma(z)$ la différence de $\Delta(z)$, $dz\Psi(z)$ la différence de $\Gamma(z)$ etc. Soit ensuite $\varphi(z+\zeta)$ une fonction de $z+\zeta$ pareille à la fonction $\varphi(z)$, ζ étant une très-petite quantité. ..."

In 1753 Euler[3] designated a function of x and t by $\Phi:(x,\ t)$. A year later he wrote:[4] "Elle sera donc une certaine fonction de a et n, que J'indiquerai par $f:(a, n)$, dont la composition nous est encore inconnüe." In 1790 Mascheroni[5] marked functions of the variable u thus, F, u and f, u.

644. A great impetus toward the more general use of the letters f, F, φ and ψ for the designation of functional relations was given by Lagrange[6] in his *Théorie des fonctions analytiques* (Paris, 1797). But such designations occur much earlier in his writings; nor are they the only symbols by which he represented functions. In one of his papers of 1759 "X et Z expriment des fonctions quelconques de la variable x."[7] In another paper of the same year he[8] wrote the equation $y=\Psi(t\sqrt{c}+x)+\Gamma(t\sqrt{c}-x)$, "$\Psi$ et Γ exprimant des fonctions quelconques des quantités $t\sqrt{c}+x$ et $t\sqrt{c}-x$." The year following he wrote[9] "$\varphi(t-ax)$." Somewhat later he[10] discussed a differential equation, "P étant une fonction quelconque de p et Q une fonction quelconque de q," and in another article considered the variation of the function

[1] D'Alembert, *Réflexions sur la cause générale des vents* (Paris, 1747), p. 164. See G. Eneström, *Bibliotheca mathematica* (2d ser., Vol. V 1891), p. 90.

[2] D'Alembert, *Recherches sur differens points importans du système du monde*, Vol. I (Paris, 1754), p. 50.

[3] L. Euler in *Mémoires d. l'acad. d. sciences et des belles lettres* (Berlin, 1753), p. 204.

[4] L. Euler, *op. cit.*, année 1754 (Berlin, 1756), p. 209.

[5] L. Mascheroni, *Adnotationes ad Calculum integralem Euleri* (Ticini, 1790). Reprinted in L. Euler's *Opera omnia* (1st ser.), Vol. XII (1914), p. 484.

[6] *Œuvres de Lagrange*, Vol. IX (Paris, 1881), p. 21, 43, 93, 99, 113, 171, 243, 383.

[7] *Miscellanea Taurinensia*, Vol. I (1759); *Œuvres de Lagrange*, Vol. I (Paris, 1867), p. 23.

[8] *Miscellanea Taurinensia*, Vol. I (1759); *Œuvres de Lagrange*, Vol. I (Paris, 1867), p. 64.

[9] *Miscellanea Taurinensia*, Vol. II (1760–61); *Œuvres*, Vol. I, p. 154.

[10] *Miscellanea Taurinensia*, Vol. IV (1766–69); *Œuvres*, Vol. II (Paris, 1868), p. 20.

φ, and took[1] Φ as a function φ, $x, y, z, \ldots$. In 1772 Lagrange[2] took u as a function of x. Another time he[3] wrote Clairaut's equation, $y - px + f(p) = 0$, "$f(p)$ dénotant une fonction quelconque de p seul," and he gave $f'(p) = \dfrac{df(p)}{dp}$. A curious symbol, $\oplus x$, for our $f(x)$ was used by W. Bolyai.[4]

645. During the early part of the nineteenth century functional notations found their way into elementary textbooks; for instance, into Legendre's *Elements of Geometry*,[5] where a function of p and q was marked $\Phi : (p, q)$.

J. F. W. Herschel[6] uses the sign $f(x)$ and says, "In general $f(f(x))$ or ffx may be written $f^2(x) \ldots$ and $f^m f^n(x) = f^{m+n}(x), \ldots$ and $f^{-1}(x)$ must denote that quantity whose function f is x."

646. G. Peano[7] writes $y = f(x)$ and $x = \bar{f}(y)$, where $\bar{f}$ means the inverse function of f. This notation is free from the objection raised to that of Herschel and Bürmann (§533, 645), and to $gd^{-1}\phi$, used as the inverse Gudermannian by some writers, for instance, by J. M. Peirce in his *Mathematical Tables* (1881, p. 42) and D. A. Murray in his *Differential and Integral Calculus* (1908, p. 422), but by others more appropriately marked $\lambda(\phi)$.

B. SYMBOLS FOR SOME SPECIAL FUNCTIONS

647. *Symmetric functions.*—Attention was directed in § 558 to Leibniz' dot notation for symmetric functions. Many writers used no special symbol for symmetric functions; the elementary functions were placed equal to the respective coefficients of the given equation, due attention being paid to algebraic signs. This course was followed

---

[1] *Loc. cit.; Œuvres*, Vol. II (Paris, 1868), p. 40.

[2] *N. Mémoires d. l'acad. r. d. scienc. et bell.-lett. de Berlin*, année 1772; *Œuvres*, Vol. III (Paris, 1869), p. 442.

[3] *N. Mémoires d. l'acad. r. d. scienc. et bell.-lett. de Berlin*, année 1774; *Œuvres*, Vol. IV (Paris, 1869), p. 30. See also p. 591.

[4] Wolfgang Bolyai, *Az arithmetica eleje* (Maros-Vásárhelyt, 1830). See B. Boncompagni, *Bullettino*, Vol. I (1868), p. 287.

[5] A. M. Legendre, *Éléments de géométrie* (éd. par J. B. Balleroy ... avec notes ... par M. A. L. Marchand; Bruxelles, 1845), p. 188.

[6] J. F. W. Herschel, *Calculus of Finite Differences* (Cambridge, 1820), p. 5. Herschel says that he first explained his notation for inverse functions in the *Philosophical Transactions* of London in 1813 in his paper "On a Remarkable Application of Cotes's Theorem," but that he was anticipated in this notation by Bürmann.

[7] G. Peano, *Lezioni di analisi infinitesimale*, Vol. I (Torino, 1893), p. 8.

by Cauchy[1] in 1821. Kramp[2] denoted these functions by $\int A$, where A is the general term.

Charles Babbage[3] wrote expressions like $F[\bar{x}, \overline{\psi(x)}]$, where the bars above the x and $\psi(x)$ were intended to specify that the function F is symmetric with respect to these two quantities. E. Meissel[4] represents by $(x|\varphi x)$ a symmetric function of the quantities x and $\varphi(x)$, where $\varphi(x)$ is determined by the equation $(x|\varphi x) = 0$.

648. J. R. Young[5] in 1843 puts "for abridgment $\Sigma_m = a_1^m + a_2^m + a_3^m + \ldots a_n^m$," "where $a_1, a_2, \ldots a_n$ are the roots of an equation." J. A. Serret[6] said: "Let $a, b, c, \ldots, k, l$ be the m roots of an equation $X = 0$ of the degree m, and let us consider a double symmetric function, of which one term is $a^\alpha b^\beta$; the function in question being determined when one term is known, we represent, for brevity, by $\Sigma a^\alpha b^\beta$, and we continue to designate by s_α the sum of the αth powers of all the roots." W. S. Burnside and A. W. Panton state: "It is usual to represent a symmetric function by the Greek letter Σ attached to one term of it, from which the entire expression may be written down."[7]

649. *Gamma and beta functions.*—Both of these functions are the creations of L. Euler, but their names and their symbolic representations originated in the nineteenth century. Euler mentions the gamma function in a letter of October 13, 1729, to Goldbach,[8] also in a paper of 1730[9] and in subsequent writings. Over sixty years later Legendre[10] gave the function its sign and corresponding name. In one place he says:[11] "... l'intégrale $\int dx \left(\log \frac{1}{x} \right)^{a-1}$, dans laquelle nous supposerons que a est positif et plus petit que l'unité. Cette quantité étant simplement fonction de a, nous désignerons par $\Gamma(a)$." It was

---

[1] A. L. Cauchy, *Cours d'analyse* (Paris, 1821), p. 70, 71.

[2] C. Kramp, *Élémens d'arithmétique universelle* (Cologne, 1808), "Notations."

[3] Gergonne, *Annales de mathématiques* (Nismes), Vol. XII (1819–20), p. 81; Charles Babbage, *Solutions of Functional Equations* [1820], p. 2.

[4] E. Meissel, *Crelle's Journal*, Vol. XLVIII (1854), p. 303.

[5] J. R. Young, *Algebraical Equations* (2d ed.; London, 1843), p. 416.

[6] J. A. Serrett, *Cours d'algèbre supérieure* (2d ed.; Paris, 1854), p. 11, 12.

[7] W. S. Burnside and A. W. Panton, *Theory of Equations* (4th ed.; Dublin and London), Vol. I (1899), p. 47.

[8] P. H. Fuss, *Correspondance math. et phys.*, Vol. I (1843), p. 3, 4.

[9] *Comment. acad. Petropolitanae ad annos 1730 et 1731*, Vol. V, p. 36–57.

[10] See our § 448. See also A. M. Legendre, *Exercices de calcul intégral*, Vol. I (Paris, 1811), p. 277.

[11] A. M. Legendre, *Traité des fonctions elliptiques*, Vol. II (Paris, 1826), p. 406.

in the same paper of 1730 that Euler gave what we now call the "beta function." He says:[1] "Sit proposita haec formula $\int x^e dx (1-x)^n$ vicem termini generalis subiens, quae integrata ita, ut fiat$=0$, si sit $x=0$, et tum posito $x=1$, dat terminum ordine n progressionis inde ortae." About a century after Euler's first introduction of this function, Binet wrote the integral in the form $\int_0^1 x^{p-1} dx (1-x)^{q-1}$ and introduced the Greek letter beta, B. Considering both beta and gamma functions, Binet said:[2] "Je désignerai la première de ces fonctions par $B(p, q)$; et pour la seconde j'adoperai la notation $\Gamma(p)$ proposée par M. Legendre." Legendre had represented the beta function by the sign $\left(\dfrac{p}{q} \right)$.

650. Gauss[3] expressed the gamma function by Πz and defined $\Pi(k, z) = \dfrac{k^z \Pi k \cdot \Pi z}{\Pi(k+z)}$; Γz means the same as $\Pi(z-1)$. Weierstrass called the reciprocal of $\Gamma(1+u)$ the *factorielle* of u and wrote it $Fc\,u$.[4]

651. *Elliptic functions.*—In his treatise on elliptic functions, Legendre[5] starts out by letting R represent the radical $\sqrt{(\alpha + \beta x + \gamma x^2 + \delta x^3 + \epsilon x^4)}$, Π^m the integral $\int \dfrac{x^m dx}{R}$, and Γ^k the integral $\int \dfrac{dx}{(1+nx)^k R}$. A little later[6] he denotes (taking the modulus $c<1$) the radical $\sqrt{(1-c^2 \sin^2 \varphi)}$ by Δ, also by $\Delta(\varphi)$ and by $\Delta(c, \varphi)$. He represents the general formula $\int \dfrac{A + B \sin^2 \varphi}{1 + n \sin^2 \varphi} \cdot \dfrac{d\varphi}{\Delta}$ by H, and special cases of it, namely (p. 15), $\int \Delta d\varphi$, by E, $\int \dfrac{d\varphi}{\Delta}$ by F. He writes also $F(\varphi)$, $F(c, \varphi)$, $E(c, \varphi)$, $\Pi(n, c, \varphi)$.

652. Jacobi in 1829 introduced a new notation, placing[7] $\int_0^\varphi \dfrac{d\varphi}{\sqrt{1 - k^2 \sin^2 \varphi}} = u$, where φ is the amplitude of the function u.

[1] Quoted by M. Cantor, *op. cit.*, Vol. III (3d ed.), p. 653.

[2] Jacques P. M. Binet in *Jour. école polyt.*, Vol. XVI (1839), p. 131.

[3] K. F. Gauss, *Werke*, Vol. III (Göttingen, 1866), p. 146.

[4] *Crelle's Journal*, Vol. LI (1856), p. 7; Weierstrass, *Mathematische Werke*, Vol. I (1894), p. 161.

[5] A. M. Legendre, *Traité des fonctions elliptiques* (Paris), Vol. I (1825), p. 4, 5, 17.

[6] *Op. cit.*, p. 11, 14, 35, 107.

[7] C. G. J. Jacobi's *Fundamenta nova*, in *Gesammelte Werke*, Vol. I (Berlin, 1881), p. 81, 82.

Hence this angle is called "ampl u," or briefly "$\varphi=$ am u." Also when

$$u=\int_0^x \frac{dx}{\sqrt{(1-x^2)(1-k^2x^2)}},$$

$x=\sin$ am u. Writing

$$\int_0^{\frac{\pi}{2}} \frac{d\varphi}{\sqrt{1-k^2\sin^2\varphi}}=K,$$

he calls $K-u$ the complement of the function u, and "coam" the amplitude of the complement, so that am$(K-u)=$coam u. In partial conformity with Legendre, Jacobi writes Δ am $u=\dfrac{d\text{ am }u}{du}$ $=\sqrt{1-k^2\sin^2\text{ am }u}$. He uses k in place of Legendre's c; Jacobi marked the complement of the modulus k by k' so that $kk+k'k'=1$. He uses the expressions "sin am u," "sin coam u," "cos am u," "cos coam u," "Δ am u," "Δ coam u," etc. The Greek letter Z was introduced by Jacobi thus:[1] "E. Cl^i. Legendre notatione erit, posito $\dfrac{2Kx}{\pi}=u$, $\varphi=$ am u:

$$Z(u)=\frac{F^I E(\varphi)-E^I F(\varphi)}{F^I}.\text{ ''}$$

This is Jacobi's zeta function.

653. If in Jacobi's expression for K, k' is written for k, the resulting expression is marked K'. Weierstrass and Glaisher denoted $K-E$ by J. With Weierstrass $J'=E'$; with Glaisher[2] $J'=K'-E'$. Glaisher writes also $G=E-k'^2 K$ and $G'=E'-k^2K'$.

Using Jacobi's zeta function, Glaisher[3] introduces three functions, ez x, iz x, gz x, defined thus: ez $x=\dfrac{E}{K}x+Z(x)$, iz $x=-\dfrac{J}{K}x+Z(x)$, gz $x=\dfrac{G}{K}x+Z(x)$. Glaisher defines also $Ae_s x$, $Ai_s x$, $Ag_s x$; for example, $Ai_s x=e^{\int_0^x iz\,x\,dx}$. The $Ai_s x$ is the same as Weierstrass' $Al_s x$.

654. To some, Jacobi's notation seemed rather lengthy. Accordingly, Gudermann,[4] taking $\varphi=$ am u and $u=$ arg am (φ), suggested the

[1] *Op. cit.*, Vol. I, p. 187.

[2] J. W. L. Glaisher in *Quarterly Journal of Mathematics*, Vol. XX (London, 1885), p. 313; see also Vol. XIX (1883), p. 145–57.

[3] J. W. L. Glaisher, *ibid.*, Vol. XX, p. 349, 351, 352.

[4] C. Gudermann in *Crelle's Journal*, Vol. XVIII (Berlin, 1838), p. 14.

more intense abbreviations[1] sn u for sin am u, cn u for cos am u, tn u for tang am u, dn u for $\sqrt{(1-k^2 \sin^2 \text{am } u)}$. The argument $K-u$, which is the complement of u, appears in the following abbreviations used by Guderman (p. 20): amc $u=$ am $(K-u)$, snc $u=$ sin amc $u=$ sn $(K-u)$, and similarly in cnc u, tnc u, dnc u. Before Gudermann, Abel[2] had marked sn u by the sign $\lambda(\theta)$. Gudermann's symbols sn u, cn u, dn u were adopted by Weierstrass in a manuscript of 1840, where he considered also functions which he then marked $A(u)$, $B(u)$, $C(u)$, $D(u)$, but which he and others designated later[3] by $\text{Al}(u)_1$, $\text{Al}(u)_2$, $\text{Al}(u)_3$, $\text{Al}(u)$ and called "Abelian functions." In 1854[4] Weierstrass called certain $2n+1$ expressions $\text{al}(u_1, u_2, \ldots)_0$, $\text{al}(u_1, u_2, \ldots)_1$, etc., Abelian functions; "it is they which correspond perfectly to the elliptic functions sin am u, cos am u, Δ am u." He shows the relation[5] $\text{al}(u_1, u_2, \ldots)_a = \text{Al}(u_1, u_2, \ldots)_a : \text{Al}(u_1, u_2, \ldots)$. In 1856 he changed the notation slightly[6] from $\text{al}(u_1, u_2, \ldots)_0$ to $\text{al}(u_1, \ldots)_1$, etc.

If $\displaystyle\int^\infty \frac{dx}{2\sqrt{X}} = \int \frac{ds}{\sqrt{S}} = u$, where X is a cubic expression in x, $x = s+f$, and f is so chosen that $S = 4s^3 - g_2 s - g_3$, then s is an elliptic function of u, of the second degree, denoted by $\wp u$ in the notation of Weierstrass.[7] He marked its derivative $\wp^1 u$. Connected with these is an analytic function which he[8] marked σu and employed in the expression of elliptic functions as quotients of integral transcendental functions. He defines it thus:

$$\sigma u = u \Pi'_w \left(1 - \frac{u}{w}\right) e^{\frac{u}{w} + \frac{1}{2}\frac{u^2}{w^2}}, \qquad w = 2\mu\omega + 2\mu'\omega',$$

$$\mu, \mu' = 0, \pm 1, \pm 2, \pm 3, \ldots \pm \infty,$$

[1] A. C. Dixon, in his *Elliptic Functions* (London, 1894), p. 70 n., reminds us that "the notation sg u, cg u, for sn $(u, 1)$, cn $(u, 1)$ is sometimes used, in honour of Gudermann."

[2] N. H. Abel, *Crelle's Journal*, Vol. IV (1829), p. 244.

[3] K. Weierstrass, *Mathematische Werke*, Vol. I, p. 5, 50; *Crelle's Journal*, Vol. LII (1856), p. 349.

[4] K. Weierstrass, *Crelle's Journal*, Vol. XLVII (1854), p. 291.

[5] K. Weierstrass, *op. cit.*, p. 300.

[6] K. Weierstrass, *Crelle's Journal*, Vol. LII (1856), p. 313.

[7] A. G. Greenhill, *Applications of Elliptic Functions* (London, 1892), p. 43; H. A. Schwarz, *Formeln und Lehrsätze zum Gebrauche der elliptischen Functionen. Nach Vorlesungen des Herrn K. Weierstrass* (Göttingen, 1885), p. 2.

[8] R. Fricke in *Encyclopädie der Math. Wissenschaften*, Band II, 2 (Leipzig, 1913), p. 268.

where the accent (') is to serve as a reminder that the value $w = 0$ is to be omitted.[1] Weierstrass obtains $\wp u = -\dfrac{d^2}{du^2} \log \sigma u$. The definitions of three Weierstrass functions $\sigma_1 u$, $\sigma_2 u$, $\sigma_3 u$ are given by Schwarz[2] and Fricke.

655. J. W. L. Glaisher introduced a notation for the reciprocals of elliptic functions by simply interchanging the order of letters; he designated[3] the reciprocals of cn u, sn u, dn u by nc u, ns u, nd u, respectively. He marked the quotient cn u/dn u by cd u. He proceeded similarly with other quotients. According to this plan, sn u/cn u would be sc u, but, according to Greenhill,[4] it is more commonly denoted by tanam u, abbreviated to tn u, while cn u/sn u or cs u was denoted by cotam u or ctn u. Glaisher marks the other four quotients thus: sn u/dn $u =$ sd u, cn u/dn $u =$ cd u, dn u/sn $u =$ ds u, dn u/cn $u =$ dc u. This notation has not met with wide adoption.

Glaisher says:[5] "The function $Z(x)$ is especially related to the group sn x, cn x, dn x, and the functions $Z_1(x)$, $Z_2(x)$, $Z_3(x)$ are similarly related to the other three groups ns x, ds x, cs x; dc x, nc x, sc x; cd x, sd x, nd x, respectively. It thus appears that a two-letter notation for the four zeta functions, in which they are distinguished from each other by the four final letters n, s, c, d, would be more representative than one in which the distinction is made by means of the suffixes 0, 1, 2, 3. I therefore denote the four zeta functions by zn x, zs x, zc x, zd x, so that zn $x = Z(x)$, zs $x = Z_1(x)$, zc $x = Z_2(x)$, zd $x = Z_3(x)$."

656. *Theta functions.*—Jacobi, in his *Fundamenta nova functionum ellipticarum* (1829), introduced not only the $Z(u)$ function in course of the representation of integrals of the "2. Gattung" in the form of Fourier series, but also the new transcendent $\Theta(u)$ which appears in the treatment of integrals of the "3. Gattung." Dirichlet in his memorial address remarked: "Strangely such an important function has as yet no other name than that of transcendent Θ, which is the chance designation by which it first appears in Jacobi, and mathematicians would only fulfill an obligation of thankfulness if they were

[1] H. S. Schwarz, *op. cit.*, p. 5; K. Weierstrass, *Mathematische Werke*, Vol. II (Berlin, 1895), p. 245–55.

[2] H. A. Schwarz, *op. cit.*, p. 21; Fricke, *op. cit.*, p. 279.

[3] J. W. L. Glaisher in *Messenger of Mathematics*, Vol. XI (London and Cambridge, 1882), p. 86.

[4] A. G. Greenhill, *op. cit.*, p. 17.

[5] J. W. L. Glaisher in *Messenger of Mathematics*. Vol. XVIII (1889), p. 2.

to unite, and attach to it the name of Jacobi."[1] Jacobi says in his *Fundamenta nova* (1829):[2] "Designabimus in posterum per characterem $\Theta(u)$ expressionem:

$$\Theta(u) = \Theta(0)e^{\int_0^u Z(u)du},$$

designante $\Theta(0)$ constantem, quam adhuc indeterminatam relinquimus, dum commodam eius determinationem infra obtinebimus." Jacobi expresses the elliptic functions by the functions Θ and H. The latter is called the "eta function." He writes,[3] for instance,

$$\sin \operatorname{am} \frac{2Kx}{\pi} = \frac{1}{\sqrt{k}} \cdot \frac{\mathrm{H}\left(\dfrac{2Kx}{\pi}\right)}{\Theta\left(\dfrac{2Kx}{\pi}\right)}.$$

Later Jacobi gave lectures (elaborated in 1838 by C. W. Borchardt) in which the theta and eta functions and series, which he had reached at the end of his *Fundamenta nova*, were made the starting-point of a new treatment of elliptic functions.[4] He uses now the following symbolism:

$$\vartheta \ (x) = 1 - 2q \cos 2x + 2q^4 \cos 4x - \ldots,$$
$$\vartheta_1(x) = 2\sqrt[4]{q} \sin x - 2\sqrt[4]{q^9} \sin 3x + \ldots,$$
$$\vartheta_2(x) = 2\sqrt[4]{q} \cos x + 2\sqrt[4]{q^9} \cos 3x + \ldots,$$
$$\vartheta_3(x) = 1 + 2q \cos 2x + 2q^4 \cos 4x + \ldots,$$

or, when necessary, the fuller designations $\vartheta(x, q)$, $\vartheta_1(x, q)$, $\vartheta_2(x, q)$, $\vartheta_3(x, q)$, where q lies between 0 and 1.

Unfortunately, the literature of theta functions shows marked diversity of notations.

657. Weierstrass[5] wrote ϑ_0 in place of ϑ, and defined $\vartheta_0(v) = 1 - 2h \cos 2v\pi + 2h^4 \cos 4v\pi - \ldots$; he defined $\vartheta_1(v) = 2h^{\frac{1}{4}} \sin v\pi - 2h^{\frac{9}{4}} \sin 3v\pi + \ldots$, and similarly for $\vartheta_2(v)$, $\vartheta_3(v)$.

When it seemed serviceable to indicate the two periods, $2\tilde{\omega}$ and $2\tilde{\omega}'$, of the functions, Weierstrass[6] wrote

$$\vartheta_\rho\left(v \left| \frac{\tilde{\omega}'}{\tilde{\omega}} \right.\right) = \vartheta_\rho(v \mid \tau), \ \rho = 1, 2, 3, 0 \ ;$$

---

[1] Leo Königsberger, *Carl Gustav Jacob Jacobi* (Leipzig, 1904), p. 98.

[2] C. G. J. Jacobi, *Gesammelte Werke*, Vol. I (Berlin, 1881), p. 198, 257.

[3] *Op. cit.*, p. 224.

[4] C. G. J. Jacobi, *op. cit.*, Vol. I, p. 501.

[5] H. A. Schwarz, *op. cit.*, p. 41.   [6] H. A. Schwarz, *op. cit.*, p. 42.

he employed also

$$\Theta_\rho(u) = \Theta_\rho(u \mid \tilde{\omega}, \omega') = 2^{2\tilde{\eta}\tilde{\omega}} \vartheta_\rho\!\left(v \,\Big|\, \frac{\tilde{\omega}'}{\omega}\right),\ u=2\tilde{\omega}v,\ \rho=1,\,2,\,3,\,0\ .$$

The notation of Weierstrass was followed by Halphen. Camille Jordan adopts the signs θ, θ_1, θ_2, θ_3 in the place of the Weierstrassian $\vartheta_1, \vartheta_2, \vartheta_0, \vartheta_3$, respectively. "Par le changement d'indices que nous avons pris la liberté de faire, on rétablit le parallélisme des notations entre ces fonctions et les ϕ correspondants."[1] Harkness and Morley, in their *Theory of Analytic Functions* (1898), page 242, represent Jordan's θ, θ_1, θ_2, θ_3 by ϑ, ϑ_1, ϑ_2, ϑ_3, respectively, so that Harkness and Morley's ϑ_3 is the same as the ϑ_3 of Weierstrass, and Harkness and Morley's ϑ, ϑ_1, ϑ_2 are equal to the Weierstrassian ϑ_1, ϑ_2, ϑ_0, respectively. Harkness and Morley remind[2] the reader that in their *Treatise on the Theory of Functions* they have a different notation from that in their *Analytic Functions*.

In 1858 Hermite[3] used a symbolism $\theta_{\mu,\,\nu}$ which, if $\omega=\tau$, yields the following equivalences with the signs of Weierstrass; $\theta_{0,\,0}(x)=\vartheta_3(x)$, $\theta_{0,\,1}(x)=\vartheta_0(x)$, $\theta_{1,\,0}(x)=\vartheta_2(x)$, $\theta_{1,\,1}(x)=i\vartheta_1(x)$.

G. H. Cresse[4] prepared two tables exhibiting the various notations in theta functions followed by authors whom he cites in the sixth chapter of the third volume of Dickson's *History*. A most exhaustive statement of notations for theta functions is given by Brill and Noether.[5]

The sign $\theta(z)$ has also been used to represent what Poincaré[6] called the "thetafuchsian series and functions"; A. R. Forsyth[7] marks these functions by a capital letter $\Theta(z)$.

658. In the development of general theta functions involving p variables (arguments) the notation $\vartheta(v_1 \mid v_2 \ldots \mid v_p)$ has been used; when the mark or characteristic $\left[{}^g_h\right]$ is introduced, and the arguments differ from each other only in their subscripts, the subscripts are

[1] Camille Jordan, *Cours d'analyse*, Vol. II (Paris, 1894), p. 411.

[2] Harkness and Morley, *Analytic Functions* (1898), p. 242.

[3] C. Hermite in *Journal de mathématiques* (Liouville; 2d ser.), Vol. III (1858), p. 26; H. A. Schwarz, *op. cit.*, p. 42.

[4] L. E. Dickson, *History of the Theory of Numbers*, Vol. III (Washington, 1923), chap. vi (written by G. H. Cresse), p. 93, 94.

[5] A. Brill and M. Noether, "Entwicklung d. Theorie d. alg. Functionen," *Jahresbericht d. d. Mathematiker Vereinigung*, Vol. III (Berlin, 1894), p. 500–520.

[6] H. Poincaré in *Acta mathematica*, Vol. I (Stockholm, 1882), p. 207, 208, 210.

[7] A. R. Forsyth, *Theory of Functions of a Complex Variable* (Cambridge, 1893), p. 642.

omitted and double parentheses are introduced so that the functions are represented[1] by the symbolism $\vartheta[^g_h]((v))$, or in case of a theta function of the nth order, by the symbolism $\Theta[^g_h]((v))$. Harkness and Morley[2] mark the p-tuple theta functions by $\theta(v_1, v_2, \ldots, v_p)$.

659. *Zeta functions.*—Reference has been made to Jacobi's zeta function, marked by the capital Greek letter Z (§ 653). The small letter zeta has been used to stand for the integral $\zeta(u) = -\int \wp(u)du$ by various writers.[3]

660. This same small letter ζ was used by B. Riemann[4] as early as 1857 to represent a function on which he based his analysis of prime numbers and which he introduced thus: "In this research there serves as point of departure the remark of Euler that the product

$$\Pi \frac{1}{1-\dfrac{1}{p^s}} = \Sigma \frac{1}{n^s} \ ,$$

if there are taken for p all prime numbers, for n all integral numbers. The function of the complex variable s which is represented by these two expressions, as long as they converge, I designate by $\zeta(s)$." This notation has maintained its place in number theory.

661. *Power series.*—Weierstrass[5] denotes by a German capital letter $\mathfrak{P}$ a power series; he marks a "Potenzreihe von x" by $\mathfrak{P}(x)$ and by $\mathfrak{P}(x-a)$, when a is the center of the circle of convergence. More commonly the Latin capital P is used. Harkness and Morley[6] remark: "The usual notation for such a series is $P(z)$."

662. *Laplace, Lamé, and Bessel functions.*—What are called Laplace's coefficients and functions was first worked out by Legendre and then more fully by Laplace.[7] If $(1-2ax+x^2)^{-\frac{1}{2}}$ is expanded

[1] A. Krazer and W. Wirtinger in *Encyklopädie der Math. Wissensch.*, Vol. II (1921), p. 637, 640, 641.

[2] Harkness and Morley, *Theory of Functions* (1893), p. 459. On p. 460 and 461 they refer to differences in notation between Riemann, and Clebsch and Gordan, with regard to the lower limits for the integrals there involved.

[3] A. R. Forsyth, *Theory of Functions of a Complex Variable* (Cambridge, 1893), p. 250, 251; R. Fricke in *Encyklopädie d. math. Wissensch.*, Vol. II, 2 (Leipzig, 1913), p. 258; R. Fricke, *Elliptische Funktionen* (1916), p. 168; H. Burkhardt, *Funktionen-theoretische Vorlesungen*, Vol. II (Leipzig, 1906), p. 49.

[4] B. Riemann, *Gesammelte Werke* (Leipzig, 1876), p. 136.

[5] K. Weierstrass, *Mathem. Werke* (Berlin), Vol. II (1895), p. 77.

[6] J. Harkness and Morley, *Theory of Functions* (New York, 1893), p. 86.

[7] *Mémoires de math. et phys., présentés à l'académie r. d. sciences par divers savans*, Vol. X (Paris, 1785).

in a series of ascending powers of a, the coefficient of a^n is a function of x, often called "Legendre's coefficient." Dirichlet[1] represents it by P_n, E. Heine[2] by $P^{(n)}$, or, when no confusion with exponents is possible, simply by P^n, while Todhunter[3] denotes it by P_n or $P_n(x)$, and says, "French writers very commonly use X_n for the same thing." When for x there is substituted the value $\cos \gamma = \cos \theta \cos \theta_1 + \sin \theta \sin \theta_1 \cdot \cos (\phi - \phi_1)$, where both θ and ϕ are regarded as variables, while θ_1 and ϕ_1 are constants, then the coefficient is called "Laplace's coefficient" and is marked Y_n by Todhunter.[4] Laplace's coefficients are particular cases of Laplace's functions which satisfy a certain partial-differential equation, and are marked by Todhunter[5] X_n or Z_n, n denoting the order of the function. Heine[6] marks them $P^n (\cos \gamma)$.

663. Analogous to Laplace's functions are those of Lamé which Heine[7] marks $E_s(\mu)$. Lamé's functions must satisfy a linear-differential equation of the second order.

664. Bessel[8] defined the function now known by his name by the following definite integral:

$$\text{"} \int \text{Cos } (h\epsilon - k \text{ Sin } \epsilon) \ d\epsilon = 2\pi I_k^h,\text{"}$$

where h is an integer and the limits of integration are 0 and 2π. His I_k^h is the same as the modern $J_h(k)$, or rather $J_n(x)$. O. Schlömilch,[9] following P. A. Hansen,[10] explained the notation $J_{\lambda, \pm n}$ where λ signifies the argument and $\pm$ the index of the function. Schlömilch usually omits the argument. Watson[11] points out that Hansen and Schlömilch express by $J_{\lambda, n}$ what now is expressed by $J_n(2\lambda)$. Schläfli[12] marked it $\overset{n}{J}(x)$. Todhunter[13] uses the sign $J_n(x)$. $J_n(x)$ is known as the "Bessel

[1] L. Dirichlet in *Crelle's Journal*, Vol. XVII (1837), p. 35, 39.

[2] E. Heine, *Kugelfunctionen* (2d ed.; Berlin, 1878), p. 3.

[3] I. Todhunter, *Laplace's Functions, Lamé's Functions, and Bessel's Functions* (London, 1875), p. 3.

[4] I. Todhunter, *op. cit.*, p. 131.

[5] I. Todhunter, *op. cit.*, p. 152.

[6] E. Heine, *op. cit.*, p. 302.

[7] E. Heine, *op. cit.*, p. 358.

[8] F. W. Bessel in *Abhandlungen Akademie der Wissensch. (Math. Kl.) 1824* (Berlin, 1826), p. 22, 41.

[9] O. Schlömilch, *Zeitschr. d. Math. u. Phys.*, Vol. II (1857), p. 137.

[10] P. A. Hansen, *Ermittelung der absoluten Störungen*, 1. Theil (Gotha, 1843).

[11] G. N. Watson, *Theory of Bessel Functions* (Cambridge, 1922), p. 14.

[12] L. Schläfli, *Mathemat. Annalen*, Vol. III (1871), p. 135

[13] I. Todhunter, *op. cit.*, p. 285.

function of the first kind of order n," while $Y^n(x)$, an allied function introduced in 1867 by Karl Neumann,[1] is sometimes called "Newmann's Bessel function of the second kind of order n." It[2] is sometimes marked $Y_n(x)$.

Watson says: "Functions of the types $J \pm (n+\frac{1}{2})^{(z)}$ occur with such frequency in various branches of Mathematical Physics that various writers have found it desirable to denote them by a special functional symbol. Unfortunately no common notation has been agreed upon and none of the many existing notations can be said to predominate over the others." He proceeds to give a summary of the various notations.[3]

In his *Theory of Bessel Functions*, pages 789, 790, Watson gives a list of 183 symbols used by him as special signs pertaining to that subject.

665. *Logarithm integral, cosine integral, etc.*—Soldner[4] uses the symbol li. a, from the initial letters of *logarithme integral*, to stand for $\int_0^a dx : \log x$. This notation has been retained by Bretschneider,[5] De Morgan,[6] and others.

As regards the sine integral and cosine integral, first considered by Mascheroni, Bessel[7] in 1812 still writes the expressions $\int \dfrac{dx}{x} \sin x$ and $\int \dfrac{dx}{x} \cos x$ without special signs to represent them. Bretschneider[8] marked the cosine integral by "ci x," and the sine-integral by "si x." He denoted the hyperbolic functions $\int \mathrm{Cos}\, x \cdot \dfrac{dx}{x}$ by "$\mathfrak{CS}$" and $\int \mathrm{Sin}\, x \cdot \dfrac{dx}{x}$ by "$\mathfrak{SS}$." Later still Schlömilch[9] denotes $\int_0^1 \dfrac{\sin w\theta}{\theta}\, db$

---

[1] Karl Neumann, *Theorie der Bessel'schen Funktionen* (Leipzig, 1867), p. 41, 52; A. Gray and G. B. Mathews, *Treatise on Bessel Functions* (2d ed.; London, 1922), p. 15; G. N. Watson, *op. cit.*, p. 67.

[2] Gray and Mathews, *op. cit.* (1922), p. 15.

[3] G. N. Watson, *op. cit.* (1922), p. 55, 56.

[4] J. Soldner, *Théorie et tables d'une nouvelle fonction transcendante* (Munich, 1809); see also F. W. Bessel, *Abhandlungen*, Vol. II (Leipzig, 1876), p. 331.

[5] C. A. Bretschneider in *Crelle's Journal*, Vol. XVII (Berlin, 1837), p. 257.

[6] A. de Morgan, *Differential and Integral Calculus* (London, 1842), p. 660.

[7] F. W. Bessel, *Abhandlungen*, Vol. II (Leipzig, 1876), p. 341.

[8] C. A. Bretschneider, *Grunert's Archiv*, Vol. III (1843), p. 30.

[9] O. Schlömilch in *Crelle's Journal*, Vol. XXXIII (1846), p. 318.

by Si(w), and writes similarly Ci(w) and the exponential integral Ei(w), where Ei(w) = li(e^w). This notation was used by J. W. L. Glaisher[1] in 1870, except that he drops the parentheses and writes simply Si x, etc.

666. Niels Nielsen in 1904[2] marked the sine integral and cosine integral by "S$_i$(x)" and "C$_i$(x)," and in 1906[3] by "ci(x)" and "si(x)." Closely allied with these transcendents are the integrals of Kramp,[4] $\int_0^x e^{-x^2}dx$, which Nielsen in 1904 represents by K(x) and in 1906 by L(x), and the integrals of Fresnel,[5] $\int dz \cos (a-z)^2$ and $\int dz \sin (az^2)$, for which he used no sign and which are variously represented in later publications. Thus Nielsen[6] writes $F_1(x)$ and $F_2(x)$; Gray and Mathews[7] use C and S as does also Nielsen[8] in 1906.

In the Alphabetical Index of Nielsen's *Cylinderfunktionen* (1904), under the word "Funktion," the reader will find a long list of signs of functions which Nielsen used in that work.

SYMBOLS IN MATHEMATICAL LOGIC

667. *Some early symbols.*—Leibniz is generally regarded as the earliest outstanding worker in the field of symbolic logic. But before him certain logical symbols were introduced by Hérigone, in his *Cursus mathematicus* (Paris, 1634, 1644), and by Rahn, in his *Teutsche Algebra* (1659). Hérigone says in his Preface: "I have invented a new method of giving demonstrations, brief and intelligible, without the use of any language. . . . . The demonstration is carried on from beginning to end by a continuous series of legitimate consequences, necessary and immediate, each contained in a short line, and they can be readily resolved into syllogisms. . . . ." Hérigone employs "hyp." for "from the hypothesis it follows"; "constr." for "from the construction one has"; "concl." for "conclusion"; "arbitr." for "taking arbitrarily"; "20.1" for "from proposition 20 of the first book it follows." Most

[1] *Philosophical Transactions* (London, 1870), p. 367.

[2] N. Nielsen, *Handbuch der Cylinderfunktionen* (Leipzig, 1904), p. 71, 72.

[3] N. Nielsen, *Theorie des Integrallogarithmus* (Leipzig, 1906), p. 6.

[4] C. Kramp in *Mémoires d. l'acad. d. sciences*, Vol. V (Paris, 1818), p. 434. Reference taken from N. Nielsen, *Cylinderfunktionen* (1904), p. 72.

[5] A. J. Fresnel, *Analyse des réfractions astronomiques et terrestres* (1798). Also *Œuvres*, Vol. I (Paris, 1866), p. 176, 177.

[6] N. Nielsen, *Cylinderfunktionen* (1904), p. 72.

[7] A. Gray and G. B. Mathews, *op. cit.* (1922), p. 219.

[8] N. Nielsen, *Integrallogarithmus* (1906), p. 8.

of these abbreviations are hardly new with Hérigone; they were used more or less before him, in editions of Euclid. Included in his scheme of symbolic writing are his ideographic symbols for algebra and geometry which we noted in § 189.[1]

The symbols of Rahn have been listed in § 194. Logical, as distinct from operational, are his symbols for "therefore" and for "impossible." The wide use of the sign for "therefore" in English and American books impels us to trace its history more fully.

668. *The sign for "therefore."*—In Rahn's *Teutsche Algebra* (Zurich, 1659) one finds both ∴ and ∵ for "therefore," the ∴ predominating. In the English translation of 1668 both signs are found, but ∵ predominates. On page 37 of the translation one reads: "Which Deduction is hereafter noted with [∵] that is as much as [*ergo*.] only [∵] is set in the work, [*ergo*] in the Margin." It should be noted that with Rahn and with many writers of the eighteenth century the three dots were used especially in connection with the process of finding the products of means and extremes of a proportion. Thus, Thomas Walter[2] says: "∵ Therefore; signifying the product of the two extremes is equal to that of the means." William Jones[3] writes the sign :· and uses it in operating with proportions and also equations.

The ·· as signifying "therefore" occurs in Oughtred's *Opuscula mathematica* (1677) (§ 181), in texts by Alexander[4] and Ward,[5] Ronayne,[6] and Birks.[7] Both forms ∵ and ∴ are found in Jones, Woodhouse,[8] and Simpson.[9] Stone[10] gives the form :·. During the eighteenth century ∵ was used for "therefore" about as often as was ∴ .

669. *The sign for "because."*—We have not been able to find the use of ∵ for "because" in the eighteenth century. This usage seems to have been introduced in Great Britain and the United States in the nineteenth century. It is found in the *Gentleman's Mathematical*

[1] See also Gino Loria's article, "La logique mathématique avant Leibniz," *Bull. d. scienc. math.* (G. Darboux; 2d ser.), Vol. XVIII (Paris, 1894), p. 107–12.

[2] Thomas Walter, *A New Mathematical Dictionary* (London, [n.d.]), p. 9.

[3] William Jones, Synopsis (London, 1706), p. 143, 145.

[4] John Alexander, *Synopsis algebraica* (London, 1693), p. 7.

[5] John Ward, *Young Mathematician's Guide* (1707), p. 189, 371, etc.

[6] Philip Ronayne, *Treatise of Algebra* (London, 1727), p. 3.

[7] John Birks, *Arithmetical Collections* (London, 1766), p. 482, etc.

[8] Robert Woodhouse, *Principles of Analytical Calculation* (Cambridge, 1803).

[9] Thomas Simpson, *Treatise of Fluxions* (London, 1737), p. 95.

[10] E. Stone, *New Mathematical Dictionary* (London, 1726, 1743), art. "Series."

Companion (1805).[1] It did not meet with as wide acceptance in Great Britain and America as did the sign ∴ for "therefore."

The signs ∴ for "therefore" and ∵ for "because" are both found in the *Elements* of Euclid edited by members of the University of Cambridge (1827), in Wright,[2] Nixon,[3] C. Smith,[4] Buddon,[5] Cockshott and Walters,[6] White,[7] Shutts,[8] Young and Schwartz,[9] Auerbach and Walsh,[10] and Macnie.[11] In mathematical publications on the European Continent the two symbols are of very rare occurrence, except in papers on symbolic logic.[12] The sign ∵ for "therefore" is used by T. Svedberg[13] in 1907.

670. *The program of Leibniz.*—The contribution of Leibniz to symbolic logic was a program, rather than actual accomplishment. He set up an ideal for others to approach. He came to regard logic "like a universal mathematics."[14] He advocated "a universal language" or "universal characteristic," or a "universal calculus." "The true method should furnish us with an Ariadne's thread, that is to say, with a certain sensible and palpable medium, which will guide the mind as do the lines drawn in geometry and the formulas for operations, which are laid down for the learner in arithmetic."[15] We quoted an extremely optimistic passage from Leibniz in § 530. His earliest publication, the *De arte combinatoria* (1666), printed when

[1] See *Mathematical Gazette*, Vol. XI (London, 1923), p. 275.

[2] J. M. F. Wright, *Euclid* (Cambridge, 1829).

[3] R. C. J. Nixon, *Geometry of Space* (Oxford, 1888), p. 25.

[4] C. Smith, *Elements of Algebra* (ed. Irving Stringham; New York, 1895), p. 18.

[5] E. Budden, *Elementary Pure Geometry* (London, 1904), p. 22.

[6] A. Cockshott and F. B. Walters, *Treatise on Geometrical Conics* (London, 1881).

[7] E. E. White, *Elements of Geometry* (New York, [1895]).

[8] George C. Shutts, *Plane and Solid Geometry* [1912], p. 13.

[9] John W. Young and Albert J. Schwartz, *Plane Geometry* (New York, [1915]).

[10] Matilda Auerbach and Charles B. Walsh, *Plane Geometry* (Philadelphia, [1920]), p. xi.

[11] John Macnie, *Elements of Geometry* (ed. E. E. White; New York, 1895), p. 10.

[12] Both ∴ ("therefore") and ∵ ("because") are found in Platon Poretsky's *Théorie des non-égalités* (Kazan, 1904), p. 22, 23.

[13] *Nova acta regiae societatis scientiarum Upsaliensis* (4th ser.), Vol. II, No. 1, p. 114.

[14] G. W. Leibniz, *Initia et Specimina scientiae novae generalis*, in C. I. Gerhardt's *Philosophische Schriften von Leibniz*, Vol. VII (Berlin, 1890), p. 65, also p. 59.

[15] Letters of 1677 or 1678 to Galloys, C. I. Gerhardt's *Philosophische Schriften von Leibniz*, Vol. VII (Berlin, 1890), p. 22.

he was only twenty years of age, is his first attempt in this direction. Leibniz himself regarded it later as of little value. The task which he outlined seemed really beyond the powers of one man. Some manuscripts of his contain later developments. The symbols he used are listed in § 538–64. Above we quoted from his *Initia et specimina scientiae novae generalis*. Recently, additional Leibnizian manuscripts on this topic have been found in the Hanover Library.[1] In a manuscript dated April, 1679,[2] Leibniz experiments on the use of number and on mathematical operations in matters of logic. He proposes that primitive concepts be symbolized by prime number and the combinations of two concepts by their product. "For example, if it be assumed that the term 'animal' be expressed by some number 2 (or more generally by a) and the term 'rational' by 3 (or more generally by r), then the term 'man' is expressed by the number 2.3, that is 6, or by the product of the multiplication of 2 and 3 (or more generally by the number ar)." The proposition—all S is P—will be true if the number which contains the concept S is exactly divisible by that which represents the concept P. Accordingly, a universal affirmative proposition may be symbolized by $\frac{S}{P}$ yielding an integral quotient without remainder. For a particular affirmative proposition it suffices that either $\frac{S}{P}$ or $\frac{P}{S}$ be an integer. For a universal negative it suffices that neither $\frac{S}{P}$ nor $\frac{P}{S}$ be an integer. Certain difficulties of this scheme led Leibniz to experiment on the use of negative numbers, but he finally abandoned the plan of representing concepts by numbers.

Later fragments are included in plans for an encyclopedia[3] on science to be developed in terms of a universal characteristic, dealing with the equivalence of concepts. C. I. Lewis says:[4] "In these fragments, the relations of equivalence, inclusion, and qualification of one concept by another, or combination, are defined and used." As

---

[1] These fragments are contained in Louis Couturat, *Opuscules et fragments inédits de Leibniz; extraits des manuscrits de la Bibliothèque royale de Hanovre* (Paris, 1903). Two fragments from Leibniz are translated into English in C. I. Lewis' *Survey of Symbolic Logic* (Berkeley: University of California Press, 1918), p. 373–87.

[2] L. Couturat, *Opuscules, etc.* (1903), p. 42, 43.

[3] Gerhardt, *Philosoph. Schriften von Leibniz*, Vol. VII (1890), XVI–XX, p. 208–47.

[4] C. I. Lewis, *op. cit.*, p. 13.

the expositions of Leibniz deal with questions of logical theory without the use of new symbols, we must pass them by.

In another place[1] Leibniz formulates the symbolism for the four propositions (using the Cartesian ∞ for equality):

1. All A is B, that is $AB \infty A$;
2. Some A is not B, that is AB not ∞A;
3. No A is B, that is AB does not exist, or AB not ∞ AB *Ens;*
4. Some A is B, that is AB exists, or $AB \infty AB$ *Ens.*

In (3) and (4) the AB represents the *possible* AB's or the AB "in the region of ideas," while the AB *Ens* represents the existing AB's, or actual members of the class AB. Here and in other fragments of Leibniz no new symbols are used, but different interpretations are given to old symbols.[2]

The breadth of view which characterized these plans precluded the possibility of their complete execution by Leibniz without the help of others. He invited the co-operation of others, but without effective response. His inability to secure promising pupils is voiced in a letter[3] of 1714: "I have spoken of my Spécieuse Générale to Mr. le Marquis de l'Hospital and to others, but they have given it no more attention than if I had related a dream. It seemed that I must back it up by some palpable example, for this purpose one would have to fabricate at least a part of my characteristic, which is not easy, especially at my age and without conversing with persons who could stimulate and assist me in researches of this sort."

A motive of disappointment at his failure to carry his plans to completion appears in another letter, written two years before his death:[4] "If I had been less distracted, or if I were younger, or assisted by well disposed young men, I could hope to give a sample of . . . . this language or universal writing, infinitely different from all hitherto projected, because the very characters . . . . would guide the reason; and errors, except those of fact, would be only errors of computation. It is very difficult to form or invent such a language or characteristic, but very easy to learn without a dictionary."

671. *The signs of J. H. Lambert.*—After Leibniz various attempts were made in the eighteenth century to develop a calculus of logic

[1] Gerhardt, *Philosoph. Schriften von Leibniz*, Vol. VII (1890), p. 214, 215. See also Lewis, *op. cit.*, p. 15.

[2] Confer C. I. Lewis, *op. cit.*, p. 5–18.

[3] *Leibnitii Opera philosophica* (ed. J. E. Erdmann; Berlin, 1840), p. 703.

[4] Letter in *Leibnitii Opera philosophica* (Erdmann, 1840), p. 701.

by J. A. Segner, Jakob Bernoulli, Gottfried Ploucquet, I. H. Tönnies, J. H. Lambert, G. J. von Holland, G. F. Castillon, and others. Peano[1] mentions also Ludovico Richeri[2] who in 1761 introduced a symbolism, for instance, $\cup$ and $\cap$, for "all" and "none." We give the symbols employed by J. H. Lambert:[3]

Gleichgültigkeit, equality	$=$	Particularity	$<$
Zusetzung, addition	$+$	Copula	$\sim$
Absonderung, abstraction	$-$	*Begriffe*, given concepts	a, b, c, d, etc.
Des Gegentheils, opposition	$\times$	Undeterminal	
Universality	$>$	concepts	n, m, l, etc.
		Unknowns	x, y, z
		The genus	γ
		The difference	δ

Lambert deals with concepts, not classes. He uses the sign $a|b$, sometimes also $a:b$, to indicate that part of concept a which is different from concept b. In syllogisms the proposition "all A is B" has two cases: (1) $A=B$ in which the converse is $B>A$; (2) $A>B$, in which the converse is $B<A$. The proposition "some A is B" also has two cases: (1) $A<B$, in which the converse is $B>A$; (2) $A<B$, in which the converse is $B<A$. Lambert translates "some A" by mA. Hence $A>B$ is equivalent to $A=mB$; $A<B$ is equivalent to $nA=B$; $mA>B$ and $A<nB$ are together equivalent to $mA=kB$ and $lA=nB$, or equivalent to $pA=qB$. For "no A is B" Lambert writes $\frac{A}{m}=\frac{B}{n}$; by $\frac{A}{m}$ he means that the peculiar properties of the subject be taken away; by $\frac{B}{n}$ he means that the peculiar properties of the predicate be taken away. Lambert transforms fractions, though he carefully refrains from doing so in cases like $\frac{A}{m}=\frac{B}{n}$, which are supposed to represent universal negatives. Lambert's treatment of the negative is a failure.[4] Lambert represents non-logical or "metaphysical" relations

[1] G. Peano in *Rivista di matematica*, Vol. IV (1894), p. 120.

[2] L. Richeri in *Miscellanea Taurinensia*, Vol. II (1761), Part 3, p. 46–63.

[3] Johann Heinrich Lambert, "Sechs Versuche einer Zeichenkunst in der Vernunftlehre," *Logische und philosophische Abhandlungen* (ed. Joh. Bernoulli), Vol. I (Berlin, 1782). For a fuller account of Lambert see C. I. Lewis, *op. cit.*, p. 19–29.

[4] C. I. Lewis, *op. cit.*, p. 26.

by Greek letters. If $f=$ fire, $h=$ heat, $a=$ cause, then he writes $f=a::h$, where $::$ represents a relation which behaves like multiplication. From this equality he derives $\frac{f}{h}=\frac{a}{\cdot}$, i.e., fire is to heat as cause is to effect. The dot represents here *Wirkung* ("effect").[1]

672. *Signs of Holland.*—G. J. von Holland in a letter[2] to Lambert points out a weakness in Lambert's algorithm by the following example: (1) All triangles are figures, $T=tF$; (2) all quadrangles are figures, $Q=qF$; (3) whence $F=\frac{T}{t}=\frac{Q}{q}$ or $qT=tQ$, which is absurd. In his own calculus Holland lets S represent the subject, P the predicate, p, π undetermined positive variable numbers. He interprets $\frac{S}{p}=\frac{P}{\pi}$ as meaning a part of S is a part of P. If $p=1$, then S/p is as many as all S. The sign $\frac{1}{\infty}$ is the same as 0; $\frac{S}{1}=\frac{P}{\infty}$ means "all S is not P." Holland's calculus does not permit his equations to be cleared of fractions.

673. *Signs of Castillon.*—The next writer on a calculus of logic was G. F. Castillon[3] who proposed a new algorithm. He lets S, A, etc., represent concepts in intension, M an indeterminate, $S+M$ the synthesis of S and M, $S-M$ the withdrawal of M from S, so that $S-M$ is a genus concept in which S is subsumed, $S+M$ a species concept. Accordingly, "all S is A" is represented by $S=A+M$; "no S is A" is represented by $S=-A+M=(-A)+M$, that is, withdrawing A from M is "no A." The real particular affirmative "some S is A" is expressed by $A=S-M$ and is the converse of $S=A+M$; the illusory particular affirmative is denoted by $S=A \mp M$ and means that in "some S is A" some S is got from A by the abstraction $S=A-M$ when in reality it is A that is drawn from S by the abstraction $S=A+M$. Hence this judgment puts $-M$ where there should be $+M$; this error is marked by $S=A \mp M$. Castillon's notation works out fairly well, but its defect lies in the circumstance that the sign $-$ is used both for abstraction and for negation.

674. *Signs of Gergonne.*—A theory of the meccanisme du raisonnement was offered by J. D. Gergonne in an *Essai de dialectique ration-*

[1] Our account of Lambert is taken from C. I. Lewis, *op. cit.*, p. 18–29.

[2] *Johann Lambert's deutscher Gelehrten Briefwechsel* (herausgegeben von J. Bernoulli), Brief III, p. 16 ff.; C. I. Lewis, *op. cit.*, p. 29, 30.

[3] G. F. Castillon, "Sur un nouvel algorithme logique," *Mémoires de l'académie des sciences de Berlin, 1803, Classe de philos. speculat.*, p. 1–14.

nelle;[1] there the symbol H stands for complete logical disjunction, X for logical product, I for "identity," C for "contains," and $\supset$ for "is contained in."

675. *Signs of Bolyai.*—Wolfgang Bolyai used logical symbols in his mathematical treatise, the *Tentamen,*[2] which were new in mathematical works of that time. Thus $A \doteq B$ signified A absolutely equal to B; $A \overline{=} B$ signified A equal to B with respect to content; $A(=B$ or $B=)A$ signified that each value of A is equal to some value of B; $A(=)B$ signified that each value of A is equal to some value of B, and vice versa.

676. *Signs of Bentham.*—The earlier studies of logic in England hardly belong to symbolic logic. George Bentham[3] in 1827 introduces a few symbols. He lets $=$ stand for identity, $\|$ for diversity, t for *in toto* (i.e., universality), p for "partiality." Accordingly, "$tX = tY$" means X *in toto* $= Y$ *in toto;* "$tX = pY$" means X *in toto* $= Y$ *ex parte;*

$$\text{"}tX \parallel \textit{,}Y\text{" means } X \textit{ in toto } \parallel Y \begin{cases} \textit{in toto} \\ \text{or} \\ \textit{ex parte} \end{cases}.$$

677. *Signs of A. de Morgan.*—The earliest important research in symbolic logic in Great Britain is that of *De Morgan.*[4] He had not seen the publications of Lambert, nor the paper of J. D. Gergonne,[5] when his paper of 1846 was published. In 1831 De Morgan[6] used squares, circles, and triangles to represent terms. He does this also in his *Formal Logic*[7] but in 1831 he did not know[8] that Euler[9] had

[1] J. D. Gergonne in *Annales de mathématiques pures et appliquées*, Vol. VII (Nismes, 1816–17), p. 189–228. Our information is drawn from G. Vacca's article in *Revue de mathématiques*, Vol. VI (Turin, 1896–99), p. 184.

[2] Wolfgangi Bolyai de Bolya, *Tentamen . . . . in elementa matheseos* (2d ed.), Vol. I (Budapest, 1897), p. xi.

[3] George Bentham, *Outline of a New System of Logic* (London, 1827), p. 133.

[4] Augustus de Morgan, *Formal Logic* (London, 1847); five papers in the *Transactions of the Cambridge Philosophical Society*, Vol. VIII (1846), p. 379–408; Vol. IX (1850), p. 79–127; Vol. X (1858), p. 173–230; Vol. X (1860), p. 331–58; Vol. X (1863), p. 428–87.

[5] J. D. Gergonne, "Essai de dialectique rationelle," *Annales de mathématique.* (Nismes, 1816, 1817).

[6] De Morgan, "The Study and Difficulties of Mathematics," *Library of Useful Knowledge* (1831), p. 71–73.

[7] De Morgan, *Formal Logic* (1847), p. 8, 9.

[8] *Ibid.*, p. 323, 324.

[9] L. Euler, *Lettres à une Princesse d'Allemagne sur quelques sujets de Physique et de Philosophie* (Petersburg, 1768–72), Lettre CV.

used circles in the same manner. In his *Formal Logic*, De Morgan uses $X)$ or $(X$ to indicate distribution, and $X($ or $)X$ no distribution. Expressing in his own language (p. 60): "Let the following abbreviations be employed:

$X)Y$ means 'every X is Y'; $X.Y$ means 'no X is Y'

$X:Y$ means 'some Xs are not Ys'; XY means 'some Xs are Ys'."

He lets (p. 60) A stand for the universal affirmative, I for the particular affirmative, E for the universal negative, and O for the particular negative. He lets x, y, z, be the negatives or "contrary names" of the terms X, Y, Z. The four forms A_1, E_1, I_1, O_1 are used when choice is made out of X, Y, Z; the forms A', E', I', O' are used whence choice is made out of x, y, z. On page 61 he writes identities, one of which is "$A_1X)Y = X.y = y)x$," that is, every A_1X is Y = no X is y = every y is x. On page 115: "P,Q,R, being certain names, if we wish to give a name to everything which is all three, we may join them thus, PQR: if we wish to give a name to everything which is either of the three (one or more of them) we may write P,Q,R: if we want to signify anything that is either both P and Q, or R, we have PQ,R. The contrary of PQR is p,q,r; that of P,Q,R is pqr; that of PQ,R is $(p,q)r$."

In his *Syllabus*,[1] De Morgan uses some other symbols as follows:

"1. $X)\circ)Y$ or both $X))Y$ and $X).)Y$.
    All Xs   and some things besides are Ys.

 2. $X||Y$ or both $X))Y$ and $X((Y$.
    All Xs   are Ys, and all Ys are Xs. . . . .

 5. $X|\cdot|Y$ or both $X)\cdot(Y$ and $X(\cdot)Y$.
    Nothing both X and Y and everything one or the other."

De Morgan also takes L^{-1} as the converse of L.

The following quotation from De Morgan is interesting: "I end with a word on the new symbols which I have employed. Most writers on logic strongly object to all symbols, except the venerable *Barbara, Celarent*, etc., ... I should advise the reader not to make up his mind on this point until he has well weighed two facts which nobody disputes, both separately and in connexion. First, logic is the only science which has made no progress since the revival of letters; secondly, logic is the only science which has produced no growth of symbols."[2]

[1] A. de Morgan, *Syllabus of a Proposed System of Logic* (London, 1860), p. 22; C. I. Lewis, *op. cit.*, p. 41, 42.

[2] A. de Morgan, *op. cit.*, p. 72; see *Monist*, Vol. XXV (1915), p. 636.

678. *Signs of G. Boole.*—On the same day of the year 1847 on which De Morgan's *Formal Logic* was published appeared Boole's *Mathematical Analysis of Logic*.[1] But the more authoritative statement of Boole's system appeared later in his *Laws of Thought* (1854). He introduced mathematical operations into logic in a manner more general and systematic than any of his predecessors. Boole uses the symbols of ordinary algebra, but gives them different significations. In his *Laws of Thought* he employs signs as follows:[2]

"1st. Literal symbols, as x, y, etc., representing things as subjects of our conceptions.

2nd. Signs of operation, as $+$, $-$, $\times$, standing for those operations of the mind by which the conceptions of things are combined or resolved so as to form new conceptions involving the same elements.

3rd. The sign of identity, $=$.

And these symbols of Logic are in their use subject to definite laws, partly agreeing with and partly differing from the laws of the corresponding symbols in the science of Algebra."

The words "and," "or" are analogous to the sign $+$, respectively. If x represent "men," y "women," z "European," one has $z(x+y) = zx+zy$, "European men and women" being the same as "European men and European women" (p. 33). Also, $x^2 = x$ for "to say 'good, good,' in relation to any subject, though a cumbrous and useless pleonasm, is the same as to say 'good' " (p. 32). The equation $x^2 = x$ has no other roots than 0 and 1. "Let us conceive, then, of an Algebra in which the symbols x, y, z, etc., admit indifferently of the values 0 and 1, and of these values alone" (p. 37).

"If x represent any class of objects, then will $1-x$ represent the contrary or supplementary class of objects" (p. 48). The equation $x(1-x) = 0$ represents the "principle of contradiction" (p. 49). "The operation of division cannot be *performed* with the symbols with which we are now engaged. Our resource, then, is to *express* the operation, and develop the results by the method of the preceding chapter (p. 89). We may properly term $\frac{0}{0}$ an *indefinite class symbol*, ... *some*, *none*, or *all* of its members are to be taken" (p. 90, 92). A coefficient $\frac{1}{0}$ shows that the constituent to which it belongs must be equated to zero (p. 91). C. I. Lewis remarks that Boole's methods of solution sometimes involve an uninterpretable stage.[3]

[1] C. I. Lewis, *op. cit.*, p. 52.

[2] George Boole, *Laws of Thought* (London and Cambridge, 1854), p. 27.

[3] C. I. Lewis, *op. cit.*, p. 57.

679. *Signs of Jevons*.—Jevons[1] endeavored to simplify Boole's algebra by discarding what has no obvious interpretation in logic. But Jevons, in the opinion of C. I. Lewis,[2] "unduly restricts the operations and methods of Boole." Jevons discards the inverse operations $A-B$ and A/B, and he interprets the sum of A and B as "either A or B, where A and B are not necessarily exclusive classes." In 1864 he uses the notation $A+B$, later[3] he uses $A \cdot | \cdot B$. Thus $A+B=A$ with Jevons, while with Boole $A+B$, is not interpretable as any relation of logical classes.

680. *Remarks of Macfarlane*.—In continuation of the researches of Boole, De Morgan, and Jevons was Alexander Macfarlane's *Principles of the Algebra of Logic* (Edinburgh, 1879), from which (p. 32) we quote: "The reason why Formal Logic has so long been unable to cope with the subtlety of nature is that too much attention has been given to *pictorial notations*. Arithmetic could never be developed by means of the Roman system of notations; and Formal Logic cannot be developed so long as Barbara is represented by

$$C, \;\blacktriangleleft\!\!\!\blacksquare\; : M, \;\blacktriangleleft\!\!\!\blacksquare\; : T \,;$$

or even by the simpler spicular notation of De Morgan. We cannot manipulate data so crudely expressed; because the nature of the symbols has not been investigated, and laws of manipulation derived from their general properties."

681. *Signs of C. S. Peirce*.—C. S. Peirce made numerous contributions to symbolic logic, advancing the work of Boole and De Morgan. His researches contain anticipations of the important procedures of his successors. Among the symbols that C. S. Peirce used in 1867[4] are "the sign of equality with a comma beneath," to express numerical identity; thus $a \overset{=}{,} b$ indicates that a and b denote the same class; $a+,b$ denoting all the individuals contained under a and b together; a, b denoting "the individuals contained at once under the classes a and b," $\bar{a}$ denoting *not-a.*, [1;0] standing for "some uninterpretable symbol"; $a;b$ being a logically divided by b; $a:b$ "being the maximum value of $a;b$"; $\overset{-}{,}$ being the sign of logical subtraction.

In 1867 C. S. Peirce gave a "Description of a Notation for the

[1] W. S. Jevons, *Pure Logic* (London, 1864).

[2] C. I. Lewis, *op. cit.*, p. 73.

[3] See C. I. Lewis, *op. cit.*, "Bibliography of Symbolic Logic," p. 395.

[4] C. S. Peirce, "On an Improvement in Boole's Calculus of Logic," *Proceedings of the American Academy of Arts and Sciences*, Vol. VII (Cambridge and Boston, 1868), p. 250–61.

Logic of Relatives"[1] which contains an enlarged and somewhat modified array of symbols. He lets $\prec$ stand for "inclusion in" or "being as small as"; $x+, y$ for addition, where the comma is written except when the inverse operation (subtraction) is determinative. In his logic of relatives multiplication xy is not generally commutative. Commutative multiplication is written with a comma, $x, y = y, x$. Invertible multiplication is indicated by a dot: $x.y$. He lets (p. 319) log abc † def ‡ ghi stand for (log $abc)ghi$, the base being def. Indeterminative subtraction (p. 320) is marked $x \overline{} y$. The two divisions are indicated as in $(x:y)y = x$, $x \dfrac{y}{x} = y$. The two inverses of involution are as in $(\sqrt[x]{y})^x = y$, $x^{\log_x y} = y$. On page 321 Peirce gives his father's notation ତ for $(1+i)^{\frac{1}{i}}$, where i is an infinitesimal, ଵ for 3.14159 . . . . , ⌣ for $\sqrt{-1}$. He denotes (p. 322) absolute terms by the Roman alphabet, relative terms by italics, and conjugate terms by a kind of type called Madisonian. He denotes also individuals by capitals, generals by small letters, general symbols for numbers in black letter, operations by Greek letters, the number of a logical term by $[t]$. In multiplication (p. 326) "identical with" is 1, so that $x1 = x$ ("a lover of something identical with anything is the same as a lover of that thing"). On page 326 he lets $\mathfrak{g}xy$ signify "a giver to x of y." He uses (p. 328) the marks of reference † ‡ ‖ § ¶, which are placed adjacent to the relative term and before and above the correlate of the term. "Thus, giver of a horse to a lover of a woman may be written,

$$\mathfrak{g} \dagger \ddagger \, {}^{\dagger}l_{\,\|} \, {}^{\|}w \ddagger h \, ."$$

Lack of space prevents the enumeration here of the symbols for infinitesimal relatives.

682. Two pupils of C. S. Peirce, Miss Christine Ladd (now Mrs. Fabian Franklin) and O. H. Mitchell, proposed new notations. Peirce[2] indicated by "Griffin $\prec$ breathing fire" that "no griffin not breathing fire exists" and by "Animal $\precsim$ Aquatic" that a non-aquatic animal does exist; Peirce adds: "Miss Ladd and Mr. Mitchell also use two signs expressive of simple relations involving existence and non-existence; but in their choice of these relations they diverge

[1] *Memoirs Amer. Acad. of Arts and Sciences* (Cambridge and Boston; N.S.), Vol. IX (1867), p. 317–78.

[2] *Studies in Logic,* by members of the Johns Hopkins University (Boston, 1883), p. iv.

both from McColl and me, and from one another. In fact, of the eight simple relations of terms signalized by De Morgan, Mr. McColl and I have chosen two, Miss Ladd two others, Mr. Mitchell a fifth and sixth." Mrs. Ladd-Franklin[1] uses the copula $\vee$, so that $A \vee B$ signifies "A is in part B," and $A \overline{\vee} B$ signifies "A is excluded from B."

683. O. H. Mitchell[2] showed how to express universal and particular propositions as Π and Σ functions. "The most general proposition under the given conditions is of the form $\Pi(F_u + \Sigma G_1)$ or $\Sigma(F, \Pi G_u)$, where F and G are any logical polynomials of class terms, Π denotes a product, and Σ denotes a sum."

684. *Signs of R. Grassmann.*—Robert Grassmann, the brother of Hermann Gunther Grassmann of the Ausdehnungslehre, wrote on logic and the logical development of elementary mathematics. He develops the laws of classes, but his development labors under the defect, as C. S. Peirce observed, of breaking down when the number of individuals in the class is infinite.[3]

In R. Grassmann's *Formenlehre*[4] of 1872 he uses as the general sign of combination (*Knüpfung*) a small circle placed between the two things combined, as $a \circ b$. Special signs of *Knüpfung* are the sign of equality ($=$), and the sign of inequality which he gives as $\gtrless$. When two general signs of combination are needed, he uses[5] $\circ$ and $\odot$. The notation[6] $\bar{a}$ means "not-a." Accordingly, a double negation $\bar{\bar{a}}$ gives a. Instead of the double sign $\leqq$ he introduces the simple sign[7] $\angle$, so that $a \angle u$ means that a is either equal to u or subordinate to it.

685. *Signs of Schröder.*—Ernst Schröder, of Karlsruhe, in 1877 published investigations on Boole's algebra and later his advances on C. S. Peirce's researches. Peirce's original ideas crossed the Atlantic to Continental Europe, but not his symbolism. It was Schröder who gave to the algebra of logic "its present form,"[8] in so far as it may be said to have any universally fixed form. In 1877 Schröder used largely the symbolism of Boole. If in $c + b = a$ and $cb = a$ the inverse operations are introduced by solving each of the two equations for the c,

---

[1] *Op. cit.*, p. 25.

[2] *Op. cit.*, p. 79; see also C. I. Lewis, *op. cit.* (1918), p. 108.

[3] C. I. Lewis, *op. cit.*, p. 107.

[4] Robert Grassmann, *Formenlehre oder Mathematik* (Stettin, 1872), p. 8.

[5] *Op. cit.*, p. 18.

[6] Robert Grassmann, *Begriffslehre der Logik* (Stettin, 1872), p. 16, 17.

[7] *Op. cit.*, p. 24.

[8] C. I. Lewis, *op. cit.*, p. 118.

and the expressions $a-b$ and $a:b=\dfrac{a}{b}$ receive complete and general solutions, he denotes them by $c=a\div b$ and $c=a::b$, respectively.[1]

Later Schröder used[2] $\subset$ for "is included in" (*untergeordnet*) and $\supset$ for "includes" (*übergeordnet*), $\prec\!\!=$ "included in or equal," in preference to the symbols $<$, $>$ and $\leqq$ used by some authors on logic, (it was introduced by Schröder[3] in 1873, three years after C. S. Peirce's $\prec$). Schröder employs also $\succ\!\!=$ (p. 167). Later, Couturat rejected Schröder's $\in\!\!=$, "parce qu'il est complexe tandis que la relation d'inclusion est simple,"[4] and uses $<$.

Schröder represents the identical product and sum (logical aggregate) of a and b in his logic by ab (or $a.b$) and $a+b$. If a distinction is at any time desirable between the logical and arithmetical symbols, it can be effected by the use of parentheses as in $(+)$, $(\cdot)$ or by the use of a cross of the form $\bullet\!\!\!+$, or of smaller, black, or cursive symbols. Schröder criticizes C. S. Peirce's logical symbols, $=$, $+$, $-$ with commas attached, as rather clumsy.[5] Schröder represents the "identical one" by 1, while Jevons designated it the "Universe U" and Robert Grassmann the "*Totalität T*." Peirce had used 1 in his earlier writings, but later[6] he and his pupils used ∞, as did also W. Wundt.[7] Schröder[8] justifies his notation partly by the statement that in the theory of probability, as developed by De Morgan, Boole, C. S. Peirce, Macfarlane, and McColl, the identical one corresponds always to the symbol, 1, of certainty. Schröder lets a_1 stand[9] for "not-a"; Boole, R. Grassmann, and C. S. Peirce had represented this by $\bar{a}$, as does later also Schröder himself.[10] Couturat[11] rejects $\bar{a}$ as typographically inconvenient; he rejects also Schröder's a_1, and

[1] Ernst Schröder, *Der Operationskreis des Logikkalkuls* (Leipzig, 1877), p. 29.

[2] E. Schröder, *Vorlesungen über die Algebra der Logik*, Vol. I (Leipzig, 1890), p. 129, 132.

[3] Schröder, *Vorlesungen*, Vol. I, p. 140, 712.

[4] Louis Couturat, "L'Algèbre de la logique," *Scientia* (Mars, 1905), No. 24, p. 5.

[5] Schröder, *op. cit.*, p. 193.

[6] C. S. Peirce in *Amer. Jour. of Math.*, Vol. III (1880), p. 32. See Schröder, *op. cit.*, p. 274.

[7] W. Wundt, *Logik*, Vol. I (1880).

[8] Schröder, *op. cit.*, p. 275.

[9] Schröder, *op. cit.*, p. 300, 301.

[10] E. Schröder, *Algebra und Logik der Relative*, Vol. I (Leipzig, 1895), p. 18, 29.

[11] Louis Couturat, *op. cit.* (Mars, 1905), No. 24, p. 22.

adopts McColl's symbol a' as not excluding the use of either indices or exponents. In his logic of relatives Schröder[1] let i and j represent a pair of ordered elements in his first domain; they are associated as binary relatives by the symbolism $i:j$, and he introduces also the identical moduli 1, 0 and the relative moduli 1', 0'. The relative product[2] is $a; b;$ the relative sum $a \not j b;$ the converse of a is $\breve{a}$.

686. *Signs of MacColl.*—Hugh MacColl worked for many years quite independently of Peirce and Schröder. His symbolic logic of 1906 contains a notation of his own. His sign[3] of equivalence is $=$, of implication is $:$, of disjunction or alternation is $+$, of a proposition is A^B where A is the subject and B the predicate, of non-existence is O. MacColl lets $A^B \times C^D$, or $A^B \cdot C^D$, or $A^B C^D$ mean that A belongs to B and that C belongs to $D;$ he lets $A^B + C^D$ mean that either A belongs to B or else C to the class D. He classifies statements not simply as "true" or "false," but as "true," "false," "certain," "impossible," "variable," respectively, denoted by the five Greek letters[4] $\tau, \iota, \epsilon, \eta, \theta$. Thus A^η means that A is impossible, $A \therefore B$ means[5] that "A is true, therefore B is true"; $B \because A$ means "B is true because A is true"; (AB) denotes the total of individuals common to A and $B;$ (AB') denotes the total number in A but not in $B;$ letting x be any word or symbol, $\phi(x)$ is any proposition containing $x;$ the symbol $\dfrac{A}{B}$ expresses the chance that A is true on the assumption that B is true."[6]

687. *Signs of Frege.*—Gottlob Frege, of the University of Jena, published several books on the logic of mathematics, one in 1879, another in 1893.[7] Two of them precede Schröder's *Vorlesungen*, yet Frege's influence on Schröder was negligible. Nor did Peano pay attention to Frege, but Bertrand Russell was deeply affected by him. This early neglect has been attributed to Frege's repulsive symbolism.

[1] Schröder, *op. cit.* (1895), p. 8, 24–36.

[2] Schröder, *loc. cit.* (1895), p. 30.

[3] Hugh MacColl, *Symbolic Logic and Its Applications* (London, 1906), p. 4, 8. One of his early papers was on "Symbolical or Abridged Language, with an Application to Mathematical Probability," *Mathematical Questions . . . . from the Educational Times*, Vol. XXVIII (1878), p. 20, 100. Before about 1884 the name was spelled "McColl."

[4] MacColl, *Symbolic Logic*, p. 6.

[5] MacColl, *op. cit.*, p. 80, 94, 96.

[6] MacColl, *op. cit.*, p. 128.

[7] G. Frege, *Begriffsschrift, eine der arithmetischen nachgebildete Formelsprache des reinen Denkens* (Halle, 1879); G. Frege, *Grundgesetze der Arithmetik* (2 vols.; Jena, 1893–1903).

In a paper on the "Fundamental Laws of Arithmetic,"[1] Frege himself admits: "Even the first impression must frighten people away: unknown signs, pages of nothing but strange-looking formulas. It is for that reason that I turned at times toward other subjects." One of the questions which Frege proposed to himself was (as P. E. B. Jourdain expressed it) "as to whether arithmetical judgments can be proved in a purely logical manner or must rest ultimately on facts of experience. Consequently, he began by finding how far it was possible to go in arithmetic by inferences which depend merely on the laws of general logic. In order that nothing that is due to intuition should come in without being noticed, it was most important to preserve the unbrokenness of the chain of inferences; and ordinary language was found to be unequal to the accuracy required for this purpose."[2]

In Figure 125 he starts out to prove the theorem: "The number of a concept is equal to the number of a second concept, if a relation images (*abbildet*) the first upon the second and the converse of that relation images the second upon the first."

In the publication of 1893 Frege uses all the symbols employed by him in 1879, except that he now writes $=$ instead of $\equiv$. He uses $\vdash$ to indicate that a statement is true. Thus[3] the symbolism $\vdash 2^2 = 4$ declares the truth of the assertion that $2^2 = 4$. He distinguishes between the "thought" and the "judgment"; the "judgment" is the recognition of the truth of the "thought." He calls the vertical stroke in $\vdash$ the "judgment stroke" (*Urtheilstrich*); the horizontal stroke is usually combined with other signs but when appearing alone it is drawn longer than the minus sign, to prevent confusion between the two. The horizontal stroke is a "function name" (*Functionsname*) such that $-\Delta$ is true when Δ is true; and $-\Delta$ is false when Δ is false (p. 9).

The value of the function $\top \xi$ shall be false for every argument for which the value of the function $-\xi$ is true, and it shall be true for all other arguments. The short vertical stroke (p. 10) in $\top$ is a sign of negation. $\top \Delta$ means the same as $\top(-\Delta)$, the same as $-\top\Delta$, and the same as $-\top(-\Delta)$.

The notation (p. 12) $\smile 2+3 \cdot a = 5 \cdot a$ declares the generality of negation, i.e., it declares as a truth that for every argument the value

[1] *Monist*, Vol. XXV (Chicago, 1915), p. 491.

[2] P. E. B. Jourdain, *Monist*, Vol. XXV (1915), p. 482.

[3] G. Frege, *Grundgesetze der Arithmetik*, Vol. I, p. 9.

of the function $\dashv 2+3 . \xi = 5 . \xi$ is true. The notation $\smile 2+3 . a = 5a$ declares the negation of the generality, i.e., it signifies the truth that

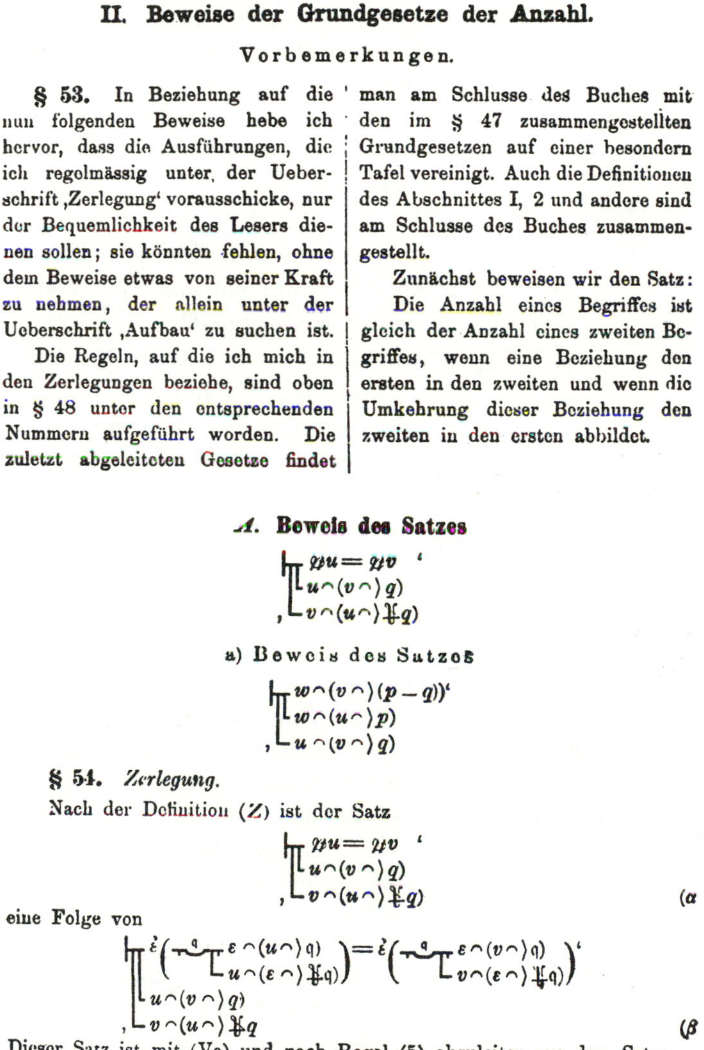

II. Beweise der Grundgesetze der Anzahl.

Vorbemerkungen.

§ 53. In Beziehung auf die nun folgenden Beweise hebe ich hervor, dass die Ausführungen, die ich regelmässig unter der Ueberschrift ‚Zerlegung‘ vorausschicke, nur der Bequemlichkeit des Lesers dienen sollen; sie könnten fehlen, ohne dem Beweise etwas von seiner Kraft zu nehmen, der allein unter der Ueberschrift ‚Aufbau‘ zu suchen ist.

Die Regeln, auf die ich mich in den Zerlegungen beziehe, sind oben in § 48 unter den entsprechenden Nummern aufgeführt worden. Die zuletzt abgeleiteten Gesetze findet man am Schlusse des Buches mit den im § 47 zusammengestellten Grundgesetzen auf einer besondern Tafel vereinigt. Auch die Definitionen des Abschnittes I, 2 und andere sind am Schlusse des Buches zusammengestellt.

Zunächst beweisen wir den Satz:

Die Anzahl eines Begriffes ist gleich der Anzahl eines zweiten Begriffes, wenn eine Beziehung den ersten in den zweiten und wenn die Umkehrung dieser Beziehung den zweiten in den ersten abbildet.

A. **Beweis des Satzes**

$$
\begin{array}{l}
\llcorner \; \mathfrak{y} u = \mathfrak{y} v \\
\quad \llcorner u \frown (v \frown) q) \\
, \; \llcorner v \frown (u \frown) \not{\mathfrak{E}} q)
\end{array}
$$

a) Beweis des Satzes

$$
\begin{array}{l}
\llcorner w \frown (v \frown) (p - q))' \\
\quad \llcorner w \frown (u \frown) p) \\
, \; \llcorner u \frown (v \frown) q)
\end{array}
$$

§ 54. *Zerlegung.*

Nach der Definition (Z) ist der Satz

$$
\begin{array}{l}
\llcorner \; \mathfrak{y} u = \mathfrak{y} v \\
\quad \llcorner u \frown (v \frown) q) \\
, \; \llcorner v \frown (u \frown) \not{\mathfrak{E}} q)
\end{array} \tag{α}
$$

eine Folge von

$$
\begin{array}{l}
\llcorner \acute{\varepsilon} \Big(\neg{}^{\mathfrak{q}} \llcorner \varepsilon \frown (u \frown) q) \Big) = \acute{\varepsilon} \Big(\neg{}^{\mathfrak{q}} \llcorner \varepsilon \frown (v \frown) q) \Big)' \\
\qquad \qquad \llcorner u \frown (\varepsilon \frown) \not{\mathfrak{E}} q) \qquad \qquad \llcorner v \frown (\varepsilon \frown) \not{\mathfrak{E}} q) \\
\quad \llcorner u \frown (v \frown) q) \\
, \; \llcorner v \frown (u \frown) \not{\mathfrak{E}} q
\end{array} \tag{β}
$$

Dieser Satz ist mit (Va) und nach Regel (5) abzuleiten aus dem Satze

Fig. 125.—Frege's notation as found in his *Grundgesetze* (1893), Vol. I, p. 70

not for every argument the value of the function $2 + 3 . \xi = 5 . \xi$ is true.

Frege (p. 20) lets τ_ζ^ξ be a function with two arguments whose value is false when the ζ-argument is assumed to be true and the ξ-argument is assumed to be false; in all other cases the functional value is true. The vertical stroke is here designated as a "condition stroke" (*Bedingungstrich*).

ι_Δ^Γ is true when, and only when, Δ is true and Γ is not true.

Hence, $\vdash_{2+3=5}^{2>3}$ affirms the statement that "2 is not greater than 3, and the sum of 2 and 3 is 5" (p. 21).

$\vdash_{2+3=5}^{3>2}$ affirms the statement that "3 is greater than 2, and the sum of 2 and 3 is 5."

$\vdash_{1^2=2^1}^{2^3=3^2}$ affirms the statement that "Neither is the 3rd power of 2 the 2nd power of 3, nor is the 2nd power of 1 the 1st power of 2."

One can pass (p. 27) from the proposition $\vdash_\Delta^\Gamma$ to $\vdash_\Gamma^\Delta$, and vice versa. These two passages are symbolized in this manner:

$$\begin{array}{ccc} \vdash_\Delta^\Gamma & & \top_\Gamma^\Delta \\ \times & \text{and} & \times \\ \vdash_\Gamma^\Delta & & \top_\Delta^\Gamma \end{array}$$

688. *Signs of Peano.*—Schröder once remarked that in symbolic logic we have elaborated an instrument and nothing for it to do. Peano proceeded to use this instrument in mathematical proof. In 1888 he published his *Calcolo geometrico*,[1] where he states that, in order to avoid confusion between logical signs and mathematical signs, he substitutes in place of the signs $\times$, $+$, A_1, 0, 1, used by Schröder[2] in 1877, the signs $\cap$, $\cup$, $-A$, O, ●.

In his *Formulaire de mathématiques*, the first volume of which appeared at Turin in 1895, Peano announces confidently a realization of the project set by Leibniz in 1666, namely, the creation of a universal script in which all composite ideas are expressed by means of conventional signs of simple ideas, according to fixed rules. Says Peano:[3] "One may change the shape of the signs, which have been

[1] G. Peano, *Calcolo geometrico secondo l'Ausdehnungslehre di H. Grassmann* (Torino, 1888), p. x.

[2] E. Schröder, *Der Operationskreis der Logikkalkuls* (1877).

[3] *Formulaire de mathématiques*, Tome I (Turin, 1895), Introd. (dated 1894), p. 52.

introduced whenever needed; one may suppress some or annex others, modify certain conventions, etc., but we are now in a position to express all the propositions of mathematics by means of a small number of signs, having a precise signification and subject to well determined rules."

The *Formulaire de mathématiques* of 1895 was not the earliest of Peano's publications of this subject. The first publication of the *Formulaire de logique* was in 1891, in the *Rivista di matematica* (p. 24, 182). This was reproduced with additions as Part I of the *Formulaire de mathématiques*, Volume I (1893–95). The historical statements in it are due mainly to G. Vailati. The first three parts of the first volume deal with mathematical logic, algebraic operations, arithmetic, and were prepared by Peano with the aid of G. Vailati, F. Castellano, and C. Burali-Forti. The subsequent parts of this first volume are on the theory of magnitudes (by C. Burali-Forti), of classes (by Peano), of assemblages (by G. Vivanti), of limits (by R. Bettazzi), of series (by F. Giudice), of algebraic numbers (by G. Fano).

689. Peano expresses the relations and operations of logic in Volume I of his *Formulaire de mathématiques* (Introduction, p. 7) by the signs ϵ, c, $\backsim$, $=$, $\cap$, $\cup$, $-$, V, Λ, the meanings of which are, respectively, "is" (i.e., is a number of), "contains," "is contained in," "is equal" (i.e., is equal to), "and," "or," "not " "all," "nothing." He adds that c and v are mentioned here for the sake of symmetry, but are not used in practice. These symbols are used for propositions as well as for classes, but receive a somewhat different signification when applied to propositions. For example, if a and b are classes, $a \backsim b$ signifies "a is contained in b," but if a and b are propositions, then $a \backsim b$ signifies "from a one deduces b" or "b is a consequence of a." As examples of the notation used for certain classes K, we quote (Introduction, p. 4):

"N signifies number, integral positive
n "   number, integral (positive, zero or negative)
N_0 "   number, integral positive or zero
R "   number, rational positive
r "   number, rational
Q "   number, real positive (quantity)
q "   number, real
Q_0 "   number, real positive or zero."

As symbols of aggregation, dots are used here in preference to parentheses.

On pages 134–39 of the first volume of his *Formulaire* Peano gives a table of the signs used in that volume. There are altogether 158 signs, of which 26 are symbols having special forms (ideographs), 16 are Greek letters, and 116 are Latin letters or combinations of them.

As specimens of the use he makes of his symbols, we give the following:

Vol. I, p. 1. $a \, \mathfrak{d} \, b \, . \, b \, \mathfrak{d} \, c \, . \, \mathfrak{d} \, . \, a \, \mathfrak{d} \, c$ [Pp.].

"If a is contained in b, and b in c, then, a is contained in c."

p. 16. $a \epsilon Q \, . \, x \epsilon q \, . \, \mathfrak{d} \, . \, a^x \epsilon Q$.

Primitive proposition: "If a is a positive real number, and x is a real number, then a^x is a positive real number."

p. 49. $A, B, C, D, U, U' \, \epsilon \, G \, . \, \mathfrak{d} : A|U \, \epsilon \, Q$.

"If A, B, C, D, U, U' are magnitudes (G=grandeur), then the fraction $A|U$ is a positive real number."

p. 85. $u, v \, . \, \epsilon q f N \, . \, m \epsilon N \, . \, \mathfrak{d} : \Sigma u_\infty \epsilon q \, . \, \mathfrak{d} \, . \, \lim u_n = 0. \quad \lim \sum_n^{n+m} u = 0.$

"If u is a real number defined in terms of positive integral numbers, and m is a positive integer, then, when the sum of an infinite number of successive u's is a real number, it follows that the limit of u_n is zero. and the limit of the sum of the u's from u_n to u_{n+m} is also zero."

Volume II of Peano's *Formulaire de mathématiques* appeared at Turin in three numbers: No. 1 in 1897, No. 2 in 1898, and No. 3 in 1899. Number 1 gives a fuller development of his symbolism of logic. Additional symbols are introduced and a few of the older notations are modified.

One of the new symbols is $\exists$ (No. 1, p. 47), $\exists_a$ signifying "il y a des a" or "there are a's." It was introduced "because the notation $a - = \Lambda$ was held by many of the collaborators as long and differing too widely from ordinary language." Peano adds: "One has thus the advantage that by the three signs $\exists$, $\frown$, $-$ one can express $\mathfrak{d}$ and Λ [viz. (p. 14, § 1), $a \mathfrak{d} b \, . = \, . - \exists (a-b)$ and $a = \Lambda \, . = \, . - \exists a \, .]$."

ιx expresses the class formed by the object x; reciprocally, if a is a class which contains a single individual x, Peano now writes (No. 1, p. 49) $\bar{\iota} a$ to indicate the individual which forms the class a, so that $x = \bar{\iota} a \, . = \, . a = \iota x$. In other words, "$x$ is the individual which constitutes the class a" is a statement identical with the statements that "a is the class formed by the object x." Example: $a, b \epsilon N \, . \, b > a \, . \, \mathfrak{d} \, . \, b - a = \bar{\iota} N \frown \overline{\chi \epsilon} (a+x=b)$ signifies: "If a and b are positive integers, and $b > a$, it follows that $b-a$ is the number which is to be added to a, to obtain b." This class exists and contains the single individual $b-a$.

Among other changes, Peano extends the use of the sign '. In the first volume[1] he had used $\smile$' for "the smallest class containing all the classes u," $\frown$' for "the greatest class contained in all the classes $\smile$." Now he lets (No. 1, p. 51) K'a stand for "the class contained in a."

In grouping symbols Peano no longer limits himself to the use of dots; now he uses also parentheses.

690. Number 2 of Volume II of Peano's *Formulaire* is devoted mainly to the theory of numbers, integral, fractional, positive, and negative. Theorems are expressed in ideographic signs, together with proofs and historical notes. Even the references to contributors are indicated by special symbols. Thus ⊃ means "Peano, *Arithmeticae principia*, 1889"; $\prec$ means "Vacca, *Form. de Math.*, t. II."

Some additional symbols are introduced; others are altered in form. Thus ɔ becomes $\supset$. By the symbolism $p . \supset x \ldots . z . q$ is expressed "from p one deduces, whatever $x \ldots . z$ may be, and q."

The new symbolism $a /^{1} m$ signifies a^{m}. Peano says on page 20: "It permits one to write all the signs in the same line, which presents a great typographical advantage in complicated formulas." It is the radical sign $\sqrt{}$ turned in its own plane through 180°.

Number III of Volume II of Peano's *Formulaire* (Turin, 1899) contains, as stated in its Preface, "the principal propositions thus far expressed by ideographic symbols. It is divided into §§. Each § has for its heading an ideographic sign. These signs follow each other in an order so that all signs appear defined by the preceding (with the exception of primitive ideas). Every § contains the propositions which one expresses by the sign heading that § and those preceding. . . . . This disposition, already partially applied in the preceding publications, is followed here with greater rigor, ..." Some new notations are given. For instance (No. 3, p. 23, § 8), the new symbol ⌐ is defined thus: "$(u \colon v)$ signifies the assemblage of couples formed by an object of the class u with an object of the class v; this new sign is to be distinguished from $(u \, ; v)$ which signifies the couple of which the two elements are the classes u and v."

$$u, v \epsilon Cls . \supset . (u \colon v) = (x \, ; y) \wp (x \epsilon u . y \epsilon v) \text{ Df.}$$

This ideographic definition is in words: "If u and v are classes, then $(u \, \colon v)$ is equal to the couple of which the two elements are x and

[1] *Formulaire de mathématiques*, Vol. I, Introd., p. 25, § 21.

y which satisfy the condition, x is u, and y is v." The symbol $℘$ may be read "which."

An edition (Vol. III) of the *Formulaire de mathématiques* of the year 1901 was brought out in Paris.

691. To give an idea of the extent to which the Peano ideographs have been used, it is stated in the seventh volume of the *Revue de mathématiques* (1901) that the symbols occur in sixty-seven memoirs published in different countries by fifteen authors. The third volume of the *Formulaire* collects the formulae previously published in the *Revue de mathématiques* and in the *Formulaire*, together with new propositions reduced to symbols by M. Nassò, F. Castellano, G. Vacca, M. Chini, T. Boggio.

In Volume IV of the *Formulaire* (Turin, 1903) seven pages are given to exercises on mathematical logic and drill in the mastery of the symbols, and seventeen pages to biographical and bibliographical notes prepared by G. Vacca.

Volume V of the *Formulaire*, the final volume which appeared at Turin in 1908, is much larger than the preceding, and is written in Latin instead of French. It gives full explanations of symbols and many quotations from mathematical writers. It is divided into eight parts devoted, respectively, to "Logica-Mathematica," "Arithmetica," "Algebra," "Geometria," "Limites," "Calculo differentiale," "Calculo integrale," "Theoria de curvas."

Peano compared[1] his symbols with those of Schröder in the following grouping:

Schröder........0 1 $+$ $\cdot$ Σ Π $\bar{a}$ $\leqq$

Formulario.......Λ V $\smile$ $\frown$ $\smile^{\prime}$ $\frown^{\prime}$ $-a$ $\epsilon\supset$

692. *Signs of A. N. Whitehead.*—A. N. Whitehead, in his *Treatise on Universal Algebra*, Volume I (Cambridge, 1898), deviates considerably from the Peano symbolism, as appears from the following comparison, made by Vacca,[2] of the symbols for addition, none, multiplication, the "universe," and "supplementary element":

Whitehead..............$a+b$, 0, ab, i, $\bar{a}$

Peano..................$a\smile b$, Λ, ab, V, $-a$

693. *Signs of Moore.*—Eliakim Hastings Moore, in his General Analysis, introduces a thorough uniformization of notations, but displays conservatism in the amount and variety of symbolism used. He

[1] *Revue de mathématiques* (Turin), Vol. VI (1896–99), p. 96.

[2] G. Vacca, *Revue de mathématiques*, Vol. VI (Turin, 1896–99), p. 102.

feels that the extreme tendencies of Peano, Schroeder, Whitehead, and Russell are not likely to be followed by the working mathematicians, that, on the other hand, writers in the abstract regions of mathematics will find that mathematical and logical ideas of constant recurrence can profitably be indicated symbolically, in the interest of brevity and clarity. Moore does not aim to reduce all mathematical representation and reasoning to ideographic notations; he endeavors to restrict himself to such symbolism as will meet the real needs of the great body of mathematicians interested in recent developments of analysis.

In general, Moore uses small Latin letters for elements, capital Latin letters for transformations or properties or relations, Greek letters for functions, and capital German letters for classes. For example, if a property P is defined for elements t of a class $\mathfrak{T}$, the notation t^P concerning an element t indicates that t has the property P. Similarly, $t_1 R t_2$ or $(t_1, t_2)^R$ indicates that the element t_1 is in the relation R to the element t_2. Naturally, t^P in the hypothesis of an implication indicates, "for every element t having the property P." In this notation Moore follows more or less closely Hugh MacColl.[1]

Consider two classes $\mathfrak{P} = [p]$, $\mathfrak{Q} = [q]$ of elements p, q. A single-valued function φ on $\mathfrak{P}$ to $\mathfrak{Q}$, $\varphi^{\text{on } \mathfrak{P} \text{ to } \mathfrak{Q}}$, is conceptually a table giving for every argument value p a definite functional value, $\varphi(p)$ or q_p, belonging to the class $\mathfrak{Q}$. The function φ as the table of functional values $\varphi(p)$ has the notation $\varphi \equiv (\varphi(p) \mid p) \equiv (q_p \mid p)$, where the "$\mid p$" indicates that p ranges over $\mathfrak{P}$. This precise notational distinction between function φ and functional value $\varphi(p)$ is of importance: it obviates the necessity of determining from the context whether $\varphi(p)$ denotes the function φ or the functional value $\varphi(p)$ for the value p of the argument. "Function φ on $\mathfrak{P}$ to $\mathfrak{Q}$" is, of course, precisely Georg Cantor's[2] "Belegung von $\mathfrak{P}$ mit $\mathfrak{Q}$"; for use in analysis the locution "function on to" is happier than "Belegung von mit."

694. We quote the list of logical signs which E. H. Moore used in his *General Analysis* in 1910. With the exception of $\neq$; $\equiv$; $\sim$; [ ], these signs are taken from Peano's *Formulario* of 1906 and are used approximately in the sense of Peano. The list is as follows:[3]

[1] Hugh MacColl in *Bibliothèque du Congrès International de Philosophie. III. Logique et Histoire* (Paris, 1901), p. 137, 167.

[2] G. Cantor, *Mathematische Annalen*, Vol. XLVI (1895), p. 486.

[3] E. H. Moore, *Introduction to a Form of General Analysis* (New Haven, 1910; The New Haven Mathematical Colloquium), p. 150. See also *Bull. Amer. Math. Society* (2d ser.), Vol. XVIII (1912), p. 344, 345; E. H. Moore, "On the Funda-

"= logical identity

≠ logical diversity

≡ definitional identity

•⊃• (for every . . . .) it is true that
( ) implies ( )
if ( ), then ( )

•⊂• ( ) is implied by ( )

•∼• ( ) is equivalent to ( )
( ) implies and is implied by ( )

⊃ ; ⊂ ; ∼ implies; is implied by; is equivalent to (as relations of
properties)

Ⅎ there exists a (system; class; element; etc.)

𝕑 such that; where

• and

·•; ::; ∴ ∴ signs of punctuation in connection with signs of implication, etc.; the principal implication of a sentence
has its sign accompanied with the largest number of
punctuation dots

⌣ or

— not

[ ] a class of (elements; functions; etc.)

[all] the class consisting of all (elements; functions; etc.,
having a specified property or satisfying a specified
condition)

∪[𝕻] the least common superclass of the classes 𝕻 of the
class [𝕻] of classes

∩[𝕻] the greatest common subclass of the classes 𝕻 of the
class [𝕻] of classes."

In further illustration of Moore's use of logical symbols, we quote
the following:

"From a class: [P], of properties of elements arises by composition

mental Functional Operation of a General Theory of Linear Integral Equations,"
Proceedings of the Fifth International Congress of Mathematicians (Cambridge,
1912); E. H. Moore, "On a Form of General Analysis with Applications to Linear
Differential and Integral Equations," *Atti del IV Congresso internazionale dei
matematici* (Roma, 1909); E. H. Moore, "Definition of Limit in General Integral
Analysis," *Proceedings of the National Academy of Sciences*, Vol. I (1915), p. 628–
32; E. H. Moore, "On Power Series in General Analysis," *Festschrift David Hil-
bert zu seinem sechzigsten Geburtstag* (Berlin, 1922), p. 355–64, and *Mathematische
Annalen*, Vol. LXXXVI (1922), p. 30–39; E. H. Moore and H. L. Smith, "A
General Theory of Limits," *American Journal of Mathematics*, Vol. XLIV (1922),
p. 102–21.

the *composite property:* $\cap[P]$, viz., the property of having every property P of the class $[P]$ of properties. . . . ."[1]

"From a class: $[P]$, of properties of elements arises by disjunction the *disjunctive property:* $\cup[P]$, viz., the property of having at least one of the properties P of the class $[P]$ of properties. The notation is so chosen that the class of all elements having the disjunctive property $\cup[P]$ is the class $\cup[\mathfrak{P}]$, viz., the least common superclass of the classes $\mathfrak{P}$ of the corresponding class $[\mathfrak{P}]$ of classes of elements."[2]

"We may write the definitions as follows:

$$x^{P_1 \cdots \cdot P_n} :\equiv : x \ni i^{1 \leqq i \leqq n} . \supset . x^{P_i} ,$$

viz., an element x having the composite property $P_1 \ldots P_n$ *is by definition* ($\equiv$) an element x *such that* ($\ni$) for every integer i having the property: $1 \leqq i \leqq n$, *it is true that* ($.\supset.$)x has the property P_i;

$$x^{\cap[P]} :\equiv : x \ni P^{[P]} . \supset . x^P ,$$

viz., an element x has the composite property $\cap[P]$ in case for every property P belonging to the class $[P]$ it is true that x has the property P;

$$x^{P_1 \smile P_2 \smile \cdots \smile P_n} . \equiv . x \ni \mathrm{\Xi} \ i^{1 \leqq i \leqq n} \ni x^{P_i} ,$$

viz., an element x has the disjunctive property $P_1 \smile \ldots \smile P_n$ in case *there exists an* ($\mathrm{\Xi}$) integer i having the property: $1 \leqq i \leqq n$, such that x has the property P_i;

$$x^{\cup[P]} . \equiv . x \ni \mathrm{\Xi} \ P^{[P]} \ni x^P ,$$

viz., an element x has the disjunctive property $\cup[P]$ in case there exists a property P belonging to the class $[P]$ such that x has the property P."[3]

695. *Signs of Whitehead and Russell.*—The *Principia mathematica* of Alfred North Whitehead and Bertrand Russell, the first volume of which appeared in 1910 at Cambridge, England, is a continuation of the path laid out in recent years mainly by Frege and Peano. Whitehead and Russell state in their Preface: "In the matter of notation, we have as far as possible followed Peano, supplementing his notation, when necessary, by that of Frege or by that of Schröder. A great deal of symbolism, however, has had to be new, not so much through dissatisfaction with the symbolism of others, as through the fact that

[1] E. H. Moore, *Introduction to a Form of General Analysis* (1910), p. 18.

[2] *Op. cit.*, p. 19.

[3] E. H. Moore, *op. cit.*, p. 19, 20.

we deal with ideas not previously symbolised. In all questions of logical analysis, our chief debt is Frege."

In the Introduction, Whitehead and Russell state the three aims they had in view in writing the *Principia:* (1) "the greatest possible analysis of the ideas with which it [mathematical logic] deals and of the processes by which it conducts demonstrations, and at diminishing to the utmost the number of the undefined ideas and undemonstrated propositions"; (2) "the perfectly precise expression, in its symbols, of mathematical propositions"; (3) "to solve the paradoxes which, in recent years, have troubled students of symbolic logic and the theory of aggregates." On matters of notation they state further:

"The use of symbolism, other than that of words, in all parts of the book which aim at embodying strictly accurate demonstrative reasoning, has been forced on us by the consistent pursuit of the above three purposes. The reasons for this extension of symbolism beyond the familiar regions of number and allied ideas are many: (1) The ideas here employed are more abstract than those familiarly considered in language. . . . . (2) The grammatical structure of language is adopted to a wide variety of usages. Thus it possesses no unique simplicity in representing the few simple, though highly abstract, processes and ideas arising in the deductive trains of reasoning employed here. . . . . (3) The adoption of the rules of the symbolism to the processes of deduction aids the intuition in regions too abstract for the imagination readily to present to the mind the true relation between the ideas employed. . . . . (4) The terseness of the symbolism enables a whole proposition to be represented to the eyesight as one whole, or at most in two or three parts." The authors say further: "Most mathematical investigation is concerned not with the analysis of the complete process of reasoning, but with the presentation of such an abstract of the proof as is sufficient to convince a properly instructed mind. For such investigations the detailed presentation of steps in reasoning is of course unnecessary, provided that the detail is carried far enough to guard against error. . . . . In proportion as the imagination works easily in any region of thought, symbolism (except for the express purpose of analysis) becomes only necessary as a convenient shorthand writing to register results obtained without its help. It is a subsidiary object of this work to show that, with the aid of symbolism, deductive reasoning can be extended to regions of thought not usually supposed amenable to mathematical treatment."

The symbolism used in the first volume of their *Principia mathematica* is as follows (1910, p. 5–38):

The small letters of our alphabet are used for variables, except p and
s to which constant meanings are assigned in the latter part of
the volume;

The capitals A, B, C, D, E, F, I, J, have constant meanings.

Also the Greek letters ϵ, ι, π and, at a later stage, η, θ, and ω have
constant meanings;

The letters p, q, r, are called 'propositional letters' and stand for
variable *propositions*.

The letters f g ϕ ψ χ θ and, at first F, are called 'functional letters'
and are used for variable functions.

Other small Greek letters and other ones of our ordinary capitals
will be used as special types of variables.

$\sim p$ means *not-p*, which means the negation of p, where p is any
proposition.

$p \vee q$ means that at least p or q is true; it is the logical sum with p and
q as arguments.

$p.q$ means that both p and q are true; it is the logical product with
p and q as arguments. Thus, $p.q$ is merely a shortened form for
$\sim(\sim p \vee \sim q)$.

$p)q$ means p implies q; it stands for $\sim p \vee q$ when p is true.

$p \equiv q$ means p implies q and q implies p.

$\vdash$ is Frege's "assertion-sign," to show that what follows is asserted.
Thus, $\vdash .(p \supset q)$ means an assertion that $p \supset q$ is true; without
the assertion-sign, a proposition is not asserted, and is merely
put forward for consideration, or as a subordinate part of an
asserted proposition.

Dots are used to bracket off propositions, or else to indicate the
logical product as shown above. Thus, $p \vee q . \supset . q \vee p$ means
'p or q' implies 'q or p.'

As an example of the authors' "primitive propositions," we quote:

$$\vdash : p \vee q . \supset . q \vee p . \qquad \text{Pp.,}$$

i.e., "If p or q is true, then q or p is true. Proposition."

The law of excluded middle is one of the simple propositions and
is indicated thus:

$$\vdash . p \vee \sim p ,$$

i.e., "p is true, or *not-p* is true."

The law of contradiction:

$$\vdash . \sim(p . \sim p) ,$$

i.e., "It is not true that p and *not-p* are both true."

The law of double negation:

$$\vdash . \, p \equiv \sim(\sim p) \, ,$$

i.e., "p is the negation of the negation of p."

ϕx designates a "propositional function"; it is not a proposition, since x is a variable, having no fixed determined meaning. "x is human" is a propositional function; so long as x remains undetermined, it is neither true nor false.

$(x) \cdot \phi x$ means "$\phi(x)$ is true for all possible values of x."

$(\exists x) . \phi x$ means "there exists an x for which ϕx is true," so that *some* propositions of the range are true.

$(x) \cdot \sim \phi x$ means "there exists no x for which ϕx is true."

$\phi x . \supset_x . \psi x$ is a notation due to Peano and means, "If ϕx is always true, then ψx is true."

$x \epsilon a$ means, as with Peano, "x is a" or "x is a member of the class a."

$\vdash : x \epsilon \hat{z}(\phi z) . \equiv . \phi x$ means " 'x is a member of the class determined by $\phi \hat{z}$' is equivalent to 'x satisfies $\phi \hat{z}$,' or to 'ϕx is true.' " Here $\hat{z}$ is individual z.

$\phi(x, y)$ is a function which determines a *relation* R between x and y. $\hat{x}\hat{y}\phi(x, y)$ is the relation determined by the function $\phi(x, y)$. xRy means "x has the relation R to y."

$$\vdash . z\{\hat{x}\hat{y}\phi(x,y)\}w . \equiv . \phi(z,w) \, ,$$

i.e., " 'z has to w the relation determined by the function $\phi(x,y)$' is equivalent to $\phi(z, w)$."

$a \cap \beta$ is the logical product of two *classes* a and β, representing the class of terms which are members of both.

$$\vdash : x \epsilon a \cap \beta . \equiv . x \epsilon a . x \epsilon \beta \, ,$$

means " 'x is a member of the logical product of a and β' is equivalent to the logical product of 'x is a member of a' and 'x is a member of β.' "

$a \cup \beta$ is the logical sum of two classes a and β, representing the class of terms which are members of either.

$- a$ is the class of terms of suitable type which are not members of class a; it may be defined thus:

$$- a = \hat{x}(x \sim \epsilon a) \qquad \text{Df.} \, ,$$

and the connection with the negation of proposition is given by

$$\vdash : x \epsilon - a . \equiv . x \sim \epsilon a \, .$$

$a \subset \beta$ means inclusion; i.e., class a is contained in class β. As an example of propositions concerning classes which are analogous of propositions concerning propositions, we give:

$$\vdash . a \cap \beta = -(-a \cup -\beta) \ .$$

The law of absorption holds in the form

$$\vdash : a \subset \beta . \equiv . a = a \cap \beta \ ,$$

as, for instance, "all Cretans are liars" is equivalent to "Cretans are identical with lying Cretans."

Whitehead and Russell[1] repeat a statement due to Peano, namely, that $x\epsilon\beta$ is not a particular case of $a \subset \beta$; on this point, traditional logic is mistaken.

Definitions and propositions similar to those for classes hold for relations. The symbol for relations is obtained by adding a dot to the corresponding symbol for classes. For example:

$R \mathbin{\dot\cap} S = \hat{x}\hat{y}(xRy . xSy)$ Df.

$\vdash : x(R \mathbin{\dot\cap} S)y . \equiv . xRy . xSy,$

i.e., " 'x has to y the relation which is common to R and S' is equivalent to 'x has the relation R to y and x has the relation S to y.' "

$R \mathbin{\dot\cup} S = \hat{x}\hat{y}(xRy . \mathrm{v} . xSy)$ Df.

$\mathbin{\dot-} R = \hat{x}\hat{y}\{\sim(xRy)\}$ Df.

$R \mathbin{\dot\subset} S . = : xRy . \supset _{x, y} . xSy$ Df.

Further symbols for the algebra of classes:

$\exists ! a$ denotes "a exists."

$\exists ! a . = . (\exists x) . x\epsilon a$ Df.

i.e., " 'a exists' is equivalent to 'x exists and x is a.' "

Λ is a class that has no members (null-class).

V is a universal class, being determined by a function which is always true. Thus, Λ is the negation of V.

The corresponding symbols for relations are:

$\mathbin{\dot\exists} ! R . = . (\exists xy) . xRy \ ,$

i.e., $\mathbin{\dot\exists} ! R$ means that there is at least one couple x, y between which the relation R holds.

$\dot\Lambda$ is the relation that never holds.

$\dot V$ is the relation that always holds.

$E ! (\imath x)(\phi x) . = : (\exists c) : \phi x . \equiv_x . x = c$ Df.

[1] Whitehead and Russell, *op. cit.*, Vol. I (1910), p. 29.

i.e., " 'the x satisfying $\phi\hat{x}$ exists' is to mean 'there is an object c such that ϕx is true when x is c but not otherwise.' "

$R'y = (\imath x)(xRy)$ Df.

The inverted comma may be real "of"; $R'y$ is read "the R of y."

For the converse of R, the authors have the notation

$\text{Cnv`} R = \breve{R} = \hat{x}\hat{y}(yRx)$ Df.

$\overrightarrow{R}'y = \hat{x}(xRy)$ Df.

$\overleftarrow{R}'x = \hat{y}(xRy)$ Df.

"If R is the relation of parent to child, $\overrightarrow{R}'y$ = the parents of y, $\overleftarrow{R}'x$ = the children of x."

$D'R = \hat{x}\{(\exists y) \cdot xRy\}$ Df.

$D'R$ is the *domain* of R, i.e., the class of all terms that have the relation R to y. The *converse domain* of R is symbolized and defined thus:

$\mathbb{Q}'R = \hat{y}\{(\exists x) \cdot xRy\}$ Df.

The sum of the domain and the converse domain is called the "field" and is represented by $C'R$,

$C'R = D'R \smile \mathbb{Q}'R$ Df.

The *relative product* of two relations R and S arises when xRy and ySz; it is written $R \mid S$.

$R \mid S = \hat{x}\hat{z}\{(\exists y) \cdot xRy \cdot ySz\}$ Df.

"Paternal aunt" is the relative product of "sister" and "father."

$R''a$ is the class of terms x which have the relation R to some member of a class a. We have,

$R''a = \hat{x}\{(\exists y) \cdot y\epsilon a \cdot xRy\}$ Df.

Thus, if R is *inhabiting* and a is *towns*, then $R''a$ = inhabitants of towns.

$\iota'x$ is the class whose only member is x. Peano and Frege showed that the class whose only member is x is not identical with x.

$a \upharpoonleft R$ is a relation R with its domain limited to members of a (Vol. I [1910], p. 278).

$R \upharpoonright \beta$ is a relation R with its converse domain limited to members of β.

$a \upharpoonleft R \upharpoonright \beta$ is a relation R with both of these limitations.

$a \uparrow \beta$ is the relation between x and y in which x is a member of a and y is a member of β; i.e.,

$a \uparrow \beta = \hat{x}\hat{y}(x\epsilon a \cdot y\epsilon\beta)$ Df.

The foregoing list represents most of the symbols used by Whitehead and Russell in the first volume of their *Principia mathematica*. This volume, after an Introduction of sixty-nine pages, is devoted to two parts; the first part on "Mathematical Logic," the second to "Prolegomena to Cardinal Arithmetic."

The authors introduce the concept of types which is intended to serve the purpose of avoiding contradictions. The two chief notations used therewith are (Vol. I, p. 419):

$t_0^\iota a$ meaning the type in which a is contained.

$t^\iota x$ meaning the type of which x is a member.

"The type of members of a" is defined thus,

$t_0^\iota a = a \cup -a$ Df.

If $\iota^\iota x$ is the class whose only member is x, we define $t^\iota x = \iota^\iota x \cup -\iota^\iota x$ Df. It follows that $\vdash \cdot t^\iota x = t_0^\iota \iota^\iota x$.

"If R is any relation (Whitehead and Russell, Vol. I, p. 429) it is of the same type as $t_0^\iota D^\iota R \uparrow t_0^\iota \mathbf{C}^\iota R$. If $D^\iota R$ and $\mathbf{C}^\iota R$ are both of the same type as a, R is of the same type as $t_0^\iota a \uparrow t_0^\iota a$, which is of the same type as $a \uparrow a$. The type of $t_0^\iota a \uparrow t_0^\iota a$ we call $t_{00}^\iota a$, and the type of $t^{m\iota} a \uparrow t^{n\iota} a$ we call $t^{mn\iota} a$, and the type of $t_m^\iota a \uparrow t_n^\iota a$ we call $t_{mn}^\iota a$, and the type of $t_m^\iota a \uparrow t^{n\iota} a$ we call $t_m^{n\iota} a$, and the type of $t^{m\iota} a \uparrow t_n^\iota a$ we call $^m t_n^\iota a$. We thus have a means of expressing the type of any relation R in terms of the type of a, provided the types of the domain and converse domain of R are given relatively to a."

If a is an ambiguous symbol (*ibid.*, p. 434) representing a class (such as Λ or V, for example),

a_x denotes what a becomes when its members are determined as belonging to the type of x, while

$a(x)$ denotes what a becomes when its members are determined as belonging to the type of $t^\iota x$. Thus,

V_x is everything of the same type as x, i.e., $t^\iota x$.

$V(x)$ is $t^\iota t^\iota x$.

$a \, sm \, \beta$ means "a is similar to β," i.e., there is a one-one relation whose domain is a and whose converse domain is β (*ibid.*, p. 476). We have

$$a \, sm \, \beta \cdot \equiv \cdot (\exists R) \cdot R\epsilon 1 \to 1 \cdot a = D^\iota R \beta = \mathbf{C} R .$$

$k \, \epsilon \, Cls^2 \, excl$ means that k is a class of mutually exclusive classes (*ibid.*, p. 540).

$k \, \epsilon \, Cls \, ex^2 \, excl$ means the foregoing when no member of k is null.

$Cl \, excl^\iota y$ means a $Cls^2 \, excl$ which is contained in a class of classes γ.

R_* means an ancestral relation to a given relation R (*ibid.*, p. 569).

$x \, R_* \, y$ means that x is an ancestor of y with respect to R (p. 576).

xBP means $x\epsilon D^\iota P - \mathbf{C}^\iota P$ (*ibid.*, p. 607).

$x \ min_p \ a$ means $x \epsilon a \frown C'P - \breve{P}''a$,

 i.e., x is a member of a and of $C'P$, and no member of a precedes
 x in $C'P$.

$\overleftarrow{R}_*'x$ means the posterity of a term with respect to the relation R
($ibid.$, p. 637).

$(\overleftarrow{R}_*'x) \dashv R$ means the relation R confined to the posterity of x.

In the second volume of the $Principia \ mathematica$ (1912) Whitehead and Russell treat of "Cardinal Arithmetic" (Part III), "Rela-
tion-Arithmetic" (Part IV), and begin the subject of "Series" (Part
V), which they continue in the third volume (which was brought out
in 1913). This third volume gives also a treatment of "Quantity"
(Part VI).

Some new notations and fresh interpretations of symbols previously used are introduced in the last two volumes. In the second
volume the theory of types is continued and applied to cardinal
number. "Contradictions concerning the maximum cardinal are
solved by this theory" (Vol. II, p. 3).

$Nc'a$ means the class of all classes similar to a, i.e., as $\hat{\beta}(\beta \ sm \ a)$. This
 definition is due to Frege and was first published in 1884. For
 formal purposes of definition, the authors put

$Nc = \overrightarrow{sm}$ Df.
$NC = D'Nc$ Df. ($ibid.$, p. 5).

$k \ \overline{sm} \ \overline{sm} \ \lambda = (1 \to 1) \frown \overleftarrow{Cl}'S'\lambda \frown \hat{T}(k = T_\epsilon''\lambda)$ Df. ($ibid.$, p. 88). If P
 and Q are generating relations of two series ($ibid.$, p. 347),
$P \nleftrightarrow Q = P \cup Q \cup C'P \uparrow C'Q$ Df.

Lack of space prevents us from enumerating all the specialized
notations adopted in the treatment of particular topics.

696. $Signs \ of \ Poretsky.$—The Russian, Platon Poretsky,[1] uses a
symbolism of his own. For example, he starts out with three known
forms of the logical equality $A = B$, which he expresses in the form
$(A = B) = (A_0 = B_0) = (AB_0 + A_0B = 0) = ((AB + A_0B_0 = 1)$ where A_0 and
B_0 are the negations of the classes A and B. Letting N stand for
$nihil$ or the complete zero, M for $mundus$ or the logical all, he re-
writes the foregoing relations thus:

$$(A = B) = (A_0 = B_0) = (0 = N) = (1 = M) \ ,$$

where $N = M_0 = AB_0 + A_0B, \ M = N_0 = AB + A_0B_0.$

---

[1] P. Poretsky, $Sept \ Lois \ fondamentales \ de \ la \ théorie \ des \ égalités \ logiques$ (Kazan,
1899), p. 1.

In logical inequalities[1] he takes $A < B$, meaning "A is contained in B" or $A = AB$; $A > B$, meaning "A contains B" or $A = A + C$; $A \neq B$, meaning "A is not equal to B." The author uses $\because$ for "since" and $\therefore$ for "therefore."

Signs of Julius König are found in his book[2] of 1914. For example, "a rel$_a$ b" stands for the declaration that "the thing a" corresponds a-wise to "the thing b"; $\cong$ is the sign for isology, $\times$ for conjunction, $\overset{\smile}{+}$ for disjunction, $(=)$ for equivalence.

697. *Signs of Wittgenstein.*—A partial following of the symbols of Peano, Whitehead, and Russell is seen in the *Tractatus Logico-Philosophicus* of Ludwig Wittgenstein. "*That a* stands[3] in a certain relation to b says *that aRb*"; $\sim p$ is "not p," p v q is "p or q" (p. 60, 61). "If we want to express in logical symbolism the general proposition 'b is a successor of a' we need for this an expression for the general term of the formal series: aRb, $(\exists x):aRx.xRb$, $(\exists x, y):aRx.xRy.yRb, \ldots$" (p. 86, 87).

"When we conclude from p v q and $\sim p$ to q the relation between the forms of the propositions 'p v q' and '$\sim p$' is here concealed by the method of symbolizing. But if we write, e.g., instead of 'p v q,' '$p \mid q \cdot \mid \cdot p \mid q$' and instead of '$\sim p$', '$p \mid p$' ($p \mid q =$ neither p nor q, then the inner connexion becomes obvious" (p. 106, 107).

Our study of notations in mathematical logic reveals the continuance of marked divergence in the symbolism employed. We close with some general remarks relating to notations in this field.

698. *Remarks by Rignano and Jourdain.*—Jourdain says:[4] "He [Rignano][5] has rightly emphasized the exceedingly important part that analogy plays in facilitating the reasoning used in mathematics, and has pointed out that a great many important mathematical discoveries have been brought about by this property of the symbolism used. Further he held that the symbolism of mathematical logic is

[1] P. Poretsky, *Théorie des non-égalités logiques* (Kazan, 1904), p. 1, 22, 23.

[2] Julius König, *Neue Grundlagen der Logik, Arithmetik und Mengenlehre* (Leipzig, 1914), p. 35, 73, 75, 81.

[3] Ludwig Wittgenstein, *Tractatus Logico-Philosophicus* (English ed.; London, 1922), p. 46, 47.

[4] P. E. B. Jourdain, "The Function of Symbolism in Mathematical Logic," *Scientia*, Vol. XXI (1917), p. 1–12.

[5] Consult Eugenio Rignano in *Scientia*, Vol. XIII (1913), p. 45–69; Vol. XIV (1913), p. 67–89, 213–39; Vol. XVII (1915), p. 11–37, 164–80, 237–56. See also G. Peano, "Importanza dei simboli in matematica," *Scientia*, Vol. XVIII (1915), p. 165–73.

not to be expected to be so fruitful, and, in fact, that it has not been so fruitful.

"It is important to bear in mind that this study is avowedly concerned with the psychological aspect of symbolism, and consequently such a symbolism as that of Frege, which is of much less service in the economy of thought than in the attainment of the most scrupulous precision, is not considered. . . . .

"In Peano's important work, dating from 1888 onwards, on logical and mathematical symbolism, what he always aimed at, and largely succeeded in attaining, was the accurate formulation and deduction from explicit premises of a number of mathematical theories, such as the calculus of vectors, arithmetic, and metrical geometry. It was, then, consistently with this point of view that he maintained that Rignano's criticisms hold against those who consider mathematical logic as a science in itself, but not against those who consider it as an instrument for solving mathematical problems which resist the ordinary methods.

"The proper reply to Rignano seems, however, to be that, until comparatively lately, symbolism in mathematics and the algebra of logic had the sole aim of helping reasoning by giving a fairly thorough analysis of reasoning and a condensed form of the analyzed reasoning, which should, by suggesting to us analogies in familiar branches of algebra, make mechanical the process of following the thread of deduction; but that, on the other hand, a great part of what modern mathematical logic does is to increase our subtlety by emphasizing *differences* in concepts and reasonings instead of *analogies*."

699. *A question.*—No topic which we have discussed approaches closer to the problem of a uniform and universal language in mathematics than does the topic of symbolic logic. The problem of efficient and uniform notations is perhaps the most serious one facing the mathematical public. No group of workers has been more active in the endeavor to find a solution of that problem than those who have busied themselves with symbolic logic—Leibniz, Lambert, De Morgan, Boole, C. S. Peirce, Schröder, Peano, E. H. Moore, Whitehead, Russell. Excepting Leibniz, their mode of procedure has been in the main individualistic. Each proposed a list of symbols, with the hope, no doubt, that mathematicians in general would adopt them. That expectation has not been realized. What other mode of procedure is open for the attainment of the end which all desire?

III

SYMBOLS IN GEOMETRY
(ADVANCED PART)

1. RECENT GEOMETRY OF TRIANGLE AND CIRCLE, ETC.

700. Lack of space prevents the enumeration of the great masses of symbols occurring in the extensive literature on the recent geometry of the triangle and circle. There exists an almost hopeless heterogeneity in symbolism. This is evident from a comparison of the publications of J. Lange,[1] A. Emmerich,[2] C. V. Durell,[3] and Godfrey and Siddons.[4] Lange represents a triangle by ABC; the feet of the altitudes by $A_.$, $B_.$, $C_.$; the intersection of the altitudes by H; the center of the circumcenter by M; the centers of the circles of contact by J, J_a, J_b, J_c; the point of contact of J with BC by X; the intersection of AJ with BC by D; the diameter of the circumcircle that is perpendicular to BC by EF; the radius of the circumcenter by r; the midpoints of the sides by A', B', C'; the midpoints of the upper parts of the altitudes by A_0, B_0, C_0; the center of gravity by S; the center of the circumcircle of $A'B'C'$ by M'; the radius of the incircle by ρ.

Emmerich's designation of terms and list of signs are in part as follows: The Brocard points (p. 23) are Ω and Ω'; the Grebe point (p. 23) is K; the circumcenter of a triangle (p. 29) is O; the collineation center of the first Brocard triangle $\triangle$ (p. 85), D; the center of the Feuerbach circle (p. 97) is F; the intersection of a median (p. 118) from the vertex A of a triangle with the opposite side g_a; the angle that this median makes with the side b is $\angle (g_{a,b})$; the centers of the Neuberg circles (p. 131) are N_a, N_b, N_c; homologous points respective to the sides a, b, c of a $\triangle ABC$ (p. 138) are J_a, J_b, J_c; the centers of McKay circles (p. 142) are M_a, M_b, M_c; antiparallelism is denoted (p. 13) by $a\|$.

[1] Julius Lange, *Geschichte des Feuerbachschen Kreises* (Berlin, 1894), p. 4.

[2] A. Emmerich, *Die Brocardschen Gebilde* (Berlin, 1891). See also the "Anhang."

[3] Clement V. Durell, *Modern Geometry. The Straight Line and Circle* (London, 1920).

[4] C. Godfrey and A. W. Siddons, *Modern Geometry* (Cambridge, 1908), p. 16.

The Ω and Ω' for Brocard points are found also in Lachlan's treatise.[1]

The symbols of Godfrey and Siddons, and Durell, are in partial agreement. Godfrey and Siddons give the following list of symbols:

A, B, C vertices of the triangle	$\triangle$ area of the triangle
D, E, F feet of the altitudes	S circumcenter
a, β, γ midpoints of the sides	H orthocenter
X, Y, Z points of contact of the incircle	G centroid
	I incenter
a, b, c lengths of the sides	I_1, I_2, I_3 excenters
s semiperimeter	N nine-points center
R circumradius	P, Q, R midpoints of HA, HB,
r inradius	HC
r_1, r_2, r_3 exradii	

701. *Geometrography.*—Lemoine[2] in his *Géométrographie* used:

C_1 to mark the placing a point of the compasses at a fixed point

C_2 to mark the placing a point of the compass at an undetermined point of a line

C_3 drawing the circle

R_1 placing a ruler at a point

R_2 drawing a straight line

He counts the number of steps in a geometric construction by the polynomial

$$l_1 \cdot R_1 + l_2 \cdot R_2 + l_3 \cdot C_1 + l_4 \cdot C_2 + l_5 \cdot C_3$$

and represents the degree of simplicity by the sum

$$l_1 + l_2 + l_3 + l_4 + l_5$$

and the degree of precision by

$$l_1 + l_3 + l_4 .$$

702. *Signs for polyhedra.*—In the expression of the relation between the edges, vertices, and faces of a convex polyhedron the initial letters of the words for "edges," "vertices," "faces" have usually been employed. Hence the designations have varied with the language used. The theorem was first enunciated by Descartes, but his

---

[1] R. Lachlan, *Modern Pure Geometry* (London, 1893), p. 65

[2] E. Lemoine, *Comptes rendus*, Vol. CVII (1888), p. 169. See also J. Tropfke, *op. cit.*, Vol. IV (2d ed., 1923), p. 91.

treatment of it was not published until 1859 and 1860.[1]   It was found in a manuscript of Leibniz in the Royal Library of Hanover and is an excerpt made by Leibniz, during his residence in Paris in the years 1675–76, from a Cartesian manuscript. Descartes says: "Ponam semper pro numero angularum solidorum a et pro numero facierum φ," that is, a is the number of vertices (*anguli solidi*) and φ is the number of faces (*facies*). He states the theorem thus:[2] "Numerus verorum angulorum planorum est $2\varphi+2a-4$." ("The number of polygonal angles is $2\varphi+2a-4$.")   As the sum of the polygonal angles is twice as great as the number of edges, it is evident that Descartes' phrasing is equivalent to the statement: faces +vertices=edges+2. Euler,[3] after whom the theorem is named, gave it the form $S+H=A+2$, where $S=$ *numerus angulorum solidorum* (vertices), $A=$ *numerus acierum* (edges), $H=$ *numerus hedrarum* (faces). Euler's letters were retained by Legendre,[4] Poinsot,[5] and Cayley,[6] the last two in their discussions introducing some half-dozen additional letters.

In the statement of the theorem, German writers[7] usually employ the letters e (*Ecken*), k (*Kanten*), f (*Flächen*); English writers,[8] e ("edges"), v ("vertices"), f ("faces").

In the morphology of polyhedra unusual symbols appear here and there. V. Eberhard[9] explains the notation used by him thus:

According as the plane a_n contains a three-, four-, or five-sided face of the n-faced A_n, it cuts from the $(n-1)$-faced A_{n-1} a three-edged vertex (a_i, a_k, a_l), or an edge $|a_k, a_l|$, or two edges meeting in a vertex $|a_l a_k,|$ and $|a_i, a_l|$. For these three operations Eberhard selects, respectively, the symbolism

$$a_n \doteq (a_i, a_k, a_l), \qquad a_n \stackrel{\bullet}{=} |a_k, a_l|, \qquad a_n \triangle [\,|a_i, a_k|, \,|a_i, a_l|\,].$$

[1] *Œuvres de Descartes* (éd. C. Adam et P. Tannery), Vol. X (Paris, 1908), p. 257.

[2] *Op. cit.*, Vol. X, p. 269.

[3] L. Euler, *Novi Comment. Acad. Petropol. ad annum 1752 et 1753* (1758), Vol. IV, p. 119, 158–60. See also M. Cantor, *Vorlesungen*, Vol. III (2. Aufl.), p. 557.

[4] A. M. Legendre, *Éléments de géométrie* (Paris, 1794), p. 228.

[5] L. Poinsot, *Jour. Polyt.*, Vol. IV (1810), p. 16–48.

[6] A. Cayley, *London, Edinburgh and Dublin Philosophical Magazine* (4th ser.). Vol. XVII (1859), p. 123–28.

[7] See, for instance, H. Durège, *Elemente der Theorie der Funktionen* (Leipzig, 1882), p. 226

[8] For instance, W. W. Beman and D. E. Smith, *New Plane and Solid Geometry* (Boston, 1899), p. 284.

[9] V. Eberhard, *Zur Morphologie der Polyeder* (Leipzig, 1891), p. 17, 29.

If, in a special case, the plane a_n passes through several vertices, a, b, ... of A_{n-1}, this is expressed by

$$(a, b, \ldots) \, a_n \,\overset{\cdot}{-}, \; \overset{\underline{\quad}}{-}, \; \triangle \, .$$

If a_i is a plane (p. 29) of the polyhedron A_n, then the m-sided polygon bounding the face is designated by $\langle a_i \rangle_m$. Two faces, $\langle a_i \rangle$ and $\langle a_k \rangle$, can be in direct connection with each other in three ways: They have one edge in common, for which we write $\langle a_i \rangle | \langle a_k \rangle$; or there exist outside of the two faces m edges such that each edge connects a vertex of the one face with a vertex of the other face, for which we

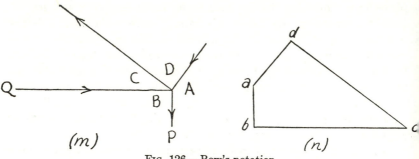

(m)   *(n)*

Fig. 126.—Bow's notation

write $\langle a_1 \rangle \rangle \overline{}_{m} \langle \langle a_k \rangle$; or both of these circumstances occur simultaneously, for which we write $\langle a_i \rangle |, \rangle \overline{}_{m} \langle \langle a_k \rangle$.

Max Brückner[1] designates the relation between three planes (p), (q), (r) without common edges by the symbols $(p) \overline{}_{x} (q)$, $(q) \overline{}_{y} (r)$, $(r) \overline{}_{z} (p)$, where x, y, z are the numbers of edges (*Scheitelkanten*) to be determined.

703. *Geometry of graphics.*—The civil engineer of Edinburgh, R. H. Bow,[2] devised the notation shown in Figure 126, known as "Bow's notation," in practical graphics. Diagram (m) shows the lines of action of a number of forces which are in equilibrium. Instead of denoting a force by a single letter, P, as is practiced in some systems, the force is here designated by two letters which are placed on opposite sides of the line of action of the force in the fram-diagram (m), and at the angular points of the polygon in force-diagram (n). Thus the force P is referred to in Bow's notation as AB.

[1] Max Brückner, *Vielecke und Vielflache. Theorie and Geschichte* (Leipzig, 1900), p. 87.

[2] Robert Henry Bow, *Economics of Construction* (London, 1873), p. 53–55. Our figure is taken from D. A. Low, *Practical Geometry and Graphics* (London, 1912), p. 92.

The angle of contact[1] of a curve $y = ax^m$ (where $a > 0$ and $m > 1$) with the axis of X as its tangent is designated by Giulio Vivanti by the symbol (a, m), so that $(a, m) > (a', m)$ when $a > a'$, and $(a, m) < (a', m')$ when $m > m'$.

2. PROJECTIVE AND ANALYTICAL GEOMETRY

704. Poncelet,[2] in his *Traité des propriétés projectives des figures*, uses little symbolism. He indicates points by capital letters A, B, C, ..., and the center of projection by S.

705. *Signs for projectivity and perspectivity.*—Von Staudt, in his *Geometry of Position* (1847), expressed himself as follows:[3] "Gleichwie ein Gebilde, welches aus den Elementen A, B, C, D besteht, gewöhnlich durch $ABCD$ bezeichnet wird, so soll ein Gebilde, welches aus den Elementen SA, SB, SC, SD besteht, durch $S(ABCD)$ bezeichnet werden." Again:[4] "Zwei einförmige Grundgebilde heissen zu einander projektivisch ($\wedge$), wenn sie auf einander bezogen sind, dass jedem harmonischen Gebilde in dem einen ein harmonisches Gebilde im andern entspricht." Also:[5] "Wenn in zwei einander projektivischen Grundgebilden je zwei homologe Elemente P, P_1 einander doppelt (abwechselnd) entsprechen, nämlich dem Punkte P des einen Gebildes das Element P_1 des anderen, und dem Elemente P_1 des ersteren das Element P des letzteren entspricht, so heissen die Gebilde involutorisch liegend ($\overline{\overline{\wedge}}$)."

Here we have introduced the two noted signs $\overline{\wedge}$ and $\overline{\overline{\wedge}}$. Neither came to be adopted by all prominent writers on projective geometry, but $\overline{\wedge}$ has met with far wider acceptance than the other. One finds $\overline{\wedge}$ employed by Hankel,[6] Reye,[7] Fiedler,[8] Halsted,[9] Aschieri,[10] Sturm,[11]

---

[1] C. Isenkrahe, *Das Endliche und das Unendliche* (Münster, 1915), p. 306.

[2] J. V. Poncelet, *Traité des propriétés projectives des figures* (2d ed.), Vol. I (Paris, 1865), p. 6. (First edition appeared in 1822.)

[3] Georg Kane Christian Von Staudt, *Geometrie der Lage* (Nürnberg, 1847), p. 32.

[4] *Op. cit.*, p. 49.

[5] *Op. cit.*, p. 118. Von Staudt uses this symbolism also in his *Beiträge*, fasc. 3, p. 332.

[6] H. Hankel, *Elemente der projectivischen Geometrie* (Leipzig, 1875), p. 79.

[7] Theodor Reye, *Die Geometrie der Lage*, Vol. I (2d ed.; Leipzig, 1882), p. 121. (First edition, 1866.)

[8] W. Fiedler, *Darstellende Geometrie*, 1. Theil (Leipzig, 1883), p. 31.

[9] G. B. Halsted, *Elementary Synthetic Geometry* (New York, 1892), p. vii.

[10] Ferdinando Aschieri, *Geometria projettiva* (Milano, 1895), p. 162.

[11] Rudolf Sturm, *Liniengeometrie* (Leipzig, 1896), p. 8.

Henrici and Treutlein,[1] Bobek,[2] Böger,[3] Enriques,[4] Amodeo,[5] Doehlemann,[6] Filon,[7] Schur,[8] Guareschi,[9] Schoenflies and Tresse,[10] Holgate,[11] and Dowling.[12] Frequently the notation $\overline{\wedge}$ is used in modern geometry to indicate a homology of points; thus, $ABC \ldots MN \ldots$ $\overline{\wedge} A'B'C' \ldots M'N'$ indicates that $ABC \ldots MN \ldots$ are homologues with $A'B'C' \ldots M'N' \ldots$ in two homographic figures of range 1, 2, or 3.[13] This same notation is used also to designate homologous elements of two planes or two sheaves, or two spaces, in homographic correspondence.

The sign $\overline{\overline{\wedge}}$ to mark perspective position is found in Böger,[14] Doehlemann,[15] Holgate,[16] and Dowling. Some prominent writers, Cremona[17] for instance, make no use either of $\overline{\overline{\wedge}}$ or $\overline{\wedge}$.

706. *Signs for harmonic and anharmonic ratios.*—Möbius[18] in 1827 marked the anharmonic ratio (*Doppelschnittsverhältniss*) of four

[1] J. Henrici and P. Treutlein, *Lehrbuch der Elementar-Geometrie*, 2. Teil (Leipzig, 1897), p. 10, 42.

[2] Karl Bobek, *Einleitung in die projektivische Geometrie der Ebene* (Leipzig, 1897), p. 7.

[3] R. Böger, *Elemente der Geometrie der Lage* (Leipzig, 1900), p. 17.

[4] Federigo Enriques, *Vorlesungen über Projektive Geometrie*, deutsch von H. Fleischer (Leipzig, 1903), p. 106.

[5] F. Amodeo, *Lezioni di geometria proiettiva* (Napoli, 1905), p. 60.

[6] Karl Doehlemann, *Projektive Geometrie*, 3. Aufl. (Leipzig, 1905), p. 78.

[7] L. N. G. Filon, *Introduction to Projective Geometry* (London, [1908]), p. 46.

[8] F. Schur, *Grundlagen der Geometrie* (Leipzig und Berlin, 1909), p. 60.

[9] G. Guareschi in E. Pascal's *Repertorium* (2d ed.), Vol. II₁ (Leipzig, 1910), p. 114.

[10] A. Schoenflies and A. Tresse in *Encyclopédie des scien. math.*, Tome III, Vol. II (1913), p. 32, 49.

[11] T. F. Holgate in J. W. A. Young's *Monographs on* .... *Modern Mathematics* (New York, 1911), p. 75.

[12] L. W. Dowling, *Projective Geometry* (New York, 1917), p. 19, 38.

[13] A. Schoenflies and A. Tresse in *Encyclopédie des scien. math.*, Tom. III, Vol. II (1913), p. 32.

[14] Rudolf Böger, *Elemente der Geometrie der Lage* (Leipzig, 1900), p. 14.

[15] Karl Doehlemann, *Projektive Geometrie*, 3. Aufl. (Leipzig, 1905), p. 132.

[16] T. F. Holgate, *op. cit.*, p. 75.

[17] Luigi Cremona, *Projective Geometry* (trans. Ch. Leudesdorf; 2d ed.; Oxford, 1893).

[18] A. F. Möbius, *Der barycentrische Calcul* (Leipzig, 1827); *Gesammelte Werke*, Vol. I (Leipzig, 1885), p. 221.

collinear points, that is, the ratio of ratios, $\dfrac{AC}{CB} : \dfrac{AD}{DB}$, in which AB is divided by C and D, by the symbolism (A,B,C,D).

Chasles[1] in 1852 did not use any special notation and denoted the anharmonic ratio, in which cd is divided by a and b, by $\dfrac{ac}{ad} : \dfrac{bc}{bd}$. Salmon[2] marks the anharmonic ratio of four points A, B, C, D by $\dfrac{AB}{BC} \div \dfrac{AD}{DC}$ and uses the abbreviation $\{ABCD\}$, and, for the anharmonic ratio of pencils, $\{O.ABCD\}$. Salmon omits the commas. Clifford[3] writes this relation for the points A,B,D,E in the form $[ADBE]$. In 1865 Chasles[4] adopts for the anharmonic ratio of four points a, b, c, d the notation (a, b, c, d), and for pencils (O_a, O_b, O_c, O_d) or $O(a, b, c, d)$. J. W. Russell[5] writes the anharmonic ratio $\dfrac{BC}{CA} \div \dfrac{BD}{DA}$ of four collinear points A, B, C, D in the form (BA, CD); and, correspondingly, $V(BA, CD)$. Others prefer braces,[6] as in $\{AB, PQ\}$. W. A. Whitworth[7] uses $\{p.abcd\}$ to denote the anharmonic ratio of the points in which the straight lines a, b, c, d intersect the straight line p.

707. *Signs in descriptive geometry.*—In this field of geometry gross diversity of notation has existed. Monge[8] had a simple symbolism. The projection of a point upon the horizontal plane he marked by a capital letter, say D, and the projection on the vertical plane by the corresponding small letter d; the projections of a line were marked AB and ab, respectively. The traces of a plane he represents by capital letters, such as AB, BC. He uses single and double accents for the marking of certain corresponding points.

Aubré[9] uses the letters h ("horizontal") and v ("vertical"), and

[1] M. Chasles, *Traité de géométrie supérieure* (Paris, 1852), p. 28, 29, 183, etc.

[2] George Salmon, *Conic Sections* (6th ed.; London, 1879), § 326, p. 297, 301; the first edition appeared in 1848; the third in 1855.

[3] W. K. Clifford in *Oxford, Cambridge and Dublin Messenger of Mathematics*, Vol. II (1864), p. 233; *Mathematical Papers* (London, 1882), p. 27.

[4] M. Chasles, *Traité des sections coniques* (Paris, 1865), p. 7, 8.

[5] J. W. Russell, *Pure Geometry* (Oxford, 1893), p. 98, 102.

[6] R. Lachlan, *Modern Pure Geometry* (London, 1893), p. 266.

[7] William Allen Whitworth, *Trilinear Coordinates* (Cambridge, 1866), p. 381.

[8] *Géométrie descriptive*, par G. Monge (5th ed., M. Brisson; Paris, 1827), p. 19, 20. The first edition bears the date of An III (1795).

[9] *Cours de géométrie descriptive, d'après la méthode de M. Th. Olivier*, par L. E. Aubré (1856), p. 6. Oliver's first edition (1843) and second (1852) contained the notations described by Aubré.

designates the projection upon the horizontal and vertical planes of a point o by o^h and o^v. Aubré marks the four right dihedral angles formed by the interesting horizontal and vertical planes by $\widehat{A,S}$, $\widehat{P,S}$, $\widehat{P,I}$, $\widehat{A,I}$, where A stands for *antérieur*, S for *supérieur*, P for *postérieur*, I for *inférieur*. A straight line in space is marked by capital letter, say D, and its projections by D^h and D^v. Church[1] denoted a point in space by, say, M; its horizontal projection by m; and its vertical by m'; a line by, say, MN. Wiener[2] represents points by Latin capitals, surfaces by black-faced Latin capitals, lines by small Latin letters, angles by small Greek letters. Berthold[3] marks a point by, say, a; its horizontal projection by a'; and its vertical by a''. He marks a line by, say, ab. Bernhard[4] prefers small Latin letters for points, Latin capitals for straight lines and surfaces, small Greek letters for angles. He uses $\times$ for intersection, so that $a = G \times L$ means that the point a is the intersection of the coplanar lines G and L; P_1 and P_2 are the two mutually perpendicular planes of projection; X the axis of projection; a' and a'' are the projections of the point a, G' and G'' of the line G; g_1 and g_2 are the traces of G; E_1 and E_2 are the traces of the plane E.

Different again is the procedure of Blessing and Darling.[5] With them, planes are represented by capital letters, the last letters of the alphabet being used, thus, P, Q, R, S, T, U; the horizontal trace of a plane has the letter H prefixed, as in HP, HQ, etc.; the vertical trace has the letter V prefixed, as in VP, VQ, etc. The ground line is marked $G - L$; a line in space $C - D$ has two projections, $c - d$ and $c' - d'$.

The illustrations cited suffice to show the lack of uniformity in notation.[6] They indicate, moreover, that the later books on descriptive geometry introduce a larger amount of symbolism than did the earlier.

[1] Albert Church, *Elements of Descriptive Geometry* (New York, 1870), p. 2, 5.

[2] Christian Wiener, *Lehrbuch der Darstellenden Geometrie*, Vol. I (Leipzig, 1884), p. 62.

[3] *Die Darstellende Geometrie*, von W. H. Behse, bearbeitet von P. Berthold, 1. Theil (Leipzig, 1895), p. 3, 11.

[4] Max Bernhard, *Darstellende Geometrie* (Stuttgart, 1909), p. 3.

[5] George F. Blessing and Lewis A. Darling, *Elements of Descriptive Geometry* (New York, 1913), p. 19, 20, 49.

[6] Consult also G. Loria, "I doveri dei simboli matematici," *Bollettino della Mathesis* (Pavia, 1916), p. 35.

708. *Signs in analytical geometry.*—Descartes introduced the letters $z, y, x, \ldots$ as variable co-ordinates (§§ 339, 340). With him a letter represented as a rule only a positive number. It was Johann Hudde who in 1659 first let a letter stand for negative, as well as positive, values (§ 392). In Descartes' *Géométrie* (1637), the co-ordinate axes are in no case explicitly set forth and drawn. He selects a straight line which he sometimes calls a "diameter" and which serves as the x-axis. The formal introduction of the y-axis is found in G. Cramer,[1] though occasional references to a y-axis are found earlier in De Gua, L. Euler, W. Murdock, and others. With Descartes an equation of a geometric *locus* is not considered valid except for the angle of the co-ordinates (quadrant) in which it was established. The extension of an equation to other quadrants was freely made in particular cases for the interpretation of negative roots.

Polar co-ordinates were introduced in a general manner by Jakob Bernoulli[2] and Pierre Varignon.[3] The latter puts $x = \rho$ and $y = l\omega$ and gets wholly different curves; the parabolas $x^m = a^{m-1}y$ become the Fermatian spirals.

An abridged notation and new systems of co-ordinates were introduced by E. Bobillier[4] and by Julius Plücker. Bobillier writes the equations of the sides of a triangle in the form $A = 0$, $B = 0$, $C = 0$, and the conic passing through the three vertices of the triangle, $aBC + bCA + cAB = 0$, where a, b, c may take any values. Bobillier proves Pascal's theorem on a hexagon inscribed in a conic in this manner:[5] Let $aAA' - bBB' = 0$ be the conic circumscribing a quadrilateral, then the six sides of the inscribed hexagon may be written $A = 0$, $\gamma A + bB = 0$, $\gamma B' + aA' = 0$, $A' = 0$, $aA' + \gamma'B = 0$, $\gamma'A + bB' = 0$. The Pascal line is then $\gamma\gamma'A - abA' = 0$. Plücker used abridged notations and formulated the principles of duality and homogeneity. The abridged notations are used extensively in several chapters of Salmon's *Conic Sections* (London, 1848), and in later editions of that work.

[1] G. Cramer, *Introduction à l'analyse des lignes courbes algébriques* (1750).

[2] Jakob Bernoulli, *Opera*, Vol. I (Geneva, 1744), p. 578–79. See also G. Eneström, *Bibliotheca mathematica* (3d ser.), Vol. XIII (1912–13), p. 76.

[3] *Mémoires de l'académie r. d. sciences*, année 1704 (Paris, 1722).

[4] E. Bobillier in Gergonne's *Annales de mathématiques*, Vol. XVIII (1828), p. 320, etc. See also Julius Plücker, *Gesammelte Mathematische Abhandlungen*, Vol. I (Leipzig, 1895), p. 599.

[5] E. Bobillier in Gergonne's *Annales*, Vol. XVIII, p. 359–67. See also E. Kötter, *Entwickelung der synthetischen Geometrie* (Leipzig, 1901), p. 27 n.

709. *Plücker's equations* on the ordinary singularities of curves exhibit considerable variety, among different authors, in the use of letters. Thus, the order (degree) of a curve is marked n by Plücker,[1] Berzolari,[2] and Kötter,[3] and m by Salmon;[4] the class of the curve is marked m by Plücker and Kötter, n' by Berzolari, n by Salmon; the number of double points is marked x by Plücker, δ by Salmon and Kötter, d by Berzolari; the number of double tangents is marked u by Plücker, τ by Salmon and Kötter, d' by Berzolari; the number of stationary points (*Rückkehrpunkte, Spitzen*) is marked y by Plücker, κ by Kötter and Salmon, τ by Berzolari; the number of stationary tangents (*Wendepunkte, Wendetangenten*) is marked v by Plücker, ι by Salmon and Kötter, r' by Berzolari.

710. *The twenty-seven lines on cubic surface.*—For the treatment of these lines, noteworthy notations have been invented. We give that of Andrew S. Hart, as described by Salmon:[5] "The right lines are denoted by letters of three alphabets, as follows:

$$A_1, B_1, C_1 \; ; \qquad A_2, B_2, C_2 \; ; \qquad A_3, B_3, C_3 \; ;$$
$$a_1, b_1, c_1 \; ; \qquad a_2, b_2, c_2 \; ; \qquad a_3, b_3, c_3 \; ;$$
$$a_1, \beta_1, \gamma_1 \; ; \qquad a_2, \beta_2, \gamma_2 \; ; \qquad a_3, \beta_3, \gamma_3 \; .$$

Letters of the same alphabet denote lines which meet if either the letters or the suffixes be the same;'for example, $A_1 A_2 A_3$ denote lines on the same plane as also do $A_1 B_1 C_1$. Letters of different alphabets denote lines which meet," according to a given table. Salmon introduced a notation of his own.

The notation most generally adopted is that of the "double-six," due to L. Schläfli,[6] of Bern. He represents a double six by

$$\begin{pmatrix} a_1, & a_2, & a_3, & a_4, & a_5, & a_6 \\ b_1, & b_2, & b_3, & b_4, & b_5, & b_6 \end{pmatrix}$$

[1] Julius Plücker, *Theorie der algebraischen Curven* (Bonn, 1839), p. 211.

[2] Berzolari in Pascal's *Reportorium der höheren Mathematik* (ed. P. Epstein und H. E. Timerding), Vol. II (2d ed.; Leipzig und Berlin, 1910), p. 286.

[3] Ernst Kötter, *Entwickelung der synthetischen Geometrie* (Leipzig, 1901), p. 465.

[4] George Salmon, *Higher Plane Curves* (3d ed.; Dublin, 1879), p. 64.

[5] G. Salmon, *Cambridge und Dublin Mathematical Journal*, Vol. IV (1849), p. 252, 253. See also A. Henderson, *The Twenty-seven Lines upon the Cubic Surface* (Cambridge, 1911).

[6] L. Schläfli, *Quarterly Journal of Mathematics*, Vol. II (London, 1858), p. 116.

and finds that "no two lines of the same horizontal row and no two lines of the same vertical row intersect, but any two lines otherwise selected do intersect."

711. *The Pascal hexagram.*—The treatment of this complicated figure exhibiting its properties has given rise to various notations. With Plücker,[1] if (a) and (b) represented two points, then (a, b) represented the line joining them, and $[(a, b), (c, d)]$ or $[a, b; c, d]$ represented the intersection of the right lines (a, b), (c, d). A corresponding interpretation was given to these symbols when (a) and (b) represented two right lines. Plücker designated the vertices of a hexagon inscribed in a conic by the numerals (1), (2), (3), (4), (5), (6), and the line joining, say, (1) and (2), by $(1, 2)$. Also, by $\begin{cases}(1,2)\ (2,3)\ (3,4)\\(4,5)\ (5,6)\ (6,1)\end{cases}$ he designated the right line passing through the three points in which the pairs of lines $(1, 2)$ and $(4, 5)$, $(2, 3)$ and $(5, 6)$, $(3, 4)$ and $(6, 1)$, intersect. He distinguishes the sixty different hexagons obtained by changing the sequence of the points (1), (2), (3), (4), (5), (6), by the order of the numerals, and arranges these hexagons into twenty groups, so that the sixth group, for example, is written

$$\left.\begin{array}{cccccc}1&2&3&4&5&6\\1&6&3&2&5&4\\1&4&3&6&5&2\end{array}\right\} \text{ VI}$$

Upon the Pascal line of 1 2 3 4 5 6 lie the three points $(1, 2; 4, 5)$, $(2, 3; 5, 6)$, $(3, 4; 6, 1)$. He found it convenient to write a group, say Group VI, also in the form

$$\begin{pmatrix}1&3&5\\2&4&6\end{pmatrix}, \qquad \begin{pmatrix}3&5&1\\2&4&6\end{pmatrix}, \qquad \begin{pmatrix}1&5&3\\4&2&6\end{pmatrix}.$$

Kirkman[2] takes the numerals 1 2 3 4 5 6 to represent the vertices of the hexagon inscribed in a conic C and lets $12=0$ be the equation of the right line joining 1 and 2; $1\ 2.3\ 4.5\ 6-\lambda2\ 3.4\ 5.6\ 1=0=CA$ is a Pascal line (A). The twenty points through which the sixty Pascal lines pass, three by three, are marked G. He denotes the line A by $A\ 1\ 2\ 3\ 4\ 5\ 6$, and the point G by $G\ 1\ 2.\ 3\ 4.5\ 6$. Pascal lines pass in addition, three together, through sixty points H, marked according to the plan H (12. 34) 56; the twenty points G lie, four together, on

[1] Plücker, *Crelle's Journal*, Vol. V (1830), p. 270–77.

[2] T. P. Kirkman, *Cambridge and Dublin Mathematical Journal*, Vol. V (1850), p. 185.

fifteen lines I, marked according to the plan I 12. 34. 56. The ninety lines, each containing two of the sixty points H, are marked J, according to the plan J 12. 43. The sixty points H lie three together on twenty lines, X.

Different notations are employed by von Staudt,[1] Cayley,[2] and Salmon.[3] A new notation was offered by Christine Ladd (Mrs. Franklin),[4] in which the vertices of the conic S are represented by A, B, C, D, E, F; the lines tangent to S at these vertices by a, b, c, d, e, f; the intersection of two fundamental lines AB, DE is called $P(AB.DE)$; the line joining two fundamental points ab, de is called $p'(ab.de)$. Here $p'(ab.de)$ is the pole of $P(AB.DE)$. The Pascal line of $ABCDEF$ is called $h(ABCDEF)$; it passes through the points $P(AB.DE)$, $P(BC.EF)$, $P(CD.FA)$. Similarly, the intersection of the lines $p'(ab.de)$, $p'(bc.ef)$, $p'(cd.fa)$ is the Brianchon point $H'(abcdef)$ of the hexagon $abcdef$. The three Pascal lines, $h(ABCFED)$, $h(AFCDEB)$, $h(ADCBEF)$, meet in a Steiner point $G(ACE.BFD)$. The four G points, $G(BDA.ECF)$ $G(EDF.BCA)$, $G(BCF.EDA)$, $G(BDF.ECA)$, lie on a Steiner-Plücker line $i(BE.CD.AF)$. The Kirkman point $H(ABCDEF)$ is the intersection of the Pascal lines $h(ACEBFD)$, $h(CEADBF)$, $h(EACFDB)$. A Cayley-Salmon line is marked $g(ACE.BFD)$; a Salmon point by $I(BE.CD.AF)$. Cayley "gives a table to show in what kind of a point each Pascal line meets every one of the 59 other Pascal lines. By attending to the notation of Pascal lines such a table may be dispensed with. His 90 points m, 360 points r, 360 points t, 360 points z, and 90 points w are the intersections each of two Pascals whose symbols can easily be derived one from another." For instance,

$$\left.\begin{array}{l} h(ABCDEF) \\ h(ABC\,FED) \end{array}\right\rangle m\,.$$

"By the notation here given," continues the author, "it is immediately evident what points are on every line and what lines pass through every point, without referring to tables, as Veronese[5] is obliged to do."

[1] Von Staudt, *Crelle's Journal*, Vol. LXII (1863), p. 142–50.

[2] A. Cayley, *Quarterly Journal*, Vol. IX (1868), p. 268–74.

[3] G. Salmon, *Conic Sections* (6th ed., 1879), p. 379–83 n.

[4] C. Ladd, *American Journal of Mathematics*, Vol. II (1879), p. 1–12.

[5] Veronese's notation and additions to the Pascal hexagram are given in *Atti della R. Accademia dei Lincei* (3d ser.), "Memorie della classe de scienze fisiche ecc.," Vol. I (Rome, 1879), p. 649–703.

IV

THE TEACHINGS OF HISTORY

A. THE TEACHINGS OF HISTORY AS INTERPRETED BY VARIOUS WRITERS; INDIVIDUAL JUDGMENTS

712. *Review of André.*—A reviewer of Désiré André's book, *Notations mathématiques* (Paris, 1909), directs attention to the fact that, before the appearance of this book, no work existed that was devoted exclusively to mathematical notations. The reviewer says:[1] "They have at all times and have now more and more a very considerable influence upon the progress of the science. One may say that without them modern mathematics could not exist. They constitute therefore, in the study of the sciences, a subject of capital importance. One is therefore amazed that there exists no work consecrated entirely to them."

713. *Augustus de Morgan,* a close student of the history of mathematics, expresses himself, in part, as follows:[2] ". . . . Mathematical notation, like language, has grown up without much looking to, at the dictates of convenience and with the sanction of the majority. Resemblance, real or fancied, has been the first guide, and analogy has succeeded."

"Signs are of two kinds,—1st. Those which spring up and are found in existence, but cannot be traced to their origin; 2ndly, Those of which we know either the origin, or the epoch of introduction, or both. Those of the first kind pass into the second as inquiry advances. . . . ."

"Mathematical marks or signs differ from those of written language in being almost entirely of the purely abbreviative character.

". . . . Too much abbreviation may create confusion and doubt as to the meaning; too little may give the meaning with certainty, but not with more certainty than might have been more easily attained. . . . ."

"With the exception of an article by Mr. Babbage in the Edinburgh Encyclopædia, we do not know of anything written in modern times on notation in general.

[1] *Bulletin des sciences mathématiques* (2d ser.), Vol. XXXIII (October, 1909), p. 228.

[2] A. de Morgan, *Penny Cyclopaedia* (1842), art. "Symbols."

"We hardly need mention a thing so well known to the mathematician as that the progress of his science now depends more than at any previous time upon protection of established notation, when good, and the introduction of nothing which is of an opposite character. . . . . The language of the exact sciences is in a continual state of wholesome fermentation, which throws up and rejects all that is incongruous, obstructive and even useless. . . . .

"RULES.—Distinctions must be such only as are necessary, and they must be sufficient. . . . . The simplicity of notative distinctions must bear some proportion to that of the real differences they are meant to represent. . . . . Pictorial or descriptive notation is preferable to any other, when it can be obtained by simple symbols. . . . . Legitimate associations which have become permanent must not be destroyed, even to gain an advantage. . . . .

"A few Cambridge writers have of late years chosen to make a purely arbitrary change, and to signify by dy, dz, etc., not increments, but limiting ratios of increments: and students trained in these works must learn a new language before they can read Euler, Lagrange, Laplace, and a host of others. Thus $d_x y$ has been made to stand for $dy:dx$, and the old association connected with dy has (in the works spoken of) been destroyed. . . . .

"Notation may be modified for mere work in a manner which cannot be admitted in the expression of results which are to be reflected upon. The mathematical inquirer must learn to substitute, for his own private and momentary use, abbreviations which could not be tolerated in the final expression of results. . . . .

"Among the worst of barbarisms is that of introducing symbols which are quite new in mathematical, but perfectly understood in common, language. Writers have borrowed from the Germans the abbreviation $n!$ to signify $1.2.3 . . . . (n-1)n$, which gives their pages the appearance of expressing surprise and admiration that 2, 3, 4, etc., should be found in mathematical results. . . . .

"The subject of mathematical printing has never been methodically treated, and many details are left to the compositor which should be attended to by the mathematician. Until some mathematician shall turn printer, or some printer mathematician, it is hardly to be hoped that this subject will be properly treated."

714. *J. W. L. Glaisher*[1] emphasizes the important rôle which notations have played in the development of mathematics: "Nothing

[1] J. W. L. Glaisher, "Logarithms and Computation" from the *Napier Tercentenary Memorial Volume* (ed. Cargill Gilston Knott; London, 1915), p. 75.

in the history of mathematics is to me so surprising or impressive as the power it has gained by its notation or language. No one could have imagined that such 'trumpery tricks of abbreviation' as writing $+$ and $-$ for 'added to' and 'diminished by,' or x^2, x^3, ... for xx, xxx, ..., etc., could have led to the creation of a language so powerful that it has actually itself become an instrument of research which can point the way to future progress. Without suitable notation it would be impossible to express differential equations, or even to conceive of them if complicated, much less to deal with them; and even comparatively simple algebraic quantities could not be treated in combination. Mathematics as it has advanced has constructed its own language to meet its need, and the ability of a mathematician in devising or extending a new calculus is displayed almost as much in finding the true means of representing his results as in the discovery of the results themselves.

"When mathematical notation has reached a point where the product of n x'es was replaced by x^n, and the extension of the law $x^m x^n = x^{m+n}$ had suggested $x^{\frac{1}{2}} \cdot x^{\frac{1}{2}} = x$ so that $x^{\frac{1}{2}}$ could be taken to denote $\sqrt{x}$, then fractional exponents would follow as a matter of course, and the tabulation of x in the equation $10^x = y$ for integral values of y might naturally suggest itself as a means of performing multiplication by addition. But in Napier's time, when there was practically no notation, his discovery or invention [of logarithms] was accomplished by mind alone without any aid from symbols."

715. *D. E. Smith*[1] directs attention to the teachings of history as aids to modern efforts to improve our notations: "In considering the improvement of our present algebraic symbolism, the possible elimination of certain signs that may have outlived their usefulness, the conservation of the best that have come down to us, the variation observed in passing from the literature of one country to that of another, and the possible unifying of the best of our present symbols, it has been found helpful to consider the early history of the subject."

716. *Alexandre Savérien* in the eighteenth century uttered a protest to the conditions then existing:[2] "Of what advantage is this multiplicity of characters? Is this expression for a square $\overline{a+b}^2$ not sufficiently simple and natural without recourse to this one $\bigodot \overline{a+b}^2$, and

---

[1] David Eugene Smith, in an Introductory Note to Suzan R. Benedict's "Development of Algebraic Symbolism from Paciuolo to Newton" in *Teachers College, Columbia University, Department of Mathematics* (1906–7), p. 11.

[2] Savérien *Dictionnaire universel de mathematique et de physique*, Vol. I (Paris, 1753), art. "Caractere."

is $\sqrt{aa+2ab+bb}$ not better than uu which some wish to substitute for it? The reader will pardon me for writing thus. I could not let this occasion pass without stating what I think on this subject. Nothing is more pernicious than diversity in expressions. To invent new characters which signify nothing else than those which have been already accepted, is a wanton embroilment of things. If only one could once reach agreement on symbols, accord would continue. For a long time we have known that 6 signifies 'six'; what would we say, if some one instructed us to make it equal to 'seven'! Oh! What can be more useless, and more capable of disgusting a beginner and of embarrassing a geometer, than the three expressions ., :, ÷ to mark division? How could we guess that this meant to divide, rather than to multiply or to indicate a continued arithmetical progression, since the first and the last characters are designed, one for multiplication and one for progression? You may say all you wish about this, but I believe that the less one uses characters, the more one learns of mathematics. The memory is less burdened and consequently mathematical propositions are more easily mastered."

717. *Colin Maclaurin.*—The obscurity of concepts which has sometimes hidden behind mathematical symbols is treated by the eighteenth-century writer, Maclaurin, in the following passage:[1]

"The improvements that have been made by it [the doctrine of fluxions] . . . . are in a great measure owing to the facility, conciseness, and great extent of the method of computation, or algebraic part. It is for the sake of these advantages that so many symbols are employed in algebra. . . . . It [algebra] may have been employed to cover, under a complication of symbols, obstruse doctrines, that could not bear the light so well in a plain geometrical form; but, without doubt, obscurity may be avoided in this art as well as in geometry, by defining clearly the import and use of the symbols, and proceeding with care afterwards.

"Hence[2] the $\sqrt{-1}$ or the square-root of a negative, implies an imaginary quantity, and in resolution is a mark or character of the impossible cases of a problem; unless it is compensated by another imaginary symbol or supposition, when the whole expression may have real signification. Thus $1+\sqrt{-1}$, and $1-\sqrt{-1}$ taken separately are imaginary, but their sum is 2; as the conditions that separately

[1] Colin Maclaurin, *A Treatise of Fluxions*. In Two Books, Vol. II (Edinburgh, MDCCXLII), p. 575, 576.

[2] C. Maclaurin, *op. cit.*, Vol. II, p. 577, 578.

would render the solution of a problem impossible, in some cases destroy each others effect when conjoined. . . . . The theorems that are sometimes briefly discovered by the use of this symbol, may be demonstrated without it by the inverse operation, or some other way; and tho' such symbols are of some use in the computations in the method of fluxions, its evidence cannot be said to depend upon any arts of this kind."

718. *Charles Babbage* stresses the power of algebraic symbolism as follows:[1] "The quantity of meaning compressed into small space by algebraic signs, is another circumstance that facilitates the reasonings we are accustomed to carry on by their aid. The assumption of lines and figures to represent quantity and magnitude, was the method employed by the ancient geometers to present to the eye some picture by which the course of their reasonings might be traced: it was however necessary to fill up this outline by a tedious description, which in some instances even of no peculiar difficulty became nearly unintelligible, simply from its extreme length: the invention of algebra almost entirely removed this inconvenience. . . . . A still better illustration of this fact is noticed by Lagrange and Delambre, in their report to the French Institute on the translation of the works of Archimedes by M. Peyrard.[2] It occurs in the ninth proposition of the second book on the equilibrium of planes, on which they observe, 'La demonstration d'Archimede a trois énormes colonnes in-folio, et n'est rien moin que lumineuse.' Eutochius commence sa note 'en disant que le theorême est fort peu clair, et il promet de l'expliquer de son mieux. Il emploie quatre colonnes du même format et d'un caractère plus serré sans reussir d'avantage; au lieu que quatre lignes d'algebre suffisent a M. Peyrard pour mettre la verité du theorême dans le plus grand jour.' "

But Babbage also points out the danger of lack of uniformity in notations: "Time which has at length developed the various bearings of the differential calculus, has also accumulated a mass of materials of a very heterogeneous nature, comprehending fragments of unfinished theories, contrivances adapted to peculiar purposes, views perhaps sufficiently general, enveloped in notation sufficiently obscure, a multitude of methods leading to one result, and bounded by the same difficulties, and what is worse than all, a profusion of notations (when we regard the whole science) which threaten, if not duly

[1] C. Babbage, "On the Influence of Signs in Mathematical Reasoning," *Trans. Cambridge Philosophical Society*, Vol. II (1827), p. 330.

[2] *Ouvrages d'Archimede* (traduites par M. Peyrard), Tom. II, p. 415.

corrected, to multiply our difficulties instead of promoting our progress," p. 326.

719. *E. Mach*[1] discusses the process of clarification of the significance of symbols resulting from intellectual experimentation, and says: "Symbols which initially appear to have no meaning whatever, acquire gradually, after subjection to what might be called intellectual experimenting, a lucid and precise significance. Think only of the negative, fractional and variable exponents of algebra, or of the cases in which important and vital extensions of ideas have taken place which otherwise would have been totally lost or have made their appearance at a much later date. Think only of the so-called imaginary quantities with which mathematicians long operated, and from which they even obtained important results ere they were in a position to assign to them a perfectly determinate and withal visualizable meaning. . . . . Mathematicians worked many years with expressions like $\cos x + \sqrt{-1} \sin x$ and with exponentials having imaginary exponents before in the struggle for adapting concept and symbol to each other the idea that had been germinating for a century finally found expression [in 1797 in Wessel and] in 1806 in Argand, viz., that a relationship could be conceived between magnitude and *direction* by which $\sqrt{-1}$ was represented as a mean direction-proportional between $+1$ and -1."

720. *B. Branford* expresses himself on the "fluidity of mathematical symbols" as follows:[2] "Mathematical symbols are to be temporarily regarded as rigid and fixed in meaning, but in reality are continually changing and actually fluid. But this change is so infinitely gradual and so wholly subconscious in general that we are not sensibly inconvenienced in our operations with symbols by this paradoxical fact. Indeed, it is actually owing to this strange truth that progress in mathematical science is possible at all. An excellent instance is the gradual evolution of algebra from arithmetic—a clear hint this for teachers."

721. *A. N. Whitehead*[3] writes on the power of symbols: "By relieving the brain of all unnecessary work, a good notation sets it free to concentrate on more advanced problems, and in effect increases the mental power of the race. Before the introduction of the

---

[1] E. Mach, *Space and Geometry* (trans. T. J. McCormack, 1906), p. 103, 104.

[2] Benchara Branford, *A Study of Mathematical Education* (Oxford, 1908), p. 370.

[3] A. N. Whitehead, *An Introduction to Mathematics* (New York and London, 1911), p. 59.

Arabic notation, multiplication was difficult, and the division even of integers called into play the highest mathematical faculties. Probably nothing in the modern world would have more astonished a Greek mathematician than to learn that, under the influence of compulsory education, the whole population of Western Europe, from the highest to the lowest, could perform the operation of division for the largest numbers. . . . . Mathematics is often considered a difficult and mysterious science, because of the numerous symbols which it employs. Of course, nothing is more incomprehensive than a symbolism which we do not understand. Also a symbolism, which we only partially understand and are unaccustomed to use, is difficult to follow. . . . . By the aid of symbolism, we can make transitions in reasoning almost mechanically by the eye, which otherwise would call into play the higher faculties of the brain. It is a profoundly erroneous truism, repeated by all copy-books and by eminent people when they are making speeches, that we should cultivate the habit of thinking of what we are doing. The precise opposite is the case. Civilization advances by extending the number of important operations which we can perform without thinking about them."

722. *H. F. Baker*[1] discusses the use of symbols in geometry: "Two kinds of symbols are employed in what follows: Symbols for geometrical elements, generally points, which we may call element-symbols; and symbols subject to laws of computation, which we may call algebraic symbols. . . . . The use of algebraic symbols for geometrical reasoning dates from very early times, ... and has prompted much in the development of Abstract Geometry. . . . . And, while it would be foolish not to employ the symbols for purposes of discovery, the view taken in the present volume is, that the object of a geometrical theory is to reach such a comprehensive scheme of conceptions, that every geometrical result may be obvious on geometrical grounds alone."

723. *H. Burkhardt*[2] utters his conviction that in geometry symbolic language has limitations: "An all-inclusive geometrical symbolism, such as Hamilton and Grassmann conceived of, is impossible."

724. *P. G. Tait*[3] emphasizes the need of self-restraint in the pro-

[1] H. F. Baker, *Principles of Geometry*, Vol. I (Cambridge, 1922), p. 70.

[2] H. Burkhardt in *Jahresbericht der Deutschen Math. Vereinigung*, Vol. V (1901, p. 52), p. xi.

[3] P. G. Tait, *An Elementary Treatise on Quaternions* (2d ed., enlarged; Cambridge. Preface dated 1873), p. xi.

posal of new notations: "Many abbreviations are possible, and some-times very useful in private work; but, as a rule, they are unsuited for print. Every analyst, like every shorthand writer, has his own special contractions; but when he comes to publish his results, he ought in-variably to put such devices aside. If all did not use a common mode of public expression, but each were to print as he is in the habit of writing for his own use, the confusion would be utterly intoler-able."

O. S. Adams, in a review of W. D. Lambert's *Figure of the Earth,*[1] dwells upon the notational diversities in geodesy:[2] "A feature of the publication that is to be commended is the table of comparison of symbols used by various authors. One of the things that is always discouraging in reading scientific literature is the use of different notations for the same thing by different authors. Not the least of such offenders against clearness are the mathematicians both pure and applied. This table is therefore of great value to those who want to look into this question without too much preliminary study."

725. *British Committee.*[3]—The attention which should be given to typography is brought out by a British Committee as follows: "The process of 'composition' of ordinary matter consists in arranging types uniform in height and depth (or 'body' as it is termed) in simple straight lines. The complications peculiar to mathematical matter are mainly of two kinds.

"First, figures or letters, generally of a smaller size than those to which they are appended, have to be set as indices or suffixes; and consequently, except when the expressions are of such frequent occur-

[1] W. D. Lambert, *Effect of Variations in the Assumed Figure of the Earth on the Mapping of a Large Area,* Spec. Pub. No. 100 of the U.S. Coast and Geodetic Survey (Washington, 1924), p. 26, 35.

[2] O. S. Adams, *Science* (N.S.), Vol. LX (1924), p. 269.

[3] "Report of the Committee, consisting of W. Spottiswoode, F.R.S., Pro-fessor Stokes, F.R.S., Professor Cayley, F.R.S., Professor Clifford, F.R.S., and J. W. L. Glaisher, F.R.S., appointed to report on Mathematical Notation and Printing with the view of leading Mathematicians to prefer in optional cases such forms as are more easily put into type, and of promoting uniformity of notation," from the *British Association Report for 1875,* p. 337.

"Suggestions for notation and printing" were issued in 1915 by the Council of the London Mathematical Society (see *Mathematical Gazette,* Vol. VIII [London, 1917], p. 172, 220) and by G. Peano in an article, "L'esecuzione tipografica delle formule matematiche," in *Atti della r. accademia delle scienze di Torino,* Vol. LI (Torino, 1916), p. 279–86. See also Gino Loria, "I doveri dei simboli matematici," *Bollettino della Mathesis* (Pavia, 1916), p. 32–39.

rence as to make it worth while to have them cast upon type of the various bodies with which they are used, it becomes necessary to fit these smaller types in their proper positions by special methods. This process, which is called 'justification,' consists either in filling up the difference between the bodies of the larger and smaller types with suitable pieces of metal, if such exist, or in cutting away a portion of the larger, so as to admit the insertion of the smaller types.

"The second difficulty arises from the use of lines or 'rules' which occur between the numerator and denominator of fractions, and (in one mode of writing) over expressions contained under radical signs. In whatever part of a line such a rule is used, it is necessary to fill up, or compensate, the thickness of it throughout the entire line. When no letters or mathematical signs occur on a line with the rule the process is comparatively simple; but when, for example, a comma or sign of equality follows a fraction, or a $+$ or $-$ is prefixed to it, the middle of these types must be made to range with the rule itself, and the thickness of the rule must be divided, and half placed above and half below the type.

"The complications above described may arise in combination, or may be repeated more than once in a single expression; and in proportion as the pieces to be 'justified' become smaller and more numerous, so do the difficulties of the workman, the time occupied on the work, and the chances of subsequent dislocation of parts augment.

"The cost of 'composing' mathematical matter may in general be estimated at three times that of ordinary or plain matter. . . . .

"Your Committee are unwilling to close this Report without alluding to the advantages which may incidentally accrue to mathematical science by even a partial adoption of the modifications here suggested. Anything which tends towards uniformity in notation may be said to tend towards a common language in mathematics; and whatever contributes to cheapening the production of mathematical books must ultimately assist in disseminating a knowledge of the science of which they treat.

"MATHEMATICAL SIGNS NOT INVOLVING 'JUSTIFICATION'

$$\times \quad - \quad + \quad = \quad \sqrt{} \quad \pm \quad :: \quad \therefore \quad \because \quad : \quad \} \quad < \quad > \quad \div$$

$$(\quad [\quad \} \quad \int \quad \sqrt{}$$

$$a, \ a_1, \ a^1, \ a^2, \ a_2 \ \& \ , \ a^{\frac{1}{2}}, \ a_{\frac{1}{2}}.$$

"EQUIVALENT FORMS

Involving Justification	Not Involving Justification
$\dfrac{x}{a}$	$x \div a$ or $x:a$
$\sqrt{x}$	$\sqrt{x}$ or $x^{\frac{1}{2}}$
$\sqrt[3]{x}$	$\sqrt[3]{x}$ or $x^{\frac{1}{3}}$
$\sqrt{x-y}$	$\sqrt{(x-y)}$ or $(x-y)^{\frac{1}{2}}$
$\sqrt{-1}$	i
$x \cdot \overline{x+a}$	$x(x+a)$
$\epsilon^{\frac{n\pi x}{a}}$	$\exp{(n\pi x)}(a)$
$\tan^{-1} x$	arc tan x
$\dfrac{x}{l}=\dfrac{y}{m}=\dfrac{z}{n}$	$x:y:z = l:m:n$ "

726. *In the United States*, the editors of several mathematical journals,[1] in 1927, issued "Suggestions to Authors" for typewriting articles intended for print. The editors recommend the use of the solidus to avoid fractions in solid lines, the use of fractional exponents to avoid root signs; they give a list of unusual types available on the monotype machines which are used in printing the mathematical journals.

B. EMPIRICAL GENERALIZATIONS ON THE GROWTH OF MATHEMATICAL NOTATIONS

727. *Forms of symbols.*—A. *Primitive forms:* (1) Many symbols originated as abbreviations of words. Later, some of them (their ancestral connections being forgotten) assumed florescent forms and masqueraded as ideographic symbols. Examples: $+$ (*et*), π (periphery of circle of unit diameter), $\int$ (*summa*), d (*differentia*), i (imaginary), log (logarithm), lim (limit), $\sim$ (similar), sin, cos, tan, sec. (2) Others are pictographic or picture symbols, such as $\triangle$ (triangle), $\bigcirc$ (circle), $\square$ (square), $\parallel$ (parallel), $\square$ (parallelogram). (3) Others still are ideographic or arbitrary symbols: $\times$ (multiplication), $\div$ (division), ( ) aggregation, $\therefore$ (therefore), $\because$ (since), $=$ (equality), and letters (not abbreviates) representing numbers or magnitudes.

[1] For instance, see *Bull. Amer. Math. Soc.*, Vol. XXXIII (May–June, 1927), the advertising page following p. 384.

B. *Incorporative forms*, representing combinations of two or more mathematical ideas and symbols, such as $\int_a^b$ in definite integrals, $_nC_r$ in the theory of combinations, $\lim\limits_{x \to a}$ in the theory of limits, $f(x)$ in the theory of functions.

C. The preference for symbols which avoid double or multiple lines of type $\left(\text{as illustrated by } \dfrac{a^n}{b^m}\right)$ is intensified by the decline of type-setting by hand and the increased use of machines.

Invention of symbols.—Whenever the source is known, it is found to have been individualistic—the conscious suggestion of one mind.

728. *Nature of symbols.*—(1) Some are merely shorthand signs which enable an otherwise long written statement to be compressed within a small space for convenient and rapid mental survey. (2) Others serve also in placing and keeping logical relationships before the mind. (3) Adaptability of symbols to changing viewpoint and varying needs constitutes superiority. The Leibnizian $\int$ has admitted of incorporative devices in integration, such as the Newtonian $\square$ or $\dot{x}$ could not admit so easily. The first derivative $\dfrac{dy}{dx}$ lends itself readily to change into differentials by elementary algebraic processes.

729. *Potency of symbols.*—Some symbols, like a^n, $\dfrac{d^n y}{dx^n}$, $\sqrt[n]{n}$, $\log n$, that were used originally for only positive integral values of n stimulated intellectual experimentation when n is fractional, negative, or complex, which led to vital extensions of ideas.

730. *Selection and spread of symbols.*—(1) While the origin of symbols has been individualistic, their adoption has been usually by non-conferring groups of mathematicians. As in other inventions, so here, many symbols are made, but few are chosen. The list of discarded mathematical signs would fill a volume. (2) In no branch of mathematics is the system of symbols now used identical with what any one inventor ever designed them to be; present notations are a mosaic of individual signs of rejected systems. Modern elementary algebra contains symbols from over a dozen different inventors. Excepting Leibniz and Euler, no mathematician has invented more than two ideographs which are universally adopted in modern mathematics. (3) Often the choice of a particular symbol was due to a special configuration of circumstances (large group of pupils friend-ships, popularity of a certain book, translation of a text) other than

those of intrinsic merit of the symbol. Thus the inferior : : was adopted in proportion after = had been introduced as the sign of equality. At times there existed other extraneous influences. Perhaps commercial relations with Hebrews and Syrians led the Greeks to abandon the Herodianic numerals and to adopt an inferior alphabetic numeral system. Friction between British and continental mathematicians prevented England, to her loss, from adopting in the eighteenth century the Leibnizian calculus notations, in which the great analytical researches of Swiss and French mathematicians were being recorded. (4) Mathematical symbols cross linguistic borders less readily than do mathematical ideas. This observation applies to abbreviates and ideographs rather than to pictographs. The Arabs adopted no symbols used by the Babylonians, Greeks, or Hindus (except probably the Hindu numerals), nevertheless the Arabs acquired the Babylonian sexagesimals, Hindu trigonometry, part of Hindu algebra, Diophantine algebra, and the Greek system of geometric axioms and proofs. The Arabic algebraic symbolism of the fifteenth century antedated the German *coss*, yet acquired no foothold in Christian Europe. In the sixteenth century the Italian solutions of cubic and quartic equation were eagerly studied in Germany, but the Italian notation was ignored by the Germans. Sometimes even national boundaries between linguistically homogeneous peoples yield diversity of symbols. The notation 2.5 means $2\frac{5}{10}$ in the United States and 2 times 5 in Great Britain; the notation 2·5 means 2 times 5 in the United States and $2\frac{5}{10}$ in Great Britain. The symbols + and − which sprang up in Germany in the fifteenth century were extremely convenient, yet they were not introduced into France, Italy, England, and Spain until after the middle of the sixteenth century. These symbols first reached France by the republication in Paris, in 1551, of an algebra written in Latin by the German author, Scheubel, and first published in Germany. These symbols first appear in Italy in 1608 in a book written by a German (Clavius) who had taken up his residence in Rome. They first appear in Spain in a book written in Spain in 1552 by a German immigrant. In some cases the symbols were actually carried across national borders by individual travelers. The book trade did not seem to stimulate the prompt spread of symbols. The continent of America offers the earliest fully developed number system (the Maya system), constructed on the principle of local value and a symbol for zero, but the knowledge of this did not spread at the time. Mathematical symbols proposed in America such as ⇌ for equivalent or ≐ for approach to a limit, or Gibbs's symbols in vector analysis, or

C. S. Peirce's signs in symbolic logic, were not adopted in Europe. In fact, no American mathematical symbol, except perhaps the dollar mark ($), has found acceptance in Europe. And yet the ideas set forth in Gibbs's vector algebra and in C. S. Peirce's symbolic logic admittedly impressed themselves upon European thought. (5) The rejection of old symbols and the adoption of new ones are greatly retarded by acquired habits. The symbol $\div$ for subtraction has persisted for four hundred years among small groups of Teutonic writers even though less desirable than the usual symbol $-$. (6) Mathematicians have uniformly resisted the general adoption of very lavish and profuse symbolism for any branch of mathematics. The eruption of symbols due to Oughtred, Hérigone, Rahn, and Hindenburg has been followed by the adoption of only certain few of them.

731. *State of flux.*—The maxim of Heraclitus that all things are in a state of flux applies to mathematical symbolism. During the last four centuries symbols of aggregation have undergone several changes. The :: in proportion is now passing into disuse; the arrow as a symbol for approaching a limit is meeting with enthusiastic adoption everywhere.

732. *Defects in symbolism.*—(1) Sometimes the haphazard selection by non-conferring groups of workers has led to two or more things being represented by the same symbol. One cannot interpret dx without first knowing whether it occurs in algebra or in the calculus. The sign $\sim$ may mean "difference" or "similar to" or "equivalent to" or "approximately equal to" or, in the case of $a_v \sim b_v$, that the ratio of a_v and b_v approaches a constant as $v \rightarrow \infty$. (2) Often two or more symbols are used for the same thing, causing waste of intellectual effort. Algebra has a calculus of radicals and a calculus of fractional exponents, both covering the same field. Millions of children have been forced to master both systems. Here Descartes and Newton missed a splendid opportunity to prevent useless duplication. However, there are fields where a certain amount of duplication is convenient and not burdensome; for instance, in forms of derivatives,

$\frac{dy}{dx}$, $f'(x)$, also $\dot{x}$ for $\frac{dx}{dt}$.

733. *Individualism a failure.*—As clear as daylight is the teaching of history that mathematicians are still in the shadow of the Tower of Babel and that individual attempts toward the prompt attainment of uniformity of notations have been failures. Mathematicians have not been profiting by the teachings of history. They have failed to adopt

the alternative procedure. Mathematical symbols are for the use of the mathematical community, and should therefore be adopted by that community. The success of a democracy calls for mass action. As long as it is not physically possible for all to gather in close conference, the political procedure of acting through accredited representatives becomes necessary. Mathematicians as a group are far less efficient than are the citizens of our leading republics as groups. The latter have learned to act through representatives; they have learned to subordinate their individual preferences and to submit themselves to the laws passed for the greatest good of all. The mathematicians have not yet reached that stage of co-operative endeavor in the solution of the all-important question of mathematical symbolism. There is need of cultivation of the spirit of organization and co-operation.

C. CO-OPERATION IN SOME OTHER FIELDS OF SCIENTIFIC ENDEAVOR

734. *Electric units.*—How does it happen that in electricity and magnetism there is perfect agreement in the matter of units of measurement in scientific books and in commercial affairs throughout the world? How was this uniformity brought about? The answer is that the scientists in this field of investigation held international meetings, that representatives of the various countries in serious conference were able to reach agreements which the separate countries afterward confirmed by legal enactments. At the International Congress of Electricians, held in Paris in 1881, the centimeter-gram-second system was adopted and the foundation was laid for the adoption also of commercial units. A uniform system was reached only by international conferences, and by the furtherance of the spirit of compromise. The millimeter-second-milligram system of Wilhelm Weber was dropped; the British Association unit of resistance was dropped. In their stead appeared slightly modified units.

It must be admitted that in the selection of electrical and magnetic units forces were at work which are not present to the same degree in attempts to unify mathematical notations. In the case of electromagnetism there existed financial considerations which made unification imperative. International commerce in electrical machinery and electric instruments rendered a common system of units a necessity. Without international means of measurement, international exports and imports would be seriously embarrassed.

735. *Star Chart and Catalogue.*—That international co-operation on the part of scientific men is possible is shown again by the gigantic enterprise to prepare an *International Star Chart and Catalogue*, which was initiated by the International Astrographic Congress that met at Paris in 1887. About eighteen observatories in different parts of the world entered upon this colossal task. In this undertaking the financial aspect did not play the intense rôle that it did in securing uniformity of electrical units. It constitutes an example of the possibility of international co-operation in scientific work even when material considerations are not involved.

It is true, however, that in this astronomical program the individual worker encountered fewer restrictions than would a mathematician who was expected in printed articles to follow the recommendations of an international committee in matters of mathematical notation.

D.   GROUP ACTION ATTEMPTED IN MATHEMATICS

736. It is only in recent years that attempts have been made to introduce uniform notations through conferences of international representatives and recommendations made by them. The success of such endeavor involves two fundamental steps. In the first place, the international representatives must reach an agreement; in the second place, individual writers in mathematics must be willing to adopt the recommendations made by such conferences. Thus far no attempt at improvement of notations can be said to have weathered both of these ordeals.

737. *Vector analysis.*—In 1889 a small international organization was formed, called the "International Association for Promoting the Study of Quaternions and Allied Systems of Mathematics." Sir Robert S. Ball was president and A. Macfarlane secretary. At the international congress of mathematicians held at Rome in 1908 a committee was appointed on the unification of vectorial notations, but at the Congress held in Cambridge in 1912, no definite conclusion had been reached (§ 509). Considerable discussion was carried on in various journals, some new symbols were proposed, but the final and critical stage when a small representative group of international representatives gather around a table and make serious attempts to secure an agreement was never reached. The mathematical congresses held at Strasbourg in 1920 and at Toronto in 1924 did not take up the study of vectorial notations. The Great War seriously interrupted all plans for co-operation in which all the great nations could participate.

738. *Potential and elasticity.*—Another failure, due chiefly to the break of friendly international relations during the Great War,

marks the work of an International Committee of sixty members, formed in 1913, on the unification of notations and terminology in the theories of potential and elasticity. Arthur Korn, of Charlottenburg, was corresponding secretary. The Committee had laid out a regular program extending over about eight years. Mathematicians, physicists, and astronomers of the world were asked to co-operate and, in the first place, to assist the Committee by answering the question, "What are the notions and notations in respect to which it is desirable to establish uniformity?" In course of the year 1914 a second circular was to be issued for suggestions as to methods by which the desired unification may be brought about. It was the original intention to send out a third circular in 1916, setting forth the points of dispute, and to arrange a discussion thereon at the International Congress of Mathematicians that was to be held in 1916. A fourth circular, to be issued in 1917, should contain a report of the discussion and provide an opportunity for those who should not have been present at the Congress to express their views in writing. All the proposals and contributions to the discussion were to be sifted and arranged, and a fifth circular was to be sent out in 1919 containing a statement of the points with regard to which agreement had been reached and to take a vote on those in dispute. The final vote was to be taken at the International Congress of Mathematicians in 1920. The year following, a sixth circular was to announce the result of the vote. This extensive scheme was nipped in the bud by the war. The present writer has seen only the first circular issued by the Committee.[1] As no truly international congress of mathematicians has been held since 1912, the proposed work of unification has not yet been resumed. The old organization no longer exists. When the work is started again, it must be *de novo*.

739. *Actuarial science.*—A third movement, which was not interrupted by war, met with greater success. We refer to the field of actuarial science. In this line the British Actuarial Society had been active for many years, devising a full system of notation, so clear and efficient as to meet with quite general approval. With some slight modifications, the English proposals were adopted by the International Congress of Actuaries in 1895 and 1898. As a result of this action, many of the symbols in a very long list have been widely adopted, but as far as we can learn, they are not used to the exclusion of all other symbols. This movement marks a partial success. Per-

[1] For an account of the movement see *Isis*, Vol. I (1913), p. 491; Vol. II (1914), p. 184.

haps some improvement in the mode of procedure might have led to a more universal acceptance of the symbols agreed upon by the international experts. Perhaps a mistake was made in endeavoring to reach agreement on too many symbols at one time. The general adoption of a small number of symbols concerning which there was the greatest unanimity, rather than a list extending over six solid pages of printed matter, might have been better.[1]

Of the three attempts at unification through international action which we have cited, the first two failed, because the first essential step was not taken; the international representatives did not reach agreement. In the third experiment agreement was reached, but, as far as can be determined, more recent individual workers of different countries have not felt themselves compelled to adopt the symbols agreed upon. In fact, the diversity of plans of insurance which have been adopted in different countries makes it often difficult for some actuaries on the European Continent to use the notation which had its origin in England.

E. AGREEMENTS TO BE REACHED BY INTERNATIONAL COMMITTEES THE ONLY HOPE FOR UNIFORMITY IN NOTATIONS

740. Uniformity of mathematical notations has been a dream of many mathematicians—hitherto an iridescent dream. That an Italian scientist might open an English book on elasticity and find all formulae expressed in symbols familiar to him, that a Russian actuary might recognize in an English text signs known to him through the study of other works, that a German physicist might open an American book on vector analysis and be spared the necessity of mastering a new language, that a Spanish specialist might relish an English authority on symbolic logic without experiencing the need of preliminary memorizing a new sign vocabulary, that an American traveling in Asiatic Turkey might be able to decipher, without the aid of an interpreter, a bill made out in the numerals current in that country, is indeed a consummation devoutly to be wished.

Is the attainment of such a goal a reasonable hope, or is it a utopian idea over which no mathematician should lose precious time? In dealing with perplexing problems relating to human affairs we are prone to look back into history, to ascertain if possible what

[1] See list of symbols in the *Transactions of the Second International Actuarial Congress*, held at London in 1898 (London, 1899), p. 618–24.

light the past may shed upon the future and what roads hitherto have been found impassable. It is by examining the past with an eye constantly on the future that the historical student hopes to make his real contribution to the progress of intelligence.

741. The admonition of history is clearly that the chance, haphazard procedure of the past will not lead to uniformity. The history of mathematical symbolism is characterized by a certain painting representing the landing of Columbus; the artist painted three flags tossed by the breeze—one east, one west, and one south—indicating a very variable condition of the wind on that memorable day. Mathematical sign language of the present time is the result of many countercurrents. One might think that a perfect agreement upon a common device for marking decimal fractions could have been reached in the course of the centuries.

No doubt many readers will be astonished to hear that in recent books, printed in different countries, one finds as many as nine different notations for decimal fractions (§ 286). O Goddess of Chaos, thou art trespassing upon one of the noblest of the sciences! The force of habit conspires to the perpetuity of obsolete symbols. In Descartes' *Geometry* of 1637 appeared the modern exponential notation for positive integral powers. Its ideal simplicity marked it as a radical advance over the older fifteenth- and sixteenth-century symbolisms. Nevertheless, antiquated notations maintained their place in many books for the following fifty or seventy-five years. Other cases similar to this could be cited. In such redundancy and obsoleteness one sees the hand of the dead past gripping the present and guiding the future.

742. Students of the history of algebra know what a struggle it has been to secure even approximate uniformity of notation in this science, and the struggle is not yet ended. The description of all the symbols which have been suggested in print for the designation of the powers of unknown quantities and also of known quantities is a huge task. Rival notations for radical expressions unnecessarily complicate the study of algebra in our high schools today. And yet ordinary algebra constitutes the gateway to elementary mathematics. Rival notations embarrass the beginner at the very entrance into the field of this science.

This confusion is not due to the absence of individual efforts to introduce order. Many an enthusiast has proposed a system of notation for some particular branch of mathematics, with the hope, perhaps, that contemporary and succeeding workers in the same field would hasten to adopt his symbols and hold the originator in grateful remembrance. Oughtred in the seventeenth century used over one hundred and fifty signs for elementary mathematics—algebra, geometry, and trigonometry. Many of them were of his own design. But at the present time his sign of multiplication is the only one of his creation still widely used; his four dots for proportion are redundant and obsolescent; his sign for arithmetical difference ($\smile$) has been employed for wholly different purposes. In France, Hérigone's writings of the seventeenth century contained a violent eruption of symbols, now known only to the antiquarian. The Hindenburg combinatorial school at the close of the eighteenth century in Germany introduced complicated symbolic devices which long since passed into innocuous desuetude. The individual designer of mathematical symbols, looking forward to enduring fame, is doomed to disappointment. The direction of movement in mathematical notation, when left to chance, cannot be predicted. When we look into the history of symbolism, we find that real merit has not always constituted the determining factor. Frequently wholly different circumstances have dominated the situation. The sign of division ($\div$) now used in England and America first appears in a Swiss algebra of 1659 (§ 194), written in the German language, which enjoyed no popularity among the Germans. But an English translation of that book attracted wide attention in England and there led to the general adoption of that sign for division. The outcome of these chance events was that Great Britain and America have at the present time a sign of division used nowhere else; the Leibnizian colon employed for division in Continental Europe is not used for that purpose in the English-speaking countries. Another example of the popularity of a book favoring the adoption of a certain symbol is Byerly's *Calculus*, which secured the adoption in America of J. E. Oliver's symbol ($\doteq$) for approach to a limit. Oughtred's pupils (John Wallis, Christopher Wren, Seth Ward, and others) helped the spread in England of Oughtred's notation for proportion which attained popularity also in Continental Europe. The Dutch admirers of Descartes adopted his sign of equality which became so popular in Continental Europe in the latter part of the seventeenth century as almost to force out of use the Recordian sign of equality now generally adopted. The final victory of the Recordian sign over the Cartesian rival appears to be due to the fact that Newton and Leibniz, who were the two brightest stars in the mathematical firmament at the beginning of the eighteenth century, both used in print the Recordian equality. Thus history points to the generalization that individual effort has not led to uniformity in mathematical language. In fact, the signs of elementary algebra constitute a mosaic composed of symbols taken from the notations of more than a dozen mathematicians.

743. In the various fields of mathematics there has been, until very recently, no method of selection, save through mere chance. Consequently, some symbols have several significations; also, the same idea is often designated by many different signs. There is a schism between the situation as it exists and what all workers would like it to be. Specialists in advanced fields of mathematics complain that the researches of others are difficult to master because of difference in sign language. Many times have we called attention to passages in the writings of specialists in modern mathematics, deploring the present status. Usually the transition from one notation to another is not easy. In the eighteenth century mathematicians framed and tried to remember a simple rule for changing from the Leibnizian dx to the Newtonian $\dot{x}$. But the rule was inaccurate, and endless confusion arose between differentials and derivatives.

744. One might think that such a simple matter as the marking of the hyperbolic functions would not give rise to a diversity of notation, but since the time of Vincenzo Riccati there have been not less than nine distinct notations (§§ 529, 530, 536). No better is our record for the inverse trigonometric functions. For these there were at least ten varieties of symbols (§§ 532–36). Two varieties, namely, John Herschel's sign looking like exponent minus one, as in $\sin^{-1} x$, and the "arc sin," "arc cos," etc., of the Continent, still await the decision as to which shall be universally adopted.

745. The second half of the eighteenth century saw much experimentation in notations of the calculus, on the European Continent. There was a great scramble for the use of the letters D, d, and δ. The words "difference," "differential," "derivative," and "derivation" all began with the same letter. A whole century passed before any general agreement was reached. Particular difficulty in reaching uniformity prevailed in the symbolism for partial derivatives and partial differentials. A survey has revealed thirty-five different varieties of notation for partial derivatives (§§ 593–619). For about one hundred and fifty years experimentation was carried on before some semblance of uniformity was attained. Such uniformity could have been reached in fifteen years, instead of one hundred and fifty, under an efficient organization of mathematicians.

746. In this book we have advanced many other instances of

"muddling along" through decades, without endeavor on the part of mathematicians to get together and agree upon a common sign language. Cases are found in descriptive geometry, modern geometry of the triangle and circle, elliptic functions, theta functions, and many other fields. An investigator wishing to familiarize himself with the researches of other workers in his own field must be able not only to read several modern languages, English, French, Italian, and German, but he must master also the various notations employed by different authors. The task is unnecessarily strenuous. A French cynic once said: "Language is for the suppression of thought." Certainly the diversity of language tends to the isolation of thought. No doubt many mathematicians of the present time are wondering what will be the fate of the famous symbolisms of Peano and of Whitehead and Russell. In the light of past events we predict that these systems of symbols will disintegrate as did the individual systems of the seventeenth and eighteenth centuries.

747. It cannot be denied that in the desire to attain absolute logical accuracy in complicated and abstract situations arising in the study of the foundations of mathematics an elaborate system of ideographs such as are used by Peano and especially by Whitehead and Russell has been found serviceable (§§ 688–92, 695). But these symbols are intended by their inventors to be used in the expression of all mathematics in purely ideographic form.[1] Thus far, such a system designed to take the place of ordinary language almost entirely has not been found acceptable to workers in the general field of mathematics.[2] All teaching of history discourages attempts to force a scheme of this sort upon our science. More promising for mathematicians in general is some plan like that of E. H. Moore. Though fully cognizant of the need of new symbols for our growing science, he uses only a restricted amount of symbolism (§ 694) and aims at simplicity and flexibility of the signs used. He aims at a system of logical signs that would be acceptable to working mathematicians. New symbols should be ushered in only when the need for them is unquestioned. Moore's program for a mathematical sign

[1] G. Peano, *Formulaire de mathématiques*, Tom. I (Turin, 1895), Introduction, p. 52; A. N. Whitehead and B. Russell, *Principia mathematica*, Vol. I (Cambridge, 1910), Preface; also Vol. III (1913), Preface.

[2] On this question see also A. Voss, *Ueber das Wesen der Mathematik* (Leipzig und Berlin, 2. Aufl., 1913), p. 28; A. Padoa, *La logique déductive* (Paris, 1912); H. F. Baker, *Principles of Geometry*, Vol. I (Cambridge, 1922), p. 62, 137, and Vol. II, p. 4, 16, 153.

language is worthy of serious study. He believes in international co-operation and agreement, in matters of notation.

748. In the light of the teaching of history it is clear that new forces must be brought into action in order to safeguard the future against the play of blind chance. The drift and muddle of the past is intolerable. We believe that this new agency will be organization and co-operation. To be sure, the experience of the past in this direction is not altogether reassuring. We have seen that the movement toward uniformity of notation in vector analysis fell through. There was a deep appreciation, on the part of the individual worker, of the transcendent superiority of his own symbols over those of his rivals. However, this intense individualism was not the sole cause of failure. Eventually, an appreciation of the ludicrous might have saved the situation, had it not been for the disorganization, even among scientific men, resulting from the Great War. Past failures have not deterred our National Committee on Mathematical Requirements from recommending that their report on symbols and terms in elementary mathematics be considered by the International Congress of Mathematicians. The preparation of standardized symbols and abbreviations in engineering is now entered upon, under the sponsorship of the American Association for the Advancement of Science and other societies.

749. The adoption at a particular congress of a complete symbolism intended to answer all the present needs of actuarial science finds its counterpart in the procedure adopted in world-movements along somewhat different lines. In 1879 a German philologian invented a universal language, called "volapuk." Eight years later a Russian invented "esperanto," containing 2,642 root-words. Since then about eight modified systems of universal language have been set up. It will be recognized at once that securing the adoption of a universal spoken and written language by all intellectual workers is an infinitely greater task than is that of a universal sign symbolism for a very limited group of men called mathematicians. On the other hand, the task confronting the mathematician is more difficult than was that of the electrician; the mathematician cannot depend upon immense commercial enterprises involving large capital to exert a compelling influence such as brought about the creation and adoption of a world-system of electric and magnetic units. Even the meteorologist has the advantage of the mathematician in securing international signs; for symbols describing the weather assist in the quick spreading of predictions which may safeguard shipping, crops, and cattle.

The movement for a general world-language is laboring under two *impedimenta* which the mathematicians do well to avoid. In the first place, there has been no concerted, united action among the advocates of a world-language. Each works alone and advances a system which in his judgment transcends all others. He has been slow to learn the truth taught to the savage by his totem, as related by Kipling, that

> "There are nine and sixty ways
> Of constructing tribal lays
> And every single one of them is right."

Unhappily each promoter of a world-language has been opposing all other promoters. The second mistake is that each system proposed represents a finished product, a full-armed Minerva who sprung from the head of some Jupiter in the realm of philology.

In the endeavor to secure the universal adoption of mathematical symbolism international co-operation is a *sine qua non*. Agreements by representatives must take the place of individual autocracy. Perhaps scientists are in a better position to cultivate internationalism than other groups.

The second warning alluded to above is that no attempt should be made to set up at any one time a full system of notation for any one department of higher mathematics. This warning has remained unheeded by some contemporary mathematicians. In a growing field of mathematics the folly of such endeavor seems evident. Who is so far sighted as to be able to foresee the needs of a developing science? Leibniz supplied symbols for differentiation and integration. At a later period in the development of the calculus the need of a symbolism for partial differentiation arose. Later still a notation for marking the upper and lower limits in definite integrals became desirable, also a mode of designating the passage of a variable to its limit. In each instance numerous competing notations arose and chaos reigned for a quarter- or a half-century. There existed no international group of representatives, no world-court, to select the best symbols or perhaps to suggest improvements of its own. The experience of history suggests that at any one congress of representative men only such fundamental symbols should be adopted as seem imperative for rapid progress, while the adoption of symbols concerning which there exists doubt should be discussed again at future congresses. Another advantage of this procedure is that it takes cognizance of a disinclination of mathematicians as a class to master the meaning of a symbol and use it, unless its introduction has become a practical necessity. Mathematicians as a class have been accused of being lethargic, easily pledged to routine, suspicious of innovation.

Nevertheless, mathematics has been the means of great scientific achievement in the study of the physical universe. Mathematical symbols with all their imperfections have served as pathfinders of the intellect. But the problems still awaiting solution are doubtless gigantic when compared with those already solved. In studying the microcosm of the atom and the macrocosm of the Milky Way, only mathematics at its best can be expected to overcome the obstacles impeding rapid progress. The dulling effect of heterogeneous symbolisms should be avoided. Thus only can mathematics become the Goliath sword enabling each trained scientist to say: "There is none like that; give it me."

750. The statement of the Spottiswoode Committee bears repeating: "Anything which tends toward uniformity in notation may be said to tend towards a common language in mathematics, and . . . . must ultimately assist in disseminating a knowledge of the science of which they treat." These representative men saw full well that God does not absolve the mathematician from the need of the most economic application of his energies and from indifference to the well-tried wisdom of the ages.

Our mathematical sign language is still heterogeneous and sometimes contradictory. And yet, whatever it lacks appears to be supplied by the spirit of the mathematician. The defect of his language has been compensated by the keenness of his insight and the sublimity of his devotion. It is hardly worth while to indulge in speculation as to how much more might have been achieved with greater symbolic uniformity. With full knowledge of the past it, is more to the point to contemplate the increasingly brilliant progress that may become possible when mathematicians readdress themselves to the task of breaking the infatuation of extreme individualism on a matter intrinsically communistic, when mathematicians learn to organize, to appoint strong and representative international committees whose duty it shall be to pass on the general adoption of new symbols and the rejection of outgrown symbols, when in their publications mathematicians, by a gentlemen's agreement, shall abide by the decisions of such committees.

ALPHABETICAL INDEX

(Numbers refer to paragraphs)

Abel, N. H.: trigonometry, 526; powers of trigonometric functions, 537; elliptic functions, 654; Abelian functions, 654

Abraham, M., vectors, 507

Absolute difference, 487. See *Difference*

Absolute value, 492

Abu Kamil, continued fractions, 422

Actuarial science, 739, 749

Adams, D., 404

Adams, J. C., 407, 419

Adams, O. S., 724

Adrain, R., calculus, 630

Aggregates, 421

Aggregation of terms, Leibniz, 541, 551

Airy, G. B., 449; sign of integration, 620

Albrecht, Theodor, 515

Alcala-Galciano, D., 488

Aleph, 421

Alexander, John: absolute difference, 487; "therefore," 668

Alfonsine tables, 511

Amodeo, F., 705

Ampère, A. M., products, 440, 446

Analogy a guide, 713

Analytical geometry, 708

Anderegg, F., and E. D. Rowe, trigonometry, 527

André, D., 712; absolute value, 492; angular measure, 514; logarithms, 470, 475, 481; use of ⌣, 490

Angle of contact, 703

Angle, sign same as for "less than," 485

Anharmonic ratio, 706

Annuities, 447

Apian, P., degrees and minutes, 512

Apollonius of Perga, letters for numbers, 388

Appell, P., 449; vectors, 502

Arbogast, L. F. A., 447, 577, 578, 580, 584, 586, 596, 602, 611, 613, 641

Archibald, R. C., 524

Archimedes, 388, 395, 398

Argand, J. R., 446, 719; sign ⌣, 489; vectors, 502

Aristotle, letters for magnitudes, 388

Arithmetic, theoretical, survey of, 483, 494

Arithmetical progression, 440

Arnauld, Antoine, sign for small parts, 570

Aronhold, S. H., 456

Arrow: in approaching a limit, 636; use of, in dual arithmetic, 457; with vectors, 502

Aschieri, A., 705

Astronomical signs, 394; in Leibniz, 560; in Bellavitis, 505

Athelard of Bath, 511

Atled, 507

Aubré, L. E., 707

Aubrey, John, 516

Auerbach, M., and C. B. Walsh, "therefore" and "because," 669

Babbage, Charles, 393, 408; article on notations, 713; calculus, 583; logarithms, 478; signs for functions, 647; quoted, 718

Bachet, C. G., 407

Bachmann, P., 406, 407, 409

Bäcklund, A. V., 598

Bails, Benito: logarithm, 476; trigonometry, 525

Bainbridge, John, 521

Baker, H. F., quoted, 722

Baker, Th., greater or less, 483

Balleroy, J. B., 645

Bangma, O. S., 421

Bardey, E., logarithms, 479

Barlow, Peter, 407, 408; calculus, 582, 605, 611; trigonometry, 526

Barrow, Isaac: absolute difference, 487; calculus, 539; greater or less, 483; sign for 3.141 . . . . , 395

Bartlett, W. P. G., 449; difference, 487

Basedow, J. B., 448

Basset, A. B., partial derivatives, 618

351

Bauer, G. N., and W. E. Brooke, 515

Because, sign for, 669

Beguelin, Nic. de, use of π, 398

Behse, W. H., 707

Bellavitis, G., vectors, 502, 504, 505

Beman, W. W., 498

Benedict, S. R., 715

Bentham, George, logic, 676

Bérard, inverse trigonometric functions, 532

Berlet, Bruno, 390

Bernhard, Max, descriptive geometry, 707

Bernoulli, Daniel (b. 1700): inverse trigonometric functions, 532; use of π, 398; use of e, 400

Bernoulli, Daniel (b. 1751), trigonometry, 525

Bernoulli, Jakob (James) I, 538; Bernoulli's numbers, 419, 447; polar co-ordinates, 708; symbol for function, 642; symbolic logic, 671; word "integral," 539, 620

Bernoulli, Johann (John) I, 425, 538, 540, 571; calculus notation, 539, 572, 577, 638; greater or less, 484; logarithms, 469; sign for 3.141 . . . ., 397; sign of integration, 539, 620; symbol for function, 642; use of ∞, 421

Bernoulli, Johann II (b. 1710), 671

Bernoulli, Johann III (b. 1744), 514 524

Bernoulli, Nicolaus (b. 1687), sign for 3.141 . . . . , 397, 436

Berry, Arthur, 394

Berthold, P., 707

Bertrand, J., 449

Bertrand, Louis, trigonometry, 525

Berzolari, 709

Bessel, F. W.: Bessel functions, 662, 664; logarithm integral, 665

Beta functions, 649

Bettazzi, R.: thesis and lysis, 494; symbolic logic, 688

Beutel, Tobias, logarithm, 475

Bézout, E.: calculus, 572, 630; determinants, 460; logarithms, 477; trigonometry, 525; use of e, 400

Binet, J., 419; beta function, 649; determinants, 461

Binomial formula, 439, 443; coefficients, 439, 443

Birkhoff, G. D., 510

Birks, John, "therefore," 668

Bischoff, Anton, logarithms, 479

Blassière, J. J., logarithms, 477

Bledsoe, A. T., sign $\frac{0}{0}$, 638

Blessing, G. F., and L. A. Darling, 707

Blundeville, Th., 517; trigonometry, 521

Bobek, Karl, 705

Bobillier, E., 708

Bôcher, M., 468

Bocx, D., 568

Böger, R., 705

Böhm, J. G., 477

Boggio, T., symbolic logic, 691

Bois Reymond, P. du: signs for limits, 637; use of $\sim$, 490

Bolyai, Wolfgang: absolute quantity, 635; approaching the limit, 633; greater or less, 486; imaginaries, 499; logic, 675; partial derivations, 609; trigonometry, 526, 527

Bombelli, Rafaele, imaginaries, 496

Boncompagni, B., 389, 422, 425, 483, 517, 526, 527, 644

Bonnet, O., 472

Bonnycastle, John, greater or less, 484

Boole, George: calculus, 592, 598; finite differences, 641; symbolic logic, 678, 680, 685

Booth, James, parabolic functions, 531

Bopp, Karl, 570

Borchardt, C. W., 591, 656

Borel, E.: inequality in series, 486; n-factorial, 449

Bortolotti, E., 496

Boscovich, R. G., 623

Bouger, P., greater or equal, 484

Bouman, J., 568

Bourdon, P. L. M., 449

Bourget, J., 481

Bow, R. H., 703

Bowditch, Nathaniel, 630

Bowser, E. A., 449; trigonometry, 526

Brancker, T., greater or less, 483

Branford, B., quoted, 720

Braunmühl, A. von, 517, 519, 523, 527

Bravais, A., 456

Bretschneider, C. A., logarithm integral, cosine integral, etc., 665

Bridge, B.: absolute difference, 487; fluxions, 630

Briggs, H.: logarithms, 469, 474; marking triangles, 516; trigonometry, 520, 521

Brill, A., and M. Noether, 499, 657

Brinkley, John, 621; sign for limit, 631

British Association, report of a committee of the, 725

Brocard points, 700; triangles, 700

Bromwich, T. J. I'A., 490; approaching a limit, 636

Brounker, W., infinite series, 435

Brückner, Max, polyhedra, 702

Bruno, Faà de, 439, 449, 456; determinants, 462

Bryan, G. H., 575

Bryan, W. W., 394

Budden, E., "therefore" and "because," 669

Bürgi, Joost, logarithms, 474

Bürmann (Burman), Heinrich: combinatorial analysis, 444; continued fractions, 428; inverse function, 533

Burali-Forti, C.: vector analysis, 506–9; symbolic logic, 688

Burja, Abel, 479

Burkhardt, H., 499, 659; quoted, 723

Burmann, H. See Bürmann

Burnside, W., 453, 454

Burnside, W. S., and A. W. Panton, symmetric functions, 648

Byerly, W. E., 449; approaching the limit, 635; equal or less than, 486; trigonometry, 526

Byrne, O., 457; dual logarithms, 482

Cagnoli, A.: absolute difference, 488; trigonometry, 525; powers of trigonometric functions, 537

Caille, De la, 410

Cajori, F., 480, 515, 568

Calculus: differential and integral, survey of, 566–639; in the United States, 630; Cambridge writers, 713; definite integrals, 626, 627; differential, 567–619; epsilon proofs, 634; fluxions, 567, 568, 579, 621; fractional differentiation, 543, 591; integral, 544, 622–30; limits, 631; partial differentiation, 593–619; quotation from Newton, 569; residual, 629; sign $\frac{0}{0}$, 638; Arbogast, 577, 602; Babbage, 583; Bolyai, 609, 633; Barlow, 582, 605; Carr, 616; Cauchy, 585, 587, 606, 610, 629; Condorcet, 596;

Crelle, 604, 624; Da Cunha, 599; Dirichlet, 637; Dirksen, 632; Duhamel, 615; Euler, 593, 625; Fontaine, 595; Fourier, 592, 626; Hamilton, 608; Herschel, 583; Hesse, 612; Jacobi, 611; Karsten, 594; Kramp, 578; Lacroix, 580, 598; Lagrange, 575, 581, 597, 601, 603; Landen, 573; Leathem, 636; Legendre, 596; Leibniz, 541, 543, 544, 570, 620, 621; Lhuilier, 600; Mansion, 619; Méray, 617; Mitchell, 582; Moigno, 610; Monge, 596; Muir, 618; Newton, 567–69, 580, 601, 621, 622; Ohm, 586, 607; Oliver, 635; Pasquich, 576; Peacock, 583, 591; Peano, 627, 637; Peirce, B., 588, 613; Reyneau, 623; Scheffer, 637; Strauch, 614; Volterra, 627; Weierstrass, 634; Woodhouse, 579; Young, W. H., 637

Camus, C. E. L., 525

Cantor, Georg, 414; transfinite numbers, 421

Cantor, Moritz, 388, 390, 422, 426, 518, 649; history of calculus, 639; use of $\sim$, 489

Caramuel, J., degrees and minutes, 514

Cardano (Cardan), Hieronimo, 390; imaginaries, 495

Carey, F. S., approaching a limit, 636

Carmichael, Robert: calculus, 592, 641; logarithms, 472; n-factorial, 449

Carmichael, R. D., 409

Carnot, L. N. M.: equipollence, 505; sign $\frac{0}{0}$, 638; use of π, 398

Carr, G. S., 589, 616

Carré, L., sign of integration, 620

Carvallo, E., 449

Casorati, F., 641

Cass, Lewis, 403

Cassany, F., 403

Castellano, Fr., logic, 688, 691

Castillon, G. F., 483; symbolic logic, 671, 673

Castle, F., 477; vectors, 502

Caswell, John, trigonometry, 522

Catalan, E., determinants, 462

Cataldi, P. A., continued fractions, 424, 425, 428

Cauchy, A. L.: calculus, 585, 587, 598, 606, 610, 611, 613, 614, 626, 631, 634; determinants, 461; imaginaries, 499, 502; logarithms, 470; "mod," 492; substitutions, 453; symmetric functions, 647; trigonometry, 526; use of Σ, 438; residual calculus, 629; $0 < \theta < 1$, 631; sign for taurus, 394

Cavalieri, B.: logarithms, 469; trigonometry, 521

Cayley, A., 725; determinants, 462, 468; n-factorial, 449; partial derivatives, 611; quaternions, 508; quantics, 439, 456; substitutions, 453; polyhedra, 702; Pascal hexagram, 711

Centen, G., 568

Chase, P. E., 449

Chasles, M., anharmonic ratio, 706

Chauvenet, W., 449

Cheney, George: Leibnizian symbols, 621; use of ∞, 421

Chessboard problem, 458

Cheyne, G., 399; fluxion, 582

Child, J. M., 571

Chini, M., symbolic logic, 691

Christoffel, E. B., 510

Chrystal, G., 411, 428, 430, 449; logarithms, 478; permutations, 452; subfactorial n, 450; use of π, 451

Chuquet, N.: survey of his signs, 390; imaginaries, 495

Church, Albert, 707

Cicero, letters for quantities, 388

Circle, recent geometry of, 700

Clairaut, A. C., signs for functions, 643

Clap, 630

Clark, George Roger, 403

Clarke, H., absolute difference, 487

Clarke, John, 567, 622

Clavius, C.: degrees and minutes, 513; letters for numbers, 388; plus and minus, 730

Clebsch, A., 449; theta functions, 658; quantics, 456

Clifford, W. K., 507, 725; anharmonic ratio, 706; hyperbolic functions, 530

Cocker, Edward, greater or less, 483

Cockshott, A., and F. B. Walters, "therefore" and "because," 669

Colburn, Warren, 403

Cole, F. N., 453

Cole, John, trigonometry, 526

Collins, John, 439, 521, 569

Combebiac, 509

Combinations: survey of, 439–58; Jarrett's signs for, 447

Combinatorial school, 443

Condorcet, N. C. de: inverse trigonometric functions, 532; partial derivatives, 596; series, 436; use of e, 400

Congruence of numbers, 408

Congruent, in Leibniz, 557, 565

Continuants, 432

Continued fractions: survey of, 422–38; ascending, 422, 423, 433; descending, 424–33; in Leibniz, 562

Continued products, 451

Copernicus, N., degrees, minutes, and seconds, 513

Cortazar, J., trigonometry, 526

Cosine integral, 665

Costard, G., 525

Cotes, Roger: his numbers, 418; binomial formula, 439; greater or less, 483; infinite series, 435

Couturat, L.: on Leibniz, 670; logic, 685

Covariants, 456

Covarrubias, F. D., 535

Craig, John: calculus, 621; sign of integration, 620, 621

Craig, Thomas, 449

Cramer, G.: analytical geometry, 708; determinants, 460; the sign $\frac{0}{0}$, 638

Crelle, A. L.: calculus, 584, 604, 611; combinatorial signs, 446; logarithms, 471; sign of integration, 624; trigonometry, 526

Cremona, L., 705

Cresse, G. H.: theta functions, 657; quadratic forms, 413

Cross-ratio. *See* Anharmonic ratio

Crozet, Claude, 630

Crüger, Peter, 514

Cubic surfaces, lines on, 710

Cullis, C. E., 468

Cunn, Samuel, finite differences, 640

Cunningham, A. J. C., 407, 416

Curtze, M., 389, 512

Da Cunha, J. A.: absolute difference, 488; calculus, 574, 596, 599, 611; trigonometry, 525

D'Alembert, J.: calculus, 595; sign for functions, 643; sign for $\frac{0}{0}$, 638; trigonometry, 525; use of e, 400

Darboux, G., 626, 667

Davies, Charles: calculus, 595; *n*-factorial, 449

Day, J., 565

Dedekind, R., 413; theory of numbers, 409; Zahlkörper, 414

Degrees, minutes, and seconds, survey of, 511, 514

De la Caille, 410

Delamain, Richard, 516, 522

Delambre, 525, 718

Del (or-Dal) Ferro, Scipio, 496

De Moivre, A.: calculus, 621; sign for 3.141 . . . . , 395; his formula, 499

De Morgan, Augustus, 388; Bernoulli's numbers, 419; imaginaries, 499, 501, 518; logarithms, 477, 480; logarithm integral, 665; *n*-factorial, 449; powers of trigonometric function, 537; quoted, 713; symbolic logic, 677, 680, 685; use of *e*, 400; partial derivative, 606

Descartes, René, 538; analytical geometry, 565, 708; equality, 670; polyhedra, 702; use of letters, 392, 565

Descriptive geometry, 707

Desnanot, P., determinants, 461

Determinants: survey of, 459–68; cubic determinants, 466; infinite, 467; suffix notation in Leibniz, 559; vertical bars, 492

Dickson, L. E., 407–10, 412, 413, 416, 417, 494, 657; multiply perfect numbers, 415; substitutions, 453

Diderot, Denys: *Encyclopédie*, 483; logarithms, 469; sign for 3.141 . . . . , 398

Difference (arithmetical): survey of, 487; symbol for, 488

Digges, Leonard and Thomas: equality, 394; use of dots, 567

Dini, Ulisse, signs for limits, 637

Diophantine analysis, 413

Dirichlet, P. G. L., 407, 414, 656; approaching a limit, 637; continued fractions, 431; Legendre's coefficient, 662; theory of numbers, 409

Dirksen, E. H., 437; infinite series, 486; sign for limit, 632

Ditton, H., 567, 622

Division: Leibniz, 541, 548; sign ÷ for, 481

Division of numbers, 407

Dixon, A. C., 499, 654

Dodgson, C. L., 462; trigonometry, 526

Doehlemann, Karl, projectivity, 705

Dollar mark, survey of, 402–5

Donkin, W. F., 464

Donn, B., absolute difference, 487

Dot: fluctions, 567; in later writers, 541, 547; small quantity, 567

Dowling, L. W., 705

Drobisch, M. W., 410

Dual arithmetic, 457

Duhamel, J. M. C., calculus, 595, 598, 615, 617, 638

Dulaurens, F., greater or less, 484

Durell, Clement V., 700

Durrande, J. B., 449

e for 2.718 . . . . , survey of, 399–401

Eberhard, V., 702

Ebert, J., 525

Echols, W. H., 635

Eddington, A. S., 510

Einstein, A.: determinants, 463; tensors, 510

Eisenstein, G., 413, 449, 490

Elasticity, 738

Electric units, 734

Elliott, E. B., 456

Elliptic functions, 651–55

Emerson, W., finite differences, 640

Emmerich, A., 700

Eneström, G., 389, 392, 400, 423, 542, 572, 591; functions, 643; greater or less, 483, 567

Engel, F., 455

Enriques, F., projectivity, 705

Epsilon proofs, 634

Epstein, P., 431

Equality: Jakob Bernoulli, 419; Digges, 394; Leibniz, 545

Equivalence, 490

Erdmann, J. E., 670

Erler, W., logarithms, 479

Escudero, J. A., 403

Esperanto, 749

Eta functions, 656

Ettingshausen, A. von, 439

Eucken, A., 637

Euclid's *Elements;* numbers, 415; use of letters for numbers, 388

Euler, L.: analytical geometry, 708; beta and gamma functions, 649; Bernoulli's numbers, 419; binomial coefficients, 439, 452; calculus, 572,

580, 591, 620, 621; continued fractions, 425, 426; diophantine analysis, 413; Euler's numbers, 415, 420; Euler's constant, 407; finite differences, 640, 641; figurate numbers, 412; imaginaries, 496, 498; inverse trigonometric functions, 532; limits of integration, 625; logarithms, 469; logic, 677; n-factorial, 448; not less than, 484, 485; number of symbols invented by Euler, 730; partial derivatives, 593, 596, 598, 607, 608, 611; partitions, 411; polyhedra, 702; powers of trigonometric functions, 537; prime and relatively prime numbers, 409; signs for functions, 643; sum of divisors, 407; trigonometry, 522, 525; use of π, 397; use of e, 400; use of Σ, 438; use of ∞, 421; zeta function, 660

Eutocius, 718

Exponents ("exp"), 725

Faber, G., 536

Factorial n: survey of, 448; De Morgan on (!), 713; Jarrett's sign, 447

Fagano, Count de, 469

Fano, G., symbolic logic, 688

Farrar, John, Leibnizian calculus symbols, 630

Fas, J. A.: logarithms, 469; use of e, 400

Fehr, Henri, vector analysis, 509

Fenn, Joseph, use of Leibnizian symbols, 621

Fermat, P., calculus notation, 539; his numbers, 417

Ferraris, G., 506

Ferro, Scipio Dal, 496

Ferroni, P.: trigonometry, 525; use of π, 398

Feuerbach circle, 700

Ficklin, J., 449

Fiedler, W.: partial derivatives, 612; projectivity, 705

Figurate numbers, 412

Filon, L. N. G., 705

Finck (or Fincke), Thomas, 517, 518, 520, 521

Fine, O., degrees, minutes, 512

Finite differences, 640, 641

Fischer, E. G., 436

Fisher, A. M., 630

Fiske, A. H., 403

Fluents 567, 622. See Calculus, integral

Föppl, A., 506

Foncenex, François Daviet de: first derivative, 575; use of π, 398

Fontaine, A., 477; fluxions, 568, 571, 574, 582; partial differentials and derivatives, 595, 598, 600, 607, 611

Fontenelle, B., use of ∞, 421

Ford, W. B., 437, 490

Forster, Mark, 521

Forsyth, A. R.: calculus, 577, 627; greater or less, 484

Fourier, Joseph: calculus, 591, 592; limits of integration, 626; series, 656; sign of integration, 623; use of Σ for sum, 438

Fournier, C. F., logarithms, 469, 477

Fractions: continued (survey of), 422–38; tiered, 434

Français, J. F., 499, 502, 641

Franklin, Mrs. Christine. See Ladd-Franklin, Mrs. C.

Freeman, A., 438

Frege, Gottlob, logic, 687, 695

Frend, John, greater or less, 483

Fresnel, A. J.: Fresnel functions, 666; limits of integration, 626

Fricke, R., 654, 659

Frischauf, J., hyperbolic functions, 529

Frisi, P.: calculus, 623; powers of trigonometric functions, 537; trigonometry, 525; use of e, 400

Frobenius, G., 454, 455

Frullani, hyperbolic functions, 529

Functions: survey of symbols for, 642–66; beta, 649; elliptic, 651–55; eta, 656; Fresnel, 666; gamma, 649; hyperbolic, 665; Lagrange, 575; Laplace's coefficient, 662; Laplace, Lamé, and Bessel, 662–64; Legendre's coefficient, 662; Leibniz' signs, 540, 558; logarithmic integral, cosine integral, 665, 666; power series, 661; $\wp$, 654; sigma, 654; symmetric, 647, 648; theta, 656–58; theta fuchsian, 657; zeta (Jacobi's), 653, 655, 656, 659

Fuss, Paul H. von, 397, 412, 484, 649

Galois, Évariste, 414; groups, 454

Gamma functions, 649, 650

Gans, R., 504

Garcia, Florentino, 403

Gardiner, W., logarithms, 469

Gauss, K. F.: calculus, 595; congruence of numbers, 408; determinants,

461; diophantine analysis, 413; hypergeometric series, 451; imaginaries, 498; logarithms, 478; n-factorial, 449, 451; "mod," 492; powers of trigonometric functions, 537; Π for product, 451; $\varphi (n)$, 409; sign for Euler's constant, 407

Gausz, F. G., 478

Gellibrand, H., 474, 516; trigonometry, 521

Gemma Frisius, degrees, minutes, 514

General analysis: symbols, 693; geodesy, 723

Geometrography, 701

Geometry: descriptive, 707; marking points, lines, surfaces by Leibniz, 561; projective, 705; lines on cubic surface, 710; spherical triangles, 524

Gerbert (Pope Silvester II), 393

Gergonne, J. D., 448, 449; *Annales*, 626; symbolic logic, 674, 677

Gerhardt, C. I., 394, 410, 538–42, 570–72, 620, 670

Gernardus, 389

Gerson, Levi ben, 389

Gherli, O.: calculus, 593; sign of integration, 620, 623; trigonometry, 525

Gibbs, J. W., 502, 506–10, 730

Gibson, Thomas, greater or less, 483

Gioseppe, Dim., 513

Girard, A.: degrees and minutes, 514; greater or less, 484; marking triangles, 516; trigonometry, 519, 521; imaginaries, 497

Giudice, F., symbolic logic, 688

Glaisher, J. W. L., 407, 516, 725; calculus, 583; cosine integral, exponential integral, 665; elliptic functions, 653, 655; quoted, 583; trigonometry, 518–20

Godfrey, C., and A. W. Siddons, 700

Goldbach, C., 649; greater or equal, 484; use of π, 397

Goodwin, H., 449, 452; absolute difference, 487

Gordan, P., 464; theta functions, 658; quantics, 456

Grammateus, Heinrich Schreiber, letters for numbers, 390

Grandi, Guido, 538

Grassmann, H. G., 494, 684; vectors, 503–9

Grassmann, Robert: combinations, 494; logarithms, 479; logic, 684, 685

Graves, John, 480

Gray, A., and G. B. Mathews, Bessel functions, 664, 666

Greater or equal, 484

Greater or less: Leibniz, 549, 564; Oughtred, 483

Greatheed, S. S., 592

Grebe points, 700

Greenhill, A. G., elliptic functions, 654, 655

Gregory, David: fluxions, 568; sign for 3.141 . . . . , 395

Gregory, Duncan F., 502, 591, 592

Gregory, James, infinite series, 435

Gregory, Olinthus, 621

Groups, survey, 454, 455

Grüson, J. P., 576, 580

Grunert, J. A., 419, 438; calculus, 596; logarithms, 469; hyperbolic functions, 529

Gua, De: analytical geometry, 708; series, 436; trigonometry, 525

Guareschi, G., 705

Gudermann, C.: elliptic functions, 654; hyperbolic functions, 529; inverse trigonometric functions, 535

Gudermannian, 536; inverse, 646

Günther, S., 425; hyperbolic functions, 529; use of $\sim$, 489

Gunter, E., 474, 516; trigonometry, 520, 522

Haan, Bierens de, 568

Halcke, P., 568

Hale, William, 622

Hales, William, 571

Hall, A. G., and F. G. Frink, 515

Hall, H. S., and S. R. Knight, 428; absolute difference, 487; logarithms, 477

Halphen, theta function, 657

Halsted, G. B., projectivity, 705

Hamilton, W. R., 480, 499, 502, 503, 504, 506; operators, 507–10; partial derivatives, 608, 611; sign of integration, 620; sign for limit, 634

Handson, R., 516

Hankel, H.: combinations, 494; projectivity, 705

Hanson, P. A., Bessel functions, 664

Hardy, A. S., 499

Hardy, G. H., 411; approaching a limit, 636; use of $\sim$, 490

Harkness, J., and F. Morley: "cis," 503; power series, 661; theta functions, 657, 658

Harmonic ratio, 706

Harriot, Th., 518; greater or less, 483

Hart, Andrew S., 710

Hatton, Edward: absolute difference, 487; greater or less, 483

Hausdorff, F., 421

Hauser, M., 421

Heaviside, P.: calculus, 591; vectors, 506, 507

Hebrew numerals, aleph, 421

Heiberg, J. L., 388

Heine, E.: continued fractions, 431; Laplace's function, 662; Legendre's coefficient, 662

Helmert, 515

Henderson, A., 710

Hendricks, J. E., 449

Henrici, J., and P. Treutlein, projectivity, 705

Henrici, O., vectors, 506

Hensel, Kurt, 463

Heraclitus, 731

Hérigone, P., 538, 730; absolute difference, 487; greater than, 484; symbolic logic, 667; trigonometry, 521

Hermann, J., 538

Hermite, C., 419; calculus, 595; theta functions, 657

Herschel, J. F. W., 411, 447; and Burmann, 444; calculus, 583; continued fractions, 428; finite differences, 641; inverse trigonometric functions, 533, 535; signs for functions, 645; symmetric functions, 647

Hesse, L. O., 464; Hessian, 465; partial derivatives, 612

Heun, K., 506

Heurlinger, Torsten, 486

Heynes, Samuel, 522

Hilton, H., 454

Hind, John, 606

Hindenburg, C. F., 428, 447, 578, 730; combinatorial school, 443, 444, 445; series, 436

History, teachings of, 712–50

Hobson, E. W.: approaching the limit, 635; hyperbolic functions, 529; powers of trigonometric functions, 537; trigonometry, 526

Hölder, P., 454

Hoffmann, J. C. V., logarithms, 479

Hohenburg, Herwart von, 514

Holgate, T. F., 705

Holland, G. J. von, logic, 671, 672

Horn-D'Arturo, G., 394

Horsley, S., 514, 567

Hospital, L', 538, 541, 571, 638

Houël, G. J., 499; hyperbolic functions, 529; inverse trigonometric functions, 535; inverse hyperbolic functions, 536

Hudde, J., analytical geometry, 708; letters for + or − numbers, 392

Hudson, R. G., 489, 500

Hudson, R. G., and J. Lipka, trigonometry, 527

Hunt, William, 522

Huntington, E. V., 421; combinations, 494

Hutton, Ch.: calculus, 596, 630; trigonometry, 525

Huygens, Chr., continued fractions, 425

Hyperbolic functions, 529, 744; hyperbolic cosine integral, 665

Hypergeometric series, 451

Ideographs, 726

Ignatowsky, tensors, 510

Imaginary $\sqrt{-1}$ or $\sqrt{(-1)}$ or $\sqrt{-1}$: survey of, 495–510; letter for, 498; Leibniz, 553; discussed by Maclaurin, 717; discussed by Mach, 719; discussed by Riabouchinski, 493

Ince, L., 428

Indeterminate form $\frac{0}{0}$, 638

Inequality (greater and less): survey of, 483–86; Leibniz, 550

Infinite products, 451; in Wallis, 395

Infinite series: survey of, 466–68; in Leibniz, 556

Infinity: survey of transfinite number, 435–38; aleph, 421; ω, 421; Nieuwentijt's m, 421; Wallis' sign, ∞, 421, 454

Inman, J., 527

Integral calculus, 539, 541, 544, 620–29

Interval of integration, 637

Invariants, 456

Isenkrahe, C., 703

Izquierdo, G., 403

Jacobi, C. G. I., 407, 411; determinants, 461; elliptic functions, 652, 654; Jacobian, 464; partial derivatives, 601, 611, 612, 617–19; Π for product, 451; theta functions, 656; trigonometry, 526; use of Σ for sum, 438; zeta function, 653, 659; n-factorial, 449

Jacquier, François, 568

Jahnke, E., 438

Jarrett, Thomas, 440, 447, 452; n-factorial, 449

Jeake, S.: absolute difference, 487; greater or less, 483; trigonometry, 522

Jefferson, Thomas, 403

Jeffrey, M., 449

Jevons, W. S., logic, 679, 680

Joachimsthal, F., determinants, 462

Jöstel, Melchior, 516, 517

Johnson, W. W., 418

Joly, C. J., 509

Jones, David, 447

Jones, Henry, 640

Jones, William: absolute difference, 487; binomial formula, 439; logarithms, 469; powers of trigonometric functions, 537; π, 396; "therefore," 668; trigonometry, 525, 527

Jordan, C., 409; groups, 454; n-factorial, 449; π, 451; signs for integration, 623, 627; substitutions, 453; theta functions, 657

Jourdain, P. E. B., 421, 500, 575; limits of integration, 626; logic, 687, 698

Jurin, J., 621

Kästner, A. G.: greater than, 485; sign for 3.141 . . . . , 398; use of astronomical signs, 394; on imaginaries, 497; trigonometry, 525

Kambly, L., 471

Karpinski, L. C., 422

Karsten, W. J. G.: calculus, 594, 595, 600, 601, 611; trigonometry, 525; use of π, 398; powers of trigonometric functions, 537

Keill, J.: sign of integration, 621; trigonometry, 525

Kelland, P., 591

Kelvin, Lord, 515, 581, 598; quaternions, 508

Kempner, A. J., 411

Kenyon, A. M., and L. Ingold: inverse trigonometric functions, 535; trigonometry, 526, 527

Kepler, J., 517; degrees and minutes, 514; Kepler's problem, 440; logarithms, 469

Kersey, John: absolute difference, 487; greater or less, 483

Kingsley, W. L., 630

Kipling, R., 749

Kirkman, T. P.: ($\pm$), 491; Pascal hexagram, 711

Klein, Felix, vector analysis, 509

Klügel, G. S., 419, 438; binomial coefficients, 443; calculus, 596; combinatory analysis, 445; sign of integration, 620; trigonometry, 525

Knott, C. G., 477, 620; vectors, 509

König, Julius, 696

Königsberger, L., on Jacobi, 656

Köpp, logarithms, 479

Kötter, E., 709

Kötterlitzsch, Th., 467

Kok, J., 568

Kooyman, J. T., 568

Korn, A., 738

Koudou, G. de, 525, 595

Kovalewski, G., 468

Krafft, G. W., use of π, 398

Kramers, H. A., 486, 627

Kramp, Christian: binomial coefficients, 439; calculus, 578; combinatorial analysis, 445, 452, 641; congruence of numbers, 408; imaginaries, 498; integral of, 666; n-factorial, 448; products, 440; use of e, 400; symmetric functions, 647

Krazer, A., 395, 658

Kresa, J., trigonometry, 525

Kronecker, L., 413; determinants, 463; divisors of numbers, 407; use of $\sim$, 490; "Kronecker symbol," 463

Labosne, A., 407

Lachlan, R., 700, 706

Lacroix, S. F., 447; calculus, 572, 576, 577, 580, 598; powers of trigonometric functions, 537; trigonometry, 526

Ladd-Franklin, Mrs. C.: logic, 682; Pascal hexagram, 711

Lagrange, J., 718; calculus, 566, 580, 582, 585, 593, 595, 597, 598, 601, 603, 611; continued fractions, 425; finite differences, 640; *Fonctions analytique,*

575, 576, 584, 601; inverse trigonometric functions, 532; *Mécanique
analytique*, 581, 623; powers of
trigonometric functions, 537; signs
for functions, 575, 644; sign $\frac{0}{0}$, 638;
trigonometry, 525; use of Σ, 438

Lalande, F. de, logarithms, 475

Lambert, J. H., hyperbolic functions, 592; inverse trigonometric
functions, 532; symbolic logic, 671,
677; use of e, 400

Lambert, W. D., 515, 724

Lamé, G., Lamé functions, 662, 663

Lampe, E., 393

Landau, E., 407, 409

Landen, John, 479; residual analysis,
573, 580

Lange, J., 700

Langevin, P., 502, 507

Langley, E. M., 477

Lansberge, Philipp van, 513, 518;
marking triangles, 516; trigonometry, 517

Laplace, P. S., 447; calculus, 593;
determinants, 460, 461; finite differences, 640; his coefficients, 662; his
functions, 662; resultant, 442; use of
π, 398

Latham, M. L., 392

Laue, M., tensors, 510

Laurent, H., 449

Leathem, J. G., approaching the limit,
636

Lebesgue, V. A., 461

Lee, Chauncey, 491; dollar mark, 404

Lefort, F., 439, 483

Legendre, A. M., 398; coefficient of,
662; congruence of numbers, 408;
elliptic functions, 651, 652; n-factori-
al, 448; partial derivatives, 593, 596,
598, 611, 617; polygonal numbers,
412; polyhedra, 702; signs for functions, 645; symbol of Legendre, 407,
413; trigonometry, 526; gamma function, 649

Leibniz, G. W.: survey of his signs,
538–65, 569, 621, 639, 702, 749;
astronomical signs, 394; binomial
formula, 439; continued fractions,
425, 435; determinants, 459; finite
differences, 640; fractional differential, 543; greater or less, 484;
integral analysis, 620, 624; logic,
688; logarithms, 470; number of
symbols invented, 730; partial de-

rivatives, 543, 593, 596; sign for
function, 642; sign for 2.718 . . . . ,
399; symbolic logic, 667, 670; symmetric functions, 647; sums of numbers, 410

Lemoine, E., 701

Lenthéric, 446

Leonardo of Pisa: continued fractions,
422, 423; letters for numbers, 389

Lerch, M., 407

Le Seur (Leseur), Thomas, 568

Leslie, John, 535

Less than, 483, 550, 564. *See* Inequality

Letronne, 394

Letters, use of: survey of, 388–94;
by C. S. Peirce in logic, 681; combinatorial analysis, 443–45; Newton, 565; Van Schooten, 565; letters
π and e, 395–401; Levi ben Gerson,
389; Nemorarius, 389; lettering of
figures, 711

Levi-Civita, T., vector analysis, 510

Levy, L., 506

Lewis, C. I.: on Boole, 678; De Morgan, 677; R. Grassmann, 684; Jevons,
679; Leibniz, 670, 671

Leybourn, William, trigonometry, 522,
525

L'Hommedieu, Ezra, 403

L'Huilier, S., 436; calculus, 600; sign
for limit, 631; trigonometry, 525

Lie, S., 455

Limits, 631–37

Liouville, Joseph, 591

Lipka, J., 489, 500

Lippmann, Edmund O. von, 511

Littlewood, J. E., 411; use of $\sim$, 490

Logarithm integral, 665

Logarithms: survey of, 469–82; Ditton,
567; Leibniz, 554; notation and discovery of log's, 714; base e, 400, 401

Logic: survey of symbols, 667–99;
Leibniz, 546; symbols, 546

Longchamps, G. de, 434

Longomontanus, C. S., 517, 519

Loomis, Elias, trigonometry, 526

Lorentz, H. A., vectors, 506, 507, 510

Lorenz, J. F., 477; equal, greater, or
less, 484; powers of trigonometric
functions, 537; trigonometry, 525

Loria, Gino, 667, 707, 725

Love, C. E., 535

Low, D. A., 703

Ludlow, H. H., trigonometry, 527

Ludolph van Ceulen, on 3.141 . . . . , 396

MacColl, H., logic, 682, 685, 686

Macfarlane, A.: symbolic logic, 680, 685; vectors, 506, 509

Mach, E., quoted, 719

Machin, John, 396; products, 440

McKay circles, 700

Maclaurin, Colin: fluxions, 568; greater or less, 483; quoted, 717

Macnie, J., "therefore" and "because," 669

Mächtigkeit, 421

Magini, G. A., 517

Maier, F. G., 525

Maillet. E., 406, 409

Mako, Paulus, 436

Manfredi, Gabriele, sign of integration, 620

Mangoldt, Hans von, 393; hyperbolic functions, 529

Mansfield, Jared, 630

Mansion, P., partial derivatives, 619

Marchand, M. A. L., 645

Marcolongo, R., vector analysis, 506, 509

Martin, John, 439

Mascheroni, L., 407; Bernoulli's numbers, 419; sign for functions, 643; sine and cosine integral, 665

Maseres, Francis, 571; trigonometry, 525

Mathematical logic, survey of symbols in, 667–99

Mathews, G. B., 407, 408

Matrix, 462, 468

Matzka, W., hyperbolic functions, 529

Mauduit, A. R.: trigonometry, 525, 527

Maurolycus, Fr., signs in trigonometry, 517

Maxwell, C., vectors, 507

Maya, 730

Medina, T. F., 402

Mehmke, R., vector analysis, 509

Meissel, E., 409; symmetric functions, 647

Melandri, Daniel, use of *e*, 400

Mendoza y Rios, José de, 527

Mengoli, P., 570

Méray, Ch., partial derivatives, 617

Mercator, N.: infinite series, 435; use of dots, 567

Mercer, James, approaching a limit, 636

Merriam, M., 449

Mersenne, M., his numbers, 416

Metzburg, Freyherr von, 475; trigonometry, 525

Milne, Joshua, 447

Minchin, G. M., hyperbolic functions, 530

Mitchell, P. H., logic, 683

Mitschell, J.: calculus, 582; trigonometry, 526

Möbius, A. F.: anharmonic ratio, 706; continued fractions, 429; vectors, 502, 505

Moigno, F., 587, 595, 610, 626

Molk, J., 421, 433, 473; partial derivatives, 617; use of $\sim$, 490

Molyneux, W., logarithms, 469

Monge, G.: descriptive geometry, 707; partial derivatives, 593, 596

Montferrier, A. S. de, 574; calculus, 593

Moore, E. H., GF[q^n], 414; attitude toward symbolism, 747; fractional matrices, 468; symbols in *General Analysis*, 693; determinants, 463; definite integrals, 628

Moore, Jonas, trigonometry, 522

Moritz, R. E., 393

Morley, F. (Harkness and Morley), "cis," 503

Morris, Robert, 402

Morton, P., absolute difference, 487

Mosante, L., 403

Muir, Thomas; determinants, 442, 453, 461–66; partial derivatives, 618

Müller, J. H. T., continued fractions, 430

Muller, John, fluxions, 568

Multiplication: dot, 541, 547; in Leibniz, 540; $\frown$ in Leibniz, 541, 546, 547, 563; star used by Leibniz, 547; circle by Leonardo of Pisa, 423

Munsterus, Sebastianus, 512

Murdock, W., 708

Murhard, F., 479

Murray, D. A., 530; approaching limit, 635

Nabla, 507
Napier, John, 522; logarithms, 469, 474
Nash's *Diary*, 630
Nassò, M., 415; symbolic logic, 691
Navier, L. M. H., sign of integration, 623
Nemorarius, Jordanus, letters for numbers, 389
Nesselmann, G. H. F., 388
Netto, E., 443; permutations and combinations, 452; substitutions, 453
Neuberg circles, 700
Neumann, Karl, Bessel functions, 664
Newcomb, Simon: calculus, 593; *n*-factorial, 449
Newton, John, 516, 520, 522
Newton, Sir Isaac, 538, 640; binomial formula, 439; equality, 639; fluxions, 567, 569, 580, 601, 621, 622; greater or less, 483; imaginaries, 497; Newton quoted, 569; trigonometry (facsimile from notebook), 522
Nicole, F., series, 436
Nielson, Niels, cosine integral, 666
Nieuwentijt, B., 571; infinity, 421
Nixon, R. C. G., "therefore" and "because," 669
Noether, Emmy, 468
Norwood, R.: logarithms, 476; trigonometry, 520, 521, 522
Notation: defects in, 732; De Morgan's rules, 664; excessive, 664; Maillet, 406; empirical generalizations on, 726–33; linguistic boundaries, 730; spread, 730; uniformity of, 740–50
Numbers, theory of: survey of, 406–20; divisors of numbers, 407; fields, 414; prime and relatively prime, 409
Numerals, used in marking geometric figures, 711
Nuñez, Pedro, degrees, minutes, 513

Ohm, Martin: binomial coefficients, 439; Bernoulli's numbers, 419; calculus, 586, 607, 611, 614; inverse trigonometric function, 534; limits of integration, 627; logarithms, 471, 479, 480; *n*-factorial, 449; permutations, 452; powers of trigonometric functions, 537; series, 437; trigonometry, 526
Oldenburgh, H., 538, 539
Oliver, J. E., approaching limit, 635
Oliver, Wait, and Jones (joint authors),

449; approaching the limit, 635; inequality, 486; trigonometry, 526
Olney, E., *n*-factorial, 449
Oostwoud, D., 568
Oresme, N., 389
Osgood, W. F., 619
Ostwald, Wilhelm, 596, 611
Oughtred, William, 538, 541, 668, 730; degrees and minutes, 514; greater or less, 483; logarithms, 469, 476, 477; marking triangles, 516; sign for 3.141 . . . ., 395; trigonometry, 518, 520, 521
Ozanam, J.: logarithms, 476; trigonometry, 525

Padoa, A., 541
Paoli, P., 525; inverse trigonometric functions, 532; powers of trigonometric functions, 537; use of *e*, 400
Pappus, letters for magnitudes, 388
Parentheses, in Gerard, 519
Partition of numbers, 411
Partridge, Seth, 522
Pascal, B., theorem on inscribed hexagon, 708, 711
Pascal, Ernst, 407, 419, 420, 431
Pascal hexagram, 711
Pasch, A., upper and lower limit, 637
Pasquich, J., 576, 580
Paterson, W. E., 515
Paugger, logarithms, 479
Pauli, W., Jr., tensors, 486, 510
Peacock, G., 452; calculus, 583; fractional derivatives, 591
Peano, G., 409, 671, 698; Bernoulli's numbers, 419; calculus, 590, 627; continued fractions, 432; *Formulaire*, 688–91; inverse trigonometric functions, 535; logarithms, 470; matrices, 468; signs for functions, 646; signs for limits, symbolic logic, 687–95, 697; trigonometry, 526; use of ∼, 489; vectors, 505–9; printing of symbols, 725
Peck, W. G., 595
Peirce, B.: calculus, 588, 613, 620; *c* for 2.718 . . . ., 400, 681; hyperbolic functions, 530; inverse trigonometric functions, 535; signs for 3.141 . . . and 2.718 . . . ., 401, 681; trigonometry, 526
Peirce, Charles S., 449; symbolic logic, 681, 682, 684, 685, 686, 730

Peirce, James Mills, 401; hyperbolic functions, 530; imaginary, 499

Peletier, Jacques, degrees and minutes, 514

Perfect numbers, 415

Permutations, survey of, 452

Perron, O., 427, 430

Perry, J., 593, 618, 619

Peyrard, 718

Pfannenstiel, E., 615

π: survey of signs for 3.141 . . . . , 395; Wallis' expression for, 451

Picard, E., 449

Pierpont, James, sign for limit, 635

Pires, F. M., trigonometry, 526

Pitiscus, B.: degrees and minutes, 514; marking triangles, 516; trigonometry, 517, 518

Plana, G. A. A., limits of integration, 626

Planets, signs for, 394

Ploucquet, Gottfried, logic, 671

Plücker, J.: abridged notation, 708; his equations, 709; Pascal hexagram, 711

Poincaré, H., 490; inequality in series, 486

Poinsot, L., 702

Poisson, S. D.: calculus, 595, 598; limits of integration, 626; trigonometry, 526

Pollock, Oliver, 403

Polygonal numbers. See Figurate numbers

Polyhedra, 702; morphology of, 702

Polynomial theorem, 443, 444

Poncelet, J. V., 704

Poretsky, Planton: logic, 696; "therefore" and "because," 669

Potential, 738

Poullet-Delisle, A. Ch. M., 498

Power series, 661

Powers, Leibniz, 552

Praalder, L., absolute difference, 488

Prändel, J. G.: fluxions, 630; infinite series, 436; letters of different alphabets, 393; logarithms, 477

Praetorius, 517

Prandtl, L., 509

Prasse, Mor. von, 478

Price, B., 593

Prime numbers, 409

Principal values, 480

Pringsheim, Alfred, 421; continued fractions, 432, 433; infinite series, 438; logarithms, 472, 473; signs in theory of limits, 486; use of $\sim$, 490; upper and lower limit, 637; use of π, 451

Pringsheim, A., and J. Molk: inverse hyperbolic functions, 536; trigonometry, 526

Printing, mathematical, 724-26

Product of terms in arithmetical progression, 440; continued and infinite products, 451; Jarrett's products, 447

Proportion: Finck, 517; Leibniz, 540, 541, 549

Prosdocimo de' Beldomandi, 389

Prym, F. A., 436

Ptolemy, signs for degrees, minutes, 512

Puryear, C., 515

Quaternions, 502-9

Raabe, J. L., binomial coefficients, 419, 439

Radians, 515

Radicals, 541

Radix, 580

Rahn, J. H., 730; greater or less, 484; symbolic logic, 667, 668; "therefore," 668

Raphson, J., 640

Rawlinson, Rich.: greater or less, 483; trigonometry, 522

Rayleigh, Lord (John William Strutt), 598

Reeson, Paul, 514

Regiomontanus, 512

Reinhold, E., 514

Reiss, M., determinants, 461, 462

Relatively prime numbers, 409

Résal, H., 506, 593

Residual calculus, 629

Reye, Theodor, projectivity, 705

Reyher, S., greater or less, 484

Reyneau, Ch.: logarithms, 472; sign of integration, 623

Rhaeticus, 514

Riabouchinski, D., different zeroes, 493

Riccati, Vincente, hyperbolic functions, 529

Ricci, G., vector analysis, 510

Richeri, Ludovico, logic, 671

Rider, P. R., and A. Davis, 515, 527

Riemann, G. F. B., 507; theta functions, 658; zeta functions, 659; →, 636

Riese, Adam, letters for numbers, 390

Riesz, F., 421

Rigaud, S. P., 520, 567

Rignano, Eugenio, symbolic logic, 698

Rivard, D. F., 525

Robins, Benjamin, Leibnizian symbols, 621

Romanus, A., trigonometry, 517, 518, 521

Ronayne, Philip: greater or less, 483; trigonometry, 525; "therefore," 668

Rondet, 622

Roots, Leibniz, 552

Rothe, A., 439; logarithms, 479

Rothe, H., 526; hyperbolic functions, 529; powers of trigonometric functions, 537

Rouché, E., and Ch. de Comberousse, vectors, 502

Rowe, J., 622

Rudio, F., 397

Rudolff, Chr., letters for numbers, 390

Ruffini, Paolo: n-factorial, 449; substitutions, 453

Runkle, J. D., 449, 487, 613

Russell, Bertrand, 421, 449; logic, 687, 694–97

Russell, J. W., 706

Saalschütz, L., 407, 419; binomial coefficients, 439

Safford, T. H., use of π, 401

Salmon, G., 449; anharmonic ratio, 706; abridged notation, 708; determinants, 463; partial derivatives, 612; quantics, 456; Plücker's equations, 709; lines on cubic surface, 710; Pascal's hexagram, 711

Sarrus, F., 625, 626

Sault, Richard: absolute difference, 487; greater or less, 483

Saurin, Abbé, 400

Saverien, A.: absolute difference, 488; quoted, 716

Scheffer, L., on limits, 637

Scheffers, G., trigonometry, 526

Schell, W., 502

Schellbach, K. H., 446, 479

Scherffer, C.: logarithms, 477; trigonometry, 525; inverse trigonometric functions, 532; powers of trigonometric functions, 537

Scherk, H. F., 419, 420; determinants, 461

Scherwin, Hen., 514

Schläfli, L.: Bessel functions, 664; double-six lines on cubic surface, 710

Schlegel, Gustave, 394

Schlegel, Victor, vectors, 508

Schlömilch, O.: Bessel function, 664; cosine integral, 665; exponential integral, 665; logarithms, 479; n-factorial, 449

Schmeisser, F., inequality, 485

Schnuse, C. H., 472

Schoen, J., 568

Schöner, Joh., degrees, minutes, 513

Schoenflies, A., 421; groups, 455; trigonometry, 526

Schoner, L., degrees, minutes, 514

Schooten, Fr. van. See Van Schooten

Schott, K., trigonometry, 521

Schouten, J. A., 507, 510

Schröder, E., symbolic logic, 685, 686, 687, 691

Schrön, L., 478

Schubert, F. T., 630

Schultz, Johann, 436

Schur, F., projectivity, 705

Schwarz, H. A., 654, 657

Schweins, F., 447; determinants, 461

Scott, R. F., 465

Seaver, E. P., 535

Segner, J. A.: logic, 671; logarithms, 477; sign for 3.141 . . . . , 398; trigonometry, 525; use of e, 400

Seidel, L., less than, 485

Seller, John, 516, 520, 522

Series: sum of, 438; signs of Jarrett, 447; hypergeometric series, 451. See Infinite series

Serret, J. A., 425; calculus, 598; hyperbolic functions, 529; powers of trigonometric functions, 537; substitutions, 453; symmetric functions, 648; trigonometry, 526

Servois, F., 447, 593, 641

Sevenoak, F. L., 428, 477

Sexagesimal system in Babylonia, 512; Ptolemy, 511

Shanks, W., 407

Shaw, J. B., vector analysis, 509

Sheppard, W. F., determinants, 463

Sherwin, H., 474, 398; logarithms, 469

Shirtcliffe, Robert, logarithms, 469

Shutts, G. C., "therefore" and "because," 669

Sigma Σ, 505

Sign $\sharp$, 493

Sign $\frac{0}{0}$, 638

Sign $\sim$ or $\backsim$, 487–89

Sign $\equiv$, 408

Similar, 488, 541; in Leibniz, 541, 557, 565

Simpson, Th.: difference, 491; no sign for integration, 622; greater or less, 483; trigonometry, 526; "therefore," 668

Simson, Robert, letters for numbers, 388

Sine integral, 665

Slaught, H. E., 535

Smith, C., 428, 449; absolute difference, 487; "therefore" and "because," 669

Smith, D. E., 392, 402; quoted, 715

Smith, H. J. S., determinants, 463, 468

Smith, W. B., 530

Snell, W., trigonometry, 519, 521

Sohncke, L. A., hyperbolic functions, 529

Soldner, J., logarithm integral, 665

Somov, J., 506

Speidell, John, 520

Spence, W., Logarithms, 472

Spottiswoode, W., 725, 750

Sprague, T. B., 458

Stäckel, P., 499, 596, 601, 611

Stampioen, J. J.: □ as an operator, 519; trigonometry, 519

Star: *Star Chart and Catalogue*, 735; in logarithmic tables, 478; for multiplication, 547

Staudt, K. G. C. von, 419

Steenstra, P., 525, 568

Stein, J. P. W., 480

Steiner, Jacob, trigonometry, 526

Steinhauser, A.: logarithms, 471, 477; nearly equal, 635

Steinmetz, C. P., 500

Stern, M. A., 419; continued fractions, 429; combinations, 452; infinite series, 437

Stifel, M., 390

Stirling, James, 638; series, 439

Stockler, B. G., 448; sign for limit, 631

Stokes, G. G., 725

Stolz, O., 494; powers of trigonometric functions, 537; trigonometry, 526

Stolz, O., and J. A. Gmeiner (joint authors): binomial coefficients, 439; on $\left[\frac{a}{b}\right]$, 407; infinite series, 437; logarithms, 470, 471; thesis, 494

Stone, E.: continued fractions, 427; greater or less, 483; finite differences, 640; sign for fluents, 622; "therefore," 668

Story, W. E., 409

Strabbe, A. B., 568

Strauch, G. W., 614

Street, Joseph M., dollar mark, 403

Stridsberg, E., 409

Stringham, I., 449, 471; "cis," 503; trigonometry, 528; vectors, 510

Strong, Th., 593, 630

Studnička, F. S., logarithms, 479

Study, E., vectors, 503

Sturm, Christoph, on sign for 3.141 . . . . , 395

Sturm, Rudolf, 705

Subfactorial n, 450

Substitutions, survey of, 453

Sums of numbers, 410; use of Σ, 438; signs of Jarret, 447

Suter, H., 422, 511

Svedberg, T., "therefore," 669

Sylvester, J. J., 410; determinants, 461, 462; partitions, 411; n-factorial, 449; totient, 409; umbral notation, 462

Symmetric functions, 647

Symmetric polynomials, 410

Tacquet, A., trigonometry, 521

Tait, P. G., 502, 507–9; calculus, 581, 598, 620; quoted, 724

Tannery, J., vectors, 502

Tannery, P., 394

Taylor, Brook: finite differences, 640; logarithms, 469; greater or less, 483; his theorem, 578, 584; signs for fluents, 622

Taylor, John, 522

Taylor, T. U., 515

Tensors, 510

Terquem, O., determinants, 462

Tetens, J. N., 444

Thacker, A., 410

"Therefore," symbol for, 668

Theta functions, 656–58

Thomson, James, absolute difference, 487

Thomson, William (Lord Kelvin), 515; calculus, 581, 598; quaternions, 508

Timerding, H. E., 431; hyperbolic functions, 529; vectors, 506, 509

Todhunter, I., 452; Bessel's functions, 664; calculus, 598; Legendre's coefficients, 662; n-factorial, 449; powers of trigonometric functions, 537; trigonometry, 526

Tönnies, I. H., logic, 671

Toledo, L. O. de: powers of trigonometric functions, 537; trigonometry, 526

Torija, Manuel Torres, inverse trigonometric functions, 535

Toro, A. de la Rosa, 403

Torporley, N., trigonometry, 518

Townsend, M., 402

Tralles, George, 577

Transfinite cardinals, 421

Transfinite ordinal number, 421

Tresse, A., 705

Treutlein, P., 389

Triangles: marking of, 516; geometry of triangle, 700

Trigonometry: survey of, 511–37; spherical, 524; hyperbolic functions, 529, 530; inverse trigonometric functions, 532–35; parabolic functions, 531; inverse hyperbolic functions, 536; haversine, 527; Leibniz, 555; Newton, 522

Tropfke, J., 390, 469, 514, 643

Tschirnhausen, E. W., 538, 539, 540

Turner, G. C., 506

Tweedie, Charles, 638

Tycho, Brahe, 514, 517

Urquhart, Thomas, 523

Ursus, Raymarus, 514

Vacca, G., 674, 691, 692

Vailati, G., symbolic logic, 688

Vandermonde, C. A.:   combinatorial symbols, 446; combinations, 452; determinants, 460; n-factorial, 448, 449; series, 436; sums, 441; use of π, 398

Van Schooten, Fr. (Sr.), 519

Van Schooten, Fr. (the Younger), 565

Varignon, P.: logarithms, 469; polar co-ordinates, 708; sign for 3.141 . . . . , 396

Vasquez, de Serna, 403

Vazquez, Maximo, 403

Vector analysis, 502–10; international movement failed, 737

Vega, G., 475; calculus, 596; sign for degree, 514; trigonometry, 525

Veronese, G., Pascal's hexagram, 711

Vessiot, E., 468

Vieta, Francis, 518, 520; infinite product, 451; trigonometry, 517

Vieth, G. U. A., 526; powers of trigonometric functions, 537

Vince, S., 622, 630

Vinculum, Ditton, 567; joined to sign of integration, 623; Leibniz, 539, 541; Nicole, 436

Vivanti, G., 571; symbolic logic, 688

Vlacq, A., 474, 516

Voigt, J. V., 477

Voigt, W., 502, 507

Volapuk, 749

Volterra, V.: matrices, 468; sign of integration, 627

Von Staudt, K. G. C., 419, 705, 711

Vryer, A., 568

Wallis, John, 538, 622; absolute difference, 487; binomial formula, 439; continued fractions, 425; expression for π, 451; greater or equal, 484; greater or less, 483; imaginaries, 497; infinite series, 435; sign for 3.141 . . . . , 395; trigonometry, 522; treatise on algebra, 567; ∞, 436

Walter, Thomas, 668

Walther, J. L., 511

Warburton, Henry, 449

Ward, John: logarithms, 476; trigonometry, 522, 525; "therefore," 668

Ward, Seth, 522; greater or less, 483

Waring, E.: calculus, 579, 593, 601; finite differences, 640; infinite series, 436; trigonometry, 525; sign of integration, 620

Watkins, Thomas, 436; logarithms. 469

Watson, G. N., Bessel functions, 664

Wattenbach, W., 421

Weatherburn, C. E., tensors, 510

Webber, S.: fluxions, 630; greater or less, 483

Weber, 507

Weber, H., and J. Wellstein, trigonometry, 526

Webster, Noah, 449

Wedderburn, J. H. M., 620

Wegener, G., 511

Weierstrass, K.: absolute value, 492; elliptic functions, 653; gamma function, 650; $\wp$ function, 654; sign for limit, 634; theta functions, 657; use of ∞, 421; n-factorial, 449; Hebrew letters, 393

Weigel, E., greater or less, 484

Weissenborn, H., 421

Weitzenböck, R.: matrices, 468; tensors, 510

Weld, L. G., 433, 464

Well-ordered aggregates, 421

Wells, E., trigonometry, 525

Wells, John, 516, 520, 521

Wentworth, G. A., trigonometry, 526, 527

Wertheim, G., 409, 425, 453

Wessel, C., 398, 497, 503, 719

West, John, 535

Weyl, H., tensors, 510

White, E. E., "therefore" and "because," 669

White, J. D., 527

Whitehead, A. N., 449; infinite series, 437; symbolic logic, 692, 693–97; quoted, 721

Whitworth, W. A., 450, 452, 468; anharmonic ratio, 706

Wiechert, E., 507

Wiegand, A., 452; equivalence, 490

Wiener, C., 707

Wilson, E. B., vector analysis, 508–10

Wilson, John, trigonometry, 522, 525

Wing, V., trigonometry, 522

Wingate, E., trigonometry, 520, 521

Winlock, J., use of π, 401

Winsor, Justin, 630

Wirtinger, W., 658

Wittgenstein, L., logic, 697

Witting, A., 567

Wolf, R., 394

Wolff (also Wolf), Chr.: finite series, 435; greater or less, 484; logarithms, 469; sign of integration, 620

Wolff, F., 586

Woodhouse, Robert, 579, 583, 603, 630; "therefore," 668

Woodward, R. S., 449

Worpitzky, J., logarithms, 479

Wright, Chauncey, use of π, 401

Wright, Edward, 516; "Appendix," 516, 518, 519, 520, 522

Wright, Elizur, 630

Wright, J. E., determinants, 463

Wright, J. M. F.: absolute difference, 487; "therefore" and "because," 669

Wronski, H., 421, 447

Wundt, W., logic, 685

Young, J. R., symmetric functions, 648

Young, J. W., and A. J. Schwartz, "therefore" and "because," 669

Young, J. W. A., 494

Young, W. H., 598; signs for limits, 637

Zahlkörper, 414

Zendrini, 538

Zermelo, E., 421

Zeta function, Jacobi's, 653, 655, 656, 659, 660

Zodiac, signs of 394

A CATALOG OF SELECTED
DOVER BOOKS
IN SCIENCE AND MATHEMATICS

A CATALOG OF SELECTED
DOVER BOOKS
IN SCIENCE AND MATHEMATICS

QUALITATIVE THEORY OF DIFFERENTIAL EQUATIONS, V.V. Nemytskii and V.V. Stepanov. Classic graduate-level text by two prominent Soviet mathematicians covers classical differential equations as well as topological dynamics and ergodic theory. Bibliographies. 523pp. 5⅜ × 8½.   65954-2 Pa. $10.95

MATRICES AND LINEAR ALGEBRA, Hans Schneider and George Phillip Barker. Basic textbook covers theory of matrices and its applications to systems of linear equations and related topics such as determinants, eigenvalues and differential equations. Numerous exercises. 432pp. 5⅜ × 8½.   66014-1 Pa. $9.95

QUANTUM THEORY, David Bohm. This advanced undergraduate-level text presents the quantum theory in terms of qualitative and imaginative concepts, followed by specific applications worked out in mathematical detail. Preface. Index. 655pp. 5⅜ × 8½.   65969-0 Pa. $13.95

ATOMIC PHYSICS (8th edition), Max Born. Nobel laureate's lucid treatment of kinetic theory of gases, elementary particles, nuclear atom, wave-corpuscles, atomic structure and spectral lines, much more. Over 40 appendices, bibliography. 495pp. 5⅜ × 8½.   65984-4 Pa. $11.95

ELECTRONIC STRUCTURE AND THE PROPERTIES OF SOLIDS: The Physics of the Chemical Bond, Walter A. Harrison. Innovative text offers basic understanding of the electronic structure of covalent and ionic solids, simple metals, transition metals and their compounds. Problems. 1980 edition. 582pp. 6⅛ × 9¼.   66021-4 Pa. $14.95

BOUNDARY VALUE PROBLEMS OF HEAT CONDUCTION, M. Necati Özisik. Systematic, comprehensive treatment of modern mathematical methods of solving problems in heat conduction and diffusion. Numerous examples and problems. Selected references. Appendices. 505pp. 5⅜ × 8½.   65990-9 Pa. $11.95

A SHORT HISTORY OF CHEMISTRY (3rd edition), J.R. Partington. Classic exposition explores origins of chemistry, alchemy, early medical chemistry, nature of atmosphere, theory of valency, laws and structure of atomic theory, much more. 428pp. 5⅜ × 8½. (Available in U.S. only)   65977-1 Pa. $10.95

A HISTORY OF ASTRONOMY, A. Pannekoek. Well-balanced, carefully reasoned study covers such topics as Ptolemaic theory, work of Copernicus, Kepler, Newton, Eddington's work on stars, much more. Illustrated. References. 521pp. 5⅜ × 8½.   65994-1 Pa. $11.95

PRINCIPLES OF METEOROLOGICAL ANALYSIS, Walter J. Saucier. Highly respected, abundantly illustrated classic reviews atmospheric variables, hydrostatics, static stability, various analyses (scalar, cross-section, isobaric, isentropic, more). For intermediate meteorology students. 454pp. 6⅛ × 9¼.   65979-8 Pa. $12.95

RELATIVITY, THERMODYNAMICS AND COSMOLOGY, Richard C. Tolman. Landmark study extends thermodynamics to special, general relativity; also applications of relativistic mechanics, thermodynamics to cosmological models. 501pp. 5⅜ × 8½. 65383-8 Pa. $12.95

APPLIED ANALYSIS, Cornelius Lanczos. Classic work on analysis and design of finite processes for approximating solution of analytical problems. Algebraic equations, matrices, harmonic analysis, quadrature methods, much more. 559pp. 5⅜ × 8½. 65656-X Pa. $12.95

SPECIAL RELATIVITY FOR PHYSICISTS, G. Stephenson and C.W. Kilmister. Concise elegant account for nonspecialists. Lorentz transformation, optical and dynamical applications, more. Bibliography. 108pp. 5⅜ × 8½. 65519-9 Pa. $4.95

INTRODUCTION TO ANALYSIS, Maxwell Rosenlicht. Unusually clear, accessible coverage of set theory, real number system, metric spaces, continuous functions, Riemann integration, multiple integrals, more. Wide range of problems. Undergraduate level. Bibliography. 254pp. 5⅜ × 8½. 65038-3 Pa. $7.95

INTRODUCTION TO QUANTUM MECHANICS With Applications to Chemistry, Linus Pauling & E. Bright Wilson, Jr. Classic undergraduate text by Nobel Prize winner applies quantum mechanics to chemical and physical problems. Numerous tables and figures enhance the text. Chapter bibliographies. Appendices. Index. 468pp. 5⅜ × 8½. 64871-0 Pa. $11.95

ASYMPTOTIC EXPANSIONS OF INTEGRALS, Norman Bleistein & Richard A. Handelsman. Best introduction to important field with applications in a variety of scientific disciplines. New preface. Problems. Diagrams. Tables. Bibliography. Index. 448pp. 5⅜ × 8½. 65082-0 Pa. $11.95

MATHEMATICS APPLIED TO CONTINUUM MECHANICS, Lee A. Segel. Analyzes models of fluid flow and solid deformation. For upper-level math, science and engineering students. 608pp. 5⅜ × 8½. 65369-2 Pa. $13.95

ELEMENTS OF REAL ANALYSIS, David A. Sprecher. Classic text covers fundamental concepts, real number system, point sets, functions of a real variable, Fourier series, much more. Over 500 exercises. 352pp. 5⅜ × 8½. 65385-4 Pa. $9.95

PHYSICAL PRINCIPLES OF THE QUANTUM THEORY, Werner Heisenberg. Nobel Laureate discusses quantum theory, uncertainty, wave mechanics, work of Dirac, Schroedinger, Compton, Wilson, Einstein, etc. 184pp. 5⅜ × 8½.
60113-7 Pa. $4.95

INTRODUCTORY REAL ANALYSIS, A.N. Kolmogorov, S.V. Fomin. Translated by Richard A. Silverman. Self-contained, evenly paced introduction to real and functional analysis. Some 350 problems. 403pp. 5⅜ × 8½. 61226-0 Pa. $9.95

PROBLEMS AND SOLUTIONS IN QUANTUM CHEMISTRY AND PHYSICS, Charles S. Johnson, Jr. and Lee G. Pedersen. Unusually varied problems, detailed solutions in coverage of quantum mechanics, wave mechanics, angular momentum, molecular spectroscopy, scattering theory, more. 280 problems plus 139 supplementary exercises. 430pp. 6½ × 9¼. 65236-X Pa. $11.95

ASYMPTOTIC METHODS IN ANALYSIS, N.G. de Bruijn. An inexpensive, comprehensive guide to asymptotic methods—the pioneering work that teaches by explaining worked examples in detail. Index. 224pp. 5⅜ × 8½.   64221-6 Pa. $6.95

OPTICAL RESONANCE AND TWO-LEVEL ATOMS, L. Allen and J.H. Eberly. Clear, comprehensive introduction to basic principles behind all quantum optical resonance phenomena. 53 illustrations. Preface. Index. 256pp. 5⅜ × 8½.
65533-4 Pa. $7.95

COMPLEX VARIABLES, Francis J. Flanigan. Unusual approach, delaying complex algebra till harmonic functions have been analyzed from real variable viewpoint. Includes problems with answers. 364pp. 5⅜ × 8½.   61388-7 Pa. $7.95

ATOMIC SPECTRA AND ATOMIC STRUCTURE, Gerhard Herzberg. One of best introductions; especially for specialist in other fields. Treatment is physical rather than mathematical. 80 illustrations. 257pp. 5⅜ × 8½.   60115-3 Pa. $5.95

APPLIED COMPLEX VARIABLES, John W. Dettman. Step-by-step coverage of fundamentals of analytic function theory—plus lucid exposition of five important applications: Potential Theory; Ordinary Differential Equations; Fourier Transforms; Laplace Transforms; Asymptotic Expansions. 66 figures. Exercises at chapter ends. 512pp. 5⅜ × 8½.   64670-X Pa. $10.95

ULTRASONIC ABSORPTION: An Introduction to the Theory of Sound Absorption and Dispersion in Gases, Liquids and Solids, A.B. Bhatia. Standard reference in the field provides a clear, systematically organized introductory review of fundamental concepts for advanced graduate students, research workers. Numerous diagrams. Bibliography. 440pp. 5⅜ × 8½.   64917-2 Pa. $11.95

UNBOUNDED LINEAR OPERATORS: Theory and Applications, Seymour Goldberg. Classic presents systematic treatment of the theory of unbounded linear operators in normed linear spaces with applications to differential equations. Bibliography. 199pp. 5⅜ × 8½.   64830-3 Pa. $7.95

LIGHT SCATTERING BY SMALL PARTICLES, H.C. van de Hulst. Comprehensive treatment including full range of useful approximation methods for researchers in chemistry, meteorology and astronomy. 44 illustrations. 470pp. 5⅜ × 8½.   64228-3 Pa. $10.95

CONFORMAL MAPPING ON RIEMANN SURFACES, Harvey Cohn. Lucid, insightful book presents ideal coverage of subject. 334 exercises make book perfect for self-study. 55 figures. 352pp. 5⅜ × 8¼.   64025-6 Pa. $8.95

OPTICKS, Sir Isaac Newton. Newton's own experiments with spectroscopy, colors, lenses, reflection, refraction, etc., in language the layman can follow. Foreword by Albert Einstein. 532pp. 5⅜ × 8½.   60205-2 Pa. $9.95

GENERALIZED INTEGRAL TRANSFORMATIONS, A.H. Zemanian. Graduate-level study of recent generalizations of the Laplace, Mellin, Hankel, K. Weierstrass, convolution and other simple transformations. Bibliography. 320pp. 5⅜ × 8½.   65375-7 Pa. $7.95

THE ELECTROMAGNETIC FIELD, Albert Shadowitz. Comprehensive undergraduate text covers basics of electric and magnetic fields, builds up to electromagnetic theory. Also related topics, including relativity. Over 900 problems. 768pp. 5⅜ × 8¼. 65660-8 Pa. $17.95

FOURIER SERIES, Georgi P. Tolstov. Translated by Richard A. Silverman. A valuable addition to the literature on the subject, moving clearly from subject to subject and theorem to theorem. 107 problems, answers. 336pp. 5⅜ × 8½. 63317-9 Pa. $7.95

THEORY OF ELECTROMAGNETIC WAVE PROPAGATION, Charles Herach Papas. Graduate-level study discusses the Maxwell field equations, radiation from wire antennas, the Doppler effect and more. xiii + 244pp. 5⅜ × 8½. 65678-0 Pa. $6.95

DISTRIBUTION THEORY AND TRANSFORM ANALYSIS: An Introduction to Generalized Functions, with Applications, A.H. Zemanian. Provides basics of distribution theory, describes generalized Fourier and Laplace transformations. Numerous problems. 384pp. 5⅜ × 8½. 65479-6 Pa. $9.95

THE PHYSICS OF WAVES, William C. Elmore and Mark A. Heald. Unique overview of classical wave theory. Acoustics, optics, electromagnetic radiation, more. Ideal as classroom text or for self-study. Problems. 477pp. 5⅜ × 8½. 64926-1 Pa. $11.95

CALCULUS OF VARIATIONS WITH APPLICATIONS, George M. Ewing. Applications-oriented introduction to variational theory develops insight and promotes understanding of specialized books, research papers. Suitable for advanced undergraduate/graduate students as primary, supplementary text. 352pp. 5⅜ × 8½. 64856-7 Pa. $8.95

A TREATISE ON ELECTRICITY AND MAGNETISM, James Clerk Maxwell. Important foundation work of modern physics. Brings to final form Maxwell's theory of electromagnetism and rigorously derives his general equations of field theory. 1,084pp. 5⅜ × 8½. 60636-8, 60637-6 Pa., Two-vol. set $19.90

AN INTRODUCTION TO THE CALCULUS OF VARIATIONS, Charles Fox. Graduate-level text covers variations of an integral, isoperimetrical problems, least action, special relativity, approximations, more. References. 279pp. 5⅜ × 8½. 65499-0 Pa. $7.95

HYDRODYNAMIC AND HYDROMAGNETIC STABILITY, S. Chandrasekhar. Lucid examination of the Rayleigh-Benard problem; clear coverage of the theory of instabilities causing convection. 704pp. 5⅜ × 8¼. 64071-X Pa. $14.95

CALCULUS OF VARIATIONS, Robert Weinstock. Basic introduction covering isoperimetric problems, theory of elasticity, quantum mechanics, electrostatics, etc. Exercises throughout. 326pp. 5⅜ × 8½. 63069-2 Pa. $7.95

DYNAMICS OF FLUIDS IN POROUS MEDIA, Jacob Bear. For advanced students of ground water hydrology, soil mechanics and physics, drainage and irrigation engineering and more. 335 illustrations. Exercises, with answers. 784pp. 6⅛ × 9¼. 65675-6 Pa. $19.95

NUMERICAL METHODS FOR SCIENTISTS AND ENGINEERS, Richard Hamming. Classic text stresses frequency approach in coverage of algorithms, polynomial approximation, Fourier approximation, exponential approximation, other topics. Revised and enlarged 2nd edition. 721pp. 5⅜ × 8½.
65241-6 Pa. $14.95

THEORETICAL SOLID STATE PHYSICS, Vol. I: Perfect Lattices in Equilibrium; Vol. II: Non-Equilibrium and Disorder, William Jones and Norman H. March. Monumental reference work covers fundamental theory of equilibrium properties of perfect crystalline solids, non-equilibrium properties, defects and disordered systems. Appendices. Problems. Preface. Diagrams. Index. Bibliography. Total of 1,301pp. 5⅜ × 8½. Two volumes. Vol. I 65015-4 Pa. $12.95
Vol. II 65016-2 Pa. $12.95

OPTIMIZATION THEORY WITH APPLICATIONS, Donald A. Pierre. Broad-spectrum approach to important topic. Classical theory of minima and maxima, calculus of variations, simplex technique and linear programming, more. Many problems, examples. 640pp. 5⅜ × 8½. 65205-X Pa. $13.95

THE MODERN THEORY OF SOLIDS, Frederick Seitz. First inexpensive edition of classic work on theory of ionic crystals, free-electron theory of metals and semiconductors, molecular binding, much more. 736pp. 5⅜ × 8½.
65482-6 Pa. $15.95

ESSAYS ON THE THEORY OF NUMBERS, Richard Dedekind. Two classic essays by great German mathematician: on the theory of irrational numbers; and on transfinite numbers and properties of natural numbers. 115pp. 5⅜ × 8½.
21010-3 Pa. $4.95

THE FUNCTIONS OF MATHEMATICAL PHYSICS, Harry Hochstadt. Comprehensive treatment of orthogonal polynomials, hypergeometric functions, Hill's equation, much more. Bibliography. Index. 322pp. 5⅜ × 8½. 65214-9 Pa. $9.95

NUMBER THEORY AND ITS HISTORY, Oystein Ore. Unusually clear, accessible introduction covers counting, properties of numbers, prime numbers, much more. Bibliography. 380pp. 5⅜ × 8½. 65620-9 Pa. $8.95

THE VARIATIONAL PRINCIPLES OF MECHANICS, Cornelius Lanczos. Graduate level coverage of calculus of variations, equations of motion, relativistic mechanics, more. First inexpensive paperbound edition of classic treatise. Index. Bibliography. 418pp. 5⅜ × 8½. 65067-7 Pa. $10.95

MATHEMATICAL TABLES AND FORMULAS, Robert D. Carmichael and Edwin R. Smith. Logarithms, sines, tangents, trig functions, powers, roots, reciprocals, exponential and hyperbolic functions, formulas and theorems. 269pp. 5⅜ × 8½. 60111-0 Pa. $5.95

THEORETICAL PHYSICS, Georg Joos, with Ira M. Freeman. Classic overview covers essential math, mechanics, electromagnetic theory, thermodynamics, quantum mechanics, nuclear physics, other topics. First paperback edition. xxiii + 885pp. 5⅜ × 8½. 65227-0 Pa. $18.95

HANDBOOK OF MATHEMATICAL FUNCTIONS WITH FORMULAS, GRAPHS, AND MATHEMATICAL TABLES, edited by Milton Abramowitz and Irene A. Stegun. Vast compendium: 29 sets of tables, some to as high as 20 places. 1,046pp. 8 × 10½. 61272-4 Pa. $22.95

MATHEMATICAL METHODS IN PHYSICS AND ENGINEERING, John W. Dettman. Algebraically based approach to vectors, mapping, diffraction, other topics in applied math. Also generalized functions, analytic function theory, more. Exercises. 448pp. 5⅜ × 8¼. 65649-7 Pa. $8.95

A SURVEY OF NUMERICAL MATHEMATICS, David M. Young and Robert Todd Gregory. Broad self-contained coverage of computer-oriented numerical algorithms for solving various types of mathematical problems in linear algebra, ordinary and partial, differential equations, much more. Exercises. Total of 1,248pp. 5⅜ × 8½. Two volumes. Vol. I 65691-8 Pa. $14.95
Vol. II 65692-6 Pa. $14.95

TENSOR ANALYSIS FOR PHYSICISTS, J.A. Schouten. Concise exposition of the mathematical basis of tensor analysis, integrated with well-chosen physical examples of the theory. Exercises. Index. Bibliography. 289pp. 5⅜ × 8½.
65582-2 Pa. $7.95

INTRODUCTION TO NUMERICAL ANALYSIS (2nd Edition), F.B. Hildebrand. Classic, fundamental treatment covers computation, approximation, interpolation, numerical differentiation and integration, other topics. 150 new problems. 669pp. 5⅜ × 8½. 65363-3 Pa. $14.95

INVESTIGATIONS ON THE THEORY OF THE BROWNIAN MOVEMENT, Albert Einstein. Five papers (1905–8) investigating dynamics of Brownian motion and evolving elementary theory. Notes by R. Fürth. 122pp. 5⅜ × 8½.
60304-0 Pa. $4.95

NUMERICAL METHODS FOR SCIENTISTS AND ENGINEERS, Richard Hamming. Classic text stresses frequency approach in coverage of algorithms, polynomial approximation, Fourier approximation, exponential approximation, other topics. Revised and enlarged 2nd edition. 721pp. 5⅜ × 8½. 65241-6 Pa. $14.95

AN INTRODUCTION TO STATISTICAL THERMODYNAMICS, Terrell L. Hill. Excellent basic text offers wide-ranging coverage of quantum statistical mechanics, systems of interacting molecules, quantum statistics, more. 523pp. 5⅜ × 8½. 65242-4 Pa. $11.95

ELEMENTARY DIFFERENTIAL EQUATIONS, William Ted Martin and Eric Reissner. Exceptionally clear, comprehensive introduction at undergraduate level. Nature and origin of differential equations, differential equations of first, second and higher orders. Picard's Theorem, much more. Problems with solutions. 331pp. 5⅜ × 8½. 65024-3 Pa. $8.95

STATISTICAL PHYSICS, Gregory H. Wannier. Classic text combines thermodynamics, statistical mechanics and kinetic theory in one unified presentation of thermal physics. Problems with solutions. Bibliography. 532pp. 5⅜ × 8½.
65401-X Pa. $11.95

ORDINARY DIFFERENTIAL EQUATIONS, Morris Tenenbaum and Harry Pollard. Exhaustive survey of ordinary differential equations for undergraduates in mathematics, engineering, science. Thorough analysis of theorems. Diagrams. Bibliography. Index. 818pp. 5⅜ × 8½. 64940-7 Pa. $16.95

STATISTICAL MECHANICS: Principles and Applications, Terrell L. Hill. Standard text covers fundamentals of statistical mechanics, applications to fluctuation theory, imperfect gases, distribution functions, more. 448pp. 5⅜ × 8½. 65390-0 Pa. $9.95

ORDINARY DIFFERENTIAL EQUATIONS AND STABILITY THEORY: An Introduction, David A. Sánchez. Brief, modern treatment. Linear equation, stability theory for autonomous and nonautonomous systems, etc. 164pp. 5⅜ × 8¼. 63828-6 Pa. $5.95

THIRTY YEARS THAT SHOOK PHYSICS: The Story of Quantum Theory, George Gamow. Lucid, accessible introduction to influential theory of energy and matter. Careful explanations of Dirac's anti-particles, Bohr's model of the atom, much more. 12 plates. Numerous drawings. 240pp. 5⅜ × 8½. 24895-X Pa. $5.95

THEORY OF MATRICES, Sam Perlis. Outstanding text covering rank, non-singularity and inverses in connection with the development of canonical matrices under the relation of equivalence, and without the intervention of determinants. Includes exercises. 237pp. 5⅜ × 8½. 66810-X Pa. $7.95

GREAT EXPERIMENTS IN PHYSICS: Firsthand Accounts from Galileo to Einstein, edited by Morris H. Shamos. 25 crucial discoveries: Newton's laws of motion, Chadwick's study of the neutron, Hertz on electromagnetic waves, more. Original accounts clearly annotated. 370pp. 5⅜ × 8½. 25346-5 Pa. $9.95

INTRODUCTION TO PARTIAL DIFFERENTIAL EQUATIONS WITH APPLICATIONS, E.C. Zachmanoglou and Dale W. Thoe. Essentials of partial differential equations applied to common problems in engineering and the physical sciences. Problems and answers. 416pp. 5⅜ × 8½. 65251-3 Pa. $10.95

BURNHAM'S CELESTIAL HANDBOOK, Robert Burnham, Jr. Thorough guide to the stars beyond our solar system. Exhaustive treatment. Alphabetical by constellation: Andromeda to Cetus in Vol. 1; Chamaeleon to Orion in Vol. 2; and Pavo to Vulpecula in Vol. 3. Hundreds of illustrations. Index in Vol. 3. 2,000pp. 6⅛ × 9¼. 23567-X, 23568-8, 23673-0 Pa., Three-vol. set $41.85

ASYMPTOTIC EXPANSIONS FOR ORDINARY DIFFERENTIAL EQUATIONS, Wolfgang Wasow. Outstanding text covers asymptotic power series, Jordan's canonical form, turning point problems, singular perturbations, much more. Problems. 384pp. 5⅜ × 8½. 65456-7 Pa. $9.95

AMATEUR ASTRONOMER'S HANDBOOK, J.B. Sidgwick. Timeless, comprehensive coverage of telescopes, mirrors, lenses, mountings, telescope drives, micrometers, spectroscopes, more. 189 illustrations. 576pp. 5⅜ × 8¼. (USO) 24034-7 Pa. $9.95

SPECIAL FUNCTIONS, N.N. Lebedev. Translated by Richard Silverman. Famous Russian work treating more important special functions, with applications to specific problems of physics and engineering. 38 figures. 308pp. 5⅜ × 8½.
60624-4 Pa. $7.95

OBSERVATIONAL ASTRONOMY FOR AMATEURS, J.B. Sidgwick. Mine of useful data for observation of sun, moon, planets, asteroids, aurorae, meteors, comets, variables, binaries, etc. 39 illustrations. 384pp. 5⅜ × 8¼. (Available in U.S. only)
24033-9 Pa. $8.95

INTEGRAL EQUATIONS, F.G. Tricomi. Authoritative, well-written treatment of extremely useful mathematical tool with wide applications. Volterra Equations, Fredholm Equations, much more. Advanced undergraduate to graduate level. Exercises. Bibliography. 238pp. 5⅜ × 8½.
64828-1 Pa. $6.95

CELESTIAL OBJECTS FOR COMMON TELESCOPES, T.W. Webb. Inestimable aid for locating and identifying nearly 4,000 celestial objects. 77 illustrations. 645pp. 5⅜ × 8½.
20917-2, 20918-0 Pa., Two-vol. set $12.00

MODERN NONLINEAR EQUATIONS, Thomas L. Saaty. Emphasizes practical solution of problems; covers seven types of equations. ". . . a welcome contribution to the existing literature. . . ."—Math Reviews. 490pp. 5⅜ × 8½. 64232-1 Pa. $9.95

FUNDAMENTALS OF ASTRODYNAMICS, Roger Bate et al. Modern approach developed by U.S. Air Force Academy. Designed as a first course. Problems, exercises. Numerous illustrations. 455pp. 5⅜ × 8½.
60061-0 Pa. $8.95

INTRODUCTION TO LINEAR ALGEBRA AND DIFFERENTIAL EQUATIONS, John W. Dettman. Excellent text covers complex numbers, determinants, orthonormal bases, Laplace transforms, much more. Exercises with solutions. Undergraduate level. 416pp. 5⅜ × 8½.
65191-6 Pa. $9.95

INCOMPRESSIBLE AERODYNAMICS, edited by Bryan Thwaites. Covers theoretical and experimental treatment of the uniform flow of air and viscous fluids past two-dimensional aerofoils and three-dimensional wings; many other topics. 654pp. 5⅜ × 8½.
65465-6 Pa. $16.95

INTRODUCTION TO DIFFERENCE EQUATIONS, Samuel Goldberg. Exceptionally clear exposition of important discipline with applications to sociology, psychology, economics. Many illustrative examples; over 250 problems. 260pp. 5⅜ × 8½.
65084-7 Pa. $7.95

LAMINAR BOUNDARY LAYERS, edited by L. Rosenhead. Engineering classic covers steady boundary layers in two- and three-dimensional flow, unsteady boundary layers, stability, observational techniques, much more. 708pp. 5⅜ × 8½.
65646-2 Pa. $15.95

LECTURES ON CLASSICAL DIFFERENTIAL GEOMETRY, Second Edition, Dirk J. Struik. Excellent brief introduction covers curves, theory of surfaces, fundamental equations, geometry on a surface, conformal mapping, other topics. Problems. 240pp. 5⅜ × 8½.
65609-8 Pa. $6.95

ROTARY-WING AERODYNAMICS, W.Z. Stepniewski. Clear, concise text covers aerodynamic phenomena of the rotor and offers guidelines for helicopter performance evaluation. Originally prepared for NASA. 537 figures. 640pp. 6⅛ × 9¼.
64647-5 Pa. $14.95

DIFFERENTIAL GEOMETRY, Heinrich W. Guggenheimer. Local differential geometry as an application of advanced calculus and linear algebra. Curvature, transformation groups, surfaces, more. Exercises. 62 figures. 378pp. 5⅜ × 8½.
63433-7 Pa. $7.95

INTRODUCTION TO SPACE DYNAMICS, William Tyrrell Thomson. Comprehensive, classic introduction to space-flight engineering for advanced undergraduate and graduate students. Includes vector algebra, kinematics, transformation of coordinates. Bibliography. Index. 352pp. 5⅜ × 8½.
65113-4 Pa. $8.95

A SURVEY OF MINIMAL SURFACES, Robert Osserman. Up-to-date, in-depth discussion of the field for advanced students. Corrected and enlarged edition covers new developments. Includes numerous problems. 192pp. 5⅜ × 8½.
64998-9 Pa. $8.95

ANALYTICAL MECHANICS OF GEARS, Earle Buckingham. Indispensable reference for modern gear manufacture covers conjugate gear-tooth action, gear-tooth profiles of various gears, many other topics. 263 figures. 102 tables. 546pp. 5⅜ × 8½.
65712-4 Pa. $11.95

SET THEORY AND LOGIC, Robert R. Stoll. Lucid introduction to unified theory of mathematical concepts. Set theory and logic seen as tools for conceptual understanding of real number system. 496pp. 5⅜ × 8¼.
63829-4 Pa. $10.95

A HISTORY OF MECHANICS, René Dugas. Monumental study of mechanical principles from antiquity to quantum mechanics. Contributions of ancient Greeks, Galileo, Leonardo, Kepler, Lagrange, many others. 671pp. 5⅜ × 8½.
65632-2 Pa. $14.95

FAMOUS PROBLEMS OF GEOMETRY AND HOW TO SOLVE THEM, Benjamin Bold. Squaring the circle, trisecting the angle, duplicating the cube: learn their history, why they are impossible to solve, then solve them yourself. 128pp. 5⅜ × 8½.
24297-8 Pa. $3.95

MECHANICAL VIBRATIONS, J.P. Den Hartog. Classic textbook offers lucid explanations and illustrative models, applying theories of vibrations to a variety of practical industrial engineering problems. Numerous figures. 233 problems, solutions. Appendix. Index. Preface. 436pp. 5⅜ × 8½.
64785-4 Pa. $9.95

CURVATURE AND HOMOLOGY, Samuel I. Goldberg. Thorough treatment of specialized branch of differential geometry. Covers Riemannian manifolds, topology of differentiable manifolds, compact Lie groups, other topics. Exercises. 315pp. 5⅜ × 8½.
64314-X Pa. $8.95

HISTORY OF STRENGTH OF MATERIALS, Stephen P. Timoshenko. Excellent historical survey of the strength of materials with many references to the theories of elasticity and structure. 245 figures. 452pp. 5⅜ × 8½. 61187-6 Pa. $10.95

GEOMETRY OF COMPLEX NUMBERS, Hans Schwerdtfeger. Illuminating, widely praised book on analytic geometry of circles, the Moebius transformation, and two-dimensional non-Euclidean geometries. 200pp. 5⅜ × 8¼.

63830-8 Pa. $6.95

MECHANICS, J.P. Den Hartog. A classic introductory text or refresher. Hundreds of applications and design problems illuminate fundamentals of trusses, loaded beams and cables, etc. 334 answered problems. 462pp. 5⅜ × 8½.   60754-2 Pa. $8.95

TOPOLOGY, John G. Hocking and Gail S. Young. Superb one-year course in classical topology. Topological spaces and functions, point-set topology, much more. Examples and problems. Bibliography. Index. 384pp. 5⅜ × 8¼.

65676-4 Pa. $8.95

STRENGTH OF MATERIALS, J.P. Den Hartog. Full, clear treatment of basic material (tension, torsion, bending, etc.) plus advanced material on engineering methods, applications. 350 answered problems. 323pp. 5⅜ × 8½.   60755-0 Pa. $7.50

ELEMENTARY CONCEPTS OF TOPOLOGY, Paul Alexandroff. Elegant, intuitive approach to topology from set-theoretic topology to Betti groups; how concepts of topology are useful in math and physics. 25 figures. 57pp. 5⅜ × 8½.

60747-X Pa. $2.95

ADVANCED STRENGTH OF MATERIALS, J.P. Den Hartog. Superbly written advanced text covers torsion, rotating disks, membrane stresses in shells, much more. Many problems and answers. 388pp. 5⅜ × 8½.   65407-9 Pa. $9.95

COMPUTABILITY AND UNSOLVABILITY, Martin Davis. Classic graduate-level introduction to theory of computability, usually referred to as theory of recurrent functions. New preface and appendix. 288pp. 5⅜ × 8½. 61471-9 Pa. $6.95

GENERAL CHEMISTRY, Linus Pauling. Revised 3rd edition of classic first-year text by Nobel laureate. Atomic and molecular structure, quantum mechanics, statistical mechanics, thermodynamics correlated with descriptive chemistry. Problems. 992pp. 5⅜ × 8½.   65622-5 Pa. $19.95

AN INTRODUCTION TO MATRICES, SETS AND GROUPS FOR SCIENCE STUDENTS, G. Stephenson. Concise, readable text introduces sets, groups, and most importantly, matrices to undergraduate students of physics, chemistry, and engineering. Problems. 164pp. 5⅜ × 8½.   65077-4 Pa. $6.95

THE HISTORICAL BACKGROUND OF CHEMISTRY, Henry M. Leicester. Evolution of ideas, not individual biography. Concentrates on formulation of a coherent set of chemical laws. 260pp. 5⅜ × 8½.   61053-5 Pa. $6.95

THE PHILOSOPHY OF MATHEMATICS: An Introductory Essay, Stephan Körner. Surveys the views of Plato, Aristotle, Leibniz & Kant concerning propositions and theories of applied and pure mathematics. Introduction. Two appendices. Index. 198pp. 5⅜ × 8½.   25048-2 Pa. $6.95

THE DEVELOPMENT OF MODERN CHEMISTRY, Aaron J. Ihde. Authoritative history of chemistry from ancient Greek theory to 20th-century innovation. Covers major chemists and their discoveries. 209 illustrations. 14 tables. Bibliographies. Indices. Appendices. 851pp. 5⅜ × 8½.   64235-6 Pa. $17.95

DE RE METALLICA, Georgius Agricola. The famous Hoover translation of greatest treatise on technological chemistry, engineering, geology, mining of early modern times (1556). All 289 original woodcuts. 638pp. 6¾ × 11.
60006-8 Pa. $17.95

SOME THEORY OF SAMPLING, William Edwards Deming. Analysis of the problems, theory and design of sampling techniques for social scientists, industrial managers and others who find statistics increasingly important in their work. 61 tables. 90 figures. xvii + 602pp. 5⅜ × 8½.
64684-X Pa. $15.95

THE VARIOUS AND INGENIOUS MACHINES OF AGOSTINO RAMELLI: A Classic Sixteenth-Century Illustrated Treatise on Technology, Agostino Ramelli. One of the most widely known and copied works on machinery in the 16th century. 194 detailed plates of water pumps, grain mills, cranes, more. 608pp. 9 × 12. (EBE)
25497-6 Clothbd. $34.95

LINEAR PROGRAMMING AND ECONOMIC ANALYSIS, Robert Dorfman, Paul A. Samuelson and Robert M. Solow. First comprehensive treatment of linear programming in standard economic analysis. Game theory, modern welfare economics, Leontief input-output, more. 525pp. 5⅜ × 8½.
65491-5 Pa. $13.95

ELEMENTARY DECISION THEORY, Herman Chernoff and Lincoln E. Moses. Clear introduction to statistics and statistical theory covers data processing, probability and random variables, testing hypotheses, much more. Exercises. 364pp. 5⅜ × 8½.
65218-1 Pa. $9.95

THE COMPLEAT STRATEGYST: Being a Primer on the Theory of Games of Strategy, J.D. Williams. Highly entertaining classic describes, with many illustrated examples, how to select best strategies in conflict situations. Prefaces. Appendices. 268pp. 5⅜ × 8½.
25101-2 Pa. $6.95

MATHEMATICAL METHODS OF OPERATIONS RESEARCH, Thomas L. Saaty. Classic graduate-level text covers historical background, classical methods of forming models, optimization, game theory, probability, queueing theory, much more. Exercises. Bibliography. 448pp. 5⅜ × 8¼.
65703-5 Pa. $12.95

CONSTRUCTIONS AND COMBINATORIAL PROBLEMS IN DESIGN OF EXPERIMENTS, Damaraju Raghavarao. In-depth reference work examines orthogonal Latin squares, incomplete block designs, tactical configuration, partial geometry, much more. Abundant explanations, examples. 416pp. 5⅜ × 8¼.
65685-3 Pa. $10.95

THE ABSOLUTE DIFFERENTIAL CALCULUS (CALCULUS OF TENSORS), Tullio Levi-Civita. Great 20th-century mathematician's classic work on material necessary for mathematical grasp of theory of relativity. 452pp. 5⅜ × 8½.
63401-9 Pa. $9.95

VECTOR AND TENSOR ANALYSIS WITH APPLICATIONS, A.I. Borisenko and I.E. Tarapov. Concise introduction. Worked-out problems, solutions, exercises. 257pp. 5⅜ × 8¼.
63833-2 Pa. $6.95

THE FOUR-COLOR PROBLEM: Assaults and Conquest, Thomas L. Saaty and Paul G. Kainen. Engrossing, comprehensive account of the century-old combinatorial topological problem, its history and solution. Bibliographies. Index. 110 figures. 228pp. 5⅜ × 8½. 65092-8 Pa. $6.95

CATALYSIS IN CHEMISTRY AND ENZYMOLOGY, William P. Jencks. Exceptionally clear coverage of mechanisms for catalysis, forces in aqueous solution, carbonyl- and acyl-group reactions, practical kinetics, more. 864pp. 5⅜ × 8½. 65460-5 Pa. $19.95

PROBABILITY: An Introduction, Samuel Goldberg. Excellent basic text covers set theory, probability theory for finite sample spaces, binomial theorem, much more. 360 problems. Bibliographies. 322pp. 5⅜ × 8½. 65252-1 Pa. $8.95

LIGHTNING, Martin A. Uman. Revised, updated edition of classic work on the physics of lightning. Phenomena, terminology, measurement, photography, spectroscopy, thunder, more. Reviews recent research. Bibliography. Indices. 320pp. 5⅜ × 8¼. 64575-4 Pa. $8.95

PROBABILITY THEORY: A Concise Course, Y.A. Rozanov. Highly readable, self-contained introduction covers combination of events, dependent events, Bernoulli trials, etc. Translation by Richard Silverman. 148pp. 5⅜ × 8¼. 63544-9 Pa. $5.95

THE CEASELESS WIND: An Introduction to the Theory of Atmospheric Motion, John A. Dutton. Acclaimed text integrates disciplines of mathematics and physics for full understanding of dynamics of atmospheric motion. Over 400 problems. Index. 97 illustrations. 640pp. 6 × 9. 65096-0 Pa. $17.95

STATISTICS MANUAL, Edwin L. Crow, et al. Comprehensive, practical collection of classical and modern methods prepared by U.S. Naval Ordnance Test Station. Stress on use. Basics of statistics assumed. 288pp. 5⅜ × 8½. 60599-X Pa. $6.95

DICTIONARY/OUTLINE OF BASIC STATISTICS, John E. Freund and Frank J. Williams. A clear concise dictionary of over 1,000 statistical terms and an outline of statistical formulas covering probability, nonparametric tests, much more. 208pp. 5⅜ × 8½. 66796-0 Pa. $6.95

STATISTICAL METHOD FROM THE VIEWPOINT OF QUALITY CONTROL, Walter A. Shewhart. Important text explains regulation of variables, uses of statistical control to achieve quality control in industry, agriculture, other areas. 192pp. 5⅜ × 8½. 65232-7 Pa. $6.95

THE INTERPRETATION OF GEOLOGICAL PHASE DIAGRAMS, Ernest G. Ehlers. Clear, concise text emphasizes diagrams of systems under fluid or containing pressure; also coverage of complex binary systems, hydrothermal melting, more. 288pp. 6½ × 9¼. 65389-7 Pa. $10.95

STATISTICAL ADJUSTMENT OF DATA, W. Edwards Deming. Introduction to basic concepts of statistics, curve fitting, least squares solution, conditions without parameter, conditions containing parameters. 26 exercises worked out. 271pp. 5⅜ × 8½. 64685-8 Pa. $7.95

TENSOR CALCULUS, J.L. Synge and A. Schild. Widely used introductory text covers spaces and tensors, basic operations in Riemannian space, non-Riemannian spaces, etc. 324pp. 5⅜ × 8¼. 63612-7 Pa. $7.95

A CONCISE HISTORY OF MATHEMATICS, Dirk J. Struik. The best brief history of mathematics. Stresses origins and covers every major figure from ancient Near East to 19th century. 41 illustrations. 195pp. 5⅜ × 8½. 60255-9 Pa. $7.95

A SHORT ACCOUNT OF THE HISTORY OF MATHEMATICS, W.W. Rouse Ball. One of clearest, most authoritative surveys from the Egyptians and Phoenicians through 19th-century figures such as Grassman, Galois, Riemann. Fourth edition. 522pp. 5⅜ × 8½. 20630-0 Pa. $10.95

HISTORY OF MATHEMATICS, David E. Smith. Nontechnical survey from ancient Greece and Orient to late 19th century; evolution of arithmetic, geometry, trigonometry, calculating devices, algebra, the calculus. 362 illustrations. 1,355pp. 5⅜ × 8½. 20429-4, 20430-8 Pa., Two-vol. set $23.90

THE GEOMETRY OF RENÉ DESCARTES, René Descartes. The great work founded analytical geometry. Original French text, Descartes' own diagrams, together with definitive Smith-Latham translation. 244pp. 5⅜ × 8½.
60068-8 Pa. $6.95

THE ORIGINS OF THE INFINITESIMAL CALCULUS, Margaret E. Baron. Only fully detailed and documented account of crucial discipline: origins; development by Galileo, Kepler, Cavalieri; contributions of Newton, Leibniz, more. 304pp. 5⅜ × 8½. (Available in U.S. and Canada only) 65371-4 Pa. $9.95

THE HISTORY OF THE CALCULUS AND ITS CONCEPTUAL DEVELOP-MENT, Carl B. Boyer. Origins in antiquity, medieval contributions, work of Newton, Leibniz, rigorous formulation. Treatment is verbal. 346pp. 5⅜ × 8½.
60509-4 Pa. $7.95

THE THIRTEEN BOOKS OF EUCLID'S ELEMENTS, translated with introduction and commentary by Sir Thomas L. Heath. Definitive edition. Textual and linguistic notes, mathematical analysis. 2,500 years of critical commentary. Not abridged. 1,414pp. 5⅜ × 8½. 60088-2, 60089-0, 60090-4 Pa., Three-vol. set $29.85

GAMES AND DECISIONS: Introduction and Critical Survey, R. Duncan Luce and Howard Raiffa. Superb nontechnical introduction to game theory, primarily applied to social sciences. Utility theory, zero-sum games, n-person games, decision-making, much more. Bibliography. 509pp. 5⅜ × 8½. 65943-7 Pa. $11.95

THE HISTORICAL ROOTS OF ELEMENTARY MATHEMATICS, Lucas N.H. Bunt, Phillip S. Jones, and Jack D. Bedient. Fundamental underpinnings of modern arithmetic, algebra, geometry and number systems derived from ancient civilizations. 320pp. 5⅜ × 8½. 25563-8 Pa. $8.95

CALCULUS REFRESHER FOR TECHNICAL PEOPLE, A. Albert Klaf. Covers important aspects of integral and differential calculus via 756 questions. 566 problems, most answered. 431pp. 5⅜ × 8½. 20370-0 Pa. $8.95

CHALLENGING MATHEMATICAL PROBLEMS WITH ELEMENTARY SOLUTIONS, A.M. Yaglom and I.M. Yaglom. Over 170 challenging problems on probability theory, combinatorial analysis, points and lines, topology, convex polygons, many other topics. Solutions. Total of 445pp. 5⅜ × 8½. Two-vol. set.
Vol. I 65536-9 Pa. $6.95
Vol. II 65537-7 Pa. $6.95

FIFTY CHALLENGING PROBLEMS IN PROBABILITY WITH SOLU-TIONS, Frederick Mosteller. Remarkable puzzlers, graded in difficulty, illustrate elementary and advanced aspects of probability. Detailed solutions. 88pp. 5⅜ × 8½.
65355-2 Pa. $3.95

EXPERIMENTS IN TOPOLOGY, Stephen Barr. Classic, lively explanation of one of the byways of mathematics. Klein bottles, Moebius strips, projective planes, map coloring, problem of the Koenigsberg bridges, much more, described with clarity and wit. 43 figures. 210pp. 5⅜ × 8½. 25933-1 Pa. $5.95

RELATIVITY IN ILLUSTRATIONS, Jacob T. Schwartz. Clear nontechnical treatment makes relativity more accessible than ever before. Over 60 drawings illustrate concepts more clearly than text alone. Only high school geometry needed. Bibliography. 128pp. 6⅛ × 9¼. 25965-X Pa. $5.95

AN INTRODUCTION TO ORDINARY DIFFERENTIAL EQUATIONS, Earl A. Coddington. A thorough and systematic first course in elementary differential equations for undergraduates in mathematics and science, with many exercises and problems (with answers). Index. 304pp. 5⅜ × 8½. 65942-9 Pa. $7.95

FOURIER SERIES AND ORTHOGONAL FUNCTIONS, Harry F. Davis. An incisive text combining theory and practical example to introduce Fourier series, orthogonal functions and applications of the Fourier method to boundary-value problems. 570 exercises. Answers and notes. 416pp. 5⅜ × 8½. 65973-9 Pa. $9.95

THE THEORY OF BRANCHING PROCESSES, Theodore E. Harris. First systematic, comprehensive treatment of branching (i.e. multiplicative) processes and their applications. Galton-Watson model, Markov branching processes, electron-photon cascade, many other topics. Rigorous proofs. Bibliography. 240pp. 5⅜ × 8½. 65952-6 Pa. $6.95

AN INTRODUCTION TO ALGEBRAIC STRUCTURES, Joseph Landin. Superb self-contained text covers "abstract algebra": sets and numbers, theory of groups, theory of rings, much more. Numerous well-chosen examples, exercises. 247pp. 5⅜ × 8½. 65940-2 Pa. $6.95

Prices subject to change without notice.
Available at your book dealer or write for free Mathematics and Science Catalog to Dept. GI, Dover Publications, Inc., 31 East 2nd St., Mineola, N.Y. 11501. Dover publishes more than 175 books each year on science, elementary and advanced mathematics, biology, music, art, literature, history, social sciences and other areas.